Retrosynthesis in the Manufacture of Generic Drugs

Selected Case Studies

RETROSYNTHESIS IN THE MANUFACTURE OF GENERIC DRUGS

Selected Case Studies

Pedro Paulo Santos

William Heggie

Wiley

This edition first published 2021

Registered Offices
John Wiley & Sons, Inc., 111 River Street, Hoboken, NJ 07030, USA
John Wiley & Sons Ltd, The Atrium, Southern Gate, Chichester, West Sussex, PO19 8SQ, UK

Editorial Office
The Atrium, Southern Gate, Chichester, West Sussex, PO19 8SQ, UK
For details of our global editorial offices, customer services, and more information about Wiley products visit us at www.wiley.com.
Wiley also publishes its books in a variety of electronic formats and by print-on-demand. Some content that appears in standard print versions of this book may not be available in other formats.

Library of Congress Cataloging-in-Publication Data
Names: Santos, Pedro Paulo, author. | Heggie, William, author.
Title: Retrosynthesis in the manufacture of generic drugs : selected case studies / Pedro Paulo Santos, William Heggie.
Description: Hoboken, New Jersey : Wiley, [2021] | Includes index.
Identifiers: LCCN 2020032121 (print) | LCCN 2020032122 (ebook) | ISBN 9781119155539 (cloth) | ISBN 9781119155546 (adobe pdf) | ISBN 9781119155553 (epub)
Subjects: LCSH: Generic drugs. | Organic compounds–Synthesis.
Classification: LCC RS55.2 .S26 2021 (print) | LCC RS55.2 (ebook) | DDC 615.1–dc23
LC record available at https://lccn.loc.gov/2020032121
LC ebook record available at https://lccn.loc.gov/2020032122

Cover Design: Wiley
Cover Image: Courtesy of Pedro Paulo Santos

Set in 10/11.5pt TimesNewRomanMTStd by SPi Global, Chennai, India

To
Pedro, Rita, Isabel,
André and Daniela

ACKNOWLEDGEMENTS

The authors wish to thank the following for their invaluable assistance in the preparation of this work:

Peter Villax and Hovione that allowed us to access the resources needed for the comprehensive literature search.

Jose Luis Pires from Hovione, for the databases patent search. Vitor Pires for the help in the production of the high-resolution photos of the author's original sketches for book cover. Nuno Cartaxo for the updating the original production draft into a professional level document.

Wiley for giving us the opportunity to write this book.

Finally, we wish to thank our families for their love, patient and support during the countless hours we needed to complete this book.

Contents

Preface

This book aims to illustrate how the retrosynthetic tool, widely used by academics in the search for synthetic routes used to prepare novel organic compounds, is applied in the Pharmaceutical Industry. However, merely finding a viable synthetic route is not sufficient. Many different options have to be considered and evaluated, as there exist numerous constraints on the industrialist in terms of what can and what cannot be reduced to practice. The reader will be guided through the various steps leading to the "best" processes and the limits encountered if these are put into practise on an industrial scale. Thereafter, an evaluation the potential each process has for implementation is presented.

A great deal of information is presented from both the patent and academic literature. The more serious reader can use this as a basis for further study and thought but for those who have neither time nor necessity the material is presented in such a way that it can be appreciated without delving into excessive detail.

William Heggie
Pedro Paulo Santos

A NOTE ABOUT THE BOOK AND ITS USE

Books on process chemistry usually use many short schemes, each referring to a particular synthetic route. This methodology makes comparison of different synthetic routes somewhat difficult and we can lose sight of the fact that frequently some intermediates are common in several processes. A unique feature of this book is the use of "reaction maps" (schemes identified with a letter coding). Whenever possible, similar strategies will be included in the same scheme.

The reaction maps collect several processes in the same scheme according to the intermediates used during the synthetic process. This allows easy comparison of the different routes which give the same molecule or intermediate. An analogy with an Underground Map can be made, where lots of lines (synthetic routes) are combined in the same chart to provide a map (our reaction map). For instance, if we travel in a city having a collection of schemes for each subway line as a guide we may be easily disoriented. Instead, the use of a map of the whole underground network where interconnections can be easily identified will be much more efficient and clear.

Each reaction in these maps will be coded with a scheme letter/reaction number (example, A22 means reaction 22 of scheme A) and the reaction conditions are included at the end of each chapter. In this way the reader can easily spot which reactants/conditions were used for the chemical transformation. In cases where the reactions have been used by several authors a summary of the conditions will be included.

The patent references in the book will be in company/year format (assignee/priority date). This system will allow the evolution of different synthetic routes to the molecule to be observed in accordance with the date of its disclosure (or discovery). If relevant, "references from peer review journals" will be also included.

Abbreviations

2-MeCBS	2-methyl-CBS-oxazaborolidine; 1-methyl-3,3-diphenyl-tetrahydro-pyrrolo[1,2*c*][1,3,2]oxazaborole
Ac	acetyl
acac	acetylacetone
AEAI	asymmetric enolate addition to imine
AHT	asymmetric hydrogen transfer
AIBN	azobisisobutyronitrile
all	allyl
alloc	allyloxycarbonyl
anh.	anhydrous
API	active pharmaceutical ingredient
aq.	aqueous
Ar	aryl
AS	asymmetric synthesis
ASA	acetylsalicylic acid
B_2pin_2	bis(pinacolato)diboron
BARF	tetrakis[3,5-bis(trifluoromethyl)phenyl]-borate
BBB	blood brain barrier
BIMAH	1-(1*H*-benzimidazol-2-yl)ethanamine
BINAP	2,2'-bis(diphenylphosphino)-1,1'-binaphthalene
BINOL	1,1'-bi-2-naphthol
BIPHEPHOS	6,6'-[(3,3'-di-*tert*-butyl-5,5'-dimethoxy-1,1'-biphenyl-2,2'-diyl)bis(oxy)]-bis(dibenzo[*d,f*][1,3,2]dioxaphosphepin)
bmim	1-butyl-3-methylimidazolium
BMS	borane dimethylsulfide
Bn	benzyl
BnEZT	benzyl ezetimibe
Boc	*t*-butyloxycarbonyl
BSA	bis(trimethylsilyl)acetamide
BTHF	borane tetrahydrofuran complex
Bu	butyl
Bz	benzoyl
CA	chiral auxiliary
CAH	catalytic asymmetric hydrogenation
CAN	ceric ammonium nitrate
cat.	catalyst
CBS	Corey-Bakshi-Shibata
Cbz	carbobenzyloxy
CDI	1,1'-carbonyldiimidazole
cod	1,5-cyclooctadiene
conc.	concentrated
Cp	cyclopentadienyl
CSA	camphor-10-sulfonic acid
Cy	cyclohexyl
DABCO	1,4-diazabicyclo[2.2.2]octane
DAIPEN	1,1-bis(4-methoxyphenyl)-3-methyl-1,2-butanediamine
dba	dibenzylideneacetone
DBSA	4-dodecylbenzenesulfonic acid
DBTA	dibenzoyl tartaric acid
DBU	1,8-diazabicyclo[5.4.0]undec-7-ene
DCC	*N,N'*-dicyclohexylcarbodiimide
DCDMH	1,3-dichloro-5,5-dimethyl hydantoin
DDQ	2,3-dichloro-5,6-dicyano-1,4-benzoquinone
de	diastereomeric excess
DEAD	diethyl azodicarboxylate
DET	diethyl tartrate
DHP	3,4-dihydro-2*H*-pyran

DIAD	diisopropyl azodicarboxylate
DIBAL	diisobutylaluminum hydride
DIC	*N*,*N*'-diisopropylcarbodiimide
dil	diluted
DIOP	2,3-*O*-isopropylidene-2,3-dihydroxy-1,4-bis(diphenylphosphino)butane
DIP-Cl	B-chlorodiisopinocampheylborane
DIPEA	*N*,*N*-diisopropylethylamine
DKR	dynamic kinetic resolution
DMA	*N*,*N*-dimethylacetamide
DMAP	4-dimethylaminopyridine
DMC	dimethylcarbonate
DME	1,2-dimethoxyethane
DMF	*N*,*N*-dimethylformamide
DMI	1,3-dimethyl-2-imidazolidinone
DMP	Dess-Martin periodinane
DMSO	dimethyl sulfoxide
DPEN	1,2-diphenylethylenediamine
DPMS	diphenylmethyl silyl
DPPA	diphenyl phosphoryl azide
DPPB	1,4-bis(diphenylphosphino)butane
dppbz	1,2-bis(diphenylphosphino)benzene
dppf	1,1'-ferrocenediyl-bis(diphenylphosphine)
DPPP	1,3-bis(diphenylphosphino)propane
DPTTA	di-*p*-toluyl-D-tartaric acid
dr	diastereomeric ratio
dtbpf	1,1'-bis(di-*tert*-butylphosphino)ferrocene
E1cB	Elimination Unimolecular conjugate Base
EAS	electrophilic aromatic substitution
EDC	*N*-(3-dimethylaminopropyl)-*N*'-ethylcarbodiimide
ee	enantiomeric excess
EP	European Pharmacopoeia
eq.	equivalent
Et	ethyl
EWG	electron withdrawing group
EZT	Ezetimibe
FGA	functional group addition
FGI	functional group interconversion
Fmoc	9-fluorenylmethoxycarbonyl
g	gas or gram
GABA	γ-aminobutyric acid
h	hour
HBTU	2-(1*H*-benzotriazol-1-yl)-1,1,3,3-tetramethyluronium hexafluorophosphate
hfacac	hexafluoroacetylacetone
HMDS	hexamethyldisilazane
HMPA	hexamethylphosphoramide
HOAt	1-hydroxy-7-azabenzotriazole
HOBt	1-hydroxybenzotriazole
HPLC	high-performance liquid chromatography
HWE	Horner-Wardsworth-Emmons
i	iso
Ipc_2BCl	diisopinocamphenylchloroborane
kg	kilogram
KRED	ketoreductase
LG	leaving group
LiHMDS	lithium bis(trimethylsilyl)amide
LTMP	lithium 2,2,6,6-tetramethylpiperidide
MCPBA	3-chloroperbenzoic acid
MCR	multicomponent reactions
Me	methyl
MEK	methyl ethyl ketone
MIBK	methyl isobutyl ketone
ML	mother liquor
MOM	methoxymethyl
morph.	morpholine
Ms	methanesulfonyl
MS	molecular sieve
MTBE	methyl *tert*-butyl ether
MW	microwave
n	normal
NAD	nicotinamide adenine dinucleotide
NaDCC	sodium dichloroisocyanurate
nbd	norbornadiene
NBS	*N*-bromosuccinimide
NCE	new chemical entity
NCS	*N*-chlorosuccinimide
NI	not isolated

NIS	*N*-iodosuccinimide
NMM	*N*-methylmorpholine
NMO	*N*-methylmorpholine-*N*-oxide
NMP	1-methyl-2-pyrrolidinone
NMR	nuclear magnetic resonance
Nu	nucleophile
op	optical purity
PCC	pyridinium chlorochromate
PEG	polyethylene glycol
PG	protecting group
Ph	phenyl
Phth	phthalimide
pin	pinacolato
Piv	pivaloyl
PLE	pig-liver esterase
PLP	pyridoxal phosphate
PMB	*p*-methoxybenzyl
PMHS	polymethylhydrosiloxane
PMP	*p*-methoxyphenyl
PNBA	*p*-nitrobenzoic acid
PPA	polyphosphoric acid
PPTS	pyridine *p*-toluenesulfonate
Pr	propyl
py	pyridine
QA	quinic acid
QN	7-chlorquinolin-2-yl
RCM	ring-closing metathesis
rt	room temperature
S/C	substrate-catalyst ratio
SA	shikimic acid in Oseltamivir chapter, salicylic acid in Sitagliptin chapter
SEGPHOS	5,5'-bis(diphenylphosphino)-4,4'-bi-1,3-benzodioxole
SES	2-(trimethylsilyl)ethanesulfonyl
SM	starting material
SMB	simulated moving bed
S_N	nucleophilic substitution
t	*tert*
t	ton
TA	tartaric acid
TBA	tetrabutylammonium
TBAF	tetra-*n*-butylammonium fluoride
TBDMS	*t*-butyldimethylsilyl
TBDPS	*t*-butyldiphenylsilyl
TCCA	trichloroisocyanuric acid
TEBA	benzyltriethylammonium chloride
TEMPO	(2,2,6,6-tetramethylpiperidin-1-yl)oxyl
TES	triethylsilyl
Tf	trifluoromethanosulfonyl
TFA	trifluoracetic acid
TFAA	trifluoracetic anhydride
THF	tetrahydrofuran
THP	tetrahydropyranyl
ThPD	bis(6,7-dihydrothieno[3,2-*c*]pyridin-5(4*H*)-yl)methane
THTP	4,5,6,7-tetrahydrothieno-[3,2-*c*]-pyridine in the Clopidogrel chapter; 3-(trifluoromethyl)-5,6,7,8-tetrahydro-[1,2,4]triazolo[4,3-*a*]pyrazine in the Sitagliptin chapter
TIPS	triisopropylsilyl
TMEDA	*N,N,N',N'*-tetramethylethylenediamine
TMS	tetramethylsilyl
Tol	toluyl
TpEA	2-(thiophen-2-yl)ethanamine
Tr	trityl
Ts	*p*-toluenesulfonyl
UK	United Kingdom
USA	United States of America
USD	United States Dollar
USP	United States Pharmacopeia
v	volume
w	weight
WW	Worldwide
Xyl-BINAP	1,1'-binaphthalene-2,2'-diylbis[bis(3,5-dimethylphenyl)phosphine]
YNR	yield not reported

1 INTRODUCTION

Retrosynthetic analysis is a powerful tool, first developed by E.J. Corey in the late 1960s, and extensively used by modern organic chemists to plan the syntheses of organic molecules. This book applies this technique to a series of important Drug Substances, also known as Active Pharmaceutical Ingredients (APIs). It is not our intention to teach, or even briefly review, the principles of retrosynthetic analysis. This is now a major part of any advanced organic chemistry course and many excellent publications, including the book: The Logic of Chemical Synthesis, New York: Wiley (1995) by E. J. Corey and X-M. Cheng, have been published. The reader is strongly advised to review the main tenets of retrosynthesis to be able to take maximum advantage of this volume.

In an academic setting the major goals of organic synthesis are:

i) to establish that the molecule (generally complex) can be synthesized;
ii) to establish new and novel synthetic routes to molecules that have already been synthesized and
iii) to demonstrate the applicability of new reactions to the synthesis of organic molecules.

There are few, if any, restrictions to carrying out these syntheses on a small scale in the laboratory. Overall yield, transformation efficiencies, potential hazards and effluent treatment, although undoubtedly important, are not showstoppers in this setting. If the synthesis produces enough material, at a milligram scale or less, which allows the end product to be characterized, then it is deemed successful. Obviously, no synthesis starts with the most rudimentary starting materials but relies on what is easily accessible. For an academic synthesis, the major source of starting materials is a chemical products catalogue such as that of Sigma-Aldrich. The vast majority of the thousands of chemicals listed have extremely limited availability and the price in many cases is enormous. Woodward's total synthesis of cephalosporin C started with L-(+)-cysteine, which was more expensive than Cephalosporin C itself, so it might appear that the undertaking of such a synthesis is illogical. However, the objective of the study was to demonstrate the conversion of a structurally simple molecule into a more complex one following a rationally designed synthesis. We return to this theme at the end of this introduction.

When carrying out a synthesis of an API on an industrial scale different constraints apply. The objective of any Pharmaceutical Company, whether originator or generic, is to get the product to market and maximise profitability for as long as possible. Hence, the design of a synthesis for industrial production has a significant number of limitations. The Process Chemist, as well as having a broad knowledge of synthetic transformations, an in-depth understanding of chemical group compatible and a well-founded acquaintance with the conditions necessary to carry out a plethora of chemical reactions, also has to appreciate what is viable on a large scale.

The retrosynthetic process will uncover a number of options for the construction of any molecule as no molecular entity has only one approach to its preparation. Nonetheless, not all of the possibilities will have the same degree of success when the chemistry is put into practice. The Process Chemist can, however, make a fairly accurate appraisal of what is more likely to work and what is not by applying basic chemical wisdom. Priorities for laboratory testing can be identified, but it should be stressed that those routes that have potential for success can only be confirmed after laboratory testing. Generally speaking, a particular route must be able to provide final product, API, in high yield and purity in addition to being feasible on a large scale. The product must be free of potentially toxic materials, some of which need to be controlled down to ppb levels. If time allows, it is prudent to do some quick trials on a route that current wisdom may have excluded, as sometimes happy surprises can occur.

Chemical considerations and laboratory verification are key in determining the success of any proposed synthetic route but additional factors have to be taken into consideration for operation in a chemical production plant. There are certain reactions that are comparatively more difficult to implement on a large scale than others. Amongst these are: highly exothermic reactions; Reactions requiring very high or low temperatures; Reactions requiring fast heating and cooling; Reactions using highly toxic reagents and that require containment; Reactions which generate large

Retrosynthesis in the Manufacture of Generic Drugs: Selected Case Studies, First Edition, Pedro Paulo Santos and William Heggie.

quantities of waste; Reactions done under high dilution; Reactions which require a high degree of control i.e. temperature, pressure and time. Whilst all of these reactions can be carried out at scale, if they can be avoided by using simpler, easier alternatives large scale manufacturing becomes more sustainable.

Again it is not our intention in this book to discuss the details of and the requirements imposed by large scale manufacturing. There are many excellent publications dealing with this topic. However, it is useful to mention some of the more salient points for each of the cases mentioned above.

Highly exothermic reactions

Highly exothermic reactions generate large amounts of heat over a very short time. On an industrial scale, the removal of this excess energy is problematic as the area to volume ratio of large vessels is relatively small. When this excess energy cannot be removed run away reactions, sometimes explosively, occur with grave consequences to both health and the environment. Many serious industrial accidents have occurred in this way.

Reactions requiring very high or low temperatures

One way to control highly exothermic reactions is to carry these out at low temperatures where the reaction rate is greatly reduced and hence the release of energy is slower and more easily removed from the reaction medium. However, expensive specialized equipment to achieve and maintain homogeneous low temperature conditions is necessary. Reactions requiring high temperatures (nominally above 150 °C) also require specialized equipment and materials resistant to chemicals and reagents at these elevated temperatures. Pyrolysis reactions at extremely high temperatures are practically never carried out in the fine chemical industry. These reactions require rapid cooling immediately following the transformation as the reaction products are subject to decomposition under these extreme conditions.

Reactions requiring fast heating and cooling

Suffice it to say that it takes a long time to heat and cool a large mass. Heat transfer is dependent on surface area. Surface area increases less (n^2 correlation) in comparison to mass/volume (n^3 correlation) on scale up. This causes problems for the Process Chemist, even to the extent that scaling up some reactions might turn out to be impractical. If the kinetics of a synthetic reaction increases greatly with temperature, and the kinetics of competing/impurity forming reactions do not, then fast heating to high temperatures so that the desired reaction goes to completion in the shortest possible time, helps keep impurities to low levels. Likewise, decomposition of the product can be controlled by fast cooling. Obviously, for large volumes of reactants, heating and cooling times are much longer than in the laboratory. It is no surprise that occasionally new impurities are observed and known impurities increase on scale up. Simulating times, temperature gradients and other variables that are expected at large scale in small scale equipment will establish whether this is true or not, and the Process Chemist will have to find alternative reaction conditions if this turns out to be the case. Flow chemistry, still in its infancy in the Pharmaceutical Industry, can resolve these types of situations and is an attractive alternative to classical batch chemistry.

Reactions using highly toxic reagents and that require containment

It is self-evident that plant personnel, the general populace and the environment need to be protected from noxious materials. Supply and transport of toxic chemicals can pose difficult logistic problems. Facilities which are designed to handle highly toxic materials are expensive to build and require constant monitoring to ensure that no harmful leakage occurs. Disastrous accidents, where large amounts are released into the environment, can have serious and even fatal consequences. Where possible companies avoid such materials and only highly specialized companies have adequate facilities to carry out this type of chemistry. In many instances, the waste products are also highly toxic and require special care in their disposal.

Reactions which generate large quantities of waste

Chemical reactions generally use stoichiometric quantities of reagents that are transformed into waste products. Few of these materials are environmentally friendly and their disposal poses problems for the chemical industry. Indeed, the major part of the bad publicity that the chemical industry, including the Pharmaceutical Sector, derives from this issue. The problem is exacerbated when more than one equivalent of reagent is required, resulting in a corresponding increase in effluent, and when toxic wastes are involved. Although recycling is an acceptable option, in the vast majority of cases recycling is not achievable. Even so, if the cost of recycling exceeds that of waste disposal, it may not be a choice worth considering. Hence, one of the principal focuses for the Process Chemist has become the application of green chemistry and catalytic reactions wherever possible.

Reactions done under high dilution

Aside from the fact that reactions carried out at high dilution require a much larger plant and bigger equipment, they severely restrict throughput, thus reducing the overall quantities that can be produced in a given time. They are also much more energy and labour intensive as the large volumes of solvent need to be heated and cooled whilst handling takes more time. This all adds to the final cost of the drug substance, which may in turn reduce the economic viability of a drug.

Reactions which require a high degree of control i.e. temperature, pressure and time

The Process Chemist takes into consideration the limitations of the equipment that will be used for the industrial process that he/she develops. The gauges that measure variables on plant have limitations (as do those in a laboratory setting) although Good Manufacturing Practices (GMP) require that these are at least adequate to the task. There is no guarantee that the probes used for measurements accurately depict real values in the reaction mass as they are usually located in the jacket of the reactor and not in body of the liquid. Another issue is that, at industrial scale, the large mass of the material gives it a certain inertia. As a consequence, overshooting of programmed reaction parameters such as temperature is anticipated when fast heating is applied. The system then takes time to settle into a steady state where temperature fluctuation is minimal. It is therefore pointless to develop a process whose success depends on tight control of such reaction parameters. The process should be as "forgiving as possible" especially when multipurpose batch reactors are involved. Tight controls are only feasible in equipment, such as flow reactors, designed specifically for the purpose. Thus, robustness and reproducibility of a processes are important factors for its success on plant.

The analyses given in this book each start with a theoretical exercise where retrosynthetic principals are applied to the molecular entities of a selected number of important drugs to ascertain viable routes of synthesis. In each case, the applicability to an industrial scenario is assessed and conclusions drawn. Following this, the original literature is examined to evaluate which options had been studied and published. An attempt is made to evaluate which of the synthetic routes uncovered are eligible for introduction into a manufacturing scenario. It is hoped that the methodology exploited will aid the reader to uncover practical synthetic routes to other drug substances whether they be NCEs (New Chemical Entities) or generic APIs (Active Pharmaceutical Ingredients).

Finally, some observations on the origin of the information that is presented in this volume is in order. The two major sources that have been used are the patent literature and peer review journals. Other sources such as books, presentations/posters given at scientific conference and personal communications exist, but are not as comprehensive and/or as easily accessible.

There exist fundamental differences between these two types of literature. Peer review journals are mainly the province of academics whose careers are largely defined by what they publish. Consequently, academics focus on publishing as many original papers as possible. If the end point of the original postulate is reached experimentally, then the prime objective has been reached. A multitude of chemical syntheses, based on original and inspired chemistry, have been published where the accumulated yields are minimal, use chromatography to separate and purify intermediates and final product and are carried out on a milligram, or even a microgram, scale. The investigator does not concern him/herself by doing further work to optimize the route further from the point where it is acceptable for publication in a peer review journal. It is challenging to assess whether these synthetic routes could be viable at an industrial scale. We have made an attempt to do so using experience and common sense but, only process development and actual hands on experimentation can validly answer this question.

Patents are legal documents and can be regarded as contracts between a company (or individuals) and the state. Pharmaceutical companies submit a patent specification and, if after examination it is granted, the state allows the company a number of years (usually 20) to commercially exploit the invention without competition. The alternative would be to keep the invention secret (referred to as a trade secret). However, this is not an option for the Pharmaceutical Industry as, once the product is commercialized, it could be reverse engineered and competitors would very quickly enter into the market. To obtain a patent is not straightforward. The very first patent of a drug (a product patent) need only describe, almost in a rudimentary fashion, how it can be prepared. Thereafter, subsequent patents which describe new and novel ways to produce the drug substance (process patents) have to establish superiority over what is known in the Prior Art (the sum of all publications related to the subject matter of the patent). Patent offices worldwide have experienced patent examiners who assess patent novelty and inventiveness and only after they have approved the patent claims, does the patent proceed to grant. However, it cannot be presumed that the work of assessing the suitability of processes therein described to be implemented industrially has been done. The patent examiner only looks at novelty and inventiveness, which may not necessarily signify that the process is industrially practicable. There is an additional constraint when trying to interpret the content of patent literature. It is required by law to include the best examples that allow a patent to be reduced to practice. This is obviously not in the patentee's interest – no-one willingly lays bare their manufacturing knowhow for competitors to copy. Hence, patents are published revealing numerous options to carry out the process, usually before real industrial development work has been done. In this way companies can "disguise" their chosen route in an avalanche of information whist still revealing the best examples at the time of patent application. Once more we have used common sense and experience to interpret which processes are worthy of application in a manufacturing scenario, but as was the case for academic syntheses, only experimentation can give definitive answers.

2 GABAPENTIN

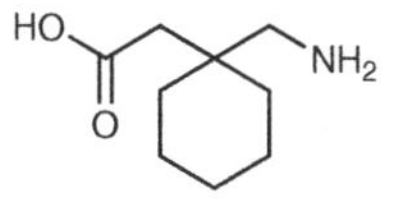

Year of discovery: 1974 Warner-Lambert Company
First WW Approval: 05 February 1993 (UK)
US approval: 30 December 1993
Drug category: Calcium channel ligand, structurally related to γ-aminobutyric acid (GABA)
Uses: Antipsychotic; Analgesic (non-narcotic); Antidepressant; Antimigraine; Anticonvulsant

Gabapentin (**1**) resulted from the search for synthetic analogues of γ-aminobutyric acid (GABA), a natural inhibitor of neurotransmission in the mammalian brain. GABA itself cannot enter the central nervous system by systemic administration; hence, a more lipophilic derivative was sought. Both GABA and Gabapentin are γ-amino acids (Figure 2.1), the higher lipophilicity of Gabapentin resulting from the presence of the cyclohexane ring. Contrary to the desired result it was found that Gabapentin does not interact with the GABA receptor in the brain but does display activity due to binding to a protein subunit that regulates a calcium ion channel.

In 1993, Gabapentin was initially approved to control seizures and was marketed by Pfizer under the trademark Neurotin®. Later, the drug was also approved (2002) for neuropathic neuralgia and some other uses anticipated, for example, in anxiety and in panic disorders.

In the aftermath of the success of Gabapentin, a new more active and potent derivative, Pregabalin, emerged and was marketed as Lyrica® by Pfizer from 2004 onwards.

Gabapentin and Pregabalin belong to a class of drugs known as gabapentinoids. Recent studies have revealed that the Large-neutral Amino acid Transporter 1 (LAT-1), which is found in the membranes of the Blood Brain barrier (BBB), transports Gabapentin across the BBB explaining the uptake of the drug to the brain. This discovery facilitated the development of gabapentin prodrugs, such as Gabapentin enacarbil (Figure 2.2), which more easily cross the BBB.

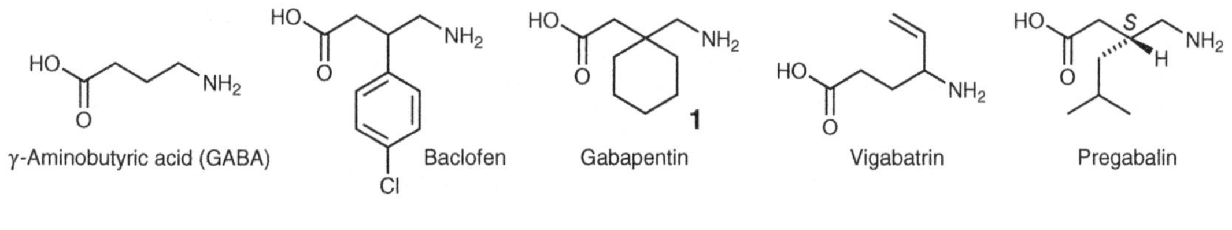

FIGURE 2.1 GABA AND SOME SYNTHETIC API ANALOGUES

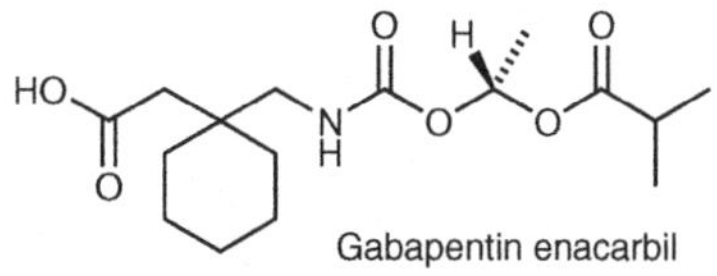

FIGURE 2.2 GABAPENTENTIN ENACARBIL CHEMICAL STRUCTURE

Retrosynthesis in the Manufacture of Generic Drugs: Selected Case Studies, First Edition, Pedro Paulo Santos and William Heggie.

2.1 WORLDWIDE SALES AND PATENT STATUS

The following graphs illustrate how research carried out in the pharmaceutical industry towards the production of the API and worldwide sales of the finished drug product evolved.

- US basic patent term 02/05/1995 (US4087544, WarnerL1974), USA Product Patent term extension 16/01/2000.
- Annual production[1]: 2019, 3,997.3 t; 2018, 3,851.6 t; 2017, 3,743.0 t.
- Annual WW sales[1]: 2019, 1377M USD; 2018, 1401.3M USD; 2017, 1346.9M USD.

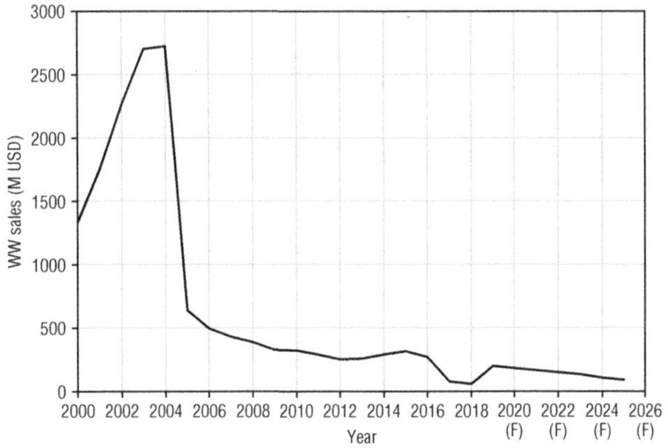

FIGURE 2.3 GABAPENTIN WW SALES EVOLUTION (IN M USD)

Source: Data from GlobalData [Online] https://globaldata.com/

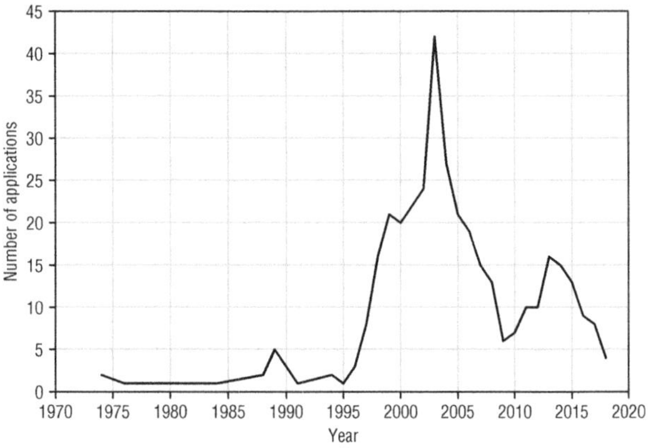

FIGURE 2.4 NEW PATENT APPLICATIONS ON GABAPENTIN FILED BY YEAR

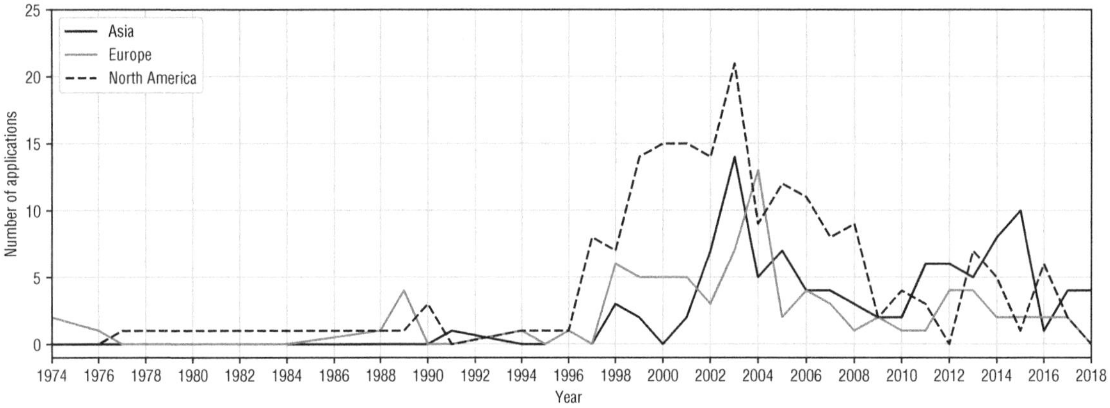

FIGURE 2.5 PATENT APPLICATIONS ON GABAPENTIN FILED BY REGION

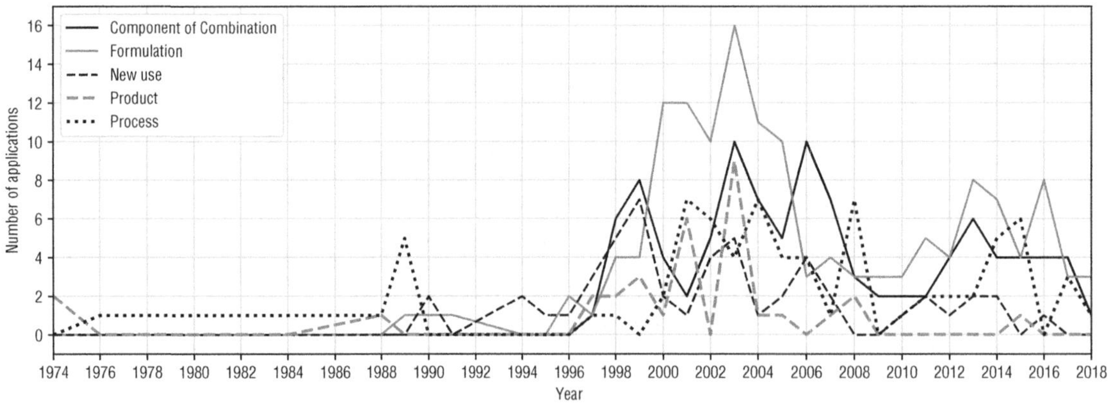

FIGURE 2.6 PATENT APPLICATIONS ON GABAPENTIN FILED BY TYPE OF PATENT

1 Newport Modular Platform – Clarivate Analytics [Online] https://clarivate.com/cortellis/ [2020, April]

2.2 GABAPENTIN STRUCTURE AND GENERAL RETROSYNTHETIC STRATEGY

Gabapentin is a relatively simple molecule containing a cyclohexane ring with two functionalized side chains linked to the same carbon atom of the ring (Figure 2.7).

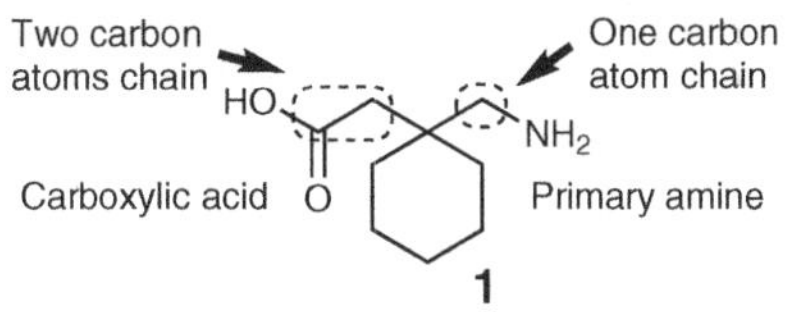

FIGURE 2.7 KEY STRUCTURE FEATURES OF GABAPENTIN

Some of the key structural aspects of the molecule are:

- It is a non-chiral molecule, containing a C6 ring.
- Gabapentin has only two functional groups and is a γ-amino acid. The drug product is sold in the neutral form, which exists as the zwitterion, although the molecule is usually represented without any charges (as above).
- The two side chains, a two-carbon and a one-carbon chain, are connected to the same cyclohexane ring atom. This feature is important for any synthetic strategy.
- The two functional groups present in the molecule may react to form a γ-lactam (Scheme 2.1).

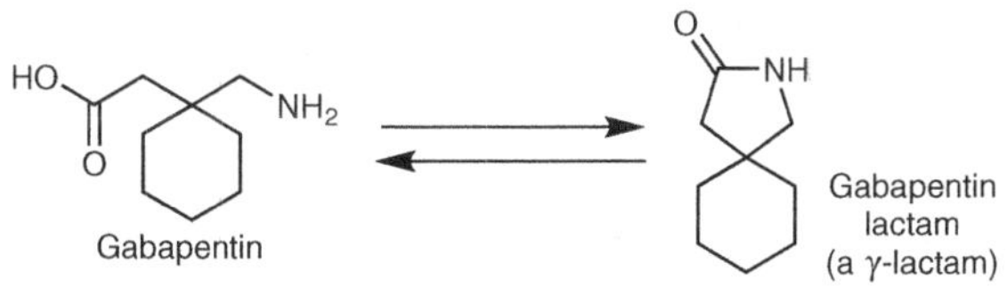

SCHEME 2.1 GABAPENTIN AND GABAPENTIN LACTAM INTERCONVERSION

Under appropriate reaction conditions, interconversion between Gabapentin and its lactam is possible. This situation can be used favourably in the industrial synthesis of Gabapentin such as using this interconversion to purify the API. The lactam is considered to be a relatively toxic impurity and only very low levels are permitted in the final drug product. Hence, special attention must be given, not only to its removal from the API, but also to avoid lactam formation during the final process steps, that is during both isolation of the API and formulation.

In the design of the retrosynthetic pathway, Gabapentin can be considered as a carbocycle with two carbon chains connected to the same carbon atom. From the synthetic point of view, it is desirable to select a starting material containing a C6 ring, hence avoiding extra synthetic steps to introduce the cyclohexane ring, which is in any case, widely available in a great number of potential starting materials. A good strategic approach is to disconnect at the branch point, removing one or both of the two chains and ending up with a monosubstituted functionalized cyclohexane starting material (Scheme 2.2). The presence of the amine and the carboxylic acid functional groups in Gabapentin are both useful for the disconnections of the molecule.

A disconnection between the carbon atoms α and β to the carbonyl function of Gabapentin leads to an unfavourable result. Enolate chemistry to promote CC bond formation is not easily applied (Scheme 2.3) and therefore, focusing first on the amine would seem to be more promising.

2.2.1 Using the primary amine for retrosynthesis

By focusing on the primary amine of Gabapentin, several retrosynthetic pathways can be developed. Some of the classical methods to obtain primary amines are reduction of primary amides, reduction of oximes and primary imines, reduction of nitro, nitrile or azide groups and rearrangement of carboxylic acid derivatives

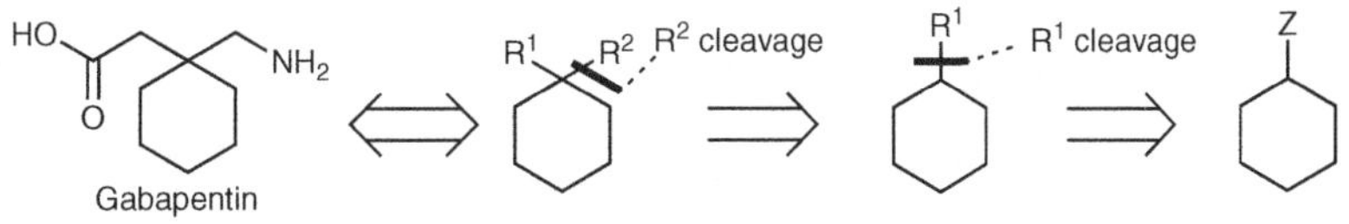

SCHEME 2.2 GENERAL RETROSYNTHESIS OF GABAPENTIN INTO A MONOSUBSTITUTED CYCLOHEXANE STARTING MATERIAL

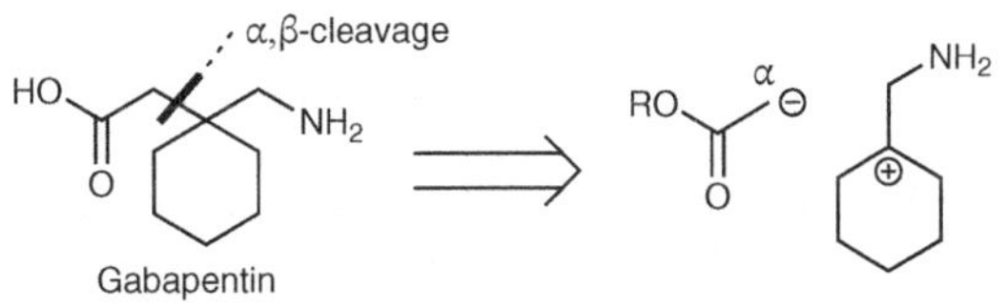

SCHEME 2.3 RETROSYNTHESIS OF GABAPENTIN BY α,β-CLEAVAGE

(acyl azides or amides *via* Curtius, Schmidt, Hofmann and Lossen rearrangements).

Producing the primary amine by an amide reduction at a late stage in the process is not a good option. Selectivity issues would make this synthetic conversion problematic as most likely the carboxylic acid group of Gabapentin will be present (either as an ester or the carboxylic acid itself) and would probably also be reduced (Scheme 2.4).

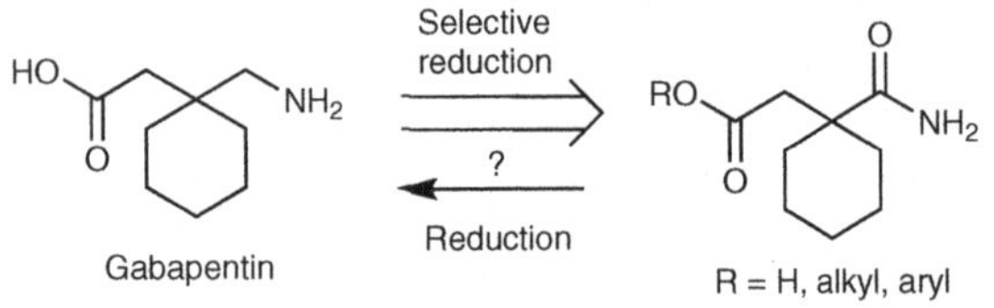

Scheme 2.4 Gabapentin by selective reduction of amide

2.2.2 Rearrangement as a key synthetic step: taking advantage of symmetrical intermediates

When a rearrangement is employed to produce the primary amine, the starting material for this rearrangement step will have one additional carbon atom in the chain destined to terminate in the amine group. The synthetic process must excise one carbon atom from this side chain *via* the rearrangement to an electron-deficient nitrogen atom (Curtius, Hofmann, Lossen or Schmidt rearrangements, Scheme 2.5).

When a rearrangement is the key step to produce the primary amine the retrosynthetic analysis indicates a symmetric spiro intermediate (derivatives of glutaric anhydride **2** or glutarimide **3**). Finding symmetric intermediates is highly desirable in retrosynthetic strategy as it often allows simpler methods to produce them.

Further analysis of the symmetric spiro compounds is shown below (Scheme 2.6). After ring opening, a retro 1,5-disconnection followed by a retroaldol gives cheap and readily available cyclohexanone as starting material, a commodity widely used in the synthesis of caprolactam and nylon 6.

Malonate or 2-cyanoacetate derivatives are useful reactants, having high predisposition for Michael additions to unsaturated carbonyl compounds. In these cases, hydrolysis and decarboxylation steps are required to remove the unwanted ester or cyano groups resulting from these synthetic processes (Scheme 2.7).

The retrosynthetic routes analysed so far, all ending up at cyclohexanone, are summarized in Scheme 2.8.

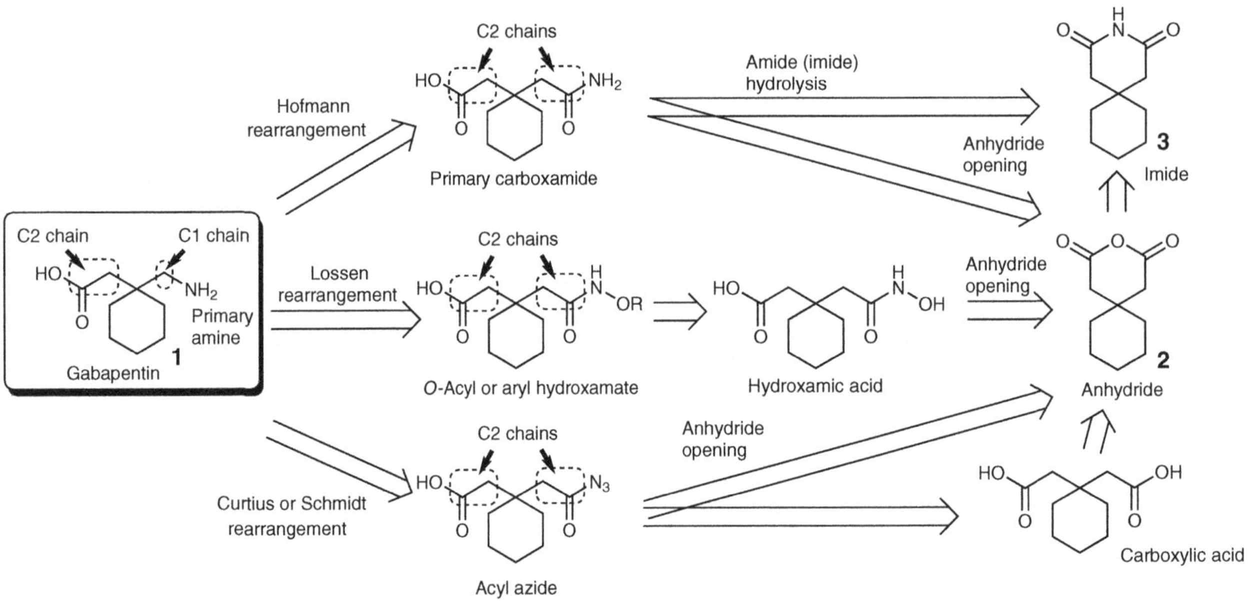

Scheme 2.5 Retrosynthetic use of a rearrangement step in Gabapentin

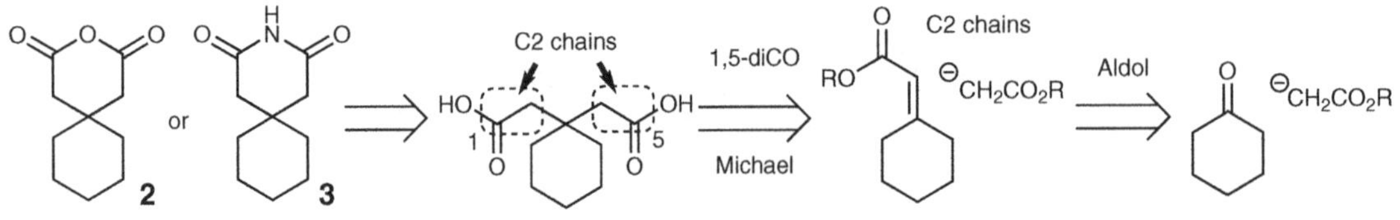

Scheme 2.6 Retrosynthesis of anhydride and imide intermediates

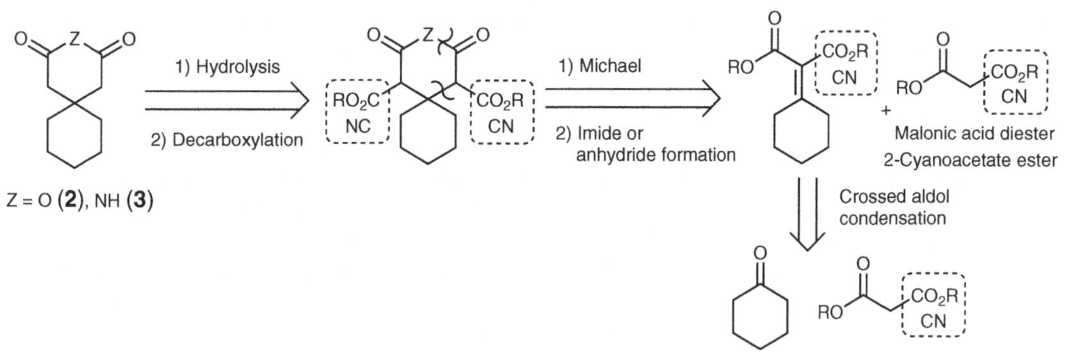

Scheme 2.7 Retrosynthesis of spiro-anhydride or spiro-imide derivatives

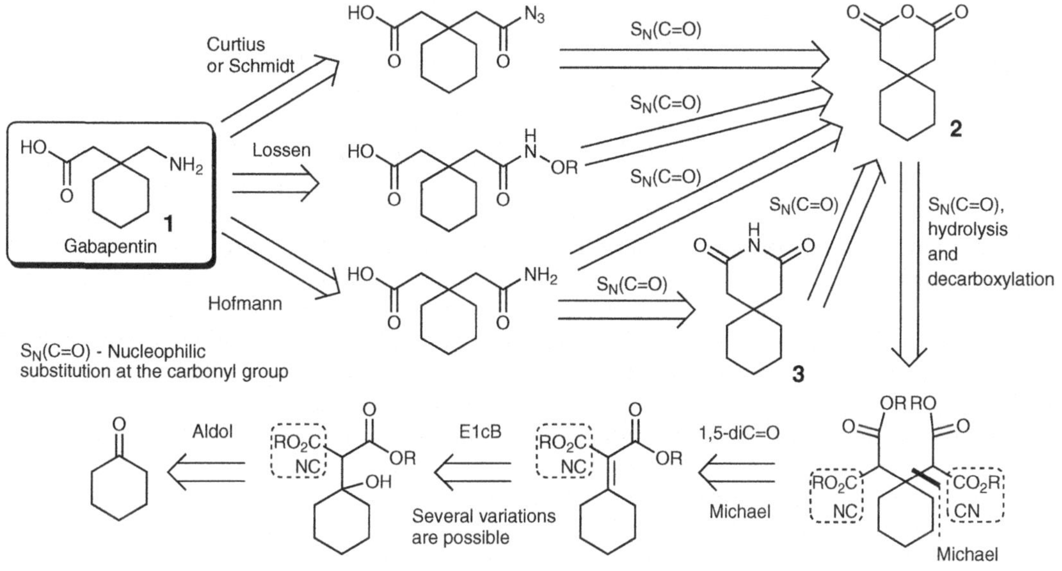

Scheme 2.8 Complete retrosynthetic pathway of Gabapentin to cyclohexanone

Yet another type of rearrangement involving migration to an electron-deficient nitrogen atom can be applied to produce Gabapentin. A retro-Beckmann rearrangement also results in a symmetric intermediate (Scheme 2.9). The Beckmann rearrangement is particularly useful at industrial scale and is extensively utilized in the production of caprolactam, as referred to above, in the production of nylon 6. Interestingly, this reaction is used in API manufacturing in the production of the antibiotic Azithromycin.

A major drawback of this strategy is the limited number of methods to produce the cyclobutanone ring. Addition of ketene to methylene cyclohexane would give the desired cyclobutanone. However, ketene although used on a multi-tonne scale in the production of acetic anhydride, is produced and consumed immediately on site in dedicated facilities because of its inherent instability. In addition, methylenecyclohexane is a problematic starting material and expensive to produce in a pure state because of its instability with respect to its isomer, 1-methylcyclohexene.

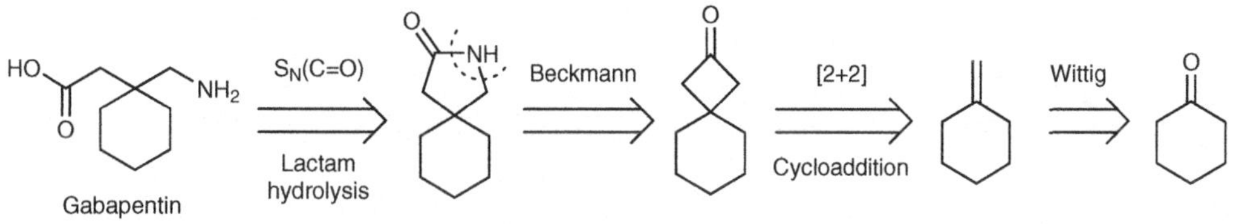

Scheme 2.9 Retrosynthesis of Gabapentin using a Beckmann rearrangement

2.2.3 Avoiding the rearrangement and obtaining the amine by a reduction step

Some of the alternative strategies to the ones that apply a rearrangement as a key step during the synthesis, mandates that two chains with different lengths must be built up thus, the symmetrical intermediates in the previous retrosyntheses cannot be exploited.

When a C1 unit (formally a d1 synthon) is used as nucleophile in a Michael addition, the rearrangement step can be avoided (Scheme 2.10).

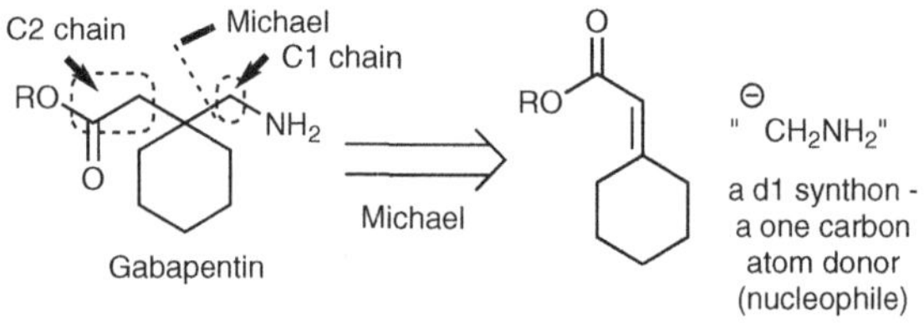

Scheme 2.10 Retrosynthesis of Gabapentin based on a $^{-}CH_2NH_2$ synthon

Cyanide and the anion produced from nitromethane (by base treatment) are useful reactants for the "$^{-}CH_2NH_2$" synthon. The desired primary amino can be conveniently produced from both the nitro and the cyano group by reduction, opening up new synthetic routes to Gabapentin (Scheme 2.11).

There are several variants of this synthetic strategy: the Michael acceptor can contain only one electron withdrawing ester or cyano group or have a second electron withdrawing group (cyano or ester). In this later case, a decarboxylation step is required to obtain the correct substitution. Several successful combinations are expected to be feasible.

2.2.4 Disconnecting only one carbon chain from the cyclohexane ring

Up until this point, all of the retrosynthetic analyses ended up with cyclohexanone wherein the two side chains were detached at different stages of the process. The removal of only one of the carbon side chains from the C6 ring is also an option, resulting in a branched cyclohexane as a plausible starting material.

The primary amine can be obtained from an imine or an oxime, both of which can be produced from an aldehyde. In turn the 1,4-dicarbonyl intermediate can be disconnected using an α-alkylation reaction resulting in cyclohexanecarbaldehyde as starting material (Scheme 2.12).

Other classic methods can be considered to yield the 1,4-dicarbonyl compound: the aldehyde function can be obtained from a nitrile or from a nitro group (Scheme 2.13).

The use of such routes to produce the 1,4-dicarbonyl compound may be troubled by selectivity issues, specifically in the aldehyde formation steps. However, more importantly, since both the cyanoester and the nitroester can give rise to Gabapentin by reduction and hydrolysis (**a** and **b**, Scheme 2.13), their conversion to an aldehyde function seems completely unnecessary, adding more steps to the synthetic route without major benefits and hence, these routes *via* the 1,4-dicarbonyl compound are devoid of interest.

In the previous retrosynthesis (Scheme 2.12), the aldehyde group was used in an α-alkylation to produce the 1,4-dicarbonyl intermediate. An alternative retrosynthetic option is to disconnect the cyanoester intermediate at the α-carbon of the nitrile group. An α-halocarbonyl or an α-haloallylic reagent is then used as electrophile for the alkylations step (Scheme 2.14).

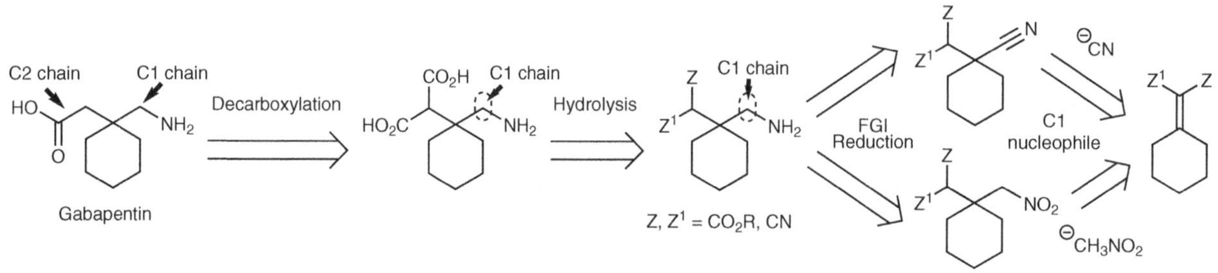

Scheme 2.11 Retrosynthesis of Gabapentin based on a C1 nucleophile

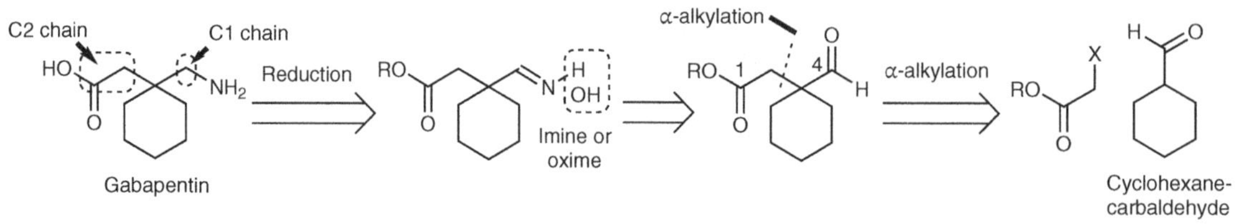

Scheme 2.12 Primary amine by reduction of imine or oxime intermediates

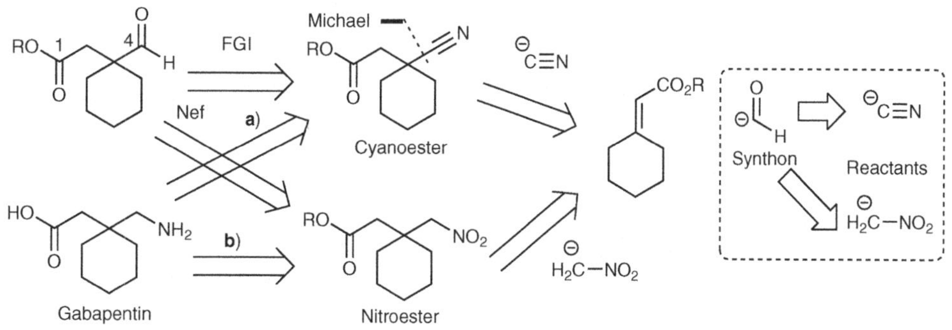

Scheme 2.13 Retrosynthesis of the 1,4-dicarbonyl intermediate

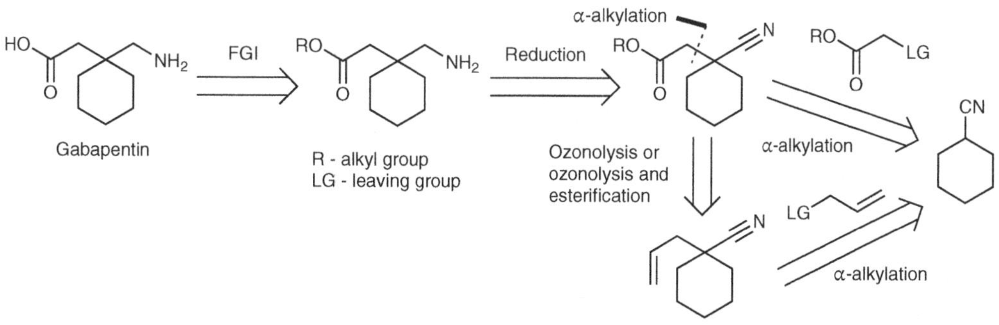

Scheme 2.14 Retrosynthesis of Gabapentin to cyclohexanecarbonitrile

2.2.5 Using an aromatic ring to produce the saturated cyclohexane ring

A cyclohexane derivative can be obtained by total reduction of a suitable phenyl derivative. For example, Birch reduction of benzonitrile can be ingeniously used in the synthesis of Gabapentin by taking advantage of the fact that the initial anionic intermediate formed has a greater negative charge at the carbon atom directly connected to the nitrile group (Scheme 2.15).

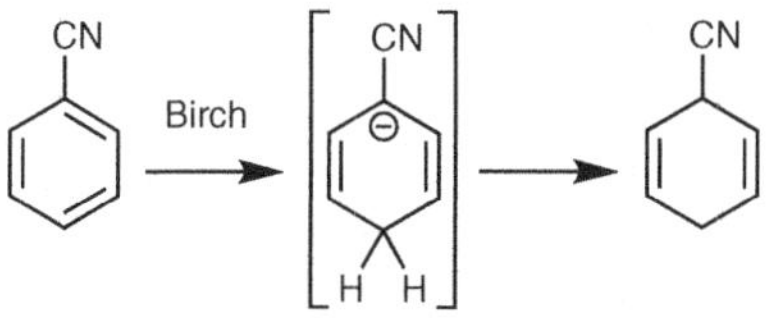

Scheme 2.15 Birch reduction of benzonitrile

A retrosynthesis based on the Birch reduction in the key step is shown in Scheme 2.16.

Although this process looks very simple and uses low cost starting materials, the Birch reduction itself is not straightforward to carry out on an industrial scale, giving rise to safety concerns. In addition, specialised equipment is required for large scale industrial production.

A global scheme for the main retrosynthetic routes of Gabapentin can be found in Scheme 2.17.

Overall, relying on a purely retrosynthetic analysis, several routes can quickly provide Gabapentin. The 1,4-dicarbonyl routes look less promising as they correspond to longer synthetic processes. The nitromethane and cyano routes are short and have the potential to produce Gabapentin quickly and efficiently if safety concerns can be overcome. However, after an in-depth assessment of the various retrosynthetic analyses, routes *via* symmetric intermediates that employ a rearrangement to obtain the correct carbon framework seem to be most preferred.

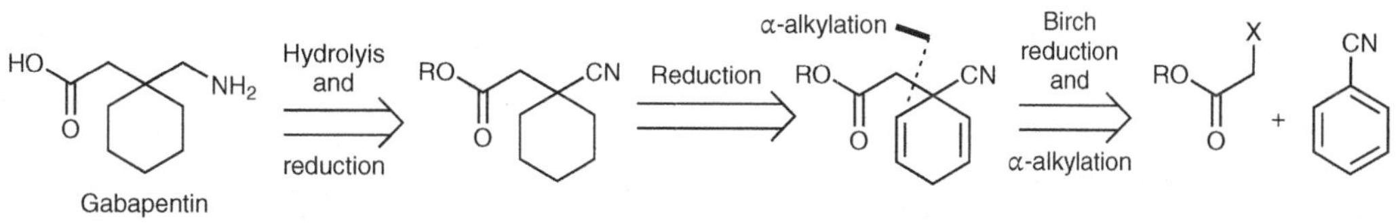

Scheme 2.16 Retrosynthesis of Gabapentin based on a Birch reduction

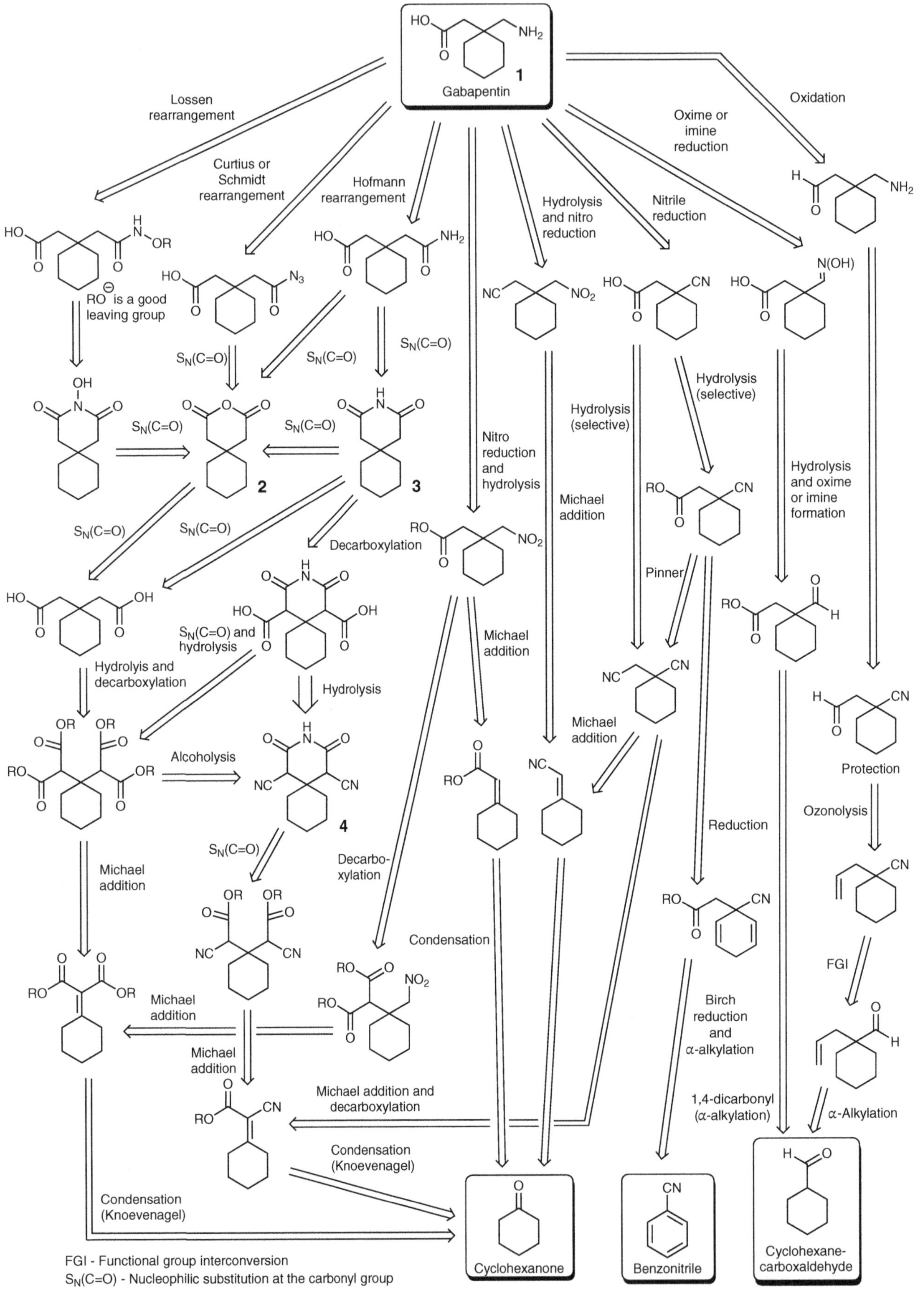

SCHEME 2.17 MAIN GABAPENTIN RETROSYNTHETIC ANALYSIS

2.3 THE FIRST REPORTED SYNTHESIS OF GABAPENTIN

The first synthesis of Gabapentin was reported by Warner-Lambert in 1974 (WarnerL1974).

In the original patent, compounds of general formula **I** were dealt with and at that time, it was not clear that the cyclohexane derivative would be the one to reach the market (Figure 2.8).

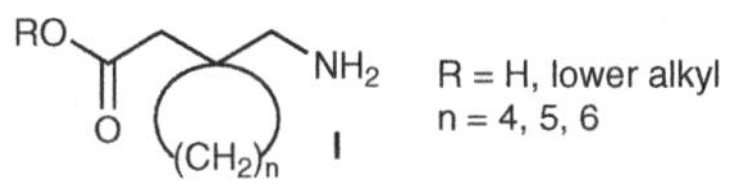

FIGURE 2.8 GENERAL FORMULA FOR COMPOUNDS INCLUDED IN THE ORIGINAL PATENT OF GABAPENTIN

The synthesis of Gabapentin (n = 5) was described in example 1 of the patent. The synthesis starts with the spiro-anhydride intermediate, a Curtius rearrangement being used as key step in the process (Scheme 2.18).

At the time of application of the original Gabapentin patent, product claims were not allowed in some countries (for example, Spain, Portugal and Greece) and only process claims permitted. In order to have a greater county patent coverage, process claims for the Curtius, Hoffman and Lossen rearrangements to produce the compounds of Figure 2.8 were included in the correspondent patent.

Although the examples themselves only included the use of the Curtius rearrangement to produce Gabapentin itself, the Hoffman and Lossen rearrangements were successfully applied to produce the cyclopentane analogue (Scheme 2.19). By analogy, these methods are valid to produce Gabapentin.

As indicated above, the first reported Gabapentin synthesis starts from the cyclic spiro-anhydride (**2**). The synthetic method to produce this compound was already known (Scheme 2.20).

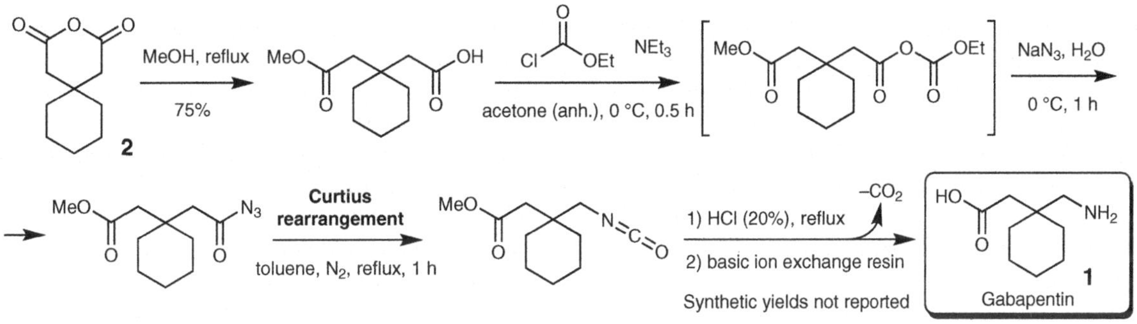

SCHEME 2.18 FIRST REPORTED SYNTHESIS OF GABAPENTIN

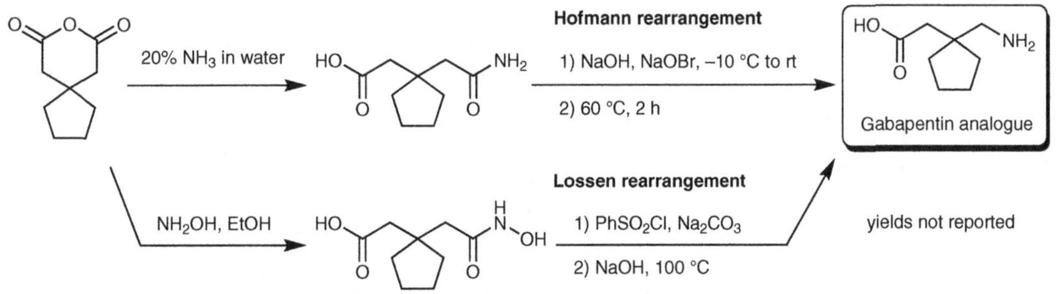

SCHEME 2.19 SYNTHESIS OF A GABAPENTIN ANALOGUE BY A HOFMANN OR LOSSEN REARRANGEMENT

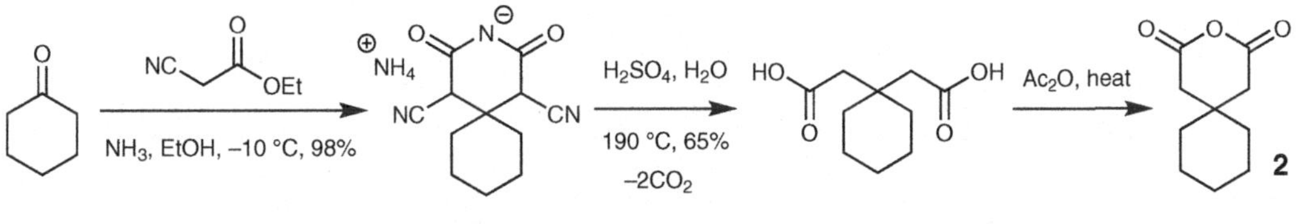

SCHEME 2.20 SYNTHESIS OF THE SPIRO-ANHYDRIDE INTERMEDIATE

Mechanistically the Lossen, Curtius, Hofmann and Schmidt rearrangements, the viable alternatives to carry out the critical rearrangement, are closely related, each transformation occurring *via* a migration to an electron deficient nitrogen atom (Scheme 2.21). Highly unstable intermediates are common to all these processes and safety may be a concern for implementation at industrial scale.

2.4 THE EVOLUTION OF THE CHEMICAL SYNTHESIS OF GABAPENTIN

Given the commercial success of Gabapentin, it is no surprise that a great deal of further work was carried out by various process development groups to improve on the synthesis. The evolution of these are described below.

2.4.1 Initial development of synthetic routes to Gabapentin using a rearrangement step

The synthetic routes to produce Gabapentin from cyclohexanone that use, as key in the process, a rearrangement step are summarised in Scheme A (the rearrangements are highlighted by a thicker reaction arrow).

As noted earlier, the Curtius rearrangement was applied in the first described Gabapentin synthesis (WarnerL1974). This reaction is fairly general and can be applied to carboxylic acids containing a wide range of functional groups. The acyl azide is usually produced by reaction of nucleophilic azide anion with an activated carboxylic acid derivative (Scheme 2.21).

Industrially the Curtius rearrangement gives rise to various safety concerns. The azide reactants required to produce the acyl azide are toxic, unstable and require careful handling. Sodium azide is a severe poison, its toxicity being comparable to that of alkali cyanides. Hydrazoic acid is also extremely toxic and easily formed by contact of sodium azide with strong acids.

The acyl azide intermediates, which undergo the key rearrangement, are inherently unstable in contact with acid, vigorously decomposing and exploding when heated (ACIE2005, PSP2004).

Because the rearrangement is triggered by heating the acyl azide any large batch scale synthesis requires special conditions and precautions.

During the conversion substantial (stoichiometric) amounts of nitrogen and carbon dioxide are produced whose release can cause problems when the reaction is carried out at large scale. Uncontrolled foaming and mass transfer of the reaction mixture out of the reaction vessel or even tank rupture are possible. Obviously, these issues have to be dealt with before any process can become an industrial reality.

In the absence of a simpler and safer alternative, implementation of the Curtius rearrangement at industrial scale implies that extensive safety tests must be undertaken (see for example OPRD1998). Instead of having all of the reactant in one pot and heating the reaction mixture, a more controlled and safer procedure is to add the dissolved acyl azide to a sufficiently large volume of a pre-heated neutral solvent such as toluene. This procedure limits the quantity of acylazide being heated at any one time to a minimum and hence, the potential for explosion or uncontrolled de-gassing is held to a minimum.

A flow chemistry approach is yet another possibility to carry out the Curtius rearrangement. Examples where flow chemistry has been applied to similar reactions can be found in the literature (JFC2012, OBC2008a,b,

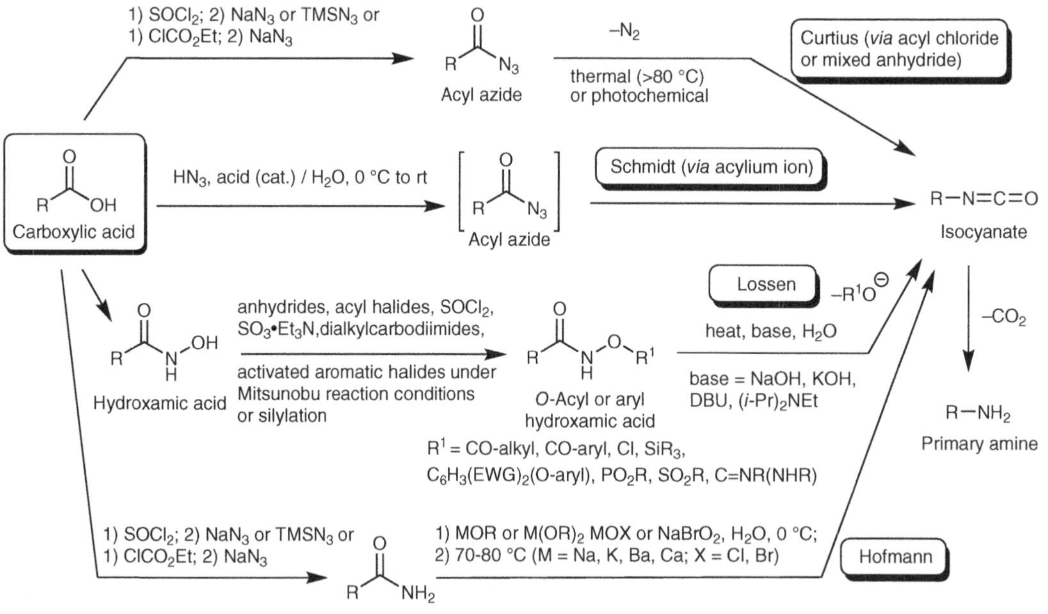

SCHEME 2.21 GENERIC REACTION CONDITIONS FOR THE CURTIUS, SCHMIDT, LOSSEN AND HOFMANN REARRANGEMENTS

EJOC2011, GC2017). The release of gases still must be contemplated and dealt with.

Given the hazardous nature of the acyl azide intermediate, an alternative synthetic solution free of such issues would be ideal.

The Schmidt reaction is not an appealing alternative to the Curtius rearrangement as the same unstable acyl azide is also the intermediate in this reaction. In addition, highly explosive free hydrazoic acid (OPRD2006), instead of an azide salt, is used to carry out the preparation of the acyl azide, in this case by addition to the free carboxylic acid.

A synthesis of Gabapentin using a Lossen rearrangement, which avoids the acyl azide and the use of the azide anion or hydrazoic acid, was disclosed in 1976 (WarnerL1976, Scheme 2.22).

Hydroxamic acids themselves do not undergo the Lossen rearrangement and activation of the oxygen atom is necessary to achieve conversion to a much better leaving group to facilitate the rearrangement. Generally, the Lossen rearrangement can be performed under mild reaction conditions giving this methodology a clear advantage over the Curtius rearrangement. Heating to reflux in a basic aqueous solution is required. Initially, ring opening takes place with formation of the unstable alkali *O*-acyl hydroxamate salt. This is followed by a rapid concerted rearrangement to the isocyanate *via* a bridged anion (compare the mechanism for the Hoffman rearrangement given below, Scheme 2.23). As in the Curtius rearrangement, concerns about the instability of the intermediates made a detailed thermochemical study mandatory prior to scale-up.

The Hofmann rearrangement is the last option in this group of related reactions. It was successfully applied to synthesis a cyclopentane analogue of Gabapentin in the original "discovery" patent. The yield was not reported but the rearrangement was carried out by slow addition of a cold basic solution of primary amide to an alkali hypobromite or hypochlorite. The resulting solution was then heated to initiate the rearrangement. Freshly prepared solutions of hypohalite are preferred for this reaction and are produced by the addition of chlorine gas or bromine to an aqueous solution of sodium or potassium hydroxide. The first step in the sequence is the formation of an *N*-halo substituted amide followed by deprotonation to the corresponding salt. This salt is quite unstable and quickly undergoes a concerted rearrangement to the isocyanate *via* a bridged anion.

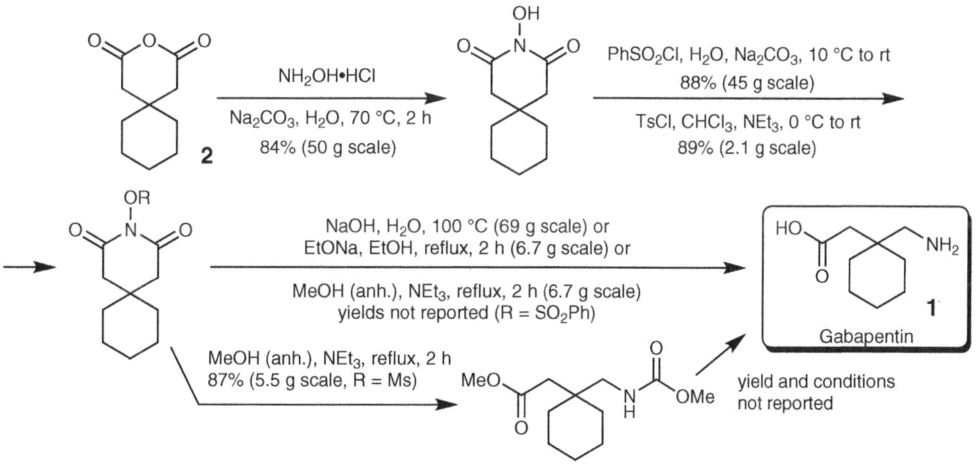

SCHEME 2.22 GABAPENTIN SYNTHESIS *VIA* A LOSSEN REARRANGEMENT STEP

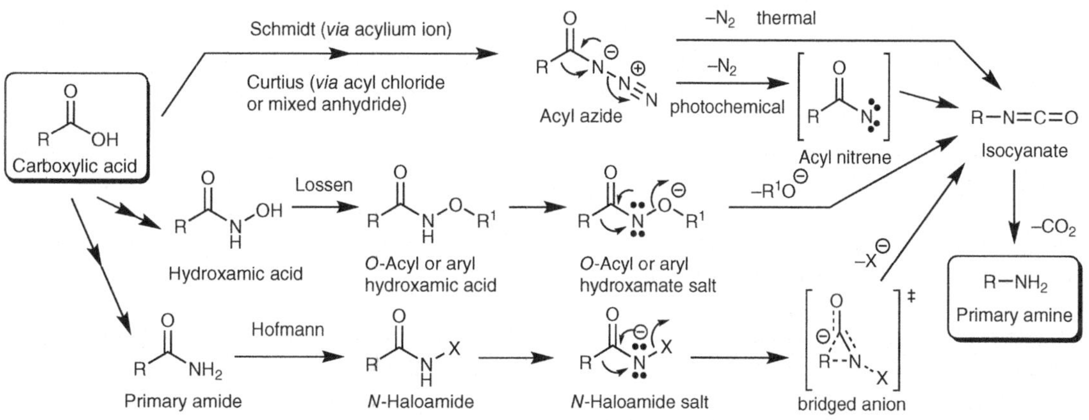

SCHEME 2.23 MECHANISM OF THE CURTIUS, SCHMIDT, LOSSEN AND HOFMANN REARRANGEMENT STEP

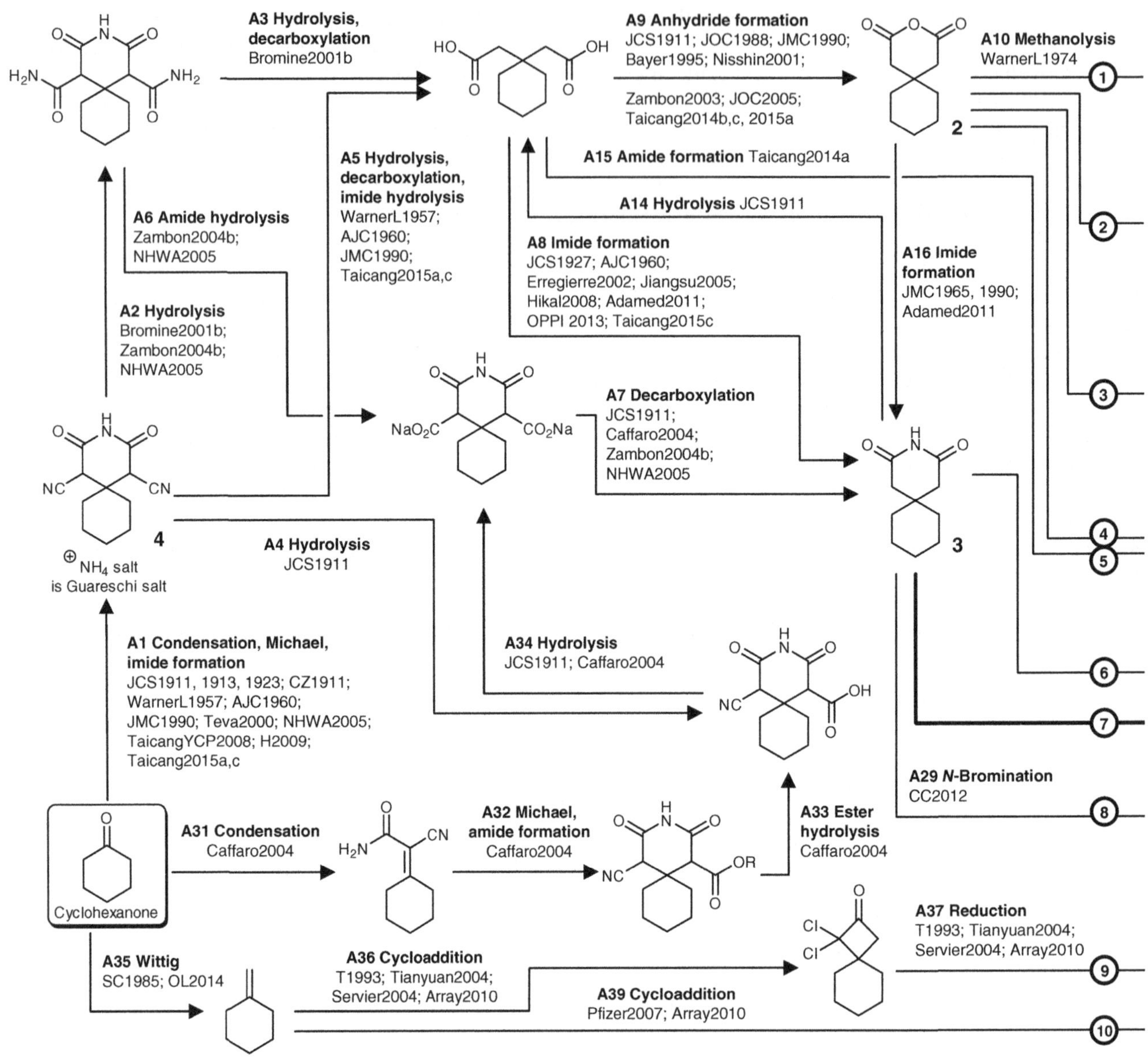

Scheme A Synthesis of Gabapentin using a rearrangement as key step (for reaction conditions see page page 31)

Once more, as for the other rearrangements, all consisting of a multistep reaction having highly energetic intermediates, extensive safety studies are required before scaling up. The Hofmann rearrangement does however, have some features that enables easier scale-up of the process, which becomes obvious after the ubiquitous process development study:

- Hypohalite is cheap and easily prepared. Development studies can usually establish conditions for utilization of commercial material.
- Formation of the *N*-halo intermediate is usually fast at room temperature.
- Only gentle heating of the *N*-halo intermediate is required for the rearrangement to take place and a rapid reaction is observed.

Appropriate control of the reaction conditions is anticipated to circumvent safety issues and allow the use of the reaction at industrial scale.

In conclusion, the Lossen, Curtius and Schmidt rearrangements are to be avoided if possible. The Hofmann rearrangement can be regarded as the most promising approach for implementation on an industrial scale but, as always, extensive safety studies have to be carried out before running even scale-up batches.

2.4.2 Routes to the spiro-anhydride or spiro-imide intermediates

Spiro-anhydride (**2**) or spiro-imide (**3**) derivatives are important intermediates in the synthesis of Gabapentin.

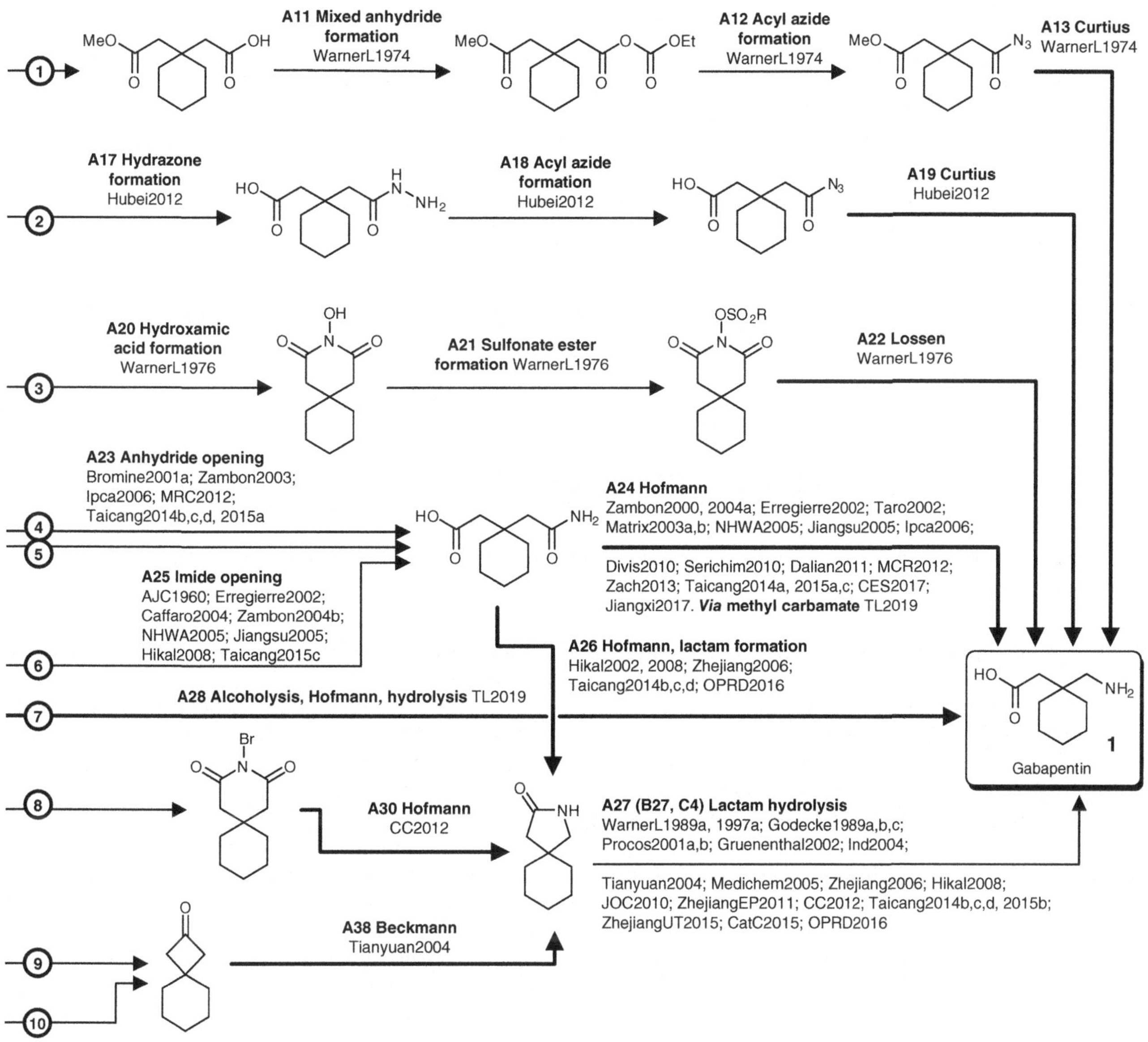

The synthesis of both had been reported well before the discovery of Gabapentin but several improved processes for their industrial preparation have since been disclosed (Scheme A).

In the first reported approach, condensation of cyclohexanone with ethyl cyanoacetate in the presence of ammonia (**A1** and Scheme 2.24) produces the ammonium salt of **4**, known as the Guareschi salt. A Knoevenagel condensation between cyclohexanone and ethyl cyanoacetate followed by a Michael addition of a second molecule of ethyl cyanoacetate produces a bis-adduct, which is then converted into the heterocyclic imide ammonium salt.

The initial process uses anhydrous conditions and requires the use of a large molar excess of gaseous ammonia. This, along with the fact that the reaction usually takes from 48 to 168 hours, stimulated the development and patenting of several alternative conditions to produce the Guareschi salt more efficiently and in a shorter time.

There was also activity focussed on different processes to convert the Guareschi salt into **2** and **3** resulting in several new patents (Scheme 2.25 and Scheme A, top-left).

These transformations require either strongly acidic and/or basic conditions to affect hydrolysis. But, common to all is the subsequent decarboxylation reaction, which releases copious quantities of carbon dioxide. Gas release and foaming is a real safety issue when carrying out large scale production batches. Heating the Guareschi salt in aqueous acidic conditions (**A5**, Scheme A) results in direct formation of the dicarboxylic acid but the rate of carbon dioxide release is impossible to control. Hence, managing the release of gas is a requirement for optimum operation at scale. Controlled hydrolysis of the cyano groups, *via* the amides, to a dicarboxylate salt, followed by decarboxylation is a good option. These conditions allow for controlled hydrolysis followed by production and release of carbon dioxide gas by precise addition of acid at a pre-defined temperature. An outline

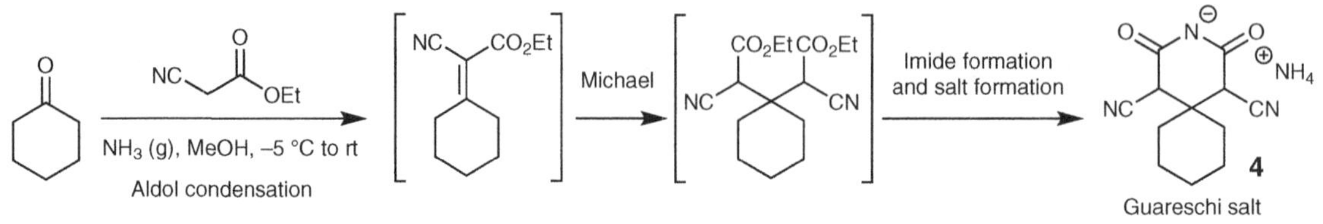

Scheme 2.24 Synthesis of the Guareschi salt from cyclohexanone

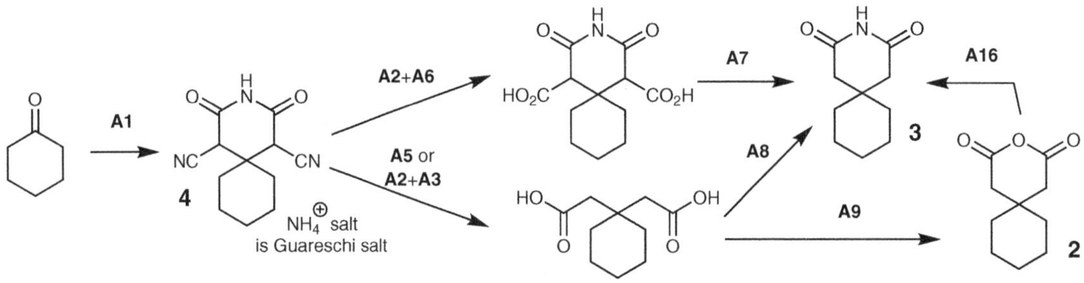

Scheme 2.25 Outline of the synthetic routes to spiro anhydride (2) and spiro imide (3)

of these routes is summarized in Scheme 2.25 and A (top/left).

The intermediates formed along the route to the spiro-imide and anhydride can be isolated, thus allowing purification steps to be introduced in the process. Not only are structurally related compounds removed but also acidic/basic residues and inorganic salts. The removal of impurities at this stage is important as they can have impact on the final conversion to Gabapentin. For example, acid trapped in the spiro intermediates can affect subsequent reactions, especially the nucleophilic ring opening of the anhydride or imide.

2.4.3 Synthetic routes of Gabapentin using a one carbon atom nucleophile (Scheme B)

Although not explicitly expressed in the patent literature, it can be inferred that safety concerns regarding the rearrangement step, in addition to the length of the then existing routes, induced several companies to develop new syntheses of Gabapentin (from 1988, Scheme B). Addition of a nucleophile with one less carbon atom (a C1 unit, formally a d1 synthon) in the Michael addition step results in an intermediate wherein the rearrangement step is avoided. Cyanide and the carbanion from deprotonated nitromethane are two good options for the nucleophile.

Knoevenagel condensations (Scheme B, **B1**, **B11**, **B35**) and Wittig-like carbonyl olefinations (Scheme B, **B21**, **B28**) are the main routes to produce the Michael acceptors for the conjugated addition (Scheme 2.26).

2.4.3.1 Using cyanide as nucleophile

Cyanide is a perfect choice for the required one carbon atom nucleophile and hence it is of no surprise to learn that several syntheses use this as a d1 synthon. Once in place the nitrile moiety can give the aminomethyl side chain directly by reduction.

Conjugate addition of cyanide introduces the carbon and the nitrogen atoms of the final CH_2NH_2 side chain of Gabapentin (Scheme 2.27). Although, in principle, electron-deficient double bonds containing either one or two electron-withdrawing groups can be used for the Michael addition, the patented processes to Gabapentin and related substances exclusively employ Knoevenagel condensation products as acceptors for the 1,4-addition, perhaps because these are most viable in an industrial scenario. The extra carbon is removed by the classical hydrolysis-decarboxylation process. Reduction of the remaining nitrile group gives the Gabapentin aminomethyl side chain.

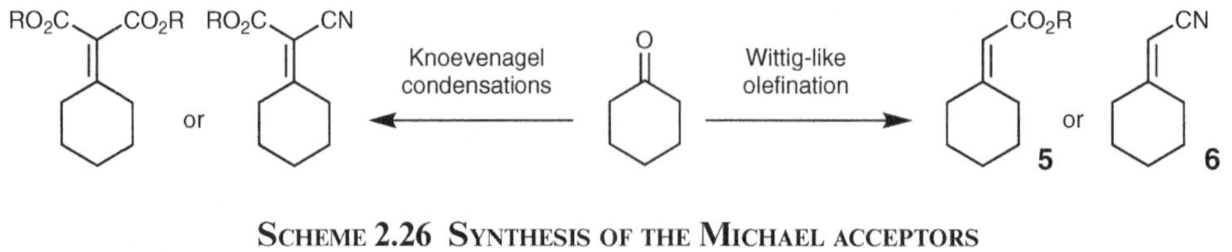

Scheme 2.26 Synthesis of the Michael acceptors

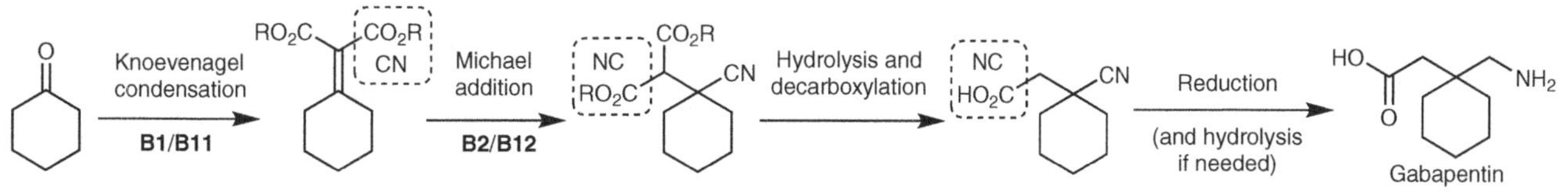

Scheme 2.27 General synthetic route using a Knoevenagel condensation-Michael addition of cyanide sequence

Based on this reaction scheme, various processes were developed and patented (Scheme B). Generally, ketones are less reactive than aldehydes as electrophilic species in a Knoevenagel condensation (OR1967) whilst malononitrile is considered to be the most reactive methylene compound for the transformation. The yields and rates of the reactions are influenced by steric factors. In the present case, the use of a cyano acetic acid derivative gives highest yields and seems preferred. This was revealed in a case-study indicating that the use of a cyano acetic acid derivative gives highest yields and is keeping with the observation that the cyano group activates the methylene best. The Knoevenagel condensations of cyclohexanone and malonic acid derivatives are reported to give yields below 62% (**B1**), whereas the use of cyano acid derivatives produce yields above 75% (**B11**). It should be noted that the Knoevenagel condensation with a cyano acetic acid ester is the first step in the two-step one-pot process to the Guareschi salt, **4** (**A1**, reported yields around 85%). Reported conditions for the Knoevenagel reactions with malonic ester require two equivalents of titanium tetrachloride and four equivalents of pyridine, making any process based on this procedure rather costly. Moreover, conditions which use stoichiometric amounts of heavy metal salts can create issues with effluent disposal and the large amounts of TiO_2-pyridine complex produced require additional treatment before disposal.

A. Using a malonic diester in the formation of the Michael acceptor

When a cyclohexylidene malonic acid ester is chosen as the Knoevenagel condensation product then conjugate addition of the cyanide moiety (**B2**) can be effected with hydrocyanic acid in the presence of a catalytic amount of alkali cyanide or with an excess of alkali cyanide (up to 5 equivalents of alkali cyanide).

Acidic conditions can be used in a pressured reactor but large excess of hydrocyanic acid is required to achieve good yields (Lonza1988). Extremely basic conditions were not viable as several side products were observed and low yields reported (27%). However, by controlling the pH at 10.5-12 excellent yields were obtained (88-89%, Lonza1988, HCA1991).

Of course, use of highly toxic cyanide salts or hydrocyanic acid creates safety concerns and requires special condition, especially at industrial scale where large amounts are in play.

Three further transformations are thereafter required convert the Michael adduct into Gabapentin: (di)ester hydrolysis, decarboxylation and nitrile reduction.

One option is to reduce the nitrile group first under conditions where the ester groups are not affected. The end product of this reaction is a lactam formed by intramolecular ring closure (Scheme 2.28, route **a**). Hydrolysis

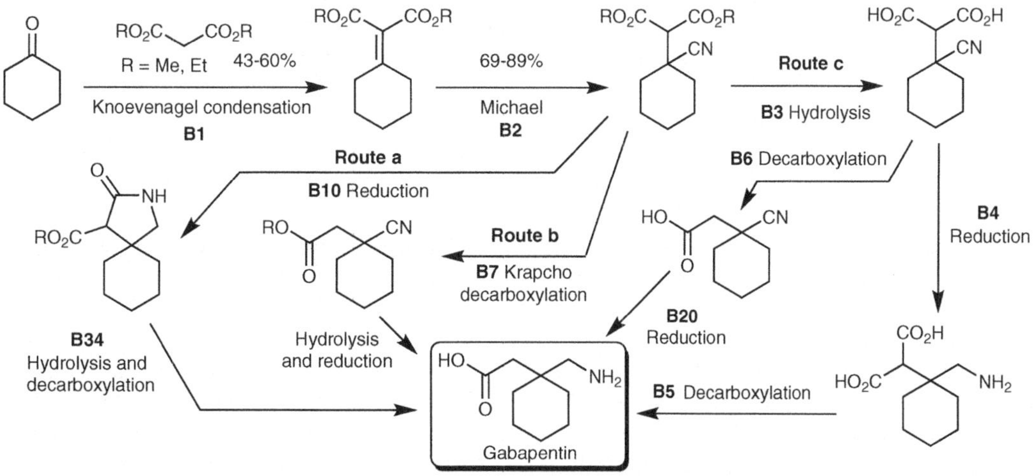

Scheme 2.28 Routes to Gabapentin by cyanide addition to a cyclohexylidene malonic acid ester derivative

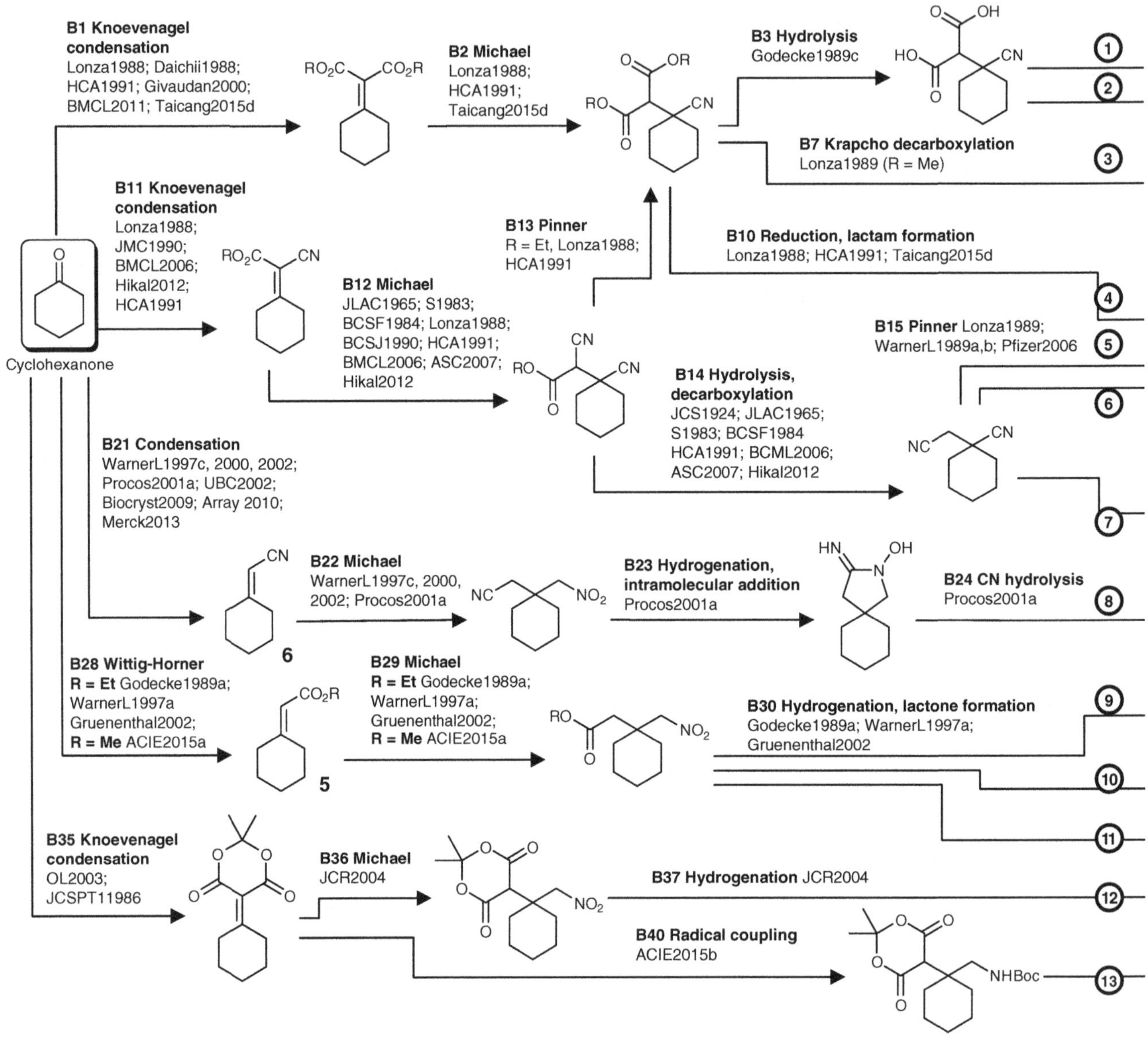

Scheme B Synthesis of Gabapentin using a conjugate addition as a key step (for reaction conditions see page page 32)

and decarboxylation by heating in an acidic medium provides Gabapentin, isolated as an ammonium salt. Once again, carbon dioxide gas liberation must be controlled to avoid industrial hazards.

Other alternative sequences are: Krapcho decarboxylation (which has an additional issue with the release of toxic methyl chloride), hydrolysis and reduction (route **b**); hydrolysis of the esters, decarboxylation and reduction; or hydrolysis of esters, nitrile reduction and decarboxylation (**c** routes). Other variations are possible and are illustrated in Scheme B.

All of these routes suffer from the fact that the first synthetic step, the Knoevenagel condensation, has a low yield.

B. Using a cyanoacetic acid ester in the formation of the Michael acceptor

When cyanoacetic acid ester is used in the Knoevenagel condensation the reaction is more facile and good yields are achieved (**B11**). An additional bonus is that the Michael reaction also proceeds more smoothly than with the malonic acid derivative. Potassium cyanide in boiling ethanol or hydrocyanic acid in the presence of a catalytic amount of base gives the dicyano intermediate in good yield (**B12**, 82-94% Lonza1988).

Fortunately, the two cyano groups, when present in the same molecule, can be differentiated and regioselective attack at the more accessible cyano group results in

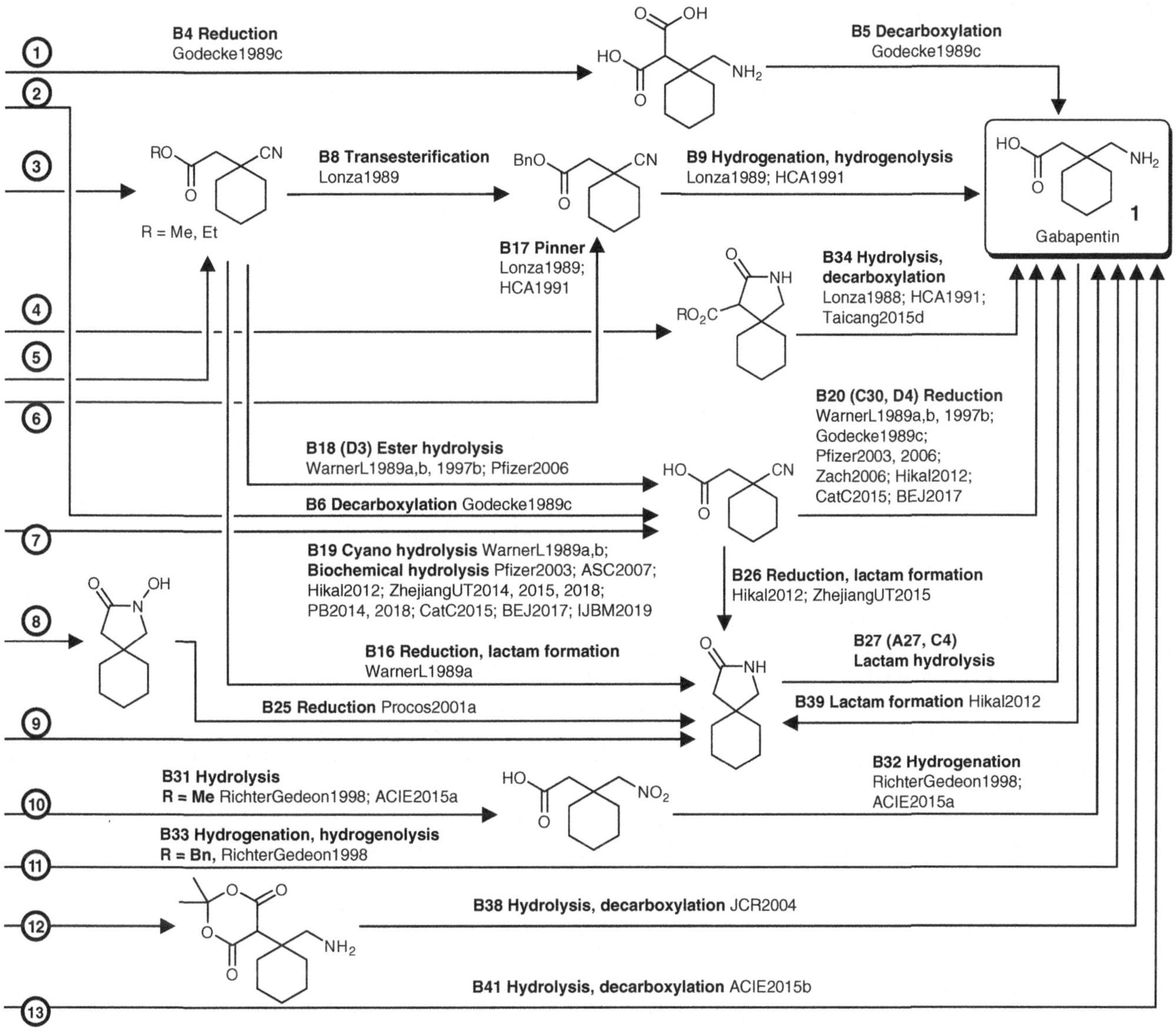

the same intermediates obtained in the malonate ester sequence. A Pinner reaction is then used on the different intermediates (**B13**, **B15** and **B17**) to transform the less hindered cyano group into an ester (Scheme 2.29).

When a benzyl ester is generated in one of the Pinner reactions (**B17**), the ester can be converted into the carboxylic acid and the nitrile into the methylamine group simultaneously, by hydrogenolysis/hydrogenation using a rhodium catalyst. Reported yields are, however, somewhat low (11 to 60% for the Pinner reaction).

As mentioned above, all of these routes require the use of cyanide as nucleophile, which is particularly toxic and may create issues with handling and effluent treatment. This may well mandate the rejection of these routes if satisfactory safety measures cannot be implemented.

2.4.3.2 Using nitromethane as reactant

The anion produced by base treatment of nitromethane (pKa~10) is the other obvious choice of a d1 synthon as this carbanion is a good nucleophile. Hydroxide is a sufficiently strong base to deprotonate nitromethane. The easy conversion of the nitro group to a primary amine by hydrogenation makes nitromethane a good synthetic equivalent to the $^{-}CH_2NH_2$ synthon.

Not surprisingly, several routes that use nitromethane have been patented (Scheme B, lower part). The nitromethane anion, which is stabilized by resonance, is particularly apt for conjugate addition and can add to Michael acceptors that incorporate only one electron-withdrawing group (ester or cyano). A possible side reaction is cyclopropanation (Scheme 2.30). This is less likely and, largely avoided when only one electron-withdrawing group is present. Hence, only this option has been the focus for patenting.

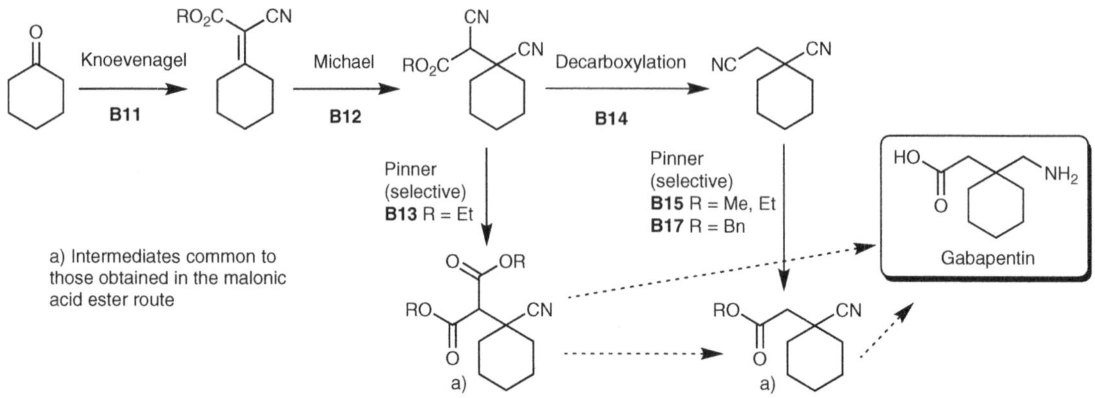

Scheme 2.29 Routes to Gabapentin by cyanide addition to a cyclohexylidene cyanoacetic acid ester derivative

After Michael addition, reduction of the nitro group to amine provides Gabapentin-lactam by ring closure at the carbonyl group. This route is very attractive as it offers a quick synthesis of the API (Scheme 2.31).

There are however, certain precautions that have to be taken. The use of strongly basic conditions can provoke deprotonation of the active nitro-methylene group, which in turn could result in the carbanion participating in various side reactions (both inter- and intra-molecular) with electron acceptors presents in the mixture.

This nitromethane route was reported in a continuous small-scale process procuring Gabapentin from cyclohexanone in a 49% overall yield (0.3 g scale, ACIE2015a).

2-Cyclohexylideneacetonitrile (**6**) can be used in a similar sequence, even though hydrolysis of the cyano group requires more vigorous conditions to accomplish. The facile reduction of the nitro group under conditions that do not affect the nitrile function has been used in an ingenious strategy by Procos, wherein the hydroxylamine intermediate, formed during the reduction of the nitro group, is used to assist concomitant hydrolysis of the cyano group (Scheme 2.32, Procos2001a).

Successful Michael addition of the nitromethane anion to the condensation product of Meldrum's acid and cyclohexanone was reported (**B36**) as a key step in another laboratory route to produce Gabapentin (JCR2004).

None of the routes mentioned in this section are very lengthy and this factor probably has considerable weight when making decisions on route selection. However, the use of nitromethane, as discussed in the section on route selection, in a similar fashion to the use of cyanide, can be major drawback.

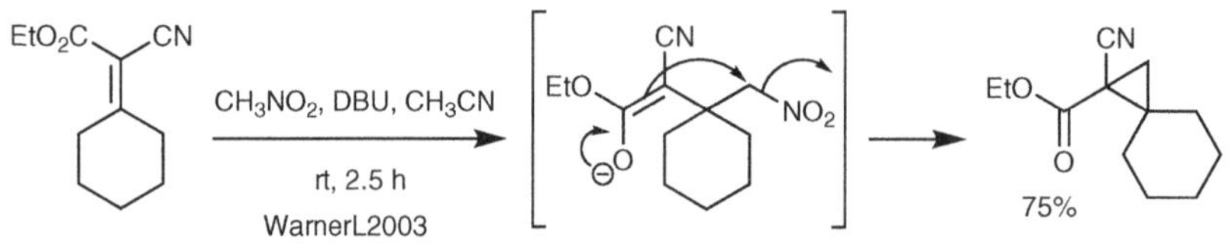

Scheme 2.30 Cyclopropanation side-reaction in the Michael addition step

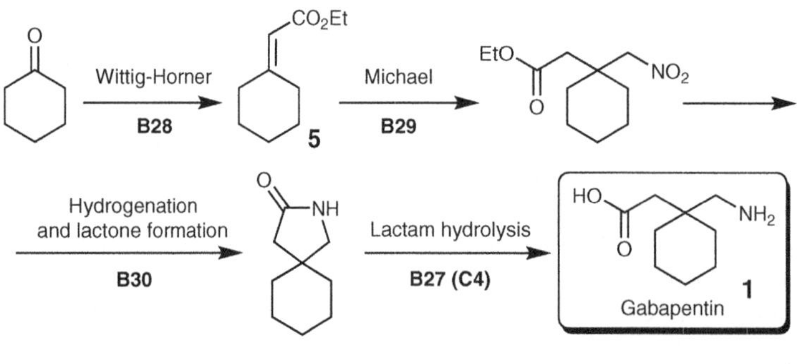

Scheme 2.31 Gabapentin synthesis by conjugate addition of the nitromethane anion to ethyl 2-cyclohexylideneacetate (5)

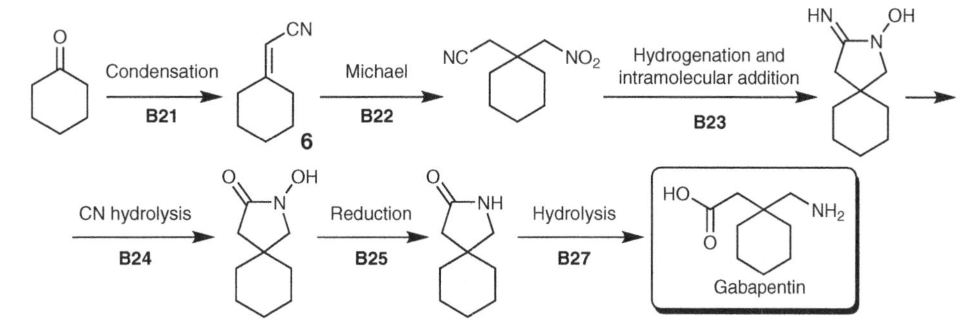

Scheme 2.32 Gabapentin synthesis by conjugate addition of the nitromethane anion to 2-cyclohexylideneacetonitrile

2.4.3.3 A radical coupling approach

A new radical-based coupling method has been developed for a single-step generation of various γ-amino acids using α-aminoacyl tellurides as reactants (ACIE2015b). Upon activation by Et_3B and oxygen at room temperature, the α-aminoacyl tellurides were readily converted into α-amino carbon radicals by decarbonylation, which then react with the CC double bond, as indicated in Scheme 2.33. After hydrolysis, decarboxylation and Boc removal, Gabapentin was obtained in 52% overall yield for the two steps.

2.4.4 The 1,4-dicarbonyl intermediate routes

Several other synthetic routes start form cyclohexanecarbaldehyde (Scheme C) wherein the aldehyde can be used to generate the required amine *via* formation of an imine or oxime followed by reduction (Scheme 2.34). The other side chain is attached making use of the nucleophilic nature of the α-carbonyl carbon. A 1,4-dicarbonyl compound is the key intermediate in this methodology and is an example of a classical laboratory approach to the synthesis of Gabapentin.

The concerns centred around the various rearrangements, and the use of cyanide and nitromethane as reactants are not present in this route.

The first step is a classic α-alkylation of an aldehyde using enamine chemistry (**C1, C14**). After the formation of the enamine, which requires water removal by azeotrope distillation, alkylation and hydrolysis results in the formation of the 1,4-dicarbonyl compound. Conversion of the newly formed aldehyde to the amine can be achieved *via* oxime or imine intermediates.

One disadvantage of this technique is the explosive nature of hydroxylamine; it is a suspected carcinogenic and can be hazardous to the environment. Hence, its use in any industrial process should be avoided where viable alternatives exist. The alternative, reductive amination of an aldehyde requires the use of ammonia to produce the

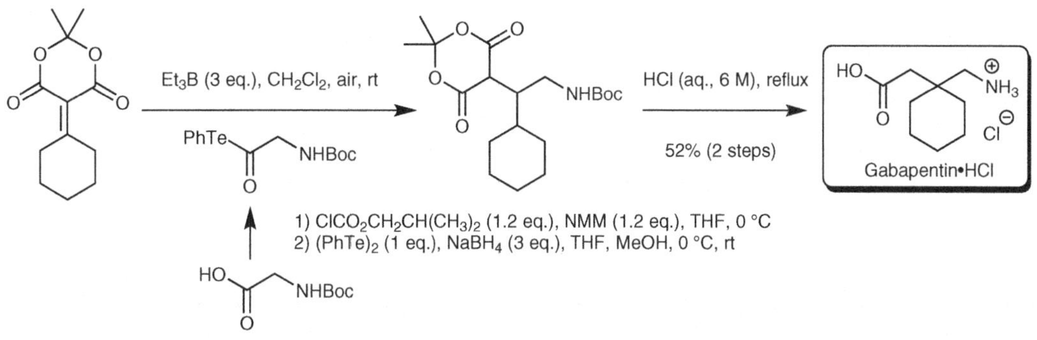

Scheme 2.33 A radical coupling approach to Gabapentin

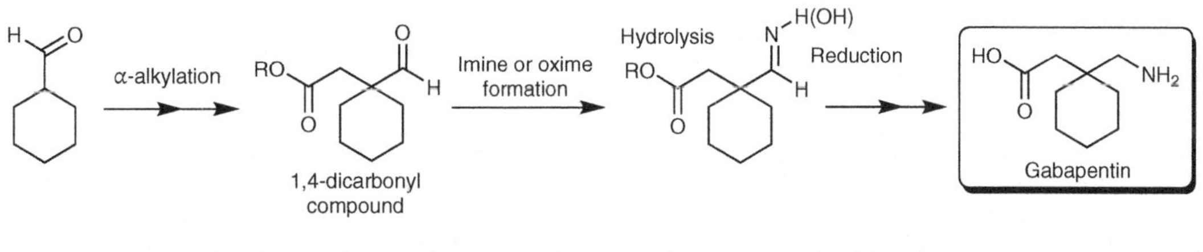

Scheme 2.34 Outline of Gabapentin syntheses *via* a 1,4-dicarbonyl intermediate

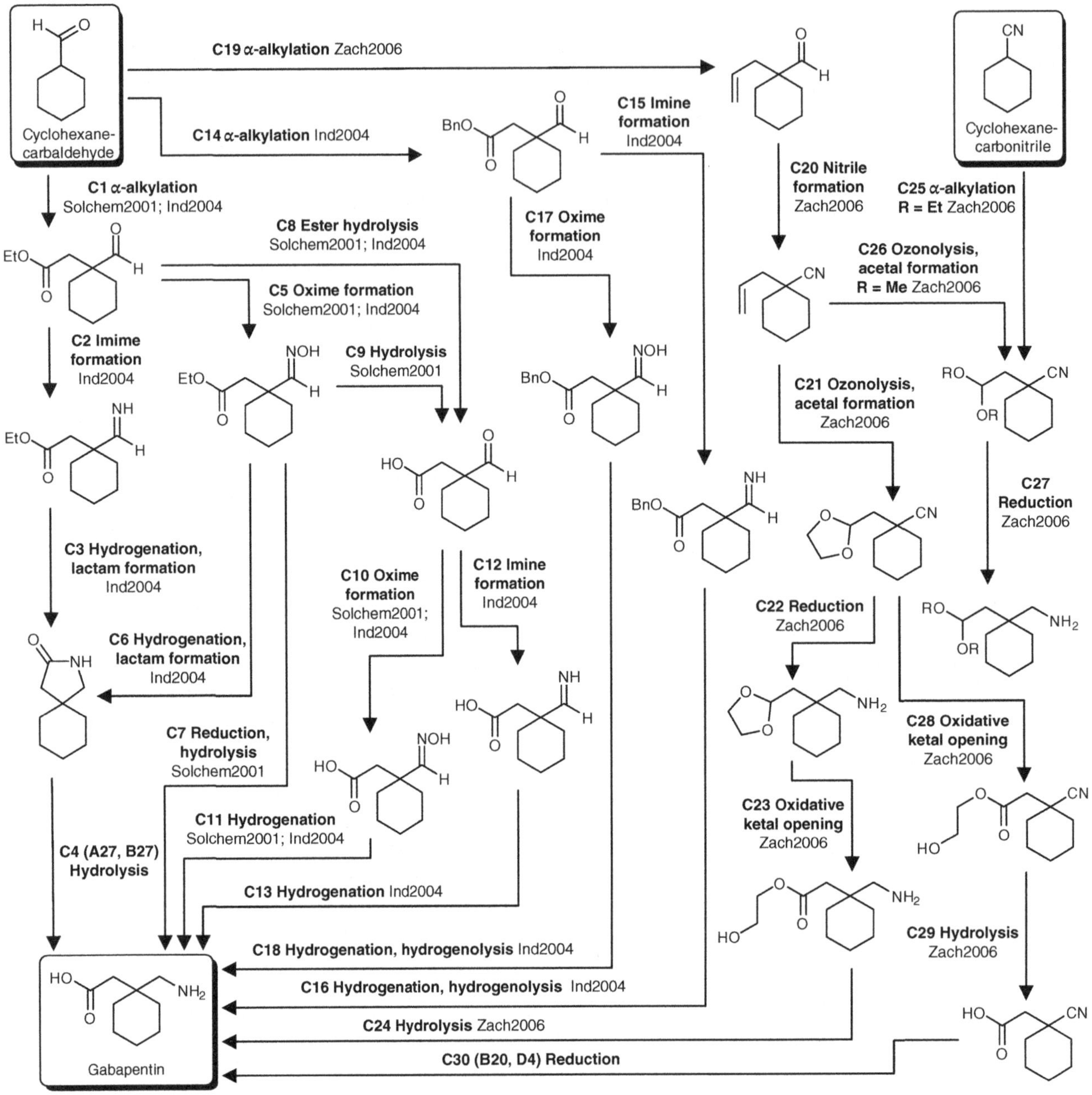

Scheme C Synthesis of Gabapentin based on a α-alkylation as a key step (for reaction conditions see page page 33)

unstable unsubstituted imine. Once formed, the primary amine can in turn form imines, which are reduced giving rise to secondary amines, and so on. This cascade terminates with the formation of quaternary ammonium salts and mixtures are to be expected unless appropriate conditions are established. Typical reaction conditions use ammonium acetate as a source of ammonia and sodium cyanoborohydride or triacetoxyborohydride as reducing agent at a controlled pH. Note that cyanoborohydride is a source of cyanide or hydrocyanic acid during reaction work-up. The alternative, catalytic hydrogenation to reduce the imine or oxime is preferable, although the use of hydrogen under pressure is not always a straightforward matter on a large scale.

In what appears to be a mere academic exercise, more recently accounts report alternate synthetic processes starting from cyclohexanecarbaldehyde and an allylic substrate (see Scheme C, right side and Scheme 2.35). Although reported yields for the various reactions in the sequence are generally high, these routes are longer than some of those previously reported and require a hazardous ozonolysis step, which requires specialized equipment and conditions for implementation on an industrial scale.

The ozonolysis reaction and conversion of the aldehyde into a cyano group can be avoided when cyclohexanecarbonitrile is alkylated directly with chloroacetaldehyde dimethyl acetal (**C25**). Alkylations at the α carbon to a cyano group are possible but require very strong bases

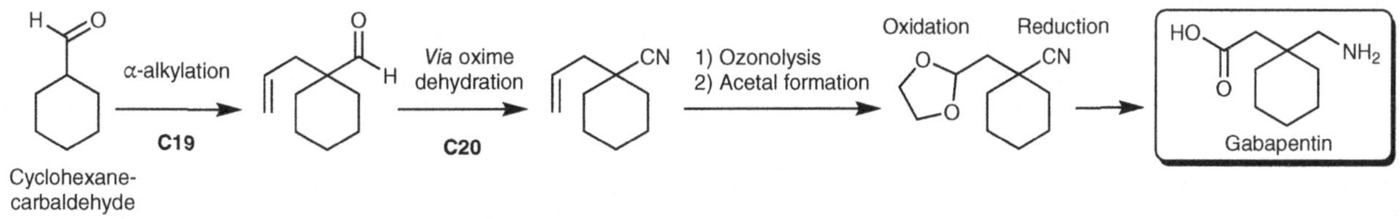

SCHEME 2.35 OUTLINE OF SYNTHETIC ROUTES OF GABAPENTIN STARTING BY CYCLOHEXANECARBALDEHYDE ALKYLATION WITH ALLYLIC HALIDE

(BuLi, NaH or similar), which in turn require specialized handling and equipment and low temperatures. The highly flammable off-gases, butane and hydrogen, also present a challenge.

2.4.5 Synthetic processes based on a Birch reduction

Warner Lambert developed a completely different synthetic methodology. In this approach, an aromatic phenyl ring is converted to a saturated cyclohexane ring (Scheme D, Scheme 2.36). Here, a Birch reduction is employed in the key step and permits a rapid access to Gabapentin. Cumulative yields of up to 73% were reported for the Birch reduction and nucleophilic substitution at a 10 g scale, which would be expected to be much improved during process development work.

Birch reductions are carried out at low temperatures (–78 °C) using lithium metal in liquid ammonia in the presence of an alcohol. This dissolving metal reduction results in a 1,4-cyclohexadiene anion which is subsequently alkylated *in situ* to give the key intermediate. The use of liquid ammonia and lithium metal, along with temperature of –78 °C, are rather demanding in an industrial setting, thus making these particular conditions, yet again, not highly favoured.

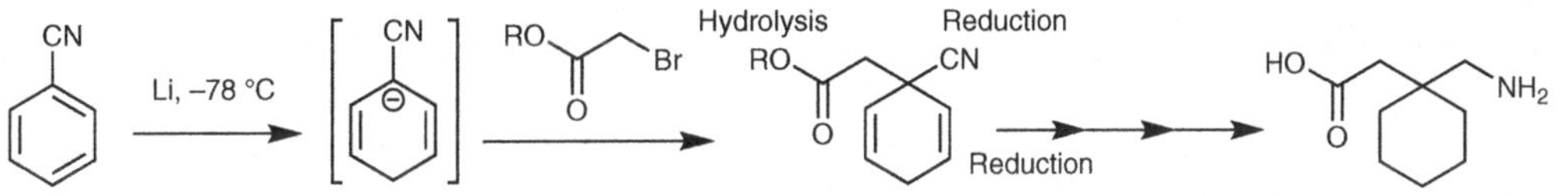

SCHEME 2.36 A BIRCH APPROACH TO GABAPENTIN

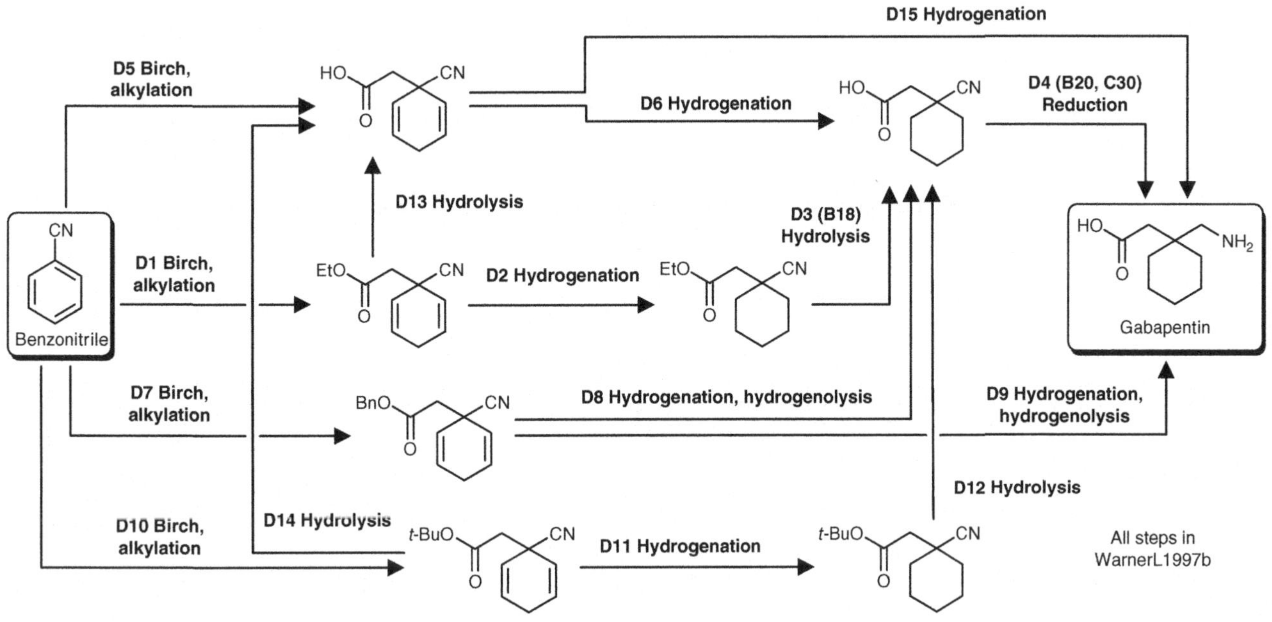

SCHEME D SYNTHESIS OF GABAPENTIN USING A BIRCH REDUCTION AS A KEY STEP (FOR REACTION CONDITIONS SEE PAGE PAGE 34)

2.4.6 Re-visiting the rearrangement as a key step

Later patent publications reveal a tendency to develop processes based on the Hoffman rearrangement (Scheme A, **A24**).

The routes that do not employ a rearrangement had proven to be efficacious but all suffer from issues related to implementation on an industrial scale (use of cyanide or nitromethane; nitro intermediates; ozonolysis and dissolving metal reduction at very low temperatures, amongst others).

As disclosed previously, the Hofmann rearrangement is the most likely of all of the alternative rearrangements to be employed in the synthesis of Gabapentin (Scheme 2.37). Hofmann rearrangements are commonly used on large scale giving results comparable to Curtius rearrangements, without the intermediacy of an acyl azide. Common reagents used to form the *N*-halide bond on commercial scale are the inexpensive and easy to handle oxidants, NaOCl and NaOBr.

As stated above, the preferred method of performing the Hofmann rearrangement is by slow addition of a basic solution of the amide to a cold sodium hypochlorite or hypobromite solution, thus allowing controlled release of the energy during the transformation, whilst the isocyanate is continuously converted to the amine under the basic aqueous conditions.

A method to carry out the Hofmann rearrangement of the monoamide at relatively high temperature (40-45 °C) using a continuous flow microreactor has been developed and patented (Jinagxi2017, CES2017). Yields above 95% of Gabapentin for relatively short residence times (5-7 min.) were reported (up to 670 g scale).

The successful use of trichloroisocyanuric acid, sodium dichloroisocyanurate (NaDCC), 1,3-dichloro-5,5-dimethylhydantoin (DCDMH) and *N*-chlorosuccinimide as chlorinating agents in promoting the Hofmann rearrangement have also been reported (100 g scale, ORPD2016). The best result was obtained with TCCA, which gives Gabapentin lactam, formed during work-up from Gabapentin, in a 90% yield.

DCDMH was used in the Hofmann rearrangement in the presence of sodium methoxide using methanol as solvent thus giving Gabapentin methyl carbamate (Scheme 2.38, TL2019). Hydrolysis then produces Gabapentin. When this methodology was extended to cyclic imides the transformation proceeds *via* a tandem reaction involving hydrolysis and rearrangement (Scheme 2.38, TL2109).

2.4.7 Other synthetic routes

Gabapentin synthesis is also an interesting topic from an academic point of view and some alternative laboratory routes have been published but have not been the subject any patent application.

In one particular case, a radical cyclization was the key step used to form Gabapentin lactone from cyclohexenylcarboxaldehyde as staring material. The C2 side chain is derived from a trichloroacetic acid residue (Scheme 2.39, T2003a,b, CC2012).

In yet another approach, a transition metal Rh(II) catalysed intramolecular C-H insertion reaction of an α-diazoacetamide produced the Gabapentin lactam ring (Scheme 2.40, T2005, Synlett2003). In this case the staring material is a secondary amine and the C2 side chain is derived from ketene (which normally exists in the form of the dimer, diketene).

Both of the processes gave good yields at laboratory scale but are not expected to be useful at industrial scale.

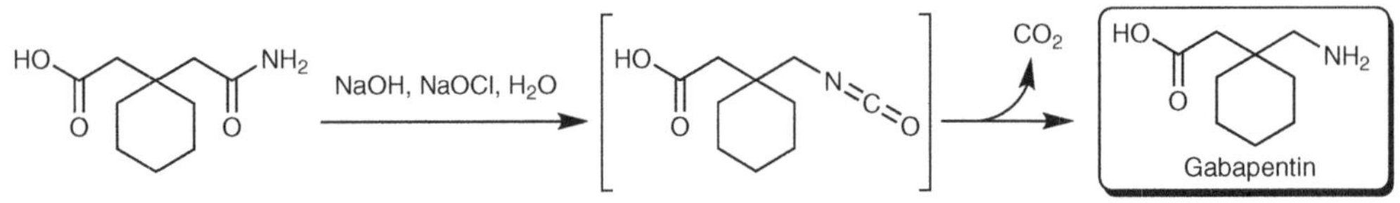

Scheme 2.37 Outline of the Hofmann rearrangement to produce Gabapentin

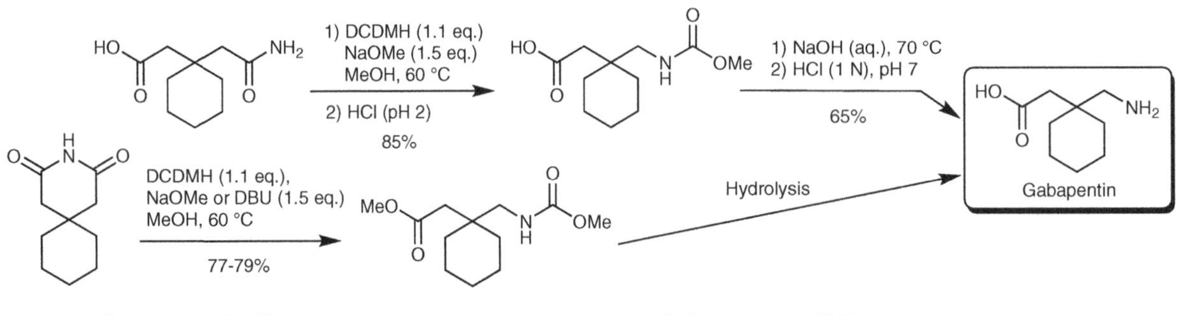

Scheme 2.38 Hofmann rearrangement induced by 1,3-dichloro-5,5-dimethylhydantoin

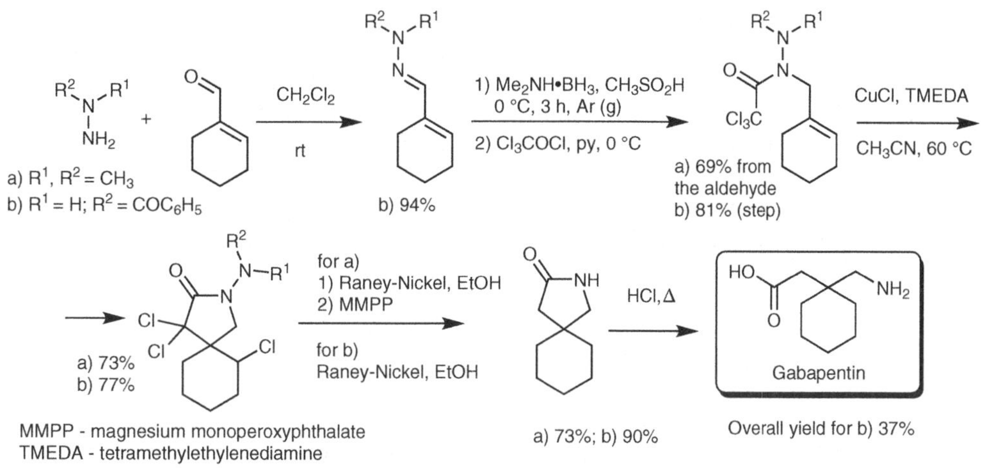

Scheme 2.39 Gabapentin synthesis using a radical cyclization a key synthetic step

2.4.8 Purifying the product

The purity and impurity profile of Active Pharmaceutical Ingredients (also referred to a Drug Substances) are rigorously controlled by the Pharmaceutical Industry and constantly monitored by the various Regulatory Authorities. As a rule, all drugs need to be produced with very high purity. This criterion makes the development of any industrial synthetic process a extremely demanding task. It is most unlikely that a synthetic sequence, *per se*, will produce an API with the required purity. Hence, various methods of purification and isolation of the final drug substance are frequently developed and patented.

It is impossible to determine the efficiency of the various purification methods available to remove impurities in any particular case without carrying out experimental work. However, a brief analysis of the type of operations, which have warranted the issuance of a patent, is relevant here.

Many of the synthesis of Gabapentin that have been described end with a reaction step under acidic conditions. An example of this is the hydrolysis of Gabapentin lactam. The resulting product is an acid salt of the API. The approved form sold in the Pharmacy is the neutral/zwiterionic compound and hence methods to produce this form need to be developed.

The acid salt can be converted to the neutral form by elution through a weak basic ionic resin followed by total removal of the water and crystallization from an alcoholic solvent. Solvents such as methanol or methanol/isopropanol or ethanol/ether mixtures have been described.

Alternatively, when alcohol is added to the aqueous eluate from the ionic resin column a slurry is obtained and the free amino acid is isolated by direct centrifugation. Thereafter, the solid is purified by crystallization as described above.

Other groups opted to solubilize Gabapentin hydrochloride in an organic solvent from which insoluble inorganic salts are removed by filtration. After an optional concentration, the Gabapentin hydrochloride solution is treated with a solution of an amine which results in the precipitation of Gabapentin in the switerionic form. This is followed by crystallization to obtain the pure compound.

Other methods have been described that do not give a hydrochloride or an alternative mineral acid salt (for example hydrogenation in the final step). Here, direct crystallization processes can be applied to the resulting crude product.

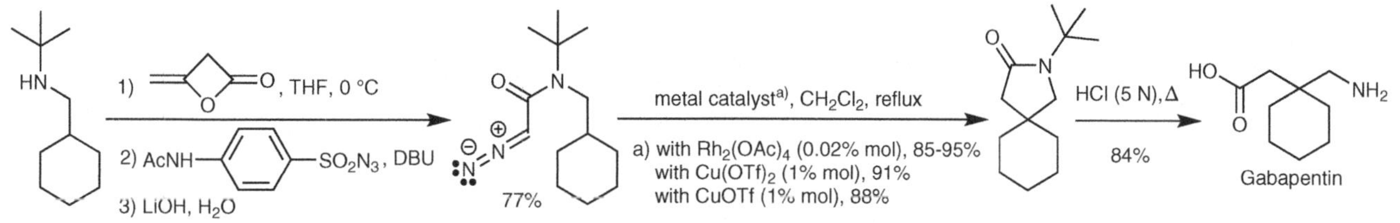

Scheme 2.40 Gabapentin synthesis by ketene chemistry

2.5 STRATEGY COMPARISON AND CONCLUSIONS

Five synthetic routes have been developed by a number of companies to produce Gabapentin at industrial scale (Scheme 2.41).

Although presumably all of these routes can be applied to industrial production of Gabapentin, some are somewhat less appropriate, for various reasons.

Without exception, each route presents hazards that have to be overcome before they can be put into practice at larger scale. The chemistry developed in the laboratory can usually be industrialized for large-scale manufacturing providing that safety and proper operating conditions are followed. This implies that extensive safety studies be carried out and significant investment in plant and equipment may be necessary.

The **1,4-dicarbonyl routes** are longer in comparison with others. The alkylation requires pre-formation of an enamine, resulting in two extra steps, i.e. enamine formation and a hydrolysis. Similarly, as to what is found in the other routes, special precautions are required, in this case entailing the handling of hydroxylamine. Hence, no simplification is bestowed from this perspective. A further drawback is that the starting material, cyclohexanecarboxaldehyde, is more expensive than both cyclohexanone and benzonitrile.

Birch routes are short and start from a simple and readily available aromatic compound, benzonitrile. Overall yields are reported to be high. However, the methodology necessitates the use of lithium (or alternatively, sodium or potassium) metal in liquid ammonia at low temperature and requires specialized equipment.

Those **routes that use cyanide** normally require the use of an excess of highly toxic sodium or potassium cyanide, which calls for careful handling and judicious treatment of the resulting waste streams. In spite of the good yields, not all companies may have the conditions in-house for industrial scale production.

Nitromethane, and the various **nitro-containing intermediates** resulting from its use, are particularly unstable and have inherently high associated decomposition energies (the decomposition energy of the nitro group is estimated to be 310-360 kJ/mol). The handling of hydrogen entails special facilities and precautions, whilst the hydrogenation of the nitro group to the amine is exceedingly exothermic and can be difficult to control. Because hydrogenation proceeds through a series of complex intermediates, there is a high potential for the formation of impurities due to various side reactions. However, the overall yield of these routes was found to be quite impressive.

The different rearrangements for the introduction of the primary amine of Gabapentin all present some hazards. These normally entail the use of toxic, unstable

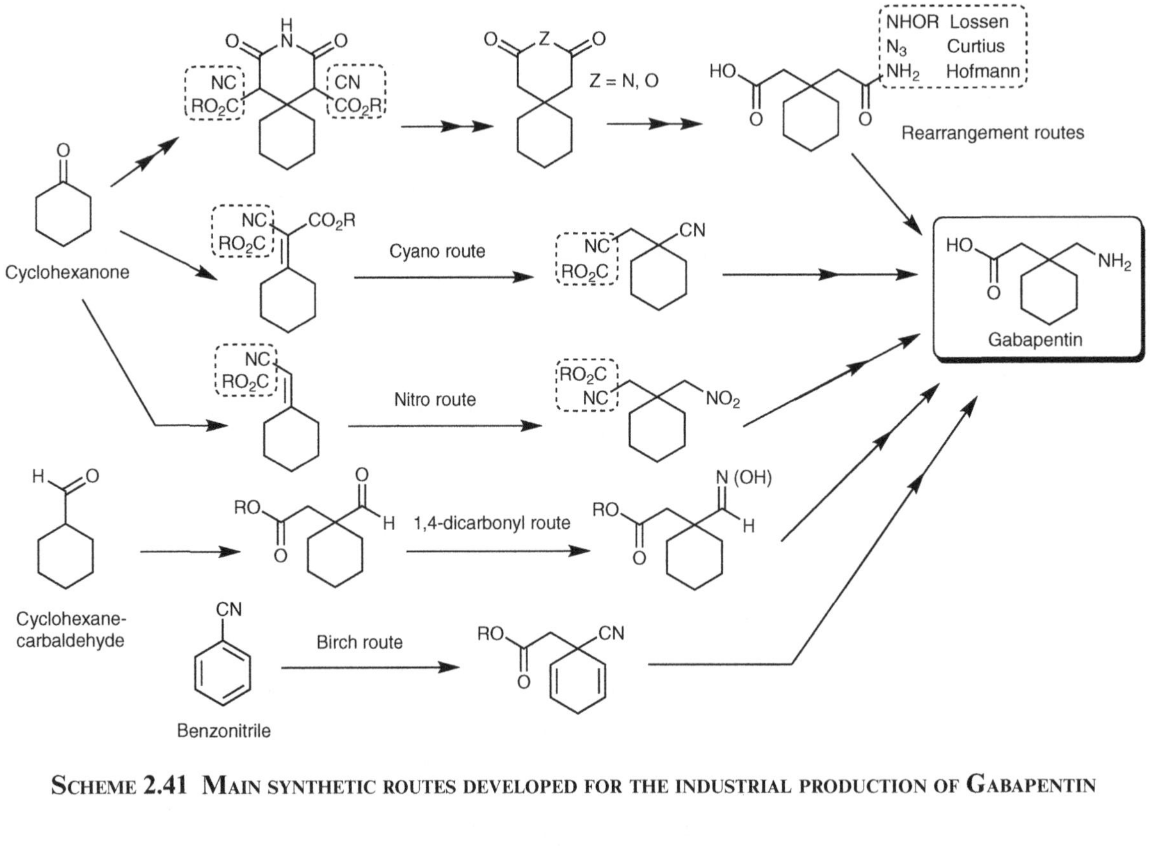

SCHEME 2.41 MAIN SYNTHETIC ROUTES DEVELOPED FOR THE INDUSTRIAL PRODUCTION OF GABAPENTIN

azide, the exception being the Hofmann route, which is easier and safer to carry out at an industrial scale.

The acyl azide[2] intermediates are known to be unstable and characteristically have the potential to decompose violently. A less hazardous practice for handling these azides is to keep the concentration and total quantity present at any one time to a minimum. The best option is to generate the acyl azide under conditions where it rearranges almost spontaneously.

The large (stoichiometric) amounts of nitrogen and carbon dioxide gases generated can create catastrophic events at large scale if they are released in an extremely short time frame. Careful monitoring of waste streams and the neutralization of deleterious substances is mandatory (for the treatment of azide waste streams see OPRD2009).

The Hofmann rearrangement is probably the rearrangement most likely to be successfully implemented at industrial scale (Scheme 2.42). Both formation of the *N*-halo intermediate and the rearrangement steps are rapid at moderate temperatures allowing good control of the reaction exotherm and gas release. In this way, the transformation can be successfully achieved efficaciously and safely.

In summary, all the described routes present some degree of concerns regarding safety. Careful safety studies are mandatory before even scale-up batches can be undertaken. The route *via* a Hofmann rearrangement is one of the most promising methods for the production of Gabapentin industrially.

2.6 LESSONS FROM THE GABAPENTIN CASE

Safety

Safety is a key and crucial issue in industrial process chemistry. When scaling up a process, the development chemist should analyze all possible hazards in great detail and put in place solutions that allow for the safe running of commercial batches. An inherently unsafe process, from which the dangers cannot be eliminated, should never be scale-up!

Approaches to eradicating the dangers include modifying the manner in which the transformations are carried out, avoiding/replacing problematic reactants or solvents and avoiding specific transformations by changing the route of synthesis. In this respect, Gabapentin is a good case study. Development chemists have used all of these approaches to run safer processes at full scale.

It is also worthy to note that in many instances a "chemical" solution to a problem, rather than an "engineering" one, will mean less investment in equipment and ultimately reduce costs.

Notwithstanding, it should be pointed out that companies that already practice and have already made the large investment in equipment and have the expertise to carry out "hazardous chemistry" in a safe mode, will have real commercial opportunities in these circumstances.

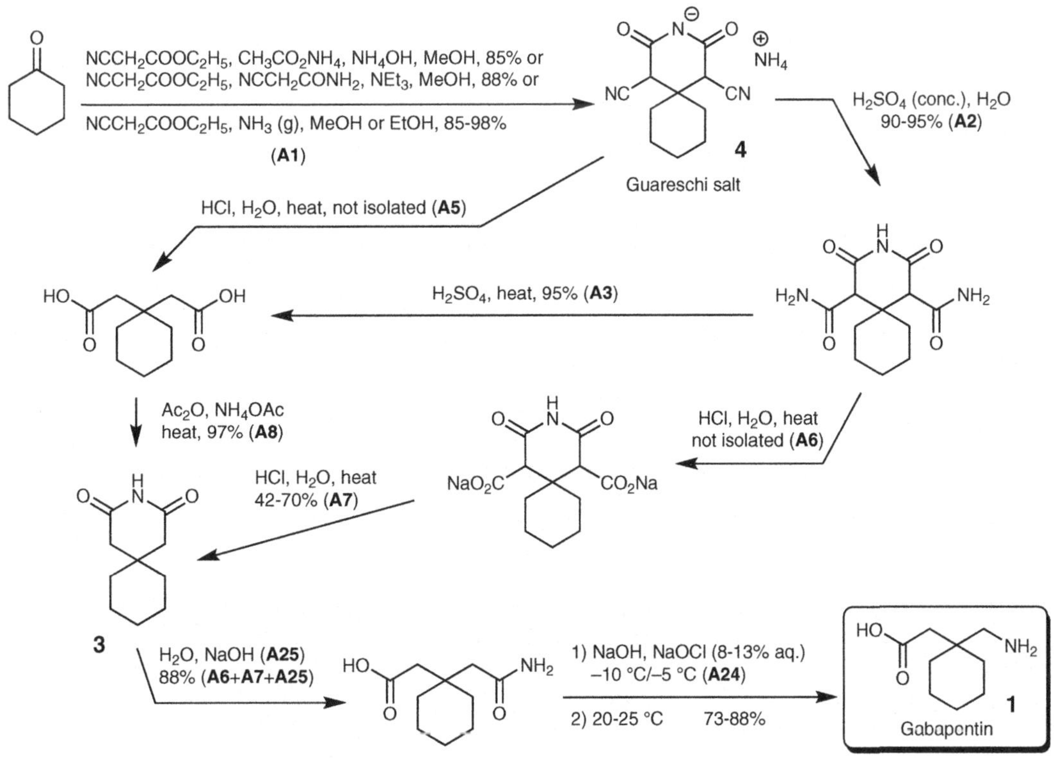

Scheme 2.42 The most promising industrial scale processes for Gabapentin synthesis

2 For a discussion of the hazards involved with carrying out these reactions at an industrial scale see: OPRD2006; OPRD2009; OA2010

Safety and process development

Most of the laboratory procedures have the potential to create dangerous situations at industrial scale and an appropriate process development is required prior to scale-up. A deep understanding of all the process (not only the reactions) and a very rigorous control of operating conditions will be needed.

In the particular case of Gabapentin, one of the most favourable routes uses a rearrangement step that might have safety concerns. Proper investigation of the transformation conditions seem to resolves this issue, confirming that correct process development is frequently key to successful scale-up.

Safety and Flow chemistry

Active pharmaceutical ingredients are almost always produced in batches. However, in some situations, flow chemistry is an effective alternative to resolve problems that may be encountered in a batch process. The Curtius rearrangement is a dangerous reaction to run at scale, both because of issues inherent with the handling of the azide reactant and the instability of the acyl azide intermediate. The acyl azide is made by reaction of the acyl chloride with azide anion, followed by heating to promote rearrangement (Scheme 2.43). Due to the high concentration of acyl azide at the beginning of the rearrangement, this reaction presents tangible hazards, even in the laboratory.

Scheme 2.43 represents how a batch process would look on an industrial scale. Each step in the train being associated with a particular hazard. The fact that a considerable quantity of material is present means that the consequences of an uncontrolled event could indeed be extremely serious. Obviously, no reputable industrial enterprise would run such a risk without having adequate controls in place.

A solution is to run the process continuously. In this approach, the acyl azide is formed by mixing the two reactants in a pipe flow reactor. When the initial formation reaction is complete, the pipe is heated to initiate the rearrangement. Reactants are continually consumed as they move along the length of the reactor, keeping the quantity of azide present to a minimum at any one time. This inherently safer process is portrayed in Scheme 2.44.

As mentioned earlier, this flow technique has been successfully applied to the less hazardous Hofmann rearrangement, giving yields of >95% on a 670 g scale (Jiangxi2017, CES2017).

Safety and gas liberation

The sudden release of gases can be extremely hazardous, especially in large industrial batches. Sudden, uncontrolled gas formation leads to build up of excessive pressure and can provoke rupture of the reaction vessel. Dangerous materials may suddenly be ejected and cause widespread contamination of the surrounding area. The rate of gas release must be appropriately controlled

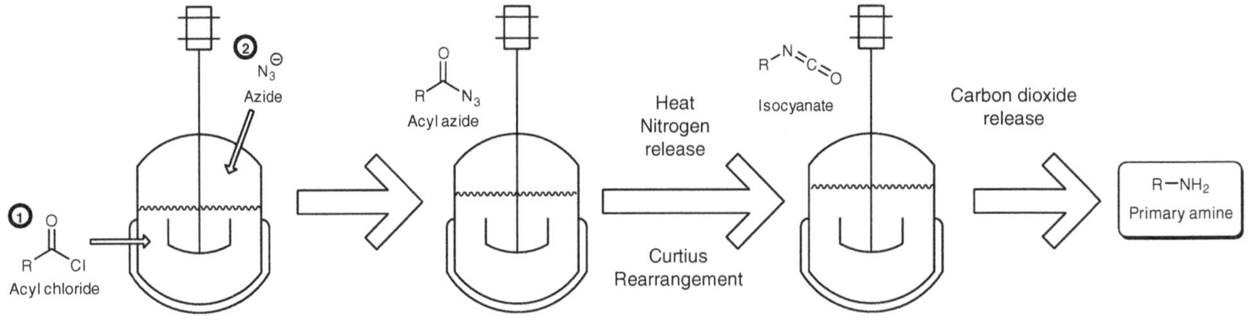

Scheme 2.43 Schematic diagram for a batch process for a Curtius rearrangement

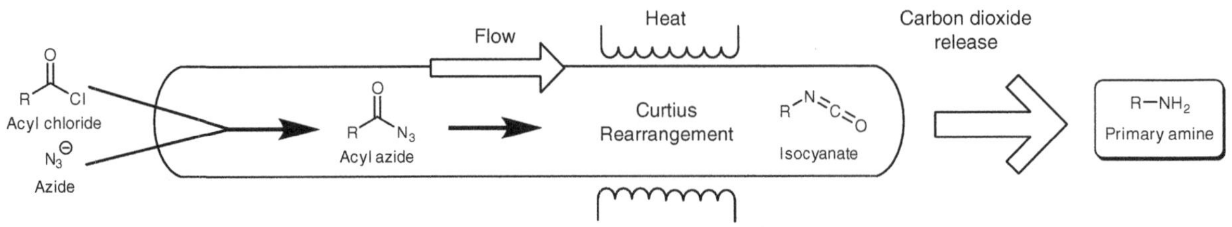

Scheme 2.44 Schematic diagram for a flow process for a Curtius rearrangement

to prevent such a hazardous event. Pipe reactors must be rated to withstand such sudden pressure surges and exiting flow streams suitably handled to avoid or at least meliorate an uncontrolled "Coca-Cola" event on depressurization.

Cost and stability of raw materials

The costs associated with the production of a generic API is of prime importance and is a key factor in deciding the final price to the consumer and the success of the product. Raw material costs make up a significant proportion of the total. The use of a commodity chemical, normally available at the lowest possible prices as controlled by the market, is clearly a preferred starting material. The general availability of cyclohexanone, its stability and long term shelf-life when shipped and stored appropriately, make it an obvious choice as a starting point for the chemical development of an industrial process to Gabapentin. However, there is a caveat: low yields and expensive processing techniques may make this scenario null and void.

The ASAP "rule"

The ASAP rule (As Simple As Possible) is a standard that all Process Chemist ought to follow. The simplest synthetic process is generally easier to control, more reliable, more robust and frequently cheaper to run. Paying attention to the simplicity of a synthetic process can facilitate route selection and help exclude less straightforward, more esoteric options, when deciding which can be scaled-up for implementation at an industrial level.

2.7 REACTION CONDITIONS FOR SCHEMES A TO D

2.7.1 Reaction conditions for Scheme A

A1 $NCCH_2CO_2NH_2$, piperidine, EtOH, YNR (JCS1911, 1913). $NCCH_2CO_2Et$, NH_3, YNR (CZ1911, JCS1923). $NCCH_2CO_2Et$, NH_3, EtOH, –10 °C, 98% (WarnerL1957), YNR (JMC1990), 0-5 °C, 73% (AJC1960). $NCCH_2CO_2Et$, $MeCO_2NH_4$, NH_4OH (25% aq.), MeOH, 8 °C to 25 °C, 85% (Teva2000). $NCCH_2CO_2Et$, $NCCH_2CONH_2$, MeOH, NEt_3, 25 °C, 88% (NHWA2005). $NCCH_2CO_2Et$, NH_3, MeOH, –5 °C to 25 °C, 85% (NHWA2005), in *i*-PrOH at –10 °C to 10 °C, YNR (Taicang2015a). $NCCH_2CO_2Me$, NH_3, $MeCO_2NH_4$ in $MeOH/H_2O$, NI (TaicangYC2008, Taicang2015c), in *i*-PrOH, NI (Taicang2015c). $LiCH_2CN$, MeOH, 20 °C, 83% (H2009).
A2 H_2SO_4 (conc.), H_2O (cat.), 60-80 °C, 90-95%.
A3 H_2SO_4 (70%), 170 °C, 95%.
A4 HCl, H_2O, reflux, YNR.
A5 H_2SO_4, H_2O, 160 °C, 65% (WarnerL1957). H_2SO_4 (60%), heat, 93% (AJC1960). HCl, AcOH, H_2O, reflux, YNR (JMC1990). *i*-PrOH or H_2O, H_2SO_4 (5-20%), 190-220 °C, YNR (Taicang2015a,c).
A6 KOH or NaOH, H_2O, reflux, NI.
A7 H_2SO_4, 180 °C, YNR (JCS1911). HCl (conc.), H_2O, 95 °C (38 kg, Zambon2004b). HCl, H_2O (pH 2), reflux, 69% (A34+A7, Caffaro2004). HCl (conc.) pH 3, MeOH, reflux, 42% (A32+A33+A34+A7, Caffaro2004). HCl (conc.), pH 1, 70 °C to reflux, NI (NHWA2005).
A8 NH_3, heat, YNR (JCS1927). NH_3 (25%), H_2O, 180 °C, YNR (Adamed2011). H_2NCONH_2 or $(NH_4)_2CO_3$ 160-190 °C, 80% (AJC1960). H_2NCONH_2, 150-180 °C, 95% (Jiangsu2005), 140° C, YNR (Hikal2008). Ac_2O, NH_4OAc, 170 °C, 97% (Erregierre2002). $HCONH_2$, 160 °C, 97% (OPPI 2013). H_2NCONH_2, H_2O, heat (Taicang2015c).
A9 AcCl, heat YNR (JCS1911), 60 °C (JOC1988, JOC2005). Ac_2O, heat, YNR (JCS1911, JMC1990, Bayer1995, Nisshin2001, Taicang2014b,c). Ac_2O, toluene, 80 °C, NI (Zambon2003). 1,2-dichloroethane, P_2O_5, 60-80 °C, YNR (Taicang2015a).
A10 MeOH, reflux, 75%.
A11 $ClCO_2Et$, NEt_3, acetone, 0 °C, NI.
A12 NaN_3, H_2O, 0 °C, NI.
A13 1) H_2O, reflux; 2) HCl (20% aq.), reflux, YNR.
A14 NaOH, H_2O, reflux, YNR.
A15 NH_3, CO_2 (60-80 bar), 80-110 °C, NI.
A16 NH_3, 180°C, YNR (JMC1965, Adamed2011). NH_3, toluene, reflux, YNR (JMC1990).
A17 $NH_2NH_2{\bullet}H_2O$, YNR.
A18 $NaNO_2$, acid, YNR.
A19 Heat, YNR.
A20 NH_2OH, H_2O, 70 °C, 84%.
A21 $PhSO_2Cl$, Na_2CO_3 (10% aq), 10 °C to rt, 88%. $TsSO_2Cl$, NEt_3, $CHCl_3$, rt, 89%.
A22 NaOH (10% aq.), 100°C, YNR (R = Ph). 1) NaOEt, EtOH, reflux; 2) HCl (aq.), reflux, YNR (R = Ph). 1) NEt_3, MeOH (anh.), reflux; 2) HCl (aq.) reflux, YNR (R = Ph). 1) NEt_3, MeOH (anh.), reflux, 87% (*N*-carbomethoxy derivative); 2) HCl (aq.), reflux, YNR. 1) NEt_3, MeOH (anh.), reflux, 83% (*N*-carbomethoxy derivative); 2) HCl (aq.), reflux, YNR.
A23 NH_3, solvent (addition of another base optional). NH_3 (28% aq.), MIBK, 20 °C, 78% (Bromine2001a). NH_3, *i*-PrOH < 5 °C to rt, 87% (Ipca2006). NH_3, H_2O, 5 °C, YNR (MRC2012). NH_3 (20% aq.), benzene, H_2O, 90 °C, NI (Taicang2014b). NH_3 (5-25%, aq.), 15-30 °C, YNR (Taicang2014c, 2015a). NH_4Cl, NaOH, H_2O, 30 °C, 96% (Ipca2006). $(NH_4)_2CO_3$, toluene, 40 °C to 70 °C, 98% (Ipca2006). NH_3 (aq.), NEt_3, dichlorotoluene, rt, 97% (Taicang2014d). NH_3 (aq.), py or NEt_3, THF, rt, 96-99% (Taicang2014d). NH_3 (28% aq.), H_2O, 30 °C, >95% (A9+A23, Zambon2003).
A24 NaOH (aq.), NaOCl (10-15%), –5 °C to rt, 65-88% (A24), 70-85% (A25+A24, Jiangsu2005), 45-50% (A6+A7+A25+A24, NHWA2005), with KOH, 84-

88% (A25+A24, Jiangsu2005), in flow conditions, YNR (Dalian2011). NaOH, NaOBr, 55 °C, YNR (Zach2013). Br_2, NaOBr, –15 °C to 70 °C, 48-51% (from cyclohexanone, A1+A5+A9+A23+A24, Taicang2015a). $NaHCO_3$ (30%), H_2O_2 (15%), YNR (Taicang2014a). **In flow conditions:** 1) NaOH (aq., 10%); 2) NaOCl (8% aq.), NaOH, 20-50 °C, flow rate 5-68 mL/min, residence time 2-5 min, >95% HPLC (up to 670 g, Jiangxi2017, CES2017). ***Via* methyl carbamate** 1) DCDMH (1,3-dichloro-5,5-dimethyl hydantoin, 1.1 eq.), NaOMe (1.5 eq.), MeOH, 60 °C, 85% (methyl carbamate, 20 g); 2) NaOH (aq.), 70 °C, 65%.

A25 NaOH, H_2O, heat, 80% (AJC1960), 50-80 °C, NI (Erregierre2002, Hikal2008), at reflux, 65% (Caffaro2004), 88% (A6+A7+A25, Zambon2004b). KOH (15-25% aq.), rt, NI (Jiangsu2005). KOH or NaOH, toluene, (20-25% aq.), 25 °C, NI (NHWA2005, Hikal2008).

A26 1) NaOH or KOH (10-20% aq.), Br_2 or NaOBr (5-10%) or Cl_2 or NaOCl (5-20%) or KOCl (20%), H_2O –10 °C to 10 °C, then heat to 50 °C; 2) toluene, 80-85 °C, 73-81% or HCl (aq.), 100 °C, 94-100%, 68-72% (A8+A25+A26, Hikal2008). NaOH (15%), NaOCl (20%), benzene, 90 °C, NI (Taicang2014b,c). 1) TCCA (trichloroisocyanuric acid), MeOH, rt; 2) NaOH, 0 °C, rt; 3) toluene, 80 °C, 90% (OPRD2016). 1) TCCA, H_2O, rt; 2) NaOH, 0 °C, rt; 3) toluene, 80 °C, 75%, with NaDCC (dichloroisocyanuric acid), 85%, with NCS (*N*-chlorosuccinimide), 46% (OPRD2016).

A27 (B27, C4) HCl (conc.), H_2O, reflux, 60-90%. HBr (conc.), H_2O, reflux, 27-78% (Medichem2005). HCl (conc.), dioxane, reflux, 40% (WarnerL1997a). H_2SO_4 (20%), 100 °C, 99% (Taicang2014d). Oxalic acid, 50-60 °C, YNR (Taicang2015b).

A28 DCDMH (1,3-dichloro-5,5-dimethyl hydantoin, 1.1 eq.), NaOMe or DBU (1.5 eq.), MeOH, 60 °C, 77-79% (of methyl carbamate), hydrolysis not shown.

A29+A30 KBr (20 eq.), *t*-BuOCl (3 eq.), *t*-BuOK (4 eq.), MeOH, 60 °C, 92%.

A31 $NCCH_2CONH_2$, $CH_3CO_2NH_4$, AcOH (cat.), toluene, reflux, 66%.

A32 $CH_2(CO_2Me)_2$, NaOMe, MeOH, 25 °C, 87% (R = Me). $CH_2(CO_2Et)_2$, NaOEt, EtOH (anh.), 25 °C, 82% (R = Et). $CH_2(CO_2Bn)_2$, NaH, toluene, 25 °C, 78% (R = Bn).

A33 H_2, Pd/C (5%), EtOAc, 12 °C, 100% (R = Bn). NaOH (aq.), MeOH, reflux, NI (R = Me).

A34 NaOH, H_2O or H_2O, EtOH or MeOH, reflux, YNR or NI.

A35 Ph_3PCH_3, *t*-BuOK, Et_2O, 40 °C, 92% (SC1985). Ph_3PCH_3, NaH, DMSO, 20 °C, 83% (OL2014).

A36 Zn, Cl_3CCOCl, Et_2O, N_2, MW, 93-97%.

A37 Zn, AcOH, H_2O, –20 °C, 77% (Array2010). Zn, AcOH, 20-30 °C, 93% (Tianyuan2004), 84% (A36+A37, T1993).

A38 H_2NOSO_3H, HCO_2H, reflux, YNR. *O*-(2,4,6-trimethylbenzenesulfonyl)hydroxylamine, CH_2Cl_2, rt, 82-92%.

A39 $ClCH_2COCl$, Zn, $CuSO_4$•$5H_2O$, Et_2O, MW, YNR or $ClCH_2COCl$, Zn, AcOH, H_2O, YNR (Array2010). 1,2-dichloroethane, DMA, Tf_2O, 2,4,6-collidine, –15 °C, YNR (Pfizer2007).

2.7.2 Reaction conditions for Scheme B

B1 R = Me: 1) $CH_2(CO_2Me)_2$, $TiCl_4$, CCl_4, THF, 0 °C; 2) py, THF, rt, 43% (Lonza1988). **R = Et:** 1) $CH_2(CO_2Et)_2$, $TiCl_4$, THF, CCl_4, 0 °C; 2) py, THF, rt, YNR (Daichii1988), 46% (BMCL2011). 1) $CH_2(CO_2Et)_2$, $TiCl_4$, THF, CH_2Cl_2; 2) py, rt, 60-62% (HCA1991, Givaudan2000). $CH_2(CO_2Me)_2$, $ZrCl_4$, THF, or CCl_4 or py –5 °C to 0 °C, YNR (Taicang2015d).

B2 R = Me: HCN (5 eq.), KCN (cat.), 120 °C, 69% (Lonza1988). **R = Et:** KCN (1.5 eq.), EtOH, HCl, reflux, 88-89% (Lonza1988, HCA1991). HCN, EtOH, 20-30 °C, YNR (Taicang2015d).

B3 NaOH, MeOH, H_2O, 40 °C, 90%,

B4 H_2 (10 bar), Raney Nickel, EtOH, 90 °C, NI.

B5 HCl (25% aq.), reflux, 57% (B4+B5).

B6 Toluene, 80-85 °C, 70%.

B7 LiCl, H_2O, DMSO, 150 °C, 74%.

B8 BnOH (5 eq.), HCl (cat.), toluene, reflux, 68%.

B9 H_2 (10 bar), Rh/C (5%), MeOH, rt, 27%.

B10 H_2 (9-10 bar), Raney Nickel, EtOH, 90 °C, 66-89% (Lonza1988; HCA1991). H_2 (9-12 bar), Pd/C, 80-110 °C, YNR (Taicang2015d).

B11 R = Et: $NCCH_2CO_2Et$, NH_4OAc (cat.), AcOH (cat.), toluene, reflux, 74% (Lonza1988), YNR (JMC1990), 87% (BMCL2006). **R = Me:** $NCCH_2CO_2CH_2Me$, piperazine (cat.), 92% (HCA1991). $NCCH_2CO_2Me$, NH_4OAc, AcOH (cat.), MTBE, reflux, 94% (Hikal2012), in toluene with $PhCO_2H$ 95% (Hikal2012), in cyclohexane 87% (Hikal2012).

B12 KCN, EtOH, H_2O, heat, YNR (JLAC1965, S1983). KCN, EtOH, reflux, 94% (Lonza1988), YNR (BMCL2006). KCN, NH_4Cl, H_2O, DMF, 100 °C, YNR (BCSF1984). KCN (cat.), HCN, EtOH, pH 11, 50 °C, 82% (Lonza1988). KCN, Me_3SiCN, 18-crown-5-ether, 25 °C, 90% (BCSJ1990). HCN, NaOEt (cat.), EtOH, H_2O, 30 °C, 62% (HCA1991). NaCN, EtOH or MeOH, H_2O, reflux, YNR (HCA1991, ASC2007, Hikal2012).

B13 1) HCl (g, cat.), EtOH, 0 °C; 2) H_2O, 75% (Lonza1988). 1) HCl (2-3 bar), EtOH, rt; 2) H_2O, 5 °C, 43% (HCA1991).

B14 NaOH, YNR (JCS1924). EtOH, H_2O, heat, 64-99% (B12+B14). MeOH, H_2O, reflux, 89-99% (B12+B14, Hikal2012). NH_4Cl, H_2O, DMF, 100 °C, YNR (BCSF1984).

B15 1) HCl (g, 3 bar), EtOH, 0 °C; 2) H_2O, 10 °C, 41% (Lonza1989). 1) HCl (g, anh., 0.7 bar), EtOH, toluene, 35 °C; 2) H_2O, 91% (WarnerL1989a,b, Pfizer2006).

B16 H_2 (21 bar), Raney Nickel, *i*-PrOH, 130 °C, 81%.

B17 1) HCl (g), BnOH, toluene, 0 °C; 2) H_2O, Et_2O, 11% (Lonza1989). 1) HCl (g), BnOH, 0 °C; 2) H_2O, 0-5 °C, 60% (HCA1991).

B18 (D3) NaOH, H_2O, MeOH, 50 °C, YNR (WarnerL1989a,b, 1997b). KOH, H_2O, 80 °C, YNR (Pfizer2006).
B19 HCl (g, anh.), EtOH, toluene, 10 °C to rt, 78% (WarnerL1989a,b). Nitrilase-catalyzed hydrolysis 50-100%.
B20 (C30, D4) H_2 (3.5 bar), Rh (10%)/Pd (1%)/C, MeOH, rt, 79% (WarnerL1989a,b, 1997b). H_2 (10 bar), Rh (5%)/C, MeOH, 30 °C, 65% (Godecke1989c). H_2 (15 bar), Rh (10%)/C, MIBK, 60 °C, 84% (Hikal2012). H_2 (15 bar), Pt or PtO_2, MIBK, 60 °C, 78% (Hikal2012). H_2 (12 bar), Raney Nickel, MeOH, H_2O, rt, 60% (WarnerL1989a,b, 1997b). H_2 (12 bar), Raney Nickel, H_2O, MeOH, NaOH (50%), rt, 60% (Pfizer2003). H_2 (3.5 bar), Raney Nickel, H_2O, 30 °C, 68-73% (from the ester, B18+B20, Pfizer2006). H_2(1.3 bar), Raney Nickel, NaOH, MeOH, H_2O, 25 °C, 60% (Zach2006). H_2 (10-15 bar), Raney Nickel, H_2O, base, 60 °C, 84-90% (Hikal2012), in MeOH, 84-91% (Hikal2012), in MIBK, 80-89% (Hikal2012), in MeOH/H_2O, 83% (Hikal2012), in THF, 80% (Hikal2012), in *i*-PrOH/H_2O, 84% (Hikal2012). H_2 (21 bar), Raney Nickel, H_2O, 110 °C, YNR (CatC2015). H_2 (20 bar), Raney Nickel, YNR (BEJ2017).
B21 1) $NCCH_2PO(OEt)_2$, NaH, THF, 0 °C; 2) Cyclohexanone, THF, 0 °C, 67% (WarnerL1997c, 2000, 2002), 40% (Biocryst2009). $NCCH_2PO(OEt)_2$, KOH, hexane, CH_3CN, reflux, 55% (Procos2001a). $NCCH_2PO(OEt)_2$, base, Et_2O or DMF (UBC2002). 1) $NCCH_2PO(OEt)_2$, NaOH, CH_3CN, 80 °C; 2) Cyclohexanone, CH_3CN, 80 °C, YNR (Merck2013). $NCCH_2PO(OEt)_2$, *t*-BuOK, THF, rt, YNR (Array 2010).
B22 CH_3NO_2 (2 eq.), Bu_4NF, THF, 70 °C, 71% (WarnerL1997c, 2000, 2002). CH_3NO_2, K_2CO_3, DMSO, 95 °C, 87% (Procos2001a).
B23 H_2 (40 bar), Pd/C (10%), EtOH (anh.), 67%.
B24 NaOH (30% aq.), reflux, 74%.
B25 H_2 (40 bar), Raney Nickel, MeOH, 80 °C, 93%.
B26 1) H_2 (8-10 bar), Raney Nickel, NaOH (1.2 eq.), H_2O, pH 8-14, 80 °C; 2) toluene, 80-85 °C, 88-97% (Hikal2012), 80-94% (B19+B26, ZhejiangUT2015).
B27 (A27, C4) see A27.
B28 R = Et: $EtOCOCH_2PO(OEt)_2$, KOH, THF, rt to 40 °C, 91-94% (Godecke1989a). $EtOCOCH_2PO(OEt)_2$, K_2CO_3, H_2O, 0 °C to 25 °C, 94% (Gruenenthal2002). 1) $EtOCOCH_2PO(OEt)_2$, NaH, THF, 0 °C; 2) SM, 0 °C to rt, 78% (WarnerL1997a). **R = Me:** $EtOCOCH_2PO(OEt)_2$, *t*-BuOK, toluene/MeOH, rt, YNR (in flow, ACIE2015).
B29 R = Et: CH_3NO_2, K_2CO_3 or KOH or Bu_4NF, DMSO, 95-110 °C, 52-91% (Godecke1989a). CH_3NO_2, Bu_4NF, THF, 70 °C, 46% (WarnerL1997a), at reflux, 98% (Gruenenthal2002). **R = Me:** CH_3NO_2, Bu_4NF, toluene, THF, 50 °C, YNR (in flow, ACIE2015).
B30 H_2 (0.6 bar), Pd/C, EtOH, 60-80 °C, 92-95% (Godecke1989a). H_2 (2.7 bar), Raney Nickel, MeOH, 35 °C, 100% (WarnerL1997a, Gruenenthal2002).
B31 KOH (10% aq.), MeOH, rt, 80% (RichterGedeon1998). LiOH, toluene, H_2O, YNR (in flow, ACIE2015).
B32 H_2 (1 bar), Pd/C, MeOH, 80% (RichterGedeon1998). H_2 (60 bar), Pd/C (10%), MeOH, H_2O, 60 °C, 49% (from cyclohexanone in flow, ACIE2015).
B33 H_2 (1 bar), Pd/C (10%), MeOH, 51%.
B34 HCl (20% aq.), reflux, 81% (Lonza1988). HCl (20% aq.), reflux, 54-75% (HCA1991). HCl (to pH 1), 15-30 °C, YNR (Taicang2015d), 46% (B1+B2+B10+B34, Taicang2015d).
B35 Meldrum's acid (2,2-dimethyl-1,3-dioxane-4,6-dione), piperidine, py, 50 °C, 52% (OL2003). Meldrum's acid , py, 4 Å MS, rt, 50% (JCSPT11986).
B36 NaOMe (1 eq.), CH_3NO_2 (1 eq.), MeOH, 0 °C to rt, 89%.
B37 H_2 (5 bar), Pt, EtOH, 130 °C, NI.
B38 HCl (20% aq.), reflux, 62% (B37+B38).
B39 Toluene, 80-85 °C, 88-97% (B20+B39, Hikal2012).
B40 $BocNHCH_2COTePh$ (0.9 eq.), Et_3B, CH_2Cl_2, air, rt, NI.
B41 HCl (aq., 6M), reflux, 52% (B40+B41).

2.7.3 Reaction conditions for Scheme C

C1 1) *i*-Bu_2NH, toluene, reflux; 2) $BrCH_2CO_2Et$ (1.5 eq.), CH_3CN, reflux; 3) AcOH, H_2O, reflux, 60% (Solchem2001), 75% (Ind2004).
C2+C3 1) NH_3, H_2 (3.5 bar), Pd/C (10%), EtOH, YNR. 1) NH_3, EtOH, rt, 97% (C2); 2) H_2 (3.8 bar), Pd/C (5%), AcOH, 100 °C, 37% (C3).
C4 (A27, B27) see A27.
C5 NH_2OH•HCl, Na_2CO_3, H_2O or MeOH/H_2O or EtOH/H_2O, rt, 93-95%.
C6 H_2 (3.5 bar), Pd/C (5%), AcOH, 80 °C, NI.
C7 1) H_2 (9 bar), Rh/Al_2O_3 (5%), EtOH, 60 °C; 2) NaOH, H_2O, reflux, 74%.
C8 NaOH (40% aq.), H_2O, rt, NI (Solchem2001). NaOH, EtOH, H_2O, rt, 94% (Ind2004).
C9 NaOH (aq.), H_2O, 10 °C, NI.
C10 1) NH_2OH•HCl, Na_2CO_3; 2) HCl (30%, pH 3), 91% (C8+C10 or C9+C10, Solchem2001).
C11 H_2 (9 bar), Rh/Al_2O_3 (5%), *i*-PrOH, H_2O, 20 °C, 87% (Solchem2001). H_2 (3.5 bar), Pd/C (5%), AcOH, 80 °C, YNR (Ind2004).
C12+C13 NH_3, H_2 (3.5 bar), Pd/C (10%), EtOH, YNR.
C14 1) *i*-Bu_2NH, toluene, reflux; 2) $BrCH_2CO_2Bn$, CH_3CN, reflux; 3) AcOH, H_2O, YNR.
C15+C16 NH_3, H_2 (3.5 bar), Pd/C (10%), EtOH, YNR.
C17 NH_2OH•HCl, Na_2CO_3, EtOH/H_2O, rt, YNR.
C18 H_2 (3.5 bar), Pd/C (5%), AcOH, 80 °C, YNR.
C19 $H_2C{=}CHCH_2OH$, benzene, TsOH, 130-180 °C, 78%.
C20 NH_2OH•HCl, HCO_2H, reflux, 81%.
C21 1) O_3, $HOCH_2CH_2OH$, *t*-BuOH, 20 °C; 2) Me_2S, <25 °C; 3) $HOCH_2CH_2OH$, MsOH (cat.), cyclohexane, heat, 75%.
C22 $LiAlH_4$, Et_2O, reflux, 85%.
C23 O_3, TFA, EtOAc, –5 °C, YNR.

C24 AcOH, HCl (18%, aq.), reflux, 62% (C23+C24).
C25 Et_2NLi, 2-bromo-1,1-diethoxyethane, –70 °C to rt, 80%.
C26 1) O_3, MeOH, –10 °C; 2) Me_2S, <25 °C; 3) $CH(OMe)_3$, MsOH (cat.), rt, 84%.
C27 $LiAlH_4$, Et_2O, reflux, 95% (R = Me). Na, EtOH, rt, 90% (R = Et).
C28 1) O_3, *t*-BuOAc, 25 °C; 2) Me_2S <35 °C, 90%.
C29 K_2CO_3, MeOH, H_2O, 57%. NaOH, H_2O, MTBE, 0 °C, 82% (C28+C29).
C30 (B20, D4) see B20.

2.7.4 Reaction conditions for Scheme D

D1 1) Li (2.5 eq.), *t*-BuOH (2.5 eq.), NH_3, THF, –78 °C, N_2; 2) $BrCH_2COOC_2H_5$, –78 °C; 3) NH_4Cl, 60%.
D2 Pd/C (5%), H_2 (1.4 bar), MeOH, 94%.
D3 (B18) see B18.
D4 (B20, C30) see B20.
D5 1) Li (2.5 eq.), *t*-BuOH (1 eq.), NH_3, THF, –78 °C, N_2; 2) $BrCH_2COONH_4$ (2 eq.), –78 °C; 3) NH_4Cl, 31%.
D6 H_2 (3.5 bar), Pd/C (5%), NH_4OH (28%, cat.), MeOH, rt, 94%.
D7 1) Li (2.5 eq.), *t*-BuOH (1 eq.), NH_3, THF, –78 °C, N_2 2) $BrCH_2COOBn$ (2 eq.), –78 °C; 3) NH_4Cl (7.7 eq.), 73%.
D8 H_2, Pd/$BaSO_4$ (5%), MeOH, rt, 35%.
D9 H_2 (3.5 bar), Pd/C (5%), NH_4OH (24%), MeOH, 50 °C, 93%.
D10 1) Li (2.5 eq.), *t*-BuOH (1 eq.), NH_3, THF, –78 °C, N_2; 2) $BrCH_2COOtBu$ (2 eq.), –78 °C; 3) NH_4Cl (7.8 eq.), 73%.
D11 H_2 (1.4 bar), Pd/C (5%), MeOH, rt, 56%.
D12 TFA, $PhOCH_3$, YNR.
D13 KOH, H_2O, EtOH, 0 °C, 61%.
D14 TFA, $PhOCH_3$ YNR.
D15 H_2 (3.5 bar), Pd/C (5%), NH_4OH (28%), MeOH, 50 °C, 100%.

2.8 REFERENCES

ACIE2005 Bräse, S.; Ggil, C.; Knepper, K.; Zimmermann, V. *Angewandte Chemie International Edition* **2005**, *44*, 5188-5240.
ACIE2015a Ghislieri, D.; Gilmore, K.; Seeberger, P. *Angewandte Chemie International Edition* **2015**, *54*, 678-682.
ACIE2015b Nagatomo, M.; Nishiyama, H.; Fujino, H.; Inoue, M. *Angewandte Chemie International Edition* **2015**, *54*, 1537-1541.
Adamed2011 Majka, Z.; Kolaczkowski, M.; Czopek, A.; Miodonski, A.; Kaminski, K.; Pawlowski, M.; Zagorska, A.; Zajdel, P.; Kowalski, P.; Mitka, K.; Jaskowska, J.; Wesolowska, A.; Mierzejewski, P.; Bienkowski, P. (Adamed Sp Zoo) WO2013001498 (prior. 29/06/2011, PL).
AJC1960 Handley, G.; Nelson, E.; Somers, T. *Australian Journal of Chemistry* **1960**, *13*, 129-144.
Array2010 Boys, M.; Burgess, L.; Groneberg, R.; Harvey, D.; Huang, L.; Kercher, T.; Kraser, C.; Laird, E.; Tarlton, E.; Zhao, Q. (Array Biopharma Inc) WO2011130146, US2013131039, US8962596, EP2558468 (prior. 14/04/2010, US).
ASC2007 Zhu, D.; Mukherjee, C.; Biehl, E. R.; Hua, L. *Advanced Synthesis & Catalysis* **2007**, *349*, 1667-1670.
Bayer1995 Wilhelm, S. M.; Vanzandt, M. C.; Kluender, H. C.; Dixon, B. R.; Wolanin, D. J.; Bullock, W. H.; Schneider, S. (Bayer Corp) US5886022 (prior. 05/06/1995, US).
BCSF1984 Picard, P.; Moulines, J.; Lecoustre, M. *Bulletin de la Societe Chimique de France* **1984**, 2, 65-70.
BCSJ1990 Chatani, N.; Hanafusa, T. *Bulletin of the Chemical Society of Japan* **1990**, *63*, 2134-2135.
BEJ2017 Xue, Y.; Zhong, H.; Zou, S.; Zheng, Y. *Biochem. Eng. J.* **2017**, *125*, 190-195.
Biocryst2009 Babu, Y. S.; Kotian, P. L.; Kumar, V. S.; Wu, M.; Lin, T. (Biocryst Pharm Inc) WO2011031554, US2012149662, EP2470537 (prior. 27/08/2009, US).
BMCL2006 Burgos-Lepley, C. E.; Thompson, L. R.; Kneen, C. O.; Osborne, S. A.; Bryans, J. S.; Capiris, T.; Suman-Chauhan, N.; Dooley, D. J.; Donovan, C. M.; Field, M. J.; Vartanian, M. G.; Kinsora, J. J.; Lotarski, S. M.; El-Kattan, A.; Walters, K.; Cherukury, M.; Taylor, C. P.; Wustrow, D. J.; Schwarz, J. B. *Bioorganic & Medicinal Chemistry Letters* **2006**, *16*, 2333-2336.
BMCL2011 Kerns, J. K.; Nie, H.; Bondinell, W.; Widdowson, K. L.; Yamashita, D. S.; Rahman, A.; Podolin, P. L.; Carpenter, D. C.; Jin, Q.; Riflade, B.; Dong, X.; Nevins, N.; Keller, P. M.; Mitchell, L.; Tomaszek, T. *Bioorganic & Medicinal Chemistry Letters* **2011**, *21*, 4409-4415.
Bromine2001a Oren, J. (Bromine Compounds Ltd) WO2003002517 (prior. 28/06/2001, IL).
Bromine2001b Oren, J. (Bromine Compounds Ltd) WO2003002504 (prior. 28/06/2001, IL).
Caffaro2004 Danieli, B.; Delogu, P.; De Rosa, S.; Fugazza, L.; Rosa, S. D. (Caffaro Srl, Serichum Srl, Caffaro Spa) WO2005090310, WO2005090310, US2007208175, US7381823, EP1732899 (prior. 17/03/2004, IT).
CatC2015 Xue, Y.-P.; Wang, Y.-P.; Xu, Z.; Liu, Z.-Q.; Shu, X.-R.; Jia, D.-X.; Zheng, Y.-G.; Shen, Y.-C. *Catalysis Communications* **2015**, *66*, 121-125.
CC2012 Moriyama, K.; Ishida, K.; Togo, H. *Chemical Communications* **2012**, *48*, 8574-8576.
CES2017 Huang, J.; Sang, F.; Luo, G.; Xu, J. *Chem. Eng. Sci.* **2017**, *173*, 507-513.
CZ1911 Guareschi, I. *Chem. Zentralbl.* **1911**, *82*, 362.
Daichii1988 Hayakawa, I.; Atarashi, S.; Imamura, M.; Kimura, Y. (Daiichi Seiyaku Co) US5508428, EP357047, EP529688 (prior. 31/08/1988, JP).
Dalian2011 Zhao, W.; Li, Y.; Song, Q.; Hu, Y. (Univ Dalian Technology) CN102093237 (prior. 04/01/2011, CN).

Divis2010 Divi, M.; Padakandla, G.; Nowshuddin, S.; Prasad, M. (Divis Lab Ltd) IN201001640, US2011306787, US8431739 (prior. 14/06/2010, IN).
EJOC2011 Ishikawa, H.; Bondzic, B. P.; Hayashi, Y. *European Journal of Organic Chemistry* **2011**, *2011*, 6020-6031.
Erregierre2002 Ferrari, M.; Ghezzi, M.; Belotti, P. (Erregierre Spa) WO2004031126, US2004063997, US6846950, EP1558564 (prior. 01/10/2002, IT).
GC2017 Marsini, M. A.; Buono, F. G.; Lorenz, J. C.; Yang, B.-S.; Reeves, J. T.; Sidhu, K.; Sarvestani, M.; Tan, Z.; Zhang, Y.; Li, N.; Lee, H.; Brazzillo, J.; Nummy, L. J.; Chung, J. C.; Luvaga, I. K.; Narayanan, B. A.; Wei, X.; Song, J. J.; Roschangar, F.; Yee, N. K.; Senanayake, C. H. *Green Chemistry 2017*, *19*, 1454-1461.
Givaudan2000 Goeke, A.; Berg-Schulz, K.; Bajgrowicz, J.; Berg-Schultz, K.; Bajgrowicz, J. A. (Givaudan Sa) WO200216307, US2004023839, EP1182190, EP1311470 (prior. 25/08/2000, EP).
Godecke1989a Geibel, W.; Herrmann, W.; Witzke, J.; Hartenstein, J. (Goedecke Ag) DE3928182, US5091567, EP414274 (prior. 25/08/1989, DE).
Godecke1989b Augart, H.; Gebhardt, U.; Herrmann, W. (Goedecke Ag) DE59007550, US6054482, EP414263 (prior. 25/08/1989, DE).
Godecke1989c Steiner, K.; Herrmann, W.; Crone, G.; Combs, C. S. (Goedecke Ag) DE59003771, US5068413, EP414275 (prior. 25/08/1989, DE).
Gruenenthal2002 Przewosny, M. T.; Puetz, C. (Gruenenthal Gmbh) DE10203122, WO2003062185, US2005043565, US7141695, EP1472210 (prior. 25/01/2002, DE).
H2009 Wu, L.; Yang, C.; Yang, L.; Yang, L. *Heterocycles* **2009**, *78*, 977 - 982.
HCA1991 Griffiths, G.; Mettler, H.; Mills, L. S.; Previdoli, F. *Helvetica Chimica Acta* **1991**, *74*, 309-314.
Hikal2002 Kuppuswamy, N.; Hariharan, S.; Mariadas, A. (Hikal Ltd) WO2004046108, US2006135789, US7632953, EP1599446 (prior. 20/11/2002, EP, WO).
Hikal2008 Gupta, R. P.; Kuppuswamy, N.; Mariadas, A. (Hikal Ltd) IN200802138, WO2010023694 (prior. 01/09/2008, IN).
Hikal2012 Yerande, S. G.; Thakur, R. M.; Sharma, S. S.; Gangopadhyay, A. K.; Rupp, H.; Kubavat, H.; Nagarajan, K.; Selvan, A.; Nunna, R.; Jalajakshi, V. I. (Hikal Ltd) WO2013190357 (prior. 18/06/2012, IN).
Hubei2012 Chen, T.; Deng, D.; Guo, H.; Li, Z.; Wang, P. (Hubei Chuyang Sci&Technology Co Ltd) CN102690207 (prior. 31/05/2012, CN).
IJBM2019 Xu, Z.; Huang, J.; Xia, C.; Zou, S.; Xue, Y.; Zheng, Y. *International Journal of Biological Macromolecules* **2019**, *130*, 117-124.
Ind2004 Gopal, D. (Individual) US2005148792 (prior. 02/01/2004, US).
Ipca2006 Kumar, A.; Soudagar, S. R.; Nijasure, A. M.; Panda, N. B.; Gautam, P.; Thakur, G. R. (Ipca Laboratories Ltd) IN200601762, US2008103334 (prior. 26/10/2006, IN).
JCR2004 Li, J.-H.; Li, Z.-G.; Chen, Q.-G. *Journal of Chemical Research* **2004**, *2004*, 758-759.
JCS1911 Thorpe, J. F.; Thole, F. B. *Journal of the Chemical Society, Transactions* **1911**, *99*, 422 - 448.
JCS1913 Thorpe, J. F.; Wood, A. S. *Journal of the Chemical Society, Transactions* **1913**, *103*, 1569-1578.
JCS1923 Birch, S. F.; Kon, G. A.; Norris, W. S.; Thorpe, J. F. *Journal of the Chemical Society, Transactions* **1923**, *123*, 1361-1374.
JCS1924 Dickens, F.; Horton, L.; Thorpe, J. F. *Journal of the Chemical Society, Transactions* **1924**, *125*, 1830-1840.
JCS1927 Sircar, S. S. G. *Journal of the Chemical Society (Resumed)* **1927**, 600-605.
JCSPT11986 Bell, V. L.; Holmes, A. B.; Hsu, S.-Y.; Mock, G. A.; Raphael, R. A. *Journal of the Chemical Society, Perkin Transactions 1* **1986**, 1507-1514.
JFC2012 Sprecher, H.; Payán, M. N. P.; Weber, M.; Yilmaz, G.; Wille, G. *Journal of Flow Chemistry* **2012**, *2*, 20-23.
Jiangsu2005 Peng, Z.; Zhu, W. (Jiangsu Nhua Pharm Group Co Ltd) CN1880299 (prior. 13/06/2005, CN).
Jiangxi2017 Xu, J.; Pang, Z.; Huang, J.; Huang, W.; Luo, G.; Zhou, L. (Jiangxi Tonghe Pharmaceutical Co Ltd) CN107011197 (prior. 21/04/2017, CN).
JLAC1965 Schafer, H. *Justus Liebigs Annalen der Chemie* **1965**, *688*, 113 - 121.
JMC1965 Grogan, C. H.; Geschickter, C. F.; Freed, M. E.; Rice, L. M. *Journal of Medicinal Chemistry* **1965**, **8**, 62-73.
JMC1990 Badger, A. M.; Schwartz, D. A.; Picker, D. H.; Dorman, J. W.; Bradley, F. C.; Cheeseman, E. N.; DiMartino, M. J.; Hanna, N.; Mirabelli, C. K. *Journal of Medicinal Chemistry* **1990**, *33*, 2963-2970.
JOC1988 Callahan, J. F.; Newlander, K. A.; Bryan, H. G.; Huffman, W. F.; Moore, M. L.; Yim, N. C. F. The *Journal of Organic Chemistry* **1988**, *53*, 1527-1530.
JOC2005 Stymiest, J. L.; Mitchell, B. F.; Wong, S.; Vederas, J. C. *The Journal of Organic Chemistry* **2005**, *70*, 7799-7809.
JOC2010 Niphakis, M. J.; Turunen, B. J.; Georg, G. I. *The Journal of Organic Chemistry* **2010**, *75*, 6793-6805.
Lonza1988 Mettler, H. P.; Griffiths, G.; Mills, L.; Previdoli, F. (Lonza Ag, Lonza Ltd) US4956473, US4958044, EP358092 (prior. 01/09/1988, CH).
Lonza1989 Mettler, H. P.; Previdoli, F. (Lonza Ag, Lonza Ltd) US5095148, US5130455, US5136091, US5149870, EP432504 (prior. 16/11/1989, CH).
Matrix2003a Satyanarayana, C.; Ramanjaneyulu, G.; Kumar, I. (Matrix Lab Ltd) IN198649, WO200411098 (prior. 12/06/2003, IN).
Matrix2003b Satyanarayana, C.; Ramanjaneyulu, G.; Kumar, I.; Chava, S.; Gorantla, S.; Indukuri, V. (Matrix Lab Ltd) IN200300480, WO2004110342 (prior. 12/06/2003, IN).
MCR2012 Mallesha, L.; Mohana, K.; Veeresh, B. *Medicinal Chemistry Research* **2012**, *21*, 1-9.

Medichem2005 Bosch, I. L.; Arnalot Aguilar, C.; Burgarolas Montero, M. (Medichem Sa) WO2006090208, US2009043126, EP1853551 (prior. 28/02/2005, EP, WO).

Merck2013 Childers, M. L.; Fuller, P.; Guerin, D.; Katz, J. D.; Pu, Q.; Scott, M. E.; Thompson, C. F.; Martinez, M.; Falcone, D.; Torres, L.; Deng, Y.; Kuruklasuriya, R.; Zeng, H.; Bai, Y.; Kong, N.; Liu, Y.; Zheng, Z. (Merck Sharp & Dohme) WO2014146491, WO2014146249, EP2976339, US2016272632, US9957264 (prior. 19/03/2013, WO).

NHWA2005 Peng, Z.; Zhu, W. (Nhwa Pharma Corp, Enhua Pharma Corp, Xuzhou Enhua Pharm Ind Group Co Ltd) CN1740161, WO2007019752, US2009099362, US7667071, EP1939170 (prior. 19/08/2005, CN).

Nisshin2001 Fukutomi, R.; Inoue, H.; Kawamura, K.; Kishimoto, T.; Suzuki, M.; Shibayama, R.; Kojima, K.; Hagihara, K. (Kyorin Pharm Co Ltd, Nisshin Pharma Inc) JP2003538161, WO2003035645, US2005085508, US2008194826, US7375115, EP1443046 (prior. 09/10/2001, JP).

OA2010 Organic azides: syntheses and applications, Ed. Bräse, S.; Banert. K. **2010** John Wiley & Sons Ltd.

OBC2008a Baumann, M.; Baxendale, I.; Ley, S.; Nikbin, N.; Smith, C.; Tierney, J. *Organic & Biomolecular Chemistry* **2008**, *6*, 1577-1586.

OBC2008b Baumann, M.; Baxendale, I.; Ley, S.; Nikbin, N.; Smith, C. *Organic & Biomolecular Chemistry* **2008**, *6*, 1587-1593.

OL2003 Fillion, E.; Fishlock, D. *Organic Letters* **2003**, *5*, 4653-4656.

OL2014 Hu, G.; Xu, J.; Li, P. *Organic Letters* **2014**, *16*, 6036-6039.

OPPI2013 Divi, M.; Rao, B.; Rao, A. *Organic Preparations and Procedures International 2013*, **45**, 504-506.

OPRD1998 am Ende, D.; DeVries, K.; Clifford, P.; Brenek, S. *Organic Process Research & Development* **1998**, *2*, 382-392.

OPRD2006 Wiss, J.; Fleury, C.; Onken, U. *Organic Process Research & Development* **2006**, *10*, 349-353.

OPRD2009 Kopach, M.; Murray, M.; Braden, T.; Kobierski, M.; Williams, O. *Organic Process Research & Development* **2009**, *13*, 152-160.

OPRD2016 Katuri, J.; Ekkundi, V.; Nagarajan, K. *Organic Process Research & Development* **2016**, *20*, 1828-1832.

OR1967 Jones, G. *Organic Reactions* **1967**, 15.

PB2014 Zhang, X.; Liu, Z.; Xue, Y.; Zheng, Y. *Process Biochemistry* **2014**, *49*, 2141-2148.

PB2018 Zou, S.; Huang, J.; Xue, Y.; Zheng, Y. *Process Biochemistry* **2018**, *65*, 93-99.

Pfizer2003 Burns, M.; Wong, J. (Pfizer Prod Inc, Pfizer Inc) WO2004111256, US2005009154, EP1639118 (prior. 19/06/2003, US).

Pfizer2006 Davies, S.; Kelleher, P.; Tully, W. E. (Pfizer Sci & Technology Ireland Ltd) WO2007129286, EP2024324 (prior. 05/05/2006, WO, EP).

Pfizer2007 Brickner, S.; Zhang, Z.; Chen, J.; Li, Z. B.; Marfat, A.; Reilly, U. D.; Mitton-Fry, M.; Plotkin, M. A.; Robinson, S.; Subramanyam, C. (Pfizer Inc, Agouron Pharma) WO2008139288, US2008280879, EP2155716, EP2481735 (prior. 09/05/2007, US).

Procos2001a Velardi, F.; Fornaroli, M. (Procos Spa) IT1324891, US2003009055, US6521788 (prior. 29/05/2001, IT).

Procos2001b Peverali, D.; Fornaroli, M.; Velardi, F. (Procos Spa) IT1327011, US6518456 (prior. 21/12/2001, IT).

PSP2004 Leggett, D. *Process Safety Progress* **2004**, *23*, 21-28.

RichterGedeon1998 Gizur, T.; Lengyel, Z.; Szalai, K. (Richter Gedeon Vegyeszeti Gyar) HU9803034, WO200039074, US6765114, EP1140793 (prior. 29/12/1998, HU).

S1983 New, J.; Yevich, J. *Synthesis* **1983**, *1983*, 388-389.

SC1985 Fitjer, L.; Quabeck, U. *Synthetic Communications* **1985**, *15*, 855-864.

Serichim2010 Delogu, P.; Cordero, L.; De Rosa, S.; Puntin, G.; Velardi, R. (Serichum Srl) EP2368872 (prior. 25/03/2010, EP).

Servier2004 De Nanteuil, G.; Benoist, A.; Combettes, M.; Harley, E. (Lab Servier Sa) FR2870538, US2005261501, US7439263, EP1598350 (prior. 19/05/2004, FR).

Solchem2001 Rossi, P.; Vecchio, E. (Solchem Italiana Spa) IT1324716, WO200274727, US2004063994, EP1373186 (prior. 16/03/2001, IT).

Synlett2003 Chen, Z.; Chen, Z.; Jiang, Y.; Hu, W. *Synlett* **2003**, *2003*, 1965-1966.

T1993 Gruhn, A. G.; Reusch, W. *Tetrahedron* **1993**, *49*, 8159-8168.

T2003a Bryans, J. S.; Chessum, N. E. A.; Huther, N.; Parsons, A. F.; Ghelfi, F. *Tetrahedron* **2003**, *59*, 6221-6231.

T2003b Cagnoli, R.; Ghelfi, F.; Pagnoni, U. M.; Parsons, A. F.; Schenetti, L. *Tetrahedron* **2003**, *59*, 9951-9960.

T2005 Chen, Z.; Chen, Z.; Jiang, Y.; Hu, W. *Tetrahedron* **2005**, *61*, 1579-1586.

Taicang2014a Zhang, W. (Taicang Yuntong Biochem. Eng.Co Ltd) CN104230735 (prior. 28/08/2014, CN).

Taicang2014b Zhang, W. (Taicang Yuntong Biochem. Eng.Co Ltd) CN104496833 (prior. 26/11/2014, CN).

Taicang2014c Zhang, W. (Taicang Yuntong Biochem. Eng.Co Ltd) CN104447372 (prior. 26/11/2014, CN.

Taicang2014d Zhang, W. (Taicang Yuntong Biochem. Eng.Co Ltd) CN104402744 (prior. 28/11/2014, CN).

Taicang2015a Zhang, W. (Taicang Yuntong Biochem. Eng.Co Ltd) CN105061239 (prior. 18/08/2015, CN).

Taicang2015b Zhang, W. (Taicang Yuntong Biochem. Eng.Co Ltd) CN105061240 (prior. 18/08/2015, CN).

Taicang2015c Zhang, W. (Taicang Yuntong Biochem. Eng.Co Ltd) CN105061241 (prior. 18/08/2015, CN).

Taicang2015d Zhang, W. (Taicang Yuntong Biochem. Eng.Co Ltd) CN105198763 (prior. 18/08/2015, CN).

TaicangYCP2008 Zhang, W. (Taicang Yuntong Chemical Plant) CN101417975 (prior. 01/12/2008, CN).

Taro2002 Bercovici, S.; Sasson, S.; Ulanenko, K.; Konstantin, U. (Taro Pharm Ind Ltd, Taro Pharm USA Inc, Taro Pharm Ltd) WO2003089403, US2004034248, EP1494996 (prior. 16/04/2002, US).

Teva2000 Montanari, S. (Teva Pharm Ind Ltd, Teva Pharm USA Inc) WO200236545, US2002107395, US2003195358, US6613904, US6881843, EP1337506 (prior. 02/11/2000, US).

Tianyuan2004 Bai, X.; Xu, X.; Wu, J. (Tianyuan Chem Tech Co Ltd) CN1727329 (prior. 28/07/2004, CN).

TL2019 Katuri, J. ; Nagarajan, K. *Tetrahedron Letters* **2019**, *60*, 552-556.

UBC2002 Abbott, F.; Karagiozov, S. (Univ British Columbia) US2006223888, WO2004054957 (prior. 16/12/2002, US).

WarnerL1957 Wessem, G.; Sakal, E. (Warner Lambert Pharmaceutical) US2960441 (prior. 17/09/1957, US).

WarnerL1974 Satzinger, G.; Hartenstein, J.; Herrmann, M.; Heldt, W. (Warner-Lambert Co) DE2460891, US4024175, US4087544 (prior. 21/12/1974, DE).

WarnerL1976 Hartenstein, J.; Satzinger, G. (Warner-Lambert Co) DE2611690, US4152326 (prior. 19/03/1976, DE).

WarnerL1989a Seamans, R.; Zeller, J.; Johnson, D.; Jennings, R. (Warner Lambert Co) US5319135 (prior. 25/08/1989, US).

WarnerL1989b Jennings, R.; Seamans, R.; Zeller, J.; Johnson, D. (Warner-Lambert Co) US5132451, EP414262 (prior. 25/08/1989, US).

WarnerL1997a Bryans, J.; Morrell, A. (Warner Lambert Co) WO9914184, US6465689, EP1015415 (prior. 18/09/1997, US).

WarnerL1997b Deering, C.; Mennen, K.; Ramage, R. (Warner Lambert Co) WO9918063, US2002010371, US6294690, US6538148 (prior. 07/10/1997, US).

WarnerL1997c Bryans, J.; Capiris, T.; Horwell, D.; Kneen, C.; Wustrow, D. (Warner Lambert Co) WO9931075, US6518289, EP1047678, EP1475371 (prior. 16/12/1997, US).

WarnerL2000 Schrier, D.; Welgus, H.; Wustrow, D. (Warner Lambert Co) US2002072533, US2004097405, US6620829, US7189747, EP1199072 (prior. 17/10/2000, US)

WarnerL2002 Dooley, D.; Wustrow, D. (Warner Lambert Co Llc) WO2003063845, US2003176504, US2006100281, US7026505, US7138542, EP1469841 (prior. 31/01/2002, US).

WarnerL2003 Schwarz, J. B.; Wustrow, D. J. (Warner-Lambert Company, Llc) WO2004065361, US2004147608, US7030267, EP1592661 (prior. 2003/01/22, US).

Zach2006 Ghelfi, F.; Cotarca, L.; Roncaglia, F.; Giovanetti, R.; Nicoli, A. (Zach System Spa) WO2008004115, EP2038249, US2010029983, US8143443 (prior. 30/06/2006, EP, US).

Zach2013 Verzini, M.; Cotarca, L.; Belluzzo, F.; Soriato, G.; Urbani, D.; Pace, E. (Zach System Spa) ITMI20131757, WO2015059150, US2016237025, US9481635, EP3060544 (prior. 22/10/2013, IT).

Zambon2000 Cannata, V.; Nicoli, A.; Corcella, F. (Zambon Groups Spa, Zach System Spa) IT1319234, WO200234709, US2004068011, US2007043237, US7199266, US7365227, EP1341749, EP1475366 (prior. 23/10/2000, IT).

Zambon2003 Arrighi, K.; Cannata, V.; Corcella, F.; Marchioro, G.; Nicoli, A.; Paiocchi, M.; Villa, M. (Zambon Groups Spa, Zach System Spa) IT1349674, WO2005044779, US2007066843, US2009326266, EP1682488 (prior. 11/11/2003, IT).

Zambon2004a Giovanetti, R.; Nicoli, A.; Verzini, M.; Soriato, G.; Cotarca, L. (Zambon Groups Spa, Zach System Spa) IT1354905, WO2005092837 US2007293700, US7417166, EP1919856 (prior. 25/03/2004, IT).

Zambon2004b Villa, M.; Paiocchi, M.; Arrighi, K.; Corcella, F.; Cannata, V.; Soriato, G.; Verzini, M. (Zambon Group Spa, Zach System Spa) IT1356528, WO2006000562, US2008097122, US7759517, EP1765770 (prior. 24/06/2004, IT).

Zhejiang2006 Liu, T.; Fan, W.; Huang, Y. (Zhejiang Chiral Medicine Chem Co Ltd) CN1861579, CN100404504, US2007287861, US7442834 (prior. 12/06/2006, CN).

ZhejiangEP2011 Zhao, J.; Wang, H.; Du, C.; Chen, W.; Li, B. (Zhejiang Excel Pharm Co Ltd) CN102363598 (prior. 25/11/2011, CN).

ZhejiangUT2014 Zheng, Y.; Xue, Y.; Liu, Z.; Su, X.; Huang, Y.; Weng, J.; Wang, Y.; Jia, D. (Univ Zhejiang Technology) CN104342463 (prior. 12/08/2014, CN).

ZhejiangUT2015 Xue, Y.; Zheng, Y.; Wang, Y.; Zou, S. (Univ Zhejiang Technology) CN104911225 (prior. 05/06/2015, CN).

ZhejiangUT2018 Xue, Y.; Zheng, Y.; Xu, Z.; Liu, Z. (Univ Zhejiang Technology) CN108486088 (prior. 14/02/2018, CN).

3 Clopidogrel

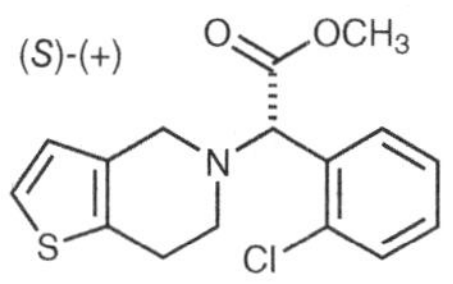

Year of discovery: Clopidogrel (*S* isomer), 1987 (Sanofi); racemic mixture, 1982 (Sanofi)
US approval: November 1997 (Plavix®, Sanofi Aventis/Bristol-Myer Squibb)
EU approval: July 1998 (Iscover®, Sanofi Aventis/Bristol-Myer Squibb)
Drug cat.: Antiplatelet agent
Uses: reduction of thrombotic events for persons with recent stroke, myocardial infarction or established peripheral vascular disease by inhibition of blood clots.

Thrombosis occurs with the formation of a blood clot inside a blood vessel, resulting in the obstruction of the flow of blood through the circulatory system. Blood clots are a major cause of heart attacks and ischemic strokes, both being a consequence of restricted blood flow. The prevention of platelet aggregation and clot formation can therefore be lifesaving and drugs for this objective are of great value.

Low dose acetylsalicylic acid (ASA), or Aspirin, is used for this purpose, the benefits largely overriding the increased probability of minor stomach bleeding at the low doses used. However, other alternatives are desirable, particularly for patients with gastric-related diseases and those who are ASA allergic or ASA sensitive. Some nucleoside/nucleotide analogues (for example: Cangrelor, Elinogrel, Ticagrelor) and some molecules containing a thienopyridine substructure were found to possess interesting activity in the field. Thienopyridine class drugs (Figure 3.1), operate *via* a different biochemical mechanism from ASA, by acting on inhibitors of the adenosine diphosphate receptor (P2Y12) on platelet cell membranes and have found significant clinical application.

Ticlopidine (Ticlid®, Sanofi-Aventis) was the first drug of this class to be marketed but, due to some adverse side effects and to its toxicity towards white blood cells, its use remained limited. Clopidogrel, an improved version of this drug, has been demonstrated to be extremely potent whilst having low levels of side effects. Unlike its predecessor, Clopidogrel achieved great commercial success.

Clopidogrel is a prodrug. It is itself not active and requires activation in the liver by cytochrome P450 enzymes, including CYP2C19 (Scheme 3.1).

The selection of the *S* isomer as the drug candidate was obvious from the beginning (CRC2012) as the first biological tests indicated that only the *S*-isomer (the eutomer) had antiplatelet and antithrombotic effects while the *R*-isomer (the distomer) was inactive and clearly less well tolerated.

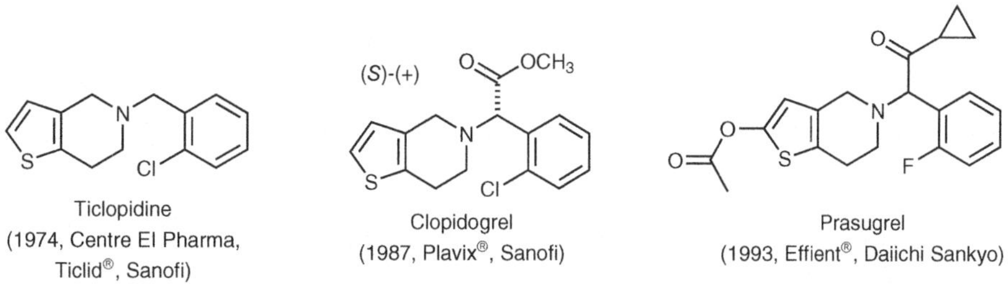

Figure 3.1 Marketed thienopyridine drugs as antiplatelet agents

Retrosynthesis in the Manufacture of Generic Drugs: Selected Case Studies, First Edition, Pedro Paulo Santos and William Heggie.

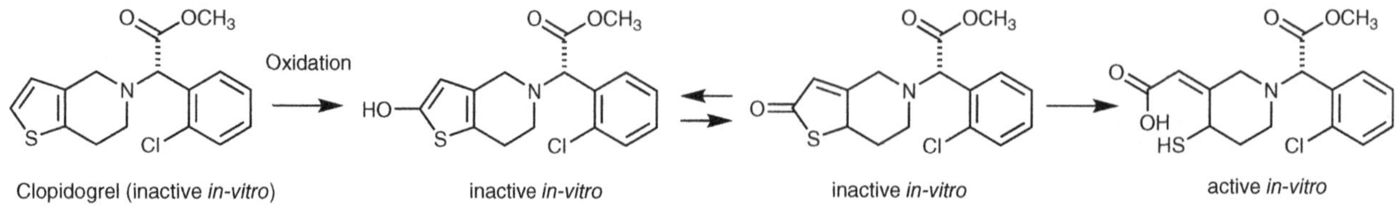

Scheme 3.1 Metabolic activation of Clopidogrel

3.1 WORLDWIDE SALES AND PATENT STATUS

The following graphs illustrate how research carried out in the pharmaceutical industry towards the production of the API and worldwide sales of the finished drug product evolved.

- US basic patent term 12/02/1998 (US4847265, Sanofi1987), USA Product Patent term extension 17/10/2001.
- Annual production[1]: 2019, 850.0 t; 2018, 808.3 t; 2017, 763.9 t.
- Annual WW sales[1]: 2019, 2883.9M USD; 2018, 3085.2M USD; 2017, 3111.8M USD.

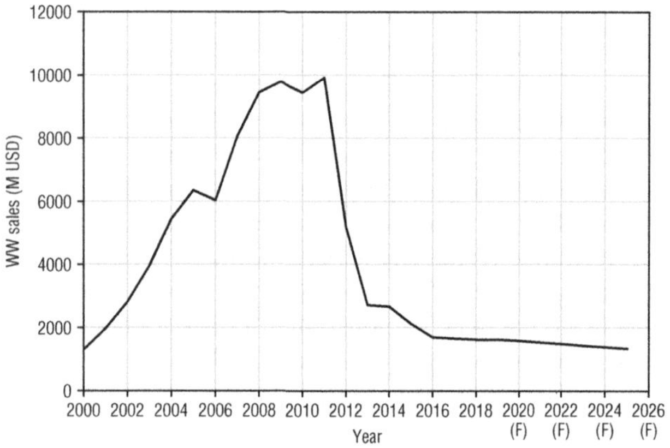

Figure 3.2 Clopidogrel WW sales evolution (in M USD)

Source: Data from GlobalData [Online] https://globaldata.com/

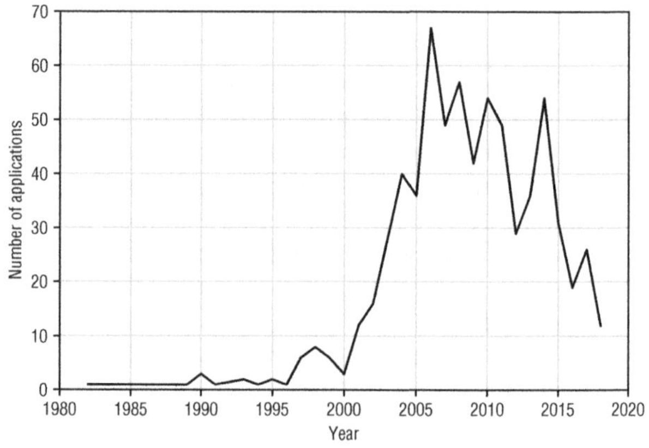

Figure 3.3 Patent applications on clopidogrel filed by year

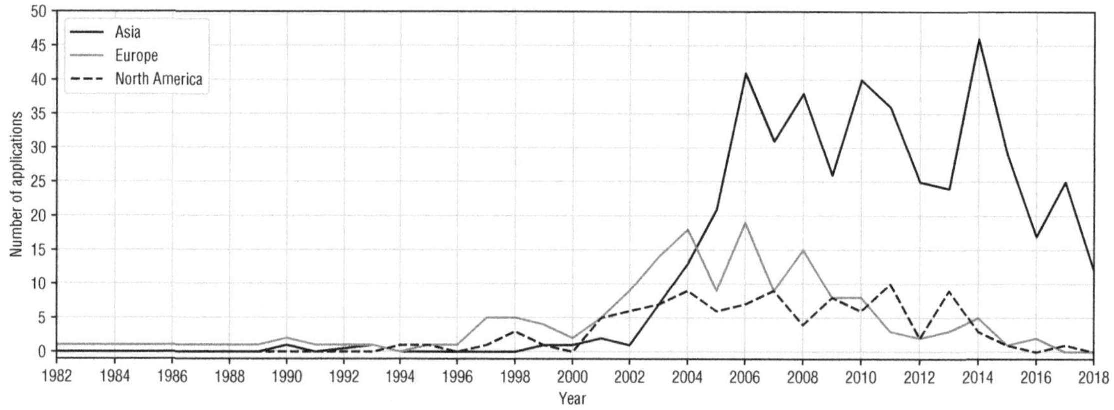

Figure 3.4 New patent applications on clopidogrel filed by region

1 Newport Modular Platform – Clarivate Analytics [Online] https://clarivate.com/cortellis/ [2020, April]

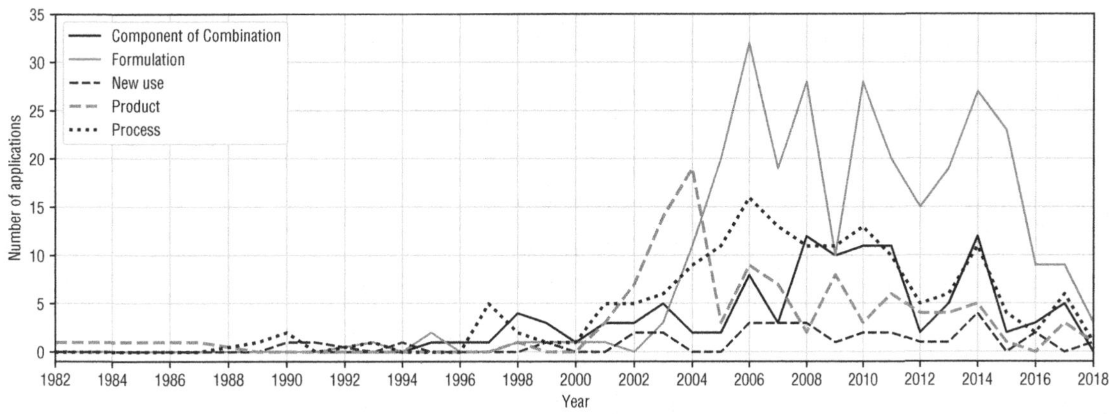

FIGURE 3.5 PATENT APPLICATIONS ON CLOPIDOGREL FILED BY TYPE OF PATENT

3.2 CLOPIDOGREL STRUCTURE AND GENERAL RETROSYNTHETIC STRATEGY

Clopidogrel is a relatively small molecule containing an α-amino ester where the amine is incorporated into a hydrothienopyridine unit. Some important structural features of the active pharmaceutical ingredient (see Figure 3.6) are:

- Contains one chiral carbon atom and the API is required to be enantiopure.
- It is formally an ester of an α-amino acid and formation of at least one of the C-N bonds is expected to be key in the assembly of the molecule.
- Construction of the hydrothienopyridine ring system can be envisaged to start from an appropriately mono-functionalized thiophene derivative. The two-ring system may be produced before connecting it to the aryl acetate group or the closure of the saturated ring may be completed at a later stage. At present, the more advanced starting material, tetrahydrothienopyridine (THTP), should be readily available commercially as its hydrochloride salt.

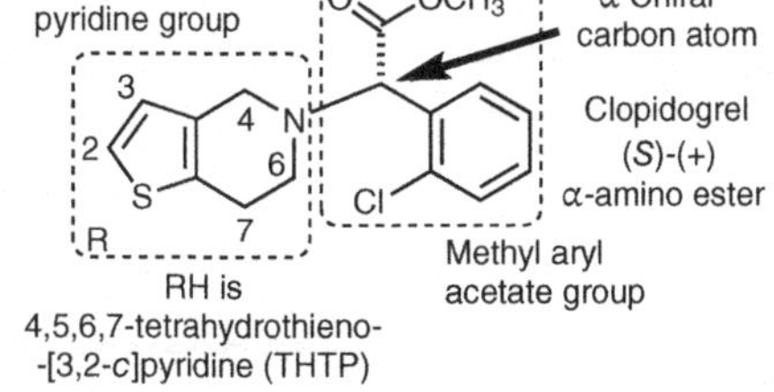

FIGURE 3.6 KEY STRUCTURE FEATURES OF CLOPIDOGREL

The nitrogen atom plays a major role in the retrosynthetic analysis of Clopidogrel. Its central position allows for disconnection of the molecule into two almost equally sized fragments.

3.2.1 Chirality and Clopidogrel

Three main strategies can be used to obtain chiral-enantiopure molecules such as Clopidogrel (illustrated in Scheme 3.2).

- Chiral-pool approach (Scheme 3.2, **a**): readily available optically active compounds (mainly natural products) are converted into the intended chiral molecule. Frequently, the chiral elements present in the starting materials remain unchanged during the whole of the synthetic route. Because of the pressure to produce new optically active compounds, the number and type of readily available chiral building blocks is constantly increasing.
- Resolution (Scheme 3.2, **b**): separation of a racemic mixture of enantiomers. This is often done by classical resolution (*via* diastereomeric salts or derivatives) but direct conglomerate crystallization, kinetic resolution and dynamic kinetic resolution (DKR) may be good alternatives. Simulated moving bed (SMB) chromatography can be very useful to some extent but its application may be limited due to throughput restrictions and high operating costs.
- Asymmetric synthesis (AS, Scheme 3.2, **c**): by definition, is a chemical reaction (or reaction sequence) in which one or more new elements of chirality are formed in a substrate molecule and which produces the stereoisomeric (enantiomeric or diastereoisomeric) products in unequal amounts. Frequently, the existence of chiral features in substrates is used to induce the selective formation of a new chiral centre in the ensuing product (asymmetric induction). Hence, a new chiral centre is formed at a remote "achiral" point in an already optically active substrate. This procedure forms preferentially one of two diastereomers as opposed to one of two enantiomers. As this kind of reaction selectively creates a new chiral centre, it still can be classified as an asymmetric synthesis. Arguably to avoid ambiguity, it would seem preferable to replace AS by stereoselective, enantiomeric or diastereomeric synthesis, in accordance with the case in hand.

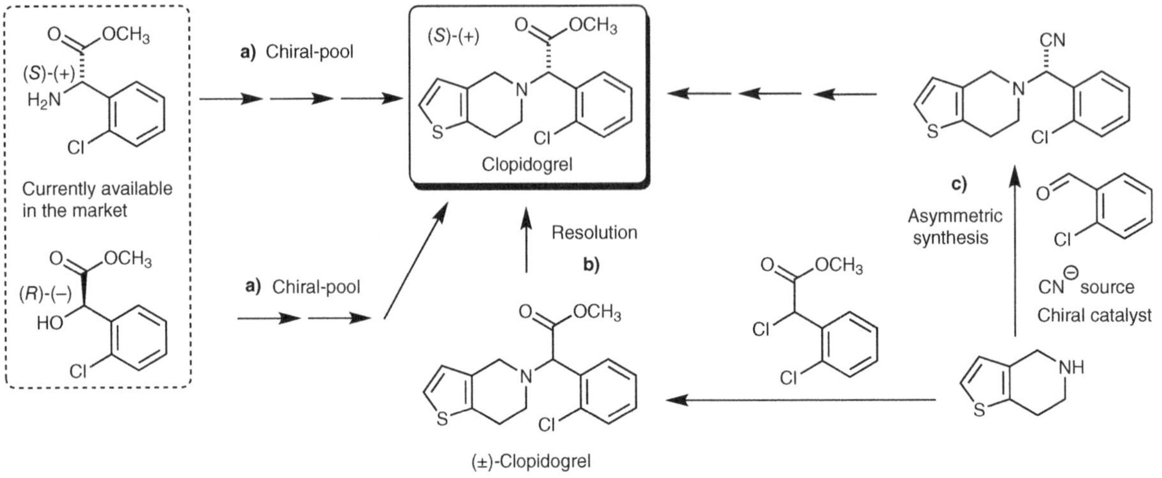

Scheme 3.2 General strategies to obtain (*S*)-Clopidogrel

As expected, all of these approaches have been used in the published synthetic routes to Clopidogrel.

3.2.1.1 The asymmetric synthesis strategy

Ideally, an asymmetric synthesis scenario would appear to be the correct choice, thus avoiding formation of the undesirable isomer. However, asymmetric reactions, which can be easily applied to industrial scale are still limited. The cost of a chiral catalyst or auxiliary, its bulk availability and recovery and reuse, in addition to the operating conditions of the chemical reaction where it is employed, can also of be a concern. Moreover, the enantiomeric purity obtained in any asymmetric reaction must be high and must be retained, preferably even increased, throughout the whole process up until the final API is isolated.

3.2.1.2 The resolution strategy

For a laboratory synthesis of a chiral enantiopure molecule with one chiral centre, a "simple" resolution process of the final molecule is often the first choice. When this separation is feasible, the synthetic routes which can be employed are generally simpler.

Resolution at the final stage also has advantages during the medicinal synthesis of possible target molecules. With the two pure enantiomers in hand, comparative activities and toxicities for both can be obtained. The distomer is considered as an impurity of the API and acceptable levels in the API may eventually be dependent on these data. Regulatory guidelines require analytical methods to control the unwanted isomer and the specification for the API includes a limit for it.

One of the most important resolution methods exploits the different solubility of the diastereomeric salts of the enantiomeric molecules and uses these properties to separate them. Since Clopidogrel is basic, a chiral enantiopure acid is the ideal choice for resolution (Scheme 3.3).

After physical separation, usually by filtration, pH adjustment regenerates the desired basic molecule. This

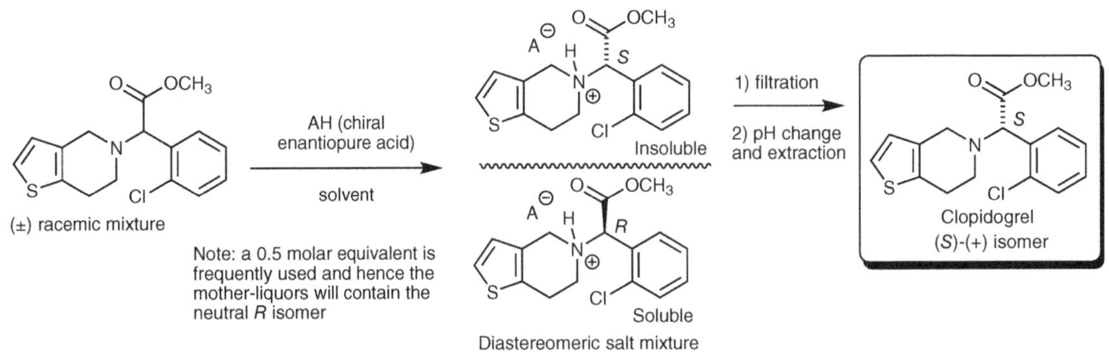

Scheme 3.3 Example of a resolution process to obtain Clopidogrel

procedure is frequently seen as a low-tech approach and hence sometimes considered as suboptimal and to be avoided. However, if properly developed, it can be very reliable and robust. Some disadvantages are inherent to the resolution process. The major drawbacks can be summarized as:

- The synthetic process generates the unwanted enantiomer in equal quantities and, unless this can be recycled, must be considered as waste.
- The intermediates of the unwanted isomer are transformed along with those of the desired product during the synthetic process. Hence, 50% extra reactants and solvents will generally be required along with a larger plant (or acceptance of reduced throughput for a smaller plant) to carry out the synthesis. This consequentially has a negative impact on the cost of the drug.
- Resolving agents are frequently expensive and sometimes difficult to find in large quantities unless a commodity type reagent can be used. Processes to recover them from mother liquors must be developed or must be accepted as waste, with the inherent cost of waste treatment. In most cases purification and reuse of the recovered resolution agents is mandatory to reduce process costs to a minimum.
- Extra processing steps are required. At least the formation of diastereomeric salts and their separation by filtration is obligatory. When the salt generated is the drug product to be marketed no further processing is required.However, in most cases, the salt that was used for the resolution is not the final drug substance. A counter ion switch must be implemented, mostly by reverting to the neutral compound and subsequent formation of the new salt.

It is sometimes possible to introduce the resolution step earlier rather than later in the synthetic route. When the unwanted isomer is considered to be process waste, an earlier resolution is generally better, thus avoiding carrying the unwanted isomer through to a later stage. Obviously, racemization of the chiral centres must be avoided during subsequent reactions, which might limit the conditions under which these are carried out.

A well-designed resolution process at the right place in the synthesis may compete effectively or even outdo an AS route. A detailed evaluation of the relative *pros* and *cons* must be carried out before a final decision is taken on which type of route is to be developed.

3.2.1.3 Recycling unwanted isomers as part of a resolution-based route

If an intermediate or final product containing the unwanted configuration was to be treated as waste, an asymmetric synthesis would ostensibly appear to be the correct choice. However, structural features inherent in the molecules may provide an opportunity to recycle the "wrong" enantiomer (the eutomer). For example, racemization of the chiral centre to generate the racemic mixture allows for a reiteration of the resolution step, thus producing additional API.

As stated above, in a classic resolution process, half of the racemic mixture is generally lost. In the absence of recycling, since Clopidogrel is a drug with particularly high worldwide sales, this would mean that huge quantities of unwanted waste material would be produced. Fortunately, the presence of a carbonyl carbon atom α to the chiral centre allows for recycling of the *R*-enantiomer. Racemization can be carried out either using base or acid to form an achiral enol or enolate, in equilibrium with the racemic carboxylic ester, resulting in the formation of a 50:50 mixture of the two enantiomers (Scheme 3.4). By putting this hypothesis into practice, the waste enantiomer scenario can be avoided, theoretically 0% enantiomer waste being possible.

Many intermediates that are encountered in various synthetic routes to Clopidogrel, share the α-amino carboxylic acid ester substructure with the API itself and hence, can be expected to be racemized using acidic or basic conditions.

In conclusion, the enormous material waste predicted in resolution-based processes can be avoided and these routes can easily compete or surpass an asymmetric synthetic route.

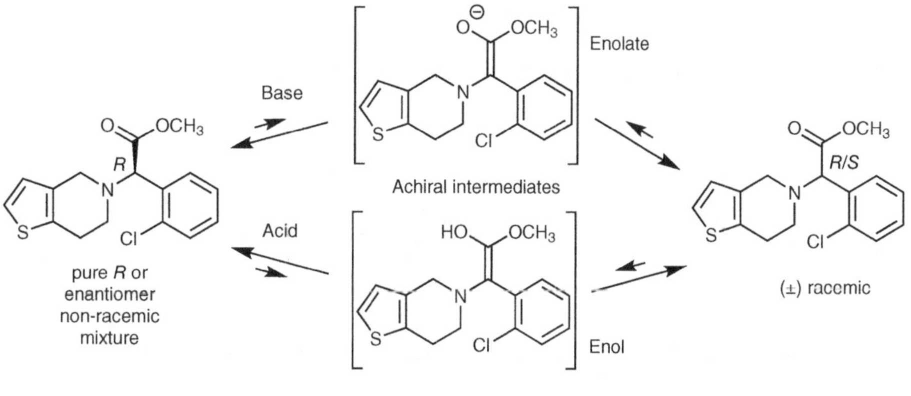

Scheme 3.4 Examples of racemizations of Clopidogrel isomer in acidic or basic conditions

3.2.2 Retrosynthetic analysis

The central nitrogen atom assumes a major role in most synthetic strategies and hence, there is a great deal of focus on the formation of C-N bonds.

3.2.2.1 Using a C-N cleavage (nucleophilic nitrogen)

Separation of the molecule into two large building blocks is a major objective in any retrosynthetic analysis of Clopidogrel. Two main disconnections are possible (Scheme 3.5) when the nucleophilic character of the nitrogen atom is taken into account. Hence, the nucleophilic nitrogen is either present in the hydrothienopyridine building block (**a**), or in the α-aryl acetate building block (**b**).

A priori, disconnection **a** looks more favourable, as the rate of S_N2 substitution reactions is extremely favourable due to the presence of the unsaturation at the two carbon atoms α to the one containing the leaving group (the carbonyl ester and the aryl ring). In addition, synthesis of a 1,2 bi-functionalized thiophene is circumvented. Nevertheless, the relatively bulky THTP nucleophile may eventually reduce the rate and yield of the S_N2 reaction. A simple retrosynthesis of the THTP building block is shown in Scheme 3.6.

A key step in the synthetic sequence to THTP is closure of the piperidine ring by the Pictet-Spengler reaction (Scheme 3.7). Initially, an imine (Schiff base) is formed which, upon protonation, is attacked by the electron-rich aromatic thiophene ring, forming the six-membered heterocycle.

3.2.2.2 Opening the saturated ring first

A different retrosynthetic strategy opens the saturated ring first (Scheme 3.8), the final step in the synthesis being the Pictet-Spengler reaction.

In all of the foregoing, the ester group was not modified during retrosynthesis but obviously, a carboxylic acid or an equivalent derivative can take the place of the ester group at several points in the retrosynthetic plan. The total of the various retrosynthetic routes considered so far are depicted in Scheme A.

In the construction of this scheme it has been assumed that resolution of the different racemic intermediates, starting materials or even racemic Clopidogrel is possible.

The scheme is divided into two parts, on left are the enantiopure compounds while the right shows only racemic mixtures. All arrows crossing the vertical dashed line represent a resolution step in the route process. Of course, chemistry is an investigational science and the feasibility and efficiency of each resolution has to be the subject of experimental studies. It should be noted that the scheme is quite symmetric and equivalent retrosynthetic steps can be seen as if reflected in the vertical dotted line.

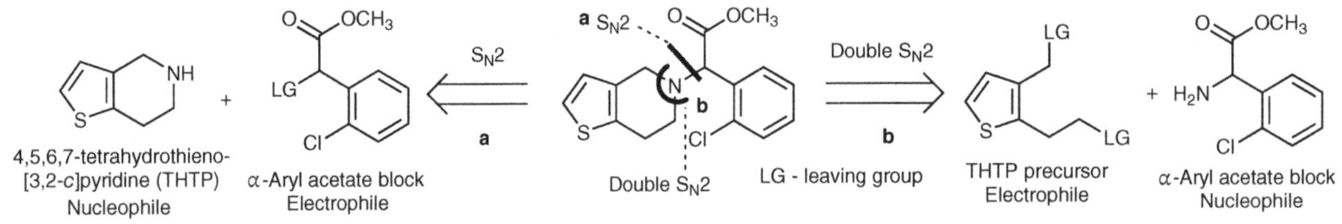

Scheme 3.5 Retrosynthesis of Clopidogrel using C-N disconnections

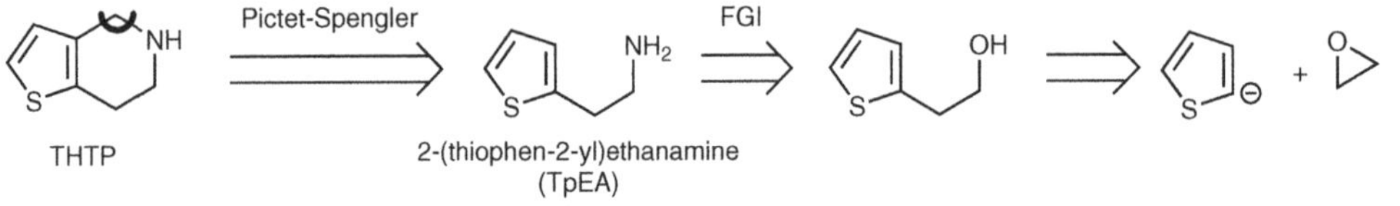

Scheme 3.6 Retrosynthetic analysis of THTP

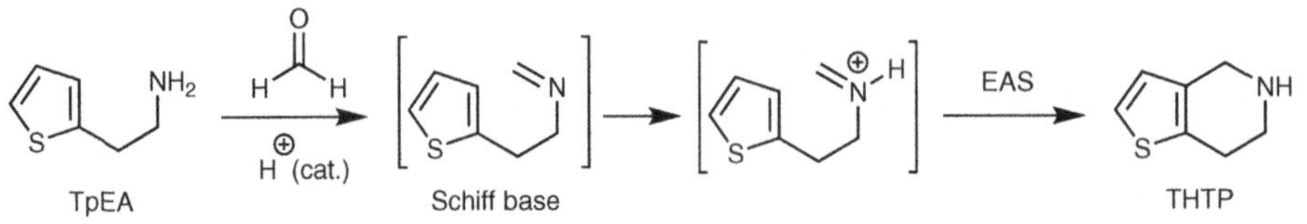

Scheme 3.7 The Pictet-Spengler reaction to produce THTP

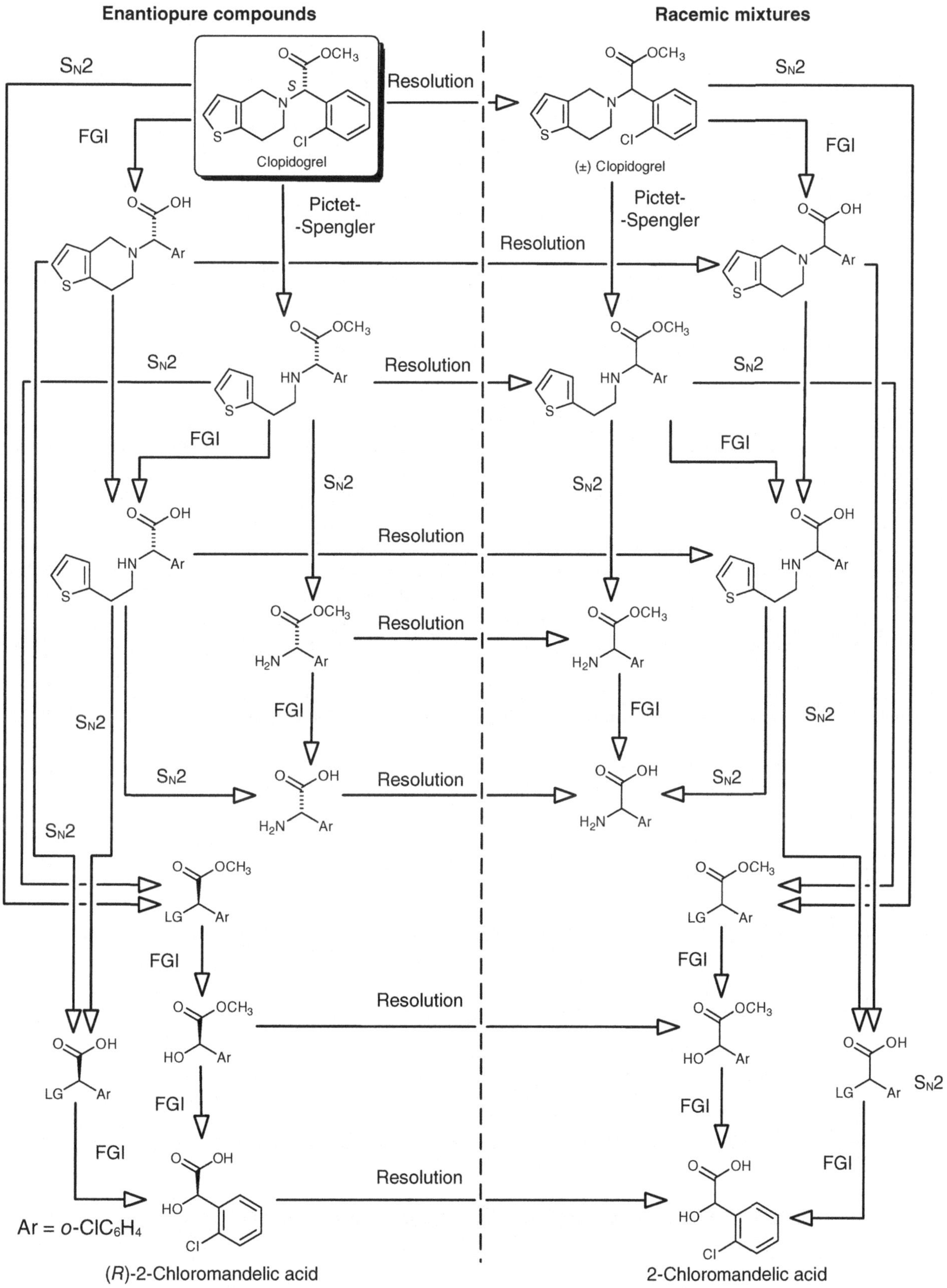

Scheme A Main retrosynthetic routes for Clopidogrel

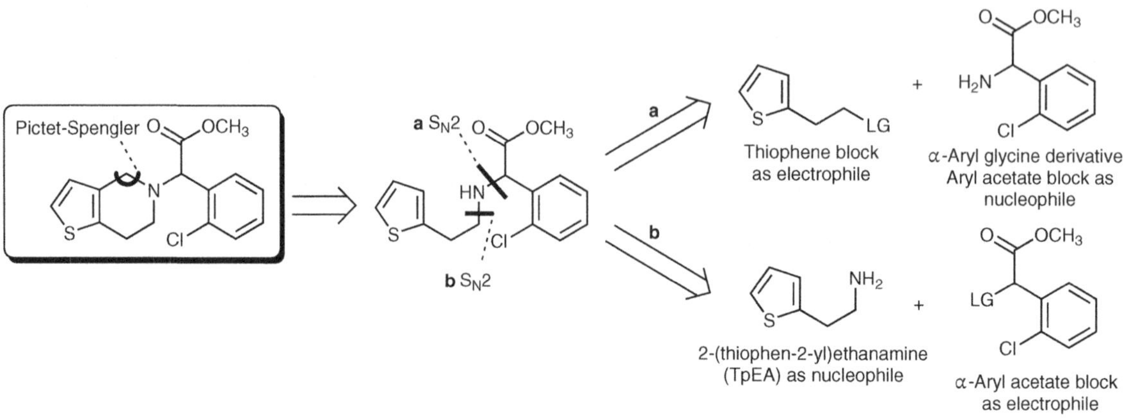

Scheme 3.8 Retrosynthesis of Clopidogrel by a retro Pictet-Spengler reaction

3.2.2.3 Deriving the ester group from a nitrile

The ester group can be generated from a nitrile, opening up further retrosynthetic routes (Scheme 3.9). The key reaction to connect the two main blocks is now a Strecker-type reaction (involving three components). These routes use 2-chlorobenzaldehyde as one of the building blocks.

Conversion of the nitrile group into the methyl ester of Clopidogrel can be performed in one step, for example by reaction of the nitrile with methanol under acidic conditions (Pinner reaction). In addition, multistep sequences can also be used to convert the nitrile group into the methyl ester *via* other carbonyl compounds such as amides and carboxylic acids.

Functional group interconversion (FGI) of various carboxylic acid derivatives allows for the design of other retrosynthetic routes (scheme B). As before, resolution can be theoretically executed at different stages along the synthetic sequence. Some of the synthetic routes may suffer drawbacks and may not be viable experimentally. Problems are expected when pure enantiomeric intermediates are heated in strongly acidic or basic conditions: partial isomerization or even complete racemization of the chiral centre is a strong possibility and any enantiomeric excess will thus be drastically reduced having a negative impact on the efficacy of that process.

The α-amino nitrile intermediates obtained from a Strecker reaction can also be obtained *via* an S_N2 reaction (Scheme 3.10). Access to large quantities of α-halo cyano aryl compounds is crucial to the use of this strategy at industrial scale.

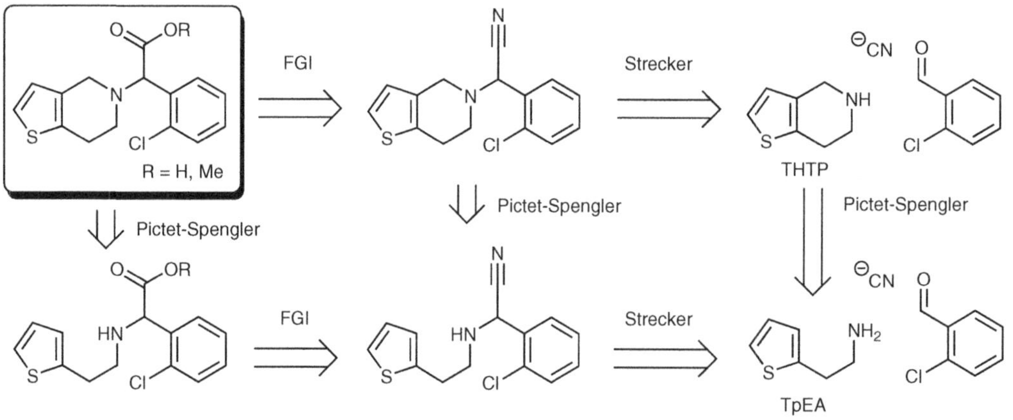

Scheme 3.9 Main retrosynthetic routes of Clopidogrel *via* a Strecker reaction

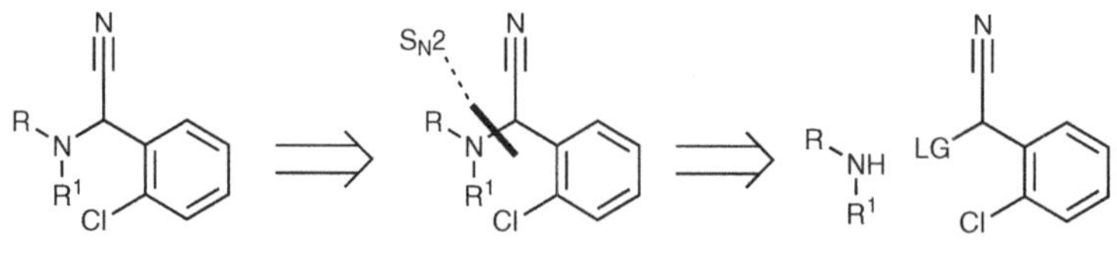

Scheme 3.10 Retrosynthesis of nitrile intermediates by S_N2 reaction

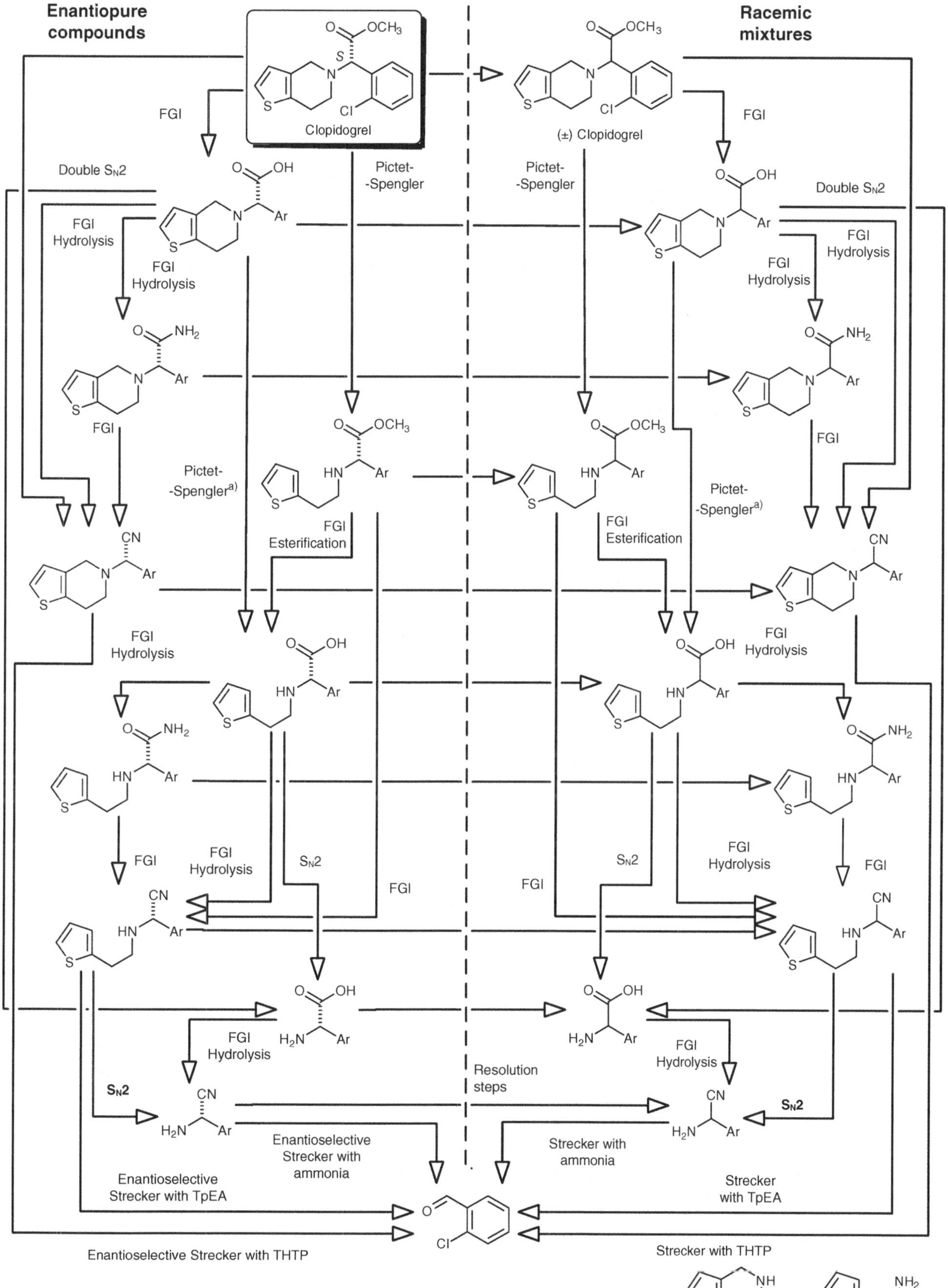

a) a protecting group may be required;
FGI – functional group interconversion

Scheme B Main retrosynthetic routes for Clopidogrel (part 2)

In conclusion, from the analysis presented above, three main strategies are expected to be most favourable in connecting the aryl ester (or carboxylic acid derivative/analogue) to the thiophene block. This is summarized in Scheme 3.11.

These strategies are:

- Bimolecular substitution using the thiophene building block as nucleophile (disconnections **a** and **d**);
- Bimolecular substitution using the thiophene building block as electrophile (disconnections **b** and **e**);
- Strecker reaction, where the thiophene building block is the nucleophile (disconnections **f** and **g**).

The use of either THTP or 2-(thiophen-2-yl)ethanamine (TpEA) - in the latter case ring closure is secured by a Pictet-Spengler reaction – is a sound choice for the formation of the hydrothienopyridine unit. In the cases that use this option, a difficult-to-synthesise disubstituted thiophene electrophilic building block is avoided (disconnection **b**).

3.2.2.4 Retrosynthetic analysis of *S* isomer (Clopidogrel)

As described above, the three options to produce enantiomerically pure *S*-Clopidogrel are: using the chiral pool; carrying out a resolution of a racemic intermediate; or developing an asymmetric synthesis.

Using a resolution or a chiral pool approach

A resolution at some stage in the process appears to be the most feasible entry into an asymmetric intermediate. Resolution of α-amino acids and esters, as well as the resolution of mandelic acid derivatives, are known and expected to be adaptable to starting materials and intermediates used in the synthesis of Clopidogrel. The API is an α-amino ester in which the chiral carbon atom adjoins the basic nitrogen and hence, there is a high possibility that the resolution can be successfully accomplished. It has to be stressed, once more, that experimentation is crucial to confirm this assumption. Variables such as resolving agent, solvent and temperature have to be investigated to arrive at a viable industrial process. As mentioned before, when a resolution is adopted in the synthetic sequence, the chirality of the enantiopure intermediate must be preserved during subsequent reactions up until the final product, Clopidogrel, is reached.

A chiral pool approach depends on the availability of chiral 2-amino-2-(2-chlorophenyl)acetic acid or

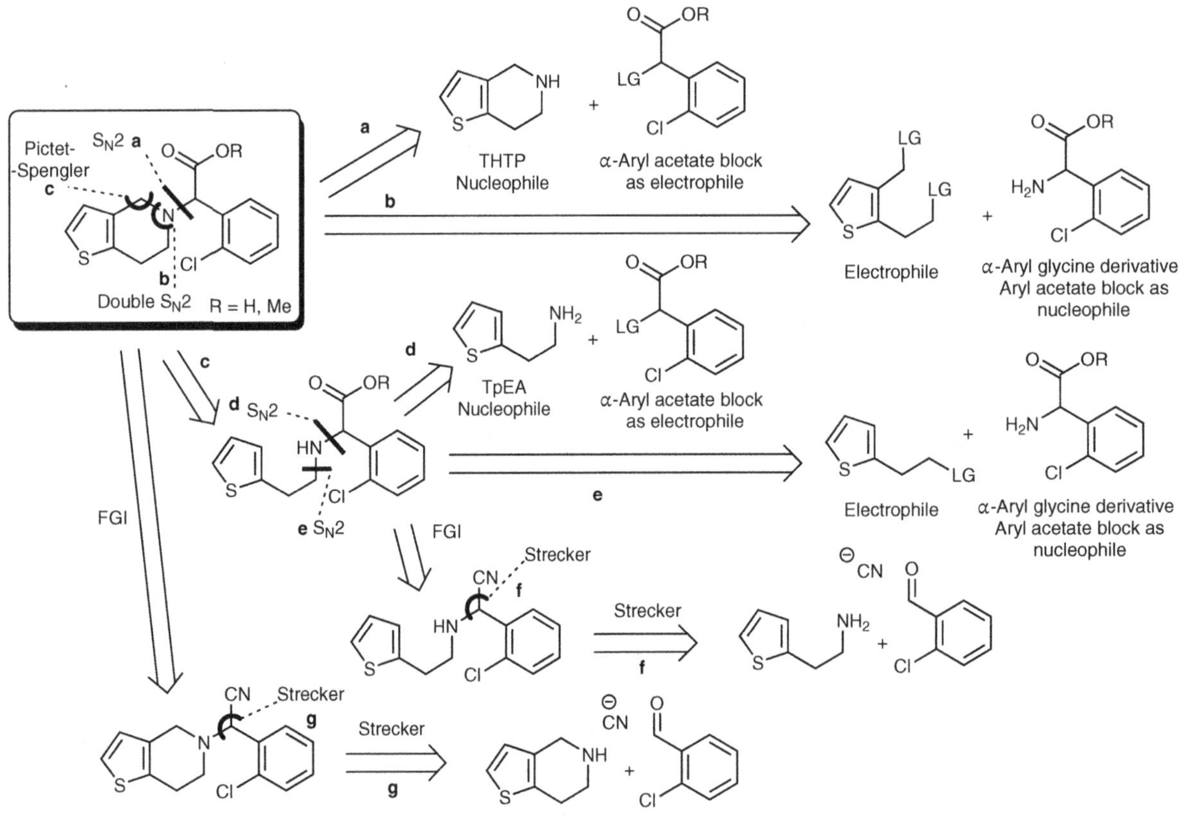

SCHEME 3.11 MAIN RETROSYNTHETIC ROUTES OF RACEMIC CLOPIDOGREL

2-hydroxy-2-(2-chlorophenyl)acetic acid (or some similar building block). Those compounds are not part of what is usually considered to belong to the chiral pool, but due to the success of Clopidogrel, are now being made available commercially and so serve as starting materials. Again, it should be noted that chiral integrity has to be maintained during chemical reactions employed and processing.

Using an asymmetric synthesis

Reactions, which create the chiral centre on formation of the C-N bond, are optimum candidates for use in the synthesis.

Asymmetric versions of the Strecker reaction are known and involve the use of chiral catalysts (CR2011). When applied to the synthesis of Clopidogrel, a variety of important pure chiral intermediates or starting materials can be produced directly. Exploration of this opportunity is anticipated to be important (Scheme 3.12).

The formation of asymmetric cyanohydrins is also expected to be of use in the production of chiral electrophiles (Scheme 3.13).

3.3 THE FIRST REPORTED SYNTHESIS OF CLOPIDOGREL

The synthesis of the racemate was described by Sanofi in 1982 without separation and isolation of the *S* enantiomer, later to become known as Clopidogrel (Scheme 3.14). The only experimental conditions that were described in the patent were those of the final step as the synthesis of methyl 2-chloro-2-(2-chlorophenyl)acetate and 4,5,6,7-tetrahydrothieno[3,2-c]-pyridine (THTP) were generated by known procedures (OSC1963). Noteworthy is the fact that THTP had been previously used by Sanofi in the synthesis of Ticlopidine.

The *S*-isomer, Clopidogrel, was first isolated by resolution of the racemic mixture in 1987 (Sanofi1987), using (–)-camphor-10-sulfonic acid (CSA) as the resolving agent (Scheme 3.15).

The disclosed resolution process required three days to complete and several crystallizations were needed to achieve an acceptable pharmaceutical enantiomeric excess. Clopidogrel base is an oil, which therefore, was converted into a pharmaceutically acceptable salt. Eventually the bisulfate salt was selected by Sanofi to market Clopidogrel as Plavix®.

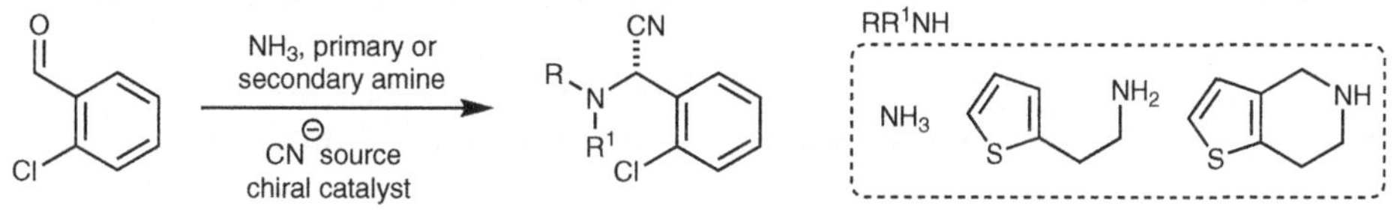

Scheme 3.12 The asymmetric Strecker reaction in the synthesis of Clopidogrel

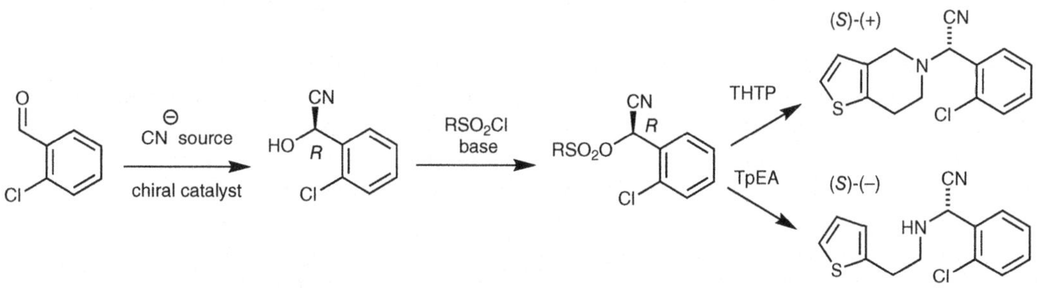

Scheme 3.13 Enantiomeric Clopidogrel synthesis *via* asymmetric cyanohydrin formation

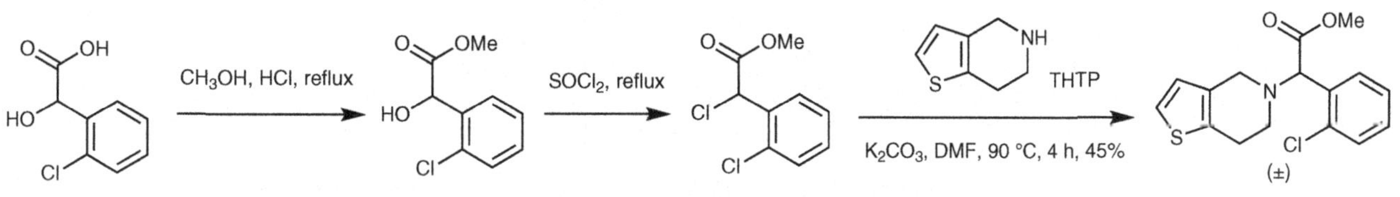

Scheme 3.14 Synthesis of racemic Clopidogrel

3.4 THE EVOLUTION OF CHEMICAL SYNTHESIS OF CLOPIDOGREL

The first synthetic approach to produce Clopidogrel was noted to be less than ideal for industrial scale production:

- α-halogenophenylacetic acid derivatives are strongly lachrymatory and are mucous membrane irritants hence, handling them at large scale is difficult. Furthermore, the yields of the known methods for their formation are rather poor (Sanofi1997c).
- The yield of the last reaction (the S_N2) is not very high, and a great deal of material is lost at this late stage in the process.
- 4,5,6,7-tetrahydrothieno[3,2-*c*]pyridine (THTP) is a low-melting solid and is difficult to obtain in pure form.
- The resolution at the end of the synthesis is associated with several inconveniences, mainly, the loss of material as unwanted isomer, the use of an expensive resolving agent and the necessity to carry out a salt exchange. The resolution process takes a relatively long time and since the initial diastereomeric salt has insufficient optical purity, further processing is required to achieve the pharmaceutically acceptable optical purity. The net outcome is a significant loss in yield.
- Costly camphorsulfonic acid resolving agent is discarded and not recovered.

As a result, improved synthetic processes for Clopidogrel were sought.

3.4.1 Synthetic routes using a resolution step

During the 90's, Sanofi, the innovator, dedicated much attention to protect the main synthetic routes to Clopidogrel, using a resolution step at some stage of the process to obtain acceptable enantiopure product. The extensive efforts by Sanofi to protect various synthetic routes through patenting was certainly a principle part of their anti-generic strategy. In this way, Sanofi obviously hoped that they could limit the market share of generic companies after expiry of the product patent by forcing them to use the highly inefficient original route. After publication of many of the innovator's synthesis many generic companies published patents using some of these same routes whilst claiming improvements to what had previously been described. New, independent synthetic routes were also explored and patented by various pharmaceutical companies.

Routes to obtain Clopidogrel that were the subject of a patent application and which used a resolution step to promote the production of enantiopure compounds are summarized in schemes C to H according to the key reaction used to couple the thiophene building block to the aryl ester or its precursor. The various resolution processes and the enantiomer recovery processes are condensed in scheme I.

Five main synthetic options are used to connect the two building blocks:

1) Using TpEA as nucleophile, in S_N2 reactions (scheme C)
2) Using TpEA as nucleophile, in Strecker reactions (scheme E)
3) Using THTP as nucleophile, in S_N2 reactions (scheme F)
4) Using THTP as nucleophile, in Strecker reactions (scheme G)
5) Using an alkylthiophene derivative as electrophile and an ester block as nucleophile, in S_N2 reactions (scheme H)

3.4.1.1 Using TpEA as nucleophile in S_N2 reactions

The synthetic routes using TpEA as nucleophile are summarized in Scheme C. The S_N2 reaction is made easier by using the primary amine, 2-(thiophen-2-yl)ethanamine, (TpEA) as the nucleophile, instead of the more bulky secondary amine THTP, in the reactions with the electrophilic aryl acetic ester block (Scheme 3.16). This alkylation is predicted to be trouble free whilst limiting polyalkylation of the amine as the desired product is less nucleophilic than the starting amine. Closing the saturated ring is carried out *via* a Pictet-Spengler reaction.

As anticipated, the S_N2 reaction was carried out on both enantiopure chiral materials and on racemic mixtures. The part of the scheme to the right of the vertical dotted line shows the transformation of enantiopure intermediates, whilst the left half indicates the conversions of racemic intermediates. Scheme C also discloses at which points resolution was undertaken (horizontal arrows from left to right). In contrast, the horizontal arrows from right to left represent racemization reactions indicating where

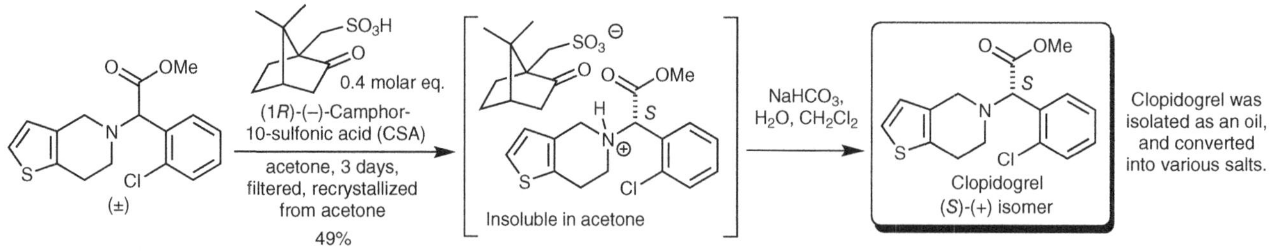

SCHEME 3.15 FIRST ISOLATION PROCESS FOR CLOPIDOGREL

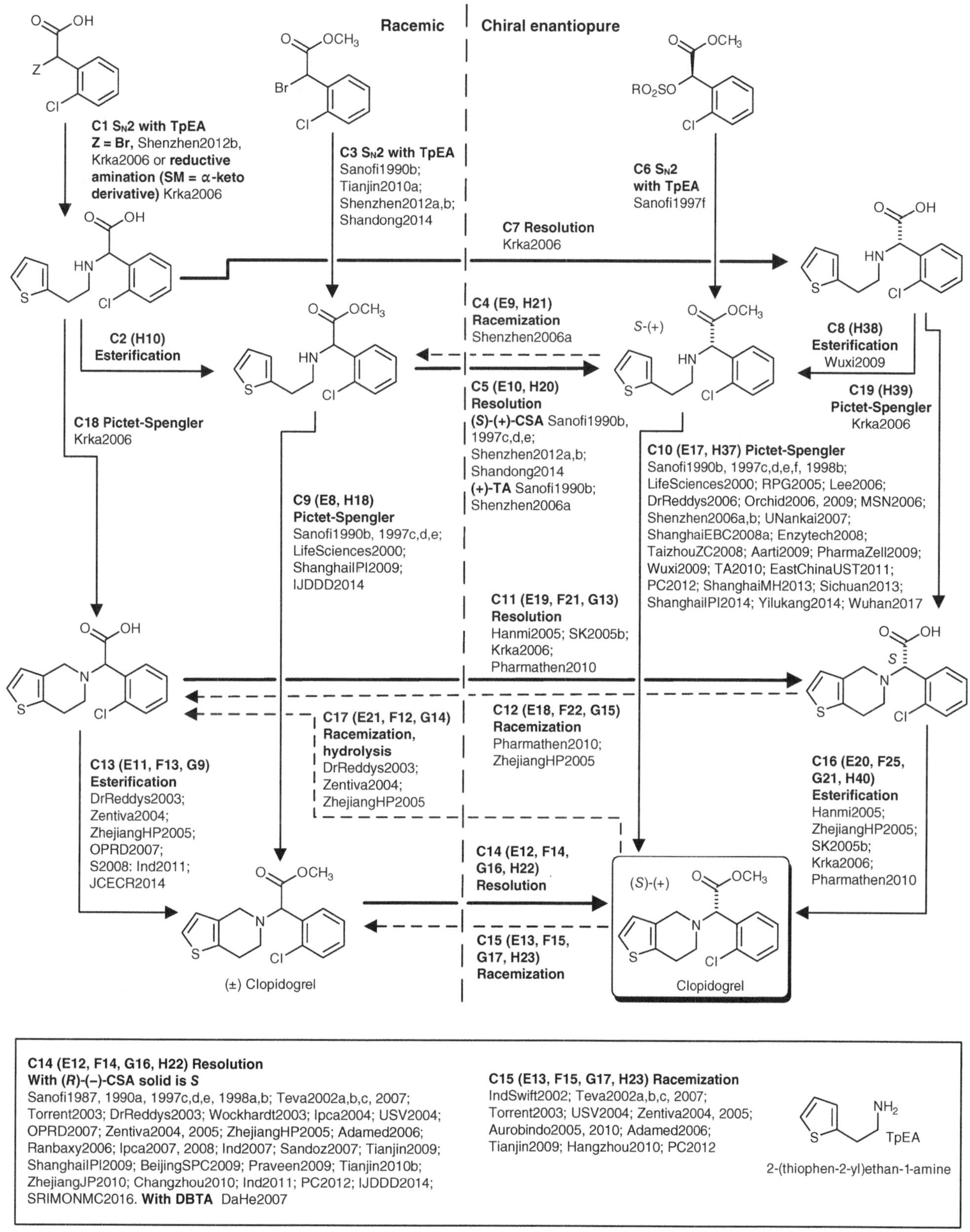

Scheme C Using 2-(thiophen-2-yl)ethan-1-amine (TpEA) as nucleophile in S_N2 reactions in Clopidogrel synthesis (for reaction conditions see page 73)

recycling of the unwanted enantiomer has been carried out, either from "pure" material or mixtures that could not be further processed conveniently. It is worth noting that most of these recycling processes were included in patents submitted during the first decade of the twenty-first century, a much later date than the resolution processes themselves. These racemization reactions have played a very important role in the development of industrially viable synthetic routes for the production of Clopidogrel.

The Pictet-Spengler reaction is ideally suited to obtain the hydropyridine ring (**C9**, **C10**). The reaction involves a Schiff base, which on activation by a protic or Lewis acid undergoes cyclization *via* an intramolecular electrophilic aromatic substitution facilitated by the activated thiophene ring. Formaldehyde, dimethyl acetal or any acid-labile surrogate are commonly used to introduce the methylene group. The reaction can be carried out in either protic or aprotic solvents. At least a slight excess of the formaldehyde source is required. In the case of Clopidogrel, formalin solutions or paraformaldehyde (de-polymerized by acid catalysis) are the most used reactants, the most common solvents reported being methylene chloride, water or an alcohol. Naturally, normal safety procedures must be followed whenever formaldehyde or one of its sources are used in the industrial setting.

Yields are generally high and the starting amine can be in the form of the free base or its hydrochloric acid salt. Importantly, since hydrolysis of the ester and racemization can be controlled, the reaction can be carried out starting with the *S*-isomer of the intermediate (**C10**), generating Clopidogrel directly.

In an alternative approach, a racemic intermediate was obtained *via* an S_N2 reaction with 2-bromo-2-(2-chlorophenyl)acetic acid or by a reductive alkylation with 2-(2-chlorophenyl)-2-oxoacetic acid (**C1**). Esterification, ring closure and resolution are required to produce Clopidogrel. Different sequences of these reactions are feasible, thus creating various possible routes.

The electrophilic ester block and the S_N2 displacement

The esters or acids, which are electrophilic at the α carbonyl carbon atom, were produced as depicted in scheme D. Preferred leaving groups at the α-carbonyl carbon atom are halides (bromide or chloride), mesylate, tosylate or analogous sulfonic acid esters. The carboxylic acid or ester can be replaced by a nitrile group. Scheme D illustrates the synthesis of the carboxylic acids, esters and nitriles (shown in the highlighted boxes) which were used in the synthesis of Clopidogrel in conjunction with either THTP or TpEA. The horizontal dotted line separates racemic and achiral compounds from enantiopure compounds and the upward vertical arrows indicate resolution steps or asymmetric transformations.

The first synthetic route of Clopidogrel used α-halo esters as electrophiles in the S_N2 reaction and produced racemic products. Production of pure chiral α-halo esters is not easy, either from chiral pure compounds (**D8**) or by resolution (**D13**, Scheme 3.17).

During the first attempts to use chirally pure 2-chloro-2-(2-chlorophenyl) acetate in reactions with THTP to produce chirally pure Clopidogrel, extensive or complete racemization was observed in several solvents and under various reaction conditions (Sanofi1997f). This result was attributed to the inherent chiral lability of these halo esters and hence the use of these halo esters was limited to S_N2 reactions resulting in racemic intermediates and a resolution step had to be applied to obtain Clopidogrel. More recently, a process using a chirally pure halo derivative was claimed to give good enantiomeric excess and yields, in reaction with THTP (ShanghaiAT2005).

The search for more stable chiral electrophiles gave rise to several patent applications that claimed the use of chirally pure electrophiles to generate chirally pure intermediates or Clopidogrel. These stable electrophiles were aryl or tosylate esters of (*R*)-2-hydroxy-2-(2-chlorophenyl)acetic acid methyl ester (**D7**).

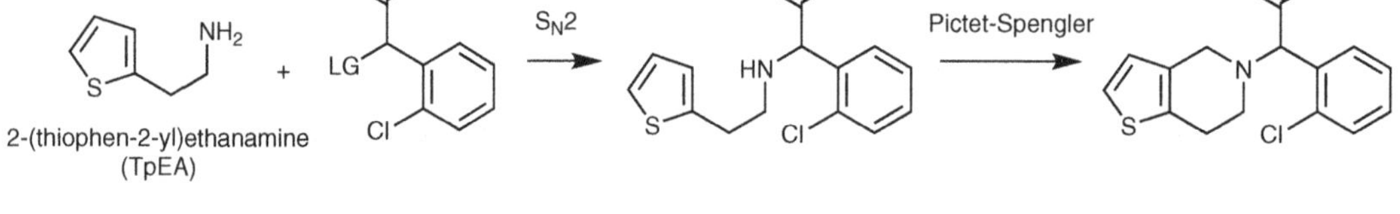

SCHEME 3.16 GENERIC SEQUENCE TO ASSEMBLE CLOPIDOGREL USING TpEA AS NUCLEOPHILE IN S_N2 REACTIONS

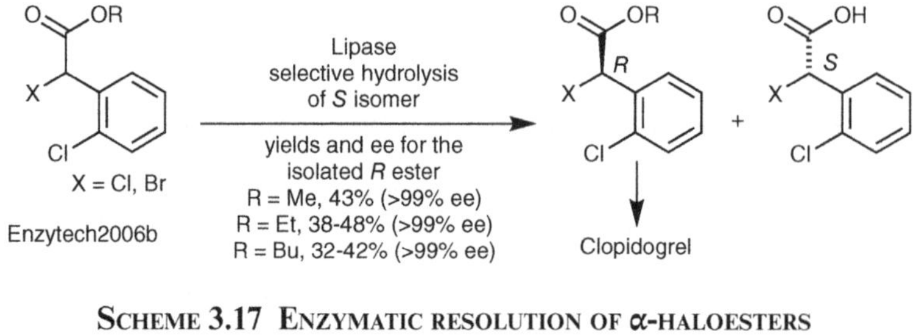

SCHEME 3.17 ENZYMATIC RESOLUTION OF α-HALOESTERS

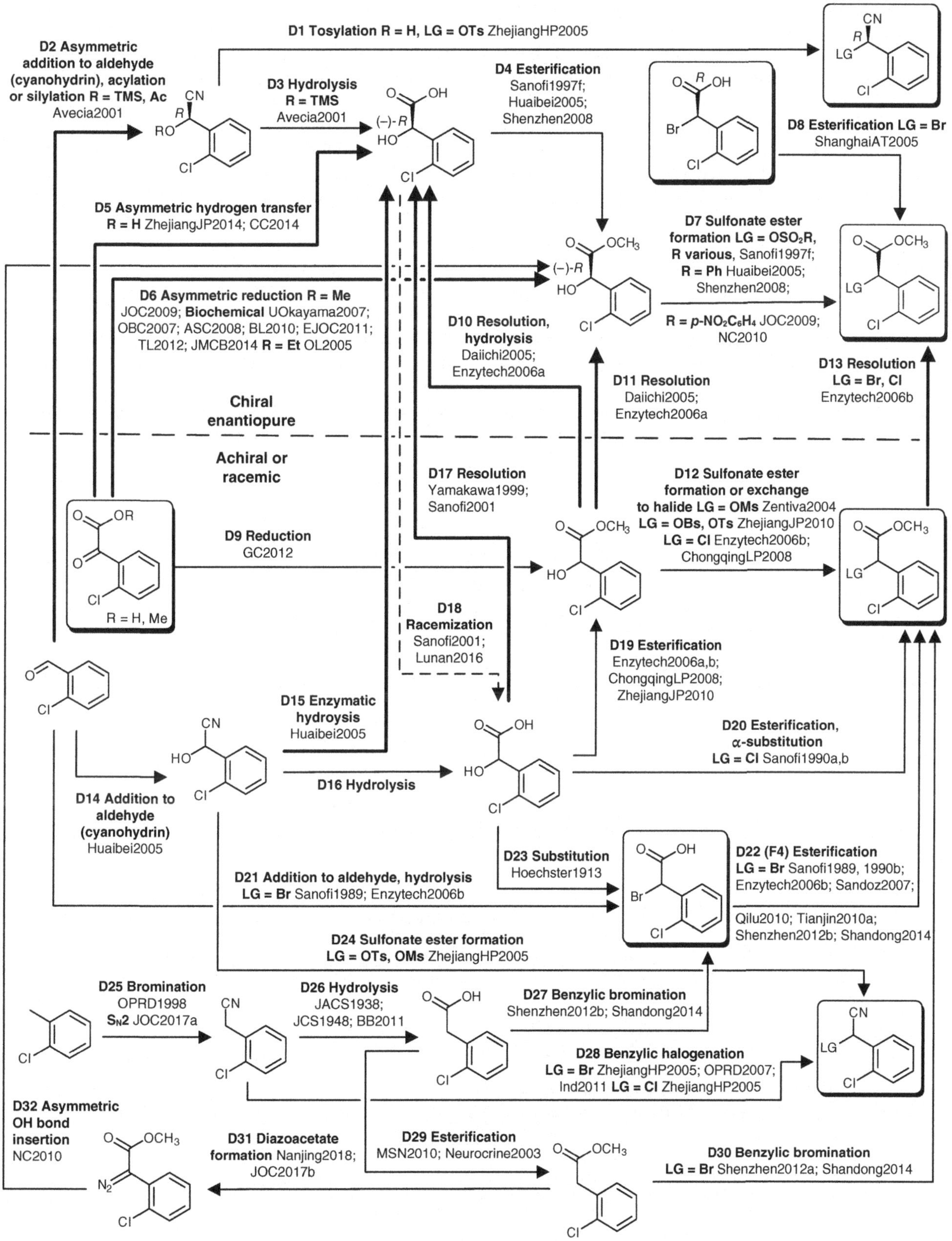

Scheme D Synthetic routes to electrophilic ester building blocks (for reaction conditions see page 74)

Racemic 2-hydroxy-2-(2-chlorophenyl)acetic acid or its methyl ester may be obtained by several different processes and several resolutions of both compounds have been developed (**D17**, **D10**, **D11**). (*R*)-2-chloromandelic acid can be obtained from the racemic cyanohydrin of 2-chlorobenzaldehyde (**D14**), hydrolysis (**D16**) and then resolution (**D17**) but a correspondent asymmetric process has also been disclosed (**D2+D3**, Avecia2001, Scheme 3.18). Selective formation of one of the enantiomers of 2-chlorobenzaldehyde cyanohydrin (**D2**), followed by hydrolysis, without loss of stereochemical purity (**D3**), is one of the methodologies employed. Thereafter, the ester can be obtained from the acid by classical esterification without racemization (**D4**). Many methods for the asymmetric addition of cyanating agents to aryl aldehydes have been reported in the literature (CAC1999).

Kinetic resolution processes have been applied to separate enantiomers. Enzymatic hydrolysis of esters or cyano groups were claimed and produce (*R*)-2-chloromandelic acid directly (**D10**, **D15**, Scheme 3.19). Selective hydrolysis of the *S*-ester and separation of the corresponding carboxylic acid to obtain the pure (*R*)-2-chloromandelic acid ester was proposed (**D11**). A simple hydrolysis produces the intended (*R*)-2-chloromandelic acid.

In yet another approach, a bio-catalytic deacylation of a racemic mixture of an *O*-acetylated mandelic acid derivative using whole cells of newly isolated *Pseudomonas sp.* ECU1011 yielded (*S*)-mandelic acid with 97.2% enantiomeric excess after 44% conversion (Scheme 3.20, AMB2010).

As shown above, (*R*)-2-chloromandelic methyl ester can be obtained by a kinetic resolution process (**D11**), but asymmetric ruthenium catalysed reduction of methyl 2-(2-chlorophenyl)-2-oxoacetate (JOC2009, **D6**, also OL2005 for an ethyl ester) is also a viable option (Scheme 3.21) and gives an enantiomerically pure mandelic acid derivative, which serves as a raw material in the

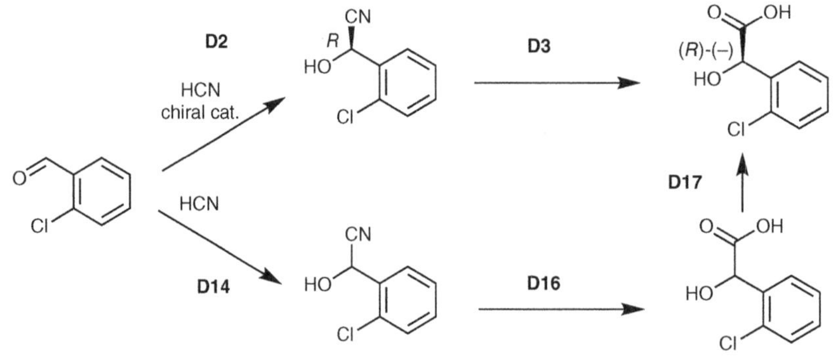

Scheme 3.18 Main routes to (*R*)-2-chloromandelic acid using a cyanohydrin as intermediate

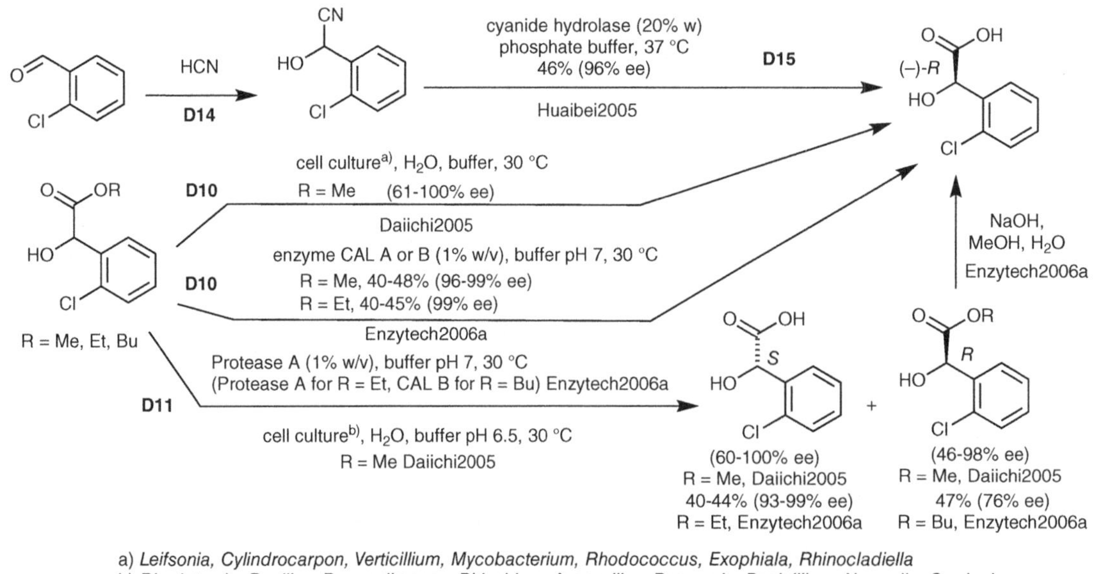

a) *Leifsonia, Cylindrocarpon, Verticillium, Mycobacterium, Rhodococcus, Exophiala, Rhinocladiella*
b) *Rhodotorula, Bacillus, Brevundimonas, Rhizobium, Aspergillus, Beauveria, Penicillium, Nocardia, Gordonia, Rhodococcus, Ramichloridium, Porphyrobacter*

Scheme 3.19 Kinetic resolution processes for the production of (*R*)-2-chloromandelic acid

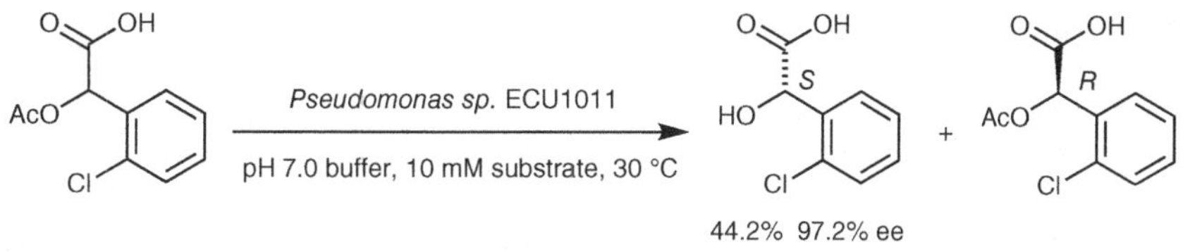

Scheme 3.20 Bio-catalytic deacylation of a racemic mixture of *O*-acetylated chloromandelic acid

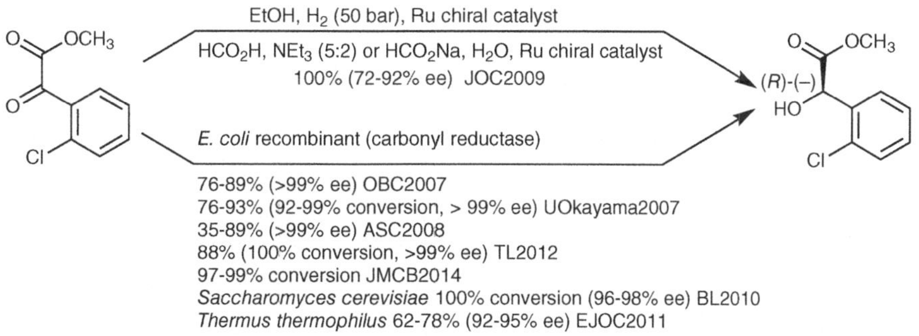

Scheme 3.21 Asymmetric reduction of methyl 2-(2-chlorophenyl)-2-oxoacetate (D6)

synthesis. Bio-reduction methods were also proposed as an alternative (**D6**, Scheme 3.21).

A similar asymmetric reduction of 2-(2-chlorophenyl)-2-oxoacetic acid has been developed and patented, giving direct access to (*R*)-2-(2-chlorophenyl)-2-hydroxyacetic acid in good yield and enantioselectivity (**D5**, Scheme 3.22, ZhejiangJP2014, CC2014).

In another asymmetric synthetic approach towards the *R*-hydroxyester, an iron complex of a spiro-bisoxazoline ligand was used as a highly efficient catalysts for an asymmetric O–H bond insertion reaction (**D32**, Scheme 3.23, NC2010).

Both the halide and the alkyl or arylsulfonate derivatives are inherently unstable compounds but their application in S_N2 reactions to give racemic racemic products appears promising.

3.4.1.2 Using TpEA as nucleophile in Strecker reactions

The Strecker reaction represents a good option to assemble the essential structural elements of the Clopidogrel molecule. Hence, condensation of 2-chlorobenzaldehyde with 2-(thiophen-2-yl)ethanamine in the presence of hydrogen cyanide (or a hydrogen cyanide equivalent), in a one-pot three component reaction gives the α-amino

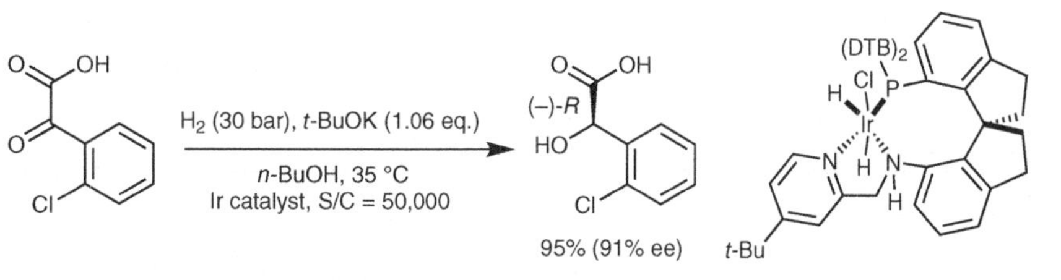

Scheme 3.22 Asymmetric reduction of 2-(2-chlorophenyl)-2-oxoacetic acid

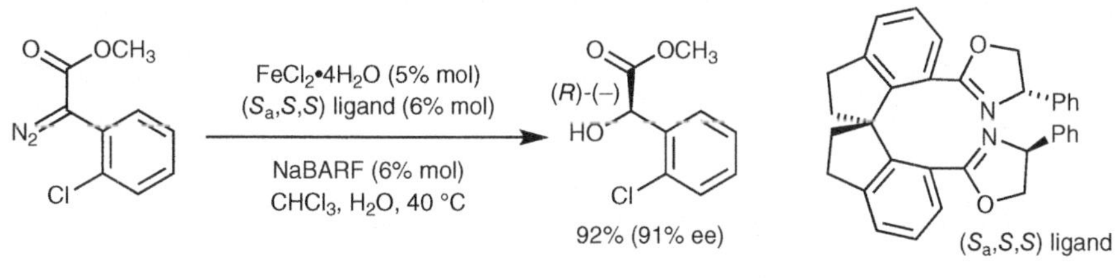

Scheme 3.23 Synthesis of the α-hydroxyester by a metal-catalysed asymmetric OH insertion

nitrile (Scheme 3.24, scheme E, **E1**). On methanolysis of the nitrile group the methyl ester of the desired carboxylic acid is obtained directly (**E3**). An alternative process proceeds through the primary amide (**E4**+**E5** or **E15**+**E16**).

The high toxicity of hydrogen cyanide is clearly a drawback to application on an industrial scale. There are several options to meliorate the hazards. Buffered aqueous alkyl cyanides can increase the level of safety but ostensibly, the safest procedure uses a bisulfite addition compound as intermediate *en route* to the cyanohydrin (Scheme 3.25, Sanofi1997c). Hydrogen sulfite ion is produced on formation of the imine from the bisulfite addition compound and is the source of the proton required for the production of the cyanohydrin during addition of cyanide to the enamine. Hence, the possibility of gaseous hydrogen cyanide release is reduced to near zero by avoiding any other acid component (proton source) in the reaction mixture.

Yet another option is to use the hydrochloric salt of the amine as a starting material (Sanofi1997c) which can act as the source of protons and hence, avoids the

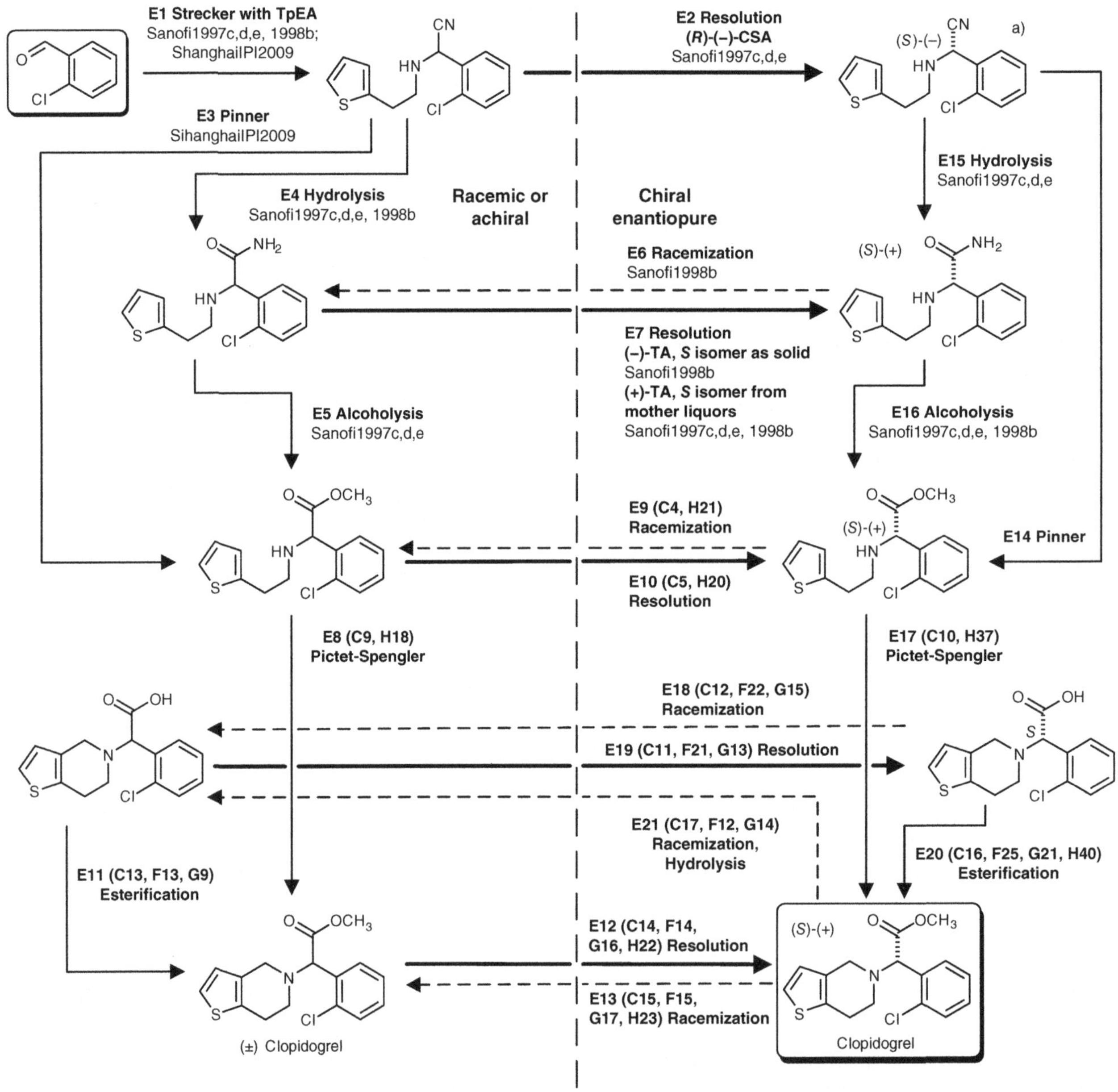

a) also produced directly by an asymmetric Strecker reaction (see section 3.4.2, Rhodia2003)

Scheme E Synthesis of Clopidogrel using 2-(thiophen-2-yl)ethan-1-amine (TpEA) as nucleophile, in Strecker reaction (for reaction conditions see page 75)

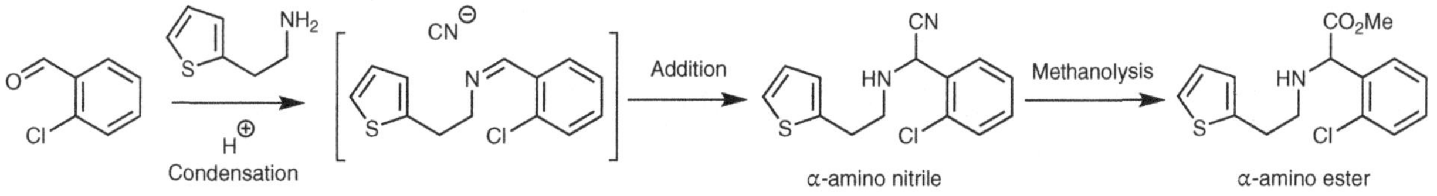

SCHEME 3.24 EXAMPLE OF A STRECKER REACTION APPLIED TO CLOPIDOGREL SYNTHESIS

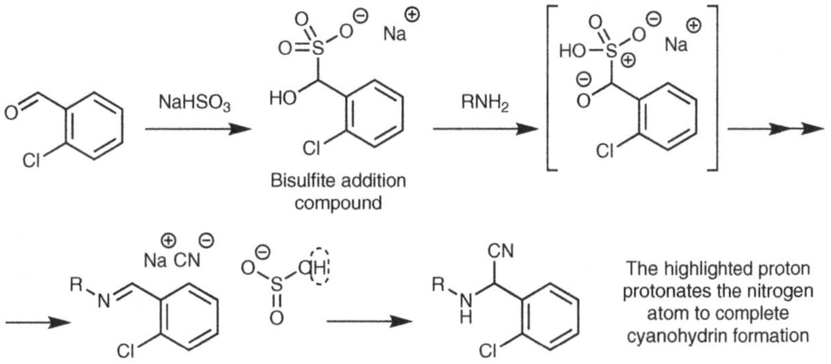

SCHEME 3.25 USE OF BISULFITE ADDITION COMPOUND IN THE CYANOHYDRIN FORMATION

use of excess acid. Other HCN carriers are available and useful (for example acetone cyanohydrin, cyanoformate or trimethylsilyl cyanide) but their use is restricted to small-scale or very expensive products because of their high cost.

Asymmetric versions of the Strecker reaction using chiral catalysts can be extremely useful in constructing the chiral Clopidogrel molecule (CR2011).

Since the nitrile group is the precursor to the ester, which can be generated at several different points *en route*, thus giving slight variations resulting in potential "new" processes. The routes using 2-(thiophen-2-yl)-ethanamine (TpEA) as the nucleophile in Strecker reactions are summarized in scheme E. The vertical dashed line and the horizontal arrows have the same connotation as before. The processes in scheme E end with a Pictet-Spengler ring-closing reaction (**E8** and **E17**).

Note that racemic Clopidogrel carboxylic acid plays an important role in recycling when a resolution (**E12**) is the final processing step of the production route.

3.4.1.3 Using THTP as nucleophile, in S_N2 reactions

4,5,6,7-tetrahydrothieno[3,2-*c*]pyridine (THTP) was used in the first process to produce Clopidogrel. Despite being reported as not easy to manipulate, having a low melting point and difficult to obtain in a pure form, it has been used over the years in various patent applications covering synthetic routes to Clopidogrel. Employing its hydrochloric salt obviates these problems, the neutral amine being released *in situ* by reaction with a base. Using THTP as staring material allows for a shorter route and avoids the Pictet-Spengler reaction in the final chemical transformation. THTP has become commercially available in industrial quantities due to the commercial success of Clopidogrel, thus making large-scale application viable. The use of THTP in S_N2 reactions in the synthesis of Clopidogrel is summarized in scheme F. The vertical dashed line and the horizontal arrows have the same connotation as before.

The α-ester electrophiles are those that were used in reactions with TpEA (see scheme D) and, as before, racemic and chiral enantiopure versions of the synthetic routes were patented.

In the sequence where the intermediates are racemic, the S_N2 reactions with the α-halogen ester (**F5**), acid (**F3**) or nitrile (**F1**) electrophiles are appropriate, as these compounds are more easily produced and because there is no need to avoid racemization. When a chiral enantiopure electrophile is involved, the sulfonate esters were almost exclusively employed since they have a reduced tendency to engender racemization (**F16**, **F17**).

At each of the possible steps after the bimolecular substitution reaction in the racemic sequence, resolutions were undertaken giving an isolated pure chiral *S*-derivative (**F6** for the nitrile, **F9** for the amide, **F21** for the carboxylic acid and **F14** for the ester, Clopidogrel). With the exception of the nitrile derivatives, yields for isolation of the *S* enantiomer are reported to be around 40% (50% maximum). Recovery processes for the unwanted isomers were reported (**F10**, **F12**, **F15**, **F18**, **F22**).

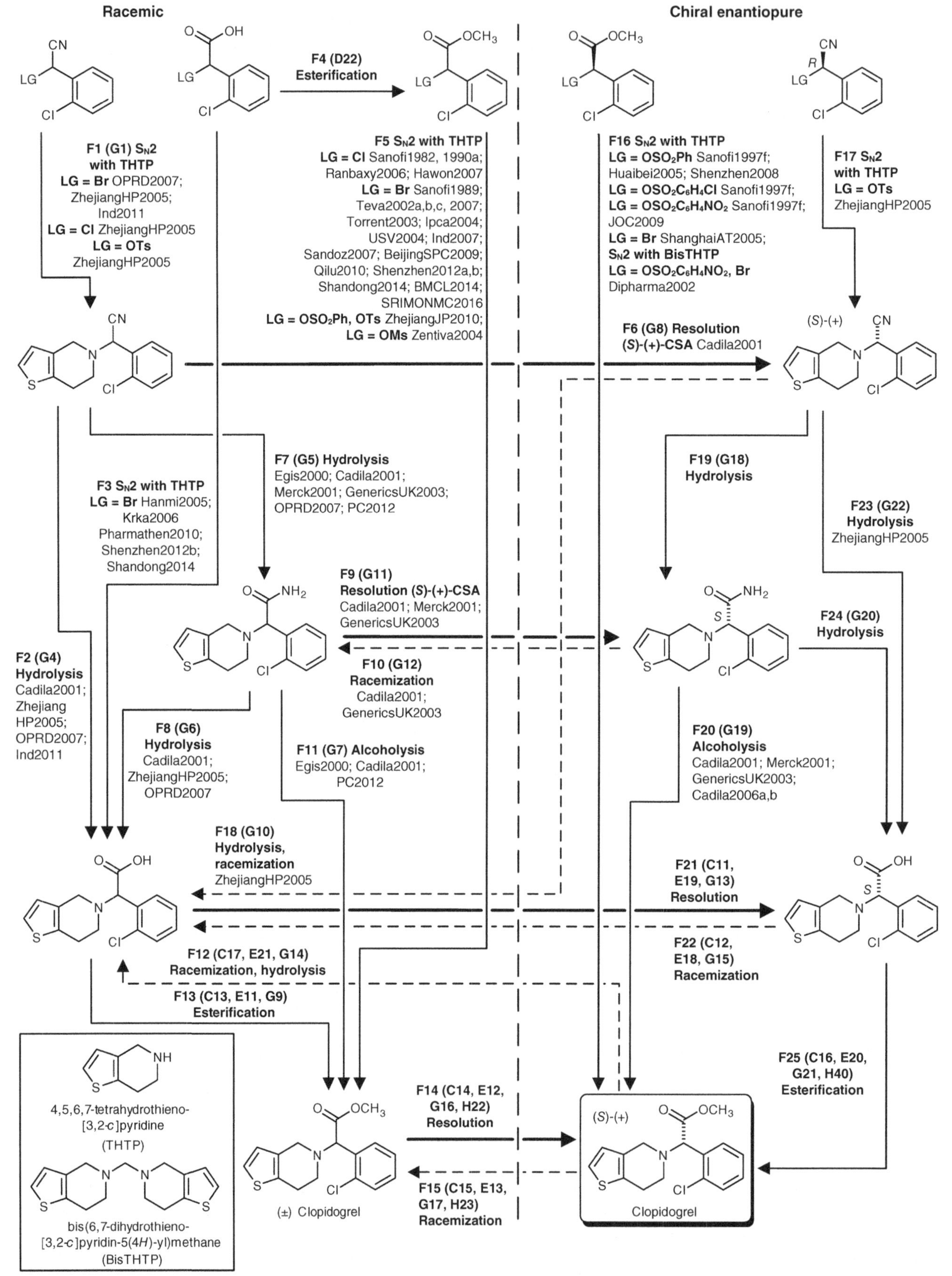

Scheme F Synthesis of Clopidogrel using 4,5,6,7-tetrahydrothieno[3,2-c]pyridine (THTP) as nucleophile in S_N2 reactions (for reaction conditions see page 76)

3.4.1.4 Using THTP as nucleophile, in Strecker reactions

The Strecker reaction has also been carried out using THTP as nucleophile. These processes are summarized in scheme G. The vertical dashed line and the horizontal arrows have the same connotation as before. Once more, this strategy avoids the use of the ring closing Pictet-Spengler reaction at the end of the synthesis.

3.4.1.5 Using the ester block as nucleophilic reactant

An alternative approach was to switch the role of the two building blocks in the S_N2 reaction, the α-aryl acetate building block acting as a nucleophile and the thiophene block as electrophile (Scheme 3.26). Two main options are illustrated in the scheme (**a** and **b**), the second (**b**), which uses a mono-substituted thiophene, is clearly pre-

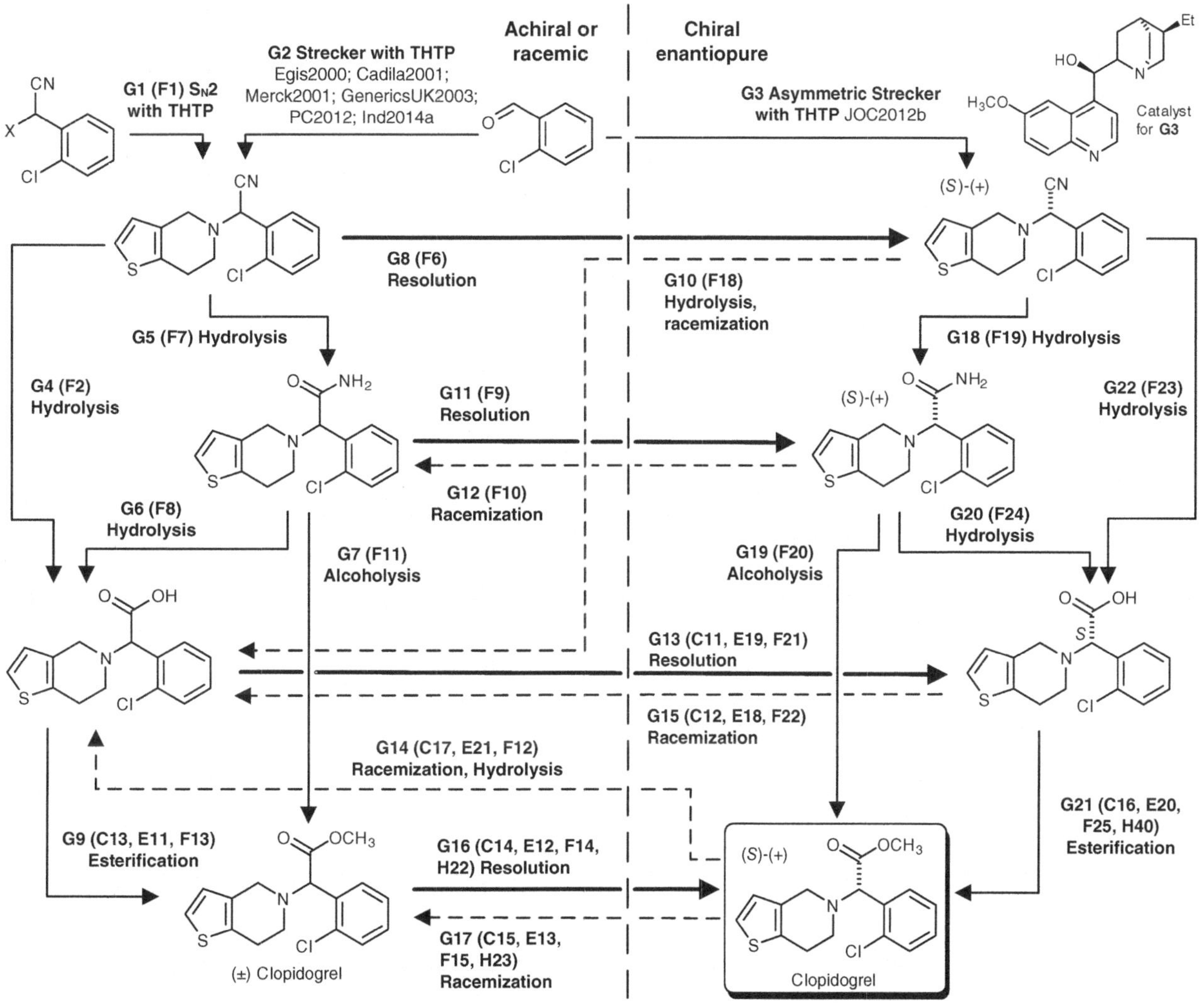

SCHEME G SYNTHESIS OF CLOPIDOGREL USING 4,5,6,7-TETRAHYDROTHIENO[3,2-*c*]PYRIDINE (THTP) AS NUCLEOPHILE IN STRECKER REACTIONS (FOR REACTION CONDITIONS SEE PAGE 76)

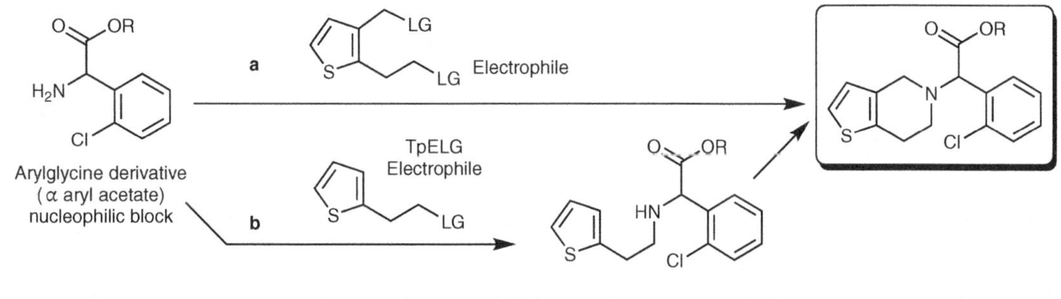

SCHEME 3.26 GENERAL ROUTES TO OBTAIN CLOPIDOGREL USING AN ARYLGLYCINE DERIVATIVE AS REACTANT

ferred due to the thiophene's easier synthesis, although (**a**) assembles the molecule in one step.

A reflection of this is that only one patent claimed a disubstituted thiophene as an electrophile in which the piperidine ring is built directly (strategy **a**, Scheme 3.27, Scheme 3.28, Hanmi2004). Only one other example of this strategy has been reported in the literature (Scheme 3.28, S2013).

In option (**b**), the preferred electrophile is 2-(thiophen-2-yl)ethyl tosylate (LG = OTs), which is more stable than the bromo or chloro analogues. A brief analysis of the methods to produce these raw materials is presented in section 3.4.4. The routes using an arylglycine amino ester as nucleophile to produce Clopidogrel are summarized in scheme H. Once more, as in all of the main synthetic routes described above, racemic and enantiopure routes were developed. As has been seen above, classical resolution can produce chiral enantiopure intermediates at different points in the synthesis, resulting in several options to produce Clopidogrel. The vertical dashed line and the horizontal arrows have the same connotation as before.

As an alternative to incorporating a resolution, an asymmetric Strecker reaction (**H1**), followed by hydrolysis (**H2**) has been proposed for the production of the (*S*)-2-amino-2-(2-chlorophenyl)acetic acid (Scheme 3.29, S2013).

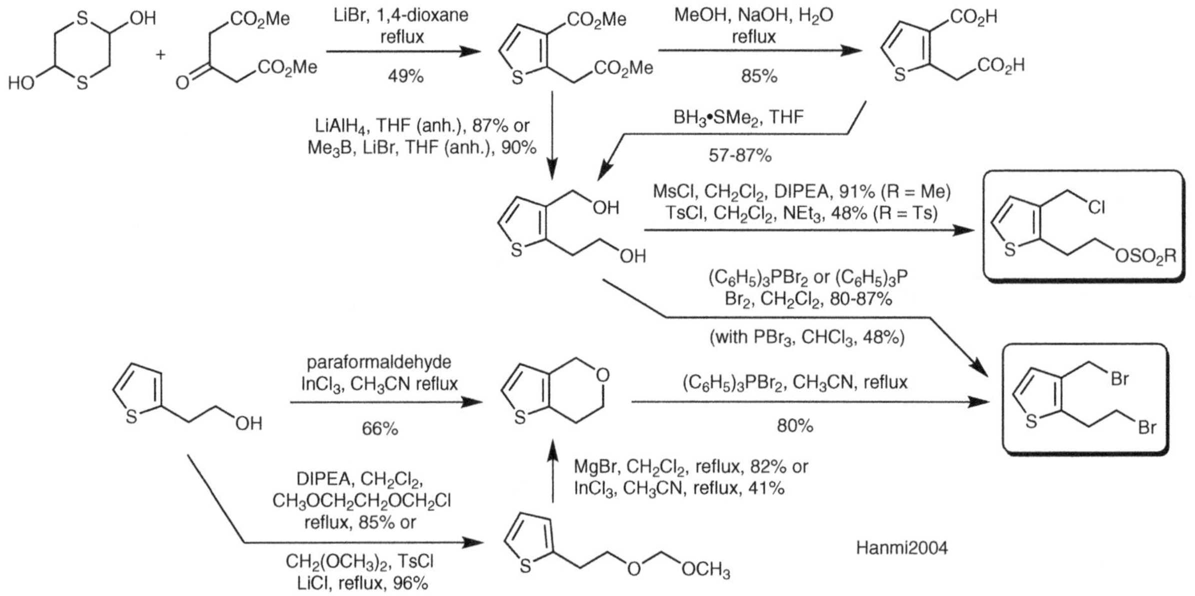

Scheme 3.27 Synthesis of thiophene disubstituted electrophiles

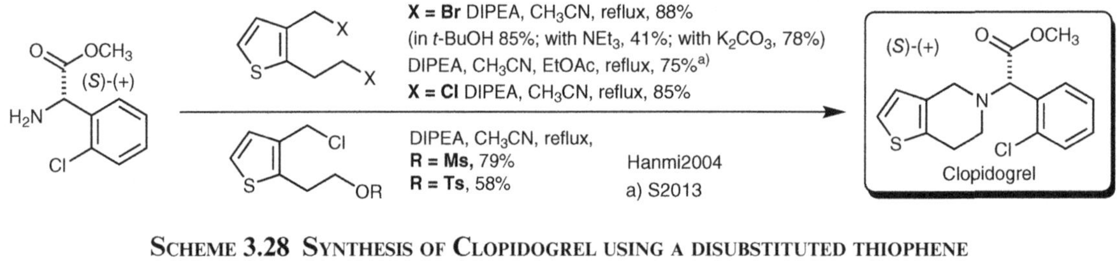

Scheme 3.28 Synthesis of Clopidogrel using a disubstituted thiophene electrophile in a double S_N2 reaction

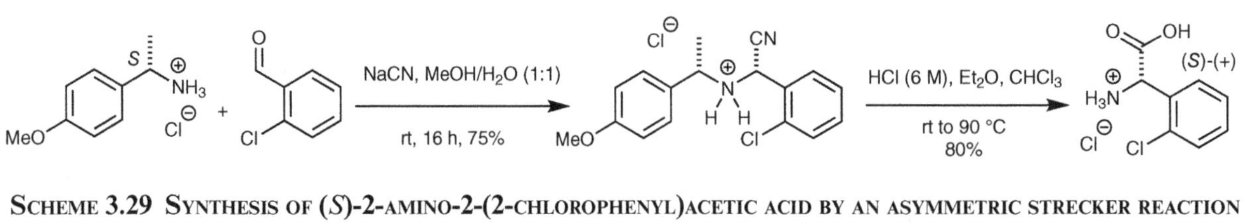

Scheme 3.29 Synthesis of (*S*)-2-amino-2-(2-chlorophenyl)acetic acid by an asymmetric Strecker reaction

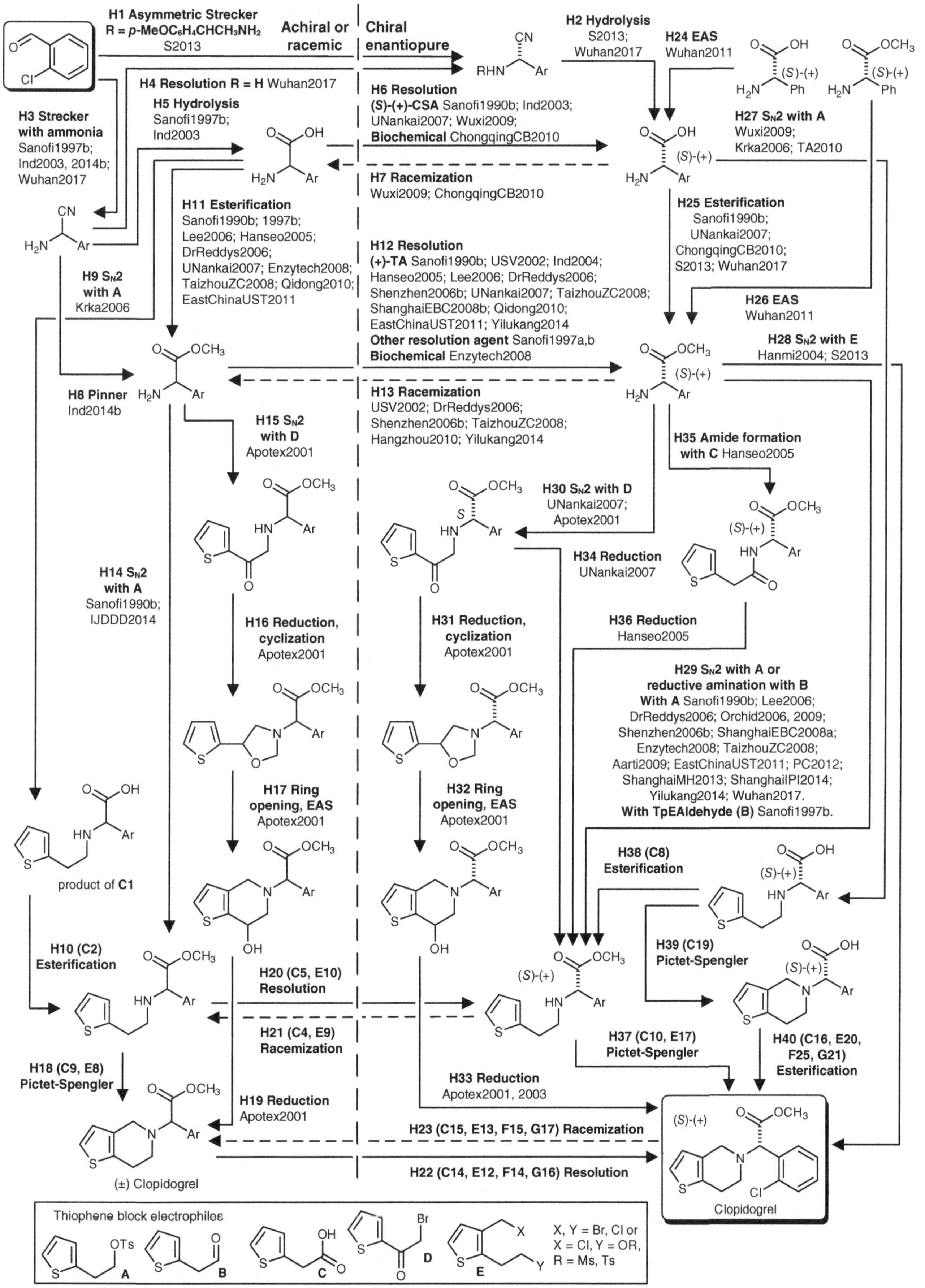

Scheme H Routes to produce Clopidogrel using an arylglycine derivative as nucleophile (for reaction conditions see page 77)

Separation processes for the both the acid [**H6**, with (+)-CSA] and the ester [**H12** with (+)-tartaric acid] by classical resolution have been patented.

The *S*-acid was isolated by selective bio-mediated hydrolysis of a racemic mixture of an *N*-acyl derivative (Scheme 3.30, ChongqingCB2010, BEJ2013) and, the *R*-isomer racemized and recycled.

Enzymatic kinetic resolutions were also successfully applied to promote selective ester hydrolysis (Scheme 3.31, TA2010, Enzytech2008).

A dynamic kinetic resolution process with simultaneous isomerization of the unwanted isomer would be an ideal scenario in this context since high yield DKR processes allow conversion of up to 100% of a racemic mixture into an enantiopure product.

In keeping with this, a DKR process was patented by Sanofi wherein racemization occurs *via* the reversible formation of an imine. Good yields of chirally pure ester were obtained when this technique was applied to several analogues, whilst the example directly relevant to Clopidogrel, although not carried out to give the desirable *S* enantiomer, gave a moderate yield (58%) of the *R* isomer (Sanofi1997a, **H12**, Scheme 3.32). The same yield of the *S* isomer was obtained in a similar experiment using the levo isomer of 2-(2,4-dinitrobenzamido)-2-phenylacetic acid as resolution agent (Sanofi 1997b).

In a novel approach, DKR was achieved by means of attrition-induced grinding of a solid conglomerate of a benzaldehyde imine in contact with its saturated solution. Racemization occurs in solution followed by preferential precipitation of the desired enantiomer (Scheme 3.33, OPRD2009, 2010).

Literature data indicate that enantiopure amino acid or its ester are stable. However, partial or total racemization has been reported during the S_N2 substitution reaction with 2-(thiophen-2-yl)ethyl tosylate (**H29**, Scheme 3.26, route **b**, with LG = OTs). Since the reaction conditions (solvent, base) play an important role, the most suitable conditions to obtain high yield and purity require a methodical and comprehensive development effort. The tosylate derivative was clearly the preferred electrophile in the alkylation of the aryl glycine ester by bimolecular substitution (**H29**). An alternative reductive alkylation of the amine, *via* an imine

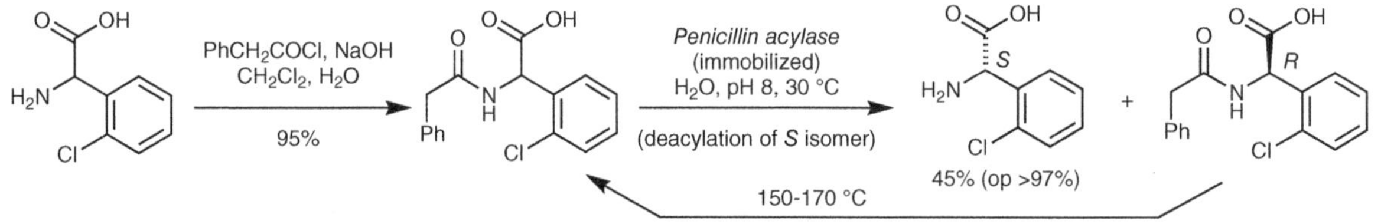

SCHEME 3.30 (*S*)-2-CHLOROPHENYL GLYCINE BY ENZYMATIC HYDROLYSIS OF AN *N*-ACYL DERIVATIVE

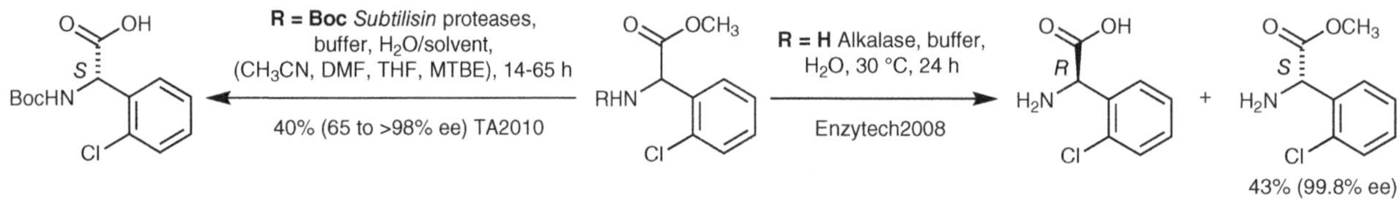

SCHEME 3.31 KINETIC RESOLUTION PROCESS FOR THE PRODUCTION OF (*S*)-2-CHLOROPHENYL GLYCINES

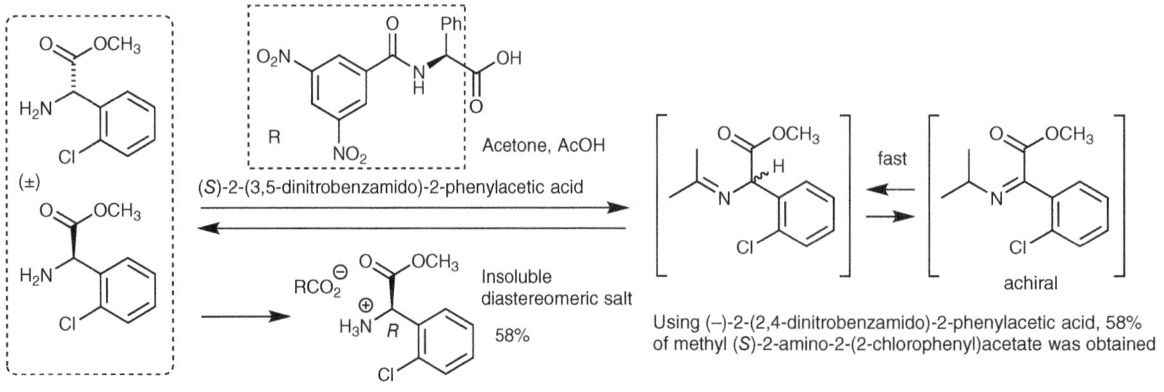

SCHEME 3.32 DKR RESOLUTION FOR METHYL 2-AMINO-2-(2-CHLOROPHENYL)ACETATE

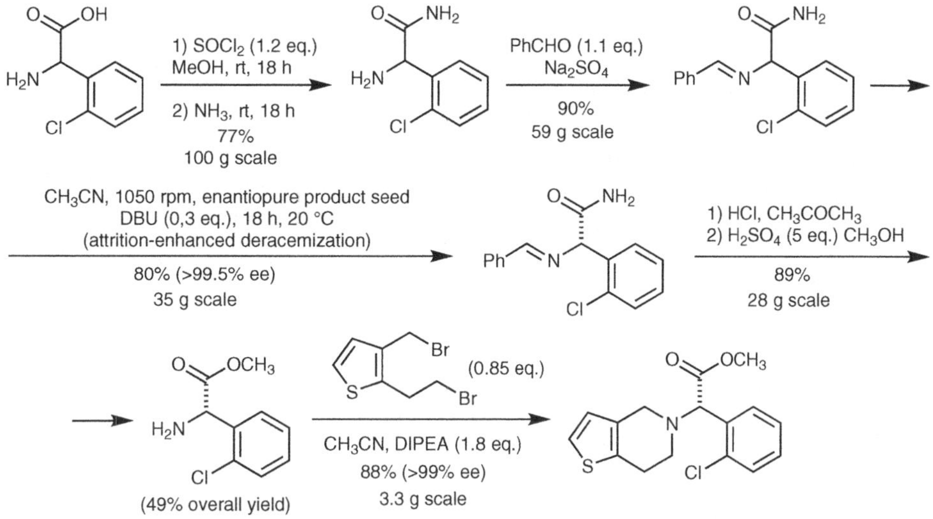

Scheme 3.33 DKR resolution for 2-amino-2-(2-chlorophenyl)acetamide and direct conversion to Clopidogrel

intermediate, has also been applied (Scheme 3.34, **H29**, Sanofi1997b).

Use of the methyl ester of 2-amino-2-(2-chlorophenyl)-acetic acid seems the obvious choice of building block to be employed in the synthesis in this approach as Clopidogrel is a methyl ester. Notwithstanding, several other esters have been used in similar routes and, of course, hydrolysis and methylation are required to arrive at the final molecule (Scheme 3.35, EastChinaUST2011).

Other electrophiles have been combined with the arylglycine ester resulting in patentable novel synthetic procedures (Scheme 3.36 and scheme H). In one case, an amide is initially formed (**H35**) and in another an α-amino ketone results from a S_N2 reaction (**H15**, **H30**).

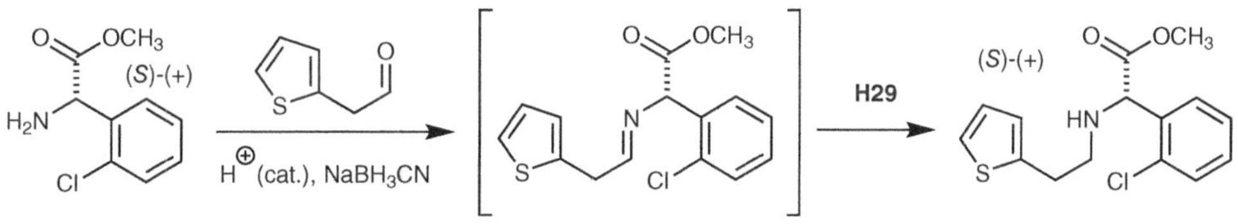

Scheme 3.34 Reductive alkylation of the aryl glycine ester

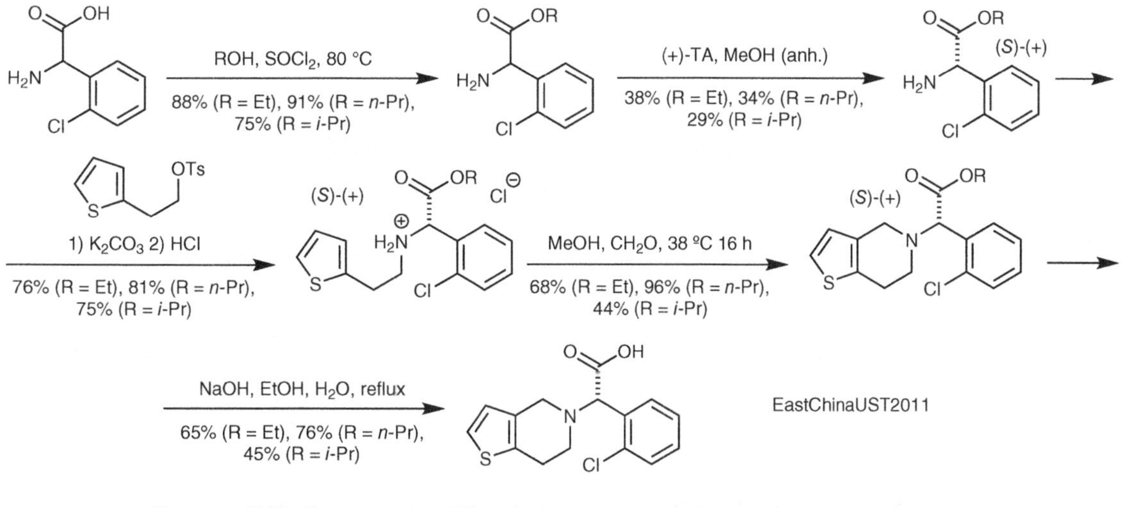

Scheme 3.35 Synthesis of Clopidogrel using ethyl and propyl esters of 2-amino-2-(2-chlorophenyl)acetic acid

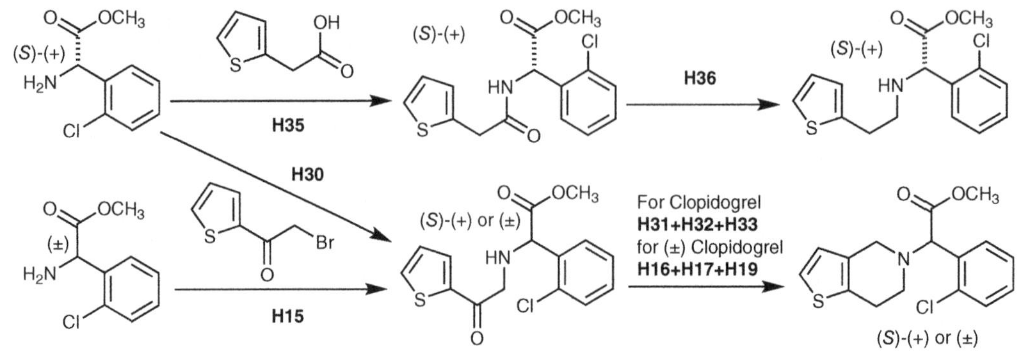

Scheme 3.36 Synthesis of Clopidogrel using other electrophiles to react with the aryl glycine ester

In both instances, a subsequent reduction is required to convert the carbonyl group into a methylene group (**H36, H16/H17/H19, H31/H32/H33**).

As seen above, the synthesis of the di-substituted thiophene building block generally requires that the thiophene ring be constructed from simple acyclic substrates. Two other synthetic routes also build up the thiophene ring, but towards the end of the synthesis (Scheme 3.37, BeijingHUTD2009, Xinfa2015).

3.4.1.6 Resolution processes in Clopidogrel synthesis

Resolution and racemization/recovery processes that have been described are summarized in scheme I. All processes involving racemic synthesis require a resolution step at some point. As already stated, a reliable and efficient resolution coupled with a recovery process for the unwanted isomer has a huge impact on the efficiency of any synthetic route. It is, therefore, not surprising that after having successfully developed synthetic methods

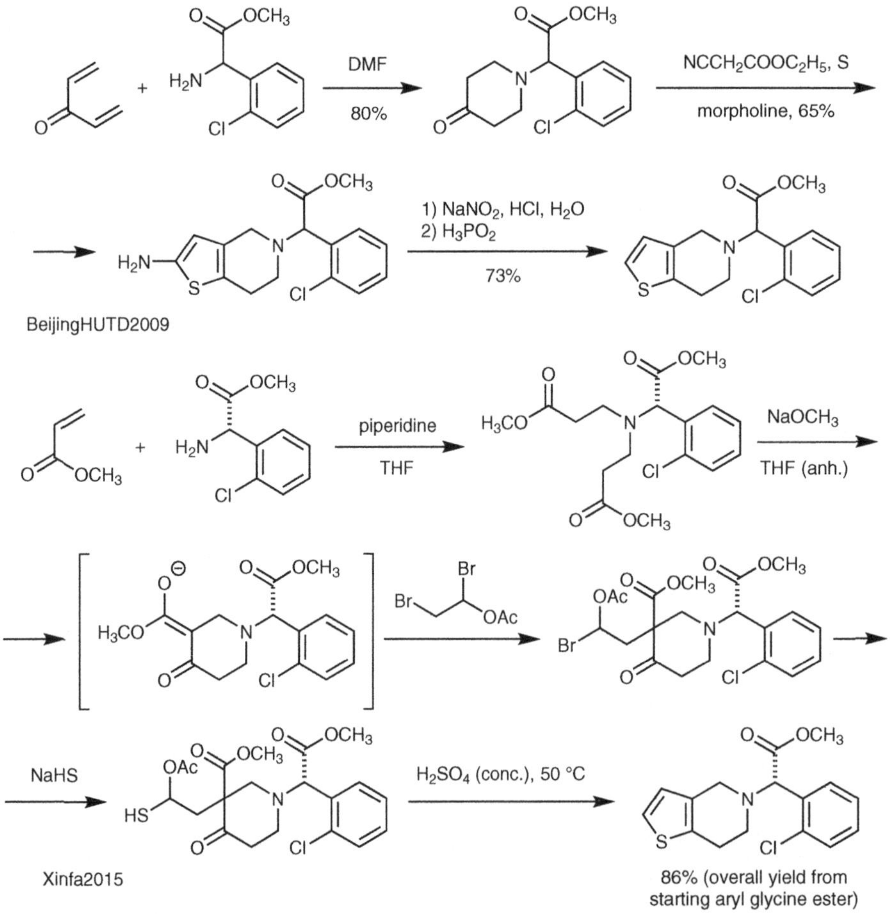

Scheme 3.37 Synthesis of Clopidogrel building the thiophene at the final steps

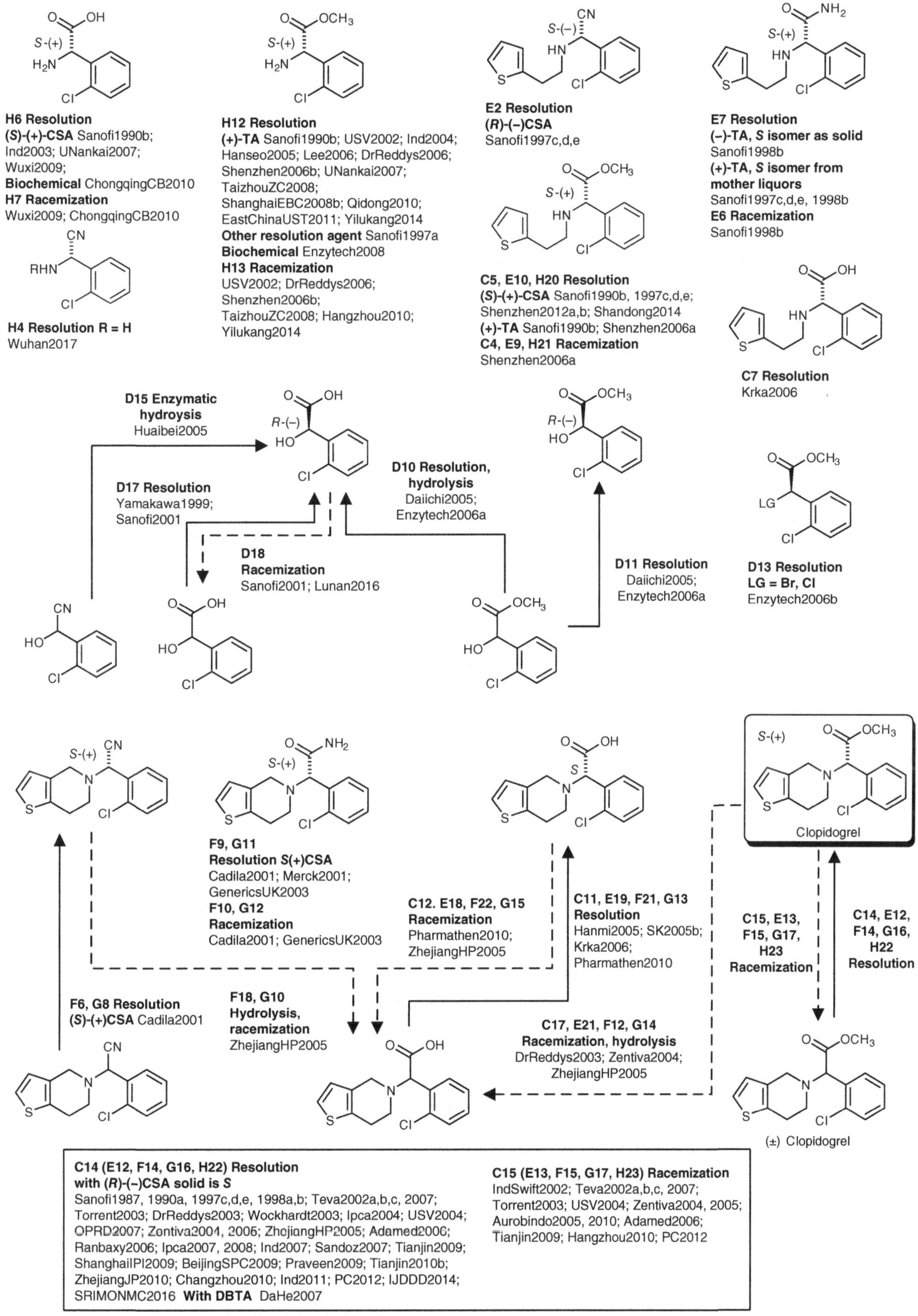

SCHEME I RESOLUTION AND RECYCLING PROCESSES APPLICABLE IN THE SYNTHESIS OF CLOPIDOGREL to produce the API, Process Chemists turned their focus on the development and improvement of resolution processes. The total effort expended in this area was considerable. Processes for the recovery and reuse of the expensive resolving agents were also the subject of a great deal of patent activity (for CSA: Ipca2007, Tianjin2010c, Zhejiang2011, ZhejiangHP2005 and for TA: EastChinaUST2010) driven by the need to bring down costs. In parallel, yet another objective was to develop raw materials with suitable enantiomeric purity.

Classical resolution *via* the formation of diastereomeric salts is the methodology most reported to produce the chirally pure compounds. Since low yields are reported in some of those processes, it is likely that these are not suitable for industrial use.

DKR processes that give enantiomerically pure compounds are obviously of great value as, theoretically, all of the racemic material is converted to Clopidogrel. The most promising of any DKR process would convert racemic Clopidogrel to the desired *S* isomer during the final isolation step, but up to date no such process has been reported. DKR processes have been reported and developed for the aryl glycine starting material, but this example stands alone as DKR has not been applied to other intermediates. Focus is then placed on an asymmetric synthesis as an alternative to avoid a resolution step.

3.4.2 Asymmetric synthesis in Clopidogrel production

Several asymmetric synthetic reactions that produce mandelic acid derivatives were touched on previously. The focus of this section is on asymmetric assembly of the two main building blocks.

The use of an asymmetric Strecker reaction turned out to be an optimum choice for application to a synthesis of Clopidogrel. Rhodia (Rhodia2003) produced an enantiomerically enriched nitrile in quantitative yield and 75-85% ee by contacting the starting amine and aldehyde in the presence of a catalyst and using a combination of trimethylsilyl cyanide (TMSCN) and methanol as a source of cyanide (Scheme 3.38, 0.55 g aldehyde scale). The catalyst employed was a Jacobsen non-metallic Strecker catalysts (OL2000). Thereafter, the nitrile derivative was converted to Clopidogrel by previously known methods.

A similar process used THTP as nucleophile, the catalyst in this case being quinine (JOC2012b, Scheme 3.39).

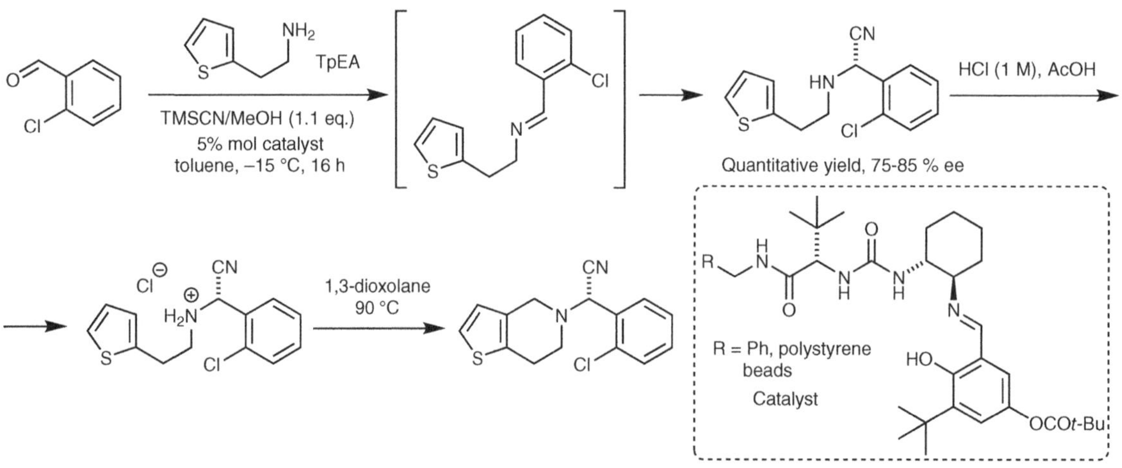

SCHEME 3.38 ASYMMETRIC STRECKER REACTION WITH TpEA IN CLOPIDOGREL SYNTHESIS

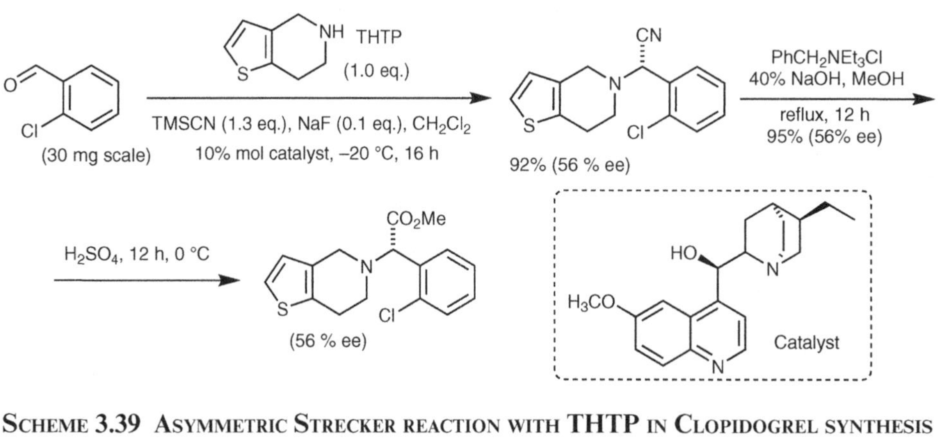

SCHEME 3.39 ASYMMETRIC STRECKER REACTION WITH THTP IN CLOPIDOGREL SYNTHESIS

S_N2 reactions are stereospecific and can provide chirally pure Clopidogrel directly when the starting materials are enantiomerically pure. This strategy was frequently adopted in the synthesis of Clopidogrel, the chirally pure intermediates frequently being obtained by classical resolution. It is, of course, necessary to preserve chiral integrity throughout the remainder of the synthesis once chirality has been established. Relayed asymmetric induction has also been applied successfully (Scheme 3.40, Ratiopharm2004).

The intended configuration at the α-halo carbon atom was fostered by a DKR during the formation of the ester with (*R*)-dimepranol in basic medium. Notably, a further increase in the diastereomeric excess was observed during the subsequent S_N2 reaction.

Another two examples of asymmetric induction by a chiral auxiliary *via* epimerization of an α-bromo aryl ester was reported in the patent literature, good yields and excellent enantiomeric excesses being observed (Scheme 3.41).

3.4.3 Other synthetic processes

Clopidogrel is an interesting target molecule to test new synthetic strategies and, as such, attracted interest from several academic groups. These syntheses, published in scientific journals, were generally not the subject of patent specifications and it is unknown whether any have been used industrially.

A concise and efficient synthesis of racemic Clopidogrel was achieved by exploiting a carbon dioxide insertion reaction to close the hydropyridine ring (Scheme 3.42). Notably, this key reaction does not use toxic transition metal catalysis. The overall yield of the sequence was reported to be 38% (SC2015).

A direct α-amination of methyl 2-(2-chlorophenyl) acetate with THTP has been accomplished using a copper catalysis (JACS2013, Scheme 3.43). Presumably, the transformation proceeds *via* the formation of an α-bromo carbonyl species that undergoes a nucleophilic

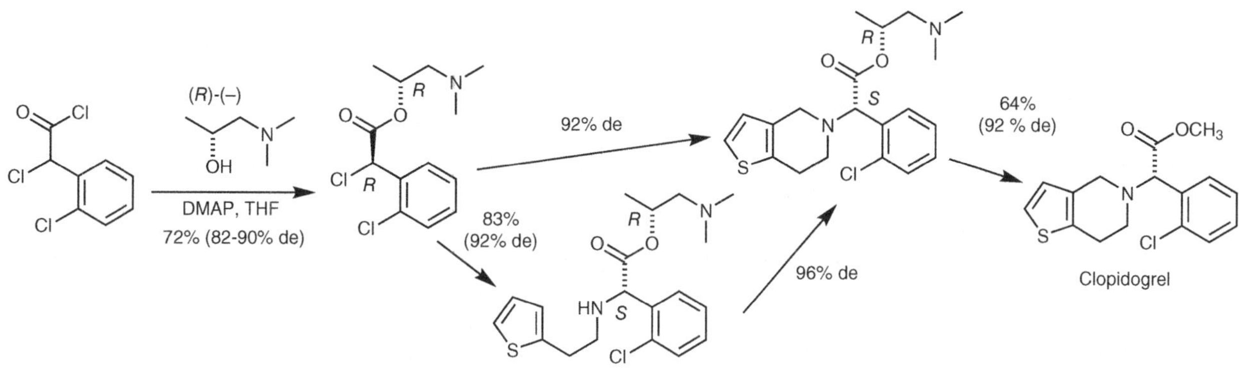

SCHEME 3.40 RELAYED ASYMMETRIC INDUCTION IN A S_N2 REACTION *VIA* A CHIRAL AUXILIARY

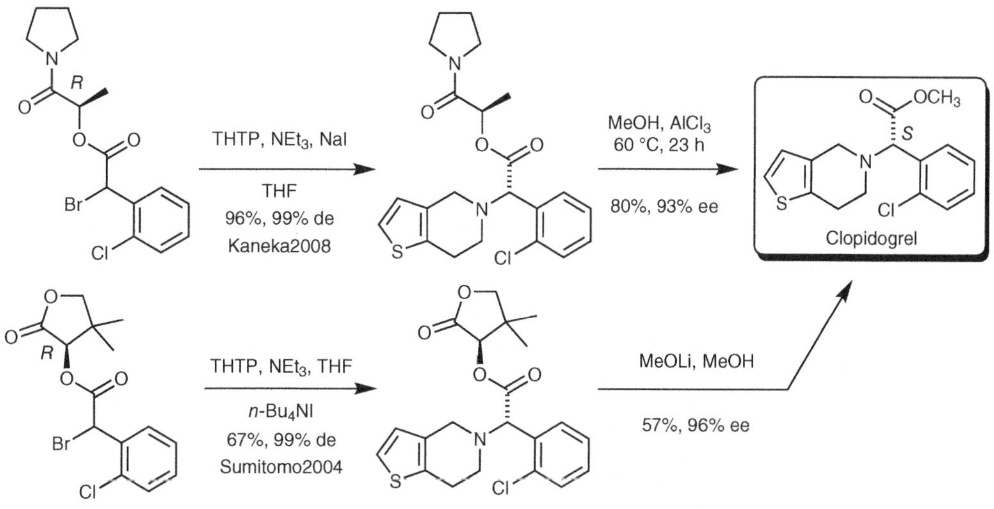

SCHEME 3.41 CLOPIDOGREL SYNTHESIS USING ASYMMETRIC INDUCTION IN A S_N2 REACTION

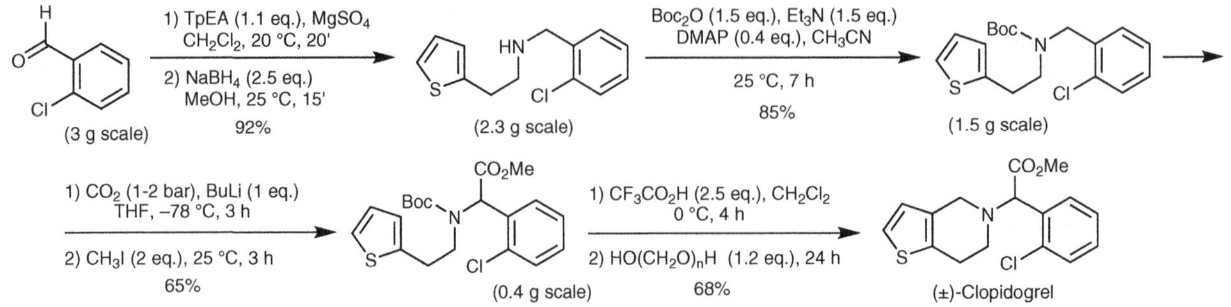

Scheme 3.42 Clopidogrel synthesis *via* a carbon dioxide insertion reaction

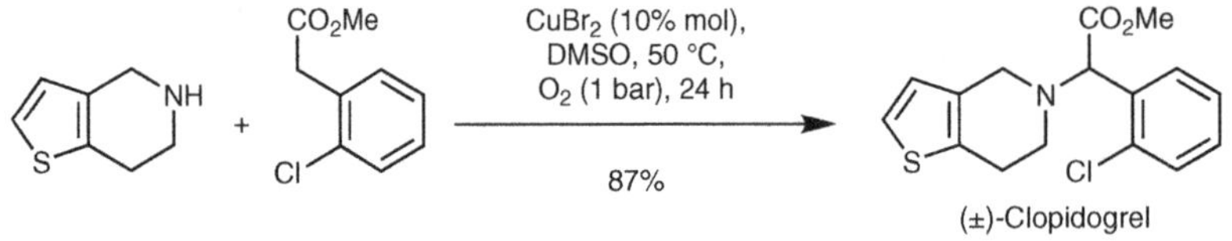

Scheme 3.43 Direct coupling of Methyl 2-(2-chlorophenyl)acetate with THTP

displacement of the bromide by the amine, yielding racemic Clopidogrel.

A solid phase synthesis of (*S*)-Clopidogrel has been reported using commercially available Wang resin and was carried out under mild reaction conditions, giving an overall yield of 55% in six steps and 98.0% ee (HCA2013, SK2005a, Scheme 3.44).

The use of multicomponent reactions (MCR) has also been explored. Kalinski *et al* applied this powerful synthetic tool in three component Ugi and Petasis reactions to produce, in two steps, racemic product (S2008) with acceptable to high global yields (Scheme 3.45). The Ugi reaction was one of the earliest examples of this type of powerful synthetic transformation and was originally applied to the preparation of both α-amino acids and to peptide couplings. The components of the reaction are combined in one pot under mild conditions, 60 °C and microwave irradiation, in the presence of an equimolar amount of formic acid, followed by the cleavage of the initially formed isocyanide by treatment with acid or base. The Petasis reaction uses different components; THTP, glyoxylic acid and 2-chlorophenylboronic acid

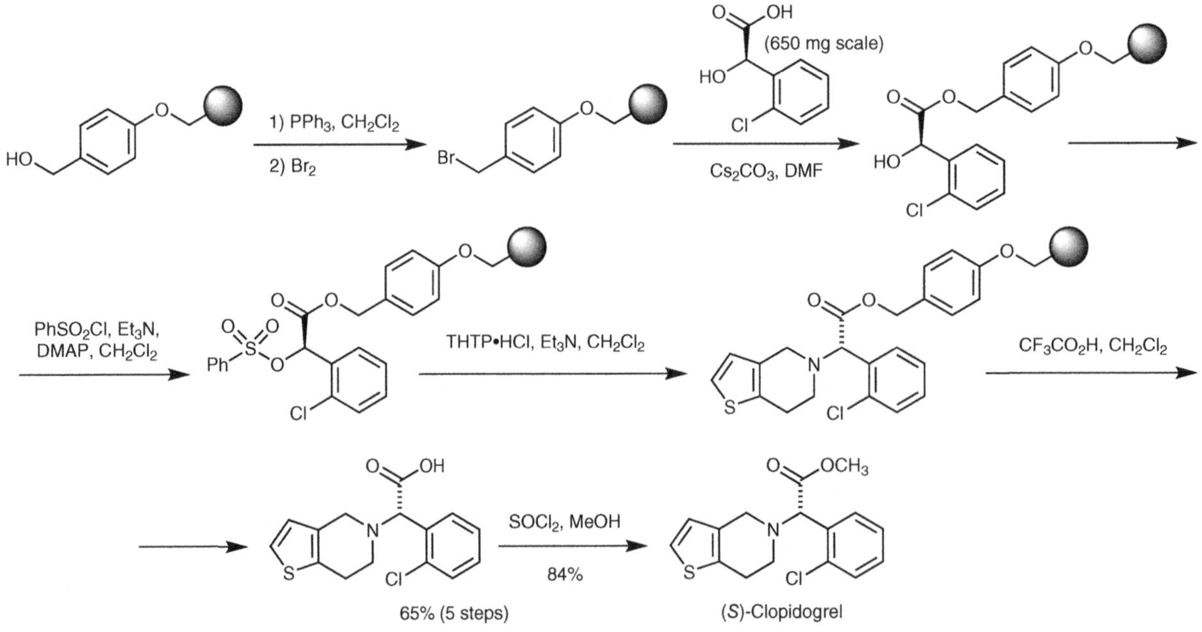

Scheme 3.44 A solid phase process for Clopidogrel

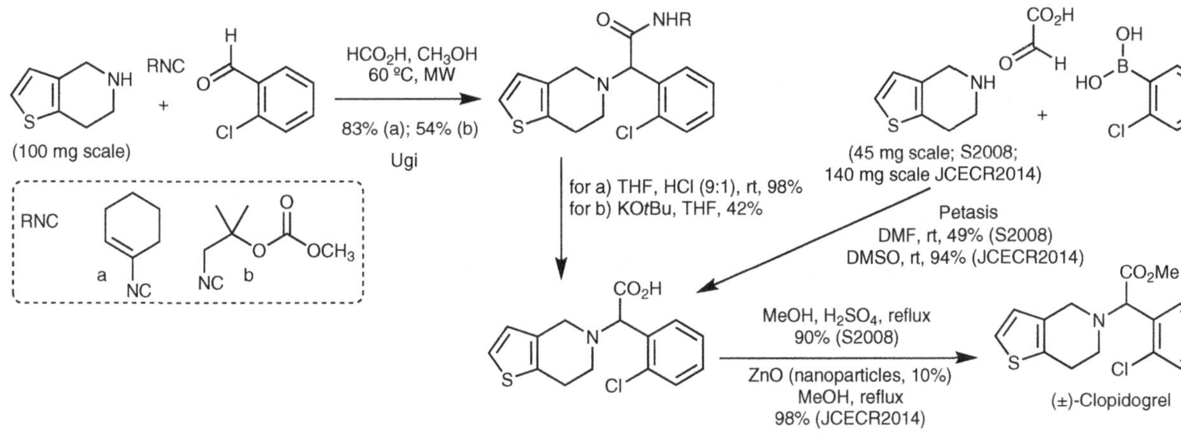

SCHEME 3.45 MULTICOMPONENT REACTIONS IN CLOPIDOGREL SYNTHESIS

to obtain the same carboxylic acid precursor. Recently, a very high yield has been reported for this reaction (JCECR2014).

In a later approach, yet another MCR reaction with an arylzinc bromide in place of the boronic acid component was developed, but the yields in this case were somewhat lower (Scheme 3.46, M2010).

3.4.4 Synthesis of the Thiophene building block

In all of the synthetic routes described two different nucleophilic thiophene derivatives were used to synthesise Clopidogrel, namely 4,5,6,7-tetrahydrothieno[3,2-*c*]-pyridine, THTP and 2-(thiophen-2-yl)ethanamine, TpEA (Scheme 3.47). As THTP and TpEA are commodities and readily available on the market, an extensive analysis of their synthesis is not given here. Several patents describe the formation of THTP from its logical precursor, TpEA, by a Pictet-Spengler reaction.

The main synthetic routes to produce TpEA are shown in scheme J. Although several methods can be used to produce TpEA, the most obvious starts with the corresponding alcohol, available by direct metalation of thiophene followed by reaction with ethylene oxide (**J3**). This route avoids the use of nitromethane chemistry, which is involved in an alternative approach from 2-formylthiophene (**J13**).

One of the most common and problematic of the impurities found in the drug substance is the isomeric compound resulting from the positional isomer in the thiophene building block. This impurity is controlled in both EP and US Pharmacopeia. Metalation of thiophene gives rise to two positional isomers and, since this is the most practical entry into these building blocks, it is to be expected that the impurity will be present in standard raw materials (Scheme 3.48). It is therefore, imperative that only highly purified thiophene building blocks be used or that adequate purification methods for either the intermediates or the final API be developed.

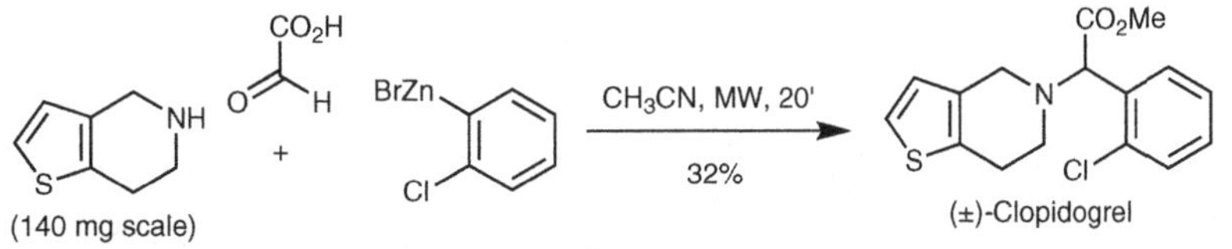

SCHEME 3.46 A MCR REACTION USING AN ARYL ZINC BROMIDE NUCLEOPHILE

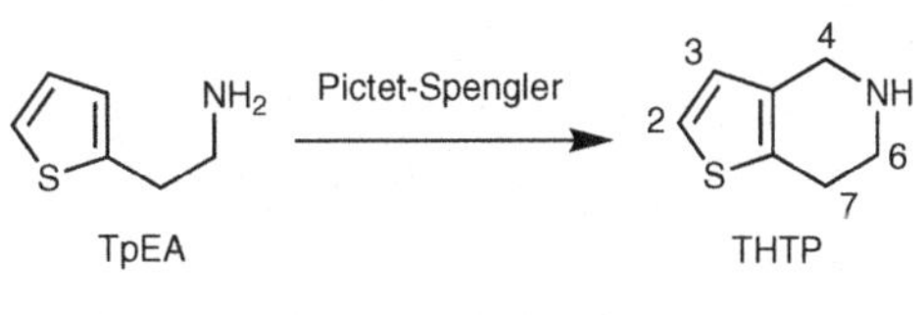

SCHEME 3.47 THE TWO AMINE-THIOPHENE BUILDING BLOCKS FOR CLOPIDOGREL SYNTHESIS

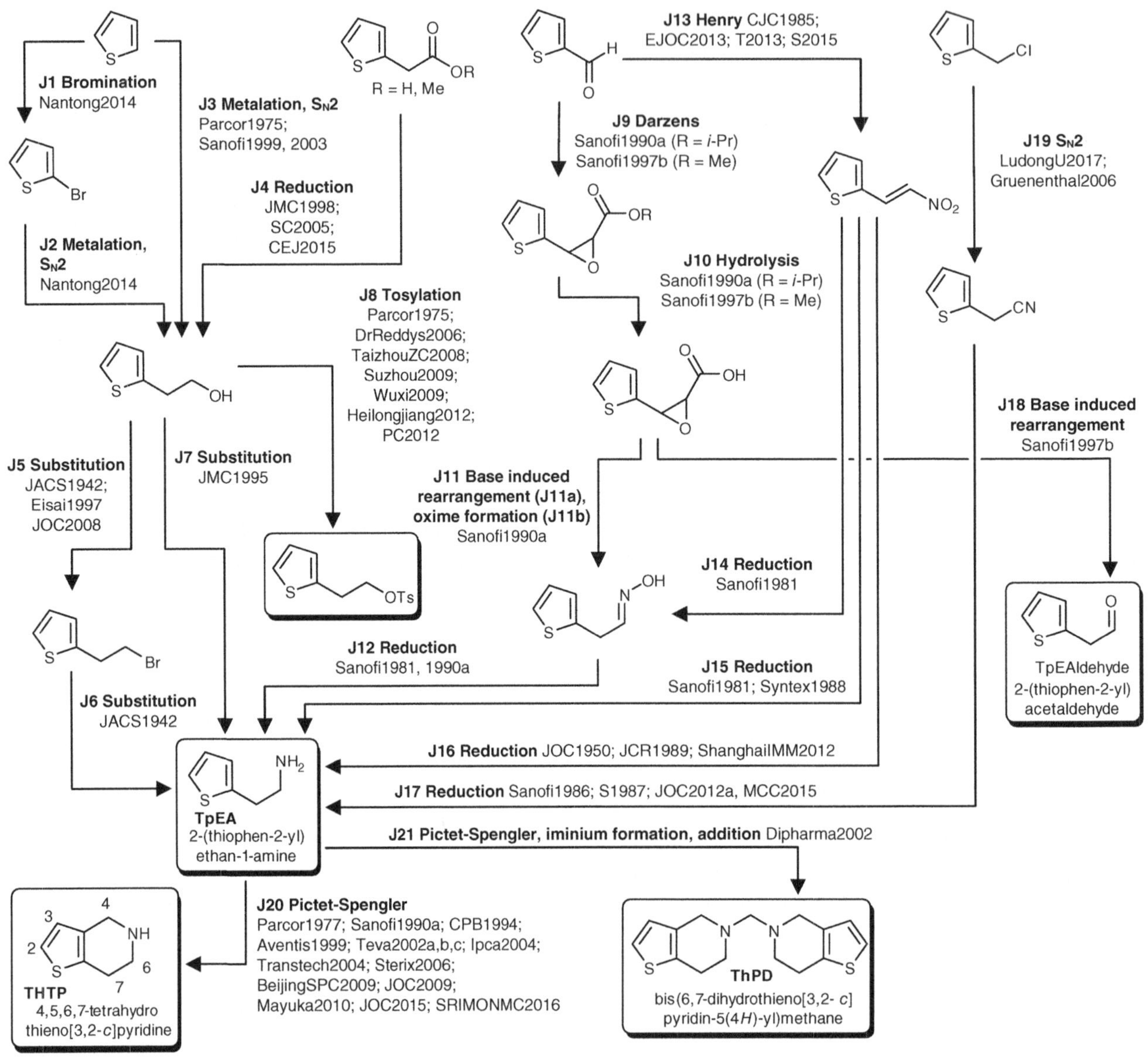

Scheme J Main synthetic routes to TpEA, THTP and ThPD (for reaction conditions see page 78)

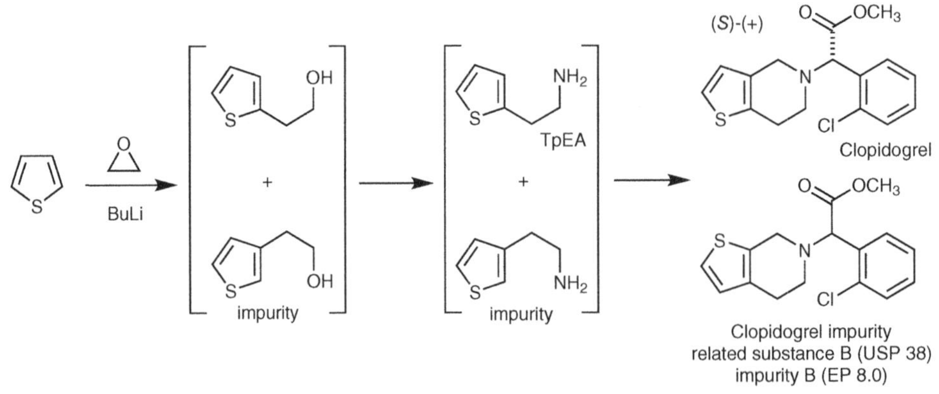

Scheme 3.48 Formation of Clopidogrel contaminant due to the carry of a starting material impurity until the final API

The alternative electrophilic thiophene derivatives are conveniently produced from 2-(thiophen-2-yl)ethanol (**J8**). The halogen derivatives are reported to be quite unstable whereas the tosylated and related derivatives are reported to be more stable. α-Thiophenylacetaldehyde (product of **J18**) can be used as a synthetic equivalent but this option is less favourable due to its comparatively more difficult synthesis and inherent instability.

3.5 STRATEGY COMPARISON AND CONCLUSIONS

For a molecule of the size and with the features of Clopidogrel, it can be expected that several synthetic routes will be applicable to its commercial production. At first glance, an asymmetric synthetic route that accesses Clopidogrel directly appears to an ideal choice because the resolution step is avoided and, along with this, there is no necessity to use a resolving agent, treat unwanted material waste and develop a recycling process. Although the prevailing belief, especially among some in the academic community, is that an asymmetric synthesis is always preferred to a resolution-based approach, there are circumstances where this is not always the case.

Several asymmetric routes have been reported, the asymmetric Strecker reaction being particularly significant as good yields and ee are obtainable. However, the cost, availability and recycling of the catalyst could be problematic for implementation on an industrial scale. Asymmetric Strecker reactions with both TpEA and THTP have been described, although when applied to the former gave better yields and ee (Scheme 3.38, Scheme 3.39, Scheme 3.49). Since ee's of this reaction are generally modest, the enantiomeric purity has to be increased during the subsequent steps. In addition, since conversion of the nitrile to the methyl ester normally requires stringent conditions, the likelihood of racemization is increased. Hence, these asymmetric type processes may not be easy to implement at industrial scale.

An alternative strategy to avoid a late resolution step in the synthesis is to use an appropriate chiral starting material. This material may be produced in-house or obtained commercially. This tactic clearly has many advantages. Two different approaches were commonly applied, each depending on the role of the building block containing the chirality (electrophilic ester block, **a**, **b** and nucleophilic ester block **c**, Scheme 3.50). Notably, when the ester group is present in an early intermediate the necessity to use vigorous conditions for its installation is negated.

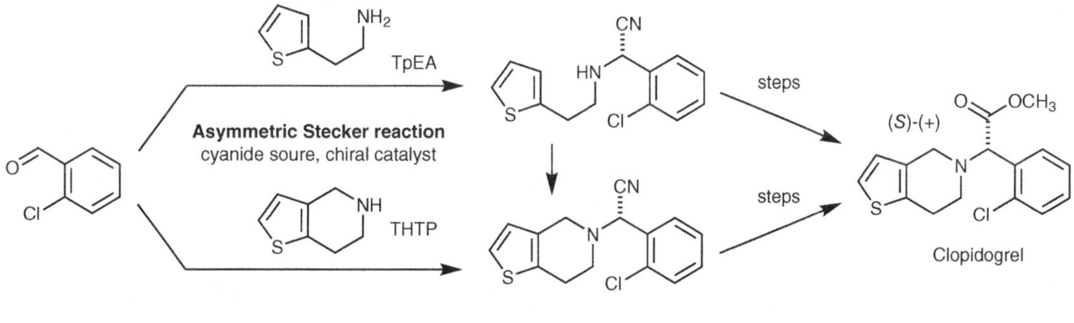

SCHEME 3.49 ASYMMETRIC STRECKER REACTION FOR CLOPIDOGREL SYNTHESIS

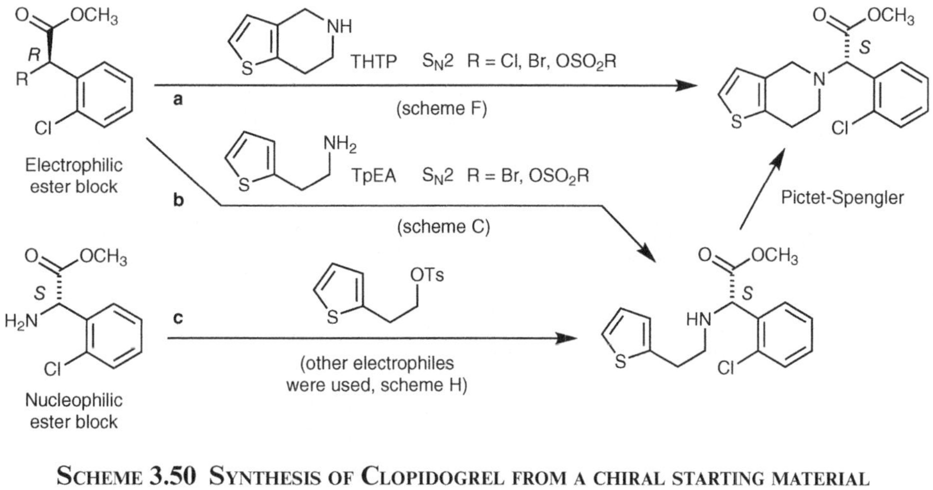

SCHEME 3.50 SYNTHESIS OF CLOPIDOGREL FROM A CHIRAL STARTING MATERIAL

The syntheses of the electrophilic ester blocks were summarized in scheme D. Generally, these electrophiles are reported to be unstable, particularly the α-halogen derivatives. Partial or complete racemization of the chiral centre in the molecule produced by the bimolecular substitution reaction was observed, thus issues may arise when Clopidogrel is produced directly using this approach. When a nucleophilic ester block is selected as a building block (strategy **c**) isomerization or racemization of the chiral centre is no longer referred to as a major problem. This stability of the intermediates in the aryl glycine route favours it use in comparison to routes **a** and **b**. In route **c**, a Pictet-Spengler reaction is the last synthetic step, requiring the implementation of adequate precautions for safe handling of formaldehyde or its surrogates. The tosyl electrophile is the most convenient but other electrophiles can be used.

In a molecule such as Clopidogrel, containing one chiral centre, a classical resolution-based process is perfectly viable. Resolution can be carried out on several intermediates (see schemes C to I) and on the final product. With some exceptions, most of the processes were able to isolate the intended enantiomer in a yield of about 45% (50% being maximum). Hence, several routes Clopidogrel can be obtained in similar overall yields *via* various routes. A crucial and key factor as to where to implement the resolution is whether the unwanted isomer can be recycled, i.e. racemized and reintroduction into the process. The principle methods to racemic Clopidogrel are shown in Scheme 3.51.

Racemization at the α position in esters is easier than in the corresponding nitrile analogues. Because of this, a resolution step at the end of the synthesis could have advantages as racemization of the unwanted isomer can be carried out under milder conditions. Scheme 3.52 summarizes what was discussed above in terms of the two most preferred routes. Both of these options allow for racemization of unwanted isomer *via* a catalytic base mediated enolate formation which can thereafter be incorporated into other material to be resolved. Recovery of the CSA resolving agent presumably is possible by routine extraction and crystallization procedures.

Some issues, which may arise when a late resolution step is employed, are:

- The resolution is required to produce acceptable pharmaceutical grade material directly to avoid the necessity for additional purification steps. In the case of Clopidogrel a salt exchange, to produce the hemisulfate, has to be carried out during which an improvement in chiral purity may be observed.
- Decomposition may occur during racemization, which may result in the formation additional impurities. These have to be closely monitored and removed by further purification before or after the resolution step.

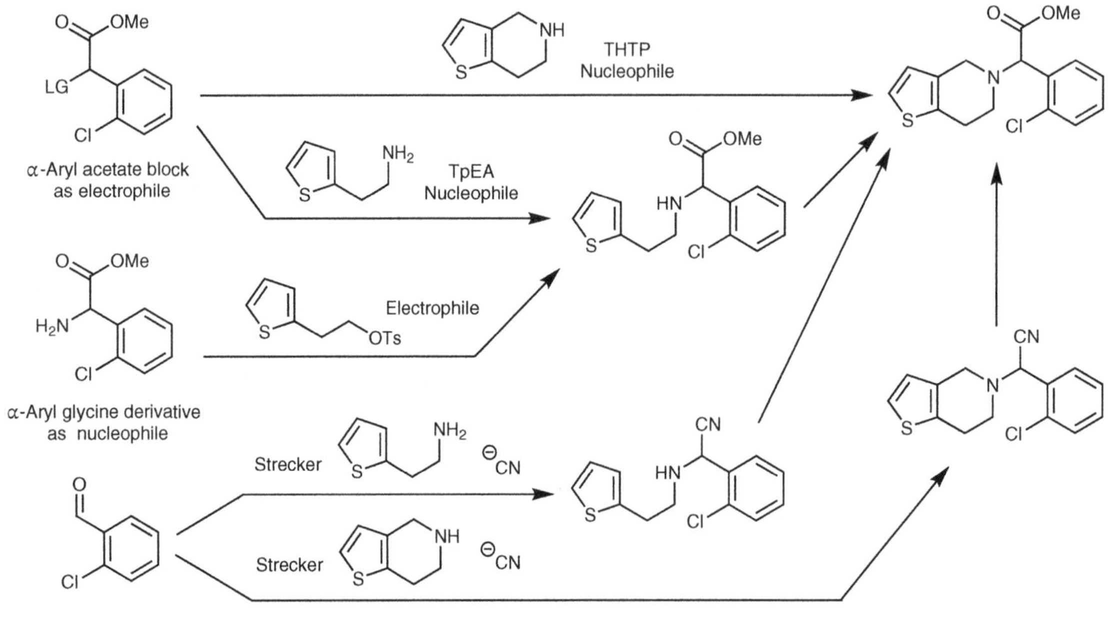

Scheme 3.51 Main routes to racemic Clopidogrel

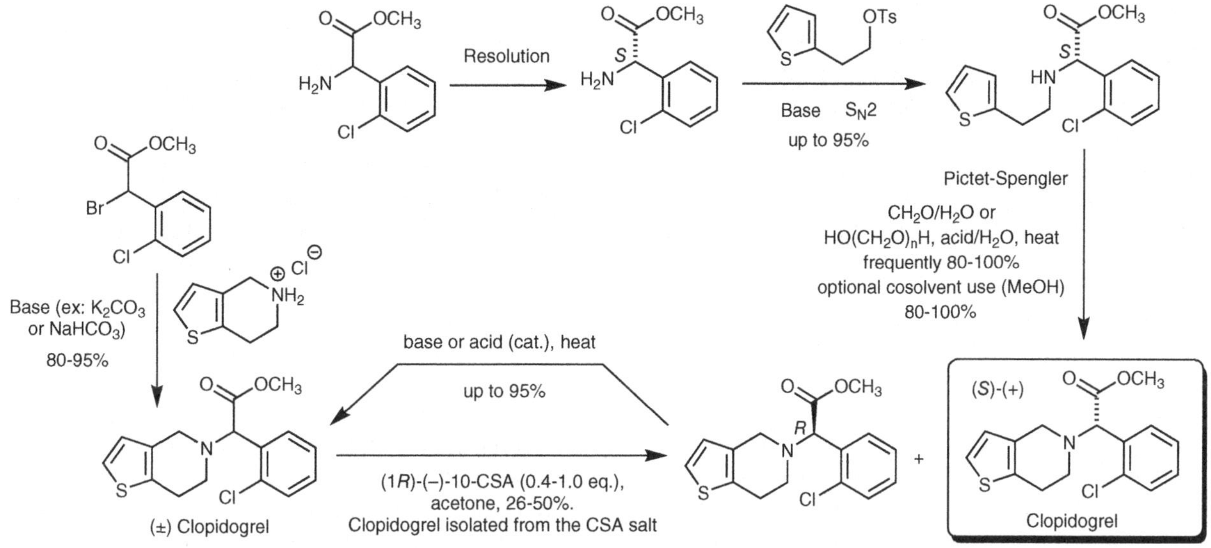

SCHEME 3.52 PREFERRED ROUTES TO CLOPIDOGREL

3.6 LESSONS FROM THE CLOPIDOGREL CASE

Asymmetric synthesis *vs* resolution-recycling

A pure asymmetric synthesis is often regarded as the "Holy Grail" in synthetic chemistry. Resolution-based process that do not contemplate recycling of the unwanted isomer would generate a huge amount of waste and hence be inherently inefficient. However, when the resolution is adequately developed and coupled with a recycling process it is often preferred to an asymmetric synthesis being routinely easier to carry out, cheaper and more robust. Clopidogrel can be considered as an excellent model where asymmetric synthesis does achieve levels of efficiency to beat out a resolution approach.

Placement of a Resolution step

In a non-recycling scenario, the resolution step should be placed as early as possible along the synthetic route. Since this is looked upon as a low yield step, its earlier placement less penalizing in terms of productivity and production costs.

Stability of Asymmetric centre

The stability of the asymmetric chiral centre is of extreme importance in route selection in a synthesis of a chiral molecule. Since reactions of the α-halo esters with nucleophiles are not completely stereoselective and partial or total racemization is observed, this is not the ideal point at which to introduce chirality and invalidates it for large-scale production.

Dynamic Kinetic Resolution *vs* Classical Resolution

A resolution process, without introduction of a process for the recovery of the unwanted isomer, limits the yield in this step to 50%. A dynamic kinetic resolution is therefore, highly desirable as the *in situ* isomerization of the unwanted isomer can theoretically increase the yield to nigh on quantitative. Some degree of success was apparent in the Clopidogrel case when a DKR technique was used in the separation of the aryl glycine esters (Scheme 3.32, Scheme 3.33).

3.7 REACTION CONDITIONS FOR SCHEMES C TO J

3.7.1 Reaction conditions for scheme C

C1 X = Br: TpEA, KOH, H_2O or MeOH, rt to 40 °C, 80-90%. **SM = α-keto derivative:** TpEA, 2-(2-chlorophenyl)-2-oxoacetic acid, H_2 (3 bar), Pd/C (5%), 100%.

C2 (H10) not reported.

C3 TpEA, $KHCO_3$, K_2CO_3 or $NaHCO_3$ or Na_2CO_3, MeOH, reflux, 78-90% (ester formation D22+C3, Sanofi1990b, Tianjin2010a, Shenzhen2012a,b). TpEA, $NaHCO_3$, CH_2Cl_2, H_2O, reflux (benzylic bromination D30+C3) 83-86% (Shenzhen2012a, Shandong2014).

C4 (E9, H21) NaOMe, MeOH (anh.), rt, YNR.
C5 (E10, H20) (*S*)-(+)-CSA: (*S*)-(+)-CSA, acetone, 34-35% (Sanofi1990b, Sanofi1997c,d,e), 26% (Shenzhen2012a, Shandong2014), 22% from 2-bromo-2-(2-chlorophenyl)acetic acid (Shenzhen2012b). **(+)-TA:** L-(+)-Tartaric acid, *i*-PrOH, YNR.
C6 TpEA, K_2CO_3 (23% aq.), CH_2Cl_2, reflux, R = Me, 65% (52% ee), R = Ph, 71% (84% ee), R = 4-ClC_6H_4, 94% (84% ee), R = 4-$O_2NC_6H_4$, 99% (90% ee), R = 2,5-$Cl_2C_6H_3$, 98% (93% ee).
C7 D-*threo*-2-amino-1-(4-nitrophenyl)-1,3-propanediol, MeOH, 37%.
C8 (H38) MeOH (anh.), $SOCl_2$, 40 °C, 90%.
C9 (E8, H18) CH_2O/H_2O (30% w/w), CH_2Cl_2, 25°C, 32% (Sanofi1990b), at reflux 85-88% (Sanofi1990b, ShanghaiIPI2009). $HO(CH_2O)_nH$, HCO_2H, 50°C, 87-92% (Sanofi1990b, 1997c,d,e). $HO(CH_2O)_nH$, HCl or H_2SO_4 (cat. conc.), *i*-PrOH, 80 °C, 80% (LifeSciences2000). 1,3-dioxolane, MeOH•HCl, 70 °C, 95% (IJDDD2014).
C10 (E17, H37) CH_2O/H_2O, rt to 80 °C or $HO(CH_2O)_nH$, acid/H_2O (acid, HCl, HCO_2H, H_2SO_4), 45-90 °C, frequently 80-100%, optional cosolvent use (MeOH).
C11 (E19, F21, G13) (1*R*,2*R*)-(–)-2-amino-1-(4-nitrophenyl)-1,3-propanediol (0.5 eq.), H_2O/MeOH, reflux to 5 °C, 42% (99% ee, Hanmi2005), in H_2O/*i*-PrOH, 29-38% (99% ee). (+)-Cinchonine (0.5-1 eq.), EtOH/CH_3CN or *i*-PrOH/acetone, rt, 29-42% (98% ee, SK2005b). (*R*)-(–)-CSA, acetone, 42-44% (Krka2006), in EtOAc 35-36% (Krka2006). L-DBTA (0.5 eq.), EtOH, rt, 26-29% (Pharmathen2010).
C12 (E18, F22, G15) KOH or NaOH, H_2O, reflux, 43-98%.
C13 (E11, F13, G9) Me_2SO_4, NaOH (aq.), Bu_4NBr (cat.), CH_2Cl_2, 27 °C, 64% (DrReddys2003), 61% (C17+C13, DrReddys2003). Me_2SO_4, NaOH, Et_3BnNCl, MeOH, H_2O, 40 °C, 87% (OPRD2007, Ind2011), in toluene/H_2O or MeOH/H_2O or *n*-BuOH/H_2O, rt to 40 °C, 44-51% (ZhejiangHP2005). MeI, Bu_4NHSO_4, CH_2Cl_2, H_2O, rt, 57% (Zentiva2004). MeI, DMF or acetone, 73% using potassium salt (Zentiva2004). MeOH, H_2SO_4, reflux, 70-90% (OPRD2007, S2008). MeOH, ZnO (nanoparticles, 10%), reflux, 98% (JCECR2014).
C14 (E12, F14, G16, H22) (*R*)-(–)-CSA (0.4–1.0 eq.), acetone 26-50%. Large scale: (*R*)-(–)-CSA (0.6 eq.), acetone, reflux to 25 °C, 33% (360 kg scale, Sanofi1998a, PC2012). Other resolution solvents: acetone/CH_2Cl_2 or acetone/toluene, 39-40% (Ipca2004), acetone/MTBE, 30-58% (USV2004, ZhejiangJP2010), toluene/DMF, 32-35% (Teva2002a,b,c, 2007), *i*-PrOH, 17-28% (Torrent2003, DrReddys2003), diglyme, 28% (Torrent2003), MTBE/EtOH, 51% (Zentiva2004), MTBE/H_2O, 34% (ZhejiangJP2010), *i*-PrOAc/MeOH, 35-59% (Zentiva2004), *n*-BuOAc/MeOH, 26% (Zentiva2004), *i*-PrOAc/EtOH, 45% (Zentiva2004), *n*-BuOAc/EtOH, 39% (Zentiva2004), EtOAc, 26-33% (Ind2007), toluene/H_2O, 46% (Ind2011). With DBTA, YNR (DaeHe2007).
C15 (E13, F15, G17, H23) With base: Li_2CO_3 or Na_2CO_3 or K_2CO_3 or LiOH or NaOH or KOH, organic solvent/H_2O, 0-60 °C, YNR (IndSwift2002). K_2CO_3 or KOH, MeOH, reflux, 20-22% (Zentiva2004). NaOH or KOH, MeOH, reflux, 100% (Hangzhou2010). *t*-BuOK, toluene (anh.), 0 °C, YNR (Teva2002a,b,c, 2007). *t*-BuOK, Et_2O, rt, 75% (USV2004). RONa, ROH, 20-50 °C, 63-92% (R = Me, Et, *i*-Pr, Tianjin2009). MeOH or *i*-PrOAc, Me_4NOH or *n*-Bu_4NOH, 5-10 °C 69-98% (Zentiva2005). **With acid:** *i*-PrOH, HCl (aq.), reflux, 63% (Torrent2003). HCO_2H, 100 °C, 90-94% (Adamed2006), in EtOH, reflux, 75% (Adamed2006). AcOH, 100 °C, 87% (Adamed2006). H_2SO_4, acetone, water, 22 °C, 45% (PC2012).
C16 (E20, F25, G21, H40) 1) $(ClCO)_2$, DMF (cat.), CH_2Cl_2, –5 °C; 2) MeOH, 95-98% (Hanmi2005). MeOH, $SOCl_2$, 60-70 °C, 88% (SK2005b), 69-80% (Pharmathen2010). MeOH (anh.), H_2SO_4, reflux, 77-72% (Hanmi2005, Krka2006). MeOH (anh.), $MeSO_3H$, reflux, 65% (Hanmi2005). $ClCOOCH_3$, NEt_3, CH_2Cl_2, 0 °C; 2) MeOH, 73% (Hanmi2005). Me_2SO_4, PEG400, $NaHCO_3$, *n*-BuOH, H_2O, 51% (ZhejiangHP2005). MeI, NEt_3, DMF, 88% (Krka2006). Me_2SO_4, NaOH, Bu_4NBr, CH_2Cl_2, rt, 89% (Krka2006).
C17 (E21, F12, G14) NaOH or KOH, H_2O or EtOH/H_2O, 85 °C to reflux, 96-100%.
C18 $HO(CH_2O)_nH$, HCO_2H, 40-70 °C, 62-90%. CH_2O/H_2O, SM as HCl salt, 40 °C, 84%.
C19 (H39) $HO(CH_2O)_nH$, HCO_2H, 40 °C, 70-77%.

3.7.2 Reaction conditions for scheme D

D1 *n*-BuOAc, TsCl, *N*,*N*-dimethylaniline, 25 °C, 97%.
D2 R = TMS: TMSCN, CH_2Cl_2, 20 °C, chiral V catalyst (10^{-3} eq.), 92-95%, 76-81% ee. **R = Ac:** KCN (3 eq.), CH_2Cl_2, H_2O, 2,6-lutidine (cat.), Ac_2O, 0 °C, chiral V catalyst (10^{-3} eq.), 51%, 62% ee.
D3 HCl (37%), H_2O, 80 °C, 100%.
D4 MeOH, H_2SO_4 (cat.), reflux, 91-94% (Sanofi1997f, Huaibei2005), at 75 °C, 90-97% (Shenzhen2008).
D5 H_2 (30 bar), catalyst (see Scheme 3.22), *t*-BuOK, *n*-BuOH, 35 °C, 95%, 91% ee (80%, 99.1% ee, after toluene crystallisation).
D6 R = Me: H_2 (50 bar), EtOH, catalyst, $CeCl_3$•7H_2O. Catalysts: [Ru(*R*)-xyl-P-Phos(benzene)Cl]Cl, 100%, 65% ee, [Ru(*R*)-BINAP(benzene)Cl]Cl, 100%, 65% ee (JOC2009). $[Ru(p\text{-cymene})Cl_2]_2$, (*R*,*R*)-Ts-DPEN ligands, HCOOH/NEt_3/DMF (5:2) or HCO_2Na/H_2O, PEG, 100%, 72-93% ee (JOC2009). *E. coli* recombinant (carbonyl reductase), 76-89%, >99% ee (OBC2007), 76-93% (92-99% conversion), >99% ee, UOkayama2007), 35-89%, >99% ee (ASC2008), 88% (100% conversion), >99% ee, TL2012), 97-99% conversion (JMCB2014). *Saccharomyces cerevisiae*, 100% conversion, 96-98% ee (BL2010). *Thermus thermophilus*, 62-78%, 92-95% ee

(EJOC2011). **R = Et:** H_2 (50-60 bar), [RuL(benzene)Cl]Cl, 70-100 °C, 100% conversion, 96% ee (L = (*S*)-(2,2,2',2'-tetramethyl-[4,4'-bibenzo[*d*][1,3]dioxole]-5,5'-diyl)bis(diphenylphosphane) (OL2005).

D7 RSO_2Cl, $LiClO_4$, py, CH_2Cl_2, R = Ph, 90%, R = *p*-MeC_6H_4, 85%, R = Me, 87%, R = *p*-ClC_6H_4, 90%, R = *p*-$O_2NC_6H_4$, 92%, all 99% ee (Sanofi1997f). RSO_2Cl, NEt_3, DMAP (cat.), CH_2Cl_2, 0 °C, R = Ph, 97%, R = *p*-$O_2NC_6H_4$, 88%, R = Ph, 97%, all 99% ee (Sanofi1997f), R = *p*-$O_2NC_6H_4$, 78-90% (JOC2009, NC2010). $PhSO_2Cl$, NEt_3, NH_3, DMAP (cat.), CH_2Cl_2 (anh), 0 °C, 89% (Huaibei2005). $PhSO_2Cl$, NEt_3, Me_3N•HCl, CH_2Cl_2, 85-91% (Shenzhen2008).

D8 H_2SO_4 (conc.), MeOH (anh.), reflux, 88%. $SOCl_2$, MeOH (anh.), rt, 93%.

D9 BH_3•NH_3, H_2O, 20 °C, 92%.

D10 Enzymatic hydrolysis *R* isomer see Scheme 3.19.

D11 Enzymatic hydrolysis *S* isomer see Scheme 3.19.

D12 LG = OMs: MsCl, NEt_3, CH_2Cl_2, YNR. **LG = OBs:** $PhSO_2Cl$, *n*-Pr_3N, 91-93% (D19+D12). **LG = OTs:** TsCl, NEt_3, 92% (D19+D12). TsCl, *n*-Pr_2NH, 88% (D19+D12). **LG = Cl:** $SOCl_2$, py, 40 °C, 76-99% (ChongqingLP2008). $SOCl_2$, 65 °C (Me, Et and Bu ester, Enzytech2006b).

D13 Lipase selective hydrolysis of *S* isomer, yields and ee for the isolated *R* ester, R = Me, 43%, >99% ee, R = Et, 38-48%, >99% ee, R = Bu, 32-42%, >99% ee.

D14 MeOH, KCN, AcOH, 0 °C, YNR. MeOH, KCN, pH 8.5 (phopshate buffer), NI.

D15 Cyanide hydrolase (20% w), CH_3CN, MeOH, pH 8.5 (phosphate buffer), 37 °C, N_2, YNR.

D16 H_2SO_4, H_2O, heat.

D17 Optically active *N*-benzyl-1-phenylethanamine (0.3-1.0 eq), H_2O or EtOH or MeOH. With (*R*)-*N*-benzyl-1-phenylethylamine, 38-45% (Yamakawa1999). (1*S*,2*S*)-2-amino-1-phenylpropane-1,3-diol (1 eq.), EtOAc saturated with H_2O, seed of *S* isomer salt, rt, salt of *S* isomer in suspension, 22%, *R*-isomer from mother liquors, 69% ee (Sanofi2001). (1*R*,2*R*)-2-amino-1-(4-nitrophenyl)propane-1,3-diol, *i*-PrOAc/MeOH/H_2O (70/8/1), seeded with *R* isomer salt, rt, 29%, *R*-isomer from solid in suspension, 92% ee (Sanofi2001). L-(+)-lysine (1 eq.), MeOH, 40-45°C, seeded with *R* isomer salt, 30-40 °C, 26% *R*-isomer from solid in suspension, 96% ee (Sanofi2001).

D18 NaOH, H_2O/DMSO, 100 °C, 95% (Sanofi2001). 1) *i*-PrOAc or EtOAc or *i*-PrOAc/EtOAc, (*R*)-1--phenylethan-1-amine, 85 °C; 2) NaOH (aq., 32%) or Na_2CO_3 (aq., 40%) or KOH (aq, 20%) or K_2CO_3 (aq. 35%), 100 °C; 3) HCl, H_2O (Lunan2016).

D19 MeOH (or EtOH or *t*BuOH), HCl or H_2SO_4 (conc, cat.), heat, YNR (Enzytech2006a,b, ZhejiangJP2010). MeOH, H_2SO_4, 60 °C, 86% (ChongqingLP2008).

D20 1) PCl_5, 100 °C; 2) MeOH, 45%.

D21 $CHBr_3$, KOH, ice, dioxane, 0-5 °C, 63% (Sanofi1989). $CHBr_3$, KOH, Me_3NBnCl (cat.), *i*-Pr_2O, H_2O, 0-5 °C, 47% (Sanofi1989). $CHBr_3$, KOH, Me_4PBr, Et_2O, H_2O, 0-5 °C, 48% (Sanofi1989). $CHBr_3$, KOH (aq.), dimethoxyethane, 44-67% (Sanofi1989, Enzytech2006b).

D22 (F4) MeOH, H_2SO_4 (conc., cat.), reflux. 92-93% (Sanofi1989, Sandoz2007).

D23 HBr, 120 °C, under pressure, YNR.

D24 TsCl, NEt_3, CH_3CN, 0-25 °C, 94%. TsCl, NaOH or KOH, toluene, H_2O, 10 °C, 91%. TsCl, *N*,*N*-dimethylaniline, *n*-BuOAc, 10 °C, 97%. MsCl, NEt_3, CH_3CN, 0-25 °C, 98%.

D25 Bromination: H_2O_2, Br_2, CH_2Cl_2, H_2O, reflux, 94%. **S_N2:** NaCN, H_2O, 30 °C, 95%.

D26 H_2SO_4, heat, YNR (JACS1938). KOH, EtOH, heat, YNR (JCS1948). HCl, H_2O, reflux, YNR (BB2011).

D27 NaBr or KBr, H_2SO_4 or HBr, H_2O_2, UV or visible light, CH_2Cl_2 or $ClCH_2CH_2Cl$, rt, 88-92%.

D28 LG = Br: Br_2, 100-110 °C, 86-86% (OPRD2007, Ind2011). Br_2, CH_3CN, 110 °C, 93-96% (ZhejiangHP2005). **LG = Cl:** Cl_2, CH_3CN, 110 °C, 93%.

D29 MeOH, H_2SO_4, reflux, 98%.

D30 KBr or NaBr or HBr, H_2SO_4, H_2O_2, acid, UV or visible light, CH_2Cl_2, H_2O, 90-93%.

D31 *p*-$TsSO_2N_3$, DBU, CH_3CN, 49 °C, 85-86%.

D32 $FeCl_2$•$4H_2O$ (5% mol), (*Sa*,*S*,*S*) ligand (6% mol), NaBARF (6% mol), $CHCl_3$, H_2O, 40 °C (see Scheme 3.23).

3.7.3 Reaction conditions for scheme E

E1 1) $NaHSO_3$ (1 eq.), H_2O, EtOH; 2) TpEA (1 eq.), 50 °C; 3) NaCN (1 eq.), H_2O, 60 °C, 94% (Sanofi1997c,d,e, 1998b, ShanghaiIPI2009). 1) NaCN (1 eq.), H_2O, EtOH, TpEA (1 eq.), 60 °C, 94% (Sanofi1997c,d,e, 1998b).

E2 (*R*)-(–)-CSA (0.4 eq.), acetone, HCO_2H (0.4x), 55 °C to rt 22%, 89% ee.

E3 1) MeOH (anh.), HCl (g), reflux; 2) H_2O, 93-94%.

E4 HCl (g), MeOH, AcOMe, 20-50 °C, 91-94%.

E5 MeOH, H_2SO_4, reflux, 82-83%.

E6 NaOH or KOH (cat), *i*-PrOH, 60 °C, 97%. NaOEt (cat), EtOH, 60 °C, 95%. NaOH (5 eq.), EtOH, H_2O, 50 °C, 90%. KOH (5 eq.), Bu_4NBr (cat), benzene or toluene, H_2O, reflux, 97%. KOH (1 eq.), EtOH, 60 °C 97%.

E7 D-(–)-TA (0.5 eq.), HCO_2H (0.7 eq.), *i*-PrOH, 50 °C to rt, 32% (*S* from the solid in suspension). L-(+)-TA (0.5 eq.), HCO_2H (0.7 eq.), *i*-PrOH, H_2O (0.2%), 50 °C to rt, 33-37% (*S* from the mother liquors, 98% ee).

E8 (C9, H18) see C9 CH_2O/H_2O or $HO(CH_2O)_nH$, acid, H_2O.

E9 (C4, H21) see C4.

E10 (C5, H20) see C5.

E11 (C13, F13, G9) see C13.

E12 (C14, F14, G16, H22) see C14.

E13 (C15, F15, G17, H23) see C15.

E14 not reported.

E15 HCl (g), MeOH, AcOMe, rt, 46% (94% ee).

E16 MeOH, H_2SO_4, reflux, 83% (99% ee).

E17 (C10, H37) see C10.

E18 (C12, F22, G15) see C12.

E19 (C11, F21, G13) see C11.

E20 (C16, F25, G21, H40) see C16.

E21 (C17, F12, G14) see C17.

3.7.4 Reaction conditions for scheme F

F1 LG = Br: THTP•HCl, $NaHCO_3$, MeOH, reflux, 85%. **LG = Cl:** THTP•HCl, $NaHCO_3$, *n*-BuOH, reflux, 99%, in MeOH, 82%. **LG = OTs:** THTP•HCl, $NaHCO_3$, MeOH, reflux, 89%.

F2 (G4) HCl, reflux, 38% (Cadila2001). NaOH (40-50% aq.), Et_3BnNCl (cat.), *n*-BuOH, reflux, 94-95% (ZhejiangHP2005, OPRD2007), in MeOH/H_2O, 89-95% (ZhejiangHP2005, Ind2011), in CH_3CN, MeOH, 93% (ZhejiangHP2005), with KOH, YNR (ZhejiangHP2005).

F3 THTP•HCl, KOH, H_2O, MeOH, 40 °C, 88% (Hanmi2005), at 5 °C to rt, 92-94% (Shenzhen2012b, Shandong2014), with NaOH at 25 °C, 94-100% (Pharmathen2010). THTP, NEt_3, CH_2Cl_2, reflux, 78% (Krka2006).

F4 (D22) see D22.

F5 THTP, base, solvent. **LG = Cl:** THTP, K_2CO_3, DMF, 90 °C, 45% (Sanofi1982), in THF, EtOAc, Et_2O or DMF, 69-81% (Sanofi1990a). THTP•HCl, $NaHCO_3$, MeOH, 65 °C (Ranbaxy2006). THTP, Na_2CO_3, NaI, 5-Sulfosalicylic acid, YNR (Hawon2007). **LG = Br:** THTP, K_2CO_3, MeOH or DMF, 80-85 °C, 72-92% (Sanofi1989, Torrent2003, USV2004). THTP•HCl, K_2CO_3 or $NaHCO_3$, acetone or MeOH/EtOAc or EtOAc or MeOH, heat, YNR (Teva2002a,b,c, Ind2007, BeijingSPC2009). THTP, CH_3CN, base, rt, 97% (BMCL2014). THTP, Na_2CO_3, acetone, toluene, rt, 90% (SRIMONMC2016). THTP•HCl, Na_2CO_3, $ClCH_2CH_2Cl$, 25 °C, 88% (Ipca2004). THTP•HCl, toluene or CH_2Cl_2 or EtOAc, K_2CO_3, H_2O, DMF, *n*-Bu_4NHSO_4, 30-35 °C, YNR (Teva2007). THTP•HCl, $NaHCO_3$, CH_2Cl_2, H_2O, reflux, 80% (Shandong2014). THTP, NEt_3, toluene, reflux, YNR (Sandoz2007). THTP•HCl, K_2CO_3 or Na_2CO_3, MeOH, 50-65 °C, 97% (F4+F5, 15-30 kg, Qilu2010). THTP•HCl, $NaHCO_3$, CH_2Cl_2, H_2O, reflux, 89% (D30 benzylic bromination+F5, Shenzhen2012a). THTP•HCl, $NaHCO_3$, MeOH, 60 °C, 83% (F4+F5, Shenzhen2012b, Shandong2014). **LG = OSO_2Ph:** THTP•HCl, K_2CO_3, *n*-PrOAc, 80 °C, 94%. **LG = OTs:** THTP•HCl, K_2CO_3, *n*-BuOAc or $CHCl_3$ or *n*-BuOAc, 60-90 °C, 90-94%. **LG = OMs:** THTP•HCl, K_2CO_3, CH_2Cl_2, H_2O, reflux, 70% (mesylation, D12+F5).

F6 (G8) (*S*)-(+)-CSA, acetone, H_2O, 8%.

F7 (G5) With base: K_2CO_3, H_2O_2 (30%), DMSO, H_2O, 50 °C, 85% (Egis2000), in DMSO, MeOH, 91-95% (Merck2001, GenericsUK2003). *t*-BuOH or *i*-PrOH, KOH, 80 °C, 85-94% (Cadila2001, PC2012). Et_3BnNCl, NaOH (aq.), MeOH, reflux, 96% (OPRD2007). **With acid:** HCl, TFA, *t*-BuOH, reflux, 38% (Cadila2001). HCO_2H, CH_3CN, 25 °C to reflux, 47% (Cadila2001). HBr, H_2O, reflux, 47% (Cadila2001). H_2SO_4, NaCl, H_2O, reflux, 57% (Cadila2001). $HClO_4$, H_2O, reflux, 47% (Cadila2001). H_2SO_4, acetone, 76% (Cadila2001).

F8 (G6) KOH, *t*-BuOH, H_2O, 38% (F7+F8, Cadila2001). Et_3BnNCl, NaOH (40-50% aq.), MeOH, reflux, 95% (ZhejiangHP2005, OPRD2007).

F9 (G11) (*S*)-(+)-CSA, acetone, 30-36% (Cadila2001). (*S*)-(+)-CSA (0.5 eq.), HCO_2H (0.5 eq.), acetone/MeOH (7:5), HCO_2H (0.5 eq.), 35 °C to 0 °C 17-19% (Merck2001, GenericsUK2003).

F10 (G12) *t*-BuOK, *t*-BuOH, 25-55 °C, 80-90% (Cadila2001), in DMSO, 25 °C, 80% (Cadila2001). NaH, THF, 30 °C, 95% (Cadila2001). NaOH (aq.), 50 °C, 93% (GenericsUK2003).

F11 (G7) MeOH, H_2SO_4, reflux, 74% (Egis2000), at 80 °C, 64% (Cadila2001), with HCl, reflux, 18% (Cadila2001). *N,N*-dimethylformamide dimethyl acetal, MeOH, reflux, 48% (Cadila2001). MeOH, MsOH, 86 °C, 20% (Cadila2001). MeOH, PPA, reflux, 30% (Cadila2001). MeOH, toluene, $TiCl_4$, 12-15% (Cadila2001). MeOH, $POCl_3$, H_2SO_4, 18 % (Cadila2001). H_2SO_4, MeOH, Me_2SO_4 (2 eq.), 70 °C, 67% (PC2012).

F12 (C17, E21, G14) see C17.

F13 (C13, E11, G9) see C13.

F14 (C14, E12, G16, H22) see C14.

F15 (C15, E13, G17, H23) see C15.

F16 LG = SO_2R, R = Ph: THTP, K_2CO_3 (aq.), toluene or CH_2Cl_2, reflux, 85-99%, 78-92% ee (Sanofi1997f), 85% (Huaibei2005), in EtOAc, 55 °C, 85-90% (Shenzhen2008). **R = *p*-$C_1C_6H_4$:** THTP, K_2CO_3 (aq.), CH_2Cl_2, reflux, 100%, 94% ee. **R = *p*-$O_2NC_6H_4$:** THTP, CH_2Cl_2, K_2CO_3 (aq.), reflux, 89-100%, 96% ee (Sanofi1997f), 70%, 90% ee (JOC2009). **LG = Br:** THTP•HCl, NEt_3, MeOH, reflux, 80%, 99% ee. THTP•HCl, K_2CO_3, MeOH, reflux, 85%, 99% ee, in acetone 78%, 99% ee. **S_N2 with BisTHTP, LG = $OSO_2C_6H_4NO_2$:** BisTHTP, CH_3CN, reflux, 55%, in MeOH, 79%, in acetone, 84-87%. BisTHTP, K_2CO_3, CH_3CN, reflux, 45%. BisTHTP, DIPEA, acetone, reflux, 65%. **LG = Br:** BisTHTP, CH_3CN, reflux, 64%.

F17 THTP, $NaHCO_3$, MeOH, TEBA, reflux, 89%.

F18 (G10) NaOH (30%, aq.), H_2O, MeOH or *n*-BuOH, TEBA, reflux, 89-97%.

F19 (G18) not reported.

F20 (G19) MeOH, H_2SO_4, 80 °C, 45-97% (Cadila2001). MeOH, $PhSO_3H$ or MsOH, reflux, YNR (Cadila2006a). 1) Acetone, $PhSO_3H$, reflux (besylate salt); 2) MeOH, H_2SO_4, 70 °C, YNR (Cadila2006b). 1) H_2SO_4 (conc.), MeOH, Me_2SO_4, reflux; 2) SM, 80 °C; 3) H_2O, 70-73% (Merck2001, GenericsUK2003).

F21 (C11, E19, G13) see C11.

F22 (C12, E18, G15) see C12.

F23 (G22) NaOH (50%), H_2O, MeOH, 18-crown-6 or TEBA, reflux, 89-94%.

F24 (G20) not used.

F25 (C16, E20, G21, H40) see C16.

3.7.5 Reaction conditions for scheme G

G1 (F1) see F1.

G2 NaCN, THTP, EtOH, H_2O, 40 °C, 81% (Egis2000), in MeOH, H_2O, 88% (Merck2001, GenericsUK2003). NaCN, THTP, MeOH, AcOH, 60 °C, 65% (Cadila2001). KCN, THTP, MeOH, HCl, rt to 60 °C, 18% (Cadila2001). KCN, THTP, MeOH/H_2O or H_2O, HCl, 30 °C, 74% (Cadila2001). 1) $NaHSO_3$, H_2O; 2) THTP, NaCN, 50°C, 97% (Cadila2001), in MeOH 72% (Cadila2001), at reflux 40% (Cadila2001). NaCN, THTP•HCl, H_2O, 60-56 °C, 100% (PC2012). 1) Cyclo[(*S*)-histidine-(*S*)-phenyl alanine], toluene, AcOH; 2) THTP, HCN, 30 °C, 43% (Cadila2001). 1) Cyclo[(*S*)-histidine-(*S*)-phenyl alanine], toluene, $NaHSO_3$; 2) THTP, TMSCN, 30 °C, 10% (Cadila2001). 1) $K_4Fe(CN)_6$, *p*-chlorobenzoyl chloride, EtOH, 180 °C; 2) SM, THTP•HCl, 80 °C, 85%, in MeOH, with benzoyl chloride, 80-86% (Ind2014a).

G3 THTP, TMSCN, NaF (cat.), catalyst (10%), –20 °C, CH_2Cl_2, 92% (56% ee).

G4 (F2) see F2.

G5 (F7) see F7.

G6 (F8) see F8.

G7 (F11) see F11.

G8 (F6) see F6.

G9 (C13, E11, F13) see C13.

G10 (F18) see F18.

G11 (F9) see F9.

G12 (F10) see F10.

G13 (C11, E19, F21) see C11.

G14 (C17, E21, F12) see C17.

G15 (C12, E18, F22) see C12.

G16 (C14, E12, F14, H22) see C14.

G17 (C15, E13, F15, H23) see C15.

G18 (F19) see F19.

G19 (F20) see G19.

G20 (F24) see F24.

G21 (C16, E20, F25, H40) see C16.

G22 (F23) see F23.

3.7.6 Reaction conditions for scheme H

H1 *p*-$MeOC_6H_4CHCH_3NH_2$•HCl (*S* isomer), NaCN, MeOH, H_2O, rt, 75%.

H2 HCl (aq., 6 M), Et_2O, $CHCl_3$, rt to 90 °C, 80% (S2013). Whole cells of the nitrilase-derived genetically engineered bacteria derived from *Pseudomonas fluorescens,* 90% (Wuhan2017).

H3 KCN, NH_4Cl, NH_3 (30%, aq.), MeOH, 45 °C, YNR (Sanofi1997b). NaCN, NH_4HCO_3, MeOH, H_2O, 70 °C, YNR (Ind2003). 1) $K_4Fe(CN)_6$, *p*-ClC_6H_4COCl, 180 °C; 2) SM, NH_3 (g), 10 bar, 80 °C, 81-84% (Ind2014b). NaCN, NH_4Cl, NH_4OH, EtOH, H_2O, 60 °C, 79% (Wuhan2017).

H4 L-Pyroglutamic acid, EtOH/H_2O, 46%.

H5 HCl (6 N), reflux, 46% (H3+H5, Sanofi1997b). NaOH (45% aq.), MeOH, H_2O, reflux, 120 °C, under pressure, 58% (H3+H5, Ind2003).

H6 (*S*)-(+)-CSA, H_2O, 21-44% (Sanofi1990b, Ind2003, UNankai2007, Wuxi2009). 1) $PhCH_2COCl$, NaOH, H_2O, 95%; 2) immobilized penicillin acylase, H_2O, pH 8, 30 °C (deacylation of *S* isomer), 45% (ChongqingCB2010).

H7 NaOH, H_2O, 85 °C, 90% (Wuxi2009). Racemization of *N*-acyl derivative: 2,6-dichlorobenzoic acid, 170 °C, 95 °C (ChongqingCB2010).

H8 1) MeOH (anh.), H_2SO_4, reflux; 2) H_2O, 81-87%.

H9 2-(thiophen-2-yl)ethyl 4-methylbenzenesulfonate (A), $NaHCO_3$, CH_3CH_2CN, reflux, 85%.

H10 (C2) see C2.

H11 $SOCl_2$, MeOH, rt to 68 °C, 74-100% (Sanofi1990b, 1997b, Lee2006, UNankai2007, Enzytech2008, EastChinaUST2011). H_2SO_4, MeOH, 68 °C to reflux, 72-97% (Hanseo2005, DrReddys2006, TaizhouZC2008, Qidong2010). **Ethyl ester:** $SOCl_2$, EtOH (anh.), reflux, 88% (EastChinaUST2011). ***n*-Propyl ester:** $SOCl_2$, *n*-PrOH, reflux, 91% (EastChinaUST2011). ***i*-Propyl ester:** $SOCl_2$, *i*-PrOH, reflux, 75% (EastChinaUST2011). ***t*-Butyl ester:** *t*-BuOH, $SOCl_2$, 50 °C, 95-97% (Enzytech2008).

H12 With (+)-TA: (+)-TA, CH_3CN, MEK, 60 °C to rt, 53% (Sanofi1990b). (+)-TA, MeOH, rt, 36% (80 kg, USV2002), 25 °C to 3 °C, 33% (140 kg, DrReddys2006), 50% (UNankai2007), 48% (Qidong2010). (+)-TA, MeOH/acetone, 50%, 99% ee (100 kg, Lee2006), 45% (ShanghaiEBC2008b), 23-43% (Ind2004, 200 g). (+)-TA, *i*-PrOH, YNR (Shenzhen2006b, Yilukang2014). (+)-TA, AcOH, salicylaldehyde, CH_3CN, 26% (TaizhouZC2008). (+)-TA, oxalic acid, salicylaldehyde, MEK, 43% (TaizhouZC2008). (+)-TA, salicylic acid, acetone, 43% (TaizhouZC2008). (+)-TA, MeOH, acetone, acetophenone or formaldehyde or benzaldehyde, 32-49% (ShanghaiEBC2008b, EastChinaUST2011). **Other resolution agent:** 1) *i*-PrOH, acetone, AcOH (cat.), hexane, (*S*)-(+)-*N*-(3,5-Dinitrobenzoyl)-α-phenylglycine (1.0 eq.), 58% of *R* isomer as salt (96% de, Sanofi1997a), with (-)-2-(2,4-dinitrobenzamido)-2-phenylacetic acid, 58% *S* isomer as salt, 98% op (Sanofi 1997b). A DKR resolution method was developed with 49% overall yield (see Scheme 3.33, OPRD2009, 2010). **Biocatalytic method:** alkalase, buffer, 30 °C, 43% (*S*).

H13 $SOCl_2$, MeOH, reflux, 75-80% (100 kg, USV2002). MeOH (ML's), reflux, YNR (DrReddys2006). NaOMe or KOMe, MeOH, >100% (Shenzhen2006b, Hangzhou2010, Yilukang2014). Acetone, acid, heat, YNR (TaizhouZC2008).

H14 2-(thiophen-2-yl)ethyl 4-methylbenzenesulfonate (A), $NaHCO_3$, CH_3CN, 80 °C, 74% (Sanofi1990b), in DMF, reflux, 15% (IJDDD2014).

H15 2-bromo-1-(thiophen-2-yl)ethan-1-one (D), K_2CO_3, toluene, DMF, 60 °C, 53%.

H16 1) $NaBH_4$, MeOH, rt, 96%; 2) CH_2O (37% aq.), EtOH, 40 °C, 90%.

H17 DMF (anh.), HCl (anh.), 0 °C, 87%.

H18 (C9, E8) see C9 CH_2O/H_2O or $HO(CH_2O)_nH$, acid/H_2O.

H19 $SnCl_2$•H_2O, HCl (conc.), AcOH, 80%.

H20 (C5, E10) see C5.
H21 (C4, E9) see C4.
H22 (C14, E12, F14, G16) see C14.
H23 (C15, E13, F15, G17) see C15.
H24 NCS, H_2SO_4, 110 °C, 90%.
H25 $SOCl_2$, MeOH, rt, YNR (Sanofi1990b), 98% (ChongqingCB2010). MeOH, H_2SO_4, reflux, 85-92% (UNankai2007, S2013).
H26 NCS, H_2SO_4, 110 °C, 89%.
H27 2-(thiophen-2-yl)ethyl 4-methylbenzenesulfonate (A), $NaHCO_3$, CH_3CN, 80 °C, 87%, with KI at reflux 70% (98% ee, TA2010). A, $NaHCO_3$, CH_2Cl_2, 35 °C, 95% (Wuxi2009). A, $NaHCO_3$, CH_3CH_2CN, 85% (Krka2006).
H28 E, base, solvent, heat. **X, Y = Br:** DIPEA, CH_3CN, reflux, 88%, in *t*-BuOH, 85%, with NEt_3, 41%, with K_2CO_3, 78%. **X, Y = Cl:** DIPEA, CH_3CN, reflux, 85%. **X = Cl, Y = Ms:** DIPEA, CH_3CN, reflux, 79%. **X = Cl, Y = Ts:** DIPEA, CH_3CN, reflux, 58% (all Hanmi2004). **X = Br:** DIPEA, CH_3CN, EtOAc, reflux, 75% (S2013).
H29 With 2-(thiophen-2-yl)ethyl 4-methylbenzenesulfonate, A: $NaHCO_3$, CH_3CN, 80 °C, 56-95% (Sanofi1990b, Yilukang2014), at reflux, 72% (Enzytech2008), with K_2CO_3, 75% (99% ee, 50 kg, Lee2006), with Na_2CO_3. 96% (Shenzhen2006b), with $NaHCO_3$, 69% (Wuhan2017), with $NaHCO_3$ or $KHCO_3$, 38-39% (TaizhouZC2008). K_2CO_3 or Na_2CO_3, EtOAc, 67-72% (ShanghaiIPI2014). K_2HPO_4, toluene, 90-100 °C, 66-67% (DrReddys2006, PC2012 130 kg), at reflux, 69% (18 kg, ShanghaiMH2013), with *t*-BuOAc, 63% (120 kg scale, DrReddys2006), with CH_3CN, reflux, 79% (18 kg, ShanghaiMH2013). K_2HPO_4 or Na_2CO_3 or Na_2HPO_4, H_2O, 98 °C, 81-83% (ShanghaiEBC2008a). K_2HPO_4, Bu_4NBr, toluene, H_2O, 90°C, 60% (Orchid2009). Na_2HPO_4, EtOAc, 70-77% (ShanghaiIPI2014). NEt_3, EtOAc, reflux, 42-64% (Orchid2006, ShanghaiMH2013, 18 kg). **With 2-(thiophen-2-yl)acetaldehyde, B** (produced from sodium 3-(thiophen-2-yl)oxirane-2-carboxylate, J18 product): sodium 3-(thiophen-2-yl)oxirane-2-carboxylate, $NaBH_3CN$ (1.5 eq.), MeOH, AcOH, 18 °C, 50-75%.
H30 2-bromo-1-(thiophen-2-yl)ethan-1-one (D), toluene, DMF, K_2CO_3, rt, 65-71%.
H31 1) $NaBH_4$, MeOH, rt, 92%; 2) CH_2O (37% aq.), EtOH, 40 °C, 99%.
H32 HCl (anh.), DMF (anh.), 35 °C, 100%.
H33 $SnCl_2$•H_2O, HCl (conc.), AcOH, rt, 79% (Apotex2001). 1) NaI, Me_3SiCl, CH_3CN or DMF, rt; 2) SM (base or HSO_4 salt), toluene, CH_3CN, rt, 86-98%, 98% ee (Apotex2003).
H34 BH_3•SMe_2, THF, 98%.
H35 2-(thiophen-2-yl)acetic acid, DCC, CH_2Cl_2, 5 °C, 96%.
H36 BH_3•THF, THF, 0 °C to rt, 69%.
H37 (C10, E17) see C10.
H38 (C8) see C8.
H39 (C19) see C19.
H40 (C16, E20, F25, G21) see C16.

3.7.7 Reaction conditions for scheme J

J1 HBr (48%), H_2O_2, 5 °C, 90%.
J2 1) Mg, THF, 40 °C; 2) ethylene oxide, 86-90%.
J3 1) BuLi, hexane, THF, <25 °C; 2) ethylene oxide, 79% (Parcor1975). 1) Na, toluene; 2) 2-methyl-1,3-butadiene, THF or toluene, 0 °C; 3) ethylene oxide, 83% (Sanofi1999, 2003).
J4 BH_3•THF, THF, 0-20 °C, 97% (JMC1998). $NaBH_4$, THF, MeOH, 70 °C, YNR (SC2005). 1) $ZnEt_2$, LiCl, THF, hexane, 20 °C; 2) NaOH, THF, hexane, H_2O, 20 °C, YNR(CEJ2015).
J5 PBr_3, $CHCl_3$, py, YNR (JACS1942). Br_2, PPh_3, CH_2Cl_2, 20 °C, 69% (JOC2008). NBS, PPh_3, CH_2Cl_2, 0 °C, 64% (Eisai1997).
J6 EtOH, NH_3,YNR.
J7 DMAP, Tf_2O, NH_3, CH_2Cl_2, YNR.
J8 TsCl, NEt_3, *i*-Pr_2O, rt, 91% (Parcor1975), in toluene, 96% (DrReddys2006), in CH_2Cl_2, 94-96% (DrReddys2006, TaizhouZC2008, Wuxi2009, Heilongjiang2012). TsCl, NEt_3, DMAP, toluene, 88-98% (Suzhou2009), with additional pyridine or piperidine, 86-92% (Suzhou2009). TsCl, py, CH_2Cl_2, 0 °C, 90% (TaizhouZC2008). TsCl, NEt_3, toluene, 30 °C, NI (100 kg, PC2012).
J9 $ClCH_2CO_2$*i*-Pr, NaO*i*-Pr, *i*-PrOH, rt, 93% (Sanofi1990a). $ClCH_2CO_2Me$, NaOMe, MeOH, rt, YNR (Sanofi1997b).
J10 + J11a *i*-PrOH, NaOH (35% aq.), 50 °C, NI (Sanofi1990a).
J11b NH_2OH•HCl, *i*-PrOH, H_2O, NaOH, 25 °C, 95% (J10+J11).
J12 H_2 (20 bar), Raney-Nickel, NH_3, *i*-PrOH, 50 °C, 92% (Sanofi1990a). H_2 (20 bar), Raney-Nickel, NH_3, MeOH, 60 °C, 65% (from (*E*)-2-(2-nitrovinyl) thiophene, J14+J12).
J13 CH_3NO_2, $FeCl_3$ (anh.), piperidine, toluene, reflux, 95% (EJOC2013). 1) CH_3NO_2, NaOH, MeOH, 10 °C; 2) HCl, MeOH, H_2O, 75% (T2013). CH_3NO_2, NaOH, MeOH, 84-96% (CJC1985, S2015).
J14 H_2 (20 bar), Pd/C (5%), EtOH, 35 °C, 100% (Sanofi1981).
J15 H_2 (20 bar), Pd/C (5%) or Rh/C (5%) or Pt/C (5%), EtOH, AcOH, 35 °C, 65% (Pd), 66% (Rh), 81% (Pt) (Sanofi1981). $NaBH_4$, BF_3•Et_2O, THF, –5 °C to 10 °C, 83% (Synthex1988).
J16 $LiAlH_4$, Et_2O or THF, 63-92%.
J17 H_2 (30 bar), Raney-Nickel, NaOH, EtOH, 50 °C, 74% (Sanofi1986). H_2 (30 bar), Raney-Nickel, K_2CO_3, EtOH or MeOH, 50 °C, 79-87% (Sanofi1986). $NaBH_4$, $InCl_3$, THF, 25 °C, 98% (JOC2012a). $LiAlH_4$, THF or Et_2O, 56-71% (S1987, MCC2015).
J18 $NaBH_3CN$, AcOH, MeOH, NI.
J19 NaCN, H_2O, CH_2Cl_2, reflux, 45% (Gruenenthal2006); $Me_2C(OH)CN$, EtOH, NEt_3, 60 °C, 92% (LudongU2017).
J20 In one step with CH_2O (35-40%, aq.), or $HO(CH_2O)_nH$, acid, H_2O, usually >90% yield. In two steps, cyclizing the intermediate by heating in DMF/HCl (5N), overall yields >90%.

J21 HCO_2H, $HO(CH_2O)_nH$, rt, 81%. CH_2O/H_2O (37%), HCl, H_2O, 50 °C, 46%.

3.8 REFERENCES

Aarti2009 Patil, P.; Suryavanshi, J.; Korgaonkar, R.; Kumbhar, S.; Furia, J. (Aarti Drugs Ltd) IN200901117 (prior. 28/04/2009, IN).

Adamed2006 Majka, Z.; Sawicki, A.; Trzepalka, A. (Adamed Sp Zoo) WO2008046792 (prior. 17/10/2006, PL).

AMB2010 Ju, X.; Yu, H.; Pan, J.; Wei, D.; Xu, J.-H. *Applied microbiology and biotechnology* **2009**, *86*, 83-91.

Apotex2001 Horne, S.; Weeratunga, G.; Comanita, B.; Nagireddy, J.; McConachie, L.; Horne, S. (Brantford Chem Inc, Apotex Pharmachem Inc) CA2352520, WO2003004502, US6495691, EP1404681 (prior. 06/07/2001, CA).

Apotex2003 Smyj, R. P.; Weeratunga, G. (Apotex Pharmachem Inc) CA2454015, US2005137401, US7060831 (prior. 23/12/2003, CA).

ASC2008 Ema, T.; Ide, S.; Okita, N.; Sakai, T. *Advanced Synthesis & Catalysis* **2008**, *350*, 2039-2044.

Aurobindo2005 Sarma, M.; Bhima, S.; Dandala, R.; Sivakumaran, M. (Aurobindo Pharma Ltd) IN200500051 (prior. 24/01/2005, IN).

Aurobindo2010 Reddy, A.V.; Srinivas, G.; Kumar K.S.; Prabhakar, K.J., Sivakumaran, M. (Aurobindo Pharma Ltd) IN201003099 (prior. 20/10/2010, IN).

Avecia2001 Blacker, A. J.; Houson, I. N. (Avecia Ltd) WO200266410, US2004068140, US7022876, EP1362024 (prior. 16/02/2001, GB).

Aventis1999 Fink, D. M.; Freed, B. S.; Hrib, N. J.; Kosley, R. W., Jr.; Lee, G. E.; Merriman, G. H.; Rauckman, B. S. (Aventis Pharma Inc) WO200119833, US2007004695, US7125903, US7511046, EP1216250 (prior. 14/09/1999, US).

BB2011 Yousefi, M.; Mohammadi, M.; Habibi, Z.; Cheraghi, Z. *Biocatalysis and Biotransformation* **2011**, *29*, 54-59.

BeijingHUTD2009 Bi, H.; Tang, G.; Wang, Y.; Zhang, K. (Beijing Huaxi United Technology Dev Co Ltd) CN101591346 (prior. 03/07/2009, CN).

BeijingSPC2009 Wang, L.; Wang, W.; Xie, S.; Yang, Y.; Zou, J. (Beijing Second Pharm Co Ltd) CN101585842 (prior. 13/07/2009, CN).

BEJ2013 Xue, Y.-P.; Jiang, T.; Liu, X.; Zheng, Y.-G. *Biochem. Eng. J.* **2013**, *74*, 88-94.

BL2010 Jeong, M.; Lee, Y. M.; Hong, S. H.; Park, S. Y.; Yoo, I.-k.; Han, M. J. *Biotechnology Letters* **2010**, *32*, 1529-1531.

BMCL2014 Farha, M. A.; Koteva, K.; Gale, R. T.; Sewell, E. W.; Wright, G. D.; Brown, E. D. Bioorganic & Medicinal Chemistry Letters 2014, 24, 905-910.

CAC1999 *Comprehensive Asymmetric Catalysis;* Eds. Jacobsen, E.N.; Pfaltz, A., Yamamoto, H, Chapter 28, Springer.

Cadila2001 Pandey, B.; Lohray, V. B.; Lohray, B. B. (Cadila Healthcare Ltd) WO02059128, EP1353928 (prior. 24/01/2001, IN).

Cadila2006a Dholakia, P. N.; Dave, M. G.; Pandey, B.; Lohray, B. B.; Lohray, V. B.; Patel, P. R.; Dave, M. (Cadila Healthcare Ltd) IN200600914, WO2007144895 (prior. 12/06/2006, IN).

Cadila2006b Shah, N.; Dwivedi, S. (Cadila Healthcare Ltd) IN200602165, WO2008081473 (prior. 29/12/2006, IN).

CC2014 Yan, P.-C.; Xie, J.-H.; Zhang, X.-D.; Chen, K.; Li, Y.-Q.; Zhou, Q.-L.; Che, D.-Q. *Chemical Communications* **2014**, *50*, 15987-15990.

CEJ2015 Kovalenko, O. O.; Adolfsson, H. *Chemistry – A European Journal* **2015**, *21*, 2785-2788.

Changzhou2010 Jin, X.; Yin, X. (Changzhou Pharm Factory) CN101863899 (prior. 21/06/2010, CN).

ChongqingCB2010 Xia, S.; Fang, G. (Chongqing Chiral Biocatalysis Technology Co Ltd) CN101864464 (prior. 29/04/2010, CN).

ChongqingLP2008 Xiaohai, T.; Xiaofang, H.; Yu, Q. (Chongqing Lummy Pharmaceutical) CN101434540 (prior. 23/12/2008, CN).

CJC1985 Bourguignon, J.; Nard, G. L.; Queguiner, G. *Canadian Journal of Chemistry* **1985**, *63*, 2354-2361.

CPB1994 Sumita, K.; Koumori, M.; Ohno, S. *Chemical & Pharmaceutical Bulletin* **1994**, *42*, 1676-1678.

CR2011 Wang, J.; Liu, X.; Feng, X. *Chemical Reviews* **2011**, *111*, 6947-6983.

CRC2012 Maffrand, J.-P. *Comptes Rendus Chimie* **2012**, *15*, 737-743.

DaHe2007 Hwang, S.; Lee, H.; Min, Y. (Da He Chem Co Ltd) KR2009053284 (prior. 23/11/2007, KR).

Daiichi2005 Sakamoto, K.; Kita, S.; Morii, A. (Daiichi Fine Chem Co Ltd) JP2007533348, WO2007026860, EP1930441, US2008311636 (prior. 02/09/2005, JP).

Dipharma2002 Castaldi, G.; barreca, G.; Bologna, A.; Graziano, C.; Giuseppe, B.; Alberto, B.; Boldgna, A. (Dinamite Dipharma Spa, Dipharma Francis Srl, Dipharma Spa) IT1339437, WO2003093276, EP1501838, US2005143414, US7329751 (prior. 03/05/2002, IT)

DrReddys2003 Reddy, M. S.; Eswaraiah, S.; Reddy, A. R.; Sampath, A. (Dr Reddy's Lab Ltd) IN200300389, US2005059696 (prior. 08/05/2003, IN).

DrReddys2006 Sajja, E.; Anumula, R. R.; Gilla, G.; Madivada, L. R. (Dr Reddy's Lab Ltd) IN200600545, US2007225320 (prior. 27/03/2006, IN).

EastChinaUST2010 Wu, F.; Zhao, M.; Wang, F.; Song, H.; Shi, G.; Wei, Q. (East China Univ Sci&Technology) CN101906032 (prior. 14/07/2010, CN).

EastChinaUST2011 Fang, F.; Song, H.; Wu, F.; Zhao, M. (East China Univ Sci&Technology) CN102336767 (prior. 11/07/2011, CN).

Egis2000 barkoczy, J.; Kotay, N. P.; Simig, G.; Gregor, T.; Nagy, K.; Vereczkeyne, D. G.; Seres, P.; Slegel, P. (Egis Gyogyszergyar, Egis Gyogyszergyar Nyilvanosan Mukoedo) HU200004116 (prior. 20/10/2000, HU).

Eisai1997 Kitazawa, N.; Ueno, K.; Takahashi, K.; Kimura, T.; Sasaki, A.; Kawano, K.; Okabe, T.; Komatsu, M.; Matsunaga, M.; Kubota, A. (Eisai Co Ltd) US2002019531, EP0976732, US2002086999, US6448243, US6579881, US7071201, WO9843956 (prior. 31/03/1997, US, JP.

EJOC2011 Pennacchio, A.; Giordano, A.; Rossi, M.; Raia, C. A. *European Journal of Organic Chemistry* **2011**, *2011*, 4361-4366.

EJOC2013 Jalal, S.; Sarkar, S.; Bera, K.; Maiti, S.; Jana, U. *European Journal of Organic Chemistry* **2013**, *2013*, 4823-4828.

Enzytech2006a Hwang, S. O.; Chung, S. H. (Enzytech Ltd) KR722155, WO2007078176 (prior. 05/01/2006, KR).

Enzytech2006b Hwang, S. O.; Chung, S. H. (Enzytech Ltd) KR2007105681, WO2007126258 (prior. 27/04/2006, KR).

Enzytech2008 Hwang, S. O.; Kim, Y. J. (Enzytech Ltd) KR2009127714, WO2009151256, US2011087028, EP2298777 (prior. 09/06/2008, KR).

GC2012 Shi, L.; Liu, Y.; Liu, Q.; Wei, B.; Zhang, G. *Green Chemistry* **2012**, *14*, 1372-1375.

GenericsUK2003 Arul, R.; Rawat, A. S.; Gadakar, M.; Rao, R.; Pise, A.; Gray, J. (Generics Uk Ltd) WO2005026174, US2007281964, US7897613, EP1618113, EP1772456, EP1837337, EP2149578 (prior. 11/09/2003, GB).

Gruenenthal2006 Zemolka, S.; Schunk, S.; Bergrath, E.; Kogel, B.-Y.; Englberger, W.; Linz, K.; Schick, H. (Gruenenthal Chemie) DE102006033114 , WO2008009416, US2009156593, EP2041142, US8399503 (prior. 2006/07/18, DE, WO).

Hangzhou2010 Jiang, W.; Ni, S.; Shi, W. (Hangzhou Hexo Chem Technology Co Ltd) CN101812071 (prior. 10/05/2010, CN).

Hanmi2004 Yun, S.; Kim, E. S.; Kim, H. S.; Ha, T. H.; Suh, K.; Lee, G. S. (Hanmi Pharm Co Ltd) KR2005091330, WO2005087779, US2007197789, US7449580, EP1723149 (prior. 12/03/2004, KR).

Hanmi2005 Kim, E. S.; Kim, H. C.; Kwon, B. S.; Yun, S.; Ko, M. Y.; Kim, C. K.; Suh, K. H. (Hanmi Pharm Co Ltd, Hanmi Holdings Co Ltd) KR2006134541, WO2006137628, US2008214821, US2009318701, US7652140, US7763730, EP1907397 (prior. 23/06/2005, KR).

Hanseo2005 Oh, M. K.; Kim, K. N.; Choi, H. (Hanseo Chem Co Ltd) KR2006098009 (prior. 08/03/2005, KR).

Hawon2007 Jong Geun, C. (Hawon Pharm Corp, Biochem Net Interactive Inc) KR2009013298 (prior. 01/08/2007, KR).

HCA2013 Jung, K.; Kim, J.-S.; Kim, T.-H.; Kim, J. *Helvetica Chimica Acta* **2013**, *96*, 326-329.

Heilongjiang2012 Wu, G. (Heilongjiang Fuhe Huaxing Pharmaceutical Group Co Ltd) CN103360363 (prior. 26/03/2012, CN).

Hoechster1913 Bruning, L. M. (Hoechster Farbw.). DE279198 (prior. 1913, DE).

Huaibei2005 Ye, H.; Sun, M.; Zhu, Z. (Huaibei Huike Pharm Co Ltd) WO2006086921, CN1778936 (prior. 18/02/2005, CN).

IJDDD2014 Kaushik, B.; Ballabh, M.; Dholakia1, P.; Dave1, M.; Sheikh1, B. *International Journal of Drug Design and Discovery*, **2014**, *5*, 1371-1380.

Ind2003 Lin, S. S.; Chen, C. (Individual) US2004176637 (prior. 03/03/2003, US).

Ind2004 Reddy, B. S.; Battula, S. R. (Individual) WO2006003671, IN200604759 (prior. 02/07/2004, WO, IN).

Ind2007 Battula, S. R. (Individual) IN200700734, EP1980563 (prior. 09/04/2007, IN).

Ind2011 Kandula, M. (Individual) WO2013024376 (prior. 16/08/2011, US).

Ind2014a Rui, Z.; Jianfeng, S. (Individual) CN104045652 (prior. 09/07/2014, CN).

Ind2014b Rui, Z.; Jianfeng, S. (Individual) CN104326934 (prior. 21/11/2014, CN).

IndSwift2002 Kumar, S.; Ramachandran, U.; Singh, H. P.; Wadhwa, L. K. (Ind-Swift Lab Ltd, Nat Inst Pharm Education&Res) IN193363 (prior. 26/06/2002, IN).

Ipca2004 Kumar, A.; Vyas, K. D.; barve, S. G.; Bhayani, P. J.; Nandavadekar, S.; Shah, C. H.; Burudkar, S. M.; Kushwaha, L. D.; barve, G. S.; Bueudkar, S. M.; Nandavadekar, S. K.; Bhayani, P. J.; Kushawaha, L. D. (Ipca Laboratories Ltd) IN200400281, WO2005104663, US2008097101, US2010160635, US7629465, EP1723141, EP1772455 (prior. 05/03/2004, IN).

Ipca2007 Kumar, A.; Bhayani, P. J.; Doshi, V. C.; Saxena, A.; Pathak, G. P.; Abhyankar, R.; Manavalan, S.; Purohit, M.; Parihar, S.; Sarvanan, M.; Bhayani, P.; Doshi, V.; Pathak, G. (Ipca Laboratories Ltd) IN200700292, WO2008093357, US2010081839, EP2114957 (prior. 29/01/2007, IN).

Ipca2008 Byju, N. T.; Kumar, A.; Purohit, M.; Saxena, A. (Ipca Laboratories Ltd) IN200800341 (prior. 15/02/2008, IN).

JACS1938 Johns, I. B.; Burch, J. M. *Journal of the American Chemical Society* **1938**, *60*, 919-920.

JACS1942 Blicke, F. F.; Burckhalter, J. H. *Journal of the American Chemical Society* **1942**, *64*, 477-480.

JACS2013 Evans, R. W.; Zbieg, J. R.; Zhu, S.; Li, W.; MacMillan, D. W. C. *Journal of the American Chemical Society* **2013**, *135*, 16074-16077.

JCECR2014 Gharib1, A.; Pesyan, N.; Fard, L. Roshani1, M. *J. Chem. Eng. Chem. Res.* **2014**, *1*, 1-5.

JCR1989 Meth-Cohn, O.; Vij, R. R.; Smalley, R. K.; Bass, R. J. *Journal of Chemical Research, Miniprint* **1989**, *5*, 1001-1018.

JCS1948 Campbell, N.; McKail, J. E. *Journal of the Chemical Society (Resumed)* **1948**, 1251-1255.

JMC1995 Bell, F. W.; Cantrell, A. S.; Hoegberg, M.; Jaskunas, S. R.; Johansson, N. G.; Jordan, C. L.; Kinnick, M. D.; Lind, P.; Morin, J. M. *Journal of Medicinal Chemistry* **1995**, *38*, 4929-4936.

JMC1998 Ornstein, P. L.; Bleisch, T. J.; Arnold, M. B.; Wright, R. A.; Johnson, B. G.; Schoepp, D. D. *Journal of Medicinal Chemistry* **1998**, *41*, 346-357.

JMCB2014 Xu, Y.-P.; Guan, Y. H.; Yu, H.-L.; Ni, Y.; Ma, B.-D.; Xu, J.-H. *Journal of Molecular Catalysis B: Enzymatic* **2014**, *104*, 108-114.

JOC1950 Gilsdorf, R. T.; Nord, F. F. *The Journal of Organic Chemistry* **1950**, *15*, 807-811.
JOC2008 Martins, A.; Lautens, M. *The Journal of Organic Chemistry* **2008**, *73*, 8705-8710.
JOC2009 Yin, L.; Shan, W.; Jia, X.; Li, X.; Chan, A. S. C. *Journal of Organometallic Chemistry* **2009**, *694*, 2092-2095.
JOC2012a Saavedra, J. Z.; Resendez, A.; Rovira, A.; Eagon, S.; Haddenham, D.; Singaram, B. *The Journal of Organic Chemistry* **2012**, *77*, 221-228.
JOC2012b Sadhukhan, A.; Saravanan, S.; Khan, N.-u. H.; Kureshy, R. I.; Abdi, S. H. R.; Bajaj, H. C. *The Journal of Organic Chemistry* **2012**, *77*, 7076-7080.
JOC2015 Shaw, S. A.; Balasubramanian, B.; Bonacorsi, S.; Cortes, J. C.; Cao, K.; Chen, B.-C.; Dai, J.; Decicco, C.; Goswami, A.; Guo, Z.; Hanson, R.; Humphreys, W. G.; Lam, P. Y. S.; Li, W.; Mathur, A.; Maxwell, B. D.; Michaudel, Q.; Peng, L.; Pudzianowski, A.; Qiu, F.; Su, S.; Sun, D.; Tymiak, A. A.; Vokits, B. P.; Wang, B.; Wexler, R.; Wu, D.-R.; Zhang, Y.; Zhao, R.; baran, P. S. *The Journal of Organic Chemistry* **2015**, *80*, 7019-7032.
JOC2017a Sela, T.; Lin, X.; Vigalok, A. *The Journal of Organic Chemistry* **2017**, *82*, 11609-11612.
JOC2017b Keipour, H.; Jalba, A.; Delage-Laurin, L.; Ollevier, T. *The Journal of Organic Chemistry* **2017**, *82*, 3000-3010.
Kaneka2008 Fujii, A.; Mitsuda, M.; Yamamoto, S. (Kaneka Corp) JP2010515872, WO2009148046, US2011166354, US8383833, EP2298729 (prior. 03/06/2008, JP).
Krka2006 Simonic, I.; Benkic, P.; Zupet, R.; Smrkolj, M.; Stukelj, M. (Krka Tovarna Zdravil) WO2008034912, EP2078025 (prior. 2006/09/22, SI).
Lee2006 Alla, V. R.; Vyakaranam, K. R.; Sirigiri, A. K.; Bodapati, S. R.; Billa, R. R.; Gudibandi, S. R.; Alla, R. (Lee Pharm Inc, Lee Pharma Ltd) IN200600223, US2007191609, WO2007094006 (prior. 13/02/2006, IN).
LifeSciences2000 Tarur, V. R.; Srivastava, R. P.; Srivastava, A. R.; Somani, S. K. (Rpg Life Sci Ltd) WO200218357, AU200128789 (prior. 29/08/2000, IN, WO, AU).
LudongU2017 Wang, W.; Ren, X.; Song, Y.; Chen, H. (Univ Ludong) CN106518839 (prior. 2017/01/11, CN).
Lunan2016 Zhang, G.; Shi, B.; Pei, H. (Lunan Pharmaceutical Group Corp) CN108069855 (prior. 18/11/2016, CN).
M2010 Aillaud, I.; Haurena, C.; Gall, E. L.; Martens, T.; Ricci, G. *Molecules* **2010**, *15*, 8144.
Mayuka2010 Khan, M.; Krishnamurthy, C.; Singh, J. (Mayuka Labs Pvt Ltd) IN201001509, WO2012001486 (prior. 28/06/2010, IN).
Merck2001 Arul, R.; Rawat, A.; Gadakar, M.; Pise, A.; Rao, R.; Raman, J.; Ajaysingh, R. (Merck Generics Aeukue Ltd, Merck Generics Uk Ltd) WO2003035652, US2005049415, EP1438314 (prior. 26/10/2001, GB).
MMC2015 Abdel-Fattah, M. A. O.; Abadi, A. H.; Lehmann, J.; Schweikert, P. M.; Enzensperger, C. MedChemComm 2015, 6, 1679-1686.
MSN2006 Satyanarayana Reddy, M.; Kishore Kumar, M.; Thirumalai Rajan, S.; Rama Subba Reddy, K.; Reddy, M.; Muppa, K.; Thirumalai, S.; Reddy, K.; Satyanarayana, R.; Kishore, K.; Thirumalai, R.; Rama, S. (MSN Lab Ltd) IN200601158, WO2008004249 (prior. 04/07/2006, IN).
MSN2010 Satyanarayana, R. M.; Thirumalai, R. S.; Eswaraiah, S.; Madhu, E. N.; Satyanarayana, K.; Seetha, R. S. P. (MSN Lab Ltd) WO2012038975 (prior. 2010/09/22, IN).
Nanjing2018 Gao, Z.; Yi, X.; Huang, F.; Huang, H.; Yu, Y.; Zhang, Z.; Xue, L.; Wang, L.; Qiao, N. (Univ Nanjing Tech) CN108727216 (prior. 2018/07/13, CN).
Nantong2014 Xu, L.; Liu, M.; Liu, J. (Nantong Chengxin Amino Acid Co Ltd) CN103965160 (prior. 16/05/2014, IN).
NC2010 Zhu, S.-F.; Cai, Y.; Mao, H.-X.; Xie, J.-H.; Zhou, Q.-L. *Nature Chemistry* **2010**, *2*, 546.
Neurocrine2003 Guo, Z.; Chen, Y.; Wu, D.; Chen, C.; Wade, W.; Dwight, W. J.; Huang, C. Q.; Tucci, F. C. (Neurocrine Biosciences Inc) WO2005007165, EP1646389, US2005038057, US2006122202, US2007191403, US7056927, US7176211, US7419983 (prior. 2003/07/07, US).
OBC2007 Ema, T.; Okita, N.; Ide, S.; Sakai, T. *Organic & Biomolecular Chemistry* **2007**, *5*, 1175-1176.
OL2000 Vachal, P.; Jacobsen, E. N. *Organic Letters* **2000**, *2*, 867-870.
OL2005 Sun, Y.; Wan, X.; Wang, J.; Meng, Q.; Zhang, H.; Jiang, L.; Zhang, Z. *Organic Letters* **2005**, *7*, 5425-5427.
OPRD1998 Amati, A.; Dosualdo, G.; Zhao, L.; Bravo, A.; Fontana, F.; Minisci, F.; Bjørsvik, H.-R. *Organic Process Research & Development* **1998**, *2*, 261-269.
OPRD2007 Wang, L.; Shen, J.; Tang, Y.; Chen, Y.; Wang, W.; Cai, Z.; Du, Z. *Organic Process Research & Development* **2007**, *11*, 487-489.
OPRD2009 van der Meijden, M.; Leeman, M.; Gelens, E.; Noorduin, W.; Meekes, H.; van Enckevort, W.; Kaptein, B.; Vlieg, E.; Kellogg, R. *Organic Process Research & Development* **2009**, *13*, 1195-1198.
OPRD2010 Noorduin, W.; van der Asdonk, P.; Bode, A.; Meekes, H.; van Enckevort, W.; Vlieg, E.; Kaptein, B.; van der Meijden, M.; Kellogg, R.; Deroover, G. *Organic Process Research & Development* **2010**, *14*, 908-911.
Orchid2006 Ramasubbu, C.; Mamidala, R.; Arjunan, D.; Ramasamy, K.; Siripragada, M.; Desinghy, A. (Orchid Chem & Pharm Ltd) IN200601018, WO2007144729 (prior. 13/06/2006, IN).
Orchid2009 Rao, J.; Nandam, H.; Redamalla, A.; Reguri, B. (Orchid Chem & Pharm Ltd) IN200901793, WO2011012961 (prior. 29/07/2009, IN).
OSC1963 Eliel, E. L.; Fisk, M. T.; Prosser, T. *Org. Synth.* 1963, Coll. Vol. *IV*, 169-172.
Parcor1975 Braye, E. (Parcor) FR2300090, US4127580 (prior. 07/02/1975, FR).

Parcor1977 Bousquet, A.; Braye, E. (Parcor) FR2397417, US4174448, EP453 (prior. 12/07/1977, FR).

PC2012 Madivada, L. R.; Anumala, R. R.; Gilla, G.; Kagga, M.; Bandichhor, R. *Der Pharma Chemica* **2012**, *4*, 479-488.

Pharmathen2010 Koftis, T.; Neokosmidis, E.; Soni, R.; Georgopoulou, I. (Pharmathen Sa) WO2012007019 (prior. 13/07/2010, WO).

PharmaZell2009 Burger, W.; Gudaparthi, O.; Reddy, K.; Kumar, S.; Ramana, V. (Pharmazell Gmbh) IN200902717, WO2011055378, EP2491044, US2012232274 (prior. 09/11/2009, IN).

Praveen2009 Talia, Y.; Parmar, S.; Chaudhary, B.; Burad, D.; Mehta, H. (Praveen Lab Pte Ltd) WO2011010318 (prior. 23/07/2009, WO, IN).

Qidong2010 Qiu, Z. (Qidong Hudong Chem Co Ltd) CN101898975 (prior. 15/07/2010, CN).

Qilu2010 Chu, L.; Fan, C.; Hu, J.; Wang, J.; Zhang, M.; Zhao, S.; Zhu, Y. (Qilu Pharm Co Ltd) CN102336766 (prior. 16/07/2010, CN).

Ranbaxy2006 Rathod, P.; Aryan, S.; Aryan, R.; Khanduri, C. (Ranbaxy Lab Ltd) IN200902031, WO2008029350, US2010016594, US8247558, EP2064217 (prior. 04/09/2006, IN).

Ratiopharm2004 Stohandl, J.; Frantisek, J.; Ness, W. (Ratiopharm Gmbh) EP1589019, WO2005113559, US2007219166, US2009149677, US7507827, US7799952, EP1737868 (prior. 20/04/2004, DE, EP).

Rhodia2003 Silva, R. (Rhodia Chirex Inc, Rhodia Pharma Solution Inc) WO2004094374, US2004214878, US6858734 (prior. 23/04/2003, US).

RPG2005 Srivastava, A.; Pawar, P.; Poojari, K.; Patil, P.; Dalvi, R. (RPG Life Sci Ltd) IN200500836, WO2007032023, US2010004453, US8063217, EP1902058 (prior. 09/07/2005, IN).

S1987 Frejd, T.; Klingstedt, T. *Synthesis* **1987**, *1987*, 40-42.

S2008 Kalinski, C.; Lemoine, H.; Schmidt, J.; Burdack, C.; Kolb, J.; Umkehrer, M.; Ross, G. *Synthesis* **2008**, *2008*, 4007-4011.

S2013 Sashikanth, S.; Raju, V.; Somaiah, S.; Rao, P. S.; Reddy, K. V. *Synthesis* **2013**, *45*, 621-624.

S2015 Lopchuk, J.; Song, M.; Butler, B.; Gribble, G. *Synthesis* **2015**, *47*, 2776-2780.

Sandoz2007 Kintali, R.; Nair, R.; Paul, A.; Salunke, S. (Sandoz Ag) IN201002594, WO2009080469 (prior. 24/12/2007, IN).

Sanofi1981 Annearchar, G.; Heymes, A.; Valette, G. (Sanofi Sa) FR2508456, EP69002 (prior. 30/06/1981, FR).

Sanofi1982 Aubert, D.; Ferrand, C.; Maffrand, J. (Sanofi Sa) FR2530247, US4529596, EP99802 (prior. 13/07/1982, FR).

Sanofi1986 Radisson, J.; Braye, E. (Sanofi Sa) FR2608607, US4876362, EP274324 (prior. 23/12/1986, FR).

Sanofi1987 Badorc, A.; Frehel, D.; Badore, A. (Sanofi Sa, Elf Sanofi, Sanofi-Aventis) FR2612929, US4847265, EP281459 (prior. 17/02/1987, FR).

Sanofi1989 Bouisset, M.; Radisson, J.; Joeel, R. (Sanofi Sa, Elf Sanofi) FR2652575, EP420706, US5036156, US5189170 (prior. 29/09/1989, FR).

Sanofi1990a Bousquet, A.; Calet, S.; Heymes, A. (Sanofi Sa, Elf Sanofi, Sanofi-Synthelabo) FR2664276, US5132435, EP465358 (prior. 04/07/1990, FR).

Sanofi1990b Descamps, M.; Radisson, J. (Sanofi Sa, Sanofi-Synthelabo) FR2664596, US5204469, EP466569 (prior. 10/07/1990, FR).

Sanofi1997a Castro, B.; Previero, A. (Sanofi Sa) FR2760452, WO9839286 (prior. 05/03/1997, FR).

Sanofi1997b Castro, B.; Dormoy, J.; Previero, A. (Sanofi Sa, Sanofi-Synthelabo, Sanofi-Aventis) FR2760456, WO9839322, US6080875, EP971915 (prior. 05/03/1997, FR).

Sanofi1997c Heymes, A.; Castro, B.; Bakonyi, M.; Csatarine Nagy, M.; Molnar, L. (Sanofi Sa, Sanofi-Synthelabo, Sanofi-Aventis) WO9851682, US6215005, EP981525 (prior. 13/05/1997, HU).

Sanofi1997d Bakonyi, M.; Csatarine Nagy, M.; Molnar, L.; Makovi, Z.; Jobb, P.; Bai, T. (Sanofi Sa, Sanofi-Synthelabo, Sanofi-Aventis) HU9700884, WO9851681, US6258961, EP981524 (prior. 13/05/1997, HU).

Sanofi1997e Bakonyi, M.; Csatarine Nagy, M.; Molnar, L.; Gajary, A.; Alattyani, E.; Nagy, M. C. (Sanofi Sa, Sanofi-Synthelabo, Sanofi-Aventis) HU9700885, WO9851689, EP981529, US6180793 (prior. 13/05/1997, HU).

Sanofi1997f Bousquet, A.; Musolino, A.; Andre, B.; Andree, M. (Sanofi-Synthelabo, Sanofi-Aventis, Sanofie Sante Lab) FR2769313, WO9918110, US2003208077, US2004260110, US6573381, US6894186, US7153969, EP1021449 (prior. 06/10/1997, FR).

Sanofi1998a Bousquet, A.; Castro, B.; Saint-Germain, J. (Sanofi-Synthelabo, Sanofi-Aventis, Sanofie Sante Lab) FR2779726, WO9965915, US2002198229, US6504030, US6429210, EP1087976, EP1087976 (prior. 15/06/1998, FR).

Sanofi1998b Bakonyi, M.; Bai, T.; Dombrady, Z.; Gaspar, K.; Supic, A. (Sanofi-Synthelabo, Sanofi-Aventis, Sanofie Sante Lab) HU9802586, WO200027840, US2004106824, US6670486, US7109356, EP1129087 (prior. 09/11/1998, HU).

Sanofi1999 Biard, M.; Bousquet, A. (Sanofi-Synthelabo, Sanofi-Aventis, Sanofie Sante Lab) FR2800070, WO200129024, US6639083, EP1226135 (prior. 22/10/1999, FR).

Sanofi2001 Balint, J.; Csatarine Nagy, M.; Dombrady, Z.; Fogassy, E.; Gajary, A.; Suba, C. (Sanofi-Synthelabo, Sanofi-Aventis, Sanofie Sante Lab) HU200401404, WO2003000636, US2004242921, US7381835, EP1397335 (prior. 21/06/2001, HU).

Sanofi2003 Biard, M.; Bousquet, A. (Sanofi-Synthelabo, Sanofi-Aventis, Sanofie Sante Lab) FR2800070, WO200129024, US6639083, EP1226135 (prior. 01/01/2003, FR).

SC2005 Boechat, N.; da Costa, J.; Mendonça, J.; Paes, K.; Fernandes, E.; de Oliveira, P.; Vasconcelos, T.; de

Souza, M. *Synthetic Communications* **2005**, *35*, 3187-3190.
SC2015 Venkataramasubramanian, V.; Sudalai, A. *Synthetic Communications* **2015**, *45*, 2099-2105.
Shandong2014 Tan, D.; Li, H. (Shandong Salubris Pharmaceuticals Co Ltd) CN104370732 (prior. 26/01/2014, CN).
ShanghaiAT2005 Mao, H.; Pan, X.; Lu, J. (Shanghai Applied Technology College) CN1775782 (prior. 15/12/2005, CN).
ShanghaiEBC2008a Chen, J.; Wu, F.; Yang, X.; Zhao, M. (Univ East China Sci&Technology, Shanghai Ecust Biomedicine Co Ltd, Shanghai Huali Biomedicine Co Ltd) CN101519401 (prior. 27/02/2008, CN).
ShanghaiEBC2008b Chen, J.; Ge, Q.; Niu, D.; Wu, F.; Zhao, M. (Shanghai Ecust Biomedicine Co Ltd) CN101747219 (prior. 01/12/2008, CN).
ShanghaiIMM2012 Liu, H.; Zhen, X.; Sun, H.; Zhu, L.; Qian, W.; Yu, L.; Li, Z.; Zhou, S.; Cai, W.; Jiang, H.; Chen, K. (Shanghai Inst Materia Medica) EP2848617 (prior. 2012-05-09, CN).
ShanghaiIPI2009 Chen, H.; Gong, L.; Xu, X.; Yang, L. (Shanghai Medicine Group Ltd, Shanghai Pharm Co Ltd) CN101544655 (prior. 05/05/2009, CN).
ShanghaiIPI2014 Zhang, F.; Ni, G.; Yang, Z.; Tan, Z. (Shanghai Inst Pharm Industry, China State Inst Pharm Ind) CN104761567 (prior. 02/01/2014, CN).
ShanghaiMH2013 Wang, Y.; Chen, H.; Liu, L.; Li, X.; Diao, W.; Tang, Q.; Liu, T. (Shanghai Modern Hasen Shangqiu Pharmaceutical Co Ltd) CN103044444 (prior. 21/01/2013, CN).
Shenzhen2006a Ye, C. (Shenzhen Xinlitai Pharmacy Co Ltd) CN101045731 (prior. 29/09/2006, CN).
Shenzhen2006b Ye, C. (Shenzhen Xinlitai Pharmacy Co Ltd, Xinlitai Pharm Co Ltd) CN100999525 (prior. 18/10/2006, CN).
Shenzhen2008 Cai, M.; Huang, C.; Li, Y.; Tang, T.; Wang, H.; Wang, Y.; Wu, X.; Zhu, D. (Shenzhen Haiwang Pharm Co Ltd) CN101348490 (prior. 18/08/2008, CN).
Shenzhen2012a Tan, D.; Li, H. (Shenzhen Salubris Pharm Co Ltd) CN103387497 (prior. 10/05/2012, CN).
Shenzhen2012b Tan, D.; Li, H. (Shenzhen Salubris Pharm Co Ltd; Shandong Salubris Pharmaceuticals Co Ltd) CN103450005 (prior. 01/06/2012, CN).
Sichuan2013 Tang, H.; Luo, H.; Shuai, S. (Sichuan Changwei Pharmaceutical Co Ltd, Chengdu Rongyao Group) CN103435631 (prior. 29/08/2013, CN).
SK2005a Lee, J.; Kim, N.; Kim, J. (SK Chem Co Ltd) KR2007066518, WO2007073095 (prior. 22/12/2005, KR).
SK2005b Kim, N.; Lee, J.; Kim, J.; Lee, N. (SK Chem Co Ltd) KR2007068043, WO2007074995, US2009275755, US7696351, EP1966219 (prior. 26/12/2005, KR).
SRIMONMC2016 Mahmoodi, N. O.; Ghavidast, A.; Ashkan, M.; Mamaghani, M.; Zanjanchi, M. A.; Tabatabaeian, K.; Arabanian, A. *Synthesis and Reactivity in Inorganic, Metal-Organic, and Nano-Metal Chemistry* **2016**, *46*, 1552-1557.
Sterix2006 Vicker, N.; Su, X.; Pradaux, F.; Reed, M.; Potter, B. (Sterix Ltd) WO2006100502, US2009042929, US2010120789, EP1861384 (prior. 24/03/2005, GB).
Sumitomo2004 Liang, T.; Ikemoto, T.; Wang, W. (Sumitomo Chem Co Ltd) JP2006124326 (prior. 28/10/2004, JP).
Suzhou2009 Bao, Z.; Shu, L.; Zhang, A. (Suzhou Lixin Pharmacy Co Ltd) CN101560202 (prior. 21/05/2009, CN).
Syntex1988 Lodewijk, E.; Khatri, H. N. (Syntex Inc) US4906756, EP0342118 (prior. 10/05/1988, US).
T2013 Xu, C.; Du, J.; Ma, L.; Li, G.; Tao, M.; Zhang, W. *Tetrahedron* **2013**, *69*, 4749-4757.
TA2010 Ferraboschi, P.; Mieri, M. D.; Galimberti, F. *Tetrahedron: Asymmetry* **2010**, *21*, 2136-2141.
TaizhouZC2008 Chen, M.; Chen, T.; Feng, Q.; Sun, S.; Xu, B.; Xu, H.; Xu, Q. (Taizhou Zhiqing Chem Co Ltd) CN101333223 (prior. 28/07/2008, CN).
Teva2002a Valeriano, M.; Daverio, P.; Bianchi, S. (Entire Interest, Teva Gyogyszergyar Reszvenytarsasag) US2005049275, US7259261 (prior. 02/08/2002, US).
Teva2002b Valeriano, M.; Daverio, P.; Bianchi, S. (Teva Pharm Ind Ltd) US2004024012, US6800759 (prior. 02/08/2002, US).
Teva2002c Valeriano, M.; Daverio, P.; Bianchi, S.; Merli, V. (Teva Pharm Ind Ltd, Teva Pharm USA Inc) WO2004013147, US2004024011, US6737411, EP1483269 (prior. 02/08/2002, US).
Teva2007 Kansal, V.; Praveen, K.; Mistry, D. (Teva Pharm Ind Ltd, Teva Pharm USA Inc) WO2008130642, EP2084164, US2008287679 (prior. 18/04/2007, US).
Tianjin2009 Meng, Q.; Song, F.; Yang, G.; Zhang, Y. (Tianjin Cent Pharm Co Ltd) CN101463038 (prior. 14/01/2009, CN).
Tianjin2010a Ma, L.; Li, Y.; Meng, Q.; Yang, G.; Song, F.; Wang, X. (Tianjin Cent Pharm Co Ltd) CN101775001 (prior. 28/01/2010, CN).
Tianjin2010b Li, Y.; Meng, Q.; Yang, G.; Wang, X.; Ji, J.; Zhang, X.; Song, F. (Tianjin Cent Pharm Co Ltd) CN101787033 (prior. 16/03/2010, CN).
Tianjin2010c Li, Y.; Meng, Q.; Yang, G.; Wang, X.; Ji, J.; Zhang, X. (Tianjin Cent Pharm Co Ltd) CN101786970 (prior. 16/03/2010, CN)
TL2012 Ni, Y.; Pan, J.; Ma, H.-M.; Li, C.-X.; Zhang, J.; Zheng, G.-W.; Xu, J.-H. *Tetrahedron Letters* **2012**, *53*, 4715-4717.
Torrent2003 Nadkarni, S.; Patel, H. (Torrent Pharm Ltd) IN200300054, WO2004074215, EP1606231 (prior. 03/02/2003, IN).
Transtech2004 Mjalli, A. M.; Gaddam, B.; Rao, M.; Bondlela, M.; Gopalaswamy, R.; Andrews, R.; Davis, S.; Simila, S.; Ren, T. (Transtech Pharma Inc) WO2006047302, US2006106089, US2009247536, US7582673, EP1809619 (prior. 21/10/2004, US).

UNankai2007 Yang, H.; Song, H. (Univ Nankai) CN101121720 (prior. 14/09/2007, CN).

UOkayama2007 Ema, T.; Sakai, T.; Okita, N. (Univ Okayama) JP2008182984, JP5103616 (prior. 31/01/2007, JP).

USV2002 Maheshwari, K.; Sarma, R.; Joshi, S.; barde, A.; Sutar, R.; Ranade, P. (Usv Ltd) US2004073057, US6812363 (prior. 15/10/2002, US).

USV2004 Tarur, V. R.; Sathe, D. G.; Mondkar, H. K.; Sawant, K. D.; Naik, T. A.; Deshpande, M. (Usv Ltd) WO2006087729, US2006074242, US7446200, EP1848720 (prior. 04/10/2004, EP, WO).

Wockhardt2003 Jaweed, S.; Merwade, A.; Khan, A. (Wockhardt Ltd) WO2005012300, US2006183907, US2008051581, US7291735, US7482453, EP1651646 (prior. 04/08/2003, WO, US, EP).

Wuhan2011 Xie, G.; Yang, B.; Yang, S.; Zhu, Y. (Wuhan Wuyao Sci&Technology Co Ltd) CN102320986 (prior. 25/07/2011, CN).

Wuhan2017 Hu, L.; Xi, C.; Peng, D. (Wuhan Yinmaote Biology Tech Co Ltd) CN107523594 (prior. 29/03/2017, CN).

Wuxi2009 Shen, L.; Zhou, Y.; Yuan, L.; Liu, F. (Wuxi Howfond Biopharma Co Ltd) CN101787032 (prior. 27/11/2009, CN).

Xinfa2015 Qi, Y.; Tang, A.; Ju, L.; Li, X. (Xinfa Pharmaceutical Co Ltd) CN104592249 (prior. 15/01/2015, CN).

Yamakawa1999 Noda, H.; Sakai, K.; Murakami, N. (Yamakawa Yakuhin Kogyo) JP2001072644 (prior. 06/09/1999, JP).

Yilukang2014 Peng, C. (Chengdu Yilukang Medical Technology & Service Co Ltd) CN103980288 (prior. 03/06/2014, CN).

Zentiva2004 Hajicek, J.; Pihera, P.; Stepankova, H. (Zentiva AS) CZ200401048, WO2006042481 (prior. 18/10/2004, CZ).

Zentiva2005 Hajicek, J.; Stepankova, H.; Kalidova, J. (Zentiva As, Zentiva KA) CZ200500150, WO2006094468, US2008182869, US7754883, EP1856127 (prior. 08/03/2005, CZ).

Zhejiang2011 Hong, X.; Lin, B.; Huang, M. (Zhejiang Xinhua Pharm Co Ltd) CN102093263 (prior. 13/01/2011, CN).

ZhejiangHP2005 Wang, L.; Tang, Y.; Cheng, Y.; Tian, F.; Chen, Y. (Zhejiang Huahai Pharm Co Ltd, Chengdu Organic Chem Co Ltd Chinese Acad) CN1927863, CN1927864, CN1927865, CN1927866, CN1951940, CN1951941, WO2007028337, US2008249311, US7932391, EP1942110 (prior. 08/09/2005, CN).

ZhejiangJP2010 Zhang, Y.; Ding, W.; Liang, S. (Shangyu Jingxin Pharm Co Ltd) CN101845050 (prior. 01/06/2010, CN).

ZhejiangJP2014 Yan, P.; Li, Y.; Che, D.; Zhang, X.; Chen, K.; Yan, Y. (Zhejiang Jiuzhou Pharm Co Ltd) CN105481677, WO2016041351, EP3196185, US2017260119, US10112884 (prior. 15/09/2014, CN).

4 Citalopram and Escitalopram

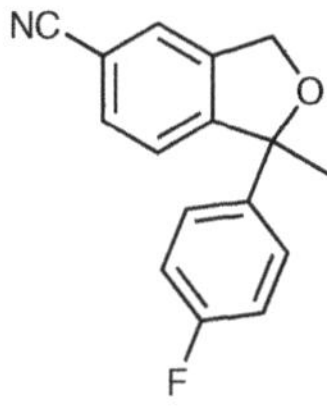

Citalopram

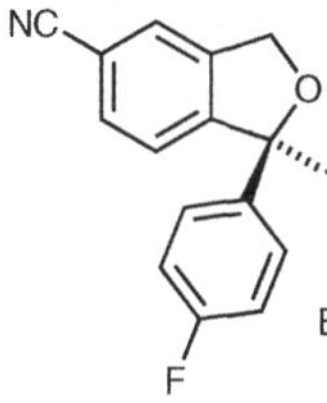

Escitalopram
(*S*)-(+)

	Citalopram	Escitalopram
Year of discovery/isolation	1976, Kefalas	1988, Lundbeck
US approval	July 1998 (Celexa®, Forest Labs)	August 2002 (Lexapro®, Forest Labs)
Trademarks	Celexa (Forest); Cipramil (Lundbeck); Elopram (Recordati); Seropram (Lundbeck)	Cipralex (Lundbeck); Gaudium (Recordati); Lexapro (Forest)

Drug category: Serotonin-specific reuptake inhibitor (SSRI)
Uses: Antidepressant, panic disorder, obsessive-compulsive disorder

Depression is a significant and common disorder. Each year about 100 million people worldwide develop some form of depression. Serotonin-specific reuptake inhibitors (SSRIs) and serotonin-noradrenaline reuptake inhibitors (SNRIs) are employed in the treatment of conditions such as depression and anxiety. SSRIs are believed to increase the extracellular level of the neurotransmitter serotonin by blocking its reabsorption. Since their introduction in the mid-80s, SSRIs have become the most widely used of all antidepressants.

The first highly successful drug in this class was Fluoxetine (Prozac®, Eli Lilly) which was approved by the FDA in 1987. A number of second-generation SSRIs were developed and approved during the 90s including Sertraline (Zoloft®, Pfizer), Paroxetine (Paxil®, GSK), Citalopram (Celexa®, Forest) and Escitalopram (Lexapro®, Lundbeck), each of which attained annual sales of billions of USD (Figure 4.1).

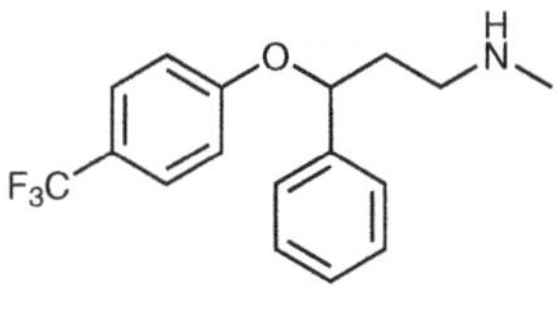

Fluoxetine (Prozac®, Eli Lilly, product patent 1974, US approval 1987)

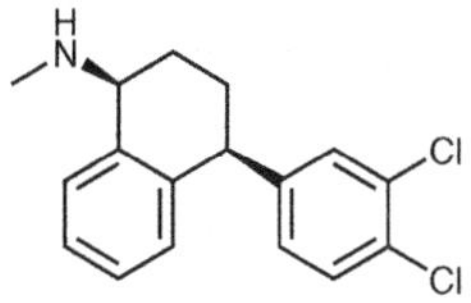

Sertraline (Zoloft®, Pfizer, product patent 1979, US approval 12/1991)

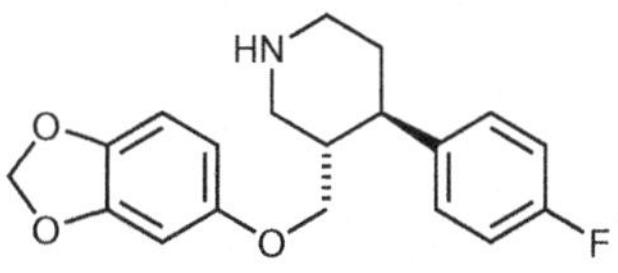

Paroxetine (Paxil®, SKB, product patent 1973, US approval 12/1992)

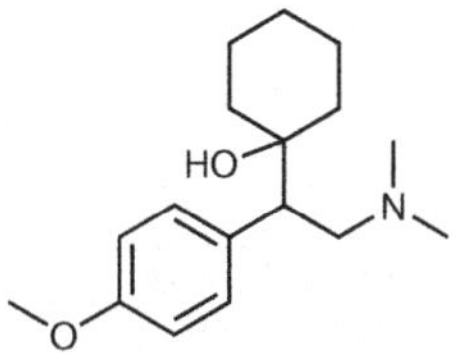

Venlafaxine (Effexor®, Wyeth, product patent 1982, US approval 12/1993)

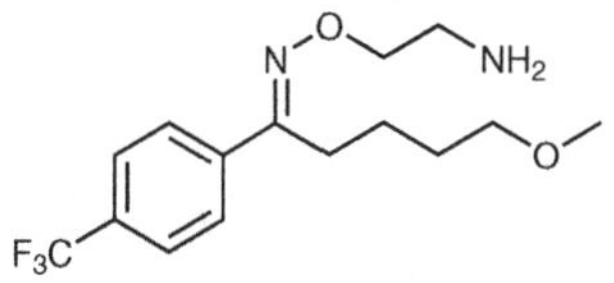

Fluvoxamine (Luvox®, Solvay, product patent 1975, US approval 12/1994)

Figure 4.1 Selected Serotonin-specific reuptake inhibitors

Retrosynthesis in the Manufacture of Generic Drugs: Selected Case Studies, First Edition, Pedro Paulo Santos and William Heggie.

Citalopram, which is a racemic mixture, was patented in 1976 (Kefalas1976) has proven to be an efficacious antidepressant compound in man (PNPBP1982). Later, it was established that the *R* isomer was inactive and a single-enantiomer (JNTG1992, EOID2002), the pure *S*-form (Escitalopram), was subsequently launched, thus effectively extending the patent life and giving Lundbeck more years of exclusivity.

4.1 WORLDWIDE SALES AND PATENT STATUS

The following graphs illustrate how research carried out in the pharmaceutical industry towards the production of the API and worldwide sales of the finished drug product evolved.

- Citalopram US basic patent term 02/08/2005 (US4650884, Lundbeck1984), USA Product Patent term extension 02/08/2010. Escitalopram US basic patent term 08/06/2009 (US34712, Lundbeck1988), USA Product Patent term extension 14/09/2011.
- Citalopram annual production[1]: 2019, 86,550 kg; 2018, 88,542 kg; 2017, 92,206 kg.
- Citalopram annual WW sales[1]: 2019, 336.5M USD; 2018, 381.6M USD; 2017, 397.4M USD.
- Escitalopram annual production[1]: 2019, 91,030 kg; 2018, 84,910 kg; 2017, 77,595 kg.
- Escitalopram annual WW sales[1]: 2019, 1541.5M USD; 2018, 1520.3M; USD; 2017, 1464.1M USD.

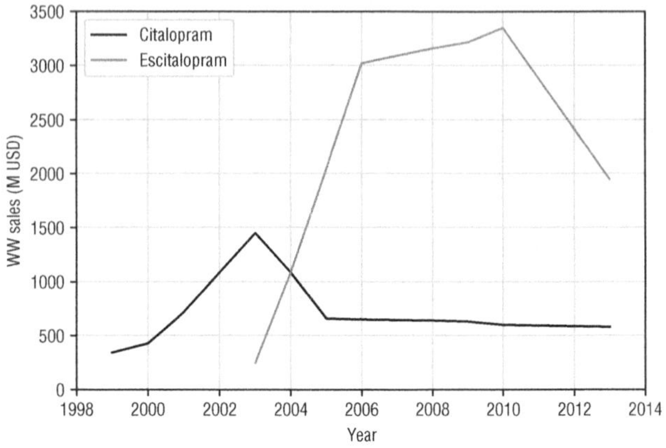

Figure 4.2 Citalopram and Escitalopram WW sales evolution (in M USD)

Source: Data from GlobalData [Online] https://globaldata.com/

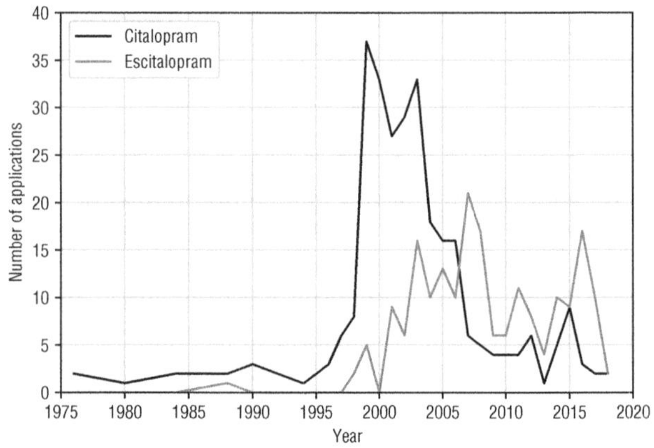

Figure 4.3 New patent applications on Citalopram and Esctitalopram filed by year

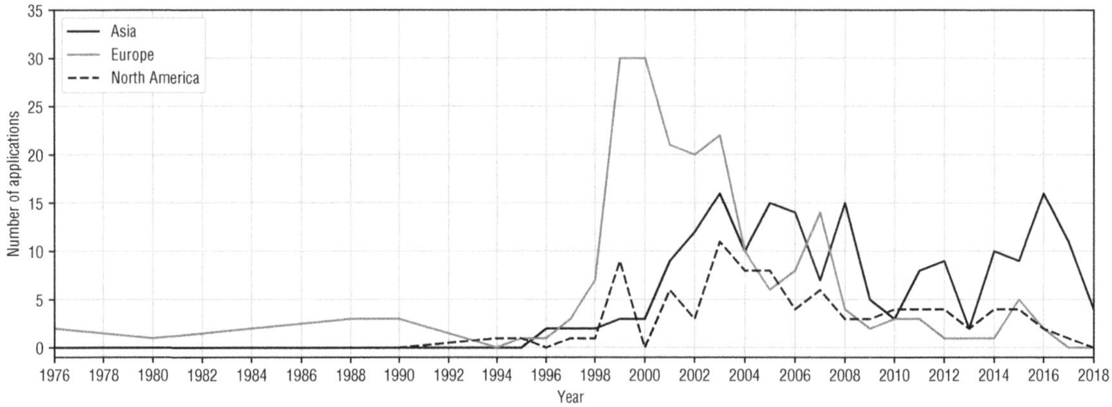

Figure 4.4 Patent applications on citalopram and escitalopram filed by region

1 Newport Modular Platform – Clarivate Analytics [Online] https://clarivate.com/cortellis/ [2020, April]

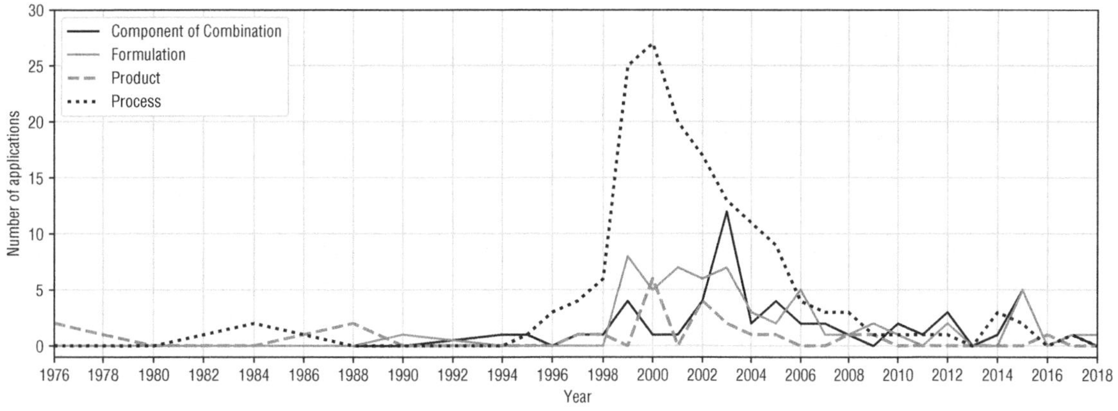

FIGURE 4.5 PATENT APPLICATIONS ON CITALOPRAM BY TYPE OF PATENT

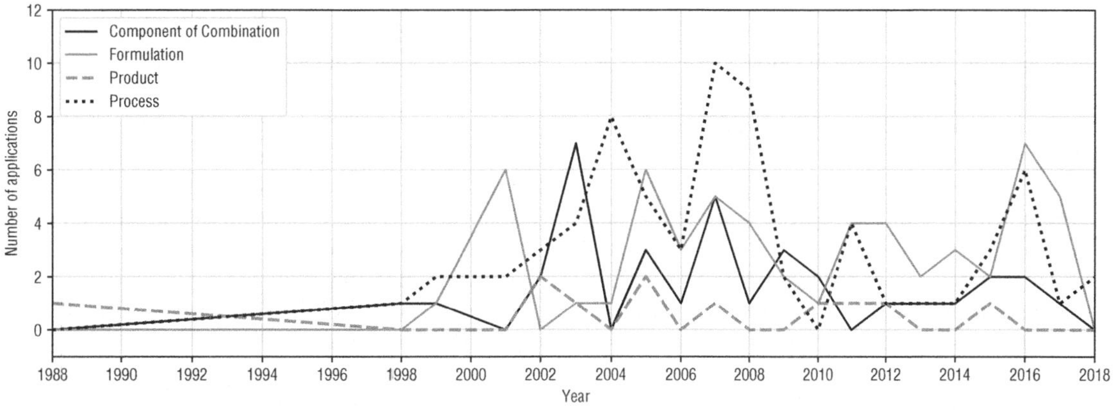

FIGURE 4.6 PATENT APPLICATIONS ON ESCITALOPRAM BY TYPE OF PATENT

4.2 ESCITALOPRAM/CITALOPRAM STRUCTURE

At first glance Escitalopram (***S*-1**) looks like a relatively simple molecule. The important structural features of this active pharmaceutical ingredient are summarized in Figure 4.7.

- The central quaternary chiral carbon atom is a key structural feature. Both in the case of Citalopram and its *S* enantiomer the formation of the quaternary carbon is certainly one of the most challenging tasks in the synthesis.
- The hydroisobenzofuran ring system may be considered the central structure element of the molecule and synthetic methods to build this hydrofuran system starting from a 1,2,4-trisubstituted aromatic ring or a 1,3 or 1,4-disubstitued aromatic ring, are the best options. When a di-substituted compound is the starting material, the remaining substituent can be introduced by an electrophilic aromatic substitution.
- The aromatic nitrile at C5 on the isobenzofuran can be introduced by functional group exchange when a convenient group is already present in the molecule.
- Introduction of a 4-fluorophenyl ring and a 3-(dimethylamino)propyl side chain at the C1 carbon atom of the isobenzofuran ring system creates the quaternary chiral centre.

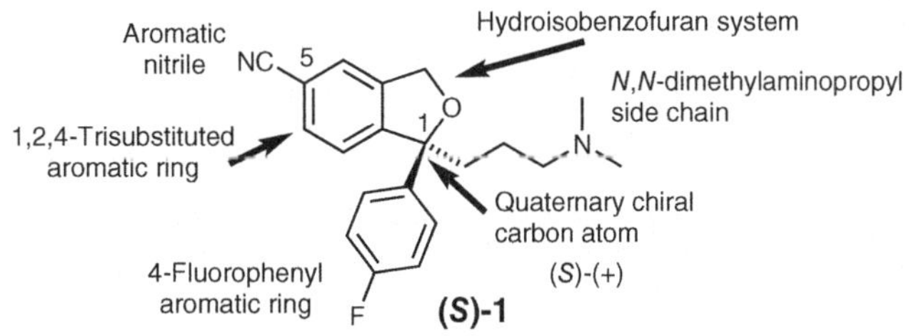

Escitalopram (*S*)-1

(*S*)-1-(3-(dimethylamino)propyl)-1--(4-fluorophenyl)-1,3--dihydroisobenzofuran-5-carbonitrile

Citalopram 1

(*RS*)-1-(3-(dimethylamino)propyl)-1--(4-fluorophenyl)-1,3--dihydroisobenzofuran-5-carbonitrile

FIGURE 4.7 KEY STRUCTURAL FEATURES OF ESCITALOPRAM

Despite its apparent simplicity, many options for the synthesis are expected to be conceivable. This chapter will focus first on the racemic molecule (Citalopram) and later on the chirally pure drug (Escitalopram).

4.3 RETROSYNTHETIC ANALYSIS OF CITALOPRAM

The centrally placed chiral carbon atom is undoubtedly a major focus of any retrosynthesis of Citalopram and having a sound strategy to build the quaternary centre is crucial. Several approaches are possible.

4.3.1 Disconnection of the propylamine side chain

Initial disconnection of the amino side chain first by cleavage of a central CC bond is one of the most effective options, resulting in a greatly simplified achiral intermediate.

4.3.1.1 Using an electrophilic C3 unit to connect the side chain

A heterolytic disconnection of the alkylamino side chain (Scheme 4.1) results in the formation of a carbanionic nucleophile. This nucleophile can potentially be formed by removal of the doubly benzylic hydrogen of **2**, resulting in a carbanion, which is doubly stabilized by the presence of the two aromatic rings. In addition, the nitrile group in the *para* position on one of the phenyl rings contributes substantially to this stabilization. Therefore, compound **2** is expected to be a significant intermediate in the synthesis of the drug.

Several electrophiles that do not contain the dimethylamino group and/or a simple leaving group (see Figure 4.8 for examples) may be envisioned as useful in reactions with the anion of **2**, but of course, in these cases additional steps are required to obtain Citalopram.

Opening the saturated furan ring of the intermediate **2** results in its accepted diol precursor. Either of the hydroxyl groups of this diol can be converted to a leaving group, thereafter a S_N2 or S_N1 displacement reaction yields the tetrahydrofuran (Scheme 4.2).

These disconnections give rise to a 1,2,4-trisubstituted aromatic ring (**3**). Strategies to obtain this compound will be discussed later in the text.

4.3.1.2 Using a nucleophilic C3 unit to connect the side chain

An alternate strategy makes use of a nucleophilic side chain. However, substitution at the quaternary benzylic centre makes this approach unviable. Nevertheless, carrying out the disconnection on an open form of the hydrofuran, i.e. the diol intermediate, renders this approach possible (Scheme 4.3).

Several nucleophilic species can be used to attach the C3 side chain (Figure 4.9).

Once more the same 1,2,4-trisubstituted aromatic intermediate (**3**) is obtained, which must be produced from simple, readily available starting materials.

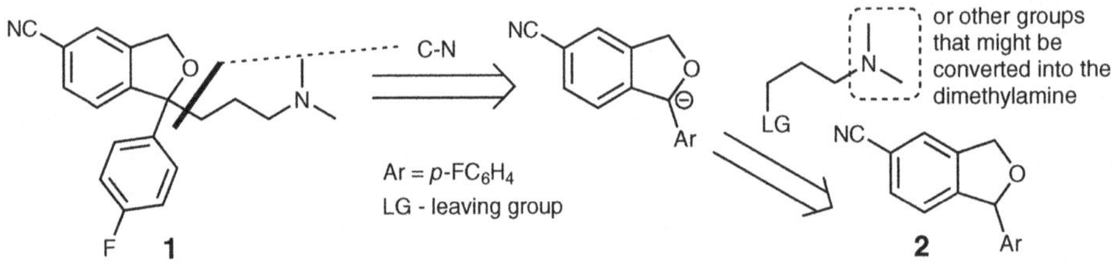

SCHEME 4.1 DIRECT DISCONNECTION OF THE C3 SIDE CHAIN OF CITALOPRAM

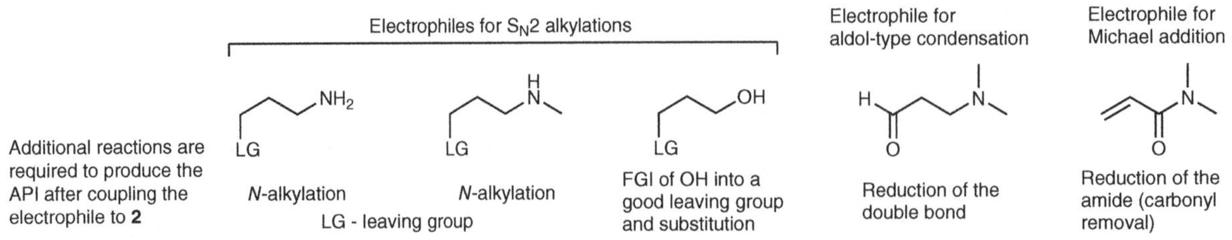

FIGURE 4.8 USEFUL ALTERNATIVE ELECTROPHILES TO REACT WITH ANIONIC **2** AND ATTACH A C3 SIDECHAIN

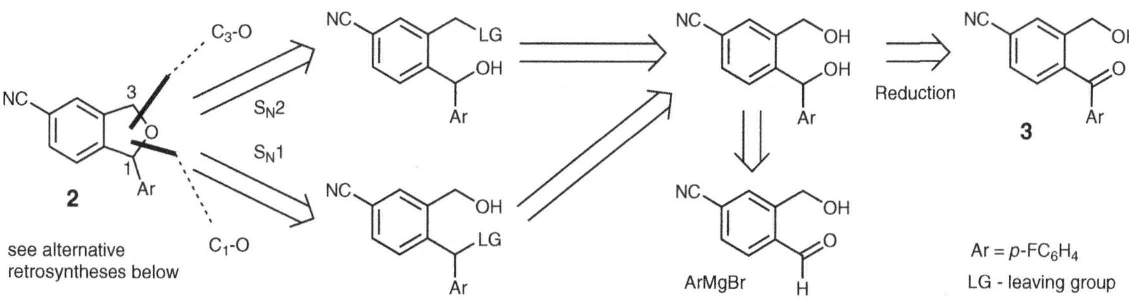

SCHEME 4.2 RETROSYNTHESIS OF THE HYDROFURAN RING

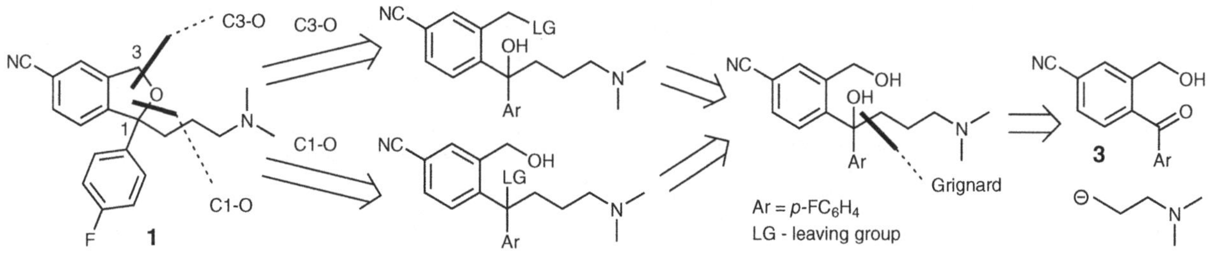

SCHEME 4.3 ALTERNATIVE DISCONNECTION OF THE C3 SIDE CHAIN

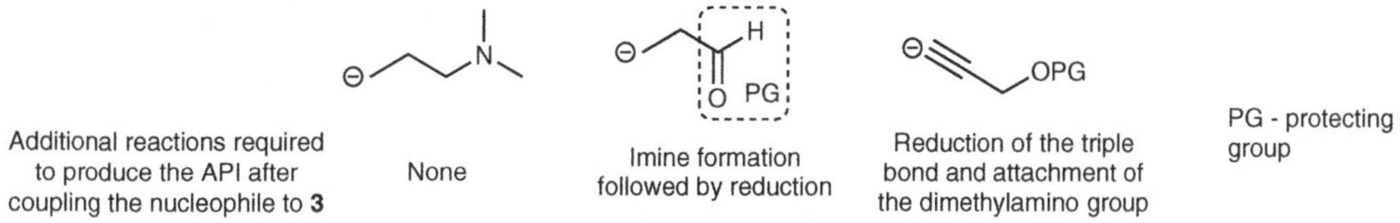

FIGURE 4.9 SOME HYPOTHETICAL USEFUL C3 NUCLEOPHILIC SPECIES FOR THE SYNTHESIS OF CITALOPRAM

4.3.2 The aromatic nitrile group

The aromatic nitrile of the benzohydrofuran may be present in the original starting material, introduced directly or produced from other functional groups during the synthetic process. Several classical strategies are feasible (Scheme 4.4).

The nitrile group can be introduced at various points in the synthesis, opening up the possibility of numerous permutations in the order of reactions, resulting in different synthetic routes to the API.

4.3.3 Retrosynthesis of the trisubstituted aromatic intermediate

As referred to above, any 1,2,4-trisubstituted intermediate, including **3**, has to be obtained from readily available starting materials. Obviously, the most useful starting materials are appropriately more elaborate than simple monosubstituted aromatic compounds. Hence, the retrosynthetic approach herein presented is directed towards commonly available targeted starting materials.

From what has been discussed previously, the raw material should ideally contain a one carbon atom substituent at C5 (preferably a carboxylic acid or a derivative thereof). Failing this, a functional group has to be introduced at this position (nitro or halogen) by an electrophilic aromatic substitution reaction (EAS). The removal of the 4-fluorophenyl ring from **3** simplifies the molecular structure further (Scheme 4.5).

At first glance, the proposed synthetic transformation seems difficult to achieve. Normally Grignard reagents add twice to esters to give tertiary alcohols, the second addition taking place at the ketone formed after the first addition. However, in the present case, the carbonyl group of the intermediate ketone can be trapped by the neighbouring alkoxide ion, formed after the initial addition, yielding a relatively stable salt and reducing the susceptibility for a second attack (Scheme 4.6). Literature precedent does indeed exist for this assumption. In addition, the electrophilic nature of the fluoro-aromatic ring assists in the stabilization of the hemiketal salt. Strict control of the quantity of the Grignard reactant is critical to avoid the second addition reaction and formation of the corresponding impurity.

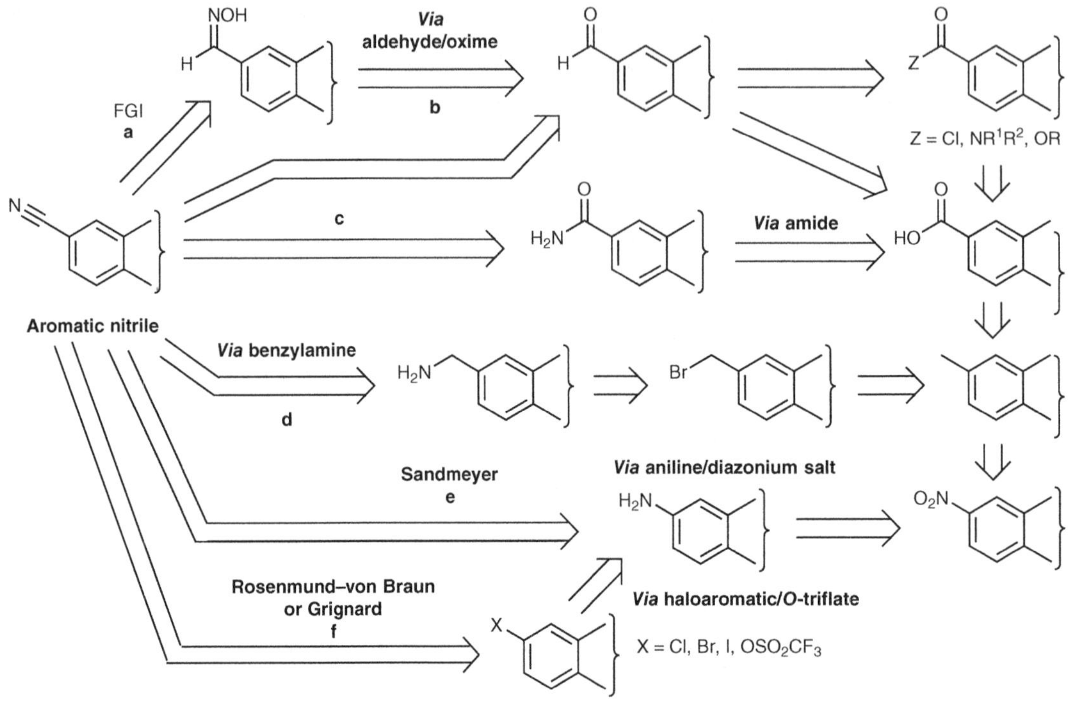

a) Ac_2O (or other dehydrating agents); b) NH_2OH, HCO_2H or NH_2OSO_2OH (*via* oxime)
c) P_2O_5 or $POCl_3$ or $SOCl_2$ or $ClSO_2NCO/NEt_3$; d) 1) NaOCl 2) H_2O; e) 1) $NaNO_2$, H^+ (cat.) 2) CuCN;
f) CuCN, heat or $Zn(CN)_2$ or Pd(II) salts for X = halogen or Ni complexes (all X) or Mg, TsCN or Mg, TMSCN

Scheme 4.4 Useful strategies to produce the aromatic nitrile group

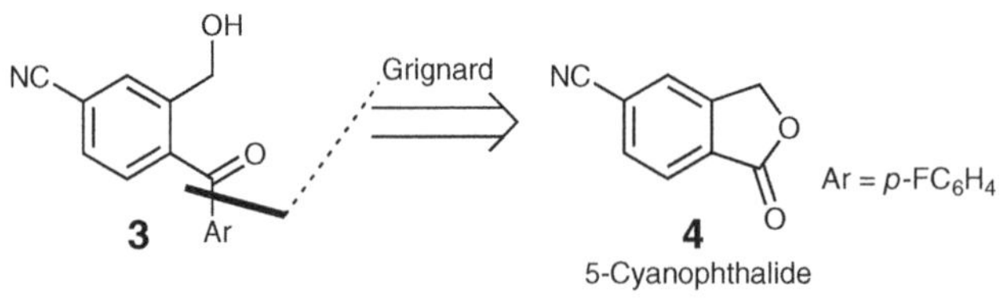

Scheme 4.5 Retrosynthesis of the 1,2,4-trisubstituted intermediate to 5-cyanophthalide

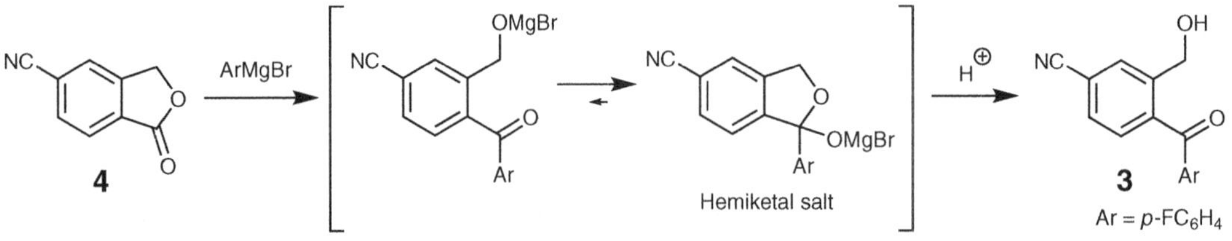

Scheme 4.6 Reaction of 5-cyanophthalide with (4-fluorophenyl)magnesium bromide

Given that the Grignard reagent may also react with the nitrile group, selectivity may be an issue. However, the lower reactivity of the nitrile compared to the ketone still makes selective transformation feasible. Phthalides containing alternative C5 groups, which can be transformed at a later stage into a nitrile, can be used to react with the Grignard.

The 5-cyanophthalide (**4**), a 1,2,4-trisubstituted aromatic compound, is obviously an important intermediate for the synthesis of Citalopram. Identification of potential precursor groups for each of the substituents allows selection of commercially available starting materials for 5-cyanophthalide (Scheme 4.7).

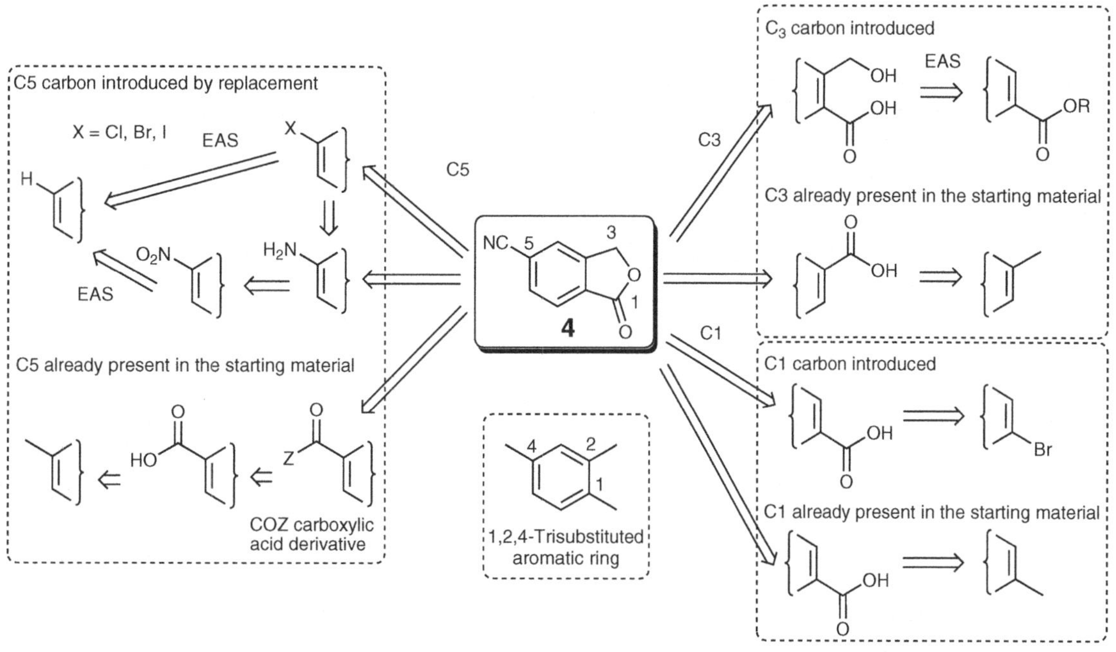

Scheme 4.7 Individual retrosynthesis of substituents in 5-cyanophthalide

Using this general scheme, a rational overall view of commercially available starting materials can be incorporated into the retrosynthetic scheme for **4** (Scheme 4.8).

A more detailed retrosynthesis plan of the 5-cyanophthalide (**4**) showing those available starting materials is presented in Scheme 4.9.

Production of the 5-cyanophthalide (**4**) from terephthalic acid (or its ester) appears to be a sound approach. No oxidations or reductions are required to produce **4**. This starting material is commercially available (obtained by oxidation of *p*-xylene) and its symmetry means that positional isomers side products are avoided. EAS with formaldehyde or equivalent, followed by lactone ring formation gives a facile entry into the hydrofuran ring. It is worth noting that the two carboxylic acid substituents deactivate the ring to EAS and severe reaction conditions are required. The alternative use of unsymmetrical trimellitic anhydride as starting material is less favoured as reduction can occur at either of the non-equivalent anhydride carbonyls yielding a mixture of isomers. Routes starting form phthalimide or phthalic anhydride suffer from the same problem. When these starting materials are used, the process must be able to both, limit the formation of the unwanted isomers and, have suitable methods for their removal from the desired products somewhere along the synthetic route.

Some alternative retrosynthetic routes showing commercially available starting materials (highlighted in boxes) are summarized in Scheme 4.10. These routes are longer than from **4** and require oxidations steps that may be difficult to carry out at industrial scale.

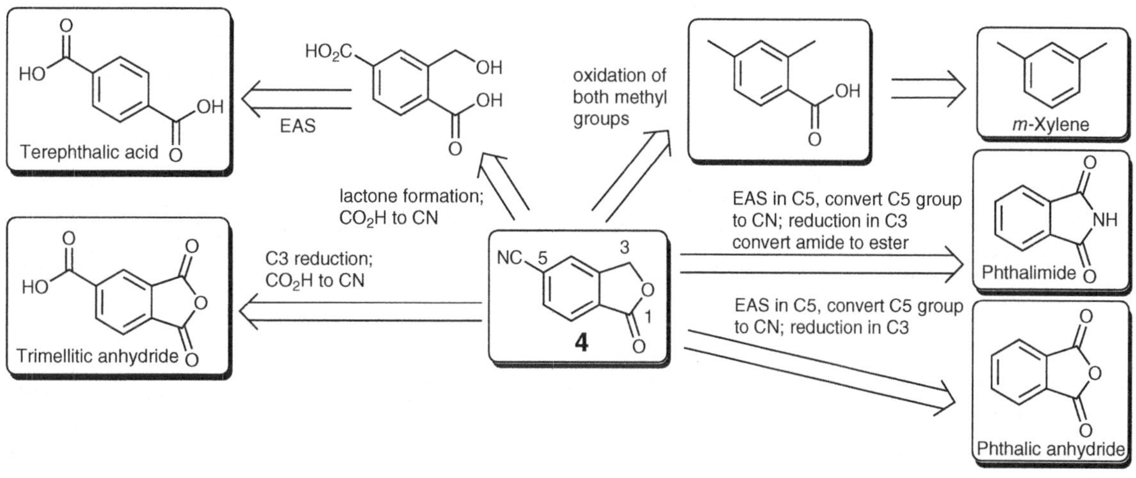

Scheme 4.8 Basic retrosynthesis of 5-cyanophthalide

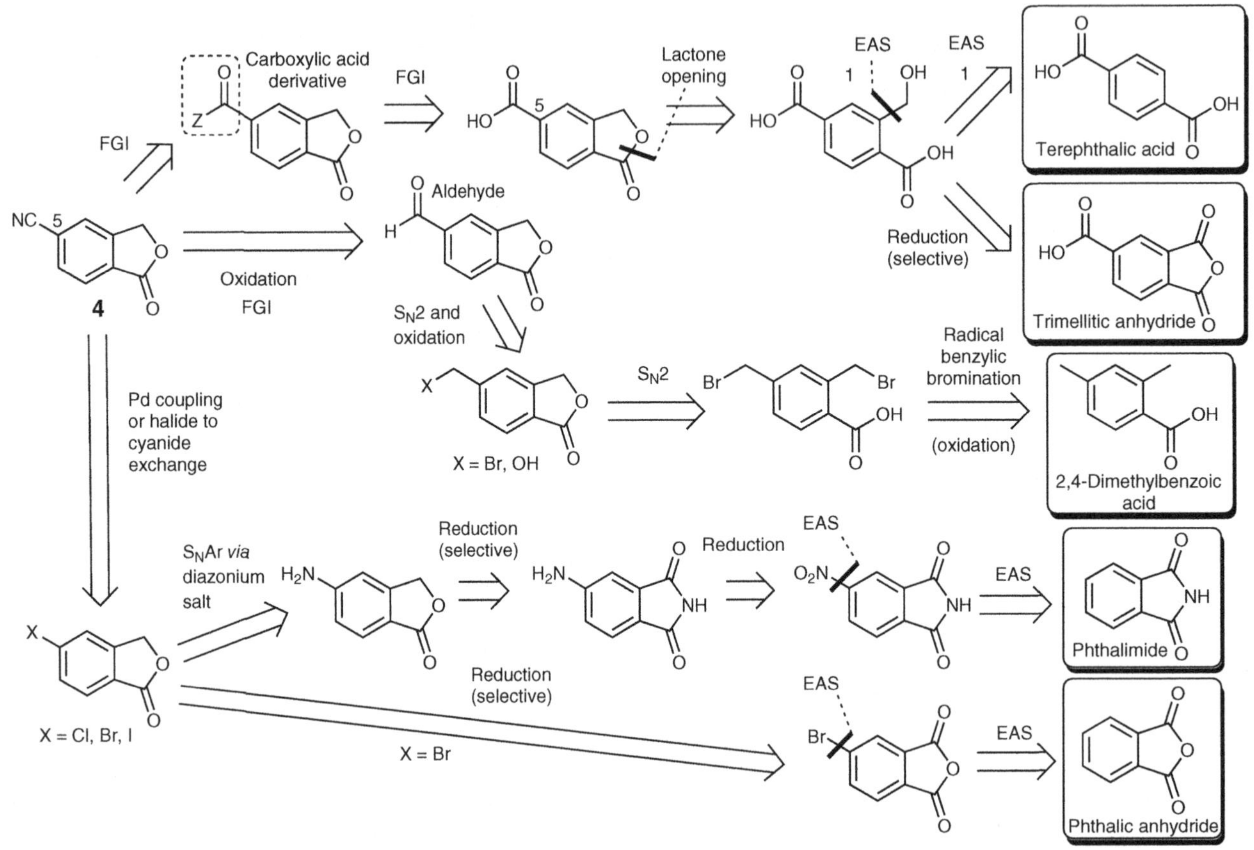

Scheme 4.9 Detailed retrosynthetic analysis of 5-cyanophyhlide

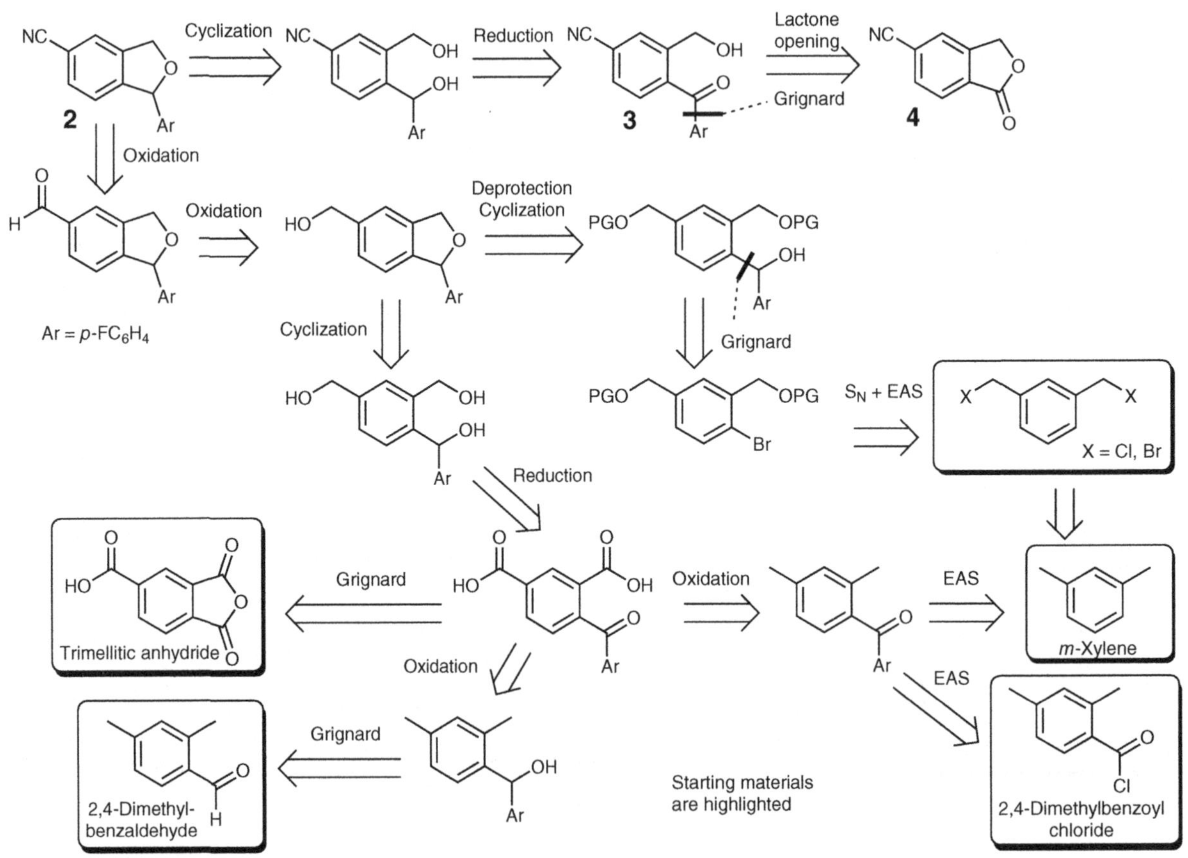

Scheme 4.10 Retrosynthesis of 1-aryl-5-cyanophthalane

From the above analysis, terephthalic acid would appear to be the starting material of choice to produce **4** as it is easily available and the synthetic route requires no oxidation steps although handling of toxic formaldehyde (or one of its equivalents) needs to be addressed. Another favourable starting material is phthalic anhydride, using an aromatic bromine as a handle to introduce the cyanide group. Scheme 4.11 summarizes those main retrosynthetic routes to Citalopram, which focus on efficiency and simplicity.

Different permutations of the order in which each individual objective in the synthesis is achieved, i.e. formation of the quaternary carbon or the introduction of the aromatic nitrile, results in a diversity of synthetic routes, as can be gathered from Scheme 4.11. Each of these routes is expected to give rise to patent publications. When the switch to Escitalopram is made some of these routes may still retain their validity through the application of additional controls, whilst others might no longer be applicable to the single enantiomer drug substance.

4.4 ESCITALOPRAM RETROSYNTHESIS

The advent of Escitalopram as a more favourable drug in comparison to Citalopram introduces a further major objective in the retrosynthetic analysis, i.e. the unearthing of appropriate and efficient methods to obtain the *S* isomer in a pure form. Since Citalopram is basic the use of a chiral acid to form a diastereomeric salt pair for a classical resolution is possible. This situation is far from ideal as the distomer is expected to be difficult to recycle because the chiral center cannot be easily racemized (Scheme 4.12). Methodologies based on opening the hydrofuran are equally impracticable.

A resolution process can still be a valid approach where the separation of diastereomers takes place at a stage that allows the recycling of the unwanted isomer. Several chiral diols are potentially viable intermediates (Scheme 4.13) which may fit this role. Of course, for resolution *via* a diastereomeric salt the basic amine side chain must already be in place. These diols have the potential for racemization under aqueous acidic conditions

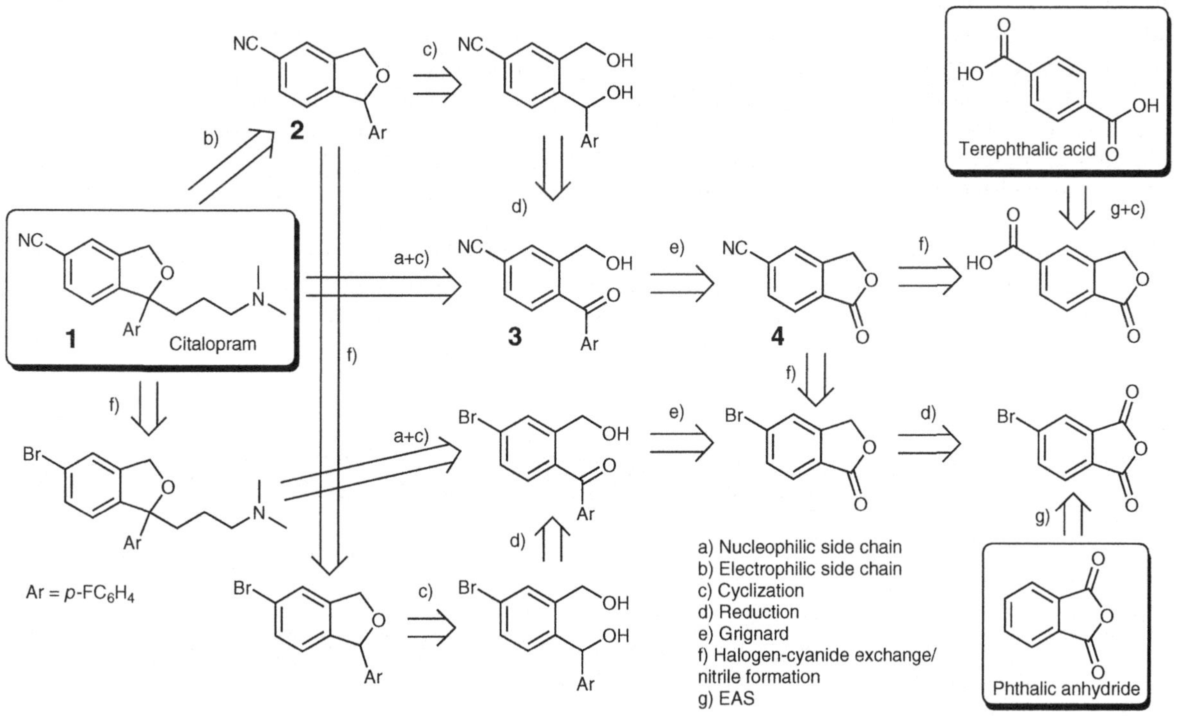

SCHEME 4.11 MAIN RETROSYNTHETIC ROUTES FOR CITALOPRAM

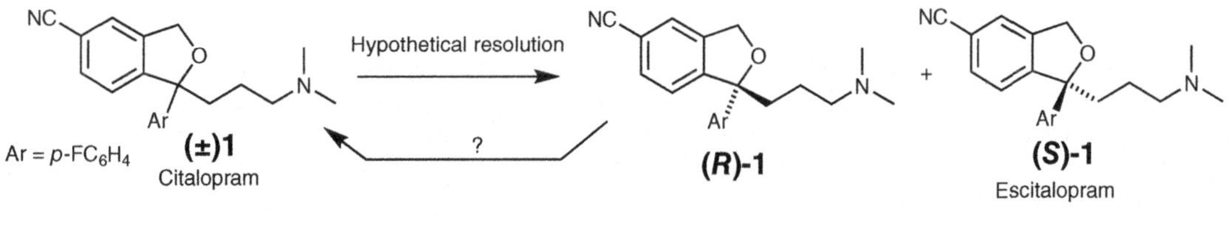

SCHEME 4.12 ESCITALOPRAM BY RESOLUTION OF CITALOPRAM AND RECYCLING OF THE *R* ISOMER

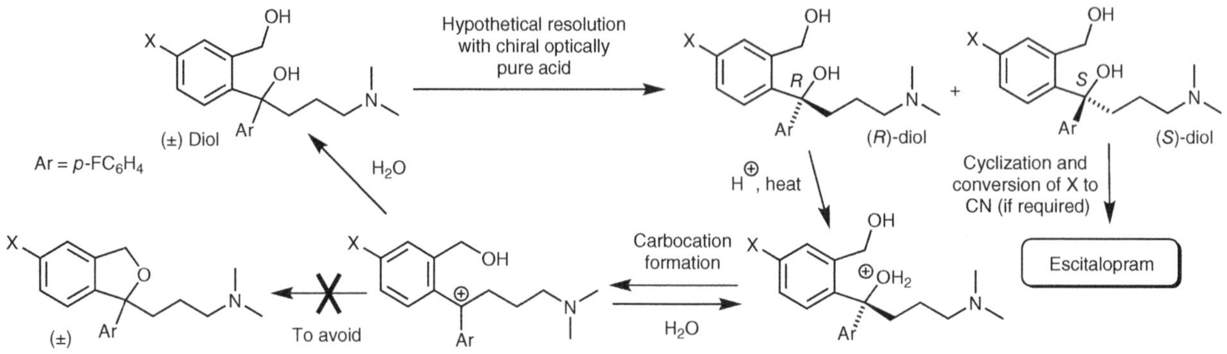

Scheme 4.13 Possible racemization of chiral basic diol intermediates

as the tertiary double-benzylic hydroxyl group can form a stable carbocation and participate in S_N1-type substitution reactions. However, conditions must be set to avoid ring closure to form the stable hydrofuran ring.

When the strategy above is employed, the ring closure reaction of pure chiral diol becomes a critical synthetic step in the synthesis of Escitalopram as conservation of chirality must be preserved. Mechanistic consideration is not important in Citalopram synthesis as a racemate is the main objective. Hence, either S_N1 or S_N2 reactions can be used (Scheme 4.14) for the ring closure, therefore the leaving group may be located on either the tertiary or primary alcoholic carbon. However, only substitution at the primary alcoholic carbon (which likely follows an S_N2 reaction path as opposed to substitution at the tertiary alcoholic carbon which likely follows an S_N1 mechanism) preserves chirality, giving Escitalopram (X = CN) or an analogous pure chiral S-intermediate (ex: X = Br, Cl, etc…) directly.

Hence, it is important that the diol intermediate with the correct final configuration of the quaternary carbon atom is selected at the resolution step and that this configuration maintained during subsequent reactions up to the point where Escitalopram is obtained. It is interesting to note that this stereochemical prediction for the S_N1 and S_N2 ring closing reactions had a very important role in the patent litigation of Escitalopram (C2016). Validity of ring closing process claims of the first Escitalopram patent (Lundbeck1988) was contested on the basis of obviousness and lack of an inventive step.

Asymmetric synthesis is another valid option. Preliminary examination indicates that reactions a and b shown below in Scheme 4.15 are difficult to achieve in good yield and enantioselectivity. Route b could have potential if the benzylic hydroxyl group is used as an anchor to connect a chiral group to produce an asymmetric environment in the vicinity of the carbonyl group.

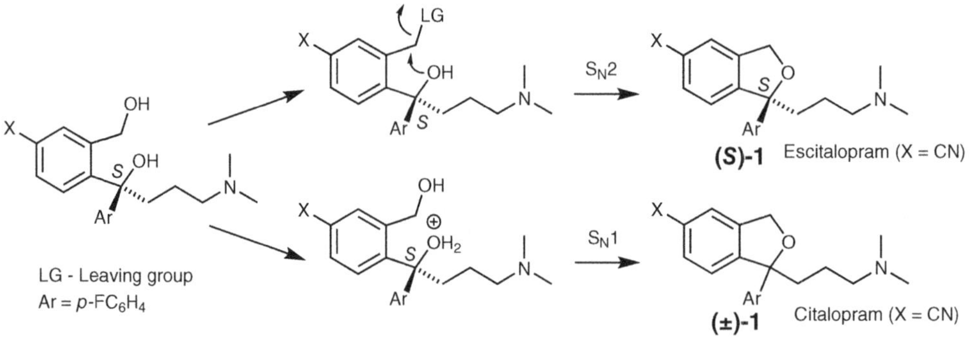

Scheme 4.14 General cyclization mechanisms of diol intermediates

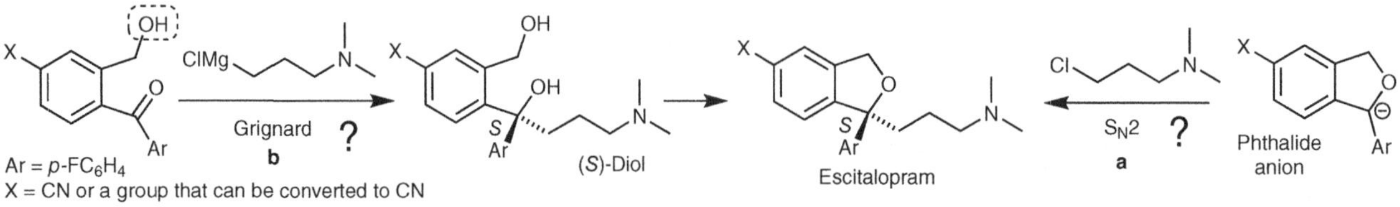

Scheme 4.15 Hypothetical asymmetric syntheses of Escitalopram

This precursory analysis reveals that the simplest and most efficient option is *via* a diol, which can be resolved and cyclize to give Escitalopram or an analogue. This assumes that recycling of the unwanted isomer of the diol is feasible.

4.5 THE FIRST REPORTED SYNTHESIS OF CITALOPRAM

Citalopram was disclosed in 1976 in the patent literature (Kefalas1976). Two different routes were presented for the synthesis (Scheme 4.16).

These synthetic routes follow the two major expected pathways to introduce the aminopropyl side chain. Route A uses two consecutive Grignard reactions, introducing the aminopropyl side chain as a nucleophilic block (Scheme 4.17). Route B introduces the C3 side chain as an electrophilic block, 3-chloro-*N*,*N*-dimethyl-propyl-1-amine. Although both of these syntheses are functional at a laboratory scale, they can present special issues at large scale.

Grignard reactions require careful analysis before implementation at industrial scale. Normal Grignard solvent, ethyl ether, is replaced with safer options such as tetrahydrofuran. The formation of Grignard reagents and their reactions are accompanied by a great release of energy and hence, thorough exothermic studies have to be undertaken and selection of adequate equipment made. Careful control of excess reagent, especially in the present case, needs to be exercised to avoid impurities.

As previously alluded to, the cyclization step uses strong reaction conditions i.e. an excess of 60% aqueous phosphoric acid at high temperatures. Under these conditions side products may be formed (Scheme 4.18).

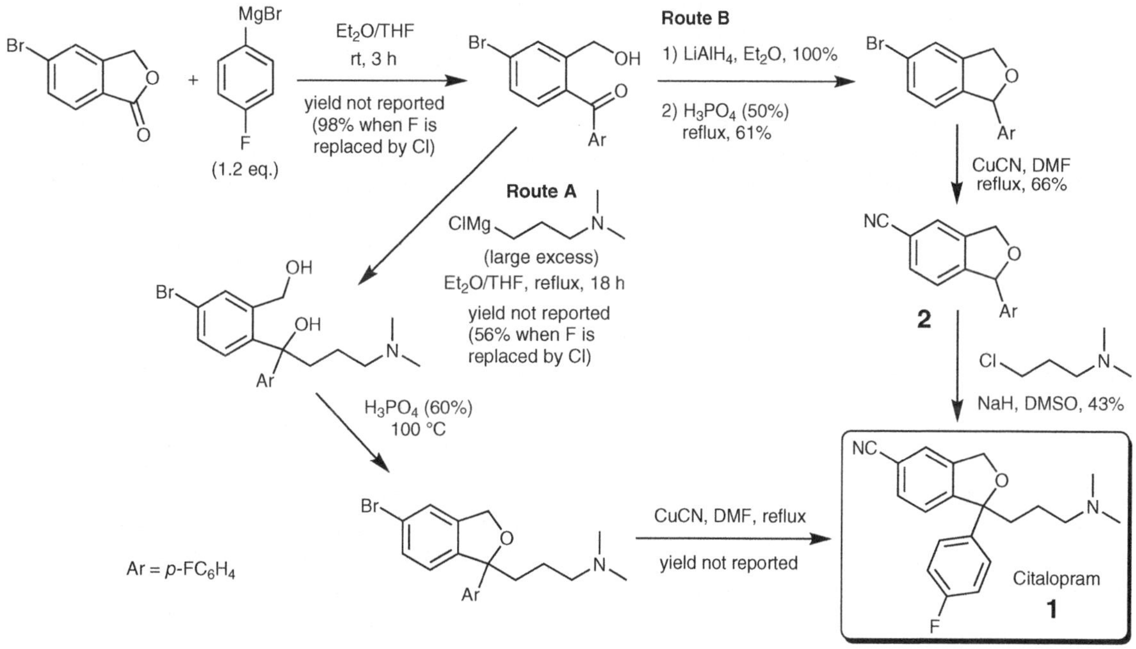

SCHEME 4.16 FIRST REPORTED SYNTHESES OF CITALOPRAM

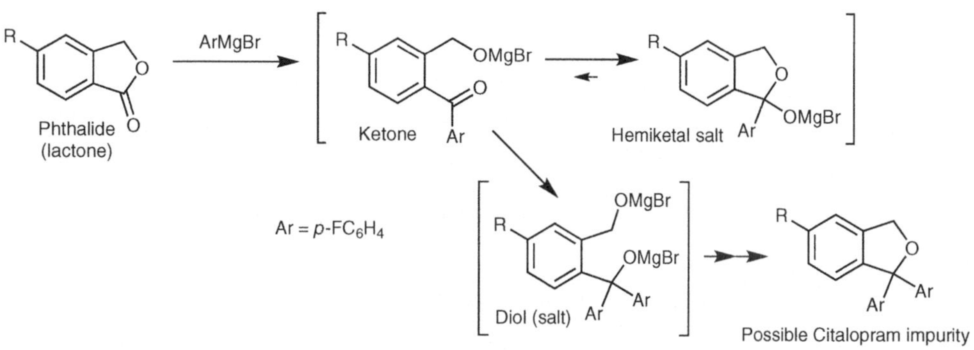

SCHEME 4.17 TWO ARYL GRIGNARD ATTACK IN A PHTHALIDE

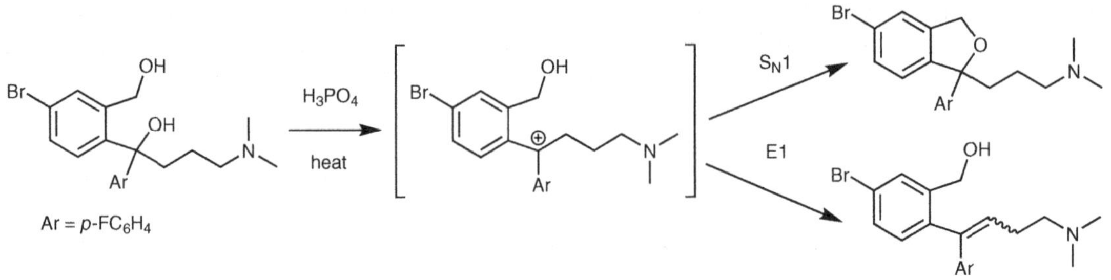

Scheme 4.18 Hydrofuran ring-closing reaction in acidic conditions and elimination side reaction

Both routes use copper cyanide in DMF to introduce the nitrile group (Rosenmund-von Braun reaction, Scheme 4.19). The toxicity of the copper cyanide and disposal of copper salts are an issue. These conditions produce a material that requires extensive purification to achieve pharmaceutical grade product. Desmethylcitalopram is found as a significant impurity at unacceptable levels, presumable from the fact that the von Braun de-methylation reaction shares similar reaction conditions. This impurity is difficult to remove and extensive and expensive purification processes have to be employed. The substitution of the bromine is reported not to go to completion and removal of the bromocitalopram is likewise cumbersome. Hence, impurity issues are of concern. Bromo-, chloro- and desmethyl impurities are controlled in the drug specification, as are heavy metals such as copper.

In concluding, both routes first reported for Citalopram have issues for scale up to industrial levels.

4.6 THE EVOLUTION OF CHEMICAL SYNTHESIS OF CITALOPRAM

Per se the industrial synthesis of Citalopram can present great challenges. Permutations of the same synthetic reactions and group transformations have given rise to various routes. However, the Process Chemist must make a given route compatible with industrial implementation. At laboratory scale, impurities can usually be easily removed using chromatographic methods. At large scale, more economical methods are crucial for commercial success and an otherwise promising synthetic route may not be useful for industrial application because of the formation of problematic, difficult to remove side products.

The product patent holder, Lundbeck, filed the vast majority of the many process patents during the years 1976 to 2000. Undoubtedly, in addition to discovering the ideal industrial process to Citalopram, a large part of this effort was directed to an anti-generic strategy to protect those processes with potential for industrial implementation.

Given that the halogen-cyanide exchange was problematic in the last synthetic step, 5-cyanophthalide was attempted as starting material for the Grignards. Two consecutive Grignard reactions, followed by acidic ring closure, afforded Citalopram (Scheme 4.20, Lundbeck1984).

Since the order of reactivity of the addition of Grignard reagents is: lactone > ketone > nitrile this sequence is feasible. The second Grignard addition being slower allows the two reactions to be carried out in one pot without isolation of the intermediate ketone. This type of telescoping opportunity has advantages industrially saving on one workup and one isolation. Nevertheless, side reactions occur and hence, this sequence is not entirely problem-free and controls are necessary to avoid high levels of impurities (Scheme 4.21).

Only a slight molar excess of the aryl Grignard is used to avoid overreaction. Three molar equivalents was found to give 14-16% of the di-aryl Grignard impurity (Adorkem2003a, Scheme 4.21). A submolar equivalent of the aryl Grignard leaves some unreacted phthalide and subsequently a di-alkyl Grignard impurity may be produced (Chengdu2014).

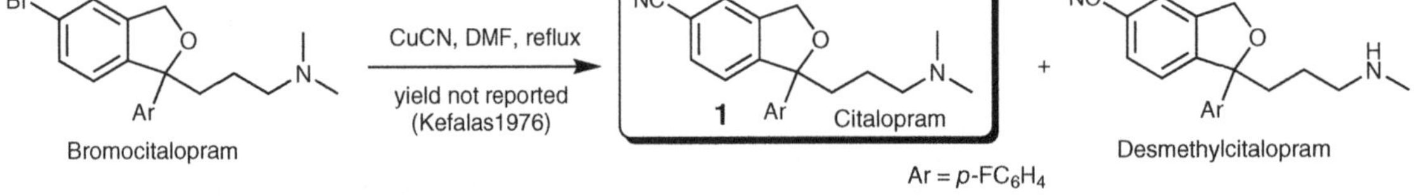

Scheme 4.19 Halide to cyanide exchange

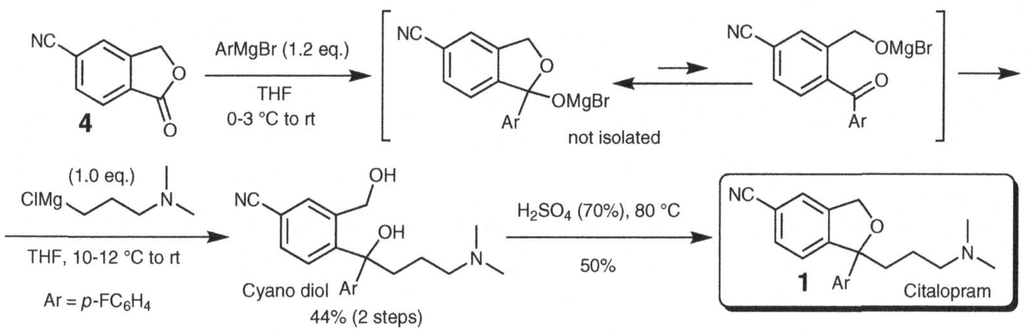

Scheme 4.20 Synthesis of Citalopram from 5-cyanophthalide

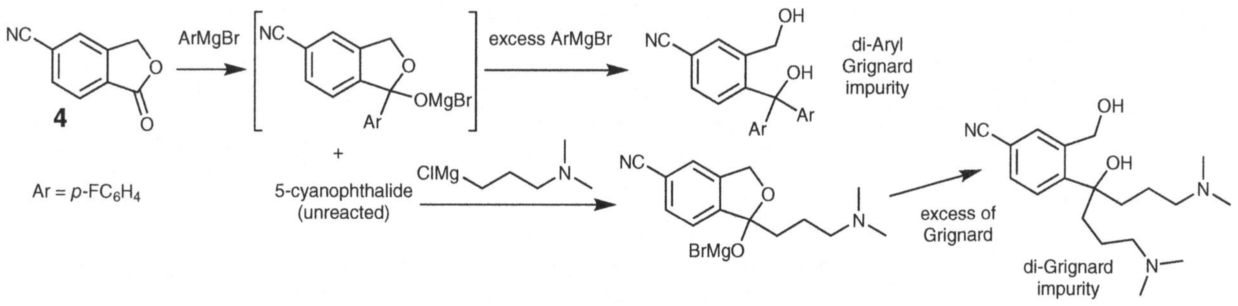

Scheme 4.21 Possible impurities in a telescoped Grignard addition

Attack at the nitrile group can also occur on addition of the second Grignard which gives, after hydrolysis, a product with a C4-acyl side chain. Up to 5% of this impurity may be formed (OPRD2013, Scheme 4.22) and both European and US pharmacopeias set limits to control this problematic impurity in the drug substance and in the drug product.

Another awkward impurity, the aromatic amide, arises during ring closure under strongly acidic conditions, where hydrolysis occurs, (Scheme 4.23, Lundbeck1984) and is controlled in EP and US pharmacopeias. Both side reactions can be inhibited by strictly controlling reaction conditions.

4.7 A QUICK GLIMPSE OF ESCITALOPRAM

Initially, Lundbeck prepared both enantiomers in a pure form to test their individual toxicities. Efforts to accomplish a chiral resolution failed and so attempts were directed at the preparation of the chiral diol.

Thus, the cyano diol was reacted with Mosher's acid chloride (α-methoxy-α-trifluoromethylphenylacetic acid chloride) and the diastereomers separated by preparative HPLC (Scheme 4.24, route **a**). In addition, the diol itself was resolved with (+)-DPTTA [(+)-*O*,*O*'-Di-*p*-toluyl-D-tartaric acid] (Scheme 4.24, route **b**, Lundbeck1988).

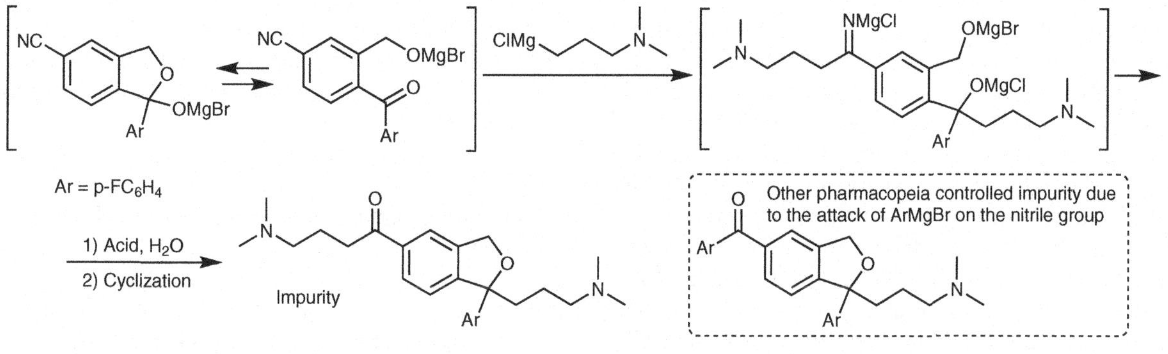

Scheme 4.22 Formation of impurities due to Grignard attack on the nitrile group

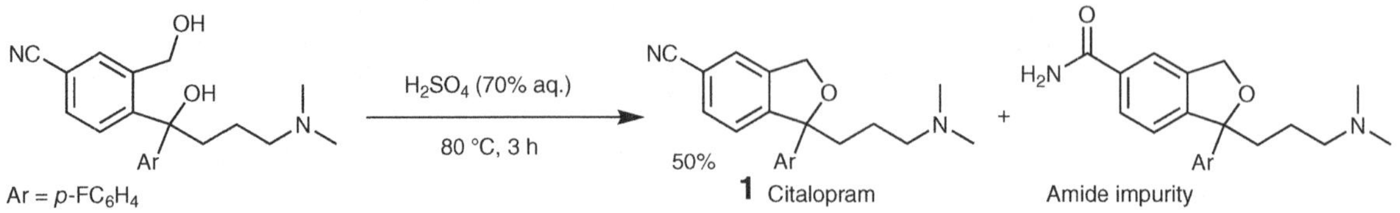

Scheme 4.23 Amide impurity formed during the aqueous acidic ring closing reaction

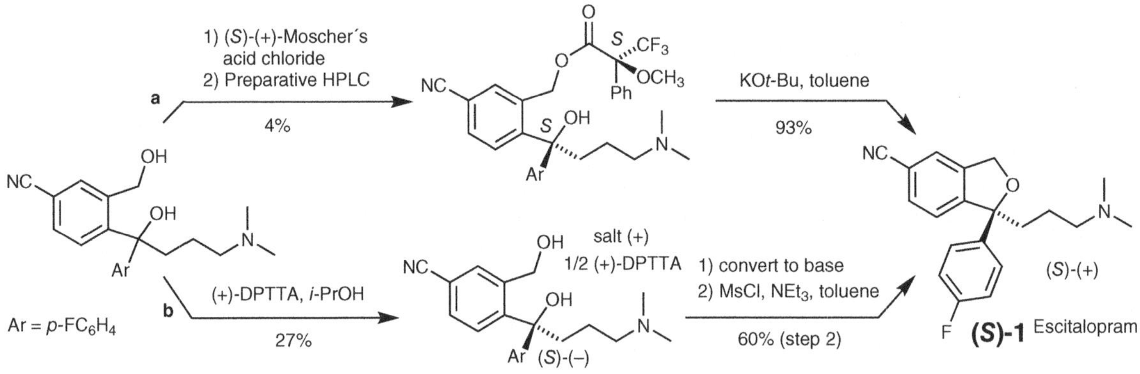

Scheme 4.24 First processes to obtain (*S*)-Citalopram

Cyclization *via* a mild S_N2 displacement on the primary mesylate conserves chirality giving pure *R* and *S* isomers. When tested it was found that drug activity was exclusively due to the *S*-(+) isomer (Lundbeck1988). The changeover to a one enantiomer version of the drug was, however, placed on-hold and synthetic efforts to produce the racemate continued.

4.8 THE TWO GRIGNARD PHASE IN CITALOPRAM SYNTHESIS

Initially, Lundbeck and other companies investigated several 5-substituted phthalides as starting materials (Scheme 4.25). These routes are summarized in scheme A. Note that many of the C5 groups which are easily converted to a nitrile, are not inert to the aryl Grignard.

Although the products from the first Grignard addition are not isolated they are indicated (as neutral molecules) in the scheme for clarity. Preparation of the Grignard reagents themselves is fairly specialized and requires experience in producing organometallic compounds. Safety considerations are key to success and effective scale up requires the ability to carry out detailed calorimetric studies. As mentioned previously, the addition reaction of the Grignard also is delicate and success depends on the ability to carry this out on plant.

In summary the following C-5 substituents in the phthalide were used:

- 5-amido or 5-ester group (**A15**, **A21**)
- 5-amino or 5-aminoacyl group (**A43**)
- 5-oxazolinyl (**A25**)
- 5-hydroxy group (**A39**, X = OH)
- 5-iodo (**A39**, X = I) or 5-bromo atom (**A39**, X = Br)
- protected 5-formyl group (**A1**)
- 5-carboxy (**A21**, R = H)
- 5-alkyloximes (**A12**).

Conversion to the nitrile compound follows traditional chemistry.

In comparison to the processes starting from the bromo and cyanophthalides, most of these alternative routes include additional steps. Sometimes these involve unstable intermediates, which often show lower conversion to Citalopram and hence are less favoured. No publications have since appeared delineating improvements, which leads to the conclusion that they are not suitable for industrial use.

4.8.1 Two Grignard route using 5-bromophthalide as starting material

Several patents have been published with improvements to the original synthetic route (Scheme 4.26).

As seen before, telescoping the two Grignard steps increased efficiency and avoided the need to isolate the first intermediate (55-70% accumulated yields for the two reactions **A39** + **A40** are common, Sekhsaria2001,

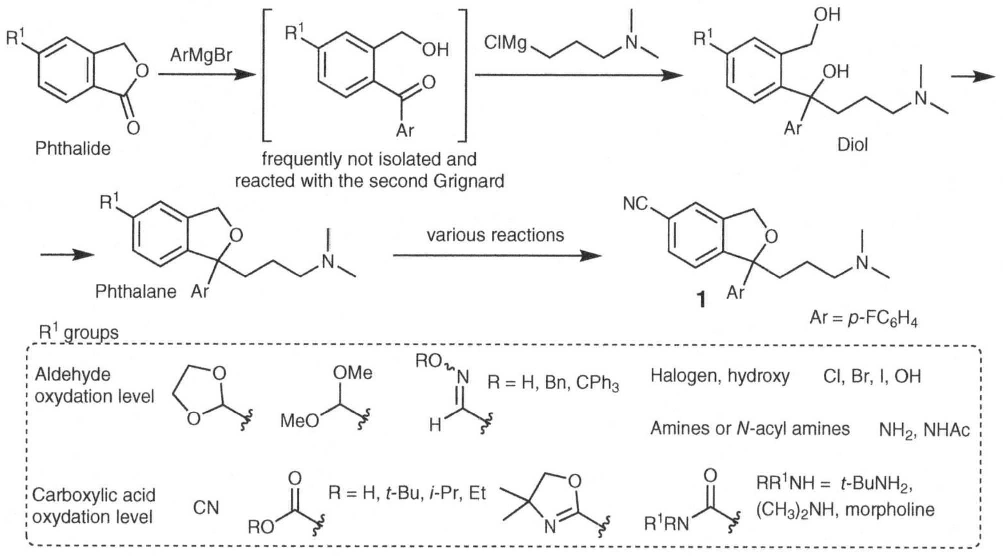

Scheme 4.25 Summary of the "two Grignard" routes

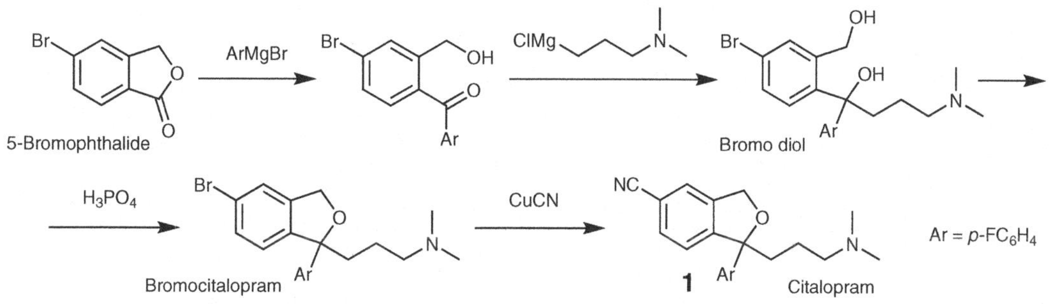

Scheme 4.26 Synthesis of Citalopram from 5-bromophthalide by the two Grignard route

2002, Torcan2002, Wockhardt2003). The direct isolation of the diol as a salt was reported to give an overall yield of 80% from the starting phthalide (Pharmachem2001, as HBr or HCl salt, Natco2003b as di magnesium halide salt) giving an opportunity for clean-up. Conversion of the salts into the free base is reported to give yields higher than 90%.

Heating in aqueous acid (without the need to use anhydrous conditions, **A41**) promotes ring formation. A "one-pot" process has been reported using these types of conditions (Sun2002, 20% global yield from the starting phthalide).

The bromine replacement (**A42**), extensively investigated using the Rosenmund-von Braun reaction, was subsequently reported to be low yielding. The issues involved have been discussed. Copper cyanide can be replaced by zinc or nickel salts in association with palladium but heavy metal removal remains a problem.

Apart from the heavy metal loading, several impurities are associated with this transformation (Scheme 4.27).

The reaction is reported to be difficult to complete and processes for purification of the final product have been proposed. Bromocitalopram can be removed by catalytic hydrogenolysis of the crude reaction mixtures, thereafter the descyanocitalopram is removed by crystallization.

As mentioned above, the desmethyl and carboxamide derivatives may be present in quantities of up to 5% and direct removal is reported to be difficult.

Methylisocyanate polystyrene resin can act as scavenger to trap the demethylated impurities (Pharmachem2001). Acetylation of the secondary amine allows extraction of the resulting neutral amide from aqueous solutions of Citalopram salt (Lundbeck2000n). Re-methylation is also an option (Wockhardt2003, Aurobindo2003a,b).

The carboxamide impurity can be converted to the nitrile by dehydration (Sun2002). Highly impure Citalopram mixtures can be hydrolysed to the amide and re-converted back to Citalopram (Ranbaxy2001b). High molecular weight impurities can be removed by thin film distillation of Citalopram (Figure 4.10, Lundbeck2000o).

When bromocitalopram oxalate salt is used for the cyanation step it is claimed that impurities associated with the dimethylamino group are controlled (Cipla2001a, oxalate, Cipla2001b oxalate, HBr, Natco2003e oxalate). However, extensive process development is likely required to make this an industrially viable alternative.

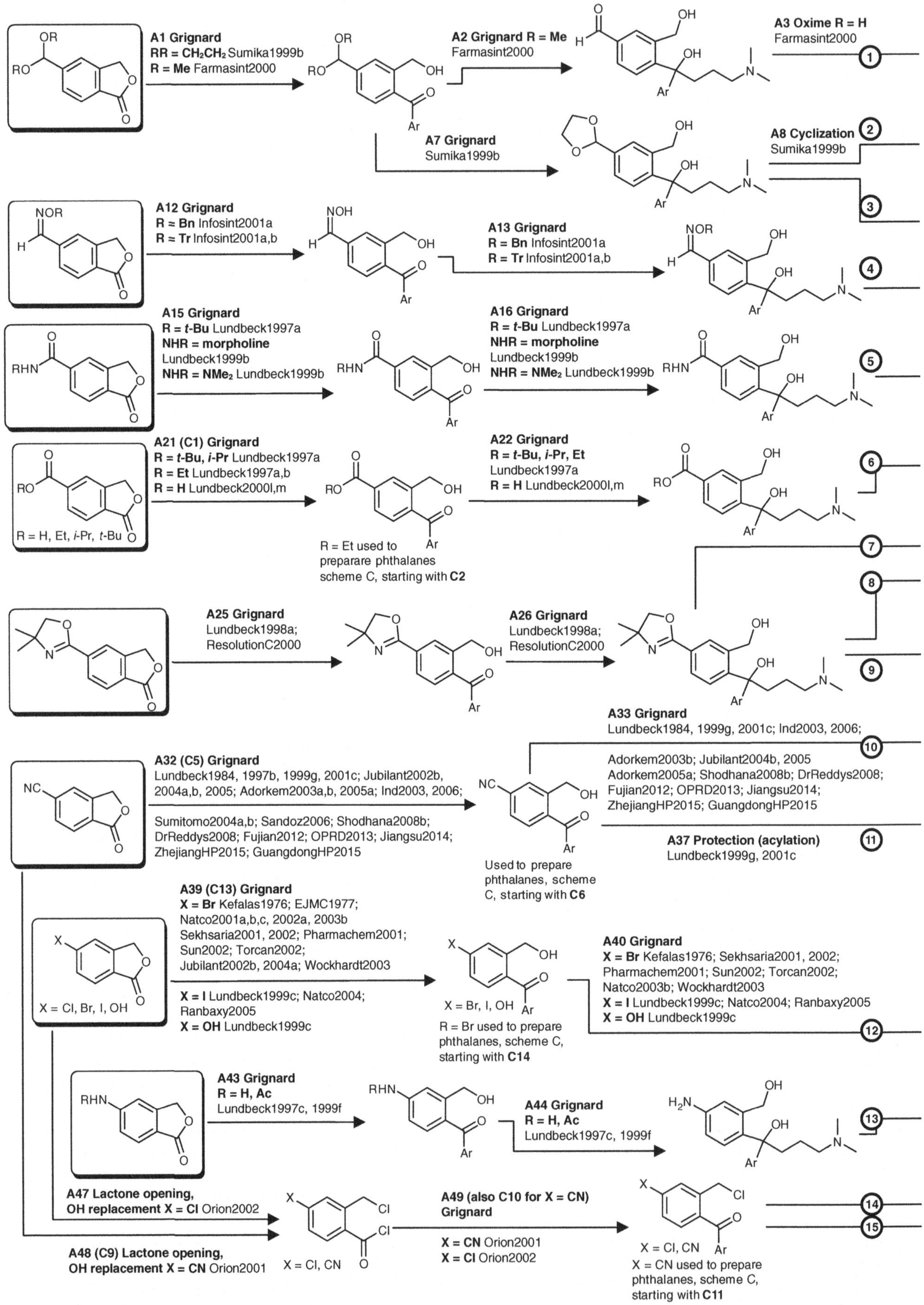

Scheme A Citalopram synthesis by the two Grignard strategy (for reaction conditions see page 128)

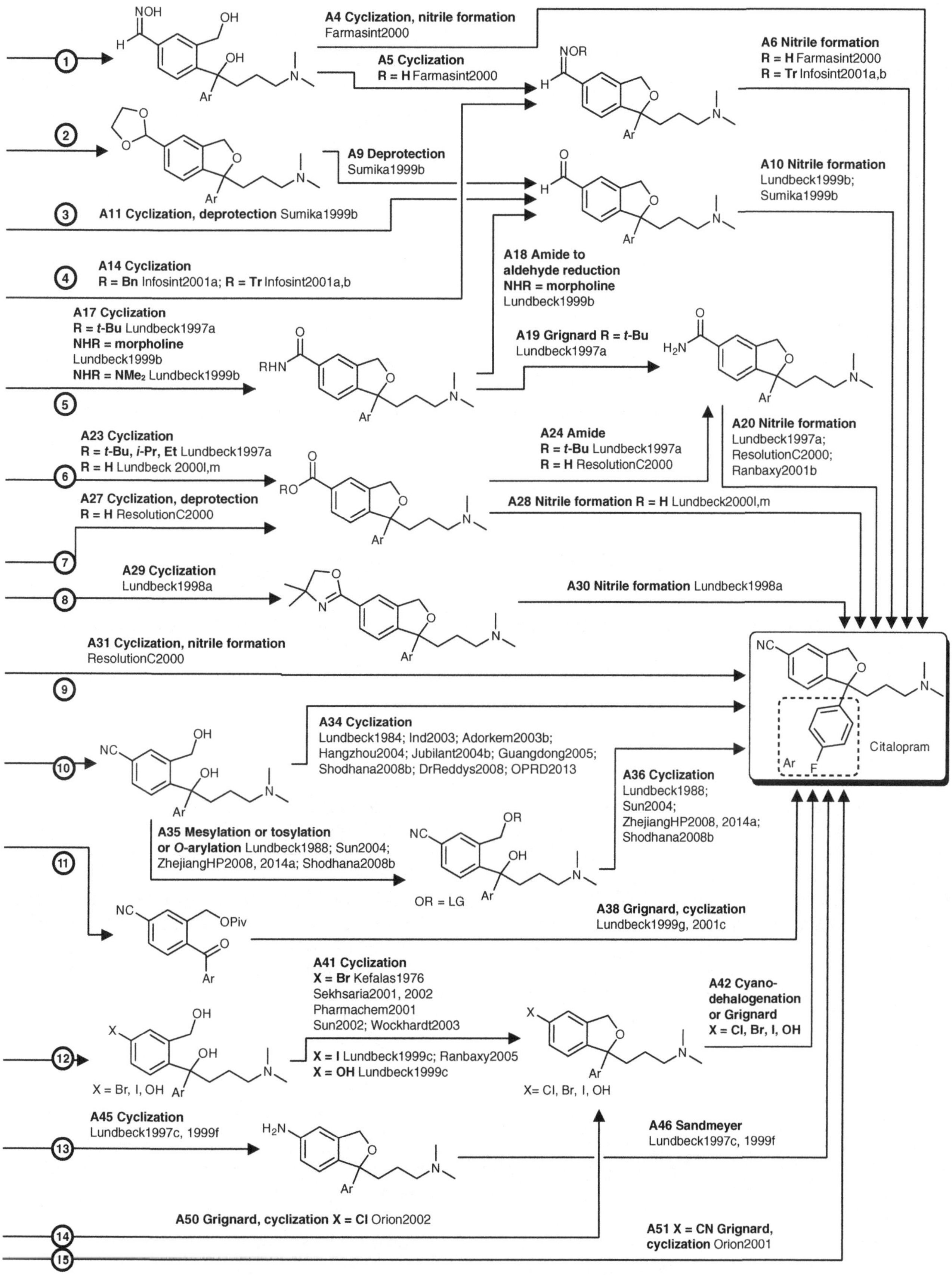

A42 Cyano-dehalogenation (X = Cl, Br, I) or Grignard
X = Cl Lundbeck1999d; Orion2002. **X = Br** Kefalas1976; Lundbeck1999c,d,e,k, 2000e,k,n,o; Cipla2001a,b; Sekhsaria2001, 2002; Pharmachem2001; Sun2002; Esteve2002; Aurobindo2003a,b; Matrix2003; Wockhardt2003; Natco2003d,e; CC2016.
X = OH Lundbeck1999c **X=I** Ranbaxy2001a, 2005.

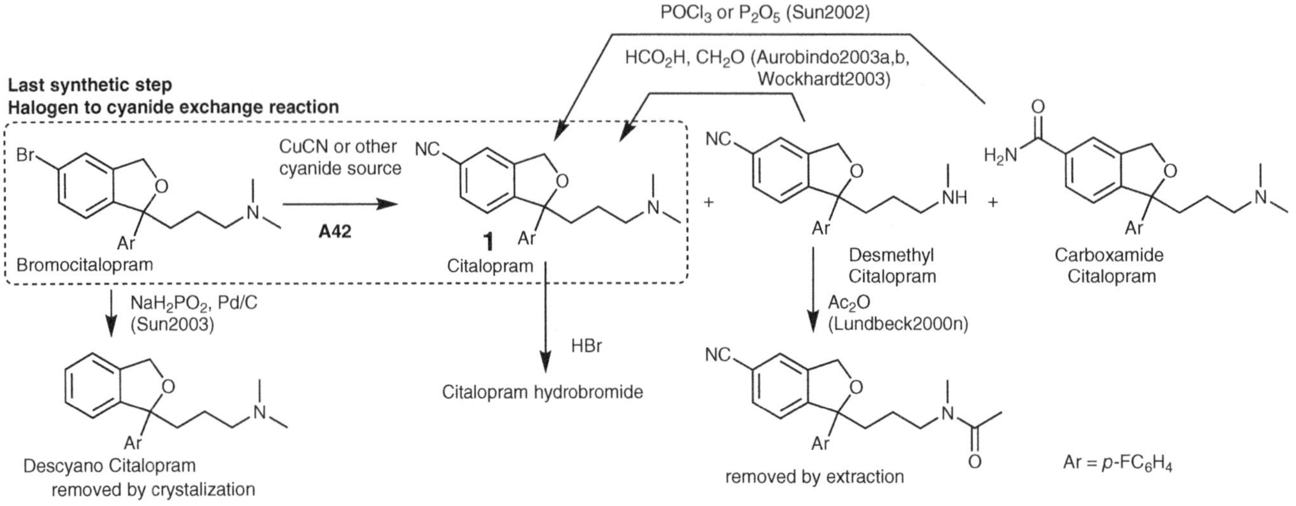

SCHEME 4.27 BROMO-CYANIDE EXCHANGE REACTION AND IMPURITIES FORMED

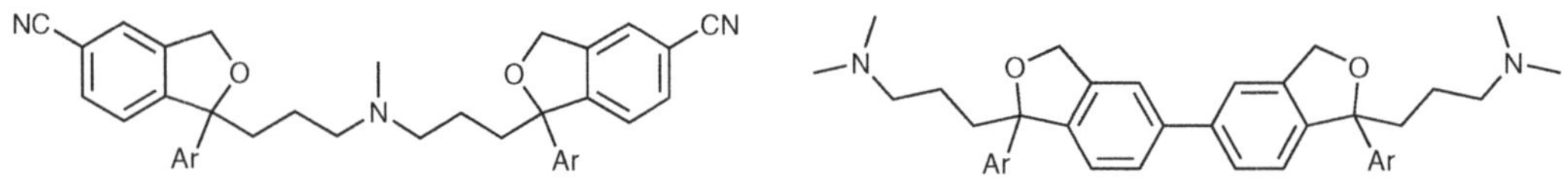

FIGURE 4.10 HIGH MOLECULAR WEIGHT IMPURITIES FORMED DURING THE ROSENMUND-VON BRAUN REACTION

4.8.2 Two Grignard route using 5-cyanophthalide as starting material

The use of 5-cyanophthalide was central to several new patents (Scheme 4.28). Since the nitrile group is in place early at the outset, the issues seen above can be avoided. This strategy represents the shortest sequence from a phthalide to Citalopram. This route although the subject of a patent in 1984, gave rise to further patent activity in the aftermath of the two-Grignard routes with other 5-substituted- and 5-bromophthalides (2002 onwards).

Some reports suggested that reaction with the propylamine Grignard was slow but other data contradicted this. Different C-5 substituents can possibly stabilize the hemiketal salt to different extents thus explaining this apparently conflicting data. It might be possible that the stability of the hemiketal salt is increased by the presence of the nitrile aromatic group, turning the second reaction slower, compared to the bromo analogue.

Formation of a C-5-acyl impurity results from attack at the nitrile group (see Scheme 4.22). Methylene chloride as a co-solvent, 2.2 molar equivalent of the Grignard and addition at –5 °C to 0 °C (OPRD2013) reduces the C-5-acyl impurity (controlled in US and EP pharmacopeias).

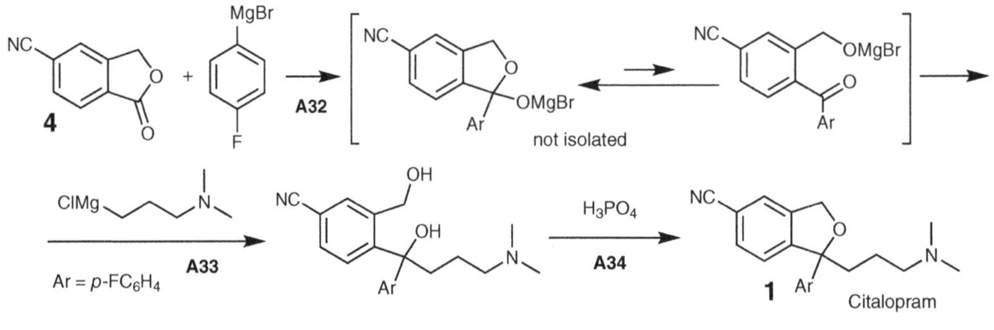

SCHEME 4.28 CITALOPRAM FROM 5-CYANOPHTHALIDE BY THE TWO GRIGNARD ROUTE

Yields of isolated cyano diol as the hydrobromide salt were increased from 44% (Lundbeck1984) to higher than 80% (DrReddys2008, Adorkem2005a, Ind2006, Jiangsu2014). Again, isolation and purification of the diol increases the the overall yield of Citalopram by about 10% (OPRD2013). The addition of an auxiliary metal salt is claimed to improve the activity and selectivity of the Grignard reagents (ZhejiangHP2015, yield of diol up to 83%).

In an interesting approach, Adorkem reacted 5-cyanophthalide with a mixture of the two Grignards (1.8 to 2.0 molar equivalents of 4-fluorophenyl magnesium halide and from 1.1 to 1.2 moles of 3-dimethylaminopropyl magnesium halide, Adorkem2003b, 2005a). Cyclization in acid was reported to produce Citalopram in an accumulated yield of 89%.

These improvements indicate that this route clearly is one of the most viable to Citalopram industrially.

Ring closure uses S_N1 conditions, heating for several hours with phosphoric, sulfuric, hydrochloric or hydrobromic acid at 70–85 °C. The carboxamide impurity has to be controlled, but does form along with other impurities, totalling 1.5% (Scheme 4.29, OPRD2013).

The alternative S_N2 ring-closures (**A35**+**A36**), carried out under anhydrous conditions, to activate the primary alcohol require more care and are more difficult to realize at an industrial scale. Azeotropic distillation with toluene is one technique to accomplish completely anhydrous conditions. Typical activating groups used were mesylate or tosylate and the ring closure proceeds at room temperature.

4.8.3 Coupling the hydrofuran ring formation with the second Grignard reaction

A slightly different approach, used by Orion and Lundbeck, (Scheme 4.30 and lower part of scheme A), employs a ketone where a leaving group at the benzylic carbon atom is in place. After reaction with the aminoalkyl Grignard cyclization occurs spontaneously yielding the corresponding phthalanes.

4.8.4 A two Grignard route not using 3-dimethylamino propyl Grignard

On one occasion, an alternative Grignard reagent to 3-(dimethylamino)propyl)magnesium chloride was used to introduce the three carbon atom chain (Scheme 4.31, Sepracor2004), thus giving a phthalane with an aldehyde at the terminal sidechain. Introduction of the nitrogen atom follows a typical reductive amination reaction hence, making this option a viable approach to produce Citalopram from an aldehyde intermediate.

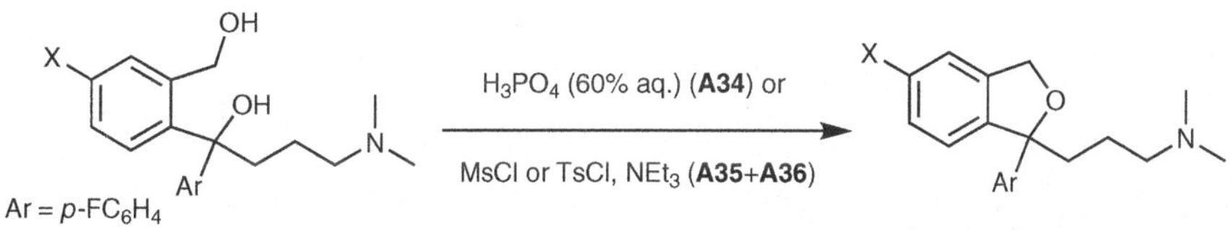

Scheme 4.29 Typical reaction conditions in the diol ring-closing reaction

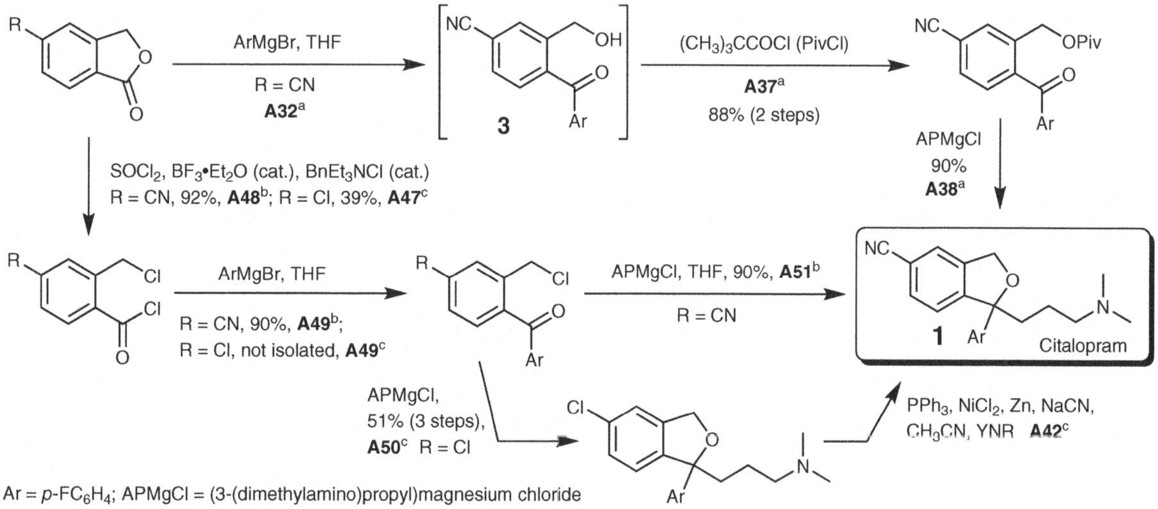

Ar = p-FC$_6$H$_4$; APMgCl = (3-(dimethylamino)propyl)magnesium chloride

a) Lundbeck1999g, 2001c; b) Orion2001 (X = CN); c) Orion2002 (X = Cl)

Scheme 4.30 Direct formation of Citalopram by a cascade nucleophilic attack in a (3-(dimethylamino)propyl)magnesium chloride reagent-cyclization process

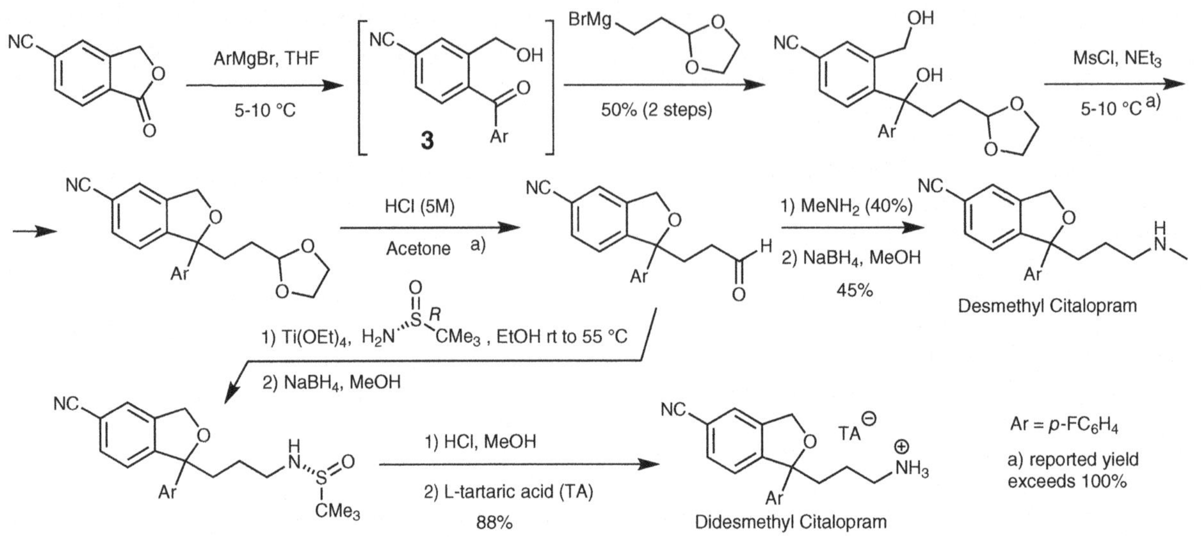

Scheme 4.31 Using (2-(1,3-dioxolan-2-yl)ethyl)magnesium bromide to introduce the three carbon side chain

This methodology was used by Sepracor to produce desmethyl analogues of Citalopram (desmethyl and didesmethyl) which can then be converted to the dimethyl by *N*-methylation. Other processes to produce the desmethyl analogues are included in scheme E (page 110).

4.8.5 Phthalide synthesis

Most 5-substituted phthalide starting materials were already known at the time Citalopram was discovered (R = CN, Cl, Br, I, JCS1931). Stimulated by the requirements to have cheap and readily available starting materials for the production of Citalopram concomitant with the developments of the "two Grignard" routes, new processes for their synthesis were developed and patented, see scheme B.

Terephthalic acid, produced by oxidation of *p*-xylene, proved itself to be the most important starting material. 5-Carboxyphthalide can be produced from terephthalic acid in one step in moderate to good yields (**B13**, Scheme 4.32) and is the preferred intermediate for most of the 5-carbon substituted phthalides, including 5-cyanophthalide (Scheme B). The preferred intermediate is 5-cyanophthalide and can be produced in one step from 5-carboxyphthalide as can the other 5-substituted phthalides.

Terephthalic acid with formaldehyde (or one of its forms) can lead to over-reaction and is problematic, but in oleum at high temperature the reaction is reported to give high-quality product on a 200 kg scale (**B13**, Scheme 4.32). Of course, there are safety concerns when carrying out this highly exothermic reaction and handling the viscous reaction mixture is challenging. Ester derivatives were similarly produced (scheme B, **B13**, Lundbeck2003).

Catalytic electrochemical reduction of trimellitic anhydride (**B6**, BASF1976) was also reported but this

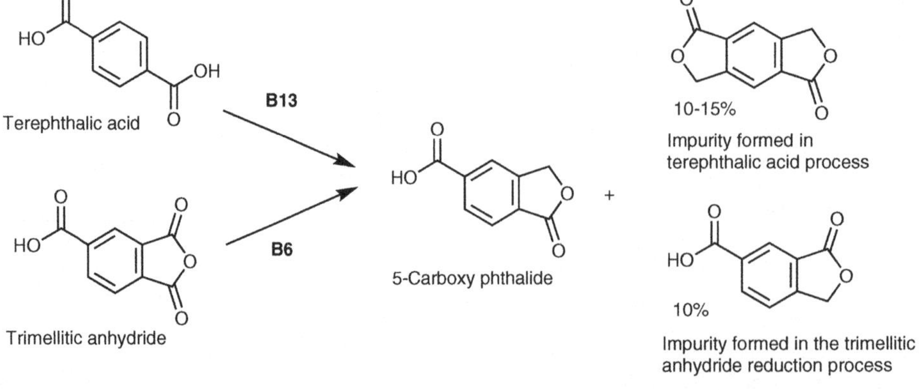

Scheme 4.32 Main synthetic reactions to produce 5-carboxyphthalide

procedure is not altogether suitable for industrial scale production. In addition, a mixture of the 5- and 6-carboxyphthalides is obtained. 5-Cyanophthalide (**4**), an important intermediate for the synthesis of Citalopram, is frequently produced from other phthalides containing a C5-carbon group, but can also be produced from 5-halo analogues using the halo-cyanide exchange (**B28**).

To produce 5-halophthalides an EAS on phthalimide or phthalic anhydride can be conveniently employed to introduce the C5 substituent (Scheme 4.33). Unwanted regioisomers are normally formed and specific reduction of the correct carbonyl groups is an issue and generally gives approximately 1:1 mixtures, for example in **B27**. In some instances (**B34**) a diazonium salt intermediate has been used to get to a chloro or iodo derivative.

A different approach, based on a low temperature *ortho*-lithiation of 4-chloro-*N*,*N*-diisopropylbenzamide and formylation with DMF, followed by reduction of the aldehyde was proposed (**B36**, OPRD2010). Ring closure (**B37**), led to 5-chlorophthalide in high overall yield.

Of the routes in scheme A, it can be stated that the cyano and bromo derivatives are preferred starting materials for the two-Grignard approach.

4.8.6 The non-Grignard C3 nucleophilic route

Other nucleophiles have been used to introduce the three carbon side-chain, for example an acetylide anion (Scheme 4.34, Sumitomo2004b) but all of these alternatives require subsequent manipulation after connecting the side chain.

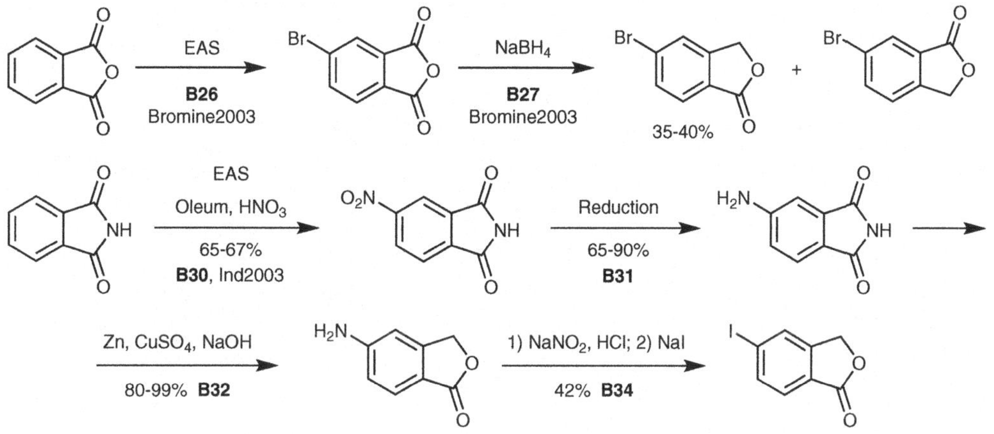

Scheme 4.33 Routes to 5-halophthalides

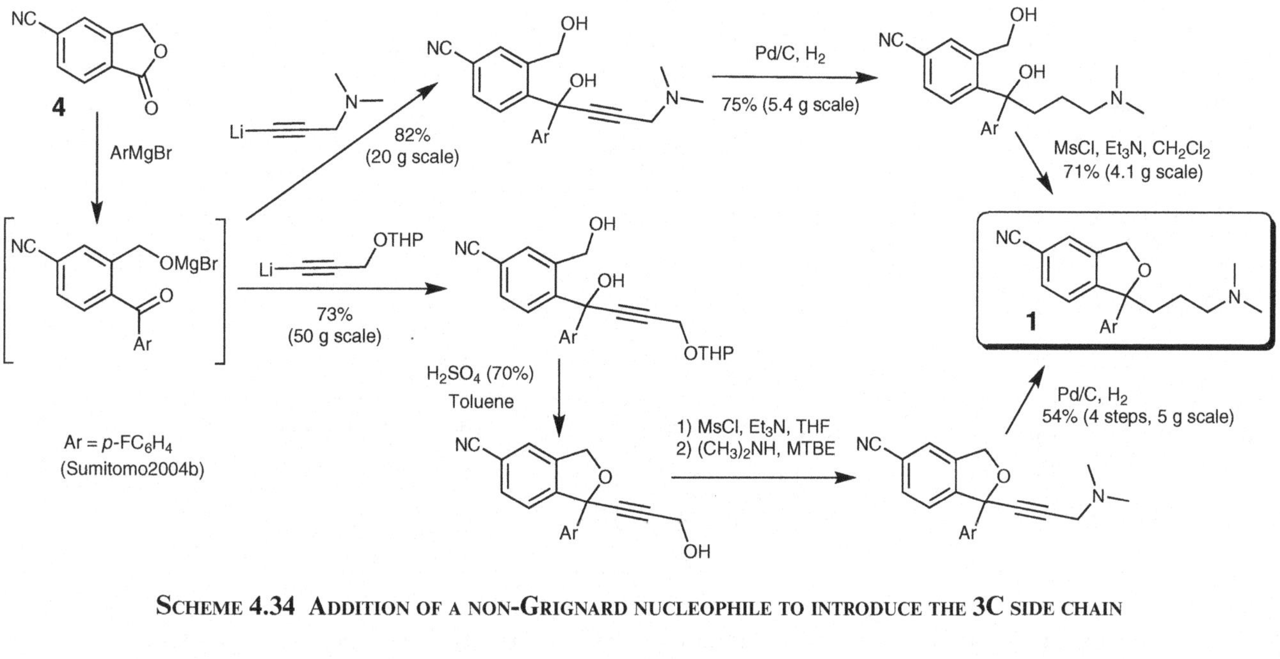

Scheme 4.34 Addition of a non-Grignard nucleophile to introduce the 3C side chain

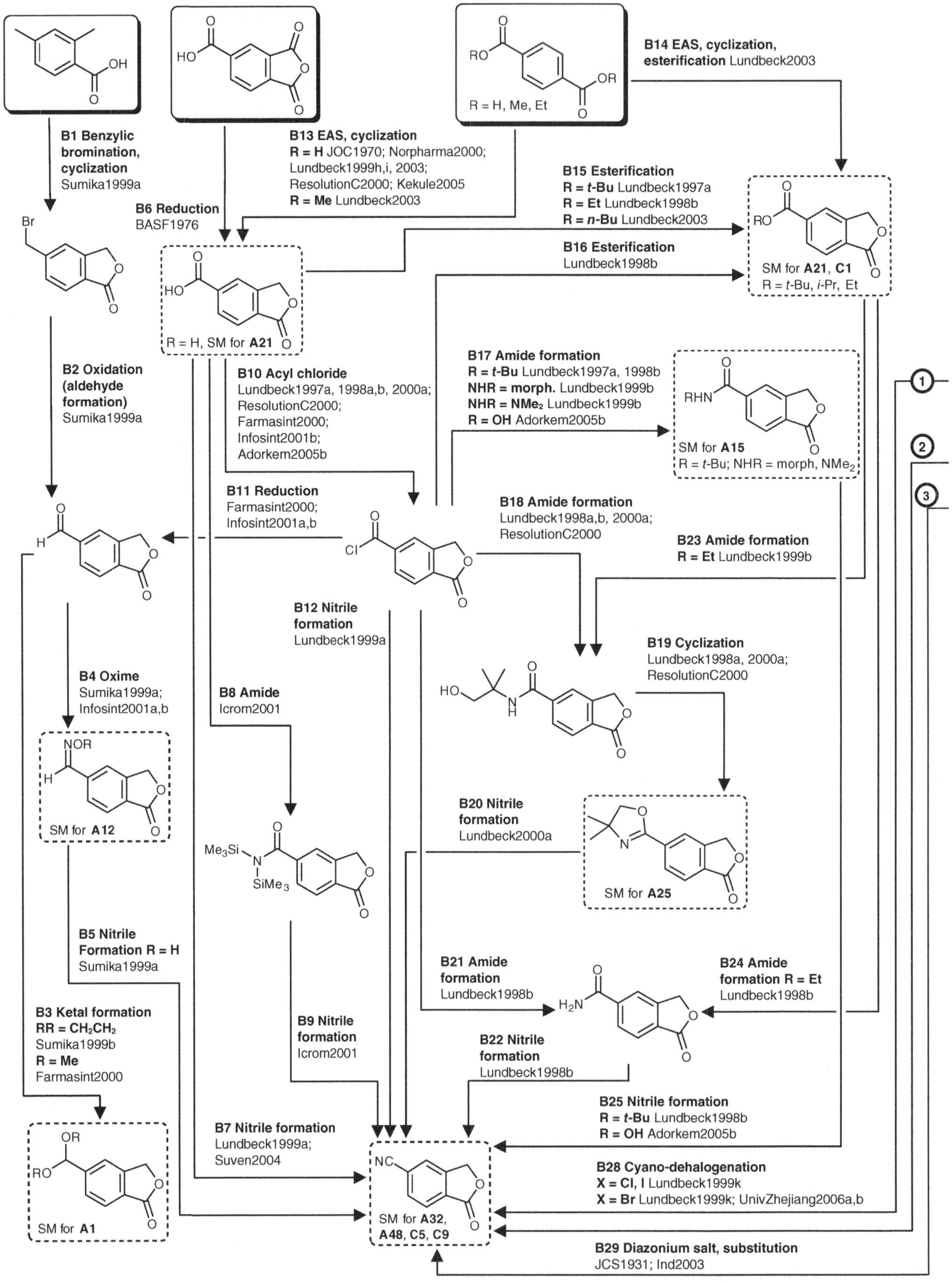

Scheme B Synthesis of phthalides (starting materials for processes in schemes A and C) (for reaction conditions see page 130)

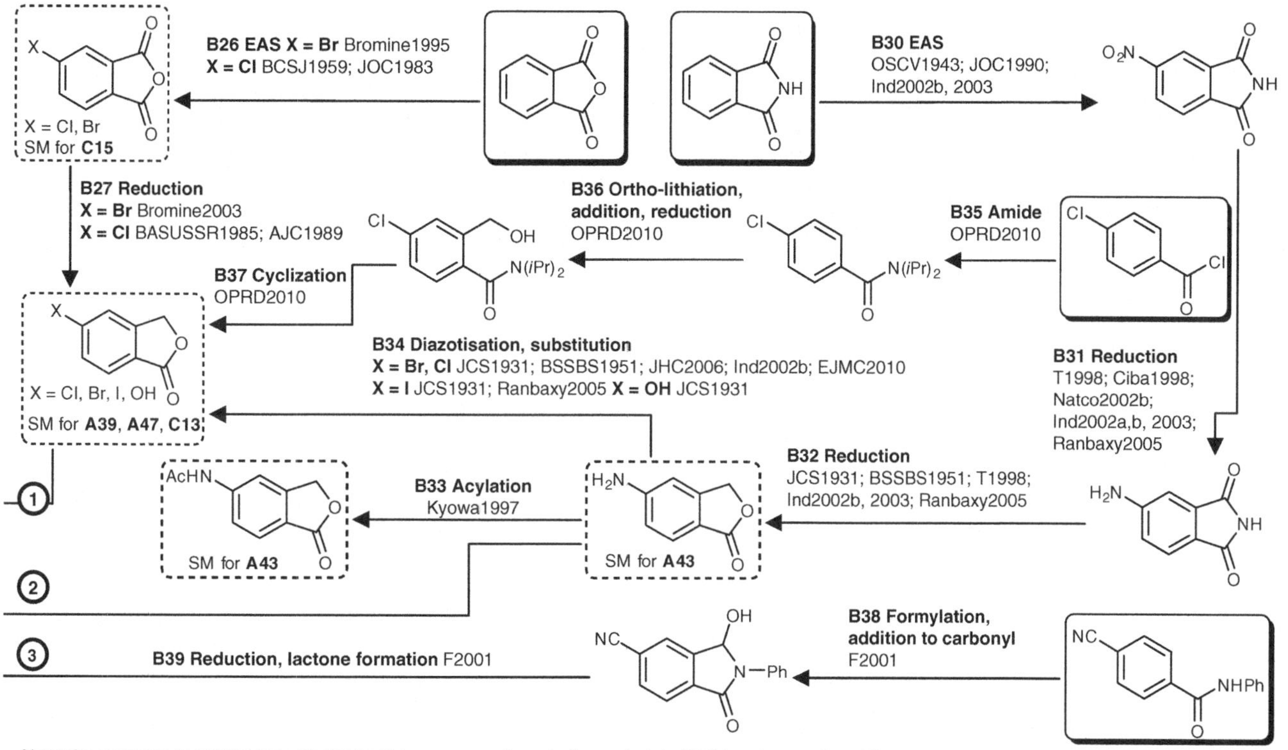

Note: the compounds highlighted with dashed boxes are used as starting materials (SM) in schemes A and C.

4.9 THE PHTHALANE ROUTE – ALKYLATION OF A PHTHALANE WITH AN ELECTROPHILIC SIDE CHAIN

Alkylation of the cyanophthalane was reinvestigated by several groups around 2000 (Scheme 4.35). This route depends on the unique stabilization of the anionic nucleophile offered by the 5-cyano substituent. A polar aprotic solvent in combination with a strong base, for example, DMSO and dimsyl anion (obtained by reacting sodium hydride or another strong base in DMSO), is used to form the anion. Note the 5-bromo derivatives would be expected to give benzynes under these conditions.

See schemes C and D for the production of 5-cyanophthalane (**2**). The most frequent approaches are to react the phthalide (5-cyano or 5 bromo) with the aryl Grignard, followed by reduction of the ketone and cyclizing the diol to the phthalane. When the starting material is 5-bromophthalide, an exchange to nitrile is required before the alkylation step (**C18**, Scheme 4.36).

The processes to produce Citalopram based on the alkylation of 5-cyanophthalane with a 3-chloropropan-1-amine are summarized in scheme E.

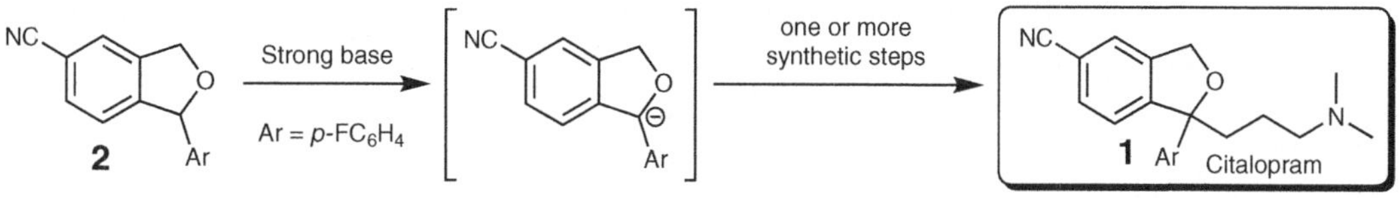

SCHEME 4.35 CITALOPRAM BY ALKYLATION OF 5-CYANOPHTHALANE

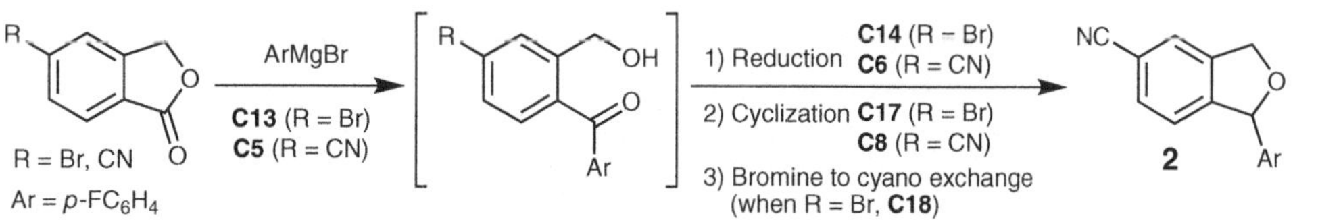

SCHEME 4.36 MAIN ROUTES TO 5-CYANOPHTHALANE

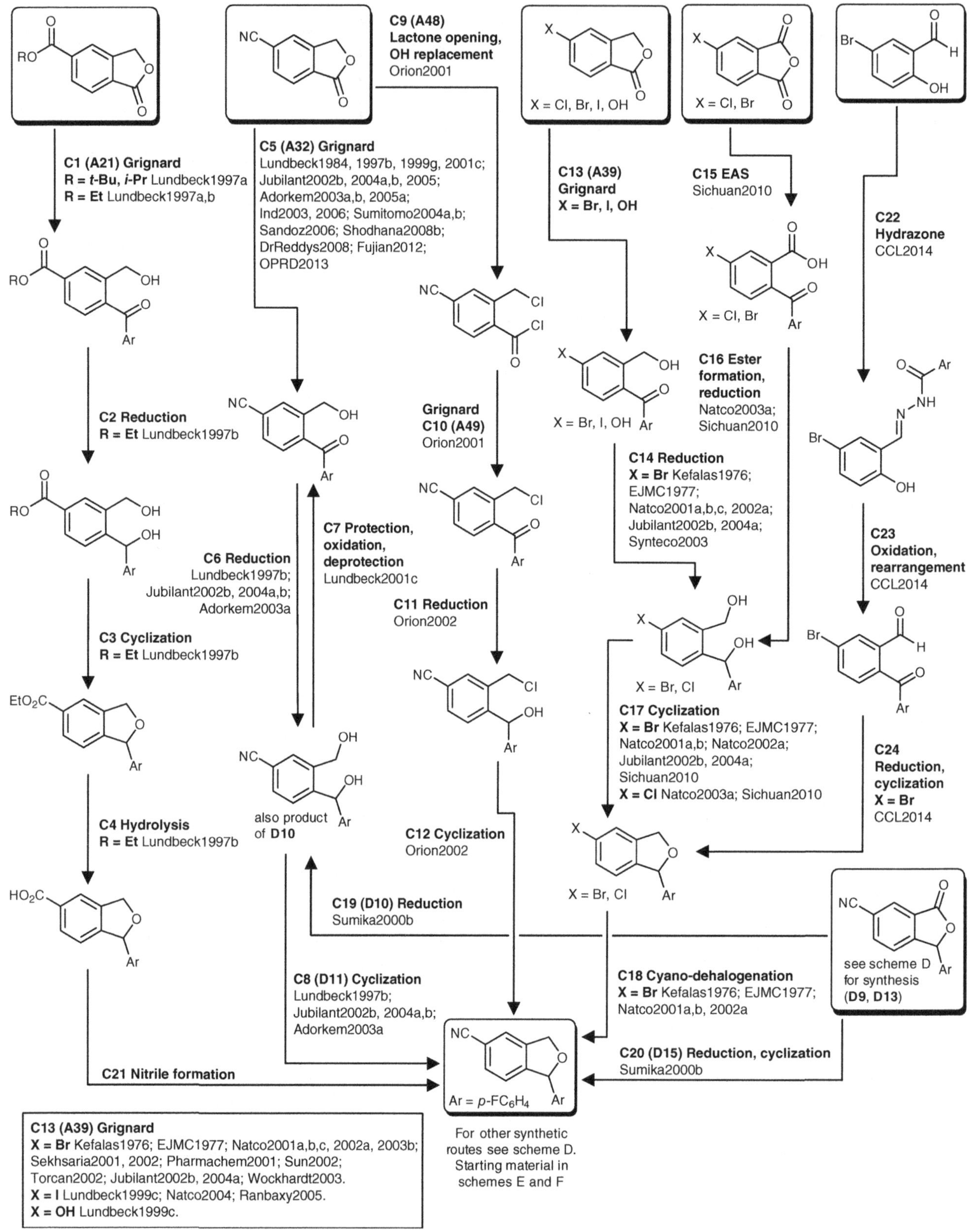

Scheme C Synthesis of 5-cyanophthalane (part 1)
(for reaction conditions see page 131)

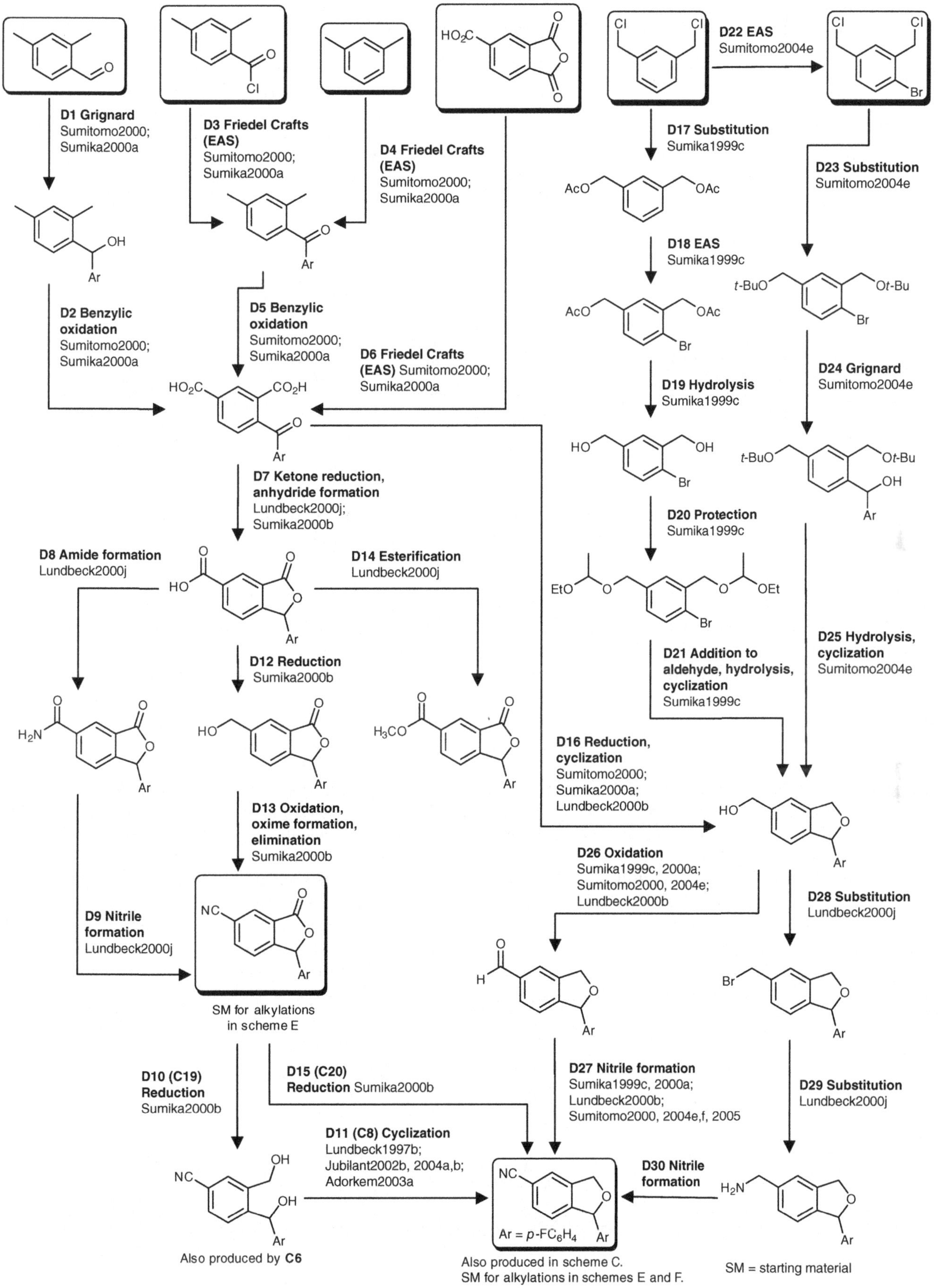

Scheme D Synthesis of 5-cyanophthalane (part 2)
(for reaction conditions see page 132)

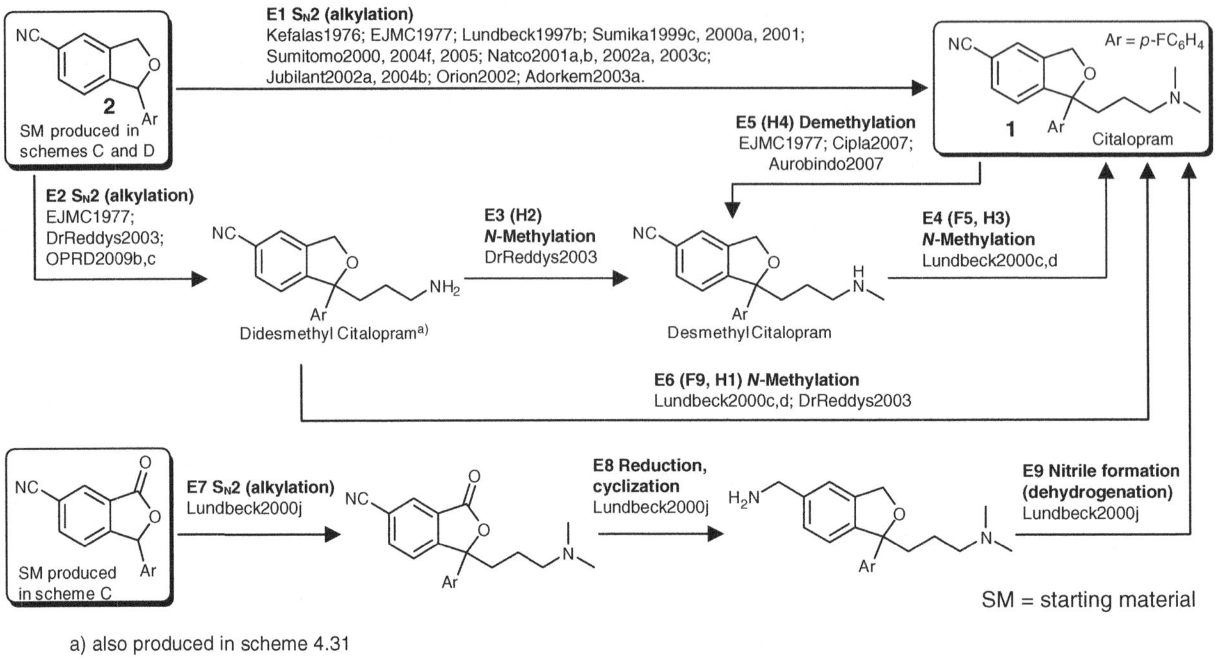

Scheme E Synthesis of Citalopram by alkylation with a 3-chloropropan-1-amine (for reaction conditions see page 133)

3-chloro-*N*,*N*-dimethylpropyl-1-amine is the simplest alkylating agent to employ, producing Citalopram directly (**E1**). The alkylating agent is produced by basification of its hydrochloric salt. Residual water is removed by azeotropic distillation or treatment with molecular sieves. Since the neutral compound polymerizes easily the reagent is used immediately.

The initially obtained yield of 43% was increased to 70-87% after development work by several groups. DMSO, DMF, 1,2-dimethoxyethane and 1,3-dimethyl-2-imidazolidinone (DMI) were used as solvents. Recovery of these solvents can be difficult, which may result in some negative impact on costs. Sodium hydride, potassium *t*-butoxide, and lithium diisopropylamide (LDA) are the most used bases.

Once more, factors such as the potential explosive decomposition of the dimsyl anion at high temperatures and the reactivity of sodium hydride with water give rise to safety concerns.

Lundbeck also published a similar alkylation process with a 3-oxophthalane analogue although the critical alkylation step (**E7**) was not exemplified. This route required further downstream processing (reduction and oxidation, **E8**, **E9**).

Use of 3-chloropropan-1-amine as alkylating agent was also described (**E2**). As expected, polymerization of the C-3 alkylating agent gives rise to side products which must controlled and removed from the final API. Di-*N*-methylation of the intermediate produces Citalopram.

The other electrophiles reported by Lundbeck, all requiring further reaction steps, were focused on the reduction of polymerization of the electrophile and are shown in Figure 4.11. Their use in the formation of Citalopram is shown in scheme F.

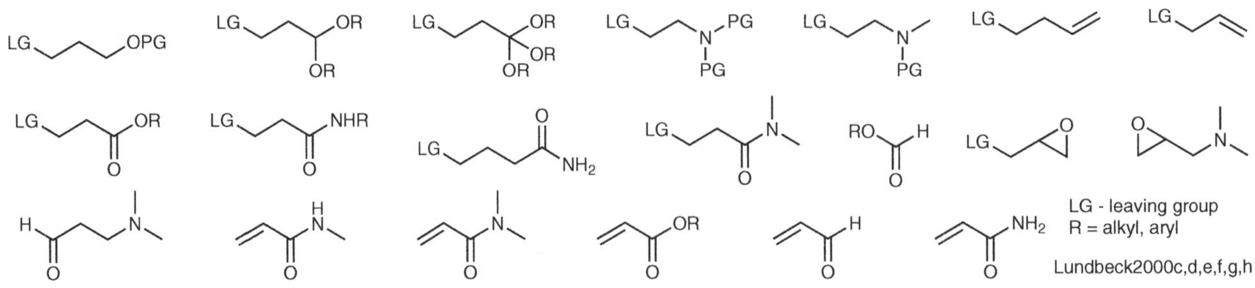

Figure 4.11 Generic structure of some electrophiles claimed for alkylation of 2

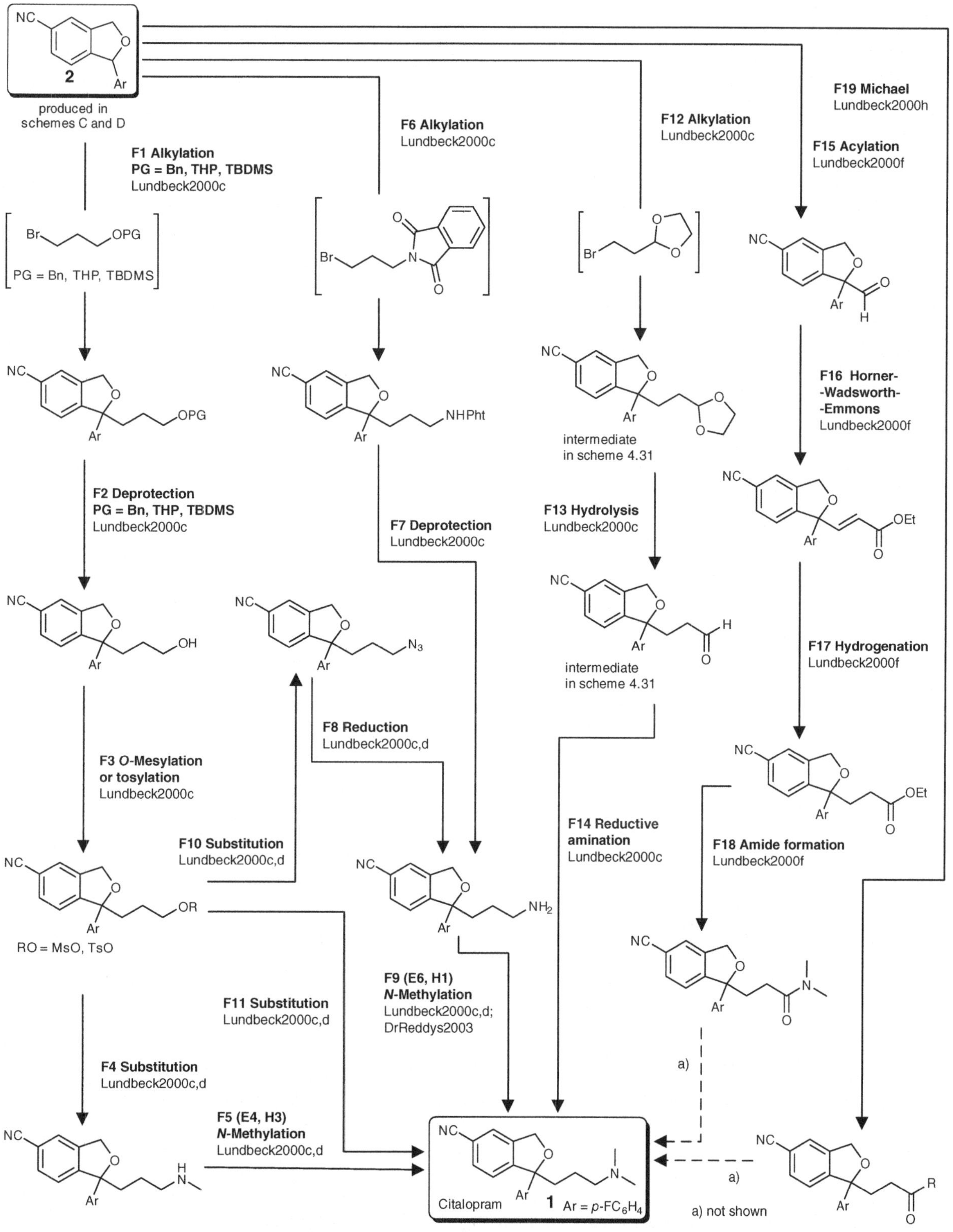

SCHEME F CITALOPRAM BY ALKYLATION OF 5-CYANOPHTHALANE NOT USING CHLOROPROPYL AMINE AS ALKYLATING AGENT (FOR REACTION CONDITIONS SEE PAGE 133)

4.10 OTHER SYNTHETIC ROUTES TO CITALOPRAM

Lundbeck patented two sequences involving cycloaddition reactions (Scheme 4.37, Lundbeck1999j) although little detail was given and no experimental examples presented. However, it seems highly unlikely that any of these routes were put into practice.

A transition-metal-free, cyanide-free, one-step direct synthesis of aryl nitriles from methyl(hetero)arenes under mild and catalytic conditions was adapted to a gram-scale Citalopram synthesis (JACS2016, Scheme 4.38). After oxidation of the remaining benzylic methyl, direct metalation of the ring with extremely reactive *t*-BuLi produces the appropriate nucleophile to couple with a ketone and allowing the access to the cyano diol intermediate.

A copper-catalysed cyanoalkylative cycloetherification of an alkene was applied as a method to obtain the basic carbon skeleton of Citalopram. Heating a solution of the substituted (2-vinylphenyl)methanol in MeCN in the presence of a catalytic amount of copper(II) tetrafluoroborate hydrate, bathophenanthroline, K_3PO_4, BnOH and $(t\text{-BuO})_2$ afforded the phthalane with the required carbon atom chain (CC2016, Scheme 4.39). Reduction, *N*-methylation and bromine replacement produced Citalopram. In another synthetic proposal, a similar cyclization of unactivated alkenes was promoted by a palladium catalyst (CS2013, Scheme 4.39).

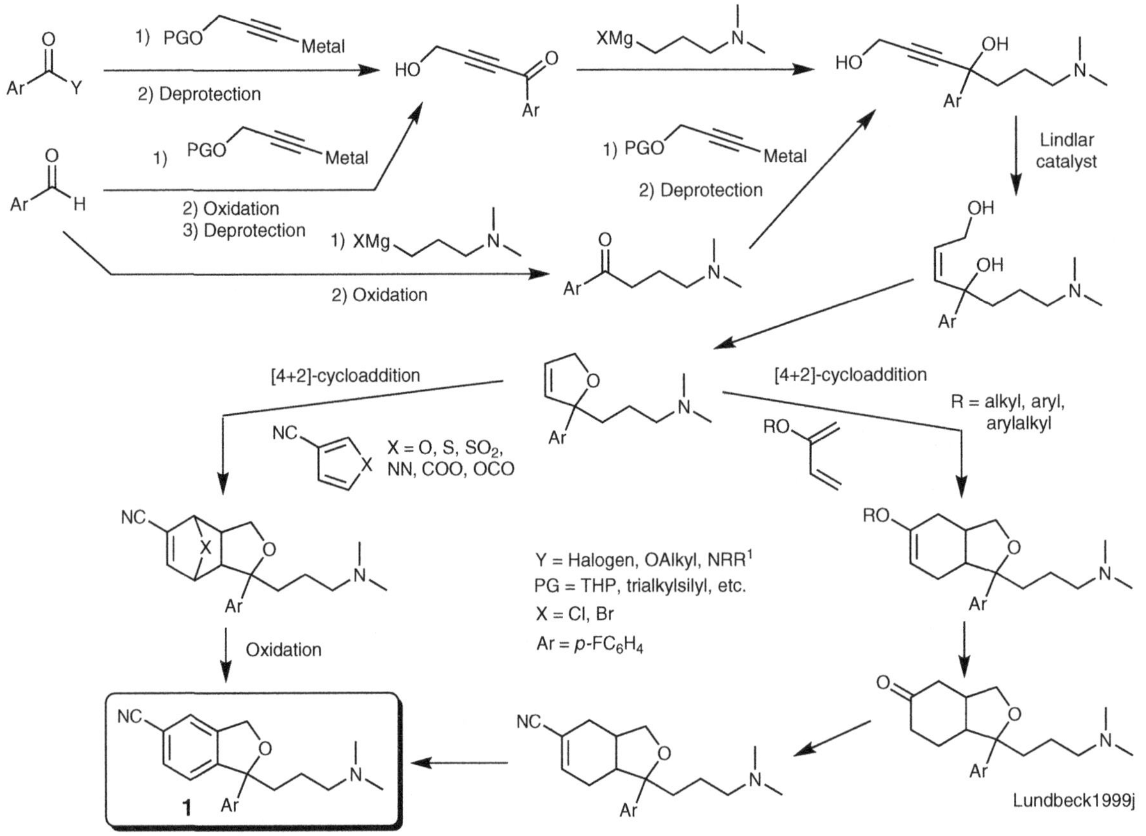

SCHEME 4.37 A SYNTHETIC PROPOSAL FOR CITALOPRAM BASED ON A DIELS-ALDER CYCLIZATION

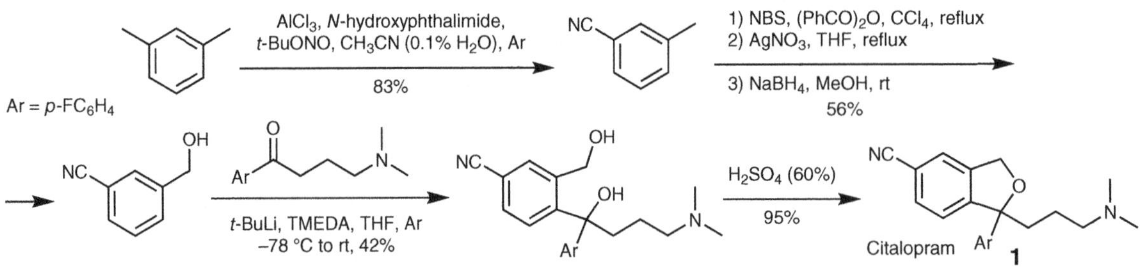

SCHEME 4.38 AN EXPEDIENT SYNTHESIS OF CITALOPRAM FROM *m*-XYLENE

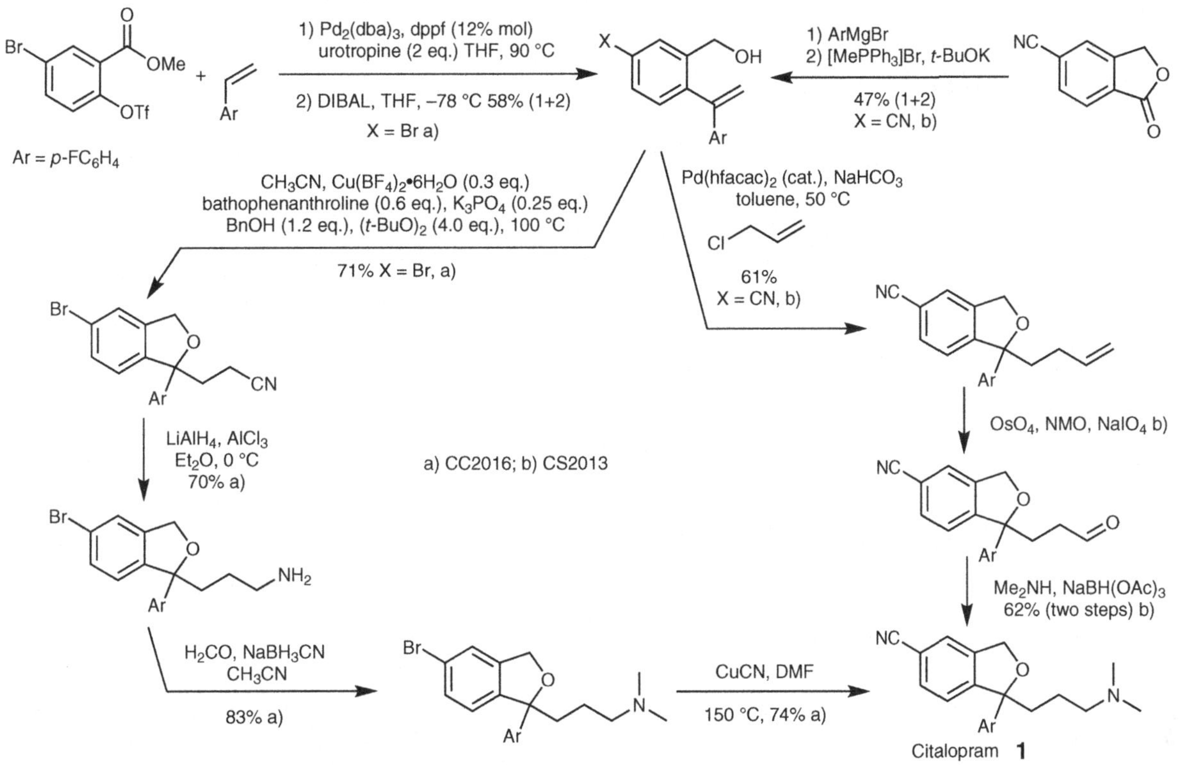

Scheme 4.39 SYNTHESIS OF CITALOPRAM BY CYCLYZATION OF UNACTIVATED ALKENES

4.11 STRATEGY COMPARISON AND CONCLUSIONS FOR THE SYNTHESIS OF CITALOPRAM

The two main strategies to produce Citalopram follow the "two Grignard" or the cyanophthalane alkylation routes. The most significant routes for each strategy are summarized in Scheme 4.40 and presumably can be used to produce Citalopram on a large scale.

Bromo and cyanophthalides are conveniently prepared from phthalic anhydride or terephthalic acid respectively (Scheme 4.41). The use of terephthalic acid (or its ester) is a better approach as there are no issues with regioselectivity and no oxidation or reduction are required. This starting material is readily available.

Phthalic anhydride has drawbacks because of the regio-selectivity of the bromination and the carbonyl reduction. The yield to pure 5-bromo compound is only 35-40%. Halogen to cyano exchange requires toxic copper cyanide. Summing up, the 5-cyanophthalide synthesis from terephthalic acid or ester look most favourable. Other less attractive routes are summarized in scheme B.

No one route is trouble-free. The reported issues in the last step, bromo to cyano exchange, as discussed previously make route D less appealing (Scheme 4.40). From the discussion above, it can be concluded that the most efficient method to produce Citalopram is that shown in Scheme 4.42. Route B (Scheme 4.40) may provide an alternative process if the safety issues related to the alkylation steps can be addressed.

4.12 THE EVOLUTION OF ESCITALOPRAM SYNTHESIS

4.12.1 The chiral switch

By the time Escitalopram was launched, the main synthetic routes to Citalopram had already been well investigated and the subject of a myriad of patent applications.

As Citalopram is still an important marketed drug, patent activity continues up to the present. As expected, the chiral switch encouraged investigations into new synthetic methods.

4.12.2 The first process for Escitalopram

Initial attempts to resolve the two Citalopram isomers were unsuccessful therefore, Lundbeck synthesised both enantiomers in a pure form by two methods (Lundbeck1988). Attention focused on the earlier cyano diol intermediate and resulted in both resolution of the diol and a diastereomeric chromatographic separation (see Scheme 4.24).

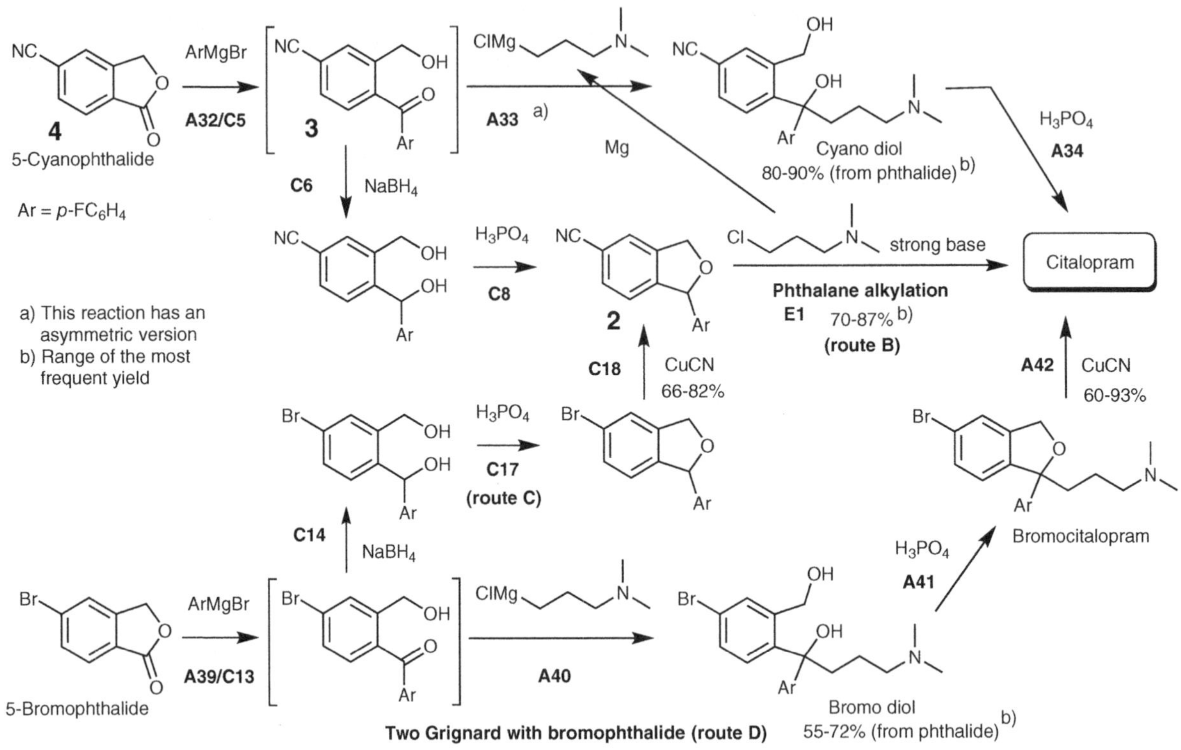

Scheme 4.40 Main routes to Citalopram

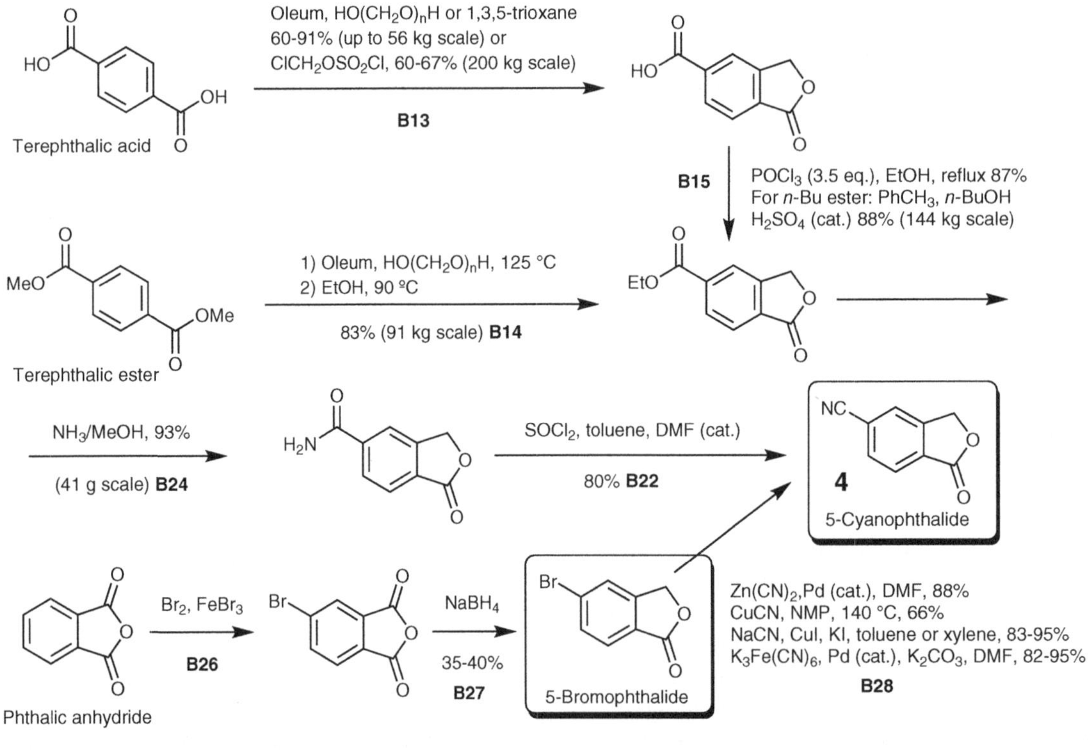

Scheme 4.41 Most important routes to produce 5-cyano and 5-bromophthalide

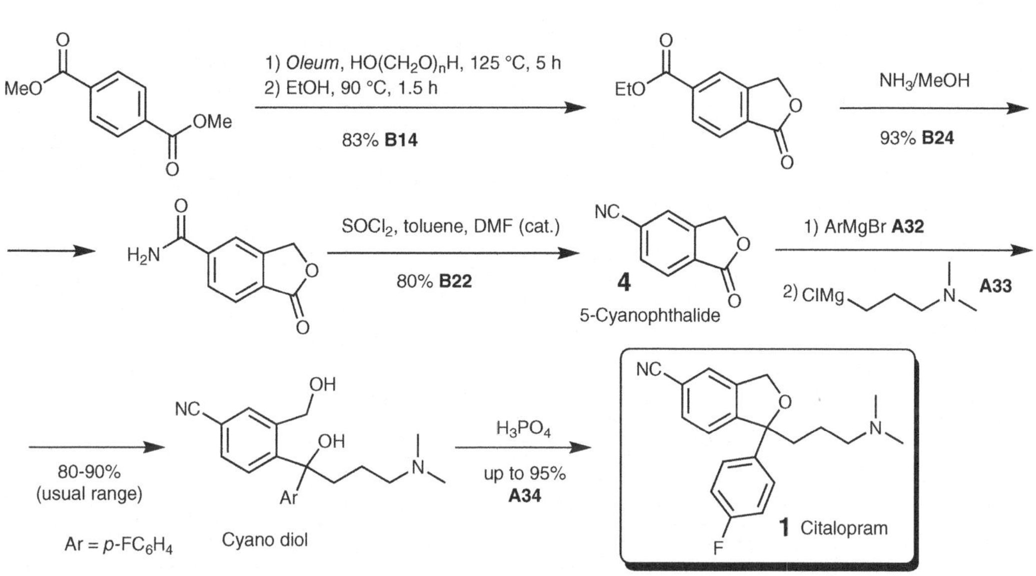

Scheme 4.42 The most efficient route to Citalopram

4.12.3 Diol as a key intermediate in Escitalopram synthesis

Being key to the synthesis of Escitalopram, investigations concentrated on the separation of a diol into pure enantiomers pursuing four main options (synthetic processes summarized in scheme G), Following isolation of enantiomerically pure diol it is cyclized by an intramolecular S_N2 to preserve the configuration of the chiral centre.

These four options to obtain an enantiomerically pure intermediate are:

- Direct **chromatographic separation** using chiral stationary phases (**G7**, **G11**, Lundbeck2001d); Lundbeck used Simulated Moving Bed (SMB) chromatography with a **chiral stationary phase** to separate chiral cyano and bromo diols (reactions **G7** and **G11**, yield for bromo derivative 50% signifying 100% isomer recovery at a 500 g scale). A similar solution was applied successfully to the separation of bromo Citalopram (**G20**). Low output, cost and the limited availability of chiral stationary phase are drawbacks for industrial scale use.
- **Enzymatic resolution**, either using a lipase to explore the different rates of hydrolysis of the monoacetate diol enantiomers or an esterase to acylate selectively one of the diol enantiomers (Scheme 4.43). In the first approach the primary benzylic hydroxyl group

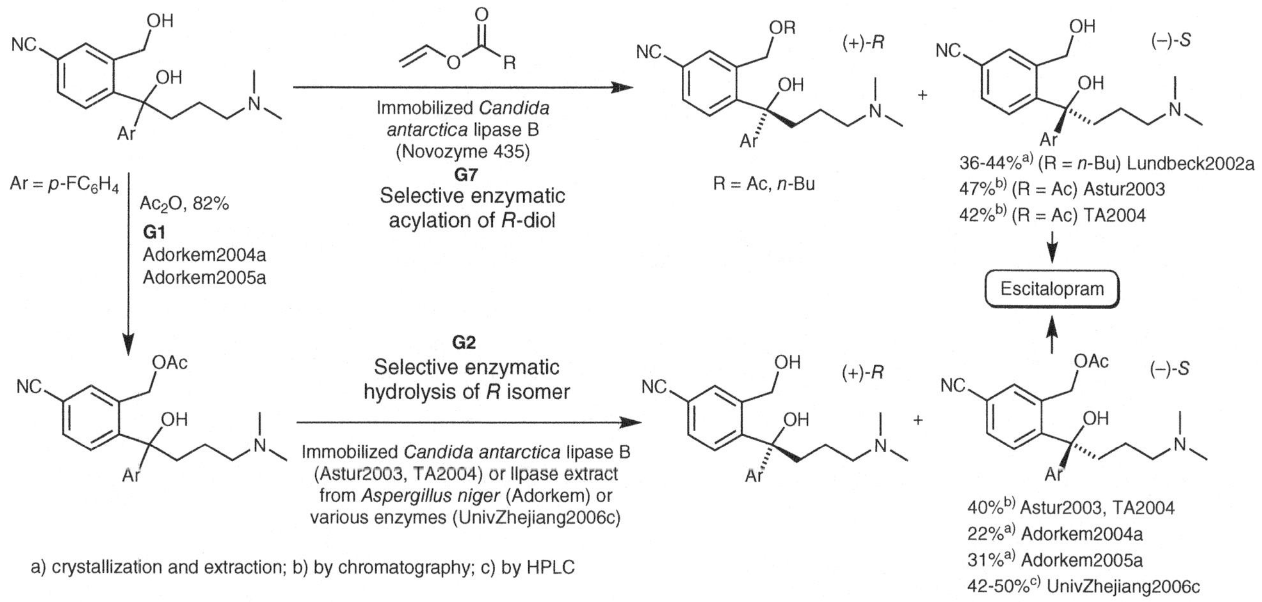

Scheme 4.43 Enzymatic resolutions of Escitalopram intermediates

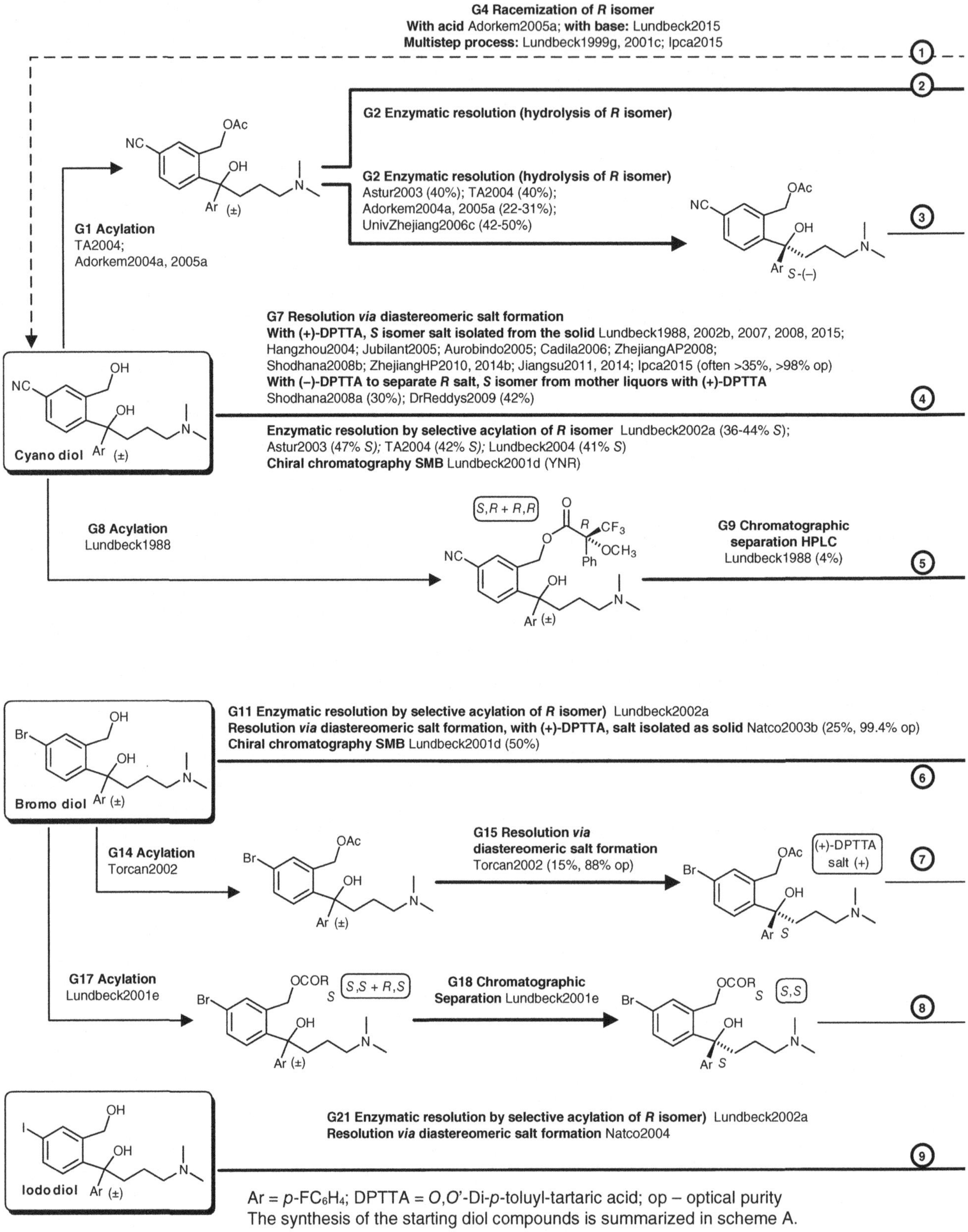

Scheme G Diol as key intermediate in Escitalopram synthesis (for reaction conditions see page 134)

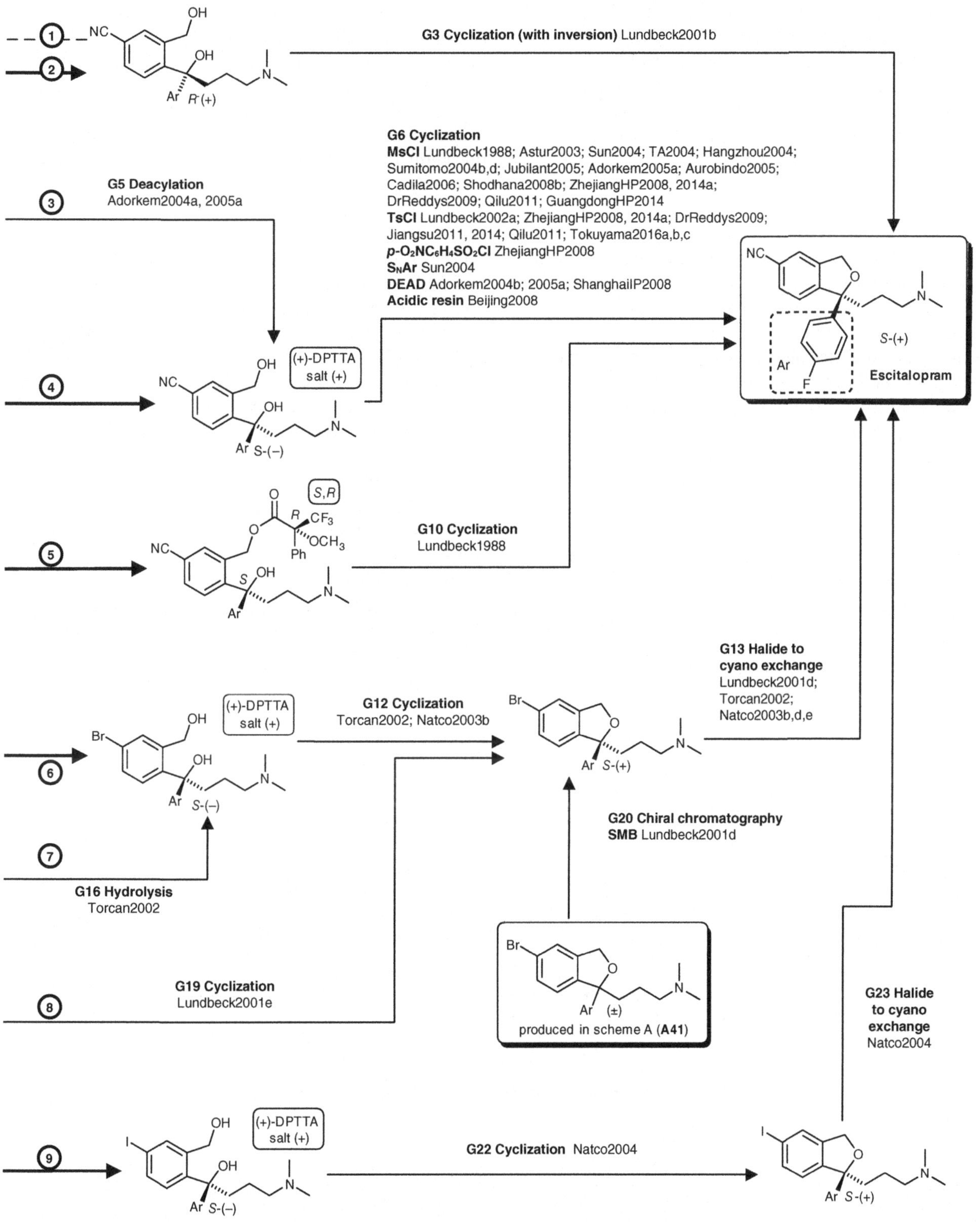
G3 Cyclization (with inversion) Lundbeck2001b
R-(+)
G6 Cyclization
MsCl Lundbeck1988; Astur2003; Sun2004; TA2004; Hangzhou2004; Sumitomo2004b,d; Jubilant2005; Adorkem2005a; Aurobindo2005; Cadila2006; Shodhana2008b; ZhejiangHP2008, 2014a; DrReddys2009; Qilu2011; GuangdongHP2014
TsCl Lundbeck2002a; ZhejiangHP2008, 2014a; DrReddys2009; Jiangsu2011, 2014; Qilu2011; Tokuyama2016a,b,c
p-O2NC6H4SO2Cl ZhejiangHP2008
SNAr Sun2004
DEAD Adorkem2004b; 2005a; ShanghaiIP2008
Acidic resin Beijing2008
G5 Deacylation
Adorkem2004a, 2005a
(+)-DPTTA salt (+)
S-(−)
S-(+)
Escitalopram
S,R
G10 Cyclization
Lundbeck1988
G13 Halide to cyano exchange
Lundbeck2001d; Torcan2002; Natco2003b,d,e
G12 Cyclization
Torcan2002; Natco2003b
G20 Chiral chromatography
SMB Lundbeck2001d
G16 Hydrolysis
Torcan2002
G19 Cyclization
Lundbeck2001e
produced in scheme A (A41)
G23 Halide to cyano exchange
Natco2004
G22 Cyclization Natco2004

of the *R* diol is preferentially acylated (for cyano diol, **G7**; for bromo diol **G11**; for iodo diol **G21**). In the second case, the primary hydroxyl group of the diol was acylated by common chemical methods to give a racemic mixture that was subjected to an enzymatic hydrolysis of the *R* acyl diol (Scheme 4.43, **G2**). To obtain Escitalopram, the *S*-acyl diol is converted to the diol (**G5**) and cyclized (**G6**).
A more recent process allows an easy separation of the *S*-diol from the *O*-acylated *R*-diol by reacting the mixture with succinic anhydride so producing an *O*-succinyl *S*-diol intermediate that can be crystallized out of the solution (Scheme 4.44, Lundbeck2004). Escitalopram is obtained by driving cyclization through the use of sodium hydride.

- Formation of **diastereomeric compounds** by reaction of the primary benzylic hydroxyl with chirally pure acyl chlorides, followed by chromatographic separation (**G8/G9**, **G17/G18**);
- Classic resolution of diol *via* **diastereomeric salt formation** (with DPTTA, **G7**, **G11**, **G21**; for *O*-acetyl derivative **G15**).

The most frequently reported process relies on the intervention of diastereomeric salts with (+)-DPTTA, which gives moderate to high separation efficiencies, the salt of the desired *S* isomer being isolated as a solid by filtration (Scheme 4.45). These separations were mainly applied to the cyano diol indicating, perhaps, that this is the most favourable intermediate for this methodology. In a more recent patent from Lundbeck (2015) a 41% yield of the *S*-isomer was obtained by a controlled crystallization of the hemi (+)-DPTTA salt from a *n*-propanol solution containing a small amount of acetic acid (99.3% *S*).

The diol or its salt, when purified by crystallization before resolution contributes greatly the success of the synthesis (Hangzhou2004, Shodhana2008b, hydrobromide salt Cadila2006, oxalic acid salt Aurobindo2005).

Crystallization and/or reslurry of the DPTT-cyano diol salt can also be carried out to improve chemical purity and enantiomeric excess (for example Jubilant2005, Lundbeck2007, Tokuyama2016a,b,c). A pure and stable new crystal form of the (*S*)-cyanodiol/(+)-DPTTA salt and a process for its formation has recently been claimed (GuangdongHP2014).

Improvement of chiral purity of the diol is possible from enantiomerically enriched mixtures. Under controlled conditions, racemic diol crystalizes from the solution and almost pure *S* isomer remains in solution (Lundbeck2002b, Scheme 4.46).

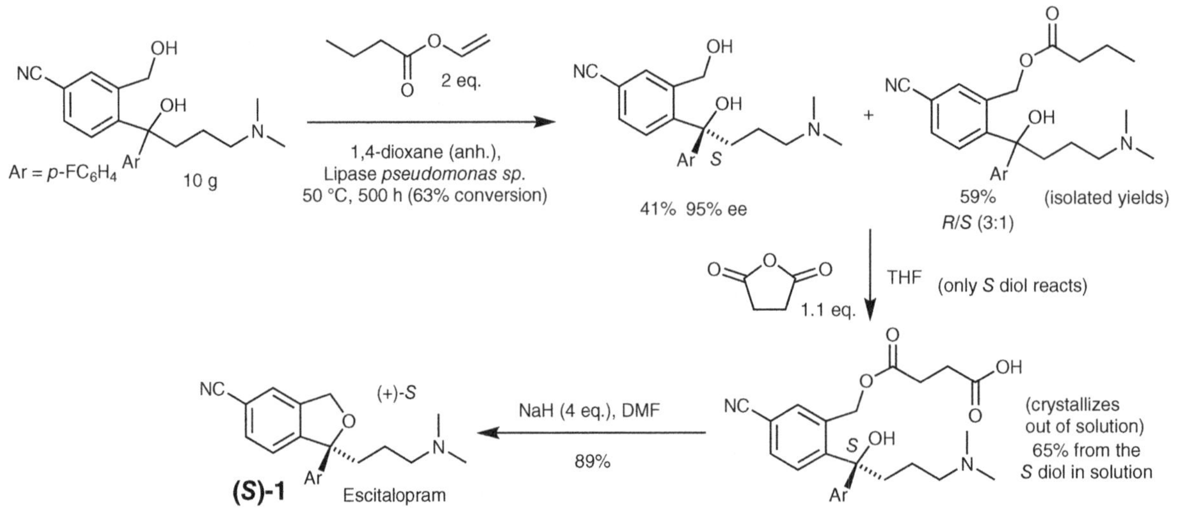

Scheme 4.44 Enzymatic resolution of 5-cyano diol

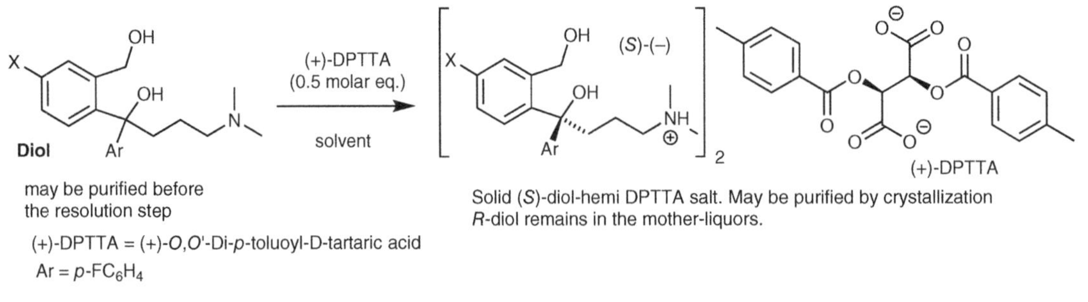

Scheme 4.45 Resolution of diol by diastereomeric salt formation

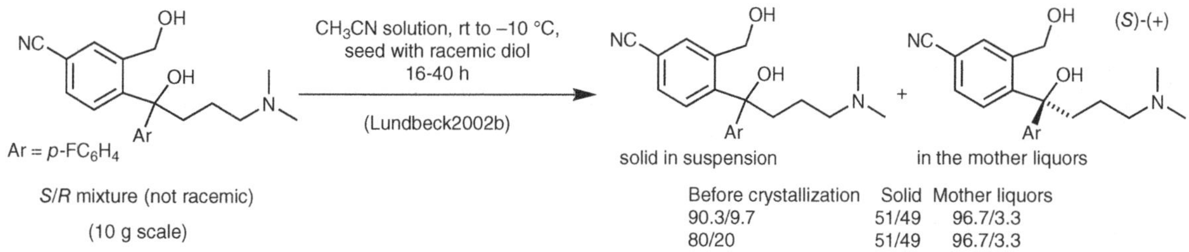

Before crystallization	Solid	Mother liquors
90.3/9.7	51/49	96.7/3.3
80/20	51/49	96.7/3.3
69/31	50/50	94.8/5.2
59/41	54/46	92.9/8.0

SCHEME 4.46 PURIFICATION OF DIOL INTERMEDIATE BY CRYSTALLIZATION

A more recent report claims that the enantiomeric purity of the cyano diol can be upgraded by selectively crystallizing the *S*-cyano diol isomer by controlled cooling and seeding of an acetonitrile solution (from 96% *S* to 99.7%, with an 81% yield in a 434 kg scale, Lundbeck2015).

Acidic conditions at the diol stage may provoke an undesirable cyclization (by a S_N1 mechanism) and give rise to other process-relevant impurities. These result from elimination of water at the double benzylic carbon atom giving an unsaturated conjugated olefin (c.f. Scheme 4.18).

Some resolution processes rely on crystallizing out the unwanted isomer with (–)-DPTTA and then recovering the *S* isomer from the mother liquors using (+)-DPTTA (Shodhana2008a).

As stressed above, the stereochemical configuration must be preserved during cyclization. The most common procedure which achieves this is:

a) Controlled addition of base to a suspension of the DPTTA salt, thus liberating the free base in a biphasic system, whereon it is extracted into the organic phase;
b) Separation of the organic phase and azeotropic drying by distillation;
c) Addition of MsCl or TsCl in the presence of a base (triethylamine is usual) at 0 to 25 °C to activate the primary hydroxyl and promote the subsequent cyclization.

Additional processes for ring closure have been patented (with DEAD/triphenylphosphine – Mitsunobu reaction – Adorkem2004b, 2005a, ShanghaiIP2008; with 2,5-dichloronitrobenzene – S_NAr followed by S_N2, Sun2004; with acidic resin, Beijing2008).

Any halogen to cyano exchange is carried out after the ring closure, no change in chirality being anticipated on carrying out this additional reaction (**G13**, **G23**).

Acid ring closure of the diol mainly follows an S_N1 mechanism with the carbocation formed giving rise to racemization. Studies indicated that in in acetonitrile the reaction proceeds with partial inversion and hence, it is not pure S_N1. The ratio of inversion versus retention is around 70:30 to 75:25, depending on the reaction conditions. Solutions of enriched Escitalopram can hence be obtained from *R*-diol (see Scheme 4.47, **G3**, Lundbeck2001b).

It is important to stress once more that when starting from a pure *S*-diol the cyclization step must be the subject of extensive development work and controls established to avoid the formation of impurities at this late stage. Two impurities are reported to be produced during this final stage at levels of less than 0.5% (HPLC) owing to elimination of water followed by nucleophilic substitution reactions on the mesylate (Scheme 4.48, OPRD2012). Processes to upgrade enantiomeric purity of Escitalopram have been patented (*via* oxalic acid salts, Tokuyama2016d).

4.12.4 Resolution of Citalopram and desmethyl analogues

Resolution of Citalopram and the desmethyl and didesmethyl derivatives has also been claimed in the patent literature as alternatives to the syntheses described above (processes summarized in scheme H).

The resolution of Citalopram reported by DrReddys (**H7**) has been controversial (OPRD2009a,b,c). The presumed multiple crystallizations and low yields make the

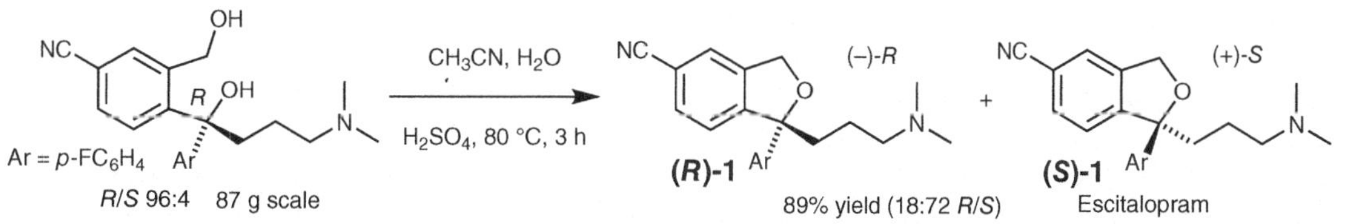

SCHEME 4.47 PARTIAL CONFIGURATION INVERSION DURING RING-CLOSING REACTION OF DIOL IN ACID

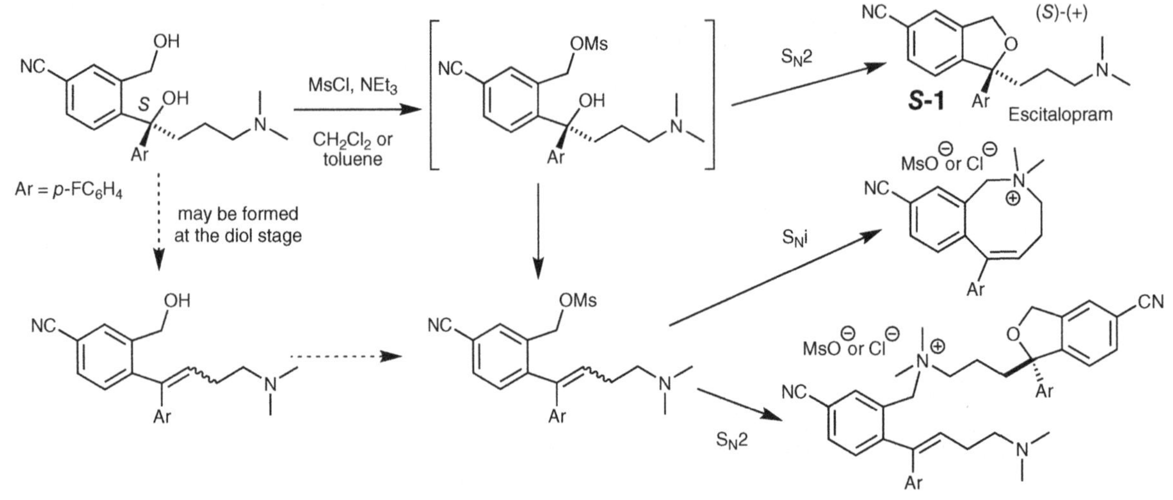

Scheme 4.48 Possible impurities formed during ring closing reaction

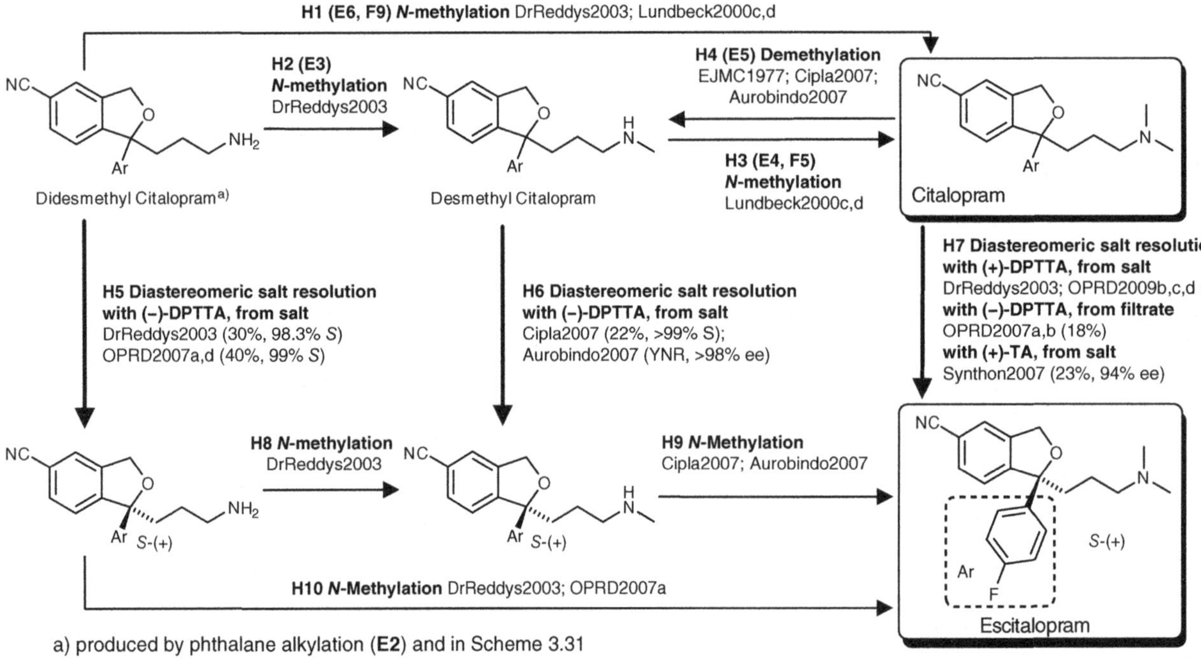

Scheme H Synthesis of Escitalopram by resolution of Citalopram, desmethyl or didesmethyl Citalopram (for reaction conditions see page 135)

process unsatisfactory at industrial scale. Another resolution process developed by Synthon (Synthon2007) using a mixture of (+)-tartaric acid and formaldehyde in water-methanol solutions gives Escitalopram after several crystallizations in 23% yield with an ee of 94%.

Resolution of didesmethyl Citalopram and desmethyl Citalopram using (–)-DPTTA were also reported (respectively **H5**, DrReddys2003 and **H6**, Cipla2007 and Aurobindo2007), a 30-40% yield given in the (*S*)-didesmethyl Citalopram case (>98.3% S). The (*S*)-desmethyl Citalopram was isolated in 22% yield (>99% *S*). In both instances, recrystallization is necessary to obtain adequately pure product. Both *S*-desmethyl derivatives can be converted to Escitalopram using the classical Eschweiler-Clarke methylation.

4.12.5 Recovery of DPTTA

For all resolution-based processes to be economically viable at industrial scale the recovery and reuse of the pure chiral resolution agent is mandatory and procedures for this to occur must be developed. In the present case, the DPTTA used is predicted to be stable and clean separation from its salts with Escitalopram should be feasible. Recovery of expensive DPTTA by extraction into a basic aqueous phase with ammonium hydroxide, adjusting the pH to 2.4 and filtering off the DPTTA-monoammonium salt gave about 93% pure product by HPLC (Sandoz2005), thus establishing the viability of the recovery of the resolution agent.

4.12.6 Recycling of unwanted *R*-isomers

For a successful industrial process, recycling of the unwanted isomer is highly desirable, if not crucial.

Because of the stability of the chiral centre in Escitalopram and its desmethyl analogues racemization is unrealistic (Scheme 4.49). At first glance opening of the furan ring may seem worthwhile, but such ring opening reactions are also untenable.

In the diol routes, an S_N2 displacement is selected to convert the *S* diol into the *S*-phthalane and control chirality. Beforehand, the *S* isomer is isolated from the racemic diol and then carried forward to Escitalopram whilst the *R* isomer is regarded as waste.

As discussed above the *R*-diol can be cyclized in acid with a high degree of inversion of configuration, 70:30 to 75:25. Hence, *R*-enriched cyano diol gives a solution with a high content of the *S*-isomer (*S*/*R* ratio can be higher than 97/3) after racemic Citalopram is removed by crystallization or precipitation (Scheme 4.50). Hence, this methodology allows for the conversion of the *R* diol isomer into Escitalopram.

Although exemplified using racemic cyano diol, a dehydration/oxidation recovery process for the *R*-isomer was patented by Lundbeck as the technique is equally applicable to the pure enantiomer (Scheme 4.51, Lundbeck1999g, Lundbeck2001c).

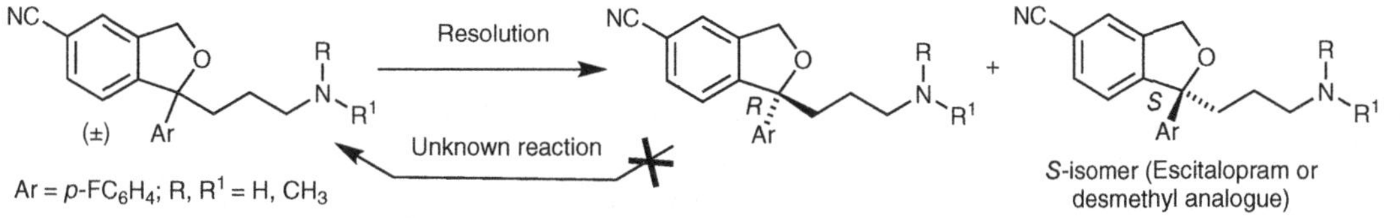

SCHEME 4.49 RESOLUTION OF CITALOPRAM AND ANALOGUES AND RECYCLING OF *R* ISOMER

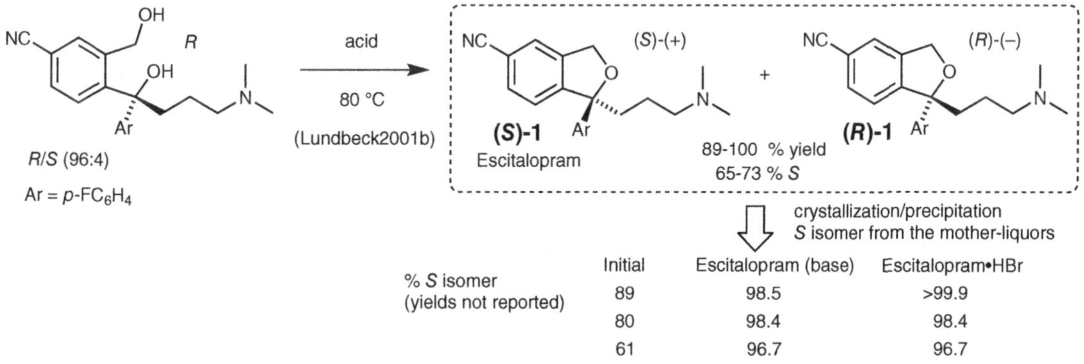

SCHEME 4.50 SYNTHESIS OF ESCITALOPRAM FROM *R*-CYANO DIOL

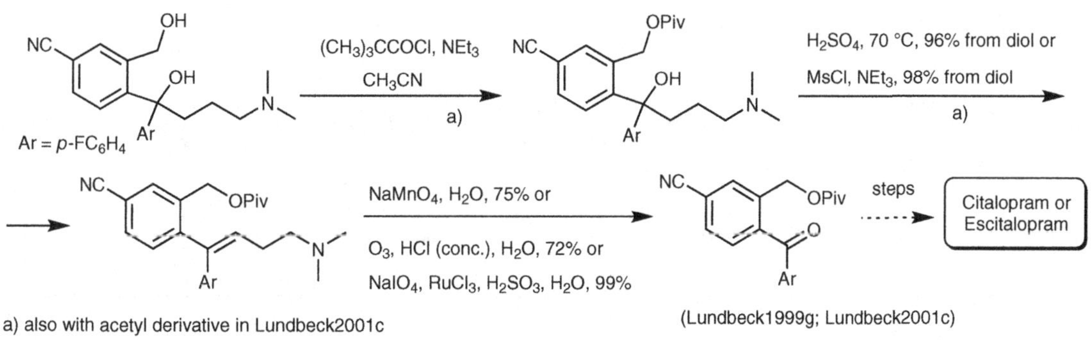

SCHEME 4.51 RECYCLING OF CYANO DIOL BY ELIMINATION/OXIDATION PROCESS

Racemization of the diol itself under conditions that avoid ring closure to the phthalane should be possible (Scheme 4.52). Racemization of the *R*-cyano diol has been reported (**G4**, Adorkem2005a) allowing for reintroduction into the resolution step.

A practical way of avoiding the possibility of cyclization is to block the nucleophilic benzylic hydroxyl group (Ipca2015), Scheme 4.53.

The presence of two aromatic rings allows a nucleophilic racemization of the diol in basic conditions, probably *via* a stabilized classical S_N1 transition state to give either *R* or *S* alcohol independently of the chirality of the initial alcohol (Scheme 4.53). Racemization of the mother-liquors of three resolution batches of cyano diol, after mixing, was disclosed by Lundbeck (Lundbeck2015). Starting from 14 tons of solution, containing a calculated amount of about 900 kg of the diol and projected to consist of 85% *R* isomer, 520 kg of racemic diol (51% yield) was obtained by heating a water-toluene solution at 55-60 °C with ammonium hydroxide, followed by a controlled crystallization.

The options to recycle and/or reduce waste *R*-cyano diol are summarized in Scheme 4.54.

4.12.7 Key Diels-Alder cycloaddition in the synthesis of Escitalopram

A novel route using a Diels-Alder cycloaddition as a key step was developed and patented (Scheme 4.55, Lundbeck2005). Isolation of the *S* isomer of a furan-alcohol intermediate uses SMB chromatography on a chiral stationary phase. This synthesis is devoid of industrial relevance and only has academic interest because of its length and uses chromatography.

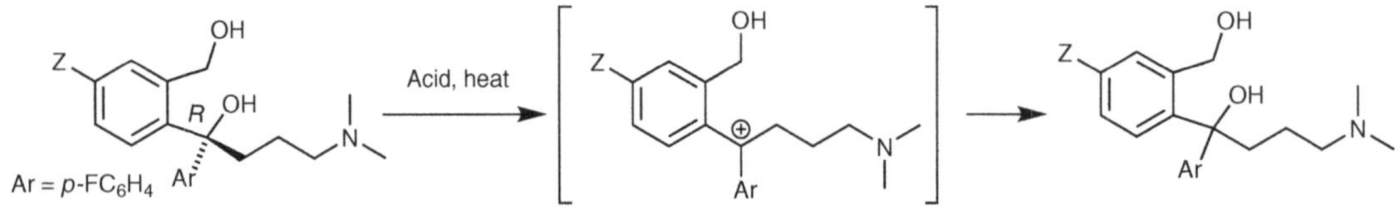

SCHEME 4.52 RACEMIZATION OF DIOL INTERMEDIATES IN ACID

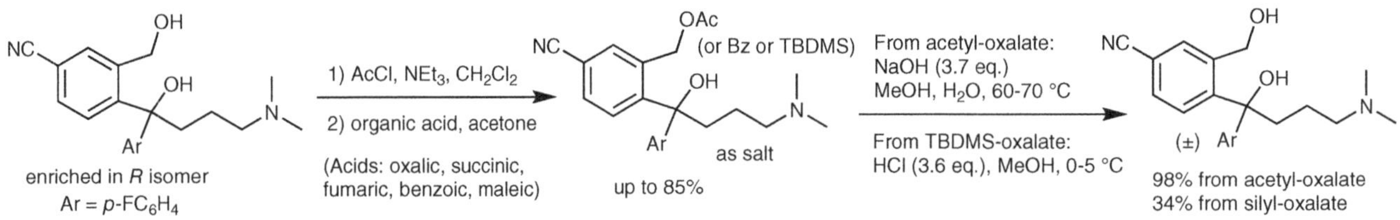

SCHEME 4.53 DIOL RACEMIZATION WITH THE USE OF A PROTECTING GROUP ON THE BENZYLIC HYDROXYL

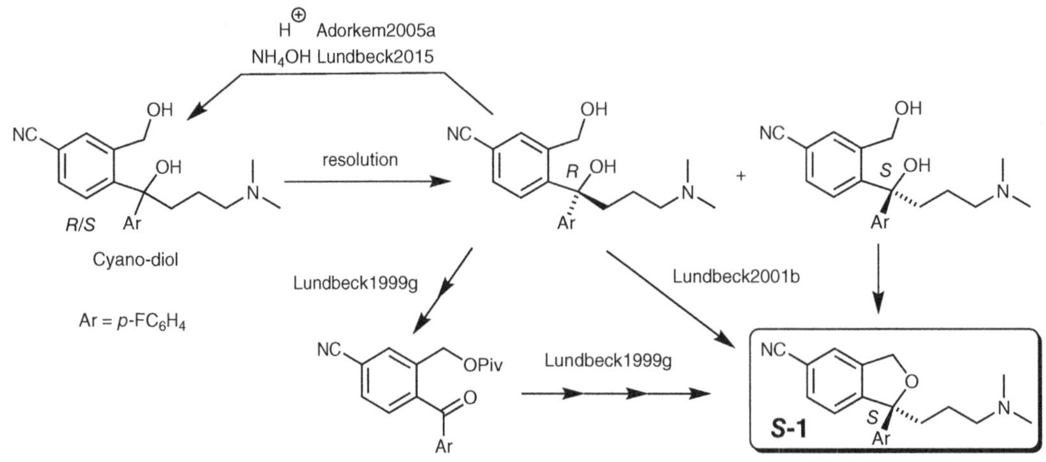

SCHEME 4.54 RECYCLING STRATEGIES FOR *R*-CYANO DIOL INTERMEDIATE

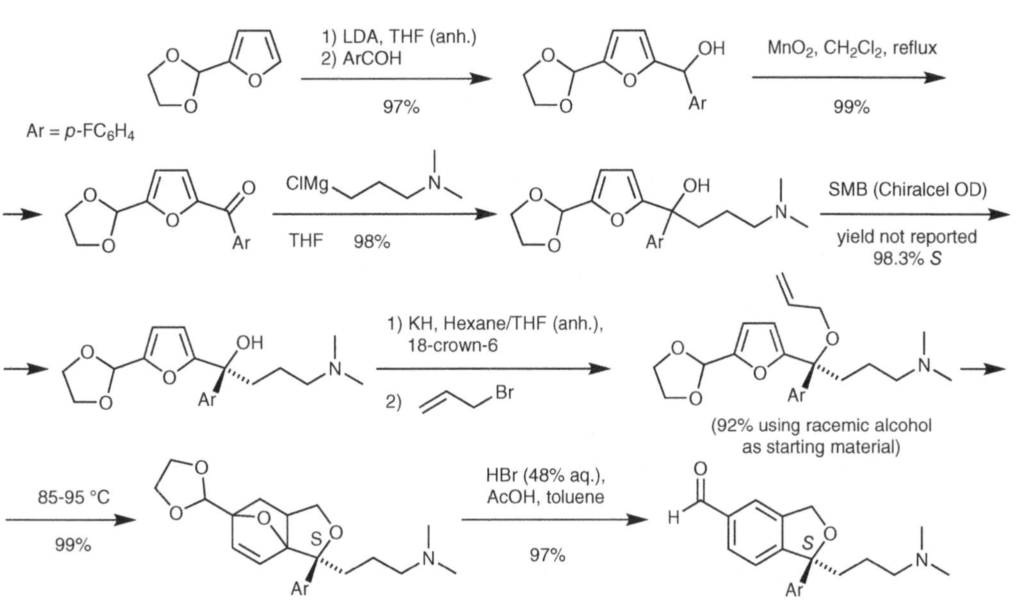

Scheme 4.55 Escitalopram using a Diels-Alder as a key step

4.12.8 Escitalopram by asymmetric synthesis

Asymmetric synthesis avoids a resolution step but construction of the asymmetric quaternary carbon atoms is not straightforward. Stereoselective addition of an organometallic reagent to a ketone in an asymmetric fashion has been patented (Sandoz2006, Scheme 4.56). The nearby hydroxymethyl group was used to anchor a borate ester of a chiral amino alcohol, thus inducing asymmetric addition. After the Grignard reaction, the borate derivative is easily hydrolysed. The best option, found after screening various amino alcohols, is shown in Scheme 4.56. An additional purification of the diol with (+)-DPTTA was required to achieve adequate ee.

Nucleophilic addition of an acetylide, mediated by an amino alcohol chiral auxiliary, was patented (Scheme 4.57, Sumitomo2004b, ee of product not reported).

In a typical laboratory route, an asymmetric addition of an acetylide to an aldehyde produced a suitably functionalized enantioenriched benzylic alcohol in good yield and selectivity (T2011, Scheme 4.58). A key lithiation-borylation reaction was used to prepare tertiary alcohol intermediates with appropriate framework to produce Escitalopram. The lithiation-borylation sequence was found to tolerate nitrile, benzylic alcohol and *N*-Boc functionalities.

In another approach, a chiral acetal was used to induce asymmetry on addition of (3-(dimethylamino)propyl) magnesium chloride to the ketonic group (Scheme 4.59, Sumitomo2004a).

An asymmetric addition of the fluoroaromatic ring to a benzylic carbon was the key step in yet another synthetic approach (Scheme 4.60, ShanghaiIOC2016, ACIE2016). A palladium catalysed process introduces two nitrile groups and promotes a specific the ring closure to the phthalide. After removal of the carbonyl group Escitalopram is obtained.

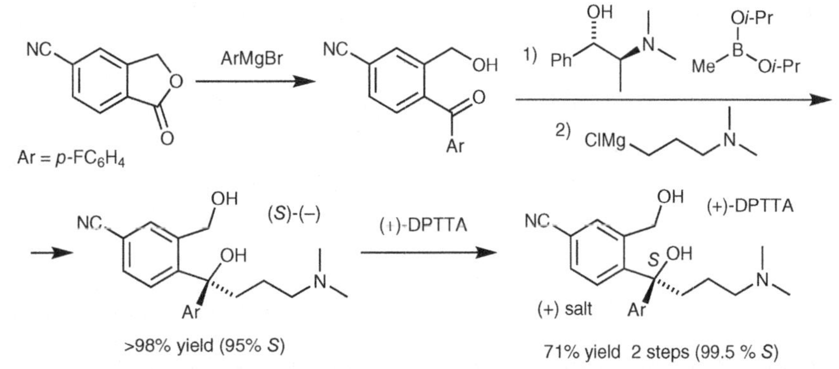

Scheme 4.56 Asymmetric synthesis of cyano diol by asymmetric addition of a Grignard

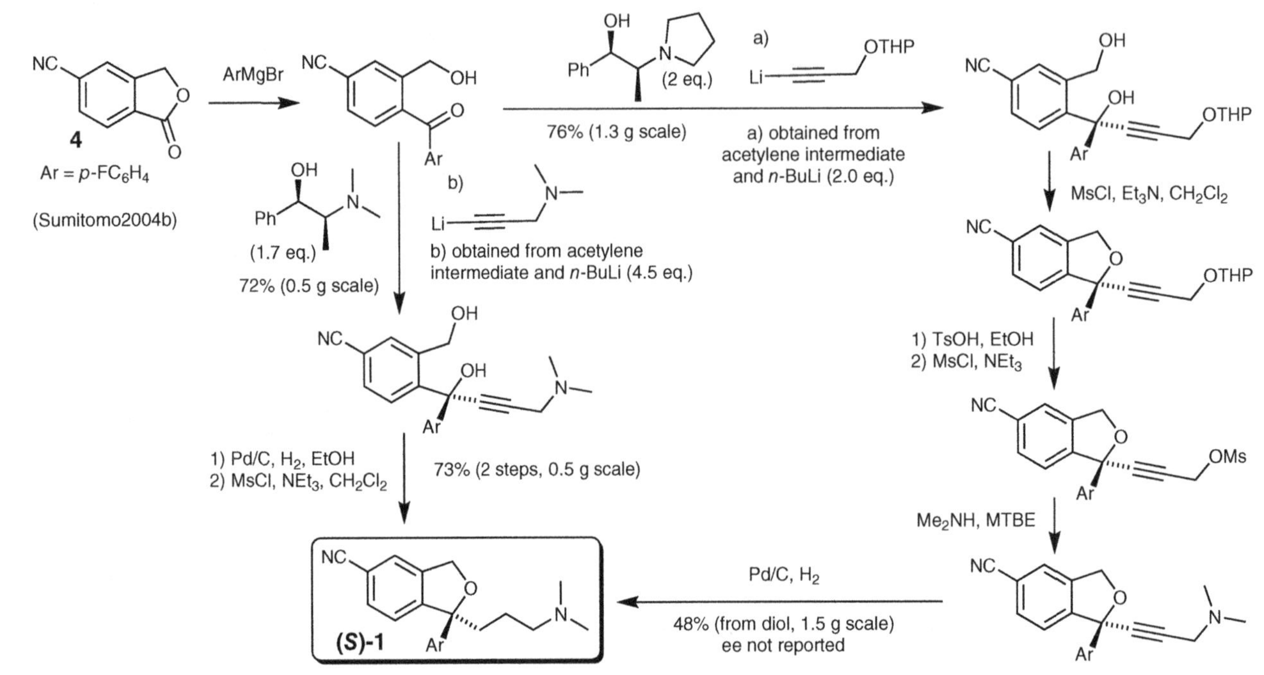

Scheme 4.57 Escitalopram synthesis by an asymmetric addition of an acetylide

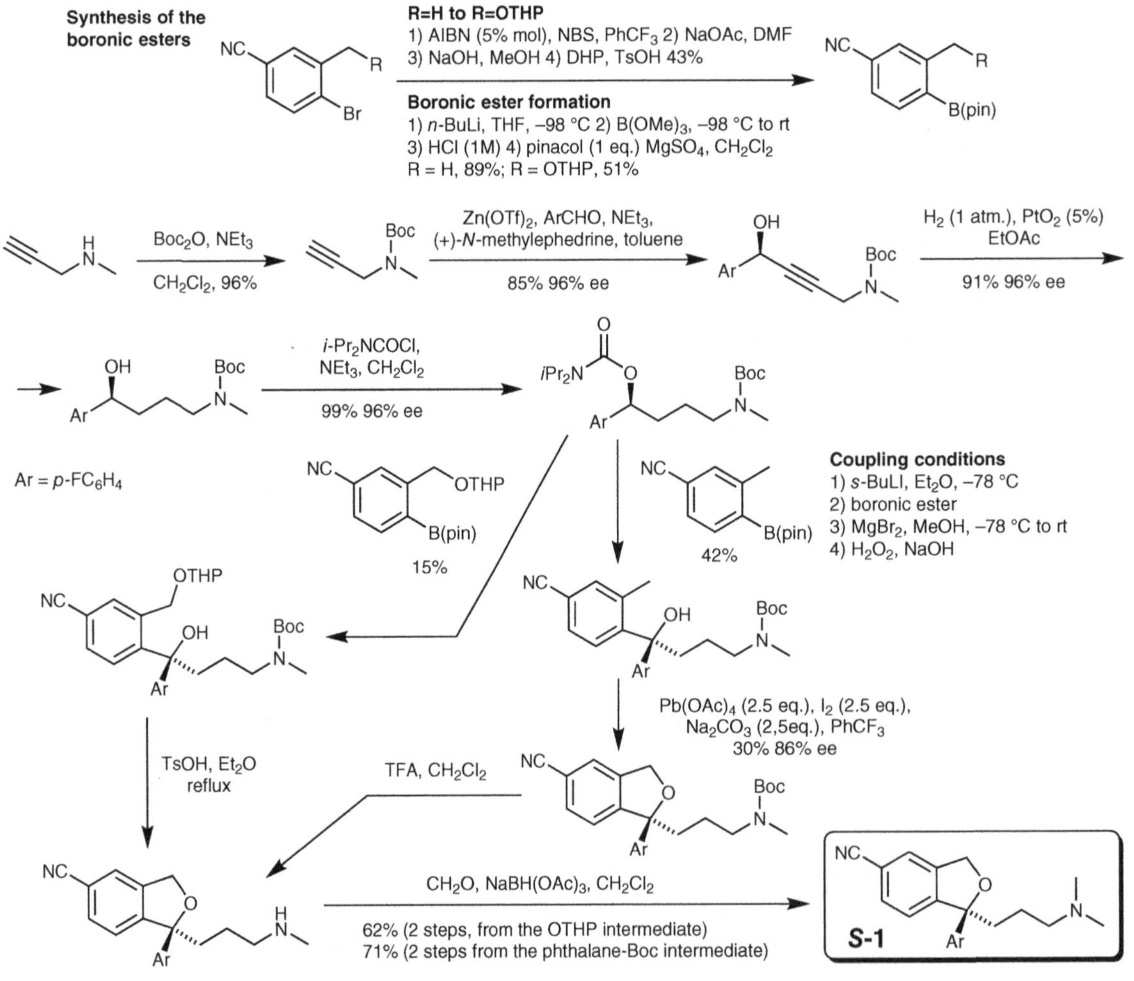

Scheme 4.58 Escitalopram synthesis by asymmetric addition of an acetylide to an aldehyde followed by a key lithiation/borylation reaction

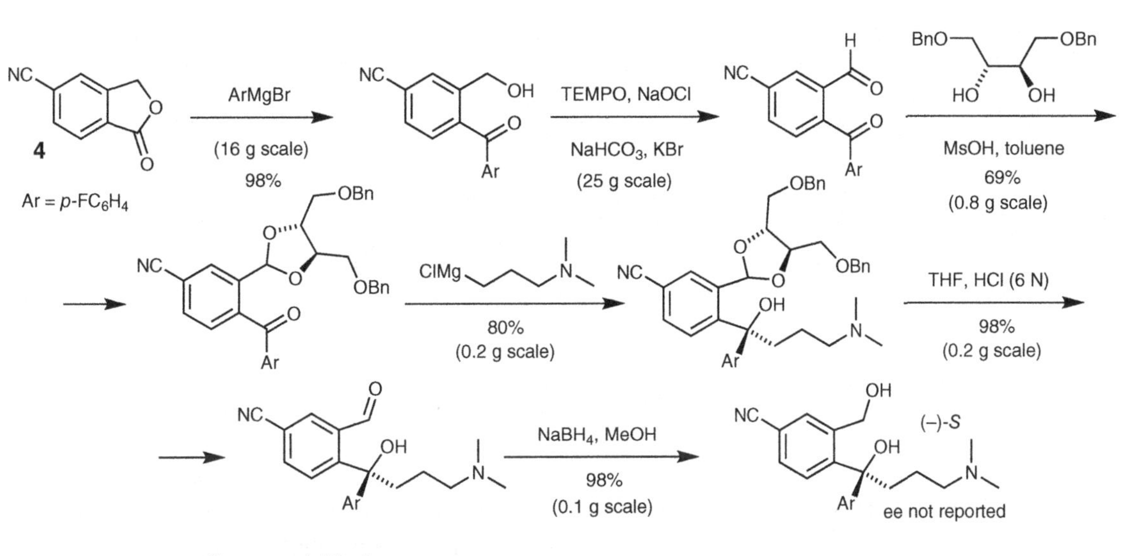

SCHEME 4.59 ASYMMETRIC SYNTHESIS OF CYANO DIOL INTERMEDIATE

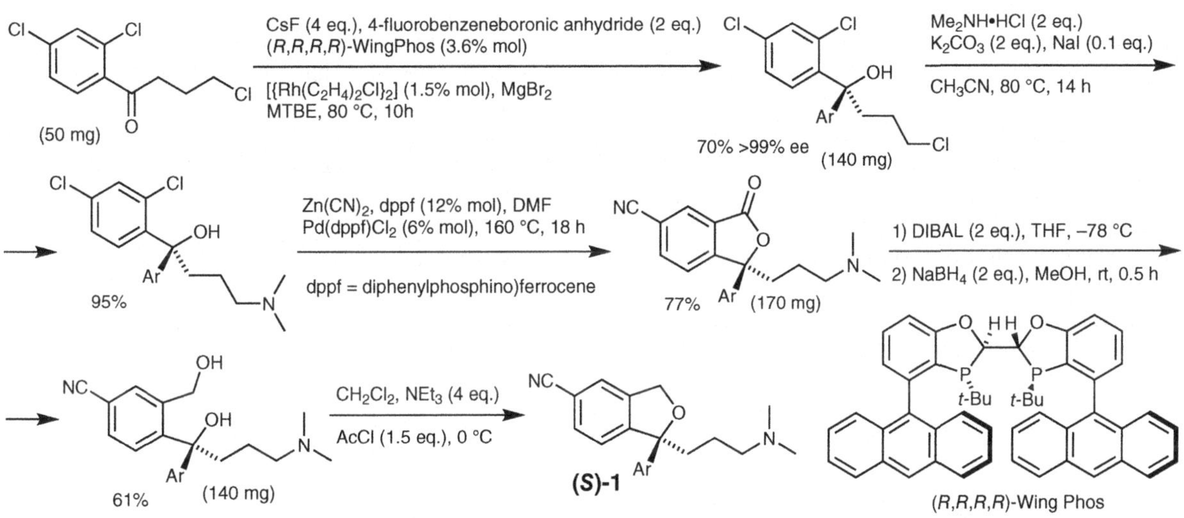

SCHEME 4.60 ESCITALOPRAM SYNTHESIS BY AN ASYMMETRIC ADDITION TO A BENZYLIC CARBONYL

(*S*)-epichlorohydrin, a versatile C-3 building block, was also used in asymmetric syntheses of Escitalopram (Scheme 4.61, Sumitomo2004c and Scheme 4.62, Sumitomo2004d), the chiral epoxide inducing asymmetry.

Neither of these processes patented by Sumitomo report chiral purities and all are rather lengthy and consist of reactions that are difficult to carry out at scale. Hence, their significance is solely academic.

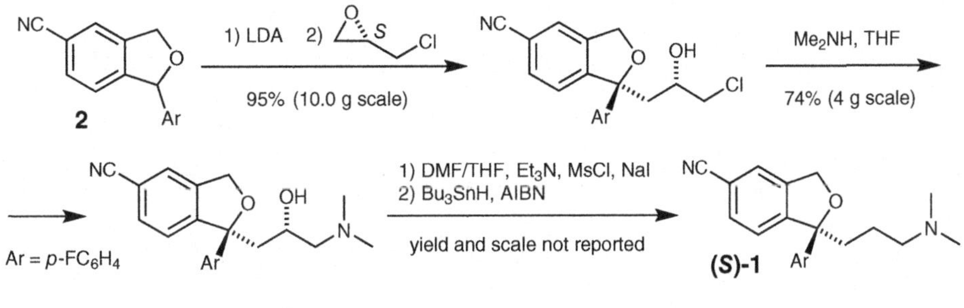

SCHEME 4.61 ASYMMETRIC SYNTHESIS WITH (*S*)-EPICHLOROHYDRIN

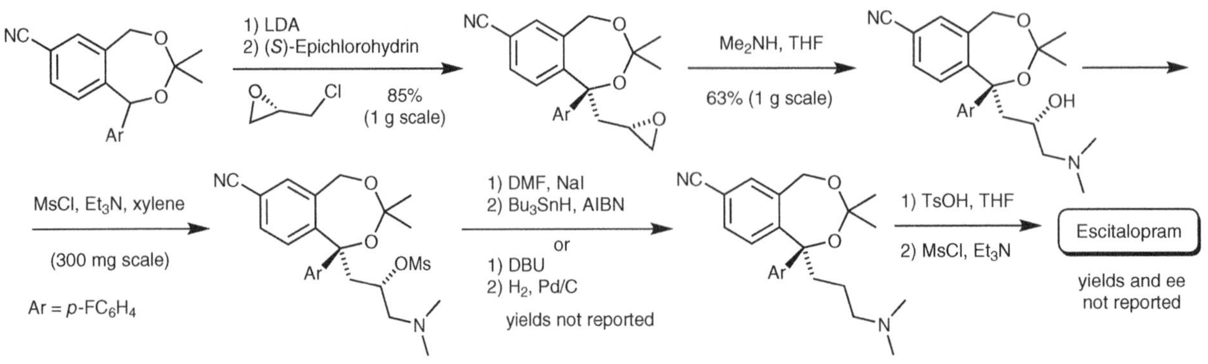

Scheme 4.62 Asymmetric synthesis with (*S*)-epichlorohydrin

4.13 BEST PROCESSES FOR ESCITALOPRAM

As with Citalopram, several processes described above could have utility for the production of Escitalopram at an industrial level. Scheme 4.40 (for Citalopram) helps indicate what might be the most useful route.

Resolution of Citalopram and desmethyl analogues takes place in low to moderate yields and importantly, no recovery for the unwanted *R*-isomer is known (Scheme 4.49) making this approach unattractive. Hence, conversion of Citalopram, or a closely related compound, to Escitalopram it is not particularly viable.

A **B**-type route (of Scheme 4.40) is also less promising as no asymmetric alkylation of the cyanophthalane is known.

As for Citalopram the cyano diol route (**A**), which avoids halogen to cyano conversion, is preferred over the bromo diol route (**D**). Yields of the two Grignard sequence also favour the 5-cyanophthalide as starting material.

An asymmetric synthesis using a Grignard addition was developed, but additional purification with (+)-DPTTA is required to achieve pharmaceutical quality. This process could be an industrial possibility providing that the cost of the asymmetric step is not high. Recycling of the unwanted isomer of the diol could have a significant impact on viability. Processes for the purification of the diol salts have been disclosed, significantly allowing for an optional clean-up.

Notably, the fact that the cyano diol racemizes both under acidic condition and in the presence of ammonium hydroxide makes this route, using the classical resolution of the intermediate diol, a most favoured one. In addition, cyclization is not expected to occur under the high pH racemization conditions.

Recovery and reuse of the (+)-DPTTA is feasible which is important in making this an industrial reality. Hence, the most efficient route to Escitalopram is shown in Scheme 4.63.

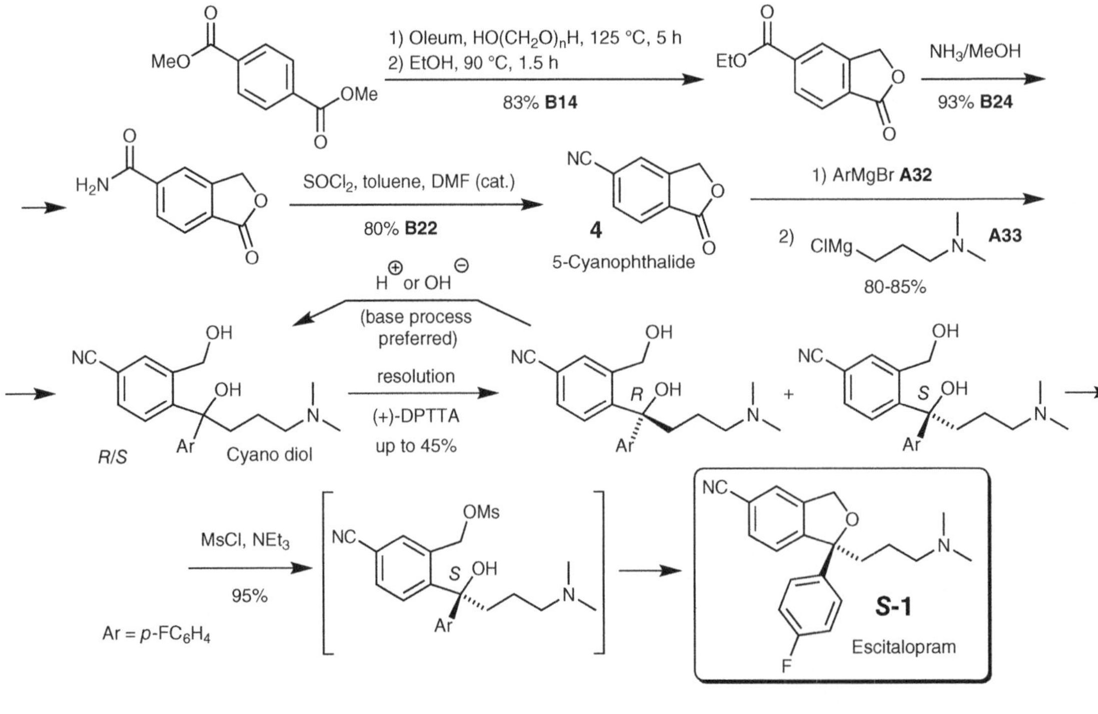

Scheme 4.63 Most efficient process to produce Escitalopram

4.14 LESSONS FROM THE CITALOPRAM/ESCITALOPRAM CASE

Chiral switch

Making the change from a racemic drug to an enantiomerically pure one allows a prolongment of the patent life for the originator. In addition, new synthetic processes have to be developed which opens up opportunities for both the original inventor and generic companies (Lancet2000).

Selectivity and unforeseen reactions

Selectivity and relative reactivities of the various functional groups largely decide the starting materials and the order of chemical reactions. In the present case, the more reactive lactone, relative to the nitrile group, allows the use of 5-cyanophthalide as starting material (Scheme 4.64).

No doubt, unforeseen outcomes to planned chemical transformation can invalidate a proposed route. However, these same unanticipated results may open up other opportunities.

At first sight, addition of the Grignard to the lactone would be expected to form a more highly reactive ketone, which would immediately react with any Grignard present in the mixture. Hence, use of the Grignard with the lactone might be expected to untenable. However, the fortuitous concomitant formation of a hydroxyl group allows for the rapid formation of a cyclic hemiketal magnesium bromide salt, which essentially protects the ketone because the reverse ring opening is slow. Thus, the addition of equimolar quantities of Grignard forms principally the ketone in the form of the hemiketal. The order of the two Grignard additions is not irrelevant. The second aryl groups stabilizes the hemiketal so that the first Grignard attack must be by the *p*-fluoroaromatic nucleophile. High selectivity and large differences in reactivity result in high yields and purer products thus increasing sustainability on an industrial scale.

Another unanticipated result is seen in the almost complete inversion of configuration on ring closure of the diol under acidic conditions. Consequently, the unwanted *R* isomer can, largely, be converted into the desired *S* isomer, Escitalopram, thus eliminating the need for racemization. This fact is key into turning this route into an unconditional industrial success (Scheme 4.65).

A take home lesson for the Synthetic and Process Chemists is: if time allows do try it out in the laboratory even if the transformations looks unreasonable at first sight.

A final example can be observed in the processes to recover the *R* diol by racemization. The racemization of *R* diol under acidic conditions is predictable as the departure of a water molecule gives rise to a highly stable double benzylic tertiary cation. Racemization thus takes place *via* a S_N1 reaction (Scheme 4.66). However, unwanted ring closure is a nuisance.

The surprising observation that racemization of the *R* diol was possible under basic conditions (Scheme 4.54, Lundbeck2015) opened up new avenues. Thus, the racemization under basic conditions may be favoured in comparison to the acidic version as ring closure is avoided and is possibly the preferred process to recover the *R* diol at industrial scale.

Racemic synthesis and chiral synthesis route

Racemic and chiral syntheses of the same compound may not necessarily follow the same route. For example, route **B** in Scheme 4.40 (by phthalane alkylation) may serve to produce Citalopram but is not useful for Escitalopram. Since a resolution of the final product was not

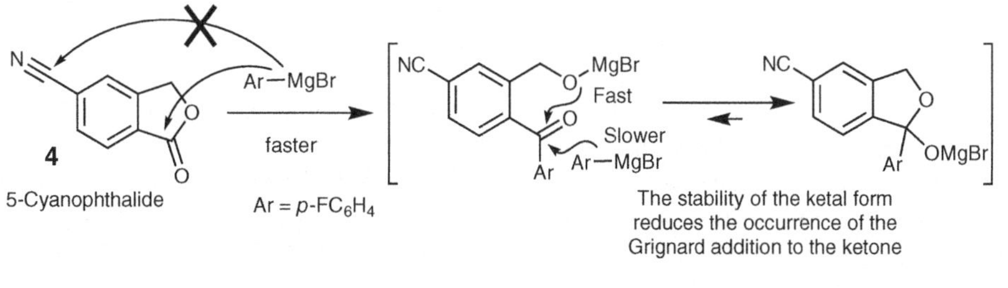

Scheme 4.64 Selectivity of the Grignard attack on 5-cyanophthalide

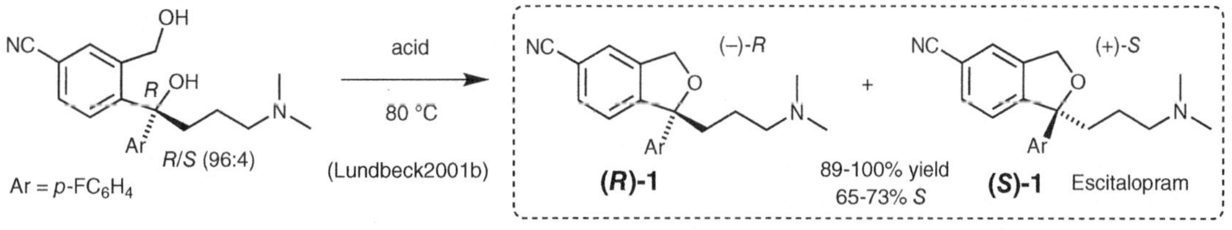

Scheme 4.65 Borderline substitution in diol ring closing reaction

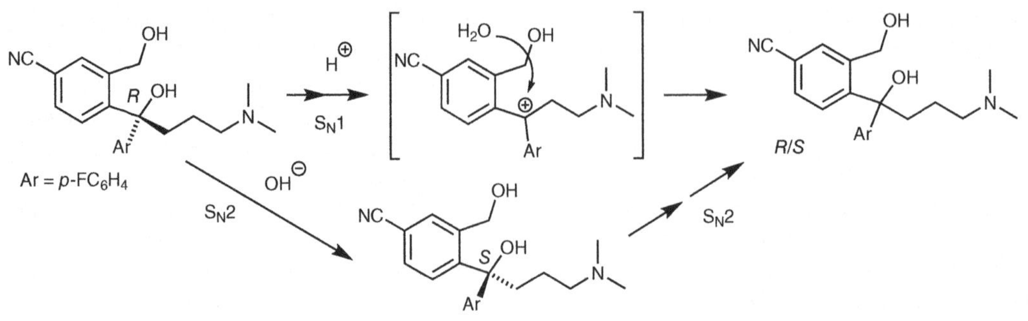

Scheme 4.66 Racemization processes for the *R* cyano diol

possible, resolution at an earlier stage was mandatory and sequential reaction conditions selected to avoid racemization. Although asymmetric syntheses were developed, these do not appear to be particularly applicable because successful chiral induction requires special circumstances and a special environment.

Asymmetric synthesis *vs* resolution

Asymmetric synthesis is often the better option to produce a chirally pure compound but the yields and enantiomeric excess in addition to operating conditions, reactant availability and the total cost of the asymmetric step are important and decisive factors for its success on an industrial scale.

Resolution may not be possible at what would ostensibly be the most favourable place and hence, the Process Chemist is obliged to explore other options at alternative points in the synthesis.

The importance of recovery and recycling in resolution-based processes

The exploitation of the "wrong" isomer can be crucial for a particular process to be successful industrially or not. Environmental concerns, associated with disposal of the unwanted isomer, are dealt with and the impact on the final cost of the API can be significant. Resolution should be carried out at the earliest opportunity, ideally at the stage where chirality is created. By recycling of the unwanted isomer early, savings can be made in the use of reagents and reaction capacities.

Starting material selection

The identification of readily available starting materials, preferably market commodities, should be given high priority by the Process Chemist when planning a synthesis. Those materials that contain functional groups, which require harsh conditions for their transformation, should be given lower priority. Suitable 1,2,4-trisubstituted, 1,3-disubstituted and 1,4-disubtituted aromatic compounds are generally available at reasonable costs as staring materials for both Citalopram and Escitalopram. Terephthalic acid is ideally suited to synthesize 5-cyanophthalides (see section 4.8.5).

Telescoping of reactions and clean-up points

Avoiding unit operations, for example work-up, isolation and drying at industrial scale, can significantly lower the overall cost and ease of carrying out the process. Carrying out several reactions in one pot, or telescoping the reactions, has significant advantages. When a clean-up of the product is necessary, because of accumulated impurities, "additional" unit operations such as crystallization have to be introduced. Ideally, a balance has to be established. In the case of Citalopram/Escitalopram, the two Grignard sequence is telescoped whilst the isolation and purification of the DPTTA diol salts is important for the final product quality.

4.15 REACTION CONDITIONS FOR SCHEMES A TO H

4.15.1 Reaction conditions for scheme A

A1 RR = CH_2CH_2: 1) 1-bromo-4-fluorobenzene (1.2 eq.), Mg (1.2 eq.), THF (anh.), 0 °C; 2) SM, rt, 77% (3 g). **R = Me:** 1) 1-bromo-4-fluorobenzene (1.1 eq.), Mg (1.8 eq.), THF (anh.), 0 °C, NI (30 g).

A2 1) $Me_2N(CH_2)_3MgCl$ (1.5 eq.), THF, rt; 2) HCl (6 N), 25-30 °C, 60% (A1+A2, 30 g).

A3 NH_2OH•HCl, EtOH (95%), H_2O, 30 °C, 88% (40 g).

A4 MsCl (2.2 eq.), NEt_3 (4 eq.), CH_2Cl_2, 0 °C, 70% (20 g).

A5 H_2SO_4 (96%), H_2O, toluene, 80 °C, 97% (2 g).

A6 R = H: Ac_2O, reflux, 83% (0.7 g). **R = Tr:** Ac_2O, HCO_2H, 120 °C, 73% (18 g).

A7 $Me_2N(CH_2)_3MgCl$ (1.3 eq.), THF, toluene, rt, 52% (A1+A7, 3 g).

A8 MsCl (1.7 eq.), NEt_3 (4.4 eq.), EtOAc, 0 °C, 65% (0.7 g).

A9 HCl (1 M), THF/H_2O, rt, 81% (1.1 g). H_2SO_4 (70%), toluene, 80 °C, 35% (4 g).

A10 1) NH_2OH•HCl (2 eq.), NaOH (28% aq.) to pH 10, EtOH, H_2O, rt, 96% (33 g); 2) Ac_2O, py, reflux, 66% (12 g, Lundbeck1999b) or 2) $SOCl_2$, toluene, reflux, 83% (1 g, Lundbeck1999b). 1) NH_2OH•HCl (1.5 eq.), NEt_3 (1.8 eq.), CH_3CN, rt, YNR; 2) Ac_2O, 140 °C, 89% (0.8 g, Sumika1999b).

A11 H_3PO_4 (60%), toluene, 80 °C, 65% (A1+A7+A11, from 6.4 g of phthalide).

A12+A13 R = Bn: 1) 4-FC_6H_4MgBr (1.7 eq.), THF, 20 °C; 2) $Me_2N(CH_2)_3MgCl$ (30% in THF, 5.4 eq.), THF, 5 °C, NI (71 g). **R = Tr:** 1) 4-FC_6H_4MgBr (1.1 eq.), THF, 15 °C; 2) $Me_2N(CH_2)_3MgCl$ (2.3 eq.), 10 °C, 65-74% (25 g).

A14 R = Bn: MsCl (2 eq.), NEt_3 (4 eq.), CH_2Cl_2, 5-25 °C, 65% (A12+A13+A14, 71 g). **R = Tr:** MsCl (2.1 eq.), NEt_3 (4.8 eq.), CH_2Cl_2, 20 °C, 86% (A14+A6, 2 g, Infosint2001a). MsCl (2 eq.), NEt_3 (4.6 eq.), CH_2Cl_2, 6 °C, 97% (A14+A6, 23 g, Infosint2001a,b).

A15 R = *t*-Bu: 4-FC_6H_4MgBr (2.4 eq.), THF (anh.), rt, NI (23 g). **NHR = morpholine:** 4-FC_6H_4MgBr (1.1 eq.), THF (anh.), rt, NI (36 g). **NHR = NMe_2:** 4-FC_6H_4MgBr (1.1 eq.), THF (anh.), rt, NI (17 g).

A16 R = *t*-Bu: $Me_2N(CH_2)_3MgCl$ (1.2 eq.), THF (anh.), rt, NI. **NHR = morpholine:** $Me_2N(CH_2)_3MgCl$ (1.1 eq.), THF (anh.), rt, NI. **NHR = NMe_2:** $Me_2N(CH_2)_3MgCl$ (1.1 eq.), THF (anh.), rt, NI.

A17 R = *t*-Bu: MsCl (1.7 eq.), NEt_3 (4.5 eq.) CH_2Cl_2, 5° C, 14% (A15+A16+A17). **NHR = morpholine:** MsCl (0.7 eq.), NEt_3 (1.3 eq.), CH_2Cl_2, 5° C, 26% (A15+A16+A17). **NHR = NMe_2:** MsCl (2 eq.), NEt_3 (2.5 eq.), CH_2Cl_2, 5° C, 69% (A15+A16+A17).

A18 1) DIBAL (1.2 eq., 1 M in toluene), toluene, 0 °C; 2) H_2O, K_2CO_3, 88% (10 g).

A19 $SOCl_2$, reflux, not isolated, converted by reaction conditions to Citalopram (A19+A20, 1 g).

A20 $SOCl_2$, reflux, 43% (A24+A20, 20 g, Lundbeck1997a). $SOCl_2$, reflux, 78% (A19+A20, 1 g, Lundbeck1997a). $POCl_3$, CH_3CN, reflux, 92% (250 g, ResolutionC2000). $SOCl_2$, toluene, 85-95 °C, 80% (60g, Ranbaxy2001b). For hydrolysis of Citalopram to the primary amide: *t*-BuOH, NaOH (2.8 eq.), 75-80°C, 71% (80g).

A21 (C1) R = *t*-Bu: 4-FC_6H_4MgBr (1.2 eq.), THF (anh.), rt, NI (35 g). **R = *i*-Pr:** 4-FC_6H_4MgBr (1.2 eq.), THF (anh.), rt, NI. **R = Et:** 4-FC_6H_4MgBr (1.2 eq.), THF (anh.), rt, NI (21 g). **R = H:** 4-FC_6H_4MgBr, $MgBr_2$, TMEDA (2.5 eq.), THF, NI (> 90% conversion, 1 g).

A22 R = *t*-Bu, *i*-Pr, Et: $Me_2N(CH_2)_3MgCl$ (1.2 eq.), THF (anh.), rt, NI (21-35 g of phthalide). **R = H:** $Me_2N(CH_2)_3MgCl$ (5 eq.), THF (anh.), rt, NI (1 g).

A23 R = *t*-Bu: MsCl, NEt_3, 5° C, 43% (A21+A22+A23), **R = *i*-Pr:** MsCl, NEt_3, 5° C, 42% (A21+A22+A23). **R = Et:** MsCl, NEt_3, 5° C, 30% (A21+A22+A23). **R = H:** HCl (10 M), YNR (>80% conversion, A21+A22+A23).

A24 R = *t*-Bu: 1) HBr (33%, AcOH), AcOH; 2) $SOCl_2$, toluene, DMF (cat.), reflux; 3) NH_4OH (25% aq.), EtOAc, NI (20 g). **R = H:** 1) $SOCl_2$, CH_2Cl_2, 25 °C; 2) NH_3 (g), 4 °C, 45% (530 g), one pot version: $POCl_3$ (1 eq.), py, 20-25 °C, 38%.

A25 4-FC_6H_4MgBr, THF, –10 °C to 10 °C, NI (135 g, Lundbeck1998a). 4-FC_6H_4MgBr (1.2 eq.), THF, 20-30 °C, NI (500 g, ResolutionC2000).

A26 $Me_2N(CH_2)_3MgCl$, THF, –5 °C to 10 °C, 71% (A25+A26, 135 g of phthalide, Lundbeck1998a). $Me_2N(CH_2)_3MgCl$ (1.3 eq.), THF, 20-30 °C, YNR (500 g of phthalide, ResolutionC2000).

A27 H_2SO_4 (12% aq.), reflux, YNR (500 g).

A28 $(NH_2)_2SO_2$ (1.1 eq.), sulfolane, $SOCl_2$, 130 °C, 78% (5 g). One pot version from 5-carboxyphthalide, 57% (10g).

A29 MsCl (1.5 eq.), NEt_3, CH_2Cl_2, 7 °C to 25 °C, 96% (140 g).

A30 1) $POCl_3$ (2 eq.), py, 10 °C; 2) NH_3; 3) HBr, reflux, 65-67% (287 g), in DMF, 48-56% (19 g).

A31 1) $POCl_3$ (1 eq.), py, 20-25 °C; 2) 60-90°C, 38%. 1) $SOCl_2$ (1 eq.), py, 20-25 °C; 2) $POCl_3$ (1.3 eq.), 48%.

A32 (C5) 4-FC_6H_4MgBr (1.2-1.4 eq.), THF (anh.), –5 °C to rt, 81-98% or not isolated (frequently), in a 25 kg, NI (OPRD2013). With 4-FC_6H_4MgBr (3.2 eq.), 14% of double-addition product is observed (95 g, Adorkem2003a). Using a mixture of Grignards: 4-FC_6H_4MgBr (1.8-2.0 eq.), $Me_2N(CH_2)_3MgCl$ (1.1 eq.), THF (anh.), –10 °C to 0 °C, NI (100 g, Adorkem2003b, 2005a).

A33 Usual conditions: $Me_2N(CH_2)_3MgCl$ (2-2.4 eq.), toluene/THF or THF or toluene, 0-5 °C (80-90%, A32+A33, product may be isolated as HBr salt). $Me_2N(CH_2)_3MgCl$ (2.3 eq.) THF/CH_2Cl_2, 0-5° C, 68% (as HBr salt, 25 kg of phthalide, OPRD2013). Using a mixture of Grignards for A32+A33: 4-FC_6H_4MgBr (1.8-2.0 eq.), $Me_2N(CH_2)_3MgCl$ (1.1 eq.), THF (anh.), –10 °C, 89% (100 g, Adorkem2005a).

A34 H_3PO_4, toluene, 80 °C, 86% (20 g, Hangzhou2004). HBr (50%), toluene, 90 °C, 91-93% (120 g, Guangdong2005). H_2SO_4, H_2O, 70-100 °C, 77 % (30-100 g, Shodhana2008b, DrReddys2008). H_2SO_4, H_2O, 100 °C, 70 % (30 kg of HBr salt, OPRD2013). H_2SO_4 (70% aq.), THF, 80 °C, 50% (A32+A33+A34, 450 g, Lundbeck1984). Ortho-phosphoric acid (85%), THF, 70-80 °C, 47% (A32+A33+A34, 100 g, Adorkem2003b). H_2SO_4, toluene, 80-90 °C, 54-67% (A32+A33+A34, 100 g, Jubilant2004b).

A35+A36 MsCl, NEt_3, CH_2Cl_2, rt, 81% (28 g, Lundbeck1988). TsCl (1.3-1.4 eq.), K_2CO_3 (3.1-3.7 eq.), toluene, H_2O, 0 °C, 71-76% (20-22 g, ZhejiangHP2008). TsCl, toluene, EtOAc, 25 °C, 74 % (50 g, Shodhana2008b). 1,4-dichloro-2-nitrobenzene, K_2CO_3 (anh.), DMSO, 100 °C, 63% (10 g, Sun2004).

A37 PivCl (1.1 eq.), 60 °C, 88% (A32+A37, 16 g).

A38 $Me_2N(CH_2)_3MgCl$ (2.2 eq.), THF (anh.), 0 °C, 90% (29 g).

A39 (C13) X = Br: Most frequent procedure, 1) 1-bromo-4-fluorobenzene (1.1-1.3 eq.), Mg (1.0-1.3 eq.), THF (anh.), –20 °C to 0 ° C; 2) SM, –5 °C to rt, product usually not isolated. 1,2-dibromoethane and/or I_2 used for activation. Co-solvents: CH_2Cl_2, $CHCl_3$, toluene, benzene, PhCl, $ClCH_2CH_2Cl$ (20 g to 1 kg).

X = I: 1) 1-bromo-4-fluorobenzene (1.1-1.3 eq.), Mg (1.2-1.4 eq.), THF (anh.), –10 ° C to 0 °C; 2) SM, –10 ° C to 0 °C, NI (26-100 g). **X = OH:** 1) 1-bromo-4-fluorobenzene (2 eq.), Mg (2.4 eq.), THF (anh.), 8 °C; 2) SM, THF, rt, YNR (10 g).

A40 X = Br: Most frequent procedure, $Me_2N(CH_2)_3MgCl$ (1.1-1.5 eq.), THF, <20 °C, 52-70% (A39+A40), 80% as diMg salt (A39+A40), in THF/toluene 80% as HBr salt (A39+A40, 1 kg, Pharmachem2001). **X = I:** $Me_2N(CH_2)_3MgCl$ (1.2 eq.), THF (anh.), rt, YNR (26 g, Lundbeck1999c). $Me_2N(CH_2)_3MgCl$ (in THF, 1.5 eq.), THF (anh.), –15 °C to 0 °C, 82% (A39+A40, 100g, Natco2004). $Me_2N(CH_2)_3MgCl$ (1.2 eq.), THF (anh.), –20 °C, YNR (50 g, Ranbaxy2005). **X = OH:** $Me_2N(CH_2)_3MgCl$ (1 eq.), THF (anh.), rt, YNR (10 g).

A41 X = Br: H_3PO_4 (20-60% aq.), 90 °C, 85-96% (160 g-1.1 kg, Sekhsaria2001, 2002, Pharmachem2001). H_3PO_4, toluene, 15-20 °C, YNR (100 g, Sun 2002). MsCl, NEt_3, CH_2Cl_2, rt, 74 % (Wockhardt2003). **X = I:** H_3PO_4, toluene, H_2O, 70 °C, 55% (A39+A40+A41, 10 g, Ranbaxy2005). H_3PO_4 (70%), rt, YNR (26 g, Lundbeck1999c). **X = OH:** H_3PO_4 (70%), rt, 48% (A39+A40+A41, 10 g).

A42 X = Cl: 1) $NiCl_2$ (0.4 eq.), PPh_3 (0.2 eq.), CH_3CN, reflux; 2) Zn (0.4 eq.), SM, NaCN (0.4 eq.), reflux, 55% (5 g, Lundbeck1999d), YNR (Orion 2002). **X = Br:** CuCN (1.1-1.8 eq.), DMF, 150 °C, 60-74% (Esteve2002, Aurobindo2003b, CC2016), in sulfolane, 88% (720 g, Lundbeck2000n,o), in diglyme, 40-51% (100g, Cipla2001a,b), in DMA, 47% (100g, Cipla2001b), with KI added, 78% (25 g, Sun2002), with KI/py/PEG-400 added, 85-93% (25 g, Sun2002), with KI/2,6-lutidine added, 82% (25 g, Sun2002). Other additives were also used. 1) $Zn(CN)_2$ (0.6 eq.), DMF, rt, Ar (g); 2) $Pd(PPh_3)_4$ (cat.), 75 °C, 92% (6 g, Lundbeck1999c). 1) NaCN (2 eq.), $Zn(CN)_2$ (cat.), THF, rt, Ar (g); 2) $Pd(PPh_3)_4$ (cat.), reflux, 82% (2.5 g, Lundbeck1999c). 1) Mg, THF; 2) 1-cyanobenzotriazol (2 eq.), THF (anh.), rt, 72% (2.4 g, Lundbeck1999e,k). 1) Mg, THF; 2) BrCN (2 eq.), THF (anh.), rt, 63% (40 g, Lundbeck1999e,k). 1) Mg (2.1 eq.), THF, EtBr (7% mol), 30-35 °C; 2) TsCN (1.6 eq.), THF (anh.) –20 °C to 20 °C, 52% (6-72 g, Lundbeck2000e,k). 1) Mg (2.1 eq.), THF, EtBr (7% mol), 30-35 °C; 2) 1-cyanobenzotriazol (2 eq.), THF (anh.), 0 °C, to rt, 71% (22g, Lundbeck2000k). 1) Mg (2.1 eq.), THF, EtBr (7% mol), 30-35 °C; 2) BrCN (2 eq.), THF (anh.), –10 °C to rt, 63% (38g, Lundbeck2000k). **X = I:** CuCN (1.5 eq.), py (3.9 eq.), DMF, 140-145 °C, 95-100% (7.5 g). **X = OH:** NaCN (2 eq.), CF_3SO_2Cl (1.2 eq.), CH_2Cl_2, rt, 33% (1.8 g). NaCN (2 eq.), CuI (0.1 eq.), $Pd(PPh_3)_4$ (0.4 mol), CH_3CN, reflux, 30% (1 g).

A43 1) 1-bromo-4-fluorobenzene (3.3 eq.), Mg (4 eq.), THF (anh.), reflux; 2) SM, 5 °C; 3) rt, NI (20-30 g).

A44 R = H: 1) $Me_2N(CH_2)_3MgCl$ (1.3 eq.), Mg (1.3 eq.), THF (anh.); 2) SM 5 °C; 3) rt, 58% (A43+A44, 30 g). **R = Ac:** 1) $Me_2N(CH_2)_3MgCl$ (1 eq.), Mg (1.2 eq.), THF (anh.); 2) SM, 5 °C; 3) rt, 63% (A43+A44, 5 g).

A45 H_3PO_4 (60%), 80 °C, YNR (40 g).

A46 1) $NaNO_2$, H_2SO_4, H_2O, 5 °C; 2) Na_2CO_3 to pH 6.5; 3) CuCN (1.1 eq.), H_2O, toluene, 55 °C, 32% (18 g).

A47 BF_3•Et_2O, $BnEt_3NCl$, $SOCl_2$, xylene, reflux, 39% (20 g).

A48 (C9) BF_3•Et_2O (cat.), $BnEt_3NCl$ (cat.), $SOCl_2$, reflux, 92% (25 g), in xylene, reflux, 73% (80 g).

A49 (C10) 4-FC_6H_4MgBr, THF, 0 °C, NI (5 g).

A50 X = Cl: $Me_2N(CH_2)_3MgCl$, 0 °C, 51% (A49+A50, 5 g).

A51 $Me_2N(CH_2)_3MgCl$ (2.5 eq.), THF, ethylene glycol dimethyl ether, –15 °C to rt, 75% (0.5 g). $Me_2N(CH_2)_3MgCl$ (1.1 eq.), THF, toluene, 0 °C to 95% (25 g), in THF, 2 °C, 56% (A49+A51, 1.8 g).

4.15.2 Reaction conditions for scheme B

B1 NBS (2.2 eq.), AIBN (cat.), *o*-dichlorobenzene, 65 °C to 160 °C, 31-45% (5.3-85 g)

B2 AcOH (80%), H_2O, hexamethylenetetramine (2 eq.), rt, 81% (13 g).

B3 RR = CH_2CH_2: Ethylene glycol (10 eq.), TsOH (cat.), $CH(OC_2H_5)_3$ (cat.), toluene, reflux, 90% (4.5 g). **R = Me:** MeOH, TsOH (cat.), 35 °C, 72% (22 g).

B4 R = H: NH_2OH•HCl (1.2 eq.), NEt_3, toluene, 65 °C, NI (1 g, Sumika1999a). NH_2OH•HCl (1.1 eq.), NEt_3 (1 eq.), DMF, 20 °C, 86% (20 g Infosint2001a). **R = Bn:** $BnONH_2$•HCl (1.4 eq.), NaOAc, EtOH (80%), 94% (81 g Infosint2001a). **R = *t*-Bu:** *t*-$BuONH_2$ (1.2 eq.), CH_2Cl_2, 25 °C, 90% (35 g, Infosint2001a). **R = Tr:** Ph_3CONH_2 (1.2 eq.), CH_2Cl_2, 25 °C, 90% (35 g, Infosint2001a,b). **Oxime to Tr oxime:** Ph_3CCl (1.5 eq.), py, 80 °C, 89% (2 g) or Ph_3CCl (1.5 eq.), K_2CO_3, *n*-Bu_4NBr (cat.), CH_3CN, reflux, 61% (2 g, Infosint2001a).

B5 Ac_2O (5 eq.), toluene, 120 °C, 69% (1 g).

B6 NH_3 (25% aq.), $(NH_4)_2CO_3$, electrolysis, 24 °C, 90% (100% with 9:1 selectivity, 250 g).

B7 $SOCl_2$ (1.2-1.4 eq.), $(NH_2)_2SO_2$ (1.1 eq.), sulfolane, 140 °C, 68-77% (50 g-6.3 kg, Lundbeck1999a). $(NH_4)_2CO_3$ (2.9-3.3 eq.), ethyl polyphosphate, 25 °C to 95 ° C, 93-95% (*via* carboxamide, 10-100 g, Suven2004).

B8+B9 1) ClCOOEt (1.2 eq.), DMA (anh.), 0 °C; 2) HMDS (1.2 eq.), rt; 3) $SOCl_2$ (2.5 eq.), 140 °C, 79-84 % (20 g), in DMF, 67%.

B10 $SOCl_2$, DMF (cat.), reflux, NI (36-525 g, Lundbeck1997a, 1998a,b), 91% (Adorkem2005b). $SOCl_2$ (2 eq.), DMF (cat.), toluene, reflux, 82-88% (53-750 g, Lundbeck1997a, 1998b, Farmasint2000, Infosint2001b), NI (900 g, ResolutionC2000).

B11 $LiAlH[OC(CH_3)_3]_3$ (1.1 eq.), THF, N_2, –70 °C, 20% (3 g, Farmasint2000). H_2 (4 bar), Pt/C (10%, cat.), quinoline, NaOAc (anh.), toluene, 70°C, 70% (75 g, Farmasint2000). H_2 (3 bar), $Pd/BaSO_4$ (5%), DMA, 60 °C, 65% (1.7 kg, Infosint2001a,b).

B12 $(NH_2)_2SO_2$ (1.2 eq.), sulfolane, 135 °C, 96% (24 g).

B13 R = H: $HO(CH_2O)_nH$, acid, 76-83%. Nitrobenzene, $HO(CH_2O)_nH$ (1.3 eq.), $ZnCl_2$ (2 eq.), 125 °C, 83% (10 kg, Lundbeck1999h). $HO(CH_2O)_nH$ (1.3 eq.), oleum, 125-140 °C, 82-83% (10-13 kg, Lundbeck1999i). 1) 1,3,5-trioxane (0.4-0.8 eq.), oleum, <35 °C to 130 °C; 2) HCl (35% aq.), 35 °C, 60-91% (50-260 g, Norpharma2000). 1) 1,3,5-trioxane (0.83 eq.), oleum, 15-20 °C to 130 °C; 2) AcOH, 20-23 °C, 73% (56 kg, Norpharma2000). $ClCH_2OSO_2Cl$ (1-1.8 eq.), 130 °C, 60-67% (200 kg, Kekule2005). **R = Me:** oleum, $HO(CH_2O)_nH$ (1.6 eq.), 150 °C, 76% (30 g, Lundbeck2003).

B14 R = Me: 1) $HO(CH_2O)_nH$, oleum, 125 °C; 2) EtOH, 90 °C, 83% (91 kg, product with **R = Et**). **R = H:** 1) $HO(CH_2O)_nH$, oleum, 155 °C; 2) BuOH, 95-100 °C, 44-54% (19 g, product with R = Bu).

B15 R = *t*-Bu: 1) Py, TsCl (2 eq.), rt; 2) *t*-BuOH (1.3 eq.), rt, 94% (100 g). **R = Et:** $POCl_3$ (3.5 eq.), EtOH, reflux, 87% (37 g). **R = *n*-Bu:** *n*-BuOH, H_2SO_4 (cat.), toluene, reflux, 88% (144 kg), without toluene, 80-85 °C, 88% (143 kg).

B16 EtOH, reflux, 88% (39 g).

B17 R = *t*-Bu: *t*-$BuNH_2$ (2.1 eq.), THF, 5 °C, 87% (B10+B17, 36 g). *t*-$BuNH_2$ (1.3 eq.), NEt_3 (1.3 eq.), THF, rt, 91% (39 g). **NHR = morph.:** morpholine (1.3 eq.), NEt_3 (1.3 eq.), THF, 0 °C, rt, 78% (39g). **NHR = NMe_2:** Me_2NH (40% aq.), THF, 90% (32 g). **R = OH:** NH_2OH•HCl (2.3 eq.), NEt_3 (2.3 eq.), THF, 10 °C, 92% (11 g).

B18 2-amino-2-methyl-1-propanol (3 eq.), THF, 20 °C, 77% (525 g, Lundbeck1998a). 2-amino-2-methyl-1-propanol (1,3 eq.), THF, NEt_3 (1.3 eq.), 0 °C, 68% (39 g, Lundbeck1999b). 2-amino-2-methyl-1-propanol (3 eq.), THF, 0 °C to rt, 76% (B10+B18, 270 g, Lundbeck2000a). 2-amino-2-methyl-1-propanol (2.2 eq.), toluene, rt, YNR (900g of acid, ResolutionC2000).

B19 $SOCl_2$, <10 °C to 30 °C, 78-80% (85-560 g, Lundbeck1998a, 2000a), one-pot methods yield for B19+B20, 60-67% (2-5 g, Lundbeck1998a). $SOCl_2$ (1.1 eq.), CH_2Cl_2, 25 °C, YNR (1kg, ResolutionC2000).

B20 $SOCl_2$, DMF, reflux, 75% (23 g).

B21 NH_4OH, THF, H_2O, 0 °C, 97% (20 g).

B22 $SOCl_2$ (1.5 eq.), toluene, DMF (cat.), reflux, 75 °C, 80% (22 g).

B23 2-amino-2-methylpropan-1-ol (1.3 eq.), toluene, reflux, 85% (82 g).

B24 NH_3 (10 M in MeOH, 10 eq.), 80 °C, 93% (41 g).

B25 R = *t*-Bu: $SOCl_2$, reflux, 93% (23 g). **R = OH:** $SOCl_2$, reflux, 91% (2 g).

B26 X = Br: 1) NaOH, heat; 2) Br_2. **X = Cl:** Cl_2, H_2SO_4, I_2, 50 °C, product mixture, YNR.

B27 X = Br: $NaBH_4$, THF, 25 °C, 61% (with 10% 6-bromophthalide), 37-40% pure (200-230 g, Bromine2003), in ethylene glycol dimethyl ether, 38% pure (Bromine2003), in DMF, 15% pure (57 g, Bromine2003), in EtOH/H_2O, 35% pure (455 g, Bromine2003). **X = Cl:** $NaBH_4$, DMF, 20 °C, 50% (AJC1989). Zn, AcOH, H_2O, 62% (BASUSSR1985).

B28 X = Cl: 1) $NiCl_2$ (0.1 eq.), PPh_3 (0.3 eq.), CH_3CN, reflux; 2) Zn (0.3 eq.), SM, NaCN (1.1 eq.), 70 °C, 79% (3 g). **X = Br:** $Zn(CN)_2$ (1 eq.), $Pd[(C_6H_5)_3P]_4$ (5% mol), DMF, 75 °C, 88% (4 g, Lundbeck1999k). $Zn(CN)_2$ (0.1 eq.), NaCN (1 eq.), $Pd[(C_6H_5)_3P]_4$ (5% mol), DMF, 75 °C, 85% (4 g, Lundbeck1999k). CuCN (1 eq.), NMP, 140 °C, 66% (4 g, Lundbeck1999k). NaCN (1-2 eq.), CuI (5-30% mol), KI (1-1.5 eq.), toluene or xylene, reflux, 83-95% (6 g, UnivZhejiang2006a). $K_3Fe(CN)_6$ (0.15-0.3 eq.), Pd catalyst (0.1-5% mol), K_2CO_3 (0.5-2 eq.), 100-150 °C, 82-95% (1.9-3.7 g, UnivZhejiang2006b). **X = I:** $Zn(CN)_2$ (1 eq.), $Pd[(C_6H_5)_3P]_4$ (5% mol), DMF, 75 °C, 75% (5 g).

B29 1) $NaNO_2$, H_2SO_4; 2) CuCN, 56%.

B30 HNO_3, oleum, 52-67%.

B31 H_2 (1 bar), Pd, Al_2O_3, NH_4CO_3, MeOH, 25°C, 100% (Ciba1998). H_2 (1.4-4.1 bar), Raney Nickel (cat.), DMF, 20-50 °C, 97% (100 g), in DMF/MeOH, 93%, using 5% Pd/C (5%, 5% w/w), 95% (Natco2002b). H_2 (2-3 bar), Pd/C (5-10%), EtOAc, 20 °C, 88-97% (T1998, Ind2002b). 1) $NaHSO_3$, Na_2SO_3, pH 3.6-4.5; 2) HCl, 65-92% (100-190 g, Ind2002a, 2003). Zn, $CuSO_4$, NaOH, H_2O, 0 °C, NI (50 g, Ranbaxy2005).

B32 Zn, NaOH, H_2O, 0 °C to 60 °C, YNR (20 g, JCS1931). Zn, $CuSO_4$, NaOH, 5 °C to 80 °C, 80-100% (BSSBS1951, T1998, Ind 2002b, Ind2003). Zn, $CuSO_4$, NaOH, H_2O, 0 °C, 88% (B31+B32, 50 g, Ranbaxy2005).

B33 Ac_2O, NEt_3, $NaHCO_3$, THF, 97% (10 g).

B34 $NaNO_2$, H_2SO_4 or HBr, CuX or heat for **X = OH**, YNR (JCS1931), for **X = Br**, 76% (Ind2002b), for **X = I**, 42% (Ranbaxy2005).

B35 *i*-Pr_2O, NEt_3, 82-84%.

B36 1) BuLi (or HexLi), THF, –78 °C; 2) DMF, 71-97% (1+2); 3) $NaBH_4$, MeOH, 82-99%.

B37 HCl (aq.), reflux 76-89% (95% from aldehyde).

B38 1) LTMP, THF, –78 °C; 2) DMF, –78 °C to 25 °C, 68% (150 mg).

B39 $NaBH_4$ (14 eq.), MeOH, 35% (50 mg).

4.15.3 Reaction conditions for scheme C

C1 (A21) see A21.

C2 $NaBH_4$ (2 eq.), EtOH, THF, rt, 98% (C1+C2).

C3 H_3PO_4 (60%), 80 °C, 94% (30 g).

C4 NaOH (aq.), EtOH, reflux, 59% (30 g).

C5 (A32) see A32.

C6 $NaBH_4$ (2 eq.), EtOH, THF, rt, 97% (Lundbeck1997b). $NaBH_4$ (0.5 eq.), MeOH, CH_2Cl_2, <25 °C, 71-74% (C5+C6, 100g, Jubilant2002b). $NaBH_4$, NaOH, THF, < 15 °C, NI (Adorkem2003a). $NaBH_4$, toluene, MeOH, 10 °C, NI (100 g, Jubilant2004b). $NaBH_4$, MeOH, CH_2Cl_2, 65-67% (as a boron complex, Jubilant2004a).

C7 1) PivCl, NEt_3, CH_3CN, 70% (9 g); 2) H_2O_2 (30%), Na_2WO_4•$2H_2O$ (cat.), (*n*-Octyl)$_3$NMe•HSO_4, EtOAc, H_2O, reflux, 98% (8 g).

C8 (D11) H_3PO_4 (60%), 80 °C, 29% (700 g, Lundbeck1997b), in EtOAc/H_2O, 66-74% (C5+C6+C8, Adorkem2003a). TsOH, toluene, 56-62 % (C5+C6+C8, Jubilant2002b), in THF, 29% (C5+C6+C8, Jubilant2002b), in THF/CH_2Cl_2, 86% (C5+C6+C8, Jubilant2002b), in THF/$CHCl_3$, 87% (C5+C6+C8, Jubilant2002b), in THF/toluene, 70% (C5+C6+C8, Jubilant2002b), in THF/PhCl, 78% (C5+C6+C8, Jubilant2002b), in THF/benzene, 78% (C5+C6+C8, Jubilant2002b), in THF/$ClCH_2CH_2Cl$, 86% (C5+C6+C8, Jubilant2002b). HCl (aq.), 65-70 °C, 100% (from the boron complex, 100 g, Jubilant2004a).

C9 (A48) see A48.

C10 (A49) see A49.

C11+C12 $NaBH_4$, EtOH, rt, 20% (1 g, C11+C12).

C13 (A39) see A39.

C14 $LiAlH_4$, Et_2O, reflux, 100% (300 g, Kefalas1976, EJMC1977). $NaBH_4$, MeOH, 25 °C, 51-54% (C13+C14, 100 g, Jubilant 2004a), in *i*-PrOH, THF, 10 °C, YNR (C13+C14, 100g, Natco2001a,b, 2002a), in EtOH, CH_2Cl_2, YNR (100g, Jubilant2002b). KBH_4, EtOH, THF, rt, 79% (150 g, Synteco2003).

C15 X = Cl: $AlCl_3$, ArF, 60 °C, 81% (182 g). **X = Br:** $AlCl_3$, ArF, 60 °C, 83% (227 g).

C16 X = Cl: 1) MeOH, H_2SO_4 (cat.), 60 °C, 91% (139 g); 2) $NaBH_4$, THF, 0 °C to rt, 93%. **X = Br:** 1) EtOH, H_2SO_4 (cat.), 80 °C, 95% (160 g); 2) $NaBH_4$, EtOH, 0 °C to rt, 90%.

C17 X = Br: H_2SO_4 (60%), 100 °C, 61-65% (300 g, Kefalas1976, EJMC1977). TsOH or H_2SO_4 (cat.), toluene, reflux, 79% (200 g, Natco2001a,b, 2002a). TsOH, toluene, reflux, 69-73% (C13+C14+C17, 100 g, Jubilant2002b), 69% (C13+C14+C17) in THF/CH_2Cl_2, 73% (C13+C14+C17), in THF/$CHCl_3$, 58% (C13+C14+C17), in THF/toluene, 58% (C13+C14+C17), in THF/benzene, 65% (C13+C14+C17), in THF/$ClCH_2CH_2Cl$ (Jubilant2002b). HCl (aq.), 65-70 °C, 92-98% (from the boron complex, 100 g, Jubilant2004a). TsOH, toluene, reflux, 66% (160 g, C16+C17, Sichuan2010). **X = Cl:** toluene, H_3PO_4 (cat.), reflux, 60% (180 g, C16+C17).

C18 CuCN (1.1-1.7 eq.), DMF, reflux, 66-82% (150-177 g).

C19 (D10) $NaBH_4$, MeOH, THF, 30-45 °C, 90% (0.5 g).

C20 (D15) 1) DIBAL (50% in THF, 1 eq.), THF, –70 °C to –10 °C; 2) H_2SO_4 80% (1 g). 1) $NaBH_4$ (2.6 eq.), MeOH, THF; 2) H_2SO_4 (60% aq.), 70-85°C, 73% (0.8 g).

C21 Not reported. Amide dehydration is a possibility. Conversion to benzylic alcohol (D16 product) opens other possible routes.

C22 *p*-$FC_6H_4CONHNH_2$, AcOH, rt, NI.

C23 $Pb(OAc)_4$, THF, 0 °C, NI.

C24 1) $NaBH_4$, EtOH; 2) TsOH (cat.), toluene, reflux, 81% (C22+C23+C24).

4.15.4 Reaction conditions for scheme D

D1 4-FC_6H_4MgBr (2.5 eq.), THF, 30 °C, 100% (80 g).

D2 $KMnO_4$ (7 eq.), *t*-BuOH, H_2O, 50 °C to 85°C, 71-75% (121 g).

D3 1) C_6H_5F (1.7 eq.), $AlCl_3$ (1.2 eq.), *o*-dichlorobenzene, 0-30 °C to 80 °C, 85% (4% side product).

D4 4-FC_6H_4COCl (1.1 eq.), $AlCl_3$ (0.1 eq.), 5 °C, 99%.

D5 $KMnO_4$ (6.5 eq.), *t*-BuOH, H_2O, 65 °C to 85 °C, 78% (10 g).

D6 C_6H_4F (2 eq.), $AlCl_3$ (3 eq.), *o*-dichlorobenzene, 70-90 °C, 75% (impure, 7:3), 23% pure (20 g).

D7 Zn (7.7 eq.), AcOH, 60 °C to reflux, 90% (21 g, Lundbeck2000j). 1) $NaBH_4$ (2 eq.), NaOH/H_2O, 20 °C; 2) H_2SO_4 (aq., 50%), 20 °C, 70% (60 g, Sumika2000b).

D8 1) $SOCl_2$, reflux; 2) NH_3, Et_2O, 80% (1 g).

D9 $SOCl_2$, DMF, reflux, 62% (14 g).

D10 (C19) see C19.

D11 (C8) see C8.

D12 BF_3•THF (1 eq.), $NaBH_4$ (1 eq.), THF, 0-20 °C, 78% (5 g).

D13 1) MnO_2 (22 eq.), EtOAc, 25-45 °C, 81% (9 g); 2) NH_2OH•HCl (1.1 eq.), NEt_3 (1.5 eq.), toluene, 80 °C; 3) Ac_2O (10 eq.) 130-140 °C, 71% (2+3, 10 g).

D14 1) $SOCl_2$, reflux; 2) MeOH, 92% (1 g).

D15 (C20) see C20.

D16 1) BF_3•THF (3.9 eq.), $NaBH_4$ (3.2 eq.), diethylene glycol dimethyl ether, 45 °C; 2) H_3PO_4 (85%), 60 °C, 43% (6 g, 70% pure SM, Sumitomo2000, Sumika2000a). 1) $NaBH_4$ (4.4 eq.), $(MeO)_2SO_2$ (4.5 eq.), THF, reflux; 2) H_3PO_4 (85%), 60 °C, 86% (26 g, Sumitomo2000, Sumika2000a). 1) $NaBH_4$ (3.3 eq.), $B(OMe)_3$ (0.25 eq.), THF, 25-35 °C; 2) BF_3•THF (3.7 eq.), 40 °C; 3) H_2SO_4 (30%), 60 °C, 81-82% (100 g-40 kg, Sumitomo2000, Sumika2000a). 1) $LiAlH_4$ (2.6 eq.), THF; 2) H_2SO_4, 25% (3 g, Sumitomo2000, Sumika2000a). 1) $LiAlH_4$ (3 eq.), toluene, 50 °C, 95% (58 g); 2) H_3PO_4 (60%), 80 °C, 90% (50 g, Lundbeck2000b).

D17 NaOAc, toluene, reflux, 99% (50 g).

D18 Br_2 (5 eq.), NaOAc (5 eq.), AcOH, 25 °C, 98% (28 g).

D19 NaOH (aq.), MeOH, rt, 84% (37 g).

D20 $EtOCH{=}CH_2$ (2.5 eq.), TsOH (cat.), toluene, 30 °C, 97% (22 g).

D21 1) BuLi (1.6 M in hexane, 1.1 eq.), 4-FC_6H_4COH (1.1 eq.), THF (anh.), –20 °C to 15 °C; 2) H_3PO_4 (60%), 85 °C, 89% (35 g).

D22 Br_2 (aq., 20%), AcOH, NaOH (aq.), 30 °C, 75% (70 g).

D23 *t*-BuOH, *t*-BuOK, 80 °C, 67% (25 g).

D24 1) Mg, THF, I_2, PrBr, 45 °C; 2) 4-FC_6H_4CHO, 0 °C, 100% (20 g).

D25 *i*-PrOH, HCl (35%), 70 °C, 43%.

D26 NaClO (13% aq. 1.1 eq.), TEMPO (1% mol), Bu_4NBr (cat.), $NaHCO_3$ (0.4 eq.), EtOAc, 84% (21 g, Sumika1999c, Sumitomo2004e). MnO_2, MTBE, 10-30 °C, 87% (300 g, Sumitomo2000, Sumika2000a). MnO_2, toluene, 20-30 °C, 88-100% (66 g-61 kg,

Sumitomo2000, Sumika2000a). MnO_2, xylene, 25-45 °C, NI (50 g, Sumitomo2000, Sumika2000a). MnO_2, CH_2Cl_2, rt, 100% (24 g, Lundbeck2000b).

D27 Oxime formation: NH_2OH•HCl (1.1 eq.), NEt_3, toluene, 85 °C, 79% (6 g, Sumika1999c). **Oxime, nitrile formation:** 1) NH_2OH•HCl (1.1 eq.), NEt_3, toluene, 90 °C; 2) Ac_2O (5 eq.), 125 °C; 3) NaOH (10% aq.), 60% (17 g, Sumika1999c, Sumitomo2004e). 1) NH_2OH•HCl (1.1 eq.), NEt_3, toluene, 70-75 °C; 2) Ac_2O (3.6 eq.), 130-140 °C, 73% (D26+D27, 50 g, Sumitomo2000, Sumika2000a). 1) NH_2OH•HCl (1.5 eq.), NaOH (1.5 eq.), EtOH, H_2O, reflux; 2) toluene, $SOCl_2$, 80 °C, 84% (2 g, Lundbeck2000b). **Nitrile formation:** DMI, $(COCl)_2$, toluene, 20-30 °C, 94-98% (14 g Sumitomo2004f). DMI, $SOCl_2$, toluene, 30 °C, 96% (14 g Sumitomo2005).

D28 $SOBr_2$ (1.3 eq.), toluene, 90% (2 g).

D29 NH_3 (aq.), –33 °C, 40% (2 g).

4.15.5 Reaction conditions for scheme E

E1 1) NaH (50%, 1.1 eq.), DMSO, 70 °C; 2) SM, $Me_2N(CH_2)_3Cl$ (1.1 eq.), 40 °C, 43% (96 g, Kefalas1976), 62% (96 g, EJMC1977), 72% (96 g, Jubilant2002a). 1) NaH (50%, 1.3 eq.), DMSO, 60-65°C; 2) SM, $Me_2N(CH_2)_3Cl$ (1.7 eq.), toluene, 25-30 °C, 89% (100 g Natco 2001a,b). 1) NaH (50%, 1.1 eq.), DMSO, 60-70 °C; 2) SM, $Me_2N(CH_2)_3Cl$ (1.1 eq.), DMSO, toluene, 25-30 °C, 74% (100 g, Natco2002a), 68-80% as HBr salt (100 g, Natco2003c). 1) NaH (50%, 1.1 eq.), DMSO, 60 °C; 2) SM, $Me_2N(CH_2)_3Cl$ (2.2 eq.), DMSO, 25 °C, 68% (30 g, Adorkem2003a). 1) *t*-BuOK (1.3 eq.), DMSO, 20-25°C; 2) $Me_2N(CH_2)_3Cl$ (1.7 eq.), toluene, 25-30 °C, 74% (100 g Natco2001a,b). 1) *t*-BuOK (1.6 eq.), DMSO, 60-70 °C; 2) SM, $Me_2N(CH_2)_3Cl$ (1.1 eq.), DMSO, 40 °C, 74% (96 g, Jubilant2002a). 1) NaH (60%, 1.2 eq.), THF, 45 °C; 2) $Me_2N(CH_2)_3Cl$ (1.3 eq.), toluene, DMSO, 70 °C, 52% (5 g, Sumika1999c, Sumitomo2004f), replacing toluene by MTBE, 61% (22 g, Sumitomo2000, Sumika2000a). 1) NaH (60%, 1.1 eq.), THF, 45 °C, 2) *n*-Bu_4NBr (cat), MTBE, $Me_2N(CH_2)_3Cl$ (1.3 eq.), DMI, 63 °C, 79% (5g, Sumitomo2000, Sumika2000a), TMEDA and DMF replacing DMI, 76%, DMF replacing DMI, 71%. 1) NaH (60%, 1.2 eq.), THF, toluene, 45 °C; 2) *n*-Bu_4NBr (cat.), toluene, DMSO, $Me_2N(CH_2)_3Cl$ (1.3 eq.), TMEDA, 63 °C, 69% (3 g, Sumitomo2000, Sumika2000a). NaH (60%, 1.2 eq.), DMI, toluene, $Me_2N(CH_2)_3Cl$ (1.3 eq.), 60-63 °C, 87% (45 kg, Sumitomo2000, Sumika2000a), 93% (120 g, Sumitomo 2004b,f). 1) NaH (60%, 1.2 eq.), DMI; 2) $Me_2N(CH_2)_3Cl$ (1.3 eq.), toluene, *n*-Bu_4NBr (3 % mol.), TMEDA (1 eq.), 60-62 °C, 86% (9 g, Sumika2001). 1) *n*-BuLi (1.6 N in hexane, 1 eq.), *i*-Pr_2NH (2 eq.), DME, – 50 °C; 2) SM, –40 °C; 3) $Me_2N(CH_2)_3Cl$ (3.2 eq.), 50 °C, 76-84% (Lundbeck1997b). 1) LDA (1.1 eq.), THF, –20 °C; 2) SM, –20 °C; 3) $Me_2N(CH_2)_3Cl$ (1.3 eq.), YNR (2 g, Orion2002).

E2 1) NaH, DMSO, 60-70°C; 2) SM, 25 °C; 3) $H_2N(CH_2)_3Cl$, 40 °C, YNR (EJMC1977). 1) *t*-BuOK (1.6 eq.), DMSO, 60-70 °C; 2) SM, DMSO, $H_2N(CH_2)_3Cl$ (CPA, 2.2 eq.), 40-45 °C, 81% (DrReddys2003), 65% using an acetone solution of CPA (10 g, DrReddys2003). 1) *t*-BuOK (1.5 eq.), DMSO, 60-70 °C; 2) SM, DMSO, $H_2N(CH_2)_3Cl$ (1.75 eq.), 40-45 °C, 67-71% (OPRD2009c). 1) *t*-BuOK (1.6 eq.), acetone, reflux; 2) SM, acetone, $H_2N(CH_2)_3Cl$ (2.2 eq.), 40-45 °C, 0% (10 g, OPRD2009b).

E3 (H2) 1) PhCHO (1.2 eq.), toluene, reflux; 2) $(MeO)_2SO_2$ (1.2 eq.), toluene, reflux; 3) H_2O, reflux 78% (10 g).

E4 (F5, H3) HCHO (37% aq., 4 eq.), HCOOH, reflux, 30% (0.7 g).

E5 (H4) 1) $ClCOOCH_2CCl_3$, benzene, reflux; 2) Zn, AcOH (90%), 25 °C, 43% (19 g, EJMC1977). 1) $ClCOOCH_2Cl$ (1.6 eq.), toluene, $ClCH_2CH_2Cl$, DIPEA (0.5 eq.), 60-65 °C; 2) MeOH, 60-65°C, 91% (150 kg, Cipla2007). 1) $ClCOOCH_2Cl$ (1.5 eq.), $ClCH_2CH_2Cl$, 85-90 °C; 2) MeOH, 60-65 °C, 75% (150 g, Aurobindo2007).

E6 (F9, H1) HCHO (5 eq.), HCO_2H (5 eq.), reflux 78% (5 g, DrReddys2003). HCHO (4 eq., 37% aq.), $NaBH_3CN$ (4 eq.), MeOH, rt, 30% (0.8 g, Lundbeck2000c,d).

E7 Base, $Me_2N(CH_2)_3X$ (X = halogen, sulfonate), YNR.

E8 1) $LiAlH_4$, THF/Et_2O (anh.), reflux; 2) H_2SO_4 (70%), 84% (5 g).

E9 1) Py, O_2, 5 Å MS, 60 °C; 2) CuCl (0.12 eq.), 14% (10 g). $KHSO_4$ ($K_2S_2O_8$ 0.2 M, 2.5 eq. in NaOH 0.4 M, 5 eq.), Ni_2SO_4 (cat.), CH_2Cl_2, H_2O, 70% (0.5 g).

4.15.6 Reaction conditions for scheme F

F1 PG = Bn: LDA (1.3 eq.), ((3-bromopropoxy)-methyl)benzene (1.3 eq.), –78 °C to rt, 60% (2.2 g). **PG = THP:** LDA (1.3 eq.), 2-(3-bromopropoxy)-tetrahydrofuran (1.3 eq.), –78 °C to rt, 60% (2.2 g). **PG = TBDMS:** LDA (1.3 eq.), (3-bromopropoxy)-(*t*-butyl)dimethylsilane (1.3 eq.), –78 °C to rt, YNR (13.4 g).

F2 PG = Bn: 1,4-cyclohexadiene (19 eq.), Pd/C (5%), EtOH, reflux, 80% (1.2 g). **PG = THP:** MeOH, TsOH•H_2O (cat.), rt, 91% (1.5 g). **PG = TBDMS:** HCl (1 M), MeOH, rt, 81% (F1+F2, 13.4 g).

F3 TsCl (1.6 eq.), NEt_3 (2.1 eq.), toluene, 0 °C to rt, 42% (2.5 g). MsCl (3 eq.), NEt_3 (0.9 eq.), THF, 0 °C to rt, 64% (14 g),

F4 $MeNH_2$ (2 M in THF, 44 eq.), EtOH, THF, 60 °C, 36% (1 g).

F5 (E4, H3) see E4.

F6 LDA (1.2 eq.), 2-(3-bromopropyl)isoindoline-1,3-dione (1.2 eq.), THF, –78 °C to rt, 36% (4.4 g).

F7 NH_2NH_2 or NaOH solvent, heat, YNR.

F8 H_2, Pd/C (5%), EtOH, 66% (1.3 g).

F9 (E6, H1) see E6.

F10 From Ms derivative: NaN_3 (7.6 eq.), DMF, 40 °C, 45% (4 g).

F11 From Ts derivative: Me_2NH_2•HCl (12.5 eq.), NEt_3 (17 eq.), DMF, 70 °C, 70% (0.2 g). **From Ms derivative:** Me_2NH_2 (33% in EtOH, 37 eq.), EtOH, THF, rt to 60 °C, 65% (1 g).

F12 LDA (eq.), 2-(2-bromoethyl)-1,3-dioxolane, THF, –78 °C to rt, 86% (4.5 g).

F13 AcOH (aq., 30%), reflux, 50% (5.3 g).

F14 Me_2NH_2•HCl (1.8 eq.), $NaBH_3CN$ (1.8 eq.), MeOH, 0 °C to rt, 82% (5 g).

F15 LDA (1.1 eq.), HCO_2Me (1.2 eq.). –78 °C to 0 °C, 50% (2.4 g).

F16 LDA (1.1 eq.), triethyl phosphonoacetate (1.1 eq.), –30 °C to rt, YNR (5.8 g).

F17 H_2, Pd/C (5%), EtOH, 57% (F16+F17, 8 g).

F18 Methylchloroaluminum *N,N*-dimethylamide (2.6 eq., prepared from Me_2NH•HCl/$AlMe_3$), toluene, 50 °C, 100 %.

F19 LDA, H_2C=CHCOR, THF, –30 °C to rt, YNR (4.8 g).

4.15.7 Reaction conditions for scheme G

G1 Ac_2O (2.0 eq.), NEt_3 (2.0 eq.), CH_2Cl_2, 85% (0.2 g, TA2004). AcCl (17 eq.), CH_2Cl_2, 30-35°C, 100% (59 g, Adorkem2004a). AcCl (3.7 eq.), CH_2Cl_2, 35-40 °C, YNR (172 g, Adorkem2005a).

G2 Enzymatic resolution (hydrolysis of *R* isomer): 22-50% *S* diol acetate (see Scheme 4.43).

G3 H_2SO_4, CH_3CN, H_2O, 80 °C (see Scheme 4.47).

G4 Racemization of *R* isomer: HCl (1 N), heat, 82% (85 g, Adorkem2005a). NH_4OH, toluene, H_2O, 55-60 °C, 51% (14 tons of *R* diol from G7, Lunbeck2015). **By *O*-acylation, water elimination, C=C cleavage**, see Scheme 4.51 (Lundbeck1999g, 2001c). **By O-acylation, racemization, hydrolysis**, see Scheme 4.53 (Ipca2015).

G5 NH_3 (30%), MeOH, 97% (10 g, Adorkem2004a). Base (exact conditions not shown), YNR (Adorkem2005a).

G6 MsCl: MsCl, NEt_3, toluene, 0 °C to 20 °C, 60-82%, with DMF as co-solvent, 57-69 %. MsCl, NEt_3, CH_2Cl_2, 0 °C to 20 °C, 62-100%. MsCl, KOH or K_2CO_3, toluene, H_2O, 0 °C, 81-83%. MsCl, NaOH, MIBK, H_2O, 3 °C to rt, 85%. **TsCl:** TsCl, NEt_3, toluene, 5 °C to rt, 78-95%. TsCl, Na_2CO_3 or K_2CO_3 or NaOH, toluene, H_2O, 0 °C, 73-86%. TsCl, NEt_3, toluene/NMP –5 °C, 99 %. TsCl, NaOH, H_2O, CH_2Cl_2 93%. TsCl, NaOH, THF, H_2O, 70%. TsCl, NaOH, MIBK, H_2O, 3 °C, 83-85%. TsCl, toluene, NaOH, H_2O, 70%, with *n*-Bu_4NCl, 75%. ***p*-$O_2NC_6H_4SO_2Cl$:** *p*-$NO_2C_6H_4SO_2Cl$ (1.4 eq.), K_2CO_3 (2.3 eq.), toluene, H_2O, 0 °C, 86% (21 g). **S_NAr:** 2,5-dichloronitrobenzene (1.1 eq.), K_2CO_3, DMSO, 100 °C, 75% (43 kg), 57% (6 g). 2,5-dichloronitrobenzene (1.1 eq.), *t*-BuOK, THF, rt, 57-59% (6 g). 1) 2,5-dichloronitrobenzene, *t*-BuOK, THF, 30 °C, 10-15 °C, 6% (6 g scale); 2) DMSO, *t*-BuOK, 95-100 °C, 100%. **DEAD:** DEAD (5.3 eq.), PPh_3 (3 eq.), *t*-BuONa (0.3 eq.), THF, 0 °C, 60% (2.8 g, Adorkem2004b, 2005a). DEAD, PPh_3, HCO_2H, THF, 20 °C, 89-97% (34 g, ShanghaiIP2008). **Acidic resin:** Strongly acidic resin (cat.), $CHCl_3/CH_2Cl_2$, $CHCl_3$/1,2-dichloroethane, $CHCl_3/CCl_4$, reflux, 92-96% (60-120 g).

G7 Resolution *via* diastereomeric salt formation. With (+)-DPTTA, *S* isomer salt isolated from the solid in suspension: (+)-DPTTA (0.4-0.5 eq.), solvent, often >35% *S* isomer obtained from the salt, >98% op. Most frequent solvents: *i*-PrOH, *n*-PrOH. Other crystallization mixtures: MeOH/*i*-PrOH, MeOH/EtOAc, *i*-PrOH/toluene, EtOH, EtOH/CH_2Cl_2, *i*-PrOH/CH_2Cl_2, acetone/*i*-PrOH, EtOAc/EtOH, EtOAc/CH_3CN, 1,2-diethoxyethane, THF, 2-MeTHF, cyclopentylmethyl ether, ethylene glycol monomethyl ether/water, ethylene glycol dimethyl ether/water, ethylene glycol dibutyl ether/propylene glycol dimethyl ether/water, propylene glycol diethyl ether/water. Example of a large scale batch: *n*-PrOH, AcOH (0.2 eq.), (+)-DTPPA (0.4 eq.) 41%, 99.3% *S* (660 kg of hemiasulfate, Lundbeck2015). **With (–)-DPTTA to separate *R* salt, *S* isomer obtained from mother liquors (ML) by crystallisation with (+)-DPTTA:** 1) *i*-PrOH, (–)-DPTTA (0.5 eq.), 40 °C to rt; 2) add (+)-DPTTA (0.3 eq.) to ML, 40 °C to rt, 30% *S* as salt (100 g of diol, Shodhana2008a). 1) *i*-PrOH, (–)-DPTTA (0.5 eq.), 40-50 °C to rt; 2) add EtOAc (3% MeOH), to ML, (+)-DPTTA (0.3 eq.), 25-30 °C, 42% *S* as salt (50 g of HBr diol salt, DrReddys2009). **Enzymatic resolution by selective acylation of *R* isomer:** with vinyl acetate or butyrate, 36-44% *S* (Lundbeck2002a), 47% *S* (Astur2003), 42% *S* (TA2004), see scheme 4.43, 41% *S* (Lundbeck2004), see Scheme 4.44. **Chiral chromatography, SMB:** Chiralpack®AD, CH_3CN, 30 °C, flow rate: 10 mL/min, YNR (Lundbeck2001d).

G8 (*R*)-(+)-α-methoxy-α-trifluoromethylacetyl chloride (1 eq.), NEt_3, CH_2Cl_2, rt, YNR (17 g scale).

G9 Chromatographic separation, HPLC, 4% (G8+G9).

G10 *t*-BuOK (2.4 eq.), toluene (anh.), 0 °C, 93%, 99.6% *S* (0.6 g).

G11 Enzymatic resolution by selective acylation of *R* isomer: Novozyme 435, vinyl butyrate, 51-54%. **Resolution *via* diastereomeric salt formation, with (+)-DPTTA, *S* isomer salt isolated from solid in suspension:** 1) *i*-PrOH, (+)-DPTTA•H_2O (0.5 eq.), 40-45 °C; 2) 20-25 °C, 25%, 99.4% *S* (100 g). **Chiral chromatography, SMB:** Chiralcel®OD, isohexane/isopropanol (98/2), rt, flow rate: 500 mL/min, detection: UV 285 nm, 48% *R* + 50% *S*, >99% ee (500 g).

G12 TsCl (1.1 eq.), Et_3N (1.2 eq.), CH_2Cl_2, rt, 96% (1.3 g, Torcan2002). MsCl (1.3 eq.), NEt_3 (2.9 eq.), toluene, –5 °C, 91-95% (46 g, Natco2003b).

G13 CuCN, DMF, 150 °C, 58-76% (AcOH, CuI, MsOH, $PhCO_2H$, used as additives). $Zn(CN)_2$

(2 eq.), $Pd(PPh_3)_4$ (5% mol), DMF or THF 80%, 99.6% ee (5 g, Lundbeck2001d).

G14 Ac_2O (1.1 eq.), Et_3N (1.2 eq.), DMAP (cat.), CH_2Cl_2, rt, 58% (21 g).

G15 1) (+)-DPTTA (1 eq.), acetone, reflux; 2) EtOAc, rt, 15%, 74% dr (13.5 g). Recrystallization of product with acetone-hexane (2:1), 65%, 94% de (4 g).

G16 NaOH (0.5 M aq., 3 eq.), EtOAc, 100% (2.8 g).

G17 1) Pure *S* acid (1.3 eq., acid not specified), $SOCl_2$ (10 eq.), DMF (cat.), toluene, reflux; 2) SM, Et_3N (1.5-2 eq.), DMAP (cat.), THF, rt, YNR.

G18 Chromatographic separation, YNR.

G19 NaH (60%, 1.1 eq.), DMF, rt, YNR.

G20 Chiralcel®OD, isohexane/isopropanol (98/2), rt, flow rate: 500 mL/min. Detection: UV 285 nm, 48% *R* + 50% *S*, >99% ee (500 g).

G21 Enzymatic resolution by selective acylation of *R* isomer: Novozyme 435, vinyl butyrate, 51-54%. **Resolution *via* diastereomeric salt formation:** (+)-DPTTA•H_2O (0.57 eq.), *i*-PrOH, 20-25 °C, 37 % (hemi DPTTA salt, 100 g).

G22 MsCl (1.3 eq.), NEt_3 (2.2 eq.), toluene, –5 °C, 97% (30 g).

G23 CuCN (2.1 eq.), DMF, 120 °C, 80 % (30 g).

4.15.8 Reaction conditions for scheme H

H1 (E6, F9) see E6.

H2 (E3) see E3.

H3 (E4, F5) see E4.

H4 (E5) see E5.

H5 (–)-DPTTA (1 eq.), CH_3CN/H_2O, 30%, 98.3% *S* (10 g, DrReddys2003). (–)-DPTTA (1 eq.), CH_3CN/H_2O, 40%, 99% *S* (136 g, OPRD2007a,d).

H6 (–)-DPTTA (0.5-1 eq.), *i*-PrOH/H_2O, 25-30 °C, 22%, >99% ee (140 kg, Cipla2007). (–)-DPTTA (1 eq.), MeOH, 25-30 °C, YNR, >98% ee (20 g, Aurobindo2007).

H7 (+)-DPTTA (1 eq.), CH_3CN/MeOH, YNR (50 g, DrReddys2003), 6%, 96% *S* (OPRD2009b,c,d). (–)-DPTTA (1 eq.), CH_3CN/H_2O, 18% (from filtrate), 99.4% *S* (25 g, OPRD2007a,b). 1) L-(+)-tartaric acid (1 eq.), CH_2O (37%, with 10-15% MeOH), H_2O, seed, rt; 2) repeated suspension of solid in CH_2O, 23%, 94% ee (22 g, Synthon2007).

H8 1) PhCHO (1.2 eq.), toluene, reflux; 2) $(MeO)_2SO_2$ (1.2 eq.), toluene, reflux; 3) H_2O, reflux, 68% (10 g).

H9 CH_2O (1 eq.), HCO_2H (0.4 eq.), 80-90 °C, YNR (28 kg Cipla2007). $HO(CH_2O)_nH$ (1.1 eq.), $NaBH_4$ (1.1 eq.), 0-5 °C, 77% (10 g, Cipla2007). CH_2O (7.2 eq.), HCO_2H (2.7 eq.), 90-95 °C, YNR (1.1 g, Aurobindo2007).

H10 CH_2O (5 eq.), HCO_2H (5 eq.), reflux, 76% (5 g, DrReddys2003). CH_2O (37%, 3.5 eq.), HCO_2H (3.5 eq.), H_2O, reflux, 80% (40 g, OPRD2007a).

4.16 REFERENCES

ACIE2016 Huang, L. W.; Zhu, J. B.; Jiao, G. J.; Wang, Z.; Yu, X. X.; Deng, W. P.; Tang, W. J. *Angew. Chem. Int. Edit.* **2016**, *55*, 4527-4531.

Adorkem2003a Cotticelli, G.; Dall'Asta, L.; Di Lernia, G. (Adorkem Technology Spa) IT1347559, WO2004080988, US2006183925, US7411077, EP1601662 (prior. 11/03/2003, IT).

Adorkem2003b Cotticelli, G.; Di Lernia, G.; Silvia, M.; Milanesi, S.; Chilcott, C.; Tagg, J. (Adorkem Technology Spa) WO2005049595, US2007060759, EP1506963, EP1682526 (prior. 28/10/2003, NZ).

Adorkem2004a Cotticelli, G.; Rocchietti, S.; Terreni, M.; Pregnolato, M.; Salvetti, R. (Adorkem Technology Spa) IT1355801, WO2005098018, US2007238887, US7470526, EP1733042 (prior. 09/04/2004, IT).

Adorkem2004b Cotticelli, G.; Salvetti, R.; Giovanni, C.; Raul, S.; Terreni, M. (Adorkem Technology Spa) WO2006037714, US2007270599, US7482477, EP1794140 (prior. 01/10/2004, IT).

Adorkem2005a Cotticelli, G.; Salvetti, R.; Bertoni, C.; Giovanni, C.; Raul, S.; Chiara, B. (Adorkem Technology Spa) WO2006136521, US2008199922, EP1893766 (prior. 22/06/2005, EP).

Adorkem2005b Cotticelli, G.; Salvetti, R.; Zappa, M.; Savetti, R. (Adorkem Technology Spa) US2007117991, US7482476, EP1777221 (prior. 14/10/2005, EP).

AJC1989 Donati, C.; Prager, R.; Weber, B. *Australian Journal of Chemistry* **1989**, *42*, 787-795.

Astur2003 Bayod, J. M.; Llorente, G. I.; Gotor, S. V.; Brieva, C. M. R.; Fernandez, S. L.; Quiros, A. M. (Astur-Pharma Sa) ES2228274 (prior. 24/09/2003, ES).

Aurobindo2003a Kaushik, V.; Divvela, V.; Meenakshisunderam, S.; Handa, V. (Aurobindo Pharma Ltd) IN200300225 (prior. 19/03/2003, IN).

Aurobindo2003b Kaushik, V.; Rao, D.; Handa, V.; Sivakumaran, M. (Aurobindo Pharma Ltd) US6781003 (prior. 09/06/2003, US).

Aurobindo2005 Kaushik, V.; Khan, M.; Meenakshisunderam, S.; Vipin, K.; Mohammed, U.; Sivakumaran, M. (Aurobindo Pharma Ltd) IN200501014, WO2007012954, US2009099375, US7939680, EP1987016 (prior. 27/07/2005, IN).

Aurobindo2007 Narsimha, R. B.; Ramesh, D.; Ranjith, K. S.; Sivakumaran, M.; Umar, K. M.; Vipin, K. K.; Kaushik, V. K.; Mohammed, U. K.; Reedy, B. N.; Srinivasan, R. K.; Dandala, R.; Meenakshisunderam, S. (Aurobindo Pharma Ltd) IN200701464, EP2017271 (prior. 06/07/2007, IN).

BASUSSR1985 Nikulin, V. I.; Pisarenko, L. M. *Bulletin of the Academy of Sciences of the USSR, Division of chemical science* **1985**, *34*, 141-144.

BASF1976 Nohe, H.; Hannebaum, H. (Basf Ag) DE2630927 (prior. 09/07/1976, DE).

BCSJ1959 Yoshiyuki, K. *Bulletin of the Chemical Society of Japan* **1959**, *32*, 690-691.

Beijing2008 Chen, L.; Shi, L. (Beijing Dezhong Wanquan Pharm Sci&Tech) CN101560199 (prior. 16/04/2008, CN).

Bromine1995 Oren, J. (Bromine Compounds Ltd) IL115814 (prior. 30/10/1995, IL).

Bromine2003 Oren, J. (Bromine Compounds Ltd) AU2003214582, WO2004089924 (prior. 11/04/2003, AU, WO).

BSSBS1951 Tirouflet, J. *Bull. Soc. Sci. Bretagne Spec.* **1951**, *26*, 35.

C2016 Michman, E.; Agranat, I. *Chirality* **2016**, *28*, 39-43.

Cadila2006 Patel, D. J.; Kumar, R.; Dwivedi, S. D. (Cadila Healthcare Ltd) IN200601210, WO2008059514 (prior. 31/07/2006, IN).

CC2016 Ha, T. M.; Wang, Q.; Zhu, J. P. *Chemical Communications* **2016**, *52*, 11100-11103.

CCL2014 Wang, P.; Zhang, R.; Cai, J.; Chen, J. Q.; Ji, M. *Chinese Chemical Letters* **2014**, *25*, 549-552.

Chengdu2014 Li, T.; Zhao, Y.; Wu, D.; Zhang, Z.; Zhi, Y. (Chengdu Organic Chemicals Co) CN105254531 (prior. 11/07/2014, CN).

Ciba1998 Lamatsch, B. (Ciba Specialty Chem Holding Inc, Ciba Specialty Chem Corp) WO9936402, US6087480, EP1047672 (prior. 13/01/1998, CH).

Cipla2001a Hamied, Y.; Kankan, R.; Rao, D. (Cipla Ltd) WO200270501, US2004092755, US6903228, EP1366034 (prior. 07/03/2001, GB).

Cipla2001b Hamied, Y.; Kankan, R.; Rao, D. (Cipla Ltd) GB2376945 (prior. 27/06/2001, GB).

Cipla2007 Singh, M.; Birari, D.; Kankan, R.; Phull, M. (Cipla Ltd) IN200902059, WO2008142379, EP2155710, US8288568, US2010204493 (prior. 18/05/2007, IN).

CS2013 Hewitt, J. F. M.; Williams, L.; Aggarwal, P.; Smith, C. D.; France, D. *J. Chem. Sci.* **2013**, *4*, 3538-3543.

DrReddys2003 Sundaram, V.; Mathad, V. T.; Venkavala, P. J.; Elati, C. R.; Kolla, N.; Govindan, S.; Chalamala, S. R.; Gangula, S. (Dr Reddy's Lab Ltd, Dr Reddy's Lab Inc) IN200400370, WO2005047274, US2009018351, EP1706394 (prior. 12/11/2003, IN).

DrReddys2008 Boini, A.; Chinta, R.; Manudhane, K.; Medisetti, R.; Sundaraneedi, V.; Vedantham, R.; Vetukuri, V. (Dr Reddy's Lab Ltd) US2010087664 (prior. 07/10/2008, US).

DrReddys2009 Bolugoddu, V.; Elati, C.; Cheemalapati, S.; Penumandla, R.; Christy, A.; Vakamudi, L. (Dr Reddy's Lab Ltd) IN200900705, US2010249438 (prior. 30/03/2009, IN).

EJMC1977 Bigler, A. L.; Bogeso, K. P.; TofT, A.; Hansen, V. *Eur. J. Med. Chem. - Chimica Therapeutica* **1977**, *May-June, 12*, 289-295.

EJMC2010 Wang, W.; Cha, X.-X.; Reiner, J.; Gao, Y.; Qiao, H.-L.; Shen, J.-X.; Chang, J.-B. *Eur. J. Med. Chem.* **2010**, *45*, 1941-1946.

EOID2002 Burke, W. J. *Expert Opinion on Investigational Drugs* **2002**, *11*, 1477-1486.

Esteve2002 Coppi, L.; Gasanz Guillen, Y.; Campon Pardo, J. (Esteve Quimica Sa) WO2003062218, US2003144534, US6635773, EP1479673 (prior. 25/01/2002, ES).

F2001 Micheli, F.; Crippa, L.; Donati, D.; Di Fabio, R.; Leslie, C. *Farmaco* **2001**, *56*, 715-718.

Farmasint2000 Guazzi, G. (Farmasint Srl) IT1319686, WO200248133 (prior. 12/12/2000, IT).

Fujian2012 Song, L.; Zong, H.; Huang, H. (Fujian Material Structure Inst) CN102675152 (prior. 26/04/2012, CN).

Guangdong2005 Liu, Z.; Huang, W.; Yuan, A.; Xiao, K.; Li, Y.; Zhuang, J. (Guangdong Xilong Chem Co Ltd) CN1740167 (prior. 12/07/2005, CN).

GuangdongHP2014 Yang, F.; Lu, C. (Guangdong Hec Pharmaceutical) CN104119248 (prior. 08/08/2014, CN).

GuangdongHP2015 Yang, F.; Qi, C.; Xiao, Y.; Yang, L.; Lu, G. (Univ Sun Yat Sen, Guangdong Hec Pharmaceutical) CN105294496) (prior. 28/09/2015, CN).

Hangzhou2004 Mei, R.; Guo, D.; Wang, S. (Hangzhou Minsheng Pharm Group Co Ltd) CN1629153, WO2005092875, US2007117992, US7435838, EP1700851 (prior. 19/12/2003, CN).

Icrom2001 Leone, M. (Icrom Spa) EP1321464 (prior. 21/12/2001, EP).

Ind2002a Szabadkai, I. (Individual) HU200204491 (prior. 23/12/2002, HU).

Ind2002b Bell, T. W.; Cline, J. I.; Cremo, C. R.; Gillett, S. L.; Frederick, J. H. (Individual) US2011077394, US7964722 (prior. 31/08/2002, US).

Ind2003 Szabadka, I. (Individual) HU200301605 (prior. 05/06/2003, HU).

Ind2006 Gao, R. (Individual) CN1830952 (prior. 10/02/2006, CN).

Infosint2001a Dall'Asta, L.; Cotticelli, G.; Dall Asta, L.; Dallasta, L.; Giovanni, C.; Leone, D. A.; Dall, A. L. (Infosint Sa) DE60108105, WO2003011846, US2004230065, US7166729, EP1281707 (prior. 02/08/2001, DE, EP).

Infosint2001b Dall'Asta, L.; Cotticelli, G.; Dall Asta, L.; Dallasta, L.; Giovanni, C.; Leone, D. A.; Dall, A. L. (Infosint Sa) DE60103530, EP1281708, WO2003011847, US2004225136, US6888010 (prior. 02/08/2001, DE, EP).

Ipca2015 Reguri, B. R.; Prasad, A.; Rawat, A. S.; Rajput, S.; Santan, R. B. (Ipca Laboratories Ltd) WO2017009866 (prior. 11/07/2015, IN).

JACS2016 Liu, J.; Zheng, H. X.; Yao, C. Z.; Sun, B. F.; Kang, Y. B. *Journal of the American Chemical Society* **2016**, *138*, 3294-3297.

JCS1931 Levy, L. F.; Stephen, H. *Journal of the Chemical Society (Resumed)* **1931**, 867-871.

JHC2006 Abaev, V. T.; Dmitriev, A. S.; Podelyakin, S. A.; Butin, A. V.; Gutnov, A. V. *Journal of Heterocyclic Chemistry* **2006**, *43*, 1195-1204.

Jiangsu2011 Yang, B.; Xiong, L.; Pan, B. (Jinagsu Hansoh Pharmaceutical, Lianyungang Hongchuang Pharm Co Ltd Jian) CN102796065 (prior. 27/05/2011, CN).

Jiangsu2014 Chen, Q.; Zhao, J.; Zhao, X.; Du, Y.; Yu, C.; Liu, Z. (Jiangsu Aosaikang Pharm Co Ltd) CN104072390 (prior. 11/03/2014, CN).

JNTG1992 Hyttel, J.; Bøgesø, K. P.; Perregaard, J.; Sánchez, C. *Journal of Neural Transmission / General Section JNT* **1992**, *88*, 157-160.
JOC1970 Forney, L. S. *The Journal of Organic Chemistry* **1970**, *35*, 1695-1696.
JOC1983 Verbicky, J. W.; Williams, L. *The Journal of Organic Chemistry* **1983**, *48*, 2465-2468.
JOC1990 Young, J. G.; Onyebuagu, W. *The Journal of Organic Chemistry* **1990**, *55*, 2155-2159.
Jubilant2002a Babu, A. N.; Goud, V. S.; Gaonkar, S. L.; Thomas, S. D.; Manjunatha, S. G.; Kulkami, A. K. (Jubilant Organosys Ltd) EP1346989, WO2003080590, US2005217562, US7255741 (prior. 21/03/2002, EP).
Jubilant2002b Narahari Babu, A.; Srinivas Goud, V.; Gaonkar, S. L.; Manjunatha, S. G.; Kulkarni, A. K. (Jubilant Organosys Ltd, Max India Ltd) GB2385051, WO2004020425, US2006116522 (prior. 29/08/2002, GB).
Jubilant2004a Narahari Babu, A.; Srinivas Goud, V.; Gaonkar, S. L.; Manjunatha, S. G.; Kulkarni, A. K. (Jubilant Organosys Ltd) WO2005066185 (prior. 12/01/2004, WO, IN).
Jubilant2004b Narahari Babu, A.; Srinivas Goud, V.; Gaonkar, S. L.; Manjunatha, S. G.; Kulkarni, A. K. (Jubilant Organosys Ltd) WO2005077927, EP1723133, US2008119662, IN200604811 (prior. 16/02/2004, IN, EP, WO).
Jubilant2005 Goankar, S. L.; Das, P. P.; Narahari Babu, A.; Manjunatha, S. G. (Jubilant Organosys Ltd, Jubilant Life Sci Ltd) IN200500856, WO2006106531, EP1877394 (prior. 04/04/2005, IN).
Kefalas1976 Bogeso, K. P.; Toft, A. S. (Kefalas as, Lundbeck as H) GB1526331, US4136193 (prior. 14/01/1976, GB).
Kekule2005 Raja Rao, K.; Muralidhar, R.; Kudaravalli, R. R.; Ramabathri, M. (Kekule Pharma Ltd) IN200500182, WO2006090409 (prior. 28/02/2005, IN).
Kyowa1997 Yoshida, M.; Seishi, T.; Aono, S.; Yamagata, T.; Atsuki, K.; Kumazawa, T.; Takai, H.; Suzuki, K.; Karasawa, A. (Kyowa Hakko Kogyo Kk) US6211227, EP0979821, WO9846587 (prior. 15/04/1997, WO, JP).
Lancet2000 Tucker, G. T. *The Lancet* **2000**, *355*, 1085-1087.
Lundbeck1984 Bogeso, K. P. (Lundbeck AS H) US4650884, EP171943 (prior. 06/08/1984, GB).
Lundbeck1988 Bogeso, K. P.; Perregaard, J. (Lundbeck AS H) US4943590, US34712, EP347066 (prior. 14/06/1988, GB).
Lundbeck1997a Petersen, H. (Lundbeck AS H) WO9819513, US6229026, EP1015416 (prior. 08/07/1997, DK, US).
Lundbeck1997b Petersen, H.; Bogeso, K. P.; Bech Sommer, M. (Lundbeck AS H) WO9819511, US6291689, EP1032566 (prior. 10/11/1997, DK, WO, EP).
Lundbeck1997c Petersen, H.; Bregnedal, P.; Bogeso, K. P. (Lundbeck AS H, Lundbeck & Co AS H) WO9819512, US6258842, EP1042310 (prior. 11/11/1997, DK, WO, EP).
Lundbeck1998a Dall'Asta, L.; Casazza, U.; Petersen, H. (Lundbeck AS H) WO200023431, US6365747, EP1123284 (prior. 20/10/1998, IT).
Lundbeck1998b Petersen, H.; Dahlberg Nielsen, P. (Lundbeck AS H) WO200039112, US2002019546, US6392060, EP1140886 (prior. 23/12/1998, DK).
Lundbeck1999a Weber, B. (Lundbeck AS H) WO200044738, US2002028956, US6441201, EP1150966 (prior. 29/01/1999, DK).
Lundbeck1999b Petersen, H. (Lundbeck AS H) WO9930548, US2002040153, US2005124817, US6849749, US7030252, EP1173431 (prior. 14/04/1999, DK, WO, EP).
Lundbeck1999c Petersen, H.; Rock, M. H.; Svane, H. (Lundbeck AS H, Lundbeck & Co AS H) DK200100060, WO200013648, US2002077353, US2009088469, EP1159274 (prior. 25/06/1999, DK).
Lundbeck1999d Petersen, H.; Rock, M. H. (Lundbeck AS H) DK200100061, WO200011926, US2002061925, US6750358, EP1105382 (prior. 25/06/1999, DK).
Lundbeck1999e Bolzonella, E.; Castellin, A.; Nicole, A. (Lundbeck AS H, Vis Pharma Spa) WO200102383, US2002128497 (prior. 06/07/1999, IT).
Lundbeck1999f (Lundbeck & Co AS H) AR025132 (prior. 22/10/1999, AR).
Lundbeck1999g Rock, M. H.; Petersen, H.; Ellegaard, P. (Lundbeck AS H, Lundbeck & Co AS H) DK200100959, WO200012044, US2002035277, US6407267, EP1228056, US6566540, EP1298124 (prior. 25/10/1999, DK, WO, EP).
Lundbeck1999h Petersen, H. (Lundbeck AS H) WO200132643 (prior. 01/11/1999, DK).
Lundbeck1999i Petersen, H.; Dahlberg Nielsen, P.; Nielsen, P. D. (Lundbeck AS H) WO200132642, US6403813, US2002165403, US6888009, EP1235819 (prior. 01/11/1999, DK).
Lundbeck1999j Petersen, H.; Felding, J. (Lundbeck AS H) WO200147909, EP1246812, US2003060641, US6762307 (prior. 28/12/1999, DK, WO, EP).
Lundbeck1999k Petersen, H.; Rock, M. H. (Lundbeck AS H) WO200149672, US2002198391, US2006079700, US7271273, EP1246813 (prior. 30/12/1999, DK, WO).
Lundbeck2000a Petersen, H. (Lundbeck AS H) WO200151477, EP1254129, US2003013895, US6911548 (prior. 14/01/2000, DK, WO, EP).
Lundbeck2000b Petersen, H.; Rock, M. H. (Lundbeck AS H) WO200162754, EP1259500 (prior. 24/02/2000, DK).
Lundbeck2000c Rock, M. H.; Ahmadian, H.; Petersen, H. (Lundbeck AS H, Lundbeck & Co AS H) WO200143525, US2001027256, US2002004604, US2003083508, US2003114692, US6420574, EP1259501 (prior. 24/02/2000, DK).
Lundbeck2000d Rock, M. H.; Ahmadian, H. (Lundbeck AS H) WO200151478, US2003092761, US6768011, EP1263750 (prior. 03/03/2000, DK).

Lundbeck2000e Petersen, H. (Lundbeck AS H) WO200168631, US2003092919, US6762308, US2004215025, US6992198, EP1265883 (prior. 13/03/2000, DK).

Lundbeck2000f Petersen, H.; Ahmadian, H. (Lundbeck AS H) WO200168629, US2003083509, US2005020670, US6864379, EP1265881 (prior. 13/03/2000, DK).

Lundbeck2000g Petersen, H. (Lundbeck AS H) WO200168628, US2003069304, US6717000, EP1265880 (prior. 13/03/2000, DK).

Lundbeck2000h Petersen, H. (Lundbeck AS H) WO200168630, US2003050484, US6806376, EP126582 (prior. 14/03/2000, DK).

Lundbeck2000i Petersen, H. (Lundbeck AS H) WO200168632, US2003060640, EP1274699 (prior. 16/03/2000, DK).

Lundbeck2000j Petersen, H.; Dancer, R. (Lundbeck AS H) WO200185712, US2003134895, US6660873, EP1296970 (prior. 12/05/2000, DK).

Lundbeck2000k Bolzonella, E.; Castellin, A.; Nicele, A. (Lundbeck AS H) WO200204435 (prior. 06/07/2000, WO, EP).

Lundbeck2000l Petersen, H.; Ahmadian, H.; Dancer, R. (Lundbeck AS H) GB2365865 (prior. 18/08/2000, DK).

Lundbeck2000m Petersen, H.; Ahmadian, H.; Dancer, R. (Lundbeck AS H) DK200101216, WO200216341, WO200216342, US2002025982, US2002026062, US6426422, US6509483, EP1309581, EP1309582 (prior. 18/08/2000, DK).

Lundbeck2000n Villa, M.; Sbrogio, F.; Dancer, R. (Lundbeck AS H) DK200100402, WO200145483, WO200145483, US2002120005, US6455710, EP1181713, EP1462447, EP1181713 (prior. 21/12/2000, DK).

Lundbeck2000o Castellin, A.; Volpe, G.; Sbrogio, F. (Lundbeck AS H) DK200100386, WO200147877, WO200147877, US2002087012, US2003178295, US6855834, EP1181272 (prior. 28/12/2000, DK).

Lundbeck2001a Dancer, R. (Lundbeck AS H) IN200100236 (prior. 15/03/2001, IN).

Lundbeck2001b Humble, R. E.; Christensen, T. V.; Rock, M. H.; Nielsen, O.; Petersen, H.; Dancer, R. (Lundbeck AS H, Lundbeck & Co AS H) WO2003000672, US2004259940, US7112686, EP1412341 (prior. 25/06/2001, DK).

Lundbeck2001c (Lundbeck & Co AS H) AT4854 (prior. 28/06/2001, AT).

Lundbeck2001d Bech Sommer, M.; Nielsen, O.; Petersen, H.; Ahmadian, H.; Pedersen, H.; Brosen, P.; Geiser, F.; Lee, J.; Cox, G.; Dapremont, O.; Suteu, C.; Assenza, S. P.; Hariharan, S.; Nair, U. (Lundbeck AS H, Lundbeck & Co AS H) WO2003006449, US2005065207, US2011065938, EP1409472 (prior. 13/07/2001, DK).

Lundbeck2001e Ahmadian, H.; Petersen, H.; Ahmedian, H. (Lundbeck & Co AS H) WO2003051861, US2005154051, EP1458701 (prior. 14/12/2001, DK, US).

Lundbeck2002a Taoka, N.; Kato, T.; Yamamoto, S.; Yoshida, T.; Takeda, T.; Ueda, Y.; Petersen, H.; Dancer, R.; Ahmadian, H.; Lyngso, L. O. (Lundbeck & Co AS H) WO2004014821, US2007129561, US2011065937, US8067640, EP1534654, EP2360150 (prior. 12/08/2002, DK, US).

Lundbeck2002b Petersen, H.; Dancer, R.; Christiansen, B.; Humble, R. E. (Lundbeck & Co AS H) WO2004056754, US2006020140, US7390913, EP1581483 (prior. 23/12/2002, DK, US).

Lundbeck2003 Pittelkow, T.; Castellin, A.; Sbrogio, F.; Dahlberg, N. P.; Zanon, J.; Sogaard, S.; Humble, R. E. (Lundbeck & Co AS H) WO2004083197, EP1611118, US2008058536, US7687645 (prior. 21/03/2003, DK, US).

Lundbeck2004 Lyngso, L. O. (Lundbeck AS H) WO2005077891, US2007190624, US7582780, EP1716108 (prior. 12/02/2004, DK, US).

Lundbeck2005 Paulon, A.; De Lucci, O.; Castellin, A.; Fabris, F.; Sbrogio, F.; Ceron, E.; Petersen, H.; Dancer, R. (Lundbeck & Co AS H) WO2007054105, US2009247772, EP1951726 (prior. 14/11/2005, DK).

Lundbeck2007 De Faveri, C.; Huber, F. A.; Dancer, R. (Lundbeck & Co AS H) DK200801265, WO2009033488, US2009069582, US8022232, EP2061753, EP2397461 (prior. 11/09/2007, DK, US).

Lundbeck2008 Dancer, R.; De Faveri, C.; Huber, F. A. (Lundbeck & Co AS H) MC200121 (prior. 11/09/2008, MC).

Lundbeck2015 Christensen, T.; Nielsen, M. A.; De Faveri, C.; Huber, F. A.; Dancer, R. J.; Diego, H. (H Lundbeck AS) DK201500520 (prior. 31/08/2015, DK).

Matrix2003 Chava, S.; Bodepudi, H.; Gorantla, S. R.; Abbineni, J.; Chunchu, V. R. (Matrix Lab Ltd) IN200300329 (prior. 21/04/2003, IN).

Natco2001a Pulla, R.; Venkaiah, C. (Natco Pharma Ltd) IN194522, WO200266453, US2004133017, EP1368330 (prior. 22/02/2001, IN).

Natco2001b Pulla, R.; Venkaiah, C. (Natco Pharma Ltd) IN193430 (prior. 23/02/2001, IN).

Natco2001c Muddasani, P.; Nannapaneni, V. (Natco Pharma Ltd) IN194950 (prior. 22/11/2001, IN).

Natco2002a Muddasani, P.; Nannapaneni, W. (Natco Pharma Ltd) AU2002330730, WO2004016602 (prior. 14/08/2002, AU, WO, IN).

Natco2002b Venkaiah Chowdary, N.; Pulla Reddy, M. (Natco Pharma Ltd) IN199096, WO2004043919 (prior. 11/11/2002, IN).

Natco2003a Muddasani, P.; Vattikuti, U.; Nannpaneni, V. (Natco Pharma Ltd) IN207524 (prior. 17/01/2003, IN).

Natco2003b Nannapaneni, V.; Talasila, S.; Nekkanti, S.; Podile, K.; Muddasani, P. (Natco Pharma Ltd) IN200300052, WO2004065375 (prior. 17/01/2003, IN).

Natco2003c Venkaiah Chowdary, N.; Khadgapathi, P.; Pulla Reddy, M.; Lingarao, D. (Natco Pharma Ltd) WO2004094399, AU2003238676 (prior. 21/04/2003, IN, AU, WO).

Natco2003d Muddasani, P.; Talasila, S.; Nekkanti, S.; Nannapaneni, V. (Natco Pharma Ltd) IN200300950 (prior. 20/11/2003, IN).

Natco2003e Pullareddy, M.; Sambasiva Rao, T.; Srinivasa Rao, N.; Venkaiah Chowdary, N. (Natco Pharma Ltd) WO2005049596, AU2003282383 (prior. 20/11/2003, IN, AU, WO).

Natco2004 Pulla Reddy, M.; Rao, T.; Chowdary, N.; Muddasani, P.; Talasila, S.; Nannapaneni, V. (Natco Pharma Ltd) IN200400885, WO2006025071 (prior. 02/08/2004, IN).

Norpharma2000 Dall'Asta, L.; Casazza, U.; Cotticelli, G. (Norpharma Spa, Infosint Sa) WO200153284, US2003009038, US2004171851, US2008249319, US6458973, US6703516, EP1118614, EP1187822 (prior. 18/01/2000, US).

OPRD2007a Elati, C.; Kolla, N.; Vankawala, P.; Gangula, S.; Chalamala, S.; Sundaram, V.; Bhattacharya, A.; Vurimidi, H.; Mathad, V. *Organic Process Research & Development* **2007**, *11*, 289-292.

OPRD2007b Elati, C.; Kolla, N.; Vankawala, P.; Gangula, S.; Chalamala, S.; Sundaram, V.; Bhattacharya, A.; Vurimidi, H.; Mathad, V. *Organic Process Research & Development* **2007**, *11*, 780-780.

OPRD2009a Pesti, J. *Organic Process Research & Development* **2009**, *13*, 22-22.

OPRD2009b Dancer, R.; Lopez de Diego, H. *Organic Process Research & Development* **2009**, *13*, 23-33.

OPRD2009c Elati, C.; Kolla, N.; Mathad, V. *Organic Process Research & Development* **2009**, *13*, 34-37.

OPRD2009d Dancer, R.; Lopez De Diego, H. *Organic Process Research & Development* **2009**, *13*, 38-43.

OPRD2010 Faigl, F.; Thurner, A.; Molnár, B.; Simig, G.; Volk, B. *Organic Process Research & Development* **2010**, *14*, 617-622.

OPRD2012 Pramanik, C.; Hivarekar, R.; Deshmukh, S.; Tripathy, N.; Kotharkar, S.; Chaudhari, A.; Gurjar, M. K. *Organic Process Research & Development* **2012**, *16*, 824-829.

OPRD2013 Vedantham, R.; Vetukuri, V.; Boini, A.; Khagga, M.; Bandichhor, R. *Organic Process Research & Development* **2013**, *17*, 798-805.

Orion2001 Hilden, L.; Huuhtanen, T.; Rummakko, P.; Grumann, A.; Pietikaeinen, P.; Pietikainen, P. (Orion Corp Fermion) FI200100175, WO200260886, US2002115872, US2003130527, US6579993, EP1355897 (prior. 30/01/2001, FI).

Orion2002 Hilden, L.; Rummakko, P.; Grumann, A.; Pietikaeinen, P.; Pietikainen, P. (Orion Corp Fermion) WO2004011450, US2005209467 (prior. 30/07/2002, FI).

OSC1943 Huntress, E.; Shriner, R. *Org. Synth., Coll. Vol. 2*, 459.

Pharmachem2001 Malik, A. A.; Palandoken, H.; Stringer, J. A.; Huang, D.; Romero, A.; Dapremont, O. (Pharmachem Technologies Ltd) WO2003029236, US2003153774, US2006074253, US6967259, EP1430044 (prior. 24/09/2001, US).

PNPBP1982 Hyttel, J. *Progress in Neuro-Psychopharmacology and Biological Psychiatry* **1982**, *6*, 277-295.

Qilu2011 Fan, C.; Leng, C.; Lin, D.; Liang, B.; Wang, J. (Qilu Pharm Co Ltd) CN102757414 (prior. 25/04/2011, CN).

Ranbaxy2001a Biswas, S.; Sharma, T.; Kumar, Y.; Sathyanarayana, S.; Vijayaraghavan, B.; Biwas, S.; Swargam, S.; Bakthavathsalan, V.; Yatendra, K. (Ranbaxy Lab Ltd) IN200100264, WO200272565, EP1370545, US2005085534 (prior. 09/03/2001, IN).

Ranbaxy2001b Sathyanarayana, S.; Kumar, Y.; Sharma, T.; Biswas, S.; Vijayraghavan, B.; Swargam, S. (Ranbaxy Lab Ltd) IN192057, WO2003007872, US2004254385, EP1412340 (prior. 19/07/2001, IN).

Ranbaxy2005 Kumar, N.; Nayyar, S.; Gupta, M. (Ranbaxy Lab Ltd) IN200500714, WO2006103550 (prior. 31/03/2005, IN).

ResolutionC2000 Greenwood, A.; McHattie, D.; Rechka, J.; Hedger, P.; Gamble, M. (Resolution Chem Ltd) WO200166536, US2004014997, US7002025, EP1268461 (prior. 07/03/2000, GB).

Sandoz2005 Albert, M.; Berger, A. (Sandoz Ag) EP1736462 (prior. 22/06/2005, EP).

Sandoz2006 Albert, M.; Sturm, H.; Berger, A.; Kremminger, P. (Sandoz Ag) WO2007082771, US2011009648, US8288569, EP1966127 (prior. 23/01/2006, GB).

Sekhsaria2001 Oak, G.; Eappan, P.; Bhobe, A.; Chodankar, N. (Sekhsaria Chem Ltd) EP1288211 (prior. 28/08/2001, US).

Sekhsaria2002 Chodankar, N.; Bhobe, A.; Oak, G.; Eappan, P. (Sekhsaria Chem Ltd) US2004077870, US6812355 (prior. 22/10/2002, US).

Sepracor2004 Barberich, T. (Sepracor Inc) WO2005079787, US2005192344, US7714023, EP1720541 (prior. 17/02/2004, US).

ShanghaiIOC2016 Tang, W.; Huang, L.; Zhu, J. (Shanghai Insititute of Organic Chemistry Chinese Acad of Sciences) CN105732249 (prior. 27/01/2016, CN).

ShanghaiIP2008 Pan, H.; Wang, Q.; Yuan, Z.; Zhu, X. (Shanghai Inst Pharm Ind, Shanghai Inst Pharm Ind Pharm Res Co Ltd) CN101607951 (prior. 20/06/2008, CN).

Shodhana2008a Gudipati, S.; Kotaru, S.; Thota, G.; Giridhar, T.; Srinivasa Rao, K.; Srinivasulu, G.; Srinivasa, R. (Shodhana Lab Ltd) IN200801358, WO2009147687, US2011230666 (prior. 03/06/2008, IN).

Shodhana2008b Giridhar, T.; Srinivasa Rao, K.; Srinivasulu, G.; Gudipati, S.; Kotaru, S. R.; Thota, G. (Shodhana Lab Ltd) IN200801450, WO2010004575, US2011092719 (prior. 16/06/2008, IN).

Sichuan2010 Li, W.; Li, C.; Cui, D. (Sichuan Kelun Pharm Res Co Ltd) CN102190641 (prior. 23/03/2011, CN).

Sumika1999a Ikemoto, T.; Kobori, K.; Iki, M. (Sumika Fine Chem Co Ltd) JP2001106681 (prior. 07/10/1999, JP).

Sumika1999b Ikemoto, T.; Kobori, K.; Iki, M. (Sumika Fine Chem Co Ltd) JP2001114773 (prior. 14/10/1999, JP).

Sumika1999c Ikemoto, T.; Igi, M.; Ikemoto, T.; Igi, M. US2001056194, US2002045770, US2002151728, US2004044232, US2005080278, US2006167285, US6310222, US6395910, US6433195, US6930211, EP1095926 (prior. 01/11/1999, JP).
Sumika2000a Ikemoto, T.; Gao, W.; Arai, N.; Igi, M. (Sumika Fine Chem Co Ltd) US2004138497, EP1428813 (prior. 17/02/2000, JP).
Sumika2000b Ko, E.; Ikemoto, T.; Iki, M. (Sumika Fine Chem Co Ltd) JP2002114770 (prior. 02/10/2000, JP).
Sumika2001 Arai, N.; Tetsuya, I.; Masami, I. (Sumika Fine Chem Co Ltd) JP2003206284 (prior. 28/12/2001, JP).
Sumitomo2000 Ikemoto, T.; Arai, N.; Iki, M.; Ko, E. (Sumitomo Chem Co Ltd) JP2005008642 (prior. 17/02/2000, JP).
Sumitomo2004a Ikemoto, T.; Watanabe, Y. (Sumitomo Chem Co Ltd) JP2005247710, WO2005082842 (prior. 01/03/2004, JP).
Sumitomo2004b Ikemoto, T.; Watanabe, Y. (Sumitomo Chem Co Ltd) JP2005263741, WO2005090290 (prior. 19/03/2004, JP).
Sumitomo2004c Ikemoto, T.; Watanabe, Y. (Sumitomo Chem Co Ltd) JP2005330249, WO2005113530 (prior. 21/05/2004, JP).
Sumitomo2004d Ikemoto, T.; Watanabe, Y. (Sumitomo Chem Co Ltd) JP2006008603, WO2006001494 (prior. 25/06/2004, JP).
Sumitomo2004e Arai, N.; Katsura, T.; Mizuno, T. (Sumitomo Chem Co Ltd) JP2006008617, WO2006001529 (prior. 28/06/2004, JP).
Sumitomo2004f Arai, N.; Katsura, T.; Mizuno, T. (Sumitomo Chem Co Ltd) JP2006176490, WO2006057261 (prior. 29/11/2004, JP).
Sumitomo2005 Arai, N.; Ikemoto, T. (Sumitomo Chem Co Ltd) JP2007045807 (prior. 13/07/2005, JP).
Sun2002 Thennati, R.; Kilaru, S.; Chinnapillai, R.; Patel, N. S.; Rajamannar, T. (Sun Pharm Ind Ltd) IN200200018, WO2003057132, US2005043550, US7148364 (prior. 07/01/2002, IN).
Sun2003 Borase, A. P.; Patel, N. S.; Kilaru, S.; Thennati, R. (Sun Pharm Ind Ltd) IN200300602, US2005004380, US7019153, EP1486492 (prior. 10/06/2003, IN).
Sun2004 Periyandi, N.; Kilaru, S.; Thennati, R. (Sun Pharm Ind Ltd, Sun Pharma Global Fze) IN200400912, WO2006021971, US2008177096, US2009326249, US2010249437, US2011251410, US7964742, US7989645, US7790935, EP1797060, EP2141156 (prior. 23/08/2004, IN).
Suven2004 Reddy, A.; Chinnapillai Rajendiran, R.; Indrasena Reddy, I.; Venakt Jasti, V. (Suven Life Sciences Ltd) WO2005111010 (prior. 13/05/2004, IN).
Synteco2003 Angeli, R.; Possenti, M. (Synteco Spa) IT1343061, EP1502907 (prior. 28/07/2003, IT).
Synthon2007 Zhu, J. (Synthon Bv) US2008281110, US7566793, EP1988086 (prior. 23/04/2007, US).
T1998 Warrener, R. N.; Liu, L.; Russell, R. A. *Tetrahedron* **1998**, *54*, 7485-7496.
T2011 Partridge, B. M.; Thomas, S. P.; Aggarwal, V. K. *Tetrahedron* **2011**, *67*, 10082-10088.
TA2004 Solares, L. F.; Brieva, R.; Quirós, M.; Llorente, I.; Bayod, M.; Gotor, V. *Tetrahedron: Asymmetry* **2004**, *15*, 341-345.
Tokuyama2016a Oba, Y.; Yokoo, Y.; Tanaka, K. (Tokuyama Corp) JP2018012653 (prior. 20/07/2016, JP).
Tokuyama2016b Yokoo, Y.; Oba, Y.; Miyaoku, T. (Tokuyama Corp) JP2018012654 (prior. 20/07/2016, JP).
Tokuyama2016c Yokoo, Y.; Oba, Y.; Miyaoku, T. (Tokuyama Corp) JP2018016568 (prior. 26/07/2016, JP).
Tokuyama2016d Yamada, K.; Miyaoku, T.; Oba, Y. (Tokuyama Corp) JP2018016569 (prior. 26/07/2016, JP).
Torcan2002 Tse, H. (Torcan Chemical Ltd) CA2381341, WO2003087081, US2006009515, EP1495013 (prior. 09/04/2002, CA).
UnivZhejiang2006a Xu, F.; Lin, X. (Univ Zhejiang) CN1962651 (prior. 21/11/2006, CN).
UnivZhejiang2006b Xu, F.; Lin, X. (Univ Zhejiang) CN1962652 (prior. 21/11/2006, CN).
UnivZhejiang2006c Wang, S.; Yang, L.; Wu, J.; Xu, G. (Univ Zhejiang) CN1974542 (prior. 07/12/2006, CN).
Wockhardt2003 Jaweed, S.; Bhargav, K.; Mishra, K.; Farooqui, M. (Wockhardt Ltd) WO2005042473, US2006293530, EP1678122 (prior. 28/10/2003, WO, EP).
ZhejiangAP2008 Zhu, X.; Ye, X.; Wu, X.; Yuan, Z.; Shi, H.; Wang, Q. (Zhejiang Aotuokang Pharm Group Co Ltd) CN101265215 (prior. 18/04/2008, CN).
ZhejiangHP2008 Li, S.; Yu, L.; Zhao, G. (Zhejiang Huahai Pharm Co Ltd) CN101538257 (prior. 21/03/2008, CN).
ZhejiangHP2010 Zhang, J.; Gan, L.; Zhou, S.; Jian, F. (Zhejiang Huahai Pharm Co Ltd) CN102190600 (prior. 13/03/2010, CN).
ZhejiangHP2014a Liang, Z.; Xiao, W.; Huang, W.; Hu, S.; Peng, C. (Zhejiang Huahai Pharm Co Ltd) CN104262306 (prior. 31/08/2014, CN).
ZhejiangHP2014b Liang, Z.; Hu, S.; Peng, C.; Huang, W.; Lu, Q.; Tu, G. (Zhejiang Huahai Pharm Co Ltd) WO2016074225, US2017240505, EP3219702 (prior. 14/11/2014, WO).
ZhejiangHP2015 Liang, Z.; Xiao, W.; Peng, C.; Huang, W.; Tu, G. (Zhejiang Huahai Pharm Co Ltd) WO2016197320, EP3309142, US2018162805 (prior. 09/06/2015, WO).

5 Sitagliptin

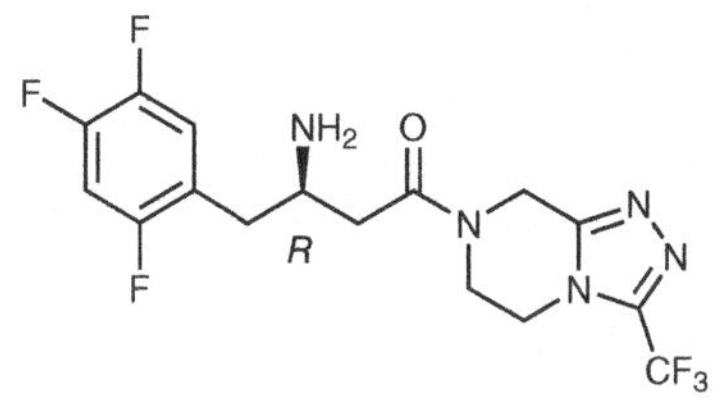

Year of discovery: 2001 (Merck & Co.)
US approval: (as phosphate salt, Januvia®, Merck) October 2006
Drug cat.: oral antihyperglycemic (antidiabetic drug), dipeptidyl peptidase-4 (DPP-4) inhibitor
Uses: Antidiabetic (for treatment of type 2 diabetes)

Type 2 diabetes has become one of the most serious public health problems at present and the search for effective and safe therapies has been a research priority for the pharmaceutical industry.

Dipeptidyl peptidase-4 inhibitors (DPP-4 inhibitors) are a recent class of drugs that indirectly regulate the secretion of insulin by inhibiting the degradation of glucagon-like peptide 1 (GLP-1), which plays an important role in glucoregulation. Sitagliptin (**1**) was the first drug of the DPP-4 inhibitor class and can be used alone or in combination with other oral antihyperglycemic agents (such as metformin or a thiazolidinedione) for treatment of diabetes mellitus type 2. Several other DPP-4 inhibitors have since been approved (Figure 5.1) but Sitagliptin is still remains a major key drug in this area.

These drugs show high selectivity for the inhibition of DPP-4 whilst showing little activity towards other structurally related peptidases and demonstrate minimal side effects.

With around 200 million people diagnosed with type 2 diabetes worldwide, this efficacious class of drugs is clearly of high pharmacological value and an especially interesting target for many pharmaceutical companies.

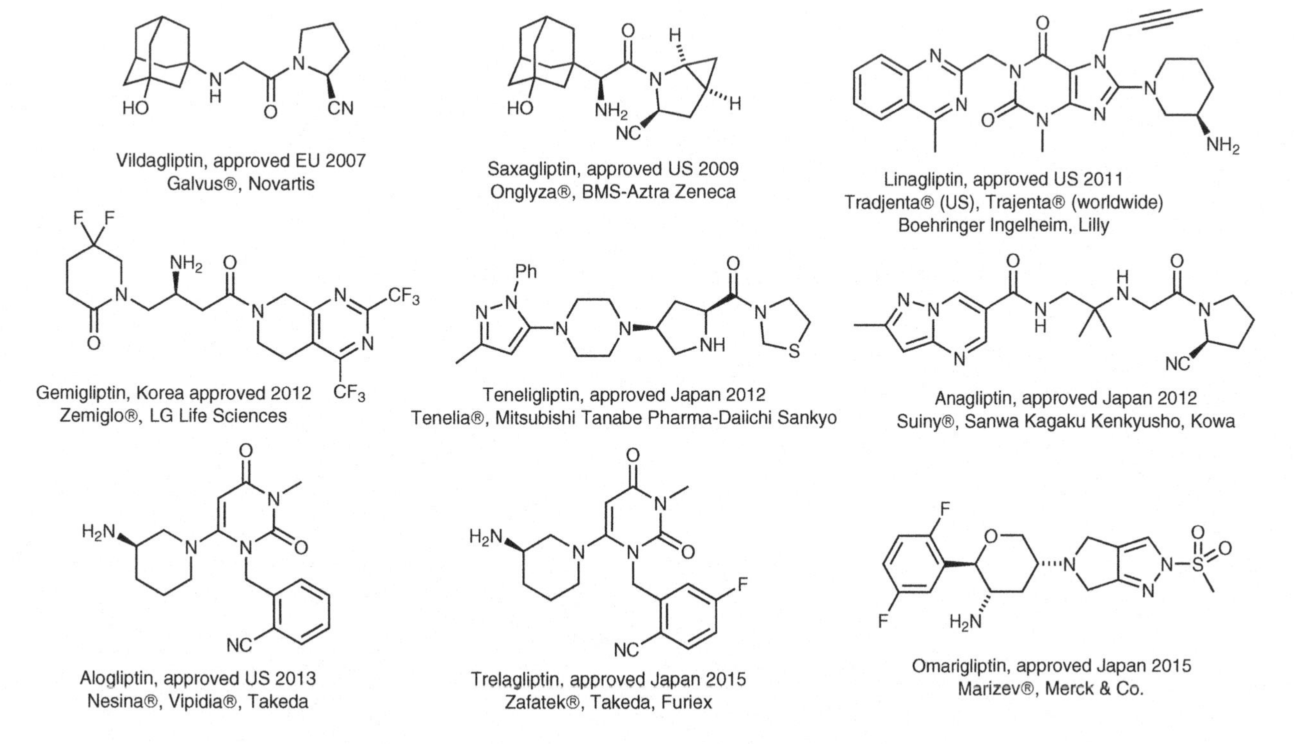

Figure 5.1 Most important approved DPP-4 inhibitors drugs

Retrosynthesis in the Manufacture of Generic Drugs: Selected Case Studies, First Edition, Pedro Paulo Santos and William Heggie.

5.1 WORLDWIDE SALES AND PATENT STATUS

The following graphs illustrate how research carried out in the pharmaceutical industry towards the production of the API and worldwide sales of the finished drug product evolved.

- US basic patent term 22/07/2022 (US6699871, Merck2001).
- Annual production[1]: 2019, 492,395 kg; 2018, 455,407 kg; 2017, 421,024 kg.
- Annual WW sales[1]: 2019, 11938.7M USD; 2018, 11474.8M USD; 2017, 10421.9M USD.

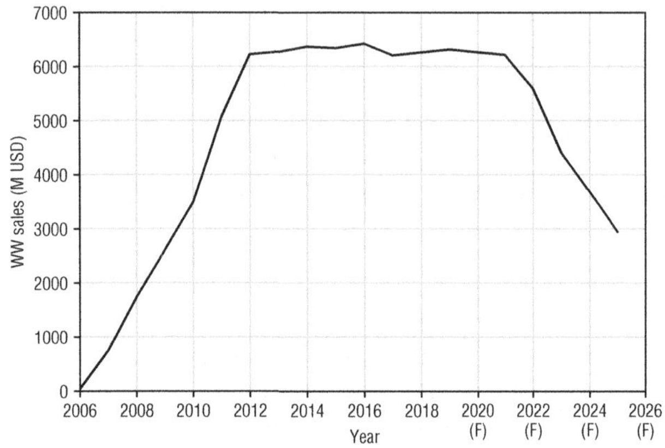

FIGURE 5.2 SITAGLIPTIN WW SALES EVOLUTION (IN M USD)

Source: Data from GlobalData [Online] https://globaldata.com/

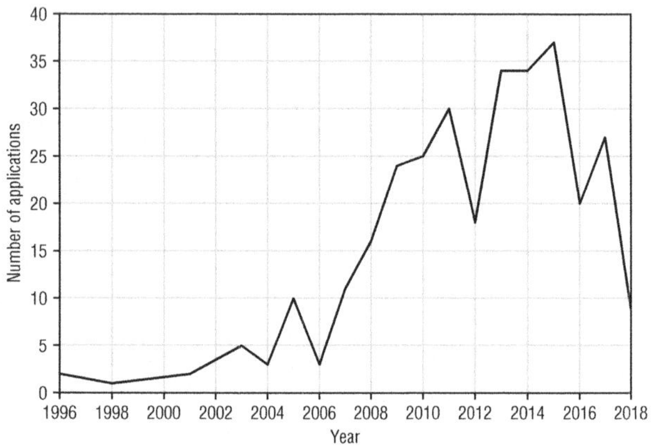

FIGURE 5.3 NEW PATENT APPLICATIONS ON SITAGLIPTIN FILED BY YEAR

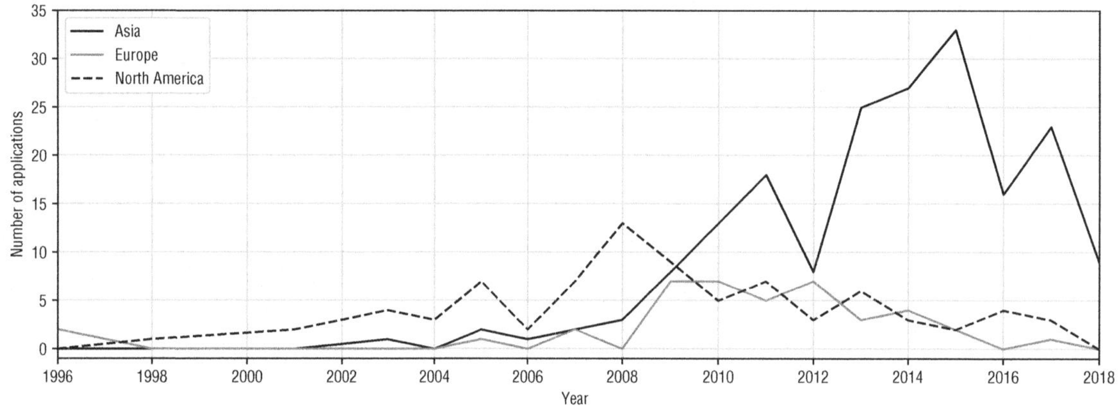

FIGURE 5.4 PATENT APPLICATIONS ON SITAGLIPTIN FILED BY REGION

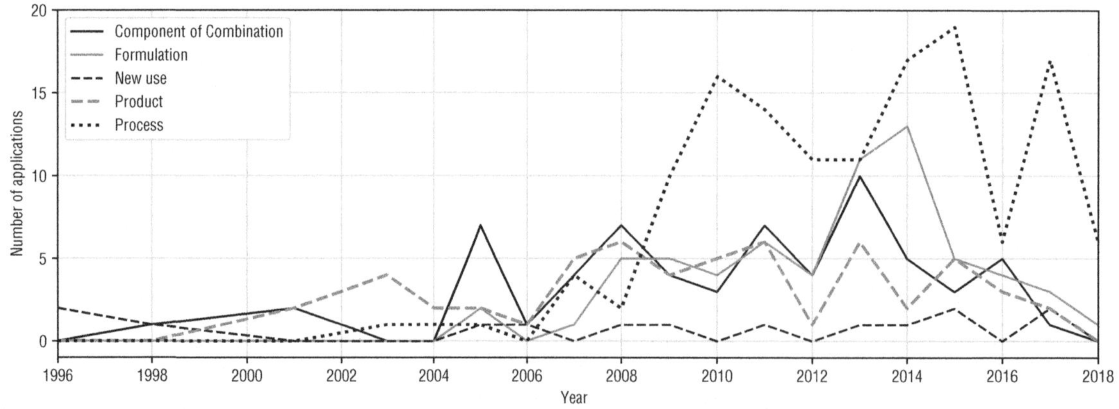

FIGURE 5.5 PATENT APPLICATIONS ON SITAGLIPTIN FILED BY TYPE OF PATENT

1 Newport Modular Platform – Clarivate Analytics [Online] https://clarivate.com/cortellis/ [2020, April]

5.2 SITAGLIPTIN STRUCTURE AND GENERAL RETROSYNTHETIC ANALYSIS

Sitagliptin is a chiral amine containing six fluorine atoms (Figure 5.6). Three structural blocks are easily recognized; a trifluoro-aromatic phenyl piece, a tetrahydrotriazolopyrazine nitrogen heterocycle (THTP, 4,5,6,7-tetrahydrothieno-[3,2-c]pyridine) and a four-atom chiral amine chain linking the two cyclic units.

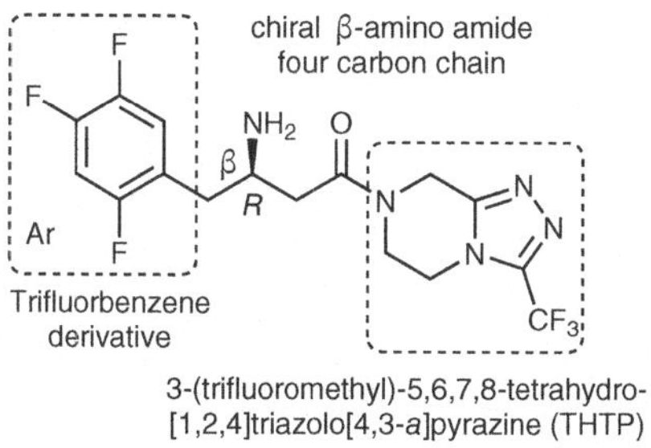

FIGURE 5.6 KEY FEATURES OF SITAGLIPTIN STRUCTURE

The fluorinated aromatic and tetrahydrotriazolopyrazine ring units can be regarded as obvious starting blocks for any synthesis of Sitagliptin (Scheme 5.1). The main tasks for the Process Chemist therefore, remains to devise a reliable and cheap process to produce the chiral amine in high yield, in high enantiomeric purity and to link the peripheral ring units to the carbon chain in a simple and robust fashion.

The construction of the chiral amine is obviously a key aspect in any synthesis design.

Conversion of a racemic mixture to chirally pure Sitagliptin by a resolution process would be expected to be viable. The primary amine attached to the chiral carbon atom can give rise to a diastereomeric salt with a chiral enantiopure acid, which in turn should be amenable to resolution to successfully produce Sitagliptin (*R*). This strategy will inevitably produce a considerable amount of waste, the enantiomeric *S*-amine. Unfortunately, direct racemization is not chemically feasible and any methodology for the recycling of the *S*-isomer involves an indirect route. A possible solution is an oxidation process to produce an unsaturated compound (with a C=C or a C=N bond) and conversion back to racemic Sitagliptin by a reductive transformation. The formation of side-products, particularly during the oxidation step, is to be expected and would have to be controlled to make the process viable. Although Sitagliptin-resolution routes cannot not be excluded, they are far from convenient to use at industrial scale. For economic sustainability, recycling of the unwanted enantiomer is desirable. In addition, the resolution itself requires two additional reactions, salt formation and re-conversion to the free amine, after the overall molecular structure has been put in place. All this results in prolonged plant occupation and a reduced overall throughput. A synthetic route based on the resolution of an early intermediate would seem to be a better option but even so, a route involving an asymmetric synthesis appears to be much more favourable.

Three main tasks have to be realised during the synthesis of Sitagliptin:

- Produce the chiral primary amine chain (*R* isomer)
- Form of the amide bond between the heterocycle and the carboxylic acid carbon chain;
- Connect the trifluoro-aromatic ring to the carbon chain.

The order of these three events may differ, according to the synthetic strategy.

5.2.1 Disconnecting the heterocycle first

Initiating the retrosynthesis of Sitagliptin by cleaving the amide C-N bond is one promising option, resulting in a convergent synthetic route with the formation of two large units (Scheme 5.2). The complex THTP heterocycle is removed, allowing for the elaboration of this building block separately. It is obvious for the backward amide-forming reaction the primary amine should be protected to avoid unwanted coupling side reactions.

A key focus of this retrosynthetic analysis is on the production of the chiral amine block. Several options for its formation can be envisaged and virtually all manner of strategies to produce a chiral pure compound may be used: resolution of chiral intermediates; chemical and biochemical asymmetric synthesis and use of a chiral pool. In a relatively simple approach, the primary amine can be obtained from a ketone by a reductive amination (Scheme 5.3). The conjugated enamine is more stable than the imine and this fact can be seen as advantageous for the synthetic process. Asymmetric catalytic versions of the enamine hydrogenation are well known – Ryōji Noyori and William S. Knowles shared a part of

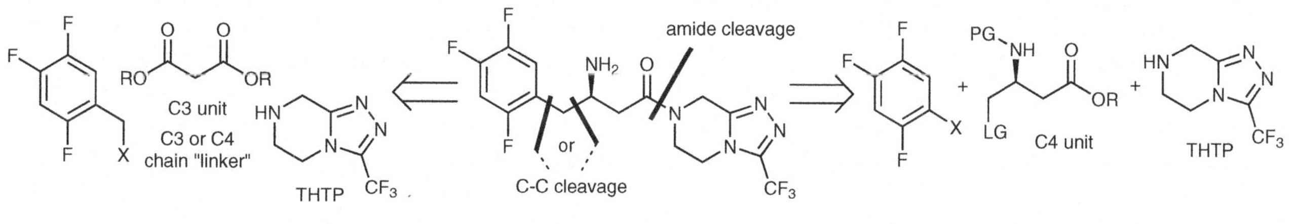

SCHEME 5.1 OUTLINE OF THE MAIN STRUCTURE CLEAVAGES FOR SITAGLIPTIN

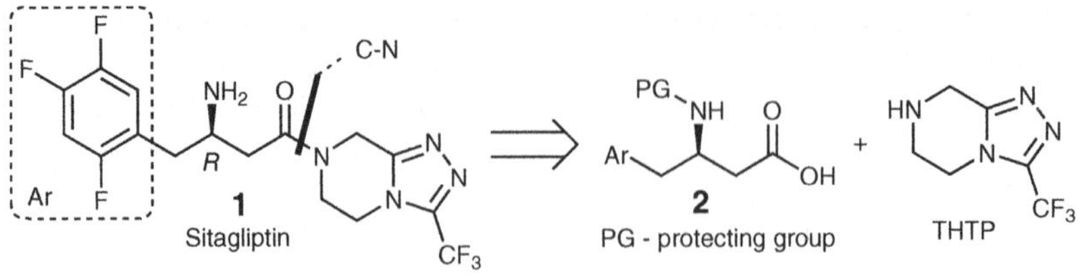

Scheme 5.2 Disconnection of Sitagliptin by amide cleavage

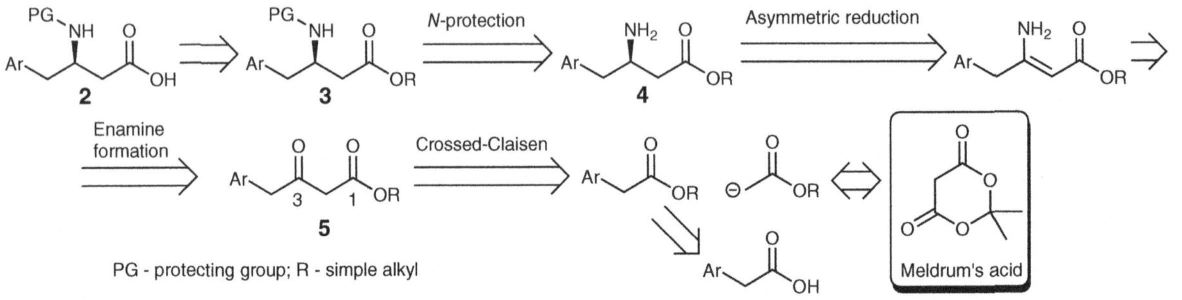

Scheme 5.3 Retrosynthesis of Sitagliptin intermediate *via* enamine

the 2001 Nobel Prize in Chemistry[2] for the study of such chirally catalysed hydrogenations.

To obtain the intended β-keto ester (or 3-keto ester, **5**) the crossed-Claisen reaction has to be fine-tuned. Meldrum's acid, which is readily available, is a good option for the nucleophile as strong basic conditions are avoided and undesirable aldol side reactions minimised.

Several retrosynthetic approaches for the chiral amine-ester intermediate are summarized in Scheme 5.4. Classical methods for the preparation of primary amines from ketones *via* their corresponding imines/enamines or oximes followed by reduction are especially applicable. Asymmetric reductive amination of the β-keto ester precursor is clearly the simplest and shortest synthetic process.

If a non-asymmetric hydrogenation is employed a resolution is necessary to obtain the desired *R* isomer.

Asymmetry can be induced by exploitation of a chiral enamine intermediate. A chiral benzyl amine presents an excellent opportunity as the benzylic group can be removed by hydrogenolysis, albeit under different conditions than those of the reductive amination. Thus, asymmetric induction is assured before cleavage of the benzyl group. Reduction of the enamine can also be achieved by using hydride reducing agents such as sodium triacetoxy borohydride.

It should be noted that although the amine can be regarded as a chiral auxiliary it is in fact a reagent as a stoichiometric amount is required. It is not recoverable as it suffers chemical transformation, i.e. after cleavage of the N-C benzylic bond an aromatic moiety is formed which does not contain the amine group. In summary, the chiral benzylamine plays two roles: placing the nitrogen on, and assuring the asymmetry at, the C3 carbon atom. The necessity for the debenzylation means that this sequence has one more step in comparison to the enantioselective version.

The amine group can also be introduced *via* a bimolecular process from an activated hydroxy ester. Azide or phthalimide are feasible nucleophiles. Side reactions are to be expected and hence strict control of the reaction conditions is required. The formation of a conjugated α,β-unsaturated derivative by an elimination process (E1 and E1cB) is one possibile side reaction, although it might be envisaged that Michael addition of the nucleophile could occur to reform the amine, but with concomitant loss of chirality. A chiral β-hydroxy intermediate can be synthesised from the corresponding β-keto ester (**5**) by chemical or enzymatic reduction. The Corey-Bakshi--Shibata reduction would be a good candidate for this transformation.

The relative positions of the primary amine and the carboxylic ester (1,3) suggests a conjugate addition should serve to achieve this key objective (Scheme 5.5). To be useful, an asymmetric addition is desirable but may not be easy to accomplish. As before, a NH_2^- synthon is required together with conditions that avoid the formation of an amide.

The primary amine can also be put in place by a rearrangement of a carboxylic acid derivative (for example by a Curtius or an analogous rearrangement, Scheme 5.6). The mono ester of a 1,4-dicarboxylic intermediate would be the key starting material in this case. However, in comparison with the previous options, this strategy is less interesting for implementation on an industrial scale as the rearrangement step raises safety concerns.

2 The other half of the prize was awarded to K. Barry Sharpless for his study in chirally catalysed oxidation reactions, the Sharpless epoxidation.

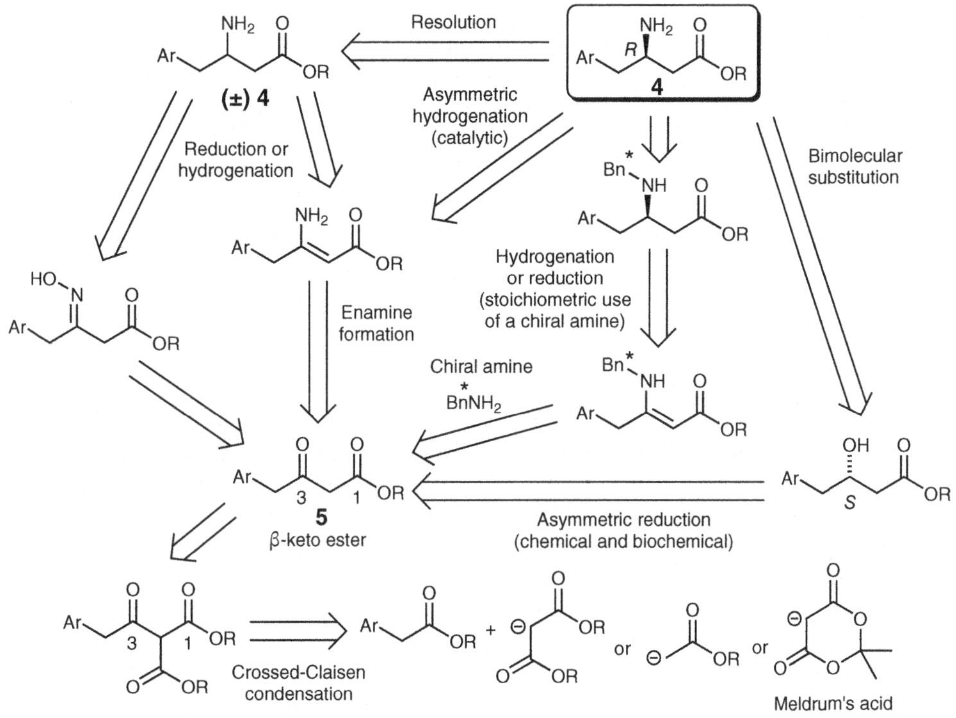

Scheme 5.4 Basic retrosynthesis of the chiral amino ester

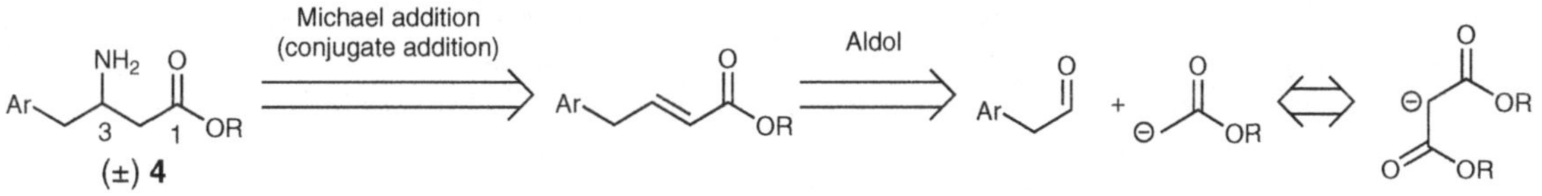

Scheme 5.5 Retrosynthesis of β-amino ester by a retro-Michael addition

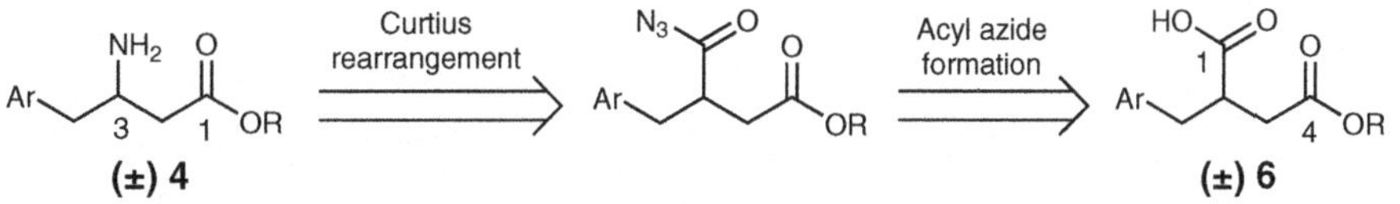

Scheme 5.6 Derivation the amino ester using a Curtius rearrangement as a key step

An alternative to creating the chiral amine by asymmetric synthesis is to use a chiral pool approach. L-Aspartic acid fulfils all the necessary requirements to produce the four-carbon atom segment (Scheme 5.7). Protection of the nitrogen and selective activation of the α-carboxylic acid, to make it amenable for a Friedel-Crafts type acylation, are necessary for this approach. The resulting ketone can be converted to the desired methylene compound by hydrogenolysis.

Many other chiral pool sources can be explored. Chirally pure *S*-epichlorohydrin is yet another obvious starting material. Retrosynthetic approaches to *S*-epichlorohydrin are shown in Scheme 5.8.

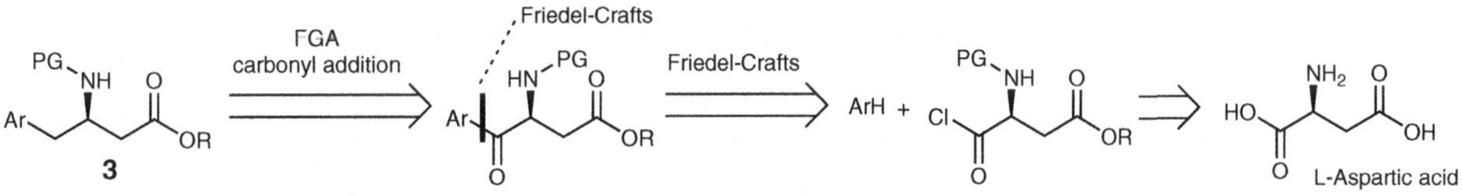

Scheme 5.7 L-aspartic acid chiral pool strategy for the Sitagliptin carbon chain

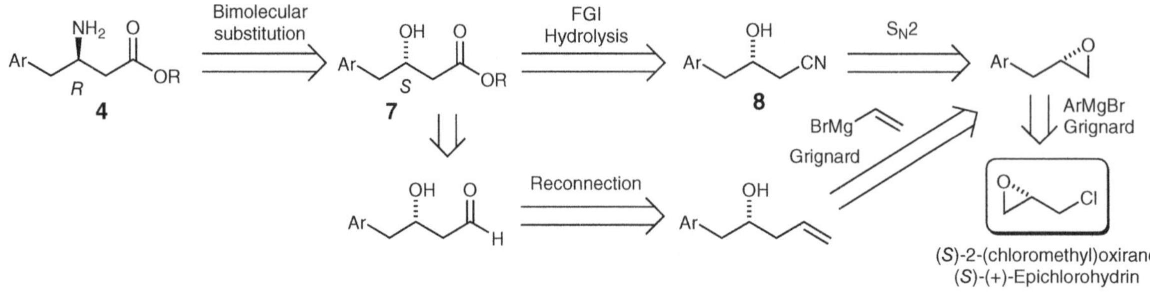

Scheme 5.8 Retrosynthesis of the chiral amino ester to (*S*)-epichlorohydrin

Tailor-made aziridines also qualify as good chiral intermediates for side chain construction. A possible retrosynthesis to L-aspartic acid is shown in Scheme 5.9.

All of the above retrosynthetic analyses have been based on the assumption of an initial disconnection of the heterocyclic block and the major focus has been on the production of the chiral amine carbon bridge. If the disconnection of the heterocycle is postponed new retrosynthetic approaches are opened up.

5.2.2 Retrosynthetic analysis starting by transformation of the chiral amine

The approach where chiral amine transformation is the first retrosynthetic step produces several other synthetic routes. Disconnecting the nitrogen to give a keto carboxylic acid derivative results in a very similar situation as to that shown in Scheme 5.4. In this case, the heterocycle, and not an ester group, is in place. Thereafter, disconnection at the amide bond gives rapid access to readily available starting materials. Obviously, the asymmetric hydrogenation is a required key synthetic step (Scheme 5.10).

This strategy avoids the use of an amine protecting group (as in 5.2.1), resulting in a shorter synthesis.

The nitro group is an effective precursor of the primary amine (Scheme 5.11). The powerful anion-stabilizing effect of the nitro group is the basis of the C-C disconnections suggesting the application of the Henry "aldol" reaction in the synthesis. Although on paper it appears easy to construct Sitagliptin in this way, the use of nitro chemistry and nitro-containing intermediates has drawbacks. There are safety hazards associated with nitro compounds and the reduction of the nitro group itself. Hence, this approach when compared to several other options is not very favourable. In addition, racemic Sitagliptin would be produced, making an additional resolution step obligatory.

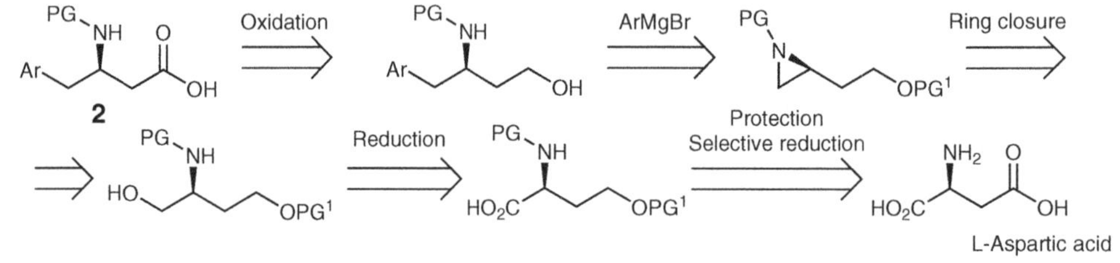

Scheme 5.9 Retrosynthesis of the chiral amino ester *via* a chiral aziridine intermediate

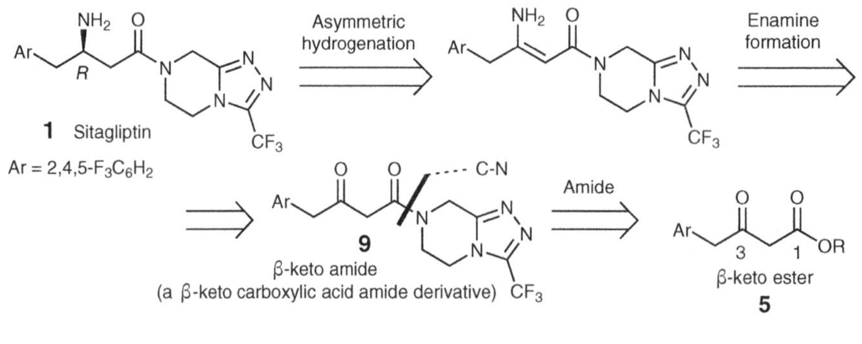

Scheme 5.10 Focusing on the chiral amine as the first retrosynthetic step of Sitagliptin

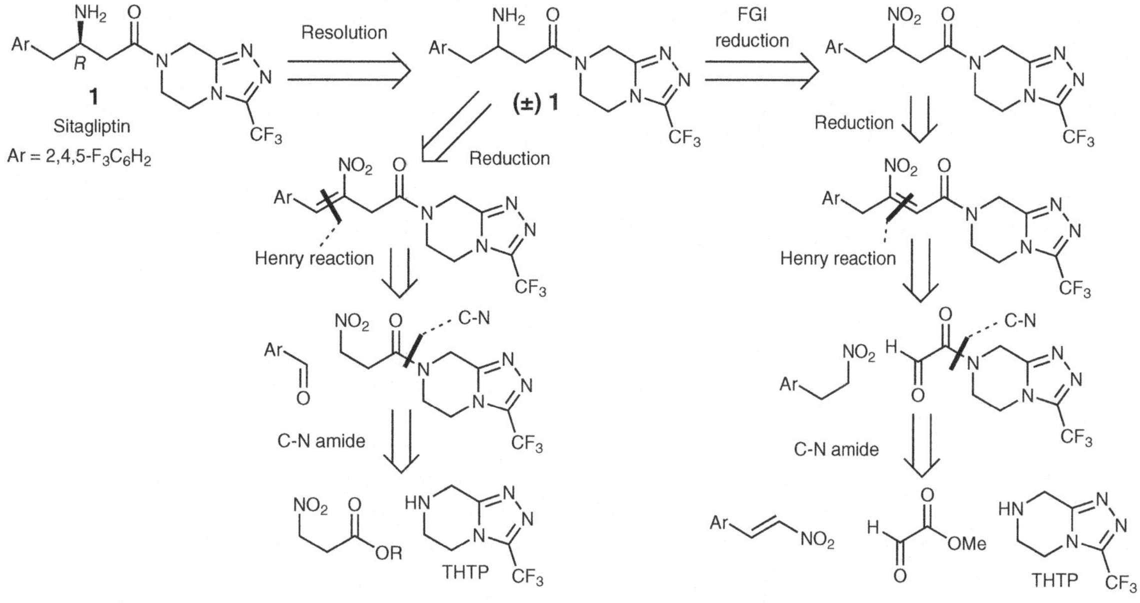

SCHEME 5.11 NITRO-APPROACH FOR SITAGLIPTIN RETROSYNTHESIS

5.2.3 Retrosynthetic analysis starting from haloaromatic ring

The last major retrosynthetic strategy starts by disconnecting the trihaloaromatic group first. Conversion of the benzylic methylene in Sitagliptin to a carbonyl group aids in the disconnection of the aromatic ring (Scheme 5.12). Grignard addition to a Weinreb amide and electrophilic aromatic substitution are two of the synthetic possibilities, with the former presenting as more favourable. This approach favours the use of a chiral pool, once more readily available L-aspartic acid being a staring material of choice. However, the conversion of the carbonyl group to a methylene group is generally less clean and the formation of impurities during this reaction may diminish the appeal of these strategies for industrial application.

5.2.4 Retrosynthesis analysis summary and most efficient routes

Disregarding the specifics of any of the synthetic chemistry published to date, it might be expected that several options are available to manufacture Sitagliptin. One of the best options to present itself is the use of a hydrogenation to produce the chiral *R*-amine.

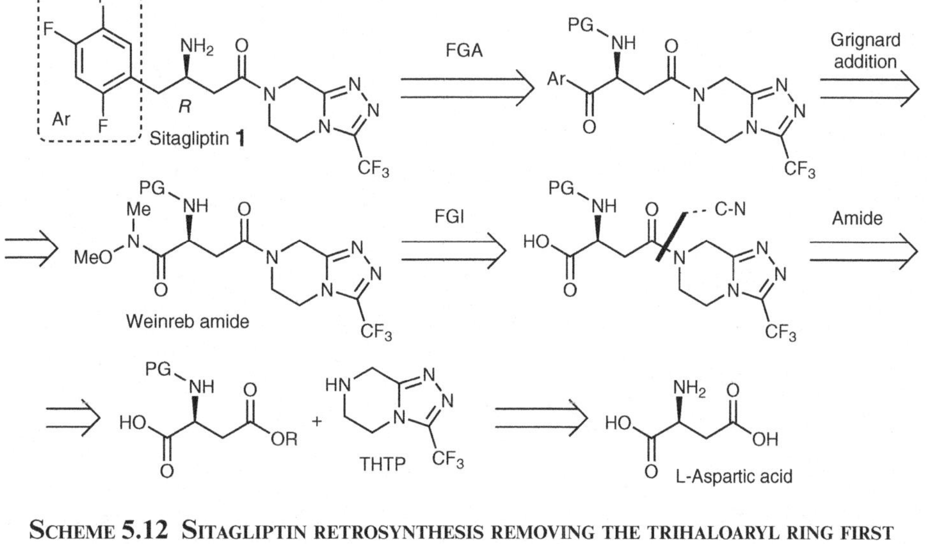

SCHEME 5.12 SITAGLIPTIN RETROSYNTHESIS REMOVING THE TRIHALOARYL RING FIRST

Regarding this key step, two main strategies are to be anticipated (Scheme 5.13):

- Classical hydrogenation of a chiral enamine intermediate (asymmetry is induced through the amine precursor of the enamine – a stoichiometric amount of this amine being required). Suitable enamines are obtained from β-keto carboxylic acid derivatives and chiral benzylamines.
- Catalytic asymmetric hydrogenation. Asymmetry is induced during reduction through the intervention of a chiral catalyst.

As noted previously, the chiral enamine approach requires a stoichiometric use of the chirally pure benzylamine, an extra step is needed to remove the benzyl group and the starting benzylamine cannot be recovered. Efficiency and costs comparisons are critical when the route to implement is chosen. However, it seems clear-cut that preferred routes involve those using either of the asymmetric options described (Scheme 5.14). It should also be remembered that key to the success of these approaches is adequate process development work to realise their true potential in terms of yield, enantiomeric purity, throughput and minimization of the catalyst and its recovery. Conceptually, the removal of the heterocycle first by cleavage of the amide bond cleaves the molecule into two large units and produces a convergent synthetic route. Selecting a preferred route from these two options, nonetheless, depends on several factors.

An outline of the two synthetic approaches is shown in Scheme 5.15. In addition to the step where asymmetry is introduced, the amide-forming reaction is critical to the success of the process. Operating conditions (the simpler the better) and low cost reactants are obviously valued. When the heterocycle is introduced early on in the sequence, no protecting group is required. On the contrary, an amine-protecting group is necessary to avoid unwanted coupling reactions when the amide formation is carried out late in the sequence and more steps are required.

However, it can be anticipated that the cost of the catalyst for each option will be significant. Hence, the point in the sequence where the asymmetric step is carried out becomes important for the cost and efficiency of an industrial synthesis. In general, because yields are never 100%, for each additional synthetic step, correspondingly less catalyst is required to obtain the same amount of final drug compared to when it is used earlier in the sequence. That is, when the asymmetric hydrogenation is postponed till the end of the synthesis lower quantities

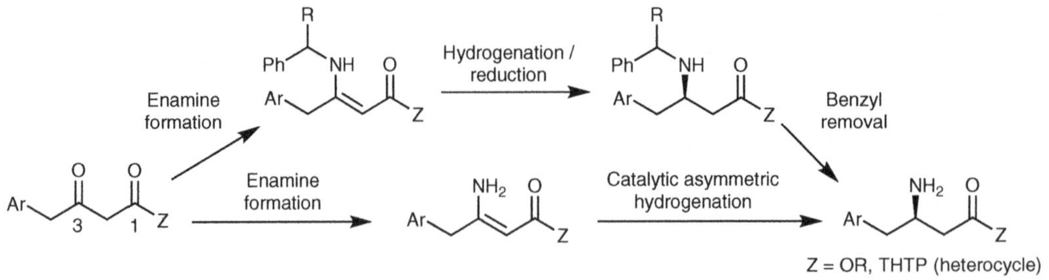

Scheme 5.13 The two main approaches to produce the chiral amine in Sitagliptin from a carbonyl group

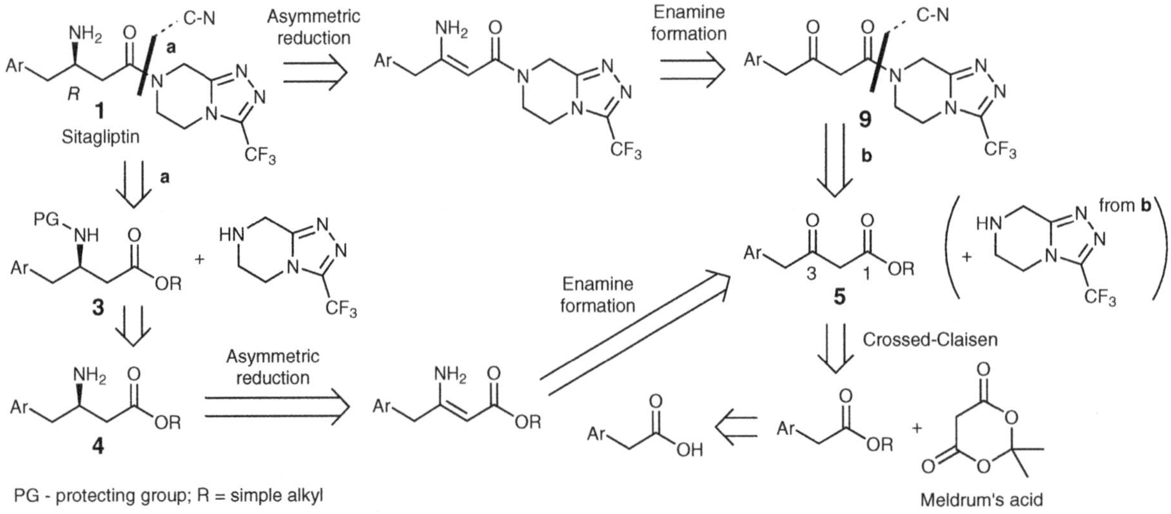

Scheme 5.14 Most promising retrosyntheses for the industrial synthesis of Sitagliptin

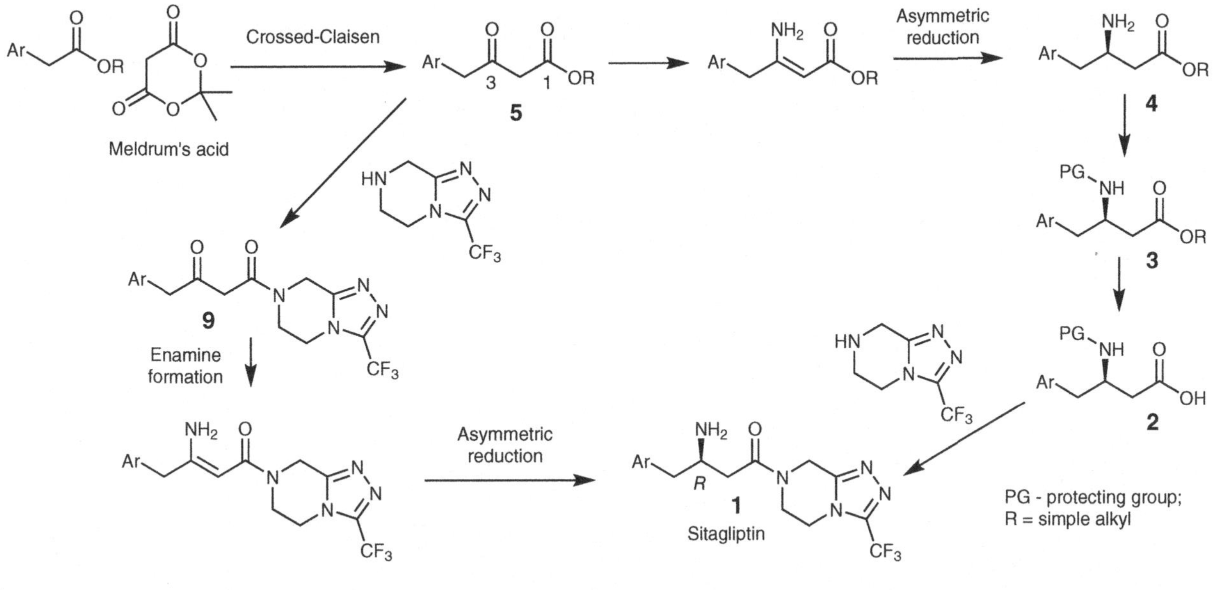

Scheme 5.15 Main synthetic routes to Sitagliptin by asymmetric hydrogenation

of catalyst are necessary, assuming the same molar ratio catalyst/product. This can be stated as a general principle: it is preferable to use an expensive reactant towards the end of a synthetic sequence to minimise the quantities required and contain costs. In practice, however, other factors must be taken into consideration to make an overall assessment and determine what the most efficient synthetic route is. For example, because catalysts are usually heavy metal complexes, their use at the end of a synthetic sequence may leave unacceptable amounts of heavy metal in the drug substance and additional processing, with attendant loss of product, may be necessary.

More than 400 tons of Sitagliptin are produced yearly, which clearly vindicates the need for an efficient streamlined process for its chemical manufacture.

5.3 THE FIRST REPORTED SITAGLIPTIN SYNTHESIS

In the first reported synthesis of Sitagliptin by Merck (Scheme 5.16, Merck2001, JMC2005), the asymmetry was generated in the very first step by a diastereoselective alkylation of a chiral dihydropyrazine auxiliary derived from L-valine (prepared in 3 steps, 75% yield, ACIE1981). Alkylation of the dihydropyrazine occurred on the β face of the heterocyclic six membered ring due to steric hindrance on the α face caused by the presence of the bulky isopropyl group, thus assuring diastereoselectivity. The configuration of the chiral centre remained unchanged up to the end of the synthesis. Arndt-Eistert homologation of the *N*-Boc protected α-amino acid into the homologous β-amino acid gives the central four-carbon unit of Sitagliptin. Coupling reaction of the *N*-Boc-protected amino acid with the triazolopyrazine building block proceeded in low yield. This asymmetric route produced Sitagliptin in 11 steps with a 12% yield from the starting dihydropyrazine (Merck2001).

The triazolopyrazine heterocycle was prepared in four steps from chloropyrazine in 23% overall yield (Scheme 5.17, JMC2005).

This first reported synthesis of Sitagliptin clearly has several drawbacks with regard to its application to large-scale synthesis. The synthesis of the chiral dihydropyrazine implies three synthetic steps in addition to 9 steps already shown in Scheme 5.16. The alkylation requires very low temperatures thus necessitating specialized and expensive equipment. *n*-Butyllithium is pyrophoric and requires special handling, especially on a large scale. A low yield in the amide formation, the penultimate step of the sequence, is unfavourable but would be expected to be surpassed by proper process development. However, an even greater concern is the use of highly toxic and explosive diazomethane. With only a few very specialized companies able to use diazomethane the elimination of the Arndt-Eistert homologation makes a change of route inevitable for application to industrial production.

5.4 THE SEARCH FOR AN INDUSTRIAL SYNTHESIS FOR SITAGLIPTIN

The enormous commercial success of Sitagliptin since its initial approval, and the accompanying demand for large quantities, has made the search for alternative and viable industrial syntheses top priority for many pharmaceutical companies.

A racemic synthesis is less attractive because the recovery of the enantiomeric *S*-amine is expected to be difficult. Thus, an asymmetric/enantiospecific syntheses becomes the preferred option giving rise to a considerable number of different routes being reported and patented.

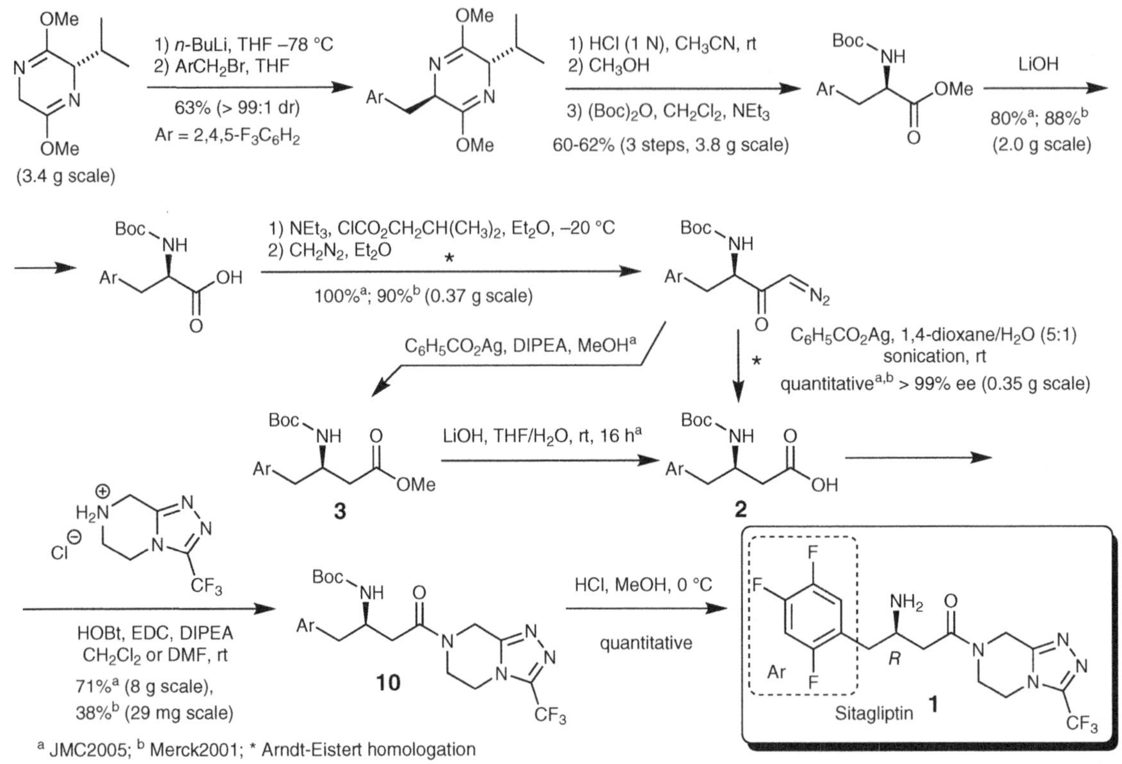

Scheme 5.16 First reported synthesis of Sitagliptin

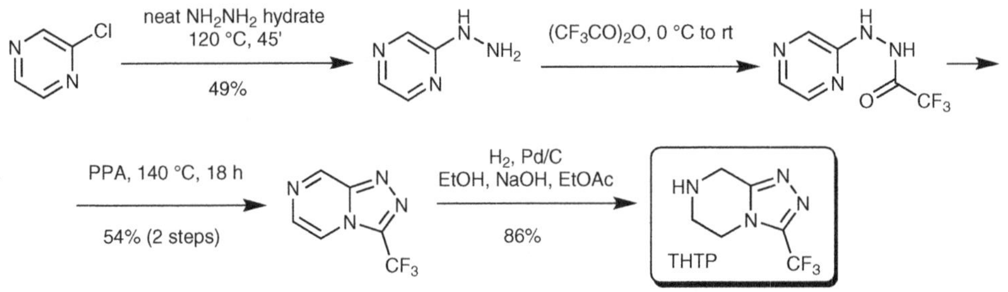

Scheme 5.17 Reported synthesis of the 3-(trifluoromethyl)-5,6,7,8-tetrahydro-[1,2,4]triazolo[4,3-*a*]pyrazine used for the first synthesis of Sitagliptin

As expected, trifluoraromatic and 3-(trifluoromethyl)-5,6,7,8-tetrahydro-[1,2,4]triazolo[4,3-*a*]pyrazine (THTP) building blocks are frequently used as starting materials. An overview reveals that three main strategies have been used (Scheme 5.18):

a) synthesising an aryl β-amino acid (**2**), preferably by asymmetric methods, followed by coupling with the triazolohydropyrazine portion, i.e. attaching the heterocyclic block at the end after formation of the chiral amine – strategy **a**;
b) producing a β-keto amide with the triazolohydropyrazine portion attached (**9**) followed by conversion to (*R*)-Sitagliptin, i.e. producing the chiral amine at the end of the synthesis after attachment of the heterocyclic block – strategy **b**;
c) building a chiral amino triazolohydropyrazine amide derivative and attaching the trihaloaromatic ring towards the end of the synthesis i.e. attaching the heterocyclic block at the start of the synthetic route – strategy **c**.

Examination of published syntheses reveal that strategies **a** and **b** are clearly preferred. Only a few examples following approach **c** have been reported. Strategy **a** provides more flexibility wherein the chiral amine can be generated by asymmetric synthesis or by the "other routes" referred to in Section 5.2.1, i.e. a chiral pool or classical resolution.

In the following sections, the synthetic routes to Sitagliptin, shown in Scheme 5.18, are analysed according to these individual strategies.

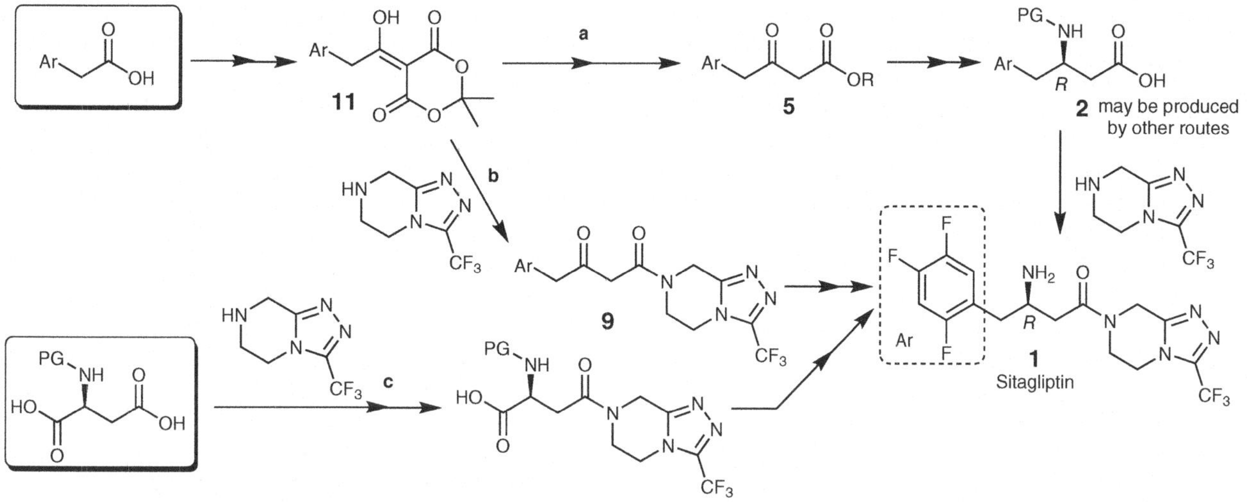

Scheme 5.18 Outline of the three main strategic routes to produce Sitagliptin

Processes exploring strategy **a** and **b** were quickly developed by Merck following the initial publication of the discovery route. After having established a scalable route using strategy **a**, Merck process research then focused attention on strategy **b**, clearly recognizing its potential for the rapid and reliable production of Sitagliptin (2003-2005). Several other companies began publishing patent protected processes from 2007 onwards, the chiral aryl β-amino acid (**2**) being a major focus of development. Because access to the chiral four-carbon chain is relatively straightforward, the majority of the new routes published follow this strategy. Hence, new methods to build chiral amino acid intermediates and to carry out the coupling reaction to the triazolohdropyrazine building block began to appear.

5.4.1 Coupling the triazolopyrazine heterocycle at the end of the synthesis: producing a key chiral β-amino acid intermediate

One of the first synthetic approaches and convergent routes to produce Sitagliptin relied upon initial synthesis of the (*R*)-amino acid (**2**) followed by coupling with the triazolohydropyrazine heterocycle.

In order to assure high productivity, the amide coupling step with THTP must be highly efficient.

Various strategies to produce the chiral β-amino acid (**2**) are conceivable and are summarized in Scheme 5.19.

Condensation of 2-(2,4,5-trifluorophenyl)acetic acid with Meldrum's acid followed by alcoholysis and loss of acetone and carbon dioxide gives the key keto-ester (**5**),

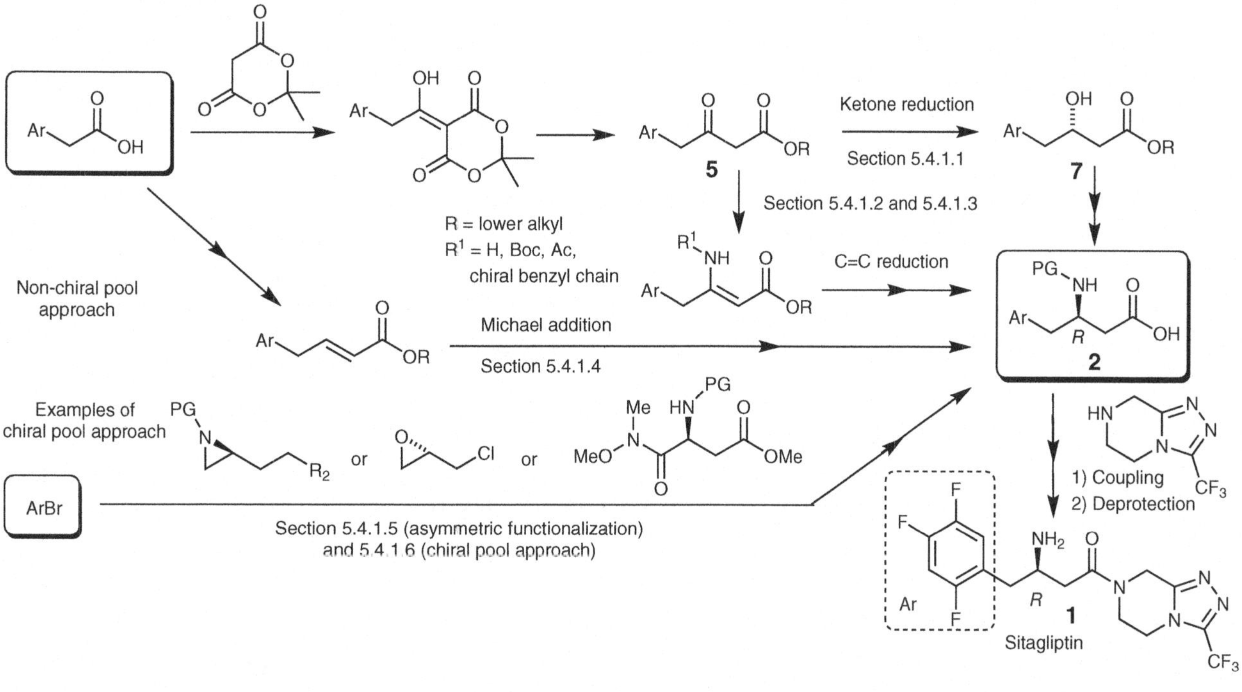

Scheme 5.19 Main routes to produce the chiral *N*-protected amino acid intermediate 2

a key intermediate on the path to the β-amino acid (**2**), which is ready for coupling with THTP. The chiral amine was synthesised alternatively by direct asymmetric reduction of the carbonyl group followed by substitution with nitrogen with inversion of stereochemistry (section 5.4.1.1), by asymmetric reductive amination of the carbonyl group (section 5.4.1.2) or by a resolution step (section 5.4.1.3).

Other routes were patented exploiting a Michael addition as key step to build the β-amino carboxylic acid (section 5.4.1.4), whilst still others start from chiral building blocks (sections 5.4.1.5 and 5.4.1.6).

Standard peptide coupling agents, with or without additives such as HOBt, (1-hydroxybenzotriazole) to suppress racemization, in the presence or absence of a base have been applied to the amidation reaction (Scheme 5.20). The hydrochloric acid salt of the THTP heterocycle is frequently used and the chiral β-amino acid is *N*-protected to avoid crossed-coupling reactions. Yields of higher than 85% are often reported for this step.

The *t*-butyloxycarbonyl group (Boc) is the standard protecting group in this situation and is easily removed by treatment with acids such as trifluoroacetic acid, either neat or in dichloromethane. Hydrochloric acid in methanol can also be used. The protecting group need not necessarily be introduced in a separate step. In many instances the nitrogen nucleophile used to attach the amine group already carries a substituent that behaves as a protecting group. Examples of these "protecting groups" are OMe, OBn and benzyl that correspond respectively to H_2NOMe, H_2NOBn and $ArCH_2NH_2$ as nucleophiles.

Common coupling reagents are *N*-(3-Dimethylaminopropyl)-*N*'-ethylcarbodiimide (EDC), *N,N*'-Dicyclohexylcarbodiimide (DCC) and 1,1'-Carbonyldiimidazole (CDI)[3]. Although the use of DIC (*N,N*'-Diisopropylcarbodiimide) has not been reported it is also an option. CDI is an attractive option, particularly at scale as it is inexpensive, available in large quantities and is a relatively safe, easy-to-handle crystalline solid. In addition, its by-product, imidazole, is easily removed during aqueous workup. All carbodiimides are skin sensitizers and should be handled with caution, thus special conditions must be implemented when used at large scale industrially. The most suggested coupling agent is EDC as the water-soluble urea by-product, easily removed from the reaction by an aqueous wash, gives it a clear advantage over DCC and DIC. However, the higher cost of EDC is a negative factor.

Although many additives have been reported to suppress racemization, 1-hydroxybenzotriazole (HOBt) is by far the most commonly used. It can be used in either stoichiometric or catalytic amounts. The explosive and shock sensitive nature of HOBt restrict its transportation and special care is required when employing it. Whilst its analogue, 1-hydroxy-7-azabenzotriazole (HOAt), is more stable, it is considerably more expensive, thus limiting its use.

As an alterantive to a direct coupling of the free carboxylic acid **2**, its activated esters have been patented as useful substrates for amide coupling (Figure 5.7, KyungDong2015). Several other solutions have been proposed *via* sulfonate-based mixed anhydrides (81-89%, Arromax2014), triazine-based coupling reagents (94%, ST2011, Zentiva2012) or *via* boron species (65-82%, Sun2014).

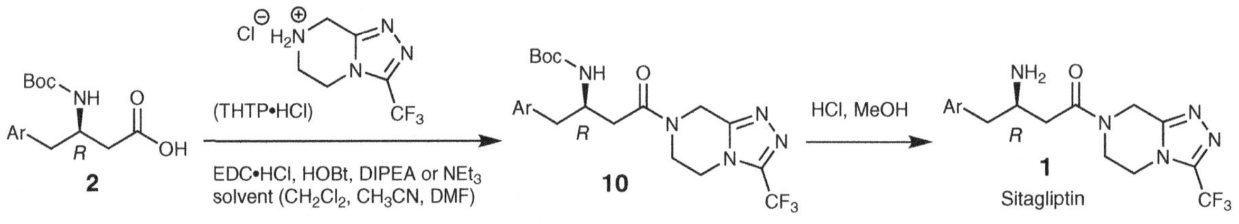

Scheme 5.20 Most frequent conditions for final amide coupling and *N*-deprotection

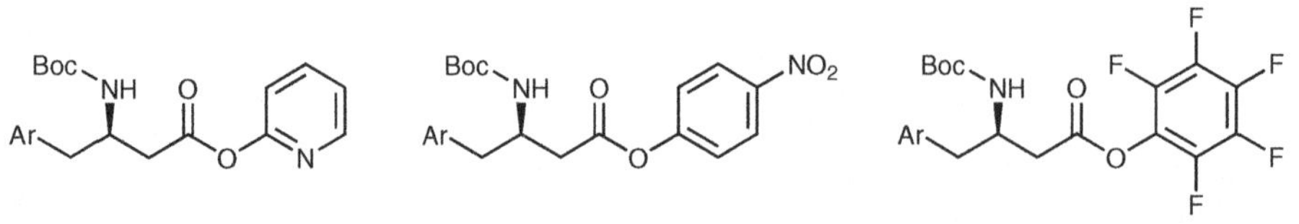

Figure 5.7 Useful esters in the coupling reaction with THTP

3 The large-scale application of amide coupling reagents in the synthesis of pharmaceuticals has been reviewed recently (OPRD2016)

5.4.1.1 Synthetic routes based on an asymmetric carbonyl reduction of a keto ester intermediate

An asymmetric carbonyl reduction as the key chiral-generating step to produce the (*R*)-amino carboxylic acid intermediate (summarized in scheme A, the asymmetric step is highlighted) was the choice of several companies.

A ruthenium-mediated asymmetric reduction of the ketone in **5** was one of the first approaches to introduce the chirality at the 3 carbon (**A5**, Merck2003c, OPRD2005, Scheme 5.21). The amine was introduced by an intramolecular displacement *via* a β-lactam using an EDC coupling/Mitsunobu reaction followed by ring opening (**A8**, **A9**). Inversion of configuration occurred during the ring closing process (**A8**) and the subsequent hydrolysis of the cyclic amide produced a suitable *N*-protected amino acid, ready for coupling with THTP. This sequence, reported by Merck, was the first-generation process for large scale synthesis of Sitagliptin. A total of eight steps from **5**, in an overall yield of 52%, was used to produce more than 100 kg of Sitagliptin for early clinical studies (ACIE2011, OPRD2005).

This route, which goes through two separate amide-forming coupling steps (**A7**, **A10**) and a Mitsunobu lactam ring formation (**A8**), generates a large amount of by-products and has poor atom economy. Control of the optical purity of the lactam was deemed to be crucial as upgrading the quality of the final product by crystallization was reported to be difficult (OPRD2005). The lactam opens easily upon heating in methanol. Appropriate conditions had to be developed to allow its isolation in high optical purity. This involved distillations during the work-up under slightly acidic conditions.

Several other companies developed similar routes *via* the *S*-hydroxy ester, enzymatic reduction of the carbonyl being proposed in one case (**A5**, Teva2007a). All these routes have a series of common disadvantages. These, in addition to the tortuous manner in which the β-amino acid moiety is created, imply that more efficient and environmentally friendly routes for the preparation of Sitagliptin are welcome.

When an asymmetric carbonyl reduction is used to produce the *R*-hydroxyester (**A16**) a double bimolecular substitution process (**A17/A18**) is required.

Intermediates with a hydroxyl or a good leaving group at the β carbon atom relative to the carbonyl group are not expected be stable, as formation of an α,β-unsaturated carbonyl compound is favoured. Hence, formation of side products is expected to be difficult to control, which may render these routes unattractive for industrial scale production. It is therefore not a surprise to find that only two routes opt to introduce the amine by a classical bimolecular substitution of a halo substituent (*via* **A18** and **A21**, MSN2009). These use azide and phthalimide nucleophiles, neither of which are problem-free, additional steps being required to produce/uncover the primary amine group.

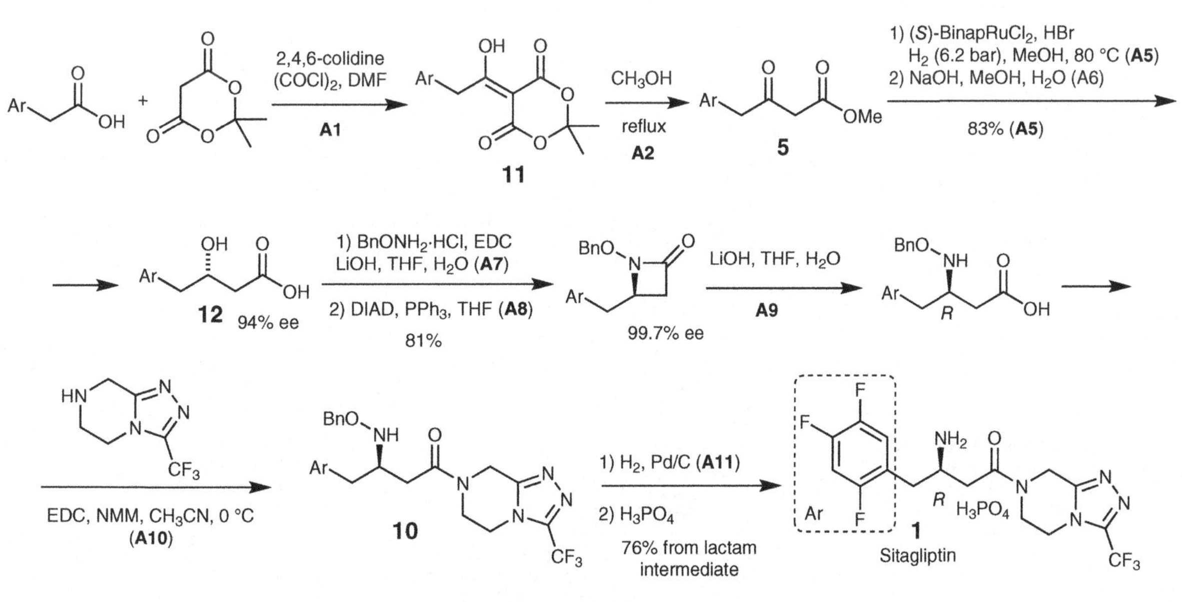

Scheme 5.21 First generation process of Sitagliptin

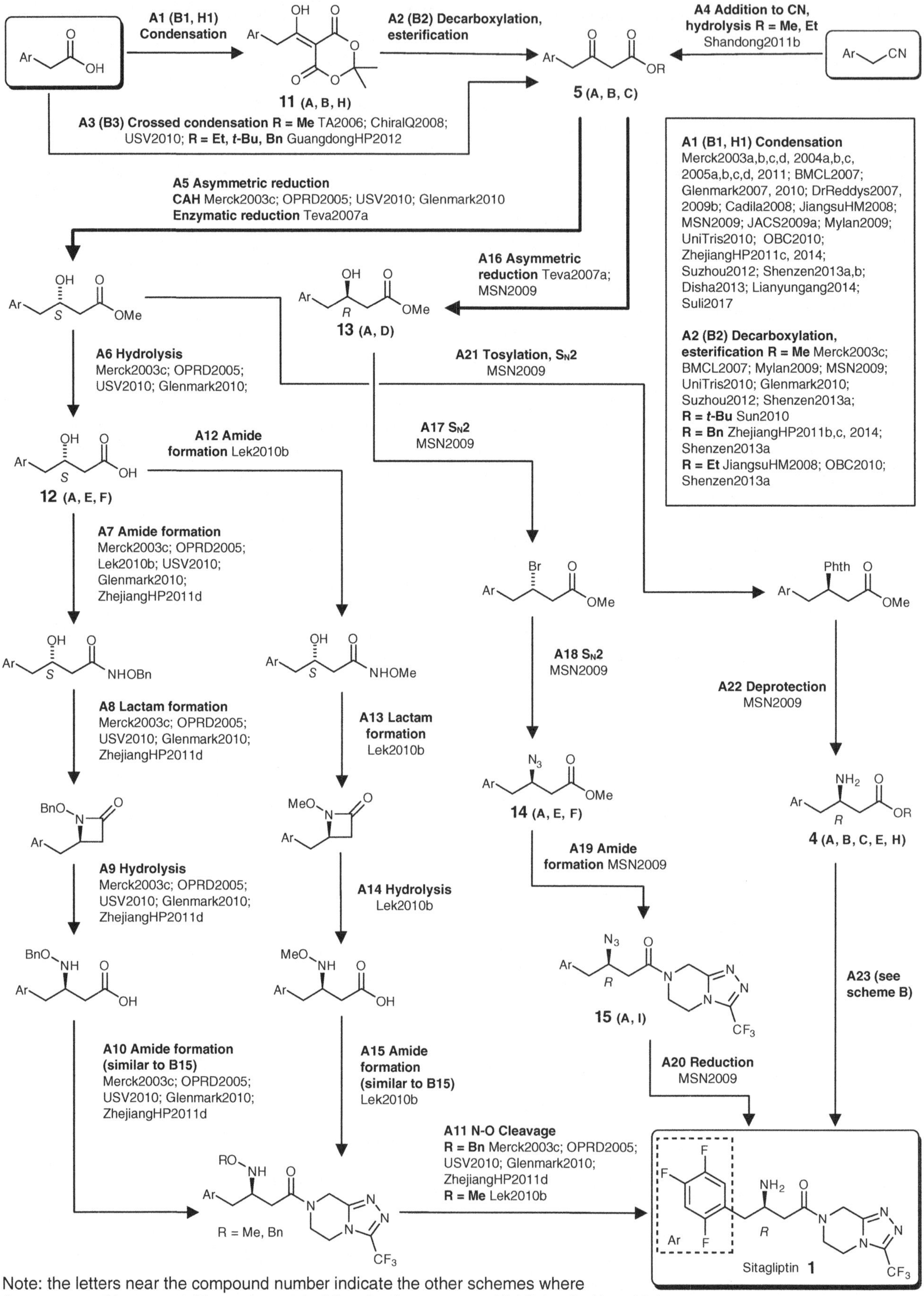

Note: the letters near the compound number indicate the other schemes where the compound is present. The asymmetric CO reduction steps are highlighted with a thick arrow.

Scheme A Synthetic routes to Sitagliptin based on an asymmetric carbonyl reduction of a β keto ester (for reaction conditions see page 191)

5.4.1.2 Synthetic routes based on an asymmetric reduction of an enamine to produce a chiral amino carboxylic acid intermediate

The direct introduction of the amino group by reductive amination of the β carbonyl of the β-keto ester (**5**) is a sound option and was exploited by several companies (Scheme 5.22).

Three main approaches were used to generate the asymmetric centre and those processes are summarized in scheme B:

a) an asymmetric induced reduction using a stoichiometric quantity of a chiral amine (a diastereoselective reduction processes, shown in the upper part of scheme B).
b) an asymmetric metal-mediated catalytic hydrogenation (central/lower part of scheme B);
c) an enzymatic transformation using a transaminase to produce the chiral amine directly from the carbonyl group (**B20**, **B27**).

In the stoichiometric approach **a**, the β-keto ester is condensed with a chiral benzylamine to give a chiral enamine (**B4**, **B17**, **B21**) which, upon reduction produces the *R*-amino isomer with good diastereoselectivity (**B5**, **B18**, **B22**, **B25**). Since the chiral centre of the benzylamine lies close to the CC double bond of the enamine, high diastereoselectivity is to be expected on hydrogenation, thus making chiral benzylamines a logical choice as a source of chirality. The C-N benzylic bond can be subsequently cleaved by simple hydrogenolysis to uncover the primary amine group (**B9**, **B19**, **B23**, **B26**, **B8**). This approach is dependent on the different reactivity of the enamine C=C double bond and the benzylic C-N bond under the hydrogenation conditions to allow the reactions to be carried out in the order desired.

The chiral β-amino ester is typically hydrolysed to the acid (**B12**) and the free amine protected as a *t*-butyloxycarbonyl derivative (**B13**) before coupling with THTP (**B15**). Alternatively, the *N*-benzylic acid, where the nitrogen is less reactive to amide formation, is coupled directly with THTP and no *N*-Boc-protecting group is required (**B7**).

The well-known catalytic asymmetric hydrogenation (CAH) of an achiral enamine derivative (option **b**) is a strong possibility. CAH of *N*-acyl enamines is well documented, the acyl group of these compounds being considered indispensable for chelation between the substrate and the metal to deliver high reactivity and selectivity. However, asymmetric reductions of primary enamines were unknown at that time. Hence, for the reduction to conform to the current understanding of the time, an acyl enamine would have been the required intermediate. However, their preparation is not trivial and efficient synthetic methods do not exist.

Likewise, the synthesis of the Boc-protected enamine was not straightforward (the yields for **B28** and **B31** are 54% and 32%, respectively). Even so, the use of a Boc-protected ester for the CAH proved to be a good choice (**B29**) as the *N*-protected free acid for the coupling step can be obtained by a simple hydrolysis of the ester after hydrogenation. The initial attempts gave low ee's and yields (41%, 33% ee, Teva2007b) but later reports indicated much better results (100%, 99.9% ee, ChiralQ2008). The asymmetric hydrogenation of the *N*-acyl ester was also reported (**B33**, 100%, 97% ee ChiralQ2008) but, even so, avoiding the protected enamine and starting from the free enamine is clearly an attractive alternative (**B35**, Figure 5.8).

A study of CAH of unprotected enamines (JACS2004a) pointed to several possibilities. Understandably, a careful screening of catalysts, reaction conditions and the correct starting material is key to delivering a process for successful application on an industrial scale.

A CAH was the key step in three other published routes to pure *R*-amino butanoic acid ester, **4** (scheme B, lower portion, CSIRIndia2014, Genchem2015, Zhejiang JP2015a). In the first approach (Genchem2015) dichloroacetyl chloride is attached to 1,2,4-trifluorobenzene *via* a Friedel-Crafts reaction. The resulting aryldichloromethyl ketone is transformed into a 2-aryl-2-chloroacetic acid, which is, in turn, converted to the key chiral β-amino carboxylic acid by a similar sequence to that

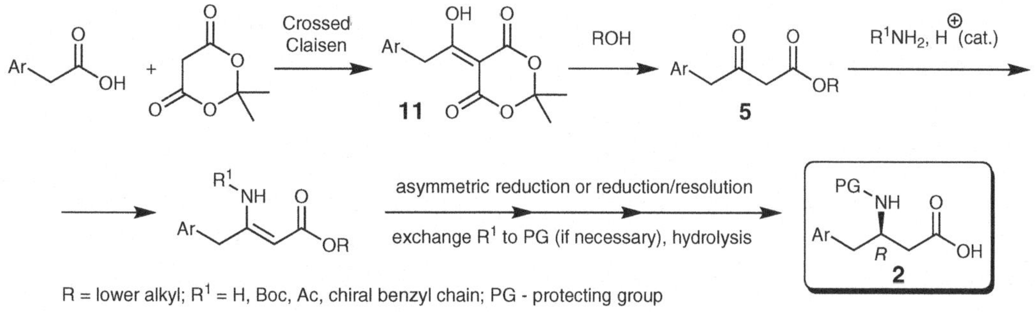

Scheme 5.22 Outline of the synthesis of the (*R*)-amino acid by reductive amination of keto ester

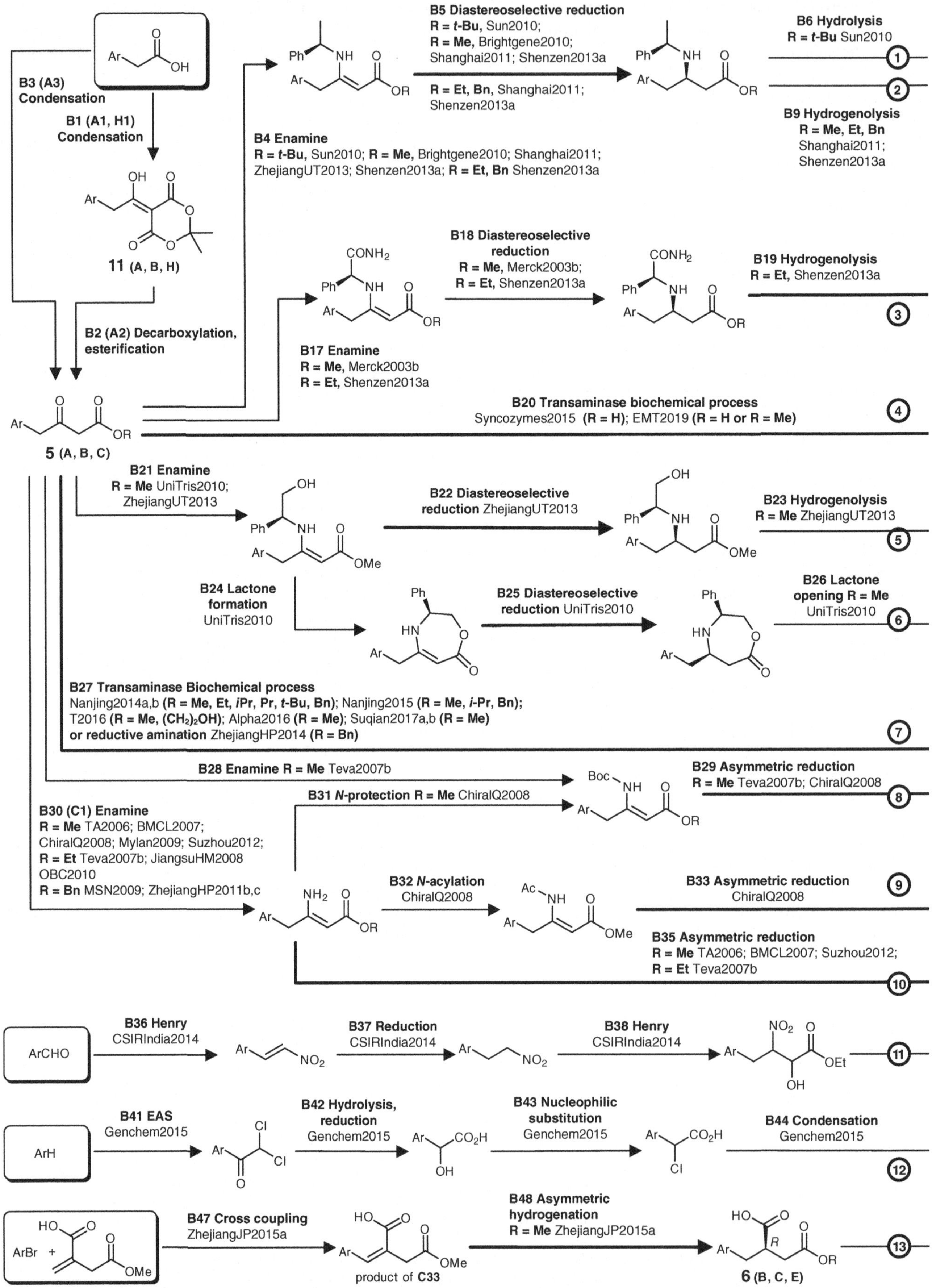

Scheme B Synthetic routes based on an asymmetric reduction of a C=C bond to produce a chiral β amino carboxylic acid intermediate (for reaction conditions see page 192)

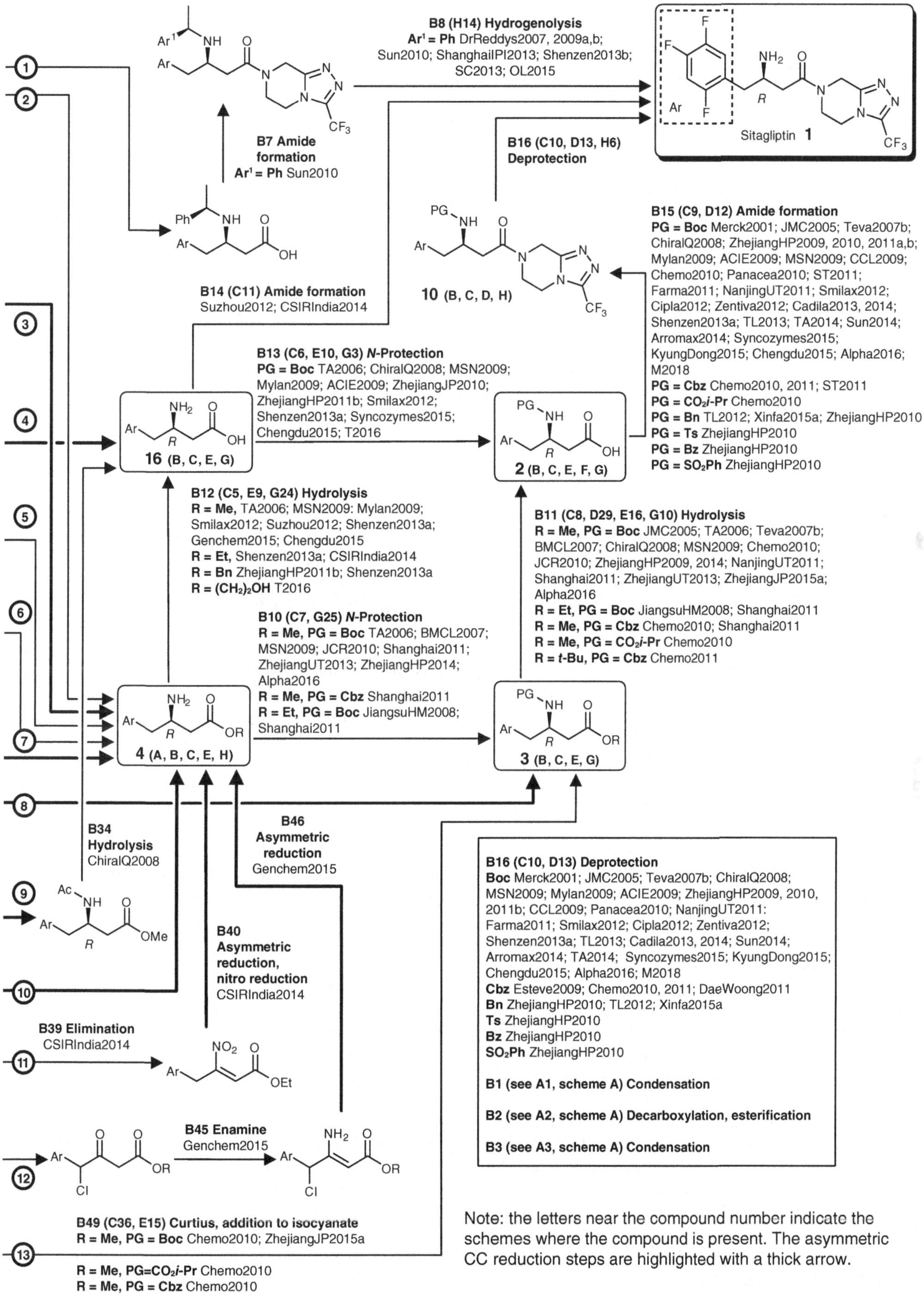

Note: the letters near the compound number indicate the schemes where the compound is present. The asymmetric CC reduction steps are highlighted with a thick arrow.

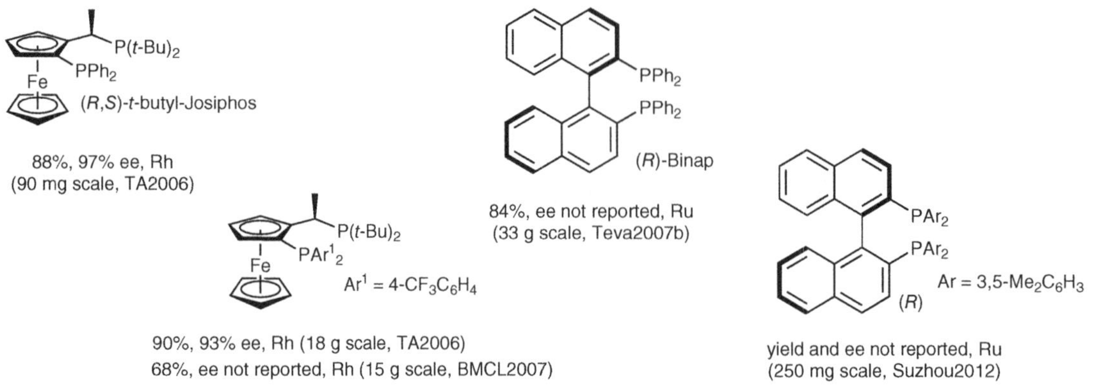

Figure 5.8 Catalysts, yields and ee in CAH of (*Z*)-3-amino-4-(2,4,5-trifluorophenyl)but--2-enoic acid ester reactions (B35)

used with the 2-aryl-acetic acid (β-keto ester, enamine formation and CAH). Note that the benzylic chlorine is removed by hydrogenolysis during CAH. In the second proposal (CSIRIndia2014), 2-arylnitroethane is converted to 4-aryl-3-nitro-2-butenoic acid ester, which in turn is reduced asymmetrically to give the desired β-amino ester in good yield.

In a different approach, (ZhejiangJP2015a), an α,β-unsaturated carboxylic acid, obtained by Pd-mediated coupling of monomethyl itaconate and 1-bromo-2,4,5-trifluorobenzene (**B47**), was subjected to an Ir-mediated CAH to produce **6** (**B48**), a β-substituted chirally pure ester. A Curtius rearrangement in the presence of *t*-butanol replaced the carboxylic acid by the Boc protected amine, giving **3**.

Enzymatic routes (option **c**) by a transaminase reaction on the β-keto acid and ester (**B20**, **B27**) were also developed.

5.4.1.3 Obtaining the chiral amino carboxylic acid intermediate using a resolution as key step

An *N*-protected chirally pure (*R*)-3-amino-4-(2,4,5-trifluorophenyl)butanoic acid (**2**) produced by resolution was fundamental in several proposals (summarized in scheme C, resolution steps are highlighted in the scheme by the use of thick arrows). Here, a simple borohydride reduction of the enamine is satisfactory.

Resolution of the racemic mixture of the β-amino ester (**±4**) was achieved by classical methodology using a chiral acid (**C4**) and by enzymatic hydrolysis of the ethyl ester to the *R*-carboxylic acid (**C3**). Both of these alternatives proved feasible, but are limited to yields no higher than 50%. For these routes to be competitive at industrial scale recycling of the waste *S*-amino ester isomer becomes critical. Hence, some processes were developed for the recycling of the *S*-isomer (scheme C, dashed lines starting in the amino ester intermediate **4**, and Scheme 5.23).

Three different approaches were used to resolve the free β-amino acid **(±)16**. The first approach (Lek2012c) accesses a racemic β-lactam from the acid and enzymatic ring opening selectively produces the *S*-isomer, leaving unreacted *R*-isomer (50 mg scale, 36% yield, 97% ee, **C20**). The second resolution process was based on an enzymatic hydrolysis of an *N*-acyl derivative (**C24**, Lek2012d). In both cases racemization by an oxidation-reduction sequence *via* an imine was claimed to recycle the *S*-isomer (95% yield, **C22**, **C26**). In a final approach, a diastereometric Ni(II) complex of the free β-amino acid and a tridentate ligand was used in a kinetic resolution (ACIE2009) (34% yield 98:2 dr, >99:1 dr after purification, **C18**). The chiral ligand was recovered quantitatively and was easily recycled. This strategy enabled asymmetric synthesis of Sitagliptin in 5 steps in 21% overall yield from racemic β-amino acid.

Once in hand, chiral (*R*)-3-amino-4-(2,4,5-trifluorophenyl)butanoic acid, or its esters, are converted into Sitagliptin as described previously (see scheme B).

Two other resolution-based processes to produce the *R* amino carboxylic acid have been patented and are shown in the lower part of scheme C. In one case, a 2-(2,4,5-trifluorobenzyl)succinic acid monoester was subjected to classical resolution *via* diastereomeric salt formation (**C35**) and the chiral *R*-acid subjected to a Curtius rearrangement (**C36**) to introduce the primary amine (Chemo2010). In the second example the separation was made on 1-(2,4,5-trifluorophenyl)pent-4-en-2-amine (**C28**) followed by an oxidative cleavage of the double bond to give the carboxylic acid (MSN2009).

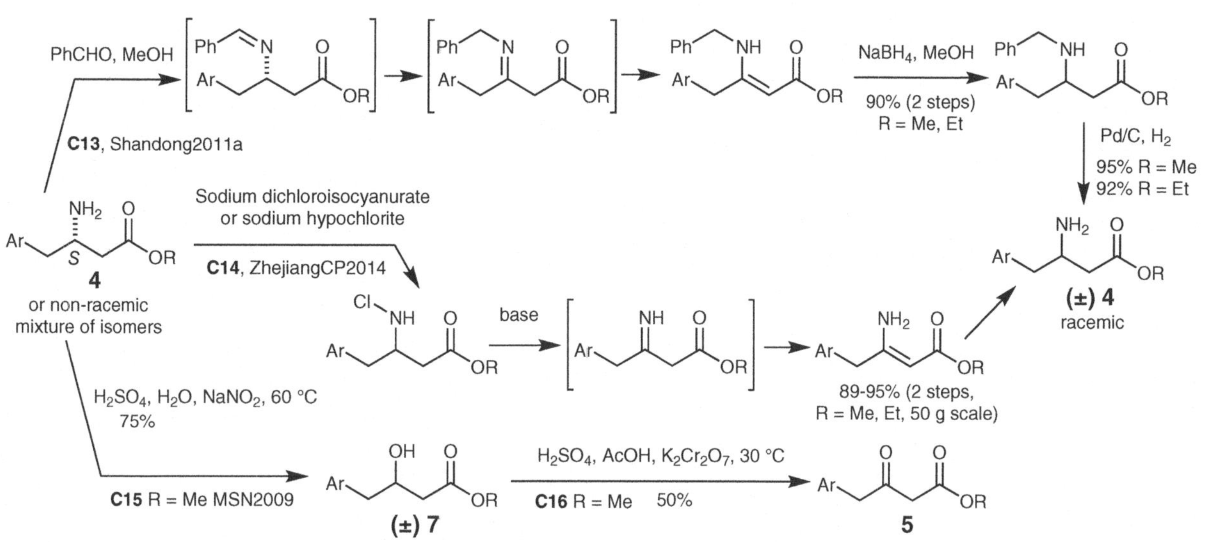

Scheme 5.23 Reported recycling processes for the (*S*)-amino ester intermediate

5.4.1.4 Obtaining the chiral β-amino carboxylic acid using a conjugate addition to install the primary amine

Two kinds of conjugate additions were applied in the construction of pure chiral β amino ester (scheme D).

A highly diastereoselective conjugate addition of an enantiopure secondary lithium amide to the α,β-unsaturated ester was the key step in a concise and efficient asymmetric syntheses of Sitagliptin (**D19**, **D23**, scheme D (JCR2010, TL2012, ZhejiangNC2014). Note that a stoichiometric amount of the chiral benzylic amine is employed to establish the desired stereochemistry. Once more, the incorporation of the nitrogen atom into the molecule of Sitagliptin on scission of the nitrogen benzylic bond means that the chiral amine is consumed in the reaction. Moderate to high yields were reported for the asymmetric conjugate addition (54-87%). In the best case, an overall yield of 42-43% from the starting aryl acetaldehyde was revealed (86-87% yield, >90% de for the asymmetric step, TL2012).

A conjugate addition of benzylamine or *O*-benzylhydroxylamine to produce a racemic intermediate directly from the α,β-unsaturated ester had been reported but low enantiomeric excesses were observed in an asymmetric version (Lek2011b, 2012a, **D2**).

Lek investigated the use of a bis(pinacolato)diboron reagent to borylate the Michael acceptor (**D3**). Oxidation of the boronic acid ester (**D4**) results in the β-alcohol which can be converted to the racemic *N*-protected β-amino ester (**D5**, Lek2011a, 2012a). Thereafter, a resolution step is required to produce the pure (*R*)-β-amino acid. A catalytic system for this reaction based on basic copper carbonate and various chiral phosphine ligands achieved asymmetric Michael addition in good yields with up to 95% ee (**D9**, Lek2011a, 2012b, AOC2013). Although the chiral boronic ester was only taken through, in good yield, to the chiral alcohol (**D10**, AOC2013). Presumably the same sequence as reported for the racemic material can be applied to synthesise Sitagliptin.

In the only example that used a conjugated aldehyde as Michael acceptor, an organocatalysed asymmetric tandem aza-Michael/hemiacetal reaction was key in producing pure chiral (*R*) β-amino acid (M2018, Scheme 5.24). A TEMPO-mediated oxidation in conjunction with sodium hypochlorite was then used to transform the primary alcohol to the carboxylic acid.

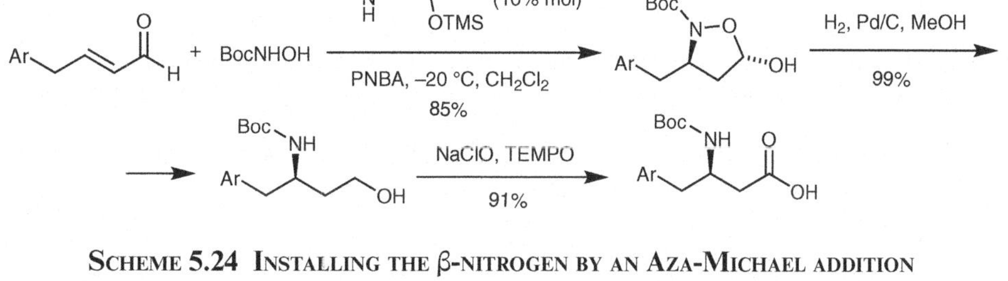

Scheme 5.24 Installing the β-nitrogen by an Aza-Michael addition to a conjugated aldehyde

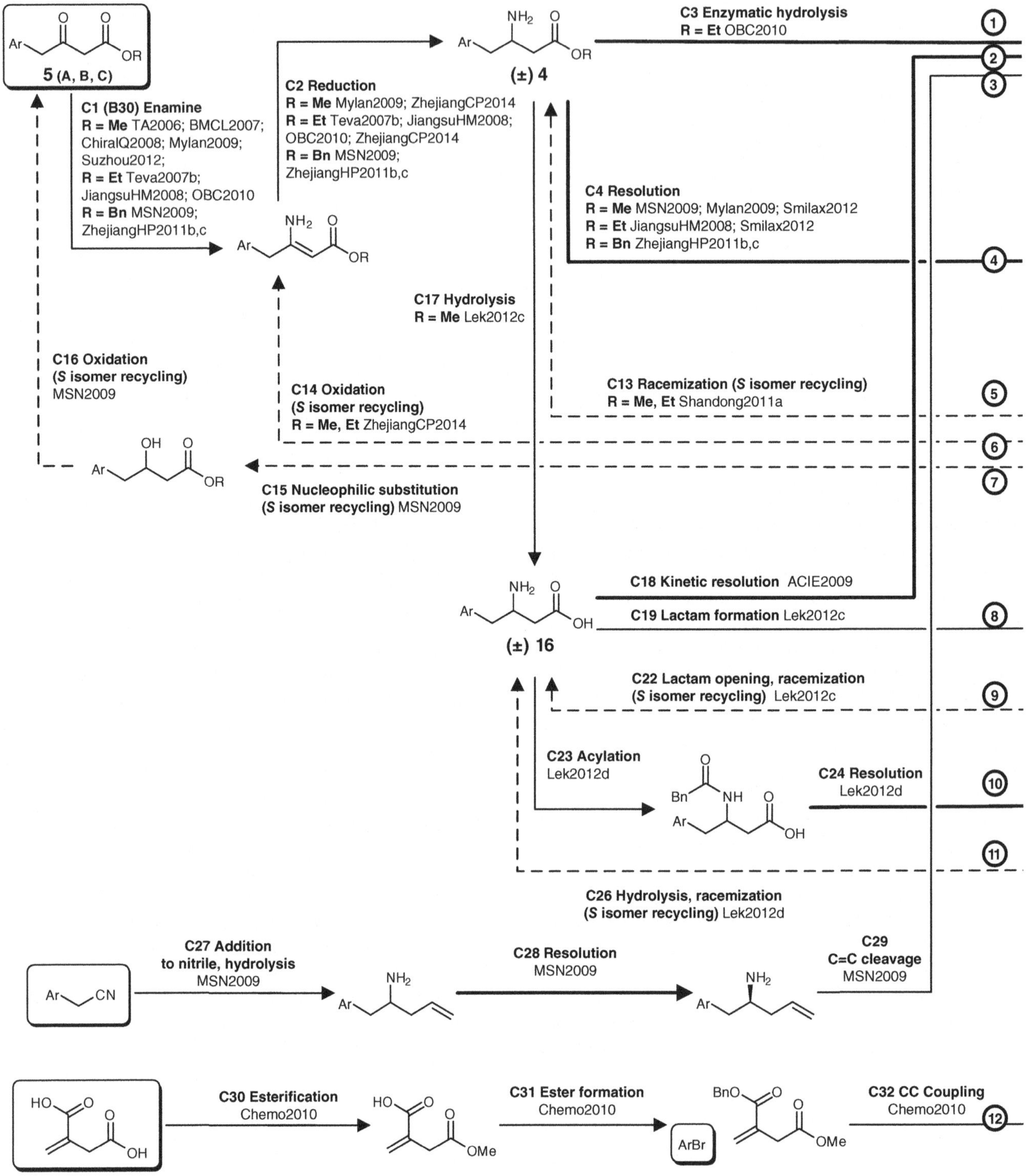

Note: the letters near the compound number indicate other schemes where the compound is present. The resolution steps are highlighted with a thick arrow; dashed lines are recycling processes of the enantiomer of the molecule shown.

Scheme C Synthetic routes obtaining the chiral β amino carboxylic acid intermediate using a resolution as key step (for reaction conditions see page 194)

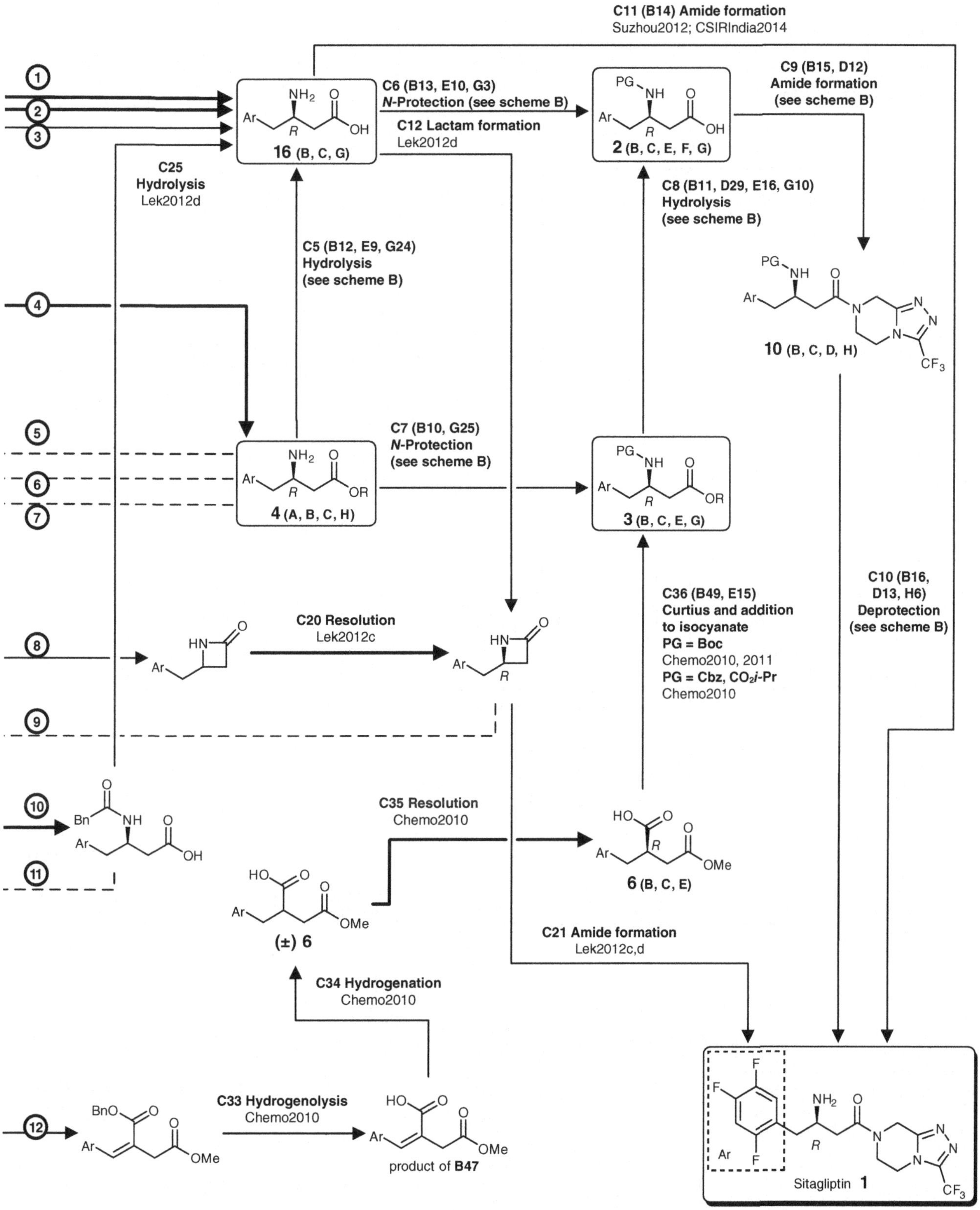

C11 (B14) Amide formation
Suzhou2012; CSIRIndia2014
C6 (B13, E10, G3)
N-Protection (see scheme B)
C12 Lactam formation
Lek2012d
C9 (B15, D12)
Amide formation
(see scheme B)
16 (B, C, G)
2 (B, C, E, F, G)
C25
Hydrolysis
Lek2012d
C8 (B11, D29, E16, G10)
Hydrolysis
(see scheme B)
C5 (B12, E9, G24)
Hydrolysis
(see scheme B)
10 (B, C, D, H)
C7 (B10, G25)
N-Protection
(see scheme B)
4 (A, B, C, H)
3 (B, C, E, G)
C36 (B49, E15)
Curtius and addition
to isocyanate
PG = Boc
Chemo2010, 2011
PG = Cbz, CO2i-Pr
Chemo2010
C10 (B16,
D13, H6)
Deprotection
(see scheme B)
C20 Resolution
Lek2012c
C35 Resolution
Chemo2010
6 (B, C, E)
(±) 6
C21 Amide formation
Lek2012c,d
C34 Hydrogenation
Chemo2010
C33 Hydrogenolysis
Chemo2010
product of B47
Sitagliptin 1

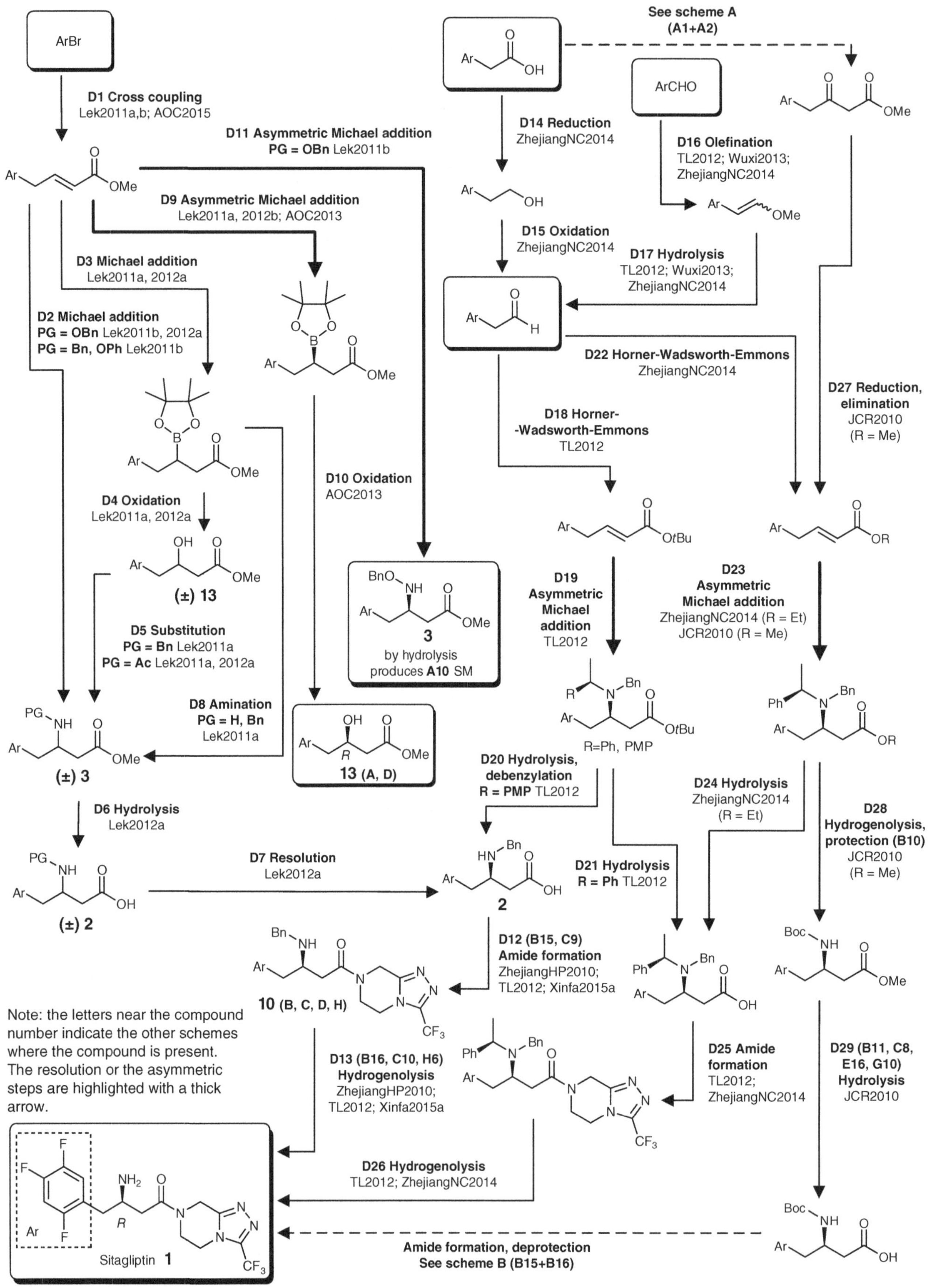

Scheme D Obtaining the chiral β amino carboxylic acid intermediate using a conjugate addition as a key reaction to install the nitrogen group (for reaction conditions see page 194)

5.4.1.5 Obtaining the chiral β-amino carboxylic acid by means of an α-functionalization or by asymmetric addition to a C=N bond

Asymmetric procedures to introduce functionality at the α-position of a carbonyl compound were the basis of several processes to produce chiral intermediates. Many of these syntheses were published in academic journals and not the subject of a patent.

In the discovery route to Sitagliptin (Merck2001, JMC2005), the asymmetry was introduced by a diastereoselective enolate alkylation (described above, Scheme 5.16). Diastereoselective enolate alkylations are found in other approaches to the stereoselective synthesis of the chiral β-amino acid (**2**).

An enantiopure 2,5-diketopiperazine template was used to introduce asymmetry, providing an (*R*)-α-amino acid (Scheme 5.25, TA2014). As before, in the first synthesis, the homologous carbon atom was introduced by Arndt-Eistert homologation. Sitagliptin was reached in eight steps and 41% overall yield from (*R*)-α-methylbenzylamine.

All processes that use an Arndt-Eistert homologation as a key step in the synthesis of the β-amino acid are summarized in Scheme 5.26. Amongst those are the discovery route synthesis (Merck2001), which starts from a chiral compound, but some routes have incorporated a classical resolution approach (CCL2009).

Evans' chiral *N*-acyloxazolidinone auxiliaries have been applied widely in a variety of reactions to elicit stereoinduction. Oxazolidinone auxiliaries are readily available, are applicable to a broad range of stereoselective reactions and are easily cleaved, thus making them ideal intermediates in asymmetric synthesis. (*S*)-4-Benzyl-oxazolidin-2-one was used as a chiral auxiliary to promote diasteroeselective alkylation of a 3-arylcarboxylic acid derivative (**E13**, scheme E, Chemo2011). A Curtius-type rearrangement substitutes the carboxylic acid unit by a primary amine with complete retention of configuration at the chiral centre (**E15**).

The remaining processes that proceed *via* an asymmetric functionalization at the α position of a carbonyl group are also indicated in Scheme E.

An enantiopure (*R*)-α-azido alcohol was prepared using the diastereoselective Evans' electrophilic azidation of the *N*-acyl (*S*)-4-benzyl-oxazolidin-2-one (88% yield and >99:1 dr) followed by reduction of the carbonyl group (**E19**, **E20**, TA2015). Further reduction of the azide and protection of the primary amine, followed by chain elongation with cyanide and hydrolysis provided the protected amino acid ready for coupling with THTP.

L-Proline catalysed addition of the enolate of the 3-arylpropanal to di-*t*-butyl azodicarboxylate followed by N-N bond cleavage produced an intermediate with the appropriate configuration at the nitrogen substituted carbon atom (**E26**, **E27**, TL2016). The same sequence as before was applied to complete the synthesis of Sitagliptin.

D-Proline catalysed enantioselective α-aminoxylation of the 3-arylpropanal with nitrosobenzene was the key step (**E28**, T2012) in yet another approach. Sodium borohydride was used to reduce the aldehyde function and subsequent treatment with Pd/hydrogen to caused scission of the N-O bond, giving chiral diol in 65% yield (>99% ee). Chain elongation and hydrolysis provided (*S*)-3-hydroxy-4-(2,4,5-trifluorophenyl)butanoic acid in 45% yield and >99% enantiomeric excess (**E30+E31**). As this acid was a key intermediate in the original industrial route (see scheme A) its conversion to Sitagliptin *via* a β-lactam had already been reported.

In an separate approach, condensation of (*R*)-2-methylpropane-2-sulfinamide with 2-(2,4,5-trifluorophenyl) acetaldehyde gave a chiral sulfinimine (**E2**) which was subjected to a Reformatsky-type reaction with ethyl bromoacetate to produce an (*R*) β-nitrogen substituted carboxylic ester (**E3**). This chiral intermediate was then

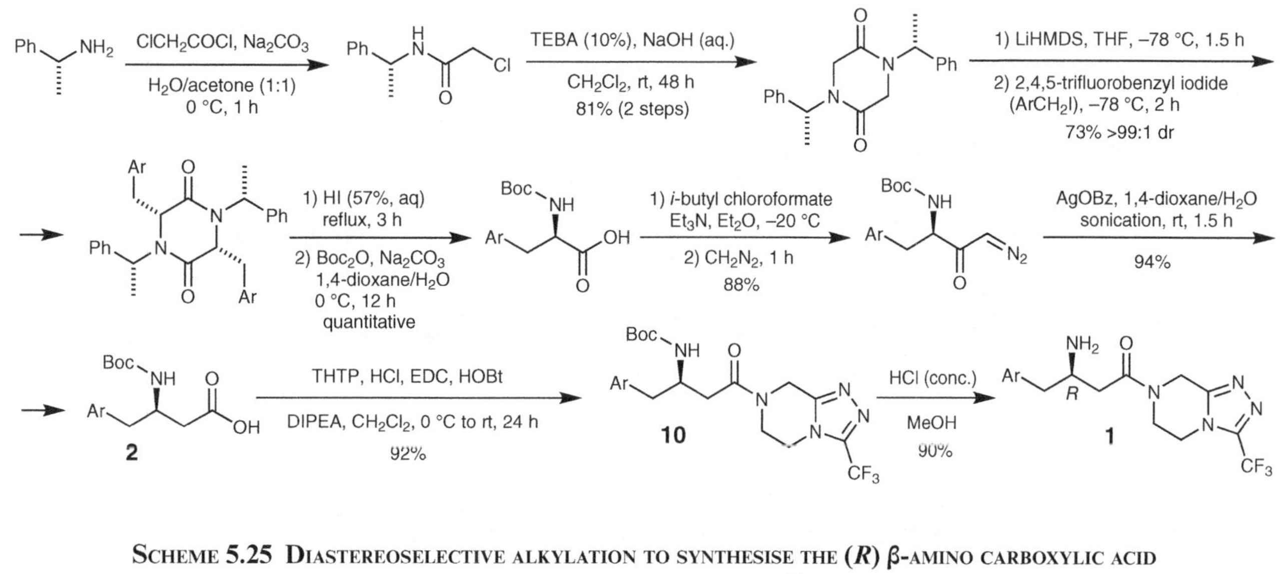

SCHEME 5.25 DIASTEREOSELECTIVE ALKYLATION TO SYNTHESISE THE (*R*) β-AMINO CARBOXYLIC ACID INTERMEDIATE AND ITS CONVERSION TO SITAGLIPTIN

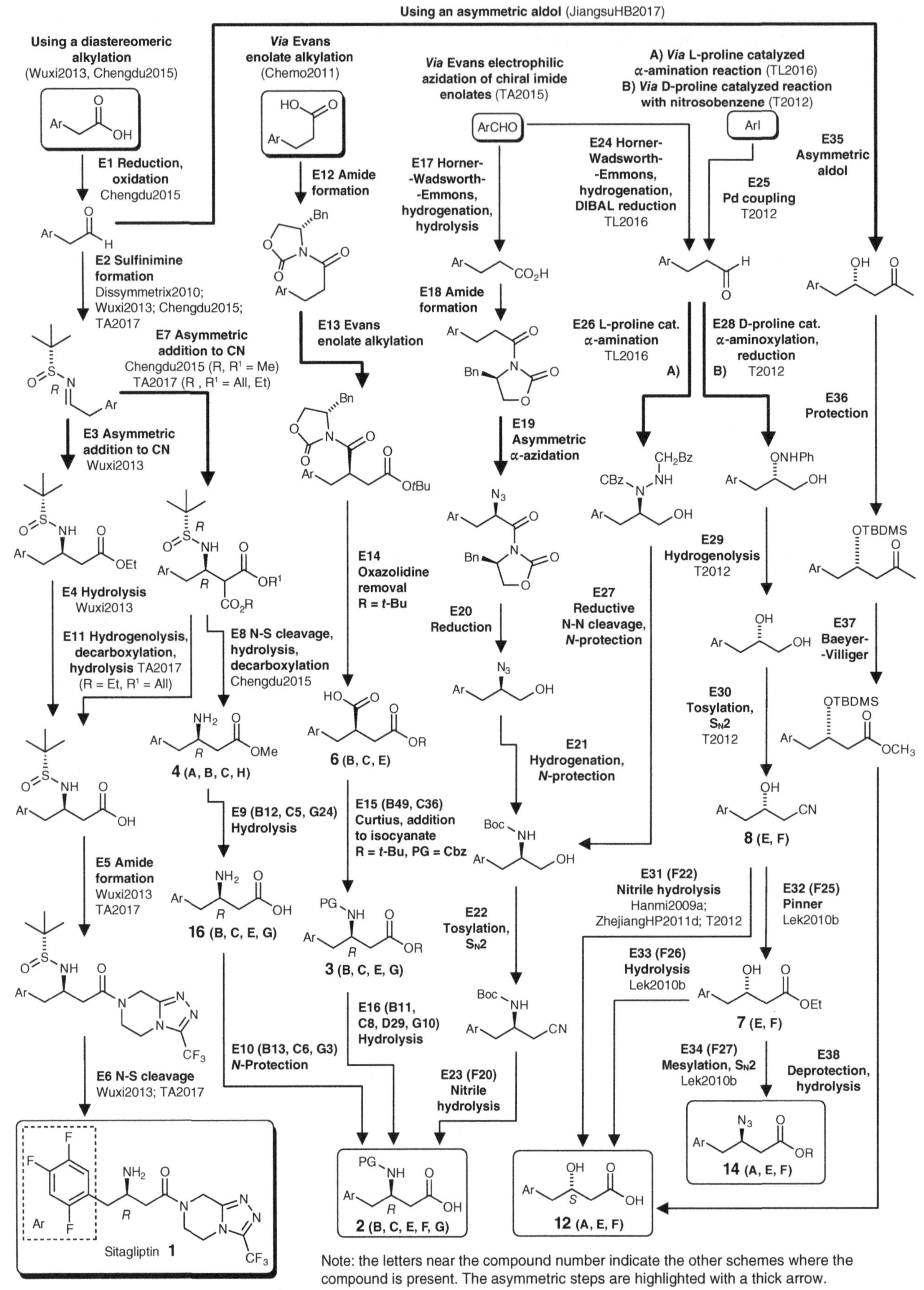

Note: the letters near the compound number indicate the other schemes where the compound is present. The asymmetric steps are highlighted with a thick arrow.

Scheme E Obtaining the chiral β-amino carboxylic acid by means an α-functionalization or by asymmetric addition to a C=N bond (for reaction conditions see page 196)

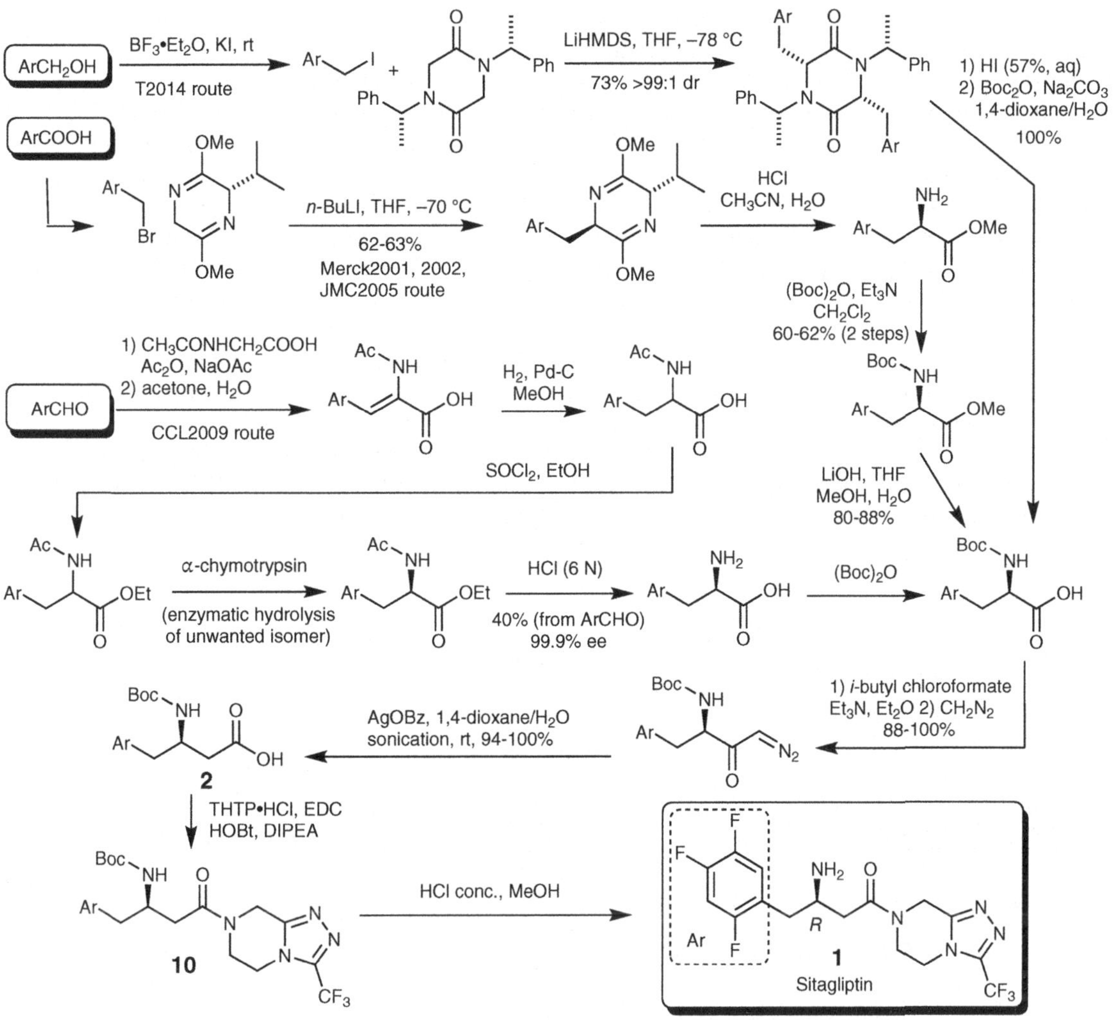

Scheme 5.26 Sitagliptin syntheses using the Arndt-Eistert homologation as a key synthetic step

converted to Sitagliptin by standard reactions (Wuxi2013, scheme E). A similar approach used malonate derivatives as nucleophiles (**E7**, Chengdu2015, TA2017). A recently disclosed route uses a proline-catalysed aldol condensation (**E35**) as the key asymmetric step (JiangsuHB2017, Scheme E).

A three-component asymmetric allylation was the key step in a patented process to produce chiral β-amino carboxylic acid (Scheme 5.27, ZhejiangUT2016). In the asymmetric step, 2-(2,4,5-trifluorophenyl)acetaldehyde, L-phenylalaninol and allyltributylstannane were combined in the presence of magnesium diiodide or magnesium dibromide in methylene chloride. After cleavage of the C-N bond of the chiral auxiliary and protection of the amine a periodate oxidation of the double bond produces the carboxylic acid.

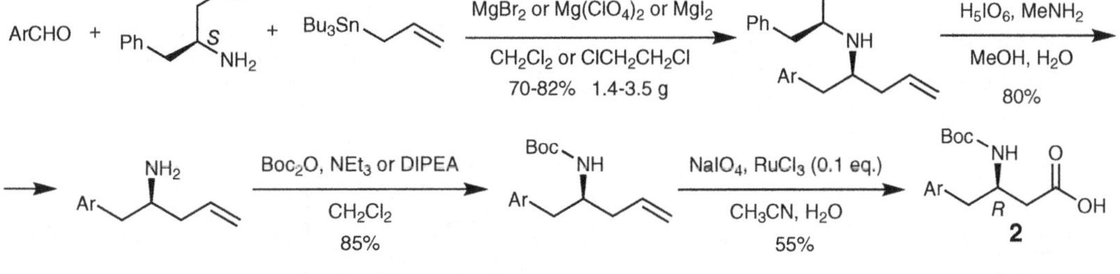

Scheme 5.27 Production of β amino carboxylic acid unit by a three-component asymmetric allylation

An asymmetric synthesis of the enantiomer (*S* isomer) of Sitagliptin *via* enantioselective allylic amination has been reported (Scheme 5.28, Synlett2013). Presumably, this methodology is applicable to the synthesis of Sitagliptin merely by using the enantiomer of the chiral ligand. Thus, catalytic enantioselective allylic amination of the terminal olefin, 4-arylbut-1-ene, *via* a sequential hetero-ene reaction followed by a [2,3]-sigmatropic rearrangement using a Pd/chiral bisoxazoline ligand catalyst system introduced the nitrogen atom with the desired *S* chirality.

5.4.1.6 Obtaining the chiral β-amino carboxylic acid using a chiral-pool approach

The chiral pool approach was also adopted by several companies to produce the chiral β-amino ester central chain. These processes are summarized in schemes F and G.

Chiral aziridines and epoxides are largely accessible and have been used as chiral pool building blocks to achieve chirality of the primary amine in Sitagliptin (scheme F).

Methionine or homoserine, with their C4 carbon chain and (*R*) primary amine at C2, are the common sources of chirality in the aziridines used as key intermediates (Scheme 5.29 and Scheme 5.30, ZhejiangHP2010, TL2013, NanjingBP2015). Opening of the aziridine ring with an aryl organometallic reagent (**F1**, **F4**, **F6**, **F8**, **F9**, **F11**, **F13**) produces intermediates with the appropriate stereochemistry and functionality for conversion to the (*R*) β-amino acid, suitable for coupling with THTP. In one case, the synthesis of (*R*)-Sitagliptin was accomplished in 12 steps and in 29% overall yield from L-homoserine (TL2013).

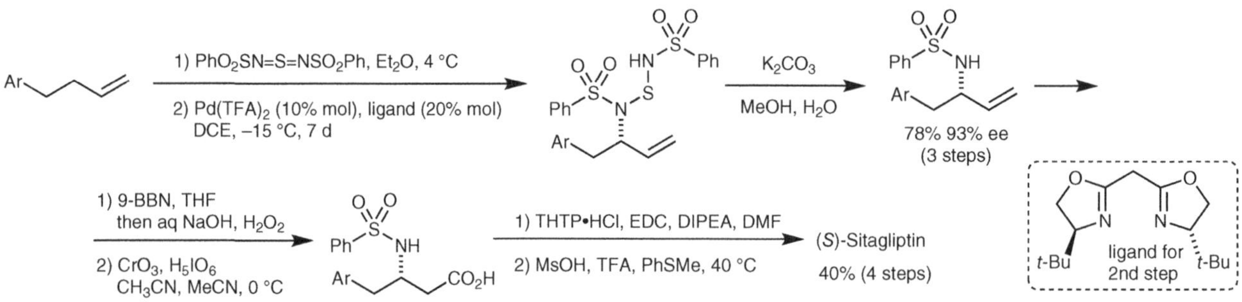

Scheme 5.28 Synthesis of (*S*)-Sitagliptin *via* enantioselective allylic amination

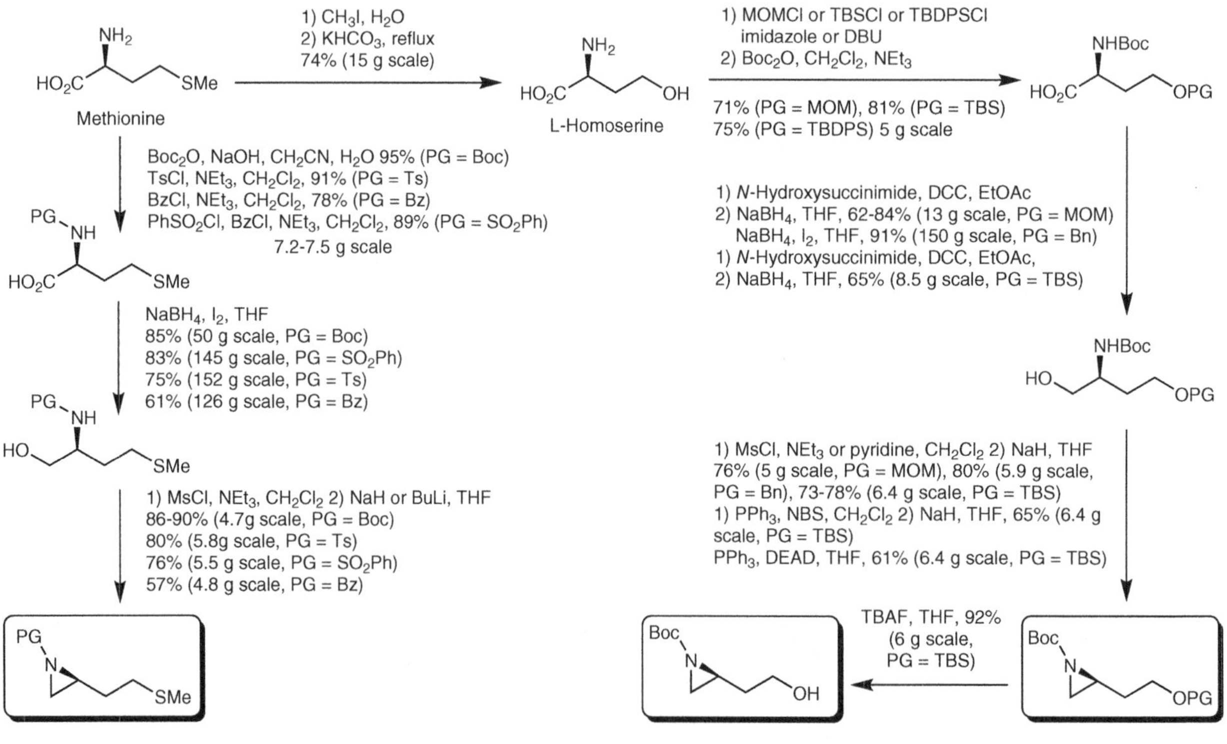

Scheme 5.29 Synthesis of electrophilic aziridines 1

Sources: ZhejiangHP2010, TL2013

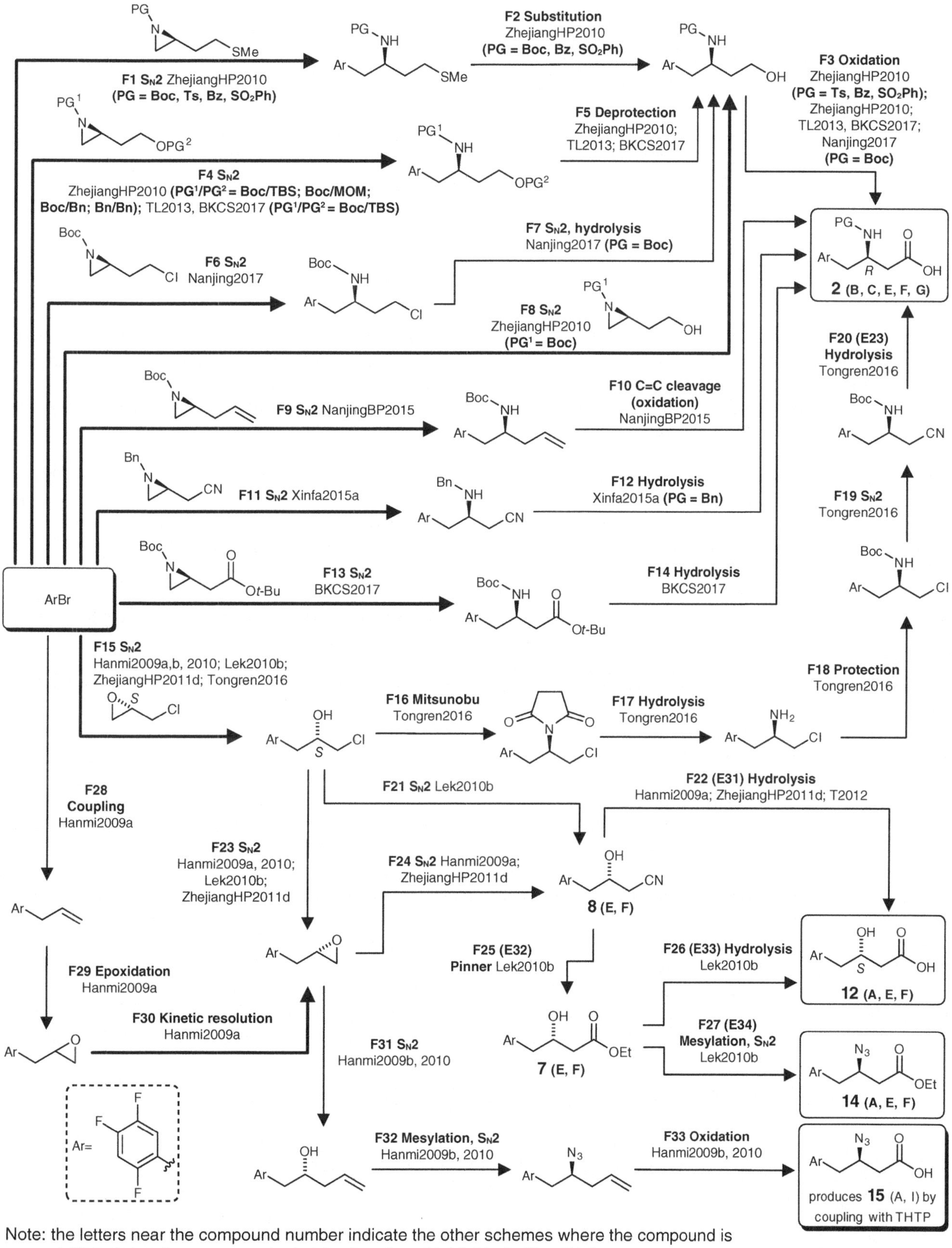

Note: the letters near the compound number indicate the other schemes where the compound is present. The steps where asymmetry is introduced are highlighted with a thick arrow.

Scheme F Aziridines and epoxides as chiral pool blocks for the synthesis of the β amino carboxylic acid intermediate (for reaction conditions see page 196)

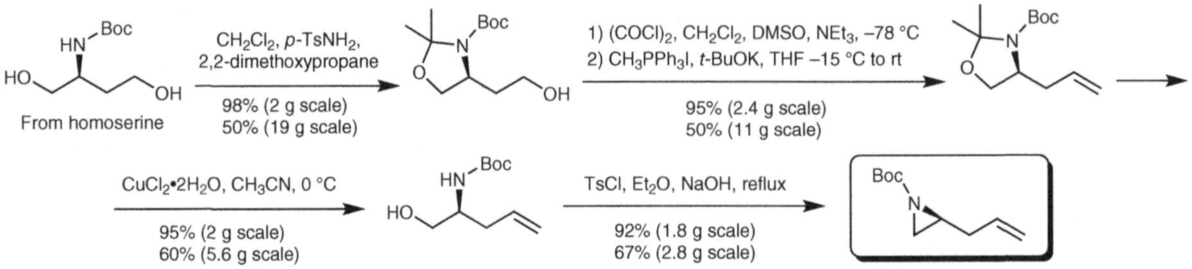

Scheme 5.30 Synthesis of electrophilic aziridines 2

Source: NanjingBP2015

In an alternative approach, a chiral *N*-benzyl protected aziridine was obtained from *S* or *R* epichlorohydrin (Scheme 5.31, Xinfa2015a).

Other routes (lower part of scheme E) use chiral epoxides to generate β substituted carboxylic acid intermediates with appropriate stereochemistry. In the first option illustrated, *S*-epichlorohydrin was the starting material (**F15**), whilst the second option relied upon a kinetic resolution to produce the chiral epoxide (**F30**).

Several routes to Sitagliptin introduced the correct chirality *via* intermediates derived from aspartic acid (processes summarized in scheme G). L-Aspartic acid is an excellent starting material as it is widely available and incorporates a C4 chain with the correct functionality and chirality. To build the molecule, the carboxylic acid of the α-amino acid has to be converted to the aryl methylene group in Sitagliptin. All the routes portrayed in scheme G build the *N*-protected chiral (*R*) β-amino acid intermediate before coupling with THTP in the final stages. A range of chemical transformations is used to attach the trifluoro aromatic ring to the central chain derived from L-aspartic acid.

Grignard additions to an aldehyde (**G8**), to a mixed anhydride (**G1**) to a lactone (**G4**) or to a Weinreb amide (**G6**) are some of the options to attach the trifluoraromatic ring. Metal mediated Negishi (**G11**, NanjingUT2011) and Suzuki-Miyaura (**G13**, ZhejiangHP2009) cross-coupling reactions with a halide, derived from L-aspartic acid, and 2,4,5-trifluorophenylboronic acid have been employed successfully. Friedel-Crafts acylations of 1,2,4-trifluorobenzene with aspartic acid derivatives have also been published (**G22**, **G26**, Ind2011 and **G30**, Guizhou2017).

These routes have the advantage that the chiral primary amine is already in place but suffer from the fact that considerable manipulation is required to introduce the trifluoraromatic moiety. Invariably, the carbon atom attached to the phenyl ring ends up functionalized and the necessity to convert this functionality to a methylene group is a disadvantage.

5.4.2 Producing the chiral amine during the final stages of the synthesis

In several published synthetic routes, the chiral amine is introduced during the final stages of the synthesis (summarized in schemes H and I).

The β-keto amide (**9**) is clearly one of the most preferred options as a precursor as it can be converted to Sitagliptin *via* various chemo- and stereoselective reactions. Several routes for the synthesis of the β-keto amide have been reported (scheme H) but by far the most convenient is a two-step process from 2-(2,4,5-trifluorophenyl)acetic acid. Condensation of Meldrum's acid and an activated mixed anhydride as electrophile (**H1**) is the first step in the sequence.

The β-keto amide (**9**) can be formed by heating a solution of the condensation product **11** with THTP. Alternatively, controlled degradation of the condensation product under catalytic acidic conditions produces

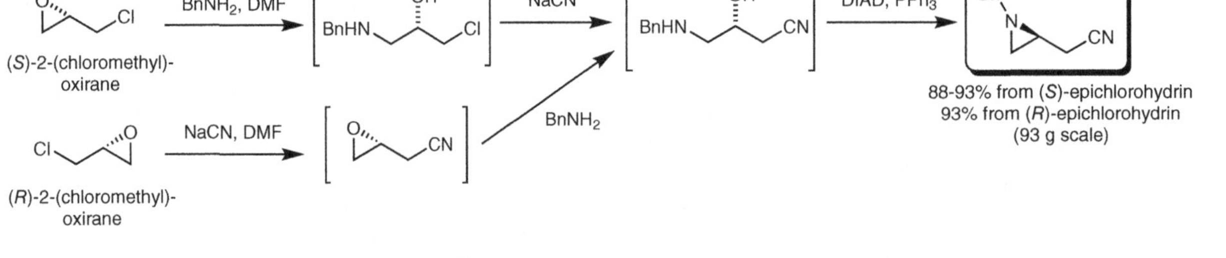

Scheme 5.31 Synthesis of electrophilic aziridines 3

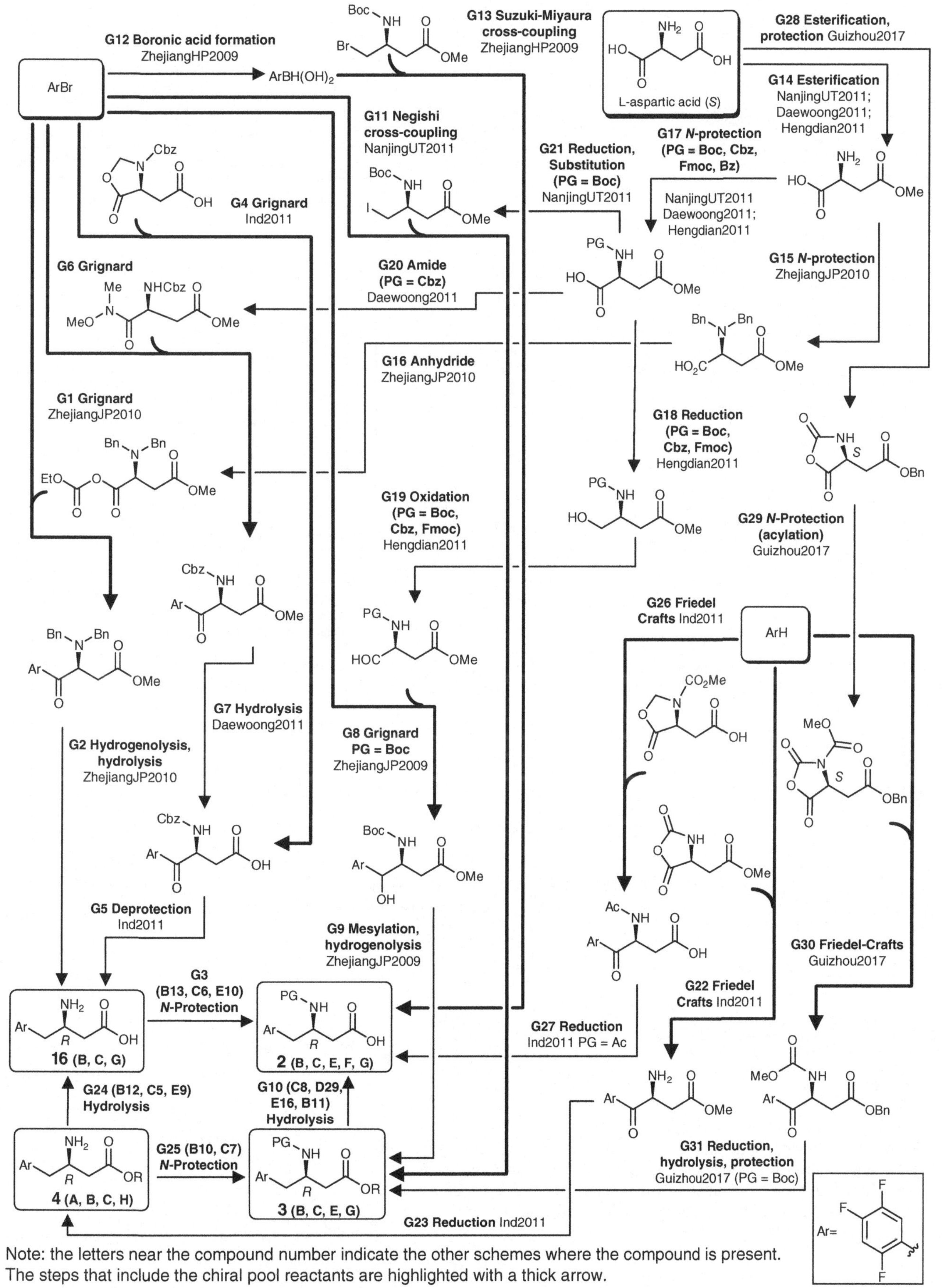

Note: the letters near the compound number indicate the other schemes where the compound is present. The steps that include the chiral pool reactants are highlighted with a thick arrow.

Scheme G Aspartic acid derivatives as chiral pool blocks for the synthesis of the β-amino carboxylic acid intermediate (for reaction conditions see page 197)

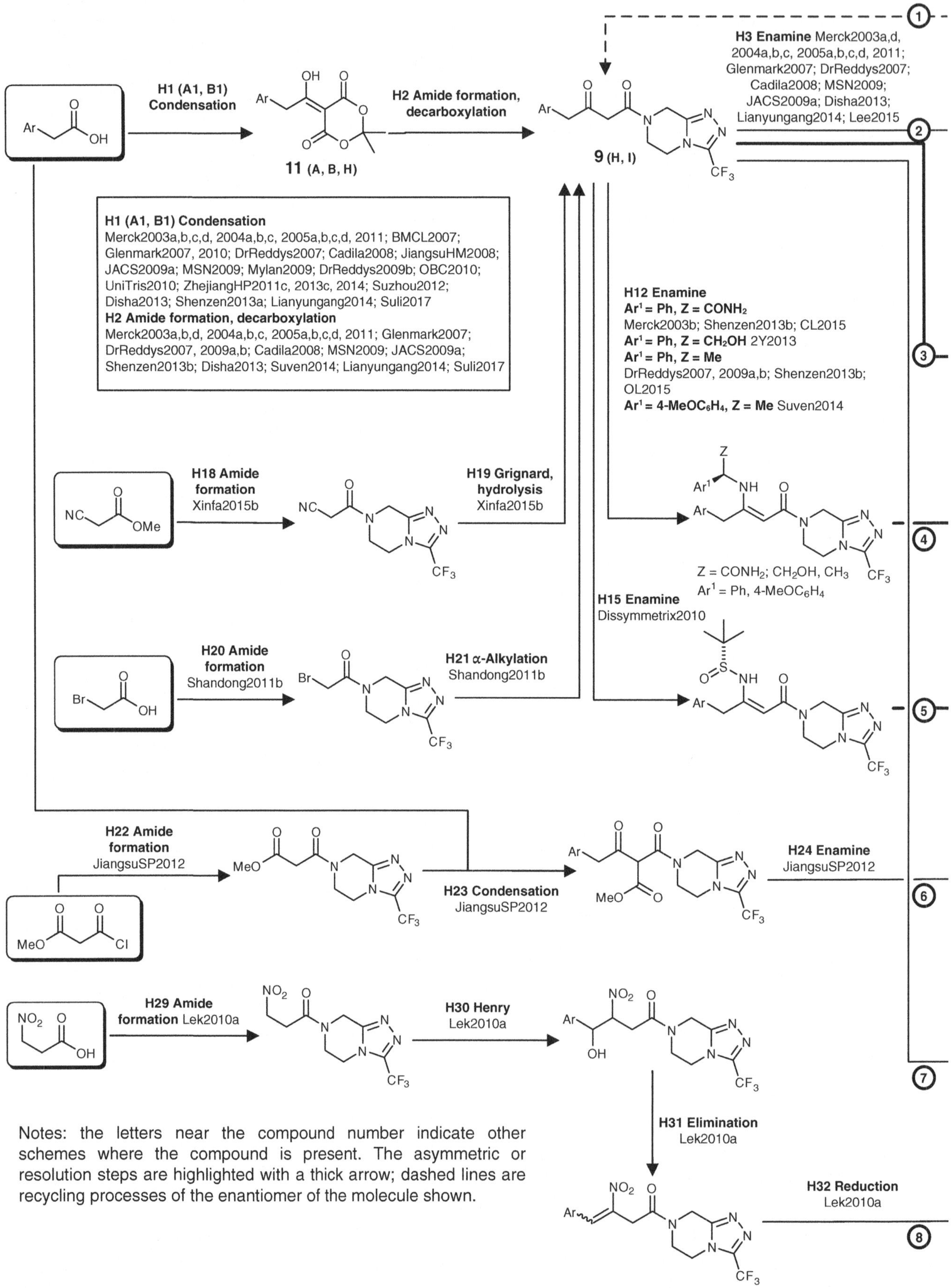

Notes: the letters near the compound number indicate other schemes where the compound is present. The asymmetric or resolution steps are highlighted with a thick arrow; dashed lines are recycling processes of the enantiomer of the molecule shown.

Scheme H Installing the primary amine by enamine-reduction and related methods (for reaction conditions see page 198)

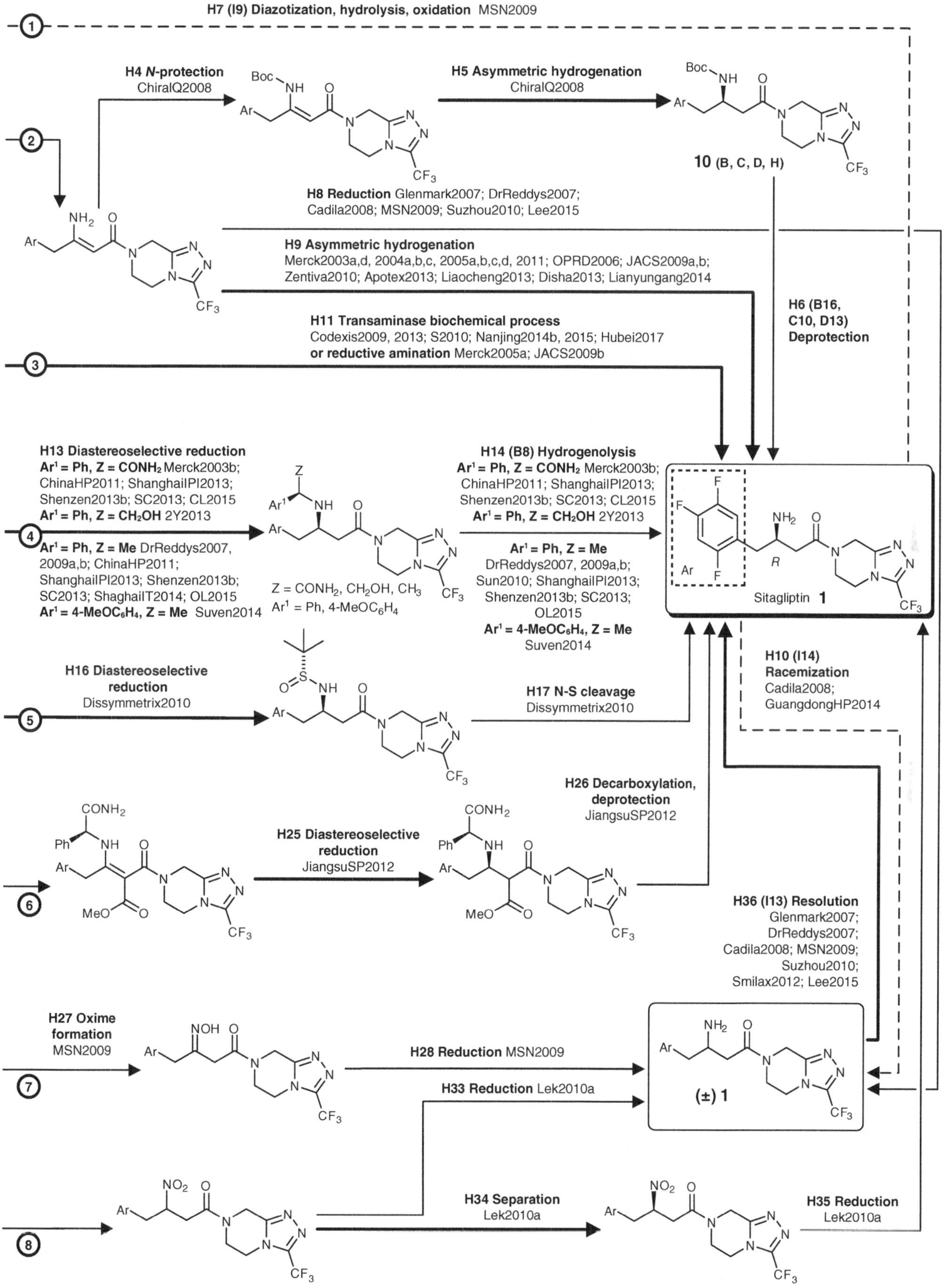

H7 (I9) Diazotization, hydrolysis, oxidation MSN2009
H4 N-protection ChiralQ2008
H5 Asymmetric hydrogenation ChiralQ2008
10 (B, C, D, H)
H8 Reduction Glenmark2007; DrReddys2007; Cadila2008; MSN2009; Suzhou2010; Lee2015
H9 Asymmetric hydrogenation
Merck2003a,d, 2004a,b,c, 2005a,b,c,d, 2011; OPRD2006; JACS2009a,b; Zentiva2010; Apotex2013; Liaocheng2013; Disha2013; Lianyungang2014
H6 (B16, C10, D13) Deprotection
H11 Transaminase biochemical process
Codexis2009, 2013; S2010; Nanjing2014b, 2015; Hubei2017
or reductive amination Merck2005a; JACS2009b
H13 Diastereoselective reduction
Ar¹ = Ph, Z = CONH₂ Merck2003b; ChinaHP2011; ShanghaiPI2013; Shenzen2013b; SC2013; CL2015
Ar¹ = Ph, Z = CH₂OH 2Y2013
Ar¹ = Ph, Z = Me DrReddys2007, 2009a,b; ChinaHP2011; ShanghaiPI2013; Shenzen2013b; SC2013; ShaghaiIT2014; OL2015
Ar¹ = 4-MeOC₆H₄, Z = Me Suven2014
Z = CONH₂, CH₂OH, CH₃
Ar¹ = Ph, 4-MeOC₆H₄
H14 (B8) Hydrogenolysis
Ar¹ = Ph, Z = CONH₂ Merck2003b; ChinaHP2011; ShanghaiPI2013; Shenzen2013b; SC2013; CL2015
Ar¹ = Ph, Z = CH₂OH 2Y2013
Ar¹ = Ph, Z = Me DrReddys2007, 2009a,b; Sun2010; ShanghaiPI2013; Shenzen2013b; SC2013; OL2015
Ar¹ = 4-MeOC₆H₄, Z = Me Suven2014
Sitagliptin 1
H10 (I14) Racemization Cadila2008; GuangdongHP2014
H16 Diastereoselective reduction Dissymmetrix2010
H17 N-S cleavage Dissymmetrix2010
H26 Decarboxylation, deprotection JiangsuSP2012
H25 Diastereoselective reduction JiangsuSP2012
H36 (I13) Resolution Glenmark2007; DrReddys2007; Cadila2008; MSN2009; Suzhou2010; Smilax2012; Lee2015
H27 Oxime formation MSN2009
H28 Reduction MSN2009
H33 Reduction Lek2010a
(±) 1
H34 Separation Lek2010a
H35 Reduction Lek2010a

an oxo-ketene, which is captured by THTP in the reaction medium (JACS2004b). This second procedure is reported to produce **9** in a 90% yield from 2-(2,4,5-trifluorophenyl)acetic acid (JACS2009a).

The β-keto amide is converted to Sitagliptin (Scheme 5.32), either by reduction of the enamine (scheme H) or reduction of the ketone to an alcohol and substitution (Scheme I). Enamine routes are the main subject of the following section.

5.4.2.1 Installing the primary amine by enamine-reduction and related methods

Reaction of the β-keto amide with an amine produces a conjugate *Z*-enamine, stabilized by hydrogen bonding the with the amide carbonyl group. The *Z* configuration is crucial for a successful outcome in the subsequent asymmetric reduction. Obviously, a formal asymmetric addition of hydrogen is preferred to a non-selective reduction with the formation of a racemate. Both asymmetric induction involving stoichiometric quantities of a chiral auxiliary (diasteroselective reduction) and catalytic asymmetric hydrogenations (enantioselective reductions) are applicable.

A. Diastereoselective reductions of an enamine intermediate: stoichiometric use of chiral amine to induce the asymmetric reduction

Stoichiometric asymmetric induction was investigated by several companies (Scheme 5.33).

Following the initial disclosure of Sitagliptin, Merck patented the use of (*S*)-phenylglycine amide to generate the chiral enamine (**H12**). Thus, hydrogenation at 6.2 bar (90 psi) mediated by Adams's catalyst, PtO_2 (**H13**, 90% yield with 97.4% de, Merck2003b) followed by hydrogenolysis at 2.8 bar (40 psi) with Pearlman's catalyst ($Pt(OH)_2$ on carbon) gave Sitagliptin in high yield (**H14**). The overall yield from β-keto amide to Sitagliptin was reported to be 76%. In summary: **9** is obtained from 2-(2,4,5-trifluorophenyl) acetic acid in two steps in 74% yield (**H1** + **H2**); β-keto enamine formation in 91% yield (**H12**); enamine reduction in 90% yield (**H13**) and hydrogenolysis in 93% (**H14**).

Several other similar processes using (*S*)-phenylglycine amide followed in addition to processes using (*R*)-1-phenylethanamine, (*S*)-2-amino-2-phenylethan-1-ol and (*R*)-1-(4-methoxyphenyl)ethan-1-amine (**H12/H13/H14**, Figure 5.9).

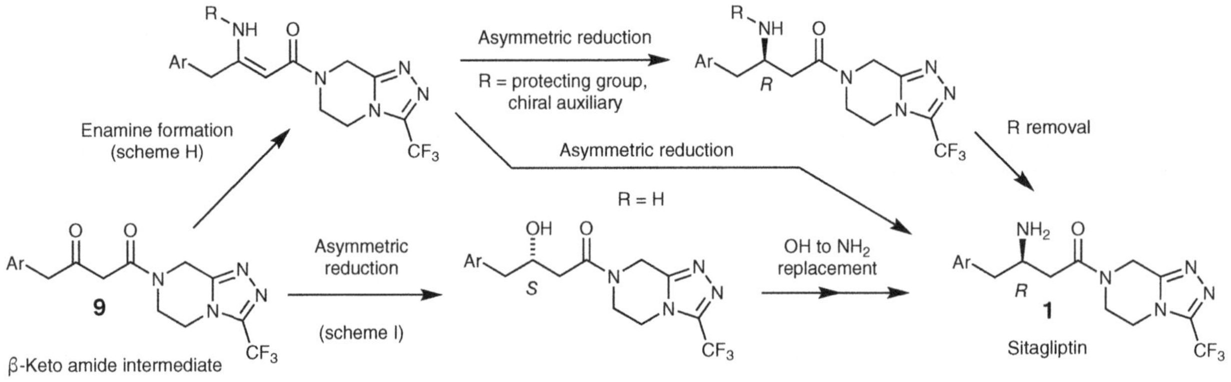

Scheme 5.32 Main routes to produce Sitagliptin from the β-keto amide

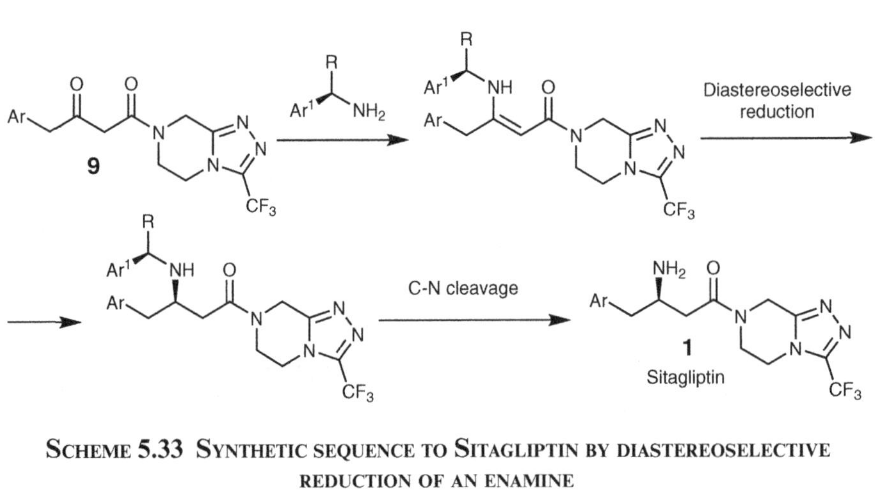

Scheme 5.33 Synthetic sequence to Sitagliptin by diastereoselective reduction of an enamine

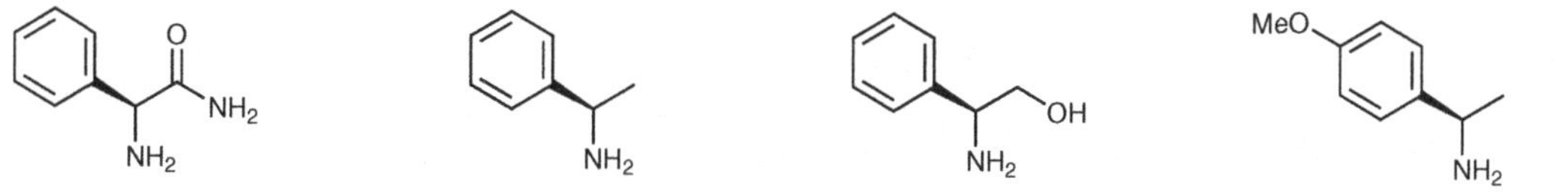

Figure 5.9 Chiral auxiliaries used to generate chiral enamines

Asymmetric reduction of the C=C double bond of the enamines was initially carried out using hydrogenation however, common borohydride reducing reagents can also be used. Good yields and diastereomeric excesses are usually obtained with borohydride reagents.

To liberate the primary amine an additional hydrogenolysis is required (**H14**). The (*R*)-1-(4-methoxyphenyl)-ethan-1-amine chiral auxiliary is the exception where strongly acidic conditions are able to sever the C-C bond and cleave off the benzyl group. Here the 4-methoxy group stabilizes the departing benzylic cation.

A high load of catalystis is required for hydrogenolysis (around 0.2-0.3 times the weight of the starting material). This heavy loading of catalyst in the last chemical transformation is not desirable in an industrial setting as removal of the heavy metal to achieve the low regulatory levels in the API may be difficult. As seen previously, the chiral amine is consumed in the process and the chiral auxiliary is sacrificed in these approaches. One drawback suffered by these routes is that the levels of des-fluoro impurities have a tendency to grow because of the harsh conditions of the final hydrogenolysis.

Similar routes use different stoichiometric auxiliaries (Dissymmetrix2010, JiangsuSP2012). In one approach a chiral sulfinamide, (Dissymmetrix2010) instead of a chiral benzylamine was used to form an asymmetric enamine (**H15**). On borohydride reduction the (*R*)-ß-amino isomer is produced (**H16**).

In second example (JiangsuSP2012), methyl 3-chloro-3-oxopropanoate is coupled with THTP to form an amide (**H22**) which is followed by a condensation reaction with the *t*-butyl ester of 2-(2,4,5-trifluorophenyl)-acetic acid (**H23**) resulting in an intermediate of similar structure to **9**, differing only in the ester group on the α carbon atom. Enamine formation with (*S*)-phenylglycine amide, reduction, decarboxylation and hydrogenolysis gives Sitagliptin.

B. Enantioselective catalytic hydrogenation of the enamine intermediate

As alluded to above, when the chiral amine is introduced at an earlier stage of the synthesis, catalytic asymmetric hydrogenation (CAH) of an enamine is obviously an attractive option (**H9**). At the time this methodology had already been demonstrated at industrial scale – an example is the synthesis of Levodopa[5,6]. Albeit only CAH of *N*-acyl amines were documented and CAH of primary amines unknown.

In 2002, Merck initiated a research programme devoted to develop methods that incorporate CAH in a synthetic route to Sitagliptin (ACR2007). High throughput screening techniques were applied to achieve an efficient chemical synthesis; its implementation; demonstration of robustness at scale; assurance of freedom of operation and surety of an unrestricted supply of catalyst supply for manufacturing.

The efficiency of the most appropriate CAH is clearly dependent on the catalyst and how the reaction is implemented. In order to select the most convenient reaction conditions hundreds of permutations (ligand/metal precursor/solvent/additives) were necessary. Extensive experimentation is required as precise prediction of the outcome of any specific combination is not possible.

The most appropriate catalyst obviously was discovered in the initial screening but its true identity only revealed at a much later date (JACS2009a). Commercially available chiral bisphosphines ligands (*R*)-BINAP, (*S*,*S*)-CHIRAPOS, (*R*,*R*)-Et BPE and (*S*,*S*)-JOSIPHOS were tested in combination with the most common metals for this application, i.e. ruthenium, rhodium and iridium. Asymmetric hydrogenation tests revealed that only JOSIPHOS in combination with Rh gave both high conversion and enantioselectivity. In these initial screening tests, using $[Rh(cod)_2]OTf$ as metal provider, a yield of >99% and an ee of 86% was obtained when the reaction was carried out in MeOH at 3.4 bar and with 5% mol catalyst loading (Merck2003b). Even better results were obtained, as seen in Scheme 5.34 (Merck2003a), on fine-tuning the reaction conditions.

A follow-up and more exhaustive screening indicated that several alternative catalysts were effective in accomplishing this hydrogenation (Figure 5.10).

The $[Rh(cod)Cl]_2$-*t*-Bu-JOSIPHOS combination was selected for further development work and optimization based on the initial exploratory results (yields and selectivity) and also on cost and availability of the ligand (JACS2009a). The serendipitous discovery that a trace amount of NH_4Cl in the isolated enamine was beneficial for the subsequent hydrogenation in terms of reaction rate and yield, led to its addition in catalytic amounts in the hydrogenation step (OPRD2006, Merck2005a).

4 Knowles, W. Asymmetric Hydrogenations – The Monsanto L-Dopa Process ch1 in in Asymmetric Catalysis on Industrial Scale: challenges, Approaches and Solutions Ed. Blaser, H.U., Schmidt E., Wiley VCH 2004

5 Selke, R. The Other L-Dopa Process, ch 2 in Asymmetric Catalysis on Industrial Scale: challenges, Approaches and Solutions Ed. Blaser, H.U., Schmidt E., Wiley VCH 2004

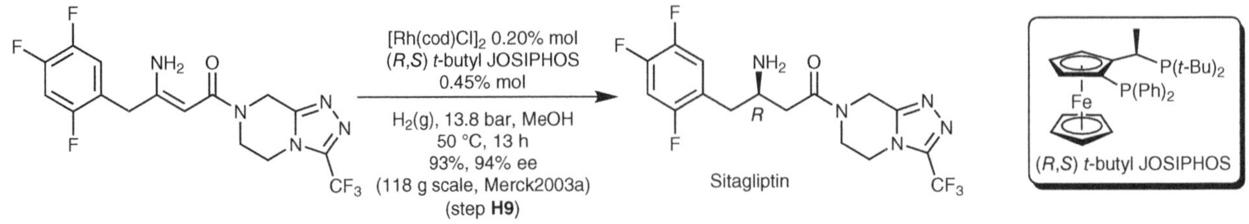

Scheme 5.34 Reaction conditions for the first reported catalytic asymmetric hydrogenation to produce Sitagliptin

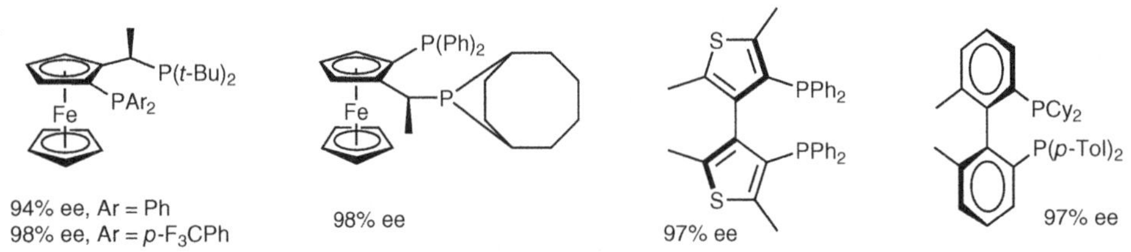

Figure 5.10 Selected results for asymmetric hydrogenations with chiral ligands and $[Rh(cod)Cl]_2$ as metal provider (reaction **H9**)

After fine-tuning of conditions such as concentration, catalyst loading, amount of ammonium and pressure (JACS2009a), optimum conditions for hydrogenation of the enamine were found to be: NH_4Cl (0.15% mol); $[Rh(cod)Cl]_2$ (0.15% mol); *t*-Bu JOSIPHOS (0.155% mol) in MeOH under 17 bar (250 psi) hydrogen at 50 °C for 16-18 h. This process proved to be extremely robust and reproducible, affording Sitagliptin in 98% yield and 95% ee (JACS2009a). Although the dimeric and conjugated olefin impurities are controlled (Figure 5.11), desfluoro impurities may be formed in undesirable amounts (Zentiva2010).

A major concern for all these routes is the use of precious metal catalysis late in the synthesis. Not only is the removal of precious metal an absolute necessity to obtain a final product that conforms to regulatory requirements, it is also imperative to recover and recycle as much as possible to keep the costs of the process within acceptable limits. Dissolved rhodium was recovered from the process stream using 10% weight of Ecosorb®, a solid polymeric absorbent impregnated with activated carbon. This resulted in near complete removal of the rhodium from the crude reaction stream (JACS2009a), the levels of both rhodium and iron in isolated Sitagliptin free base typically being below 20 ppm. Thus, recovery of extremely costly rhodium was made possible, making the process sufficiently economical to be put into practice on an industrial scale.

After publication of these initial findings many more catalyst systems were made known, attesting to the fact that extensive catalyst screening had been undertaken by numerous companies. As illustrated in scheme H, many processes to produce Sitagliptin have a key asymmetric hydrogenation of an enamine derivative and several catalysts have been cited (scheme H, **H9**). The catalyst systems and the results obtained can be seen in Figure 5.12.

Ruthenium-based systems were explored in detail as Ru is much cheaper than Rh. A $Ru(OAc)_2$((*R*)-dm-SEGPHOS) catalyst was found to be particularly efficient and produced Sitagliptin from the enamine in 96% yield and 99.5% ee in the presence of salicylic acid and ammonium salicylate (JACS2009b).

In an interesting variation, a borohydride reduction with 2.5 eq. of L-DPTTA [(–)-di-*p*-toluoyl-L-tartaric acid] to induce asymmetry was reported (86-91% yield, 98.2-99.6% HPLC, 99.4-99.7% ee, 80 g scale, Disha2013). Use of these conditions avoids the requirement for specialized hydrogenation equipment.

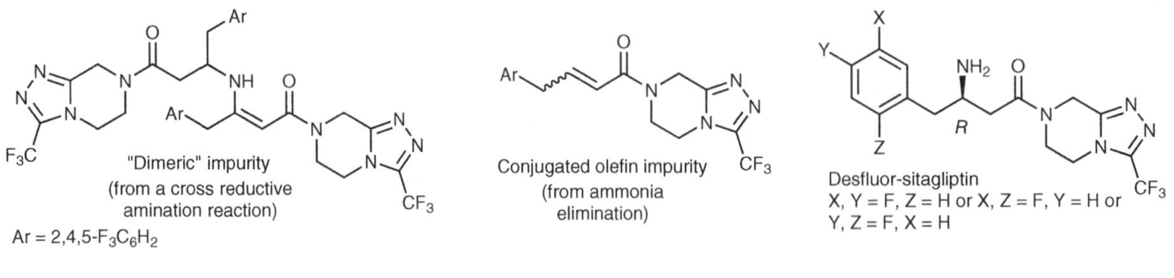

Figure 5.11 Main impurities formed in the asymmetric hydrogenation

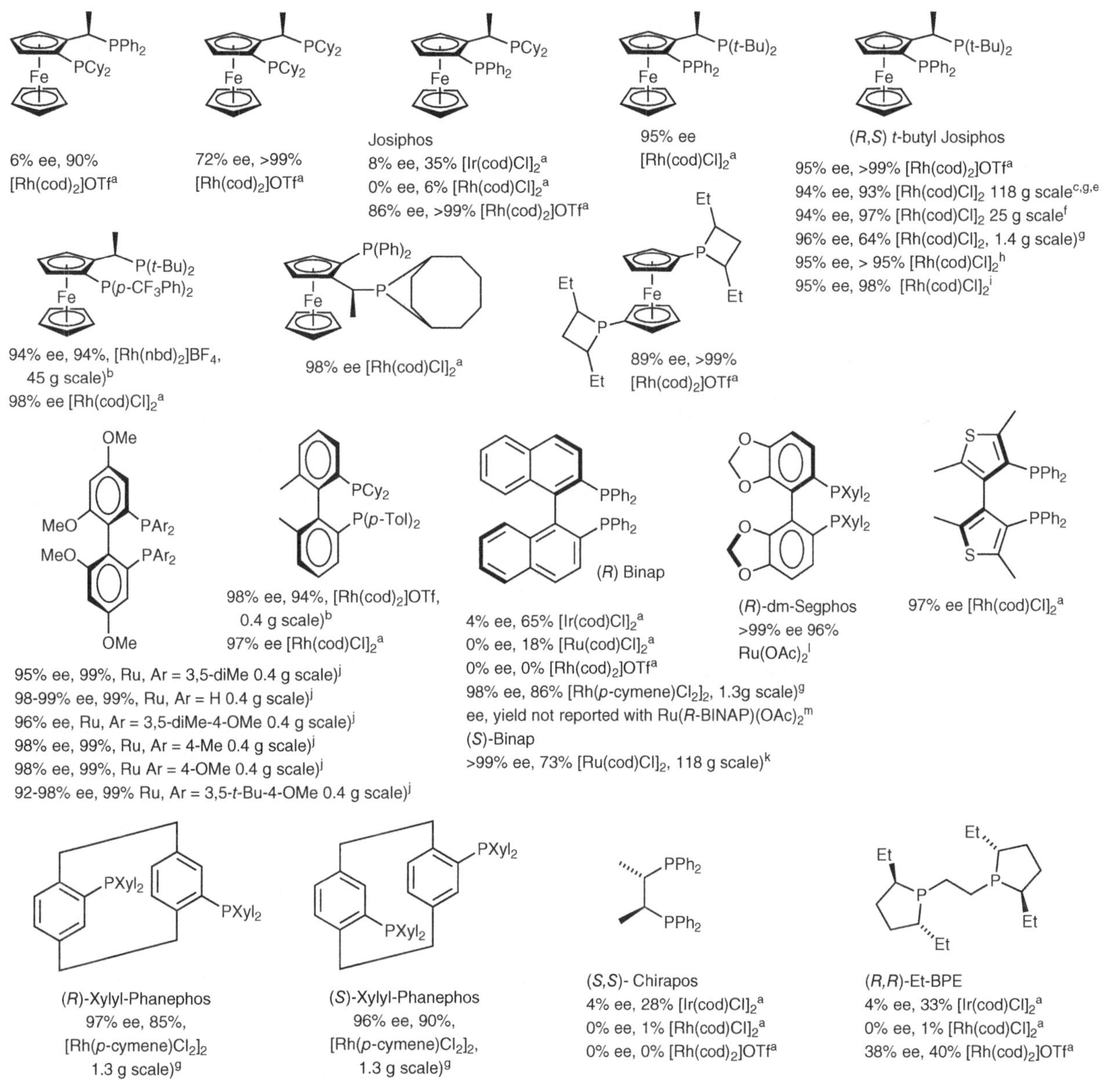

Data: ee, conversion, metal provider, scale

a) JACS2009a; b) Merck2004b; c) Merck2003a; d) Merck2003d; e) Merck2004a; f) Merck2005a;
g) Zentiva2010; h) ACR2007; i) ACIE2011; j) Apotex2013; k) Liaocheng2013; l) JACS2009b; m) Lianyungang2014

FIGURE 5.12 CATALYST LIGANDS EMPLOYED IN THE CAH OF THE β KETO ESTER ENAMINE

C. Obtaining Sitagliptin directly from the β-keto amide by reductive amination or transamination

One-pot processes for reductive amination of the β-keto amide were also developed (scheme H, **H11**). Telescoping avoids isolation and purification of the enamine and can increase throughput. However, even small changes in the impurity profile of raw materials could have an adverse effect on the CAH step, as detrimental impurities are not purged in a prior work-up/isolation procedure in this scenario.

The hydrogenation of the β-keto amide in the presence of a variety of ammonium salts using $Ru(OAc)_2$((*R*)-dm-SEGPHOS) or $[Rh(cod)Cl]_2$ and (*R*,*S*)-*t*Bu-JOSIPHOS (1:1 molar) was investigated (JACS2009b). Optimized conditions produced Sitagliptin in 91% with 99.5% ee, whilst the β-hydroxy amide, from direct reduction of the β-ketone, was formed in less than 3% (Scheme 5.35). Subsequent condensation and reduction of the starting β-ketone with Sitagliptin to give a dimeric by-product was avoided by careful control of reaction conditions.

The β-keto amide was converted into the final API using a biocatalytic approach with transaminases (Codexis2009, 2013, S2010). Since transaminases have a limited range of substrates and most of them do not accept substituents larger than methyl at the α position on the ketone, a mutated enzyme from an *Arthrobacter sp.* was engineered for this approach (Codexis2009, S2010).

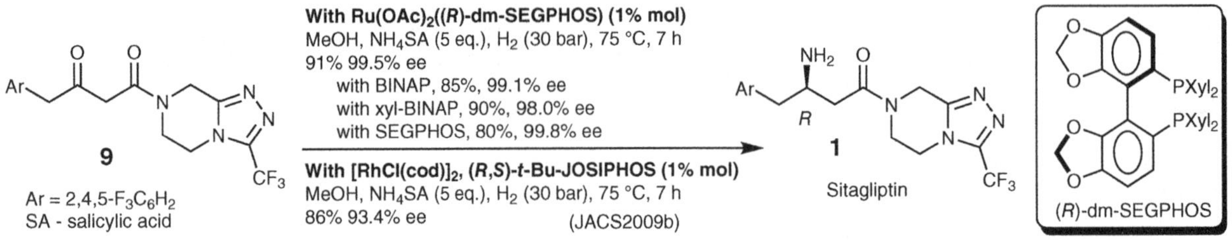

Scheme 5.35 Chemical asymmetric reductive amination of the β-keto amide

This evolved enzyme contained 27 mutations after going through 11 rounds of evolution to become optimised. As expected from these enzymatic approaches, very high ee is achieved (>99.9%, Codexis2013).

35 kg pilot plant feasibility batches gave conversions of 88-93% (Codexis2009, Scheme 5.36). The final optimized process resulted in conditions where 6 gL^{-1} of the evolved transaminase enzyme tolerates a 200 gL^{-1} concentration of the starting material in aqueous DMSO at 45 °C and produces (*R*)-Sitagliptin in 92% yield and >99.95% ee (S2010). The Sitagliptin produced by this method is reported to meet or exceed all purity specifications regarding content of enzymes (<18 ppm), PLP co-factor (<0.1 ppm) and endotoxin levels (<0.05 ng) (Codexis2009).

Further development of this route gave a 74% overall from crude enamine (77% in the isolated enamine) on 700 kg scale[6]. Immobilisation of the enzyme greatly simplifies the process and such a solution was revealed by Merck in 2015[7]. Lyophilised enzyme and PLP (pyridoxal phosphate) co-factor were incubated with a buffer and a variety of resins as support for 24 hours to produce the immobilised enzyme. Reaction with this immobilised enzyme, when run in isopropyl acetate as a solvent and isopropylamine as the amine donor in a plug flow reactor during a 180 hours run did not show any drop off in activity. Isolation is simple: water wash of the organic layer; a solvent exchange to isopropanol; addition of phosphoric acid and isolation of the salt by filtration gives the final API[2].

D. Other late-process asymmetric hydrogenation routes

The previous sections summarize the huge amount of work and process development carried out in the catalytic asymmetric reduction of the enamine intermediate to produce an *N*-protected derivative or Sitagliptin directly (see also scheme H). In contrast, only one process using a CAH of a non-enamine substrate was patented (Scheme 5.37, ZhejiangJP2015b). The reported hydrogenation is quite impressive, using only 0.03% mol of an Iridium catalyst to promote a quantitative and completely enantioselective hydrogenation. The main drawback of this route is the use of a hazardous Curtius rearrangement to replace the carboxylic acid by the primary amine group.

E. Production of Sitagliptin by a classical resolution approach

A number of companies reported routes to Sitagliptin with a resolution toward the end of the synthesis (Scheme H, lower part and Scheme I, right part). Those routes where the primary amine was produced by reduction of an enamine, oxime or nitro group are shown in the lower part of Scheme H. There are also routes that

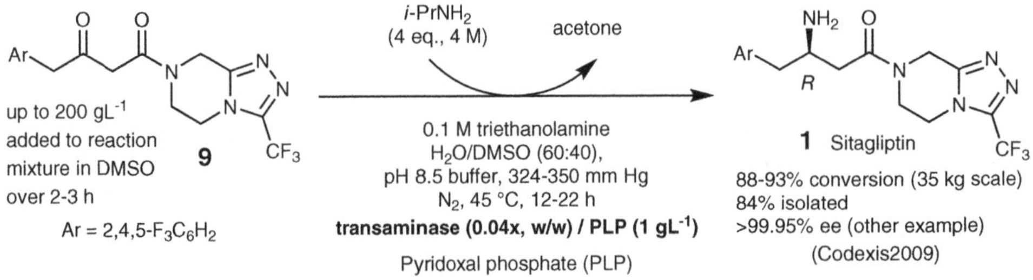

Scheme 5.36 Production of Sitagliptin by an enzymatic transaminase process

6 M. Vink (Codexis) presentation at New Horizons in Catalysis, Cologne, December 2009
7 I. Mergelsberg (Merck) presentation at The 33rd International Conference on Organic Process Research and Development, 97 Toronto, July 2015

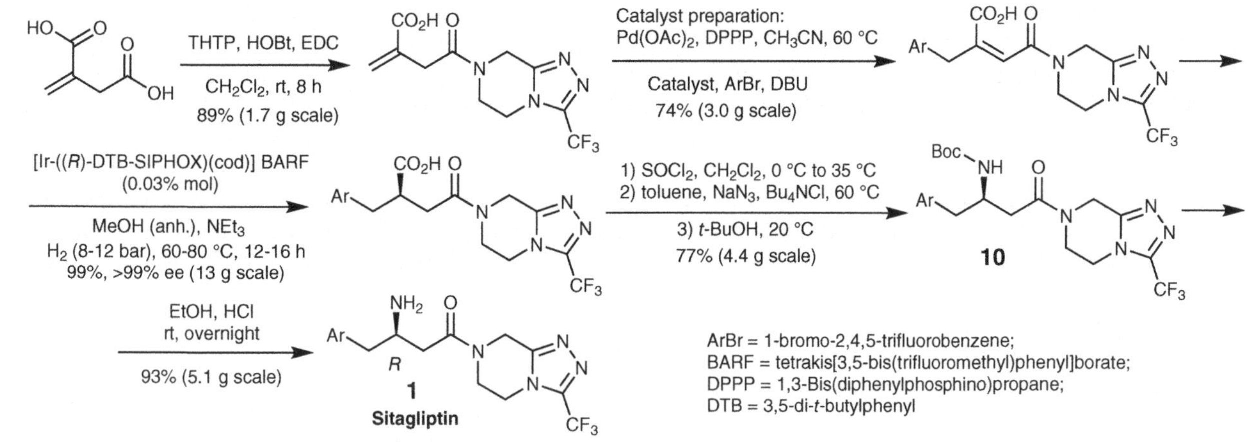

Scheme 5.37 Sitagliptin by a CAH of a non-enamine substrate

go through a β-hydroxy amide, derived from the β-keto amide (**9**) and these are shown in right half of Scheme I (see page 180). These techniques obviously give racemic Sitagliptin (scheme H, **H3+H8** or **H27+H28**). However, when a nitro group is the immediate precursor to the amine, epimerisation can be brought about by simple base treatment because of the acidic nature of the α-hydrogen. In this case, resolution can be carried out on the β-nitro amide (**H34**) and the unwanted isomer recycled (scheme H, Lek2010a).

The approach which opts to convert **9** in a non-stereoselective fashion into the correspondent β-hydroxy derivative (**I10**) is challenging and only a few instances have been reported (scheme I, right part). The instability of the β-hydroxy derivative of **9** toward dehydration to give an α,β-unsaturated amide and its susceptibility to undergo a retro-aldol reaction is anticipated to have a negative impact on yield and generate high levels of impurities.

Key to the exploitation of racemic Sitagliptin as a synthetic goal is finding an efficient method to separate the enantiomers and, just as importantly, recycling of the distomer. Classical resolution processes *via* diastereomeric salts give yields lower than 50% and, in many instances, low enantiomeric excesses after the initial crystallisation (**H36**, **I13**). In one case, an interesting dynamic kinetic resolution was reported wherein a reversible imine/enamine reaction was employed (Scheme 5.38, Suzhou2010).

Of course, a procedure based on the above is applicable to racemize the distomer, but other strategies have also been reported (Scheme 5.39). One method uses the strongly electron withdrawing trifluoroacetamide group to make the α-hydrogen sufficiently acidic

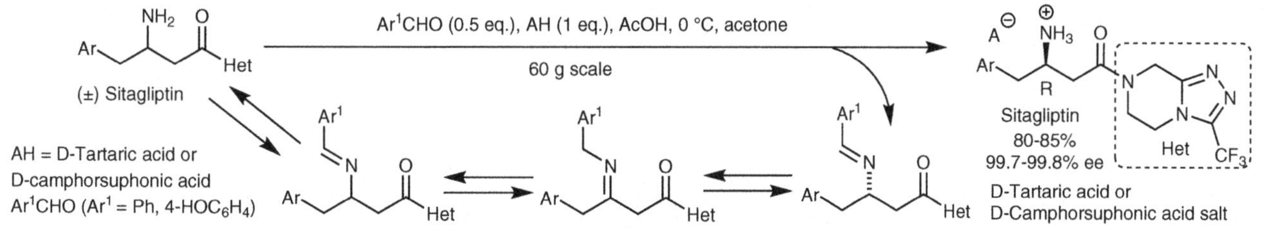

Scheme 5.38 Dynamic kinetic resolution process for Sitagliptin

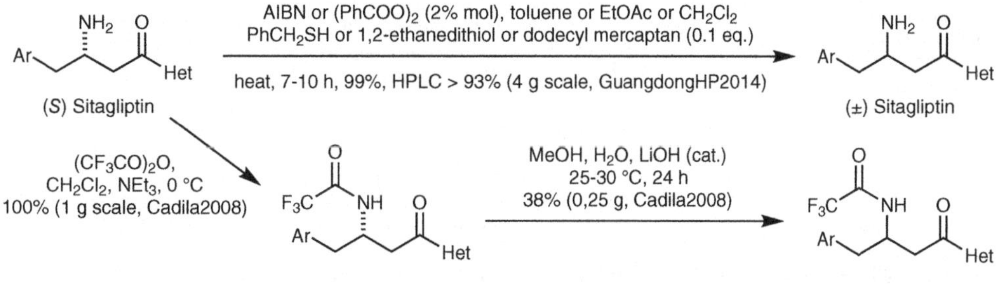

Scheme 5.39 Isomerization processes for the recovery of (*S*)-Sitagliptin

to provoke racemization under basic conditions giving a yield of 38% and a purity of 83% (by HPLC) on a 1 g scale. Heating (*S*)-Sitagliptin with a thiol in the presence of a free radical initiator produces the racemate (99% yield, HPLC purity higher than 93%, 4 g scale, GuangdongHP2014).

Other procedures opted to revert to the sp² carbon atom of the β keto amide *via* the sequence: C-NH_2 / C-N_2^+ / C–OH / C=O (MSN2009, Scheme 5.40). Both the β-keto ester (**5**) and the β-keto amide (**9**) can be reintroduced into the synthetic sequence.

F. Other approaches with a resolution late in the process

In addition to those synthetic routes wherein racemic Sitagliptin is produced and resolved, processes that rely on an enantiomer separation of a late stage intermediate have been proposed (Scheme 5.41, Dipharma2009).

Here the enantiomers are separated by enzymatic ester hydrolysis of an advanced intermediate. Diethylmalonate is the source of the C3 atom of the central chain. The acidity of the methylene hydrogens of the malonate unit facilitate the sequential aldol condensation and α-alkylation. Hydrolysis, decarboxylation and esterification gives an ester, which is then followed by selective enzymatic hydrolysis. The resulting (*R*)-acid is subjected to a hazardous Curtius-type rearrangement that proceeds with complete retention of configuration to produce the (*R*)-primary amine. Recycling of the unwanted *S* isomer (not hydrolysed in the process), is feasible by base treatment because of the acidity of the α hydrogen atom.

G. Installing the amine group at the end of the sequence by a Michael addition

Michael addition to an α,β unsaturated carboxylic acid or ester to introduce the amine function has been discussed in 5.4.1.4. This methodology can also be applied at the end of the synthesis (Scheme 5.42, Stereokem2014).

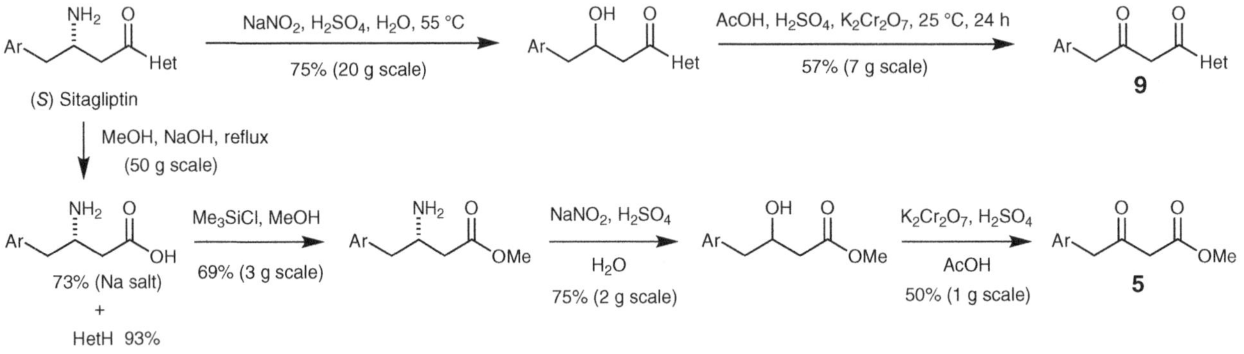

Scheme 5.40 Recycling processes for (*S*)-Sitagliptin *via* diazonium salts to a β-keto derivative

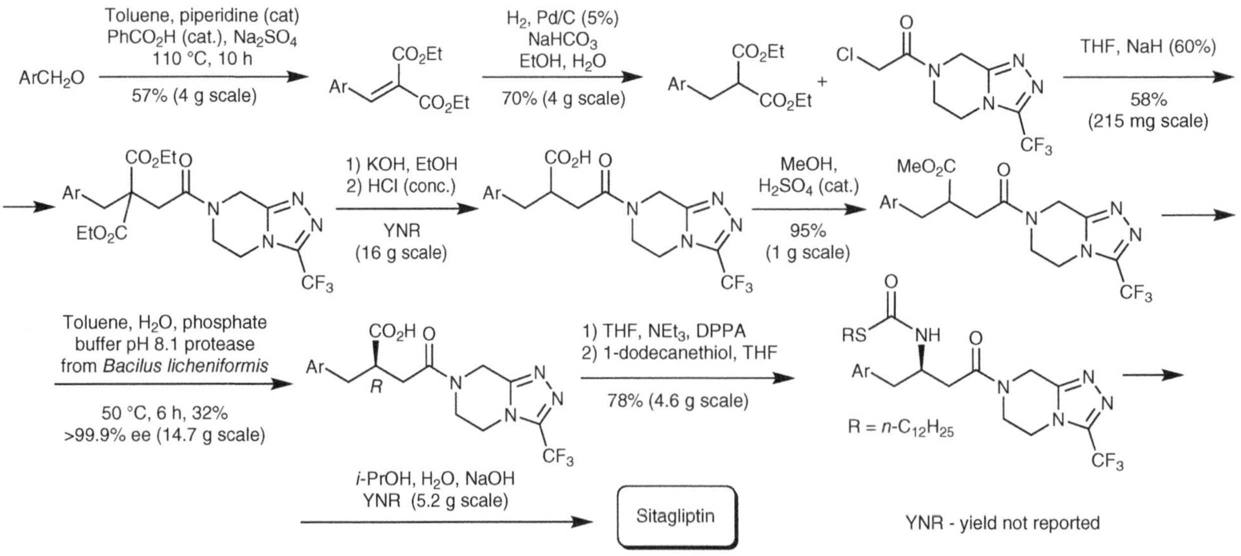

Scheme 5.41 Synthesis of Sitagliptin *via* enzymatic resolution-Curtius rearrangement

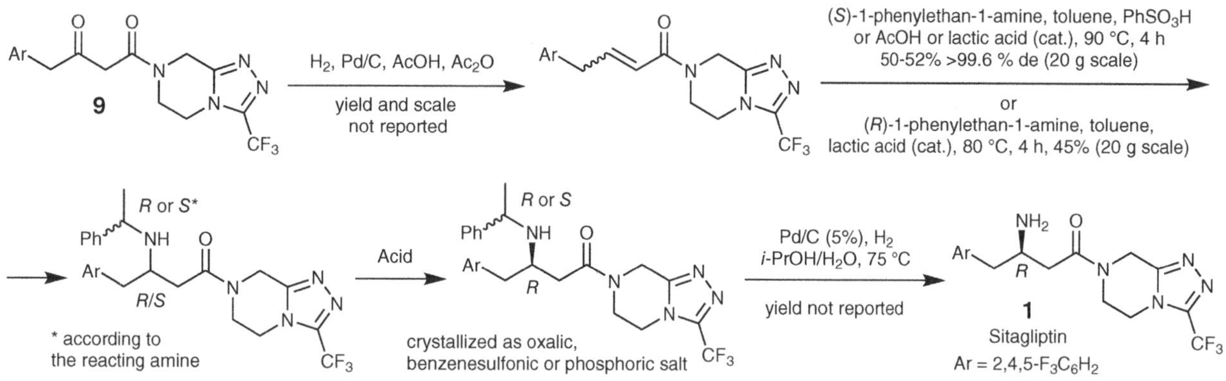

SCHEME 5.42 SITAGLIPTIN SYNTHESIS BY A LATE CONJUGATE ADDITION OF 1-PHENYLETHAN-1-AMINE

When a chiral benzyl amine is used as nucleophile an intermediate with the desired chirality can be separated by crystallisation from the resulting mixture of diastereomers. As seen previously, hydrogenolysis unmasks the primary amine. Although the separation of the diastereomers by a crystallisation is relatively facile, this procedure suffers from the fact that the yields of the *R*-isomer is only around 50% and recovery of the unwanted isomer is, as before, difficult.

5.4.2.2 Installing the chiral primary amine *via* carbonyl reduction-hydroxyl replacement at the final stage of the synthesis

Scheme I summarizes the processes from the β-keto amide intermediate (**9**) to Sitagliptin that use a carbonyl to hydroxyl reduction. The non-chiral routes of scheme I (right part) to Sitagliptin have already been mentioned above, resolution of Sitagliptin being the final step.

Other patented routes incorporate an enantioselective transformation before the amine function is in place, such as asymmetric reduction of the β-keto amide (**9**) and replacement of the hydroxyl group (left portion of scheme I). The *S*-alcohol (from **I1**) requires a single bimolecular substitution with inversion of configuration (**I2**) to install the nitrogen group with the correct chirality. In one case, the hydroxyl group was activated to give a suitable leaving group (e.g. tosyl) and an S_N2 reaction with azide (or some other nitrogen nucleophile) introduced the nitrogen with the correct configuration.

From the *R*-alcohol, a double inversion of configuration is required. This was achieved by halogenation (inversion of stereochemistry) followed by bimolecular substitution with a nitrogen nucleophile, either phthalimide or azide (**I7** or **I5**). Notably, all these synthetic sequences described in Scheme I require an additional step to uncover the primary amine after introducing the nitrogen.

5.4.2.3 Building the *R*-primary amine at the last synthetic step by an asymmetric alkylation

Yet another strategy where the primary amine is created in the final stages of the synthesis *via* an asymmetric alkylation can be seen in Scheme 5.43 (Dissymmetrix2010). The key step is reacting the chiral sulfinimide from (*R*)-2-methylpropane-2-sulfinamide and 2-(2,4,5-trifluorophenyl)acetaldehyde with the lithium enolate of *N*-acetyl THTP, which gives the *R*-isomer of the sulfinamide end product. Yields of several of the steps are not reported.

5.4.3 Synthesis of Sitagliptin by attaching the 2,4,5-trifluorophenyl ring at the end of the synthesis

This, the last major strategy derived from the retrosynthetic analysis, is the one of the least studied routes to Sitagliptin. Only two examples are found in the literature and both use a chiral-pool approach, introducing the asymmetry of the primary amine employing L-aspartic acid.

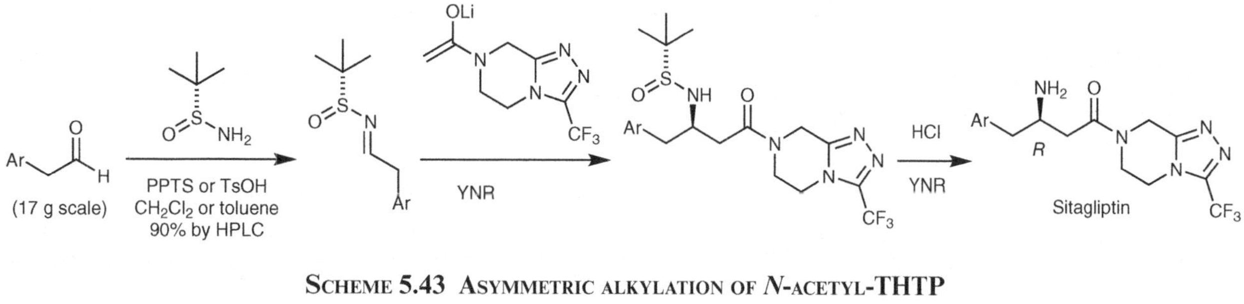

SCHEME 5.43 ASYMMETRIC ALKYLATION OF *N*-ACETYL-THTP

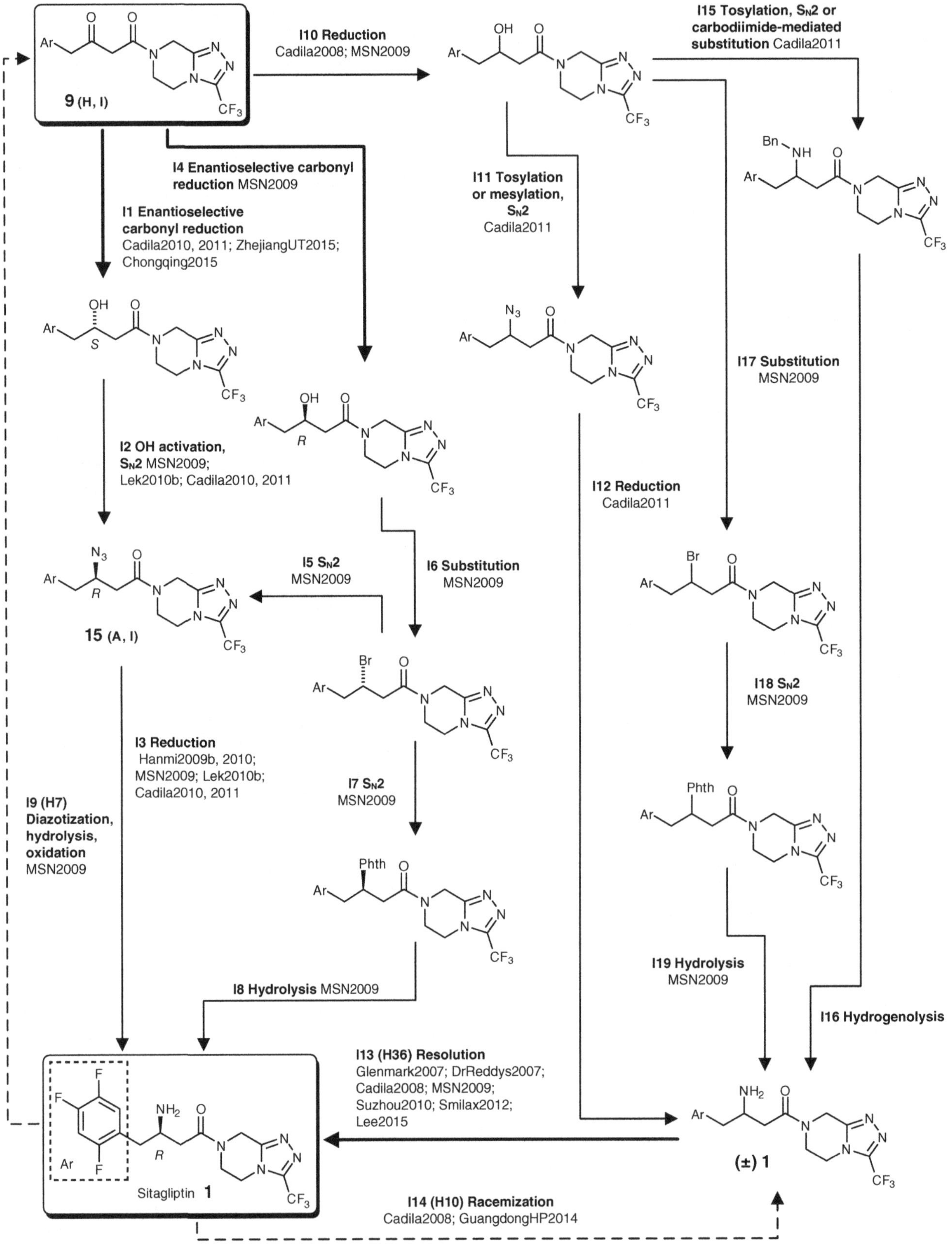

Note: the letters near the compound number indicate other schemes where the compound is present. The asymmetric or resolution steps are highlighted with a thick arrow. Dashed lines are recycling processes of the enantiomer of the molecule shown.

Scheme I Installing the primary amine *via* carbonyl reduction-hydroxyl replacement at the final synthetic steps of the route (for reaction conditions see page 199)

In one case (Esteve2009) THTP is attached by formation of the amide *via* nucleophilic attack on a lactone derivative (Scheme 5.44). Metal-mediated coupling of 1-bromo-2,4,5-trifluorobenzene follows to produce the Sitagliptin framework.

In the second of these routes (Scheme 5.45, Daewoong2011), THTP is coupled with a carboxylic acid in a classical mode using 2-(1*H*-benzotriazol-1-yl)-1,1,3,3-tetramethyluronium hexafluorophosphate (HBTU) as a mild coupling agent designed to reduce racemization. The ester at the opposite end of the central chain is transformed into a Weinreb amide and then reacted with the Grignard reagent from 1-bromo-2,4,5-trifluorobenzene. Reductive removal of the benzylic group and deprotection affords Sitagliptin.

5.4.4 Triazolopyrazine synthesis

Due to the huge success of Sitagliptin, several building blocks have become freely available in the market i.e. the THTP unit and several variations for the trifluoroaromatic ring moiety.

THTP is a required building block in all Sitagliptin syntheses. A quick structure analysis leads to several synthetic routes (Scheme 5.46). A major inconvenience to overcome in all of the synthetic routes to THTP is the necessity to use hazardous hydrazine.

Two of the preferred approaches to THTP have been developed, either starting with 2-chloropyrazine or by route **b** in Scheme 5.46. The routes to produce THTP are summarized in scheme J.

In the first synthesis of Sitagliptin synthesis of THTP used 2-chloropyrazine as starting material (Scheme 5.47, Merck2001, JMC2005). A nucleophilic aromatic substitution reaction with neat hydrazine hydrate under reflux at 120 °C was the first step in THTP synthesis (**J1**).

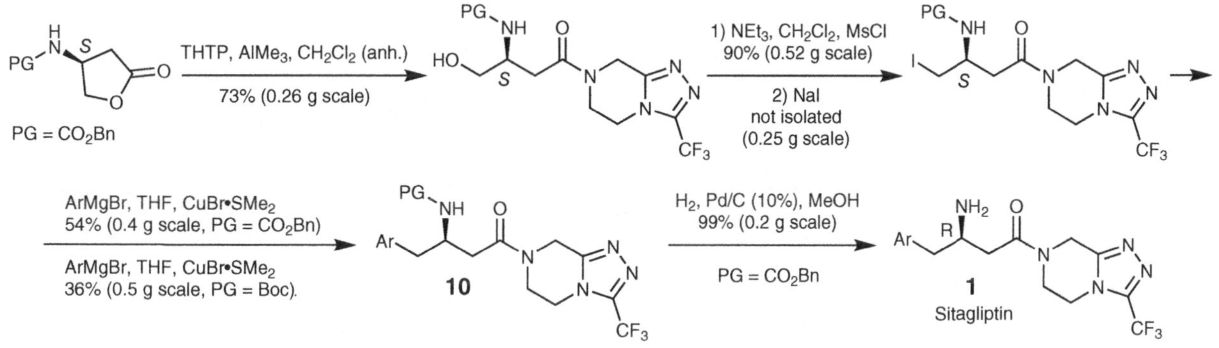

SCHEME 5.44 ATTACHING THE FLUORAROMATIC RING AT THE END BY A METAL MEDIATED COUPLING REACTION

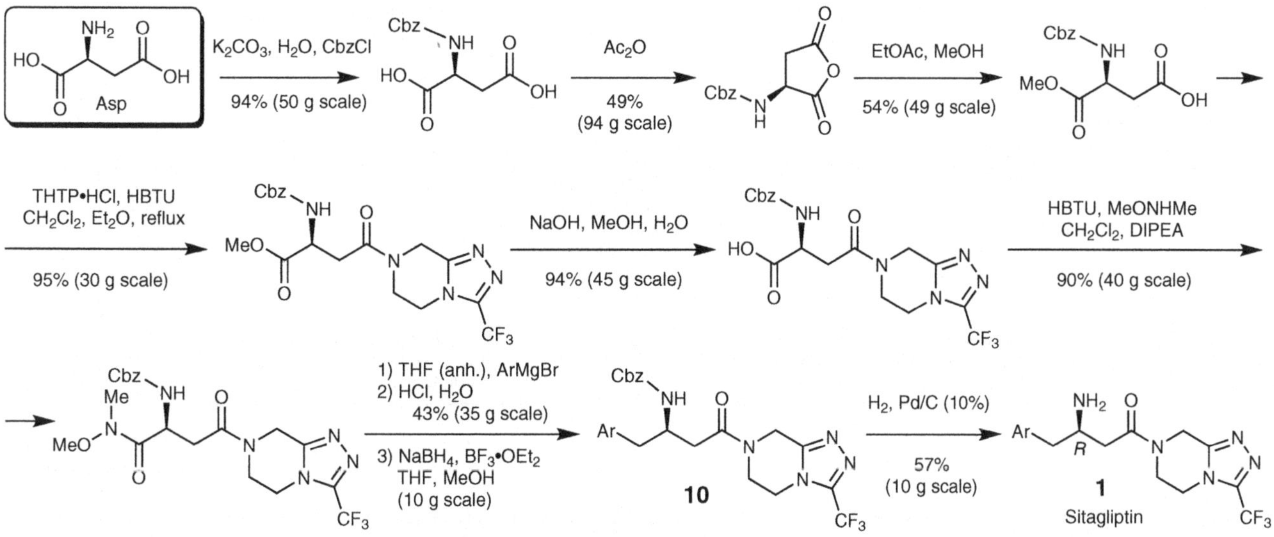

SCHEME 5.45 SYNTHESIS OF SITAGLIPTIN ATTACHING THE TRIFLUORAROMATIC RING AT THE END OF THE ROUTE BY A WEINREB AMIDE REACTION

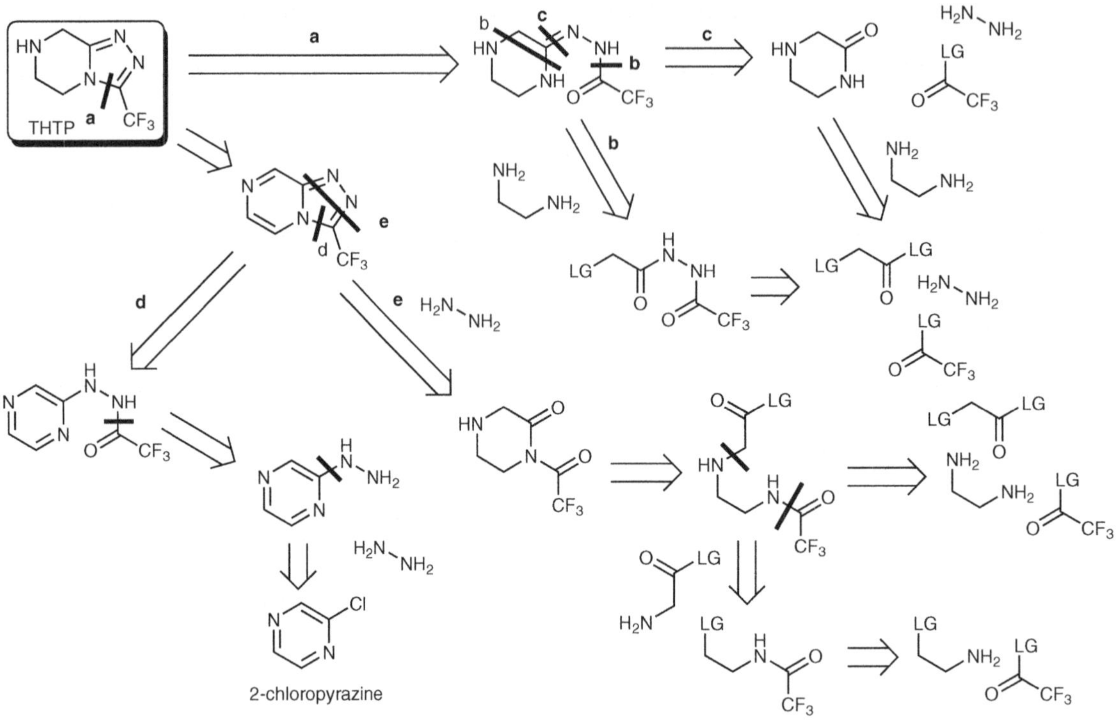

SCHEME 5.46 RETROSYNTHETIC OUTLINE FOR THTP

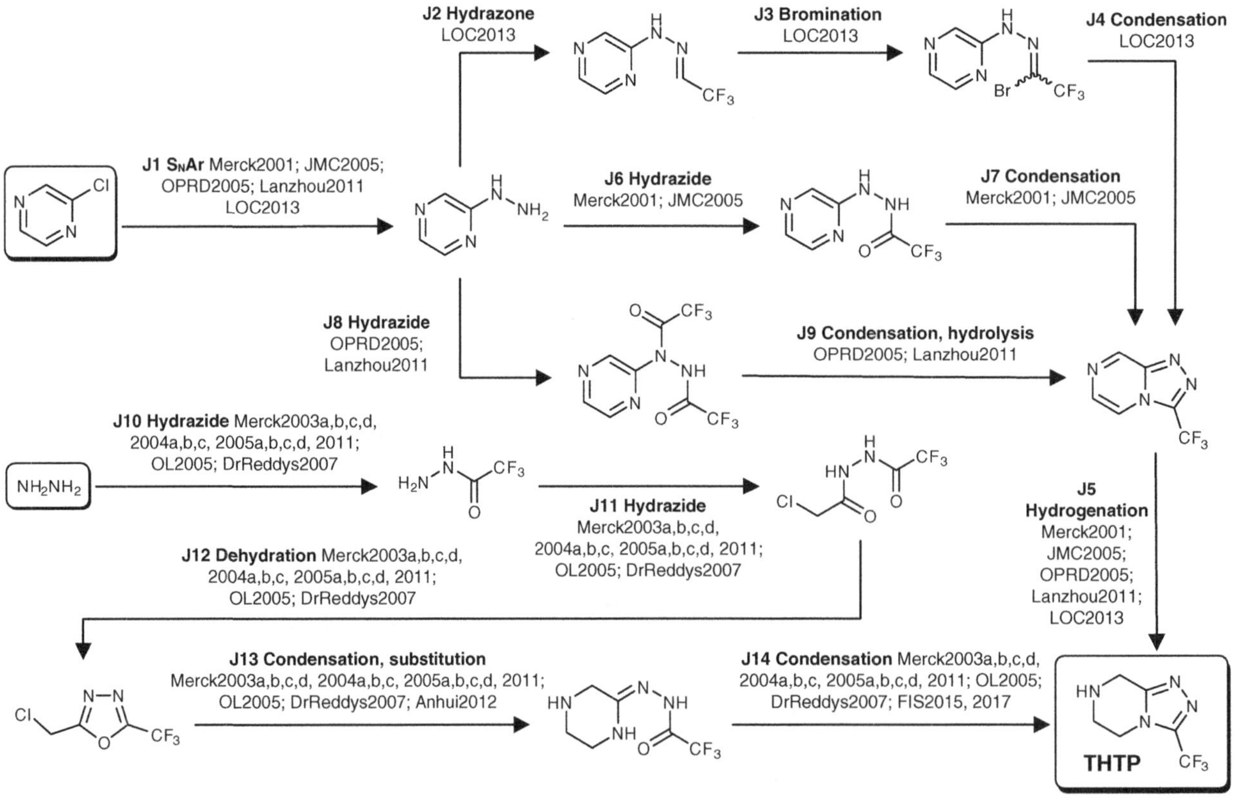

SCHEME J SYNTHETIC ROUTES TO 3-(TRIFLUOROMETHYL)-5,6,7,8-TETRAHYDRO-[1,2,4]TRIAZOLO[4,3-*a*]PYRAZINE (THTP) (FOR REACTION CONDITIONS SEE PAGE 200)

Given the dangers associated with neat hydrazine hydrate, a later laboratory procedure used 2-propanol as solvent with 5 molar equivalents of hydrazine. The product was isolated by crystallization upon cooling and filtration. Scale-up of this procedure was untenable as anhydrous explosive hydrazine could well be produced in the vapour phase, especially when azeotropic distillation removes water from the reaction mixture. Unfortunately, aqueous conditions required prolonged heating above 60 °C and calorimetric studies detected an uncontrollable exothermic event above 85 °C, relatively close to the 60 °C of reaction.

A safe procedure was developed by slowly adding the chloropyrazine to an aqueous solution of hydrazine (5.6 equiv.) at 60-65 °C (Scheme 5.47, OPRD2005). The intermediate was removed by an extraction procedure giving low levels of hydrazine in the organic extract, thus allowing safe solvent removal by distillation. Superphos was used to catalyse cyclisation (**J7**) circumventing the use of highly viscous PPA and avoiding its highly exothermic quenching with water. After hydrogenation (**J5**) the THTP was obtained in a 26% overall yield from the stating pyrazine (OPRD2005).

The large excess of hydrazine required to accomplish a complete reaction with 2-chloropyrazine was successfully avoided using a bishydrazide (Scheme 5.48). This was prepared by reacting hydrazine sequentially with equimolar amounts of ethyl trifluoroacetate and chloroacetyl chloride (**J10+J11**). Dehydration (**J12**) produced a key reactive chloromethyloxadiazole, which on condensation with 1,2-phenylenediamine followed by ring-closure afforded THTP. Investigation uncovered efficient dehydration conditions, which used of a substoichiometric amount of phosphorus oxychloride with a catalytic amount of DMAP (OL2005). The best yield was obtained with phenylphosphonic dichloride (88%) but phosphorus oxychloride is preferred as it produces similar yield, is cheaper and creates more benign waste streams. This synthesis, after adequate development work, has an overall yield of 52%. However, it should be noted that the use of hydrazine is not avoided.

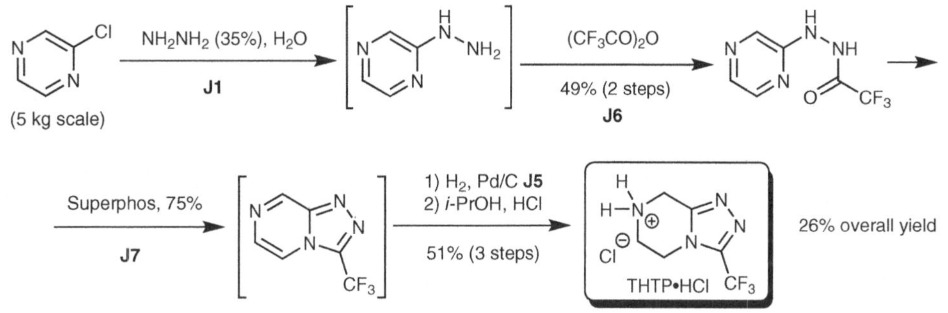

SCHEME 5.47 SCALED-UP PROCESS FOR THE SYNTHESIS OF THTP FROM 2-CHLOROPYRAZINE

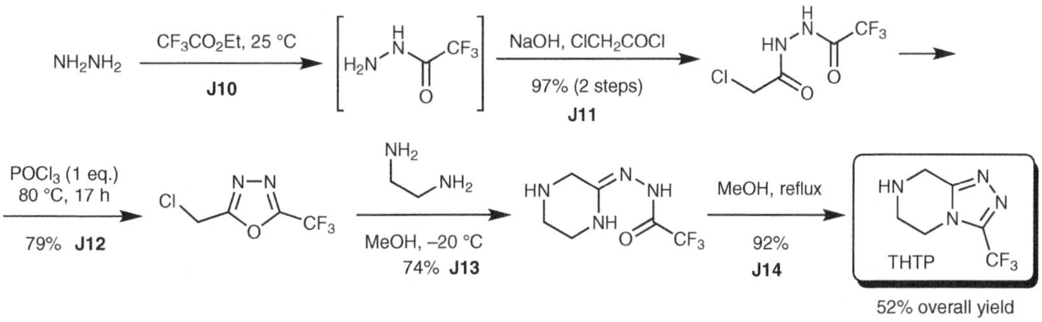

SCHEME 5.48 SYNTHESIS OF THTP USING A REACTIVE CHLOROMETHYLOXADIAZOLE AS KEY INTERMEDIATE

5.5 STRATEGY COMPARISON AND CONCLUSIONS

As is to be expected for a highly successful drug, many approaches have been published for the synthesis of Sitagliptin and many of these appear to be suitable for implementation at industrial scale. The principal factor in the selection of the most suitable route is an efficacious introduction of the chiral primary amine. Several routes use a chiral pool approach to deliver the primary chiral amine (schemes F and G). However, none of these processes is direct and the chiral pool material has to be converted into suitably reactive intermediates to accomplish the subsequent synthetic transformations. In some cases, as when L-Aspartic acid is used, the amine is protected as a benzylamine and this protecting group later removed to unveil the final product. The outcome is that an additional number of synthetic steps are required to reach the final product (Scheme 5.49).

A resolution approach at either the (*R*)-β-amino acid stage or separation of enantiomers in the final product is also far from being an ideal solution as the recycling of the unwanted isomer is challenging. Consequently, the competitiveness of resolution-based routes is questionable. Processes based on an asymmetric transformation are able to deliver the final product in fewer steps, with high purity and high ee. Hence, asymmetric induction has clear advantages over chiral pool or resolution-based routes.

5.5.1 Main strategies for asymmetric synthesis

The two main strategic options, as shown in Scheme 5.50, were the subject of much investigation and process development. A major decision to be made when selecting a route is where to place the asymmetric step. The impact of this decision on the requirements for implementation in an existing plant or, on the design and cost of new plant, clearly influence the final choice.

Of the various routes presented in schemes A, B, D, E, H and I, asymmetric enamine reduction or the use of a biochemical transaminase-based approach appear to be the most favoured. Either a chiral β-amino acid or Sitagliptin itself are the end products of these transformations (schemes B and H).

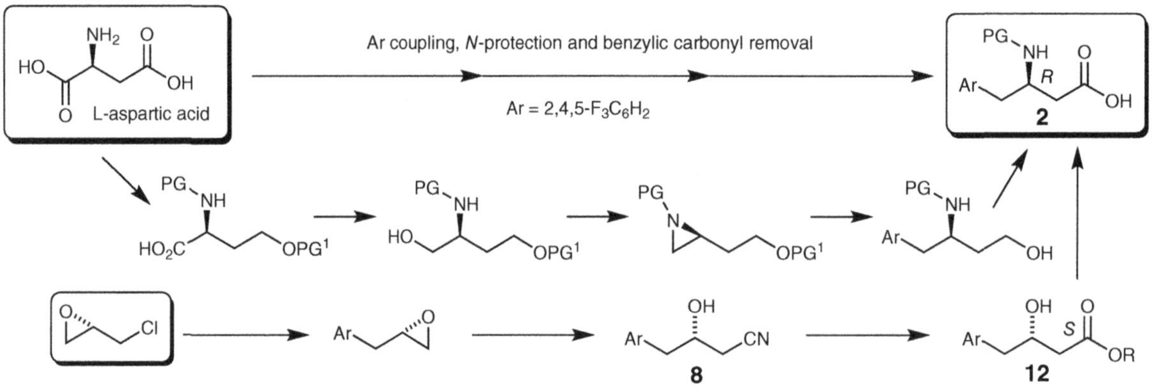

SCHEME 5.49 ILLUSTRATIVE EXAMPLES OF SITAGLIPTIN SYNTHESIS USING A CHIRAL-POOL STRATEGY APPROACH

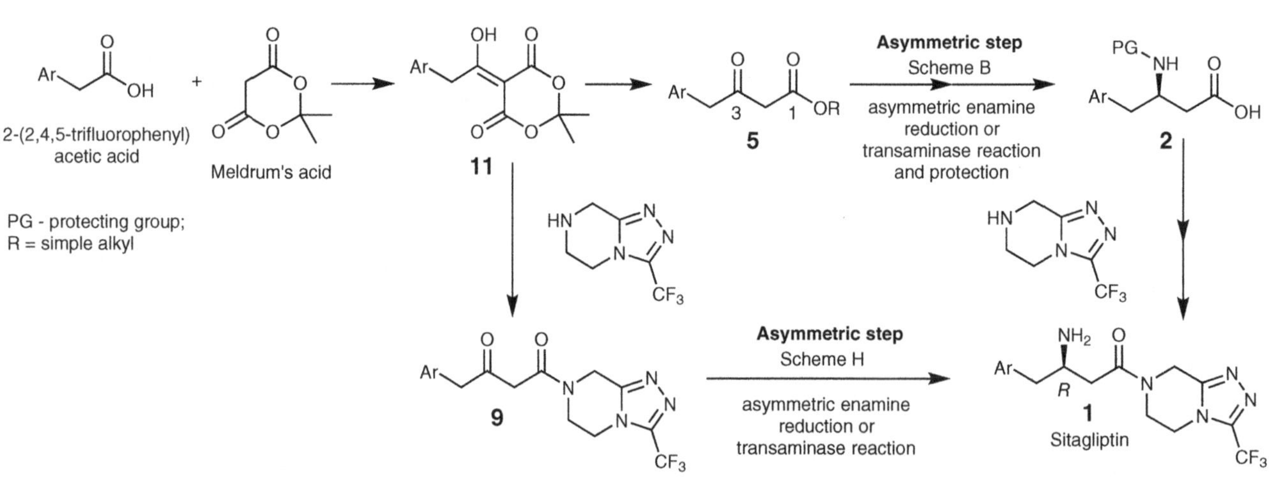

SCHEME 5.50 MOST USED ASYMMETRIC ROUTES TO SITAGLIPTIN

Coupling the chiral β amino carboxylic acid (**2**) with THTP is a sound option. The β-keto ester (**5**) is easily obtained from 2-(2,4,5-trifluorophenyl)acetic acid and asymmetry may be introduced by a chiral reduction of the ketone in **5** or of a derived enamine.

The first large scale synthesis opted to use the ketone reduction route to produce more than 100 kg of Sitagliptin, chirality being introduced by a ruthenium-mediated asymmetric reduction of the carbonyl function (Merck2003c). This first-generation process, implemented at pilot plant scale, has eight steps and an overall yield of 52% (Scheme 5.51, ACIE2011).

One main concern associated with this route is the EDC coupling/Mitsunobu/EDC coupling sequence where carbodiimide coupling agents are used three times (DIAD once and EDC twice). Poor atom economy is inherent in these coupling but no alternatives are easily available. The large amounts of by-products renders this route costly and environmentally unfavourable. Achieving the high optical purity of the lactam requires carefully controlled conditions and has a direct impact on the final quality of the Sitagliptin. Notably, the β-keto ester was transformed into the key, protected chiral β-amino acid after five chemical steps. Nonetheless a simpler, greener route, is desirable for industrial manufacture.

A direct replacement of the (*S*)-hydroxy group of **12** by a nitrogen nucleophile is challenging as the substrate is prone to undergo competing E1cB elimination to form the α,β-conjugated carbonyl compound. Therefore, development of the internal S_N2 (S_Ni) reaction was key to reduce elimination (OPRD2005).

The best options for introducing the chiral primary amine, no matter where this occurs in the synthetic sequence, are by reduction of a chiral enamine, asymmetric reduction of a non-chiral enamine or by a transaminase enzymatic process (Scheme 5.52).

Generation of the primary *R*-amine from an enamine utilizes two principle procedures (Scheme 5.52):

- In the first of these, chirality already exists in the enamine as a chiral amine has been used in its formation and a classical hydrogenation or hydride reduction induces chirality on the carbon. This diastereoselective approach uses a stoichiometric amount

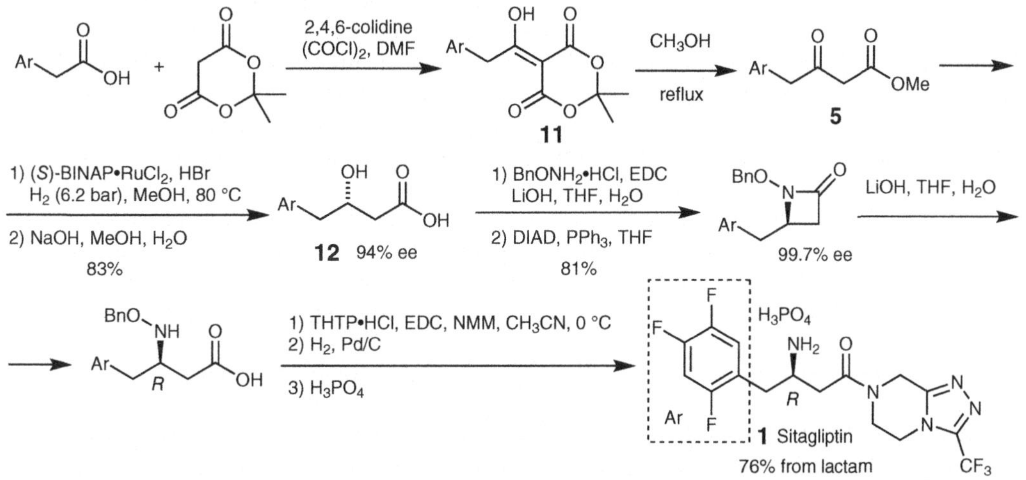

SCHEME 5.51 OUTLINE OF THE FIRST LARGE SCALE ROUTE TO SITAGLIPTIN HAVING A CARBONYL ENANTIOSELECTIVE REDUCTION AS KEY STEP

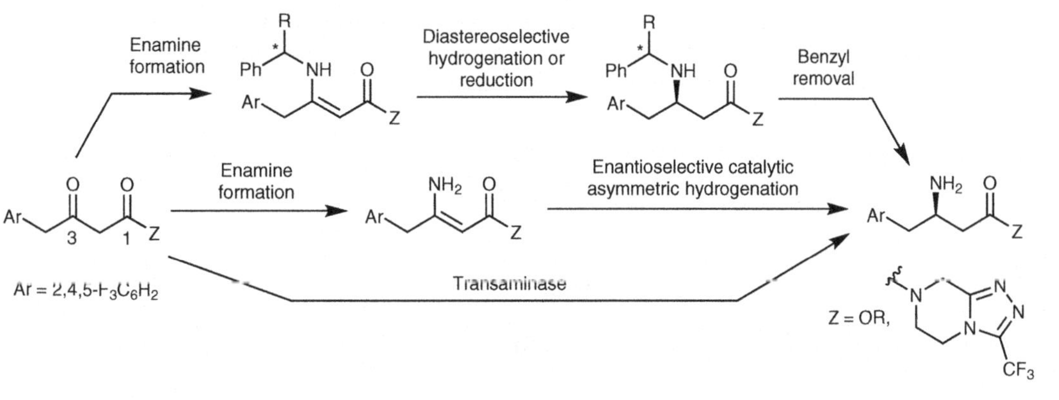

SCHEME 5.52 THE MAIN APPROACHES TO PRODUCE THE CHIRAL AMINE STARTING FROM A CARBONYL GROUP

of the chiral amine. Enamines from β-keto carboxylic acid derivatives and chiral benzylamines are the preferred intermediates for these routes. As annotated before, a stoichiometric amount of chirally pure benzylamine, which cannot be recovered, is a prerequisite and an extra chemical step, hydrogenolysis with an Adam's (platinum) catalyst, is required.
- The second approach uses an asymmetric catalytic hydrogenation. In this case the catalyst is chiral and the reduction enantioselective.

Although efficiency and cost comparisons are critical when selecting the route of choice, from this brief analysis it seems that routes that induce chirality through use of an enantioselective hydrogenation catalyst are favoured as the use of stoichiometric amounts of chiral amines is avoided. However, for such a route to be successful, painstaking development of the hydrogenation step is critical wherein the following factors are imperative: yield; enantiomeric purity; throughput; optimisation of catalyst loading and its recovery. Biochemical approaches circumvent the hydrogenation step, and although this might have value, tailor-made enzymes, which can come with a high price tag, are normally required.

5.5.2 The relative positions of the asymmetric step and the amide forming coupling reaction in route selection

After opting for an enamine reduction (or a transaminase reaction), the next decision is where to locate the asymmetric step in the synthetic sequence. As a corollary, the commercially available THTP building block will be introduced either near the end or at an earlier point in the sequence (see Scheme 5.50 above).

Reflections on the amide-forming step:

Both approaches add the THTP heterocyclic later rather than at the beginning, which makes for a convergent synthesis.

The option to connect the THTP building block during the very final stages of the process requires that the amine of the chiral β-amino-carboxylic acid be protected (Boc is preferred) to avoid unwanted coupling reactions (Scheme 5.53). Various highly efficient amidation reactions were reported, but the use of carbodiimide coupling agents is almost omnipresent. Yields above 85% and acceptable API purity were obtained in all cases. Of the many carbodiimides investigated only a few are routinely used on a large scale (DCC, DIC and EDC). Notably, all of these compounds are skin sensitizers and must be handled with care.

A downside of carbodiimide couplings is the large amount of urea by-products that are produced, making their use environmentally unfriendly. Waste dicyclohexylurea from DCC has limited solubility in most organic solvents and is typically removed by filtration, which generates a solid for disposal. However, DIC and EDC urea by-products are water soluble and normally removed by an aqueous extraction process, which can lead to ground water contamination if the aqueous waste stream is not carefully treated.

Auxiliary nucleophiles used improve reaction efficiency in the amide-forming step, reduce the formation of *N*-acylurea side products and epimerization of the α-stereocenter whilst increasing the overall rate of coupling are also of concern. HOBt is by far the most common used either catalytically or in stoichiometric amounts but is shock sensitive and subject to travel regulations. The more stable analogue, HOAt, considerably more expensive, can be an alternative.

The other approach, where the amide is introduced earlier in a one-pot process, acidic decomposition of the Meldrum's acid aldol product (**11**) in the presence of THTP hydrochloride gives amide in good yield and quality (90% overall yield from the starting acid for the two steps, JACS2009a, Scheme 5.54). Thorough investigation of the mechanism of the amide-forming step revealed that the reaction proceeded *via* a ketene intermediate. This method of creating the amide appears to be superior in an industrial situation to the carboxylic acid/amine coupling with carbodiimides.

On the *N*-protecting group:

Protection of the amine nitrogen atom is necessary to assure chemoselectivity in reactions where the free amine could take part. Of course, the use of a protecting group mandates a deprotection step. The Boc protecting group is most commonly used as it is easily removed by a simple acid treatment.

Those routes that introduces the heterocycle at an earlier stage before the free amine do not require the use of a protecting group and the end-product can be reached in fewer steps.

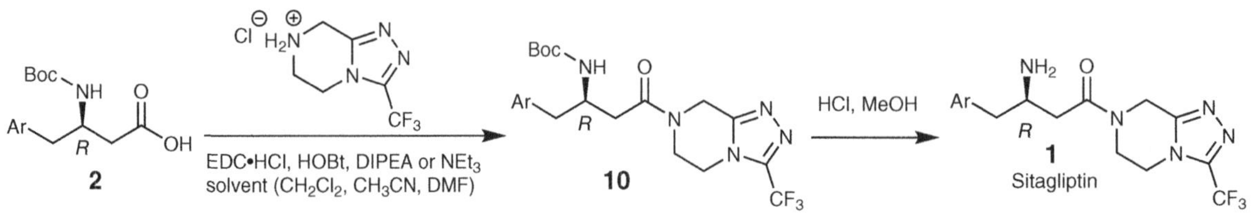

Scheme 5.53 Most frequent conditions for final amide coupling and *N*-deprotection

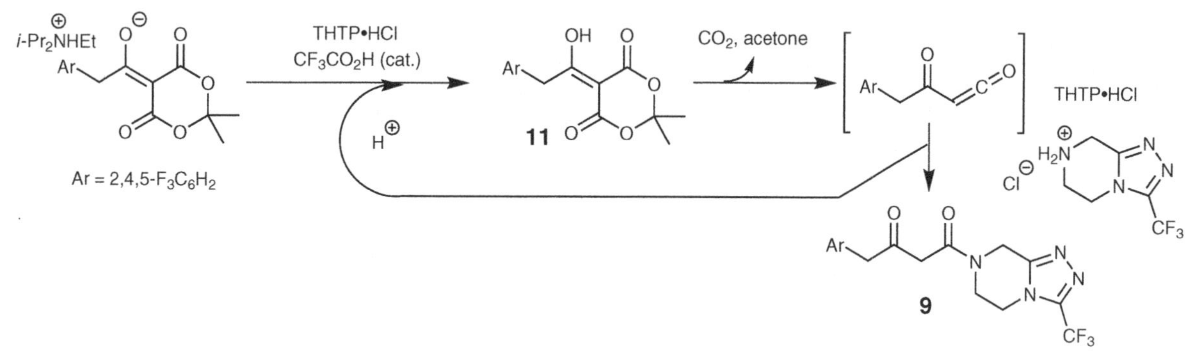

SCHEME 5.54 AMIDE-FORMING MECHANISM FOR 9

On the location of asymmetric step:

One of the decisive factors when selecting the synthetic route for further elaboration is undoubtedly the cost involved in the asymmetric catalytic reaction.

Generically speaking, to obtain the same amount of final drug more catalyst will be required when it is used earlier in the synthesis. This is a natural consequence of the fact that synthetic yields are practically never quantitative. By delaying the asymmetric hydrogenation to towards the end of the synthesis, and assuming the use of same molar ratio catalyst/product, less amounts of the expensive catalyst is required. As a general conclusion, the use of an expensive reactant near the end of the synthesis is preferred.

5.5.3 Preferred routes for the synthesis of Sitagliptin

Productivity is clearly enhanced when the catalytic asymmetric hydrogenation step is delayed until late in the synthesis. Despite the fact that the literature at the time reported only successful CAH of *N*-protected enamines, development of an asymmetric hydrogenation of the unprotected enamine was achieved. Summarising what has been discussed above, many combinations of ligands, solvents, temperatures, pressures and the metals Rh, Ru and Ir were tested. High enantioselectivity was reached with low catalyst loading and almost quantitative yield (98%) and high ee (95%, JACS2009a).

The enamine formation and CAH can be telescoped and carried out in one process step (JACS2009b, ACIE2011). Optimized conditions produced Sitagliptin in 91% with 99.5% ee containing less than 3% of β-hydroxy amide derived from direct hydrogenation of the β-ketone (Scheme 5.55). The other major impurity, which is produced by condensation of the initially formed Sitagliptin amine and the starting β-ketone, followed by hydrogenation, is avoided by judicious selection of reaction conditions. A higher loading of the catalyst was used in comparison to the hydrogenation as applied to the enamine on its own (1% versus 0,15%).

Low rhodium catalyst loading, combined with ease of recovery of both the precious metal (Merck reported than more than 95% of the rhodium is recycled) and the expensive chiral ligand, make this process highly cost-effective.

Both β-keto amide and enamine intermediates are crystalline solids, allowing detailed separate investigation of their transformation, thus facilitating the development of a robust and reliable CAH step. In addition, crystallization of one or both of these compounds can be inserted into the process as a "clean-up point", assuring success of the CAH, which can be susceptible to impurities in the substrate. This process delivers final product of appropriate chemical and enantiomeric purity indicating that this may be the most efficient and the preferred chemical synthesis of Sitagliptin (Scheme 5.56).

This process, as a testimony its success and applicability, was the winner of a coveted Presidential Green Chemistry Challenge in 2006[8].

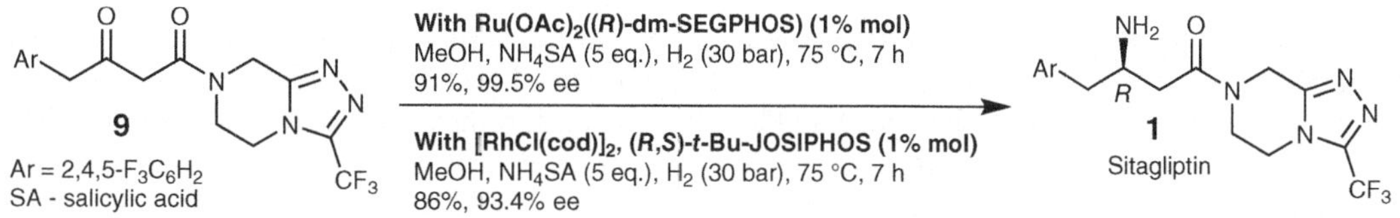

SCHEME 5.55 SELECTED CONDITIONS FOR THE Rh AND Ru CATALYTIC ASYMMETRIC REDUCTIVE AMINATION OF THE β-KETO AMIDE

8 In Presidential Green Chemistry Challenge: 2006 Greener Synthetic Pathways Award, United States Environmental Protection Agency

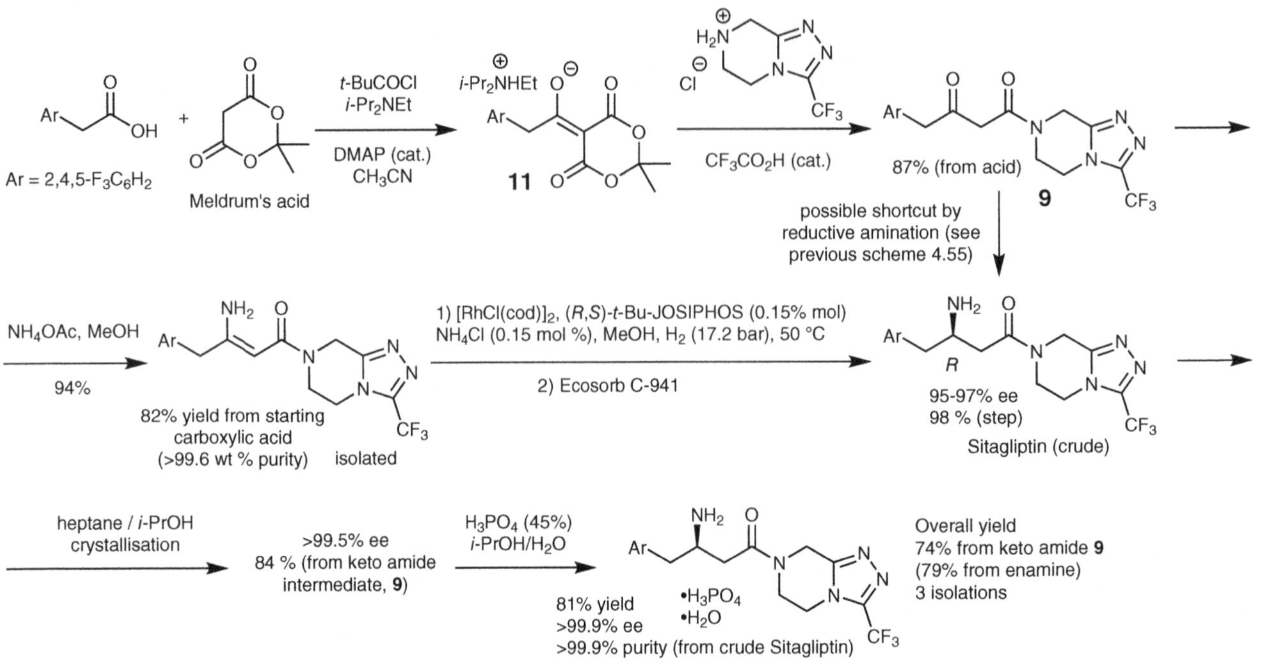

Scheme 5.56 Best industrial chemical synthesis of Sitagliptin (with isolation of the enamine intermediate)

In comparison to the first-generation process, this route is environmentally friendly ('green') and was able to significantly reduce the total waste generated per kilogram of Sitagliptin, from 250 kg to 50 kg (ACIE2011). The potentially environmentally harmful aqueous waste streams were also eliminated. It is reported that this efficient/cost-effective process, after scale up, is carried out in batches of hundreds of kilograms with an overall yield of about 65% (JACS2009a). Thereafter, in the year following its introduction over 1 MT of Sitagliptin was produced and a total of over 20 MT was produced up until the end of 2007 (ACR2007).

The only apparent issue that this process might face is the quantity of heavy metal, in this case rhodium or iron, that could remain from the use of a CAH in the last step. Since the levels of heavy metals in an API are strictly controlled, special precautions have to be taken to make sure that these metals are not retained in the final product. Use of the catalyst early in the synthesis would have presented an opportunity to purge the heavy metals during subsequent reactions and work-up. Fortunately, as indicated above, treatment with of 10% weight of Ecosorb®, a polymer impregnated with activated carbon, resulted in near complete removal of heavy metals (to levels of below 20 ppm) from the crude reaction stream (JACS2009a).

Even so, there are some downsides to the process. The CAH entails the use of high hydrogen pressure (7-34 bar, 100-500 psi), which requires special plant equipment. The limited availability and high cost of rhodium, coupled with the high cost of the ligand are inconveniencies. Even with commercially available JOSIPHOS ligand, costs remain high. For these reasons, it was still considered worthwhile to explore other alternatives for the industrial production of Sitagliptin.

Recapitulating on the discussion of section 5.4.2 a biocatalytic approach, exploiting tailor-made transaminases, converts the β-keto amide directly into the final API (Codexis2009, S2010, Codexis2013). Mutated transaminase from an *Arthrobacter sp.* delivered the desired *R* isomer (S2010, Codexis2009) in excellent yield, with an ee of 99.9% (Codexis2013) and was demonstrated on a 45 kg scale (Codexis2009). Reports indicate that the overall yield increased by 13%, productivity by 53% (ACIE2011, S2010) and the total waste reduced by 19%. Contamination of the product which could have resulted from the biological process was extremely well controlled, enzyme (<18 ppm), PLP co-factor (<0.1 ppm) and endotoxin levels (<0.05 ng) (Codexis2009, Scheme 5.57, Scheme 5.58).

Advantages of the biochemical process are:

- high yield, 85-87% from the β-keto amide
- only one isolated intermediate *vs.* three in the chemical process
- crystallisation to increase ee is not necessary
- no need to recover and recycle rhodium and catalyst ligand; removal of heavy metals from the product is unnecessary; uses multipurpose reactors - no need for specialised high pressure equipment
- immobilized enzyme simplifies operating conditions and facilitates reuse; was scaled up to 700 kg batches.

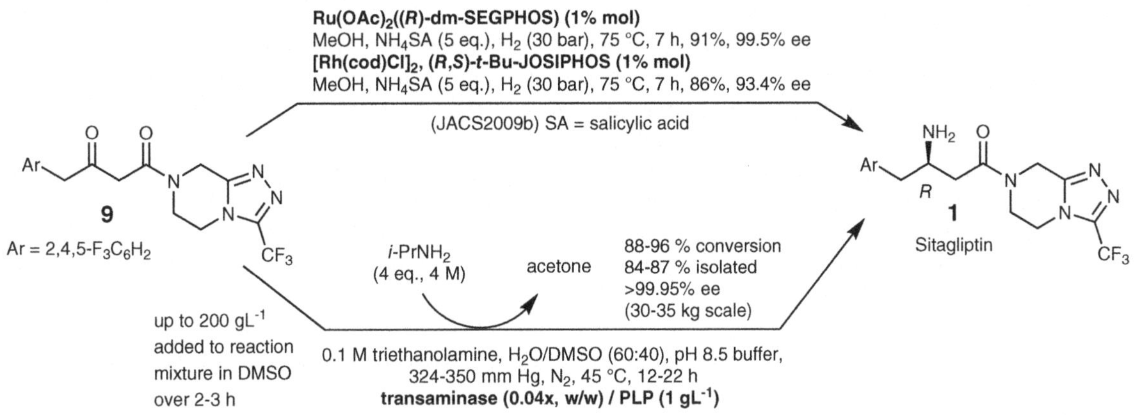

Scheme 5.57 Chemical *vs* enzymatic transaminase process of Sitagliptin production directly from the keto ester intermediate

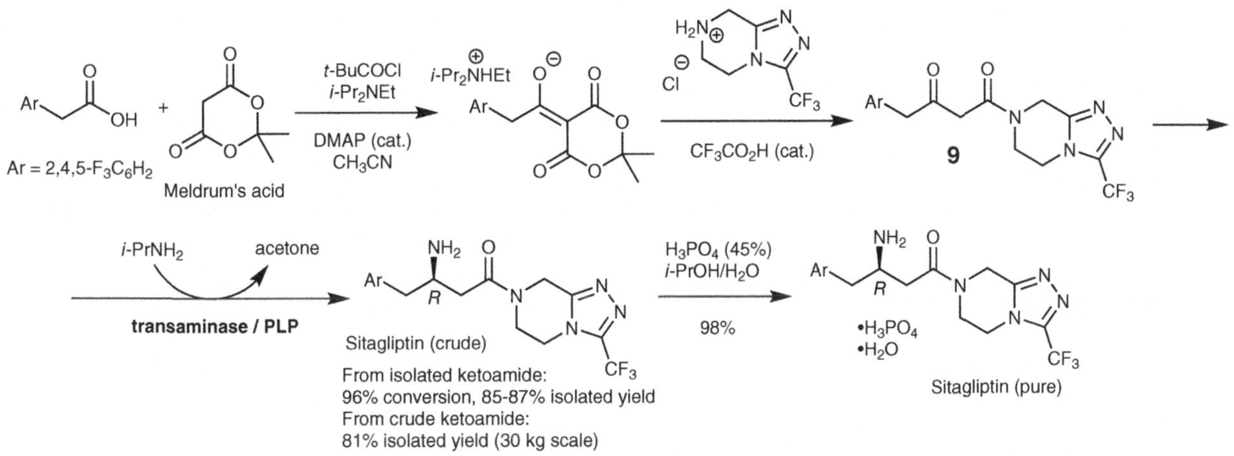

Scheme 5.58 Best industrial synthesis of Sitagliptin

5.6 LESSONS FROM THE SITAGLIPTIN CASE

Asymmetric synthesis and process chemistry

Asymmetric synthesis is an extremely powerful tool when applied to synthetic strategies to produce chemical entities and its relevance to manufacture APIs is far-reaching. Many asymmetric reactions have been the subject of scientific publications. However, transforming these into industrial reality is not straightforward. In the case of Sitagliptin, even though precedent for an analogous reaction, asymmetric hydrogenation of acyl enamines, already had been described, this did not fit the exact requirements for the transformation that Process Chemists wished to bring about. Hence, investigations were initiated into making the catalytic step fit the process rather than altering the chemistry to follow published precedent. Adoption of new technologies can play a major role in the evolution of process chemistry. However, these new technologies are not always adaptable to the problem at hand, and a great deal of effort on the part of the Process Chemist is necessary to come up with variations to make implementation practicable.

In the Sitagliptin case, the starting point was the CAH of *N*-acyl enamines. The use of a protected enamine entails additional protection/de-protection steps and compatibility with the chemical process. Process development resulted in the discovery of techniques to use unprotected enamines as substrates in CAH, thus giving rise to considerable simplification of the synthesis.

The factors, which assure successful implementation of a catalytic asymmetric reaction at industrial scale, are amongst others:

- use of high throughput screening techniques to find the best catalyst and reaction conditions
- the demonstration of performance and robustness at scale
- a steady supply of catalyst and freedom to operate to ensure manufacturing with no disruptions

Given that catalysts used in the asymmetric hydrogenation are extremely sensitive to foreign substances, it is imperative that the quality of the substrate is consistent, especially with regard to its impurity profile. If constant quality cannot be guaranteed, purification/crystallisation prior to the catalytic reaction should be contemplated. For a process, which relies upon precious metal catalysis and the use of expensive chiral ligands, their recovery is vital for economic viability. Finally, complete removal or reduction to below the regulatory limits of all heavy metals from the final API is obligatory.

Placement of the catalytic asymmetric step

The choice of where an asymmetric step in the synthesis of any API can have an enormous impact on its cost. Delaying the catalytic step to the end of a synthesis reduces the total amount of catalyst used to produce correspondingly the same amount of product and consequently the global cost. Incorporation of the asymmetric reaction late in the route offers fewer opportunities to improve the chiral purity as well as removal of heavy metal impurities. These issues are usually addressed during the course of process development. In general, the use of an expensive reactant or building block late in a synthetic sequence can be relied upon to reduce costs.

One-pot processes and isolation steps

One-pot processes are particularly beneficial in an industrial synthesis, allowing sequential reactions to be carried out in one vessel. Efficiency is improved; plant occupancy and resources are decreased whilst yield frequently increases. Many separation and purification operations can be avoided, but this is offset by the tendency to carry side- and by-products through to a later stage and unwanted build-up of these materials could adversely affect subsequent reactions. Therefore, it is convenient to identify a clean-up point whereat, a crystallisation may be performed to reduce impurity carry-over and so guarantee that the final product quality is not compromised.

In the "best" Sitagliptin chemical synthesis indicated above, there are two possible one-pot sequences. In one of these, the β-keto amide is the first isolated crystallisable intermediate and in the other, the sequence is carried up to the crystalline enamine (Scheme 5.59).

Telescoping a reaction in a synthesis and carrying out the process in one-pot, in conjunction with the isolation of intermediates at carefully selected points, is a valuable strategy to increase efficiency and achieve process intensification.

Biochemical reactions

Biochemical approaches for certain transformations may have positive benefits in a chemical synthesis. Enzymes can be highly chemo, regio and stereoselective and many exhibit high activity under mild aqueous conditions. However, their application is constrained by several factors: high specify for their native substrate may mean little activity for the intended starting material; they normally function in dilute aqueous conditions within a narrow range of operating parameters; the solubility of the substrate might be problematic; inhibition of the enzyme frequently occurs; small quantities of biological materials, such as endotoxins, may remain in the final product and can cause adverse reactions in patients. However, when this type of technology is applicable, even given the hurdles that are normally encountered in process development, results can be outstanding.

A biocatalytic process was found appropriate to the conversion of the β-keto amide to Sitagliptin and a genetically engineered transaminase was developed for the purpose. The process was onerous, the first "hit" showed only 1% conversion after 24 hours, 36,480 enzymes being screened before an adequate enzyme was found. Pursuit of increased activity and stability of the enzyme to the operating conditions required 11 cycles of directed

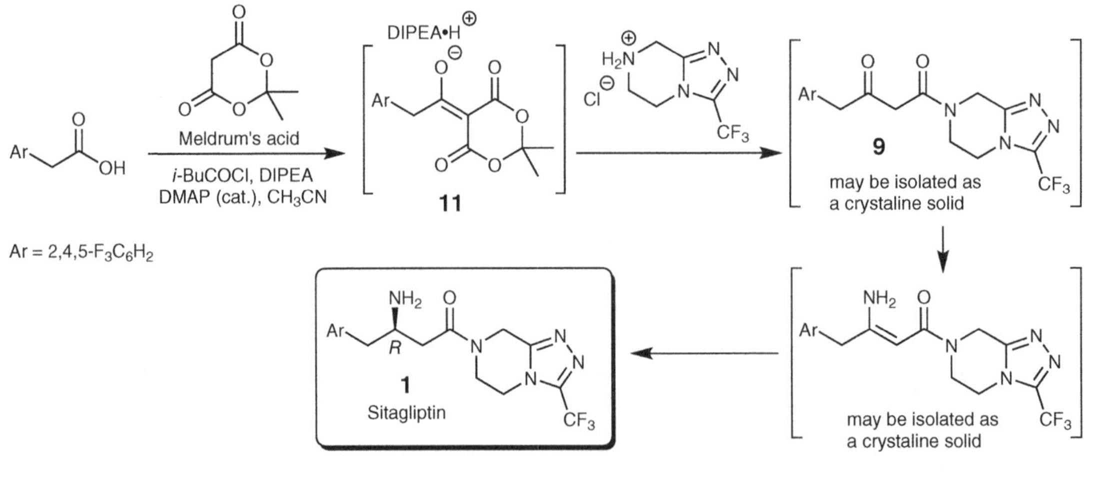

Scheme 5.59 One-pot sequences and the possibility isolation of an intermediate in one of the best synthetic proposals for Sitagliptin

evolution. The use of DMSO as co-solvent overcame the solubility issues and prevented oiling out of β-keto amide. As anticipated for enzyme reactions, the pH had to be controlled within strict limits. Acetone, which is produced continuously from the nitrogen donor, isopropyl amine, was conveniently removed using a nitrogen sweep and the controlled application of vacuum.

Even though the CAH process is exceptionally efficient for an industrial process, alternatives, which overcome the inherent drawbacks associated with its use, are welcome. The driving force behind the investigations into an enzyme-catalysed transformation are: elimination of the enamine as an intermediate in the chemical process (either *in situ* or isolated); elimination the high-pressure hydrogenation step and the specialized equipment required: elimination of the use of a noble metal and expensive chiral ligand along with their respective recoveries; no heavy metal clean up at the end. The biocatalytic process gives, after 96% conversion, a high isolated yield (87%) and high ee (96%) on a 30 kg scale from β-keto amide. No crystallisation was necessary to improve ee[9].

5.7 REACTION CONDITIONS FOR SCHEMES A TO J

5.7.1 Reaction conditions for Scheme A

A1 (B1, H1) Most frequent: Meldrum's acid (1.1 eq.), DMAP (cat.), DMA or CH_3CN, DIPEA (2 eq.), PivCl (1.1 eq.), 0-5 °C, usually not isolated (2.5 kg, JACS2009a), 85-95% (BMCL2007, OBC2010, Shenzen2013a), with amyl chloride replacing PivCl (Shenzen2013b), with NEt_3 at 50 °C, 81% (100 g, ZhejiangHP2014). CDI (1 eq.), Meldrum's acid (1 eq.), THF, 50 °C, 60-96 % (11 g Merck2003b, DrReddys2007, JiangsuHM2008. Mylan2009). 1) ClCOCOCl, CH_2Cl_2, DMF, 20-30 °C; 2) Meldrum's acid (1.6 eq.), DMAP or colidine (2 eq.), –5 °C to 0 °C, 72-95% (Glenmark2007, 2010, Cadila2008). 1) $SOCl_2$, CH_2Cl_2, DMF, 45 °C; 2) Meldrum's acid, NEt_3, CH_2Cl_2, 5 °C, 78% (50 g, MSN2009).

A2 (B2) R = Me: MeOH, reflux, 85-94% [base ($NaHCO_3$), or acid (HCl, MsOH) may be added]. **R = *t*-Bu:** *t*-BuOH, reflux, NI. **R = Bn:** BnOH, toluene or CH_3CN, reflux, 82-90%. **R = Et:** EtOH, reflux, 86-94% (cosolvent may be used).

A3 (B3) R = Me: 1) CDI (1.1 eq.), CH_3CN; 2) methyl malonic acid magnesium salt (1.3 eq.), NEt_3 (2.8 eq.), $MgCl_2$, 50 °C, 84% (3.5 kg, TA2006), 77% (43 g, ChiralQ2008), 83% (100 g, USV2010). **R = Et, Bn or *t*-Bu:** ethyl or benzyl or *t*-butylmalonic acid potassium salt, NEt_3, $MgCl_2$ (anh.), CDI, EtOAc or *i*-PrOAc, rt to 60 °C, 99-100% (10 g).

A4 1) Mg, $BrCH_2CO_2R$ (R = Me, Et), Et_2O or THF, I_2 (cat.), reflux or 40 °C; 2) SM, 15 °C to rt; 3) HCl (3 N), 87-90% (17-22 g).

A5 CAH: MeOH, HCl (2 N) or AcOH, (*S*)-BINAP•$RuCl_2$ (0.1-0.4% mol), H_2 (4.8-10 bar), 60-85 °C, YNR (1.5 kg, Merck2003c, 100 g, USV2010, 400 g, Glenmark2010). MeOH, HBr (48% aq.), (*S*)-BINAP•$RuCl_2$, NEt_3 (0.2% mol), H_2 (6 bar), 80 °C, 99%, 91% ee (2.5 kg, OPRD2005). **Enzymatic reduction:** Ketoreductase enzyme (KRED, 10% w), DMSO, YNR, >98% ee (50 mg, Teva2007a, R = Me).

A6 LiOH (4 eq.), THF/H_2O, 20-25 °C, NI (1.5 kg, Merck2003c), with NaOH, 83%, 94% ee (5 kg, OPRD2005), 86% (A5+A6, 100 g, USV2010), 64% (A2+A5+A6, 400g, Glenmark2010).

A7 $BnONH_2$•HCl, EDC•HCl, H_2O, HCl (37%), 20-22 °C, 100%, 92% ee (for a difluoride analogue, 2.8 kg, Merck2003c). $BnONH_2$•HCl, LiOH, EDC•HCl, THF, H_2O, 20 °C, NI (4 kg, OPRD2005), 88% (200g, Glenmark2010), YNR (4.3 g, ZhejiangHP2011d), with DCC, NI (80 g, USV2010), with DIC, CH_2Cl_2, 100% (15 g, Lek2010b).

A8 PPh_3 (1.2 eq.), DIAD (1.2 eq.), toluene, 20-30 °C, 79%, 99.7% ee for a diF analogue (4.1 kg, Merck2003c). PPh_3 (1.1 eq.), DIAD (1.1 eq.), MTBE, 10 °C to 20 °C, 82% (A7+A8), 99.7% ee (4 kg, OPRD2005), in THF, 76-82% (A7+A8, 4-80 g, USV2010, ZhejiangHP2011d).

A9 LiOH (1.5-2.6 eq.), THF/H_2O, rt, NI (2.7-4.7 kg, Merck2003c, OPRD2005). NaOH, MeOH/H_2O, rt, 92% (4.5 g, ZhejiangHP2011d).

A10 1) THTP•HCl (1.3 eq.), NMM (1.1 eq.), DMF/H_2O or CH_3CN, 0-18 °C; 2) EDC•HCl (1.5 eq.), 0 °C, YNR (2.7-4.7 kg, Merck2003c, OPRD2005), in CH_3CN, 86% (80 g, USV2010). 1) THTP•HCl, NEt_3, CH_3CN; 2) EDC, HOBt, CH_3CN, rt, 91% (3.4 g, ZhejiangHP2011d), in CH_2Cl_2 with EDC•HCl or DCC, NI (100 g, Glenmark2010).

A11 R = Bn: H_2 (2.7 bar), Pd/C (10%), MeOH, 60 °C, 75%, (4.4 kg for a difluoride analogue, Merck2003c). H_2 (2.7 bar), Pd/C (10%), EtOH/H_2O, 50 °C, 86%, (6.9 kg, OPRD2005), at 30-40 °C, 63-66% (100 g, USV2010). H_2 (3.5 bar), Pd/C (10%), MeOH, 50-55 °C, 61% (A10+A11), 99% ee (100 g, Glenmark2010). H_2, Pd/C (5%), MeOH, rt, 87% (A10+A11, 5.1 g, ZhejiangHP2011d). **R = Me:** H_2 (10 bar), Pd/C (5%), MeOH, 55 °C, 97% (1.4 g).

A12 1) $MeONH_2$•HCl, LiOH, H_2O; 2) EDC, HOBt, H_2O, 25 °C, 90% (30 g).

A13 Mesylation, cyclization: 1) MsCl, DIPEA, CH_2Cl_2, 0 °C, 68% (8.8g), with 2,4,6-trimethylpyridine, 20 °C, 93% (2.6 g); 2) NaH, CH_2Cl_2, DMF, 23 °C, 100% (2.2 g). **Cyclization:** PPh_3, DIAD, toluene, 19 °C, 100% (20 g).

A14 LiOH, THF, H_2O, 35 °C, 38% (40 g).

A15 THTP, DEAD, HOBt, CH_2Cl_2, 23 °C, 100% (6 g, Lek2010b).

9 M. Vink (Codexis) presentation at New Horizons in Catalysis, Cologne, December 2009

A16 Biochemical: Ketoreductase enzyme (KRED, 10% w), DMSO, H_2O, pH 7, >98% ee (50 mg, Teva2007a). **CBS:** (*R*)-2-MeCBS, BH_3•Me_2S, toluene, 5 ° C, 80% (5 g, MSN2009).

A17 PPh_3, NBS, CH_2Cl_2, 5 °C to 25 °C, 80% (4 g).

A18 NaN_3, DMF, H_2O, 30 °C, 71% (4 g). NaN_3, NMP, H_2O, 30 °C to reflux, 65% (4 g).

A19 THTP•HCl, NaOH, MeOH, reflux, 76% (3.4 g). 1) LiOH, THF, 80 °C; 2) THTP•HCl, DCC, DMAP, NEt_3, DMF, 5 °C, 47% (6 g).

A20 H_2, Pd/C, MeOH, 71% (3 g).

A21 1) TsCl, py, toluene, 22 °C, 56% (3 g); 2) Potassium phthalimide, DMF, 110 °C, 65% (2 g).

A22 $MeNH_2$, H_2O, toluene, 40 °C, 65% (3 g).

A23 (see scheme B).

5.7.2 Reaction conditions for Scheme B

B1 (A1, H1) see A1.

B2 (A2) see A2.

B3 (A3) see A3.

B4 R = *t*-Bu: *t*-BuOH, (*R*)-1-phenylethan-1-amine (1.1 eq.), 30 °C, 83% (B2+B4, 50 g). **R = Me:** (*R*)-1-phenylethan-1-amine (1 eq.), MeOH or *i*-PrOH, AcOH (cat.), 40 °C, 87-90% (200 g, Brightgene2010). (*R*)-1-phenylethan-1-amine (2 eq.), AcOH (cat.), MeOH, reflux, 99% (1.7 kg, Shanghai2011). (*R*)-1-phenylethan-1-amine (1 eq.), MeOH, rt, 92% (3 g, ZhejiangUT2013). (*R*)-1-phenylethan-1-amine (1 eq.), Ti(O*i*-Pr)$_3$Cl, CH_2Cl_2, NI (25 g, Shenzen2013a). **R = Et, Bn:** (*R*)-1-phenylethan-1-amine (1 eq.), Ti(O*i*-Pr)$_3$Cl, CH_3CN, NI (26 g, R = Et). (*R*)-1-phenylethan-1-amine (1 eq.), Ti(O*i*-Pr)$_3$Cl, THF, NI (32 g, R = Bn).

B5 R = *t*-Bu: H_2 (4 bar), Pt/C (5%), *i*-PrOH, 35 °C, YNR, 99.7% de (300 g). **R = Me:** H_2 (16-18 bar), Pt/Al_2O_3 (2%) or Pd/C (10%), MeOH, 35 °C, 93-98%, >98% ee (200 g, Brightgene2010). $(AcO)_3BHNa$, AcOH, rt, 83% (B4+B5), >99% de (25 g, Shenzen2013a). $NaBH_4$, THF, BF_3•OEt_2, 100% (70 g-1.7 kg, Shanghai2011). $NaBH_4$, MsOH, 0 °C, 100% (69 g, Shanghai2011). **R = Et, Bn:** $(AcO)_3BHNa$, AcOH, rt, 83% >98% de (B4+B5, 26-32 g, Shenzen2013a). $NaBH_4$, THF, BF_3•OEt_2, 100% (170 g, R = Et, Shanghai2011).

B6 H_2SO_4, H_2O, 70 °C, 95% (50 g).

B7 PivCl, HOBt, THTP•HCl, NEt_3, CH_2Cl_2, –8 °C to 25 °C, 100% (25 g).

B8 (H14) HCO_2H, $Pd(OH)_2/C$ (20%), H_2O, THF, MeOH, 60 °C, 61-93% (2-22 g, DrReddys2007, 2009a,b, Shenzen2013b). H_2 (10 bar), $Pd(OH)_2/C$ (20%), AcOH, MeOH, H_2O, 50 °C, 94% (13 g, SC2013). H_2 (2-15 bar), Pd/C (5-10%), AcOH, MeOH or H_2O or MeOH/H_2O, 50-55 °C, 55-85% (3-25 g, DrReddys2009b, Sun2010, ShanghaiIPI2013). H_2 (7 bar), Pd/C, *i*-PrOH, H_2O, 70 °C, 88% (300 g, OL2015).

B9 HCO_2H, $Pd(OH)_2$ (20%), MeOH, THF, reflux, NI (29-33 g, R = Me, Et, Bn, Shenzen2013a). H_2, Pd/C, NI (R = Me, Shanghai2011).

B10 (C7, G25) R = Me, PG = Boc: Boc_2O, CH_2Cl_2, rt, 93% (6.7 g, BMCL2007). Boc_2O, base (NEt_3, Na_2CO_3, LiOH), solvent (EtOAc, CH_3CN, dioxane), 0 °C to 25 °C, 90-99% (1 g to 350 g). **R = Me, PG = Cbz:** $ClCO_2CH_2C_6H_5$, NEt_3, CH_2Cl_2, 96% (34 g). **R = Et, PG = Boc:** $(Boc)_2O$ (1.2 eq.), Na_2CO_3, MeOH, H_2O, 30 °C, NI (1 g, JiangsuHM2008), 95% (350 g, Shanghai2011).

B11 (C8, E16, G10) R = Me, PG = Boc: LiOH or NaOH, THF/MeOH/H_2O or EtOH, or MeOH, rt, 95-99%. **R = Et, PG = Boc:** LiOH, MeOH, H_2O, 40 °C, 95% (65 g, Shanghai2011). MeOH, H_2O, NaOH, 40 °C, 89% (B10+B11, 1 g, JiangsuHM2008). **R = Me, PG = Cbz:** LiOH, MeOH, H_2O, 40 °C, 95% (38 g, Shanghai2011). **R = Me, PG = CO_2*i*-Pr:** NaOH, H_2O, rt, 87-100% (5.9 g). **R = *t*-Bu, PG = Cbz:** TFA, 0 °C, 80% (2 g).

B12 (C5, E9, G24) R = Me: LiOH or NaOH, H_2O, THF or MeOH, NI. LiOH, THF, 10 °C, 98% (12 g, Genchem2015). **R = Et:** LiOH, MeOH, H_2O rt, NI (Shenzen2013a). HCl (2 M), reflux, 92% (ClSIRIndia2014). **R = Bn:** LiOH, MeOH, H_2O, rt, NI (Shenzen2013a). H_2, Pd/C (7%), MeOH, 25-30 °C, NI (ZhejiangHP2011b). **R = $(CH_2)_2OH$:** LiOH, DMSO, 45 °C, NI (0,3 g).

B13 (C6, E10, G3) Boc_2O, LiOH or Na_2CO_3, THF, H_2O, 80-95% (B12+B13), in MeOH/H_2O, 87-89% (B12+B13), in EtOH, 77% (B12+B13), with NEt_3 in CH_2Cl_2, 82% (B12+B13), with NEt_3 in CH_2Cl_2/H_2O, 100% (B12+B13).

B14 (C11) THTP, EDC (1.2 eq.), HOBt (cat.), NMM, EtOH, 25 °C, 86% (2.4 g, CSIRIndia2014).

B15 (C9, D12) PG = Boc: Most common conditions, THTP or THTP•HCl (1.1-1.5 eq.), (base, when THTP•HCl is used, for example, NEt_3, py, NMM, 2,6-lutidine or DIPEA), EDC (1.1-1.5 eq.), HOBt, DMF, 0 °C to rt, 72-95%, in CH_2Cl_2 38-95%, with TBTU in DMF or CH_2Cl_2 95-99% (ZhejiangHP2011a, Cadila2013). THTP•HCl, DMF or CH_2Cl_2, DCC (1.3 eq.), NEt_3 or 2,6-lutidine, DMAP, 0 °C to rt, 41-85%. THTP, CDI, DIPEA, CH_3CN, rt to 40-70°C, 47-99%. Alternatives: *via* activated esters (KyungDong2015), *via* sulfonate-based mixed anhydrides, 81-89% (Arromax2014), triazine-based coupling reagents, 94%, (ST2011, Zentiva2012) or *via* boron species, 65-82% (Sun2014). **PG = Cbz:** THTP•HCl , CDI, NEt_3, THF, reflux, YNR (Chemo2010). THTP, EDC, HOBt, CH_2Cl_2, rt, 86% (Chemo 2011). THTP•HCl, 2-chloro-4,6-dimethoxy-1,3,5-triazine, NMM, THF, 5 °C, 90% (ST2011). **PG = CO_2*i*-Pr:** THTP, THF, CDI, NEt_3, reflux, 91% (Chemo2010). **PG = Bn:** THTP•HCl, EDC•HCl, HOBt, NEt_3, CH_3CN, 94%. 1) $SOCl_2$, DMF, THF; 2) THTP, YNR (Xinfa2015a). EDC•HCl, HOBt, DIPEA, $CHCl_3$, rt, YNR. **PG = Ts:** THTP•HCl, EDC•HCl, NEt_3, HOBt, CH_3CN, 91%. **PG = Bz:** THTP•HCl, EDC•HCl, NEt_3, HOBt, CH_3CN, 92%. **PG = SO_2Ph:** THTP•HCl , EDC•HCl, NEt_3, HOBt, CH_3CN, 85%.

B16 PG = Boc: HCl (conc. or g), MeOH or EtOH, *i*-PrOH, 80-100%, other solvents/acids used. **PG = Cbz:** H_2, Pd/C (10%), MeOH or THF, rt, 57-100%. **PG = Bn:** H_2, Pd/C (10%) or $Pd(OH)_2$/C (30% w/w), THF or DMF or MeOH, rt, 87-96%. **PG = Ts;** AcOH, HCl (conc.), reflux, 65%. **PG = Bz;** MeOH, NaOEt, rt, 75%. **PG = SO_2Ph:** AcOH, HCl (conc.), reflux, 70%.

B17 R = Me: (*S*)-2-amino-2-phenylacetamide, AcOH, MeOH, reflux, YNR (35 g, Merck2003b). **R = Et:** (*S*)-2-amino-2-phenylacetamide, *i*-PrO_3TiCl, 1,2-dichloroethane, NI (26 g, Shenzen2013a).

B18 R = Me: H_2 (6 bar), PtO_2 (cat.), AcOH, THF, rt, 84% (B17+B18), 100% de (60 g). **R = Et:** $(AcO)_3$BHNa, AcOH, rt, 80% (B17+B18), >99% de (26 g).

B19 HCO_2H, $Pd(OH)_2$ (20%), MeOH, THF, reflux, NI (33 g).

B20 Dehydrogenase, HCO_2NH_4, phosphate buffer, NAD, DMSO, 30 °C, 88%, >99% ee (19 g, Syncozymes2015). ω-Transaminases from *Ilumatobacter coccineus,* 82%, 99% ee (R = H, 23 g, EMT2019). Hydrolase + ω-Transaminases from *Ilumatobacter coccineus,* YNR (R = Me, 23 g, EMT2019).

B21 (*S*)-2-amino-2-phenylethan-1-ol, THF (anh.), MS, reflux, 100% (92 g, UniTris2010). (*S*)-2-amino-2-phenylethan-1-ol, MeOH, rt, 90% (3 g, ZhejiangUT2013).

B22 $NaBH_3CN$/AcOH or BH_3•THF, THF, rt, 83-87% (3 g).

B23 H_2 (1 bar), Pd/C (cat.), MeOH, rt, 90% (1.8 g).

B24 NaH, THF (anh.), rt, 70% (60 g).

B25 $NaBH_3CN$, AcOH (glac.), CH_2Cl_2, 1,4-dioxane, rt, 78% (19 g). $NaBH_4$, AcOH, CH_3CN, rt, 64% (3.3 g). H_2 (3.4 bar), $Pd(OH)_2$/C (10%), HCl (37%), MeOH, 50 °C, 89% (2 g).

B26 H_2 (3.4 bar), $Pd(OH)_2$/C (10%), MeOH, HCl (cat.), 50 °C, 89%, 99.7% ee (2 g).

B27 Transaminase Biochemical process: Aminotransferase derived from *Mycobacterium vanbaalenii* PYR1, >99%, 99% ee (20 g, R = Me, Et, *i*-Pr, Pr, *t*-Bu, Nanjing2014a). Immobilized transaminase, >90%, >92% ee (5 g, R = Me, *i*-Pr, Bn, Nanjing2015), >99%, >99% ee (5 g, R = Me, *i*-Pr, Bn, Nanjing2014b). Aminotransferase ATA117-rd11, from *Arthrobacter sp*, 80% conversion (0,3 g, R = CH_2CH_2OH, T2016). Recombinant transaminase, 98% conversion, 99.5% ee (32.6 kg, R = Me, Alpha2016, Suqian2017b). Recombinant *Escherichia coli* transaminase, 98%, 99% ee (139 kg, R = Me, Suqian2017a,b). **Reductive amination:** H_2 (36 bar), $Ru(OAc)_2$((*R*)dm-Segphos) (1% mol), SA (2.9 eq.), NH_4OAc (5 eq.), MeOH, 40-50 °C, 95%, 94.2% ee (100 g), at 20 bar, 90-100 °C, 89% (100 g), at 90-100 bar, 40-50 °C, 96% (100 g).

B28 *t*-Butyl carbamate (1.5 eq.), TsOH, MS 4 Å, CH_2Cl_2, 25 °C, 54% (2.5 g).

B29 H_2 (5 bar), [Ru(cod)Cl_2] (1 % mol), (*S*)-BINAP (1% mol), MeOH, 80 °C, 41%, 33 % ee (1.5 g, Teva2007b). H_2 (10 bar), TangPhosRh(cod)BF_4 (1 % mol), THF, rt, 100%, 99.9% ee (1.3 g, ChiralQ2008).

B30 (C1) R = Me: NH_4OAc (5-6 eq.), MeOH, reflux, 91% (25-350 g). **R = Et:** NH_4OAc (5 eq.), EtOH, reflux, 89% or NI (Teva2007b, OBC2010). NH_4OAc, HCO_2H, EtOH, reflux, NI (JiangsuHM2008). **R = Bn:** NH_4OAc, CH_3CN (or MeOH), reflux, 90% (29 g, ZhejiangHP2011b).

B31 Boc_2O (2.2 eq.), DMAP (1.1 eq.), CH_3CN, rt, 32% (2 g).

B32 Ac_2O, py, THF, reflux, 70% (4.9 g).

B33 H_2 (10 bar), TangPhosRh(cod)BF_4 (cat.), THF, rt, 100%, 97% ee (2.3 g).

B34 HCl (6 N, aq.), reflux, NI (2.5 g).

B35 R = Me: H_2 (13.7 bar), $[Rh(cod)Cl]_2$, JOSIPHOS II (0.1 mol %), CF_3CH_2OH, 40 °C, 90%, 93% ee (18 g, TA2006). H_2 (6.2 bar), $[Rh(cod)Cl]_2$ (0.6 mol %), JOSIPHOS I, MeOH, 88%, 97% ee (90 mg, TA2006). H_2 (6.8 bar), $[Rh(cod)Cl]_2$, JOSIPHOS (*R*)-(–)-1-[(*S*)-2-(bis(4-trifluoromethylphenyl)-phosphino)ferrocenyl ethyl-di-*t*-butylphosphine, TFA, 68% (15 g, BMCL2007). H_2 (25 bar), (*R*)-2.2'-bis[bis(3.5-dimethylphenyl)phosphino]-1,1'-binaphthyl]ruthenium diacetate [(S)-RUCY(R)-XylBINAP]$Ru(OAc)_2$ (2% mol), MeOH, 50 °C, YNR, ee not reported (0.25 g, Suzhou2012). **R = Et:** H_2 (6 bar), (*R*)-BINAP (0.5 % mol), Ru(cod)Cl_2 (1% mol), CF_3CH_2OH, 80 °C, 84%, ee not reported (33 g). $NaBH_4$, (*S*)-mandelic acid (4.2 eq.), THF, rt, 17%, 84% ee (1 g) (see figure 5.8).

B36 CH_3NO_2 (1.9 eq.), NaOH, MeOH, 93% (5 g).

B37 $NaBH_4$, *i*-PrOH, silica gel, $CHCl_3$, 27 °C, 90% (5 g).

B38 $HCOCO_2Et$ (4 eq.), DBU (cat.), CH_2Cl_2, 27 °C, 88% (4.5 g).

B39 1) Ac_2O, I_2 (cat.), 27 °C; 2) basic Al_2O_3, CH_2Cl_2, 27 °C, 95% (6 g).

B40 1) (*R,S*)-JOSIPHOS (3% mol), CuF_2 (3% mol), $C_6H_5SiH_3$ (1.1 eq.), PMHS, H_2O, toluene, 88% (4 g); 2) H_2 (3.4 bar), Pd/C (10%), EtOAc, 27 °C, 89% (40 mg).

B41 $Cl_2CHCOCl$, $AlCl_3$, 40-45 °C, 95% (13 g).

B42 NaOH (10%), H_2O, 0-5 °C, 86% (24 g).

B43 $SOCl_2$, reflux, 85% (21 g).

B44 $MeCOCH_2CO_2Me$ (1 eq.), CDI (1 eq.), THF, 73% (11 g).

B45 NH_4OH (25%), rt, 93% (14 g).

B46 H_2, $[Rh(cod)Cl]_2$ (0.2 % mol), (*R,S*) *t*-butyl JOSIPHOS (0.4 % mol), MeOH, 65 °C, 87% (32 g).

B47 1,3-Bis(diphenylphosphino)propane, $Pd(OAc)_2$, CH_3CN, DBU, 83% (1.4 g monomethyl itaconate).

B48 H_2 (8-12 bar), [Ir-((*R*)-DTB-SIPHOX) (cod)] BARF, MeOH, NEt_3, 60-80 °C, 99%, >99% ee (8.2 g).

B49 (C36, E15) PG = Boc: *t*-BuOH, DPPA, NEt_3, toluene, reflux, 69% (0.5 g, Chemo2010). 1) $SOCl_2$, CH_2Cl_2, 40 °C (2.8 g); 2) NH_2NH_2•H_2O, heat, 83% (ZhejiangJP2015a). **PG = CO_2*i*-Pr:** *i*-PrOH, DPPA, NEt_3, reflux, 89% (5.5 g). **PG = Cbz:** BnOH, DPPA, NEt_3, toluene, 80 °C, 79% (1 g).

5.7.3 Reaction conditions for Scheme C

C1 (B30) see B30.
C2 R = Me: $NaBH_3CN$ (2 eq.), H_2SO_4 (cat.), MeOH, 0-10 °C, 75-80% (350 g, Mylan2009). $NaBH_4$, THF, 90% (46 g, ZhejiangCP2014). **R = Et:** $NaBH_4$ (3 eq.), AcOH, 20-25 °C, 56% (2.6 g, Teva2007b). $NaBH_4$, THF, 91% (47 g, ZhejiangCP2014). $NaBH_3CN$ (1.3 eq.), EtOH, reflux, 91% (C1+C2, JiangsuHM2008). $NaBH_3CN$ (3 eq.), AcOH (3 eq.), EtOAc, 0 °C, 54% (1.6 g, OBC2010). **R = Bn:** 1) Me_3CNH_2•BH_3, THF, 0-5 °C; 2) H_2SO_4 (conc.), *i*-PrOH, 87% (32 g, ZhejiangHP2011b). $NaBH_4$, H_2SO_4, THF, 5 °C, 70% (50 g, ZhejiangHP2011c). $NaBH_4$, AcOH, MeOH, 5 °C 48% (C1+C2, 178 g, MSN2009).
C3 Lipase PS, *i*-Pr_2O, H_2O, 45 °C, 39-47%, 97% ee (0.8 g).
C4 R = Me: (*R*)-mandelic acid (2 eq.), *i*-PrOH, 28%, >99% ee (260 g, Mylan2009), in CH_3CN, H_2O, 30 °C, 11% (60 g, MSN2009). (*R*)-CSA, H_2O, CH_3CN, 30 °C, 10% (80 g, MSN2009). *N*-acetyl-D-Phenylalanine (0.8 eq.), acetone, reflux to 25 °C, 32-33%, 90-96% ee (5-20 g, Smilax2012). *N*-pentanoyl-L-phenylalanine (0.5 eq.), EtOAc reflux to 25 °C, 30%, 100% ee (10 g, Smilax2012). **R = Et:** D-DPTTA (0.5 eq.), MeOH or EtOH, in MeOH, 23% (>99.7% ee), with a 2nd crop, 32-36%, >99% ee, in EtOH, 21% (5 g, JiangsuHM2008). (Phenoxycarbonyl)-L-proline (0.4 eq.), acetone, 11%, 75% ee (1.3 g, Smilax2012). Pentanoyl-L-phenylalanine (0.5 eq.), acetone, 25%, 84% ee (2 g, Smilax2012), in EtOAc, 26-33%, 96.6-99.3% ee (5-20 g, Smilax2012). **R = Bn:** D-DBTA (1 eq.), CH_3CN, H_2O, 26-42% (32 g), in CH_3CN, 29%, >99% ee (32 g).
C5 (B12, E9, G24) see B12.
C6 (B13, E10, G3) see B13.
C7 (B10, G25) see B10.
C8 (B11, D29, E15, G10) see B11.
C9 (B15, D12) see B15.
C10 (B16, D13, H6) see B16.
C11 (B14) see B14.
C12 MsCl, $NaHCO_3$, CH_3CN, 130 °C, 96% (1.5 g).
C13 1) PhCHO (1 eq.), MeOH, reflux; 2) $NaBH_4$, rt, 90% (1+2, 25 g); 3) H_2, Pd/C (10%), NEt_3 (1 eq.), MeOH, rt, 95% (R = Me), 92% (R = Et).
C14 R = Me 1) Sodium dichloroisocyanurate (0.6 eq.), EtOAc, –5 °C; 2) K_2CO_3 or DBU or py (2 eq.) or *t*-BuOK (3 eq.), 15-25 °C, 89-95% (50 g). 1) NaOCl (37% aq., 1 eq.), toluene or EtOAc; 2) K_2CO_3 (2 eq.), 40 °C, 93% (50 g). **R = Et:** 1) Sodium dichloroisocyanurate (0.6 eq.), EtOAc, –5 °C; 2) NEt_3 or DABCO (2 eq.), 25 °C, 90-93% (50 g). 1) NaOCl (37% aq., 1 eq.), toluene or EtOAc; 2) ethylene diamine or DIPEA (2 eq.), 40 °C, 92-95% (50 g).
C15 $NaNO_2$, H_2SO_4, H_2O, 60 °C, 75% (2 g).
C16 $K_2Cr_2O_7$, H_2SO_4, AcOH, 30 °C, 50% (1 g).
C17 NaOH, MeOH, H_2O, reflux, 64% (6 g).
C18 1) (*R*)-catalyst (0.5 eq.), $Ni(OAc)_2$ (1 eq., anh.), *t*-BuOK (10 eq.), MeOH, reflux, 34% (0.1 g); 2) HCl (3 N, aq.), EtOH, NI.

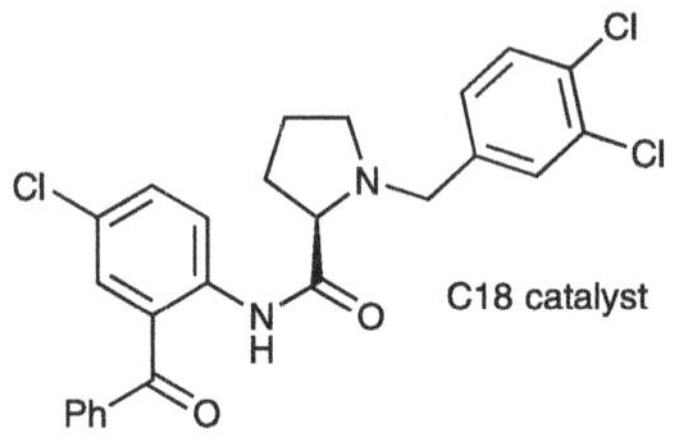

C18 catalyst

C19 MsCl, $NaHCO_3$, CH_3CN, 80 °C, 96% (1.6 g).
C20 *Candida antarctica* lipase B (CaLB), enzyme in recombinant form immobilised on polyacrylic resin, MTBE, 50 °C, 36%, 97% ee (50 mg, opens *S* isomer).
C21 THTP•HCl, THF, reflux, 90%, 99.5% ee (2.2 g, Lek2012c), in CH_3CN, 100 °C, 92%, 99.5% ee (0.1 g, Lek2012c), in MeOH, 100 °C, 83%, 99.5% ee (0.1 g, Lek2012c). THTP•HCl, NEt_3, THF, 75 °C, 95%, 99.5% ee (0.5 g, Lek2012c), in CH_3CN, H_2O, 100 °C, quantitative (0.1 g, Lek2012c), in THF, 2-ethylhexanoic acid, 75 °C, 95% (0.5 g, Lek2012d). THTP•HCl, THF, $CuCl_2$, 100 °C, 74-100%, 99.5% ee (100 mg, Lek2012c). THTP•HCl, HCl, THF, 100 °C, 90%, 99.5% ee (0.1 g, Lek2012c), with TFA or AcOH or boric acid, 100 °C, quantitative (70-100 mg, Lek2012c).
C22 1) Oxone, acetone, H_2O, pH=7-9; 2) Zn/HCl (conc.), MeOH, 95% (2 g).
C23 BnCOCl, NEt_3, acetone, H_2O, rt, 76% (1 g).
C24 Immobilized *E. coli* penicillin amidase, NaOH, H_2O, pH 8, 26 °C, 44-47% *R* isomer of lactone (94-99.5% ee), *S* beta amino acid, 32-41%, 80-84% ee (1 g); Immobilized *Providencia rettgeri* penicillin amidase, NaOH, H_2O, pH 7.5, 25 °C, 37% *R* isomer of lactone (97.3% ee), 36%, *S* beta amino acid (54% ee, 1 g).
C25 HCl (6 M), reflux, 50%, 99.5% ee (1 g).
C26 1) Oxone, phosphate buffer (pH 7-8), acetone, H_2O, 5° C; 2) Zn, HCl (37%), MeOH, rt, 95% (2 g).
C27 1) DIBAL, THF, 0 °C; 2) CH_2=$CHCH_2MgBr$, THF, MeOH; 3) NaOH, H_2O, 47% (5 g).
C28 D-(+)-Mandelic acid, H_2O, *i*-PrOH, 34% (4 g).
C29 $NaIO_4$, $RuCl_3$, CH_3CN, 20 °C, 46% (3 g).
C30 $SOCl_2$, MeOH, rt, 84% (250 g).
C31 BnBr, Na_2CO_3, DMA, rt, 84% (100 g).
C32 $Pd(OAc)_2$, Cy_2NMe, LiBr, DMA, H_2O, 71% (100 g, 124g of ArBr), 6% impurities.
C33+C34 H_2 (3 bar), Pd/C (10%), MeOH, 20 °C, 94-95% (135 g), 6% overreduction.
C35 (*S*)-1-phenylethan-1-amine, CH_3CN, 35%, 30% ee (5 g). (+)-cinchonine, acetone, 50%, 60% ee (5 g).
C36 (B49, E15) see B49.

5.7.4 Reaction conditions for Scheme D

D1 1) *i*-PrMgBr, THF (anh.), –10 °C; 2) $CoBr_2$ (0.06 eq.), TMEDA (0.06 eq.), hexamethylenetetramine (0.1 eq.), $BrCH_2CH$=$CHCO_2Me$, 0 °C, 49% (4.6 g, Lek2011a),

93% (9 g, Lek2011b). 1) *i*-PrMgBr, THF (anh.), –10 °C; 2) $BrCH_2CH{=}CHCO_2Me$, $Fe(acac)_3$ (0.1 eq.), TMEDA (0.18 eq.), 0 °C, 93% (14 g, Lek2011a). 1) THF (anh.), *i*-PrMgBr, –10 °C; 2) $BrCH_2CH{=}CHCO_2Me$, $Co(acac)_3$ (0.09 eq.), TMEDA (0.09 eq.), 0 °C, 45-62% (5.2 g, Lek2011a). 1) Grignard formation (not described); 2) $Fe(acac)_2$/D-Glucosamine•HCl/NEt_3 (5% eq.), $BrCH_2CH{=}CHCO_2Me$, THF, –20 °C, 75% (4.7 g, AOC2015).

D2 PG = OBn: $BnONH_2$•HCl, NaOH, H_2O, catalyst (DBSA, Lewis acid, acid, 2,2,2-trifluoroethanol), 60 °C, 42-94%. **PG = Bn:** $BnNH_2$, H_2O, catalyst (Lewis acid, acid, 2,2,2-trifluoroethanol, 1,1,1,3,3,3-hexafluoro-2-propanol), 60 °C, 50-92%. **PG = OPh:** $PhONH_2$.HCl, NaOH, DBSA, 60-70 °C, 40% (115 mg).

D3 B_2pin_2, *t*-BuONa, CuCl, THF, rt, 86% (0.23 g, Lek2011a), in the presence of 1,2-bis(diphenylphosphino)benzene (dppbz, 0.1 eq.) or 1,1'-bis(di-*t*-butylphosphino)ferrocene (dtbpf, 0.1 eq.), 82% (0.23 g, Lek2011a), in the presence of PPh_3/MeOH or PPh_3/MeOH/toluene, 86% (0.23 g, Lek2012a).

D4 $NaBO_3$•$5H_2O$, THF, H_2O, rt, 85-88% (0.3-1.6 g, Lek2011a, 2012a). NaOCl (aq.), CH_3OH, 91% (360 mg, Lek2011a, 2012a). H_2O_2 (30%), NaOH, THF, H_2O, rt, 81% (360 mg, Lek2011a). Oxone, THF, H_2O, rt, 84% (180 mg, Lek2011a).

D5 PG = Bn: $BnNH_2$, $[CpIrCl_2]_2$ (5% mol), $NaHCO_3$, 120 °C, 22-34% (250 mg). $BnNH_2$, $Ru_3(CO)_{12}$ (5% mol), 2-[dicyclohexyl]phosphine)-1-phenyl--1*H*-pyrole (10% mol), toluene, 120 °C, 30% (50 mg). $BnNH_2$, $FeCl_3$, L-pyroglutamic acid, 1,2,3,4,5-pentamethylcyelopenta-1,3-diene (10 % mol), $PhCF_3$, 160 °C, 29% (124 mg). **PG = Ac:** TFA, H_2O, CH_3CN, 80 °C, 64% (115 mg).

D6 NaOH, MeOH, not shown.

D7 (*R*)-*N*-benzyl-1-phenylethan-1-amine, EtOAc or *i*-PrOH, 34-37%, >96% ee (0.5-0.6 g).

D8 PG = H: 1) CH_3MgCl, THF; 2) H_2NOSO_3H, 45% (360 mg). 1) CH_3MgCl, THF; 2) NH_4OH, NaOCl, 54% (360 mg). **PG = Bn:** 1) BCl_3, toluene; 2) BnN_3, 47% (180 mg).

D9 B_2pin_2, (*R*)-BINAP (5% mol), $CuCO_3$ (cat.), H_2O, 84%, 70% ee (115 mg, Lek2011a). B_2pin_2, (*R*)-BINAP (5% mol), CuCl (10% mol), *t*-BuONa, H_2O, 84%, 86% ee (115 mg, Lek2011a). B_2pin_2, JOSIPHOS (5% mol), $CuCO_3$ (4% mol), H_2O, 87%, 86% ee (115 mg, Lek2011a). B_2pin_2, JOSIPHOS (5% mol), CuCl (10% mol), *t*-BuONa, H_2O, 72%, 81% ee (115 mg, Lek2011a). B_2pin_2, WALPHOS (5% mol), $CuCO_3$ (4% mol), H_2O, 84%, 95% ee (115 mg, Lek2011a). B_2pin_2, (1*R*,2*R*)-DPEN (10% mol), $CuCO_3$ (4% mol), H_2O, 84%, 62% ee (115 mg, Lek2011a). B_2pin_2, (*R*)-1-{(R_P)-2-[2--(Diphenylphosphino)phenyl]ferrocenyl}ethylbis[3,5--bis-(trifluoromethyl)phenyl]phosphine (5% mol), $CuCO_3$, H_2O, rt, 84%, 95% ee (115 mg, Lek2012b). B_2pin_2, chiral phosphine ligand (10% mol), $CuCO_3$, H_2O, rt, 70-100%, up to 95% ee. **Best result:** (*R*)-1-{(R_P)-2-[2-(Diphenylphosphino)phenyl]ferrocenyl} ethylbis[3,5-bis-(trifluoromethyl)phenyl]phosphine, 100%, 95% ee (230 mg, AOC2013).

D10 NaOCl (aq.), EtOAc or MeOAc, 25 °C, 94% (36 g).

D11 $BnONH_2$•HCl, NaOH, DBSA, H_2O, L-proline (0.1 eq.), 80 °C, 54%, 20% ee (230 mg). $BnONH_2$•HCl, NaOH, 2,2,2-trifluoroetanol, H_2O, L-proline (0.1 eq.), 80 °C, 57%, 15% ee (230 mg). $BnONH_2$•HCl, NaOH, DBSA, H_2O, *N*-[3,5-bis(trifluoromethyl)phenyl]-*N*'-[6'-methoxy-9-cinchonanyl]thiourea (0.2 eq.), 80 °C, 56%, 25% ee (115 mg). $BnONH_2$•HCl, NaOH, DBSA, H_2O, 2-[[3,5-bis(trifluoromethyl)phenyl]thioureido]-*N*-benzyl-*N*,3,3-trimethylbutanamide (0.2 eq.), 80 °C, 71%, 38% ee (115 mg). $BnONH_2$•HCl, NaOH, DBSA, H_2O, quinidine chiral catalyst (0.2 eq.), $PhCO_2H$, 80 °C, 68%, 15% ee (115 mg). $BnONH_2$•HCl, NaOH, DBSA, H_2O, 6'-methoxycinchonan-9-amine trihydrochloride (0.2 eq.), $PhCO_2H$, 80 °C, 51%, 22% ee (115 mg). $BnONH_2$•HCl, NaOH, DBSA, H_2O, 2,2,3-trimethyl-5-benzyl-4-imidazolidinone•HCl (0.2 eq.), $PhCO_2H$, 80 °C, 54%, 18% ee (115 mg).

D12 (B15, C9) see B15.

D13 (B16, C10, H6) see B16.

D14 DIBAL or BH_3•THF, THF, rt, 81-90% (1.9 g). DIBAL, toluene, rt, 65% (1.9 g), in dioxane 84% (1.9 g).

D15 2-Iodoxybenzoic acid, CH_3CN, reflux, 76% (1.8 g). MnO_2, benzene, reflux, 75% (1.8 g).

D16 $MeOCH_2PPh_3Cl$, NaH, THF, 80% (30 g, Wuxi2013), with *t*-BuOK, NI (TL2012).

D17 HCl, H_2O/THF, 78% (23 g, Wuxi2013), in 1,4-dioxane, NI (TL2012, ZhejiangNC2014). HCl, THF, H_2O, 78% (23 g, Wuxi2013).

D18 *t*-butyl diethylphosphonoacetate, MeMgBr, THF, reflux, 74% (D16+D17+D18), >99:1 dr.

D19 Lithium (*R*)-benzyl(1-phenylethyl)amide, THF, –78 °C, 87%, >99:1 dr. Lithium (*R*)-benzyl(1-(4-methoxyphenyl)ethyl)amide, THF, –78 °C, 86%, >99:1 dr.

D20 HCl (2 M), reflux, NI.

D21 HCl (2 M), reflux, NI.

D22 $Ph_3PCH_2CO_2Et$, NaH, THF, reflux, 60% (7 g).

D23 (*R*)-*N*-benzyl-1-phenylethan-1-amine, BuLi, THF, –40 °C, 59% (1.2 g, ZhejiangNC2014), at –78 °C, 76% (JCR2010), with NaHMDS, 54-56% (1.2 g, ZhejiangNC2014).

D24 HCl (2 N), H_2O, reflux, NI (1.5 g).

D25 EDC•HCl, HOBt, DIPEA, THTP, $CHCl_3$, rt, 70% (D21+D25, TL2012), in CH_2Cl_2, 67% (D24+D25, 0.6 g, ZhejiangNC2014).

D26 H_2 (5 bar), $Pd(OH)_2/C$ (30%), MeOH, rt, 96% (TL2012). H_2 (25 bar), Pd/C (5%), MeOH, 50 °C, 95% (1g, ZhejiangNC2014).

D27 1) $NaBH_4$, MeOH, 0 °C to rt, 92%; 2) Ac_2O, py, 80 °C, 90%; 3) *t*-BuOK, Et_2O, –60 °C, 80%.

D28 1) H_2 (5 bar), $Pd(OH)_2$, MeOH, rt; 2) Boc_2O, NEt_3, CH_2Cl_2, rt, 83%.

D29 (B11, C8, E16, G10) see B11.

5.7.5 Reaction conditions for Scheme E

E1 1) $NaBH_4$, THF, MsOH, –15 °C, 98% (25 g); 2) 2-Iodoxybenzoic acid, CH_2Cl_2, rt, NI (produced in the presence of (*R*)-*t*-BuS(O)NH_2, reaction E2 also occurs).
E2 (*R*)-*t*-BuS(O)NH_2 (1 eq.), TsOH, toluene, reflux, YNR (17 g, Dissymmetrix2010). (*R*)-*t*-BuS(O)NH_2 (1 eq.), PPTS, $MgSO_4$, CH_2Cl_2, 90% (by HPLC, 17 g, Dissymmetrix2010), YNR (Wuxi2013), 96% (45 g, TA2107). (*R*)-*t*-BuS(O)NH_2, $MgSO_4$, CH_2Cl_2, 93% (E1 step 2+ E2, 215 g, Chengdu2015).
E3 Zn, $BrCH_2CO_2Et$, CuCl, TMSCl, THF, 35% (36 g).
E4 LiOH, MeOH, H_2O, 91% (12 g).
E5 THTP, EDC, HOBt, DIPEA, CH_2Cl_2, 95% (5 g, Wuxi2013). THTP, HBTU, NEt_3, CH_2Cl_2, 0 °C to rt, 84% (48 g, TA2107).
E6 HCl (4 M), MeOH, 88% (1 g, Wuxi2013). HCl (conc.), MeOH, rt, 89% (59 g, TA2107).
E7 **R^1 = Me:** Dimethyl malonate, $NaHCO_3$, rt, 98% (9 g, Chengdu2015). **R^1 = Et:** Diethyl malonate, $NaHCO_3$ or $KHCO_3$ or Na_2CO_3 or K_2CO_3, 25-40 °C, 80-94% (1 g, TA2017). **R = Et, R^1 = All:** Allyl ethyl malonate, K_2CO_3, NaI, $NaHCO_3$, 0 °C (1 g, TA2017).
E8 HCl (aq., 6 M), heat, 26% (19.7 g).
E9 (B12, C5, G24) see B12.
E10 (B13, C6, G3) see B13.
E11 1) $Pd(OAc)_2$, PPh_3, HCO_2H, NEt_3, EtOAc, reflux, YNR (114 g); 2) NaOH (5 M, aq.), THF, 40 °C, 57% (E7+E11).
E12 1) $SOCl_2$ (2 g); 2) Lithium (*S*)-4-benzyl-2-oxooxazolidin-3-ide, THF, toluene, –5 °C to 0 °C, 60% (1+2). (*S*)-4-benzyloxazolidin-2-one, PivCl, LiCl, THF, –20 °C, 72% (5 g).
E13 LDA, THF, $BrCH_2CO_2$*t*-Bu –78 °C to –15 °C, 30%, >98% dr (1 g), with NaHMDS, 52% (2 g), with LiHMDS, 72% (2 g).
E14 LiOH, H_2O_2, THF, 0 °C, 90% (0.65 g).
E15 (B49, C36) see B49. 1) DPPA, NEt_3, toluene, 80 °C; 2) BnOH, reflux, 80% (0.8 g, Chemo 2011).
E16 (B11, C8, D29, G10) see B11.
E17 1) $Ph_3P{=}CHCO_2Et$, benzene, reflux, 98%; 2) H_2 (1 bar), Pd/C (10%), MeOH, 98%; 3) LiOH, THF, MeOH, H_2O 96% (5 g).
E18 1) PivCl, NEt_3, THF, –20 °C; 2) (*R*)-4--benzyloxazolidin-2-one, LiCl, –20 °C to 25 °C, 94% (4 g).
E19 1) KHMDS, –78 °C, THF (anh.); 2) 2,4,6-triisopropylbenzenesulfonyl azide, AcOH, –78 °C to 25 °C, 88% (3.8 g).
E20 $NaBH_4$, THF, H_2O, 0–25 °C, 98% (3 g).
E21 H_2 (1 bar), Pd/C (10%), Boc_2O, MeOH, 98% (2 g).
E22 1) TsCl, NEt_3, CH_2Cl_2; 2) NaCN, DMF, 80 °C, 65% (1.5 g).
E23 (F20) NaOH (3 M, aq.), H_2O_2, 100 °C, 75% (0.5 g). NaOH, EtOH, H_2O, 80 °C, 81% (2 g, Tongren2016).
E24 1) $Ph_3P{=}CHCO_2Et$, benzene, reflux; 2) H_2 (1 bar), Pd/C (10%), MeOH; 3) DIBAL, toluene, –78 °C, 92%.
E25 $CH_2{=}CHCH_2OH$, $Pd(OAc)_2$ (cat.), $NaHCO_3$, $BnEt_3NCl$, DMF, 45 °C, 95% (1,5 g).
E26 1) L-proline (10 mol%), di-*t*-butyl azodicarboxylate (0.9 eq.), CH_3CN, 0 °C; 2) $NaBH_4$, MeOH, 90%.
E27 $PdCl_2$ (5 mol %), PMHS, Boc_2O, H_2O/MeOH, 25 °C, 83%.
E28 1) PhNO, D-proline (cat.), DMSO, H_2O, 30 °C; 2) $NaBH_4$, MeOH, 0 °C, NI (0.3 g).
E29 H_2, Pd/C, $CHCl_3$, EtOH, 65% (E28+E29, 0.4 g).
E30 1) TsCl, NEt_3, CH_2Cl_2, 0 °C; 2) NaCN, DMF, 75 °C, 65%, >99% ee (0.5 g).
E31 (F22) NaOH (6 N), EtOH, reflux, 60% (13 g, ZhejiangHP2011d). NaOH, H_2O_2, MeOH or H_2O, reflux, 90% (0.4-5 g, Hanmi2009a, T2012), in CH_2Cl_2, 50 °C, 80% (13 g, ZhejiangHP2011d).
E32 (F25) 1) HCl (g), EtOH abs., –10 °C to 35 °C; 2) H_2O, 100% (75 g).
E33 (F26) NaOH, THF, H_2O, 40 °C, 75% (93 g).
E34 (F27) 1) MsCl, CH_2Cl_2, DIPEA, 10 °C to 20 °C, 99% (68 g); 2) NaN_3, DMF, sulfolane, 50 °C, 70% (with 20% olefinic products, 10 g).
E35 L-Proline, acetone, EtOH, –40 °C, 94% (174 g).
E36 TBDMSCl, py, 5 °C, 95%.
E37 H_2O_2 (30%), 50 °C.
E38 HCl (15%, aq.), acetone, 55 °C, 48%.

5.7.6 Reaction conditions for Scheme F

F1 **PG = Boc:** CuBr, THF, 82% (4.6 g). **PG = Ts:** CuBr, THF, 87% (4.6 g). **PG = Bz:** CuBr, THF, 77% (4.6 g). **PG = SO_2Ph:** CuBr, THF, 85% (4.6 g).
F2 **PG = Boc:** MeI, $NaHCO_3$, MeOH, 72% (4.3 g). **PG = Ts:** not shown. **PG = Bz:** MeI, $NaHCO_3$, MeOH, 77% (4 g). **PG = SO_2Ph:** MeI, $NaHCO_3$, MeOH, 76% (5 g).
F3 **PG = Boc:** TEMPO, NaClO, CH_2Cl_2. 90% (5.6 g, ZhejiangHP2010). NaClO, TEMPO, 90% (TL2013, BKCS2017). HClO, TEMPO, $NaHPO_4$, Na_2PO_4, H_2O, 10 °C. 98% (Nanjing2017). **PG = Ts:** TEMPO, NaClO, CH_2Cl_2, 85% (6.5 g). **PG = Bz:** TEMPO, NaClO, CH_2Cl_2, 92% (5.6 g). **PG = SO_2Ph:** TEMPO, NaClO, CH_2Cl_2, 88% (6.3 g).
F4 **PG1/PG2 = Boc/TBS:** CuBr, THF, 78% (4.6 g, ZhejiangHP2010). 1) *i*-PrMgBr, THF, 90%; 2) CuBr•Me_2S, SM, 0 °C, 78% (TL2013, BKCS2017). **PG1/PG2 = Boc/MOM:** CuBr, THF, 73% (4.6 g). **PG1/PG2 = Boc/Bn:** CuBr, THF, 88% (4.6 g). **PG1/PG2 = Bn/Bn:** CuBr, THF, 77% (4.6 g).
F5 **PG1/PG2 = Boc/TBS:** Bu_4NF, THF, 95% (5 g). Bu_4NF, MeOH, 95% (TL2013, BKCS2017). **PG1/PG2 = Boc/MOM:** AcOH, CH_2Cl_2, 97% (4 g). **PG1/PG2 = Boc/Bn:** H_2, Pd/C, MeOH, 98% (5.4 g). **PG1/PG2 = Bn/Bn:** 1) H_2, Pd/C, MeOH; 2) Boc_2O, NEt_3, CH_2Cl_2, 88% (5 g).
F6 1) PrBr, THF, *i*-PrMgCl, 70 °C; 2) ArBr, toluene, rt; 3) CuCl, aziridine, –5 °C, 85% (680 g).
F7 1) NaOAc, Bu_4NBr, DMA, 70 °C; 2) NaOH, 5 °C, 93% (660 g).
F8 CuBr, THF, 77% (4.6 g).
F9 1) *i*-PrMgCl, CuCl, THF, MTBE, –30 °C; 2) Aziridine, –30 °C to 0 °C, 40-90% (11-21 g).

F10 $KMnO_4$, acetone, rt, 32-89% (14-15 g).
F11 1) Zn, $BrCH_2CH_2Br$, THF, 50 °C; 2) Aziridine, –5 °C to rt, NI (49 g).
F12 NaOH (10%), THF, H_2O, 100 °C, 87-88% (F11+F12, 49 g).
F13 CuBr•SMe_2, rt, NI.
F14 1) TFA, 0 °C; 2) Boc_2O, $NaHCO_3$, 0 °C to rt, 61% (F13+F14).
F15 Mg, THF, 1,2-dibromoethane, CuI, (*S*)-epichlorohydrin, YNR (9 g, Hanmi2009a,b). 1) *i*-PrMgCl, THF, –20 °C; 2) CuI, –10 °C; 3) (*S*)-epichlorohydrin, –10 °C to 0 °C, 100% (25 g, Hanmi2010). 1) *i*-PrMgCl, THF, 0 °C; 2) CuCl, (*S*)-epichlorohydrin, 25 °C, 100% (120 g, Lek2010b). 1) Mg, THF, I_2, 1,2-dibromoethane, BuCl, –10 °C; 2) SM, –5 °C; 3) CuI, (*S*)-epichlorohydrin, 0 °C, 95% (22 g, ZhejiangHP2011d). 1) Mg, THF; 2) (*S*)-epichlorohydrin, CuI, 0 °C, 100% (94 g, Tongren2016).
F16 PPh_3/DEAD or Bu_3P/DIAD, succinimide, toluene, 0 °C, YNR (45 g).
F17 H_2O, HCl (conc.), reflux, 86% (32 g).
F18 Boc_2O, NEt_3, CH_2Cl_2, 25 °C, 85% (13.8 g).
F19 NaCN, PEG400, DMSO, 80 °C, 85% (34 g).
F20 (E23) NaOH, EtOH, H_2O, 80 °C, 81% (2 g).
F21 KCN, DMF, H_2O, 95 °C, 83% (128 g).
F22 (E31) see E31.
F23 NaOH, MeOH, 80-85% (F15+F23, 9-9.5 g, Hanmi2009a,b), in MeOH/H_2O, NI (26.6 g, Hanmi2010). NaOH or KOH, THF, H_2O, 60 °C, 87-90% (22 g, ZhejiangHP2011d).
F24 NaCN, NH_4Cl, MeOH, H_2O, 50 °C, 92% (5 g, Hanmi2009a). NaCN, Na_2CO_3 or NaOH or $NaHCO_3$, MeOH, H_2O, 40 °C, 86-92% (5 g, ZhejiangHP2011d).
F25 (E32) see E32.
F26 (E33) see E33.
F27 (E34) see E34.
F28 Mg, CH_2=CHCH_2Br, THF, 1,2-dibromoethane, YNR (25 g).
F29 MCPBA, CH_2Cl_2, rt, 75% (25 g).
F30 (*S,S*)-Co-salen (0.5% mol), AcOH (2.5%), THF, H_2O, rt, 41% (19 g).
F31 CH_2=CHMgBr, CuBr, THF, –78 °C to 0 °C, 100% (30 g, Hanmi2009b). CH_2=CHMgBr, CuI, THF, –20 °C to 0 °C, 93% (F23+F31, 26.6 g, Hanmi2010).
F32 1) MsCl, NEt_3, DMAP, CH_2Cl_2 or EtOAc, 0 °C; 2) NaN_3, DMF, H_2O, 70-75 °C, 94% (24-35 g).
F33 $NaIO_3$, $RuCl_3$, CH_3CN, 0 °C, 100% (32 g, Hanmi2009b). $KMnO_4$, acetone, H_2O, –20 °C, 98% (25 g, Hanmi2010).

5.7.7 Reaction conditions for Scheme G

G1 1) Mg, THF, I_2 (cat.), 25 °C; 2) Anhydride, 25 °C, 86% (9 g).
G2 H_2, Pd/C (10%) EtOH, 25 °C, 93% (3-4 g).
G3 (B13, C6, E10) see B13.
G4 *i*-PrMgBr, THF, –20 °C, 84% (21 g).
G5 H_2 (4 bar), Pd/C (10%), HCl, MeOH, rt, 70% (31 g).
G6 1) *i*-PrMgBr, THF, rt (19 g); 2) Amide, –10 °C, 43%.
G7 NaOH (5%), H_2O, MeOH, rt, 97% (20 g).
G8 1) Mg, THF, 30-40 °C; 2) Aldehyde, THF, 66% (116 g).
G9 1) MsCl, CH_2Cl_2, –20 °C to rt, 73% (36 g); 2) H_2 (5 bar), Pd/C (5%), MeOH, 60 °C, 76% (50 g).
G10 (B11, C8, D29, E16) see B11.
G11 1) Zn or Zn/CuBr, 1,2-dibromoethane, DMF, iodoester; 2) ArBr, $(Ph_3P)_2PdCl_2$, YNR (1.2 g).
G12 1) Mg (1 eq.), Et_2O, I_2 (cat.); 2) $B(OMe)_3$ (1 eq.), –78 °C, 91% (32 g).
G13 $Pd(OAc)_2$ (1.6% mol), *t*-BuOK (2.5 eq.), 1,4-dioxane, 50 °C, 75% (11 g).
G14 $SOCl_2$, MeOH, 0 °C to rt, 82-95% (20-40 g).
G15 BnCl, Na_2CO_3, MeOH, 25 °C, 91% (10 g).
G16 EtOCOCl, NEt_3, THF, 25 °C, NI (10 g).
G17 PG = Boc: Boc_2O, Na_2CO_3, 1,4-dioxane or 1,4-dioxane/H_2O, 75-88% (29-37 g, NanjingUT2011, Hengdian2011). **PG = Cbz:** $ClCO_2Bn$, Na_2CO_3, H_2O, 1,4-dioxane, 10 °C, 63% (2.2 g, Hengdian2011), with K_2CO_3 in Et_2O, 94% (31 g, Daewoong2011). **PG = Fmoc:** FmocCl, Na_2CO_3, H_2O, 1,4-dioxane, 50 °C, 89% (3.5 g, Hengdian2011). **PG = Bz:** PhCOCl, Na_2CO_3, H_2O, 1,4-dioxane, 50 °C, 81% (4.7 g, Hengdian2011).
G18 PG = Boc: $ClCO_2$*i*-Bu, NMM, MeOH, $NaBH_4$, THF, 0-80 °C, 76-88% (12-15 g). 1) *N*-hydroxysuccinimide, DCC, EtOAc, 10-30 °C; 2) B_2H_6 or $LiBH_4$, THF, 50-80 °C, 75-84% (5.9 g). **PG = Cbz:** $ClCO_2$*i*-Bu, NMM, MeOH, $ZnBH_4$, THF, 50 °C, 68% (5.3 g). 1) *N*-hydroxysuccinimide, DCC, EtOAc, 10 °C; 2) KBH_4, THF, 89% (12.9 g). **PG = Fmoc:** $ClCO_2$*i*-Bu, NMM, MeOH, KBH_4 or $LiBH_4$, THF, 90-100 °C, 77-91% (3.7-11 g).
G19 PG = Boc: TEMPO, KBr, NaClO, CH_2Cl_2, H_2O, 90 °C, 85% (10 g). $(COCl_2)_2$, DMSO, DIPEA, CH_2Cl_2, rt, 79% (5.1 g). DMP, CH_2Cl_2, 40 °C, 66% (0.5 g). PCC, CH_2Cl_2, 40 °C, 72% (2.3 g). PDC, CH_2Cl_2, reflux, 80% (3.1 g). **PG = Cbz:** TEMPO, KBr, NaClO, CH_2Cl_2, H_2O, 80 °C, 75% (2.5 g). **PG = Fmoc:** TEMPO, KBr, NaClO, CH_2Cl_2, H_2O, 10 °C, 91% (7.1 g).
G20 MeNHOMe, CDI, DIPEA, CH_2Cl_2, reflux, 75% (9.6 g).
G21 1) *N*-hydroxysuccinimide, EDC, EtOAc, YNR (30 g) or *i*-butylchloroformate, *N*-hydroxysuccinimide, EtOAc, YNR; 2) $NaBH_4$, THF, 30 °C; 3) PPh_3, I_2, imidazole, CH_2Cl_2, I_2, 25-30 °C.
G22 $AlCl_3$, CH_2Cl_2, 30 °C, 84% (75 g).
G23 H_2 (4 bar), Pd/C (10%), MeOH, 71% (25 g).
G24 (B12, C5, E9) see B12,
G25 (B10, C7) see B10,
G26 $AlCl_3$, CH_2Cl_2, rt, 73% (21 g).
G27 H_2 (4 bar), Pd/C (10%), HCl, MeOH, rt, 66% (22 g).
G28 1) H_2SO_4, Et_2O, BnOH, rt, 90% (25 g); 2) triphosgene, EtOAc, 90 °C, 83% (4.5 g).
G29 ClCOOMe, NMM, THF, –25 °C to rt, 88% (2.5 g).
G30 $AlCl_3$, CH_2Cl_2, <30 °C, 78% (30.7 g).
G31 1) H_2 (5 bar), Pd/C, EtOH (abs.), rt (6.2 g); 2) HCl (12 M), reflux; 3) LiOH, Boc_2O, THF, rt, 66%.

5.7.8 Reaction conditions for Scheme H

H1 (A1, B1) see A1.

H2 THTP•HCl (1 eq.), DMA, DIPEA, 70 °C, 89% (150 g Merck2003a,d, 2004a,b,c, 2005a,b,c,d, 2011), in *i*-PrOAc, 85 °C, 50-92% (2-16 g, Merck2003b, Cadila2008), in toluene, reflux, 85% (500 g, Glenmark2007), in toluene/*i*-PrOAc, 85 °C, 93% (10 g, DrReddys2007), in CH_3CN, 64% (25 g, Cadila2008), in EtOAc, reflux, 92-98% (50-200 g, DrReddys2009a, Disha2013), in EtOAc with NMM, reflux, 90% (100 g, Suven2014). THTP•HCl (1 eq.), TFA (0.3 eq.) CH_3CN, 50-55 °C, NI, 90% solution assay (2.5 kg, JACS2009a), with MsOH, 85 °C, 86% (50 g, MSN2009), in DMA with TFA, 55 °C, 84% (150 g, Shenzen2013b).

H3 NH_4OAc (2.9 eq.), NH_4OH (aq,), MeOH, reflux, 90% (200 g, Merck2003a,d, 2004a,b,c, 2005a,b,c,d, 2011), 40-55°C, 72-77% (540 g, Glenmark2007). NH_4OAc (1 eq.), MeOH, CH_3CN, 45 °C, 91% (4.8 kg, JACS2009a). $HCOONH_4$, NH_4OH (aq,), MeOH, reflux, 79% (2 g, Cadila2008). $(NH_4)_2CO_3$ (aq,), AcOH, MeOH, reflux, 53% (2 g, Cadila2008). NH_3 (g, anh.), MeOH, 25-30°C, reflux, 80% (427g scale, Lee2015).

H4 Boc_2O (2 eq.), NEt_3 (2 eq.), CH_2Cl_2, reflux, 50% (2 g).

H5 H_2 (50 bar), (*Sp*,*Rc*)-DuanPhosRh(cod)BF_4, CF_3CH_2OH, 90 °C, 100%, 44% ee (0.1 g).

H6 (B16, C10, D13) see B16.

H7 (I9) 1) $NaNO_2$ (aq.), H_2SO_4, H_2O, 55-60 °C, 76% (20 g); 2) $K_2Cr_2O_7$, AcOH, 25 °C, 57%.

H8 Most frequent: $NaBH_3CN$ or $NaBH_4$, in alcoholic solvent, acid may be added, >95%. $NaBH_3CN$ (6.7 eq.), AcOH, CH_2Cl_2, MeOH, 25-30 °C, 92% (50 g, Glenmark2007). $NaBH_3CN$ (1.2 eq.), MeOH•HCl, pH 4.5, 30 °C, 100% (14 g, DrReddys2007). KBH_4, MeOH, 25 °C, 96% (60 g, Suzhou2010). $NaBH_4$, AcOH, toluene, 10 °C, 96% (25 g, MSN2009). $NaBH_4$ (2-2.7 eq.), HCO_2H (11--36 eq.), PhCl or CH_3CN/PhCl, 0-15 °C, NI (100 g, Lee2015). H_2 (1 bar), Pd/C (3-5%), MeOH, 25 °C, 96-99% (60 g, Suzhou2010).

H9 AHT. An example: H_2 (13.7 bar), $[Rh(cod)Cl]_2$ (0.2% mol), (*R*,*S*) *t*-butyl JOSIPHOS (0.45 % mol), MeOH, 50 °C, 93%, 94% ee (118 g, Merck2003a,d, 2004a,c, 2005b,c,d, 2011). Many catalysts used: see section 5.4.2.1B and Figure 5.12.

H10 (I14) 1) $(CF_3CO)_2O$, NEt_3, 0 °C, CH_2Cl_2, 100% (1 g, Cadila2008); 2) LiOH, MeOH, H_2O, 25-30 °C, 38% (0,25 g, Cadila2008). AIBN or $(PhCOO)_2$ (2% mol), toluene or EtOAc or CH_2Cl_2, $PhCH_2SH$ or 1,2-ethanedithiol or dodecyl mercaptan (0.1 eq.), heat, 99% (4 g, GuangdongHP2014).

H11 Transaminase biochemical process: Transaminase enzyme derived from *Arthrobacter sp.*, 89%, ee not reported, 93% conversion, (20 g, Codexis2009), 90-95% conversion (1 kg, Codexis2009). Engineered enzyme (DSP), 84%, 99.9% ee (5 mg, Codexis2013). Immobilized transaminase, 50 °C, >98%, >92% ee (5 g, Nanjing2015), >99%, >99% ee (5 g, Nanjing2014b). Transaminase enzyme derived from prositagliptin transaminase 93%, >99.9% ee (S2010). Immobilized transaminase, 72% (Hubei2017). **Reductive amination:** H_2 (17 bar), $[Rh(cod)Cl]_2$ (1% mol), (*R*,*S*)-*t*-butyl JOSIPHOS (0.2 % mol), HCO_2NH_4 (1% mol), MeOH, 55 °C, 91%, 95% ee (6 mg, Merck2005a). H_2 (30 bar), $Ru(OAc)_2$((*R*)-dm-segphos) (1% mol), NH_4•SA (5 eq.), MeOH, 75 °C, 91%, 99.5% ee, with BINAP, 85%, 99.1% ee, with Xyl-BINAP, 90%, 98.0% ee, with Segphos, 80%. 99.8% ee. H_2 (30 bar), $[RhCl(cod)]_2$, (*R*,*S*)-*t*-Bu-JOSIPHOS (1% mol), NH_4•SA (5 eq.), MeOH, 75 °C, 86%, 93.4% ee (JACS2009b).

H12 Ar^1 = Ph, Z = $CONH_2$: (*S*)-2-amino-2--phenylacetamide (1.1 eq.), *i*-PrOH, AcOH, 40 °C, 91% (711 g, Merck2003b), 65% (CL2015) in CH_3CN, NI (41 g, Shenzen2013b). **Ar^1 = Ph, Z = CH_2OH:** (*S*)-2-amino-2-phenylethan-1-ol, AcOH (cat.), MeOH, 60-65 °C, 86% (2.5 g), in *i*-PrOH, 81% (23 g), in toluene, 100% (10 g). **Ar^1 = Ph, Z = Me:** (*R*)-1-phenylethan-1-amine, AcOH (cat.), *i*-PrOH, 40 °C, 100% (2 g, DrReddys2007), in toluene, reflux, 100% (10 g, OL2015). (*R*)-1-phenylethan-1-amine (1.2 eq.), toluene, reflux, 92% (25 g, DrReddys2009a), in *i*-PrOH, 40-50 °C, 43% (H1+H2+H12, 53 g, DrReddys2009b), in 1,2-dichloroethane or THF, NI (41 g, Shenzen2013b). **Ar^1 = 4-MeOC_6H_4, Z = Me:** (*R*)-1-(4-methoxyphenyl)ethan-1-amine (1.1 eq.), toluene, AcOH, reflux, 98 % (115 g).

H13 Ar^1 = Ph, Z = $CONH_2$: H_2 (6 bar), PtO_2 (6%), THF, MeOH, 22 °C, 90%, 97.4% de (20 g, Merck2003b). $NaBH_4$, HCO_2H, THF, –30 °C, 65%, >99.5% de (4 g, ChinaHP2011). KBH_4, TFA, MTBE, –30 °C, 66%, >99.5% de (4 g, ChinaHP2011). $NaBH_4$, $PrCO_2H$ or $BuCO_2H$, CH_3CN, –30 °C, 58-64%, >99.5% de (4 g, ChinaHP2011). $NaBH_4$, $ZnCl_2$, *i*-PrOH, 58%, >99% de (CL2015). $(AcO)_3BHNa$ (1.4-2.0 eq.), CH_3CN, TFA (1.1 eq.) or AcOH (1.6 eq.), rt, 85-87%, >99% de (41 g, Shenzen2013b). CH_3SiHCl_2, $EtCO_2H$, EtOAc, 30 °C, 72%, >99.5% de (7 g, ShanghaiIPI2013). $(CH_3)_2SiHCl$, $BuCO_2H$, CH_3CN, toluene, 20 °C, 74% >99.5% de (7 g, ShanghaiIPI2013). $NaBH_4$, HCO_2H, THF, –30 °C, 50% de (SC2013). **Ar^1 = Ph, Z = CH_2OH:** $NaBH_4$ or KBH_4, AcOH, CH_2Cl_2 or toluene, 0 °C, 98-99% (2.5-25 g). **Ar^1 = Ph, Z = Me:** H_2 (13 bar), PtO_2, THF, MeOH, 40 °C, 100% (8 g, DrReddys2007). $NaBH_4$ (2.5 eq.), MsOH, 1,2-dimethoxyethane or 1,2-dimethoxyethane/*i*-PrOH, –40 °C, 83-98%, >98% ee (5-33 g, DrReddys2009a,b). $NaBH_4$, HCO_2H, THF, –30 °C, 50%, 99.2% de (34.8 g, SC2013). $NaBH_4$ (2 eq.), HCO_2H (6 eq.) or TFA (3 eq.) or *t*-BuCO_2H (3 eq.) or AcOH (6 eq.), THF, –20 °C, 62-65%, >99.5% de (3 g, ChinaHP2011). $NaBH_4$ (1.5 eq.), AcOH or TFA, $ZnCl_2$ (2.8 eq.), THF, 82-98%, 59-90% de (0.5 g, ShaghaiIT2014), with $CoCl_2$ 93%, 23% de. $NaBH_4$, MsOH, *i*-PrOH, toluene, 1,2-dimethoxyethane, –65 °C, 71%, 99.2% de (21.5 kg, OL2015). $(AcO)_3BHNa$ (1.6 eq.), TFA

(1.3 eq.), 1,2-dimethoxyethane, rt, 95%, >99.5% de (H12+H13, 41 g, Shenzen2013b). $NaBH_3CN$ (1.3 eq.) TFA (1 eq.), THF, rt, 89% >99% de (H12+H13, 41 g, Shenzen2013b). $SiHCl_3$, AcOH, THF, MTBE, 0 °C, 75% >99.5% de (8 g, ShanghaiIPI2013). SiH_2Cl_2, HCO_2H, THF, 20 °C, 71%, >99.5% de (8 g, ShanghaiIPI2013). $SiHCl_3$, TFA, toluene, 25 °C, 73%, >99.5% de (8 g, ShanghaiIPI2013). **Ar¹ = 4-MeOC_6H_4, Z = Me:** $NaBH_4$, MsOH, 1,2-dimethoxyethane, –40 °C, 73% (150 g).

H14 (B8 when Z = Me) Ar¹ = Ph, Z = $CONH_2$: HCO_2H, $Pt(OH)_2/C$ (20%), THF, MeOH, H_2O, 60 °C, 90%, 97% ee (42 g, Merck2003b), 61-68%, 99% ee (22 g, Shenzen2013b). H_2 (2.7 bar), $Pt(OH)_2/C$ (20%), THF/MeOH/H_2O, AcOH, 40 °C, 93%, 96% ee (83 g, Merck2003b). H_2 (10 bar), $Pt(OH)_2/C$ (20%), AcOH, MeOH, H_2O, 50 °C, 81-85%, >99.5% ee (3-4 g, ChinaHP2011, ShanghaiIPI2013), 94%, 99.9% ee (SC2013). **Ar¹ = Ph, Z = CH_2OH:** HCO_2H, $Pd(OH)_2$, THF, MeOH, H_2O, 55-65 °C, 92%, 80% ee (5 g), with Pd/C (2%), 92%, 84% ee (5 g), with Pd/C (2%) in H_2O, 90%, 84% ee (2 g). H_2 (10 bar), Pd/C, AcOH, MeOH, THF, 60-65°C, 96%, 84% ee (2 g). **Ar¹ = Ph, Z = Me:** HCO_2H, $Pd(OH)_2/C$ (20%), H_2O, THF, MeOH, 60 °C, 61-93% (2-22 g, DrReddys2007, 2009a,b, Shenzen2013b). H_2 (10 bar), $Pd(OH)_2/C$ (20%), AcOH, MeOH, H_2O, 50 °C, 94% (13 g, SC2013). H_2 (2-15 bar), Pd/C (5-10%), AcOH, MeOH or H_2O or MeOH/H_2O, 50-55 °C, 55-85% (3-25 g, DrReddys2009b, Sun2010, ShanghaiIPI2013). H_2 (7 bar), Pd/C, *i*-PrOH, H_2O, 70 °C, 88% (300 g, OL2015). **Ar¹ = 4-MeOC_6H_4, Z = Me:** ortho-phosphoric acid or polyphosphoric acid, 95-100 °C, 68-85% (10-100 g).

H15 *t*-BuS(O)NH_2 (1 eq.), TsOH (cat.), toluene, reflux, YNR (41 g).

H16 $NaBH_4$ (1.25 eq.), CH_2Cl_2, AcOH, YNR (51 g).

H17 HCl, YNR (not shown).

H18 THTP•HCl, Na_2CO_3 or K_2CO_3, toluene or xylene, 95 °C, YNR (23 g).

H19 1) $ArCH_2MgBr$, THF, 45 °C; 2) NH_4OH, H_2O, 88-90% (H18+H19, 23 g), 85% (starting from the ethylester, 11 g).

H20 THTP•HCl, $SOCl_2$, CH_3CN, 60 °C, 93% (14 g). THTP•HCl, $POCl_3$, DMF, 100 °C, 95% (28 g).

H21 1) Mg, Et_2O, 35 °C; 2) $ArCH_2CN$, 0 °C to rt; 3) HCl (2 N), 83% (29 g). 1) Mg, THF, I_2 (cat.), 40 °C; 2) $ArCH_2CN$, 15 °C to rt; 3) HCl (2 N), 87% (58 g).

H22 THTP, MeOH, –10 °C to rt, 98% (66 g).

H23 $ArCH_2CO_2H$, CH_3CN, *t*-BuCl, DIPEA, DMAP, rt to 75 °C. 95% (92 g).

H24 *i*-PrOH, AcOH, (*S*)-2-amino-2-phenylacetamide (1.3 eq.), rt to reflux, 98% (51 g).

H25 H_2, PtO_2, THF, MeOH, rt, 73% (60 g).

H26 1) *N,N*-dimethylaniline, 150 °C, 93% (20 g); 2) H_2, $Pd(OH)_2$, EtOH, 60 °C, 90% (10 g).

H27 NH_2OH, EtOH, py, rt, 70% (5 g).

H28 Zn, NH_4Cl, MeOH, reflux, 67% (5 g).

H29 1) THTP•HCl (1.3 eq.), CH_3CN, NMM (1 eq.), rt; 2) EDC•HCl (1.5 eq.), 79% (20 g).

H30+H31 ArCHO, K_3PO_4 (0.1 eq.), CH_3CN, rt, 77% (29 g). ArCHO, morpholine (0.6 eq.), EtOH, reflux, 14% (1.1 g).

H32 $NaBH_4$ (1.2 eq.), THF, 0 °C, 39% (2.2 g).

H33 Zn, HCl (conc.), MeOH, rt, 93% (0.1 g).

H34 HPLC separation, 43% (0.1 g).

H35 Zn, HCl (conc.), MeOH, rt, 87% (40 mg).

H36 (I13) Diastereomeric sal formation: DPTTA (1 eq.), MeOH/H_2O (4.9:1), 33% as salt, 22% free base (13 g, DrReddys2007). 1) L-DBTA (0.9 eq.), MeOH, 70 °C to 20-25 °C; 2) MeOH/H_2O crystallizations, 27%, 92-95% ee; 3) 16% (free base), 99.8% ee (212 g, Glenmark2007). L-DBTA (0.9-1.2 eq.), *i*-PrOH/acetone or MeOH, YNR, 80-92% ee (Glenmark2007, Cadila2008). *i*-PrOH/MeOH, YNR, 40% ee (Cadila2008). L-DBTA, MeOH, 23%, 54% ee (2 g, Cadila2008), in *i*-PrOH/MeOH, 35-45%, 50-62% ee (2-50 g, Cadila2008), in MeOH/toluene, 32%, 54, % ee (2 g, Cadila2008), in *t*-BuOH/MeOH, 58%, 34% ee (2 g, Cadila2008), in dioxane/MeOH, 35%, 56% ee (2 g, Cadila2008), in MIBK/MeOH, 35%, 50% ee (2 g, Cadila2008), in MIBK/MeOH/H_2O, 35%, 58% ee (2 g, Cadila2008). L-TA (1 eq.), MeOH, 26-32%, 93.6-94.8% ee (5 g, Cadila2008). (*S*)-(+)-*O*-acetyl mandelic acid (1 eq.), *i*-PrOH, 4%, 50% ee (1 g, Cadila2008). *N*-acetyl-D-phenylalanine (1 eq.), MeOH, reflux to 25 °C, 11%, >98% ee (3 g, Smilax2012). (*S*)-3-(2-amino-2-oxoethyl)-5-methylhexanoic acid (0.6 eq.), EtOH, 27%, 99.4% ee (25 g, Lee2015), in CH_3CN, 38%, 98.8% ee (25 g, Lee2015), in EtOAc, 41%, 91% ee (25 g, Lee2015), in *i*-PrOH, 35-38%, 98.8-99.5% ee (50-100 g, Lee2015). (–)-MA, H_2O, *i*-PrOH, 20-25% (3-8 g, MSN2009). **Dynamic kinetic resolution:** D-TA (1 eq.), acetone, *p*-HOC_6H_4CHO or C_6H_5CHO (0.5% eq.), AcOH, 0 °C, 82-85%, 99.7-99.8% ee (60g, Suzhou2010). D-CSA (1 eq.), acetone, *p*-HOC_6H_4CHO (0.5% eq.), AcOH, 0 °C, 80%, 99.8% ee (60 g, Suzhou2010).

5.7.9 Reaction conditions for Scheme I

I1 Biochemical: Oxidoreductase in an enzyme coupled cofactor regeneration system, lysate of wholecell catalyst co-expressing oxidoreductase and glucose dehydrogenase, 60-72%, >99% ee, >90% HPLC (10-25 g, Cadila2010, 2011). Wet cell bacteria, 56-96%, 99.9% ee (0.4 g, ZhejiangUT2015). *Pseudomonas pseudoalcaligenes* XW-40, 57-99%, >99% ee (Chongqing2015).

I2 Alcohol activation: MsCl, DIPEA, DMAP, CH_3CN, 0-5 °C, 100% (3 g, MSN2009). MsCl, DIPEA, CH_2Cl_2, 0-20 °C, 99-100% (0.4-13 g, Lek2010b, Cadila2010). TsCl, py, 0-5 °C, 85%, >99% ee (80 g, Cadila2011), with DABCO, toluene, 0-5 °C, 92% (0.2 g, Cadila2011), with DABCO, CH_2Cl_2, 0-5 °C, 100% (0.2 g, Cadila2011). $(EtO)_2P(O)Cl$, py, CH_2Cl_2, 0-5 °C, 92% (0.5 g, Cadila2011). *p*-$O_2NC_6H_4SO_2Cl$, py, 0-5 °C, 72%

(4 g, Cadila2011). Ac_2O, DMAP, CH_2Cl_2, 0-5 °C, 99% (3 g, Cadila2011). **N_3, displacement, on a Ms derivative:** NaN_3, DMF, CCl_4, NI (3 g, MSN2009), in DMSO, DMF or sulfolane, 50 °C, 100% (8 g, Lek2010b), in DMF, 40 °C, 18-20% (0.3-4 g, Cadila2010). **N_3, displacement, on a Ts derivative:** NaN_3, DMSO, 24-30 °C, 91-100% (0.7-93 g, Cadila2011). Bu_4NN_3, DMSO, toluene, 24-30 °C, 80% (0.2 g of *p*-nitroTs derivative, Cadila2011).

I3 1) PPh_3, THF, 50 °C; 2) NH_4OH, H_2O, 86-90% (6-34 g, Hanmi2009b, 2010). PPh_3, AcOH, THF, EtOH, H_2O, 23 °C, 62% (12 g, Lek2010b). PPh_3, NH_4OH, 100% (0.3 g, Cadila2011). H_2 (3 bar), Pd/C, MeOH, 71% (3 g, MSN2009). $NaBH_4$, Pd/C (5%), MeOH, 0 °C to 30 °C, 65-68%, 99% ee (1.4 g, Cadila2010, 2011). Zn, $AlCl_3$•$7H_2O$, EtOH, 25 °C, 63% (0.3 g, Cadila2011). Zn, HCl (conc.), 85 °C, 37% (0.3 g, Cadila2011). $NaBH_4$, $NiCl_2$•$6H_2O$, MeOH, 30 °C, 67% (0.3 g, Cadila2011). $SnCl_2$•$2H_2O$, MeOH, 25 °C to 40 °C, 60% (0.3 g, Cadila2011).

I4 (*R*)-2-MeCBS, BH_3•Me_2S, toluene, 0-5 °C, 60% (5 g).

I5 NaN_3, CCl_4, DMF, 90 °C, 60% (2 g).

I6 PPh_3, NBS, CH_2Cl_2, 0 °C to 25 °C, 87% (4 g).

I7 Potassium phthalimide, DMF, 110 °C, 75% (1.4 g).

I8 $MeNH_2$, H_2O, toluene, 53% (2 g).

I9 (H7) see H7.

I10 $NaBH_4$, MeOH, THF, –78 °C, 100% (2 g, Cadila2008). $NaBH_4$, AcOH, 15 °C, NI (5 g, MSN2009).

I11 Tosylation or mesylation: TsCl or MsCl, py, 0-5 °C, 54-92%, 99.6% ee (5-10 g). *p*-$O_2NC_6H_4SO_2Cl$, py, 0-5 °C, 85% (0.5 g). **N_3 displacement:** DMSO, NaN_3, 24-30 °C, 100%, 63% purity by HPLC (0.5 g of Ts derivative). DMF, NaN_3, 24-30 °C, 100%, 23% purity by HPLC (0.1 g of Ms derivative). DMF, tributyl hexadecylphosphonium bromide, NaN_3, 24-30 °C, 100%, 44% purity by HPLC (0.2 g of Ts derivative).

I12 $NaBH_4$, Pd/C (5%), MeOH, 25-30 °C, 65-67%, >99% ee (1.4 g).

I13 (H36) see H36.

I14 (H10) see H10.

I15 1) TsCl, py; 2) BnOH, toluene, 80 °C, 100% (0.5 g). PPh_3, DIAD, THF, $BnNH_2$, 25-30 °C, very low yield (1 g).

I16 Not shown.

I17 PPh_3, NBS, 25 °C, 76-77% (4-5 g, MSN2009).

I18 Phthalimide, DMF, 110 °C, 64% (3 g).

I19 $MeNH_2$, H_2O, toluene, 60% (2 g).

5.7.10 Reaction conditions for Scheme J

J1 NH_2NH_2•xH_2O (neat), 120 °C, 49% (25 g, JMC2005). NH_2NH_2 (35%), H_2O, 65 °C, 81% (5 kg, OPRD2005), at 58 °C, YNR (Lanzhou2011), at 100 °C, 67% (LOC2013).

J2 F_3CCHO, EtOH, 90 °C, 72%.

J3 NBS (1 eq.), DMF, 25 °C, 92%.

J4 NEt_3, toluene, 25 °C, 90%.

J5 H_2, Pd/C (10%), EtOH, rt to 45 °C, 86-87%. H_2 (3.5 bar), Pd/C (10%), EtOH, 45 °C, 87% (5 kg, OPRD2005).

J6 $(CF_3CO)_2O$, 0 °C to rt, YNR (17 g, JMC2005).

J7 PPA, 140 °C, 54% (J6+J7, 17 g, JMC2005).

J8 CH_3COO*i*-Pr, TFAA (4 eq.), 61% (5 kg, OPRD2005). PhCl, TFAA, 50 °C, NI (Lanzhou2011).

J9 Superphosphoric acid, 75 °C, 71% (4.8 kg, OPRD2005). PhCl, MsOH (cat.), 50 °C, YNR (Lanzhou2011).

J10 CF_3CO_2Et (1 eq.), CH_3CN, 14-25 °C, NI (15-20 g).

J11 $ClCH_2COCl$ (1 eq.), NaOH (1 eq.), CH_3CN, H_2O, 15 °C, 97-98% (J10+J11, 15-20 g).

J12 $POCl_3$ (1 eq.), CH_3CN, 80 °C, 86-92% (43 g, Merck). $POCl_3$, 79% (OL2005). $POCl_3$/DMAP, 79% (OL2005). $SOCl_2$/DMAP, 55% (OL2005). $SOCl_2$, 56% (OL2005). $MePOCl_2$, 72% (OL2005). $PhPOCl_2$ (OL2005).

J13 $H_2NCH_2CH_2NH_2$ (3.5 eq.), MeOH, –20 °C, 72-74% (30 g, Merck, OL2005), 47% (79 g, DrReddys2007), 0 °C to 35 °C, 85-94% (flow reactor, Anhui2012).

J14 HCl (37%, 1.1 eq.), MeOH, 55 °C, 89% (27 g, Merck). HCl (37%, 1.1 eq.), MeOH, 55 °C, 91% (41 g, DrReddys2007). HCl (g), EtOH, H_2O, 60 °C, 90-91% (180-500 g, FIS2015). MeOH, reflux, 92% (OL2005).

5.8 REFERENCES

2Y2013 Zhu, J.; Wang, Y.; He, X.; Wang, Y.; Gao, F. (2Y Chem Ltd) CN103319487 (prior. 10/01/2013, CN).

ACIE1981 Schöllkopf, U.; Groth, U.; Deng, C. *Angewandte Chemie International Edition in English* **1981**, *20*, 798-799.

ACIE2009 Zhou, S.; Wang, J.; Chen, X.; Aceña, J. L.; Soloshonok, V. A.; Liu, H. *Angewandte Chemie International Edition* **2014**, *53*, 7883-7886.

ACIE2011 Desai, A. A. *Angewandte Chemie International Edition* **2011**, *50*, 1974-1976.

ACR2007 Shultz, C.; Krska, S. *Accounts of Chemical Research* **2007**, *40*, 1320-1326.

Alpha2016 Shi, L.; Chen, J.; Cai, J.; Qi, Z.; Zhang, W.; Ye, Y.; Gong, S.; Xu, C. (Alpha Pharmaceutical Co Ltd) CN106801043 (prior. 28/12/2016, CN).

Anhui2012 Li, J.; Lv, H.; Xia, J.; Su, W.; Xie, Y.; Ge, J.; Chen, H. (Univ Zhejiang Technology, Anhui Changming Pharmaceutical Co Ltd) CN102898387 (prior. 26/09/2012, CN).

AOC2013 Stavber, G.; Casar, Z. *Applied Organometallic Chemistry* **2013**, *27*, 159-165.

AOC2015 Sova, M.; Frlan, R.; Gobec, S.; Stavber, G.; Casar, Z. *Applied Organometallic Chemistry* **2015**, *29*, 528-535.

Apotex2013 Abdur-Rashid, K.; Guo, R.; Chen, X.; Horne, S. E. (Apotex Technologies Inc) WO2014113869, US2015353564 (prior. 22/01/2013, US).

Arromax2014 Hong, J.; Liu, G.; Xu, X. (Arromax Pharmaceutical Technology Suzhou Co Ltd) CN104693207 (prior. 17/03/2015, CN).

BKCS2017 Kwak, W. Y.; Mok, H. Y.; Yoon, S. H. *Bull. Korean Chem. Soc.* **2017**, *38*, 976-979.

BMCL2007 Ahn, J. H.; Shin, M. S.; Jun, M. A.; Jung, S. H.; Kang, S. K.; Kim, K. R.; Rhee, S. D.; Kang, N. S.; Kim, S. Y.; Sohn, S.-K.; Kim, S. G.; Jin, M. S.; Lee, J. O.; Cheon, H. G.; Kim, S. S. *Bioorganic & Medicinal Chemistry Letters* **2007**, *17*, 2622-2628.

Brightgene2010 Yuan, J.; Xu, A. (Brightgene Bio-Medical Technology Co Ltd) CN102126976 (prior. 20/12/2010, CN).

Cadila2008 Kothari, H.; Dave, M.; Shukla, B.; Pandey, B. (Cadila Healthcare Ltd) IN200801798, WO2010032264, US2011213149, US8476437, EP2331545 (prior. 27/08/2008, IN).

Cadila2010 Mendiratta, S. K.; Pandey, B.; Joshi, R.; Trivedi, U.; Dave, M. G.; Kothari, H. M.; Shukla, B. (Cadila Healthcare Ltd) IN201002805, WO2012046254, EP2625179, US2013289276 (prior. 08/10/2010, IN).

Cadila2011 Pandey, B.; Dave, M. G.; Kothari, H. M.; Shukla, B. S.; Mendiratta, S. K.; Joshi, R.; Trivedi, U. (Cadila Healthcare Ltd) WO2013065066 (prior. 02/11/2011, IN).

Cadila2013 Dwivedi, S. P.; Singh, K. K.; Narode, S. D.; Patel, D. J.; Solanki, K. S. (Cadila Healthcare Ltd) US2015087834, US9409912 (prior. 20/03/2013, IN).

Cadila2014 Dwivedi, S. P.; Khera, B.; Singh, K. K.; Solanki, K. S. (Cadila Healthcare Ltd) IN212MU2014, WO2015114657 (prior. 21/01/2014, IN).

CCL2009 Zeng, L. L.; Ding, Y. J.; Zhang, G. C.; Song, H. R.; Hu, W. H. *Chinese Chemical Letters* **2009**, *20*, 1397-1399.

Chemo2010 Colombo, L.; Rasparini, M.; Tufaro, R.; Venegoni, S.; D'Arienzo, G. (Chemo Iberica Sa) EP2423178 (prior. 28/07/2010, EP).

Chemo2011 Rasparini, M.; Tufaro, R. R.; Minelli, C. (Chemo Iberica Sa) IT1405253, WO2012150328, US2014081026, US8912327, EP2705026 (prior. 05/05/2011, IT).

Chengdu2015 Chen, S.; Yang, R. (Chengdu Qianxilai Pharmaceutical Co Ltd) CN105017099 (prior. 15/07/2015, CN).

ChinaHP2011 Zhang, S.; Li, Y.; Cai, Z.; Lin, K.; Yao, Z.; Zhou, W. (China Head Pharm Ind Res Inst, Shanghai Inst Pharm Ind) CN102838603 (prior. 24/06/2011, CN).

ChiralQ2008 Delice, A.; Wang, Y.; Wu, S.; Yu, B.; Zhu, J.; Dris, A. (Chiral Quest Inc, Chiral Quest Co Ltd) WO2010078440, US2010280245, US8278486, EP2381772 (prior. 31/12/2008, US).

Chongqing2015 He, C.; Wei, Y.; Xiong, W.; Xia, S.; Xu, H. (Univ Chongqing Posts & Telecom) CN105274027 (prior. 05/11/2015, CN).

Cipla2012 Rao, D. R.; Kankan, R. N.; Ghagare, M.; Kadam, S. A. (Cipla Ltd) WO2014023930, US2015197523, US9233967, EP2882752 (prior. 08/08/2012, IN).

CL2015 Pan, X.; Wang, K.; Yu, W.; Zhang, R.; Xu, L.; Liu, F. *Chemistry Letters* **2015**, *44*, 1170-1172.

Codexis2009 Savile, C.; Mundorff, E.; Moore, J. C.; Devine, P. N.; Janey, J. M. (Codexis Inc) WO2010099501; US2010285541; US2012329108; US2015037869; US2015329838; US2016264948; US2017088868; US8293507; US8889380; US9133445; US9353355; US9550982; US9944963; EP2401366 (prior. 26/02/2009, US).

Codexis2013 Quintanar-Audelo, M.; Eberhard, E.; Nazor, J.; Smith, D.; Wang, C. (Codexis Inc) WO2014133960, US2016002688, US2017175090, US2018223265, US9617573, US9963685, EP2961844 (prior. 28/02/2013, US).

CSIRIndia2014 Barua, N. C.; Saikia, B.; Borah, P.; Baishya, G. (Council Scient Ind Res) WO2015189862 (prior. 10/06/2014, IN).

DaeWoong2011 Choi, S. J.; Lee, B. G.; Oh, S. S.; Kim, Y. T.; Eo, J. Y.; Kim, H. S. (Dae Woong Pharm Co Ltd) KR2012122988, WO2012148246 (prior. 29/04/2011, KR).

Dipharma2009 Allegrini, P.; Attolino, E.; Dellanna, G.; Michieletti, M.; Dell'Anna, G. (Dipharma Francis Srl) IT1395596, US2010331541, EP2270009, US8097724 (prior. 30/06/2009, IT).

Disha2013 Wang, W. (Shandong Disha Pharmaceutical Co Ltd, Disha Phamaceutical Group) CN103387577 (prior. 29/07/2013, CN).

Dissymmetrix2010 Basaveswara Rao, M.; Janagani, S.; Kumar, V.; Ravisankar, V. (Dissymmetrix P Ltd) WO2011142825 (prior. 12/05/2010, US).

DrReddys2007 Padi, P.; Ireni, B.; Polavarapu, S.; Nerella, K.; Ramasamy, V.; Padamata, S.; Vangala, R. (Dr Reddy's Lab Ltd, Dr Reddy's Lab Inc) WO2009085990, US2010274017, US2013035486, US2014128607, US8309724, US8969558, EP2220093, EP2599781, EP2679590 (prior. 20/12/2007, IN).

DrReddys2009a Bandichhor, R.; Gudimalla, N.; Dwivedi, N.; Chetluru, K. K.; Gade, S. R.; Muvva, V.; Srivastava, B. (Dr Reddy's Lab Inc, Dr Reddy's Lab Ltd) IN201201509, WO2011025932, EP2470542 (prior. 28/08/2009, IN).

DrReddys2009b Gade, S.; Muvva, V.; Srivastava, B. (Dr Reddy's Lab Inc, Dr Reddy's Lab Ltd) WO2011060213 (prior. 12/11/2009, IN).

EMT2019 Kim, G. H.; Jeon, H.; Khobragade, T. P.; Patil, M. D.; Sung, S.; Yoon, S.; Won, Y.; Choi, I. S.; Yun, H. *Enzyme Microb. Technol.* **2019**, *120*, 52-60.

Esteve2009 Bartra Sanmarti, M.; Rustullet Oliver, A.; Fernandez Hernandez, S.; Monsalvatje Llagostera, M. (Esteve Quimica Sa) EP2223923 (prior. 25/02/2009, EP).

Farma2011 Zupancic, S.; Zupet, R. (Farma Grs D O O) SI2736909, WO2013013833, EP2736909 (prior. 27/07/2011, SI).

FIS2015 Galvagni, M.; Grendele, E.; Lora, G. (F I S - Fabbrica Italiana Sintetici S P A) EP3159343 (prior. 22/10/2015, EP).

FIS2017 Galvagni, M.; Grendele, E.; Lora, G. (F I S Fabbrica Italiana Sintetici S P A) US2018230151 (prior. 10/02/2017, US).

Genchem2015 Wang, Y.; Zhan, Y. (Genchem & Genpharm Changzhou Co Ltd) CN105254519 (prior. 25/11/2015, CN).

Glenmark2007 Ahmed Khan, M.; Arjunan, S.; Raghunath Patil, M.; Khan, M. A.; Patil, M. R.; Sankar, A. (Glenmark Generics Ltd) IN200702190, WO2009084024, US2010317856, US8334385 (prior. 02/11/2007, IN).

Glenmark2010 Badgujar, S.; Gharpure, M. M.; Yadav, R. S. (Glenmark Generics Ltd) IN201002698, WO2012042534 (prior. 28/09/2010, IN)

GuangdongHP2012 Lai, J.; Liang, Z.; Yang, F. (Guangdong Hec Pharmaceutical) CN103910634 (prior. 29/12/2012, CN).

GuangdongHP2014 Lai, J.; Kou, J. (Guangdong Hec Pharmaceutical) CN104130264 (prior. 14/08/2014, CN).

Guizhou2017 Liu, X.; Tian, M.; Yao, Z.; Liu, X.; Zhang, M.; Zhou, Y.; Feng, T.; Yu, Z. (Univ Guizhou) CN107129501 (prior. 19/05/2017, CN).

Hanmi2009a Kim, N.; Chang, J.; Jung, J.; Lee, H.; Kim, D.; Chang, Y.; Lee, G. (Hanmi Holdings Co Ltd, Hanmi Sci Co Ltd) KR2011036308, WO2011040717 (prior. 01/10/2009, KR).

Hanmi2009b Kim, N. D.; Chang, J. Y.; Kim, D. J.; Lee, H. S.; Jung, J. H.; Chang, Y. K.; Lee, G. S. (Hanmi Holdings Co Ltd, Hanmi Sci Co Ltd) KR2011043288, WO2011049344, US2012214997, US8624026, EP2491040 (prior. 21/10/2009, KR).

Hanmi2010 Chang, J.; Lee, H. S.; Kim, D. J.; Ko, M.; Kim, N. D.; Chang, Y. K.; Kim, M. S. (Hanmi Holdings Co Ltd, Hanmi Sci Co Ltd) WO2011102640, KR2011095672 (prior. 19/02/2010, KR).

Hengdian2011 Zhang, X.; Guo, J.; Zhang, Y.; Zhu, Y.; Zhang, L.; Wu, Y.; Xu, H. (Hengdian Group Jiayuan Chem Co Ltd) CN102304064 (prior. 21/06/2011, CN).

Hubei2017 Ye, Z.; Zhou, Y.; Deng, Z.; E, S. (Hubei Hongyuan Pharmaceutical Tech Co Ltd) CN106834376 (prior. 05/01/2017, CN).

Ind2011 Soukup, M. (Individual) WO2012156888, US2012123144, US8524936 (prior. 18/05/2011, US).

JACS2004a Hsiao, Y.; Rivera, N. R.; Rosner, T.; Krska, S. W.; Njolito, E.; Wang, F.; Sun, Y.; Armstrong, J. D.; Grabowski, E. J. J.; Tillyer, R. D.; Spindler, F.; Malan, C. *Journal of the American Chemical Society* **2004**, *126*, 9918-9919.

JACS2004b Xu, F.; Armstrong, J. D.; Zhou, G. X.; Simmons, B.; Hughes, D.; Ge, Z.; Grabowski, E. J. J. *Journal of the American Chemical Society* **2004**, *126*, 13002-13009.

JACS2009a Hansen, K. B.; Hsiao, Y.; Xu, F.; Rivera, N.; Clausen, A.; Kubryk, M.; Krska, S.; Rosner, T.; Simmons, B.; Balsells, J.; Ikemoto, N.; Sun, Y.; Spindler, F.; Malan, C.; Grabowski, E. J. J.; Armstrong, J. D. *Journal of the American Chemical Society* **2009**, *131*, 8798-8804.

JACS2009b Steinhuebel, D.; Sun, Y.; Matsumura, K.; Sayo, N.; Saito, T. *Journal of the American Chemical Society* **2009**, *131*, 11316-11317.

JCR2010 Liu, F.; Yu, W. S.; Ou, W. H.; Wang, X. K.; Ruan, L. B.; Li, Y. M.; Peng, X. J.; Tao, X. H.; Pan, X. H. *Journal of Chemical Research* **2010**, 230-232.

JiangsuHB2017 Pan, Q.; Zhou, X. (Jiangsu Hankuo Biological Co Ltd) CN107011138 (prior. 18/04/2017, CN).

JiangsuHM2008 Chen, Y.; Sun, P.; Yu, G.; Sun, P. Y. (Jiangsu Hengrui Medicine Co Ltd) CN101633625, WO2010009630, US2011130587, US8580997, EP2308829 (prior. 23/07/2008, CN).

JiangsuSP2012 Liu, Y.; Wang, J.; Wang, Q.; Wang, X. (Jiangsu Shimeikang Pharm Co Ltd) CN102627648 (prior. 14/04/2012, CN).

JMC2005 Kim, D.; Wang, L.; Beconi, M.; Eiermann, G. J.; Fisher, M. H.; He, H.; Hickey, G. J.; Kowalchick, J. E.; Leiting, B.; Lyons, K.; Marsilio, F.; McCann, M. E.; Patel, R. A.; Petrov, A.; Scapin, G.; Patel, S. B.; Roy, R. S.; Wu, J. K.; Wyvratt, M. J.; Zhang, B. B.; Zhu, L.; Thornberry, N. A.; Weber, A. E. *Journal of Medicinal Chemistry* **2005**, *48*, 141-151.

KyungDong2015 Lee, B. S.; Shin, S. H.; Kilan, Y.; Chun, E. J. (Kyung Dong Pharm Co Ltd) WO2016204376, EP3262025, US2018086765 (prior. 16/06/2015, KR).

Lanzhou2011 Qi, B. (Lanzhou Jiushengya Biological Technology Co Ltd) CN102796104 (prior. 25/05/2011, CN).

Lee2015 Alla, V. R.; Alla, R.; Mallepalli, S. R.; Nandam, S. B.; Guda, M. R.; Alluri, R. R. (Lee Pharma Ltd) WO2016110750, US2018265513, EP3242879 (prior. 08/01/2015, IN).

Lek2010a (Lek Pharmaceuticals) EP2392575-A1 (prior. 04/06/2010, EP).

Lek2010b (Lek Pharmaceuticals) EP2397141 (prior. 16/06/2010, EP).

Lek2011a Casar, Z.; Stavber, G.; Sova, M. (Lek Pharmaceuticals) WO2012136383, EP2694466, US2014213810, US9174930, EP2508506 (prior. 08/04/2011, EP).

Lek2011b Casar, Z.; Stavber, G. (Lek Pharmaceuticals) WO2012163815, US2014187558, US9359385, EP2527320, EP2714646, (prior. 27/05/2011, EP).

Lek2012a Stavber, G.; Casar, Z. (Lek Pharmaceuticals) WO2013104774, EP2615080, EP2802554 (prior. 12/01/2012, EP).

Lek2012b (Lek Pharmaceuticals) EP2647640 (prior. 06/04/2012, EP).

Lek2012c Starcevic, S.; Mrak, P.; Kopitar, G. (Lek Pharmaceuticals) WO2013186326, US2015148533, US9388188, EP2674432, EP2861598 (prior. 14/06/2012, EP).

Lek2012d Starcevic, S.; Mrak, P.; Kopitar, G. (Lek Pharmaceuticals) WO2014096375, US2016194680, EP2935205 (prior. 21/12/2012, EP).

Lianyungang2014 Sang, G.; Zhao, R.; Zhang, A.; Zhang, X.; Xia, C. (Lianyungang Runzhong Pharmaceutical Co Ltd) CN105315286 (prior. 30/07/2014, CN).

Liaocheng2013 Liu, Q.; Chen, S.; Wang, S.; Zhang, C.; Liu, M.; Guo, B. (Liaocheng Jiuzhou Hegu Biotechnology Co Ltd) CN103232455 (prior. 14/05/2013, CN).

LOC2013 Lakshmikant, H. N.; Sunil, N. D.; Devanand, B. S. *Letters in Organic Chemistry* **2013**, *10*, 348-352.

M2018 Gao, H. L.; Yu, J. G.; Ge, C. S.; Jiang, Q. *Molecules* **2018**, *23*, 12.

Merck2001 Edmondson, D.; Fisher, M.; Kim, D.; Maccoss, M.; Parmee, E.; Weber, A. E.; Xu, J.; Fisher, H.; Dooseop, K.; Jinyou, X.; Biftu, T.; Liang, G.; Qian, X.; Feng, D. (Merck & Co Inc, Merck Sharp & Dohme Corp) WO2003004498, US2003100563, US2004167133, US2006270679, US2010130504, US2011190308, US2013217695, US2015359793, US2017348309, US6699871, US7125873, US8168637, US8440668, EP1412357, EP1625847, EP2226324, EP2292232 (prior. 06/07/2001, US).

Merck2002 Duffy, J. L.; Edmondson, S. D.; Kim, D.; Kirk, B. a.; Wang, L.; Weber, A. E. (Merck & Co Inc) US20030469315P, US20020435389P, US20050540283, US2006052382, WO2003US40114 (prior. 20/12/2002, US, WO).

Merck2003a Xiao, Y.; Armstrong, J.; Krska, S. W.; Njolito, E.; Rivera, N.; Sun, Y.; Rosner, T.; Yi, X.; Eugenia, N.; Yongkui, S.; Thorsten, R. (Merck & Co Inc, Merck Sharp & Dohme Corp) WO2004085378, US2006194977, US7468459, EP1606243 (prior. 19/03/2003, US).

Merck2003b Dreher, S.; Ikemoto, N.; Njolito, E.; Rivera, N.; Tellers, D.; Xiao, Y. (Merck & Co Inc) WO2004085661 (prior. 24/03/2003, US).

Merck2003c Angelaud, R.; Armstrong, J.; Askin, D.; Balsells, J.; Hansen, K.; Lee, J.; Maligres, P.; Rivera, N.; Xiao, Y.; Zhong, Y. (Merck & Co Inc) WO2004087650 (prior. 27/03/2003, US).

Merck2003d Cypes, S.; Chen, A.; Ferlita, R.; Hansen, K.; Lee, I.; Vydra, V. K.; Wenslow, R.; Howard, C.; Ivan, L.; Karl, H.; Minhua, C.; Pokropinski Jr, H. (Merck & Co Inc, Merck Sharp & Dohme Corp, Merck Sharp & Dohme Ltd) WO2005003135, US2005032804, US7326708, EP1654263 (prior. 24/06/2003, US).

Merck2004a Ferlita, R.; Hansen, K.; Vydra, V.; Wang, Y.; Lindemann, C. (Merck & Co Inc) WO2005072530, US2008227786, EP1708571 (prior. 16/01/2004, US).

Merck2004b Xiao, Y.; Sun, Y.; Rosner, T.; Rivera, N.; Krska, S.; Clausen, A.; Armstrong, J.; Spindler, F.; Malan, C. (Merck & Co Inc) WO2005097733, US2008058522, US7495123, EP1735269 (prior. 05/04/2004, US).

Merck2004c Ferlita, R.; Wenslow, R. (Merck & Co Inc) WO2006033848, US2007281941, US7612072, EP1796671 (prior. 15/09/2004, US).

Merck2005a Xiao, Y.; Armstrong, J.; Krska, S.; Njolito, E.; Rivera, N.; Sun, Y.; Rosner, T.; Clausen, A. (Merck & Co Inc) WO2006081151, EP1856028 (prior. 24/01/2005, US).

Merck2005b Amatruda, J.; Fong, T.; Moller, D.; Thornberry, N. (Merck & Co Inc) WO2006119260, US2009306037, EP1879582 (prior. 02/05/2005, US).

Merck2005c Thornberry, N.; Kaufman, K. (Merck & Co Inc) US2006270722 (prior. 31/05/2005, US).

Merck2005d Hasegawa, P. (Merck & Co Inc, Merck Sharp & Dohme Corp) WO2007050485, US2009156579, EP1942921 (prior. 25/10/2005, US).

Merck2011 Atwood, J. (Merck Sharp & Dohme Corp) WO2012166420, US2014336196 (prior. 27/05/2011, US).

MSN2009 Satyanarayana Reddy, M.; Eswaraiah, S.; Satyanarayana, R.; Kondal Reddy, B.; Srinivas, A. (MSN Lab Ltd) IN201001084, WO2010122578 (prior. 20/04/2009, IN).

Mylan2009 Gore, V. G.; Gadakar, M.; Bhosle, P.; Shinde, S. (Generics Uk Ltd, Mylan India Private Ltd) IN200900722, WO2010131025, US8846916, EP2430025 (prior. 11/05/2009, IN).

Nanjing2014a Luo, Y.; Ding, S.; Qu, X. (Nanjing Abiochem Biomedical Technology Co Ltd) CN105018440 (prior. 24/04/2014, CN).

Nanjing2014b Luo, Y.; Ding, S.; Qu, X. (Nanjing Abiochem Biomedical Technology Co Ltd) CN105219745 (prior. 12/06/2014, CN).

Nanjing2015 Luo, Y.; Ding, S.; Qu, X.; Guo, F. (Nanjing Abiochem Biomedical Technology Co Ltd) CN104805069 (prior. 06/02/2015, CN).

Nanjing2017 Li, W. (Heading Nanjing Pharmtechnologies Co Ltd) CN107540575 (prior. 13/10/2017, CN).

NanjingBP2015 Li, W. (Heding Nanjing Medicine Technology Co Ltd) CN105017081 (prior. 08/07/2015, CN).

NanjingUT2011 Xu, H.; Shi, C. (Univ Nanjing Technology) CN102153559 (prior. 28/02/2011, CN).

OBC2010 Tasnadi, G.; Forro, E.; Fulop, F. *Organic & Biomolecular Chemistry* **2010**, *8*, 793-799.

OL2005 Balsells, J.; DiMichele, L.; Liu, J.; Kubryk, M.; Hansen, K.; Armstrong, J. *Organic Letters* **2005**, *7*, 1039-1042.

OL2015 Gutierrez, O.; Metil, D.; Dwivedi, N.; Gudimalla, N.; Chandrashekar, E.; Dahanukar, V. H.; Bhattacharya, A.; Bandichhor, R.; Kozlowski, M. C. *Organic Letters* **2015**, *17*, 1742-1745.

OPRD2005 Hansen, K. B.; Balsells, J.; Dreher, S.; Hsiao, Y.; Kubryk, M.; Palucki, M.; Rivera, N.; Steinhuebel, D.; Armstrong, J. D.; Askin, D.; Grabowski, E. J. J. *Organic Process Research & Development* **2005**, 9, 634-639.

OPRD2006 Clausen, A.; Dziadul, B.; Cappuccio, K.; Kaba, M.; Starbuck, C.; Hsiao, Y.; Dowling, T. *Organic Process Research & Development* **2006**, *10*, 723-726.

OPRD2016 Dunetz, J.; Magano, J.; Weisenburger, G. *Organic Process Research & Development* **2016**, *20*, 140-177.

Panacea2010 Jain, R.; Rao, J.; Rao, S. (Panacea Biotec Ltd) IN201002164, WO2012035549 (prior. 13/09/2010, IN).

S2010 Savile, C.; Janey, J.; Mundorff, E.; Moore, J.; Tam, S.; Jarvis, W.; Colbeck, J.; Krebber, A.; Fleitz, F.; Brands, J.; Devine, P.; Huisman, G.; Hughes, G. *Science* **2010**, 329, 305.

SC2013 Lin, K.; Cai, Z.; Zhou, W. *Synthetic Communications* **2013**, *43*, 3281-3286.

Shandong2011a Zhang, G.; Zhao, X.; Zhang, L.; Chen, G. (Shandong Boyuan Chem Co Ltd, Shandong Platinum Source Pharm Co Ltd) CN102391142 (prior. 30/09/2011, CN).

Shandong2011b Chen, G.; Zhang, G.; Zhang, L.; Zhao, X. (Shandong Boyuan Chem Co Ltd, Shandong Platinum Source Pharm Co Ltd) CN102503829 (prior. 08/10/2011, CN).

Shanghai2011 Zhao, C.; Yin, J.; Wu, Y.; Tao, H.; Wei, F.; Chen, Y. (Shanghai Sun Sail Pharmaceutical Science & Technology Co Ltd) CN103172539 (prior. 23/12/2011, CN).

ShanghaiIPI2013 Lin, K.; Cai, Z.; Pan, J.; Zhou, W.; Wu, G.; Yue, L.; Shen, D. (Zhejiang Med Xinchang Pharm; Shanghai Inst Pharm Industry) CN104418861, WO2015035735, US2016229859, US9745309, EP3045459 (prior. 10/09/2013, CN).

ShanghaiIT2014 Liu, F.; Pan, X.; Wang, K.; Yu, W.; Zhou, Y.; Zhang, R.; Xu, N.; Xu, L.; Chen, Y. (Shanghai Inst Technology) CN104193744 (prior. 15/07/2014, CN).

Shenzen2013a Yao, Z.; Zhang, W.; Liu, J.; Ma, Y.; Yuan, J. (Shenzhen Hybio Pharmaceutical) CN104447374 (prior. 17/09/2013, CN).

Shenzen2013b Yao, Z.; Zhang, W.; Liu, J.; Ma, Y.; Yuan, J. (Shenzhen Hybio Pharmaceutical) CN104447753 (prior. 17/09/2013, CN).

Smilax2012 Mallela, S.; Havale, S.; Boddepalli, N.; Gunji, N.; Mokkapati, B. (Smilax Lab Ltd) WO2013114173 (prior. 30/01/2012, IN).

ST2011 Lim, G.; Chang, S.; Byeon, C. (St Pharm Co Ltd) KR2012084622, WO2012099381, US2014005394, US9006436, EP2665728 (prior. 20/01/2011, KR).

Stereokem2014 Janagani, S.; Thaduri, V.; Vamaraju, R. (Stereokem Inc Usa) WO2015120111, US2017183351, US9938282, EP3102580 (prior. 05/02/2014, US).

Suli2017 Liang, C.; Wang, J. (Suli Pharmaceutical Tech Jiangyin Co Ltd) CN106831783 (prior. 10/01/2017, CN).

Sun2010 Jadav, K.; Bhatt, R.; Borkhataria, K.; Chitturi, T.; Thennati, R. (Sun Pharm Ind Ltd) WO2011135586, US2013041150, US8471016, EP2566843 (prior. 28/04/2010, IN).

Sun2014 Azad, A.; Pandey, G.; Singh, K.; Prasad, M. (Sun Pharm Ind Ltd) WO2015145333 (prior. 26/03/2014, IN).

Suqian2017a Shi, L.; Chen, J.; Gong, S.; Xu, C.; Du, J.; Chen, X.; Chen, Y.; Zhu, W.; Yang, Y.; Ma, J.; Zhou, D. (Suqian Aerfa Tech Co Ltd) CN106995807 (prior. 27/04/2017, CN).

Suqian2017b Shi, L.; Chen, J.; Gong, S.; Xu, C.; Chang, B.; Zhong, Z.; Liu, X.; Pang, X.; Peng, Y.; He, Y.; Liu, S. (Suqian Aerfa Tech Co Ltd) CN106995808 (prior. 27/04/2017, CN).

Suven2014 Arava, V.; Siripalli, U.; Gorentla, L.; Jasti, V. (Suven Life Sciences Ltd) WO2015162506 (prior. 21/04/2014, IN).

Suzhou2010 Song, T.; Li, M.; Zeng, D. (Suzhou Leader Chem Co Ltd) CN101824037 (prior. 18/03/2010, CN).

Suzhou2012 Zhang, S. (Suzhou Xinkai Biomedical Technology Co) CN102702205 (prior. 16/05/2012, CN).

Syncozymes2015 Zhu, W.; Gao, X.; Wang, B. (Syncozymes Shanghai Co Ltd) CN105331651 (prior. 25/11/2015, CN).

Synlett2013 Bao, H.; Bayeh, L.; Tambar, U. *Synlett* **2013**, *24*, 2459-2463.

T2012 Fistikci, M.; Gundogdu, O.; Aktas, D.; Secen, H.; Sahin, M. F.; Altundas, R.; Kara, Y. *Tetrahedron* **2012**, *68*, 2607-2610.

T2016 Hou, A.; Deng, Z.; Ma, H.; Liu, T. *Tetrahedron* **2016**, *72*, 4660-4664.

TA2006 Kubryk, M.; Hansen, K. B. *Tetrahedron: Asymmetry* **2006**, *17*, 205-209.

TA2014 Subbaiah, C.; Haq, W. *Tetrahedron: Asymmetry* **2014**, *25*, 1026-1030.

TA2015 Dey, S.; Sudalai, A. *Tetrahedron: Asymmetry* **2015**, *26*, 67-72.

TA2017 Kang, S. K.; Cho, G. H.; Leem, H. J.; Soh, B. K.; Sim, J.; Suh, Y. G. *Tetrahedron-Asymmetry* **2017**, *28*, 34-40.

Teva2007a Niddam-Hildesheim, V. (Teva Pharm Ind Ltd, Teva Pharm USA Inc) WO2009045507, US2009123983 (prior. 03/10/2007, US).

Teva2007b Abramov, M.; Etinger, M.; Niddam-Hildesheim, V.; Perlman, N. (Teva Pharm Ind Ltd, Teva Pharm USA Inc) WO2009064476, US2009192326 (prior. 13/11/2007, US).

TL2012 Davies, S.; Fletcher, A.; Lv, L.; Roberts, P.; Thomson, J. *Tetrahedron Letters* **2012**, *53*, 3052-3055.

TL2013 Pan, X.; Li, X.; Lu, Q.; Yu, W.; Li, W.; Zhang, Q.; Deng, F.; Liu, F. *Tetrahedron Letters* **2013**, *54*, 6807-6809.

TL2016 Dey, S.; Gadakh, S.; Ahuja, B.; Kamble, S.; Sudalai, A. *Tetrahedron Letters* **2016**, *57*, 684-687.

Tongren2016 Luo, H.; Chen, S.; Ma, Y.; Hou, H. (Univ Tongren) CN106748888 (prior. 22/11/2016, CN).

UniTris2010 Zhang, F.; Tao, W.; Sun, P.; Li, P.; Liu, B. (Unitris Biopharma Co Ltd, Shanghai Yuanli Biotech Co Ltd) CN102212015, WO2011127794 (prior. 12/04/2010, CN).

USV2010 Sathe, D.; Damle, S.; Arote, N.; Ambre, R.; Sawant, K.; Naik, T.; Want, K. (Usv Ltd) WO2012025944, US2013158265, EP2609099 (prior. 27/08/2010, IN).

Wuxi2013 Zhang, H.; Zhang, S.; Bian, W. (Wuxi Biortus Biosciences Co Ltd) CN103483340 (prior. 29/07/2013, CN).

Xinfa2015a Qi, Y.; Ju, L.; Niu, W.; Zhang, M.; Chen, J.; Li, X. (Xinfa Pharmaceutical Co Ltd) CN105061434 (prior. 30/07/2015, CN).

Xinfa2015b Qi, Y.; Ju, L.; Zhang, M.; Chen, J.; Li, X. (Xinfa Pharmaceutical Co Ltd) CN104987338 (prior. 30/07/2015, CN).

Zentiva2010 Richter, J.; Jirman, J. (Zentiva KS) CZ201000198, WO2011113399, EP2547683 (prior. 16/03/2010, CZ).

Zentiva2012 Richter, J.; Halama, A.; Jirman, J. (Zentiva KS) WO2014086325 (prior. 04/12/2012, CZ).

ZhejiangCP2014 Xia, J.; Li, J.; Lai, X.; Zhou, Z.; Zhang, N. (Zhejiang Changming Pharmaceutical Co Ltd; Univ Zhejiang Technology) CN104130149 (prior. CN20141344250 20140718, CN).

ZhejiangHP2009 Pan, X.; Zhang, Q. (Zhejiang Haizheng Pharm Co Ltd) CN102093245 (prior. 14/12/2009, CN).

ZhejiangHP2010 Pan, X.; Li, W.; Zhang, Q.; Ruan, L.; Yu, W.; Deng, F.; Ma, T.; Huang, M.; He, M. (Zhejiang Hisoar Pharm Co Ltd, Zhejiang Haizheng Pharm Co Ltd) CN102485718, WO2012072036, US2013281695, US9187419, EP2647624 (prior. 03/12/2010, CN).

ZhejiangHP2011a Ge, J.; Yan, F. (Zhejiang Huahai Pharm Co Ltd) CN102603749 (prior. 27/10/2011, CN).

ZhejiangHP2011b Fan, Q.; Ruan, H.; Zhu, W. (Zhejiang Huahai Pharm Co Ltd, Shanghai Aobo Bio-Pharm Technology Co Lt) CN102757431 (prior. 24/04/2011, CN).

ZhejiangHP2011c Ge, J.; Yan, F. (Zhejiang Huahai Pharm Co Ltd) CN102363599 (prior. 09/10/2011, CN)

ZhejiangHP2011d Wang, L.; Zhang, Q.; Pan, X.; Ruan, L.; Yu, W.; Peng, X.; Deng, F.; Ma, T.; Liu, F. (Zhejiang Haizheng Pharm Co Ltd) CN102838511 (prior. 24/06/2011, CN).

ZhejiangHP2014 Hong, J.; Wang, Y.; Wang, B.; Huang, L.; Gu, E. (Zhejiang Huahai Pharm Co Ltd; Shanghai Syncores Technologies Inc Ltd) CN105566138, CN106458853, WO2016055015, US2017305822, EP3156391 (prior. CN201410532331 20141011, CN).

ZhejiangJP2009 Zhu, G.; Zhang, J.; Yang, L.; Yao, Q.; Ying, J. (Zhejiang Jiuzhou Pharm Co Ltd) CN102030683, WO2011035725, US2012178957, US8471057, EP2481722 (prior. 27/09/2009, CN).

ZhejiangJP2010 Gao, H.; Li, M. (Zhejiang Jiuzhou Pharm Co Ltd) CN102199102, WO2011116686, US2013012735; EP2551259, US8710255 (prior. 25/03/2010, CN).

ZhejiangJP2015a Zhu, G.; He, Q.; Qian, L.; Xu, Y.; Yan, P. (Zhejiang Jiuzhou Pharm Co Ltd) CN106146340 (prior. 15/04/2015, CN).

ZhejiangJP2015b Zhu, G.; He, Q.; Qian, L.; Xu, Y.; Yan, P. (Zhejiang Jiuzhou Pharm Co Ltd) CN106146514 (prior. 15/04/2015, CN).

ZhejiangNC2014 Zhang, Y.; Yao, J.; Qian, H.; Lin, C.; Lu, G.; Tang, J.; Shen, C. (Zhejiang Nhu Co Ltd) CN103819475 (prior. 11/02/2014, CN).

ZhejiangUT2013 Zhang, X.; Lyu, C.; Sun, X.; Shen, F. (Univ Zhejiang Technology) CN103755596 (prior. 30/09/2013, CN).

ZhejiangUT2015 Wang, P.; Huang, J.; Ding, X. (Univ Zhejiang Technology) CN104893989 (prior. 29/05/2015, CN).

ZhejiangUT2016 Zhang, X.; Xie, X. (Univ Zhejiang Technology) CN105968030 (prior. 10/05/2016, CN).

6 Ezetimibe

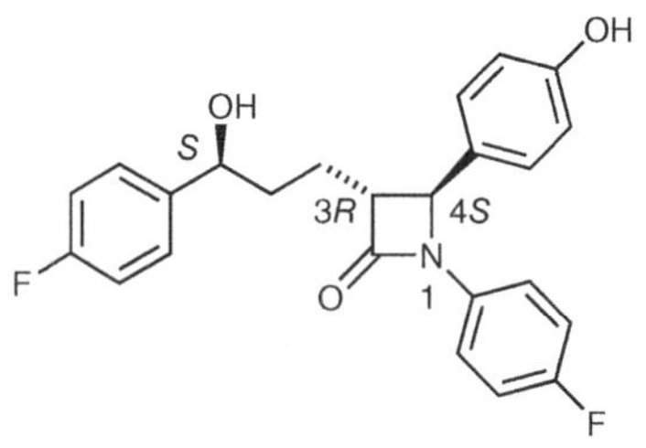

Year of discovery 1993 (Schering Corporation)
US approval (Zetia®, Ezetrol®, Schering-Plough & Merck) October 2002
Drug cat.: Antihypolipidemic agent/selective cholesterol absorption inhibitor
Uses: For the reduction of high cholesterol levels (LDL, low-density lipoprotein) as a monotherapy, or in combination with statins

Cholesterol is a vital structural component of all animal cell membranes, essential for maintenance of structural integrity and fluidity. Cholesterol also serves as a precursor for the biosynthesis of steroid hormones, bile acids and vitamin D. Hypocholesteraemia has been shown to have a high correlation with the development of cardiovascular disease and appropriate control of the amount of cholesterol in the body is essential for long-term human health. Clinical trials and extensive epidemiological studies support the proposal that reduction of the levels of low-density plasma lipoproteins (LDLs) is a major goal in the treatment and prevention of coronary heart disease. Regulation of cholesterol levels is complex, the steady state amount of cholesterol resulting from a balance of internal biosynthesis, dietary intake, metabolism and clearance (mainly in the form of oxidized sterols – the bile acids). The accumulation of cholesterol, a typical consequence of the modern western diet, is one of the root causes of high cholesterol levels. Whilst a healthy dietary regime is the preferred option to maintain low cholesterol levels, reducing cholesterol absorption is an attractive strategy for therapy.

An early phase discovery program under investigation at Schering-Plough in the 1990's targeted non-fused azetidin-2-ones (β-lactams) in a search for new monobactams. Most of these monocyclic β-lactam compounds were devoid of antibiotic activity but several showed significant inhibitory activity in a cholesterol-fed hamster model. One of the first promising prototypes was SCH48461 (Scheme 6.1). Blocking potential sites for metabolism in these prototypes gave rise to the design of new and more potent drugs. A series of analogues were prepared incorporating strategically placed hydroxyl groups and fluorine atoms to enhance oxidation to the final active drug and to block sites of potential detrimental/deactivation metabolic oxidation in the molecule. Through a SAR analysis SCH58235 (Ezetimibe) was targeted and experimentally found to exhibit remarkable efficacy. Hence, Ezetimibe at an ED50 of 0.04 mg/kg/day showed reduction of liver cholesteryl esters in a 7-day cholesterol-fed hamster model (JMC1998).

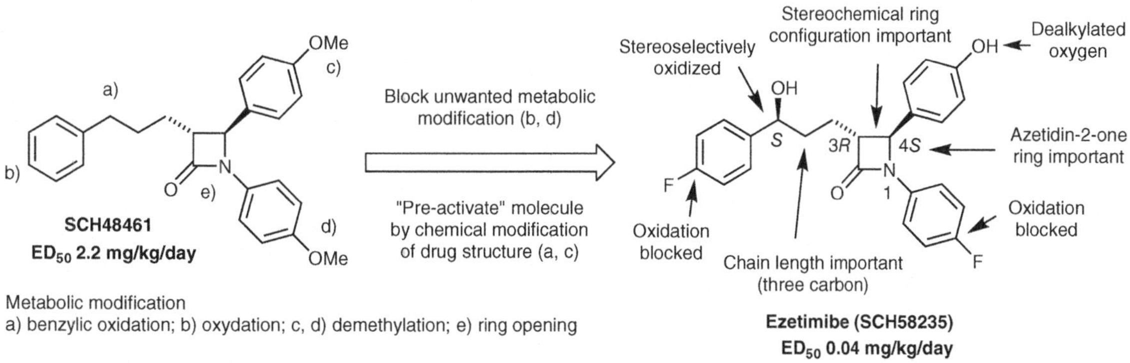

Scheme 6.1 SAR from SCH48461 to Ezetimibe

Retrosynthesis in the Manufacture of Generic Drugs: Selected Case Studies, First Edition, Pedro Paulo Santos and William Heggie.

Ezetimibe selectively inhibits the absorption of dietary cholesterol in the intestines by specifically binding to a protein that transports cholesterol through the intestinal wall. Several naturally occurring plant sterols can also inhibit the absorption of cholesterol but show significantly less potency than Ezetimibe.

This approach can be applied in the clinic as a complement to the attenuation of *in vivo* cholesterol biosynthesis, for example by the use of HMG-CoA reductase inhibitors (Figure 6.1). Ezetimibe is marketed as a monotherapy but has also been approved as combination therapy with statins (e.g. example: Ezetimibe/Simvastatin, Vytorin®) since their cholesterol-lowering effects are additive. Ezetimibe is also recommended as second line therapy for patients that are intolerant to statins or unable to achieve target LDL cholesterol levels by statins therapy alone.

The discovery and development of more efficacious, well-tolerated agents and combinations which regulate cholesterol homeostasis by complementary mechanisms of action remains of great interest.

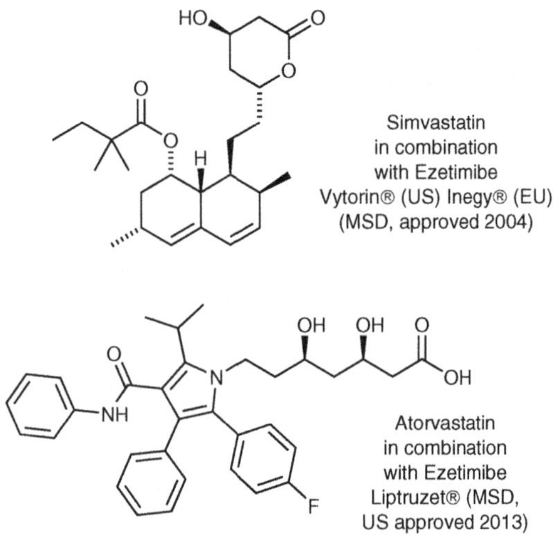

FIGURE 6.1 HMG-CoA REDUCTASE INHIBITORS (STATINS) APPROVED FOR HUMAN USE IN COMBINATION DRUGS WITH EZETIMIBE

6.1 WORLDWIDE SALES AND PATENT STATUS

The following graphs illustrate how research carried out in the pharmaceutical industry towards the production of the API and worldwide sales of the finished drug product evolved.

- US basic patent term 16/06/2015 (USRE37721, US5631365, Schering1993), patent term extension 25/10/2016.
- Annual production[1]: 2019, 33,292 kg; 2018, 29,334 kg; 2017, 26,585 kg.
- Annual WW sales[1]: 2019, 2390.7M USD; 2018, 2663.5M USD; 2017, 4326.9M USD.

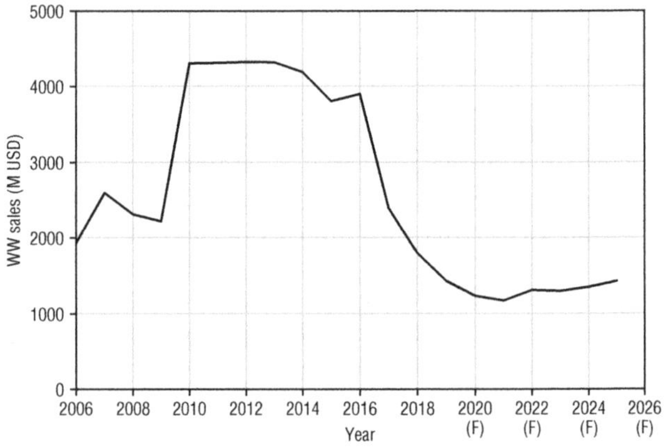

FIGURE 6.2 EZETIMIBE WW SALES EVOLUTION (IN M USD)

Source: Data from GlobalData [Online] https://globaldata.com/

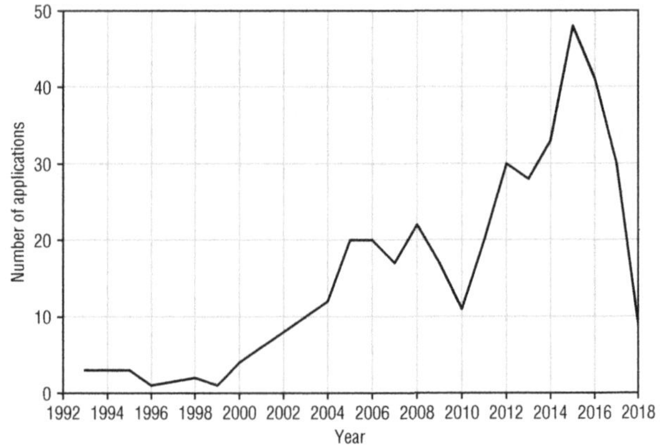

FIGURE 6.3 NEW PATENT APPLICATIONS ON EZETIMIBE FILED BY YEAR

1 Newport Modular Platform – Clarivate Analytics [Online] https://clarivate.com/cortellis/ [2020, April]

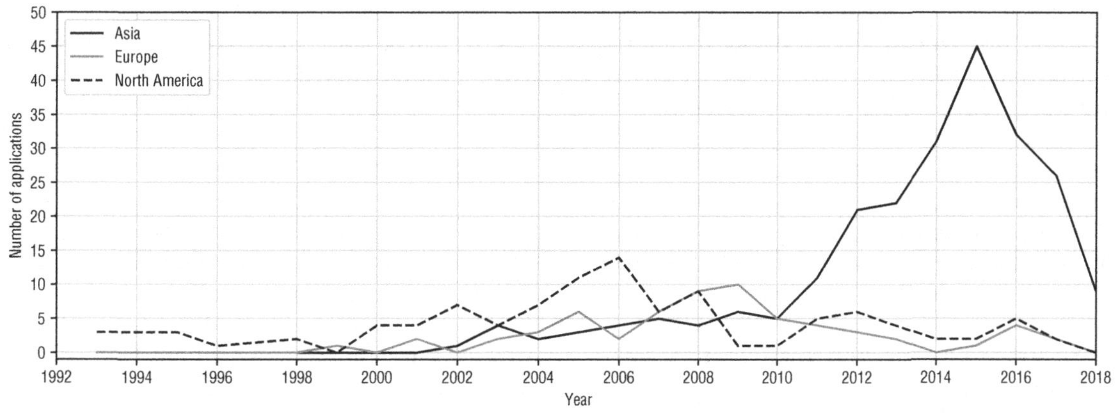

FIGURE 6.4 PATENT APPLICATIONS ON EZETIMIBE FILED BY REGION

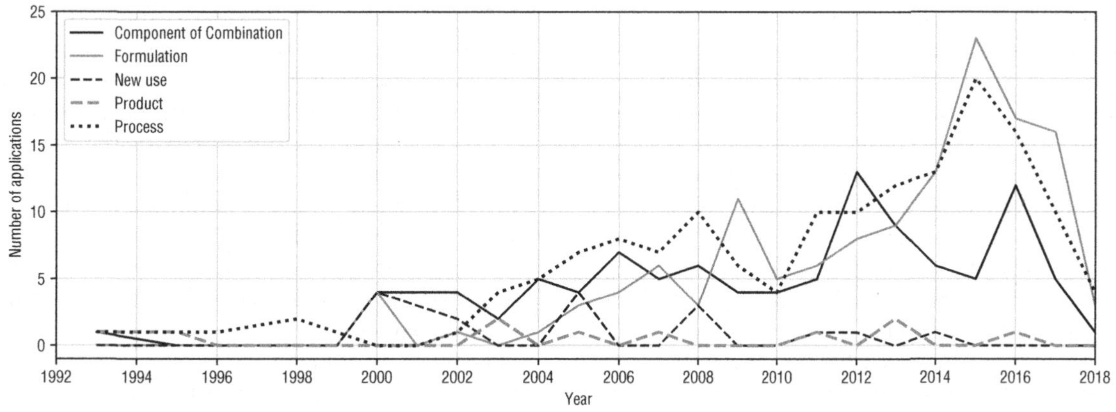

FIGURE 6.5 PATENT APPLICATIONS ON EZETIMIBE FILED BY TYPE OF PATENT

6.2 EZETIMIBE STRUCTURE AND GENERAL RETROSYNTHETIC ANALYSIS

Ezetimibe is an enantiopure azetidin-2-one (β-lactam) with three *p*-substituted aromatic rings. Two fluorine atoms block the oxidation of two of the aromatic rings. Some key features of Ezetimibe are highlighted in Figure 6.6, several of which are extremely important for the design of a reliable and efficient synthetic process.

- Ezetimibe is a single enantiomer and has three chiral centres, two of them at the sp^3 carbon atoms of the four-membered ring. The remaining chiral atom is at the benzylic carbon of the side chain.
- The synthesis of the β-lactam ring is central in any Ezetimibe synthetic process. *Trans* isomers are clearly more stable than the *cis*, but only one of the *trans* arrangement must be produced and the build-up of a *trans* 3*R*,4*S* isomer is critical. Isomerization at C3, α to the carbonyl atom, should be possible allowing a *cis* to *trans* epimerization. Appropriate planning should allow for the selective formation of the correct configuration, thus avoiding the use of chromatographic separation, which is, as always, a major challenge for any synthetic

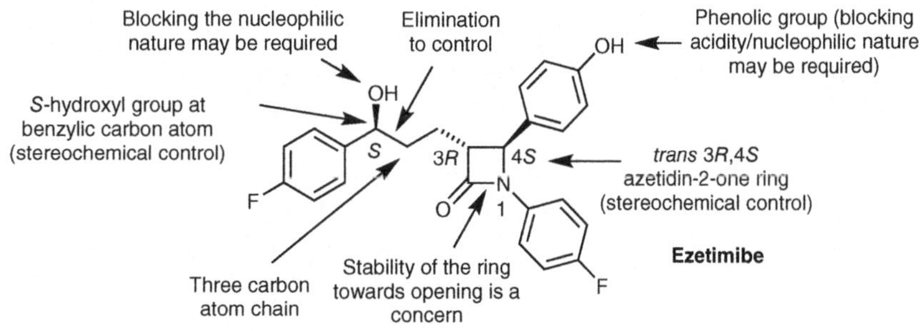

FIGURE 6.6 KEY FEATURES OF EZETIMIBE STRUCTURE

route. As is true for any β-lactam compound, avoidance of ring opening by C-N cleavage is a significant test for any synthetic route.

- The third chiral centre, at the benzylic position, ideally ought to be produced selectively and an asymmetric reduction of a carbonyl derivative should prove to be a good option. Whether this reduction be carried out before or after β-lactam ring formation is a subject to be analysed during process selection. It is to be expected that selective formation of the benzylic hydroxyl group is the second major challenge to be addressed during route selection to arrive at a viable process for the production of Ezetimibe.
- The benzylic hydroxyl group has the potential to generate a variety of side products. Elimination of water and benzylic reduction are two plausible occurrences (Scheme 6.2).
- Because of the lability and reactivity of the hydroxyl function, the use of protecting groups may be necessary to avoid side reactions when this group is present in an intermediate undergoing chemical transformation. However, protecting group strategies are to be avoided whenever possible, as these entail two "non-productive" steps. When **absolutely** necessary, these protecting group strategies should involve the use of the cleanest, cheapest and most accessible reagents, whilst using the simplest operating conditions in order to reduce any impact on final costs.
- Starting materials, which have the fluorine atoms already incorporated, are the primary option because of their ready availability and because, in general, the fluorine atoms are stable under a variety of conditions. However, under reductive conditions, the fluorine atom can be labile and hence, such reactions should be carefully monitored.

6.2.1 The azetidin-2-one ring formation

The formation of the four-membered ring is central to any synthesis of Ezetimibe. Azetidin-2-ones are a very important class of compounds due to their well-known biological activity and also because of their utility in organic synthesis. Several methods have been developed for ring formation, mainly [2+2]-cycloadditions, intramolecular cyclization reactions, carbene insertion reactions and rearrangement of heterocyclic compounds (CR2008).

Two types of disconnections are possible for [2+2]-cycloaddition (Scheme 6.3) and both options have proven to be useful. The imine-ketene combination (**a**), known as the Staudinger [2+2] ketene-imine cycloaddition, is clearly favoured as the precursors are easily accessible and the regioselectivity of the ring formation is unambiguous.

The Staudinger synthesis, in which a ketene reacts with a carbon-nitrogen double bond of an imine, is probably the most important synthetic tool to access β-lactams and its mechanism and stereoselectivity has been intensively studied (Scheme 6.4).

The reaction can be performed thermally using an acid chloride-trialkylamine base combination or an α-diazoketone as ketene precursors, the former combination is easier to use in large scale batches. The formation of

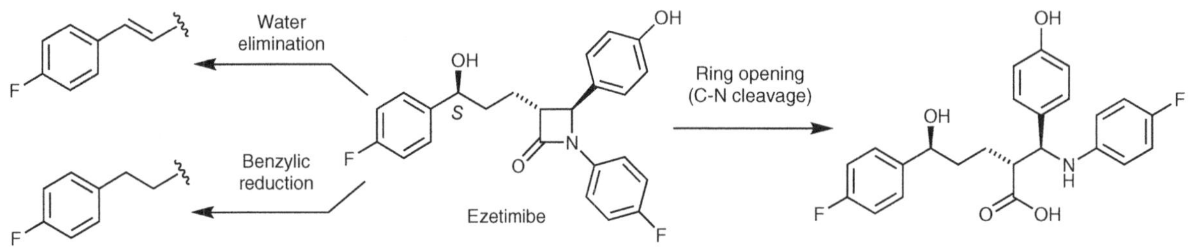

SCHEME 6.2 SOME PLAUSIBLE IMPURITIES DERIVING FROM THE STRUCTURE OF EZETIMIBE

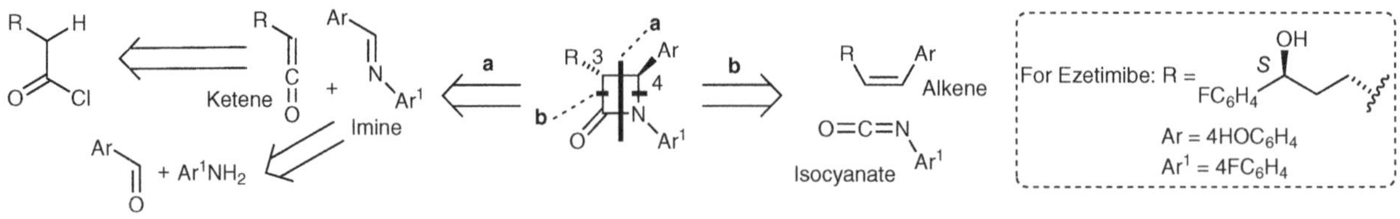

SCHEME 6.3 DISCONNECTION OF THE AZETIDIN-2-ONE RING

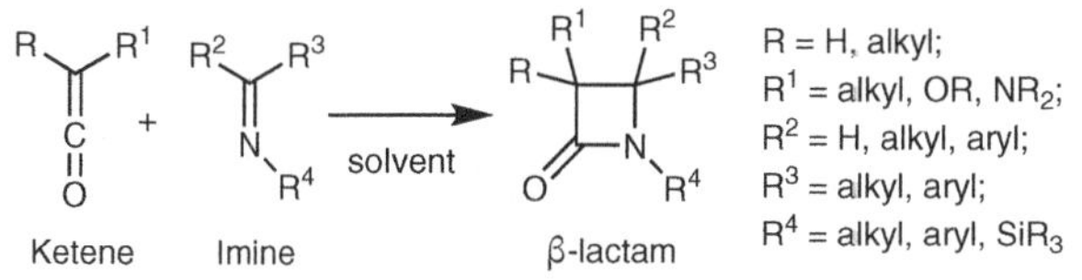

Scheme 6.4 The Staudinger [2+2]-ketene-imine cycloaddition

the highly reactive ketene is of prime importance and is almost always carried out *in situ* under mild conditions (–10 °C to room temperature). Obviously, the hydroxyl groups of the final Ezetimibe molecule must, either be protected, or introduced post cyclization to avoid reaction between the alcohol and ketene. An aryl substituent on the nitrogen atom stabilizes the imine and makes a positive contribution to the efficiency of the cycloaddition. A C-monosubstituted imine gives two new stereogenic centres at C3 and C4 of the azetidine-2-one ring when the reaction is applied in the synthesis of Ezetimibe. The reaction follows a stepwise mechanism *via* a zwitterionic intermediate, which undergoes an electrocyclic conrotatory ring closure to give the β-lactam ring. *Ab initio* calculations substantiated the presence of a zwitterionic intermediate and the electrocyclic conrotatory ring closure, which is ostensibly the rate-determining step (JACS1992). The established mechanism, (CR2008, Scheme 6.5), predicts that when the ketene and the imine have each one substituent on the relevant carbon atom and that on the imine is able to stabilize a positive charge, a *trans* four-membered ring is the preferred product.

Initial *exo* addition is followed by rapid rotation, made possible by the weakening of the double bond character of the CN bond by the stabilization of the positive charge on the carbon atom, and subsequent cyclization, gives the more stable *trans* substituted azetidin-2-one. Obviously, without additional control, a mixture of enantiomers of the *trans* product is obtained. At the discovery stage of drug development, this is desirable as it provides a quick access to both enantiomers for use in biological studies. However, for routine industrial production this is undesirable.

Asymmetric versions of the cycloaddition were developed by employing chiral auxiliaries attached to the imine or the ketene (EJOC1999). Asymmetric catalytic methods with chiral amine bases were also developed. A completely stereoselective cycloaddition to generate the Ezetimibe ring would have considerable benefit, but has been difficult to achieve.

Another approach to β-lactams is *via* the Kinugasa reaction. In this reaction a copper acetylide, formed by the reaction of a Cu catalyst and a terminal alkyne, undergoes a [3+2] dipolar cycloaddition with a nitrone (Scheme 6.5), which is followed by rearrangement to produce an azetidine-2-one (see EJOC2013 for a short review). Kinetic control usually favours the *cis*-3,4-disubstituted lactam, but epimerization to the more stable *trans* isomer can be effected. However, as in the Staudinger cycloaddition, formation of a single enantiomer is difficult to achieve.

The condensation of an enamine and an enolate, derived from an α-bromoester and a metal such as zinc (c.f. the Reformatsky reaction) is another useful synthetic approach to β-lactams (Scheme 6.7, JACS1943). The Gilman-Speeter reaction employs various metals other than zinc: lithium; aluminium; tin; boron; zirconium and titanium enolates having been described, although most of these approaches give the β-aminoesters rather than the azetidin-2-one.

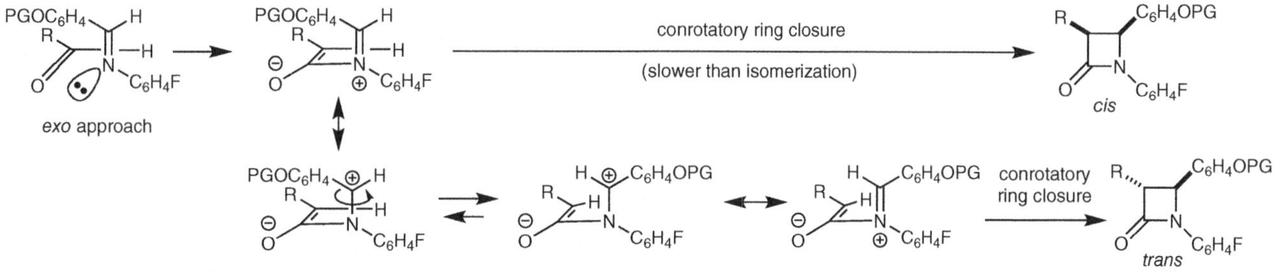

Scheme 6.5 Staudinger [2+2]-ketene–imine cycloaddition in the synthesis of Ezetimibe

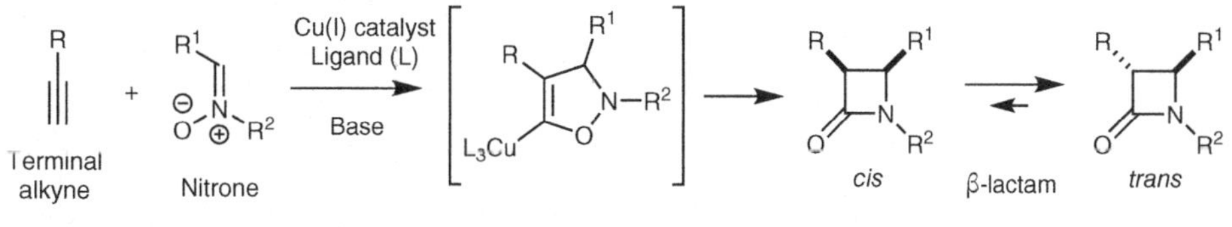

Scheme 6.6 General outline of the Kinugasa reaction

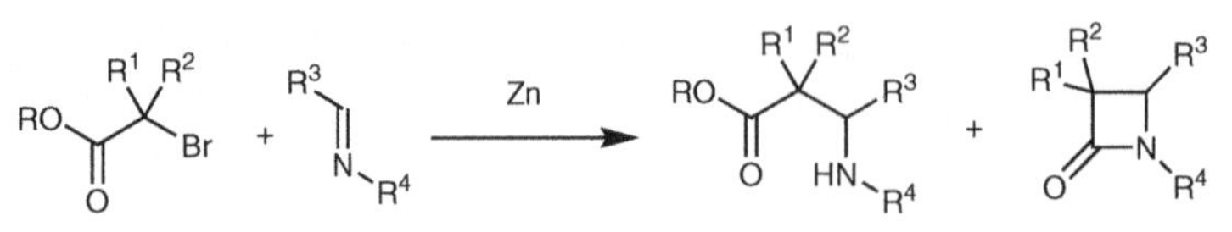

SCHEME 6.7 GENERAL OUTLINE OF A GILMAN-SPEETER REACTION

A common approach to produce the azetidin-2-one ring is cyclization of a β-amino carboxylic acid, derived from the ester or an alternative derivative (Scheme 6.8). The cyclization can be achieved using conventional amide coupling reagents such as the Mukaiyama reagent (2-chloro-1-methylpyridinium iodide) or carbodiimides.

Although other strategies can be envisioned as approaches to the synthesis of the Ezetimibe ring the four shown above are expected to prove most useful.

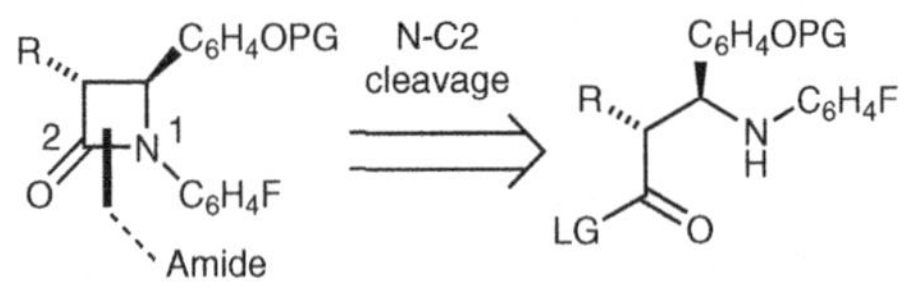

SCHEME 6.8 β-LACTAM SYNTHESIS BY INTRAMOLECULAR AMIDE FORMATION

6.2.2 Key structural features for the retrosynthesis

Apart from stereoselective formation of the lactam ring, three other features are particularly important in the design of any synthesis of Ezetimibe (Figure 6.7).

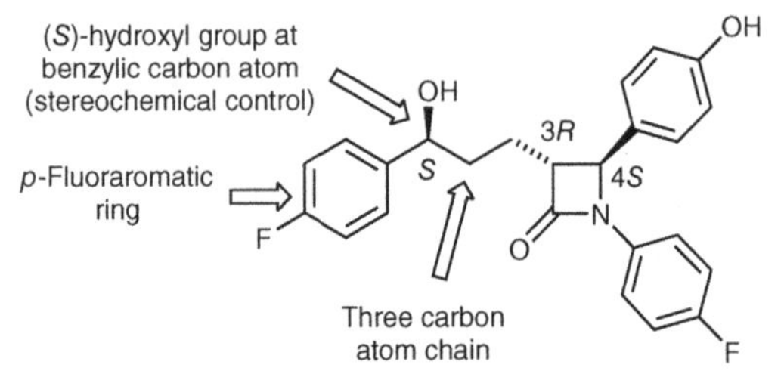

FIGURE 6.7 IMPORTANT STRUCTURAL FEATURES (IN ADDITION TO THE LACTAM) FOR THE SYNTHESIS OF EZETIMIBE

- The asymmetric benzylic hydroxyl group is likely to be produced by reduction of a benzylic carbonyl group, either by a catalytic chemical process (for example by Corey-Bakshi- Shibata oxazaborolidine reduction or by a metal mediated catalytic asymmetric hydrogen transfer) or an enzymatic process.
- The *p*-fluoroaromatic ring can be attached using the terminal oxygen functionality of the side chain as a handle. A range of options can be foreseen, from an electrophilic aromatic substitution to Grignard-type additions to carbonyl groups.
- Several strategic options can be anticipated for the three-carbon atom chain. This fragment can be obtained directly from a starting material or be produced by chain elongation.

In summary, four main tasks must be undertaken to construct the Ezetimibe molecule: stereoselective β-lactam synthesis, benzylic hydroxyl formation, *p*-fluoroaromatic ring attachment and installation of the three-carbon atom chain. Depending on the manner in which the β-lactam is constructed, the order in which the other components are assembled can vary.

6.2.3 Retrosynthesis using the Staudinger cycloaddition for the azetidin-2-one synthesis

Several retrosynthetic pathways are conceivable from Ezetimibe using the Staudinger cycloaddition to produce the azetidin-2-one ring (Scheme 6.9). A completely selective Staudinger reaction would appear to be inaccessible, meaning this reaction has its drawbacks for any industrial application. A chiral separation, with concomitant loss of 50% of product, is anticipated after formation of the azetidin-2-one ring; hence, it is desirable to place the Staudinger cyclisation at the beginning of the synthesis. Thus, large quantities of materials, which are destined to be discarded, are not carried forward into subsequent reactions, resulting in savings on staring materials and better volume efficiency. Accordingly, strategy **a)** should be avoided, **b)** and **c)** being better choices. Nonetheless, strategy **a)** looks to be an expedited way to Ezetimibe in this paper chemistry approach.

Placing the ring formation at the beginning of the route is a concern and requires mild reaction conditions in the steps following formation of the β-lactam. Wherever harsh conditions are required, at least partial hydrolysis of the strained four-member ring would ensue with consequent loss of yield and an increase in the number of impurities.

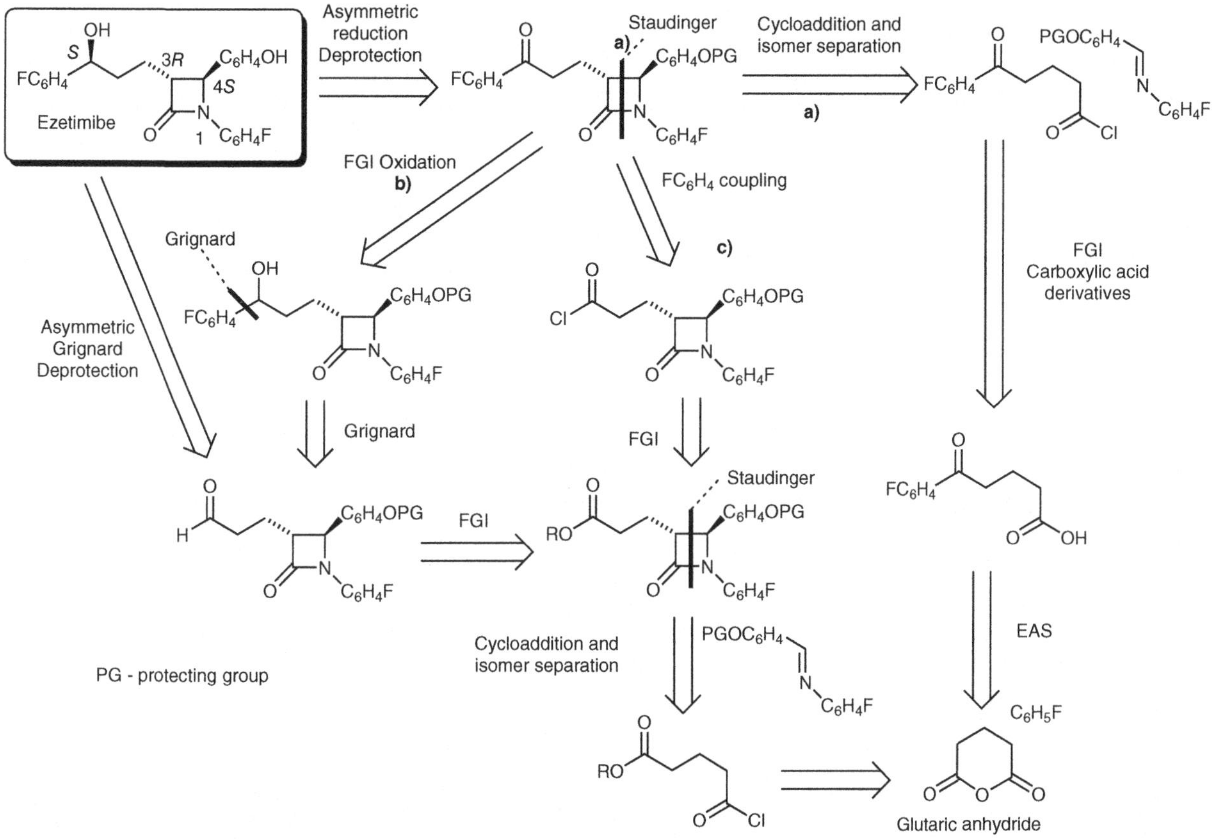

Scheme 6.9 Retrosynthesis of Ezetimibe using a Staudinger cyclisation for ring disconnection

In addition to the inherent instability of the β-lactam ring, latent lability at the asymmetric centres may present problems. Acidic and basic conditions can potentially cause epimerization at the carbon atom α to the carbonyl group. However, since the *trans* isomer is more stable and equilibrium conditions favour the *trans* 3*R*,4*S* isomer, the quantity of the unwanted 3*S* diastereomer would likely be kept under control (Scheme 6.10).

Direct attachment of the *p*-fluoroaromatic ring to the 3-carbon chain by nucleophilic addition to the carbonyl group of an ester is, predictably, difficult to achieve and the efficiency expected to be low because of the tendency for a second nucleophilic addition to the initially formed ketone. To circumvent this problem, several options are foreseeable, for example making use of a Weinreb amide or an aldehyde (Scheme 6.11). The addition of a relatively unreactive nucleophile to a reactive acyl chloride is also an option worth considering.

Asymmetric nucleophilic addition to the aldehyde would be an elegant solution and would produce Ezetimibe directly, avoiding the oxidation-asymmetric reduction sequence to introduce the (*S*)-benzylic hydroxyl group.

6.2.4 Retrosynthesis using the Kinugasa reaction to produce the azetidin-2-one

The Kinugasa Cu(I)-mediated cycloaddition-rearrangement cascade, starting with a nitrone and a terminal alkyne is an appropriate alternative to produce the azetidin-2-one ring. Moderate enantio and diastereoselectivities were often observed experimentally, whilst some chiral copper catalysts produced good yields and selectivities. *Cis* isomers are frequently the major products however; a non-nucleophilic base causes epimerization to the more stable *trans* isomer. The outline of a retrosynthetic analysis based on the Kinugasa reaction to produce the β-lactam is shown below (Scheme 6.12). An efficient cyclization to produce the correct C4 *S* configuration is key to the successful implementation of this reaction.

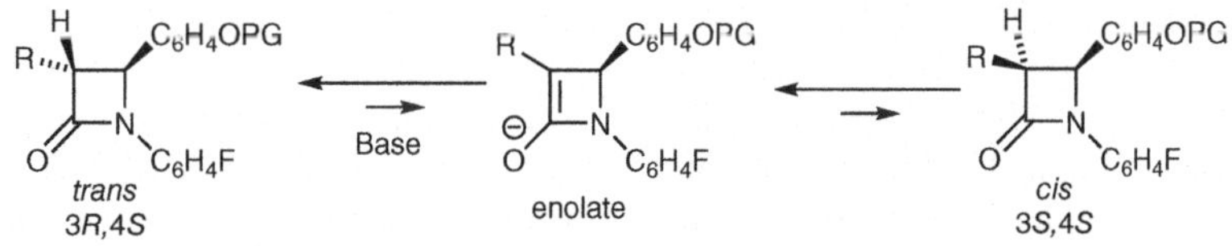

Scheme 6.10 Isomerization of C3 centre by enolization

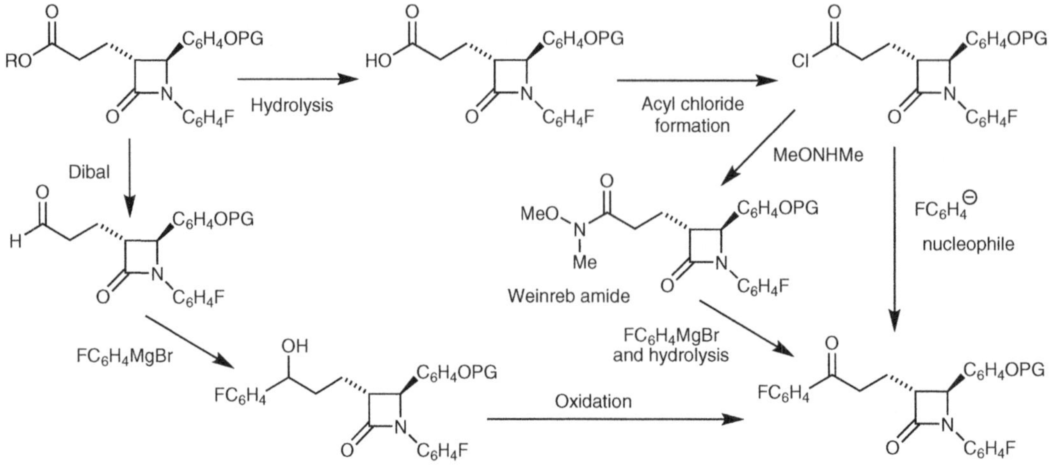

SCHEME 6.11 MAIN PROCESSES TO COUPLE THE *p*-FLUOROAROMATIC RING TO AN ESTER INTERMEDIATE

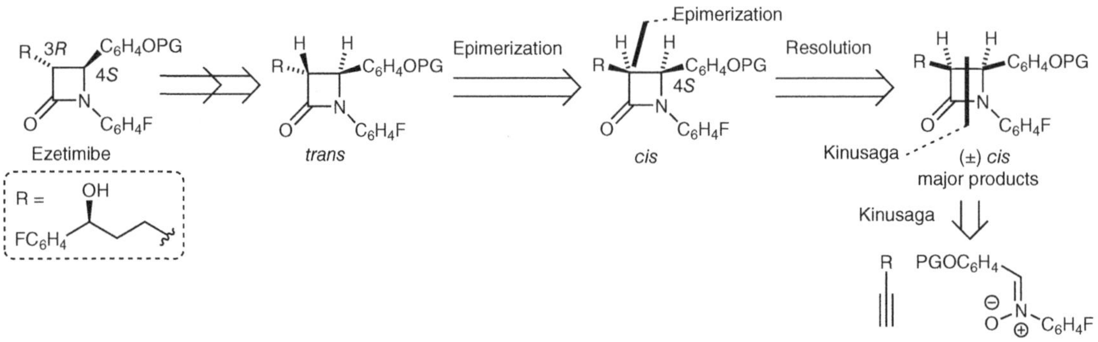

SCHEME 6.12 OUTLINE OF A RETROSYNTHETIC PATHWAY OF EZETIMIBE USING A KINUGASA REACTION FOR THE β-LACTAM FORMATION

6.2.5 Retrosynthesis using a N to C2 ring closure to produce the azetidin-2-one

An N to C2 ring closure strategy is an undoubtedly a pertinent option to take into consideration. A retrosynthesis based on an N-C2 ring cleavage is shown in Scheme 6.13. From this, it is clear that an enantiopure open chain precursor is mandatory for an efficacious synthesis.

The enantioselective enolate addition of the open chain carboxylic acid derivative to the imine CN double bond is key to the success of this strategy. The use of a Gilman-Speeter type reaction to access to the β-amino-amide and the use of a chiral auxiliary would, in theory, produce the two chiral centres on addition to the CN double bond. Following Evans pioneering use of chiral oxazolidinones in asymmetric alkylation (PAC1981, AA1982), this tool has become widespread in several separate diastereoselective synthetic applications, among them: α-alkylation; aldol reaction; Michael addition and Diels-Alder cyclisation. Several chiral auxiliaries of this type were subsequently developed and many of these are now easily accessible. Most of the oxazolidinones are conveniently prepared from the chiral pool, amino acids being the major source. An example is shown below using an Evans oxazolidine auxiliary produced from widely available and cheap L-valine (*S*) or from (+)-norephedrine (Scheme 6.14).

Since the most versatile chiral auxiliaries are available as both enantiomers, both stereoisomers can be produced, depending on the stereochemistry of the auxiliary. An important feature of this strategy is that the chiral auxiliary can be cleaved and recycled. Frequently the diastereoselectivity of these alkylations is very high (>95:5). The ee can normally be improved by a classic crystallization albeit with concomitant loss of product.

During the 90's asymmetric aldol additions using Evans chiral auxiliaries were the subject of intense synthetic and mechanistic study because of their synthetic importance. The configuration of the intermediate enolate (*E* or *Z*) and the Lewis acid used for its formation are crucial to the selectivity observed for the products (*syn*/*anti* ratio and diastereoselectivity). (*Z*)-Boron enolates with aldehydes give the *syn* aldol product whilst (*E*)-boron enolates lead to formation of the *anti* aldol product with high diastereoselectivity (Scheme 6.15).

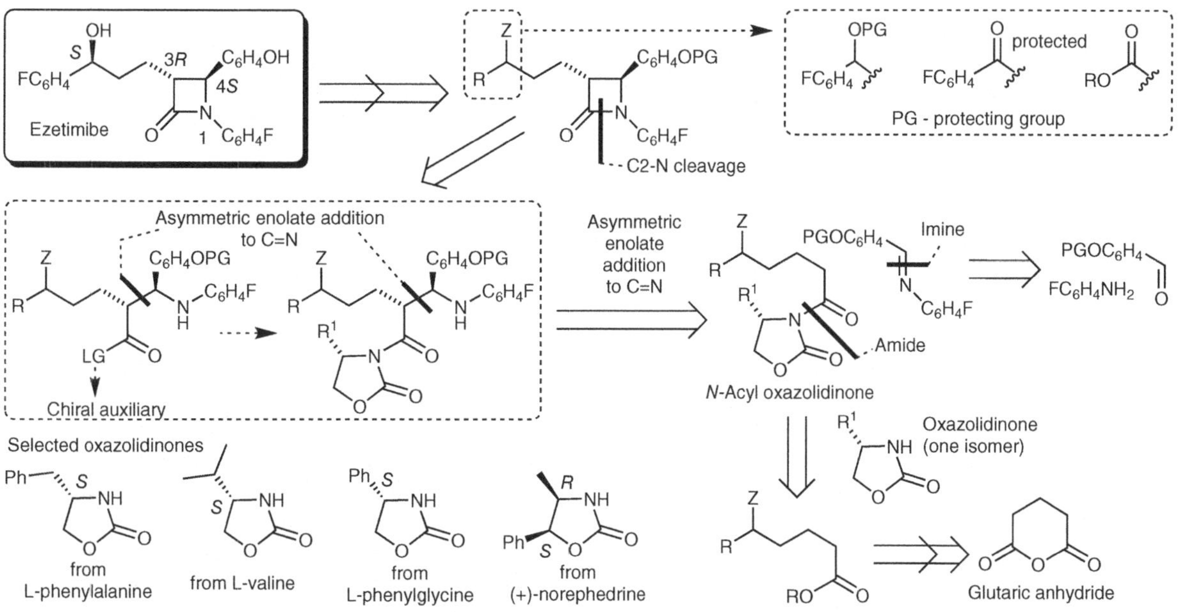

Scheme 6.13 Retrosynthesis of Ezetimibe by a N-C2 ring cleavage

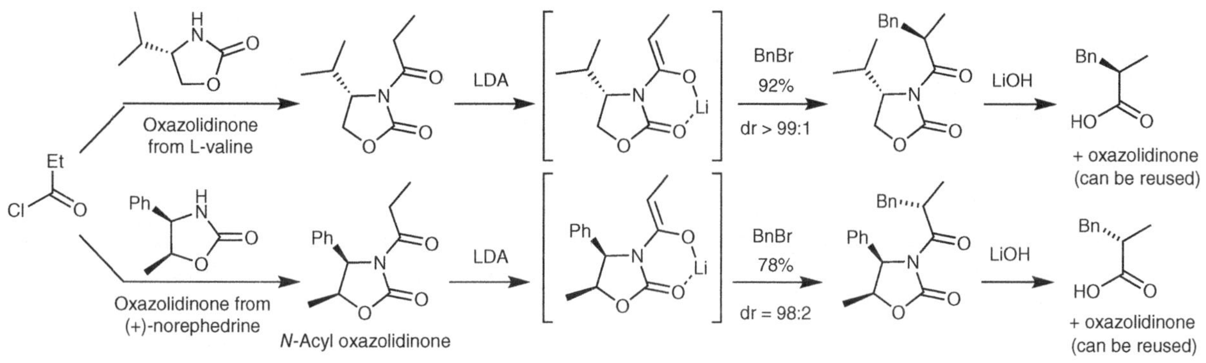

Scheme 6.14 Example of a diastereoselective enolate alkylation using an Evans'-type auxiliary

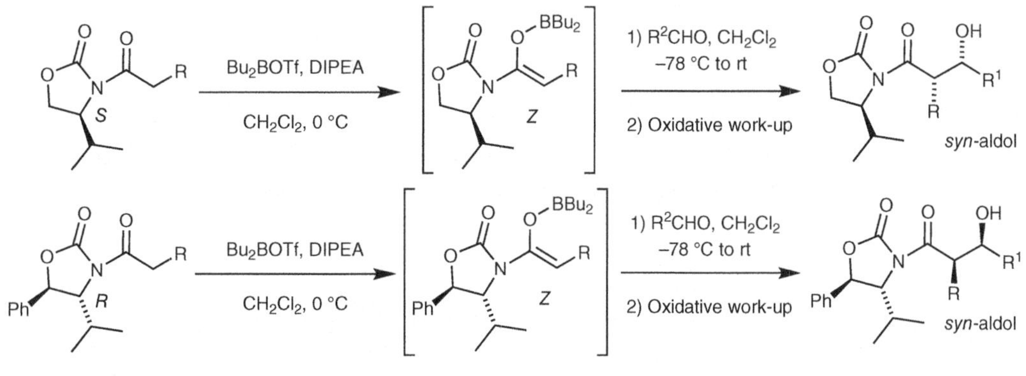

Scheme 6.15 Outline of a diastereoselective Evans aldol condensation

When applied to enolate-imine addition the stereochemistry of the two chiral centres at C3 and C4 of the azetidin-2-one can be predicted (Scheme 6.16, T1994).

Various terminal groups can be present on the 3-carbon chain when addition to the CN bond is undertaken. Esters will not intervene during the reaction and it is

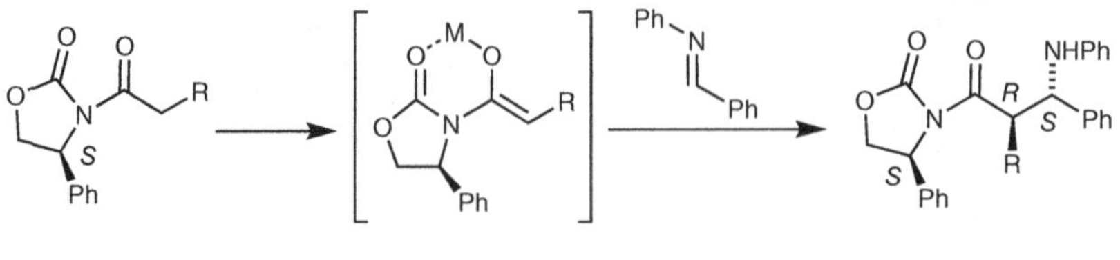

Scheme 6.16 Outline of a diastereoselective enolate-imine addition

possible to introduce the *p*-fluoroaromatic moiety after formation of the azetidine-2-one. When the *p*-fluoroaromatic group is already in place, a benzylic carbonyl or hydroxyl group will normally also be present. However, the presence of a reactive benzylic group may be detrimental to a regiospecific ring closure and this fact has to be taken in consideration when the synthetic process is designed.

6.2.5.1 The nature of the benzylic chain group and the enolate addition to C=N and cyclization steps

As alluded to above, when a benzylic function (ketone or alcohol) is present during enolate-imine addition, protection will be necessary. In the case of the ketone where undesirable enolization must be avoided introduction of a ketal protecting group, stable under the conditions of the reaction, is mandatory (Scheme 6.17). The protection of the carbonyl group also circumvents any unwanted intramolecular attack of the product amine nitrogen on the ketone.

An intermediate containing a benzylic alcohol requires protection to block the acidity and nucleophilic nature of the of the hydroxyl group. Silyl groups are ideal protecting groups in these situations (Scheme 6.18).

As discussed above, benzylic hydroxyl or carbonyl groups must be protected to avoid side reactions. Since the protecting groups are already in place at the alkylation stage, it is important that any unit operation thereafter, i.e. during work-up, does not cause even partial deprotection. If this were to occur an unnecessary re-protection step would be required. Finding adequate conditions for ring closure is also key in this approach.

Since the carbonyl group of the *N*-acyl group of the oxazolidinone is not particularly electrophilic the β-nitrogen atom requires activation to close the β-lactam ring. Non-nucleophilic bases are ideal for this purpose as they only provoke removal of the nitrogen proton and do not cause ring opening of the β-lactam once it is formed. However, isomerization is a real concern as the carbon atom α to the acyl carbonyl group (or of the β-lactam) contains an acidic proton, giving rise to racemization under the reaction conditions (Scheme 6.19). Fortunately, the configuration of the remaining chiral centre(s) should not be affected.

Other methods to activate the nitrogen nucleophile to close the β-lactam ring exist. One approach relies on the formation of *N*-silyl derivatives. In summary, the N-C2 ring closure appears to be particularly suitable to synthesis

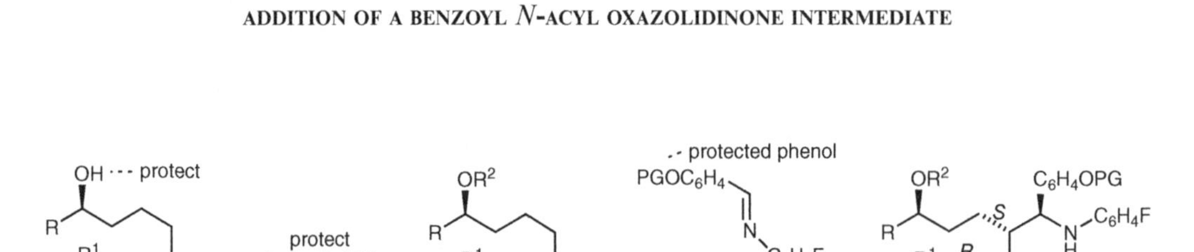

Scheme 6.17 Proposed sequence for the diastereoselective enolate-imine addition of a benzoyl *N*-acyl oxazolidinone intermediate

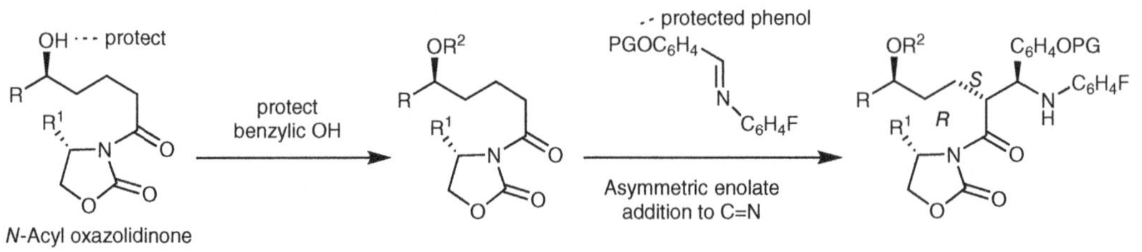

Scheme 6.18 Proposed sequence for a diastereoselective enolate-imine addition of a *N*-acyl oxazolidinone intermediate with a benzylic alcohol group

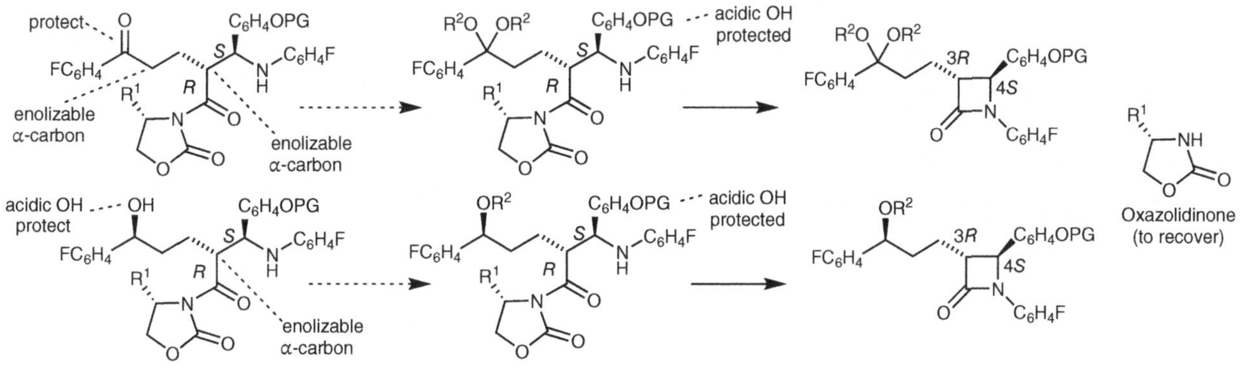

Scheme 6.19 Cyclization of the β-aminoamides

the central ring of Ezetimibe provided that the diastereoselective enolate-imine addition is efficient.

6.2.6 The (*S*)-hydroxyl benzylic group of Ezetimibe

The benzylic (*S*)-hydroxyl group can be produced *via* a number of options, allowing for its introduction at several stages of the synthesis. Indeed, exactly when the (*S*)-benzylic alcohol is created has a significant bearing on route selection. In the following analysis, disconnection of Ezetimibe centred on the benzylic hydroxyl group will be depicted. The most likely method of generating the chiral benzylic hydroxyl is by an asymmetric reduction of a ketone (Scheme 6.20). However, there are several other options providing alternative approaches for retrosynthetic analysis. Amongst these, asymmetric epoxidation-rearrangement or asymmetric addition to an aldehyde present interesting methodology.

Radical halogenation is probably not the most indicated approach to introduce functionality at the benzylic position as this method does not create asymmetry (Scheme 6.21).

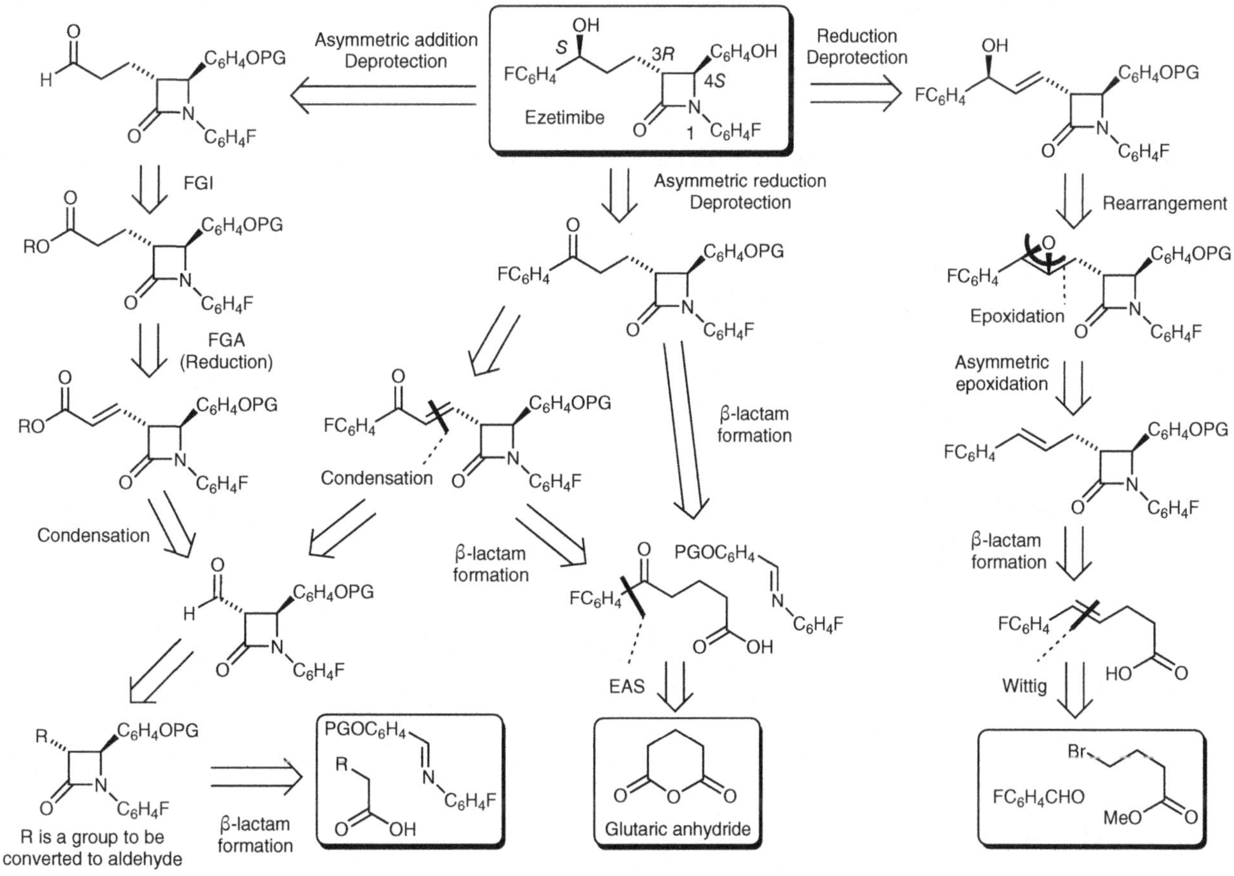

Scheme 6.20 Retrosynthesis of Ezetimibe initially focusing on the benzylic alcohol

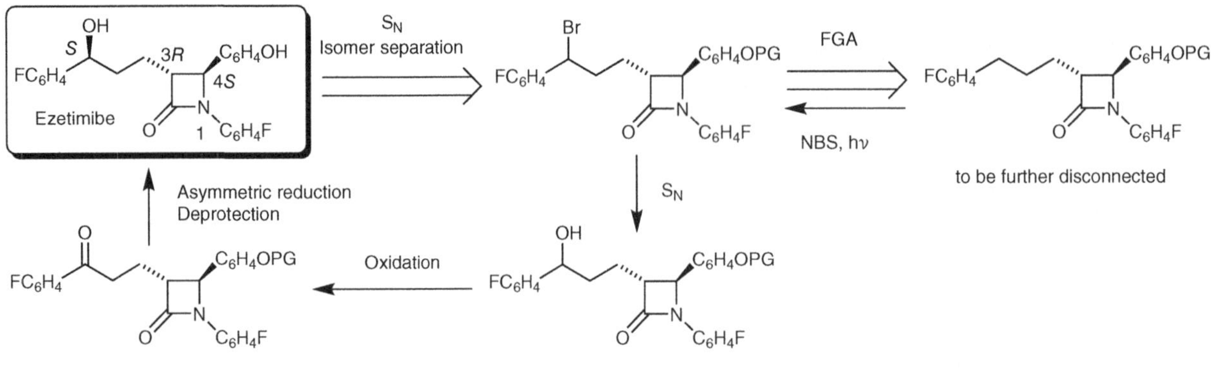

Scheme 6.21 Retrosynthesis of Ezetimibe introducing the benzylic hydroxyl group *via* oxidation

In addition, elimination to a conjugated olefin is expected to be a significant side reaction during substitution by hydroxide and additional steps are necessary to reach chirally pure Ezetimibe.

6.2.7 Summary of the retrosynthesis analysis

Ideally, each of the three chiral centres of Ezetimibe should be assembled in a selective manner. The benefits from a truly asymmetric synthesis are greater efficiency and a significant reduction of wastes. A corollary of this is that resolution-based processes are to be avoided as in addition to the fact that efficacy is low; recycling of unwanted isomers would appear to be extremely difficult.

With respect to the formation of the β-lactam, diastereoselective enolate-imine addition to give a β-amino amide followed by cyclization is the most promising option. Hence, a combination of high selectivity in the formation of a β-amino acid derivative (amide) and an efficient ring closure is a key target for development.

Attachment of the fluoroaromatic ring should be carried out before the β-lactam ring is produced as the conditions to introduce this fragment have the potential to open the lactam ring.

Likewise, the opportunity to introduce the (*S*)-benzylic alcohol using an asymmetric reduction of a carbonyl group makes the two strategies shown in Scheme 6.22 particularly promising. It is to be noted that all five carbon atoms of glutaric anhydride are elegantly incorporated into the Ezetimibe molecule in this approach.

The *S* configuration of the benzylic alcohol can be created by known asymmetric reducing methods. Amongst the most prominent of these are the Corey-Bakshi-Shibata (CBS) protocol, an asymmetric transfer hydrogenation process or a biochemical-enzymatic transformation. The positioning of the reduction step in the synthetic sequence requires careful analysis as it has considerable

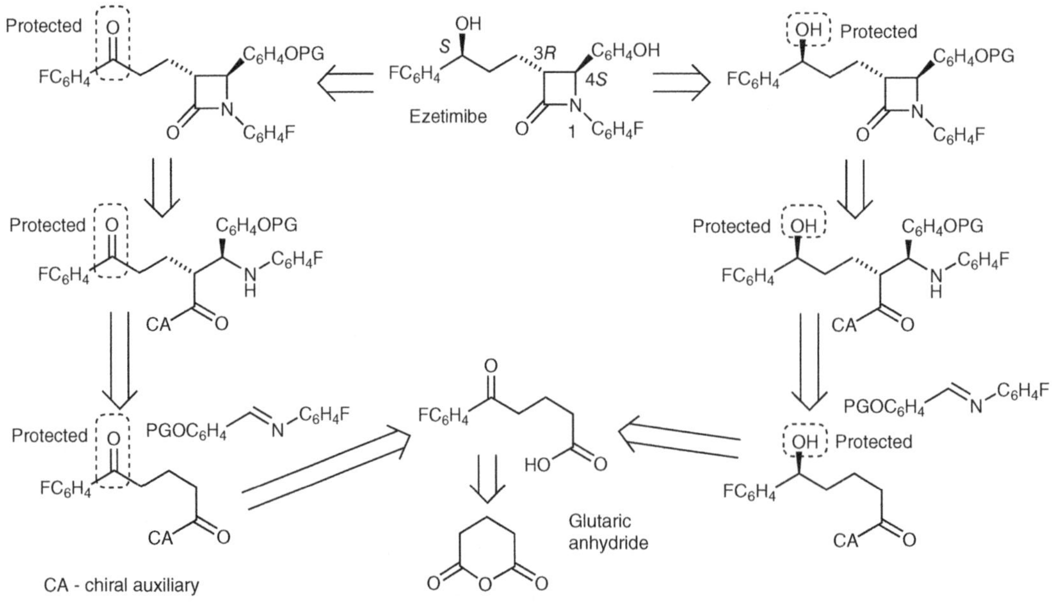

Scheme 6.22 Most promising retrosynthetic pathways for an industrial synthesis of Ezetimibe

influence on the choice of protecting group for whatever benzylic functionality is present and will, inevitably, influence the whole synthesis. Although the use of protecting groups has distinct inconveniencies, the other possible routes demonstrate diverse problems that make them less attractive. Hence, the use of a protecting group strategy ought to be regarded as a minor drawback.

Finally, it should be noted that evaluation of the different synthetic routes can only be made after initiation of development work.

6.3 THE FIRST EZETIMIBE SYNTHESES

Ezetimibe was first patented in 1993 by Schering (Schering1993, priority date 21/09/1993) but the discovery route was only reported five years later (JMC1998). That first synthesis is shown below (Scheme 6.23). Notably, this synthetic route was flexible enough to produce several analogues for biological evaluation. The non-selective nature of several of the steps, namely the Staudinger β-lactam formation and the ketone reduction permitted isolation of the various stereoisomers of Ezetimibe whilst changing the nature of the organozinc nucleophile gave easy access to a series of analogues.

Clearly, from several points of view, the synthesis is not viable as an industrial process. Chiral chromatography is used twice, the first to separate the two *trans* β-lactam enantiomers resulting from the Staudinger cycloaddition. The non-selectivity of the ring formation is a significant shortcoming of this route as the racemic mixture is difficult to resolve using conventional procedures. The β-lactam has neither acidic nor basic properties, thus separation *via* formation of diastereomeric salts is not an option. The penultimate step produces a diastereomeric mixture, also separated by the use of chiral chromatography (although a non-chiral stationary phase could potentially be used). This less than ideal situation could theoretically be resolved by applying an asymmetric reduction of the carbonyl group. This route produces a large amount of waste material and the use of chromatography drastically limits the scale of production, although more recent techniques such as Simulated Moving Bed chromatography might help meliorate this bottleneck. A total of three synthetic steps are required to convert the terminal methyl ester to the *p*-fluroaromatic ring ketone, the first of these being a chemoselective hydrolysis of the ester, which has the potential to hydrolyse the β-lactam ring resulting in a reduction in yield and purity. A palladium-mediated arylzinc-acid chloride cross-coupling Negishi reaction is used to attach the aromatic unit and is the last of these three reactions. The final product is reported to be produced in high enantiomeric purity but a scalable stereospecific synthesis based on this sequence would require considerable development work to make it suitable for industrial use.

Although the route shown above was the first approach to Ezetimibe, the first patent (Schering1993) reported a different synthesis (Scheme 6.24). The principal starting material, methyl glutaryl chloride, was easily derived from glutaric anhydride. Schering, following their expertise acquired in studies of similar molecules, used (*S*)-4-phenyloxazolidin-2-one as chiral auxiliary in the asymmetric addition of an enolate to the CN double bond in the key synthetic step.

A titanium containing Lewis acid ($TiCl_3Oi$-Pr) in the presence of a tertiary amine base, was used to improve

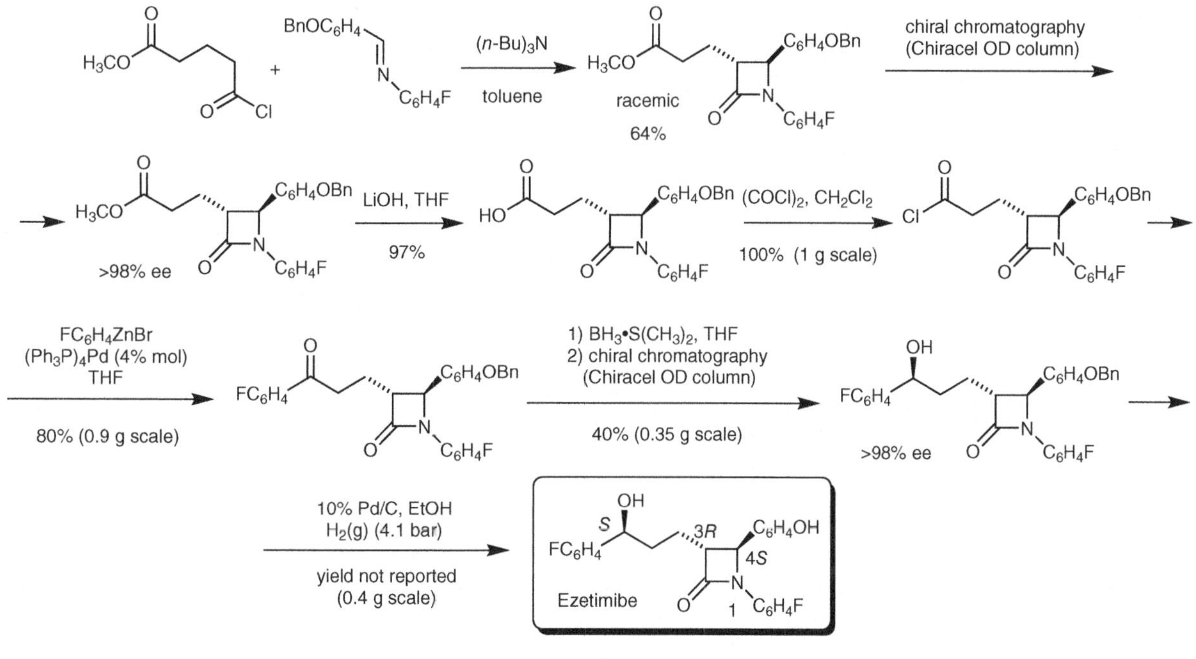

SCHEME 6.23 FIRST SYNTHESIS OF EZETIMIBE

Source: JMC1998

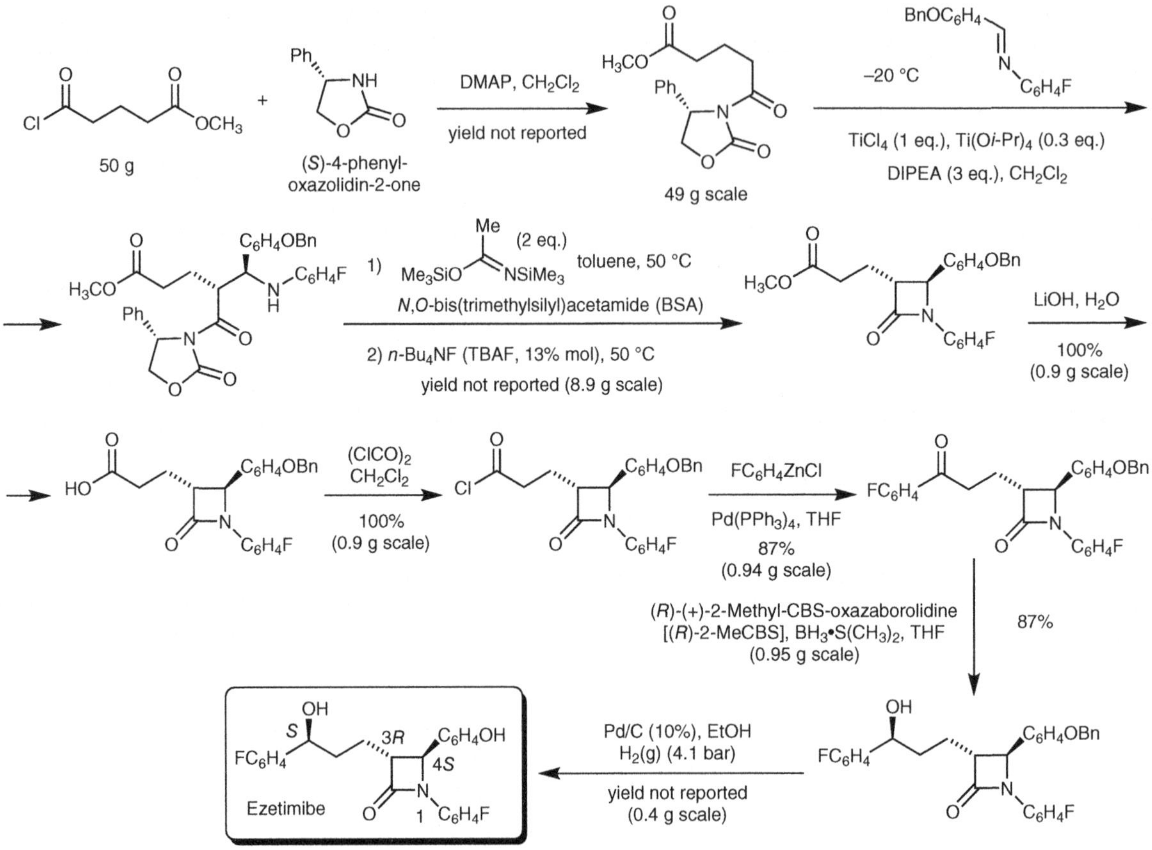

Scheme 6.24 Ezetimibe synthesis as described in the first patent

the diastereoselectivity of the reaction by coordinating with the oxygen atoms of the carbonyl groups, thus bestowing rigidity to the substrate (Scheme 6.25). A simple crystallization produced the intended isomer without the requirement for chromatography.

The β-lactam ring was produced *via* a BSA-TBAF mediated N-C2 cyclization, thus avoiding the use of a strong base and circumventing racemization of the α-carbon atom. The use of a mild-silylating agent at 50 °C effects *N*-silylation and cyclization is accomplished by treatment with a catalytic amount of a source of fluoride ion (TBAF). The asymmetric reduction of the carbonyl group was achieved by the CBS protocol without formation of the unwanted isomer (yield not reported).

This represents a huge improvement over the first synthesis with regard to stereoselectivity and operating procedures. However, the yields for several steps in this synthesis were not reported. Although it is to be assumed that problems still remain, the use of a chiral auxiliary, whilst not ideal, can be considered an effective solution and it is to be expected that it can be recovered from the cyclization step and recycled.

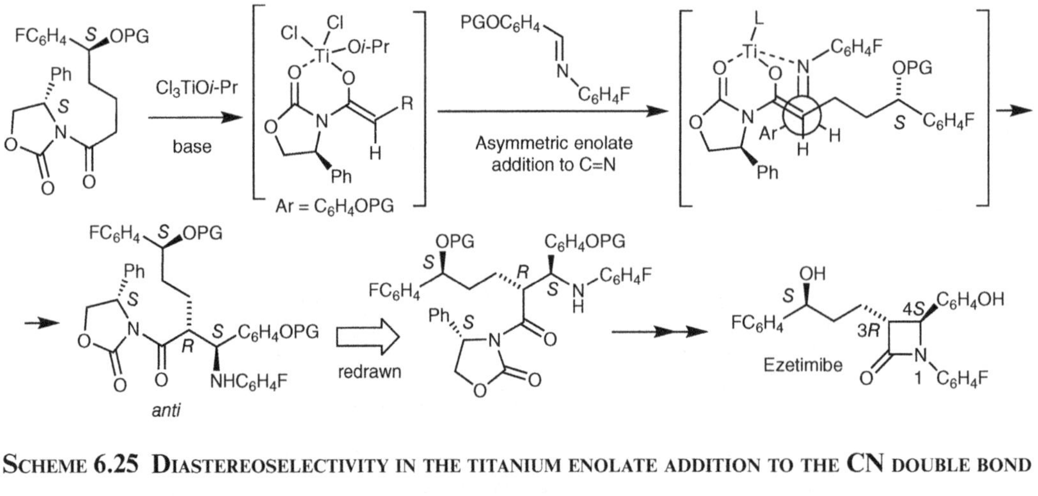

Scheme 6.25 Diastereoselectivity in the titanium enolate addition to the CN double bond

The terminal fluoroaromatic ring is introduced *via* an unstable acyl chloride as in the initial synthesis and it may be that this step imparts a negative impact on the yield and quality of the final product. Once more, the hydrolysis of the ester step is of some concern because of to the inherent instability of the β-lactam ring.

6.4 THE SEARCH FOR AN INDUSTRIAL SYNTHESIS OF EZETIMIBE

Due to the commercial success of Ezetimibe, many different companies filed a considerable number of process patents. The academic community has also published some alternative synthetic approaches. One prominent effort focused on the development of direct formation of the asymmetric ring (Schering1996b). However, the majority of the patented processes opted to build the core β-lactam ring by an N-C2 cyclization.

6.4.1 The synthesis of the C5 chain intermediates and the imines

A five-carbon atom chain compound is frequently used as starting material for both the Staudinger cyclisation or for the enolate addition to imine/cyclization sequence. All five carbon atoms are most conveniently obtained from glutaric anhydride (Scheme 6.26). It is clear from a perusal of the previous routes mentioned that compounds containing a five-carbon atom chain with a carboxylic acid derivative at one terminal and an oxygen substituent at the other are also desirable starting materials, which in turn, are obtainable from glutaric acid.

Scheme A summarizes the conversion of glutaric acid or its anhydride into the five-carbon chain compounds used in the synthesis. The highlighted compounds on the right side of scheme A were effectively used in the formation of β-lactam rings.

At C1, acyl chloride functionality was used in the Staudinger cyclization and the Evans oxazolidinone chiral auxiliaries in the N-C2 ring closing approach. The acyl chloride is typically produced using oxalyl chloride at room temperature but the use of thionyl chloride has also been reported. The oxazolidinone is linked by formation of the amide through the intervention of the mixed anhydride with pivalic acid.

The most frequent terminal C5 group in the *N*-acyl oxazolidinone enolate addition reaction is an ester, a protected aryl ketone or a protected benzylic alcohol. 5-(4-Fluorophenyl)-5-oxopentanoic acid, **1** is an important compound and its *N*-acyl oxazolidinone derivative, with a protected carbonyl, or the corresponding protected chiral C5 hydroxyl derivative, are significant intermediates in the synthesis of Ezetimibe (Scheme 6.27 and G to M).

Most of the transformations indicated in scheme A follow classical chemistry but the asymmetric reduction of the carbonyl to produce the chiral alcohol deserves special attention (thick arrows). Mostly applied methods

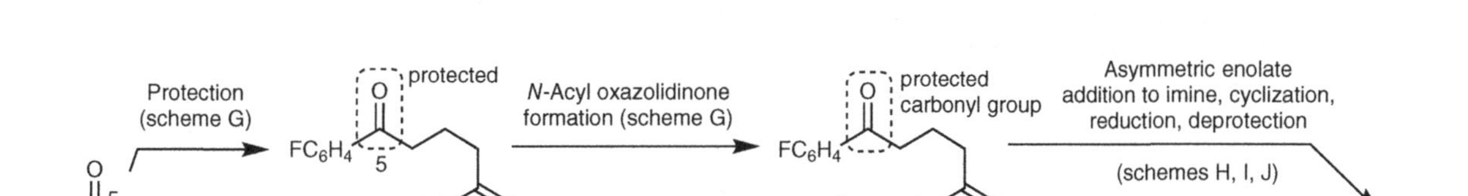

Scheme 6.26 Use of glutaric anhydride as starting material in the synthesis of Ezetimibe

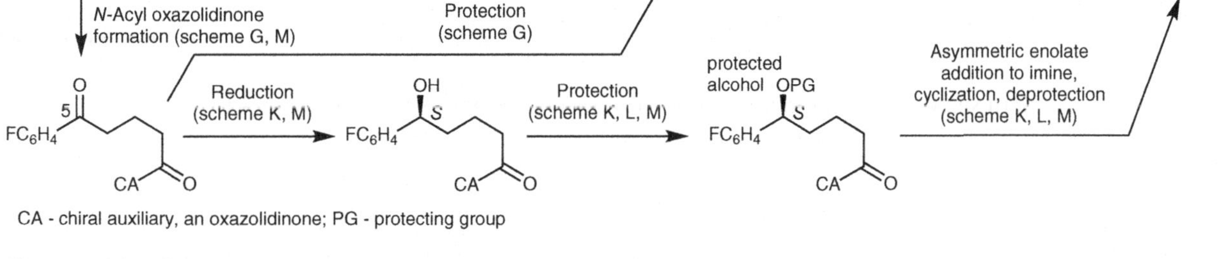

Scheme 6.27 Outline of the synthesis of *N*-acyl oxazolidinones and their use in the synthesis of Ezetimibe

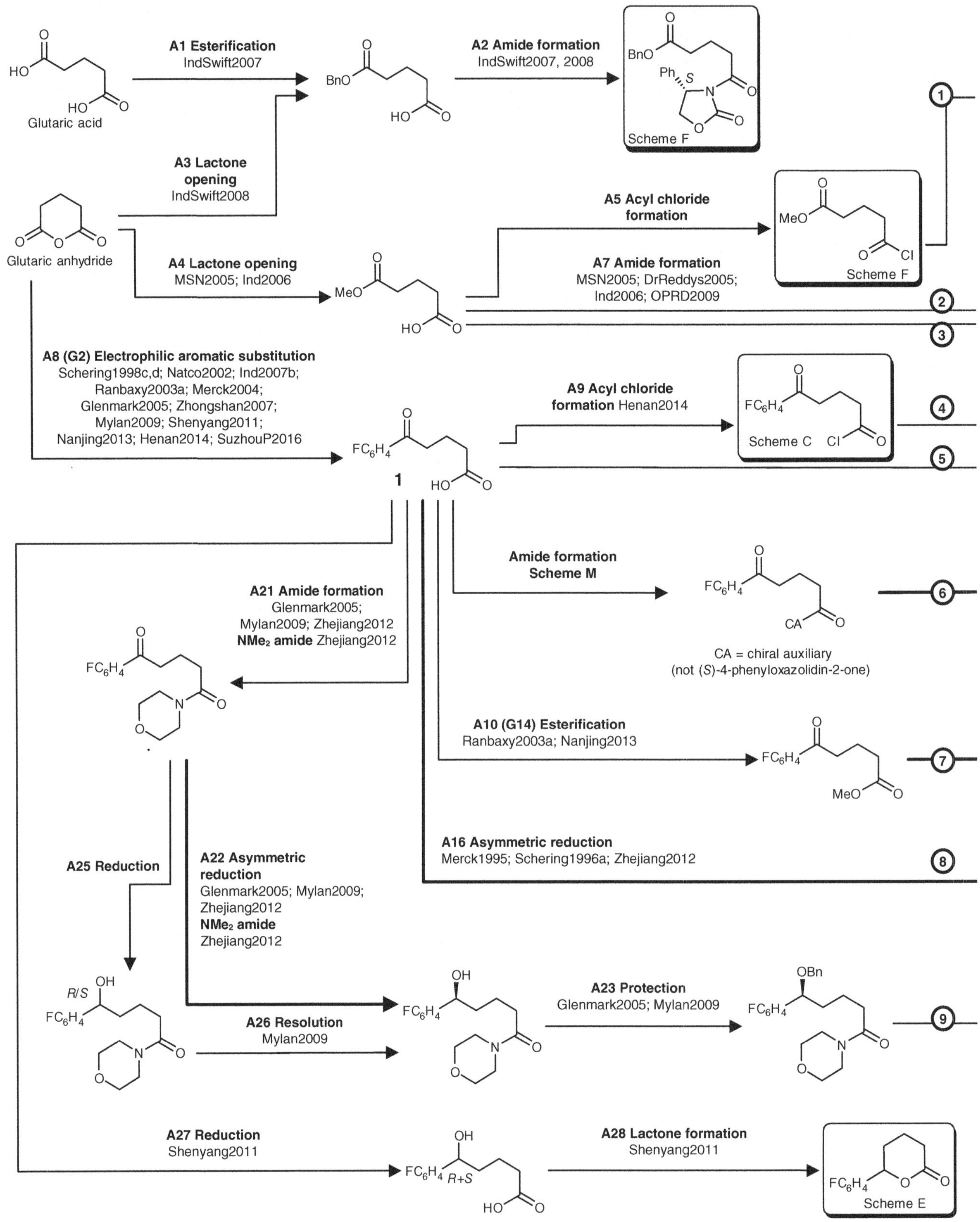

Scheme A Synthetic routes to C5 intermediates for Ezetimibe production (for reaction conditions see page 262)

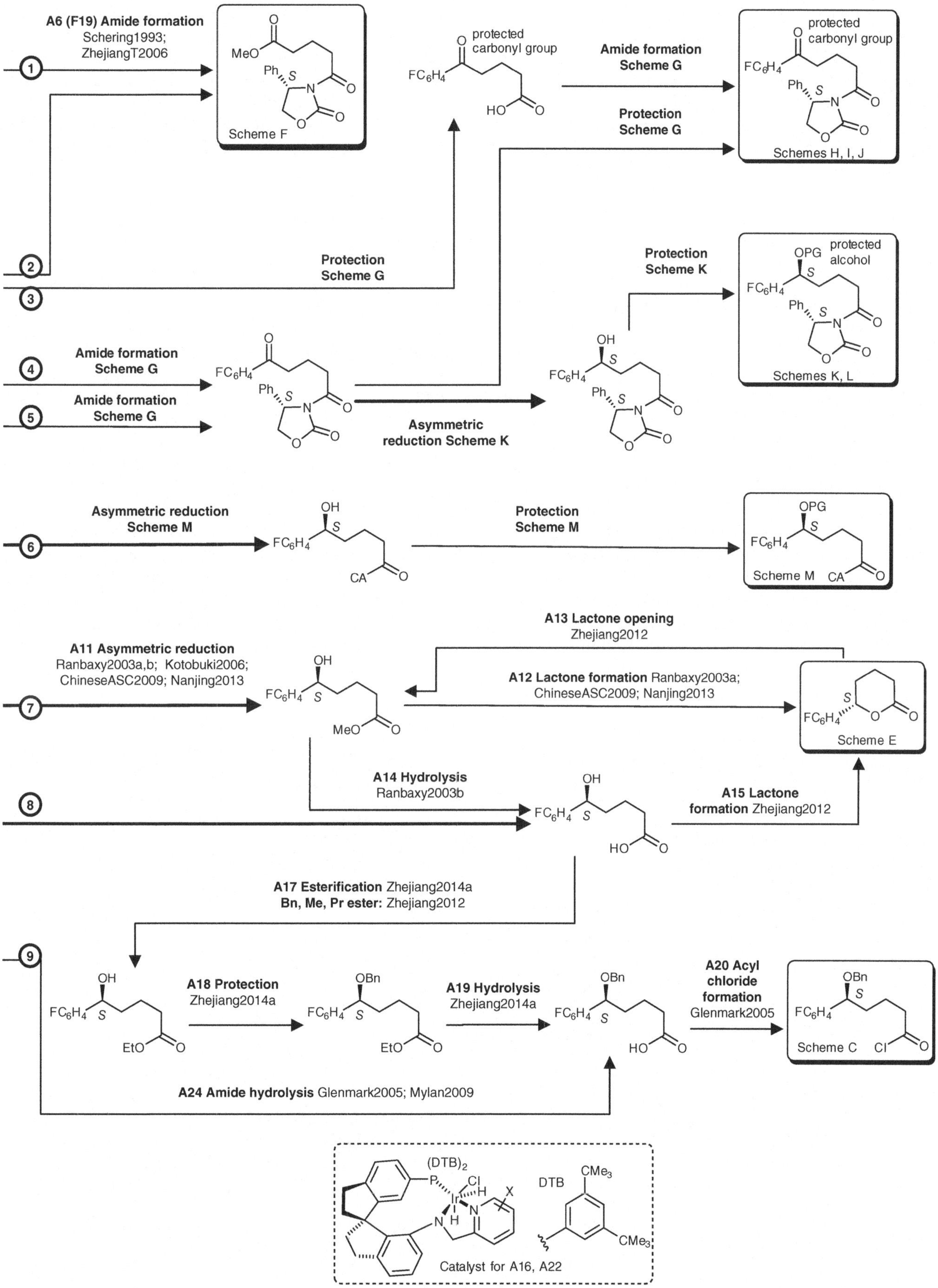
A6 (F19) Amide formation
Schering1993;
ZhejiangT2006
Scheme F
protected carbonyl group
Amide formation Scheme G
Protection Scheme G
Schemes H, I, J
Protection Scheme G
Protection Scheme K
protected alcohol
Schemes K, L
Amide formation Scheme G
Amide formation Scheme G
Asymmetric reduction Scheme K
Asymmetric reduction Scheme M
Protection Scheme M
Scheme M
A13 Lactone opening Zhejiang2012
A11 Asymmetric reduction
Ranbaxy2003a,b; Kotobuki2006;
ChineseASC2009; Nanjing2013
A12 Lactone formation Ranbaxy2003a;
ChineseASC2009; Nanjing2013
Scheme E
A14 Hydrolysis
Ranbaxy2003b
A15 Lactone formation Zhejiang2012
A17 Esterification Zhejiang2014a
Bn, Me, Pr ester: Zhejiang2012
A18 Protection
Zhejiang2014a
A19 Hydrolysis
Zhejiang2014a
A20 Acyl chloride formation
Glenmark2005
Scheme C
A24 Amide hydrolysis Glenmark2005; Mylan2009
DTB
Catalyst for A16, A22

are the CBS or DIP-Cl reduction, a catalytic asymmetric hydrogenation or an enzymatic reduction. Less frequently, a δ-lactone was used as intermediate for the β-lactam. The following sections will analyse the use of the compounds highlighted in scheme A as key intermediates in the formation of Ezetimibe.

In most cases, a protected 4-hydroxybenzaldehyde was the starting point for the imines that were incorporated into the β-lactam ring of Ezetimibe. Their synthesis is summarized in scheme B. An imine oxide was used in a few limited instances, obtained by reaction of a benzaldehyde with hydroxylamine.

6.4.2 Using the Staudinger reaction in the azetedin-2-one ring formation

In the very first synthesis of Ezetimibe, a Staudinger cyclization was used to form the azetidine-2-one ring (JMC2008, Scheme 6.23). The relatively low selectivity of the Staudinger cyclization and the formation of enantiomers of the β-lactam, along with the necessity to generate a highly unstable ketene, makes this reaction less appealing to the Process Chemist.

Even so, new proposals based on the Staudinger cyclisation were the subject of several routes (summarized in scheme C).

In two of these, a carboxylic acid derivative is built up, which upon cycloaddition, directly produces mixtures with the carbon framework of Ezetimibe in place (Glenmark2005, Henan2014). It was not possible to control the formation of undesired isomers during cyclization hence, the efficacy and yields were low. Two more synthetic routes use metal mediated coupling reactions (UShenyang2010, ACIE2010), after the Staudinger cyclization to produce the β-lactam, to elaborate the 3-carbon chain fragment. Once more mixtures were obtained in the cyclization reaction, confirming that these types of routes obviously are problematic for industrial application.

6.4.3 The Wacker oxidation route

As referred to previously, the presence of the alcohol at the benzylic position can interfere with the enolate-imine addition, thus necessitating the use of a protecting group. Previous investigations, on related compounds, indicated that direct introduction of the benzylic functionality towards the end of the synthesis was not viable (JMC1998) i.e. the yields for the sequence, benzylic bromination, S_N2 hydroxide displacement and oxidation were fairly low.

In a different approach which, introduces the benzylic hydroxyl function at a later stage, (*E*)-5-(4-fluorophenyl) pent-4-enoic acid was tested in the enolate-imine addition/cyclization sequence (Schering1995, scheme D). The Wacker oxidation (cat. $Pd(OAc)_2$/ benzoquinone/ in the presence of perchloric acid) of the alkene proceeds in a regioselective manner to give the benzylic ketone (**D5**, 70-80% yield from the *E* alkene). Ezetimibe is then produced by chiral CBS reduction and removal of the benzyl protecting group on the phenol.

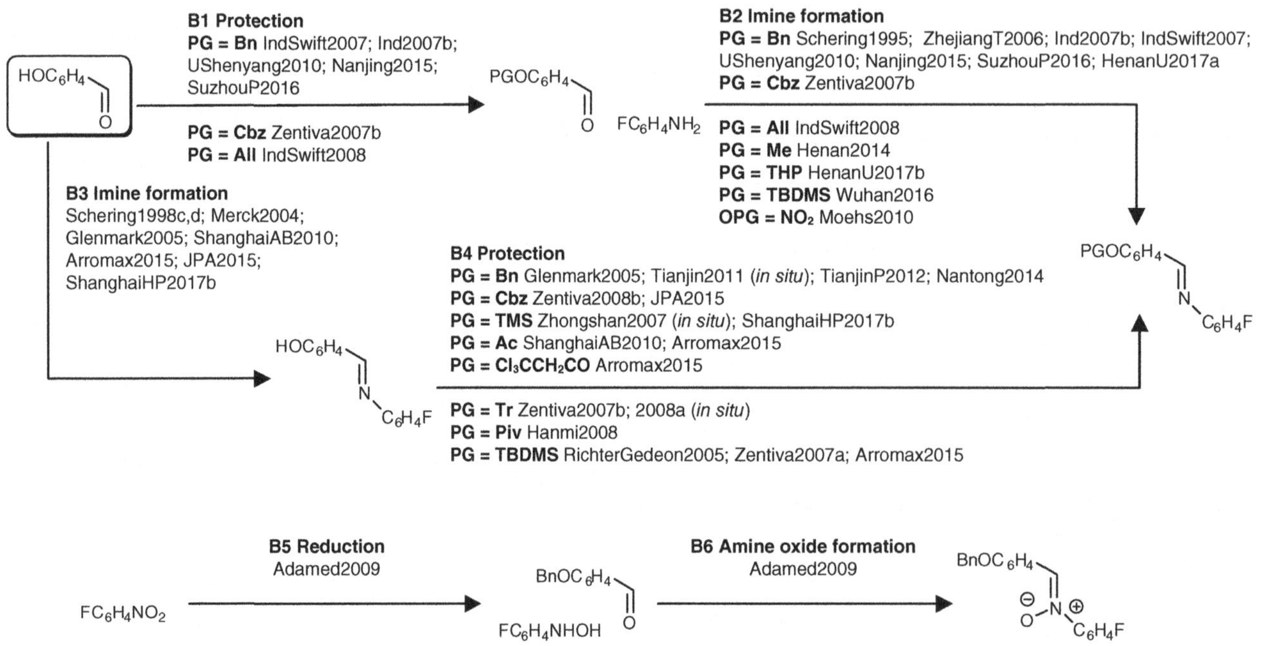

Scheme B Synthesis of imines and imine oxides (for reaction conditions see page 262)

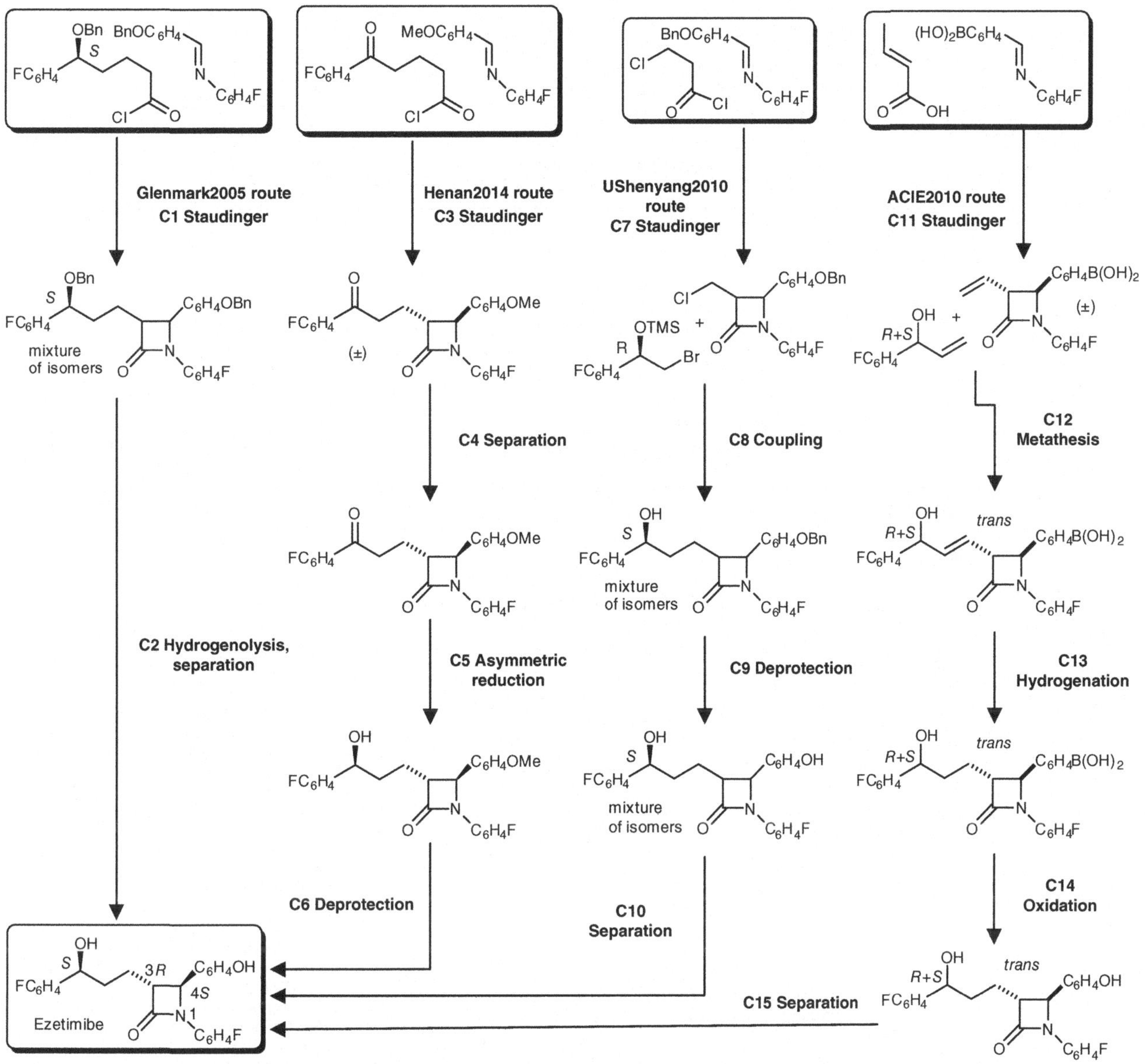

For the synthesis of starting C5 carboxylic acid derivatives see scheme A; for imines see scheme B
Note: the first synthesis of Ezetimibe also used a Staudinger cycloaddition is shown in Scheme 5.23 and also included in scheme F

SCHEME C USE OF THE STAUDINGER CYCLOADDITION IN THE SYNTHESIS OF EZETIMIBE (FOR REACTION CONDITIONS SEE PAGE 263)

This route avoids the late introduction of the terminal chain *p*-fluoroaromatic group from the ester, which is the "Achilles' heel" of these syntheses (for more detailed analysis see section 6.4.7). In addition, the route is much shorter. Similar routes using 5-(4-fluorophenyl)pent-4-enoic acid were also investigated (summarized in scheme D). (*Z*)-5-(4-fluorophenyl)-pent-4-enoic acid, conveniently prepared by a Wittig reaction in high yields (**D1**), when used as starting material gave an increased global yield (Lek2008b, 86% for step **D5** and 20% more for the global process) when compared to the *E* isomer.

In addition to the Wacker oxidation/reduction procedure, Shi epoxidation of the *Z*-alkene-lactam (**D10**, 54-56% yield) followed by ring opening (**D11**, 38% ShanghaiSP2012a) was used to introduce the benzylic (*S*)-hydroxy group. Hydrogenation of the double bond and removal of the benzyl protecting group of the phenol gave Ezetimibe.

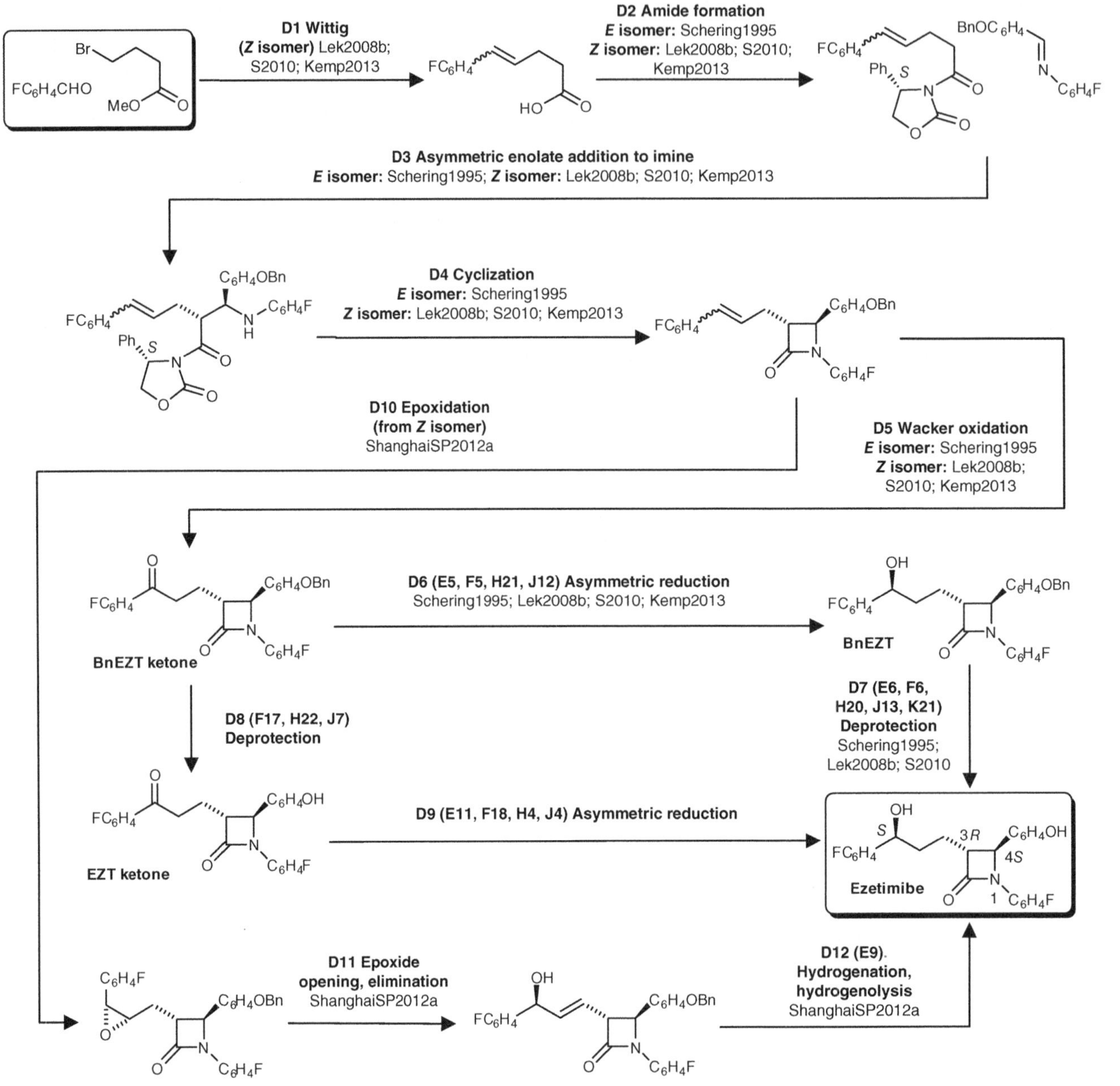

D6 (E5, F5, H21, J12) Asymmetric reduction
CBS Schering1993, 1995, 1996b; JOC1999; MSN2005; DrReddys2005; Teva2005, 2006a; Medichem2005; Ind2006; KRKA2007; IndSwift2007; Lek2008b; Adamed2009; Lupin2009a; OPRD2009; S2010; ChangzhouFP2011, Suzhou2012; ShanghaiIOC2012; Kemp2013; Chengdu2014a,b; Tokuyama2015; SuzhouP2016; ShanghaiIP2016; Shangdong2016; Sichuan2017; HenanU2017a.
$NaBH_4$ Kotobuki2006; Unichem2006.
CAH Medichem2006; KRKA2007; ChiralQ2011; Sichuan2013; Nissan2014; ZhejiangA2015; Anhui2015. **Microbial reduction** Teva2007b.
D7 (E6, F6, H20, J13, K21) Deprotection
Schering1993, 1995, 1996b; JMC1998; JOC1999; Ranbaxy2003a; MSN2005; DrReddys2005; Teva2005; Medichem2005, 2006; Unichem2006; Ind2006, 2007b; KRKA2007; IndSwift2007: Lek2008a,b; Adamed2009; Lupin2009a; OPRD2009; S2010; PolishAS2011; ChangzhouFP2011; Jiangsu2011; ChiralQ2011; Suzhou2012; ShanghaiIOC2012; TianjinP2012; ShanghaiSP2013; Nissan2014; Chengdu2014a,b; Tokuyama2015; Anhui2015; SuzhouP2016; Shangdong2016; Sichuan2017; HenanU2017a
D8 (F17, H22, J7) Deprotection
ZhejiangT2006; IndSwift2007; HenanU2017a
D9 (E11, F18, H4, J4) Asymmetric reduction
CBS Schering1996b; JOC1999; Teva2006a; Medichem2005; IndSwift2007, 2008, 2016; Zentiva2007a; Zhongshan2007; SC2016; ChangzhouFP2016; ShanghaiIP2016. **DIP-Cl** Ranbaxy2003b; Ind2007b. **CAH** Teva2005; Medichem2006; ZhejiangA2015. **Biochemical reduction** Schering1998a, 1999; Teva2007a; Ind2007a; Codexis2010; JMCB2010.

Scheme D Synthesis of Ezetimibe using a styrene benzylic oxidation step (for reaction conditions see page 263)

6.4.4 The search for a selective β-lactam formation

The development of techniques to generate the β-lactam stereoselectively led to several alternative syntheses of Ezetimibe (scheme E).

Commercially available (*S*)-β-hydroxy-γ-butyrolactone, when used in an enolate-imine/cyclisation sequence, gave the *trans*-β-lactam in a highly diastereoselective fashion in one-step (**E1**, Schering1996b, JOC1999). After optimization, the reaction was easily scaled up to 300 g and gave 70% conversion with a 9:1 diastereomeric *cis-trans* ratio. The β-lactam was subsequently isolated in a 64% yield (diastereomeric ratio, dr > 95:5). Sodium periodate oxidation of the diol (**E2**) gave a relatively unstable aldehyde at the same time as the *cis* isomer (*R* configuration at the 4 position of the β-lactam) epimerized to the enantiomer giving only a mixture of *trans* β-lactams (90% ee). A Mukaiyama aldol addition with the TMS enol ether of 4-fluoroacetophenone completed formation of the Ezetimibe carbon skeleton (**E3**). During the subsequent conversion to the final product the unwanted isomers were removed by crystallization yielding the final product in excellent ee (>99.9%). This one-step process for β-lactam formation was highly diastereoselective (and eventually enantioselective) as testified by the high yields.

This key aldehyde was also produced (Shanghai2015b) *via* a cyclization of a β-amino amide in a more recent variation (**E29/E30/E31/E32**).

Three other patents report formation of the β-lactam by an enolate-imine cyclocondensation in a similar vein (lower part of scheme E, **E23** and **E28**, Shenyang2011, Ranbaxy2003a, Zhejiang2013). A mixture of stereoisomers is obtained on ring formation using a racemic β-lactone (**E23**, Shenyang2011). An oxidation and asymmetric reduction sequence converts the unwanted *R* benzylic alcohol to the *S* isomer (**E24**, **E25**). Classical chromatography was then able to separate the diastereomeric *trans* ring isomers. However, since the diastereoisomers are present at a 1:1 ratio this process obviously suffers from the low yielding final separation step (**E27**, 39%).

Notably, this technique would provide Ezetimibe (with a protected phenolic group) in one step by using the chiral (*S*)-δ-lactone. However, it must be noted that the chiral lactone has to be produced (see previous scheme A). On a 5.5 g lactone scale, the LDA-activated cyclization at −50 °C produced BnEZT in 67% yield after chromatographic purification (**E28**, Ranbaxy2003a, Scheme 6.28). However, a very low yield was reported on conversion to pure Ezetimibe after deprotection (HCO_2NH_4, Pd/C 10%, HCO_2H, 55 °C) and isolation (20%).

An improved process for step **E28** was recently reported (Zhejiang2013) where the reaction was carried out in polar solvents using hindered lithium amide bases and an anhydrous lithium salt to increase the percentage of chiral induction (chiral purity >96%; yield >80%).

6.4.5 Using Kinugasa cycloaddition

Routes starting with a Kinugasa nitrone-olefin [3+2] cycloaddition to establish the β-lactam ring have been reported (scheme E).

In one approach, the Cu(I)-mediated cycloaddition between a terminal acetylene derived from the acetonide of L-glyceraldehyde and a suitable C,N-diarylnitrone derivative produced a *cis* 3*R*,4*S* configuration of the azetidin-2-one ring, with high stereoselectivity in 68-70% yield (**E14**, Adamed2009, JOC2011). Removal of the acetonide protecting group followed by glycolic cleavage and base induced isomerization at the carbon atom α to the carbonyl group afforded the (3*S*,4*S*) aldehyde (*cis*:*trans* = 1:6.2 to 1:8.8, JOC2011), which had already been converted to Ezetimibe (see Schering1996b). Again, the aldehyde is reported as being unstable and prone to decomposition.

In an approach reported by an academic group, an α-methylidene-β-lactam was produced by the Kinugasa protocol (**E18**, scheme E, 16% of racemic product, JOC2016). A Ru-catalysed cross metathesis reaction (**E19**) using a second generation Hoveyda-Grubbs catalyst served to attach the side chain, giving the *Z* isomer in 65% yield, and a corrected yield of 10.4% for the 4*S* isomer. Hydrogenation gave a mixture of *cis* β-lactam isomers, which were isomerized by the action of a non-nucleophilic base to the more stable *trans* compounds. Chromatographic separation of the *trans* diastereomers and deprotection furnished Ezetimibe.

Formation of isomers during the cyclization reaction proved difficult to control, thus reducing the utility of the Kinugasa reaction for a synthesis of Ezetimibe in an industrial setting.

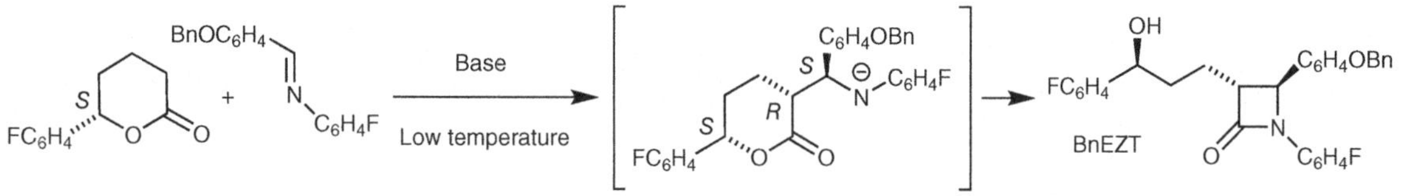

Scheme 6.28 Synthesis of BnEZT starting from a chiral (*S*)-δ-lactone

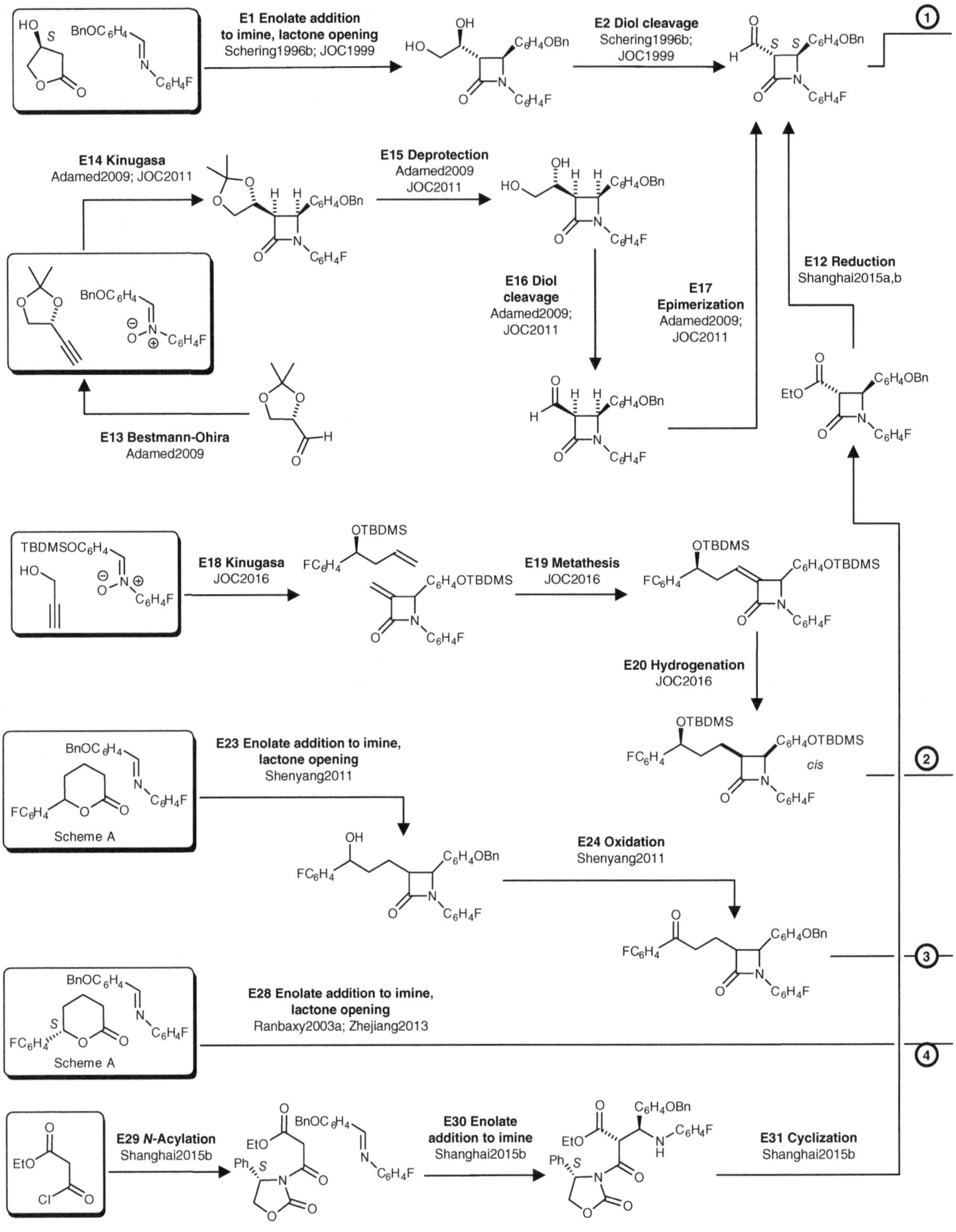

Scheme E Synthesis of Ezetimibe using an enolate addition of a lactone to an imine or a Kinugasa reaction as key steps (for reaction conditions see page 263)

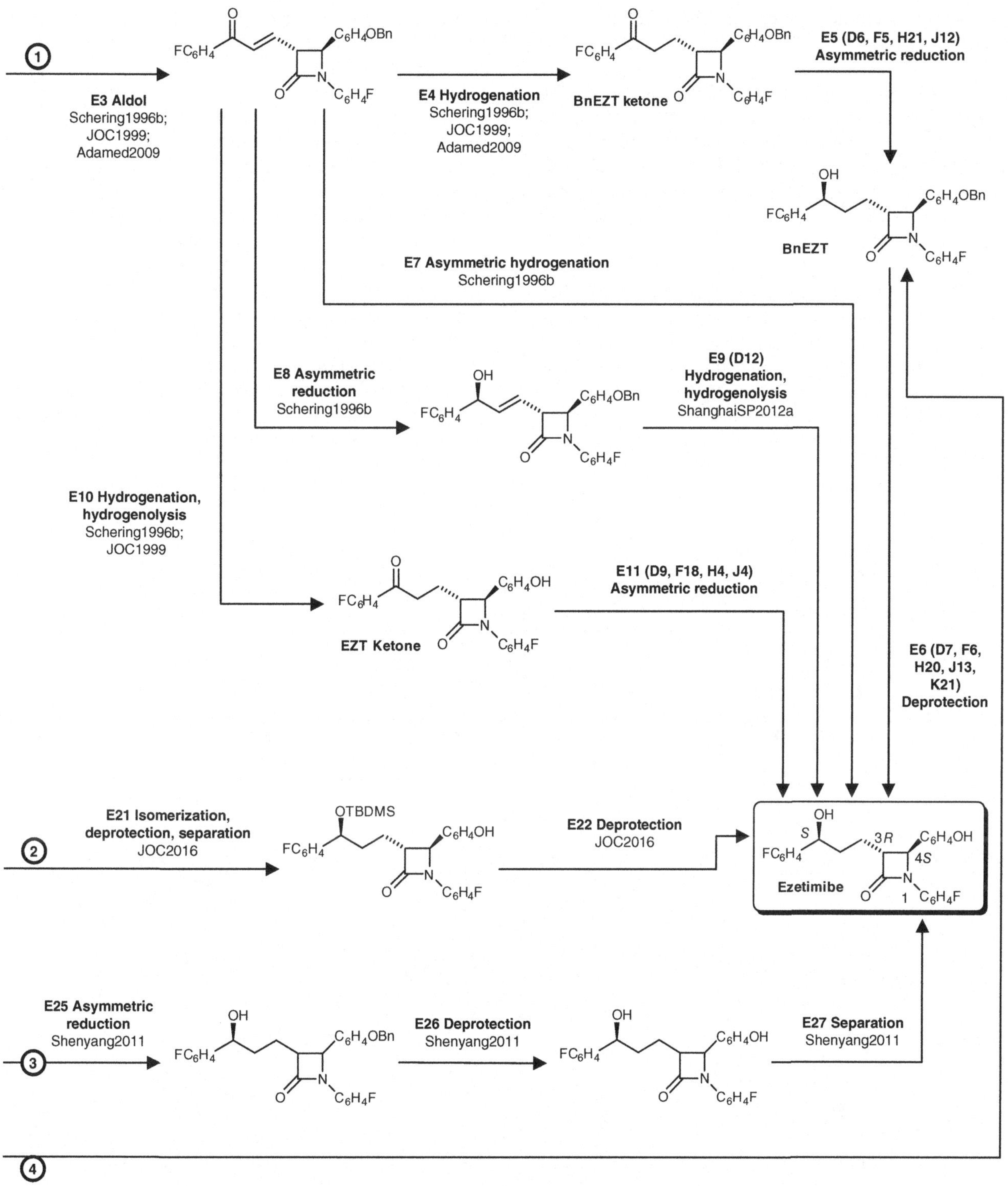

E3 Aldol
Schering1996b; JOC1999; Adamed2009
E4 Hydrogenation
Schering1996b; JOC1999; Adamed2009
BnEZT ketone
E5 (D6, F5, H21, J12) Asymmetric reduction
BnEZT
E7 Asymmetric hydrogenation
Schering1996b
E8 Asymmetric reduction
Schering1996b
E9 (D12) Hydrogenation, hydrogenolysis
ShanghaiSP2012a
E10 Hydrogenation, hydrogenolysis
Schering1996b; JOC1999
E11 (D9, F18, H4, J4) Asymmetric reduction
EZT Ketone
E6 (D7, F6, H20, J13, K21) Deprotection
E21 Isomerization, deprotection, separation
JOC2016
E22 Deprotection
JOC2016
Ezetimibe
E25 Asymmetric reduction
Shenyang2011
E26 Deprotection
Shenyang2011
E27 Separation
Shenyang2011

6.4.6 N-C2 ring closing strategy and the build-up sequence

The majority of the processes reported to synthesise Ezetimibe use an enolate addition to a C=N bond followed by N-C2 cyclization to build the azetedin-2-one ring.

To form the entire molecule of Ezetimibe, four tasks have to be accomplished: i) the asymmetric enolate addition to C=N (Mannich-type aldol reaction); ii) the N-C2 cyclization; iii) the asymmetric carbonyl reduction and iv) the attachment of the *p*-fluoroaromatic ring. The order in which these tasks are carried out can be varied, thus giving rise to different synthetic routes (Scheme 6.29). The use of protecting groups is a prerequisite when making changes in the sequence of the chemical reactions. Interestingly, the asymmetric reduction of the carbonyl group can be carried out both chemically or enzymatically and numerous processes have been reported.

The following sections analyse Ezetimibe synthesis according to the most commonly used combinations of the sequence of steps:

- 1) asymmetric enolate addition to imine (AEAI); 2) N-C2 cyclization; 3) *p*-fluoroaromatic ring coupling; 4) asymmetric carbonyl reduction (section 6.4.7)
- 1) *p*-fluoroaromatic ring coupling; 2) asymmetric enolate addition to imine (AEAI); 3) N-C2 cyclization; 4) asymmetric carbonyl reduction (section 6.4.8)
- 1) *p*-fluoroaromatic ring coupling; 2) asymmetric enolate addition to imine (AEAI); 3) asymmetric carbonyl reduction; 4) N-C2 cyclization (section 6.4.8)
- 1) *p*-fluoroaromatic ring coupling; 2) asymmetric carbonyl reduction; 3) asymmetric enolate addition to imine (AEAI); 4) N-C2 cyclization (section 6.4.9).

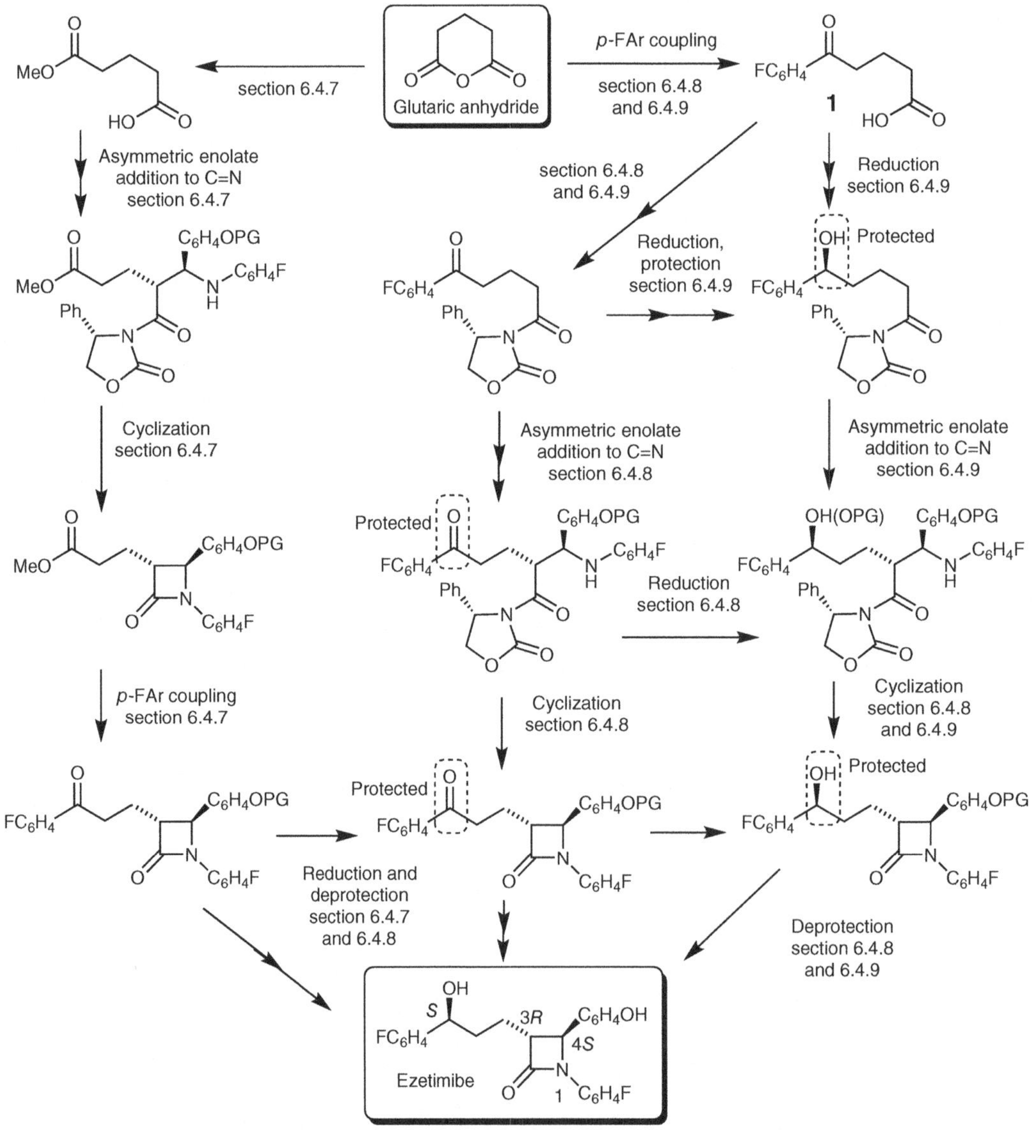

SCHEME 6.29 DIFFERENT SEQUENCES AND COMBINATIONS OF THE KEY TASKS TO PRODUCE EZETIMIBE FROM GLUTARIC ANHYDRIDE *VIA* ENOLATE ADDITION TO C=N AND CYCLIZATION

6.4.7 N-C2 ring closure strategy: late attachment of the *p*-fluoroaromatic ring

Several processes with similar strategies to that disclosed in the first patent on Ezetimibe (Schering1993) have been proposed aiming to improve efficiency, purity and operating conditions (summarized in scheme F).

An enzymatic resolution of the racemic *trans* β-lactam mixture was designed to avoid chromatographic separation (**F8**, Schering1998b). Not surprisingly, these routes reject the use of the Staudinger cyclisation to avoid the formation of enantiomeric mixtures, opting for the N-C2 cyclization of Schering1993.

Despite its high yield, the ester hydrolysis step (**F9**) produces a measurable amount of impurities from opening of the β-lactam ring. Unfortunately, these impurities can be carried forward to the final product. Purification of the intermediate carboxylic acid (ChangzhouPF2011) offers an attractive method to resolve this problem.

The Palladium-Negishi coupling (**F11**) is reported to have yields ranging from 37 to 87%, which would indicate its lack of robustness and making this step a candidate for further development work. The reaction, between an acid chloride and *p*-fluorophenyl zinc chloride, is carried out in the presence of a palladium catalyst. This combination poses problems and side products are reported. Transmetalation of the Grignard with an excess of anhydrous zinc chloride is yet again another issue, as the zinc chloride is anhydrous and difficult to handle on an industrial scale. The resulting product is closely related to Ezetimibe (BnEZT ketone) and is normally converted to the final product by reduction and deprotection (**F5/F6**) although, this order may be inverted. These final steps will be analysed in more detail below.

The carboxylic acid can also be converted into the BnEZT ketone *via* a Weinreb amide (**F14+F15**, Unichem2006, ZhejiangT2006). The asymmetric reduction step was avoided in another reported route. Here, the carboxylic acid was reduced to a primary alcohol and then oxidized to an aldehyde (**F22+F23**, Pharmathen2011; ShanghaiSP2013) followed by (*R*)-BINOL mediated asymmetric addition of a Grignard (**F24**, Pharmathen2011; ShanghaiSP2013) to produce BnEZT directly in good yield (77-89%, for **F24**) and diastereoselectivity (82-89%, for **F24**).

Formation of an ester enolate and reacting this with *p*-fluorobenzoyl chloride followed by removal of the ester group was the subject of yet another report (**F3+F4**, Jiangsu2015a) but the harsh conditions used probably limit its usefulness.

A modification of the initial 1993 route (Schering1993) used a benzyl ester and an allyl phenol-protecting group in the enolate addition step (**F25**, IndSwift2008). Here, the patent focus was on diminishing side-products formed by over reduction on deprotection of the benzylic group by hydrogenolysis in the final stages of the synthesis.

6.4.7.1 On the final reduction-deprotection sequence to produce Ezetimibe

The BnEZT ketone, or a similarly protected phenol, is a frequent intermediate in many of the routes mentioned in this chapter. Not surprisingly, the two final steps in the synthesis of the API (reduction and deprotection) must undergo extensive development work if impurity forming side reactions are to be controlled (Figure 6.8), thus assuring production of a high quality final product.

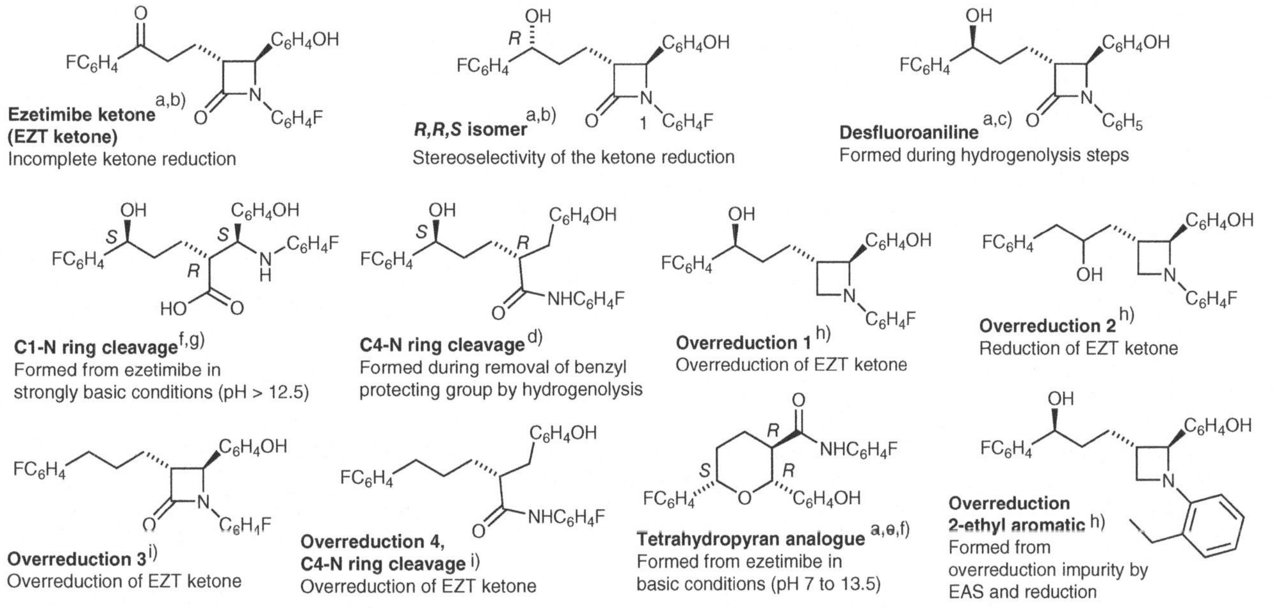

a) USP39 b) JMS2011c) JPA2015 d) JPBA2010 e) JPBA2012 f) JPS2014 g) JPBA2011a h) JPBA2015 i) JPBA2014

Figure 6.8 Selected process impurities of Ezetimibe related to the reduction and deprotection steps of a phenol-protected EZT ketone intermediate

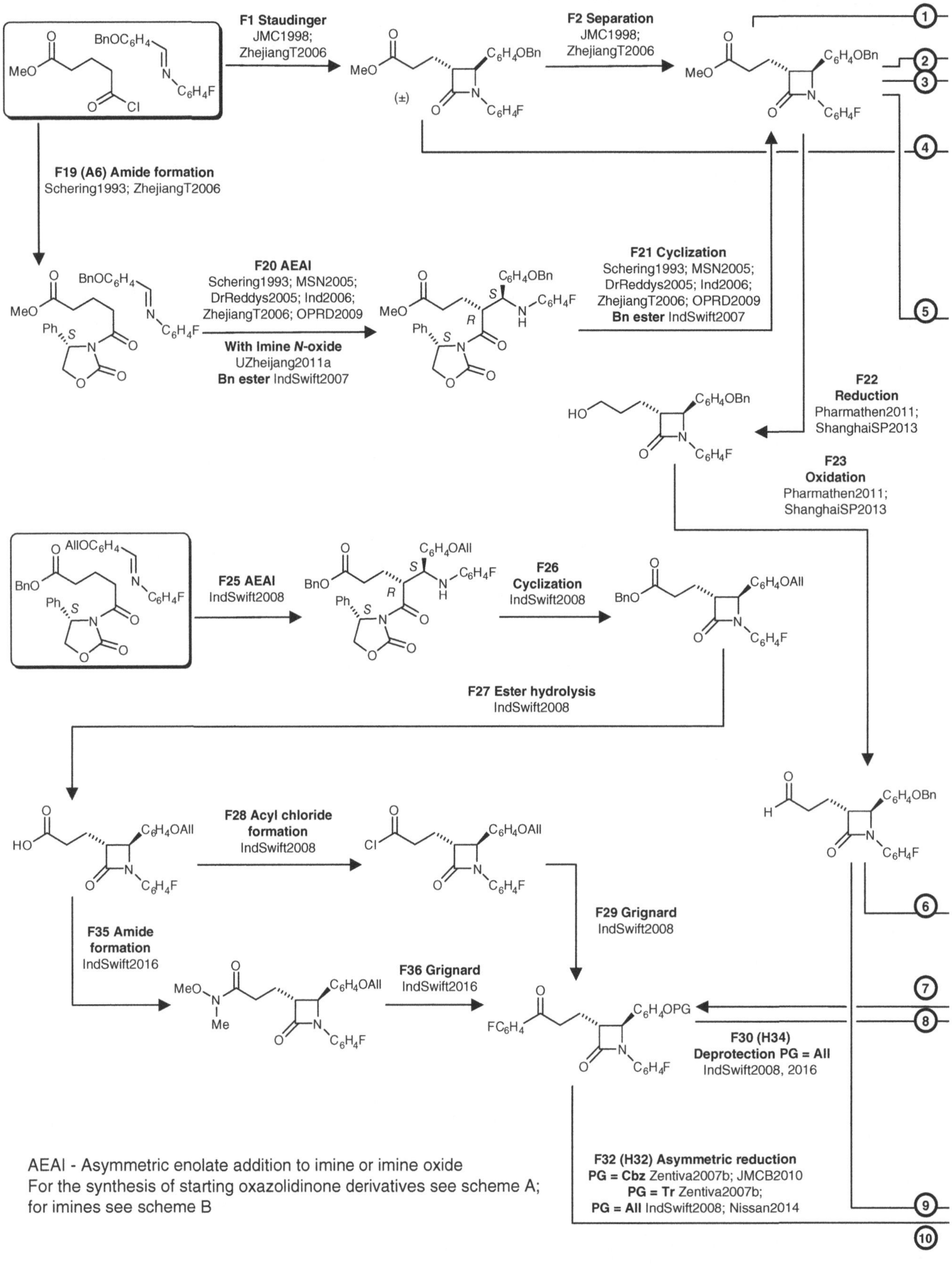

SCHEME F SYNTHESIS OF EZETIMIBE BY ATTACHING THE *p*-FLUORARОMATIC RING AT LATER ROUTE STAGES (FOR REACTION CONDITIONS SEE PAGE 264)

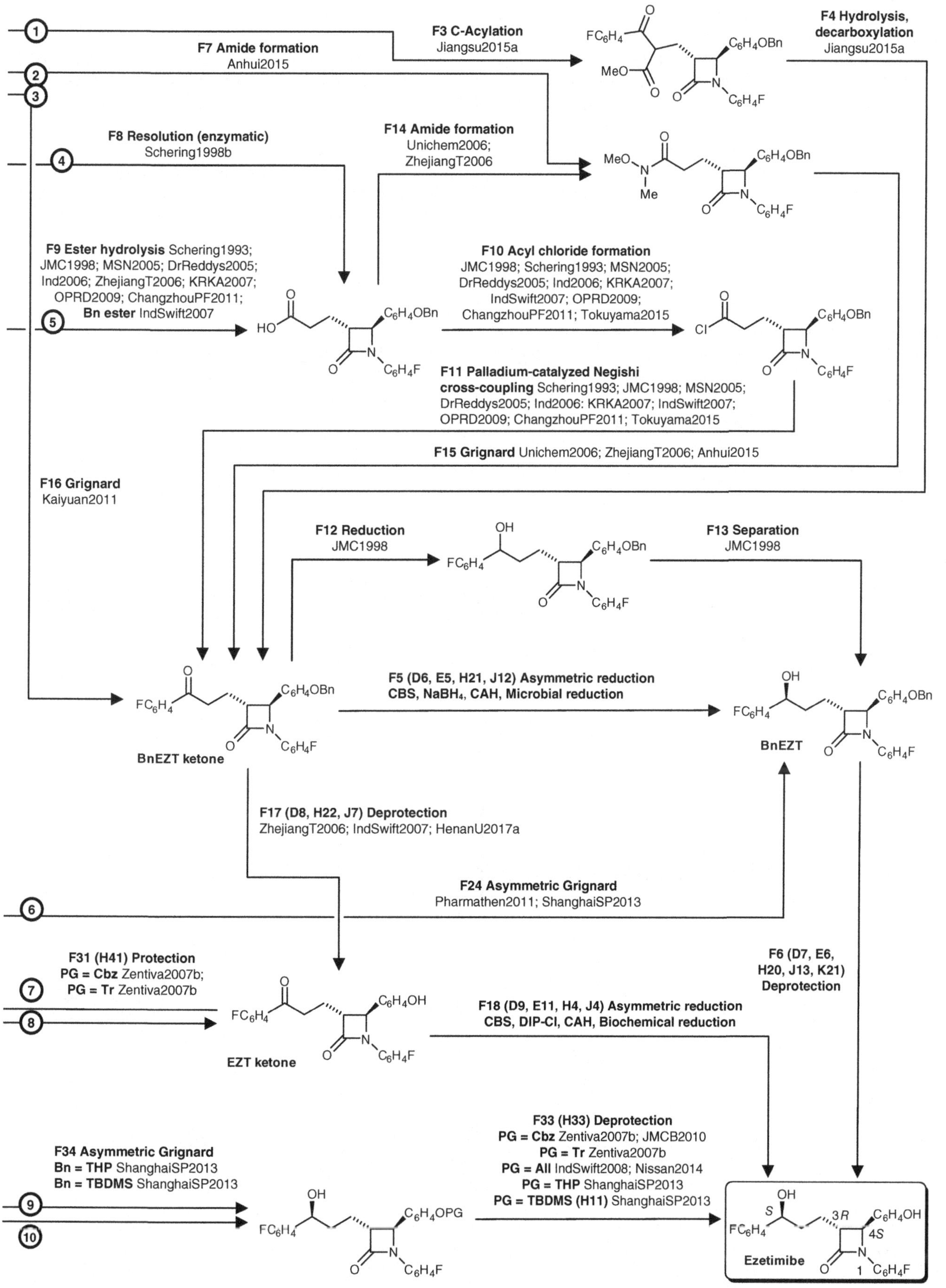
F3 C-Acylation
Jiangsu2015a
F4 Hydrolysis, decarboxylation
Jiangsu2015a
F7 Amide formation
Anhui2015
F8 Resolution (enzymatic)
Schering1998b
F14 Amide formation
Unichem2006; ZhejiangT2006
F9 Ester hydrolysis Schering1993; JMC1998; MSN2005; DrReddys2005; Ind2006; ZhejiangT2006; KRKA2007; OPRD2009; ChangzhouPF2011;
Bn ester IndSwift2007
F10 Acyl chloride formation
JMC1998; Schering1993; MSN2005; DrReddys2005; Ind2006; KRKA2007; IndSwift2007; OPRD2009; ChangzhouPF2011; Tokuyama2015
F11 Palladium-catalyzed Negishi cross-coupling Schering1993; JMC1998; MSN2005; DrReddys2005; Ind2006: KRKA2007; IndSwift2007; OPRD2009; ChangzhouPF2011; Tokuyama2015
F15 Grignard Unichem2006; ZhejiangT2006; Anhui2015
F16 Grignard
Kaiyuan2011
F12 Reduction
JMC1998
F13 Separation
JMC1998
F5 (D6, E5, H21, J12) Asymmetric reduction
CBS, NaBH4, CAH, Microbial reduction
BnEZT ketone
BnEZT
F17 (D8, H22, J7) Deprotection
ZhejiangT2006; IndSwift2007; HenanU2017a
F24 Asymmetric Grignard
Pharmathen2011; ShanghaiSP2013
F6 (D7, E6, H20, J13, K21) Deprotection
F31 (H41) Protection
PG = Cbz Zentiva2007b;
PG = Tr Zentiva2007b
F18 (D9, E11, H4, J4) Asymmetric reduction
CBS, DIP-Cl, CAH, Biochemical reduction
EZT ketone
F34 Asymmetric Grignard
Bn = THP ShanghaiSP2013
Bn = TBDMS ShanghaiSP2013
F33 (H33) Deprotection
PG = Cbz Zentiva2007b; JMCB2010
PG = Tr Zentiva2007b
PG = All IndSwift2008; Nissan2014
PG = THP ShanghaiSP2013
PG = TBDMS (H11) ShanghaiSP2013
Ezetimibe

A number of impurities reported in the literature and designated as "to be controlled" in the US Pharmacopeia Monograph are obviously related to the chemistry of these final steps. The plausible origin of these impurities is included in the following scheme (Scheme 6.30).

The presence of EZT ketone reflects incomplete reduction and the *R,R,S* isomer is due to a carbonyl reduction which is not completely stereoselective (JMS2011). Surprisingly, the Ezetimibe impurity resulting from the N-C2 cleavage is not referred to in the monograph, despite having been identified as a major basic degradation product of Ezetimibe (JPBA2011a, JPS2014).

The desfluoroaniline derivative and C4-N cleavage (in itself a benzylic hydrogenolysis) result from the catalytic conditions to remove the benzyl protecting group. The N-C2 ring opening, as mentioned above, not included in Scheme 6.30, is typical of the facile hydrolysis of β-lactams and can be controlled by avoiding highly basic conditions. The tetrahydropyran derivative results from an intramolecular S_N2 reaction, wherein the C4 of the lactam ring suffers nucleophilic attack by the (*S*)-hydroxy benzylic oxygen and is accompanied by inversion of configuration.

Most of the impurities shown above result from the conditions inherent to the chemical transformations used. Appropriate process development is expected to minimise the side reactions that give rise to these impurities. Many patents focus on these two steps and aim at maintaining the impurities at very low levels.

6.4.7.2 On the reduction of BnEZT ketone

The asymmetric reduction of BnEZT ketone (**F5**, scheme F) has been the focus of several patents, confirming that this step is crucial to the quality of the final product. The formation of the *R,R,S* isomer is reported to be problematic as it is difficult to remove from the final product (JMS2011).

The Corey-Bakshi-Shibata reduction, with (*R*)-(+)-2-Methyl-CBS-oxazaborolidine (Figure 6.9, 5% to 22% mol) and borane dimethylsulfide (BMS) in THF, toluene or methylene chloride, frequently at low temperature (–20 °C) is the most used method for this transformation. Some of the yields reported are too low to be useful at an industrial scale (around 50%) but values above 90% have been published in the patent literature. Diastereoselectivity ratios are frequently omitted but, in some

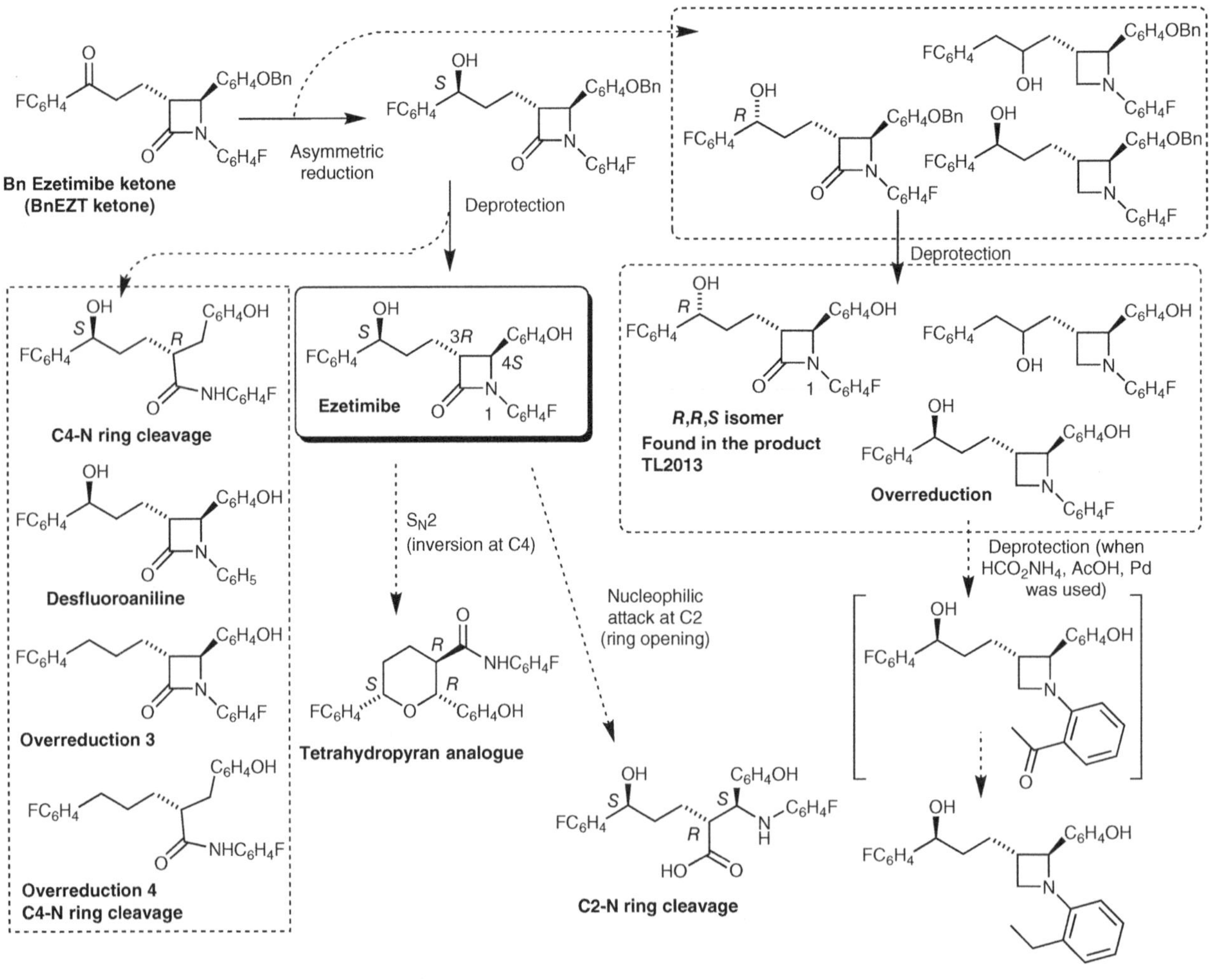

Scheme 6.30 Outline of the mechanistic pathways for the formation of impurities during the reduction and deprotection steps

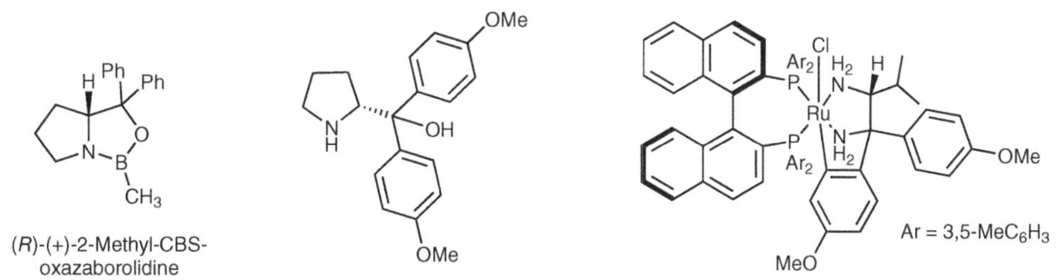

(*R*)-(+)-2-Methyl-CBS-oxazaborolidine (with BH_3•$S(CH_3)_2$)

(*R*)-bis(4-methoxyphenyl)(pyrrolidin-2-yl)methanol (with $NaBH_4$)

(*R*)-RUCY-XylBINAP (with H_2)

FIGURE 6.9 SELECTED CATALYSTS FOR ASYMMETRIC REDUCTION OF THE BNEZT KETONE

instances, values above 95% have been described and the reaction scaled up to 100 g. Diastereoselectivity depends on the rate and mode of addition of the borane-complex, the presence of moisture in the reaction medium and the reaction temperature. In addition, the amount of BMS reducing agent affects the reaction process and the impurity profile (JPBA2015). Hence, a variety of borane reagents have been employed, including borane tetrahydrofuran complex (BTHF), which, being somewhat unstable, gives rise to some concerns for use at large scale. BMS is associated with environmental concerns due to release of methylsulfide and it can be seen that implementation of processes that use either of these two borane reducing reagents is challenging. Finally, it should also be noted that his reaction generates a stoichiometric amount of borate waste salts.

The use of $NaBH_4/Me_3SiCl$ catalysed by (*R*)-2-[bis-(4-methoxyphenyl)hydroxymethyl]pyrrolidine (10% mol, Figure 6.9) has also been shown to be possible (Kotobuki2006) at a 1 g scale giving an 89% yield and a 90% de.

Ruthenium catalytic asymmetric hydrogen transfer processes, developed from 2006 onwards, were shown to give good yields (>90%) and diastereoselectivities (>99%). For example, a [(*R*)- RUCYXylBINAP] catalyst at a loading of 0.1% mol gave 100% conversion and 99.9% de on a 54 g scale (Nissan2014, Figure 6.9).

A microbial reduction was tested with limited success (10% conversion, 99.5% de, Teva2007b).

6.4.7.3 On the reduction of EZT ketone

Similar methods to those shown above were shown to be applicable to the asymmetric reduction of the ketone of an unprotected phenol (**F18**, **D9**).

However, from the reports it can be inferred that the free hydroxyl phenolic group interferes with the selectivity of the CBS reduction. Direct chiral reduction gave only 84% diastereomeric purity whereas a TMS protected phenol after reduction and removal of the TMS during the acidic work-up produced Ezetimibe in 79% overall yield and 99.7% ee (JOC1999). When EZT ketone was used as the starting material, i.e. with no protection of the phenol, in most instances no diastereomeric ratios are reported after CBS reduction, although yields of up to 94% are cited. In a series of experiments designed to optimize the CBS reduction step (10% mol catalyst being used), several reducing agents were compared (SC2016). The best isolated yields ranged from 51% to 62% with the de of the products higher than 99.6%, whilst the de of the crude product was between 78% to 82%:

- BH_3•Me_2S: 61% yield, 81% de crude and 99.7% final
- BH_3•THF: 56% yield, 78% de crude and 99.7% final
- BF_3, *N*,*N*-diethylaniline: 61% yield, 82% de crude and 99.6% final
- $NaBH_4/I_2$: 55% yield, 76% de crude and 98.9% final

were amongst the best results obtained.

(–)-DIP Chloride [(–)-*B*-Chlorodiisopinocampheylborane] reduction (Ranbaxy2003b and Ind2007b) and catalytic asymmetric hydrogen (CAH) transfers mediated by ruthenium catalyst have also been reported. In some examples, CAH transfer showed good efficiency, for example: Ru catalyst (0.2% mol, see Figure 6.10), HCO_2H, NEt_3, CH_3CN, 10 °C gave 92% yield and 99.1% optical purity at a 41 g scale (ZhejiangA2015).

Microbial reductions at high dilution have also been described, giving a 28% to 38% yield on a 450 to 600 mg scale (Schering1999; Schering1998a). Biocatalytic microbial reducing systems using *Rhodococcus fascians* or fungal isolate from *Geotrichum candidum* were also cited (Schering1998a). Later, improved processes based on *Rhodococcus fascians* were reported to give up to 95% conversion with 99.9% de (15 mg scale JMCB2010).

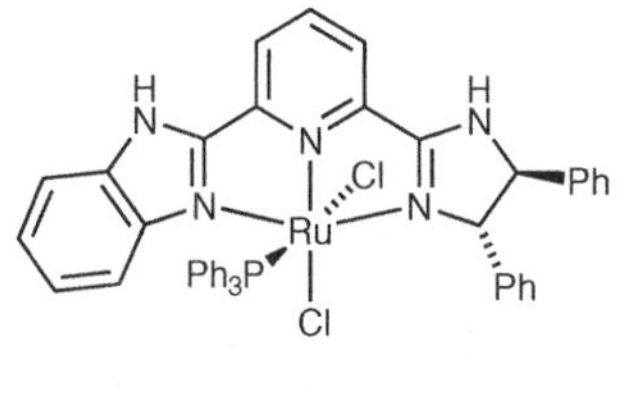

FIGURE 6.10 EXAMPLE OF AN EFFECTIVE CATALYST FOR CAH OF EZT KETONE

Enzymatic processes with ketoreductase enzymes were also developed, yielding up to 89% of Ezetimibe (20 mg scale, 65% at 600 mg scale Teva2007a, Ind2007a). A subsequent biocatalytic process using an engineered ketoreductase polypeptide was able produce crude Ezetimibe in 90% yield, 99.9% de at a 20 g scale (Codexis2010).

6.4.7.4 On the debenzylation

Removal of the benzyl protecting group (**D7**, **F6**) can have an impact on the impurity profile of the final product. The hydrogenolysis is generally carried out using palladium on carbon as catalyst and hydrogen gas at a pressure of 4 to 60 bar. The hydrogen source can also be varied, for example ammonium formate/acetic acid in an alcohol as solvent is often used. The formation of the impurity from N-C4 scission by hydrogenolysis is reported to be problematic when hydrogen gas is the hydrogen source (Lek2008a). The use of alternative hydrogen sources at atmospheric pressure and low catalyst loading (0.7% w/w of Pd) presents lower risk and high yield (Lek2008a). A design space study established ideal reaction conditions (JPBA2015). Impurities were controlled at low levels. These options are viable with the caveat that removal of palladium from the final API must also be accomplished.

6.4.8 N-C2 ring closure strategy: attaching the *p*-fluoroaromatic ring at the beginning of the synthesis

Introducing the *p*-fluoroaromatic ring at the end of the aliphatic side-chain is associated with certain drawbacks, principally manipulation of the ester-lactam intermediate and its conversion to Ezetimibe is challenging. On hydrolysis to the carboxylic acid, impurities, which are hard to remove are generated, whilst the Negishi coupling produces significant amounts of waste, thus causing concern. One method of resolving these issues is to reschedule the attachment of the *p*-fluoroaromatic and bring it nearer to the beginning of the synthesis.

The *p*-fluoroaromatic ring can conveniently be attached *via* an EAS reaction with glutaric anhydride and *p*-fluorobenzene. The product of this reaction is the key intermediate 5-(4-fluorophenyl)-5-oxopentanoic acid (**1**), see Scheme 6.31. An *o*-fluorosubstituted analogue of Ezetimibe is mentioned as a possible impurity in the US Pharmacopeia Monograph, which clearly can result from *o*-attack of the acylating agent in an EAS substitution reaction. In the late *p*-fluoraromatic attachment routes, the same impurity may be present due to the presence of an *o*-fluoro-impurity in the aryl metal reagent.

To convert 5-(4-fluorophenyl)-5-oxopentanoic acid to Ezetimibe a similar sequence to that used in Schering1993 can be adopted. This standard sequence, used by several companies, is outlined in Scheme 6.32.

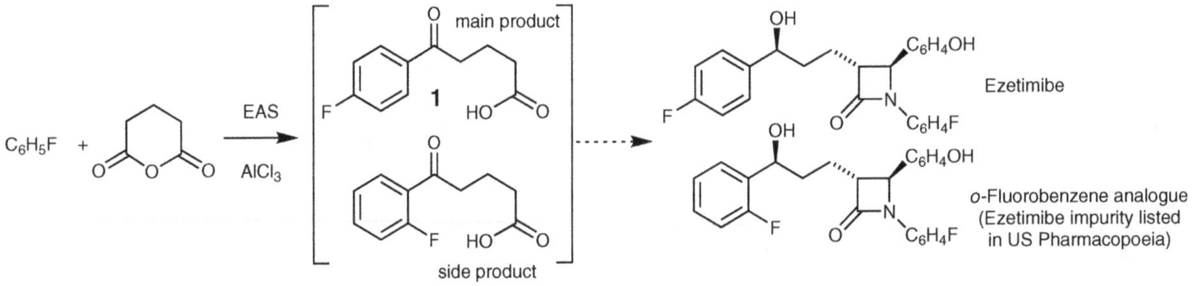

Scheme 6.31 Formation of 5-(2-fluorophenyl)-5-oxopentanoic acid as side product in the EAS reaction, generating an impurity in Ezetimibe

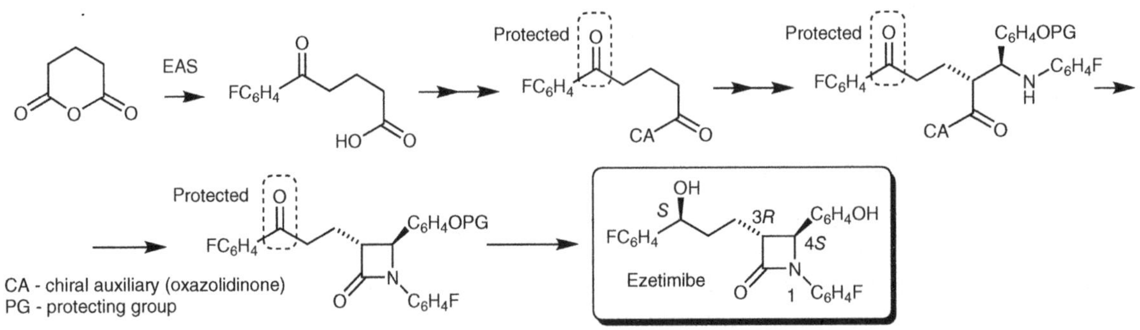

Scheme 6.32 *p*-Fluoroaromatic ring coupling/Mannich-type aldol reaction/N-C2 cyclization and asymmetric carbonyl reduction sequence

Coupling to a chiral oxazolidinone and carbonyl protection, followed by an asymmetric enolate addition to imine and then N-C2 ring closure, is routinely utilised. For success, any potential reaction involving the benzylic ketone must be prevented. Deprotection of the ketone, asymmetric reduction and deprotection of the phenol completes the synthesis. Processes following this strategy are summarized in the schemes G to J.

Scheme G shows the various carbonyl protected intermediates used in this synthetic strategy. Formation of the *N*-acyl oxazolidinone carbonyl protected intermediates is trouble free and good yields have generally been reported. The most frequently applied sequence is amide formation (**G5**) followed by protection of the ketone. The amide formation can be carried out using typical coupling reagents such as DCC, although reaction with a pivaloyl mixed anhydride in the presence of a tertiary base is a commonly employed procedure. After formation of the mixed anhydride it is reacted with (*S*)-4-phenyloxazolidin-2-one to form the desired amide. The benzylic ketone of the *N*-acyl oxazolidinone derivative is then protected as a cyclic or non-cyclic ketal, a cyclic thioketal, a cyanohydrin *O*-silyl ether or an *N*-protected oxime. These derivatives are highlighted in the lower part of Scheme G. Their use in the asymmetric enolate addition to the imine is shown in schemes H, I and J.

6.4.8.1 Asymmetric enolate addition of ketal protected *N*-acyl oxazolidinones to the imine

The enolate-imine addition is carried out using titanium-trichloride-isopropoxide, usually prepared *in-situ* by mixing Ti(IV)-isopropoxide and $TiCl_4$ in a 1:3 ratio in the reaction solvent, typically methylene chloride, at low temperatures (usually –30 °C to –40 °C). The reagent is then added to a mixture of the protected imine, the ketal derivative of the *N*-acyl oxazolidinone and a tertiary amine base in an inert water-free solvent at low temperature. Since the enolate addition is reversible (RichterGedeon2005), any isomeric intermediate in the methanolic mother liquors after filtration of the product can be converted to the desired product by subjecting it once more to the reaction conditions, thereby increasing the global yield of the conversion (**H7**, 78% yield for first crop; increased to 89% with the product recovered from the mother liquors, Scheme 6.33).

Understandably, this step is key for this strategy to manufacture Ezetimibe as both asymmetric centres of the ring are created. Notably, in most of the synthetic routes using this strategy, this step has the lowest yields. Diastereoselectivity resulting from the possible combinations of the numerous protecting groups in the imine piece and in the *N*-acyl oxazolidinone piece is not easily evaluated and this aspect remains mostly unreported in the literature.

In a study to synthesise the stereoisomers of Ezetimibe (Scheme 6.34, TL2013) an 8 to 1 diastereoselectivity in the enolate-imine addition step for the 5,5-dimethyl-1,3-dioxane/TBDMS protecting group combination was reported whilst the *S*,*R*,*R* isomer was isolated from the mother-liquors.

Not surprisingly, the same relative diastereomeric ratio was reported for the same protecting group combination when the enantiomeric (*R*)-4-phenyloxazolidin-2-one was used as chiral auxiliary to produce a mixture of the *S*,*S*,*S* and *R*,*S*,*R* isomers (CCL2014, Scheme 6.35).

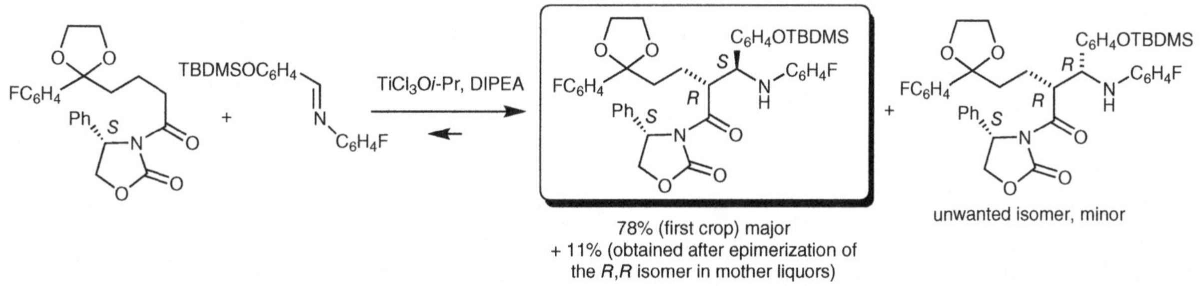

Scheme 6.33 Diastereoselectivity in the enolate-imine addition step

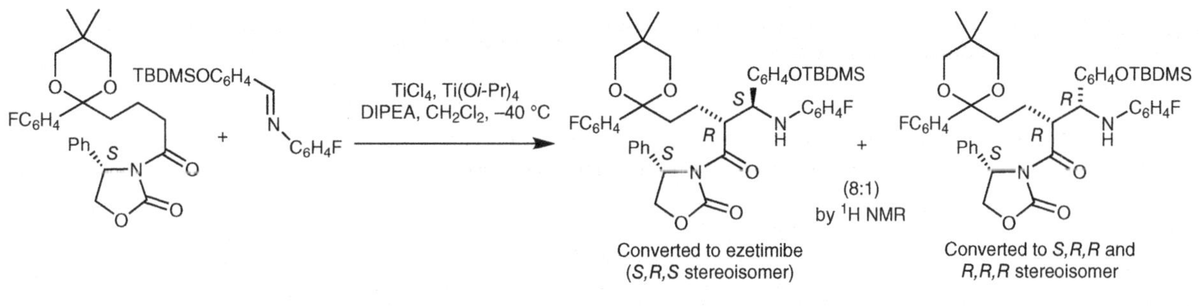

Scheme 6.34 Diastereoselective ratio in a enolate-imine addition step

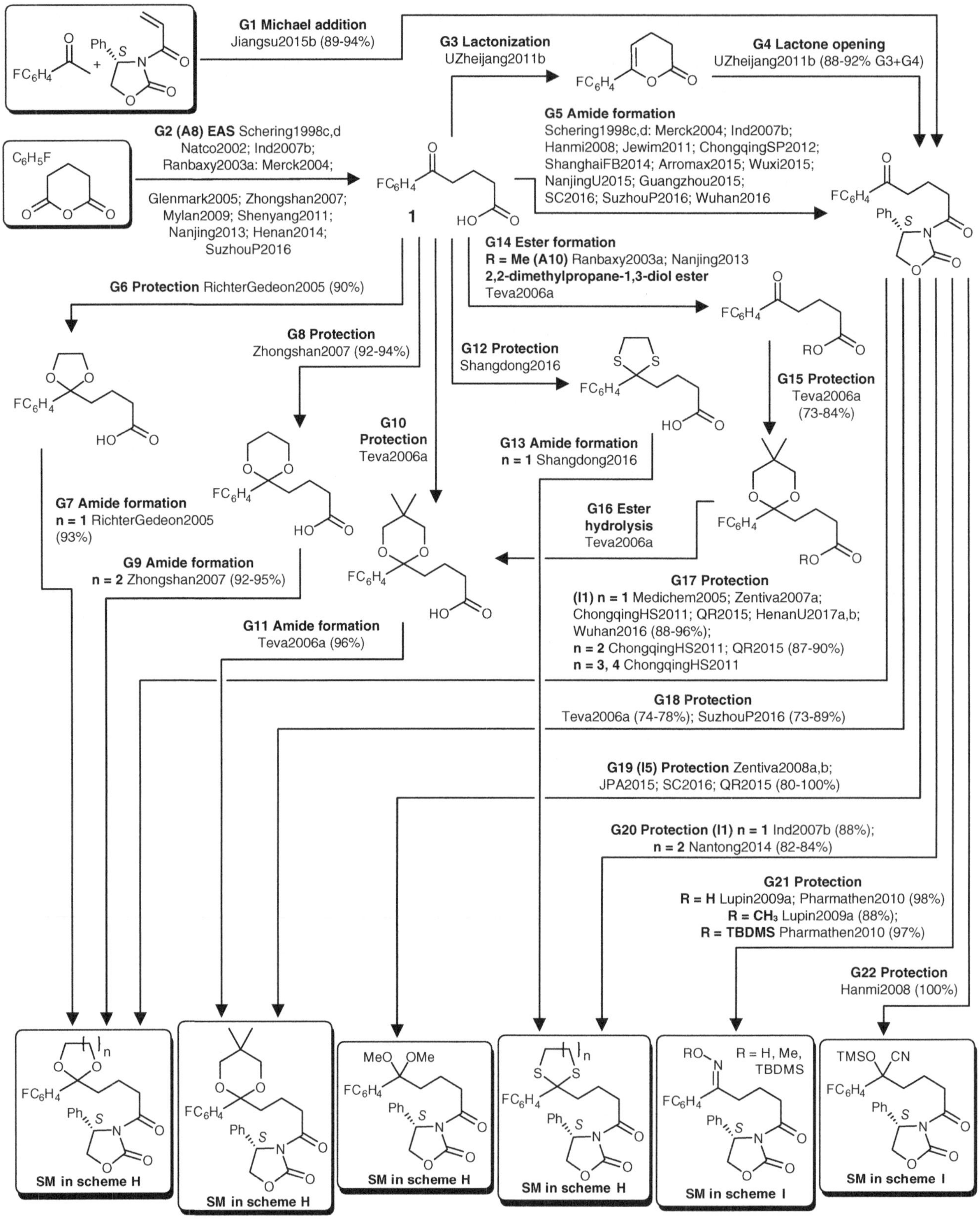

Scheme G Synthesis of carbonyl protected *N*-acyl oxazolidinones intermediates (for reaction conditions see page 265)

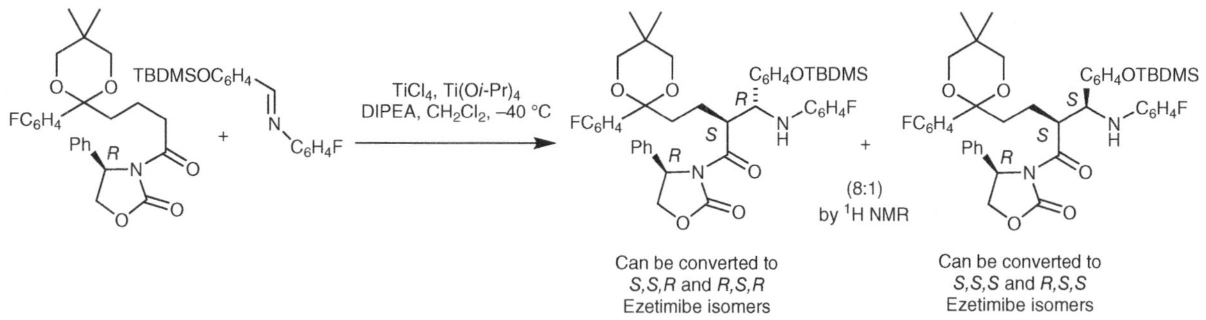

SCHEME 6.35 DIASTEREOSELECTIVE RATIO IN THE ENOLATE-IMINE ADDITION STEP USING (*R*)-4-PHENYLOXAZOLIDIN-2-ONE AS CHIRAL AUXILIARY

Source: Modified from CCL2014

Low levels of the unwanted isomer (<1%) were detected in the isolated intermediate. Hence, it can be seen that the excellent stereodirecting power of the 4-phenyloxazolidin-2-one in this crucial enolate condensation goes on to generate the desired β-lactam isomer efficiently.

A word of caution is in order when analysing the data given in the patent literature. Frequently, moderate yields quoted can be greatly improved by appropriate process development.

The first of the four enolate-imine additions shown Scheme H (top) use a 1,3-dioxolane to protect the ketone along with TMS, TBDMS, Bn and acyl protecting groups for the hydroxyl function, hence allowing a comparison of these different OH protecting groups: TMS 32-41% (**H1**); TBDMS 34-89% (**H7**); Bn 49-65% (**H13**) and COR 74-87% (**H23**). When an imine *N*-oxide was used, the yield increased from 49-63% to 84% (**H13**) but this route suffers from the fact that treatment with metallic zinc is necessary to convert the *N*-oxide to the NH group.

In general, yields vary from moderately low to high. Of note is that the use of a 1,3-dioxane as protecting group along with a TMS-protected imine gave 87-91% in the condensation step (**H27**).

The TMS group is introduced by reaction with TMSCl, which is cheap in comparison to *t*-butyl-dimethyl-silyl chloride (TBDMSCl). However, the TMS group is extremely labile and prone to cleavage even under mild condition hence, re-silylation may be required prior to isolation. Because of this ease of removal, a TMS protected phenol is always de-protected before the ketone. Therefore, in the final step of the synthesis, EZT ketone is reduced asymmetrically to provide Ezetimibe directly (**H4**).

The *t*-butyl-dimethyl-silyl (TBDMS) protecting group is much more stable under the reaction conditions used during the synthesis hence, there is no need for re-silylation before isolation of the intermediate but, the expense of TBDMSCl should be taken into consideration. Removal of silyl protecting groups are easily achieved by treatment with acid.

The other hydroxyl protecting group used in conjunction with 1,3-dioxane ketone protection is benzyl, which has the advantage that it is stable thoroughout the whole sequence. The removal of the benzyl group by hydrogenolysis was discussed above (section 6.4.7.4), wherein the use of hydrogenation conditions is of concern due to the formation of unwanted impurities.

The TBDMS and benzyl protecting groups are customarily removed in the final step of the synthesis, after carbonyl deprotection and reduction, allowing the asymmetric carbonyl reduction to take place on a protected phenol.

1,3-Dithiolanes and 1,3-dithianes have both similarly been used to protect the carbonyl group of the *N*-acyl oxazolidinone, but the yields in the enolate-imine addition step were slightly lower (**H13** and **H27**).

The enolate-imine addition product is usually cyclized by reacting with a silylating agent (BSA, bis(trimethylsilyl)acetamide) followed by the addition of a catalytic amount of a fluoride source (commonly TBAF). Observed yields are frequently above 85%. During the cyclization, (*S*)-4-phenyloxazolidin-2-one is released and hence, this expensive key raw material can be recovered and recycled. Reports indicate that 85% or more of the (*S*)-4-phenyl-2-oxazolidinone can be recovered from the mother liquors (85% RichterGedeon2005, >95% Qilu2016).

One patent claimed a process for the purification of BnEZT ketone *via* formation the neopentyl ketal (also used as a protecting group in its own right), followed by acidic hydrolysis (**H38+H37**, Teva2006b).

Up until now cyclization always precedes the reduction-deprotection sequence, but the asymmetric reduction of the ketone before cyclization (scheme I) has also been reported. In these few cases, the sequence adopted was: *p*-fluoroaromatic ring coupling; asymmetric enolate addition to the imine; asymmetric carbonyl reduction and N-C2 cyclization. This however, requires that the newly formed (*S*)-hydroxy group be protected before β-lactam formation.

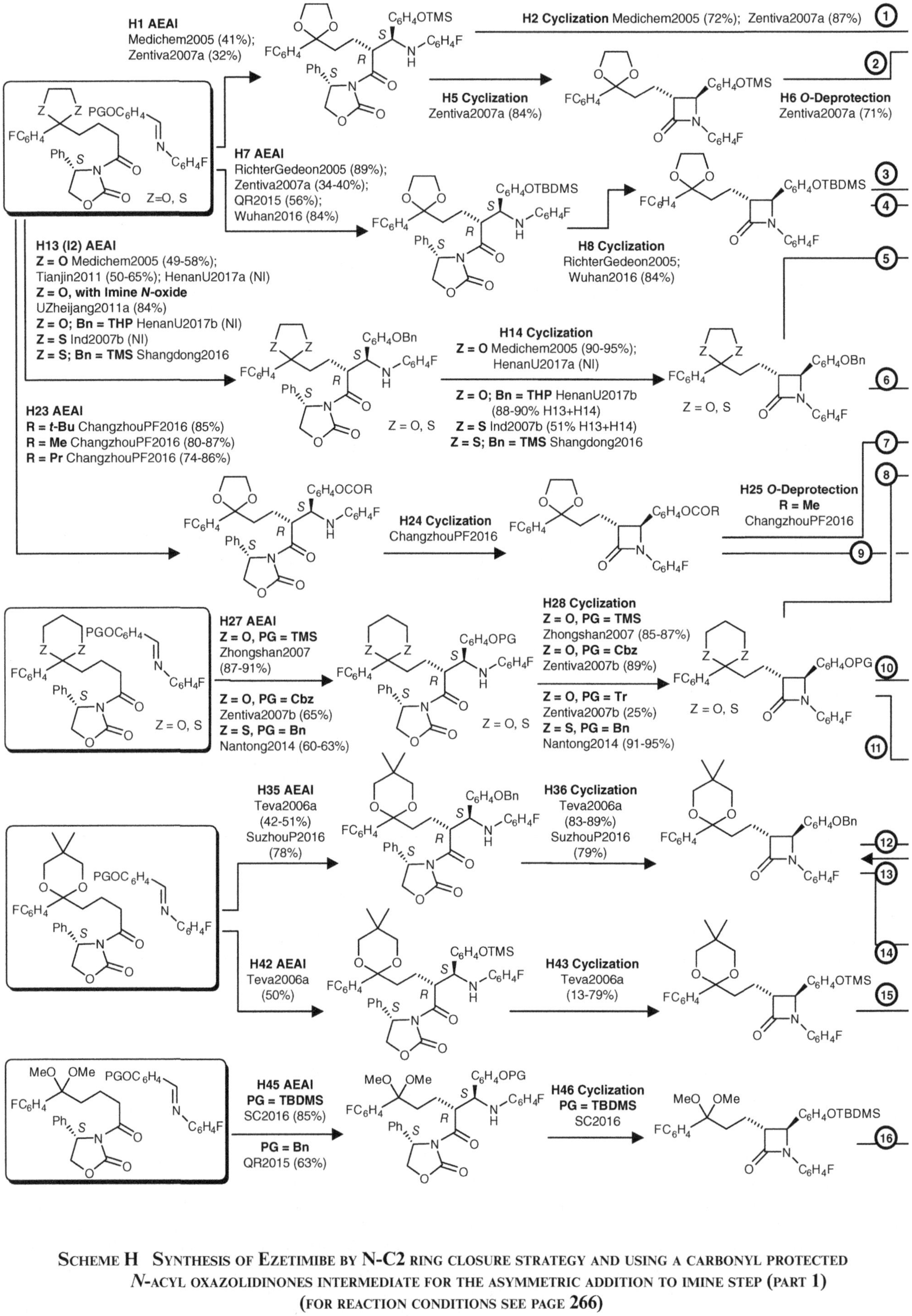

SCHEME H SYNTHESIS OF EZETIMIBE BY N-C2 RING CLOSURE STRATEGY AND USING A CARBONYL PROTECTED *N*-ACYL OXAZOLIDINONES INTERMEDIATE FOR THE ASYMMETRIC ADDITION TO IMINE STEP (PART 1) (FOR REACTION CONDITIONS SEE PAGE 266)

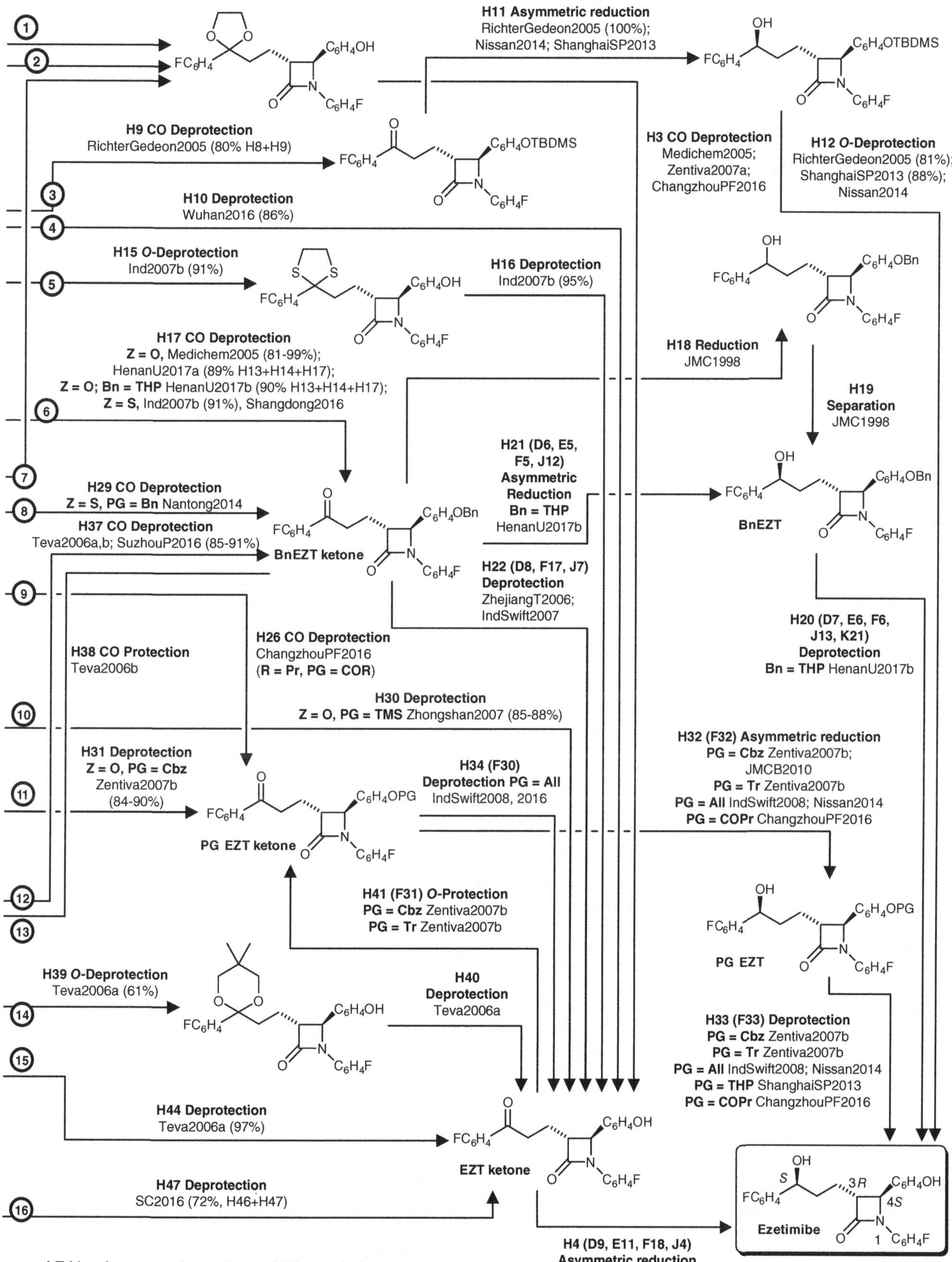

AEAI = Asymmetric enolate addition to imine or amine oxide

NI = not isolated

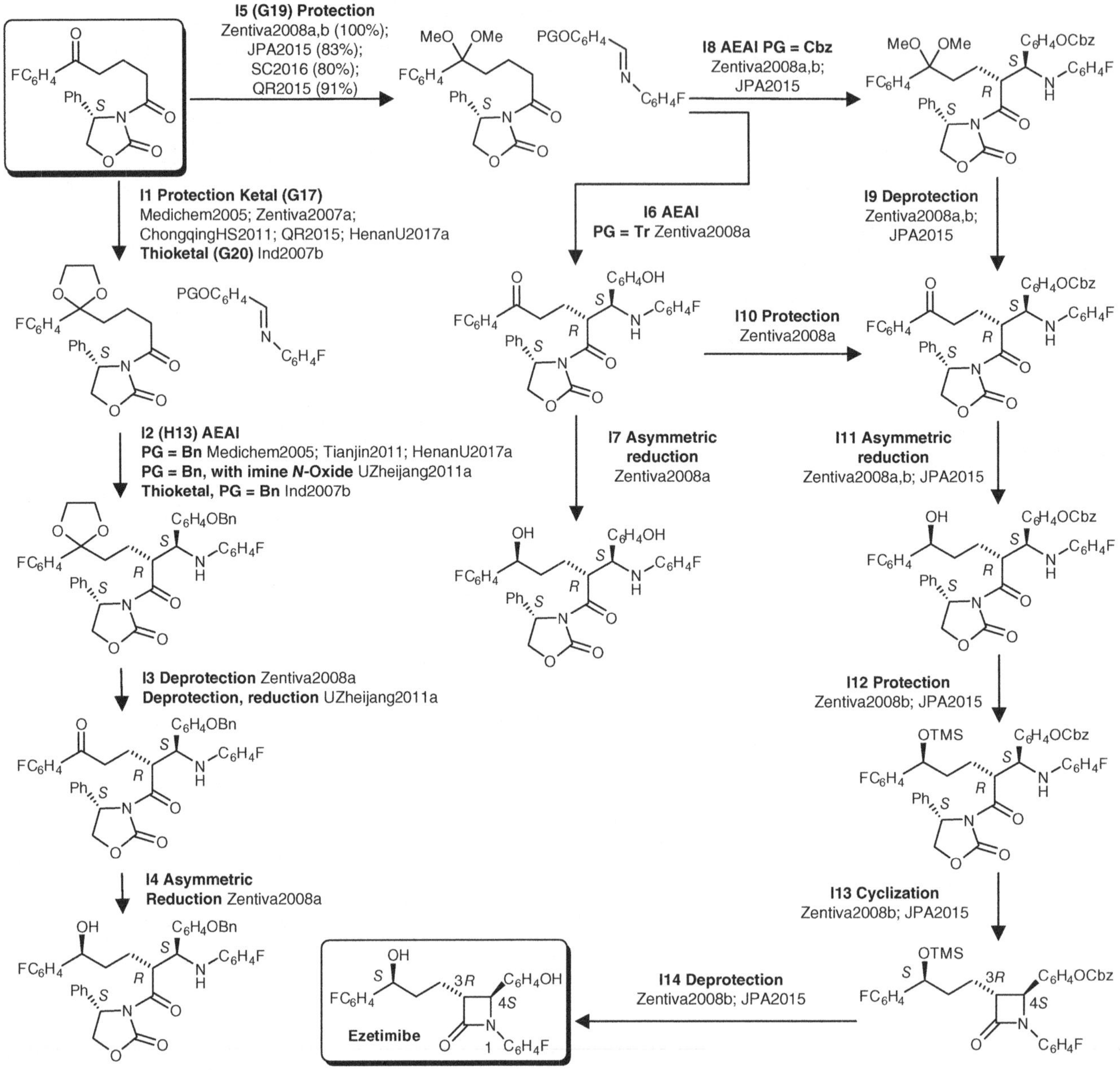

AEAI - Asymmetric enolate addition to imine
For the synthesis of starting oxazolidinone derivatives see scheme A, G; for imines or imine oxides see scheme B

Scheme I Synthesis of Ezetimibe by N-C2 ring closure strategy and using a carbonyl protected *N*-acyl oxazolidinone intermediate for the asymmetric addition to imine step (part 2) (for reaction conditions see page 267)

6.4.8.2 Enolate-imine addition using protecting groups other than a ketal for the ketone

Further processes where the key asymmetric step is the enolate-imine addition between a carbonyl protected (not ketal) *N*-acyl oxazolidinone and a protected imine are shown in scheme J.

The keto protecting groups in these cases are *N*-substituted oximes and an *O*-silyl cyanohydrin. Some of the oxime-protected compounds show similar results to those observed in their ketal-protected counterpart. The *O*-TBDMS oxime affords high yields when cyclization is carried out using the BSA/TBAF protocol by taking advantage of the difference in the reactivity of the two silyl groups (**J10**). When an *O*-TMS oxime protection is employed, its lability would be expected to create problems during the cyclization (**J2**, yield not reported). In the case of the *O*-silyl cyanohydrin, cyclization is carried out using a hindered strong nitrogen base (LiHMDS), under which conditions the protecting group is anticipated to be stable. This is followed by ketone regeneration during work-up (**J15**).

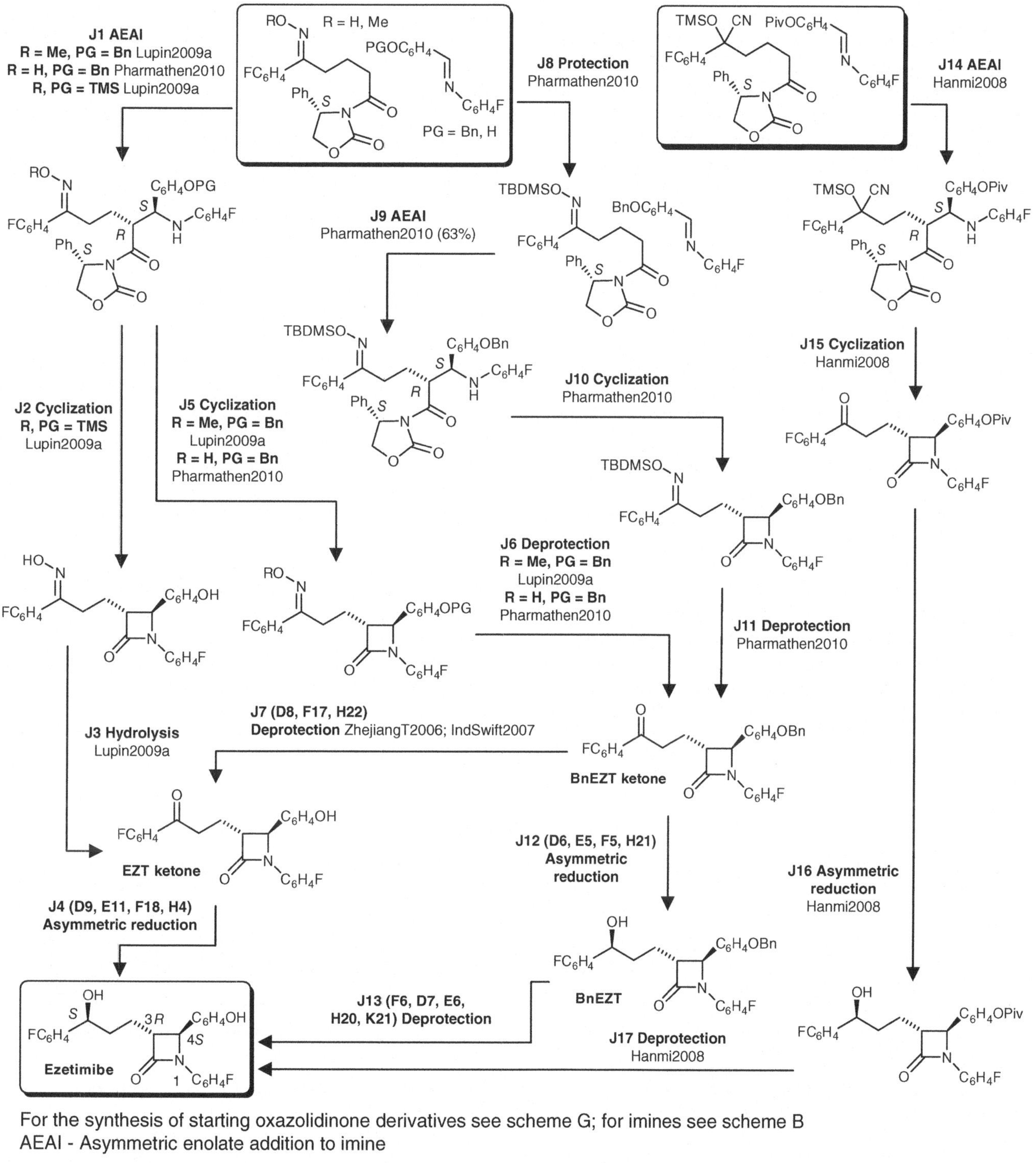

Scheme J Synthesis of Ezetimibe by N-C2 ring closure strategy and using a carbonyl protected *N*-acyl oxazolidinone intermediate for the asymmetric addition to imine step (part 3) (for reaction conditions see page 267)

Several companies have proposed Ezetimibe synthetic routes following this strategy and it may be that some are possibly used at industrial scale. The selection of the appropriate protecting group for the ketone and the phenol groups is clearly a key issue to be defined after a convenient process development program.

6.4.9 N-C2 ring closing strategy: implementing the reduction at the beginning of the route

In all the approaches described in the previous section, the *p*-fluoroaromatic ring was in place at an early stage of the synthesis thus avoiding the problems associated with the late attachment of this piece (see 6.4.7). The presence of a benzylic carbonyl group was fundamental and necessitated protection. In addition, diastereoselective reduction of the carbonyl group to the (*S*)-benzylic hydroxyl group was carried out towards the end of the sequence. The final reduction and deprotection steps require extensive development work to control impurities resulting from the conditions employed in these two reactions.

A strategy that avoids the problems associated with a late stage reduction is to bring this step forward to the beginning of the sequence, see the outline shown in Scheme 6.36. The processes following this strategy are summarized in schemes K, L and M.

The chemistry has already been discussed in detail at section 6.4.8, the sequence being initiated by attachment of *p*-fluoroaromatic piece to glutaric anhydride by an EAS.

6.4.9.1 The asymmetric reduction of the *N*-acyl oxazolidinone intermediate

(*S*)-(+)-4-phenyloxazolidin-2-one was, with few exceptions, used as the chiral auxiliary to induce asymmetry in the Ezetimibe molecule (see following section 6.4.9.4). This gave rise to extensive investigation of the asymmetric reduction of the *N*-acyl derivative (**K3**) resulting in several chemical and enzymatic conversions being reported.

A Corey-Bakshi-Shibata reduction with chiral oxazaborolidine catalysts is a favoured option, generally using (*R*)-(+)-2-Methyl-CBS-oxazaborolidine (Figure 6.11). High yields and diastereoselectivites have been reported (frequently >95% yield, >98% de) using 5% or less of the chiral catalyst. Solutions of borane tetrahydrofuran complex (BTHF) or borane dimethylsulfide complex (BMS), both of which had become commercially available at the time, or a *N*,*N*-diethylaniline complex (BH_3•DEA, BDEA), prepared *in situ* from sodium borohydride ($NaBH_4$), diethylaniline and dimethylsulfate, were used as reducing agents..

BMS is generally considered to be more stable, although less reactive than BTHF, and has been widely used in CBS catalysed ketone reductions. Each of these reagents has issues: BMS releases dimethyl sulphide, a flammable, low boiling point liquid (BP 37 °C) with an extremely disagreeable odour; BTHF decomposes to tributyl borate and hydrogen and at temperatures above 50 °C decomposes to give pyrophoric diborane which forms explosive mixtures with air. BTHF requires careful handling, must be stored between 0 and 5 °C and reactions carried at below 35 °C. Due to safety issues, BTHF has now been discontinued and is no longer a valid option.

Boron-based methodologies using borane-pyrrolidine complex have also been reported (97% yield, 93% de, 50 g scale, Kotobuki2006). A particularly selective reduction using a stoichiometric amount of (–)-DIP-Cl was described to give 100% yield and 99.2% de on a 100 g scale (HeteroDrugs2003). Several similar approaches have been reported (Ranbaxy2003b Enantiotech2015, Jewim2011, Arromax2015).

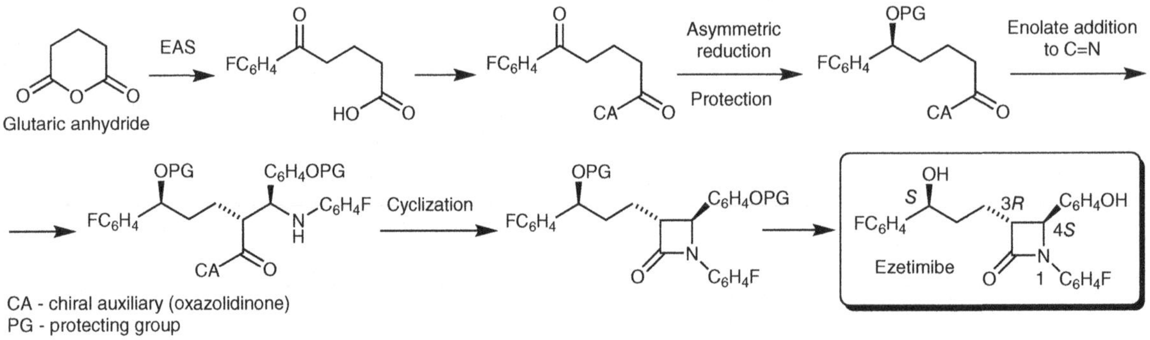

Scheme 6.36 The *p*-fluoroaromatic ring coupling/asymmetric carbonyl reduction/asymmetric enolate addition to imine/N-C2 cyclization sequence

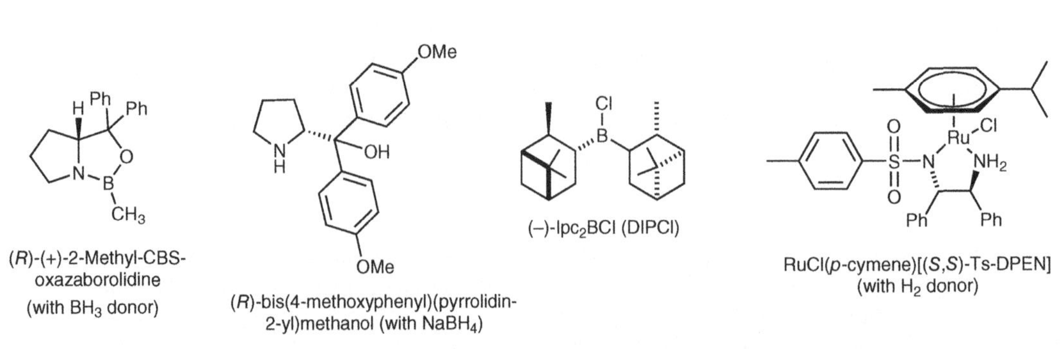

Figure 6.11 Selected catalysts for the asymmetric reduction of the *N*-acyl oxazolidinone

A ruthenium-catalysed asymmetric transfer hydrogenation was the subject of yet another patent, giving 70-93% conversion, 80-92% de (on an undisclosed scale) using 0.1% to 0.3 % mol of the metal catalyst (Pharmathen2013).

Microbial reductions using broth cultures of *Schizosaccharomyces octosporus* ATCC 2479 and *Zygosaccharomyces bailii* ATCC 38924 have also been patented (Merck1995, Schering1996a, 1999). Bioengineered ketoreductase enzymes were employed and patented in this asymmetric reduction (Codexis2008, HenanFP2011, EnzymeWorks2016). Quantitative conversion, with >99% de, has been observed on a 120 g scale (Codexis2008).

6.4.9.2 On the asymmetric enolate addition to imine reaction

In the asymmetric addition to imine step, the (*S*)-hydroxy *N*-acyl oxazolidinone intermediate is protected with a suitable hydroxy-protecting group, preferably a silyl protecting group such as that derived from chlorotrimethylsilane (TMSCI, **K4**) or *t*-butyldimethylsilyl chloride (TBDMSCI, **K27**), benzyl (Bn, **K39**), methoxymethyl (MOM, **K32**) and acetyl (Ac, **K24**) protecting groups have also been used (scheme K). Protection may be carried out *in situ* as with the TMS or MOM protecting groups or may require a separate step, as is the case with acetyl and TBDMS protection. Only two reports of an unprotected oxazolidine intermediate in the asymmetric addition have been published (**K22**, Jiangsu2011, Nanjing2015).

As mentioned previously, the phenol group in the imine intermediate also requires protection. Several options are available. When one protecting group is used for both functionalities, some degree of simplification is expected. In one application the unprotected imine and the (*S*)-hydroxy *N*-acyl oxazolidinone were mixed in an inert solvent, followed by addition of TMSCl to produce a mixture of the two protected reactants (Scheme 6.37). Thereafter, the enolate-imine addition/cyclization sequence was carried out to give the β-lactam (**K5+K6**). A similar strategy was used when MOM is used as a unique protecting group (**K32+K33**).

The following generic procedure was applied in these cases: the unprotected imine (usually 1 to 3 equivalents) and the (*S*)-hydroxy *N*-acyl oxazolidinone intermediate are mixed together in an anhydrous solvent (for example CH_2CI_2) in the presence of a tertiary amine base (frequently DIPEA, preferably 2-4 equivalents) and TMSCl (or MOMCl) added to protect the free hydroxyl groups of each substrate. After protection is complete the mixture is treated with at least one equivalent of a Lewis acid (usually $TiCI_4$ or $TiCl_3O$*i*Pr), in the presence of a tertiary amine base such as DIPEA (preferably 1-3 equivalents) for 2-4 hours at –20 °C to –35 °C (steps **K4+K5**).

The reaction is quenched by treating with an acid (glacial acetic acid – which must be kept above 20 °C to keep it liquid – is typically used) followed by addition of an aqueous solution of tartaric acid and sodium bisulfite. In the case of the TMS protecting group, re-silylation (with BSA) is necessary during the work-up as some hydrolysis of the OTMS groups occurs. The product can be isolated as solid di-TMS Ezetimibe by crystallisation.

In a report of a pilot plant batch (54 kg scale of hydroxy *N*-acyl oxazolidinone, Jiangxi2015), the di-hydroxy compound was isolated (CH_2Cl_2, H_2O, $NaHSO_3$) in 65-70% yield (protection of both OH functions, enolate-imine addition and deprotection, **K4+K5+K8**). Conversion to the cyclized product is carried out by treatment with an excess of BSA, followed by TBAF (**K9+K8**). The uncyclised

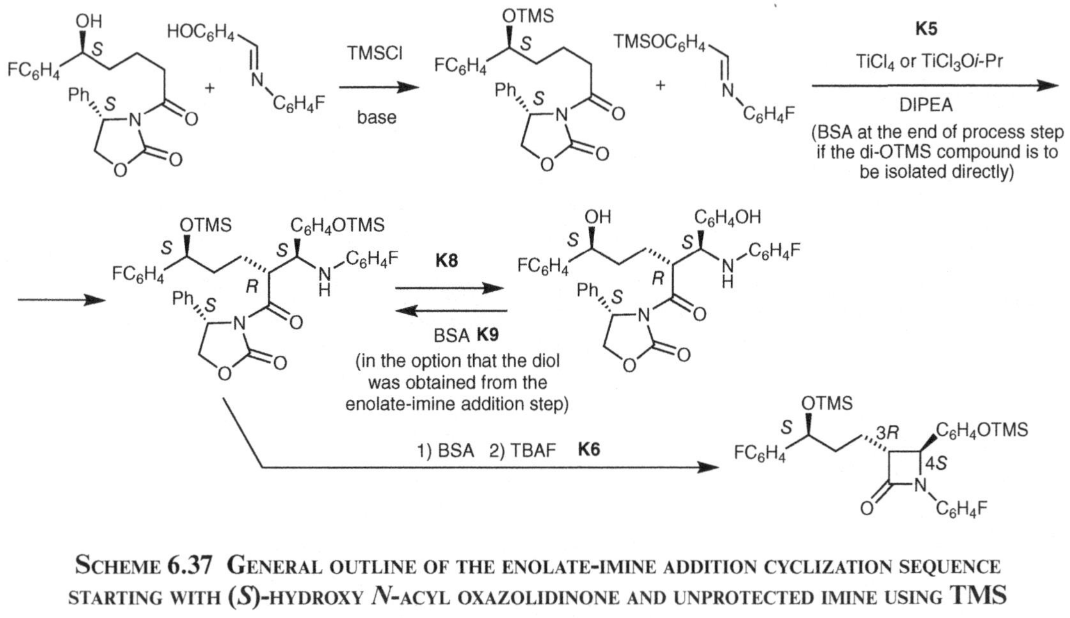

Scheme 6.37 General outline of the enolate-imine addition cyclization sequence starting with (*S*)-hydroxy *N*-acyl oxazolidinone and unprotected imine using TMS as sole protecting group

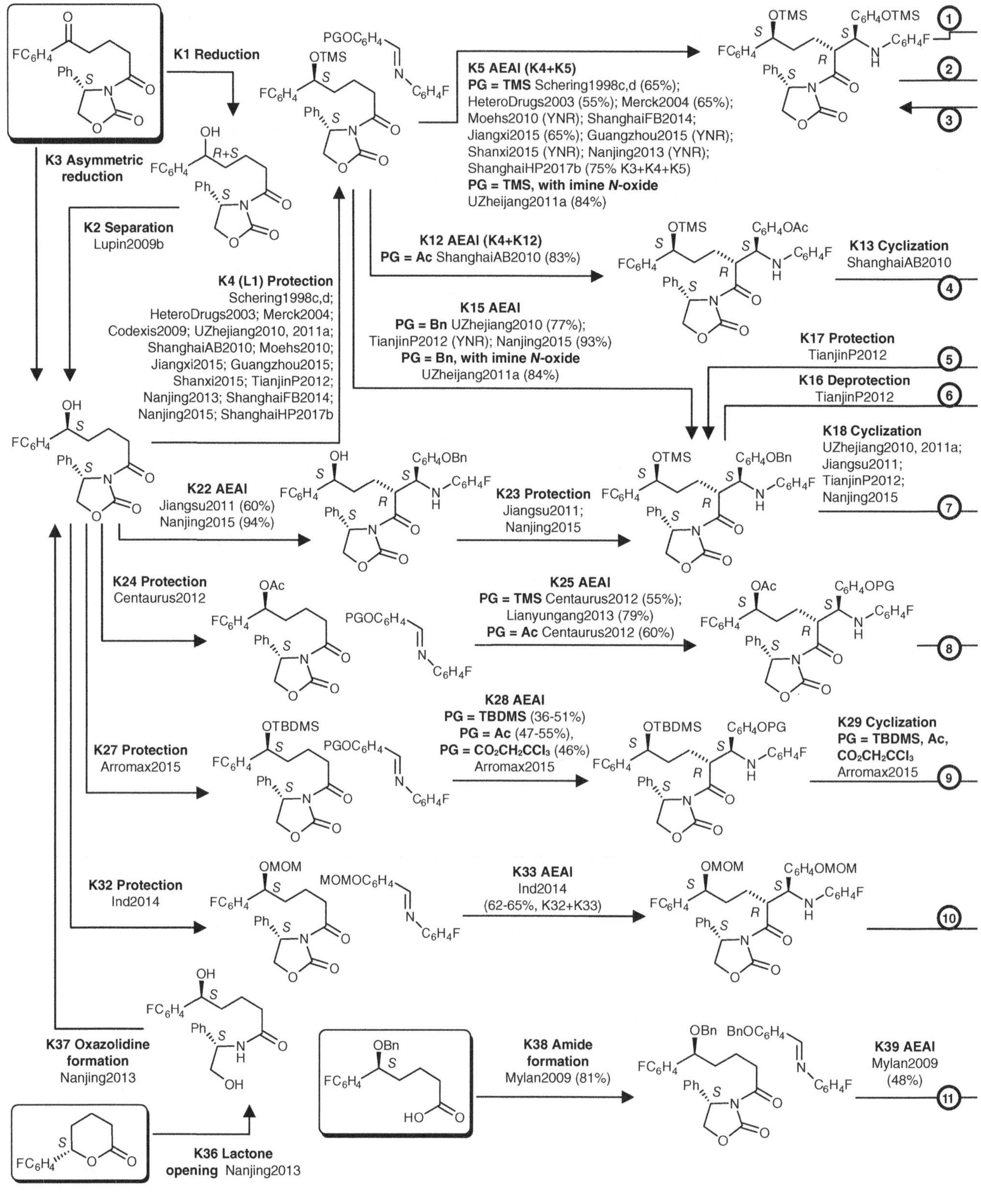

Scheme K Synthesis of Ezetimibe by N-C2 ring closure strategy and using a hydroxyl protected *N*-acyl oxazolidinone intermediate for the asymmetric addition to imine step (part 1) (for reaction conditions see page 267)

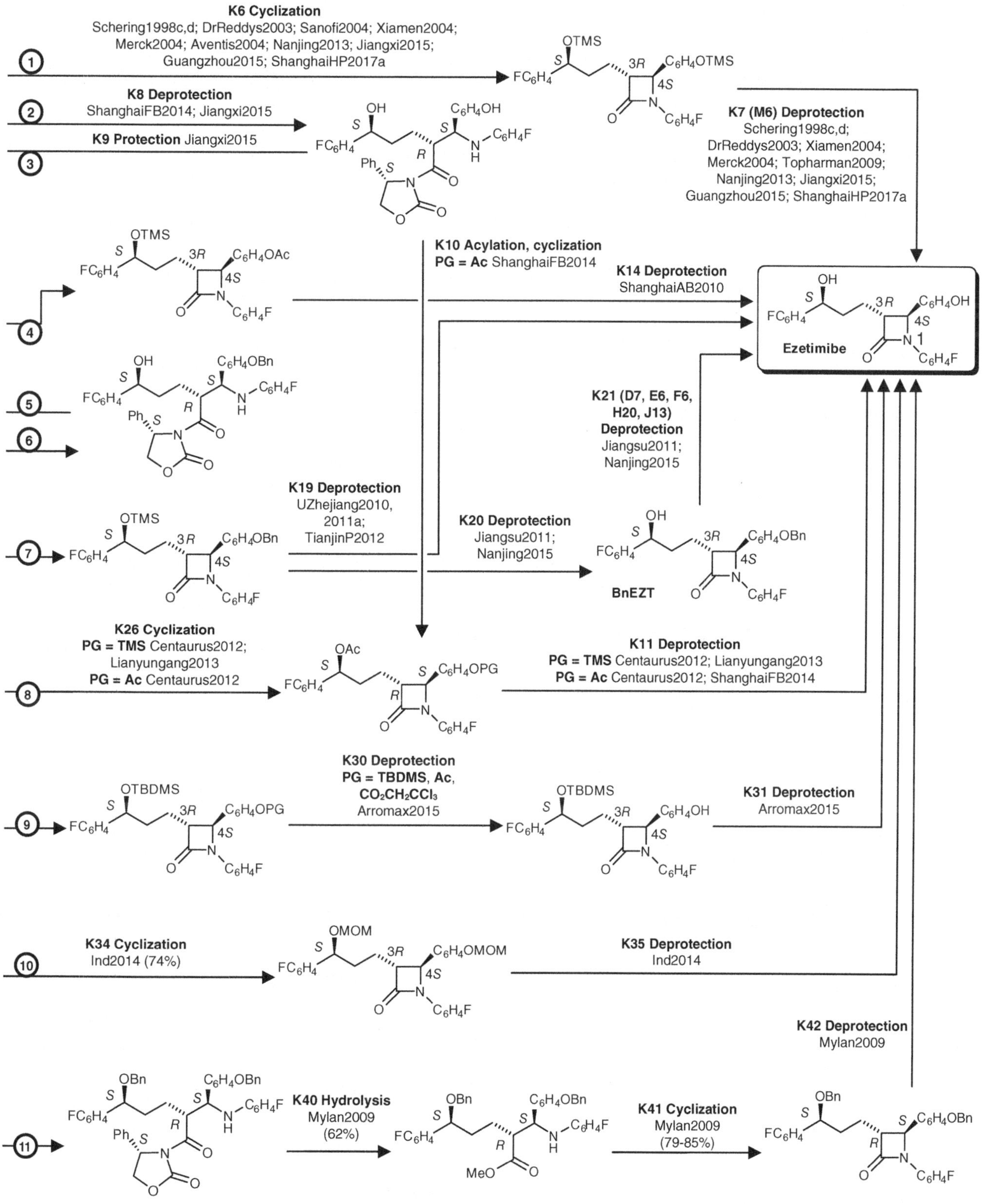

K3 Asymmetric reduction

Biochemical reduction Merck1995, Schering1996a, 1999; Codexis2008; HenanFP2011; EnzymeWorks2016.
CBS Schering1998c,d; Schering2001; TL2003, 2007; Merck2004; Zach2006; TianjinP2012; Ind2014; ShanghaiFB2014; Arromax2015; Wuxi2015; Guangzhou2015; ShanghaiHP2017b.
$NaBH_4$ Kotobuki2006. **DIPCl** HeteroDrugs2003; Ranbaxy2003b; Jewim2011; Enantiotech2015; Arromax2015.
CAH Pharmathen2013.

S,R,S,S di-hydroxy intermediate is stated to be unstable elsewhere (JMS2011).

Nitrones have been used to increase the electrophilic character of the CN double bond to facilitate the enolate-imine addition, (UZheijang2011a, **K5** and **K15**). Zinc reduction is then used to substitute the oxygen for hydrogen, giving the product in higher yield.

In most of the publications, the diastereoselectivity of the enolate addition to the imine is not revealed, but the majority of the yields for the isolated intermediate are generally in the high sixties. One example that was reported, (Moehs2010), the synthesis employed di-TMS protection and produced a 2.6 to 1 mixture of stereoisomers in the crude reaction mixture (by NMR). This translates to a maximum yield of 72% and the product being isolated in 55-65% yield.

6.4.9.3 On the cyclization-deprotection

As in the case of the carbonyl protected intermediate (section 6.4.8), the protected benzylic alcohol is cyclized to the azetidine-2-one by the use of a silylating agent, such as bis(trimethylsilyl) acetamide (BSA), in inert solvent followed by the addition of silyl releasing reagent. The cyclization step occurs readily to give the 3*R*,4*S* azetidine-2-one (*trans* isomer), usually in high yields (sometimes greater than 90%).

The fluoride source is typically a quaternary ammonium fluoride salt, TBAF being the most common choice. The reagent is used in catalytic amounts (5% - a hydrated form being usual). In some examples both reactants (BSA and TBAF) were added simultaneously. Caesium fluoride or potassium fluoride have also been used. The removal of the protecting groups after cyclization is carried out using conventional methods.

In a study aimed at synthesising various ring stereoisomers (Scheme 6.38, JMS2011), cyclization of the 4*R* isomer (the major impurity formed in the condensation step), was not achievable, presumably because transition state to the *cis* ring is too sterically congested. This implies that small amounts of this impurity remaining in the imine-aldol intermediate will not to be cyclized.

Removal of the protecting groups, most of which are readily cleaved, is achieved by use of standard methods. The TMS and MOM groups are obviously the easiest to cleave. Hydrogenolysis of a benzyl protecting group was discussed previously (see section 6.4.7) and requires strict control to minimise side reactions.

6.4.9.4 Asymmetric enolate addition to other imines and other chiral auxiliaries use

The route shown in Scheme 6.36, wherein an asymmetric enolate addition of a TMS protected (*S*)-hydroxy *N*-acyl oxazolidinone to a CN double bond is the key step (scheme K, **K5**, **K12**, **K15**), has been generically applied using modified imines as electrophiles (Figure 6.12 and scheme L). In these syntheses, one of the aromatic ring of the imine is modified chemically at a later stage to introduce the substitution pattern of Ezetimibe.

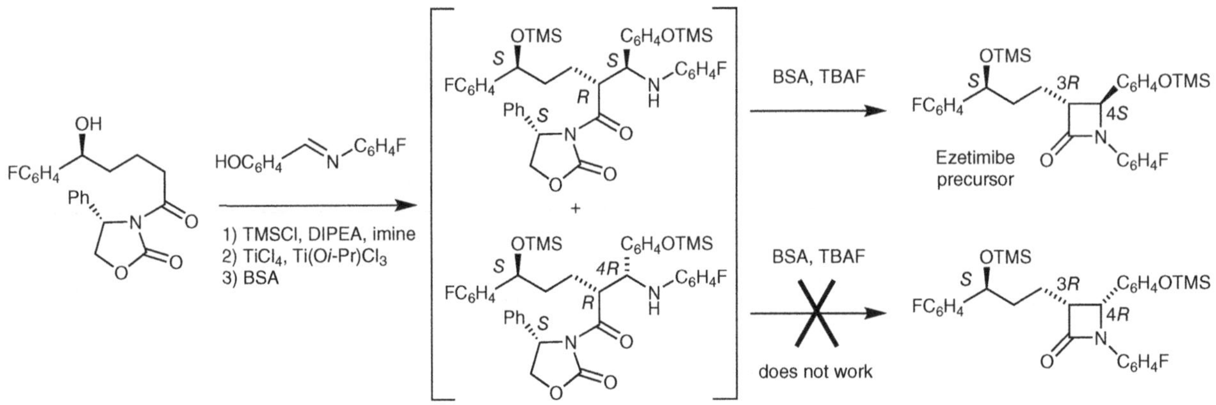

Scheme 6.38 Non-cyclization of 4*R* intermediate under usual reaction conditions

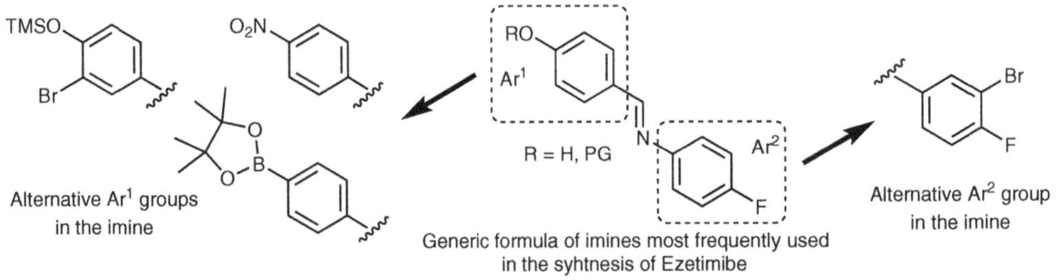

Figure 6.12 Imines used in the Ezetimibe synthesis

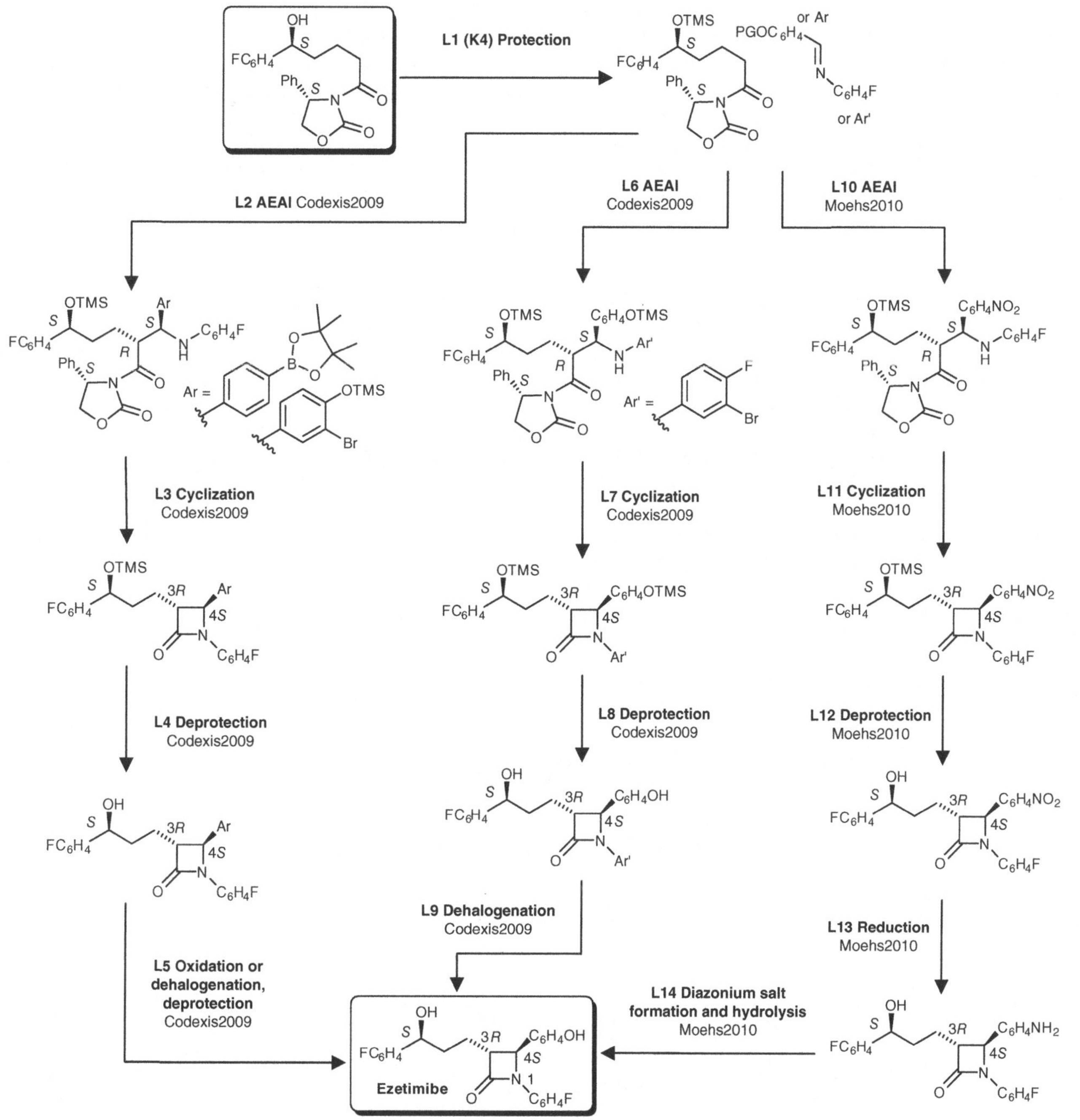

AEAI - Asymmetric enolate addition to imine

For the synthesis of starting oxazolidinone see scheme K; for imines see scheme B

Scheme L Synthesis of Ezetimibe by N-C2 ring closure strategy and using a hydroxyl protected *N*-acyl oxazolidinone intermediate for the asymmetric addition to imine step (part 2) (for reaction conditions see page 269)

Use of the aryl boronate imine (**L2**, Codexis2009) was reported to give a 54% yield for the condensation step. Here a final oxidation is necessary to replace the boron with oxygen (**L5**). When the *para* substituent of the aromatic group of the imine is nitro, a reduction/diazotization/hydrolysis sequence is applied to introduce the phenol group (**L13+L14**), a yield of 27% being reported for the condensation step (**L10**). A 3-bromo-4-fluoro substituted aromatic ring of the imine has also been used (Codexis2009, yields not reported). The removal of the bromine atom was carried out by catalytic hydrodehalogenation at the end of the synthesis. The yield (37%) was reported in only one case (**L9**). Although these sequences can be seen as being independent from all previously patented routes, they introduce a higher degree of complexity to the synthesis and are hence, probably not viable on an industrial scale.

Collating all the data published on the synthesis of Ezetimibe, it can be inferred that (*S*)-4-phenyloxazolidin-2-one is the chiral auxiliary of choice. Any impact its price may have on the cost of the API is diminished as it is conveniently recovered (>85%) from the mother liquors of the cyclization step. Three other chiral auxiliaries have been described to synthesise Ezetimibe applying the synthetic strategies shown in this section (Figure 6.13 and scheme M).

Unfortunately, none offered any significant improvement. In addition, their higher costs substantially reduces their attractiveness for industrial application.

6.4.9.5 Other N-C2 cyclizations

A number of alternative routes which still maintain a N-C2 cyclization have been reported.

In the first example, the key step is an asymmetric allylic amination, catalysed by a chiral allyl palladium complex, of a Morita-Baylis-Hillman adduct by an aromatic amine (ShanghaiIOC2012, ACIE2012, Scheme 6.39). The (*R*)-β-amine ethyl ester was obtained in 91% yield and 95% ee using 1.5 mol% of $[Pd(all)Cl]_2$/(*S*,*S*,*S*)-spiroketal bisphosphine ligand, thus setting up the configuration for the future C4 carbon of the β-lactam ring. After Michael addition of an aryl β-keto ester the cyclization step is triggered by LiHMDS in THF at –20 °C, the base isomerising any *cis* product to the more stable *trans*. Hence, a single isomer (3*R*,4*S*) of the β-lactam was obtained in 77% yield and 95% ee. The conversion to Ezetimibe then followed known procedures.

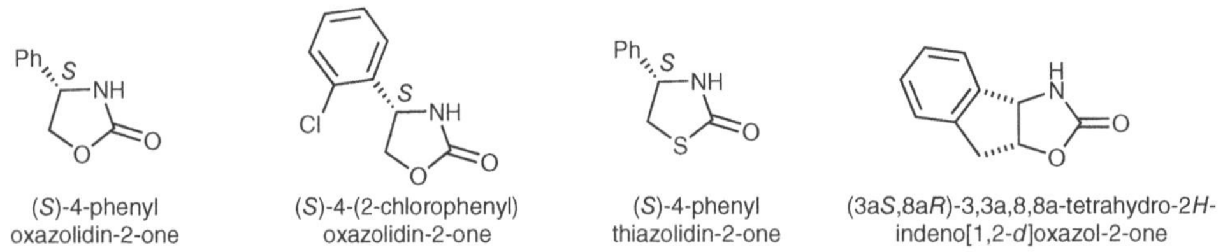

Figure 6.13 Chiral auxiliaries used in the asymmetric enolate addition to imine reaction

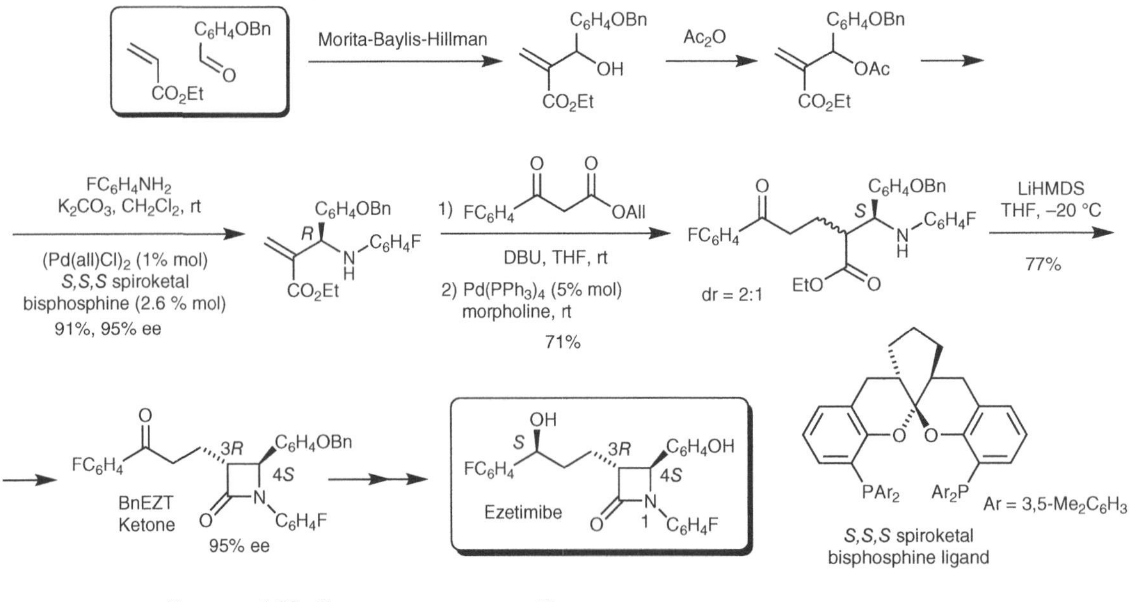

Scheme 6.39 Synthetic route to Ezetimibe using an asymmetric allylic amination of Morita-Baylis-Hillman adduct as key step

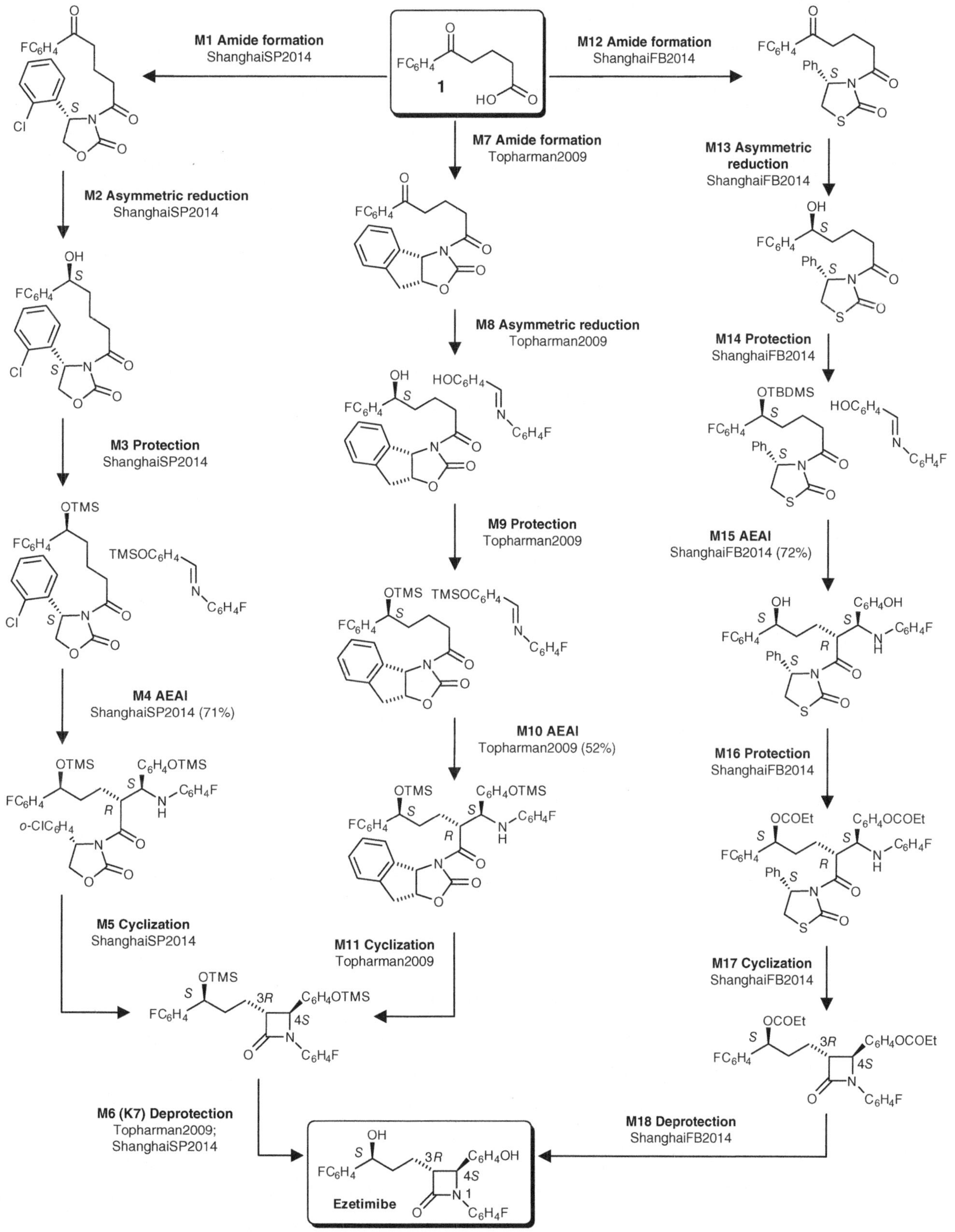

AEAI - Asymmetric enolate addition to imine

For the synthesis of starting C5 carboxylic acid see scheme A; for imines see scheme B

SCHEME M SYNTHESIS OF EZETIMIBE BY N-C2 RING CLOSURE STRATEGY AND USING A HYDROXYL PROTECTED *N*-ACYL OXAZOLIDINONE INTERMEDIATE FOR THE ASYMMETRIC ADDITION TO IMINE STEP (PART 3) (FOR REACTION CONDITIONS SEE PAGE 269)

An improved process for the Michael addition-cyclization steps using a flow process at room temperature has been claimed for this route (Desano2016, Scheme 6.40).

In another example (PolishAS2011), a 1,3-dipolar cycloaddition of a nitrone with a chiral lactone afforded a bicyclic isoxazolidine (Scheme 6.41). In this way, the configuration of both asymmetric carbon atoms of the β-lactam are set up. After N-O bond cleavage, removal the hydroxyl group (by elimination and reduction, by L-Selectride or by hydrogenation) a lactone intermediate was obtained. Inversion of configuration a C6 atom sets up the benzylic oxygen function with the correct chirality. A Grignard induced rearrangement then generates the β-lactam to give BnEZT.

6.4.10 Ezetimibe by alkylation of a C3 unsubstituted β-lactam

The feasibility of using a C3 unsubstituted β-lactam as an intermediate in the synthesis of Ezetimibe has been tested. This type of β-lactam was claimed as an intermediate in the synthesis (Ind2003, Opus2005, Scheme 6.42). The nucleophilicity of the *p*-fluoroaniline intermediate prior to ring closure would, predictably, be reduced making ring closure more difficult. Note that the starting material with the *S* configuration gives the same *S* chirality at C4 ring in the ring, which is in conformity with a double inversion in the presence of sodium iodide.

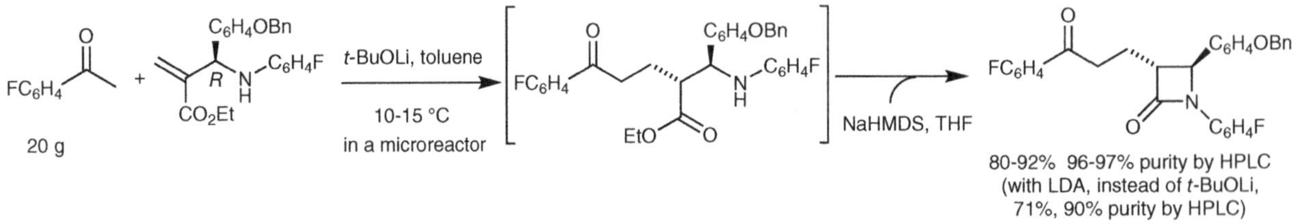

Scheme 6.40 Improved process to access BnEZT ketone by a Michael addition-cyclization strategy

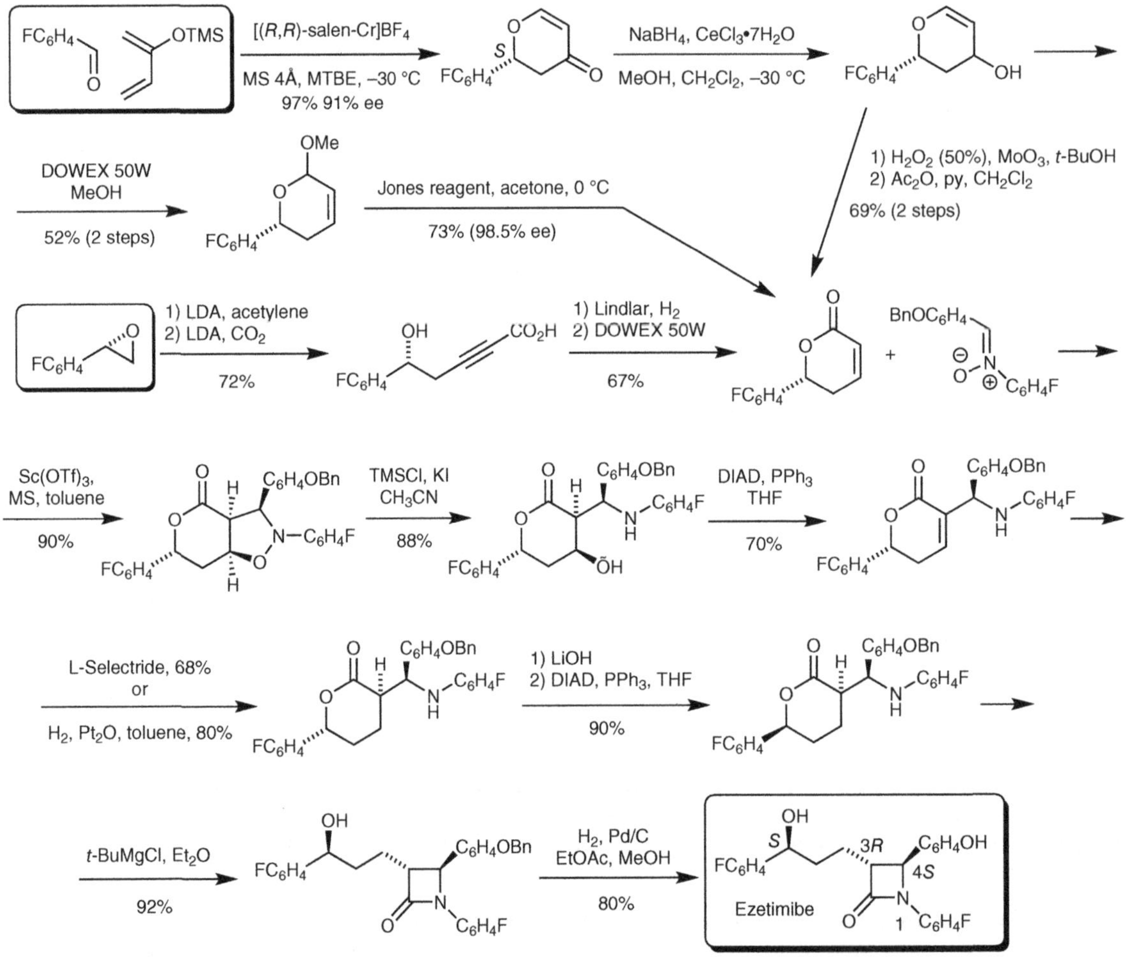

Scheme 6.41 Synthesis of Ezetimibe using a 1,3-dipolar cycloaddition as key asymmetric step

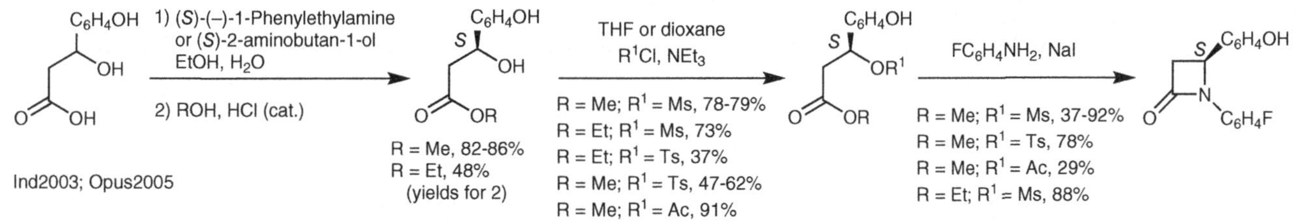

Scheme 6.42 Synthesis of a C3 unsubstituted β-lactam

Another account narrates a different tactic to construct the chiral lactam. In a Reformatsky type reaction a zinc enolate is added to an imine (TL1996, Scheme 6.43), the new chiral centre being induced by using the bromoacetate of a chiral alcohol as staring material. Despite modest yields with both chiral auxiliaries employed [(–)-*trans*-2-phenylcyclohexanol gave 45% and (–)-phenyl menthol gave 55%], high ee's (99:1) were obtained.

With the desired C3 unsubstituted β-lactam in hand, alkylation of the C3 carbon was successfully achieved through the intervention of a non-nucleophilic base at low temperatures. This gave rise to known intermediates that had already been successfully converted to Ezetimibe. As to be expected, the base effects isomerization of any *cis* product to the desired *trans* isomer (Scheme 6.44, TL1996).

Of note is the fact that application of Mitsunobu type conditions were unsuccessful in producing the β-lactam ring. The predominant product was the conjugated amide produced by elimination of water (TL1996, Scheme 6.45).

In a related approach, the chirality was obtained by an asymmetric enzymatic reduction of a β-keto ester, which

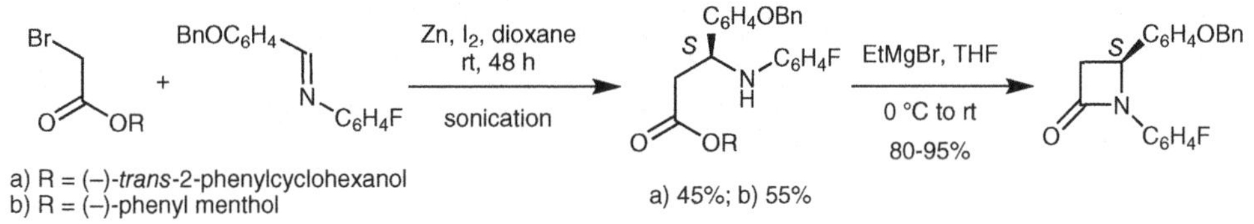

Scheme 6.43 Synthesis of a C3 unsubstituted β-lactam by a C2-N closure

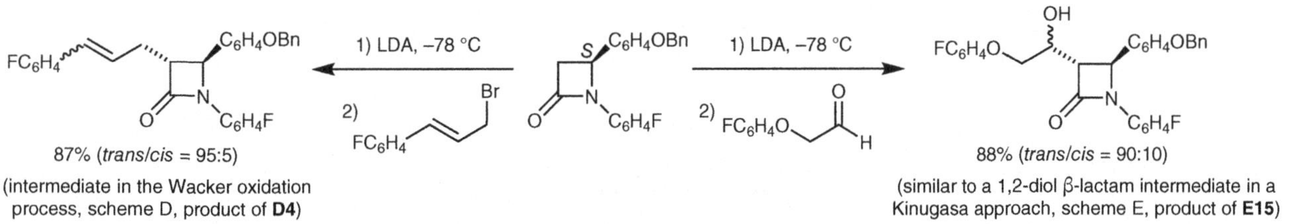

Scheme 6.44 C3 alkylation of a β-lactam towards Ezetimibe synthesis

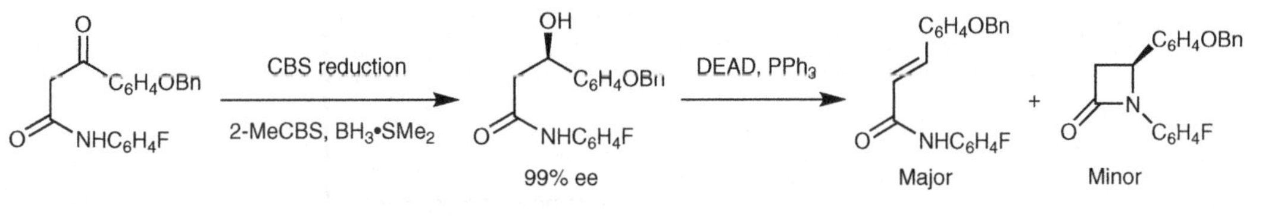

Scheme 6.45 Attempts to use Mitsunobu-type conditions in producing the β-lactam ring of Ezetimibe

was then cyclized in good yields to the β-lactam with 4-fluoroaniline (NanjingAB2015, Scheme 6.46). Alkylation of C3 is reported to proceed with high yield and diastereoselectivity.

6.4.11 Other β-lactam ring formations by N-C4 cyclization

The previous section described the preparation of a C3-unsubstituted β-lactam by N-C4 cyclization. The N-C4 ring closure strategy has also been employed in other variations to prepare Ezetimibe.

Two consecutive biochemical reductions, using a ketoreductase (KRED), have been applied to introduce the chirality of the Ezetimibe molecule (AsymchemL2013, Scheme 6.47). The second KRED reduction introduces the chirality of the two carbon atoms of the four-membered ring in high yield (86%) and stereoselectivity (99% ee). An N-C4 ring closure is used to produce the β-lactam before deprotection.

Although the use of a Mitsunobu reaction was unsuccessful in the case cited above, it has been used in the synthesis of a racemic β-lactam intermediate utilizing an intramolecular N-C4 ring closure (CCL2010, Scheme 6.48). However, a yield of only 33% was achieved.

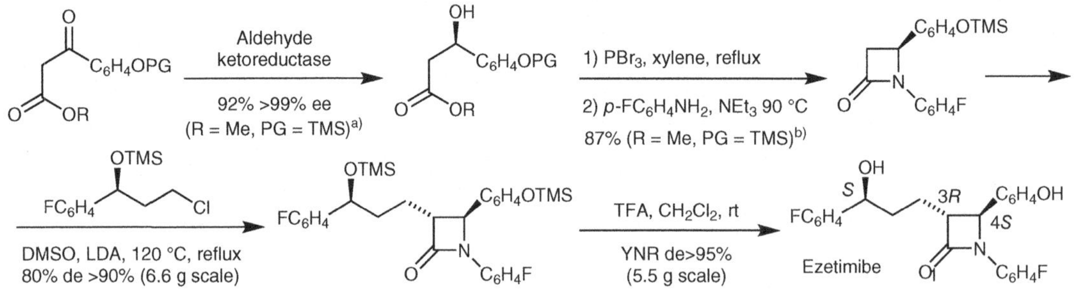

Other substrates: a) R = Et, PG = MOM or Ac or Me, 90-91%, >93% ee; R, PG = Me, 91%, >96% ee;
b) R = Et, PG = MOM or Ac, 80-83%, 83-91% de; R, PG = Me, 78%, >86% de

Scheme 6.46 C3 alkylation of a β-lactam towards Ezetimibe synthesis

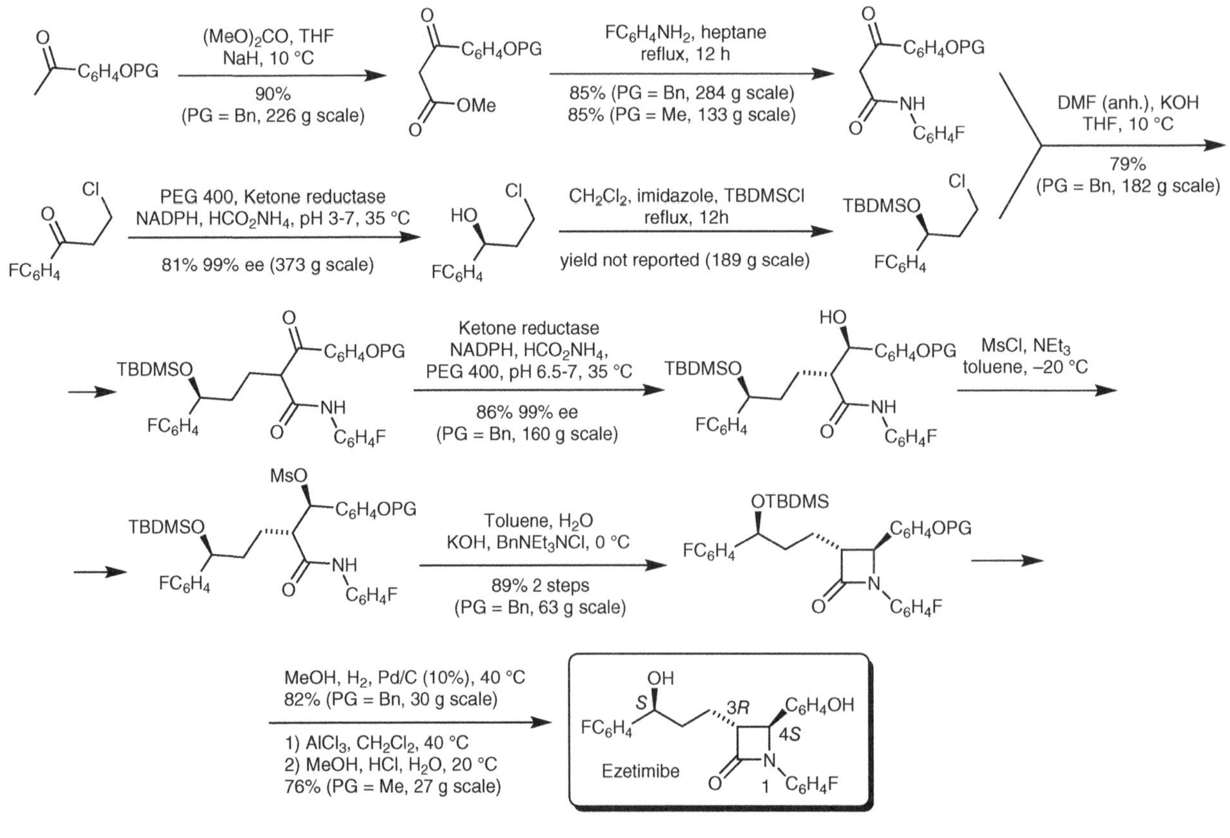

Scheme 6.47 Ezetimibe by a N-C4 cyclization using enzymatic reduction

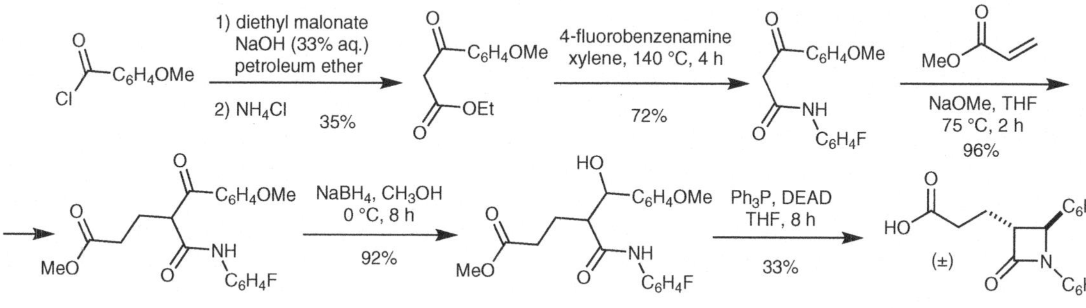

SCHEME 6.48 SYNTHESIS OF RACEMIC EZETIMIBE CORE USING AN N-C4 CYCLIZATION

6.5 COMPARISON OF STRATEGIES AND CONCLUSIONS

Several synthetic routes to Ezetimibe are expected to be applicable for implementation on an industrial scale to manufacture the API. Formation of the β-lactam ring is a key feature when deciding which route to adopt.

The pathways employing Evans chemistry to create the ring enjoy a clear advantage in terms of yield and diastereoselectivity and, when the cost of the appropriate chiral reagent is manageable, are definitely preferred. A key factor for implementation at industrial scale hinges on whether the chiral auxiliary can be recycled. In most cases, this is expected to be feasible. Both Staudinger and Kinugasa protocols for formation of the β-lactam are hampered by poor selectivity.

In the Evans-based routes, the order of each individual transformation has significant impact on the synthesis and can be decisive when selecting a route. An in-depth analysis of the published routes indicates that three options hold most promise (Scheme 6.49). Each option will be discussed in detail below.

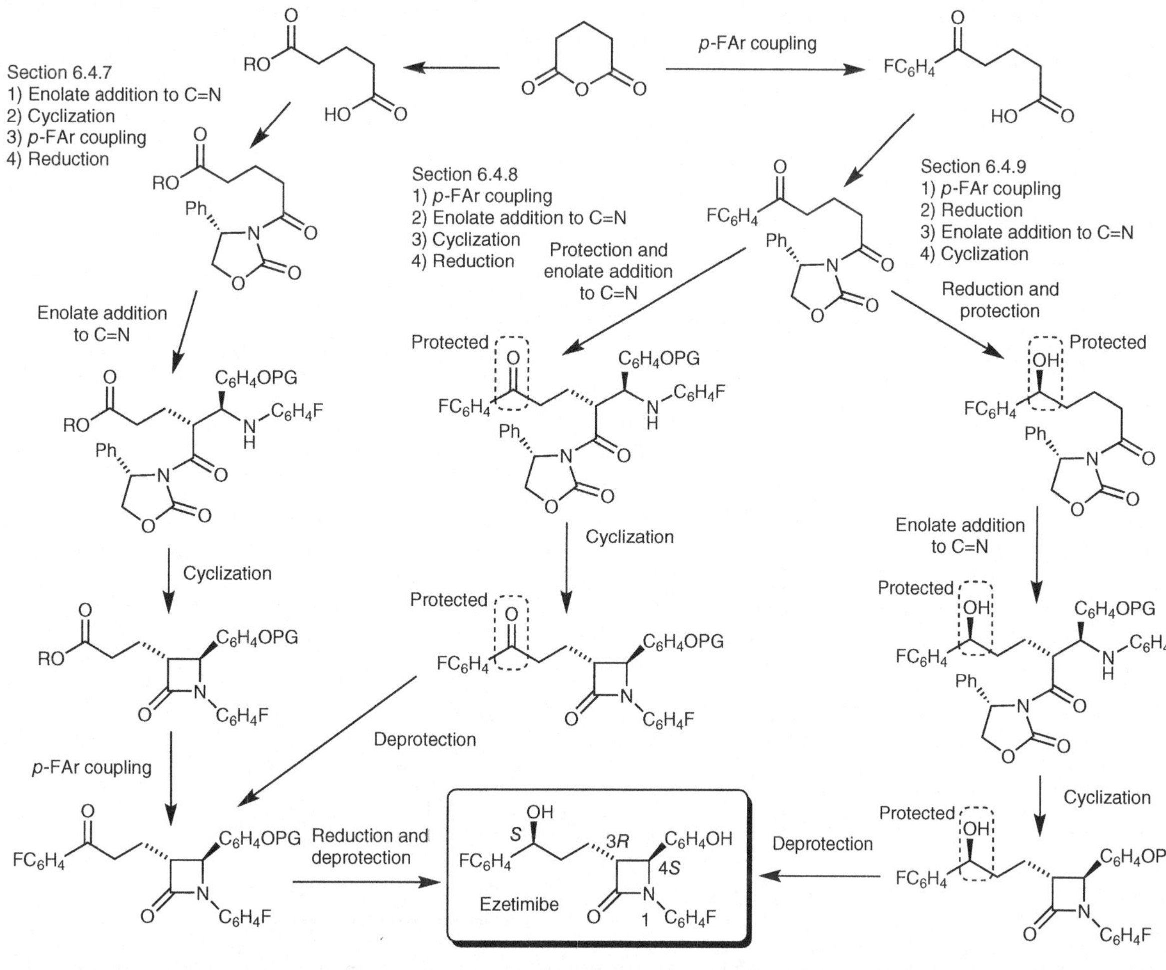

SCHEME 6.49 MOST PROMISING EVANS-BASED ROUTES TO EZETIMIBE

Late coupling of the *p*-fluoroaromatic ring avoids the use of protecting groups for the benzylic oxygen function during enolate imine addition but this is far from ideal, as coupling of the aromatic ring is not straightforward. One favoured sequence was hydrolysis of the terminal carboxylic acid ester, acyl chloride formation and palladium catalysed Negishi coupling (Scheme 6.50). Hydrolysis of an ester to a carboxylic acid in an intermediate that already has a β-lactam ring is reported to produce difficult-to-remove impurities that result from ring opening side reactions. Activation of the acid by formation of the acid chloride frequently employs oxalyl chloride at room temperature, resulting in the release of carbon monoxide and carbon dioxide gases. At large scale, measures to control this gas release must be taken. The Negishi coupling step requires extensive process development since reported yields are extremely variability (ranging from 37% to 87%). The final steps involve an asymmetric reduction of the carbonyl group and deprotection. The order of these final steps may be inverted but the reduction in the presence of a free phenol group appears to have issues. It is to be noted that, both of these late synthetic steps are generally carried out using a metal catalyst, whose removal from the API can be problematic.

Benzyl is the most commonly used protecting group for the phenolic function. However, its hydrogenolysis when carried out at the end of the synthesis can be correlated with the formation of a number of process impurities. The allyl protected version (see lower part of scheme F, IndSwift2008) avoids these issues but the coupling reaction is still carried out *via* a problematic metal catalysed coupling (**F29**).

The issues associated with late attachment of the *p*-fluoroaromatic group to terminal position of the linear chain are avoided in the other two routes in Scheme 6.49. Introduction of the aromatic group at an early stage is easily achieved *via* an electrophilic aromatic acylation, conveniently coupling this moiety to the Evans auxiliary by using a mixed anhydride to give the required *N*-acyl oxazolidinone. An additional advantage is that there is no need to produce the more active and more difficult to handle acyl chloride.

The two routes differ in the placement of the asymmetric reduction but in contrast with the late *p*-fluoroaromatic coupling strategy, the benzylic function (alcohol or ketone) requires protection before the enolate imine addition.

Various protecting groups were successfully applied when late stage reduction of the ketone is employed, with a ketal being the simple and logical choice (scheme G, H, I, Scheme 6.51). However, this requires additional steps to prepare and isolate the ketal prior to the enolate-imine addition. After formation of the β-lactam, two different options are available depending on which phenol protecting group is used.

When a TMS is used to protect the phenol, it is removed before or during the hydrolysis of the ketal. The asymmetric reduction of the ketone is carried out in the last step on the unprotected intermediate. The positioning of the asymmetric reduction at the end implies that less catalyst is required to produce the same amount of API in comparison to carrying out this reaction earlier in the sequence, but the possibility to improve the enantiopurity of the product, which might occur during subsequent processing, is lost.

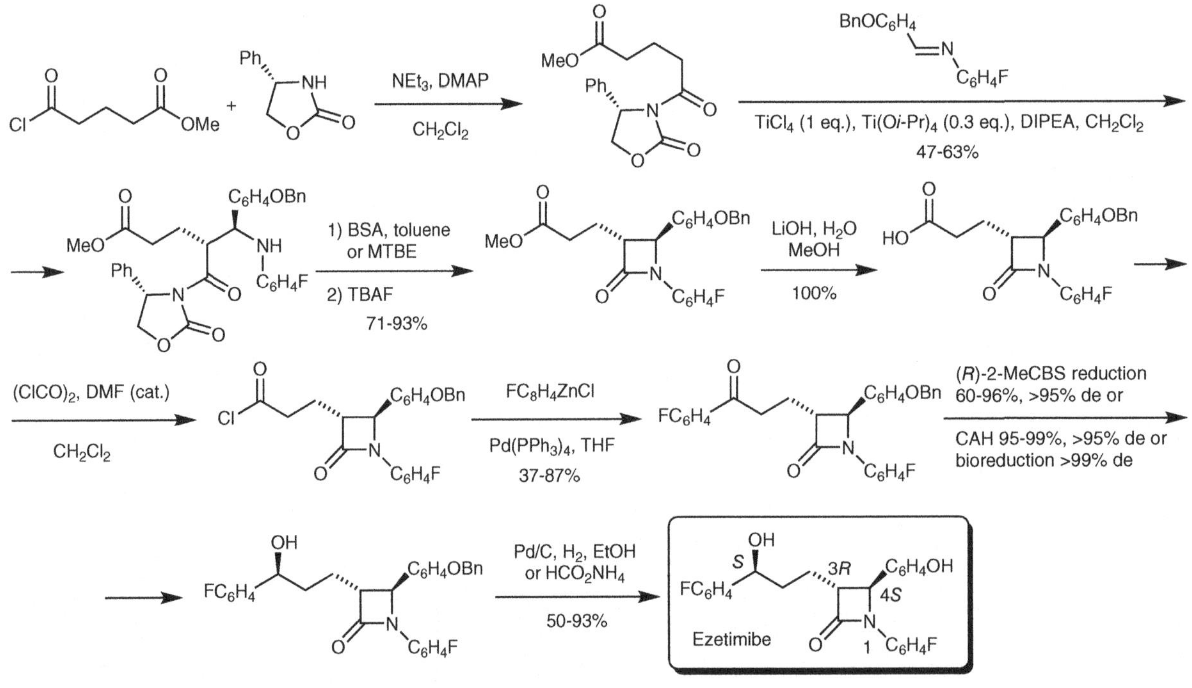

SCHEME 6.50 EZETIMIBE ROUTE USING AN EVANS METHODOLOGY AND LATE COUPLING OF THE FLUOROAROMATIC RING

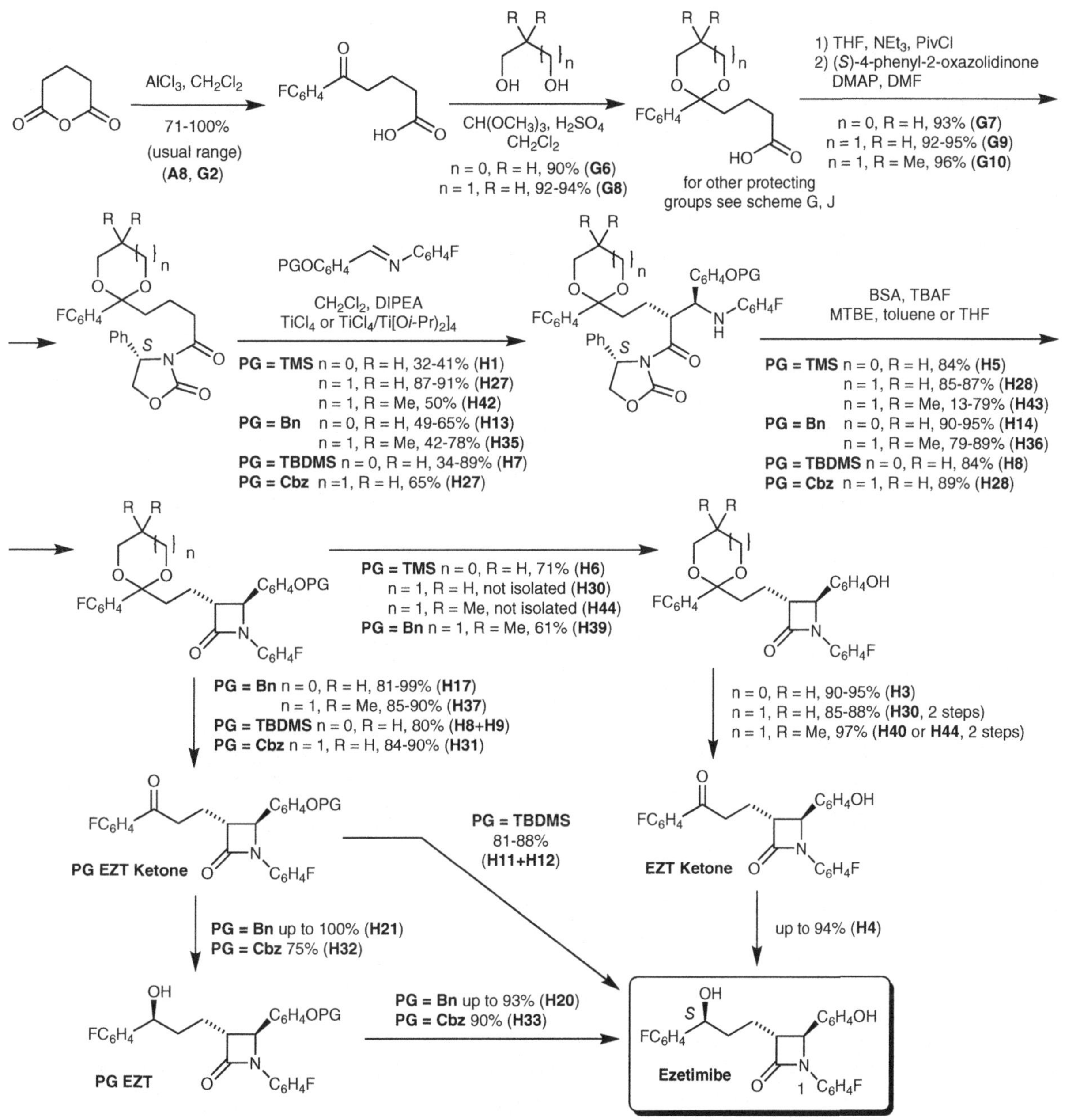

Scheme 6.51 Examples of Ezetimibe routes using an Evans chiral auxiliary enolate addition to an imine, an early coupling of the *p*-fluoroaromatic ring and late asymmetric carbonyl reduction

The use of a benzyl ether, frequently reported as the phenol protecting group, means that hydrolysis of the ketal is accomplished without phenol deprotection. The final step then becomes hydrogenolysis to remove the benzyl ether. Once more, this step requires comprehensive process development to control, or even avoid, the formation of impurities. Residual heavy metal from the catalyst (e.g. Pd) in the final product can be a concern and methods for its adequate removal have to be developed.

The last option is to place the asymmetric reduction at an earlier stage, for example, immediately after the formation of the *N*-acyl oxazolidinone. Compared to the other options, a larger quantity of the intermediate must be transformed (succeeding steps rarely give yields of 100 %!), and more catalyst is required. However, since more processing is required, there is an opportunity for improvement in enantiopurity.

As previously, two functional groups require protection to avoid interference at the enolate addition/cyclization steps. Several combinations have been proposed (Scheme 6.52).

The obligatory protection of the hydroxyl group before enolate-imine addition can be carried out in a separate step or *in situ*. The last option is of particular interest for industrial application.

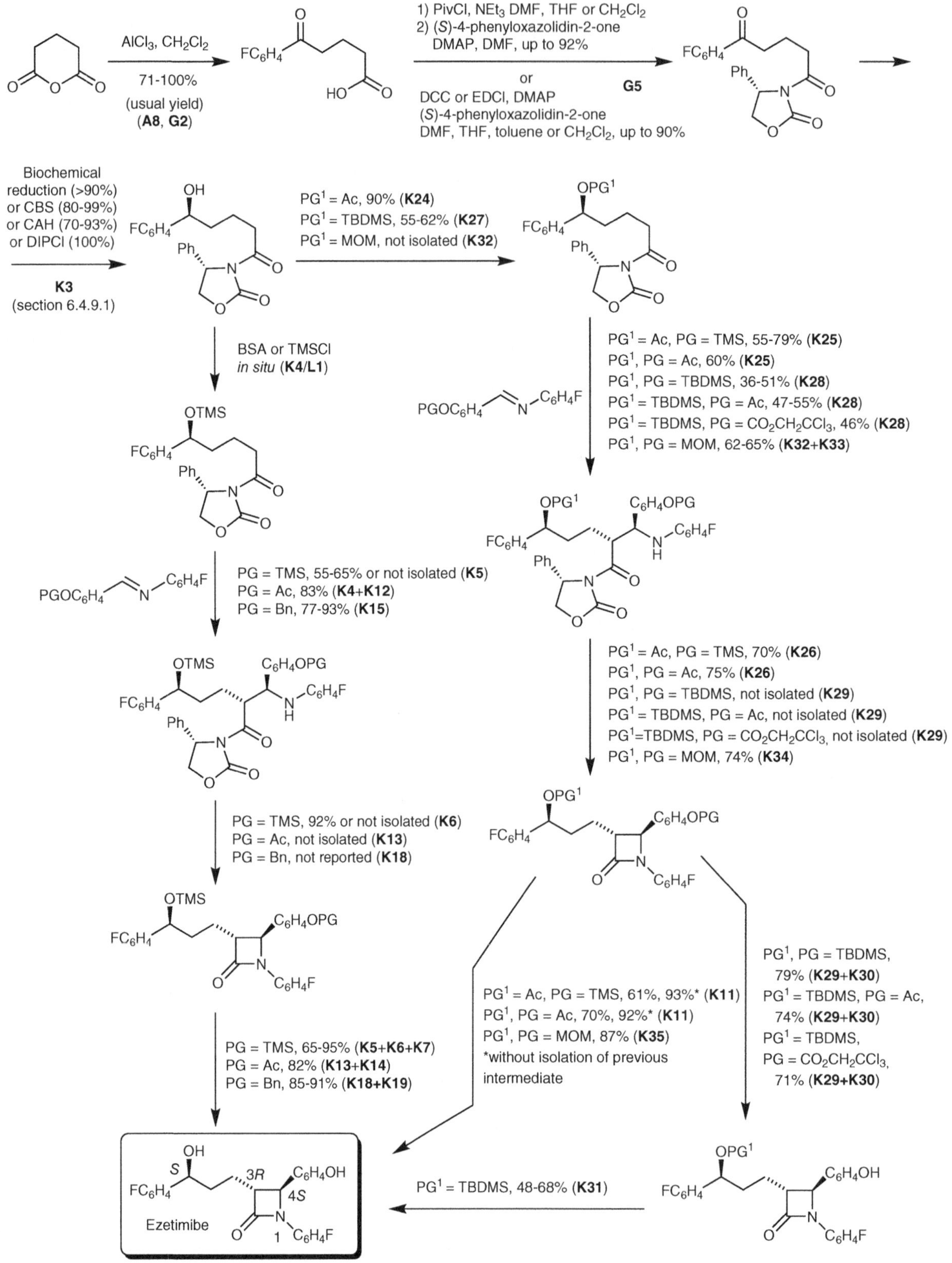

Scheme 6.52 Outline of Ezetimibe routes using an Evans chiral auxiliary enolate addition to an imine, early coupling of the *p*-fluoroaromatic ring and an asymmetric reduction of the carbonyl group

6.6 THE BEST SYNTHESES

Overall, the option to attach the *p*-fluoroaromatic ring to the linear chain at the beginning of the synthesis and introducing the chiral alcohol *via* an asymmetric reduction of a ketone in an *N*-acyl oxazolidinone, looks to be an excellent choice for industrial manufacture of Ezetimibe drug substance.

The choice of whether the protecting group is introduced at the ketone or a previously formed chiral alcohol stage, eventually comes down to cost considerations when applied to an industrial scenario. The high cost of TBDMSCl makes the use of TMS or benzyl protecting groups preferable for protection of the alcohol. Two of the most promising routes, according to this scenario, are shown below (Scheme 6.53).

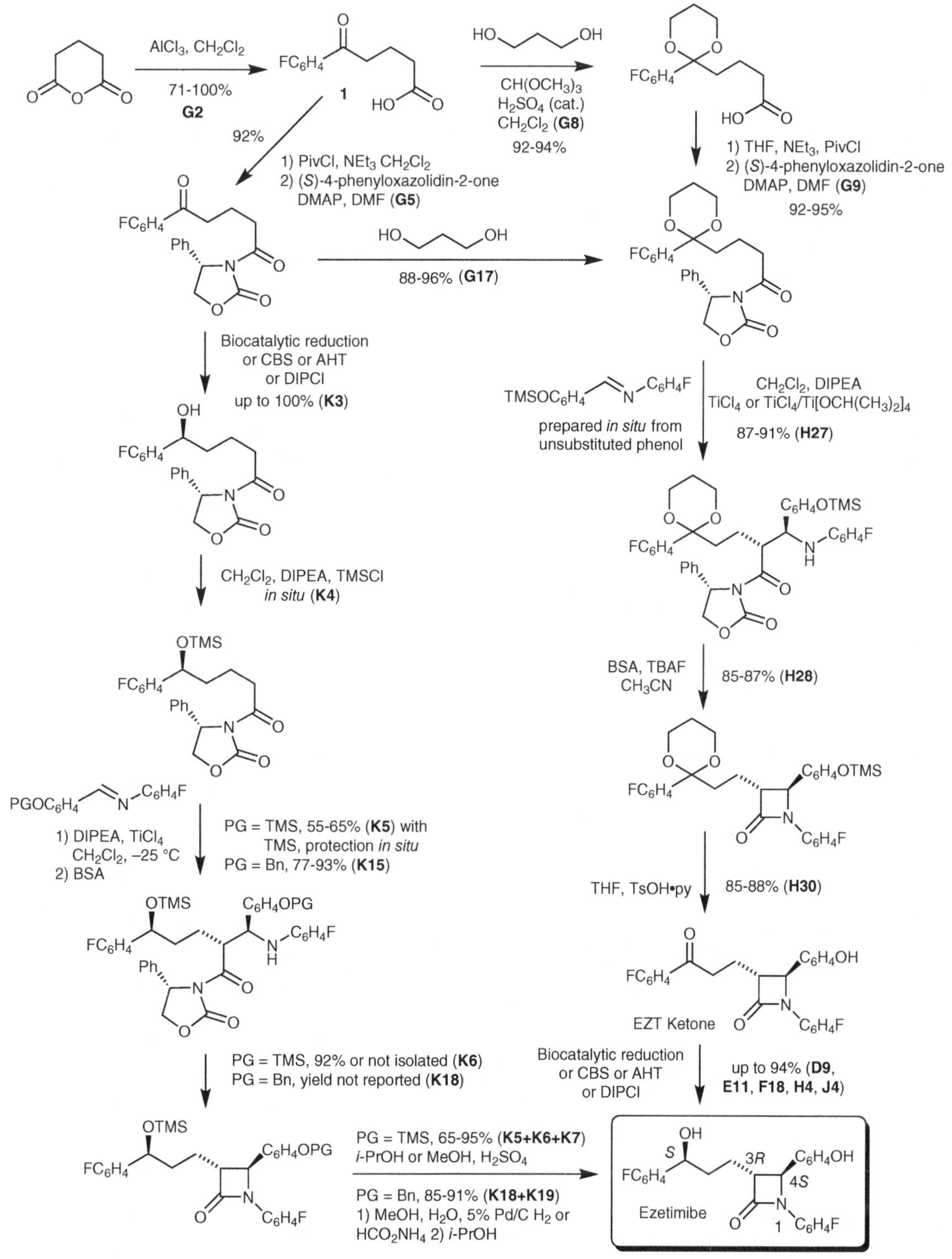

Scheme 6.53 Best routes to Ezetimibe

The use of a ketal protecting group and deferred reduction of the ketone has two main inconveniences. The belated reduction gives Ezetimibe directly and leaves little opportunity for improvement of enantiomeric purity by further processing, as would have been the case with an earlier reduction. A separate step for the protection of the carbonyl group is required and heavy metal residues from the catalyst used in the reduction may contaminate the final API and mandate additional purification.

Conversion of the ketone to the chiral alcohol can be carried out at an early stage by a CBS or DIP-Cl reduction or by an Asymmetric Hydrogen Transfer (AHT) using a ruthenium catalyst. Reduction with (*R*)-2-Methyl-CBS-oxazaborolidine [(*R*)-2-MeCBS] (5% mol) and BH_3•SMe_2 in CH_2Cl_2 afforded the best results (>95% 98% de) but AHT (70-93% conversion, 80-92% de, Pharmathen2013) exhibited high potential for further development work.

With an early carbonyl reduction, an intermediate with two OH groups is produced, a phenol and a benzylic alcohol. Both these hydroxyl functions require protection to avoid complications during enolate addition/cyclization. When the phenol-protecting group is benzyl, a separate step is essential for its introduction. The final step is removal of the benzyl group by hydrogenolysis, reported to be problematic because of the formation of impurities. In addition, potential contamination from residues from the metal catalyst is again a concern.

However, advantage can be taken from the fact that two similar groups (hydroxyl) require protection (Scheme 6.53 and Scheme 6.54). By selecting TMS as the sole protecting group, the route can be streamlined considerably since both hydroxyl functions can be protected *in situ* at the same time. This simply involves mixing of both substrates together, protecting them simultaneously and then proceeding to the enolate-imine addition, all in one pot.

The lability of the TMS group is of concern because of its potential to undergo partial cleavage during work up. However, re-silylation with BSA before isolation resolves this issue. *N*-silylation followed by cyclization catalysed by fluoride (using TBAF) produces the β-lactam ring. Deprotection of both hydroxyl groups *in situ* is thereafter, easily executed to produce Ezetimibe in high yield.

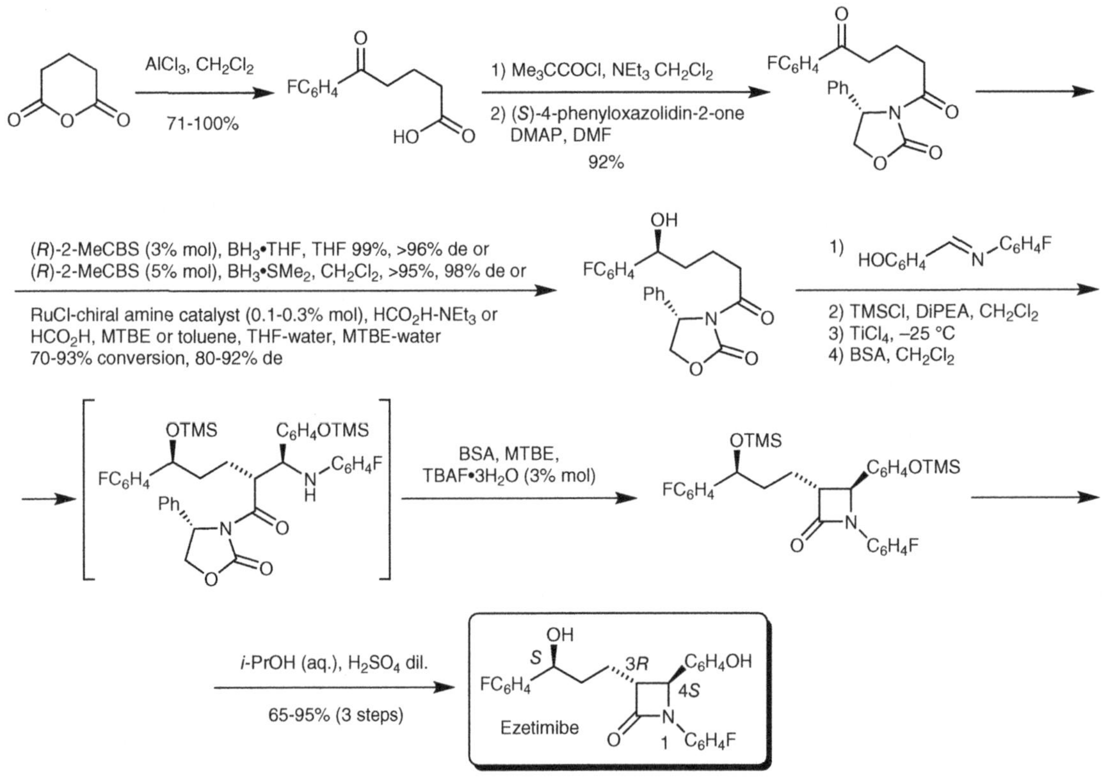

Scheme 6.54 Most practical and expedient route to Ezetimibe

6.7 LESSONS FROM THE EZETIMIBE CASE

Resolution *vs* asymmetric synthesis

The Ezetimibe molecule has three chiral centres and presents a total of eight stereoisomers. Clearly a racemic synthesis is to be avoided both because the yield is greatly diminished in these approaches and recovery of unwanted isomers is seen to be unrealistic. Hence, an appropriate industrial process should involve stereoselective transformations. Consequently, both the Kinugasa and the Staudinger approaches to construct the β-lactam ring can be excluded. An asymmetric enolate imine addition/cyclization strategy is therefore, a preferred option.

Chiral auxiliary asymmetric synthesis *vs* catalytic asymmetric synthesis

The chiral auxiliary asymmetric synthesis approach is generally regarded as less than ideal because of a stoichiometric amount of the chiral auxiliary is necessary. A catalytic asymmetric process is, therefore, favoured. Evans-based routes when applied to the synthesis of Ezetimibe use a stoichiometric amount of a chiral oxazolidinone to establish the configuration of two of the three chiral centres present in the final molecule. At first glance, this option looks have real drawbacks because of the cost of the chiral auxiliary. Fortunately, it was demonstrated that more than 85% of the chiral auxiliary can be recovered. Hence, the contribution of an oxazolidinone auxiliary to the total cost of the API is correspondingly diminished. making its use for industrial application feasible.

Protecting groups in laboratory and industrial synthesis

The use of protecting groups is regarded as unsophisticated and undesirable as the number of steps of the synthesis increases by two, one for protection and another for deprotection. Therefore, on paper, the best industrial route is that that employs the fewest protecting groups. However, this is not always feasible as the non-selectivity of numerous chemical transformations requires that certain functional groups, which are reactive in the conditions of the reaction, be removed from the sphere of activity. This is done, typically, through the use of protecting groups. This can be seen in a variety of approaches to Ezetimibe. Although the cost associated with the use of a protecting group (increased number of process steps, loss in yield associated with more reaction steps and work-ups, cost of additional reagents, etc.) is significant, this may be outweighed by the overall yield of the synthesis as a whole and purity of the final product.

Two types of functionalities may require protection in an Evans-based Ezetimibe synthesis, hydroxyl (a phenol and benzylic alcohol) and a benzylic ketone. Protection of the benzylic ketone as a ketal is easily and efficiently achieved and hydrolytic deprotection is quick and problem free. Use of a labile TMS group for protection of either of the hydroxyl groups may be problematic, as it frequently does not survive even simple synthetic manipulations. This is particularly true in the presence of protic solvents under acidic conditions. What are generally considered mild conditions, for example, citric acid in methanol at 20 °C (10 min), K_2CO_3 in methanol at 0 °C or TBAF in THF at 0 °C provoke cleavage of the TMS group in a matter of seconds. When an undesirable cleavage of this type occurs, loss of material with attendant reduction of productivity ensues, unless preventive measures are taken, such as re-protection (see below), before proceeding to the next step.

Due to the instability of the OTMS group, the much more stable TBDMS derivative has become the most popular protecting group of this family. TBDMS ethers are stable to chromatography and to mild base but are labile to mild acid. TBDMSCl is expensive but often the choice in a laboratory synthesis.

When the use of protecting groups is unavoidable, the fortuitous situation wherein several functionalities can be protected in one step, is highly beneficial. In the case of Ezetimibe this scenario is indeed applicable. Hence, a synthesis involving di-TMS or di-MOM protection of the hydroxyl functions (Scheme 6.52) has definite advantages when compared to the use of two distinct protecting groups (for example TMS combined with Bn). This also gives this approach an advantage over those procedures involving both a keto- and a hydroxy-protecting group (Scheme 6.51).

As discussed to above, the stability of TMS is a concern for a route involving di-TMS protection. Hence, it would be thought that a more stable protecting group would be more appropriate. This is particularly true at laboratory scale. At industrial scale, despite all efforts to avoid it by keeping manipulation to a minimum after the enolate-imine addition, partial hydrolysis of the TMS protecting group occurred. Re-silylation prior to isolation resolved this issue, thus making the use of the cost effective di-TMS protection possible. After cyclization to the β-lactam, simultaneous removal of both TMS groups was easily achieved. Here, the advantage over the use of a TMS/Bn protecting group strategy, which requires different reaction conditions to remove each of the two protecting groups, is clear.

In the laboratory, it is desirable to use highly stable protecting groups as this allows a variety of chemical transformation to be carried out under a wide range of conditions so that molecular modifications can be introduced without impediment. An industrial synthesis focuses on using protecting groups that will survive a well-defined set of reaction conditions over a specific number of reactions. Protecting groups should be easy to introduce and remove without altering the rest of the molecule. When applied to the synthesis of Ezetimibe, the TMS group might have been seen to be less than ideal because of its reduced stability but, it was found to perform effectively as a protecting group, being easily introduced and removed. In addition, the low cost associated with the use of the TMS protecting group makes it a good option for this particular application when compared to the other possibilities.

On unstable structural units or groups

The central β-lactam unit in Ezetimibe confers instability to the molecule. Ring opening occurs easily under a variety of conditions. Strategies that opt for an early ring formation (see, for example, routes indicated in scheme F), require subsequent steps to be carried out under mild conditions under which the β-lactam is stable. This is extremely important, as it has been reported in the literature that, when ring opening occurs, impurities that are difficult to purge are formed. Consequently, it is desirable to delay the formation of the ring until further downstream. The ideal situation places ring formation at the end of the sequence, thereby reducing manipulation of unstable intermediates to a minimum. As a general recommendation, the least stable structural features of a compound should be generated during the final stages of a synthesis.

6.8 REACTION CONDITIONS FOR SCHEMES A TO M

6.8.1 Reaction conditions for Scheme A

A1 BnOH, TsOH, toluene, reflux, YNR (200 g).
A2 1) PivCl, NEt_3, CH_2Cl_2, 30 °C; 2) (*S*)-4-phenyloxazolidin-2-one, DMAP, DMF, 55°C to reflux, 56-66% (100-330 g).
A3 1) Na, BnOH, 85 °C; 2) SM, 85 °C, NI (100 g).
A4 NaOMe, MeOH, 30 °C, 98% (100 g).
A5 $SOCl_2$, heat, 80-100%.
A6 (F19) (*S*)-4-phenyloxazolidin-2-one, DMAP, NEt_3, CH_2Cl_2, 0 °C to rt, YNR (50-58 g).
A7 1) PivCl, NEt_3, CH_2Cl_2, rt to 30 °C; 2) (*S*)-4-phenyloxazolidin-2-one, DMAP, DMF, reflux, 56% (45 g, MSN2005, Ind2006). 1) PivCl, NEt_3, CH_2Cl_2, rt; 2) (*S*)-4-phenyloxazolidin-2-one, DMAP, 45 °C, 75% (300 g, DrReddys2005, OPRD2009).
A8 (G2) C_6H_5F, $AlCl_3$, 0 °C to 30 °C, 54-100% (up to 300 g).
A9 $SOCl_2$, CH_2Cl_2, reflux, 81-94% (0.7-3 g).
A10 (G14) MeOH, MeOH•HCl, 80% (100 g, Ranbaxy2003a). MeOH, H_2SO_4, reflux, 100% (3 g, Nanjing2013).
A11 (–)-DIP-Cl (1.7 eq.), THF, –25 °C, NI (50 g, Ranbaxy2003a,b). $NaBH_4$ (1.2 eq.), TMSCl (1.2 eq.), THF, (*R*)-2-[bis(4-methoxyphenyl) hydroxymethyl]pyrrolidine (10% mol), 25 °C, 93%, 75% de (1 g, Kotobuki2006), with (*R*)-2-(diphenylhydroxymethyl)pyrrolidine (10% mol), 90%, 57% de (1 g, Kotobuki2006). H_2, $[RhCl_2Cp^*]_2$, (*R,R*)-Ts-DPEN, H_2O, NaOH, 40 °C, 98%, 92% ee (0.2 g, ChineseASC2009). Other ligands tested, for example, (*S,S,R*)-Cs-DPEN, 100%, 95% ee. $NaBH_4$, α-pinene, BCl_3, $CH_3OCH_2CH_2OCH_3$, –15 °C to 4 °C, 81% (31 g, Nanjing2013).
A12 1) NaOH (5 M), THF, H_2O; 2) PPTS, toluene, CH_2Cl_2, rt, 75% (A11+A12, 50 g, Ranbaxy2003a). HCl, CH_2Cl_2, rt, 55-73%, 80-98% ee (A11+A12, 90 mg, ChineseASC2009). TFA, CH_2Cl_2, rt, 95% (25 g, Nanjing2013).
A13 NaOMe, MeOH, rt, 90% (28 g).
A14 NaOH, H_2O, THF, 0 °C, 84% (A11+A14, 20.5 g).
A15 TFA, CH_2Cl_2, rt, 95% (14 g).
A16 Microbiological reduction, up to 100% (ee >95%, Merck1995, Schering1996a). H_2 (100 bar), Ir catalyst (0.02% mol), *t*-BuOK, EtOH, 50 °C, 100%, 99% ee (X = 3-Me, 53 g, Zhejiang2012), 96%, 97.5% op (X = H or 4-Me, 1 g, Zhejiang2012), with X = H, base = KOH in DMF, 97%, 98.6% op (1 g, Zhejiang2012), with X = 2-OEt, base = NEt_3, 100%, 98.8% op (1 g, Zhejiang2012).
A17 R = Et: EtBr, K_2CO_3, DMF, rt, 99% (21 g). **R = Me:** MeI, K_2CO_3, DMF, rt, 92% (10.6 g). **R = Bn:** BnBr, K_2CO_3, DMF, rt, 95% (10.6 g). **R = Pr:** PrCl, K_2CO_3, DMF, 60 °C, 86% (10.6 g).
A18 $CCl_3C(=NH)OCH_2C_6H_5$, MsOH, cyclohexane, CH_2Cl_2, 0 °C, 94% (4.8 g).
A19 NaOH, MeOH, H_2O, rt, 93% (5 g).
A20 $SOCl_2$, DMF, CH_2Cl_2, 0 °C to rt, NI.
A21 1) CDI, DMF, 25 °C; 2) Morpholine, 45 °C, 82% (50 g, Glenmark2005). 1) PivCl, NEt_3, toluene, rt; 2) Morpholine, 0 °C, 75% (10 g, Mylan2009), in THF 92% (10 g, Zhejiang2012). **For *N,N*-dimethyl amide derivative:** 1) PivCl, NEt_3, THF, –5 °C; 2) $HNMe_2$, rt, 93% (10 g).
A22 (*R*)-2-MeCBS (0.05 eq.), BH_3•SMe_2, CH_2Cl_2, toluene, –5 °C to 0 °C, 95% (40 g, Glenmark2005). (–)-DIP-Cl, toluene, 0-5°C, 55% (10 g, Mylan2009). Bioreduction with ES-KRED-119, 98%, >99% ee (50 g, Mylan2009). H_2 (100 bar), Ir catalyst, *t*-BuOK, EtOH, 50 °C, 100%, 98% op (X = 3-Me, 1.4 g, Zhejiang2012). **For NMe_2 amide:** H_2 (100 bar), Ir catalyst, *t*-BuOK, EtOH, 50 °C, 100%, 98.2% op (X = 3-Me, 1.2 g).
A23 NaH, BnBr, THF, 0 °C to rt, 84-93% (5-30 g).
A24 KOH, EtOH, reflux, 51% (40 g, Glenmark2005), in *i*-PrOH, 80% (10 g, Mylan2009).
A25 $NaBH_4$, MeOH, not reported.
A26 *Candida antartica* lipase A (10 % w/w), vinyl acetate, toluene, 35-40 °C, 31-36%, >99% ee (50-750 g), with succinic anhydride, 40-46%, 95-99% ee (5 g).
A27 $LiBH_4$, NaOH, H_2O, rt, 90% (55 g).
A28 $SOCl_2$, CH_2Cl_2, reflux, 80% (47 g).

6.8.2 Reaction conditions for Scheme B

B1 PG = Bn: Acetone or H_2O or DMF, NaOH or K_2CO_3 or Na_2CO_3, BnBr (frequently not isolated). Acetone, Na_2CO_3, BnBr, reflux, 98% (5 g, UShenyang2010), with K_2CO_3, DMF, 91-97% (12-100 g, Nanjing2015, SuzhouP2016). **PG = Cbz:** CbzCl, NEt_3, DMAP, CH_2Cl_2, 0 °C to rt, 78% (20 g). **PG = All:** AllBr, NaOH, Bu_4NBr, H_2O, CH_2Cl_2, 20-30 °C, not isolated (500 g).

B2 PG = Bn: *p*-$FC_6H_4NH_2$ (1 eq.), *i*-PrOH, reflux, 90-95%, (B2 or B1+B2, 2-170 g aldehyde, Schering1995, ZhejiangT2006, IndSwift2007, UShenyang2010, Nanjing2015), in acetone, reflux, 92% (B2, SuzhouP2016), 87% (B1+B2, 10 g aldehyde, Ind2007b). Toluene, reflux, 93% (2 g aldehyde, HenanU2017a). **PG = Cbz:** *p*-$FC_6H_4NH_2$ (1 eq.), *i*-PrOH, 55 °C, 96% (33 g aldehyde). **PG = All:** *p*-$FC_6H_4NH_2$ (1 eq.), *i*-PrOH, 60-65 °C, 84% (B1+B2, 500 g aldehyde). **PG = Me:** *p*-$FC_6H_4NH_2$ (1 eq.), *i*-PrOH, 50 °C, 89-90% (0.15-7.3 g). **PG = THP:** toluene, reflux, 90% (2 g aldehyde). **PG = TBDMS:** CH_2Cl_2, TBDMSCl, 79%. **OPG = NO_2:** *p*-$FC_6H_4NH_2$ (1 eq.), MeOH, rt, 100% (27 g aldehyde).

B3 *p*-$FC_6H_4NH_2$, *i*-PrOH, 50-70 °C, 80-97%.

B4 PG = Bn: BnBr, base (K_2CO_3 or NaOH or DIPEA or NEt_3), acetone or toluene or CH_2Cl_2 or *i*-Pr_2O, 83-91%. **PG = Cbz:** CbzCl, DIPEA, CH_2Cl_2, 0-5 °C to rt, NI. **PG = TMS:** TMSCl, imidazole, THF, toluene, 92%. **PG = Ac:** Ac_2O, py, rt, 71% (B3+B4, 144 g, ShanghaiAB2010). Ac_2O, NEt_3, CH_2Cl_2, rt, 98% (6 g, Arromax2015). **PG = Cl_3CCH_2CO:** Cl_3CCH_2COCl, NEt_3, CH_2Cl_2, rt, 95% (5.6 g). **PG = Tr:** TrCl, DIPEA, CH_2Cl_2, rt, 98% (9.8 g). **PG = Piv:** PivCl, NEt_3, CH_2Cl_2, 93% (180 g). **PG = TBDMS:** TBDMSCl, imidazole, THF, 25 °C, 85% (22 g, RichterGedeon2005), in DMF, 66-98% (6-10 g, Zentiva2007a, Arromax2015).

B5 Zn, NH_4Cl, H_2O, 50 °C, 90% (3.5 g).

B6 EtOH (abs), reflux, 88% (2.5 g).

6.8.3 Reaction conditions for Scheme C

C1 DIPEA, NEt_3 (cat.), toluene, reflux, 64% (5 g).

C2 H_2, Pd/C (10%), EtOAc, MeOH, YNR (3 g).

C3 CH_2Cl_2, reflux, YNR (0.6-11 g).

C4 Chiral chromatography, 11% (C3+C4, 0.6-11 g).

C5 (*R*)-2-MeCBS, BH_3•SMe_2, CH_2Cl_2, 90-93% (0.5-2.3 g).

C6 LiOH (10%), H_2O, 0-5 °C, 71-72% (0.4-2.0 g).

C7 DIPEA, toluene, reflux, 85% (11 g).

C8 1) (*R*)-(2-bromo-1-(4-fluorophenyl)ethoxy)-trimethylsilane, Mg, I_2, THF (anh.); 2) CuCl, THF; 3) TBAF, THF, YNR (5 g).

C9 HCO_2NH_4, HCO_2H, Pd/C, EtOH (abs.), 60 °C, 79% (2 g).

C10 EtOAc, Et_2O, 46% (1.3 g).

C11 $POCl_3$, NEt_3, CH_2Cl_2, rt, 83% (*trans*).

C12 Grubbs II (10% mol), CH_2Cl_2, reflux, 85%.

C13 H_2, Pd/C (5%), CH_2Cl_2, rt, 91%.

C14 H_2O_2, NaOH, H_2O, THF, 0 °C to rt, 70%.

C15 Not described.

6.8.4 Reaction conditions for Scheme D

D1 1) PPh_3, toluene (anh.), reflux, 93-97% (51-60 g of ester); 2) NaHMDS, *p*-FC_6H_4CHO, THF (anh.), 10 °C to rt; 3) THF, H_2O, NaOH, rt, 91-94% (10 g).

D2 1) $(COCl)_2$, DMF (cat.), CH_2Cl_2; 2) (*S*)-4-phenyloxazolidin-2-one, DIPEA, DMAP, 93-95% (Schering1995, Lek2008b, S2010). 1) PivCl, NEt_3, CH_2Cl_2, reflux; 2) (*S*)-4-phenyloxazolidin-2-one, LiCl (anh.), 95% (2.3 g, Kemp2013).

D3 1) $TiCl_4$, CH_2Cl_2, DIPEA, –20 °C; 2) $BnOC_6H_4CH{=}NC_6H_4F$, –20 °C, 67% (1.3 g, Schering1995), 59-65% (1-1.3 g, Lek2008b, S2010, Kemp2013).

D4 1) BSA, toluene, 80 °C; 2) TBAF, 50°C, 90% (1 g, Schering1995). 1) BSA, toluene, rt; 2) TBAF, rt to 50 °C, 71-75% (0.5 g, Lek2008b, S2010, Kemp2013).

D5 $Pd(OAc)_2$ (2% mol), benzoquinone (1.5 eq.), $HClO_4$ (cat.), CH_3CN, H_2O, 70-80% (2.2 g, Schering1995). $Pd(OAc)_2$ (3% mol), benzoquinone (1.5 eq.), $HClO_4$ (cat.), CH_3CN, H_2O, 86% (0.7 g, Lek2008b, S2010, Kemp2013).

D6 (E5, F5, H21, J12) Most frequent method, CBS reduction: (*R*)-2-MeCBS (7-20% mol), BH_3•SMe_2 (0.95-1.1 eq.), THF or CH_2Cl_2, 70-95%, >97% de, acids proposed as additives. **CAH** with Ru complexes, 94-99%, >98% de (see 6.4.7.2).

D7 (E6, F6, H20, J13, K21) Most frequent conditions: H_2 (4-60 bar) or HCO_2NH_4/HCO_2H, Pd/C, EtOH, 50-92%. MeOH, CH_2Cl_2 or *i*-PrOH also used as solvents; Raney-Nickel also used as catalyst. See section 6.4.7.4.

D8 (F17, H22, J7) H_2 (4 bar), Pd/C (10%), EtOH or MeOH/EtOAc, 92% (20-118 g).

D9 (E11, F18, H4, J4) Most frequent method, CBS reduction: (*R*)-2-MeCBS (5-20% mol), BH_3•SMe_2 (0.95-1.1 eq.), THF. toluene or CH_2Cl_2, 48-94%, >97% de reported. See section 6.4.7.3.

D10 MCPBA, CH_2Cl_2, 0-5 °C, 30% (5 g), with THP or TBDMS as protecting group 27-30%. Shi catalyst, oxone, $CH_2(OMe)_2$, K_2CO_3, CH_3CN, buffer, Me_4NHSO_4, 20 °C, 54-56% (5 g).

D11 Ph_2Se, $NaBH_4$, EtOH, 0 °C, 38% (0.4 g).

D12 (E9) H_2 (7 bar), Pd/C (10%), EtOH, 89%, 99.1% de (4.2 g).

6.8.5 Reaction conditions for Scheme E

E1 1) LDA, DMPU, DMF, –45 °C; 2) $BnOC_6H_4CH{=}NC_6H_4F$, –45 °C, 64% (47-300 g).

E2 $NaIO_4$, CH_3CN, or CH_3CN/H_2O, –10 °C to 20 °C, 90% (100 g).

E3 1) *p*-$FC_6H_4C(OSiMe_3){=}CH_2$, BF_3•OEt_2, toluene, –30 °C; 2) TsOH, MS, 75% (10-100 g).

E4 H_2 (4 bar), $(Ph_3P)_3RhCl$ (3% mol), CH_2Cl_2, 71% (40 g, Schering1996b, JOC1999b). H_2 (4.5 bar), $(Ph_3P)_3RhCl$, CH_2Cl_2, 8% (E3+E4, 10 g, Adamed2009).

E5 (D6, F5, H21, J12) see D6.

E6 (D7, F6, H20, J13, K21) see D7.
E7 H_2 (59 bar), [(*R*)-Ru(BINAP)Cl_2]$_2$•NEt_3 (<10% mol), MeOH, YNR (50 mg).
E8 (*R*)-2-MeCBS (0.1 eq.), BH_3•SMe_2, CH_2Cl_2, –15 °C to 0 °C, YNR, 75% ee (0.3 g).
E9 (D12) see D12.
E10 H_2 (2 bar), Pd/C (10%), $NaHCO_3$, EtOAc, MeOH, 90% (16 g).
E11 (D9, F18, H4, J4) see D9.
E12 1) DIBAL, THF, –45 °C to rt; 2) HCl, H_2O, 75-78% (14-16 g).
E13 Dimethyldiazomethylphosphonate (DAMP, Bestmann-Ohira reagent), K_2CO_3, MeOH, 0 °C to rt, 70% (1 g).
E14 CuI, CH_3CN, NEt_3, 0 °C to rt, 68-70% (0.3-0.4 g).
E15 TFA, THF, H_2O, 76-90% (0.5-0.9 g).
E16+R17 $NaIO_4$, $NaHCO_3$, CH_3CN/H_2O or CH_2Cl_2, 0-5 °C, 82-86%, *cis*/*trans* = 1:6.2-8.8 (0.1-0.3 g).
E18 CuI, L-Selectride proline, DMSO, rt, 16% (0.52 g of nitrone).
E19 Synthesis of OTBDMS reactant: (*S*)-[1-(4-fluorophenyl)but-3-en-1-ol, TBDMSCl, imidazole, DMF, rt, 75% (550 mg). **Methatesis:** Hoveyda--Grubbs II (10% mol), solvent, 100 °C, 65% (1:3.5 mixture of diastereoisomers, 4*S*,7*S* major/4*R*,7*S* minor, 35 mg).
E20 H_2, Pd/C, MeOH, rt, 98% (1:3.5 *cis* β-lactam mixture, 18 mg).
E21 DBN, DMSO, rt, 42% (30 mg).
E22 HF•py, THF (anh.), py (anh.), rt, 93% (20 mg).
E23 LDA, THF, DMF, –30 °C, 72% (33 g).
E24 CrO_3, py•HCl, CH_2Cl_2, 0 °C to rt, 100% (50 g).
E25 (*R*)-2-MeCBS, BH_3•SMe_2, CH_2Cl_2 (anh.), 10 °C, 92% (25 g).
E26 HCO_2H, NEt_3, Pd/C (10%), rt, NI (25 g).
E27 EtOAc, petroleum ether, rt, 39% (E26+E27, 25 g).
E28 LDA (2 eq.), hexane, THF, DMPU, DMF, –50 °C, 67% (5.5 g, Ranbaxy2003a). 1) LMTP (2 eq.) or lithium dicyclohexylamine or LDA, DMI or DMPU or toluene or MTBE, THF, DMF, –30 °C; 2) LiCl or LiOAc (anh.), –20 °C to –10 °C, 82-90% (5 g, Zhejiang2013).
E29 1) (*S*)-4-phenyloxazolidin-2-one, TMSCl, NEt_3 or DIPEA, CH_2Cl_2 or toluene, 0-10 °C; 2) Acyl chloride, TBAF, rt 85-93%.
E30 $BnOC_6H_4CH{=}NC_6H_4F$, $TiCl_4$, DIPEA, CH_2Cl_2, –30 °C, 76%.
E31 BSA, TBAF, CH_2Cl_2, 50 °C, 75%.

6.8.6 Reaction conditions for Scheme F

F1 *n*-Bu_3N, toluene, 55-64% *trans*, racemic (6-10 g of imine).
F2 Chiralcel OD column, YNR, >98% ee (9.7 g).
F3 *p*-FC_6H_4COCl, LiHMDS or NaH, THF, –10 °C, YNR (0.4-2.9 g).
F4 1) NaOH (10% aq.) H_2O, 80 °C; 2) AcOH, HCl, 90 °C, YNR (3 g).
F5 (D6, E5, H21, J12) see D6.
F6 (D7, E6, H20, J13, K21) see D7.
F7 MeNHOH, toluene, YNR.
F8 Direct resolution of racemic *trans* lactam methyl esters using microorganisms, up to 35% (91% ee) or subtractive resolution of racemic *trans* lactam methyl esters using microorganisms, up to 36% of 3*R*,4*S* ester, >99% ee.
F9 LiOH, NaOH or KOH, H_2O, acetone or MeOH or THF, rt, 89-100% (1-80 g). **Bn ester:** LiOH, THF, H_2O, rt, 97% (100 g).
F10 $(COCl)_2$, DMF (cat.), CH_2Cl_2, rt, 100% (1-100 g).
F11 *p*-FC_6H_4ZnBr (from *p*-FC_6H_4MgBr + Zn), $(Ph_3P)_4Pd$ (cat.) THF, 0 °C to rt, 80-87% (1 g, Schering1993, JMC1998). *p*-FC_6H_4ZnBr (from *p*-FC_6H_4MgBr + $ZnCl_2$ (anh.)), $Pd(OAc)_2$ (cat.), toluene, rt, 60-69% (25 g, MSN2005, DrReddys2005, Ind2006), in CH_2Cl_2, 10 °C, 69% (125 g, OPRD2009). *p*-FC_6H_4ZnBr (from *p*-FC_6H_4MgBr + $ZnCl_2$ (anh.)), $(Ph_3P)_4Pd$ (cat.), THF, 0 °C to rt, 37% (32 g, KRKA2007), in CH_2Cl_2, 60%, in THF, 55%, in CH_3CN, 51% (15 g, ChangzhouPF2011). *p*-FC_6H_4MgBr, $Fe(acac)_3$, THF, –78 °C, 92% (F10+F11, 25 g, IndSwift2007).
F12 BH_3•SMe_2, THF, 22 °C, YNR (0.35 g).
F13 Chiral chromatography (Chiracel OD column), 40%, >98% ee (F12+F13).
F14 1) $(COCl)_2$, CH_2Cl_2 (anh.), rt; 2) CH_3ONHCH_3•HCl, py, 0 °C to rt, 91% (16 g, ZhejiangT2006). CH_3ONHCH_3•HCl, CDI, NEt_3, THF, 30 °C, YNR (12 g, Unichem2006). CH_3ONHCH_3•HCl, 2-chloro-4,6-dimethoxy-[1,3,5] triazine (CDMT), NMM, THF, rt, 83-85% (16 g, ZhejiangT2006).
F15 1) *p*-FC_6H_4MgBr, THF, 5 °C; 2) HCl, H_2O, 85-100% (12-19 g).
F16 *p*-FC_6H_4MgBr, NEt_3, toluene, –30 °C, 18-50% (0.6-0.9 g).
F17 (D8, H22, J7) see D8.
F18 (D9, E11, H4, J4) see D9.
F19 (A6) see A6.
F20 $TiCl_4$ (1 eq.), Ti(O*i*-Pr)$_4$ (0.3 eq.), DIPEA (2.2 eq.), CH_2Cl_2, –20 °C, 63% (50 g, MSN2005, Ind2006), 47% (100 g, DrReddys2005), 60% (14 g, ZhejiangT2006), 48% (100 g, OPRD2009). **With imine *N*-oxide:** 1) $TiCl_2$(O*i*-Pr)$_2$, DIPEA, CH_2Cl_2, –25 °C; 2) Zn, AcOH, THF, 40 °C; 3) BSA, CH_2Cl_2, reflux, 89% (4.5 g). **With Bn ester:** $TiCl_4$, Ti(O*i*-Pr)$_4$, DIPEA, CH_2Cl_2, –5 °C to 0 °C, 81%.
F21 1) BSA, toluene, 45-60 °C; 2) TBAF (cat.), 50-60 °C, 71-93% (9-200 g). **With Bn ester:** 1) BSA, MTBE, rt; 2) TBAF, 40 °C, 73% (186 g).
F22 $LiBH_4$, THF, H_2O, –10 °C, 86% (2 g, Pharmathen2011). $NaAlH_2(OCH_2CH_2OMe)_2$ (Red-Al®), toluene, –20 °C, 80% (4 g ShanghaiSP2013). $LiAlH_4$, THF, –20 °C, 75% (4.5 g, ShanghaiSP2013).
F23 1) $(COCl_2)_2$, CH_2Cl_2, DMSO –78 °C; 2) NEt_3, 25 °C, 100 % (2 g, Pharmathen2011). TEMPO, NaOCl, $NaHCO_3$, CH_2Cl_2, 5° C to rt, 93% (0.43 g, Pharmathen2011). PCC, CH_2Cl_2, 95% (2 g, ShanghaiSP2013).

F24 1) p-FC_6H_4MgBr, bis[2-(N,N-dimethylamino)-ethyl] ether (BDMAEE), Et_2O, 0 °C; 2) Ti(O*i*-Pr)$_4$, (*R*)-BINOL, 25 °C; 3) SM, –20 °C to rt, 80% (2 g, Pharmathen2011). 1) (*R*)-BINOL (1 eq.), Et_2O, Ti(O*i*-Pr)$_4$, –10 °C; 2) *p*-FC_6H_4MgBr; 3) SM, –20 °C to rt, 85-89%, 89% de (0.2 g, ShanghaiSP2013), in MTBE, 77-83%, 82-89% de (0.2 g, ShanghaiSP2013).

F25 $TiCl_4$, Ti(O*i*-Pr)$_4$, DIPEA, CH_2Cl_2, –25 °C, 59-81% (125-500 g).

F26 BSA, TBAF, MTBE, 40 °C, YNR (500 g).

F27 LiOH, acetone, H_2O, 30 °C, 100% (F26+F27, 500 g).

F28 $(COCl)_2$, DMF, CH_2Cl_2, 15°C, NI (10-300 g).

F29 *p*-FC_6H_4MgBr, Fe(acac)$_3$, THF, –90 °C, 41-74% (10-300 g).

F30 (H34) $(Ph_3P)_4Pd$ (cat.), morpholine, CH_2Cl_2, reflux, YNR (PG = All, 250 g).

F31 (H41) PG = Cbz: CbzCl, DMAP, NEt_3, CH_2Cl_2, toluene, 5 °C to 15 °C, 83% (12 g). **PG = Tr:** TrBr, DIPEA, NEt_3, CH_2Cl_2, rt, 97% (6 g).

F32 (H32) PG = Cbz: (*R*)-2-MeCBS (15% mol), BH_3•SMe_2, THF or CH_2Cl_2, –25 °C, 99% (3 g, Zentiva2007b). Bioreduction with *Rhodococcus fascians* MO22, NI (JMCB2010). **PG = Tr:** (*R*)-2-MeCBS (15% mol), BH_3•SMe_2, THF, –25 °C, 63% (2.6 g). **PG = All:** $RuCl_2$[(*R*)-DAIPEN][(*R*)-XylB] (0.1% mol), DBU, H_2 (5 bar), EtOH, 20 °C, 99%, 99.9% de (2 g, Nissan2014), in THF with *t*-BuOK 100%, 99.9% de (2 g, Nissan2014). (*R*)-2-MeCBS (25% mol), BH_3•SMe_2, CH_2Cl_2, 0 °C, YNR (10 g, IndSwift2008).

F33 (H33) PG = Cbz: H_2 (1 bar), Pd/C (10%), MeOH, 76% (2.4 g, Zentiva2007b). Bioreduction with *Rhodococcus fascians* MO22 63% (F32+F33, JMCB2010). **PG = Tr:** H_2 (1 bar), Pd/C (10%), MeOH, 73% (1.6 g). **PG = All:** HCO_2H, morpholine, $Pd(PPh_3)_4$, EtOAc, 60 °C, 83%, 100% de (2 g, Nissan2014). $Pd(PPh_3)_4$, 5,5-dimethyl-1,3-cyclohexanedione, THF, reflux, YNR (10 g, IndSwift2008). **PG = THP:** Acetone, HCl (conc.), 97% (150 mg). **PG = TBDMS:** Bu_4NF, THF, rt, 88% (150 mg).

F34 PG = THP or TBDMS: 1) (*R*)-BINOL (1 eq.), Et_2O, Ti(O*i*-Pr)$_4$, –10 °C; 2) *p*-FC_6H_4MgBr; 3) SM –20 °C to rt, YNR.

6.8.7 Reaction conditions for Scheme G

G1 Synthesis of the Michael acceptor: 1) (*S*)-4-phenyloxazolidin-2-one, TMSCl, CH_2Cl_2, 10 °C; 2) H_2C=CHCOCl, TBAF, 10 °C to rt, 90-93% (20 g of oxazolidinone). **Synthesis of the silyl ether of 1-(4-fluorophenyl)ethan-1-one:** TMSCl, NEt_3, DMF 100°C 90% (20-22 g), in toluene 85-88%. **Michael addition:** I_2, toluene, 40 °C, 89-94% (10 g of oxazolidine SM).

G2 (A8) see A8.

G3 $(COCl)_2$ or $SOCl_2$, CH_2Cl_2 or toluene, 15 °C to 40 °C, YNR (4 g).

G4 (*S*)-4-phenyloxazolidin-2-one, NEt_3, toluene, 88-92% (G3+G4, 4 g).

G5 1) PivCl, NEt_3, CH_2Cl_2 or DMF; 2) (*S*)-4-phenyloxazolidin-2-one, DMAP, DMF, reflux, 80-92%, in THF, 20 °C, 89% (Wuhan2016). (*S*)-4-phenyloxazolidin-2-one, DCC, DMAP, CH_2Cl_2, 86-90%, in toluene, 85%, in DMF, 71%, in acetone, 82%, with EDC, 47%.

G6 $HOCH_2CH_2OH$, $CH(OMe)_3$, H_2SO_4, CH_2Cl_2, 25 °C, 90% (21 g).

G7 1) PivCl, NEt_3, THF, –20 °C; 2) (*S*)-4-phenyloxazolidin-2-one, –20 °C to rt, 93% (42 g).

G8 $HO(CH_2)_3OH$, $CH(OMe)_3$, H_2SO_4 (cat.), CH_2Cl_2, 30-40 °C, 92-94% (21 g).

G9 1) PivCl, NEt_3, THF; 2) (*S*)-4-phenyloxazolidin-2-one, DMAP, DMF, 25 °C, 92-95% (51 g).

G10 Neopentyl glycol, TsOH or py•HBr or TsOH, toluene, reflux, 96% (100 g).

G11 1) PivCl, NEt_3, CH_2Cl_2, 25 °C; 2) (*S*)-4-phenyloxazolidin-2-one, DMAP, DMF, 25 °C, 96% (100 g, Teva2006a).

G12 $ZnCl_2$, HCl, $HSCH_2CH_2SH$, 95%.

G13 PivCl, NEt_3, (*S*)-4-phenyloxazolidin-2-one, CH_2Cl_2, 95% (38 g).

G14 (A10) see A10 **for R = Me. For 2,2-dimethylpropane-1,3-diol ester, G14+G15:** Neopentyl glycol, TsOH or py•TsOH, toluene, reflux, 100% (G14+G15).

G15 Neopentyl glycol, TsOH or py•TsOH or py•HBr, toluene, reflux, 73-84% (100 g, R = Me, Et). **For 2,2-dimethylpropane-1,3-diol ester see G14.**

G16 2,2-dimethylpropane-1,3-diol ester: LiOH, THF, H_2O, rt, 75%. NaOH, MeOH, H_2O, rt 75%.

G17 (I1) n = 1: $HOCH_2CH_2OH$, TMSCl (4 eq.), 25 °C, 96% (Medichem2005). $HOCH_2CH_2OH$, TsOH, TMSCl (3 eq.), toluene, 25 °C, 89% (20 g, Medichem2005). $HOCH_2CH_2OH$, TsOH, toluene, reflux, 94% (15 g, Zentiva2007a). $HOCH_2CH_2OH$, TsOH, $CH(OMe)_3$, *n*-BuOH, rt, 88% (12 g, ChongqingHS2011). $Me_3SiOCH_2CH_2OSiMe_3$ (1.2 eq.), $CF_3SO_3SiMe_3$ (10% mol), TsOH, toluene, 50 °C, 89 % (69 g, Zentiva2007a). 1) $HOCH_2CH_2OH$, NEt_3, THF, TMSCl, 0 °C to rt; 2) SM, toluene, $CF_3SO_3SiMe_3$ (5% mol), 50 °C, 89% (57 g, Zentiva2007a). $HO(CH_2)_2OH$, $CH(OMe)_3$, NBS, 70 °C, 95% (20 g, QR2015). $HO(CH_2)_2OH$, TsOH, *n*-hexane, 25 °C, 91% (3.6 g HenanU2017a). $CH(OMe)_3$, NBS, $HOCH_2CH_2OH$, 100 °C (Wuhan2016). **n = 2:** $HO(CH_2)_3OH$, $CH(OMe)_3$, H_2SO_4 (cat.), EtOH, 90% (71 g, ChongqingHS2011). $HO(CH_2)_2OH$, $CH(OMe)_3$, NBS, 70 °C, 87% (10 g, QR2015). **n = 3, 4:** $HO(CH_2)_4OH$, AcOH (cat.) $MgSO_4$ (anh.), EtOAc, 100% (50 g). $CH(OMe)_3$, $HO(CH_2)_5OH$, HNO_3 (cat.), THF, rt, 94% (35 g).

G18 Neopentyl glycol, py•HBr or py•TsOH or TsOH, toluene, reflux, 77-78% (100 g, Teva2006a), with TsOH/TMSCl, 74%, with TsOH/BF_3•OEt_2 73%, with TsOH/TMSCl, 74%, in cyclohexane/py•TsOH, 79%. Neopentyl glycol, P_2O_5, H_2SO_4, CH_2Cl_2, reflux, 89% (36 g, SuzhouP2016).

G19 (I5) MeOH, $CH(OMe)_3$, TsOH, rt, 83-100% (15-50 g, Zentiva2008a,b, JPA2015). MeOH, $CH(OMe)_3$, 80% (SC2016). $CH(OMe)_3$, MeOH, NBS, 50 °C. 91% (20 g, QR2015).
G20 (I1) n = 1: $HSCH_2CH_2SH$, $TiCl_4$, toluene, reflux, 88% (200 g). **n = 2:** $HS(CH_2)_3SH$, TsOH, TMSCl, toluene, rt, 82% (22 g), in CH_3CN, 84% (33 g), in THF, 84% (21 g).
G21 R = H: NH_2OH•HCl, K_2CO_3 (anh.), *i*-PrOH, reflux, YNR (50g, Lupin2009a). NH_2OH•HCl, py, 0 °C, rt, 98% (30 g, Pharmathen2010). **R = Me:** CH_3NHOH•HCl, NEt_3, EtOH, 80 °C, 88% (*E*/*Z*, 93: 7, 100g). **R = TBDMS:** TBDMSCl, imidazole, DMAP, CH_2Cl_2, rt, 97% (3 g).
G22 TMSCN, CH_2Cl_2, I_2, 100% (100 g).

6.8.8 Reaction conditions for Scheme H

H1 1) $HOC_6H_4CH{=}NC_6H_4F$, DIPEA, TMSCl, CH_2Cl_2, –15 °C; 2) $TiCl_4$, –15 °C; 3) BSA, 41% (11 g, Medichem2005). 1) $HOC_6H_4CH{=}NC_6H_4F$, DIPEA, TMSCl, CH_2Cl_2, 0 °C; 2) $TiCl_4$, –30 °C, 32% (2.2 g, Zentiva2007a).
H2 1) BSA, toluene, 50 °C; 2) TBAF•H_2O, 50 °C, 72% (4 g, Medichem2005). 1) BSA, TBAF, THF, 10 °C to 20 °C; 2) AcOH, H_2O, 75%, 87% with 2nd crop (3.5 g, Zentiva2007a).
H3 TsOH, H_2O, acetone, reflux, 90-95% (60 mg-8 g, Medichem2005, Zentiva2007a, Changzhou2016). AcOH, H_2O, THF, 60 °C, 92% (2.6 g, Zentiva2007a).
H4 (D9, E11, F18, J4) see D9.
H5 1) BSA, MTBE, rt; 2) TBAF, THF, 0 °C, 84% (1.4 g).
H6 TBAF, THF, rt, 71% (0.9 g).
H7 1) $HOC_6H_4CH{=}NC_6H_4F$, TBDMSCl, DIPEA, CH_2Cl_2, 25 °C; 2) SM, $TiCl_4$, $Ti(Oi\text{-}Pr)_4$, –30 °C, 78%, 89% with 2nd crop (40 g, RichterGedeon2005), 56% (50 g, QR2016). 1) $HOC_6H_4CH{=}NC_6H_4F$, TBDMSCl, CH_2Cl_2, DIPEA, 0 °C; 2) SM, $TiCl_4$, –30 °C, 34-40% (2-3.2 g, Zentiva2007a). $TBDMSOC_6H_4CH{=}NC_6H_4F$, $TiCl_4$, CH_2Cl_2, $Ti(Oi\text{-}Pr)_4$, –40 °C to 0 °C, 84% (30 g, Wuhan2016).
H8 1) BSA (2 eq.), CH_3CN, 25 °C; 2) TBAF•$3H_2O$ (cat.), 25 °C, YNR (20 g, RichterGedeon2005). BSA, CH_3CN, TBAF, 84% (40 g, Wuhan2016).
H9 Montmorillonite K10, CH_2Cl_2, 25 °C, 80% (H8+H9, 17 g).
H10 *i*-PrOH, H_2SO_4, 70 °C, 86% (30 g).
H11 (*R*)-*o*-tolyl-CBS-oxazaborolidine (10% mol), BH_3•SMe_2, CH_2Cl_2, –5 °C, 100%, > 98% de (5 g, RichterGedeon2005). H_2 (5 bar), $RuCl_2$[(*R*)-daipen][(*R*)-XylBINAP] (0.1 % mol), *t*-BuOK, EtOH, 26 °C, YNR, 99% de (2 g, Nissan2014). (*R*)-2-MeCBS (0.1 eq.), $NaBH_4$ (2 eq.), I_2 (1 eq.), THF, –5 °C, 80% (1 g, ShanghaiSP2013).
H12 H_2SO_4 (2 M), H_2O, *i*-PrOH, 70 °C, 81% (H11+H12, 5 g, RichterGedeon2005). TBAF, THF, 88% (0,15 g, ShanghaiSP2013). H_2SO_4 (20%), H_2O, 60 °C, YNR (2 g, Nissan2014).
H13 (I2) Z = O: $BnOC_6H_4CH{=}NC_6H_4F$, $TiCl_4$, $Ti(Oi\text{-}Pr)_4$, DIPEA, CH_2Cl_2, –15 °C, 49-58% (4-52 g, Medichem2005). 1) $HOC_6H_4CH{=}NC_6H_4F$, BnBr, NaOH, CH_2Cl_2; 2) $TiCl_4$, $Ti(OEt)_4$, DIPEA, CH_2Cl_2,; 3) SM, –15 °C, 50-65% (36 g, Tianjin2011). 1) $BnOC_6H_4CH{=}NC_6H_4F$, BSA, toluene, NI (4 g-4 kg, HenanU2017a). **Z = O, with imine *N*-oxide:** 1) $HOC_6H_4CH{=}NC_6H_4F$ *N*-oxide, TMSCl, DIPEA, CH_2Cl_2, –3 °C; 2) $TiCl_4$, –25 °C; 3) THF, Zn, AcOH; 4) BSA, CH_2Cl_2, reflux, 84% (7.3 g, *N*-oxide). **Z = O; Bn = THP:** 1) $THPOC_6H_4CH{=}NC_6H_4F$, BSA, toluene, NI (4 g-4 kg, HenanU2017b). **Z = S:** $BnOC_6H_4CH{=}NC_6H_4F$, $TiCl_4$, DIPEA, CH_2Cl_2, –20 °C, YNR (10 g, Ind2007b). **Z = S, Bn = TMS:** unintelligible (Shangdong2016).
H14 Z = O: 1) BSA, toluene, 50 °C; 2) TBAF•H_2O, 50 °C, 90-95% (4-10 g, Medichem2005). 1) BSA, toluene; 2) TBAF, 60 °C, NI (4 g-4 kg, HenanU2017a). **Z = O, Bn = THP:** 1) BSA, toluene, 60 °C; 2) TBAF, 60 °C, 88-90% (H13+H14, 4 g-4 kg, Henan2017b). **Z = S:** 1) BSA, CH_2Cl_2, reflux; 2) TBAF, 51% (H13+H14, 10 g). **Z = S, Bn = TMS:** unintelligible (Shangdong2016).
H15 H_2, Pd/C (5%), EtOH, reflux, 91% (10 g).
H16 $HgCl_2$, benzene, reflux, 95% (8 g).
H17 Z = O: TsOH•H_2O, acetone, reflux, 81-99% (2-8 g). AcOH (H12+H14+H17, HenanU2017a). **Z = O; Bn = THP:** AcOH, 90% (H12+H14+H17, 4 g-4 kg, Henan2017b). **Z = S:** Pd/C (5%), EtOH, reflux, 91% (10 g, Ind2007b).
H18 BH_3•SMe_2, THF, 22 °C YNR (0.4 g).
H19 Chiral chromatography (Chiracel OD column), 40% (H18+H19), >98% ee.
H20 (D7, E6, F6, J13, K21) see D7. **Bn = THP:** PPTS, EtOH, reflux, YNR (HenanU2017b).
H21 (D6, E5, F5, J12) see D6. **Bn = THP:** (*R*)-2-MeCBS, BH_3•SMe_2, –5 °C, CH_2Cl_2, 92% (HenanU2017b).
H22 (D8, F17, J7) see D8.
H23 R = *t*-Bu: $TiCl_4$, $Ti(Oi\text{-}Pr)_4$, $BocOC_6H_4CH{=}NC_6H_4F$, DIPEA, CH_2Cl_2, –40 °C, 85% (4 g). **R = Me:** $TiCl_4$, $Ti(Oi\text{-}Pr)_4$, DIPEA, $AcOC_6H_4CH{=}NC_6H_4F$, DIPEA, CH_2Cl_2, –40 °C, 80-87% (4 g). **R = Pr:** $TiCl_4$, $Ti(Oi\text{-}Pr)_4$, DIPEA, $PrCOOC_6H_4CH{=}NC_6H_4F$, DIPEA, CH_2Cl_2, –40 °C, 74-86% (4 g).
H24 R = Boc, Me, Pr: 1) BSA, CH_3CN, 20-25 °C; 2) TBAF, 20-25 °C, YNR (20 g).
H25 NaOH, EtOH (anh.), 20-25 °C, 80% (H24+H25, R= Me).
H26 R = Pr: TsOH, acetone, 20-25 °C, 95%.
H27 Z = O, PG = TMS: 1) $TMSOC_6H_4CH{=}NC_6H_4F$, DIPEA, $TiCl_4$, $Ti(Oi\text{-}Pr)_4$, CH_2Cl_2, 87-91% (4-5 g). **Z = O, PG = Cbz:** 1) $CbzOC_6H_4CH{=}NC_6H_4F$, DIPEA, $TiCl_4$, $Ti(Oi\text{-}Pr)_4$, CH_2Cl_2, 0°C, 65% (10 g). **Z = S, PG = Bn:** 1) $TiCl_4$, $Ti(Oi\text{-}Pr)_4$, NEt_3, CH_2Cl_2 or Et_2O, –15 °C; 2) $BnOC_6H_4CH{=}NC_6H_4F$, 60-63% (30-31 g).
H28 Z = O, PG = TMS: 1) BSA, CH_3CN (anh.); 2) TBAF, 85-87% (24 g). **Z = O, PG = Cbz:** 1) BSA, THF; 2) TBAF, 0 °C to rt, 89% (2.4 g). **Z = O,**

PG = Tr: 1) BSA, THF; 2) TBAF, 0 °C to rt, 25% (1.3 g). **Z = S, PG = Bn:** 1) BSA, THF, 50 °C; 2) TBAF, 50 °C, 91-95% (13 g).

H29 Z = S, PG = Bn: TsOH, THF, H_2O, 45 °C, 86% (8.5 g). TFA, CH_3CN, H_2O, 45 °C, 83% (6.8 g). HCO_2H, toluene, H_2O, 45 °C, 85% (17 g).

H30 py•TsOH, THF, 0 °C, 85-88% (10g).

H31 AcOH, THF, H_2O, 60 °C, 84% (3.4 g). TsOH, acetone, H_2O, 60 °C, 90% (2.7 g).

H32 (F32) see also F32. **PG = COPr:** (*R*)-2-MeCBS (15% mol), BH_3•SMe_2, THF, 20-25 °C, 75% (12 g, ChangzhouPF2016).

H33 (F33) see F33. **PG = COPr:** NaOH, EtOH, 20-25 °C, 90% (8 g, ChangzhouPF2016).

H34 (F30) see F30.

H35 $BnOC_6H_4CH{=}NC_6H_4F$, $TiCl_4$, Ti(O*i*-Pr)$_4$, DIPEA, CH_2Cl_2, –30 °C, 42-51% (100 g, Teva2006a), at –5 °C 78% (44 g, SuzhouP2016).

H36 1) BSA, THF, reflux; 2) TBAF•H_2O, reflux, 83-89% (100 g, Teva2006a). 1) BSA, toluene, reflux; 2) TBAF, reflux, 79% (75 g, SuzhouP2016).

H37 HCO_2H, optional use of CH_2Cl_2, 30 °C, 85-90% (20-100 g, Teva2006a,b). HCO_2H, CH_2Cl_2, reflux, 90% (58 g, SuzhouP2016).

H38 Neopentyl glycol, py•HBr or py•TsOH or TMSCl or BF_3•OEt_2, toluene, reflux, 68-81% (50-207 g).

H39 H_2 (4 bar), Pd/C (10%), 30 °C, 61% (100 g).

H40 HCO_2H, 25 °C, 74% (100g, Teva2006a).

H41 (F31) see F31.

H42 1) $HOC_6H_4CH{=}NC_6H_4F$, TMSCl, DIPEA, CH_2Cl_2, –10 °C; 2) $TiCl_4$, –25 °C, 50% (100 g).

H43 1) BSA, THF, reflux; 2) TBAF•H_2O, reflux, 79% (100 g). 1) BSA, toluene, 50 °C; 2) TBAF•H_2O, 50 °C, 13% (100 g).

H44 HCO_2H, 25 °C, 97% (100g, Teva2006a).

H45 PG = TBDMS: $TBDMSOC_6H_4CH{=}NC_6H_4F$, $TiCl_4$, Ti(O*i*-Pr)$_4$, DIPEA, CH_2Cl_2, 85% (SC2016). **PG = Bn:** $BnOC_6H_4CH{=}NC_6H_4F$, $TiCl_4$, Ti(O*i*-Pr)$_4$, DIPEA, CH_2Cl_2, –50 °C, 63% (8 g, QR2015).

H46 TBAF•$3H_2O$, YNR.

H47 HCl (4 N), *i*-PrOH, 72% (H46+H47).

6.8.9 Reaction conditions for Scheme I

I1 for ketal see G17, for thioketal see G20.

I2 (H13) see H13.

I3 TsOH, acetone, H_2O, reflux, 95% (8.5 g, Zentiva2008a). 1) TA, AcOH, CH_2Cl_2; 2) Zn, 83% (7.2 g, UZheijang2011a).

I4 (*R*)-2-MeCBS (25% mol), BH_3•SMe_2, THF, rt, 78% (2 g),

I5 (G19) see G19.

I6 1) $HOC_6H_4CH{=}NC_6H_4F$, DIPEA, TrCl, CH_2Cl_2, 10 °C; 2) SM, $TiCl_3$(O*i*-Pr), –30 °C; 3) MeOH, HCl, 25 °C, 51% (6.6 g).

I7 (*R*)-2-MeCBS (50% mol), BH_3•SMe_2, THF, rt, 75% (1 g).

I8 1) $HOC_6H_4CH{=}NC_6H_4F$, CbzCl, DIPEA, toluene or toluene/CH_2Cl_2, 10 °C; 2) SM, $TiCl_3$(O*i*-Pr), –40 °C to 10 °C, YNR (6-13 g).

I9 TsOH, MeOH, reflux, 59% (I8+I9, 6 g, Zentiva2008a). Citrate buffer, rt, 52% (I8+I9, 12.6 g, Zentiva2008b). AcOH, CH_2Cl_2, 49% (I8+I9, JPA2015).

I10 CbzCl, NEt_3, DMAP, CH_2Cl_2, toluene, 10-15 ° C, 91-96% (1.5-6 g).

I11 (*R*)-2-MeCBS (5-25% mol), BH_3•SMe_2 (1.4 eq.), THF, rt, 83-93% (3-14 g, Zentiva2008a,b), 79% (JPA2015).

I12 $LiClO_4$, HMDS or TMSCl, CH_2Cl_2, 0-20 °C, 89-91% (5-6 g, Zentiva2008b). $LiClO_4$, HMDS, 90% (JPA2015).

I13 BSA, TBAF, CH_2Cl_2, MeOH, –10 °C to 0 °C, YNR (2-7 g).

I14 H_2 (1 bar), Pd/C (3%), MeOH, 70% (I13+I14, 6.5 g, Zentiva2008b), 60% (I13+I14, JPA2015).

6.8.10 Reaction conditions for Scheme J

J1 R = Me, PG = Bn: 1) $TiCl_4$, Ti(O*i*-Pr)$_4$, DIPEA, CH_2Cl_2, 0 °C; 2) $BnOC_6H_4CH{=}NC_6H_4F$ –25 °C, 65% (60 g). **R = H, PG = Bn:** 1) $TiCl_4$, Ti(O*i*-Pr)$_4$, DIPEA, CH_2Cl_2, 0 °C; 2) $BnOC_6H_4CH{=}NC_6H_4F$, –15 °C, 9% (2 g). **R, PG = TMS:** 1) $HOC_6H_4CH{=}NC_6H_4F$, DIPEA, TMSCl, CH_2Cl_2, –5 °C; 2) $TiCl_4$, –25 °C; 3) BSA, CH_2Cl_2, reflux, YNR (25 g).

J2 R, PG = TMS: MTBE, BSA, TBAF, rt, YNR (25 g).

J3 HCO_2H, $NaHSO_4$, EtOH, H_2O, reflux, YNR (2 g).

J4 (D9, E11, F18, H4) see D9.

J5 R = Me, PG = Bn: 1) BSA, toluene, 50 °C; 2) TBAF, 50 °C, 86% (23 g). **R = H, PG = Bn:** 1) BSA, toluene, 60°C; 2) TBAF, 60 °C, 70% (0.18 g).

J6 R = Me, PG = Bn: HCl, H_2O, acetone, 1,4-dioxane, 95 °C, YNR (500 g). **R = H, PG = Bn:** CH_2O (35%), HCl, THF, rt, 65% (0.5 g), 87% (J5+J6, 2 g).

J7 (D8, F17, H22) see D8.

J8 TBDMSCl, imidazole, DMAP, CH_2Cl_2, rt, 97% (3 g).

J9 1) CH_2Cl_2, $TiCl_4$, Ti(O*i*-Pr)$_4$, DIPEA, 0 °C; 2) $BnOC_6H_4CH{=}NC_6H_4F$, –15 °C, 63% (10 g).

J10 1) BSA, toluene, 60°C; 2) TBAF, 60 °C, 95% (2 g).

J11 CH_2O (37%), HCl, THF, H_2O, rt, 93% (0.41 g).

J12 (D6, E5, F5, H21) see D6.

J13 (D7, E6, F6, H20, K21) see D7.

J14 $TiCl_4$, DIPEA, CH_2Cl_2, –40 °C, 58% (128 g).

J15 1) LiHMDS, toluene, CH_2Cl_2, 0 °C; 2) PivCl, NEt_3, 80% (10 g).

J16 (*R*)-2-MeCBS, BH_3•SMe_2 or BH_3•SMe_2, THF, –25 °C, 88% (10 g).

J17 NaOH, MeOH, THF, –15 °C, 92% (10g).

6.8.11 Reaction conditions for Scheme K

K1 $NaBH_4$ (not shown).

K2 Lipozyme (*Thermomyces lanuginosus*), vinyl acetate, toluene, 40°C, 37% (1 g).

K3 Most frequent chemical method, CBS reduction: (*R*)-2-MeCBS (5-20% mol), BH_3•SMe_2 (0.95-1.1 eq.),

THF or CH_2Cl_2, usual yield range 80-95%, >98% de. Biochemical methods with >90% conversion, >99% de reported. See section 6.4.9.1.

K4 (L1) When PG = TMS in also the imine for K5 step, simultaneous silylation of the two hydroxyl groups is frequent: $HOC_6H_4CH{=}NC_6H_4F$, TMSCl, DIPEA, CH_2Cl_2, –10 °C/–20 °C, NI (toluene or heptane). Alternatively, the amide may be silylated in a separate step using the same reaction conditions. At the end of AEAI, BSA may be used to assure complete sylilation. 75-78% (K3+K4+K6, 55 kg, ShanghaiHP2017b).

K5 1) $TiCl_4$ or $TiCl_4$, $Ti(O\textit{i}\text{-}Pr)_4$, DIPEA (optional), CH_2Cl_2, –25 °C to –5 °C; 2) BSA, CH_2Cl_2. The product is frequently NI. Without DIPEA, 65% (K4+K5, 10 g, Schering1998c,d, Merck2004a,b), in toluene/heptane; without DIPEA, 55% (K4+K5). 1) $TiCl_4$, $Ti(O\textit{i}\text{-}Pr)_4$, –25 °C; 2) $NaHSO_3$; 3) BSA, CH_2Cl_2, 64% (K4+K5, 10 g, Jiangxi2015). **With imine *N*-oxide:** 1) $TiCl_4$, –25 °C; 2) Zn, AcOH, THF; 3) BSA, CH_2Cl_2, reflux, 84% (4.8 g amide, 8 g imine *N*-oxide).

K6 BSA, $TBAF•3H_2O$ (cat.), MTBE or toluene, rt, YNR (50 g, Schering1998c,d), with Bu_4POMe, YNR (Sanofi2004, Aventis2004), with TBAOAc (cat.), (1 g, Merck2004a,b). 1) BSA/KOAc or CsF or KOAc 2) TBABr (cat.) or TBAOH, THF, rt, YNR (5 g, Merck2004a,b). BSA, TBAF, CH_2Cl_2, rt, 92% (50 kg, Jiangxi2015).

K7 (M6) H_2SO_4, *i*-PrOH, rt, 90-92% (K6+K7, Schering1998c,d, Topharman2009, Shanghai2014, Jiangxi2015), in MeOH, 65% (25 g, DrReddys2003), in EtOAc, 65% (K6+K7, Nanjing2013). LiF, *i*-PrOH, rt, 90-95% (K6+K7, 48 kg, Jiangxi2015).

K8 $NaHSO_3$, CH_2Cl_2, H_2O, 65-70% (K4+K5+K8, 54 kg, Jiangxi2015), 75% (K5+K8, 28 g, ShanghaiFB2014).

K9 BSA, CH_2Cl_2, reflux, NI (48 kg).

K10 1) Ac_2O, NEt_3, CH_2Cl_2, rt; 2) BSA, TBAF, MTBE, rt, NI (12 g).

K11 PG = TMS: NaOH (1M), MeOH, rt, 61% (52 g, Centaurus2012). H_2SO_4 (2 N), *i*-PrOH, rt, 93% (K26+K11, 20 g, Lianyungang2013). **PG = Ac:** HCl (1 M), MeOH, rt, 70% (49 g, Centaurus2012). H_2SO_4 (2 N), *i*-PrOH, rt, 92% (K10+K11, 12 g, ShanghaiFB2014).

K12 $AcOC_6H_4CH{=}NC_6H_4F$, $TiCl_4$, NEt_3, CH_2Cl_2, 25 °C, 83% (K4+K12, 30 g).

K13+K14 1) BSA, toluene, reflux; 2) TBAF, reflux; 3) H_2O, H_2SO_4 rt, 82% (23 g).

K15 PG = Bn: 1) $BnOC_6H_4CH{=}NC_6H_4F$, $TiCl_4$, CH_2Cl_2, –25 °C; 2) BSA, 77-78% (K4+K15, 9-13 g, UZhejiang2010). $BnOC_6H_4CH{=}NC_6H_4F$, $TiCl_4$, $Ti(O\textit{i}\text{-}Pr)_4$, DMA, DIPEA, –20 °C, NI (30 g, TianjinP2012), 93% (Nanjing2015). **PG = Bn, with imine *N*-oxide:** 1) $BnOC_6H_4CH{=}N(O)C_6H_4F$, $TiCl_4$, CH_2Cl_2, –25 °C; 2) Zn, AcOH, THF; 3) BSA, CH_2Cl_2, reflux, 84% (6 g, UZheijang2011a).

K16 HCl or HOAc, DMA, H_2O, 80-81% (K4+K15+K16, 30 g), 76% (K3+K4+K15+K16).

K17 TMSCl, NEt_3, CH_2Cl_2, 97% (46 g).

K18 1) BSA, toluene, 45-60 °C; 2) TBAF, 45-60 °C, YNR (4-10 g, UZhejiang2010, 2011a, Jiangsu2011), in MTBE, NI (20 g, Nanjing2015), in CH_3CN, NI (37 g, TianjinP2012).

K19 1) H_2 (3-4 bar), Pd/C (5%), MeOH, H_2O, rt; 2) *i*-PrOH, 45°C, 85-91% (K18+K19, 4-8 g, UZhejiang2010, 2011a). Pd/C, HCO_2NH_4, MeOH, THF, 0 °C, 85% (K18+K19, 37 g, TianjinP2012).

K20 H_2SO_4 (2 N), *i*-PrOH, rt, NI (10 g, Jiangsu2011). TBAF, MTBE, 96% (20 g K18+K20, Nanjing2015).

K21 (D7, E6, F6, H20, J13, K21) see D7.

K22 $BnOC_6H_4CH{=}NC_6H_4F$, $TiCl_4$, $Ti(O\textit{i}\text{-}Pr)_4$, DIPEA, CH_2Cl_2, –25 °C, 60% (10 g, Jiangsu2011), 94% (Nanjing2015).

K23 1) BSA, toluene, 60 °C; 2) TBAF, 60°C, NI (10 g).

K24 AcCl, py, CH_2Cl_2, 90% (100 g).

K25 PG = TMS: 1) $HOC_6H_4CH{=}NC_6H_4F$, TMSCl, DIPEA, CH_2Cl_2, –15 °C; 2) SM, $TiCl_4$, –30 °C; 3) BSA, reflux, 79% (20 g, Lianyungang2013). $TMSOC_6H_4CH{=}NC_6H_4F$, $TiCl_4$, $Ti(O\textit{i}\text{-}Pr)_4$, DIPEA, CH_2Cl_2, –15 °C, 55% (40 g, Centaurus2012). **PG = Ac:** $AcOC_6H_4CH{=}NC_6H_4F$, $TiCl_4$, $Ti(O\textit{i}\text{-}Pr)_4$, DIPEA, CH_2Cl_2, –15 °C, 60% (40 g).

K26 PG = TMS: 1) BSA, toluene, rt; 2) $TBAF•3H_2O$, rt, 70% (69 g, Centaurus2012). 1) BSA, MTBE, 10 °C; 2) TBAF, NI (20 g, Lianyungang2013). **PG = OAc:** 1) BSA, toluene, rt; 2) $TBAF•3H_2O$, rt, 75% (66 g).

K27 TBDMSCl, imidazole or py or DBU, DMF, 40-80 °C, 55-62% (5.2 g).

K28 PG = TBDMS: 1) $TBDMSOC_6H_4CH{=}NC_6H_4F$, $TiCl_4$, DIPEA, CH_2Cl_2, –30 °C, 51% (0.8 g), in $CH_2Cl_2/CHCl_3$, 36% (0.8 g). **PG = OAc:** $AcOC_6H_4CH{=}NC_6H_4F$, $TiCl_4$, DIPEA, CH_2Cl_2, –30 °C, 47-55% (0.8 g). **PG = $CO_2CH_2CCl_3$:** $Cl_3CCH_2OCOOC_6H_4CH{=}NC_6H_4F$, $TiCl_4$, DIPEA, CH_2Cl_2, –30 °C, 46% (0.8 g).

K29 1) BSA, toluene, 30 °C; 2) TBAF, 30 °C, NI (0.2-0.3 g).

K30 PG = TBDMS: TBAF, MeOH, toluene, 79% (K29+K30, 0.3 g). **PG = OAc:** Na_2CO_3, MeOH, toluene, rt, 74% (K29+K30, 0.27 g). **PG = $CO_2CH_2CCl_3$:** Na_2CO_3, MeOH, toluene, rt, 71% (K29+K30, 0.25 g).

K31 TBAF, THF, 30 °C, 51% (0.1 g), in CH_2Cl_2, 48%, in $CHCl_3$, 57%, in THF/H_2O, 68%.

K32+K33 1) $HOC_6H_4CH{=}NC_6H_4F$, DIPEA, MOMCl, CH_2Cl_2 (anh.), –10 °C; 2) $TiCl_4$, $Ti(O\textit{i}\text{-}Pr)_4$, –30 °C, 62-65% (10 g).

K34 BSA, TBAF, CH_3CN, 74% (13.5 g).

K35 HBr (48%), *i*-PrOH, rt, 87% (11.3 g).

K36 (*S*)-(+)-phenylglycinol, DMAP, dioxane, 60 °C, 91% (1 g).

K37 1) Boc_2O, NEt_3, DMAP, THF, 4 °C, 98% (1 g); 2) NaH, DMF, 0 °C to rt, 75% (1.3 g).

K38 1) PivCl, NEt_3, toluene, 25 °C; 2) (*S*)-4-phenyloxazolidin-2-one, Me_2NH_2, DMF, reflux, 81% (10 g, Mylan2009).

K39 1) $BnOC_6H_4CH{=}NC_6H_4F$, $TiCl_4$, Ti(O*i*-Pr)$_4$, DIPEA, CH_2Cl_2, –35 °C, 48% (10 g).
K40 NaOMe, $(MeO)_2CO$, CH_2Cl_2, 62% (10 g).
K41 LiHMDS, THF, –25 °C, 85% (10 g). 1) BSA, toluene, 50 °C; 2) TBAF, 79% (2 g).
K42 H_2, Pd/C (5%), MeOH, 25 °C, 72% (10 g).

6.8.12 Reaction conditions for Scheme L

L1 (K4) see K4.
L2 1) $TiCl_4$, DIPEA, CH_2Cl_2, –30 °C; 2) BSA, toluene, TBAF, 54% (1 g, imine/boron intermediate). 1) TMSCl, DIPEA, CH_2Cl_2; 2) $TiCl_4$, –30 °C; 3) BSA, 40 °C, 37% (1 g, imine/silyl intermediate).
L3 BSA, TBAF•$3H_2O$, toluene, NI (0.9 g, silyl/boron derivative). BSA, TBAF•$3H_2O$, toluene, NI (0.9 g, silyl/silyl derivative).
L4 H_2SO_4 (2 N), *i*-PrOH, rt, NI (0.9 g, silyl/boron and silyl/silyl derivatives).
L5 H_2O_2, H_2O, rt, 79% (L3+L4+L5, 0.9 g, boron derivative). H_2, Pd/C (10%), MeOH, YNR (0.9 g, silyl derivative).
L6 1) TMSCl, DIPEA, CH_2Cl_2; 2) $TiCl_4$, –30 °C, YNR.
L7 BSA, 40 °C, YNR.
L8 H_2SO_4 (2 N), *i*-PrOH, rt, NI.
L9 H_2, Pd/C (10%), MeOH, YNR.
L10 $O_2NC_6H_4CH{=}NC_6H_4F$ (0.25 eq.), $TiCl_4$ (0.5 eq.), –40 °C, 27% (to OTMS SM), 100% (to imine SM) (43 g).
L11 1) BSA, toluene, 90 °C; 2) TBAF•$3H_2O$, 50 °C, 60% (3.5 g).
L12+L13 Fe, HCl (6 N), AcOH, EtOH, H_2O, 0 °C to rt, 85% (3.2 g).
L14 $NaNO_2$, H_2SO_4, H_2O, DMF, 0 °C to rt, 81% (0.3-0.4 g).

6.8.13 Reaction conditions for Scheme M

M1 DCC, (*S*)-4-phenyloxazolidin-2-one, CH_2Cl_2, rt, 84% (20 g).
M2 (*R*)-2-MeCBS (5% mol), BH_3•SMe_2, CH_2Cl_2, 0 °C, >95%, 98% de (10 g).
M3 TMSCl, DIPEA, CH_2Cl_2, –5 °C, NI (10 g).
M4 1) $TiCl_4$, DIPEA, CH_2Cl_2, –25 °C; 2) BSA, CH_2Cl_2, reflux, 71% (10 g).
M5 BSA, TBAF, MTBE, rt, NI (10 g).
M6 (K7) see K7.
M7 1) PivCl, DIPEA, CH_2Cl_2; 2) (3a*S*,8a*R*)-3,3a,8,8a-tetrahydro-2*H*-indeno[1,2-*d*]oxazol-2-one, DMAP, reflux, 85% (30 g).
M8 (+)-DIP-Cl, THF, 86% (39 g).
M9 TMSCl, Et_2NH, CH_2Cl_2, NI (2.4 g).
M10 1) Ti(O*i*-Pr)$_4$ (1.1 eq.), –25 °C; 2) BSA, 52% (2.4 g).
M11 1) BSA, toluene, 30 °C; 2) TBAF, NI (0.2 g).
M12 1) PivCl, NEt_3, CH_2Cl_2, rt; 2) (S)-4-phenylthiazolidin-2-one, DMAP, DMF, reflux, 90% (20g).
M13 (*R*)-2-MeCBS (5% mol), BH_3•SMe_2, CH_2Cl_2, 0 °C, 100%, 96% ee (10g).
M14 TBDMSCl, DIPEA, CH_2Cl_2, –10 °C, NI (10 g).
M15 1) $TiCl_4$, DIPEA, CH_2Cl_2, –25 °C; 2) HCl, H_2O, 0 °C; 3) $NaHSO_3$, H_2O, 72%, 99% ee (10 g).
M16 $EtCO_2Cl$, NEt_3, CH_2Cl_2, rt, NI (12 g).
M17 BSA, TBAF, MTBE, rt, YNR (12 g).
M18 *i*-PrOH, H_2SO_4, rt, 91%, 99.8% op (M16+M17+M18, 12 g).

6.9 REFERENCES

AA1982 Evans, D. A. *Aldrichimica Acta* **1982**, *15*, 23-32.
ACIE2010 Mothana, S.; Grassot, J. M.; Hall, D. G. *Angewandte Chemie International Edition* **2010**, *49*, 2883-2887.
ACIE2012 Xiaoming, W.; Fanye, M.; Yan, W.; Zhaobin, H.; Yong-Jun, C.; Li, L.; Zheng, W.; Kuiling, D. *Angewandte Chemie International Edition* **2012**, *51*, 9276-9282.
Adamed2009 Chmielewski, M.; Furman, B.; Stecko, S.; Mames, A.; Panfil, I. (Adamed Sp Zoo) WO2010097350 (prior. 24/02/2009, PL).
Anhui2015 Zou, Q.; Shi, J.; Xiong, S. (Anhui Menovo Pharmaceuticals Co Ltd) CN105503686 (prior. 31/12/2015, CN).
Aventis2004 Lindenschmidt, A.; Will, D. W.; Jaehne, G.; Wollmann, T. A.; Frick, W.; Junker, B.; Rigal, D.; Billen, G.; Jendralla, H. (Aventis Pharma Gmbh) WO2005113496, EP1756049, US2007149776, US7488818 (prior. 21/05/2004, DE).
Arromax2015 Hong, J.; Wang, J.; Xu, X.; Liu, G. (Arromax Pharmaceutical Technology Suzhou Co Ltd) CN104513187 (prior. 09/01/2015, CN).
AsymchemL2013 Hong, H.; James, G.; Ma, J.; Li, J. (Asymchem Lab Tianjin Co Ltd, Asymechem Life Science Technology Tianjin Co Ltd) CN103435525 (prior. 25/07/2013, CN).
CCL2010 Ji, J. F.; Zhang, H. B.; Huang, W. L.; Qian, H.; Zhou, J. P. *Chinese Chemical Letters* **2010**, *21*, 67-69.
CCL2014 Ren, Y.; Duan, Y.-J.; Li, R.-J.; Deng, Y.; Hai, L.; Wu, Y. *Chinese Chemical Letters* **2014**, *25*, 1157-1160.
Centaurus2012 Sun, D.; Yang, L.; Ji, C.; Zhang, X.; Xiao, D.; Han, Y. (Beijing Centaurus Biopharma Technology Co Ltd) CN103102297 (prior. 28/09/2012, CN).
ChangzhouFP2016 Zhang, X.; Zhou, T.; Xiong, Z. (Changzhou Fangnan Pharmaceuticals Ltd) CN107098841, WO2017140251 (prior. 19/02/2016, CN).
ChangzhouPF2011 Yin, X.; Wang, B.; Sun, G. (Changzhou Pharm Factory) CN102675177 (prior. 28/06/2011, CN).
Chengdu2014a Tian, X.; Yan, L.; Li, J.; Zhao, Z.; Luo, D. (Chengdu Senke Pharmaceutical Co Ltd) CN104356041 (prior. 06/11/2014, CN)

Chengdu2014b Tian, X.; Yan, L.; Li, J.; Zhao, Z.; Luo, D. (Chengdu Senke Pharmaceutical Co Ltd) CN104447473 (prior. 06/11/2014, CN).

ChineseASC2009 Deng, J.; Zhu, J.; Liao, J.; Chen, Y.; Che, G.; Tang, Y. (Chinese Acad Sci Chengdu Organic Chem Co) CN102126954 (prior. 15/12/2009, CN).

ChiralQ2011 Bai, J.; Zhang, X.; Xiong, F.; Li, W. (Chiral Quest Suzhou Co Ltd) CN102952055 (prior. 16/08/2011, CN).

ChongqingHS2011 Yang, Y.; Luo, B.; Zhou, X.; Gao, T.; Wu, Y. (Chongqing Huabang Shengkai Pharmaceutical Co Ltd) CN103159751 (prior. 13/12/2011, CN).

ChongqingSP2012 He, G.; Peng, Y. (Chongqing Shenghuaxi Pharma Co; Chongqing Changjie Medical Chemical Co Ltd) CN103373970 (prior. 16/04/2012, CN).

Codexis2008 De Vries, E.; Mundorff, E. (Codexis Inc) WO2010025085; US2010062499; US2012322136; US2013210098; US2015132806; US2015337272; US2016376565; US2018023059; EP2329014; US8273554; US8415126; US8956840; US9139820; US9422530; US9796964 (prior. 29/08/2008, WO, US).

Codexis2009 Collier, S. J.; Liang, J.; Fu Fan, J.; Wilson, R. J. (Codexis Inc) WO2010141494 (prior. 02/06/2009, US).

Codexis2010 Crowe, M. A.; Alvizo, O.; Behrouzian, B.; Bong, Y. K.; Collier, S. J.; Gohel, A.; Mavinahalli, J.; Modukuru, N.; Mundorff, E.; Smith, D.; Song, S.; Yeo, W. L. (Codexis Inc) WO2011140219, US2013052699, US2015218527, US2016281068, US2017204378, US2018312818, US9040262, US9388391, US9644189, US10053673, EP2566497 (prior. 04/05/2010, US).

CR2008 Brandi, A.; Cicchi, S.; Cordero, F. M. *Chemical Reviews* **2008**, *108*, 3988-4035.

CR2014 Pitts, C. R.; Lectka, T. *Chemical Reviews* **2014**, *114*, 7930-7953.

Desano2016 Li, J.; Zhao, N.; Lou, Y. (Shanghai Desano Pharmaceuticals Co Ltd) CN108264472 (prior. 30/12/2016, CN).

DrReddys2003 Sundaram, V.; Rajan, S.; Ramayya, V.; Vardhan, S.; Subrahmanyam, B.; Sasikala, C. (Dr Reddy's Lab Ltd, Dr Reddy's Lab Inc) WO2005062897, US2005171080 (prior. 23/12/2003, IN).

DrReddys2005 Uppala, V.; Vaddadi, P.; Sunkara, V.; Cheemalapati, V.; Padaga, K. (Dr Reddy's Lab Ltd) IN200501187, US2007049748 (prior. 26/08/2005, IN).

DSM2008 Busscher, G.; Lefort, L.; Vries, D. (DSM Ip Assets Bv) WO2010006954 (prior. 15/07/2008, EP).

EJOC1999 Palomo, C.; Aizpurua, J. M.; Ganboa, I.; Oiarbide, M. *European Journal of Organic Chemistry* **1999**, *1999*, 3223-3235.

EJOC2013 K., K. R.; P., K. K. *European Journal of Organic Chemistry* **2013**, *2013*, 7664-7677.

Enantiotech2015 Yang, W.; Huang, C.; Chen, R.; Xu, L.; Muhammad, M. (Enantiotech Corp Ltd) CN104693136 (prior. 02/02/2015, CN).

EnzymeWorks2016 Fu, M.; Wei, X. (Enzymeworks Inc) CN106011095 (prior. 27/07/2016, CN).

Glenmark2005 Sawant, M. S.; Pillai, B. K. G.; Bhirud, S. B.; Sharma, K. (Glenmark Pharm Ltd) WO2007017705 (prior. 09/08/2005, WO, IN).

Guangzhou2015 Deng, C.; Feng, Y.; Lin, Z. (Guangzhou Nanxin Pharmaceutical Co Ltd) CN104829511 (prior. 18/05/2015, CN).

Hanmi2008 Kim, C.; Kim, G.; Kim, N.; Lee, G.; Chang, J.; Chang, Y. (Hanmi Pharm Co Ltd) KR2010070305, WO2010071358, WO2010071358 (prior. 17/12/2008, KR).

Henan2014 Fan, Y.; Li, W.; Zhao, J.; Xue, B.; Duan, X.; Xu, X. (Inst Chemistry Co Ltd Henan Academy of Sciences) CN104072398 (prior. 03/07/2014, CN).

HenanFP2011 Zhang, Q.; Huang, P.; Xu, G.; Chen, B. (Hunan Fangsheng Pharmaceutical Co Ltd) CN102978253 (prior. 28/11/2012, CN).

HenanU2017a Hao, E.; Li, G.; Wang, S.; Zhang, Q.; Huang, J.; Hao, Z.; Fu, D.; Liu, Y.; Guo, H.; Shao, W. (Univ Henan Normal) CN107793339 (prior. 07/11/2017, CN).

HenanU2017b Hao, E.; Li, G.; Wang, S.; Zhang, Q.; Huang, J.; Hao, Z.; Fu, D.; Liu, Y.; Guo, H.; Shao, W. (Univ Henan Normal) CN107903200 (prior. 07/11/2017, CN).

HeteroDrugs2003 Parthasaradhi Reddy, B.; Rathnakar Reddy, K.; Raji Reddy, R.; Muralidhara Reddy, D.; Subash Chander Reddy, K. (Hetero Drugs Ltd) IN200301840, WO2005049592, US2006069137, US7067675, EP1687287 (prior. 24/11/2003, WO, EP, IN).

Ind2003 Framroze, B. P. (Individual) US2004254369, US7002008 (prior. 16/06/2003, US).

Ind2006 Satyanarayana Reddy, M.; Sahadeva Reddy, M. (Individual) IN200601648, WO2008032338 (prior. 11/09/2006, IN).

Ind2007a Perlman, N.; Fishman, A. (Individual) US2009047716 (prior. 07/06/2007, US).

Ind2007b Chen, X.; Du, H.; Zheng, L. (Individual) CN101423515 (prior. 30/10/2007, CN).

Ind2014 Yan, B. (Individual) CN104402790 (prior. 28/12/2014, CN).

IndSwift2007 Srinivasan, C. V.; Saxena, R.; Gupta, P.; Wadhwa, L.; Chidambaram, V. S. (Ind-Swift Lab Ltd) IN200700234, WO2008096372 (prior. 06/02/2007, IN).

IndSwift2008 Chidambaram Venkateswaran, S.; Gupta, P.; Sarin Gurdeep, S.; Wadhwa, L.; Chidambaram, V. S.; Sarin, G. S. (Ind-Swift Lab Ltd) IN200801498, WO2009157019 (prior. 23/06/2008, IN, WO).

IndSwift2016 Arul, R.; Sarin, G. S.; Vyas, S. (Ind-Swift Laboratories Ltd) WO2017168438 (prior. 31/03/2016, IN).

JACS1943 Gilman, H.; Speeter, M. *Journal of the American Chemical Society* **1943**, *65*, 2255-2256.

JACS1992 Sordo, J. A.; Gonzalez, J.; Sordo, T. L. *Journal of the American Chemical Society* **1992**, *114*, 6249-6251.

Jewim2011 Ge, Z.; Wang, D.; Wu, L.; Zhang, Y. (Jewim Beijing Pharmaceutical Inst Co Ltd) CN103172585 (prior. 26/12/2011, CN).

Jiangsu2011 Xie, J.; Yang, Y.; Jin, L. (Jiangsu Hansoh Pharmaceutical) CN102850390 (prior. 01/07/2011, CN).

Jiangsu2015a Wang, Z.; Wang, X.; Wu, H.; Ding, Z.; Chen, Z. (Jiangsu Hansyn Pharmaceutical Co Ltd) CN105541690 (prior. 16/12/2015, CN).

Jiangsu2015b Wang, X.; Ding, Z.; Wu, H.; Wang, Z.; Chen, Z. (Jiangsu Hansyn Pharmaceutical Co Ltd) CN105541742 (prior. 16/12/2015, CN).

Jiangxi2015 Jiang, H.; Fu, C.; Wei, J. (Jiangxi Shimei Pharmaceutical Co Ltd, Guangzhou Shimei Pharmaceutical Technology Co Ltd) CN104892537 (prior. 15/05/2015, CN).

JMC1998 Rosenblum, S. B.; Huynh, T.; Afonso, A.; Davis, H. R.; Yumibe, N.; Clader, J. W.; Burnett, D. A. *Journal of Medicinal Chemistry* **1998**, *41*, 973-980.

JMCB2010 Kyslíková, E.; Babiak, P.; Marešová, H.; Palyzová, A.; Hájíček, J.; Kyslík, P. *Journal of Molecular Catalysis B: Enzymatic* **2010**, *67*, 266-270.

JMS2011 Filip, K.; Bankowski, K.; Sidoryk, K.; Zagrodzka, J.; Łaszcz, M.; Trzcinska, K.; Szyprowska, A.; Cmoch, P.; Maruszak, W. *Journal of Molecular Structure* **2011**, *991*, 162-170.

JOC1999 Wu, G.; Wong, Y.; Chen, X.; Ding, Z. *The Journal of Organic Chemistry* **1999**, *64*, 3714-3718.

JOC2011 Michalak, M.; Stodulski, M.; Stecko, S.; Mames, A.; Panfil, I.; Soluch, M.; Furman, B.; Chmielewski, M. *The Journal of Organic Chemistry* **2011**, *76*, 6931-6936.

JOC2016 Humpl, M.; Tauchman, J.; Topolovčan, N.; Kretschmer, J.; Hessler, F.; Císarová, I.; Kotora, M.; Veselý, J. *The Journal of Organic Chemistry* **2016**, *81*, 7692-7699.

JPA2015 Bellur Atici, E.; Karlıga, B. *Journal of Pharmaceutical Analysis* **2015**, *5*, 356-370.

JPBA2010 Raman, B.; Sharma, B. A.; Butala, R.; Ghugare, P. D.; Kumar, A. Journal of Pharmaceutical and Biomedical Analysis 2010, 52, 73-78.

JPBA2011a Gajjar, A. K.; Shah, V. D. *Journal of Pharmaceutical and Biomedical Analysis* **2011**, *55*, 225-229.

JPBA2011b Barhate, C. R.; Mohanraj, K. *Journal of Pharmaceutical and Biomedical Analysis* **2011**, *55*, 1237-1238.

JPBA2012 Sánta, Z.; Kóti, J.; Szöke, K.; Vukics, K.; Szántay, C. *Journal of Pharmaceutical and Biomedical Analysis* **2012**, *58*, 125-129.

JPBA2014 Guntupalli, S.; Ray, U. K.; Murali, N.; Gupta, P. B.; Kumar, V. J.; Satheesh, D.; Islam, A. *Journal of Pharmaceutical and Biomedical Analysis* **2014**, *88*, 385-390.

JPBA2015 Zhang, D.; Su, J. *Journal of Pharmaceutical and Biomedical Analysis* **2015**, *107*, 355-363.

JPS2014 Batová, J.; Imramovský, A.; Hájíček, J.; Hejtmánková, L.; Hanusek, J. *Journal of Pharmaceutical Sciences* **2014**, *103*, 2240-2247.

Kaiyuan2011 An, R.; Cai, Z.; Dong, X.; Jiang, S.; Li, Y.; Peng, S. P.; Wang, W. (Shanghai Hongbo Zhiyuan Medical Technolo, Kaiyuan Hengtai Pharm Co Ltd) CN102531985 (prior. 25/04/2011, CN).

Kemp2013 Zeng, P.; Wu, F.; Wu, J.; He, L.; Zhang, J.; Zhang, H.; Zhou, J.; Liu, J. (Kamp Pharmaceuticals Co Ltd) CN104163783 (prior. 16/05/2013, CN).

Kotobuki2006 Tomiyama, H.; Yokota, M. (Kotobuki Pharm Co Ltd, Kotobuki Seiyaku Co Ltd) JP2008500579, WO2007094480, EP1988071, US2009062527, US7956182 (prior. 16/02/2006, JP).

KRKA2007 Stimac, A.; Mohar, B.; Stephan, M.; Bevc, M.; Zupet, R.; Gartner, A.; Kroselj, V.; Smrkolj, M. (Krka Dd Novo Mesto, Krka Pharm & Chem Works) WO2008089984, US2013190487, EP1953140, EP2125715 (prior. 24/01/2007, EP).

Lek2008a Selic, L. (Lek Pharmaceuticals, Lek Pharm Dd) WO2009150038, US2011130378, EP2128133, EP2294052 (prior. 26/05/2008, EP).

Lek2008b Casar, Z.; Gobec, S.; Kovac, A.; Mravljak, J.; Sova, M. (Lek Pharmaceuticals, Lek Pharm Dd) WO2010012775, US2011183956, EP2326616, EP2149547 (prior. 30/07/2008, EP).

Lianyungang2013 Yang, B.; Pan, B.; Fu, Y.; Chen, X. (Lianyungang Hengyun Medicine Science & Technology Co Ltd) CN103739537 (prior. 24/12/2013, CN).

Lupin2009a Srivastava, D.; Shakya, R.; Chaudhari, N.; Ansari, I.; Singh, G. (Lupin Ltd) IN200900562, WO2010113182, US2012029185, US9040688, EP2414326 (prior. 31/03/2009, WO, IN).

Lupin2009b Lathi, P.; Roy, B.; Singh, G.; Shrivastava, D. (Lupin Ltd) IN200900577, WO2010113184, EP2414339, US2012028340 (prior. 02/04/2009, WO, IN).

Medichem2005 Gavalda, I.; Bosch, I.; Garcia, G. (Medichem Sa) WO2007119106, US2009227786, EP1971573 (prior. 22/12/2005, US).

Medichem2006 Escude, A.; Llado, J.; Nettekoven, U.; Bosch, I. (Medichem Sa) WO2007144780, US2010168414, EP2007718 (prior. 29/03/2006, US).

Merck1995 Homann, M.; Previte, E. (Schering Corp) WO9712053, US5618707, EP862645 (prior. 27/09/1995, US).

Merck2004 Thiruvengadam, T.; Chiu, J.; Fu, X.; McAllister, T. (Schering-Plough Corp, Merck Sharp & Dohme Corp) US2012083601, US8383810 (prior. 20/12/2004, US).

Moehs2010 Sanchez Casals, C.; Dobarro Rodriguez, A.; Bosch Cartes, J.; Garcia Diaz, D. (Moehs Iberica Sl) ES2372460, WO2012004382, EP2590940 (prior. 09/07/2010, ES).

MSN2005 Satyanarayana Reddy, M.; Kishore Kumar, M.; Sahadeva Reddy, M. (MSN Lab Ltd) IN200500781, WO2006137080, US2009048441, US8013150, EP1893570 (prior. 22/06/2005, IN).

Mylan2009 Husain, M.; Gorantla, S. C.; Thorpunuri, S.; Debashish, D.; Srikanth, G. S.; Gorantla, S.; Datta, D. (Matrix Lab Ltd, Mylan Lab Ltd) IN200900760, WO2010113175, US2012028316, US9388440, US2016280642, EP2414529 (prior. 01/04/2009, IN).

Nanjing2013 Li, W.; Liu, L.; Tang, A.; Wang, Y.; Wang, H.; Yu, W. (Heading Nanjing Pharmaceutical Tech Co Ltd) CN103450065, US2015018565 (prior. 15/07/2013, CN).

Nanjing2015 Lee, W.-S. (Heading Nanjing Pharmaceutical Tech Co Ltd) CN105985275, JP2016147843, JP6153209 (prior. CN2015183806 2015/02/15, CN).

NanjingAB2015 Luo, Y.; Ding, S.; Qu, X.; Sun, C. (Nanjing Abiochem Biomedical Technology Co Ltd) CN104860980 (prior. 23/04/2015, CN).

NanjingU2015 Sun, J.; Wang, W.; Wang, L.; Xiao, Y. (Univ Nanjing Normal) CN104744390 (prior. 19/03/2015, CN).

Nantong2014 Li, Z.; Huang, H.; Liu, M.; Zhao, Y. (Nantong Changyou Pharmaceutical Science & Technology Co Ltd) CN104003921 (prior. 06/05/2014, CN).

Natco2002 Pulla Reddy, M.; Muddasani, P.; Nannapaneni, V. (Natco Pharma Ltd) IN200200427, WO2003104180, US2005250961 (prior. 05/06/2002, IN).

Nissan2014 Naitou, S.; Kitsuyama, H.; Nakamura, S. (Nissan Chemical Industries Ltd) JPWO2015133405 (prior. 06/03/2014, JP).

OPRD2009 Sasikala, C.; Reddy Padi, P.; Sunkara, V.; Ramayya, P.; Dubey, P.; Bhaskar Rao Uppala, V.; Praveen, C. *Organic Process Research & Development* **2009**, *13*, 907-910.

Opus2005 Framroze, B. (Opus Organics Pvt Ltd) US2007129540, US7498431 (prior. 01/12/2005, US)

PAC1981 Evans, D.; Takacs, J.; McGee, L.; Ennis, M.; Mathre, D.; Bartroli, J. *Pure Appl. Chem.* **1981**, *53*, 1109-1127.

Pharmathen2010 Koftis, T.; Andreou, T.; Zitrou, A. (Pharmathen Sa) WO2012076030 (prior. 10/12/2010, WO).

Pharmathen2011 Koftis, T.; Andreou, T.; Gioti, E.; Zitrou, A. (Pharmathen Sa) WO2012155932 (prior. 17/05/2011, WO).

Pharmathen2013 Koftis, T. V.; Neokosmidis, E.; Gioti, E.; Alexandraki, E.; Mandalou, P.; Menisiou, A.; Andreou, T. (Pharmathen Sa) WO2015039675 (prior. 23/09/2013, IN).

PolishAS2011 Chmielewski, M.; Stecko, S.; Furman, B.; Panfil, I.; Sniezek, M.; Michalak, M.; Jurczak, M.; Grzeszczyk, B.; Soluch, M.; Mikoaoajczyk, P. (Polish Academy of Sciences Organic Chem) PL395262, WO2012173504 (prior. 15/06/2011, PL).

Qilu2016 Zuo, J.; Li, F.; Wu, Z.; Li, W.; Zhou, X.; Li, Z. (Qilu Tianhe Pharmaceutical Co Ltd) (prior. 15/01/2016, CN).

QR2015 Ma, J.; Yu, P.; Fang, X.; Wang, C. (QR Pharmaceuticals Ltd) CN106279240 (prior. 11/06/2015, CN).

Ranbaxy2003a Karooti, K.; Rathod, P.; Aryan, R.; Kumar, Y.; Gangakhedkar, K.; Ram, C. A. (Ranbaxy Lab Ltd) IN200505651, WO2004099132, EP1626954 (prior. 05/05/2003, IN).

Ranbaxy2003b Kumar, Y.; Meeran, H.; Singh, S.; Rathod, P.; Ganagakhedkar, K.; Bose, P.; Kumar, P. (Ranbaxy Lab Ltd) IN200301643, WO2005066120 (prior. 30/12/2003, IN).

RichterGedeon2005 Bodi, J.; Eles, J.; Szoeke, K.; Vukics, K.; Gati, T.; Temesvari, K.; Kiss-Bartos, D. (Richter Gedeon Vegyeszeti Gyar, Richter Gedeon Nyrt) HU200501164, WO2007072088, US2009216009, US8178665, EP1963260 (prior. 20/12/2005, HU).

S2010 Sova, M.; Mravljak, J.; Kovac, A.; Pecar, S.; Casar, Z.; Gobec, S. *Synthesis* **2010**, *2010*, 3433-3438.

Sanofi2004 Lindenschmidt, A.; Will, D. W.; Jaehne, G.; Wollmann, T. A.; Frick, W.; Junker, B.; Rigal, D.; Billen, G.; Jendralla, H. (Aventis Pharma Deut Gmbh, Sanofi-Aventis Deut Gmbh) DE102004025071, WO2005113496, US2007149776, US7488818, EP1756049 (prior. 21/05/2004, DE).

SC2016 Zhu, Y.; Pan, J.; Zhang, S.; Liu, Z.; Ye, D.; Zhou, W. *Synthetic Communications* **2016**, *46*, 1687-1693.

Schering1993 Dugar, S.; Clader, J.; Rosenblum, S.; Burnett, D.; McKittrick, B. (Schering Corp, Schering-Plough Corp) WO9508532, US5631365, US42461, EP720599 (prior. 21/09/1993, US).

Schering1995 Shankar, B. (Schering Corp) WO9716424, US5856473 (prior. 02/11/1995, US).

Schering1996a Homann, M.; Previte, E. (Schering Corp) WO9712053, US5618707, EP862645 (prior. 04/01/1996, US).

Schering1996b Wu, G.; Chen, X.; Wong, Y.; Schumacher, D.; Steinman, M. (Schering Corp) WO9745406, US5739321, EP906278 (prior. 31/05/1996, US).

Schering1998a Previte, E.; Homann, M. (Schering Corp) US6133001 (prior. 23/02/1998, US).

Schering1998b Homann, M.; Morgan, W. (Schering Corp) WO200020623, US5919672, EP1117823 (prior. 02/10/1998, US).

Schering1998c Thiruvengadam, T. K.; Fu, X.; Tann, C.; McAllister, T. L.; Chiu, J. S.; Colon, C. (Schering Corp) WO200034240, EP1137634 (prior. 07/12/1998, US).

Schering1998d Thiruvengadam, T.; Fu, X.; Tann, C.; McAllister, T.; Chiu, J.; Colon, C. (Schering Corp) US620782 (prior. 07/12/1998, US).

Schering1999 Homann, M.; Previte, E. (Schering Corp) WO200060107, EP1169468, EP2281899 (prior. 05/04/1999, WO, EP).

Schering2001 Fu, X.; McAllister, T.; Thiruvengadam, T.; Tann, C. (Schering Corp, Merck Sharp & Dohme Corp) WO200279174, US2002193607, US6627757, US2003204096, EP1373230 (prior. 28/03/2001, US).

Shangdong2016 Zhang, G.; Zang, C.; Xia, M. (Shangdong New Time Pharm Prod) CN107488190 (prior. 10/06/2016, CN).

Shanghai2015a Wang, X.; Ding, Z.; Wu, H.; Zhang, Q. (Shanghai Hengsheng Pharmaceutical Co Ltd) CN105439929 (prior. 16/12/2015, CN).

Shanghai2015b Ding, Z.; Wang, X.; Wu, H.; Zhang, Q. (Shanghai Hengsheng Pharmaceutical Co Ltd) CN105524010 (prior. 16/12/2015, CN).

ShanghaiAB2010 Chen, Y.; Fan, Q.; Gan, L.; Gan, M.; Ma, D.; Zhu, W. (Shanghai Aobo Bio-Pharm Technology Co Lt, Zhejiang Huahai Pharm Co Ltd) CN102234246 (prior. 23/04/2010, CN).

ShanghaiFB2014 Zhang, X. (Shanghai Fangnan Pharmaceuticals Ltd) CN103965089, WO2015158191 (prior. 18/04/2014, CN).

ShanghaiHP2017a Chen, L.; Xu, Y.; Peng, Z.; Zhu, X.; Liu, Z.; Wang, J.; Zheng, Y. (Shanghai Huayuan Pharmaceutical Tech Dev Co Ltd) CN106892851 (prior. 09/03/2017, CN).

ShanghaiHP2017b Zheng, Y.; Peng, Z.; Zhu, X.; Xu, Y.; Wang, J.; Liu, M.; Chen, L.; Liu, Z. (Shanghai Huayuan Pharmaceutical Tech Development Co Ltd) CN106967106 (prior. 24/04/2017, CN).

ShanghaiIOC2012 Ding, K.; Wang, X.; Wang, Z. (Shanghai Inst Organic Chem) CN103570574, EP2876102, US2015166479, WO2014012372, US9518014 (prior. 20/07/2012, CN).

ShanghaiIP2016 Zhu, Y.; Pan, J.; Zhang, S.; Liu, Z.; Zhou, W. (Shanghai Inst of Pharmaceutical Industry; China State Inst Pharm Ind) CN107118144 (prior. 24/02/2016, CN).

ShanghaiSP2012a Wei, B.; Yu, Z.; Wang, G. (Shangshai Shyndec Pharmaceutical Co Ltd, Shanghai Shyndec Pharmaceutical Co Ltd) CN103012383 (prior. 17/12/2012, CN).

ShanghaiSP2012b Yu, Z.; Wang, G. (Shangshai Shyndec Pharmaceutical Co Ltd, Shanghai Shyndec Pharmaceutical Haimen Co Ltd) CN102993077 (prior. 17/12/2012, CN).

ShanghaiSP2013 Yu, Z.; Wei, B.; Wang, G.; Liu, S.; Hou, J. (Shangshai Shyndec Pharmaceutical Co Ltd, Shanghai Shyndec Pharmaceutical Co Ltd) CN103044305 (prior. 24/01/2013, CN).

ShanghaiSP2014 Hou, J.; Xu, G. (Shangshai Shyndec Pharmaceutical Co Ltd) CN104230978 (prior. 22/09/2014, CN).

Shanxi2015 Wang, Y.; Li, L.; Su, Z.; Zan, J.; Wu, J.; Yao, H. (Shanxi Pude Pharmaceutics Co Ltd) CN105622661 (prior. 31/12/2015, CN).

Shenyang2011 Zhao, D.; Song, S.; Miao, J.; Li, W.; Gao, L. (Univ Shenyang Pharmaceutical) CN103086938 (prior. 28/10/2011, CN).

Sichuan2013 Yao, Q.; Deng, X.; Zhu, H.; Zhang, Y. (Sichuan Jinhui Pharmaceutical Co Ltd) CN104059009 (prior. 21/03/2013, CN).

Sichuan2017 Mao, X.; Zhang, Z.; Fan, S.; Du, Z.; Xu, X.; Xu, J.; Mao, Z.; Huang, T. (Sichuan Zhiqiang Medicine Science and Tech Development Limited Company) CN106966942 (prior. 07/04/2017, CN).

Suzhou2012 Yan, W. (Suzhou Langke Biotech Co Ltd) CN102531986 (prior. 23/02/2012, CN).

SuzhouP2016 Luo, R. (Suzhou Puluoda Biological Science and Tech Co Ltd) CN106397292 (prior. 20/09/2016, CN).

T1994 Abrahams, I.; Motevalli, M.; J. Robinson, A.; B. Wyatt, P. *Tetrahedron* **1994**, *50*, 12755-12772.

Teva2005 Kansal, V. K.; Ahmad, S.; Tyagi, B.; Gupta, N. (Teva Pharm Ind Ltd, Teva Pharm USA Inc) WO2007030721, US2007259845, US2010010212, EP1922304 (prior. 08/09/2005, US).

Teva2006a Kansal, K.; Ahmad, S.; Mariappan, S.; Tyagi, B.; Perlman, N.; Le Paih, J.; Zanotti-Gerosa, A. (Teva Pharm Ind Ltd, Teva Pharm USA Inc) WO2007120824, US2008032964, EP2004639 (prior. 10/04/2006, US).

Teva2006b Ahmad, S.; Sethumadhavan, S.; Mariappan, S.; Kansal, V.; Kansal, K. (Teva Pharm Ind Ltd, Teva Pharm USA Inc) WO2008027081, US2008058305, EP1937636 (prior. 29/08/2006, US).

Teva2007a Perlman, N.; Fishman, A. (Teva Pharm Ind Ltd, Teva Pharm USA Inc) WO2008151324, EP2155667 (prior. 07/06/2007, US).

Teva2007b Perlman, N.; Szabo, L.; Toth, L. (Teva Pharm Ind Ltd, Teva Pharm USA Inc) WO2009032264, US2009093627 (prior. 30/08/2007, US).

Tianjin2011 Cai, J.; Li, L.; Luo, Z.; Zhao, S.; Wu, X. (Tianjin Inst Pharm Res, Jiangsu Deyuan Pharm Co Ltd) CN102731489 (prior. 11/04/2011, CN)

TianjinP2012 Wang, Y.; Meng, H.; Zhao, P.; Liu, N.; Su, Y.; Li, Y.; Wang, L. (Tianjin Pharmaceutical Tech Dev Co Ltd) CN103864708 (prior. 12/12/2012, CN).

TL1996 Shankar, B.; Kirkup, M.; McCombie, S.; Clader, J.; Ganguly, A. *Tetrahedron Letters* **1996**, *37*, 4095-4098.

TL2003 Fu, X.; McAllister, T.; Thiruvengadam, T.; Tann, C.-H.; Su, D. *Tetrahedron Letters* **2003**, *44*, 801-804.

TL2007 Bertrand, B.; Durassier, S.; Frein, S.; Burgos, A. *Tetrahedron Letters* **2007**, *48*, 2123-2125.

TL2013 Ren, Y.; Li, R.; Deng, Y.; Guan, M.; Wu, Y.; Hai, L. *Tetrahedron Letters* **2013**, *54*, 6443-6446.

Tokuyama2015 Shimizu, T. (Tokuyama Corp) JP2017002008 (prior. 15/06/2015, JP).

Topharman2009 Zhu, F.; Wu, M.; Tang, G.; Li, H. (Topharman Shanghai Co Ltd, Topharman Shandong Co Ltd) CN101935309, WO2011000212 (prior. 29/06/2009, CN).

Unichem2006 Ramakrishnan, A.; Shrigadi, N.; Prabhavalkar, T.; Thorat, A.; Tirtha, S. (Unichem Lab Ltd) IN200600412, WO2007108007 (prior. 23/03/2006, IN).

UShenyang2010 Li, Y.; Miao, J.; Sun, L.; Zhao, D.; Song, S.; Xiong, X. (Univ Shenyang Pharm) CN102477008 (prior. 22/11/2010, CN).

UZhejiang2010 Yang, J.; Yu, C. (Univ Zhejiang) CN101845010 (prior. 10/02/2010, CN).

UZheijang2011a Yang, J.; Yu, C. (Univ Zhejiang) CN102219803 (prior. 20/04/2011, CN).

UZheijang2011b Yang, J.; Yu, C. (Univ Zhejiang) CN102285932 (prior. 01/09/2011, CN).

Wuhan2016 Zhu, G.; Wan, Y.; Chen, Y.; Song, A. (Qr Pharmaceuticals Ltd; Wuhan Zhongyou Pharmaceutical Co Ltd) CN108059609 (prior. 07/11/2016, CN).

Wuxi2015 Wang, Q.; Wang, T.; Wang, B.; Mi, J.; Xu, J.; Zhang, Y.; Yu, J.; Cheng, L. (Wuxi Fortune Pharmaceutical Co Ltd) CN105237492 (prior. 29/10/2015, CN).

Xiamen2004 Gu, L.; Ruan, J.; Huang, Z. (Xiamen Mchem Pharma Group Ltd) WO2006050634 (prior. 15/11/2004, WO).

Zach2006 Burgos, A.; Frein, S. (Zach System Spa) FR2909671, WO2008078041, US2008139851, US7586015, EP2102218 (prior. 12/12/2006, FR).

Zentiva2007a Slavikova, M.; Vesilek, P.; Obadalova, I. V. A.; Hajicek, J. (Zentiva AS) CZ200700170, WO2008106900 (prior. 02/03/2007, CZ).

Zentiva2007b Hajicek, J.; Slavikova, M.; Stepankova, H.; Zezula, J. (Zentiva AS) CZ200700843, WO2009067960 (prior. 30/11/2007, CZ).

Zentiva2008a Hajicek, J.; Slavikova, M.; Stepankova, H.; Zezula, J. (Zentiva KS, Zentiva AS) CZ200800107, WO2009106021, US2011046389, EP2276753 (prior. 25/02/2008, CZ).

Zentiva2008b Hajlcek, J.; Slavikova, M.; Stepankova, H.; Zezula, J.; Hajicek, J. (Zentiva KS, Zentiva AS) CZ200800317, WO2009140932 (prior. 21/05/2008, CZ).

Zhejiang2012 Yan, P.; Che, D.; Li, Y.; Zhu, G.; Zhang, X.; Qu, X. (Zhejiang Jiuzhou Pharm Co Ltd) CN103896700, WO2014101690 (prior. 25/12/2012, CN).

Zhejiang2013 Li, Y.; Che, D.; Zhu, G.; Yan, P.; Yang, L.; Chen, X.; Luo, L.; Sun, L. (Zhejiang Jiuzhou Pharm Co Ltd) CN104744331 (prior. 31/12/2013, CN).

Zhejiang2014a Yan, P.; Li, Y.; Che, D.; Zhang, X. (Zhejiang Jiuzhou Pharmaceutical Technology Co Ltd) CN105272852 (prior. 16/07/2014, CN).

Zhejiang2014b Zhu, G.; Li, Y.; Yang, L.; Sun, L.; Chen, X. (Zhejiang Jiuzhou Pharm Co Ltd) CN105461649 (prior. 08/08/2014, CN).

ZhejiangA2015 Zhu, F.; Zhao, S.; Ma, X.; Li, K.; Chen, L.; Chen, Z.; Jin, L. (Zhejiang Apeloa Tospo Pharmaceutical Co Ltd; Shandong Apeloa Pharmaceutical Co Ltd) CN104860862 (prior. 09/04/2015, CN).

ZhejiangT2006 Tu, Y. (Zhejiang Tianyu Pharm Co Ltd) CN1931838 (prior. 20/10/2006, CN).

Zhongshan2007 Dai, L.; Han, Y.; Huang, Z. (Zhongshan Yiantai Medical Technology Co Ltd) CN101423511 (prior. 05/11/2007, CN).

7 Montelukast

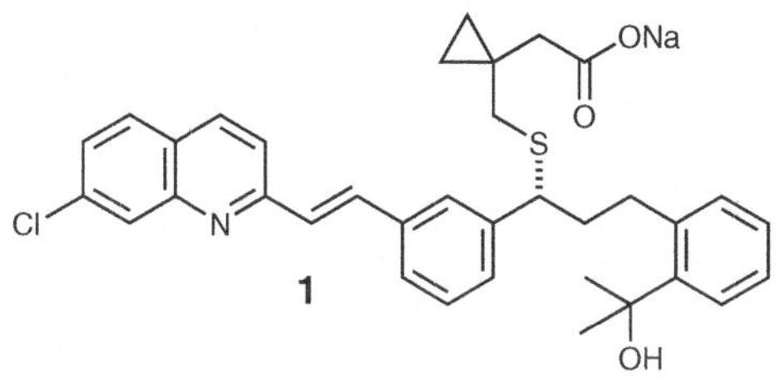

Year of discovery: 1991 (Merck Frosst)
First WW Approval: Finland 1997; US approval 1998 (Singulair®, Merck & Co.)
Drug cat: Oral leukotriene antagonist
Uses: Asthma or to relieve seasonal allergies

Montelukast sodium (**1**) is a synthetic antagonist of the cysteinyl leukotriene receptor used in the treatment of asthma. Asthma is a chronic inflammatory disorder of the airways, frequently triggered by allergens and its effects can be moderate to life threatening. Currently, asthma is one of the most prevalent of chronic diseases and its occurrence has been linked to the degree of urbanization of modern society. Genetic factors have also been appointed as contributing to the development of the disease.

The global market for asthma and chronic obstructive pulmonary disease (COPD) prescription drugs was valued at USD 18 billion in 2017. This figure was projected to reach 27.6 billion in 2026, whilst the segment made up of asthma drugs was projected to increase from 12.0 billion in 2017 to 18.9 billion in 2026.

Three categories of drugs are usually used for the treatment of asthma. First-line drugs are inhaled corticosteroids, for example, Fluticasone, Budesonide and Prednisolone. Combination of corticosteroids with short or long acting β-2-adrenoceptor agonists (for example, Salbutamol, Terbutaline, Fenoterol, Formoterol, Salmeterol and Indacaterol) is considered as a second line of treatment. The more recent third-line drugs are the leukotriene antagonists and anticholinergics. Pranlukast (Onon®, launched 1995), Zafirlukast (Accolate®, launched 1996) and Montelukast (Singulair®, launched late 1997) belong to this class, the latter being the most successful of this drug type (Figure 7.1).

7.1 WORLDWIDE SALES AND PATENT STATUS

The following graphs illustrate how research carried out in the pharmaceutical industry towards the production of the API and worldwide sales of the finished drug product evolved.

- US basic patent term 30/10/2010 (US5565473, MerckFrosst1990), USA Product Patent term extension 03/02/2012.
- Annual production[1]: 2019, 66,056 kg; 2018, 60,854 kg; 2017, 54,406 kg.
- Annual WW sales[1]: 2019, 1778.8M USD; 2018, 1799.5M USD; 2017, 1812.9M USD.

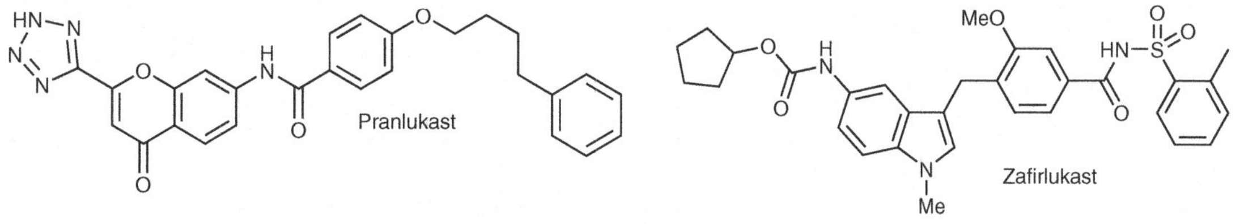

Figure 7.1 Selected leukotriene antagonists

1 Newport Modular Platform – Clarivate Analytics [Online] https://clarivate.com/cortellis/ [2020, April]

Retrosynthesis in the Manufacture of Generic Drugs: Selected Case Studies, First Edition, Pedro Paulo Santos and William Heggie.

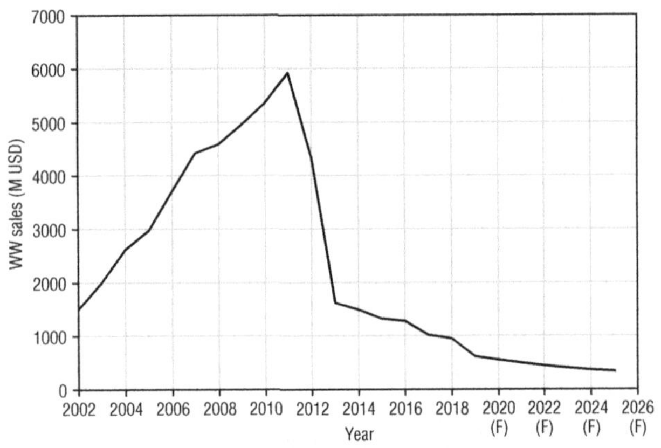

FIGURE 7.2 MONTELUKAST WW SALES EVOLUTION (IN M USD)

Source: Data from GlobalData [Online] https://globaldata.com/

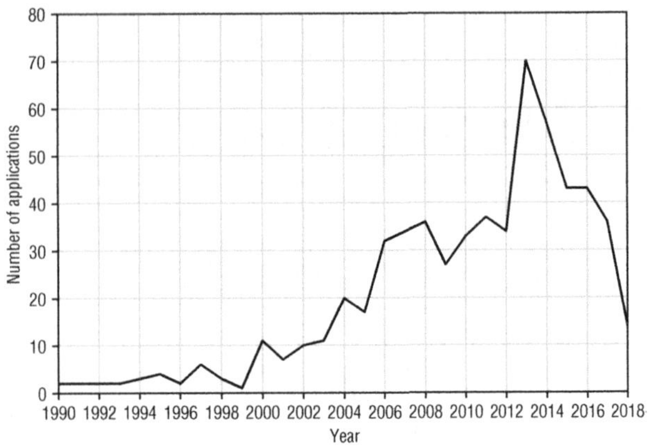

FIGURE 7.3 PATENT APPLICATIONS ON MONTELUKAST FILED BY YEAR

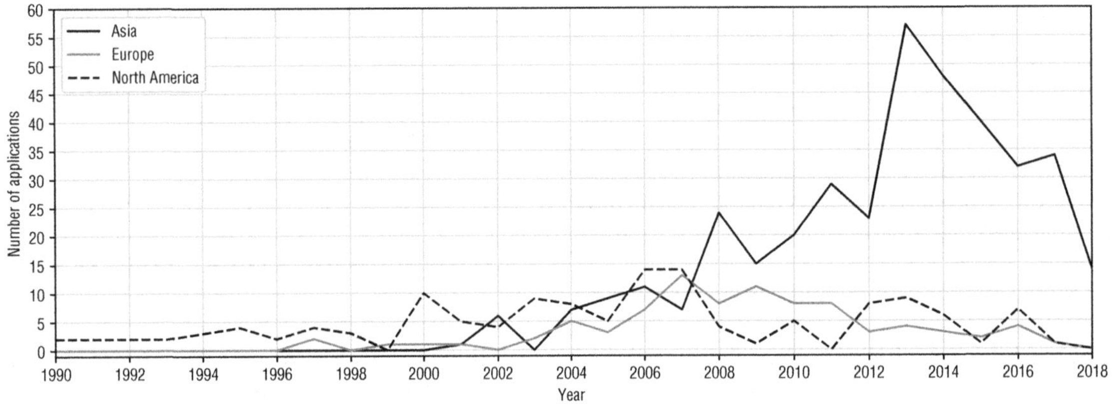

FIGURE 7.4 NEW PATENT APPLICATIONS ON MONTELUKAST FILED BY REGION

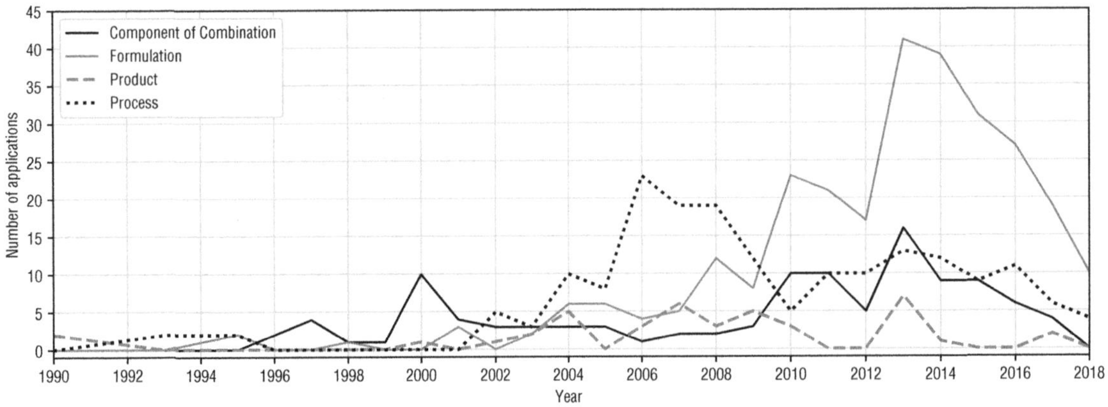

FIGURE 7.5 PATENT APPLICATIONS ON MONTELUKAST FILED BY TYPE OF PATENT

7.2 MONTELUKAST STRUCTURE AND GENERAL RETROSYNTHETIC ANALYSIS

The industrial synthesis of Montelukast in very high purity is a challenging task as the molecule contains various functional groups that influence its chemical stability, which in turn, makes the synthetic process very arduous.

The drug can be considered as being assembled from four main structural blocks: a quinoline unit (**A**), a central *meta*-disubstituted aromatic ring (**B**), a second disubstituted aromatic ring containing the tertiary alcohol (**C**) and a thiol-carboxylic acid side-chain (**D**) (Figure 7.6).

Obviously, a considerable number of Montelukast process patents focus on key reactions that enable these four building blocks to be connected. Three "main" connections have to be established during the synthetic process. Some key aspects of molecule structure are:

- Montelukast is chiral and an enantiopure drug. The correct chirality of the carbon atom connected to the sulfur atom can, in most circumstances, be achieved by a nucleophilic substitution (bimolecular) at the benzylic carbon atom during the synthetic process, thus allowing for the selective production of the intended isomer.
- A three carbon bridge links two aromatic disubstituted benzene rings. This bridge can be made by taking advantage of the fact that functionality is required at the benzylic atom to introduce the thiol side chain.
- The ethylene bridge that connects the quinoline and the central aromatic ring can be made directly by Heck or Wittig-like reactions. Aldol "type" reactions are also a good alternative.
- The presence of tertiary benzylic alcohol may cause some structural instability, particularly under acidic conditions. The departure of the hydroxyl group, catalysed by acid, is somewhat favoured by the formation of a relatively stable tertiary benzyl carbocation and a methylstyrene degradation impurity is predictable (this impurity is listed both in US end EP pharmacopeia monographs). The tertiary benzylic alcohol can be produced preferably using Grignard reactions on a ketone or an ester.
- 7-chloro-2-methylquinoline (**2**) can be regarded as the starting material for all synthetic processes and can be produced by a Doebner-Von Miller reaction.

Scheme 7.1 summarizes some of these.

An overview of the main retrosynthetic strategies to produce Montelukast will be carried out in the next section. The corresponding processes (synthetic methods) that were the subject of a patent application are condensed in schemes A to F. A listing of the main reactants for each transformation is also included.

7.2.1 Chirality issues

Montelukast is a chiral enantiopure compound and directing the formation of the intended isomer is a key issue for the development of an efficient industrial process. In principle, a resolution step at the end of a non-chiral synthetic process will be very unfavourable and costly. The unwanted isomer is projected to be very difficult to recycle and hence half of the material produced would be wasted.

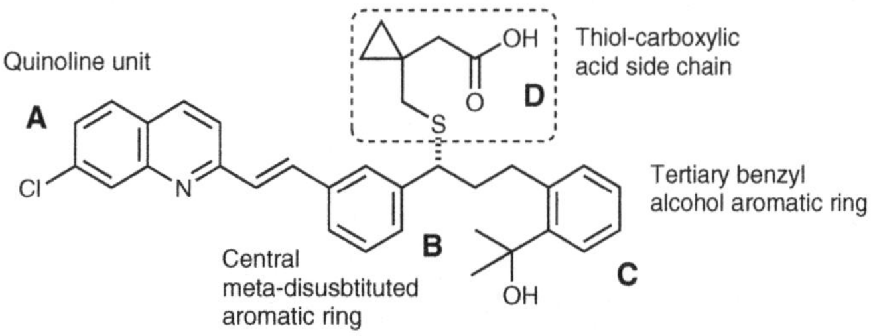

FIGURE 7.6 KEY STRUCTURE BLOCKS OF MONTELUKAST

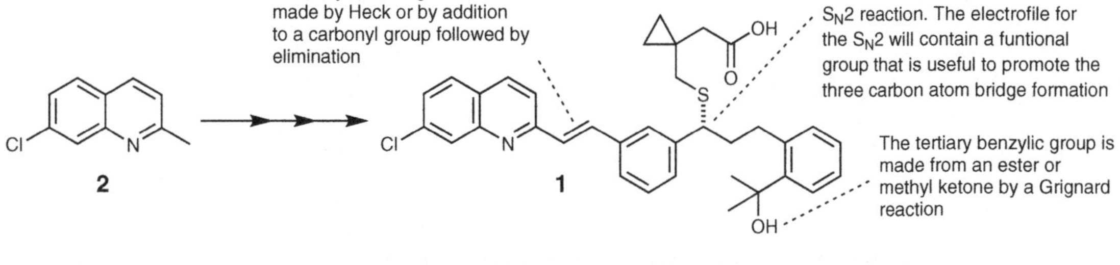

SCHEME 7.1 KEY STRUCTURE FEATURES OF MONTELUKAST AND STARTING 7-CHLORO-2-METHYLQUINOLINE

A stereoselective synthetic process is therefore preferable and, because of the particular structural elements of Montelukast, one option is that the appropriate chirality can be introduced by an asymmetric reduction of a benzylic carbonyl group (Scheme 7.2). This strategy has proven to be applicable on various substrates and usable at industrial scale.

Therefore, by synthesising an alcohol with appropriate configuration and using a bimolecular nucleophilic substitution to connect the thiol side chain, it is possible to produce the intended isomer exclusively.

7.2.2 The thiol side chain and the enantioselectivity of the synthetic process

It is useful to use a nucleophilic thiol (**3**) already containing the appropriate carbon framework as a building block (Scheme 7.3, strategy **a**). Obviously, this reactant can contain a precursor to the carboxylic acid such as a nitrile, an ester or other such group that can be hydrolysed to the carboxylic acid present in the Montelukast molecule. To obtain the correct *R*-configuration found in the product, the substrate reacting with the nucleophilic thiol will need to have the opposite *S*-configuration.

The electrophile for the S_N2 reaction can be effectively obtained from the corresponding alcohol **(*S*)-4**, by suitable activation as a mesylate or some similar ester or by substitution by a halide. The alcohol **4** should be enantiomerically pure (*S* form) in order to guarantee the desired configuration of the final molecule (*R*). Note that if an S_N2 type substitution reaction at the chiral carbon atom is used to activate the substrate for a second and subsequent S_N2 substitution, the starting alcohol will have the same chirality (*R*) as the final product. As mentioned previously, it is desirable to generate this *S* isomer by an asymmetric reduction of the benzylic carbonyl precursor (Scheme 7.2). The diol (key-intermediate, **(*S*)-4**) in Scheme 7.3 is clearly central to several processes for the synthesis of Montelukast.

An alternative to a substitution reaction at the benzylic carbon results from a disconnection of the other C-S bond in the Montelukast molecule (Scheme 7.3, strategy **b**). The thiol-hydroxy intermediate (**5**) can also be obtained from the diol by similar methods to those described above (for example, by alcohol mesylation followed by nucleophilic substitution). Obviously, these two approaches to introduce the thiol side chain can be applied at several different stages of a synthetic route and not necessarily during the final synthetic steps. This can be clearly observed when comparing several alternate synthetic routes which have been developed.

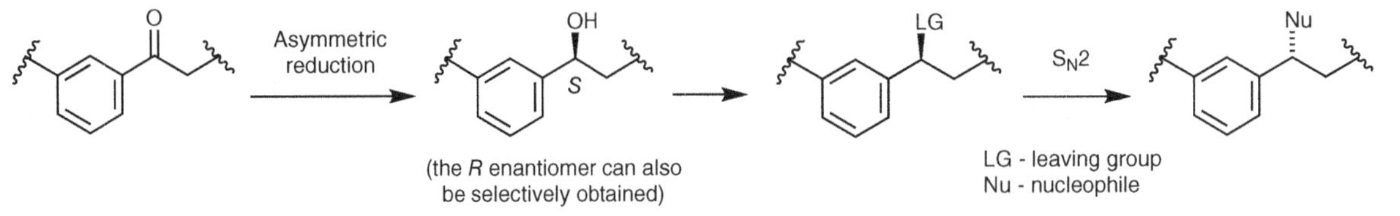

Scheme 7.2 Key synthetic sequence to produce Montelukast

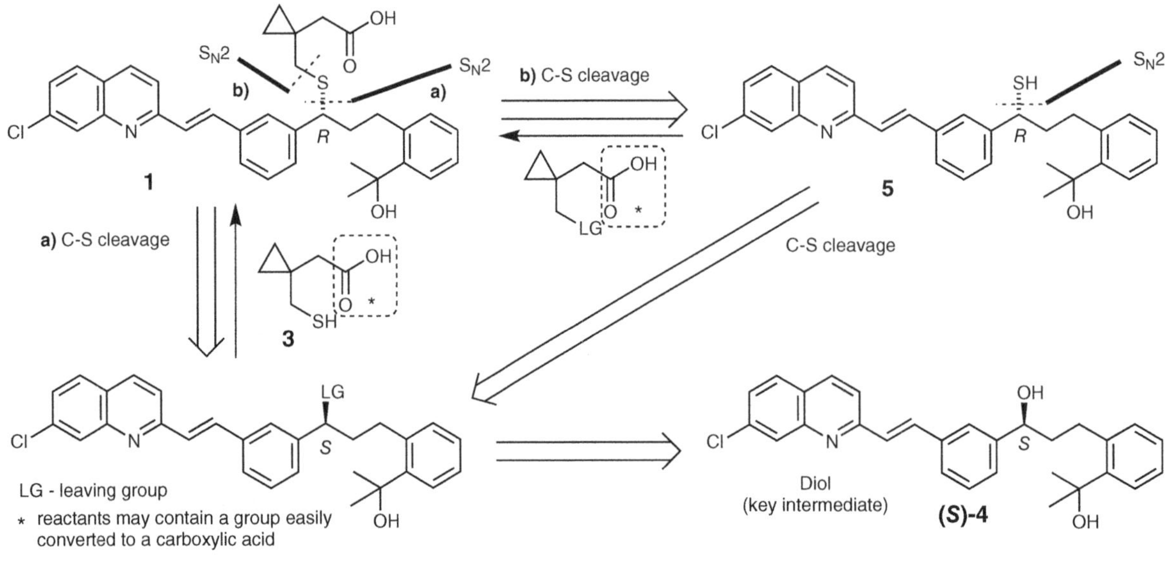

Scheme 7.3 Retrosynthesis I: thiol side chain cleavage

7.2.3 Retrosynthetic analysis of the thiol side chain

The appropriate thiol moiety itself has been the focus of several synthetic process patents. The methylene thiol and the methylene carboxylic acid groups connected to the same carbon of a cyclopropyl ring, points to a malonate ester as one probable starting material (Scheme 7.4). The carboxylic acid is conveniently obtained by hydrolysis of a nitrile group and which is in turn easily introduced by a S_N2 reaction in a substrate containing an appropriate leaving group, using cyanide as nucleophile. The introduction of the thiol group is much easier: a nucleophilic displacement using thiourea on a carbon containing a leaving group will do the job.

During the synthetic process, the two equivalent hydroxyl groups in (1-hydroxymethyl-cyclopropyl)-methanol (**6**) have to be differentiated in order to produce the two different geminal side chains on the cyclopropyl ring. Several alternative reaction sequences are possible. All of these processes produce the nucleophilic thiol building blocks.

When an electrophilic side chain is required, for example, for strategy **b** in Scheme 7.3, the functionalization of one of the two CH_2OH groups is suspended and a leaving group incorporated into this methylene group, which is then used to produce the final thioether at a later stage (Scheme 7.5).

Another interesting possibility to construct the cyclopropyl moiety is *via* cyclization of a di-halide compound. Wurtz-type coupling has been used in this case and has proved particularly valuable for the synthesis of strained small ring systems.

7.3 RETROSYNTHETIC OUTLINE

A retrosynthetic outline of Montelukast is shown in Scheme 7.6. The thiol side chain is a good candidate for initial disconnection. During construction of the Montelukast framework, this group would then be introduced at the end of the synthetic process. The three-carbon bridge between the two aromatic di-substituted benzene rings can be made by taking advantage of the functionality required to be present at the benzylic atom for attachment of the thiol side chain.

A disconnection using the double bond of the ethylene bridge is the next logical step in the retrosynthetic analysis. Either a retro-Wittig or retro-Heck are obvious choices (Scheme 7.7). The double bond can also be obtained from an alcohol precursor by dehydration, in turn, being easily obtained from the addition of a carbanion to an aromatic aldehyde.

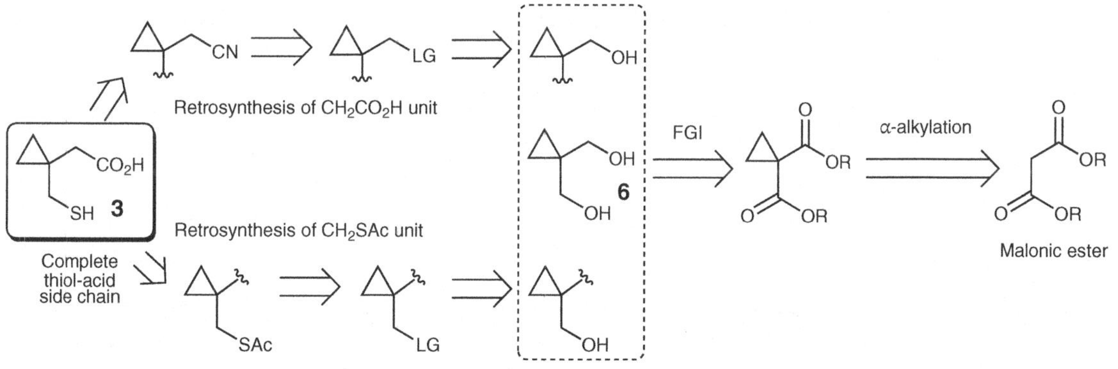

SCHEME 7.4 RETROSYNTHESIS OUTLINE OF A NUCLEOPHILIC THIOL SIDE CHAIN

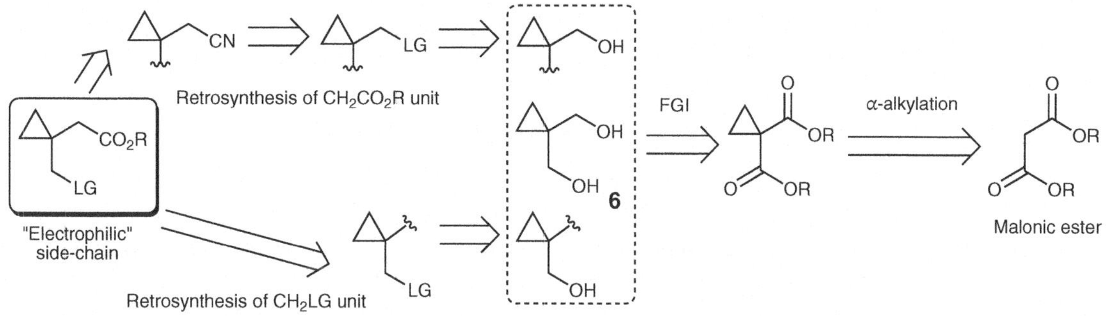

SCHEME 7.5 RETROSYNTHESIS OF AN ELECTROPHILIC THIOL SIDE CHAIN

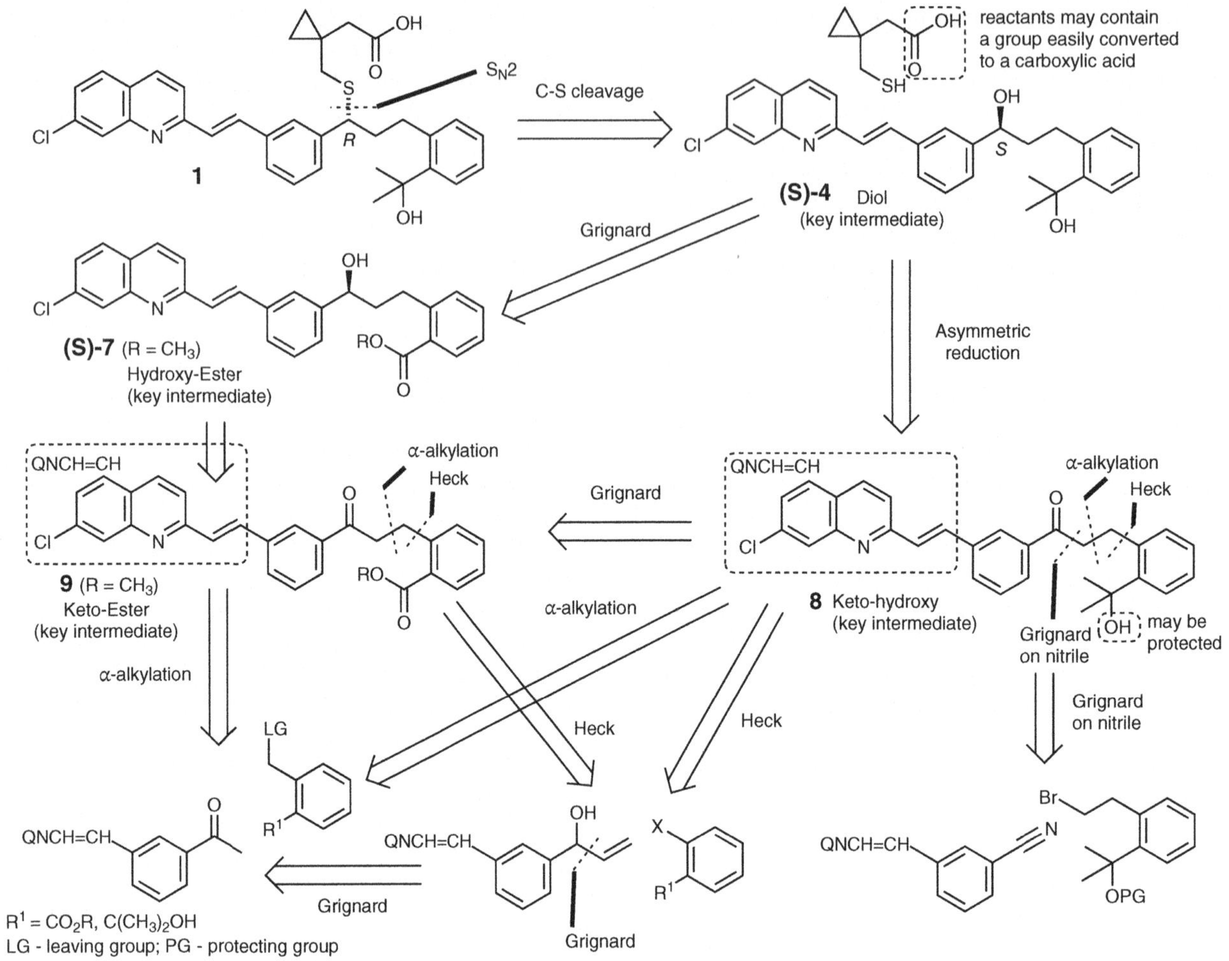

Scheme 7.6 Retrosynthetic outline of Montelukast

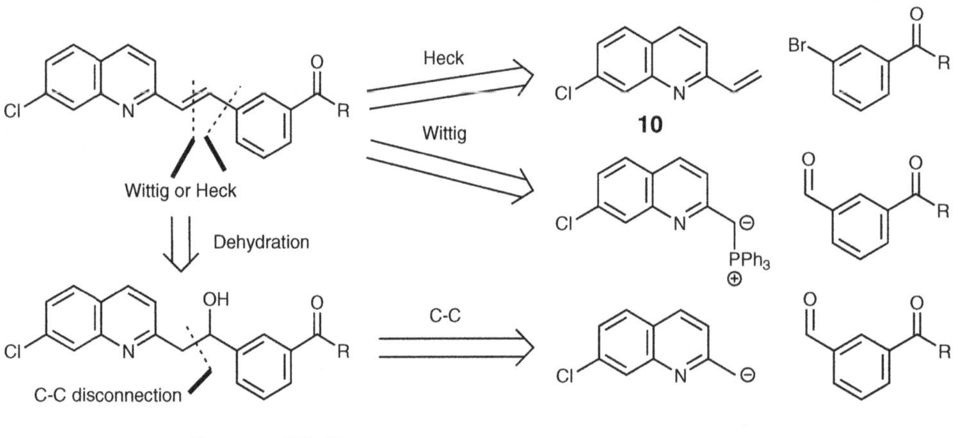

Scheme 7.7 Retrosynthesis of the ethylene bridge

Not surprisingly, due to the inherent complexity of the Montelukast molecule, many different processes have been developed with industrial synthesis in mind, some of them resulting from different sequencing of the same reactions. However, as several incompatible groups are present in some intermediates, different selectivity problems arose during the development of the various synthetic methods. Given that changes in the order of the synthetic steps may give origin to several difficulties, selectivity problems should be well assessed. Process impurities that form can be different and some synthetic routes more difficult to develop. An example of a complete retrosynthesis of Montelukast resulting in easily available raw materials is shown in Scheme 7.8.

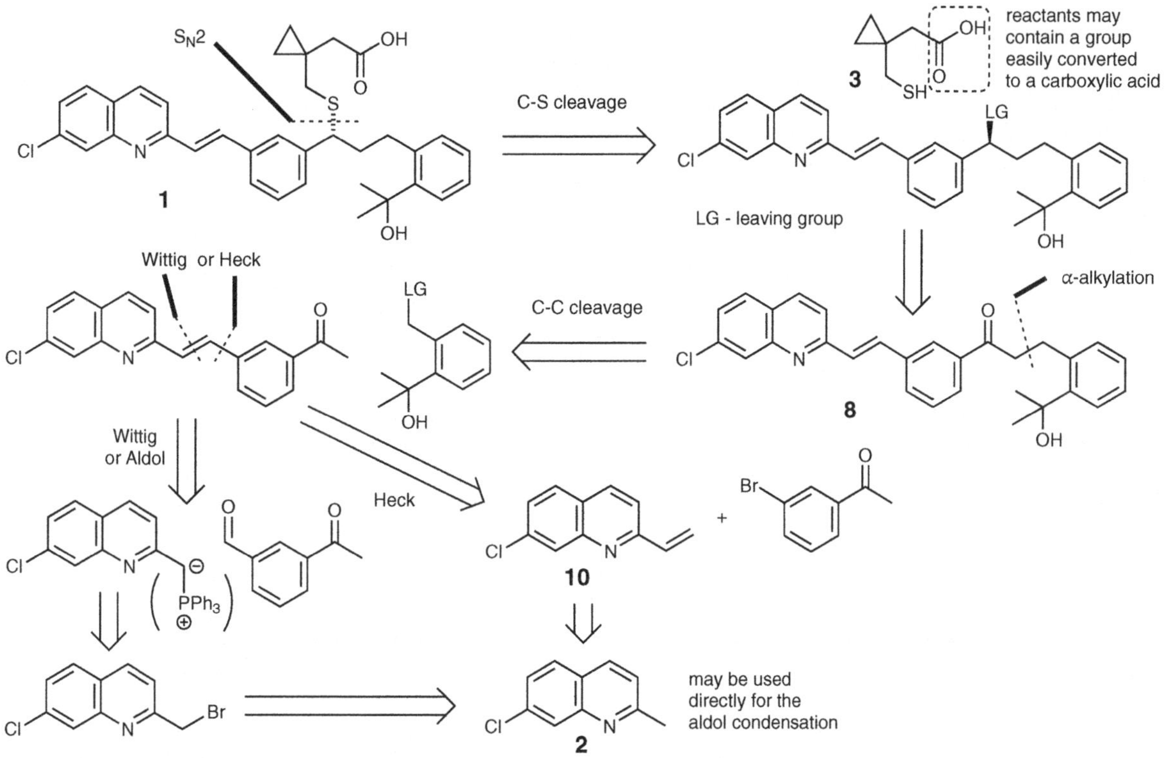

Scheme 7.8 Complete retrosynthesis of montelukast resulting in available materials

Tackling the ethylene bridge of Montelukast first will lead to other retrosynthetic approaches. A retro-Heck, a retro-Wittig or a retro aldol can be used, producing in all cases, two large fully functionalized blocks. Retrosynthetic analyses for these three opportunities are shown below (Scheme 7.9 and Scheme 7.10).

The use of a Wittig or an aldol-type disconnection as first retrosynthetic step (Scheme 7.10) means that an aldehyde (or its precursor) must be present up until the latter stages of the synthetic process. Usually, being a very reactive group, the aldehyde almost inevitably requires the use of a protecting group, particularly during the α-alkylation step.

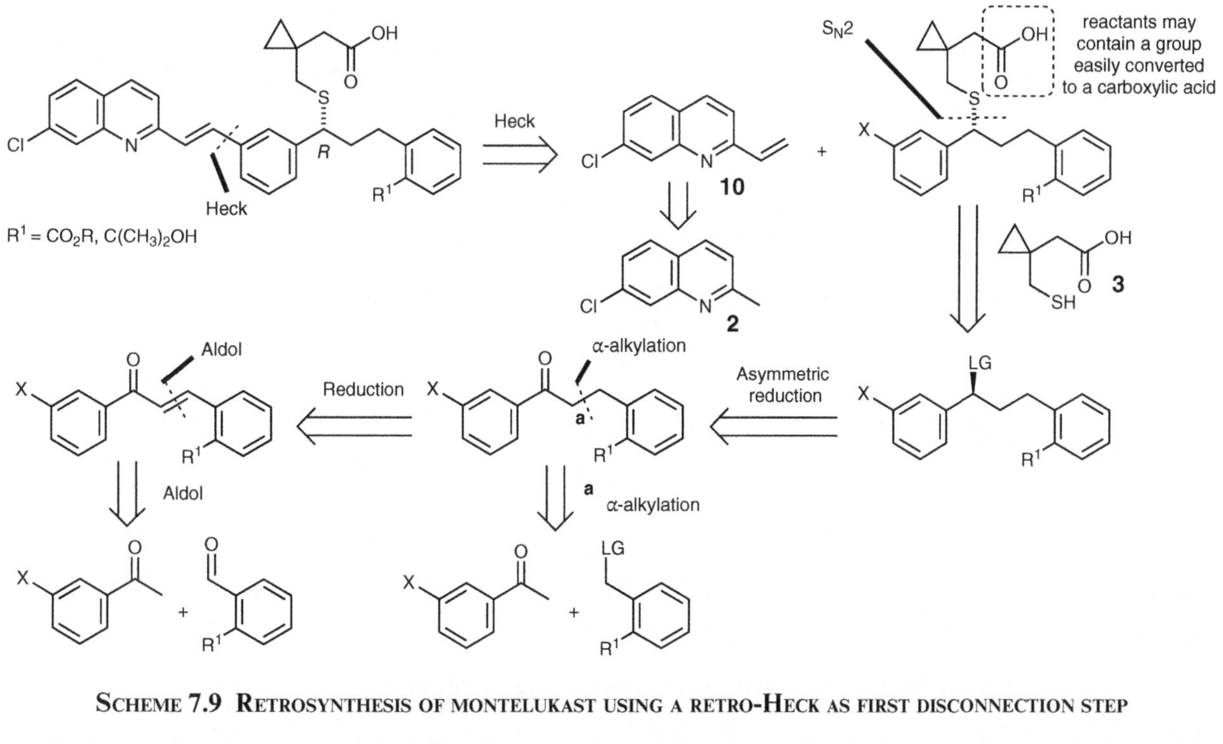

Scheme 7.9 Retrosynthesis of montelukast using a retro-Heck as first disconnection step

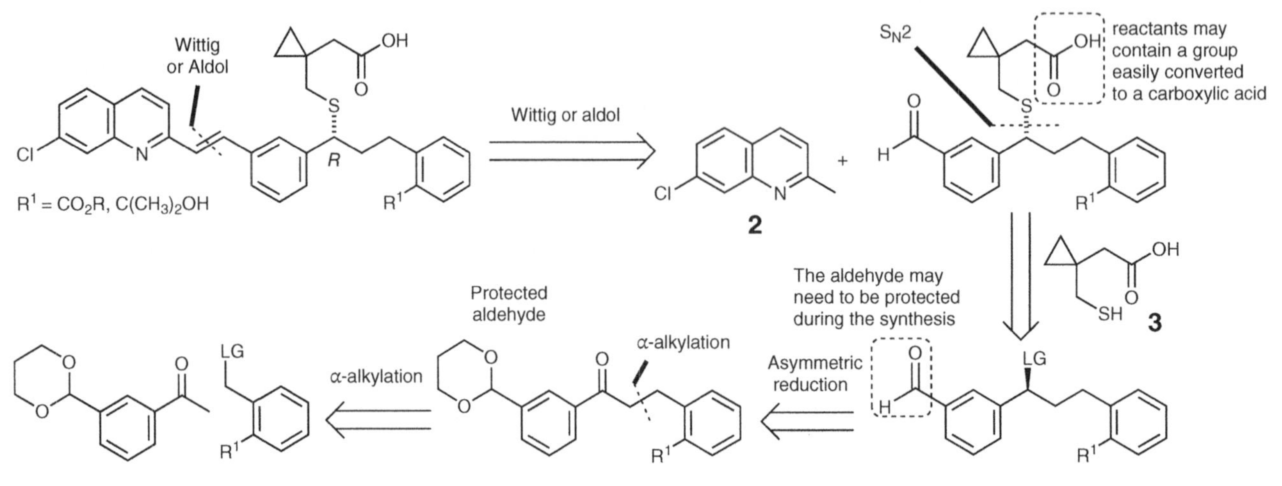

Scheme 7.10 Retrosynthesis of montelukast using a retro-Wittig or aldol as first disconnection step

7.4 THE FIRST REPORTED MONTELUKAST SYNTHESIS

The synthesis starts with an acidic condensation of isophthalaldehyde with 7-chloro-2-methyl-quinoline (**2**), establishing the ethylene bridge between the quinolone and the central aromatic ring (MerckFrosst1990, Scheme 7.11). The resulting intermediate was converted to the key keto-ester intermediate (**9**) using either a Heck reaction or a ketone α-alkylation.

In the first case, the addition of vinyl Grignard gave the vinyl alcohol used directly in the palladium catalysed Heck coupling. In the alternative processes, the aldehyde was converted to a methyl ketone by a Grignard reaction-oxidation sequence. Direct alkylation with the appropriate halo-benzylic aromatic building block produced the desired keto-ester intermediate (**9**). The third alternative converts the methyl ketone to a 3-keto-ester with dimethylcarbonate, in a smooth α-alkylation reaction. Hydrolysis, decarboxylation and esterification produced the keto-ester (**9**).

Asymmetric reduction of the benzylic ketone with (−)-diisopinocamphenylchloroborane [(−)-Ipc_2BCl or (−)-DIPCl] gave the appropriate stereochemical configuration (*S*) (Scheme 7.12). The diol intermediate **(*S*)-4** was easily obtained by a Grignard reaction on the ester group of **(*S*)-7**. At this stage, the thiol side chain requires only attachment to the secondary benzylic carbon to complete the synthesis.

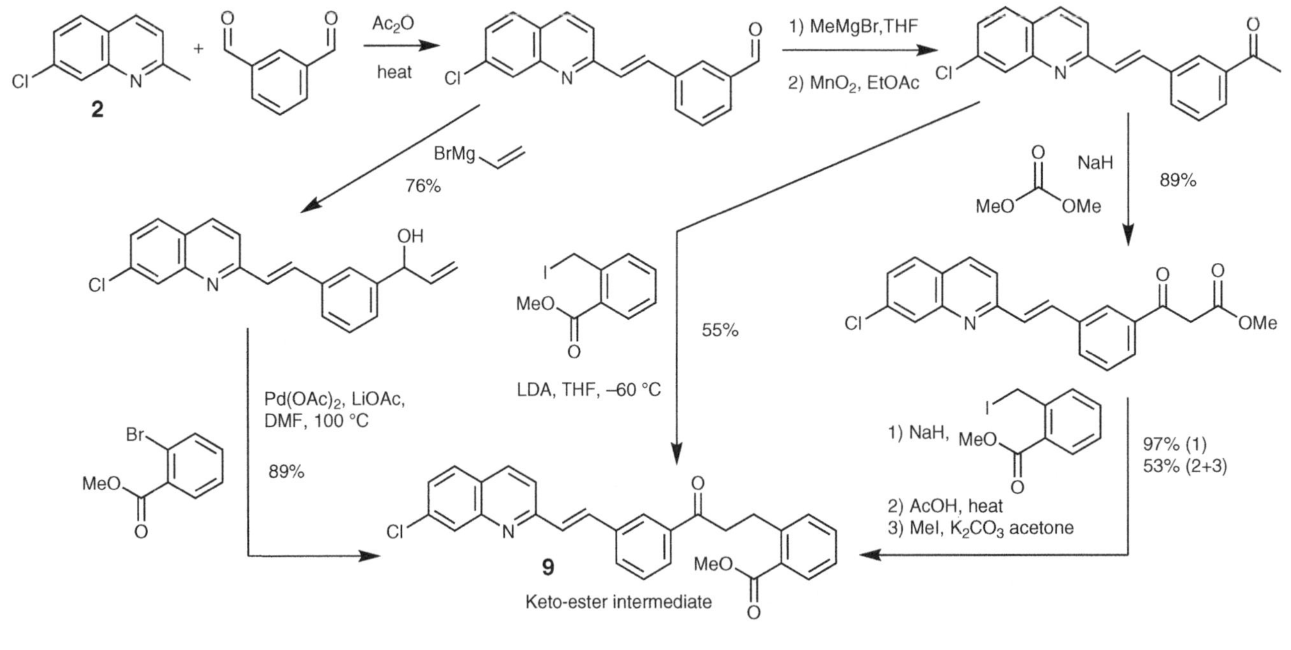

Scheme 7.11 Initial steps of the first reported Montelukast synthesis

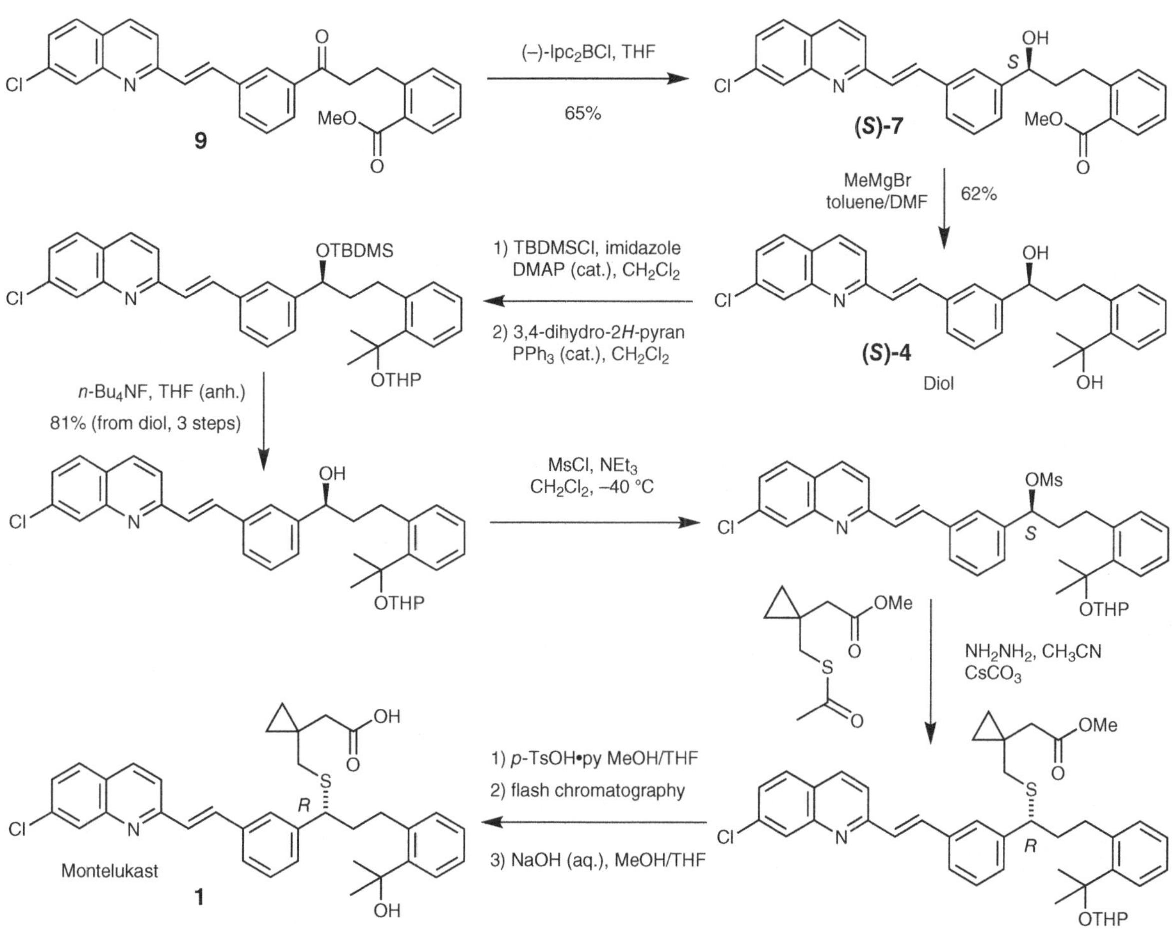

Scheme 7.12 First Montelukast synthesis: conversion of keto-ester intermediate into Montelukast

Leaving nothing to chance, Merck researchers opted to protect the tertiary alcohol for this critical S_N2 reaction and used a thiol-ester as nucleophile. Firstly, they protected both alcohols using different protecting groups and then selectively removed the protecting group on the least hindered secondary alcohol. Mesylation of the benzylic secondary alcohol set the scene for a displacement of mesylate by the thiol to give the methyl ester of Montelukast. Montelukast itself was obtained after chromatographic purification and hydrolysis.

Although the first Montelukast patent described and claimed several other synthetic routes, only the one described above was exemplified.

7.5 THE EVOLUTION OF INDUSTRIAL CHEMICAL SYNTHESES OF MONTELUKAST

The extended schemes (A to E) summarize the various methods to synthesis Montelukast that have been the subject of a patent applications.

Scheme A deals with the synthesis of the key keto-ester (**9**) and keto-hydroxy (**8**) intermediates from 7-chloro-2-methyl-quinoline (**2**). Schemes B to E are separated according to the key synthetic strategy to produce Montelukast. The various processes to convert the key intermediates, keto-ester (**9**) and keto-hydroxy (**8**), into the final API (**1**) are condensed in schemes B (strategy **a**, from Scheme 7.3) and C (strategy **b**, from Scheme 7.3). Processes that apply a Heck or a Wittig-like reaction during the latter stages of the synthesis to produce Montelukast directly, or a closely related compound, are summarized in schemes D and E (these syntheses correspond to the retrosynthetic analyses shown in Scheme 7.9 and Scheme 7.10).

The methods to produce the complete thiol side chain or an electrophilic equivalent are the subject of scheme F.

Merck's first Montelukast synthesis, decribed above, can also be found in schemes A and B. The conversion of 7-chloro-2-methyl-quinoline (**1**) into the keto-ester intermediate (**9**) (Scheme 7.11) are embodied in scheme A in reaction sequences **A2/A3/A4/A5**, **A2/A3/A4/A14/A15/A17/A18** or **A2/A19/A20**.

The conversion of the keto-ester to Montelukast in the original synthesis, as reported by Merck Frosst, (Scheme 7.12) can be followed in reactions **B1/B2/B3/B4/B5/B6/B7a** in scheme B.

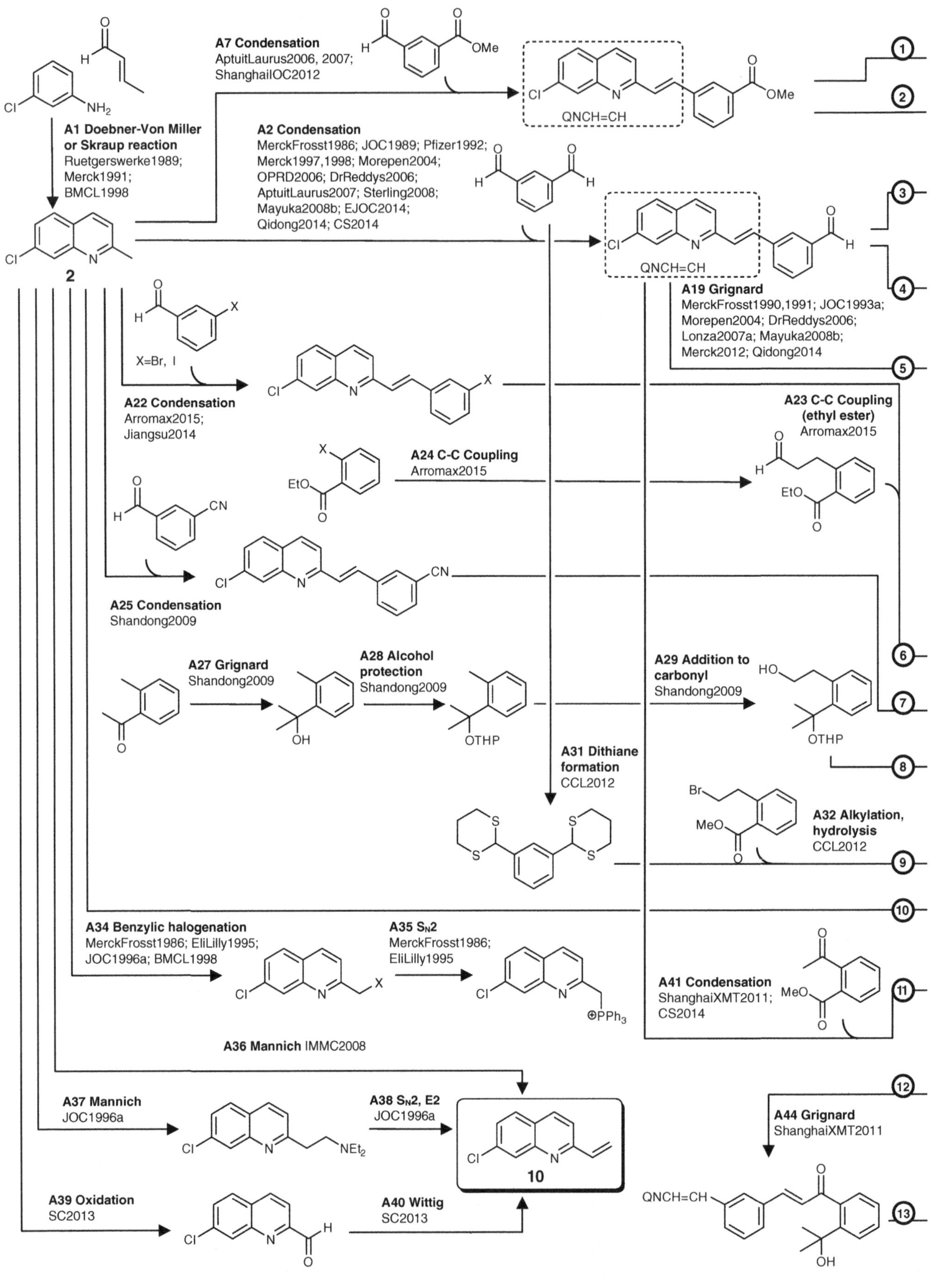

SCHEME A QUINOLINE NUCLEUS SYNTHESIS AND INTERMEDIATE KETONE PRODUCTION ROUTES (FOR REACTION CONDITIONS SEE PAGE 307)

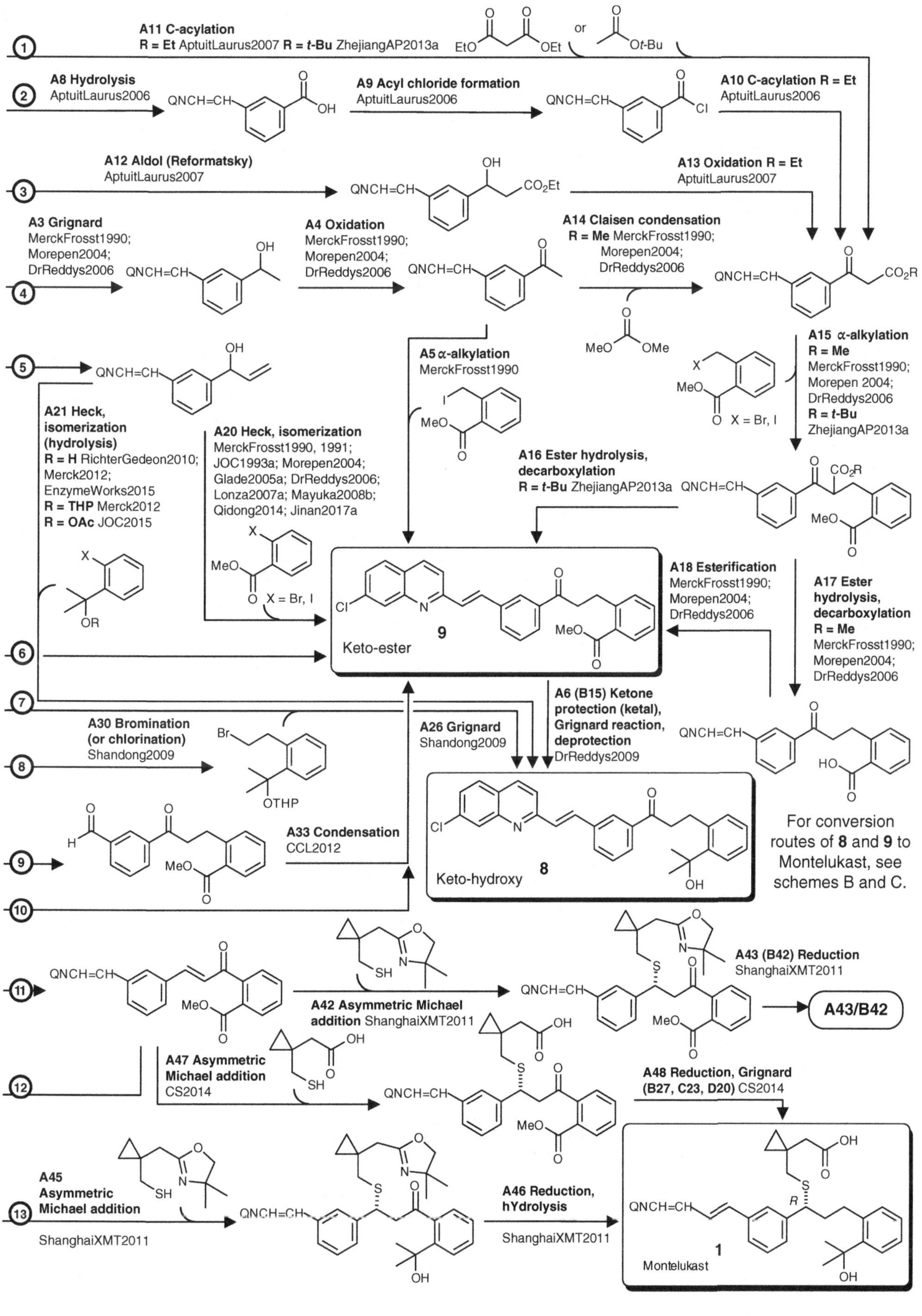

A11 C-acylation
R = Et AptuitLaurus2007 R = t-Bu ZhejiangAP2013a
EtO
OEt
or
Ot-Bu
A8 Hydrolysis
AptuitLaurus2006
QNCH=CH
OH
A9 Acyl chloride formation
AptuitLaurus2006
Cl
A10 C-acylation R = Et
AptuitLaurus2006
A12 Aldol (Reformatsky)
AptuitLaurus2007
CO2Et
A13 Oxidation R = Et
AptuitLaurus2007
A3 Grignard
MerckFrosst1990;
Morepen2004;
DrReddys2006
A4 Oxidation
MerckFrosst1990;
Morepen2004;
DrReddys2006
A14 Claisen condensation
R = Me MerckFrosst1990;
Morepen2004;
DrReddys2006
CO2R
MeO
OMe
A5 α-alkylation
MerckFrosst1990
A15 α-alkylation
R = Me
MerckFrosst1990;
Morepen 2004;
DrReddys2006
R = t-Bu
ZhejiangAP2013a
X = Br, I
A21 Heck,
isomerization
(hydrolysis)
R = H RichterGedeon2010;
Merck2012;
EnzymeWorks2015
R = THP Merck2012
R = OAc JOC2015
A20 Heck, isomerization
MerckFrosst1990, 1991;
JOC1993a; Morepen2004;
Glade2005a; DrReddys2006;
Lonza2007a; Mayuka2008b;
Qidong2014; Jinan2017a
A16 Ester hydrolysis,
decarboxylation
R = t-Bu ZhejiangAP2013a
OR
A18 Esterification
MerckFrosst1990;
Morepen2004;
DrReddys2006
A17 Ester
hydrolysis,
decarboxylation
R = Me
MerckFrosst1990;
Morepen2004;
DrReddys2006
Cl
N
9
Keto-ester
A6 (B15) Ketone
protection (ketal),
Grignard reaction,
deprotection
DrReddys2009
A26 Grignard
Shandong2009
A30 Bromination
(or chlorination)
Shandong2009
Br
OTHP
HO
A33 Condensation
CCL2012
H
8
Keto-hydroxy
For conversion routes of 8 and 9 to Montelukast, see schemes B and C.
A43 (B42) Reduction
ShanghaiXMT2011
SH
A42 Asymmetric Michael
addition ShanghaiXMT2011
A43/B42
A47 Asymmetric
Michael addition
CS2014
A48 Reduction, Grignard
(B27, C23, D20) CS2014
A45
Asymmetric
Michael addition
ShanghaiXMT2011
A46 Reduction,
hYdrolysis
ShanghaiXMT2011
R
1
Montelukast
1 2 3 4 5 6 7 8 9 10 11 12 13

7.5.1 Processes to produce keto-ester intermediate

As it can be observed from scheme A, several companies opted to use production routes similar to the original one, starting with isophthalaldehyde as raw material (**A2** is the first reaction when these routes are followed). A different route is seen when the dialdehyde is replaced with 3-formyl-benzoic acid methyl ester (routes starting at **A7**). Using 3-formyl-benzonitrile for the initial condensation (**A25**) will give, after Grignard addition and hydrolysis (**A26**), a useful keto-hydroxy intermediate, (**8**). In another approach, the condensation of 7-chloro-2-methyl-quinoline (**1**) with a 3-halobenzaldehyde (**A22**) produces an ethylene bridge compound with a *meta*-halogen substituted aromatic ring. Palladium mediated C-C coupling with an aldehyde provides the keto ester (**9**) directly (**A23**).

7.5.2 Processes to convert the keto-ester intermediate to Montelukast

Three main tasks have to be accomplished to convert the keto-ester (**9**) to Montelukast, namely the asymmetric reduction of the ketone, the conversion of the ester to the tertiary alcohol and the connection of the thiol side chain.

Many companies decided to elaborate methodologies for the asymmetric reduction of the keto-ester (**9**) to give the hydroxy-ester (**7**) (steps **B1** or **B10**). Cost effectiveness of the asymmetric reduction process is critical for of the route to be competitive. The yield of this step, and the enantiomeric excess of the product, are undoubtedly very important parameters when assessing whether to implement on an industrial scale or not. Some microbiological process were originally tested with limited success (Merck1994a,b). For industrial application, three chemical methods were developed to produce the *S*-alcohol (Scheme 7.13). More recently biocatalytic processes were claimed to be usable at a >200 kg scale.

Although it was less used, probably due to some instability of the tertiary benzylic alcohol, an asymmetric reduction of **8** (**B16**) is a convenient alternative method to produce the *S*-benzylic alcohol (**4**). The starting material for the reduction was accessed with good convergence by a tandem Heck reaction and double-bond isomerization (**A21**).

A detailed screening of [(Diphosphine)(diamine)$RuCl_2$] catalysts led to the identification of ((*R*)-Xyl-BINAP)((*R*,*R*)-DPEN)$RuCl_2$ as a catalyst that afforded an enantioselectivity of 99% ee for the asymmetric hydrogenation of **8** on a multigram lab scale (substrate:catalyst loading of 5000:1, 0.02% mol, JOC2015, Scheme 7.14).

Several other patents describe a similar strategy to that portrayed in the first reported synthesis to produce the diol intermediate and used a protected tertiary alcohol as a forerunner to the S_N2 reaction (Synthon2004b, Aurobindo2006, Pliva2006, RichterGedeon2010, Lunan2013, scheme B, on the left).

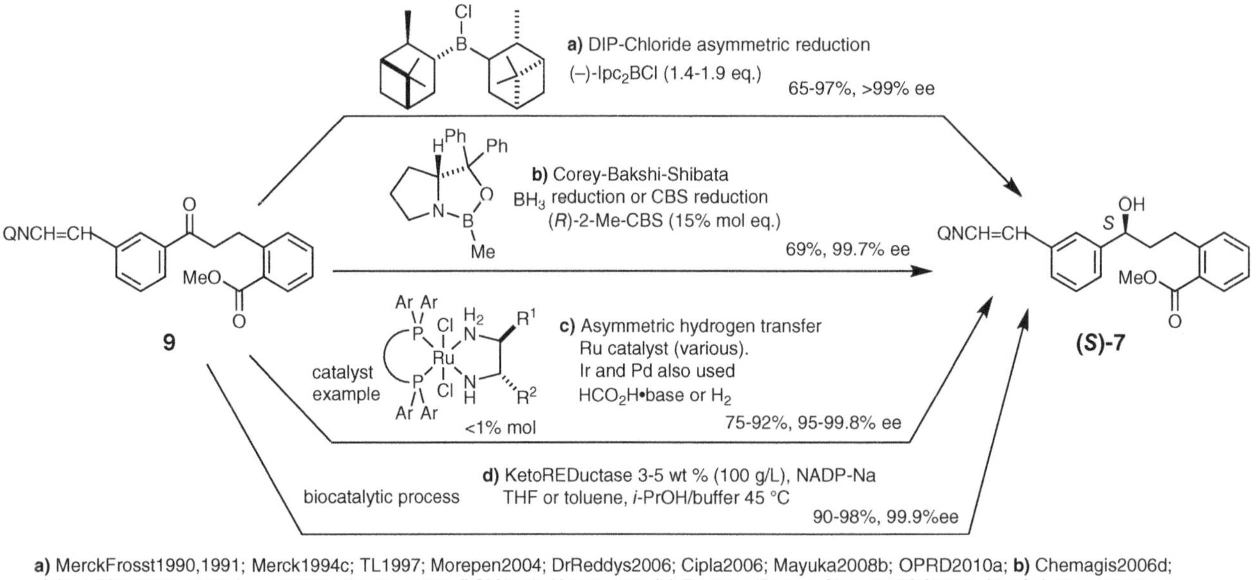

a) MerckFrosst1990,1991; Merck1994c; TL1997; Morepen2004; DrReddys2006; Cipla2006; Mayuka2008b; OPRD2010a; **b)** Chemagis2006d; **c)** (Ru) Pliva2004, 2006; Lonza2007a; Synthon2008; DSM2009; Kanata2010 (Ir) ZhejiangJP2012; ShanghaiIOC2013; (Pd-Cu) Jinan2017b; **d)** Merck1994a,b; Codexys2007; Qidong2014; ZhangjiagangXC2014; EnzymeWorks2015, 2016

SCHEME 7.13 EXISTING ASYMMETRIC REDUCTIONS OF THE KETO-ESTER INTERMEDIATE (**B1**)

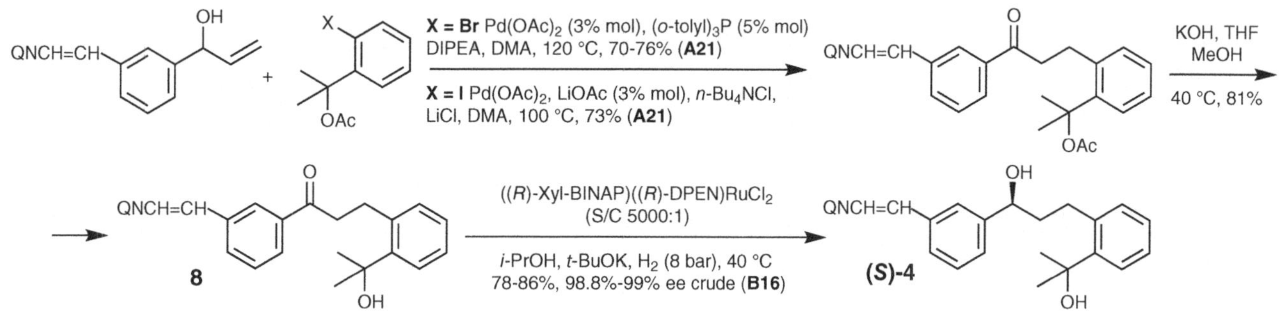

Scheme 7.14 Convenient asymmetric hydrogenation of the keto-alcohol intermediate

An innovative approach applies an enzymatic biochemical asymmetric Baeyer-Villiger reaction to generate the chiral centre. High yields and selectivities were reported for this transformation (Scheme 7.15). Opening of the lactone with a methyl Grignard in the presence of cerium (III) trichloride produces the diol (**4**), subsequently converted to Montelukast using standard procedures (mesylation and S_N2).

Different thiols have been used for the final displacement reaction (scheme B, on the left, **B7a** to **B7d**). Avoiding the use of protecting groups is a clear improvement for any industrial synthesis: less synthetic steps are required and obviously, less reagents and solvents used hence, generating less waste. A shorter synthesis usually also means a higher yielding route. Merck (1993) was able to avoid undesirable protecting groups (scheme B, center left, starting from the diol **(*S*)-4**). Direct mesylation of the key diol was possible (**B8a**) and displacement of the mesylate with thiol-acid (**3**) feasible, giving good results (**B9a**, Merck1993). Many companies developed processes that incorporate this key modification in the synthesis, thus proving its viability and, together with various other improvements, justifying the grant of a patent. The key displacement S_N2 reaction (**B9**) is clearly tricky and high levels of impurities are reported to be formed during this step (Scheme 7.16, OPRD2010b).

Many process patents focus on improvements on the key S_N2 reaction, with the principle objective being a reduction of impurities, both in number and quantity and to provide a more robust high yielding process (scheme B, **B8** and **B9**). As chromatographic purification is impractical for large-scale batches, a great deal of attention has been attributed to the development alternative purification methods for several intermediates. In most cases, the intermediate to be purified already contains a carboxylic acid group on the thiol side chain and hence, it is straightforward to produce a crystalline salt with an amine. Crystallizations of these salts generally remove unwanted impurities. Crude montelukast itself can be improved by this process.

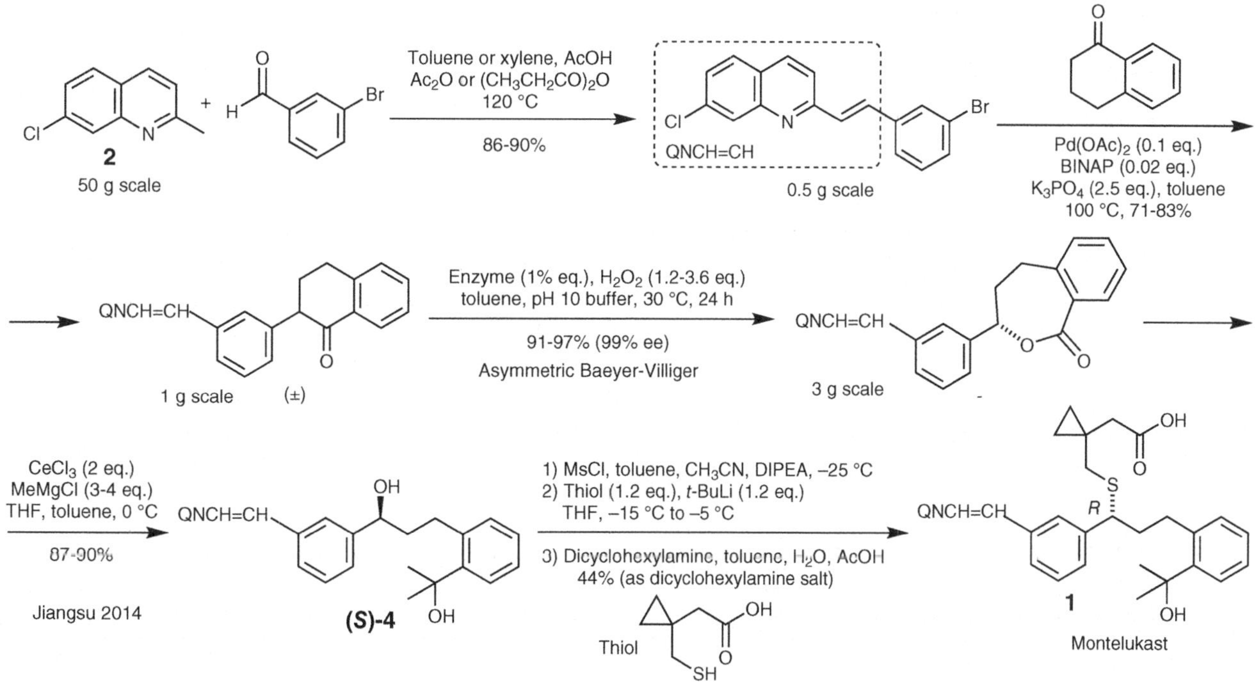

Scheme 7.15 Montelukast synthesis using an asymmetric Baeyer-Villiger reaction

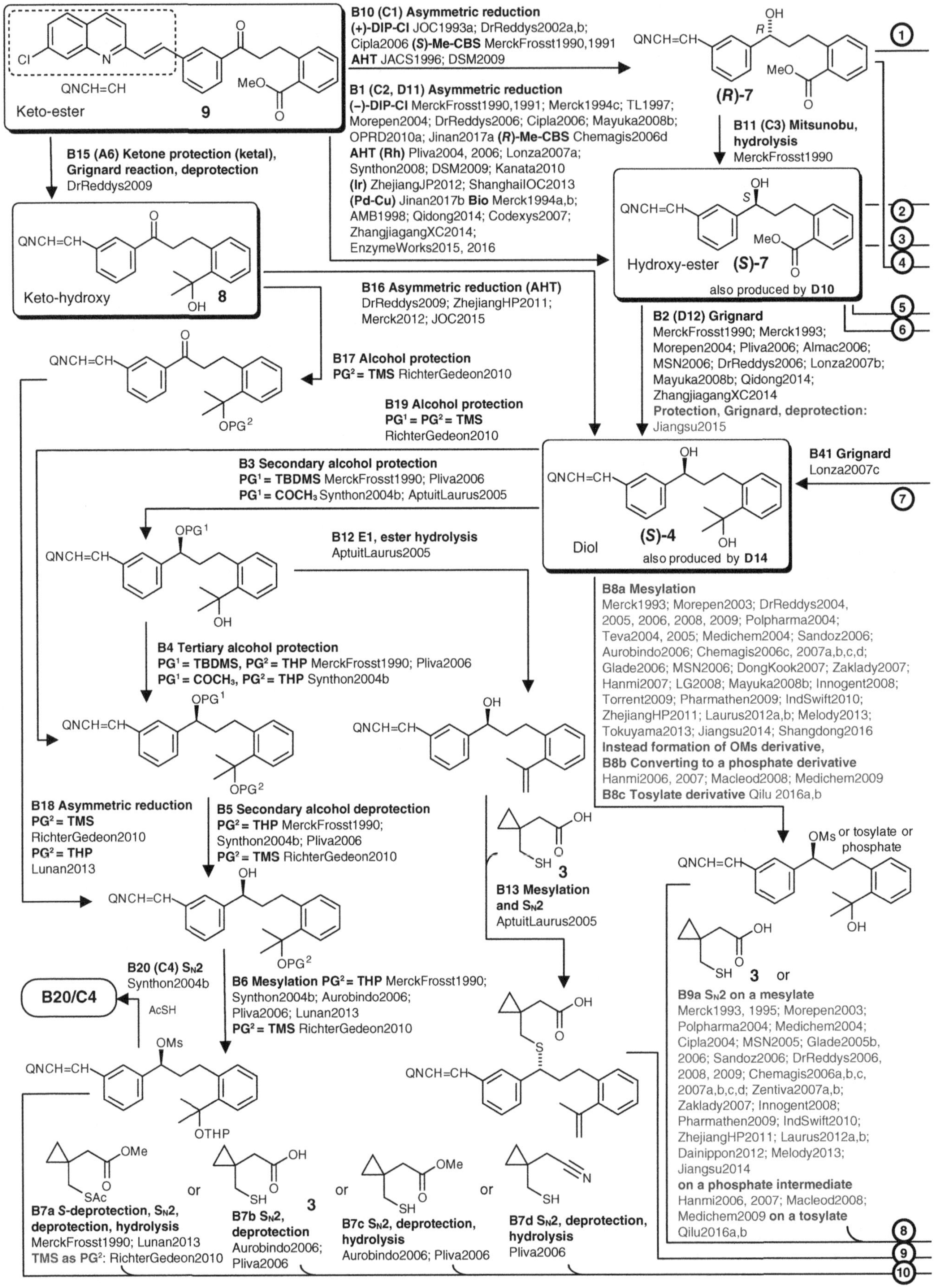

Scheme B Montelukast synthesis by introducing thiol side chain at final process stages (for reaction conditions see page 308)

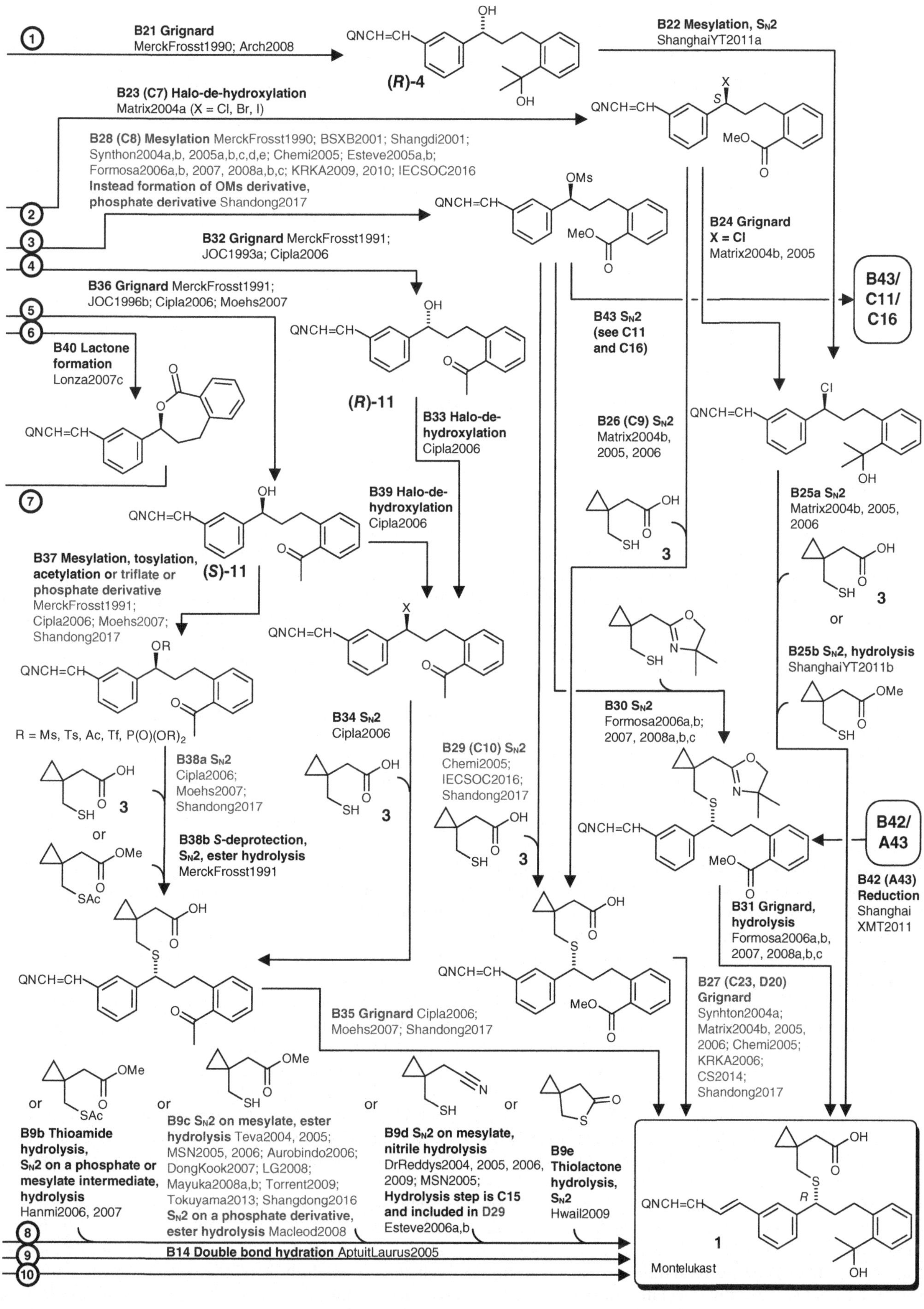
B21 Grignard
MerckFrosst1990; Arch2008
(R)-4
B22 Mesylation, SN2
ShanghaiYT2011a
B23 (C7) Halo-de-hydroxylation
Matrix2004a (X = Cl, Br, I)
B28 (C8) Mesylation MerckFrosst1990; BSXB2001; Shangdi2001; Synthon2004a,b, 2005a,b,c,d,e; Chemi2005; Esteve2005a,b; Formosa2006a,b, 2007, 2008a,b,c; KRKA2009, 2010; IECSOC2016
Instead formation of OMs derivative, phosphate derivative Shandong2017
B32 Grignard MerckFrosst1991; JOC1993a; Cipla2006
B24 Grignard
X = Cl
Matrix2004b, 2005
B43/ C11/ C16
B36 Grignard MerckFrosst1991; JOC1996b; Cipla2006; Moehs2007
B43 SN2 (see C11 and C16)
B40 Lactone formation
Lonza2007c
(R)-11
B33 Halo-de-hydroxylation
Cipla2006
B26 (C9) SN2
Matrix2004b, 2005, 2006
B39 Halo-de-hydroxylation
Cipla2006
B25a SN2
Matrix2004b, 2005, 2006
B37 Mesylation, tosylation, acetylation or triflate or phosphate derivative
MerckFrosst1991; Cipla2006; Moehs2007; Shandong2017
(S)-11
3
or
B25b SN2, hydrolysis
ShanghaiYT2011b
R = Ms, Ts, Ac, Tf, P(O)(OR)2
B34 SN2
Cipla2006
B30 SN2
Formosa2006a,b; 2007, 2008a,b,c
B29 (C10) SN2
Chemi2005; IECSOC2016; Shandong2017
B38a SN2
Cipla2006; Moehs2007; Shandong2017
B42/ A43
B38b S-deprotection, SN2, ester hydrolysis
MerckFrosst1991
B42 (A43) Reduction
Shanghai XMT2011
B31 Grignard, hydrolysis
Formosa2006a,b, 2007, 2008a,b,c
B27 (C23, D20) Grignard
Synhton2004a; Matrix2004b, 2005, 2006; Chemi2005; KRKA2006; CS2014; Shandong2017
B35 Grignard Cipla2006; Moehs2007; Shandong2017
B9b Thioamide hydrolysis, SN2 on a phosphate or mesylate intermediate, hydrolysis
Hanmi2006, 2007
B9c SN2 on mesylate, ester hydrolysis Teva2004, 2005; MSN2005, 2006; Aurobindo2006; DongKook2007; LG2008; Mayuka2008a,b; Torrent2009; Tokuyama2013; Shangdong2016
SN2 on a phosphate derivative, ester hydrolysis Macleod2008
B9d SN2 on mesylate, nitrile hydrolysis
DrReddys2004, 2005, 2006, 2009; MSN2005;
Hydrolysis step is C15 and included in D29
Esteve2006a,b
B9e Thiolactone hydrolysis, SN2
Hwail2009
B14 Double bond hydration AptuitLaurus2005
1
Montelukast

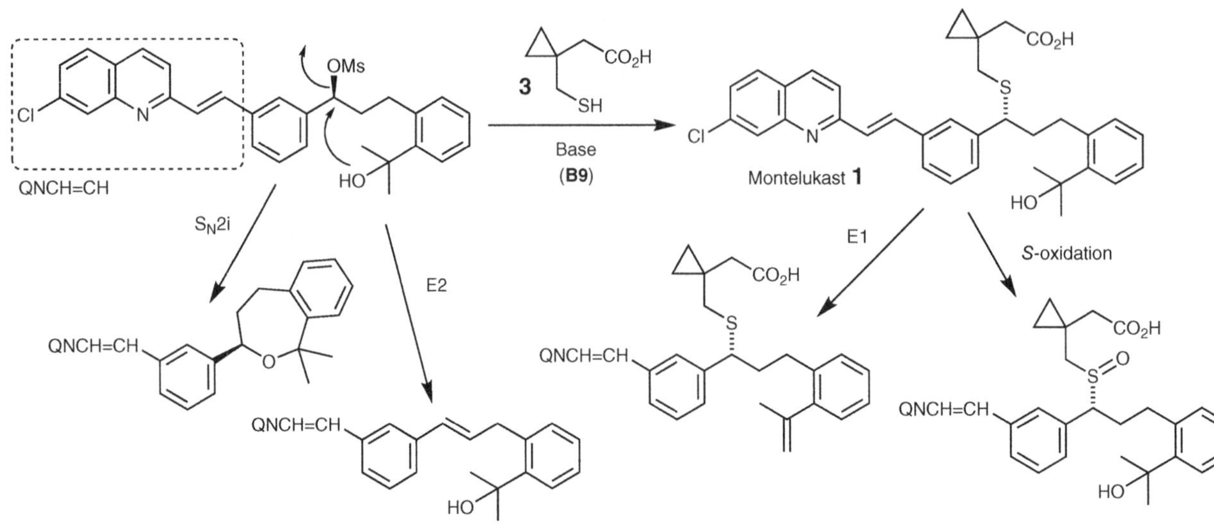

Scheme 7.16 Common impurities formed during the S_N2 displacement

Instead of using the thiol-acid (**3**), which requires double deprotonation for the S_N2 reaction to work well, thiol-ester, thiol-nitrile or thiol-amide can also be used as a nucleophile. Montelukast is thereafter obtained after a final hydrolysis step (scheme B, **B9b**, **B9c** and **B9d**).

Leaving groups at the benzylic position for the S_N2 reaction other than the ubiquitous mesylate have been reported. These include phosphate and tosylate groups (**B8b**, **B8c**).

For most of the processes referred to up till now the diol intermediate **(*S*)-4** was key and frequently obtained from the corresponding keto-ester (**5**) by an asymmetric reduction of the benzylic ketone followed by a double carbanionic addition to the aryl ester group (top-left scheme B, **B1** and **B2**) to form the tertiary alcohol. As the ketone is more electrophilic than the ester, the reduction followed by a Grignard sequence is fairly common but the Grignard followed by reduction has also been used and the subject of a patent application (DrReddys2009, scheme B, **B15**, **B16**). As expected, the ketone is protected prior to the Grignard addition.

In addition to the methods referred to above that use the diol **(*S*)-4** as a key intermediate, the hydroxy-ester intermediate **(*S*)-7** can be converted into Montelukast by several alternative processes using two main strategies (scheme B, right and centre).

The first option, involves converting the ester to a ketone to obtain a hydroxy-ketone intermediate (**11**, Scheme 7.17, route **a**), transforming the hydroxyl group into a suitable leaving group, performing the S_N2 displacement with the thiol side chain and finally reacting with the Grignard reagent to produce Montelukast (scheme B, center-mid-right, MerckFrosst1991, JOC1993c, 1996b, Cipla2006, Moehs2007). Several different leaving groups have been used (for example, acetate, various sulfonates and chloride). The addition of a Grignard to an ester usually produces the corresponding alcohol, *via* the initially formed ketone. However, in the present case, the use of a strongly basic medium promoted the formation of a stable conjugated lithium enolate that circumvented the second Grignard attack and allowed the isolation of the ketone intermediate (**11**). The use of a Weinreib amide was also used as a solution to obtain **11**.

The second option (Scheme 7.17, route **b**, scheme B, right) converts the hydroxyl into mesylate (**B28**) or halide (**B23**) followed by the S_N2 displacement with the thiol side chain and finally transforming the ester into the tertiary alcohol in the last step by a Grignard addition. In the conversion to the tertiary alcohol, a Cerium (III) salt is frequently added to prevent the reaction stopping at the ketone stage (as in **B36**).

7.5.3 Montelukast synthesis using an incomplete thiol for the S_N2 displacement

Up to now, all processes described have used a thiol side chain with the appropriate structural elements to produce Montelukast or a closely related compound directly from the S_N2 displacement. Other investigators opted not to use a completely elaborated side chain thiolate as a nucleophile. Introducing an "incomplete" thiol side chain first and building the cyclopropylacetic acid side chain during the latter steps of the synthetic process is yet another feasible choice (scheme C). Two principle options were developed to implement this strategy (Scheme 7.18).

The first of these strategies introduces a thiol group by a bimolecular displacement on a mesylate using thioacetate as nucleophile (Scheme 7.18, route **a**; **C4** or **C16**, scheme C). Hydrolysis of the thioester gives a thiol-alcohol or a thiol-ester that can be used as nucleophiles in another S_N2 reaction with a cyclopropylic electrophilic fragment, **C20** and **C22**).

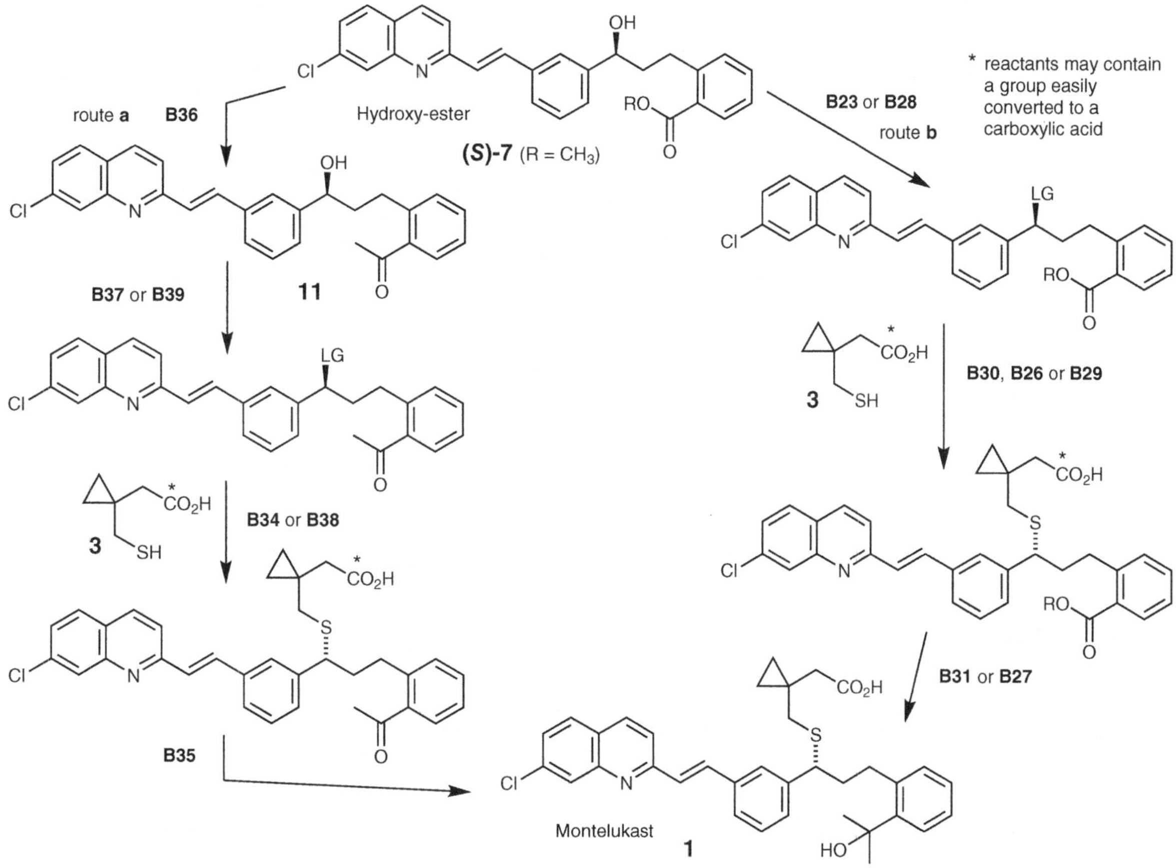

Scheme 7.17 Other routes from hydroxy-ester to Montelukast

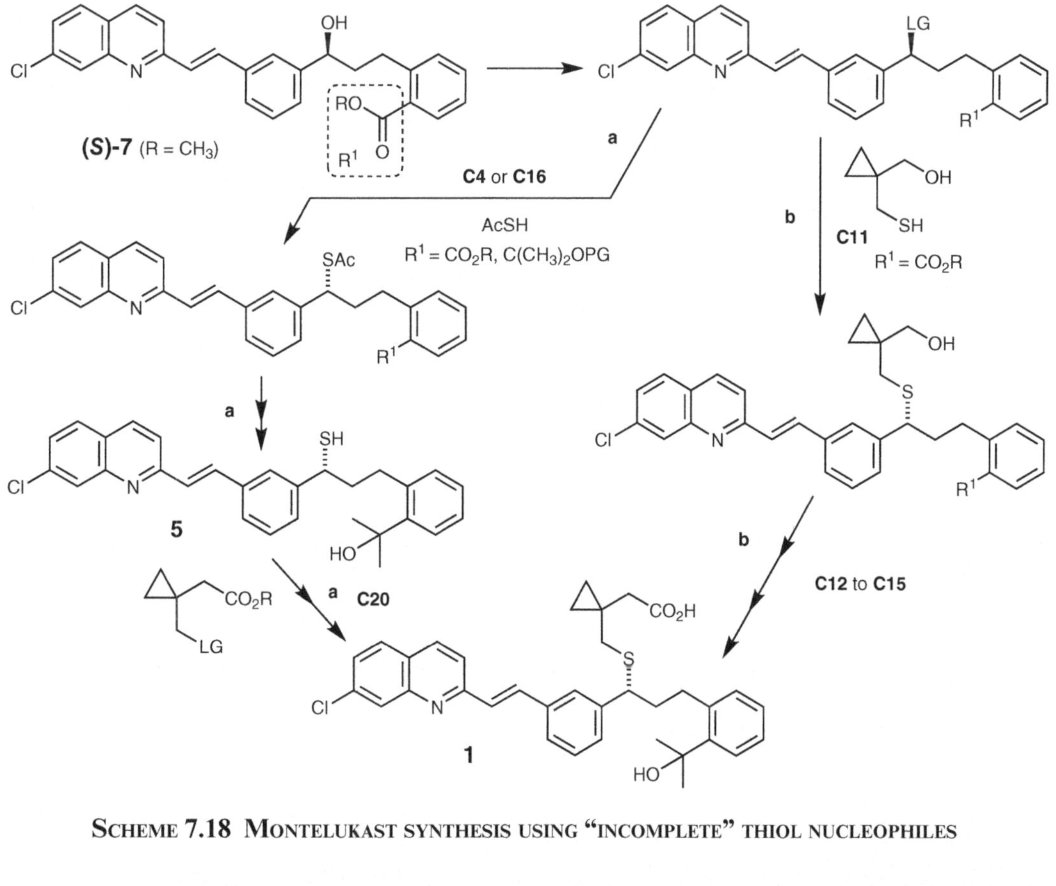

Scheme 7.18 Montelukast synthesis using "incomplete" thiol nucleophiles

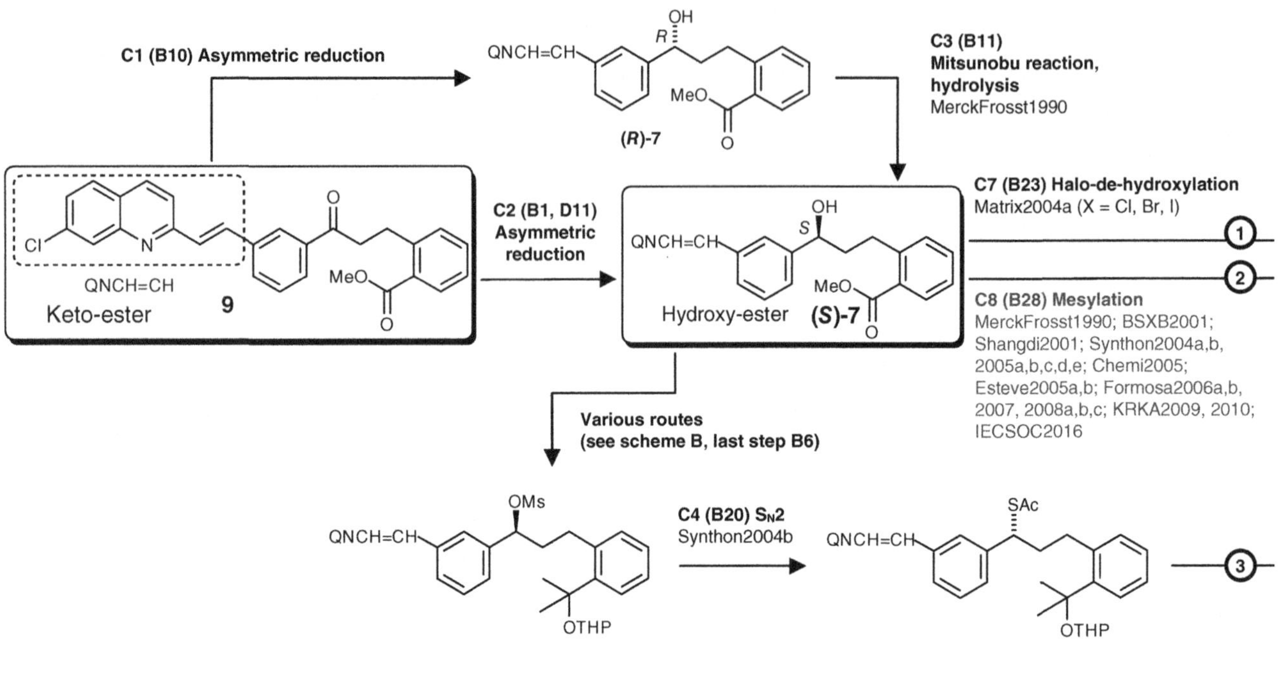

Scheme C Montelukast synthesis by building thiol side-chain at final process stages (for reaction conditions see page 310)

The second strategy uses a thio-alcohol cyclopropylic nucleophile for a S_N2 displacement (scheme C and Scheme 7.18, route **b**, Esteve2005a,b, **C11**). The resulting intermediate is converted to Montelukast by conversion of the ester into the tertiary alcohol, mesylation, reaction with cyanide and then hydrolysis (**C12** to **C15**).

7.5.4 Montelukast synthesis using a Heck reaction as a key step

Almost all the previous methods use a bimolecular nucleophilic substitution near the end of the process as the key step to assemble the Montelukast molecule. Yet another synthetic approach was developed, focusing on a Heck coupling as a final key synthetic step (Scheme 7.19 and D). This convergent synthetic process can be elaborated by coupling 7-chloro-2-vinyl-quinoline (**10**) with an appropriate haloaromatic building block that already has the complete thiol-acid side chain assembled.

Montelukast (KRKA2006, Inke2007a,b) or a closely related compound (KRKA2006, SC2013) are the products of these reactions (**D19**, **D24** and **D29**).

The Heck coupling can also be useful in producing some key intermediates: **A20**, **A23** and **D7-D8** for the keto-ester, **A21** for keto-hydroxy, **D10** for hydroxy-ester and **D14** for the diol. These intermediates are converted to Montelukast using the processes previously discussed (scheme B and C).

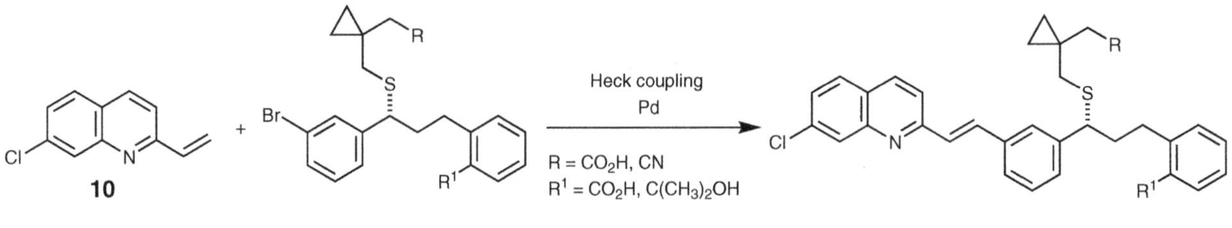

Scheme 7.19 Synthetic strategy using a Heck-type coupling approach as a final key synthetic step

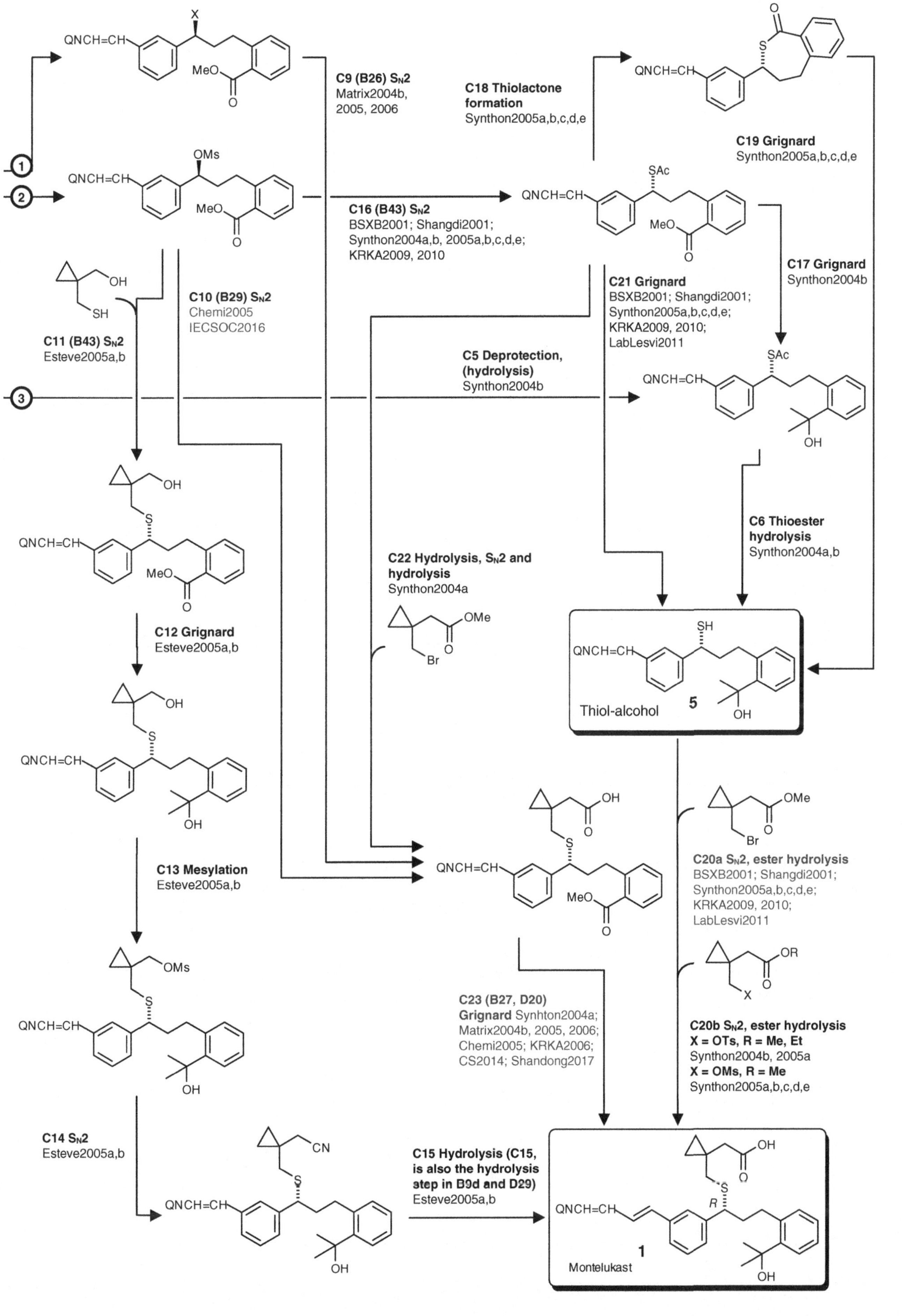

QNCH=CH
MeO
X
OMs
SAc
SH
OH
Br
OMe
OR
CN
C9 (B26) SN2
Matrix2004b, 2005, 2006
C18 Thiolactone formation
Synthon2005a,b,c,d,e
C19 Grignard
Synthon2005a,b,c,d,e
C16 (B43) SN2
BSXB2001; Shangdi2001; Synthon2004a,b, 2005a,b,c,d,e; KRKA2009, 2010
C17 Grignard
Synthon2004b
C10 (B29) SN2
Chemi2005
IECSOC2016
C21 Grignard
BSXB2001; Shangdi2001; Synthon2005a,b,c,d,e; KRKA2009, 2010; LabLesvi2011
C11 (B43) SN2
Esteve2005a,b
C5 Deprotection, (hydrolysis)
Synthon2004b
C6 Thioester hydrolysis
Synthon2004a,b
C22 Hydrolysis, SN2 and hydrolysis
Synthon2004a
Thiol-alcohol
5
C12 Grignard
Esteve2005a,b
C20a SN2, ester hydrolysis
BSXB2001; Shangdi2001; Synthon2005a,b,c,d,e; KRKA2009, 2010; LabLesvi2011
C13 Mesylation
Esteve2005a,b
C23 (B27, D20) Grignard Synhton2004a; Matrix2004b, 2005, 2006; Chemi2005; KRKA2006; CS2014; Shandong2017
C20b SN2, ester hydrolysis
X = OTs, R = Me, Et
Synthon2004b, 2005a
X = OMs, R = Me
Synthon2005a,b,c,d,e
C14 SN2
Esteve2005a,b
C15 Hydrolysis (C15, is also the hydrolysis step in B9d and D29)
Esteve2005a,b
R
1
Montelukast

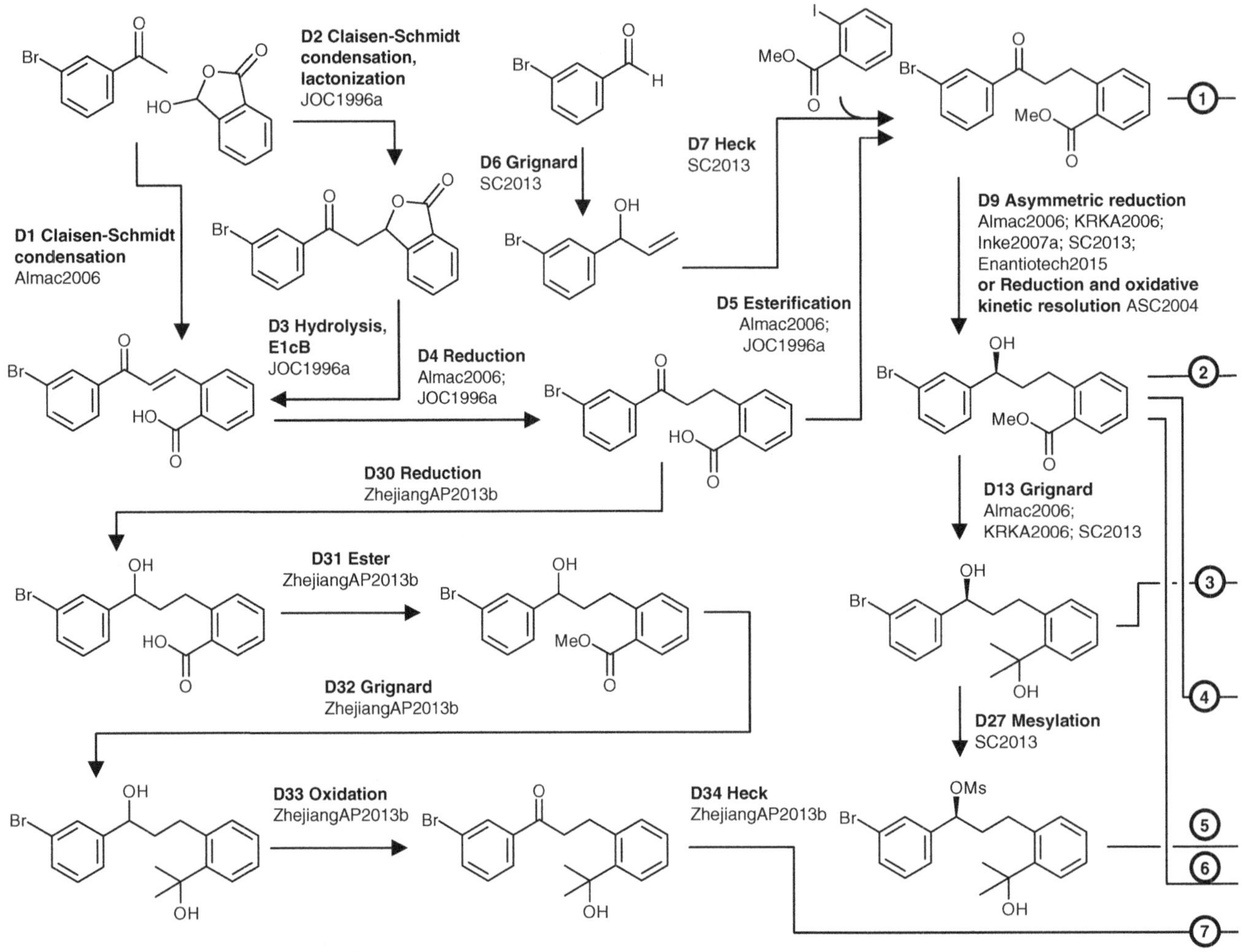

Scheme D Montelukast synthesis by a Heck "strategy" (for reaction conditions see page 310)

7.5.5 Montelukast synthesis using a carbonyl olefination reaction as a key step

Another patented synthetic method uses a carbonyl olefination reaction to produce Montelukast directly from fully functionalized intermediates (Scheme 7.20 and E, Farmaprojects2006). The methyl group on the quinoline is activated and hence, a very strong base is not required to accomplish the condensation with the aldehyde.

A similar process was claimed in the first patent on Montelukast (MerckFrosst1990), wherein an appropriate phopsphorus ylide was described in a Wittig synthesis of a Montelukast analogue, but no example was given where Montelukast itself was synthesised.

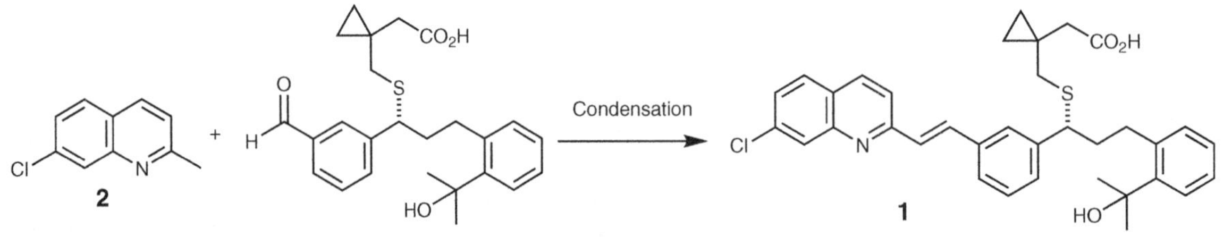

Scheme 7.20 Montelukast synthesis using a carbonyl olefination reaction as a key synthetic step

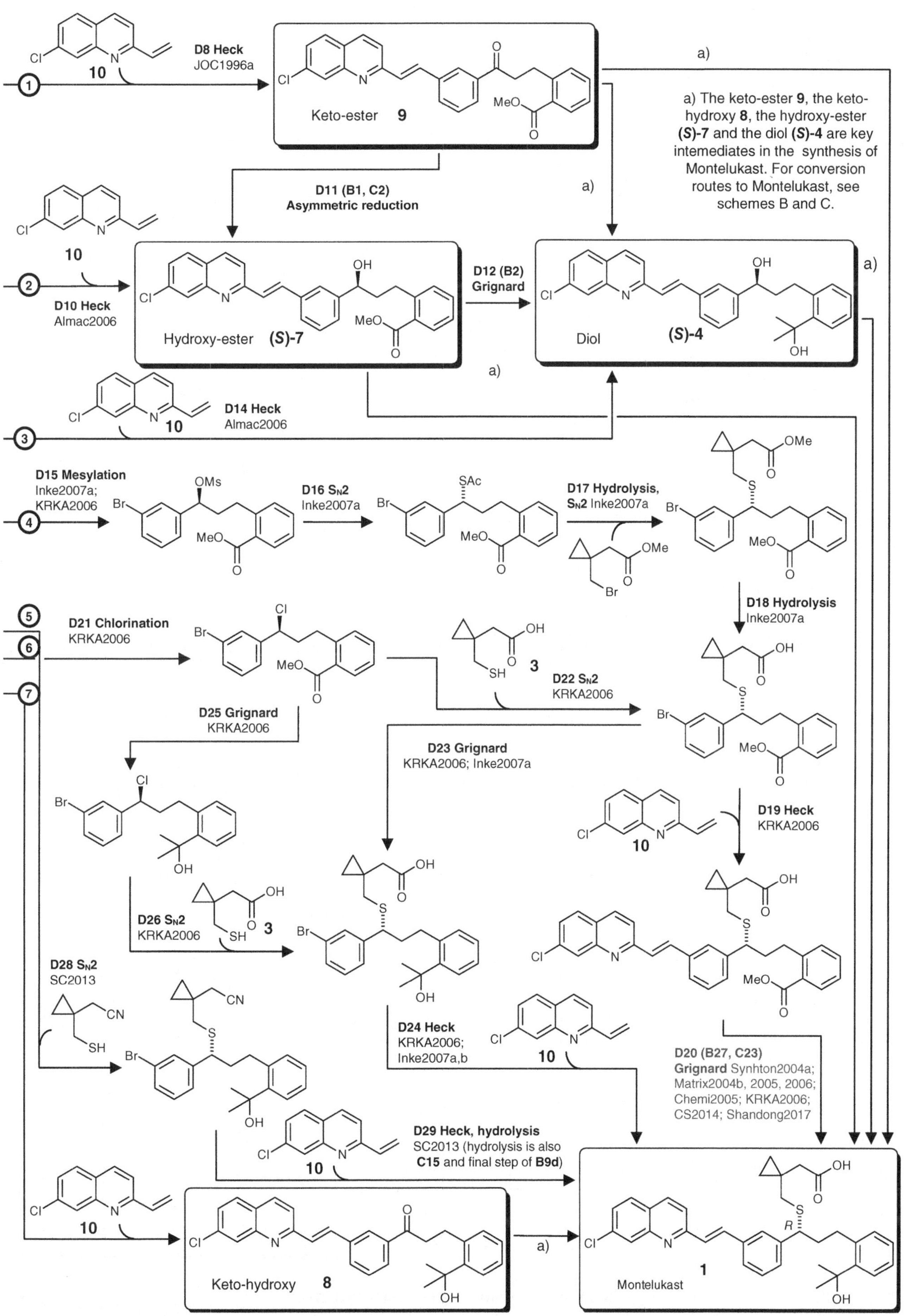
D8 Heck
JOC1996a
Keto-ester 9
a) The keto-ester 9, the keto-hydroxy 8, the hydroxy-ester (S)-7 and the diol (S)-4 are key intemediates in the synthesis of Montelukast. For conversion routes to Montelukast, see schemes B and C.
D11 (B1, C2)
Asymmetric reduction
D10 Heck
Almac2006
Hydroxy-ester (S)-7
D12 (B2)
Grignard
Diol (S)-4
D14 Heck
Almac2006
D15 Mesylation
Inke2007a;
KRKA2006
D16 SN2
Inke2007a
D17 Hydrolysis, SN2 Inke2007a
D18 Hydrolysis
Inke2007a
D21 Chlorination
KRKA2006
D22 SN2
KRKA2006
D25 Grignard
KRKA2006
D23 Grignard
KRKA2006; Inke2007a
D19 Heck
KRKA2006
D26 SN2
KRKA2006
D28 SN2
SC2013
D24 Heck
KRKA2006;
Inke2007a,b
D20 (B27, C23)
Grignard Synhton2004a; Matrix2004b, 2005, 2006; Chemi2005; KRKA2006; CS2014; Shandong2017
D29 Heck, hydrolysis
SC2013 (hydrolysis is also C15 and final step of B9d)
Keto-hydroxy 8
Montelukast 1

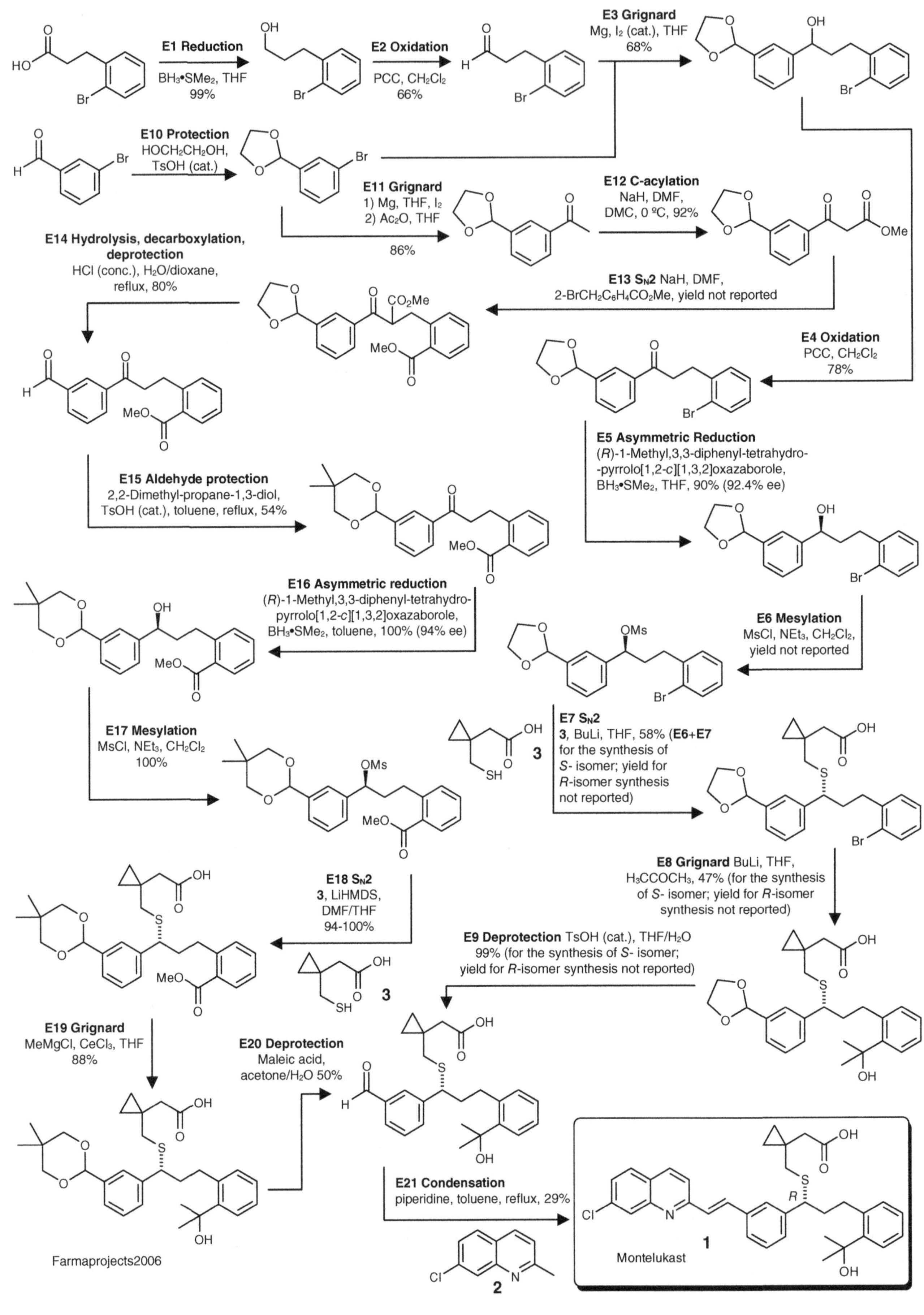

SCHEME E CARBONYL OLEFINATON STRATEGY

7.5.6 Montelukast synthesis using a Michael addition as a key step

Some later process patents apply an asymmetric Michael reaction as the key step to introduce the thiol side chain (Scheme 7.21 and A, **A42**, **A45** and **A47**, ShanghaiXMT2011, CS2014). An additional carbonyl group must be present in the intermediate to allow the Michael addition to take place. The extra carbonyl function was removed using a Wolff-Kischner reduction. Understandably, the rigorous reaction conditions of the Wolff-Kischner reaction could give rise to high levels of impurities at this late stage in the process.

7.6 SYNTHESIS OF THE THIOL SIDE CHAIN

The thiol side chain itself has been the focus of various synthetic processes (scheme F). The formation of the cyclopropane ring by a double S_N2 reaction (**F1**, α-alkylation) by the activated methylene of a malonate ester followed by reduction (**F2**) produces the key diol-cyclopropane intermediate (**6**).

In the first reported synthesis of this side chain (Scheme 7.22, MerckFrosst1990) one of the hydroxyl groups was protected as an ester (**F3**) while the other was replaced by cyanide following the sequence, mesylation and S_N2 displacement with cyanide ion (**F5/F6**). Hydrolysis of the nitrile group, with concomitant hydrolysis of the ester, gives rise to the carboxylic acid (**F7**). Esterification of the carboxylic acid function (**F8**) and replacement of the remaining hydroxyl by thiol by a sequence similar to that already described (mesylation and S_N2 displacement with thiolacetate anion) gives the desired thiol side chain (**F9/F10**).

This *S*-acyl carboxylic ester was used directly in the first Montelukast synthesis (MerckFrosst1990), the free thiol being obtained *in situ* by hydrazinolysis. Merck Frosst also described alternative processes but no examples were given.

Merck Frosst scientists probably did not consider that processes for the production of the thiol side chain were critical and hence, certain processes were not protected in the patent (MerckFrosst1990). Consequently, several other patents describe methods to obtain the thiol side chain (scheme F) but most of them also do not claim protection, the focus being on other critical key steps near the end of sequence.

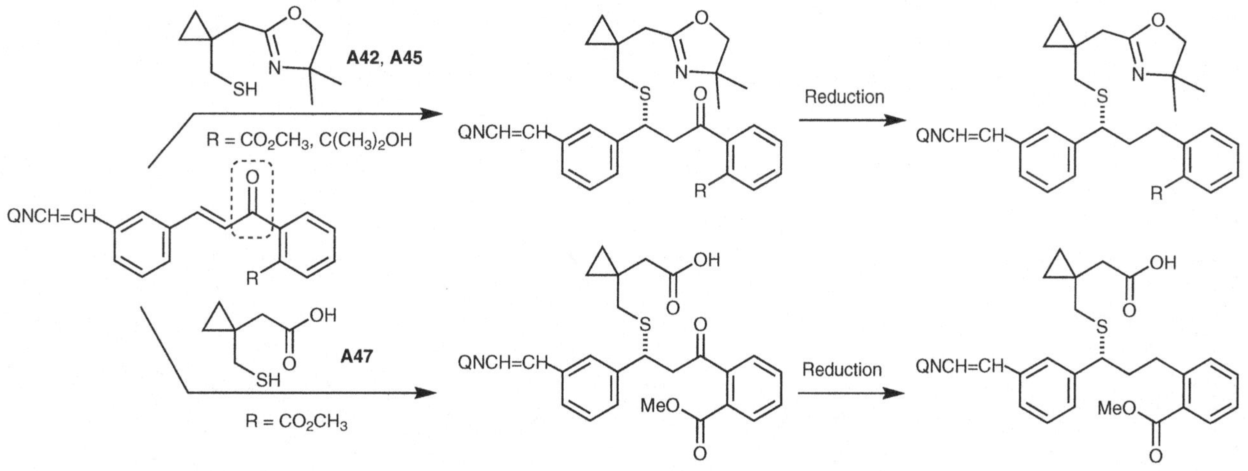

SCHEME 7.21 MONTELUKAST SYNTHESIS USING AN ASYMMETRIC MICHAEL ADDITION REACTION AS KEY SYNTHETIC STEP

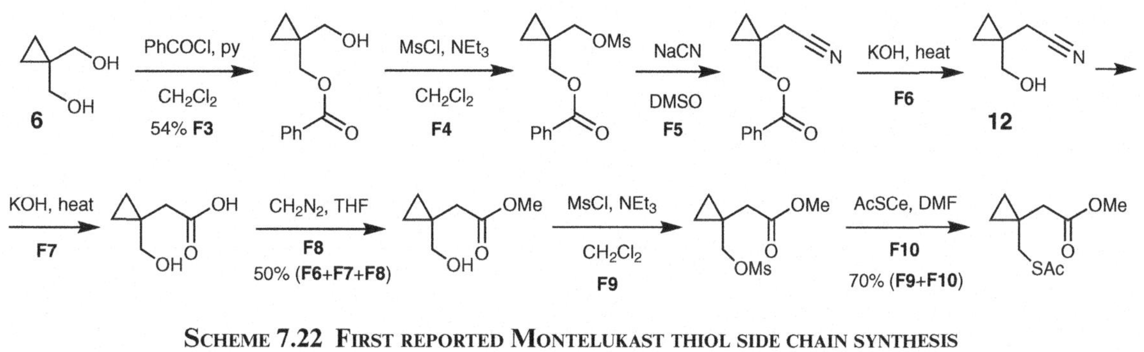

SCHEME 7.22 FIRST REPORTED MONTELUKAST THIOL SIDE CHAIN SYNTHESIS

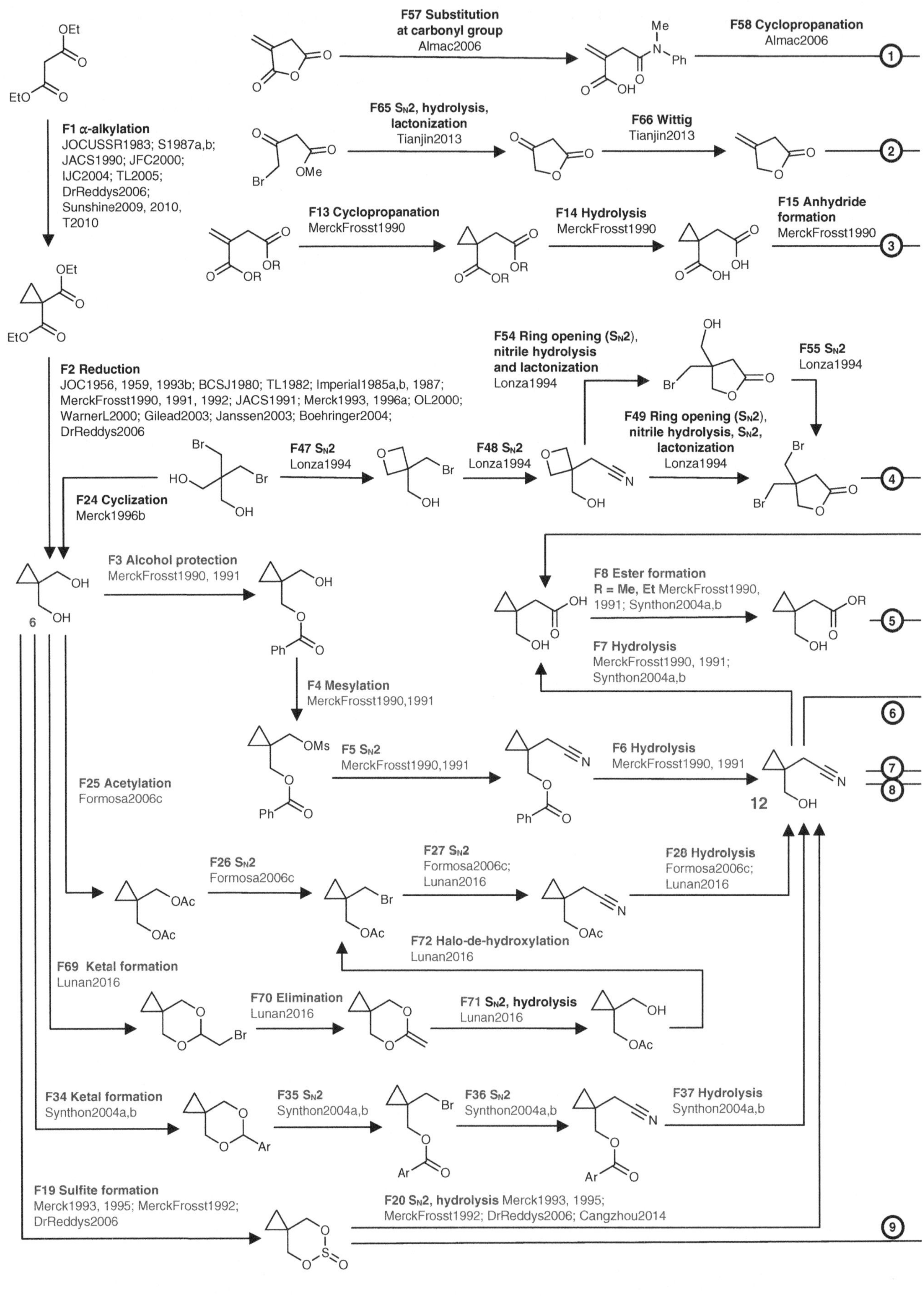

Scheme F Thiol side chain synthesis (for reaction conditions see page 311)

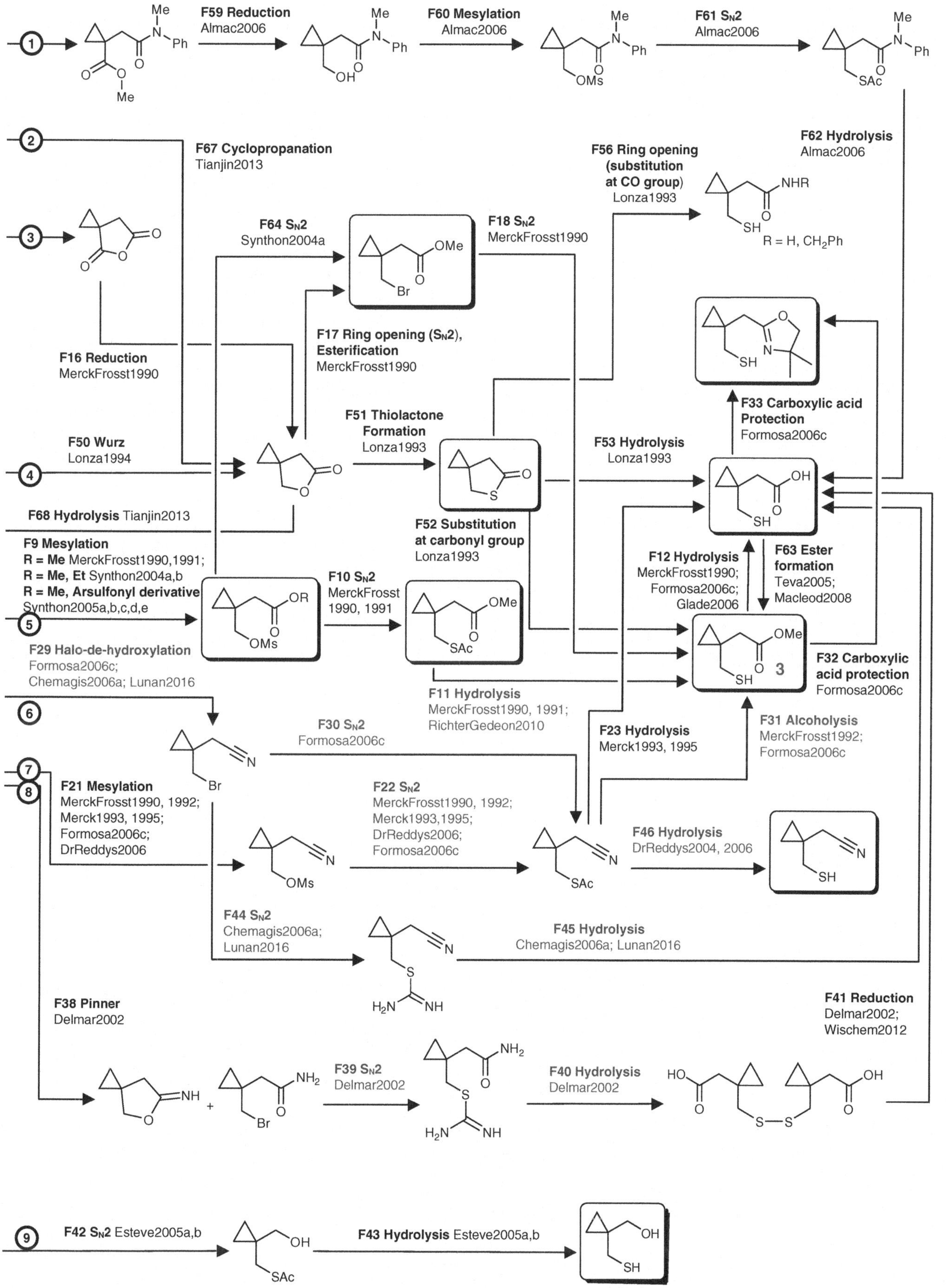

Highlighted structures are used in S_N2 reactions during Montelukast's synthetic processes (schemes A, B, C, D, E).

Other thiol nucleophiles have been used to produce Montelukast, most of them being conveniently produced from the hydroxy-nitrile intermediate **12** (Scheme 7.23 and F). Apart from the thiol-ester, a thiol-acid, a thiol-nitrile, a thiolactone and thiol-oxazolines (highlighted in scheme F) have been used as reactants in the S_N2 displacement. Their use in the synthesis of Montelukast are regarded as fundamental and hence, process patents have been granted for their use. Use of these substrates in any manufacturing process therefore, help avoid infringement of other existing patents.

Other processes to produce the hydroxy-nitrile (**12**) from the diol have been reported:

a) a double acetylation of the di(hydroxymethyl)cyclopropane (**6**) (**F25**) followed by an S_N2 reaction with bromide, substituting only one of the acyl groups (**F26**) differentiates the two hydroxyl groups. Bromide, being the most efficient leaving group, was replaced by cyanide (**F27**) followed by hydrolysis of the ester (**F28**);
b) a cyclic acetal was produced from the diol and then opened with bromide in a S_N2 reaction (**F34** and **F35**). As in **a)**, a S_N2 substitution with cyanide (**F36**) and hydrolysis of the ester (**F37**) produced the hydroxy-nitrile intermediate (**12**);
c) the diol was converted to a 2-methylene cyclic ketal (**F69** + **F70**) that was subjected to a ring opening under acidic conditions (**F71**) giving a mono-acyl intermediate. Replacement of the OH by bromine (**F72**) gives the bromo alcohol of route **a)**.
d) a cyclic sulphite (**F19**) was opened directly with cyanide (**F20**). Thereafter, hydrolysis produces the desired compound (**12**).

All of these processes were focused on differentiating the two hydroxyl groups, one to be replaced by cyanide and the other by a thiol.

So far, all the processes discussed convert the hydroxy-nitrile into the thiol-acid or the thiol-ester side chain by first hydrolysing the nitrile and introducing the thiol group afterwards (Scheme 7.24, strategy **a**). Inverting the order of the transformations gave additional opportunities to develop new processes (scheme F, right-bottom and Scheme 7.24, strategy **b**). The hydrolysis of the nitrile group is postponed until the thiol moiety is linked to the main Montelukast fragment, thus the nitrile-thiol is used directly in the S_N2 reaction (for example **B7d** and **B9d**).

The methods referred to previously are based on the diol-cyclopropane synthon, wherein the ring was conveniently produced by a double S_N2 α-alkylation reaction (**F1**). Other methods were used to build up the cyclopropane ring. Wurz-type cyclization of a 1,3-dihalocompound (**F50**) or a carbene addition using diazomethane to a double bond (**F13**, **F58** and **F67**) are alternatives. The former is more suitable for industrial scale because diazomethane use at industrial scale is always to be avoided unless specialized equipment is available. The Lonza process (Lonza1994, Scheme 7.25) appears attractive, producing first a γ-butyrolactone (**F47** to **F49**

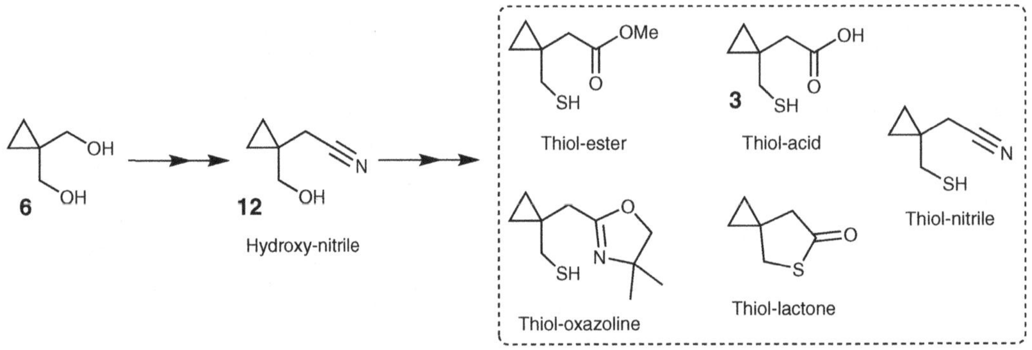

Scheme 7.23 Thiol nucleophiles used in Montelukast synthesis

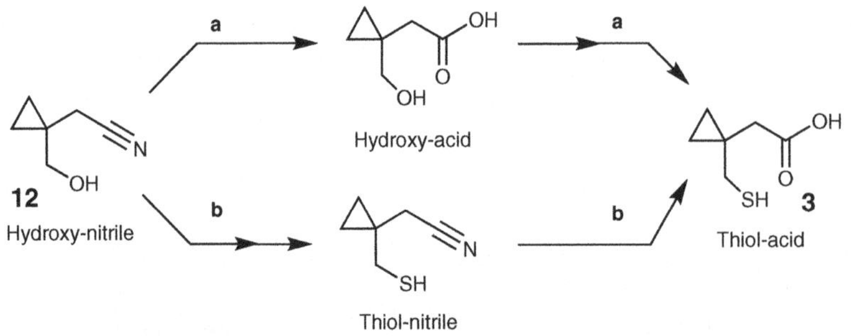

Scheme 7.24 Generic sequences to convert hydroxy-nitrile into thiol-acid side chains

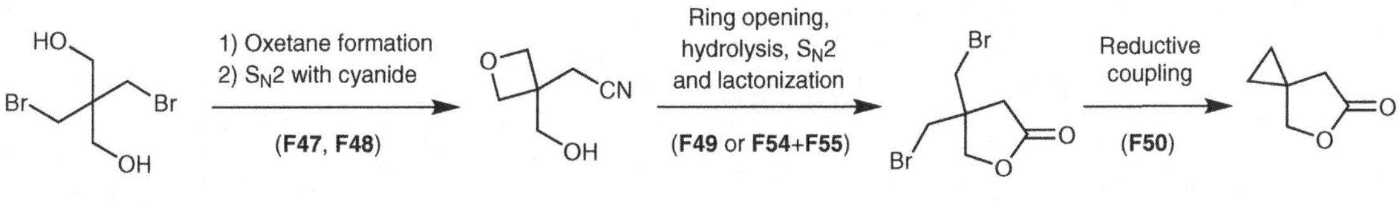

SCHEME 7.25 OUTLINE OF CYCLOPROPANE RING FORMATION IN LONZA'S SIDE CHAIN PROCESS

and **F54**, **F55**) and promoting the ring closure using a zinc-mediated reductive process (**F50**). However, the industrial use of zinc metal is not without problems as highly active and pyrophoric zinc dust, formed in the reaction, needs to be handled and disposed of.

7.6.1 Electrophilic side chain reactants

All synthetic methods previously referred to for the thiol side chain are focused on the production of nucleophilic thiols to be used in bimolecular substitutions, mainly with mesylated or halogenated electrophilic substrates (the exceptions are the use of the thiol-oxazoline or thiol-acid as nucleophiles in asymmetric Michael additions, reactions **A42**, **A45**, **A47**). From schemes B, C, D and E, it is obvious that the use of these nucleophilic thiols is preeminent in the various processes to produce Montelukast.

In an alternative strategy, some manufacturers opted to introduce a thiol group first into the main Montelukast framework using thioacetate as nucleophile, producing a nucleophilic thiol intermediate (scheme C). To convert these nucleophilic intermediates into the final API, an electrophilic reactant containing the cyclopropyl ring and a carboxylic acid derivative is required. In order to avoid intra or intermolecular substitution reactions, the nucleophilic nature of carboxylic acid must inhibited, an ester suffices for this purpose. Halogenated, mesyl and tosyl derivatives of this kind were applied in the production of Montelukast (scheme C and Figure 7.7).

Mesyl and tosyl derivatives can be obtained from the hydroxy-nitrile intermediate by a hydrolysis, esterification and mesylation or tosylation sequence (**F7**, **F8**, **F9**). Further substitution of mesylate by bromine yields the bromo electrophilic derivative (**F64**).

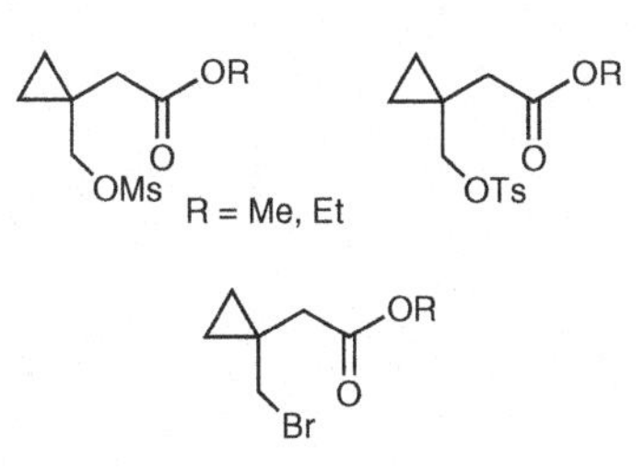

FIGURE 7.7 ELECTROPHILIC SIDE CHAINS FOR MONTELUKAST SYNTHESIS

7.7 STRATEGY COMPARISON AND CONCLUSIONS

After an examination of the many different process patents applications filed over the years it can be concluded that the relatively complex structure of Montelukast was a true source of inspiration for industrial Process Chemists. As to be expected, some of the reported processes are projected to be difficult to scaled-up successfully to an industrial scale, while others may look more promising to develop.

Most of the routes introduce the chirality by an asymmetric reduction of a benzylic carbonyl group. An asymmetric synthesis is probably essential for cost optimization as a process by enantiomer separation at the end of the synthetic process is usually inefficient in that it generates large quantities of waste and 50% of the material, the unwanted enantiomer, would be certainly difficult to recycle. From an overall viewpoint, a convergent synthetic route, where two highly functionalized intermediates are combined near the end of the process is highly desirable. This concept was applied in three of the major options to produce Montelukast (schemes B, D, E and Scheme 7.26).

The first of these strategies attaches the thiol side chain at the end of the process, while the others opted to attach a quinolone building block at the end. From the data presented in the literature, the thiol side chain looks more difficult and expensive to produce. Therefore, this fragment should logically be added nearer the end of the synthesis and, if all other factors are constant, give rise to the cheapest process.

In the "carbonyl olefination" route, the synthesis of the aldehyde fragment (scheme E) is linear and protracted. In addition, the aldehyde function must be protected up until the final coupling reaction. This situation appears to be unfavourable for large scale industrial processing. A further drawback is that the reported yield for the final condensation step is low (29%).

Compared to the "carbonyl olefination" route, the use of the Heck reaction to complete the synthesis is more favoured, but suffers from the potential contamination of the final API by palladium and the necessity to carry out further processing to reduce the level to that permitted in the Pharmacopoeias.

Collectively, all of the above factors point to a route *via* the key diol intermediate **(*S*)-4**, where the thiol side chain is added towards the end of the synthesis, as the most favoured (scheme B). The large number of filed

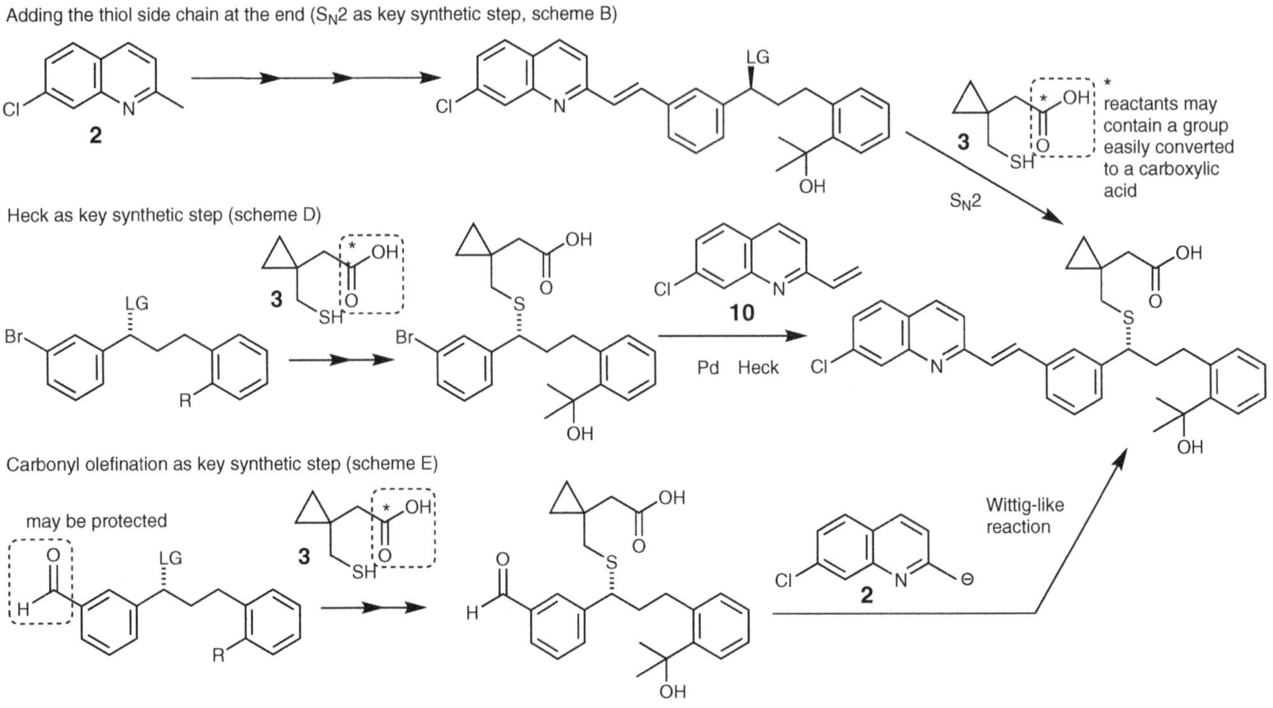

Scheme 7.26 Main convergent synthetic routes to Montelukast

process patents that use the diol **(*S*)-4** as intermediate supports this premise. Additionally, it is reported that the diol intermediate **(*S*)-4** can be easily recrystallized and therefore, purification of this intermediate serves as a "clean-up-point" in the process.

As referred to previously, the use of protecting groups are to be avoided. In line with this, a direct mesylation of the secondary, less hindered hydroxyl group of the key diol intermediate **(*S*)-4**, followed by a bimolecular substitution was tested and shown to be feasible (Scheme 7.27).

To achieve high yield and purity, conditions for the S_N2 reaction with the thiol side chain appear to be critical hence, solvent, base, temperature and other variables must be adequately fine-tuned for success. The presence of both a carboxylic acid and a thiol in (**3**) signifies that the nucleophile for the S_N2 reaction is a bi-anionic species. Protecting the carboxylic acid as an oxazoline or using

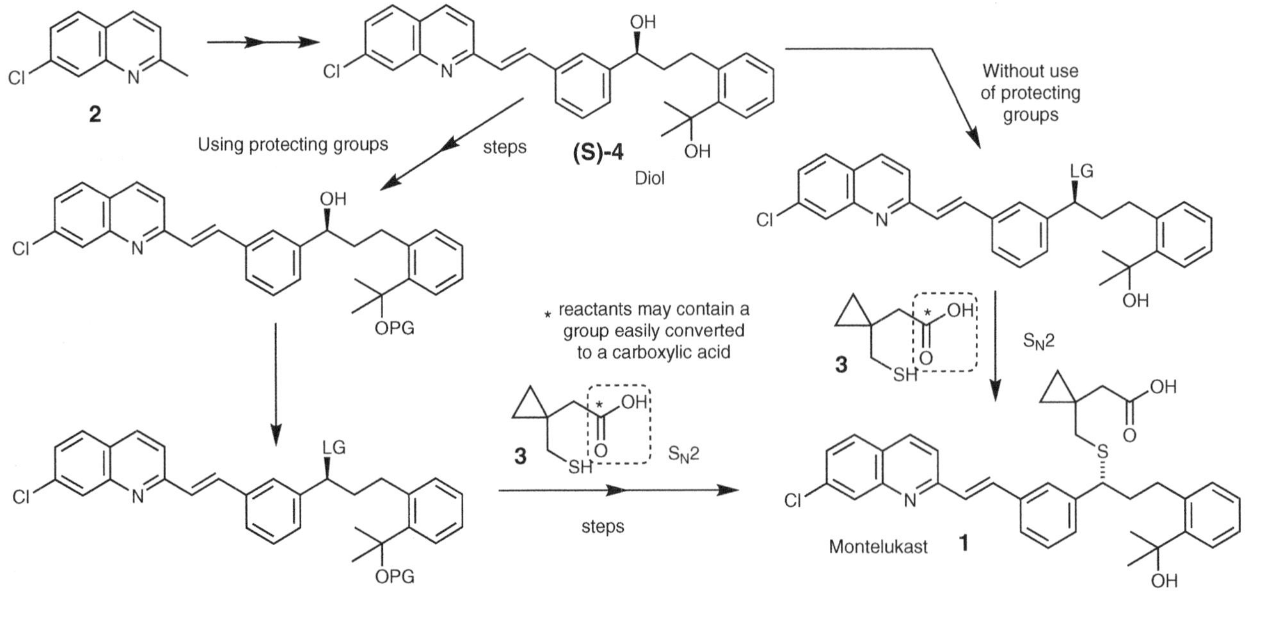

Scheme 7.27 Main synthetic routes from diol to Montelukast

non-acidic analogues (ex: ester or nitrile) avoids the formation of the bi-anionic nucleophile and results in the use of less base. However, additional processing is required to reach the end product. In some cases, these additional steps may even require harsher reaction conditions (for example hydrolysis of the nitrile) and have the potential to produce different impurities and/or increase the levels of previously known degradation compounds. Needless to say, this type of situation is to be avoided, particularly near the end of a synthetic process, especially where the final purity of the product is critical.

Most of the processes conveniently use a mesylate derivative as electrophile for the S_N2 reaction but, although less frequent, other leaving groups have been tested. Replacing the original benzylic hydroxyl group by a chlorine atom can be challenging, particularly due to the high enantioselectivity required (**B22**, **B23**, **B33**, **B39**). In some instances, the cleavage of the C−O bond can give rise to undesirable racemization. However, the patent literature claims that the chloro derivatives are much more stable compared to the mesylate, which is a potential advantage in an industrial setting.

One report indicates that the diol **(*S*)-4**, when reacted with mesyl chloride at low temperature in the presence of a hindered base (**B8a**) smoothly forms the mesylate, which can be crystalized from the reaction mixture in moderate to very good yields (64-96%). As before, a crystallization allows for an opportunity to purify and clean-up an advanced intermediate. It is also reported that the mesylate can be stored under wet conditions at low temperatures for some days without visible degradation. These factors, taken together, indicate that direct mesylation and a S_N2 reaction looks a to be a more favourable sequence.

The introduction of "incomplete" side-chains (Scheme 7.28 and C), produces intermediates that require further modification to reach Montelukast. These alternatives add more synthetic steps to a linear route and, at first sight, appear somewhat less favourable.

In conclusion, it appears that a direct mesylation of the diol **(*S*)-4** followed by a bimolecular substitution reaction would be one of the best ways to produce Montelukast from the diol. This route introduces the thiol near the end of the synthetic process, minimizes the number of steps and is, perhaps, the simplest process (Scheme 7.29).

Obviously, this does not mean that other routes cannot be used and implemented for industrial synthesis.

Also of relevance are the synthetic routes to obtain the key diol intermediate **(*S*)-4** and the thiol side chain.

Several routes can be used to produce the diol **(*S*)-4**, but one particular route that seems more favourable starts with the condensation of 2-methyl-7-chloroquinoline (**2**) with isophthalaldehyde (**A2**), followed by addition of vinyl Grignard reagent and then a Heck coupling to produce the keto-ester intermediate (**9**) (**A19** and **A20**, Scheme 7.30, route **a**). Asymmetric reduction of the benzylic ketone and converting the ester into the tertiary alcohol by addition of methyl Grignard gives the diol **(*S*)-4** (reactions **B1** and **B2**). Alternatively, (**8**) can be produced by a Heck coupling, which then gives the diol **(*S*)-4** by an asymmetric reduction of the carbonyl group.

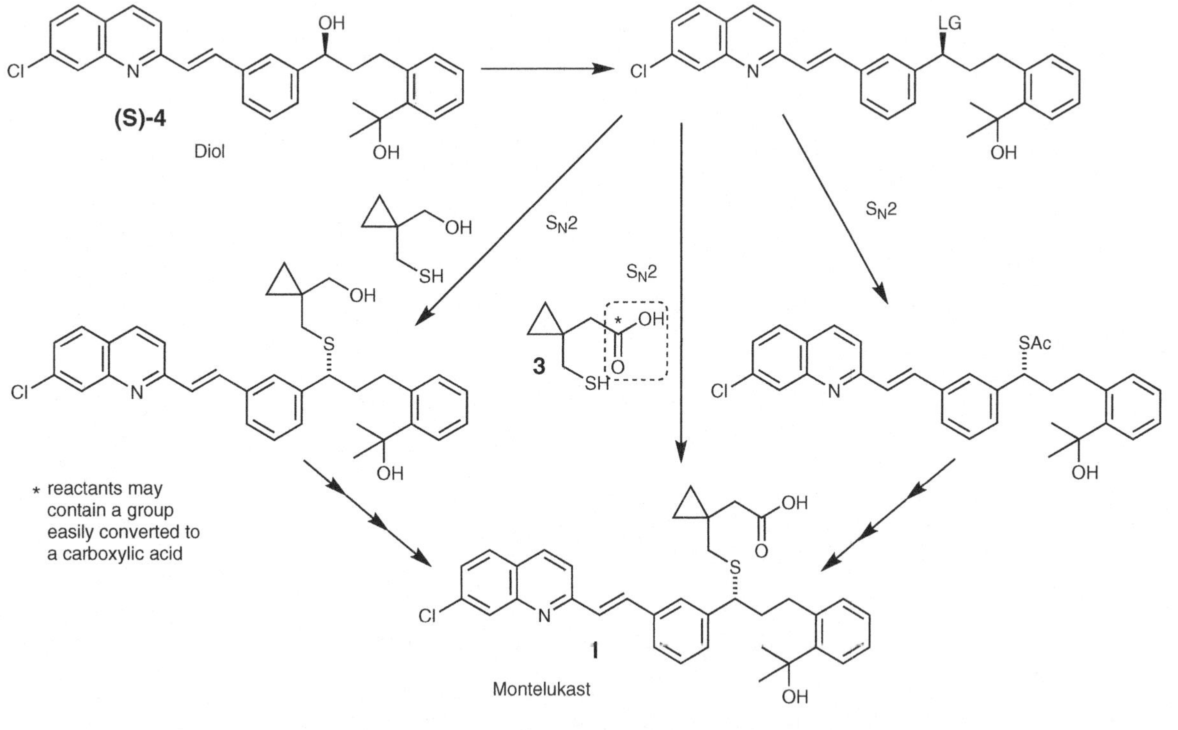

Scheme 7.28 Main synthetic routes from diol to Montelukast

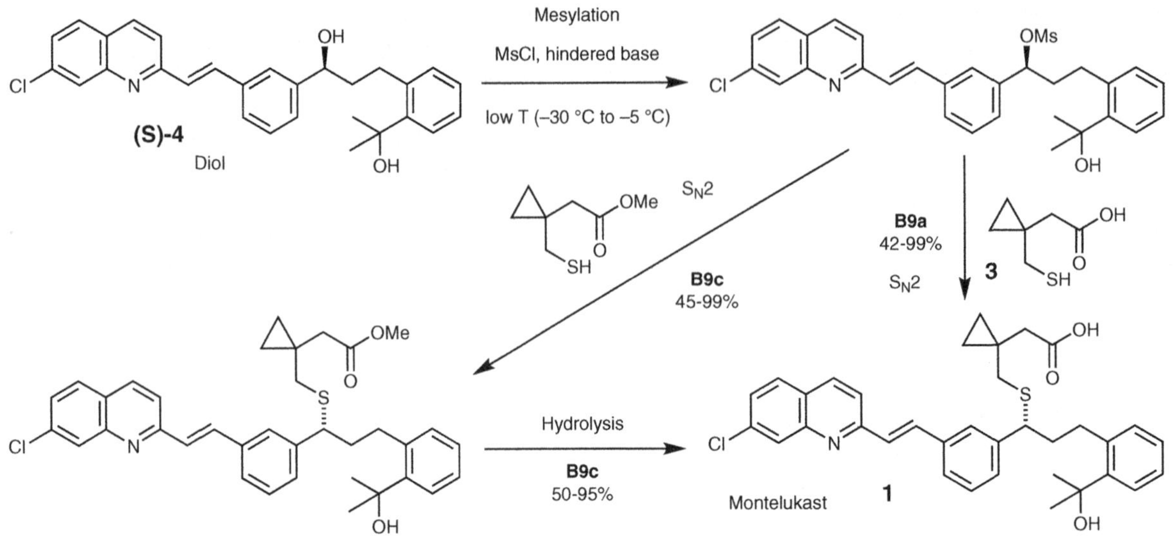

Scheme 7.29 Best routes to convert diol intermediate into Montelukast

Another option that appears appropriate for development employs an α-alkylation step to attach the *o*-disubstituted aromatic ring **A15**, Scheme 7.30, route **b**) thus avoiding undesirable addition/reduction sequences (**A3/A4** or **A12/A13**).

The nucleophilic thiol building block required for these routes can be produced from a malonic acid diester or from 2,2-bis-bromomethyl-propane-1,3-diol. Two worthy propositions are shown in Scheme 7.31. For route **a**, the reduction of the diester (**F1**) required the use of lithium aluminium hydride, which may have disadvantages in an industrial setting. When sodium borohydride was used instead of lithium aluminium hydride, lower yields were observed.

The low cost of the starting material for route **b**, may favour the process that promotes cyclopropyl ring forma-

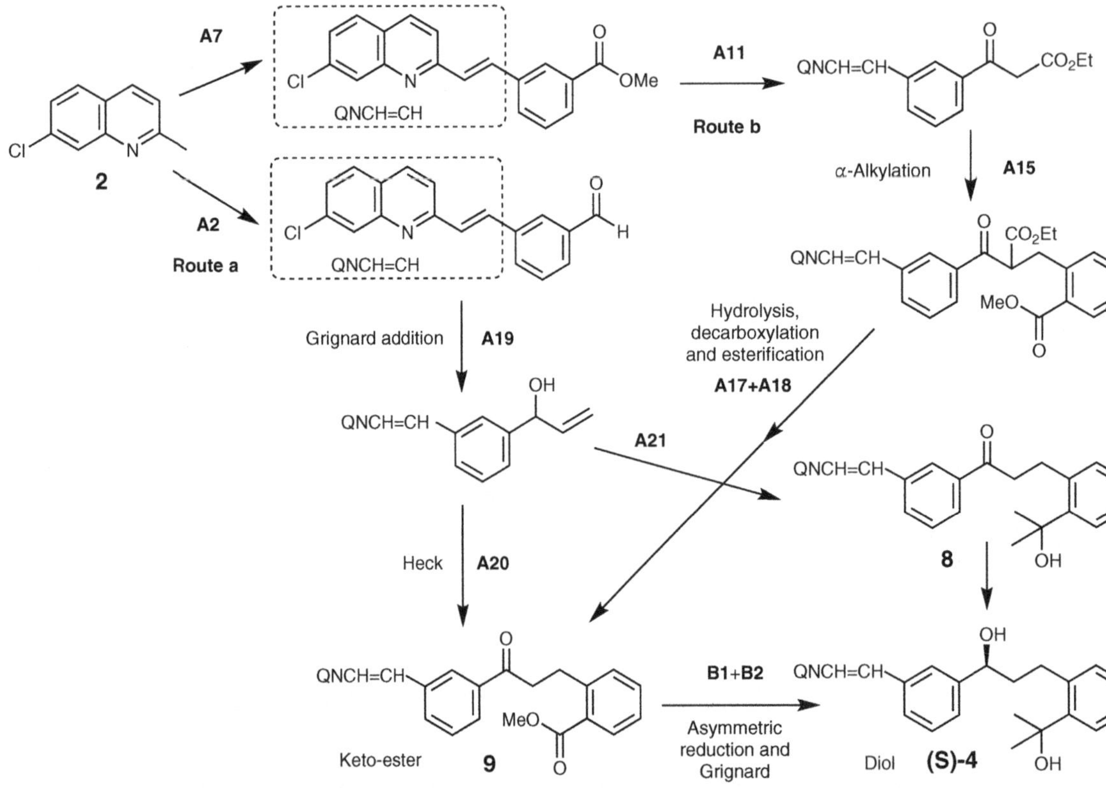

Scheme 7.30 Some of the most preferred routes to the diol intermediate

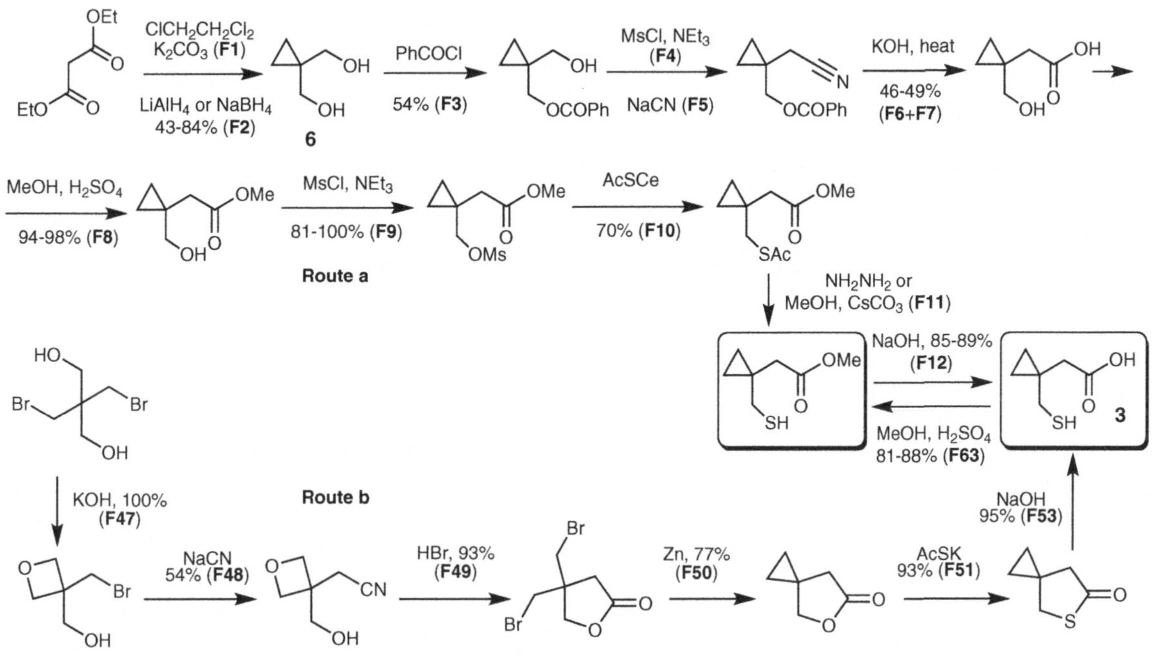

SCHEME 7.31 BEST ROUTES TO THE THIOL SIDE CHAIN SYNTHESIS

tion *via* a Wurz-type coupling process (Lonza1993), but again, the use of metallic zinc does require special handling.

Both of these methods can be used to produce (1-mercaptomethyl-cyclopropyl)-acetic acid (**3**) or its methyl ester.

In conclusion, the potentially most favoured routes for the production of Montelukast are possibly those shown in Scheme 7.29 and Scheme 7.30, the latter incorporating a synthesis of the diol intermediate **(*S*)-4** and the former concentrating on the conversion of the diol into the final API. The thiol side chain used can be advantageously produced by the routes shown in Scheme 7.31.

7.8 LESSONS FROM THE MONTELUKAST CASE

Asymmetric synthesis and enantiomeric purity of API

Montelukast is an enantiomerically pure drug. Synthetizing the racemic mixture and obtaining Montelukast by a classical resolution step or some other separation of enantiomers at the end of the synthesis, is, in principle unfavourable and costly. Because it is difficult to see how the unwanted isomer could be easily recycled, half of the synthetic effort would be wasted. Therefore, a viable industrial process for Montelukast requires an asymmetric synthesis.

Asymmetric reduction of the benzylic carbonyl group is the logical choice to introduce the chirality. Four different approaches have been developed and the selection of the most appropriate catalyst depends on a careful analysis of the cost and output of this step. Mesylation of the *S*-intermediate and bimolecular substitution is the simplest way to produce the *R*-isomer.

Some routes pursue the formation of the alkyl halide from the hydroxy derivative, before carrying out the S_N2 substitution reaction. This option can be somewhat more complicated as both the formation of the C-O bond and the formation of the C-Cl bond of the asymmetric carbon must be close to 100% specific, whereas isomerization during the chlorination is possible. That is, going through the alkyl halide introduces an additional step with the potential to perturb the asymmetric centre. Direct mesylation followed by an S_N2 substitution reaction is simpler and more robust and follows the rule: try to keep the process ASAP (as simple as possible).

Stability of intermediates

In any industrial synthesis, it is desirable that all process intermediates be relatively stable to allow for storage, at least for short to intermediate periods of time. This situation is not always possible and sometimes an unstable compound must be produced and processed as soon as possible up until a point where a stable compound is obtained. The diol mesylate is reported to be relatively unstable but still can be stored for short periods at low temperature without noticeable decomposition. In contrast, the halide analogues are reported to be stable at higher temperatures, which may help mitigate those arguments outlined above against their use. However, despite the apparent drawback due to its inherent reactivity, good results were achieved with the *O*-mesyl derivative. With appropriate production planning, it is possible to overcome this inconvenient.

Purity and quality

Regulatory authorities worldwide require all modern synthetic drugs to be produced to very high levels of purity. Recrystallization of advanced intermediates are important "clean-up-points" and it is desirable to have this type of processing step present in all efficient, robust and forgiving processes. In the synthesis of Montelukast the diol **(*S*)-4** and its mesylate can be crystallized and hence, favours routes that have these intermediates as part of the process. Several other processes have been developed wherein crystalline ammonium salts of various intermediates, and even of Montelukast itself, are produced. Many of the processes reported focused on a purification of the final product by crystallization of an ammonium salt.

Linear route *vs* convergent route

Whenever possible, an industrially applicable synthetic route (this is also generally true for a laboratory route) should be convergent, connecting two highly functionalised intermediates near the end of the process, thus avoiding a lengthy linear sequence. A convergent route reduces the number of sequential steps from the raw materials to the final product and results in a higher global yield. This concept disfavours the routes shown in scheme C.

Use of protecting groups

Whenever possible, the use of protecting groups is always to be avoided. Their use introduces at least two more steps in the synthesis, the protection and de-protection reactions, along with the additional solvent and reagents inherent to carrying out more synthetic transformations. In Merck Frosst's first synthesis, the tertiary hydroxyl group of the diol was protected before mesylation of the secondary OH and the S_N2 reaction with the thiol side chain (Scheme 7.12) resulting in four additional steps. A clear improvement in the industrial synthesis (Scheme 7.16) was introduced by elimination of this protection/deprotection sequence.

Heavy metal use

Heck coupling, using a palladium catalyst, is reported in several syntheses to establish one or both of the bridges between the three aromatic rings. The palladium allowed in the final product is strictly controlled at very low levels. This is obviously a concern for those synthetic routes that utilize this type of coupling reaction at the end of the process because opportunities to purge residual palladium are reduced. Conversely, the use of a Heck coupling earlier in the synthesis and distant from the final product results in reduced concern. Processes that use a heavy metal towards the end of the synthesis require appropriate methods to remove the metal efficiently from reaction streams. Accurate analytical quantification methods have to be implemented to determine trace amounts of the metal in the final product (usually to low ppm levels). Exceptions may be allowed for metals of low toxicity, for example iron and zinc.

Low yield steps

The formation of the ethylene bridge by reaction of an aromatic aldehyde and 2-methyl-7-chloroquinoline (**2**) (**A2**, **A7**, **A25** and **E21**) is reported to have low to moderate yields. Using this reaction as a key step to produce Montelukast (scheme E), especially in the last synthetic step, is certainly not favoured. If a low yield step has to be used in a synthetic process it should be, if possible, placed at the beginning of the synthesis rather than at a later stage. In addition, those reactions that involve expensive raw materials should aim at having as high yields as possible. Using a low yielding step in either of these situations means that a large fraction of an intermediate is not converted to the final API or that a significant part of the expensive raw material is wasted. Furthermore, solvent, reagent and equipment usage are not economically utilized, resulting in low efficiency and higher production costs.

The asymmetric reduction step and the cost of the API

Four different approaches were developed for the benzylic asymmetric reduction and a careful analysis of the costs and outputs of each is required to select the most appropriate catalyst. In most cases, the asymmetric transformation has long ranging consequences on production cost, as the chiral reactants and the metal (ruthenium) used in the process are usually the most expensive individual raw materials. Details of the experimental conditions, reactions yields and long term pricing for the steady supply of raw materials in large quantities all need to be considered when making an evaluation of each transformation. Recovery of the catalyst may play a very important role and a successful recovery protocol from the reaction mixture or mother-liquors can dramatically decrease the production cost.

The influence of the cost of a synthetic block in route selection

The "complete" thiol side chain (**3**) and 2-methyl-7-chloroquinoline (**2**) are key synthetic building blocks in the synthesis of Montelukast and are used in almost all routes previously reported. Options are to purchased one or both or opt for in-house production.

One possibility is to add the "complete" thiol side chain at the end of the synthesis (most of the routes in scheme B use this strategy), while a process using a carbonyl olefination reaction as final step uses 2-methyl-7-chloroquinoline (**2**) as a reactant in the last step (**E21**). Given that the thiol side chain is more complex and requires more manipulations to construct than the quinolone unit, it is highly probable that this building block is more expensive. Logically, it is advantageous to introduce the most costly reactant later and hence, routes shown in scheme B are to be preferred over the route of scheme E. A similar comparison can be made between routes shown in schemes B and D.

7.9 REACTION CONDITIONS FOR SCHEMES A TO F

7.9.1 Reaction conditions for scheme A

A1 H_2SO_4, heat or 2-BuOH/HCl, *p*-chloranil, heat, 46-69%.

A2 Isophthaldehyde, Ac_2O, *n*-BuOAc or toluene or *n*-heptane, or EtOAc, heat, piperidine (cat. optional), 45-75%. Isophthaldehyde, $La(Pfb)_3$ (Pfb = pentafluorobenzoate), toluene, air, 120 °C, 97%. Ac_2O, NEt_3, 100 °C, 95%. $TsNH_2$, toluene, reflux, NI.

A3 MeMgBr or MeMgCl, THF or toluene, 76-89%.

A4 MnO_2, EtOAc or toluene or CH_2Cl_2, 74-82%. 1) DMSO, $(ClCO)_2$, CH_2Cl_2; 2) NEt_3, 92%.

A5 2-$ICH_2C_6H_4CO_2Me$, DMPU, LDA, THF, 55%.

A6 (B15) 1) TsOH (cat.), $HOCH_2CH_2OH$, toluene, 110 °C; 2) MeMgCl, toluene, −5 °C, 96% (1+2); 3) HCl (10% aq.), THF, 28 °C, 62%.

A7 Methyl 3-formylbenzoate, toluene or xylene, pyridine or piperidine (cat.), reflux, 60-89%.

A8 NaOH, MeOH/H_2O, reflux, 95%.

A9 $SOCl_2$, reflux, YNR.

A10 1) $CH_2(CO_2Et)_2$, NEt_3, $MgCl_2$, CH_3CN; 2) Acyl chloride, 90% (A9+A10).

A11 R = Et: $CH_2(CO_2Et)_2$, NaH, EtOAc, THF, 96%. **R = *t*-Bu:** *t*-BuOAc, LiHMDS or NaHMDS, THF, –78 °C, 86-93%.

A12 Zn, $BrCH_2CO_2Et$, toluene/Et_2O, 46%.

A13 CrO_3/py, CH_2Cl_2, 40%.

A14 NaH, DMC, THF, MeOH (cat.) 89%. NaOMe, DMC, 1,4-dioxane, 80-87%.

A15 R = Me: 2-$ICH_2C_6H_4CO_2Me$, NaH, DMF, 97%. 2-$BrCH_2C_6H_4CO_2Me$, K_2CO_3, DMF, 85-89%. **R = *t*-Bu:** 2-$BrCH_2C_6H_4CO_2Me$, DMF, K_2CO_3, rt to 50 °C, 89%.

A16 $PhSO_3H$•H_2O, toluene, reflux, 88%.

A17 AcOH, HCl (conc.), 60-70 °C, 97%.

A18 MeI, K_2CO_3, acetone, 55 °C, 53-69% (A17+A18).

A19 CH_2=CHMgBr, toluene or ethylbenzene/THF, 76-100%.

A20 2-$IC_6H_4CO_2Me$ or 2-$BrC_6H_4CO_2Me$, $Pd(OAc)_2$, DMF or toluene or CH_3CN or xylene, LiOAc or NEt_3, 58-89%, with $Pd(OAc)_2$, toluene, NEt_3, *n*-Bu_4NF or Ph_4PBr or $MePPh_3Br$, 82-86%. 2-$BrC_6H_4CO_2Me$, graphene-supported palladium/Co catalyst (10% w Pd), DMF, DIPA, YNR.

A21 1) 2-$BrC_6H_4CMe_2OH$, $Pd(OAc)_2$, NaOAc, *n*-Bu_4NBr, toluene/NMP, 120 °C; 2) $NaHCO_3$ 100%. 1) 2-$BrC_6H_4CMe_2OH$, $(NH_3)_2PdBr_2$, NaOAc, Cy_2NMe, $(o\text{-tol})_3P$, DMF or DMF/toluene, 100 °C; 2) DBU, 120 °C, 82-100%. 1) 2-$BrC_6H_4CMe_2OTHP$, $(CH_3CN)_2PdCl_2$, Cy_2NMe, NaOAc, $(o\text{-tol})_3P$, DMF, 85-120 °C; 2) DBU, 120 °C; 3) TsOH•H_2O, YNR. 1) 2-$BrC_6H_4CMe_2OAc$, $Pd(OAc)_2$, DMA, $(o\text{-tol})_3P$, DIPEA, 100-130 °C, 70-76%; 2) KOH, THF, MeOH, 40 °C, 81%. 1) 2-$IC_6H_4CMe_2OAc$, $Pd(OAc)_2$, LiOAc, *n*-Bu_4NCl, LiCl, DMA, 100 °C, 73%; 2) KOH, THF, MeOH, 40 °C, 81%.

A22 X = I: Ac_2O, 50 °C, 80%. **X = Br:** Tf_2O, 140 °C, 77%. Ac_2O or $(CH_3CH_2CO)_2O$, toluene or xylene, AcOH, 120 °C, 86-90%.

A23 X = Br, I, R = Et: *n*-Bu_4NBr, $MgSO_4$, DMF or toluene, or DMSO, $Pd(OAc)_2$, pyrrolidine, 50 °C, 71-74%.

A24 X = I: Prop-2-en-1-ol, $NaHCO_3$, *n*-Bu_4NBr, $MgSO_4$, $Pd(OAc)_2$, 60 °C, 93%, with $Pd(PPh_3)_4$, DMSO, 140 °C, 90%, with $Pd_2(dba)_3$, dioxane, 80 °C, 87%. **X = Br:** Prop-2-en-1-ol, $NaHCO_3$, *n*-Bu_4NBr, $MgSO_4$, $Pd_2(dba)_3$, 80 °C, 88%.

A25 3-formylbenzonitrile, Ac_2O, NaOAc, 130 °C, 83-92%.

A26 Mg, THF, 70%.

A27 MeMgBr, THF, 83-88%.

A28 3,4-Dihydro-2*H*-pyran, PPh_3•HBr (cat.), CH_2Cl_2, reflux, 92%.

A29 Paraformaldehyde, $C_6H_5CH_2NMe_3OH$, , DMSO, 72-78%.

A30 Bromination: PBr_3, CH_2Cl_2, Et_2O, 0-5 °C to rt, 98-99%. HBr, toluene, 5 °C, 95%. **Chlorination:** $SOCl_2$, CH_2Cl_2, 5 °C to rt, 95%.

A31 1,3-propanedithiol, BF_3•Et_2O, CH_2Cl_2, 0 °C, 99%.

A32 1) *n*-BuLi, THF, –20 °C, 71%; 2) NBS, CH_3CN/H_2O, 10-20 °C, 94% or H_5IO_6, THF, 10 °C, 89% or DMP, CH_2Cl_2, 0-20 °C, 85%.

A33 Ac_2O, 140 °C, 93%.

A34 NBS, $(PhCO_2)_2$, CCl_4, reflux, h*ν*, 47%. NBS, AIBN, 78 °C, 85%. 1) *i*-Pr_2NH, *n*-BuLi; 2) Me_3SiCl, Br_2, 76%. PhCl, *t*-BuOCl, 88%.

A35 PPh_3, CH_3CN, 78-80%.

A36 HCHO, Et_2NH•HCl, NEt_3 (cat.), EtOH, 67%.

A37 HCHO, Et_2NH•HCl, NEt_3 (cat.), EtOH, 56 %.

A38 1) MeI, EtOH; 2) NEt_3, reflux, 81%.

A39 SeO_2, THF, 78%.

A40 CH_3PPh_3Br, *n*-BuLi, THF, 71%.

A41 EtOH, H_2O, NaOH, 25 °C, 83%. $TsNH_2$, toluene, reflux, 86% (A2+A41).

A42 [1-(4,4-Dimethyl-4,5-dihydro-oxazol-2-ylmethyl)--cyclopropyl]-methanethiol, catalyst (various), CH_2Cl_2 or THF or Et_2O or toluene, –78 °C to –40 °C, 44-89%, (44-93% ee). [1-(4,4-Dimethyl-4,5-dihydro-oxazol-2-ylmethyl)-cyclopropyl]-methanethiol, catalyst, THF, 88%, 93% ee.

A43 (B42) $(HOCH_2CH_2)_2O$, KOH, NH_2NH_2•H_2O (80%), reflux, 91%.

A44 MeMgBr, THF, 92%.

A45 [1-(4,4-Dimethyl-4,5-dihydro-oxazol-2-ylmethyl)--cyclopropyl]-methanethiol, catalyst, THF, 83-89%, 81-96% ee. (1-Mercaptomethyl-cyclopropyl)-acetic acid methyl ester or (1-mercaptomethyl-cyclopropyl)-acetic acid, catalyst, THF, 67-89%, 48-92% ee.

A46 1) $(HOCH_2CH_2)_2O$, KOH, $NH_2NH_2{\cdot}H_2O$ (80%), reflux, 89%; 2) EtOH, H_2O, NaOH, reflux, 90%.

A47 (1-Mercaptomethyl-cyclopropyl)-acetic acid, 1,2-dichloroethane, chiral iron(III)-salen catalyst (20%), –5 °C, 96%, 98% ee.

A48 1) $TsNHNH_2$, MeOH, rt; 2) $NaBH_4$, rt; 3) MeMgBr, THF, 0 °C to rt, 87% (step 3 is B27, C23, D20).

7.9.2 Reaction conditions for scheme B

B1 (C2, D23) (–)-DIP-Cl: (–)-Ipc_2BCl, THF or CH_2Cl_2, 65-97% (>99% ee). α-pinene, $NaBH_4$/BCl_3 (97% ee) or $BH_3{\cdot}Me_2S$ or BH_2Cl/Me_2S. **(*R*)-2-MeCBS:** (*R*)-2-MeCBS; BH_3, THF, 69% (99.7% ee). Graphene Pd-Co catalyst with chiral diphosphine, BBr_3, α-pinene, DMF or *i*-PrOH, 71-76%, >99.1% ee. **AHT:** 1) $[RuCl_2(mesitylene)]_2$ (1% mol), (1*S*,2*S*)-piperidyl-*N*-sulfamoyl)-1,2-diphenylethylenediamine, HCO_2H/NEt_3, DMF, 75-83%, 99.5% ee. $RuCl_2$[(*R*)-DM-BINAP][(R)-DAIPEN], H_2 (5-10 bar), $PhCH_2NEt_3Cl$, PhCl/*i*-PrOH, NaOH, 86-92%, 96.4% ee. *N*-[(1*S*,2*S*)-1,2-diphenyl-2-(3-phenylpropylamino)-ethyl]-4'-methylbenzenesulfonamide dichlororuthenium hydrochloride]dimer (cat.), NEt_3/HCO_2H, THF, 40 °C, YNR. RuCl[Ts-DPEN](*p*-cymene), HCO_2H/NEt_3, *i*-PrOH or H_2 in CH_2Cl_2, 40-80 °C. $RuCl_2$((*R*)-Xylyl-Garphos)((*R*)-DAIPEN), *n*-Bu_4NBr, H_2 (9 bar), toluene, *i*-PrOH, H_2O, NaOH, rt, YNR, 99% ee. Ir-catalyst, H_2 (15 bar), *t*-BuOK, EtOH, 50 °C >90%, >98% ee. Graphene Pd-Cu catalyst with chiral diphosphine, $BH_3{\cdot}THF$, DMF, 82%, >99.5% ee. **Bioconversion:** Micobacterium MB5614, 0.1 M phosphate buffer, pH 7.2, 60%. Immobilized enzyme from *Microbacterium campoquemadoensis* (MB5614), NaDPH, 30 °C, pH 8 YNR (>95% ee). *Mucor hiemalis* IFO 5834, 35%, 90% ee. KRED CDX-026, pH 8.0, triethanolamine·HCl buffer, NADP-Na, *i*-PrOH/toluene, 86-98%, 99% ee. Bioreductase, toluene, *i*-PrOH, TEBA, NADP, 40-50 °C, 95%. Ketoreductase, *i*-PrOH, H_2O, NADP, $MgCl_2$, triethanolamine hydrochloride buffer, 45 °C, 99%, 99.4% *S*.

B2 (D12) MeMgBr, toluene, rt, 62%. MeMgCl, $CeCl_3$, THF, 5 °C, 85-100%. MeMgCl, THF/CH_2Cl_2 or toluene or CH_2Cl_2, −5 °C to 0 °C, 65-75%. $LaCl_3$, LiCl, THF, MeMgCl, −5 °C, 68-99%. MeMgCl, $TiCl_4$, THF, 5 °C, 84%. **Protection, Grignard, Deprotection:** 1) THF, NMM, TMSCl, –5 °C, 97%; 2) THF, MeMgCl, –10 °C , 95%; 3) MeOH, MsOH, –20 °C to rt, 98%. 1) CH_2Cl_2, NMM, TBDMSCl, –5 °C, 94%; 2) toluene, MeMgCl, –5 °C, 94%; 3) acetone, H_2O, HCl (20%), 93%. 1) THF, NEt_3, TBDPSCl, –5 °C, 96%; 2) THF, MeMgCl, –5 °C, 95%; 3) py, HF (20%), H_2O, 93%. 1) CH_3CN, K_2CO_3, TIPSCl, –10 °C, 91%; 2) THF, MeMgCl, –5 °C, 97%; 3) CH_2Cl_2, HF (20%), H_2O, 94%.

B3 PG[1] = TBDMS: TBDMSCl, imidazole, DMAP, CH_2Cl_2 or toluene, 76%. **PG[1] = Ac:** Ac_2O, py or NEt_3, DMAP, CH_2Cl_2, 99%.

B4 3,4-dihydro-2*H*-pyran, PPh_3, CH_2Cl_2, reflux, YNR, $MePPh_3Br$ (optional), 100%.

B5 PG[1] = TBDMS: *n*-Bu_4NF, THF, 0 °C, 81% (B3+B4+B5). *n*-Bu_4NF (2.8 eq.), THF/CH_2Cl_2, 40 °C, 100%. **PG[1] = Ac:** NaOH, THF, MeOH, rt, 85% (B3+B4+B5). **PG[1] = TMS:** NH_4OH (aq.), MeOH, 50 °C, 94%.

B6 MsCl, NEt_3 or DIPEA, CH_2Cl_2, −40 °C/−10 °C, 93-95%.

B7a 1) (1-Acetylsulfanylmethyl-cyclopropyl)-acetic acid methyl ester, NH_2NH_2, CH_3CN; 2) Mesylate SM, Cs_2CO_3; 3) TsOH•py, MeOH/THF, YNR. 1) (1-Acetylmercaptomethyl-cyclopropyl)-acetic acid methyl ester, $CeCO_3$, MeOH, rt, solvent replaced by NMP; 2) Mesylate SM, 5 °C; 3) NaOH (aq.), MeOH, THF or 2-MeTHF, 75 °C, 87%. **TMS as PG[2]:** 1) (1-Acetylsulfanylmethyl-cyclopropyl)-acetic acid methyl ester, Cs_2CO_3, NMP; 2) Mesylate SM, 0 °C to rt, 94% (1+2); 3) 2-MeTHF, NaOH, H_2O, MeOH, 70 °C, 87%.

B7b 1) (1-Mercaptomethyl-cyclopropyl)-acetic acid, NaOMe/MeOH or *n*-BuLi/THF, −5 °C to rt; 2) Mesylate SM; 3) PPTS, THF/MeOH/DMF, quantitative.

B7c 1) (1-Mercaptomethyl-cyclopropyl)-acetic acid methyl ester, K_2CO_3/CH_3CN or NaOMe/MeOH, −5 °C to rt; 2) Mesylate SM, DMF, 30-40 °C; 3) NaOH, 88%.

B7d 1) (1-Mercaptomethyl-cyclopropyl)-acetonitrile, NaOMe, DMF/MeOH, rt; 2) Mesylate SM, DMF 98% (1+2); 3) PPTS (cat.), THF/MeOH, 79%; 4) EtOH, NaOH, reflux, 94%.

B8a MsCl, DIPEA or NMM or NEt_3, toluene or toluene /CH_3CN or CH_3CN or CH_2Cl_2 or toluene/CH_2Cl_2 or THF or MTBE/CH_3CN, −40 °C - 0 °C, 64-96%.

B8b $(Ph_2O)_2POCl$, NEt_3, DMAP (cat.), CH_2Cl_2/toluene or CH_2Cl_2/DMF or CH_2Cl_2, 68-98%.

B8c 1) TsCl, NEt_3, DMF or DMSO, YNR.

B9a S_N2 on a mesylate: 1) (1-Mercaptomethyl-cyclopropyl)-acetic acid, base, solvent (base: *n*-BuLi, *t*-BuOK, *t*-BuOLi, *t*-BuONa, *t*-AmylOK, NaOH, NaH, NaOMe; solvent: THF, NMP, DMF, DMSO, MeOH, NMP/H_2O, MIBK), −15 °C to 20 °C; 2) Mesylate SM 43-99% (note: PEG-600 or PEG-1500 or 18-crown-4 or 18-crown-5 or 18-crown-6 may be added). **S_N2 on a phosphate derivative:** 1) (1-Mercaptomethyl-cyclopropyl)-acetic acid, DMF, NaH, 5 °C; 2) Phosphate SM, DMF, rt, 87%. 1) (1-Mercaptomethyl-cyclopropyl)-acetic acid, DMSO, 1-ethyl-3-methylimidazolium bromide or 1-butyl-1-methylpyrrolidium bromide, *t*-BuOK, 10 °C; 2) Phosphate SM, 15 °C, 70-93%.

(1-Mercaptomethyl-cyclopropyl)-acetic acid, THF, aminoguanidine, 40 °C, 88%. **S_N2 on a tosylate:** 1-Mercaptomethyl-cyclopropyl)-acetic acid, NaOH or NaOMe (microreactor use), YNR.

B9b on a phosphate derivative: 1) (1-Acetylsulfanylmethyl-cyclopropyl)-acetic acid methyl ester, base (*t*-BuOK, NaH), solvent (DMSO, DMF), 5 °C to rt; 2) Phosphate SM, rt; 3) NaOH (aq.), THF/MeOH, rt, 71-90%. **On a mesylate derivative:** 1) (1-Acetylsulfanylmethyl-cyclopropyl)-acetic acid methyl ester, *t*-BuOK, 1-ethyl-3-methylimidazolium bromide, DMSO, rt; 2) Mesylate SM, rt; 3) NaOH (aq.), THF/MeOH, rt, 83%.

B9c on a mesylate: 1) (1-Mercaptomethyl-cyclopropyl)-acetic acid methyl ester, base (NaOH, NaH, LDA, NaOMe, K_2CO_3, Bu_4NOH, Cs_2CO_3, $LiN(SiMe_3)_2$, $NaN(SiMe_3)_2$, $KN(SiMe_3)_2$, Me_4NOH, *t*-BuOK), solvent (DMF, DMA, NMP, THF, DMSO/MeOH, CH_3CN); 2) Mesylate SM, THF or THF/DMF or CH_3CN/toluene or CH_3CN, −10 °C to 20 °C; 3) NaOH or LiOH (aq.), THF, MeOH, THF/MeOH, toluene /H_2O, *i*-PrOH, rt to 60 °C, 36-81% (1+2+3), 46-99% (1+2), 50-95% (3). **On a phosphate compound:** (1-Mercaptomethyl-cyclopropyl)-acetic acid methyl ester, NaOH, MeOH/THF, 30 °C, 19%, in DMF, 25 °C, 63%.

B9d 1) (1-Mercaptomethyl-cyclopropyl)-acetonitrile, base (*n*-BuLi, NaOMe), solvent (DMF, DMSO/MeOH), −15 °C to 0 °C; 2) Mesylate SM, −15 °C to 0 °C; 3) NaOH or KOH (aq.), diethyleneglycol/H_2O, toluene/NaOH/H_2O, reflux, 85-95% (1+2), 27-35% (3).

B9e 1) 5-Thia-spiro[2.4]heptan-6-one, KOH or NaOH or (C1-C4)alkoxide; 2) Mesylate SM, YNR.

B10 (C1) (*S*)-2-MeCBS: (*S*)-2-MeCBS, BH_3, THF, −20 °C, YNR. **(+)-DIP-Cl:** (+)-Ipc_2BCl, CH_2Cl_2, 0 °C to 5 °C, 71-92%. (−)-α-pinene, BH_3•Me_2S, hexane/THF, DIPEA, −20 °C to 5 °C, 87%. **AHT:** RuCl[(*R*,*R*)-TsDPEN](mesitylene), $NaCO_2H$, $MeOct_3NCl$, H_2O/CH_2Cl_2, 40 °C, 70-97% or HCO_2H/NEt_3, *i*-PrOH, 80 °C, 68%.

B11 (C3) Mitsunobu: DEAD, PPh_3, (*R*)-(−)-α-methoxyphenylacetic acid, THF, 0 °C, 98%. **Hydrolysis:** NaOH, MeOH, THF, rt, 74%.

B12 1) TsOH, CH_2Cl_2, reflux, 94%; 2) NaOH, MeOH/H_2O, 30 °C, 95%.

B13 Mesylation: MsCl, DIPEA, CH_2Cl_2, −25 °C or 5 °C, 89%. **S_N2:** 1) (1-Mercaptomethyl-cyclopropyl)-acetic acid, base/solvent (base: NaH, NaOMe, solvent: DMF, MeOH), 0 °C; 2) Mesylate SM, DMF, 0 °C, 70%.

B14 H_2SO_4 (80%), 10 °C, 76%.

B15 (A6) 1) TsOH (cat.), $HOCH_2CH_2OH$, toluene, 110 °C; 2) MeMgCl, toluene, −5 °C, 96% (1+2); 3) HCl (aq. 10%), THF, 28 °C, 62% (3).

B16 $RuCl_2$[(*R*)−DM−BINAP][(*R*,*R*)−DPEN], H_2 (8 bar), *t*-BuOK, *i*-PrOH, 40 °C, 78%. $RuCl_2$[(*R*)−DAIPEN][(*R*)−DM-SEGPHOS], H_2 (3 bar), *t*-BuOK, THF/EtOH/toluene, 40 °C, 94%, >99.5% ee. $RuCl_2$[(*R*)-DAIPEN][(*R*)-Xylyl-BINAP], *i*-PrOH, *t*-BuOK, H_2 (8 bar), 40 °C, 78-86%, 99.6-99.8% ee.

B17 Me_3SiCl, imidazole, THF, rt, 65%

B18 (*R*)-(+)-3,3-Diphenyl-1-*o*-tolyl-tetrahydropyrrolo[1,2-c][1,3,2]oxazaborole, catecholborane, CH_2Cl_2, −40 °C, 88%. (−)-Ipc_2BCl, 90%, 99.5% ee.

B19 Me_3SiCl, imidazole, THF, rt, 100%.

B20 (see C4) KOAc, toluene/DMF, YNR.

B21 MeMgCl, toluene, 0 °C, YNR. 1) MeMgCl, $CeCl_3$ (anh.), THF; 2) toluene, SM, –5 °C, 77%.

B22 1) MsCl, DIPEA, DMF; 2) LiCl 92%. 1) MsCl, DIPEA, toluene; 2) *n*-Bu_4NCl, 92%.

B23 (C7) X = Cl: $SOCl_2$, CH_2Cl_2/DMF, 25 °C, 70%. **X = Br:** 1) MsCl, NEt_3, CH_2Cl_2; 2) LiBr, CH_3CN, 37%. **X = I:** 1) NaI, Me_3SiCl, CH_3CN, 45 °C, 35%.

B24 MgMeCl, $CeCl_3$, THF, 5 °C, 65%.

B25a 1) (1-Mercaptomethyl-cyclopropyl)-acetic acid, base (NaH, Cs_2CO_3), DMF, 0 °C to 35 °C; 2) Chloro derivative SM, 0 °C, 41-71%.

B25b 1) (1-Mercaptomethyl-cyclopropyl)-acetic acid methyl ester, base, DMF, 10 °C; 2) NH_2NH_2, DMF, H_2O, rt, 73%.

B26 (C9) 1) (1-Mercaptomethyl-cyclopropyl)-acetic acid, base (NaH, Cs_2CO_3), DMF, 0 °C; 2) Chloro derivative SM, 0-35 °C, 31-81%.

B27 (C23, D20) Me ester as SM: MgLi•LiBr, THF, −40 °C, 40%. MeMgI, toluene, 10 °C, 72%. MgMeCl, $CeCl_3$, THF or toluene, 0-5 °C, 52-98%. **Other esters as SM: ethyl ester,** MeMgBr, toluene, 85%, ***i*-Pr ester,** MeMgCl, $ZnCl_2$, THF, 92%, **Ph ester,** MeMgI, $LaCl_3$, toluene, 92%.

B28 (C8) Mesylation: MsCl, NEt_3 or DIPEA or py, CH_2Cl_2 or toluene or 2-BuOH or THF, −40 °C to 20 °C, quantitative. **Phosphate derivative (diPh, Me or Et carboxylic ester):** Diphenyl chlorophosphate, NEt_3 or DIPEA or KOH, CH_2Cl_2 or toluene or 2-butanone, 97-99%; **(di *i*-Pr, Me or *i*-Pr carboxylic ester):** diisopropyl chlorophosphate, NEt_3, DMAP, THF, 94%.

B29 (C10) S_N2 in mesylate: 1) (1-Mercaptomethyl-cyclopropyl)-acetic acid, LiHMDS, DMF or THF, 5 °C; 2) Mesylate SM, DMF, 0 °C, 75-80%. (1-Mercaptomethyl-cyclopropyl)-acetic acid, toluene, 70 °C, 88%. **S_N2 in diphenylphosphate derivative:** (1-Mercaptomethyl-cyclopropyl)-acetic acid with NaOMe, MeOH, DMSO 98%, with *t*-BuOK, MeOH, DMF 92-95%, with DBU, EtOAc, 90%, with NaH, CH_3CN 90%, **with similar conditions in other phosphate derivatives (Et, *i*-Pr)** 90-95%.

B30 1) [1-(4,4-Dimethyl-4,5-dihydro-oxazol-2-ylmethyl)-cyclopropyl]-methanethiol, NaH, DMF/toluene, 0 °C; 2) Mesylate SM, 0-20 °C, 80-91%.

B31 1) MeMgCl, $CeCl_3$, THF, rt, 80-95%; 2) NaOH, ethylene glycol, 120-150 °C, 60-80%.

B32 MeMgBr, THF, 0 °C, YNR. MeMgCl, LiHMDS, THF/toluene, 0 °C, 78%.

B33 $POCl_3$, rt, 96%.

B34 1) (1-Mercaptomethyl-cyclopropyl)-acetic acid, Cs_2CO_3, DMF; 2) Chloro SM, DMF, 40 °C, 34%.

B35 MeMgCl, $CeCl_3$, THF, 0 °C-15 °C, 70-81%. MeMgBr, THF, LiCl, 90%.

B36 Weinreb amide, Grignard: 1) CH_3NHOCH_3, THF, 0 °C; 2) MeMgBr, THF, 0 °C, YNR. **Grignard:** MeMgCl, LiHMDS, THF or THF/toluene, 0 °C, 74-80%.
B37 Mesylation: MsCl, NEt_3, CH_2Cl_2 or toluene −40 °C to −10 °C, YNR. **Tosylation:** TsCl, py, toluene, 60 °C, 77%. **Triflate derivative:** $(CF_3SO_2)_2O$, py, CH_2Cl_2, 30 °C, 94%. **Acetylation:** Ac_2O, AcOH (glacial), $HClO_4$, 30 °C, 88%. **Phosphate derivative:** CH_3CN, NaOH, Diphenyl chlorophosphate, 99%.
B38a 1) (1-Mercaptomethyl-cyclopropyl)-acetic acid, Cs_2CO_3, DMF; 2) *O*-acetyl, *O*-tosyl or *O*-triflate SM, DMF, rt, 40 °C, 30-35%. 1) (1-Mercaptomethyl-cyclopropyl)-acetic acid, NaH, DMF, 5 °C; 2) *O*-mesyl SM, toluene, 75%. 1) (1-Mercaptomethyl-cyclopropyl)-acetic acid, KOH, CH_3CN, 5 °C; 2) *O*-phosphate SM, toluene, 92%.
B38b 1) (1-Acetylsulfanylmethyl-cyclopropyl)-acetic acid methyl ester, NH_2NH_2, CH_3CN; 2) Mesylate SM, $CsCO_3$, 0 °C; 3) NaOH (aq.), MeOH/THF, rt, 72%.
B39 $SOCl_2$, CH_2Cl_2, 5 °C, 96%.
B40 MgMgCl, toluene/THF, −5 °C, 52%. *t*-BuOK, toluene, 4Å MS, 20 °C, 62%.
B41 MgMgCl, $LaCl_3$/LiCl, toluene /THF, 5 °C, 70%.
B42 (A43) see A43.
B43 see C11/C16.

7.9.3 Reaction conditions for scheme C

C1 (B10) see B10.
C2 (B1, D11) see B1.
C3 (B11) see B11.
C4 (B20) KOAc, toluene/DMF, YNR.
C5 TsOH, MeOH/THF, 63% (C4+C5).
C6 $NH_2NH_2•H_2O$, CH_3CN, YNR.
C7 (B23) see B23.
C8 (B28) see B28.
C9 (B26) see B26.
C10 (B29) see B29.
C11 (B43) 1-(Mercaptomethyl)cyclopropyl)methanol, Cs_2CO_3, DMF, 86%.
C12 MeMgBr, THF, 81%. MeMgCl, $CeCl_3$, THF, 80%.
C13 MsCl, DIPEA, CH_2Cl_2, 100%.
C14 NaCN, DMF, 75%. NaCN, *n*-Bu_4NHSO_4, toluene/H_2O, >100%.
C15 NaOH, *n*-Bu_4NBr, toluene/H_2O, 83%. NaOH, EtOH/H_2O, 91%.
C16 (B43) AcSK, toluene/DMF or toluene/THF, 57-68% (C8+C16). AcSH, NEt_3, toluene, 104% (C8+C16).
C17 MeLi, THF, 92%.
C18 MeMgI, toluene, 96%. MeMgCl, toluene 64%.
C19 MeMgCl, $CeCl_3$, THF/toluene, 93%.
C20a 1) (1-Bromomethyl-cyclopropyl)-acetic acid methyl ester, NaOMe, THF/MeOH or toluene or DMF or DMF/toluene or 2-MeTHF/acetone/toluene/DMF; 2) NaOH, THF/MeOH, 40-52% (1+2).
C20b 1) (1-Methanesulfonyloxymethyl-cyclopropyl)-acetic acid ethyl ester, Cs_2CO_3/CH_3CN or NaOMe, toluene, 71%.
C21 MeMgCl, $CeCl_3$, toluene or THF/toluene 50-98%. MeMgI or MeMgBr, $LaCl_3$/LiCl, toluene, YNR.
C22 1) (1-Bromomethyl-cyclopropyl)-acetic acid methyl ester, LiOH, CH_3CN/MeOH; 2) NaOH, THF/MeOH/H_2O, 62%.
C23 (B27, D20) see B27.

7.9.4 Reaction conditions for scheme D

D1 2-carboxybenzaldehyde, NaOH, 106%.
D2 1) NaOH, EtOH; 2) HCl or H_2SO_4, 82%.
D3 NaOH, EtOH, YNR.
D4 Zn, NH_4OAc, EtOH/H_2O, NaOH, 70% (D3+D4). $(Ph_3P)_3RhCl$, H_2 (2.7 bar), NaOH, EtOH, 96% (D3+D4).
D5 MeOH, H_2SO_4, reflux, 89-92%.
D6 CH_2=CHMgBr, toluene, 95%.
D7 $Pd(OAc)_2$, PPh_3, NEt_3, 75%.
D8 7-Chloro-2-vinyl-quinoline, $Pd(OAc)_2$, P(*o*-tolyl)$_3$, NEt_3, DMF, 91%.
D9 Asymmetric reduction: RuCl[(*S*,*S*)-TsDPEN](mesitylene), $H_2CO_2H•NEt_3$, EtOAc, 95%. TarB-NO_2, THF, $NaBH_4$, 97% (TarB-NO_2 preparation: 3-nitrophenylboronic acid, D-tartaric acid, CaH_2, THF, reflux). Ru-complex, H_2CO_2H/NEt_3, CH_2Cl_2, YNR (Ru-complex preparation: $[RuCl_2(mesitylene)]_2$, (1*S*,2*S*)-*N*-piperidylsulfamoyl-1,2-diphenylethylenediamine, CH_2Cl_2). (−)-Ipc_2BCl, THF or CH_2Cl_2, 76-90%, >95.5% ee. (*R*,*R*)-DIOP, $RuCl_2$(*R*)-Me-BIMAH, *t*-BuOK, toluene, H_2 (35 bar), 35 °C, 90% (10 kg scale). **Reduction and kinetic resolution (oxidation):** 1) $NaBH_4$, CH_2Cl_2, MeOH, 91%; 2) O_2, Pd(nbd)Cl_2, (−)-sparteine, Cs_2CO_3, *t*-BuOH, MS 3 Å, toluene, 37%, 93% ee.
D10 7-Chloro-2-vinyl-quinoline, DMF, $Pd(OAc)_2$, NEt_3, P(*o*-tolyl)$_3$, 75%, >99% ee.
D11 (B1, C2) see B1 (–)-DIP-Cl, (*R*)-2-MeCBS, AHT or bioconversion.
D12 (B2) see B2.
D13 MeMgCl or MeMgBr, toluene or toluene/THF, 72-93%.
D14 7-Chloro-2-vinyl-quinoline, DMF, $Pd(OAc)_2$, NEt_3, P(*o*-tolyl)$_3$, not exemplified.
D15 MsCl, DIPEA, CH_2Cl_2, 100%.
D16 KSAc, toluene/DMF, 78%.
D17 Hydrolysis: LiOH, MeOH, CH_3CN, 89%. **Hydrolysis and S_N2:** (1-Bromomethyl-cyclopropyl)-acetic acid methyl ester, LiOH, MeOH, CH_3CN, 69%.
D18 NaOH (aq.), MeOH, THF, 61%.
D19 7-Chloro-2-vinyl-quinoline, DMF, $Pd(OAc)_2$, NEt_3, P(*o*-tolyl)$_3$, 62%.
D20 (B27, C23) see B27.
D21 $SOCl_2$, DMF (cat.), CH_2Cl_2, 92%.
D22 (1-Mercaptomethyl-cyclopropyl)-acetic acid, *t*-BuOK, DMF, 88-92%.

D23 MeLi, Et_2O, toluene, 90%. MeLi, LiBr, Et_2O, THF, 69%.
D24 7-Chloro-2-vinyl-quinoline, DMF, $Pd(OAc)_2$, NEt_3, $P(o\text{-tolyl})_3$, 68-85%.
D25 MeMgBr, toluene, 74%.
D26 (1-Mercaptomethyl-cyclopropyl)-acetic acid, *t*-BuONa, DMF, 68%.
D27 MsCl, DIPEA, CH_2Cl_2, NI.
D28 (1-Mercaptomethyl-cyclopropyl)-acetonitrile, *n*-BuLi, THF, −15 °C; 2) Mesylate starting material, CH_2Cl_2, −15 °C, 52%.
D29 1) $Pd(OAc)_2$, PPh_3, NEt_3, 110 °C, 58%; 2) NaOH/H_2O, reflux, 65%.
D30 $NaBH_4$, $CoCl_2•7H_2O$, MeOH, rt, 79%.
D31 DMF, K_2CO_3, MeI, rt, 82%.
D32 1) MeMgCl, ZnCl (anh.), THF, rt; 2) SM, 0 °C, 81%.
D33 MnO_2, CH_2Cl_2, rt, 84%.
D34 7-Chloro-2-vinyl-quinoline, $Pd(OAc)_2$, DMF, NEt_3, 100 °C, 89%.

7.9.5 Reaction conditions for scheme E (see scheme E)

7.9.6 Reaction conditions for scheme F

F1 $(CH_2)_2X_2$ (X = Cl, Br), K_2CO_3, $n\text{-}Bu_4NX$ or [bmim]Br, DMF or DMSO. Aliquat in toluene or benzene/H_2O, 70-94%.
F2 $LiAlH_4$, THF or Et_2O, 33-99%. H_2 (1-14 bar), Cu/ZnO, 150 °C, 26-39%. $NaBH_4$, *i*-PrOH, MeOH, 75 °C, 43%. $BH_3•THF$ 75-77%.
F3 PhCOCl, CH_2Cl_2, 54%.
F4 MsCl, NEt_3, CH_2Cl_2, YNR.
F5 NaCN, DMSO, rt, YNR.
F6+F7 KOH (8 N, aq.), EtOH, reflux, 46-49%.
F8 R = Me: CH_2N_2, THF, 50% (F6+F7+F8). MeOH, H_2SO_4 (cat.), rt, 94-98%. **R = Et:** EtOH, H_2SO_4 (conc.), reflux, 37-41%.
F9 Mesylation: MsCl, NEt_3, CH_2Cl_2, 81-100%. **Arenesulfonylation:** $p\text{-}MeOC_6H_4SO_2Cl$, py, 102%.
F10 AcSCe, DMF, 70%.
F11 NH_2NH_2, CH_3CN, NI. MeOH, Cs_2CO_3, YNR.
F12 NaOH (aq.), MeOH, 85-89%.
F13 $CH_2N_2/Pd(OAc)_2$. Me_3SI/*n*-BuLi, not exemplified.
F14 NaOH (aq.), not exemplified.
F15 Ac_2O, not exemplified.
F16 $LiAlH_4$ or $NaBH_4$, not exemplified.
F17 HBr, MeOH, not exemplified.
F18 KSH or AcSNa, not exemplified.
F19 $SOCl_2$, DIPEA, CH_2Cl_2 or toluene, 85-90%. $SOCl_2$, NEt_3, CH_2Cl_2, 69%. Diisopropyl sulfite, toluene, 89%.
F20 NaCN, NaI (cat.), DMF or DMF/*i*-PrOAc or DMF/toluene or THF, 70 °C, 70-88%. NaCN, NaI (cat.), DMF, 90-100 °C, 58-65%.
F21 MsCl, NEt_3, *i*-PrOAc/DMF or toluene /DMF or DMF or CH_2Cl_2, NI.
F22 AcSK, *i*-PrOAc or DMF, 70-81% (F21+F22). AcSH, NEt_3, toluene/DMF/EtOH or toluene/DMF, 49-93% (F21+F22).
F23 NaOH, toluene/H_2O, reflux, 82-83%.
F24 DMF/H_2O, CF_3SO_3Li, electrolysis (210 mA), 60%.
F25 Ac_2O, py, CH_2Cl_2, NI.
F25 + F26 HBr (33% in glacial AcOH), CH_2Cl_2, 97%.
F26 HBr (33% in glacial AcOH), 81% (F25+F26).
F27 NaCN, DMF, >100%; $Zn(CN)_2$, CuI, DMF, 50-60 °C, 93%.
F28 K_2CO_3, MeOH, 50 °C, 79-85%.
F29 PBr_3, CH_2Cl_2, 41-96%. PPh_3, Br_2, CH_3CN, 87%.
F30 AcSH, K_2CO_3, DMF, 94%.
F31 HCl (conc.) or H_2SO_4 (conc.), MeOH, reflux, 80-93%.
F32 2-Amino-2-methyl-1-propanol, TsOH (cat.), xylene, reflux, 80%.
F33 2-Amino-2-methyl-1-propanol, TsOH (cat.), xylene, reflux, 76%.
F34 $p\text{-}MeOC_6H_4CHO$, TsOH (cat.), cyclohexane, reflux, 9%.
F35 CuBr, $n\text{-}Bu_4NBr$, DDQ, $ClCH_2CH_2Cl$, YNR.
F36 KCN, EtOH/H_2O, 18-20%.
F37 KOH (aq.), EtOH, 91%.
F38 H_2SO_4, LiBr, *i*-PrOAc, YNR.
F39 H_2NCSNH_2, acetone, reflux, 83%.
F40 1) NaOH (50%), H_2O, reflux; 2) H_2O_2 (30%), H_2O; 3) HCO_2H, 81%.
F41 Zn, NH_4OH, H_2O, 88%.
F42 KSAc, DMSO, 78%.
F43 HCl (conc.), MeOH, rt, 100%.
F44 H_2NCSNH_2, acetone or CH_3CN, reflux, 91-93%.
F45 NaOH (aq.), reflux, 77-91%.
F46 NaOMe, MeOH, 71%. 1) NaOMe, MeOH; 2) H_2O, YNR.
F47 KOH, EtOH, rt, 100%.
F48 NaCN, EtOH, reflux, 54%.
F49 HBr, 100 °C, 93%.
F50 Zn, DMA, 120 °C, 77%.
F51 AcSK, hydroquinone (cat.), DMA, 155 °C, 93%.
F52 MeOH, $BF_3•Et_2O$ (cat.), reflux, NI.
F53 1) NaOH, H_2O, reflux; 2) HCl (6 N), 95%.
F54 HBr (48% aq.), DMA, 90 °C, 39%.
F55 HBr (48% aq.), 120 °C, 63%.
F56 $PhCH_2NH_2$, dioxane, reflux, 91%. NH_3(g), DMA, 50 °C, YNR.
F57 PhNHMe, toluene, 70 °C, 93%.
F58 1) CH_2N_2, Et_2O, THF, 0 °C to 130 °C, 20-30%. 1) CH_2N_2, Et_2O, THF; 2) Benzophenone, MeOH, hν (Hg lamp), 87% (1+2). CH_2N_2, $Pd(OAc)_2$, Et_2O, THF, 23%.
F59 $LiAlH_4$, *t*-BuOH, THF, 81%.
F60 MsCl, NEt_3, CH_2Cl_2, 100%.
F61 AcSH, Cs_2CO_3, MeOH/DMF, 80%.
F62 KOH, H_2O, reflux, 80%.
F63 MeOH, H_2SO_4 (conc.), 81-88%. $SOCl_2$, MeOH, 82%.
F64 LiBr, THF, 99%.
F65 KOAc, AcOH, CH_2Cl_2.
F66 PPh_3, CH_3Br, NaH.

F67 Zn, CH_2N_2.
F68 NaOH, H_2O.
F69 2-bromo-1,1-dimethoxyethane, 50 °C, TiO_2 or perfluorosulfonic acid resin or ZrO_2, THF, 89-92%.
F70 *t*-BuONa, THF, –20 °C, YNR.
F71 AcOH, H_2O, 82-93% (F70+F71).
F72 PBr_3, CH_2Cl_2, 0-10 °C, 96%.

7.10 REFERENCES

Almac2006 Wiffen, J.; Wilson, I. (Almac Sciences Ltd) WO2008035086 (prior. 22/09/2006, GB).

AMB1998 Shafiee, A.; Motamedi, H.; King, A. *Applied Microbiology and Biotechnology* **1998**, *49*, 709-717.

AptuitLaurus2005 Chava, S.; Simhadri, S.; Indukuri, V.; Kumar, I. (Aptuit Laurus Pvt Ltd) IN200801284, WO2009144742, US2011112300, US2012232276, US8207343, US8367834, EP2288595 (prior. 26/05/2005, IN).

AptuitLaurus2006 Gottumukkala, V. S.; Chava, S.; Casukhela, S. R.; Casukhel, S. R. (Aptuit Laurus Pvt Ltd) IN200601710, WO2008035379 (prior. 19/09/2006, IN).

AptuitLaurus2007 Gottumukkala, V. S.; Casukhela, S. R.; Chava, S.; Gorantla, S. R. (Aptuit Laurus Pvt Ltd) IN200701353, WO2009001374 (prior. 25/06/2007, IN).

Arch2008 Pai, G. G.; Audi, A. A.; Nair, R. V.; Bhagoji, N. R. (Arch Pharmalabs Ltd) IN253845, WO2010032256, US2011014107 (prior. 17/09/2008, IN).

Arromax2015 Hong, J.; Wang, J.; Xu, X. (Arromax Pharmaceutical Technology Suzhou Co Ltd) CN104496899 (prior. 15/01/2015, CN).

ASC2004 Caspi, D. D.; Ebner, D. C.; Bagdanoff, J. T.; Stoltz, B. M. *Advanced Synthesis & Catalysis* **2004**, *346*, 185-189.

Aurobindo2006 Ray, U. K.; Boju, S.; Pathuri, S.; Meenakshisunderam, S.; Sivakumaran, M. (Aurobindo Pharma Ltd) IN200601084, WO2008001213, US2009326232 (prior. 26/06/2006, IN).

BCSJ1980 Osamu, I.; Yoshiro, K.; Yasunori, I.; Makoto, U.; Tsutomu, O.; Katsuhiko, I. *Bulletin of the Chemical Society of Japan* **1980**, *53*, 146-153.

BMCL1998 von Sprecher, A.; Gerspacher, M.; Beck, A.; Kimmel, S.; Wiestner, H.; Anderson, G. P.; Niederhauser, U.; Subramanian, N.; Bray, M. A. *Bioorganic & Medicinal Chemistry Letters* **1998**, *8*, 965-970.

Boehringer2004 Deroy, P.; Faucher, A.-M.; Gagnon, A.; Landry, S.; Morin, S.; O'Meara, J.; Simoneau, B.; Thavonekham, B.; Yoakim, C. (Boehringer Ingelheim Int, Boehringer Ingelheim Pharma) WO2005118575, US2005282907, US2009143370, US7517998, US8198458, EP1756091 (prior. 01/06/2004, WO, US)

BSXB2001 Wang, D.; Zhang, Y.; Li, J. (Beijing Shangdi Xinshiji Biolo) CN1428335 (prior. 26/12/2001, CN).

Cangzhou2014 Zhang, S.; Zhang, J.; Wang, P.; Zhou, W.; Zhang, W. (Cangzhou Senary Chemical Science Tec Co Ltd) CN104860845 (prior. 21/02/2014, CN).

CCL2012 He, L.; Guo, Y. H.; Wang, Y. P.; Wang, X. J.; Zhang, J.; Xiang, W. S. *Chinese Chemical Letters* **2012**, *23*, 518-520.

Chemagis2006a Wang, Y.; Brand, M.; Kaspi, J. (Chemagis Ltd) WO2007088545, US2007208177, US7572930 (prior. 02/02/2006, US).

Chemagis2006b Adin, I.; Deutsch, Z.; Brand, M.; Shookrun, M.; Arad, O.; Kaspi, J. (Chemagis Ltd) WO2007096875, US2007213365, EP1986649 (prior. 21/02/2006, US).

Chemagis2006c Brand, M.; Shookrun, M.; Arad, O.; Kaspi, J. (Chemagis Ltd) WO2007096889, US2007208178, US7528254 (prior. 27/02/2006, US).

Chemagis2006d Shen, J.; Dai, Y.; Kaspi, J. (Chemagis Ltd) US2006223999 (prior. 10/05/2006, US).

Chemagis2007a Brand, M.; Weisman, A.; Gafni, Y.; Zelikovitch, L.; Davidi, G.; Manoff, E. (Chemagis Ltd) US2008188664 (prior. 15/01/2007, US).

Chemagis2007b Brand, M.; Weisman, A.; Gafni, Y.; Zelikovitch, L.; Davidi, G.; Manoff, E. (Chemagis Ltd) WO2008087628 (prior. 15/01/2007, US).

Chemagis2007c Noiman, M.; Salman, A.; Levin, I.; Arad, O.; Kaspi, J.; Weisman, A.; Gafni, Y. (Chemagis Ltd) WO2008126075 (prior. 12/04/2007, US).

Chemagis2007d Salman, A.; Gafni, Y.; Weisman, A.; Perelmuter, D.; Levin, I.; Noiman, M.; Antler, O. (Chemagis Ltd) WO2008135966 (prior. 02/05/2007, US).

Chemi2005 Turchetta, S.; Tuozzi, A.; Ullucci, E.; De Ferra, L.; Ciambecchini, U. (Chemi Spa) IT1365128, US2006194839, US7417149, EP1693368 (prior. 18/02/2005, IT).

Cipla2004 Srinivas, P.; Rao, D.; Kankan, R.; Relekar, J.; Pathi, S. (Cipla Ltd) IN200401357, WO2006064269 (prior. 17/12/2004, IN).

Cipla2006 Pathi, S.; Kankan, R.; Rao, D.; Phull, M. (Cipla Ltd) IN200900718, WO2008032099, US2010081688, EP2059509 (prior. 15/09/2006, IN).

Codexis2007 Liang, J.; Borup, B.; Mitchell, V.; Mundorff, E.; Lalonde, J.; Huisman, G. (Codexis Inc) EP2203557 (prior. 28/09/2007, WO, US).

CS2014 White, J. D.; Shaw, S. *Chem. Sci.* **2014**, *5*, 2200-2204.

Dainippon2012 Osawa, K.; Suenaga, T.; Oyama, T. (Dainippon Printing Co Ltd) JP2014005233 (prior. 25/06/2012, JP).

Delmar2002 Liu, L. W.; Wang, Y.; Marrugo, H.; Harper, S.; Quan, D. D. C.; Zhou, Z. N.; Bydlinski, G. (Delmar Chemicals Inc) CA2371048, WO2003066582, US6512140, EP1474386 (prior. 06/02/2002, CA).

DongKook2007 Cheol Kwon, H.; Ju Hong, Y.; Dong Noh, M.; Hoe Cha, G.; Kwon, H.; Hong, Y.; Rho, M.; Cha, K. (Dong Kook Pharm Co Ltd) KR990046, WO2012015255, US2013184464, US8686151, EP2598485 (prior. 30/07/2007, KR).

DrReddys2002a Aavula, S.; Bollikonda, S.; Chandra, S.; Kasturi, R.; Reguri, B. (Dr Reddy's Lab Ltd, Dr Reddy's Lab Inc) US2010063291 (prior. 30/12/2002, IN).

DrReddys2002b Reguri, B.; Bollikonda, S.; Bulusu, V.; Kasturi, R.; Aavula, S. (Dr Reddy's Lab Ltd) IN200200993, US2005107612 (prior. 30/12/2002, IN).

DrReddys2004 Sundaram, V.; Rajan, S.; Bulusu, V.; Srivastav, A.; Kasturi, R.; Aavula, S. (Dr Reddy's Lab Ltd) IN200400342, US2005234241, US7189853 (prior. 15/04/2004, IN).

DrReddys2005 Padi, P.; Bollikonda, S.; Srivastava, A.; Kasturi, R.; Jinna, R.; Mopidevi, N. (Dr Reddy's Lab Ltd, Dr Reddy's Lab Inc) WO2007012075, US2007022123, US2008214823, EP1912499 (prior. 20/01/2005, IN).

DrReddys2006 Padi, P.; Bollikonda, S.; Naidu, N.; Ghosh, D.; Venkata Kandirelli, K.; Jinna, R.; Kasturi, R.; Mohanarangam, S. (Dr Reddy's Lab Ltd, Dr Reddy's Lab Inc) IN200902532, WO2008058118, EP2094665 (prior. 06/11/2006, IN).

DrReddys2008 Bollikonda, S.; Kandirelli, K.; Medisetti, R.; Kopparapu, J.; Manudhane, K. (Dr Reddy's Lab Ltd, Dr Reddy's Lab Inc) IN201005596, WO2009117381, US2011040095, EP2265586 (prior. 17/03/2008, IN).

DrReddys2009 Fox, M.; Bollikonda, S.; Mohanarangam, S.; Jinna, R.; Kandirelli, V. (Dr Reddy's Lab Inc, Dr Reddy's Lab Ltd) IN201109433, WO2010148209 (prior. 19/06/2009, IN).

DSM2009 Lefort, L. (DSM Ip Assets Bv) WO2011042416, US2012215002, EP2486015 (prior. 09/10/2009, EP)

EJOC2014 Mao, D.; Hong, G.; Wu, S.; Liu, X.; Yu, J.; Wang, L. *European Journal of Organic Chemistry* **2014**, *2014*, 3009-3019.

EliLilly1995 Bach, N. J.; Baker, S. R.; Gilmore, J.; Lewthwaite, R. A.; McKillop, A.; Sawyer, J. S.; Stephenson, G. R.; Urquhart, M. W. J. (Lilly & Co Eli, Lilly Ind Ltd, Univ East Anglia) US5807866, US6160120, US6509476, EP775694 (prior. 23/11/1995, GB).

Enantiotech2015 Xu, L.; Huang, Z.; Li, Y.; Mao, B. (Enantiotech Corp Ltd) CN105330540 (prior. 01/12/2015, CN).

EnzymeWorks2015 Tao, J.; Le, Y.; Jiang, X. (Enzymeworks Inc) CN104745652 (prior. 25/03/2015, CN).

EnzymeWorks2016 Yuan, H. (Enzymeworks Inc) CN106011092 (prior. 20/06/2016, CN).

Esteve2005a Coppi, L.; bartra Sanmarti, M.; Gasanz Guillen, Y.; Monsalvatje Llagostera, M.; Talavera Escasany, P. (Esteve Quimica Sa) EP1783117 (prior. 04/11/2005, EP).

Esteve2005b Coppi, L.; bartra Sanmarti, M.; Gasanz Guillen, Y.; Monsalvatje Llagostera, M.; Talavera Escasany, P. (Esteve Quimica Sa) WO2007051828, US2008306270, EP1948611 (prior. 04/11/2005, EP, US).

Esteve2006a Gasanz Guillen, Y.; Montsalvatje Llagostera, M.; Talavera Escasany, P. (Esteve Quimica Sa) WO2008017669, US2009247759, US8188285, EP1886998, EP2077996 (prior. 09/08/2006, EP, US).

Esteve2006b Gasanz Guillen, Y.; Montsalvatje Llagostera, M.; Talavera Escasany, P. (Esteve Quimica Sa) WO2008017667, US2010029945, EP1886997, EP2066638 (prior. 09/08/2006, EP, US).

Farmaprojects2006 Bessa Bellmunt, J.; Rafecas Jane, L.; Pasto Aguila, M.; Verdaguer Espaulella, X.; Gordo Campon, E. (Farmaprojects Sa) WO2007101841, US2009099367, EP2004609 (prior. 06/03/2006, EP).

Formosa2006a Hung, J.; Wei, C. (Formosa Lab Inc) US2008097104, US7700776 (prior. 24/04/2006, US).

Formosa2006b Hung, J.; Wei, C. (Formosa Lab Inc) US2010152453, US8399675 (prior. 24/10/2006, US).

Formosa2006c Suen, R.-T.; Liu, Y.-L.; Wei, C.-P. (Formosa Lab Inc) US7271268 (prior. 22/12/2006, US)

Formosa2007 Hung, J.; Wei, C. (Formosa Lab Inc) DE102007061630 (prior. 20/12/2007, DE).

Formosa2008a Hong, R.; Wei, Q. (Formosa Lab Inc) CN101323589 (prior. 06/03/2008, CN).

Formosa2008b Wei, Q.; Hong, R. (Formosa Lab Inc) CN101817818 (prior. 06/03/2008, CN).

Formosa2008c Wei, Q.; Hong, R. (Formosa Lab Inc) CN101845024 (prior. 06/03/2008, CN).

Gilead2003 Krawczyk, S. H. (Gilead Sciences Inc) WO2005012324, US2005059637, US2006252729, US7579332, EP1656387 (prior. WO2004US24922 20040730 ; US20030491123P 20030730, US).

Glade2005a Parekh, S.; Chawla, H.; Chowdhary, A.; Patel, A. (Glade Organics Private Ltd) WO2006131782 (prior. 07/06/2005, IN).

Glade2005b Chawla, H.; Chowdhary, A.; Patel, A.; Joshi, V.; Patel, M. (Glade Organics Private Ltd) IN200501613, WO2007072114, US2009182148, EP1968942 (prior. 23/12/2005, IN).

Glade2006 Chawla, H.; Chowdhary, A.; Patel, A.; Patel, M. (Glade Organics Private Ltd) IN200600574, WO2007116240, US2009281323, EP2004608 (prior. 12/04/2006, IN).

Hanmi2006 Lee, G.; Chang, Y.; Lee, J.; Park, C.; Park, E.; Yoo, J. (Hanmi Pharm Co Ltd, Hanmi Holdings Co Ltd) KR774088, WO2008072872, US2010099876, US8217174, EP2114889 (prior. 14/12/2006, KR).

Hanmi2007 Park, C. H.; Park, E. J.; Kim, C. H.; Lee, G.; Jang, S. M.; Chang, Y. K.; Lee, J.; Lim, E. (Hanmi Pharm Co Ltd, Hanmi Holdings Co Ltd, Korea Inst Sci&Technology) KR2009036349, WO2009048236, US2011105757, US8426599, EP2195295 (prior. 09/11/2007, KR).

Hwail2009 Ho Jeong, Y.; U, L. J.; Cheol Kim, G.; Jung, Y. H.; Lee, J. W.; Kim, K. C. (Hwail Pharma Co Ltd) KR2010135130 (prior. 16/06/2009, KR).

IECSOC2016 Aliabad, J. M.; Hosseini, N.; Karimian, K.; Rouhani, M. *20th International Electronic Conference on Synthetic Organic Chemistry*; Mdpi: Basel, 2016.

IJC2004 Khan, A.; Bhawal, D. *Indian Journal of Chemistry – Section B Organic and Medicinal Chemistry* **2004**, *43*, 420.

IMMC2008 Deng, Y.; Feng, W.; Li, Q.; Song, Z. (Inst Medical Materials Chinese Acad Medi) CN101591291 (prior. 26/05/2008, CN).

Imperial1985a Whittle, A. J. (Imperial Chem Ind Plc) GB2180838, US4705900, EP220810 (prior. 23/09/1985, GB).

Imperial1985b Whittle, A. J. (Imperial Chem Ind Plc) GB2184439, US4812483, US4904677, EP227369 (prior. 23/12/1985, GB).

Imperial1987 Whittle, A. J. (ICI Plc) US4788348 (prior. 12/08/1987, US).

IndSwift2010 Bhirud Shekhar, B.; Sarin, G. S.; Tripathi Ratnakar, M.; Bhirud, S. B.; Mani, T. R.; Bhaskar, B. S.; Singh, S. G.; Ratnakar, M. T. (Ind-Swift Lab Ltd) IN201101655, WO2012077133 (prior. 07/12/2010, IN).

Inke2007a Perez, A. J.; Huguet, C. J.; Dalmases, B. P.; Perez Andres, J. A.; Huguet Clotet, J.; Dalmases barjoan, P. (Inke Sa) WO2008098965, US2008194825, EP1958936 (prior. 14/02/2007, EP).

Inke2007b Huguet Clotet, J.; Peirats Masia, J.; Huguet, C. J.; Peirats, M. J. (Inke Sa) WO2009053424, US2009111849, EP2053043 (prior. 26/10/2007, EP).

Innogent2008 Rammohan, K.; Ray, P. C.; Reddy, G. M.; Tirmalaiah, G. (Innogent Lab Private Ltd) WO2010064257 (prior. 01/12/2008, IN).

JACS1990 Newkome, G. R.; Baker, G. R.; Arai, S.; Saunders, M. J.; Russo, P. S.; Theriot, K. J.; Moorefield, C. N.; Rogers, L. E.; Miller, J. E. *Journal of the American Chemical Society* **1990**, *112*, 8458-8465.

JACS1991 Wade, P. A.; Kondracki, P. A.; Carroll, P. J. *Journal of the American Chemical Society* **1991**, *113*, 8807-8811.

JACS1996 Fujii, A.; Hashiguchi, S.; Uematsu, N.; Ikariya, T.; Noyori, R. *Journal of the American Chemical Society* **1996**, *118*, 2521-2522.

Janssen2003 Kuo, G.-H.; Zhang, R.; Wang, A.; Deangelis, A. R. (Janssen Pharmaceutica Nv) WO2005030694, US2005107469, US2010004470, US2011294875, US2012322875, US7598292, US8008513, US8258177, US8450371, EP1670744, EP2243766 (prior. 19/09/2003, US).

JFC2000 Dmowski, W.; Wolniewicz, A. *Journal of Fluorine Chemistry* **2000**, *102*, 141-146.

Jiangsu2014 Ding, Z.; Wang, X. (Jiangsu Hansyn Pharmaceutical Co Ltd) CN104293850 (prior. 12/08/2014, CN).

Jiangsu2015 Chen, B.; Zhou, C.; Qiu, L.; Dai, J. (Jiangsu Ocean Pharmaceutical, Chemical Engineering Co Ltd) CN104592111 (prior. 13/02/2015, CN).

Jinan2017a Gao, L.; Zheng, G.; Han, X.; Fu, K.; Wang, Q. (Univ Jinan) CN106928136 (prior. 23/02/2017, CN).

Jinan2017b Zheng, G.; Gao, L.; Han, X.; Fu, K.; Wang, Q.; Feng, X. (Jinan Univ) CN106928137 (prior. 23/02/2017, CN).

JOC1956 House, H. O.; Lord, R. C.; Rao, H. S. *The Journal of Organic Chemistry* **1956**, *21*, 1487-1491.

JOC1959 Searles, S.; Nickerson, R. G.; Witsiepe, W. K. *The Journal of Organic Chemistry* **1959**, *24*, 1839-1844.

JOC1989 McNamara, J. M.; Leazer, J. L.; Bhupathy, M.; Amato, J. S.; Reamer, R. A.; Reider, P. J.; Grabowski, E. J. J. The *Journal of Organic Chemistry* **1989**, *54*, 3718-3721.

JOC1993a King, A. O.; Corley, E. G.; Anderson, R. K.; Larsen, R. D.; Verhoeven, T. R.; Reider, P. J.; Xiang, Y. B.; Belley, M.; Leblanc, Y. *The Journal of Organic Chemistry* **1993**, *58*, 3731-3735.

JOC1993b Wright, M. E.; Allred, G. D.; Wardle, R. B.; Cannizzo, L. F. *The Journal of Organic Chemistry* **1993**, *58*, 4122-4126.

JOC1996a Larsen, R. D.; Corley, E. G.; King, A. O.; Carroll, J. D.; Davis, P.; Verhoeven, T. R.; Reider, P. J.; Labelle, M.; Gauthier, J. Y.; Xiang, Y. B.; Zamboni, R. J. *The Journal of Organic Chemistry* **1996**, *61*, 3398-3405.

JOC1996b Dufresne, C.; Gallant, M.; Gareau, Y.; Ruel, R.; Trimble, L.; Labelle, M. *The Journal of Organic Chemistry* **1996**, *61*, 8518-8525.

JOC2015 Bollikonda, S.; Mohanarangam, S.; Jinna, R. R.; Kandirelli, V. K. K.; Makthala, L.; Sen, S.; Chaplin, D. A.; Lloyd, R. C.; Mahoney, T.; Dahanukar, V. H.; Oruganti, S.; Fox, M. E. *J. Org. Chem.* **2015**, *80*, 3891-3901.

JOCUSSR1983 Zefirov, N.; Kuznetsova, T.; Kozhushkov, S.; Surmina, L.; Rashchupkina, Z. *Journal of Organic Chemistry USSR (English Translation)* **1983**, *19*, 474; *Zhurnal Organicheskoi Khimii* **1983**, *19*, 541.

Kanata2010 Abdur-Rashid, K.; Jia, W.; Lu, S.; Guo, R.; Chen, X.; Amoroso, D. (Kanata Chemical Technologies Inc) WO2012031358, US2013184479, US9062085, EP2614069 (prior. 10/09/2010, US).

KRKA2006 Kidemet, D.; Kroselj, V.; Osolnik, R. (Krka Tovarna Zdravil Dd) WO2008049922, EP2094664 (prior. 26/10/2006, SI).

KRKA2009 Kidemet, D.; Benkic, P.; Kljajic, A.; Stefane, B. (Krka Novo Mesto Dd) EP2287154 (prior. 14/07/2009, EP).

KRKA2010 Kidemet, D.; Bogdan, S.; Kljajic, A.; Benkic, P.; Stropnik, T.; Plaper, I. (Krka Novo Mesto) WO2011121091, EP2552892 (prior. 31/03/2010, EP).

LabLesvi2011 Dalmases barjoan, P.; Ozores Viturro, L. (Lab Lesvi Sl) WO2012123506, EP2502910, EP2686301 (prior. 15/03/2011, EP).

Laurus2012a Simhadri, S.; Mohammad, Y.; Indukuri, V. S.; Gorantla, S. R.; Chava, S. (Laurus Labs Private Ltd) WO2014001860, US2015299127, US9487487, EP2850064 (prior. 18/05/2012, IN).

Laurus2012b Simhadri, S.; Mohammad, Y.; Indukuri, V. S.; Gorantla, S. R.; Chava, S. (Laurus Labs Private Ltd) WO2014001860, US2015299127, US9487487, EP2850064 (prior. 07/11/2012, IN).

LG2008 Kang, J. E.; Lee, I. S.; Lee, S. J. (Lg Life Sci Ltd) KR2010035597, WO2010036048, US2011166356, US8426600 (prior. 26/09/2008, KR).

Lonza1993 Quittmann, W.; McGarrity, J. (Lonza Ag, Lonza Ltd) US5534651, EP641775 (prior. 03/09/1993, CH).

Lonza1994 Hanselmann, P. (Lonza Ag) US5486622, EP678515 (prior. 21/04/1994, CH).

Lonza2007a McGarrity, J.; Bappert, E. (Lonza Ag, Lonza Ltd) WO2008131932, US2010130745, EP1988079, EP2150533 (prior. 25/04/2007, EP).

Lonza2007b McGarrity, J.; Djojo, F.; Yon, M. (Lonza Ag, Lonza Ltd) WO2009010231, US2010217004, EP2014633, EP2178812 (prior. 13/07/2007, EP).

Lonza2007c McGarrity, J.; Djojo, F.; Yon, M. (Lonza Ag, Lonza Ltd) WO2009010230, US2010190987, EP2014650, EP2178841 (prior. 13/07/2007, EP).

Lunan2013 Wang, H.; Zhao, G.; Zhang, S. (Lunan Pharm Group Corp) CN103012261 (prior. 14/01/2013, CN).

Lunan2016 Zhang, G.; Wang, Z.; Ma, Q. (Lunan Better Pharmaceutical Co Ltd) CN105541786 (prior. 06/01/2016, CN).

Macleod2008 Agarwal, R. M.; Bhirud, S. B.; Khude, G. D.; Pillai, B. G. (Macleod Pharm Inc) WO2009138993 (prior. 15/03/2008, WO, IN).

Matrix2004a Chava, S.; Gorantla, S.; Indukuri, V.; Simhadri, S.; Satyanarayana, C.; Ramanjaneyulu, G.; Kumar, I.; Srinivas, S.; Sunil, K.; Krishna, K. (Matrix Lab Ltd) WO2006008750, US2008275243, US2011137039, US7858793, EP1812393 (prior. 19/07/2004, EP, WO, IN).

Matrix2004b Chava, S.; Gorantla, S.; Indukuri, V.; Simhadri, S.; Jammula, V. (Matrix Lab Ltd) DE602004031678, WO2006008751, EP1812394, US2009143590 (prior. 19/07/2004, DE, EP, WO).

Matrix2005 Chava, S.; Gorantla, S.; Indukuri, V.; Kumar, I.; Ramanjaneyulu, G.; Satyanarayana, C. (Matrix Lab Ltd) IN200500872, WO2007004237, US2010168432, EP1940793 (prior. 05/07/2005, IN).

Matrix2006 Gorantla, S.; Bandari, M.; Indukuri, V.; Simhadri, S.; Ketavarapu, N.; Jammulaveera, V.; Mohammed, Y.; Jammula, V.; Jammulaveera, V. (Matrix Lab Ltd) IN200601400, WO2008015703, US2011092708, EP2069307 (prior. 04/08/2006, IN).

Mayuka2008a Kashyap, T.; Nakka, K.; Singh, J. (Mayuka Labs Pvt Ltd) IN200800782 (prior. 27/03/2008, IN).

Mayuka2008b Kashyap, T.; Nakka Krishnamurthy, C.; Singh, J.; Chandra, S. (Mayuka Labs Pvt Ltd) IN200802732, WO2010064109 (prior. 02/12/2008, IN).

Medichem2004 Chamorro Gutierrez, I.; Bosch, I.; Molins, I. (Medichem Sa) WO2006058545, US2008214822, EP1817289 (prior. 30/11/2004, EP, WO).

Medichem2009 Dong Ock, C.; Youn Ja, Z.; Cho, D.; Kim, B.; Kim, S. (Medichem Korea Co Ltd) KR893756, WO2010082714 (prior. 14/01/2009, KR).

Melody2013 Patel, D.; Ghogare, A.; Vishe, K. (Melody Healthcare Pvt Ltd) WO2014118796 (prior. 31/01/2013, IN).

Merck1991 Song, Z.; Hughes, D. (Merck & Co Inc) US5126456 (prior. 17/10/1991, US).

Merck1993 Bhupathy, M.; McNamara, J.; Sidler, D.; Volante, R.; Bergan, J.; Silder, D.; Tung, H.; Beran, J. (Merck & Co Inc, Merck Sharp & Dohme Corp) WO9518107, EP737186, US5614632, US6320052 (prior. 28/12/1993, US).

Merck1994a Chartrain, M.; Chen, S.; Garrity, G.; Heimbuch, B.; Roberge, C.; Shafiee, A.; Chen, S. (Merck & Co Inc) WO9602657, US5491077 (prior. 20/07/1994, US).

Merck1994b Chen, S.-S.; Shafiee, A. (Merck & Co Inc) US5427933 (prior. 20/07/1994, US).

Merck1994c King, A.; Larsen, R.; Verhoeven, T.; Zhao, M.; King, A.; Ping, A.; Mangzhu, Z. (Merck & Co Inc) WO9605206, US5545758, EP775146, US5693816 (prior. 11/08/1994, US).

Merck1995 Pipik, B.; King, S.; Conlon, D.; Colon, D. (Merck & Co Inc) US5523477, WO9622987, EP805808 (prior. 23/01/1995, US).

Merck1996a Blackmond, D.; Sun, Y.; Wang, J. (Merck & Co Inc) GB2312896 (prior. 10/05/1996, US).

Merck1996b Blackmond, D.; Sun, Y. (Merck & Co Inc) GB2313119 (prior. 14/05/1996, US).

Merck1997 Hobbs, D.; Tung, H. (Merck & Co Inc) GB2322624, US5869673 (prior. 28/02/1997, US).

Merck1998 Amato, J.; McNamara, J.; Leazer, J. J.; Reider, P.; Reamer, R. (Merck & Co Inc) US4883878, EP328188 (prior. 02/08/1998, US).

Merck2012 Albaneze-Walker, J.; Chen, Y.; Humphrey, G.; Krska, S.; Qi, J.; Tan, L.; Itoh, T.; Funane, S.; Yokozawa, T.; Kobayashi, T. (Merck Sharp & Dohme) WO2014081616 (prior. US201261729026P 20121121, US).

MerckFrosst1986 Young, R.; Zamboni, R.; Leger, S. (Merck Frosst Canada Inc) EP233763, US4851409, US5428171, CA1337196 (prior. 14/02/1986, US, CA).

MerckFrosst1990 Belley, M.; Leger, S.; Roy, P.; Labelle, M.; Xiang, Y. B.; Guay, D. (Merck Frosst Canada Inc, Merck Sharp & Dohme Pharm Srl) US5565473, US5856322 (prior. 12/10/1990, US).

MerckFrosst1991 Belley, M. (Merck Frosst Canada Inc) US5212180 (prior. 21/02/1991, US).

MerckFrosst1992 Zamboni, R.; Guay, D.; Labelle, M. (Merck Frosst Canada Inc, Merck Frosst Canada & Co) WO9321159, US5270324, EP565185 (prior. 10/04/1992, US).

Moehs2007 Camps Ramirez De Cartagena, H.; Sanchez Casals, C. (Moehs Iberica Sl) ES2320077, WO2009016191, US2010204476, US8163924, EP2185517 (prior. 31/07/2007, ES).

Morepen2003 Suri, S.; Singh, J.; Sarin, G.; Tanwar, M.; Mahendru, M.; Sarin, G.; Tanwar, M.; Singh, G.; Tamvar, M.; Mahedru, M. (Morepen Lab Ltd) AU2003253247, WO2004108679, US2007082925, US8450491, EP1631550 (prior. 06/06/2003, AU, EP, WO).

Morepen2004 Suri, S.; Sarin, G.; Mahendru, M. (Morepen Lab Ltd) IN200401561, WO2006021974 (prior. 23/08/2004, IN).

MSN2005 Satyanarayana, R.; Kihsore, K.; Thirumalai, R.; Ramasubba, R.; Reddy, M.; Muppa, K.; Karamal, R.; Satyanarayana, R.; Kihsore, K.; Thirumalai, R.; Ramasubba, R.; Kumar, M.; Rajan, S.; Reddy, K.; Reddy, K. (MSN Lab Ltd) IN200501818, WO2007069261, US2009171092, US8178680, EP1968943 (prior. 13/12/2005, IN).

MSN2006 Satyanarayana Reddy, M.; Kumar, M.; Reddy, K.; Prasad, D. (MSN Lab Ltd) IN200602152, WO2008062478, US2010056793, US8115004 (prior. 20/11/2006, IN).

OL2000 Wender, P.; Zhang, L. *Organic Letters* **2000**, *2*, 2323-2326.

OPRD2006 Merschaert, A.; Boquel, P.; Van Hoeck, J.-P.; Gorissen, H.; Borghese, A.; Bonnier, B.; Mockel, A.; Napora, F. *Organic Process Research & Development* **2006**, *10*, 776-783.

OPRD2010a Liang, J.; Lalonde, J.; Borup, B.; Mitchell, V.; Mundorff, E.; Trinh, N.; Kochrekar, D.; Nair Cherat, R.; Pai, G. G. *Organic Process Research & Development* **2010**, *14*, 193-198.

OPRD2010b Halama, A.; Jirman, J.; Boušková, O.; Gibala, P.; Jarrah, K. *Organic Process Research & Development* **2010**, *14*, 425-431.

Pfizer1992 Marfat, A.; Eggler, J.; Fray, M.; Cooper, K. (Pfizer) US5322847 (prior. 05/11/1992, US).

Pharmathen2009 Koftis, T.; Soni, R. (Pharmathen Sa) AU2009357276, WO2011076237, US2012259121, EP2516401 (prior. 23/12/2009, AU, EP, WO).

Pliva2004 Avdagic, A.; Mohar, B.; Sterk, D.; Stephan, M. (Pliva Istrazivanje Razvoj Doo) WO2006008562, US2008009629, US7868174, EP1781615 (prior. 23/07/2004, US).

Pliva2006 Avdagic, A.; Jerkovic, J.; Pavlicic, D. (Pliva Istrazivanje Razvoj Doo, Pliva Hrvatska Doo) WO2008009970, EP2046750 (prior. 21/07/2006, GB).

Polpharma2004 Achmatowicz, O.; Wisniewski, K.; Ramza, J.; Szelejewski, W.; Szechner, B. (Inst Farmaceutyczny, Zaklady Farm Polpharma Sa) WO2006043846, US2009005413, EP1853563 (prior. 22/10/2004, PL).

Qidong2014 Song, J.; Zhang, C.; Long, Z.; Liu, X.; Cai, S. (Qidong Dongyue Pharmaceutical Co Ltd, Shanghai Dongyue Biochem Co Ltd) CN103936671 (prior. 06/05/2014, CN).

Qilu2016a Zhou, X.; Yang, Q.; Wu, K.; Zhang, Z.; Dong, T.; Gao, D.; Jiang, H.; Zhou, X.; Yang, B. (Qilu Tianhe Pharmaceutical Co Ltd) CN105541710 (prior. 22/02/2016, CN).

Qilu2016b Zhou, X.; Yang, Q.; Wu, K.; Zhang, Z.; Dong, T.; Gao, D.; Jiang, H.; Zhou, X.; Yang, B. (Qilu Tianhe Pharmaceutical Co Ltd) CN105541711 (prior. 22/02/2016, CN).

RichterGedeon2010 Bodi, J.; Farago, J.; Szoeke, K.; Ujvari, V.; Temesvari, K.; Aranyi, A.; Santa, Z.; Nagy, M.; Szoke, K. (Richter Gedeon Nyrt) HU201000425, WO2012020271 (prior. 11/08/2010, HU).

Ruetgerswerke1989 Pastorek, E.; Orth, W.; Weiss, W.; Kleffner, H. (Ruetgerswerke Ag) DE3932136, US5066806 (prior. 27/09/1989, DE).

S1987a Szabó, G.; Aranyosi, K.; Csiba, M.; T ke, L. *Synthesis* **1987**, *1987*, 565-566.

S1987b Heiszman, J.; Bitter, I.; Harsányi, K.; Töke, L. *Synthesis* **1987**, *1987*, 738-739.

Sandoz2006 Bendre, D.; Holkar, A. (Sandoz Ag) WO2008023044 (prior. 23/08/2006, EP).

SC2013 Saravanan, M.; Satyanarayana, B.; Reddy, P. *Synthetic Communications* **2013**, *43*, 2050-2056.

Shandong2009 Wang, H.; Wang, Z.; Zhao, Z. (Lunan Pharm Co Ltd , Shandong New Time Pharm Co Ltd) CN101638381, WO2011026398, US2012165535, US8569508, EP2474535 (prior. 02/09/2009, CN).

Shandong2017 Zhen, Y.; Chen, J.; Zhao, X.; Zhang, Z.; Qiu, X. (Shandong Bestcomm Pharmaceutical Co Ltd) CN106831863, WO2018133533 (prior. 20/01/2017, CN).

Shangdi2001 Wang, D.; Zhang, Y.; Li, J. (Shangdi Xinshiji Bio Medical I) CN1420113 (prior. 16/11/2001, CN).

Shangdong2016 Zhang, G.; Shi, J.; Zhang, S. (Shangdong New Time Pharm Prod) CN105585525 (prior. 29/02/2016, CN).

ShanghaiIOC2012 Gao, T.; Ma, W.; Ye, Z.; Zhang, J.; Zhang, Y.; Zhao, G. (Shanghai Organic Chem Inst) CN102617460 (prior. 29/03/2012, CN).

ShanghaiIOC2013 Ye, Z.; Zhang, Y.; Zhang, J.; Ma, W.; Gao, T.; Zhao, G. (Shanghai Inst Organic Chem) CN103073492 (prior. 04/02/2013, CN).

ShanghaiSPT2016 Zhou, S.; Feng, Z.; Song, B.; Chen, L. (Shanghai Stepup Pharmaceutical Tech Co Ltd) CN106431863 (prior. 12/09/2016, CN).

ShanghaiXMT2011 Liang, Y.; Xiong, J.; Zeng, Z.; Zhang, J. (Shanghai Xianghui Medical Sci & Technology Co Ltd) CN102690229 (prior. 23/03/2011, CN).

ShanghaiYT2011a Lin, Z.; Pan, L.; Wang, H.; Wang, T.; Yu, Q.; Yuan, B.; Zhang, F. (Shanghai Yingli Technology Co Ltd) CN102442947 (prior. 01/11/2011, CN).

ShanghaiYT2011b Lin, Z.; Pan, L.; Wang, H.; Wang, T.; Yu, Q.; Yuan, B.; Zhang, F. (Shanghai Yingli Technology Co Ltd) CN102442948 (prior. 01/11/2011, CN).

Sterling2008 Chatterjee, S.; Chavda, M.; Kadu, R.; Kumar, N.; Rawat, A. (Sterling Biotech Ltd) IN200801564, US2010022777 (prior. 23/07/2008, IN).

Sunshine2009 Xi, N. (Sunshine Lake Pharma Co Ltd) WO2010111063, US2010239576, US8232294, EP2408300 (prior. 21/03/2009, US).

Sunshine2010 Xi, N. (Sunshine Lake Pharma Co Ltd) WO2010045095, US2010093727, US8293897, EP2340244, EP2532657 (prior. 14/10/2008, US).

Synthon2004a Overeem, A.; Gieling, R.; Zhu, J.; Lambertus, T.; Gieling, R. (Synthon Bv, Synthon Pharm Inc) WO2005105750, US2005245568, US7829716, EP1740545 (prior. 30/04/2004, US).

Synthon2004b Overeem, A.; Gieling, R.; Zhu, J.; Thijs, L.; Gieling, R. (Synthon Bv, Synthon IP Inc) WO2005105749, US2005245569, US7501517, EP1740546, EP2181986 (prior. 30/04/2004, US).

Synthon2005a Overeem, A.; Thijs, L.; Benovsky, P.; Castulik, J.; Zhu, J.; bartos, P.; Skoumal, R. (Synthon Bv) WO2007057228, EP1954679 (prior. 18/11/2005, US).

Synthon2005b Benovsky, P.; Thijs, L.; Overeem, A.; Castulik, J.; Zhu, J.; bartos, P.; Skoumal, R. (Synthon Bv) US2007135480, US7786138 (prior. 18/11/2005, US).

Synthon2005c Overeem, A.; Thijs, L.; Benovsky, P.; Castulik, J.; Zhu, J.; bartos, P.; Skoumal, R. (Synthon Bv) WO2007057225, EP1948610 (prior. 18/11/2005, US).

Synthon2005d Benovsky, P.; Thijs, L.; Overeem, A.; Castulik, J.; Zhu, J.; bartos, P.; Skoumal, R. (Synthon Bv) US2007135642, US7476748 (prior. 18/11/2005, US).

Synthon2005e Benovsky, P.; Thijs, L.; Overeem, A.; Castulik, J.; Zhu, J.; bartos, P.; Skoumal, R. (Synthon Bv) WO2007057225, WO2007057227, WO2007057228, US2007135480, US2007135642, US2007135643, US2009093636, US2010099879, US7476748, US7601741, US7786138, EP1948610, EP1954679, EP1960362 (prior. 18/11/2005, US).

Synthon2008 Thijs, L.; Vinent, H.; Tejedor, V. (Synthon Bv) WO2009130056, US2009270628, EP2276739 (prior. 25/04/2008, US).

T2010 Chang, X.-W.; Han, Q.-C.; Jiao, Z.-G.; Weng, L.-H.; Zhang, D.-W. *Tetrahedron* **2010**, *66*, 9733-9737.

TA1992 Wallbaum, S.; Martens, J. *Tetrahedron: Asymmetry* **1992**, *3*, 1475-1504.

Teva2004 Shapiro, E.; Yahalomi, R.; Niddam-Hildesheim, V.; Sterimbaum, G.; Chen, K. (Teva Pharm Ind Ltd, Teva Pharm USA Inc) WO2005105751, US2005256156, US7547787, EP1646612 (prior. 21/04/2004, US).

Teva2005 Sterimbaum, G.; Shapiro, E.; Chen, K.; Evgeny, S.; Kobi, C. (Teva Pharm Ind Ltd, Teva Pharm USA Inc) WO2007005965, US2007078158, US2010076195, US7812168, EP1904448 (prior. 05/07/2005, US).

Tianjin2013 Zhang, S.; Guo, Z.; Wang, P.; Zhang, Y.; Liu, W.; Liu, Y.; Zhang, W. (Tianjin Suyan Medical Technology Co Ltd) CN103467276 (prior. 13/09/2013, CN).

TL1982 Evans, D.; Mitch, C. *Tetrahedron Letters* **1982**, *23*, 285-288.

TL1997 Zhao, M.; King, A.; Larsen, R.; Verhoeven, T.; Reider, P. *Tetrahedron Letters* **1997**, *38*, 2641-2644.

TL2005 Muthusamy, S.; Gnanaprakasam, B. *Tetrahedron Letters* **2005**, *46*, 635-638.

Tokuyama2013 Shimizu, T.; Kamizono, H.; Kobayashi, K. (Tokuyama Corp) JP2015010049 (prior. 27/06/2013, JP).

Torrent2009 Lakkad, M.; Nadkarni, S.; Shimpi, N.; Rao, A. (Torrent Pharm Ltd) IN200800038, WO2009113087, US2010292251, EP2231613 (prior. 07/01/2009, IN).

WarnerL2000 barrett, S.; Biwersi, C.; Kaufman, M.; Tecle, H.; Warmus, J.; Chen, M. (Warner Lambert Co) WO200206213, US2004054172, US2005176820, US6960614, US7411001, EP1301472 (prior. 19/07/2000, WO, US).

Wischem2012 Cho, Y.-R.; Jung, S.-Y.; Heor, J.-H.; Yu, H.-R.; Son, H.-B.; Han, N.-R.; Ji, K.-W.; Park, J.-B. (Wischem Co Ltd) KR20130082017, WO2013105688 (prior. 10/01/2012, KR).

Zaklady2007 Zyla, D.; Rynkiewicz, R.; Krzyzanowski, M.; Ramza, J. (Zaklady Farm Polpharma Sa) WO2008136693, US2010069641, EP2142508 (prior. 02/05/2007, PL).

Zentiva2007a Halama, A.; Jirman, J. (Zentiva KS, Zentiva AS) CZ200700020, WO2008083635, US2010174083, EP2118063 (prior. 09/01/2007, CZ)

Zentiva2007b Halama, A.; Jirman, J.; Petrickova, H.; Brusova, H. (Zentiva AS) CZ200700455, WO2009006861, US2010267958, EP2173718 (prior. 09/07/2007, CZ).

ZhangjiagangXC2014 Lou, X.; Wang, X. (Zhangjiagang Xinyi Chemical Co Ltd) CN104326976 (prior. 20/10/2014, CN).

ZhejiangAP2013a Zhiguo, Z. (Zhejiang Ausun Pharmaceutical Co Ltd) CN104109122 (prior. 16/04/2013, CN).

ZhejiangAP2013b Zhiguo, Z. (Zhejiang Ausun Pharmaceutical Co Ltd) CN104109123 (prior. 16/04/2013, CN).

ZhejiangHP2011 Cai, Q.; Chai, J.; Gu, R.; Li, J.; Luo, H.; Ma, J.; Zhu, G. (Zhejiang Hisun Pharm Co Ltd) CN102424673 (prior. 14/09/2011, CN).

ZhejiangJP2012 Yan, P.; Zhu, G.; Zhang, X.; Li, Y.; Che, D. (Zhejiang Jiuzhou Pharm Co Ltd) CN103420908 (prior. 25/05/2012, CN).

8 OSELTAMIVIR

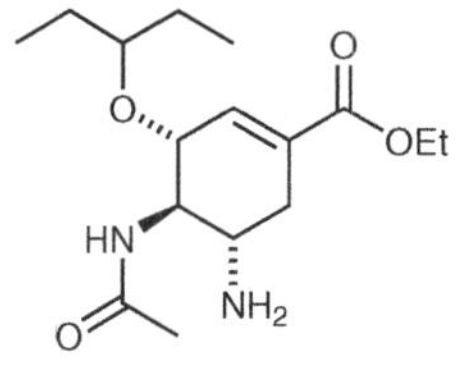

Year of discovery early 1995 (Gilead)
First WW approval US 1999
US approval (as phosphate salt, Tamiflu®, Roche) July 1999
Drug cat.: antiviral, orally active inhibitor of Neuraminidase
Uses: treatment and prevention of influenza viral infections

An influenza pandemic can have devastating effect on world populations and is a global concern. There have been three pandemics per century for the last 300 years. The Spanish flu outbreak of 1918 is estimated to have killed between 50 and 100 million people worldwide. The other two pandemics of the 20th century were less catastrophic: the Asian flu pandemic of 1957 causing 1-4 million deaths and the Hong Kong flu pandemic of 1968 causing over 1 million deaths. These flu outbreaks are caused by recombinant viruses of avian origin, although pigs are also infected with flu viruses, which can also be transmitted to humans as is the case with the 2009 H1N1 swine flu pandemic which, the WHO estimates caused between 151,700 and 575,400 deaths.

With the emergence of Fujian flu, caused by a mutation of the H5N1 strain of avian flu (A type) in 2004–2005, it quickly became apparent that, in the face of this new and virulent virus, resistant to known treatments and with a high potential for further mutation, new defensive medicinal weaponry was needed. The high mortality caused by this virus and its quick spread between poultry and migratory waterfowl caused great alarm. Fortunately, individuals who had contracted the disease invariably had been in contact with infected birds and human-to-human transmission was not observed. The outbreak was therefore contained, but the dangers associated with a mutation, which would allow spread from person-to-person, is of considerable concern.

Immunization by vaccine is an effective approach to fight flu epidemics with a protective efficacy of 70–90% in healthy individuals under 65 years of age. However, new vaccines are continuously required to be effective against the constantly mutating flu virus. For this reason, the Pharmaceutical Industry makes great efforts to reduce the time to develop, manufacture and place these new vaccines on the market. Small-molecule antiviral agents represent genuine potential for flu therapy and perhaps even prevention. However, antiviral drugs should be considered important adjuncts to vaccines, but not substitutes.

Influenza A, B and C are caused by RNA viruses. Matrix-2 (M2) protein, a proton ion channel protein present in the type A virus viral envelope, is crucial for viral replication but is not present in B and C type viral envelopes. Inhibitors of the M2 proton pump (Amantadine and Rimantadine, Figure 8.1) have been known for decades but resistance to these drugs and toxic side effects indicated the necessity to find new and more efficient drugs.

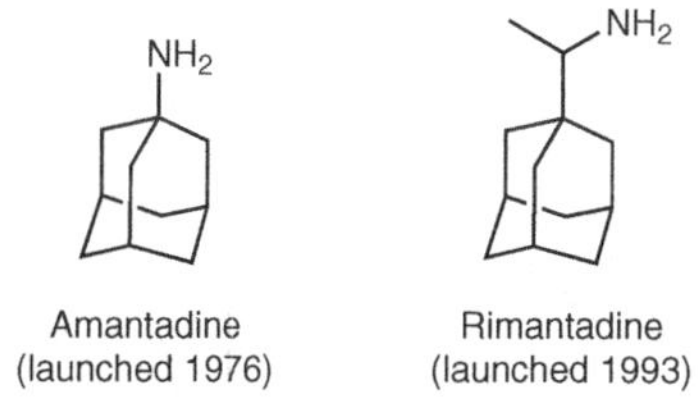

FIGURE 8.1 AMANTADINE AND RIMANTADINE CHEMICAL STRUCTURES

Neuraminidase is an essential enzyme that plays a decisive role in viral replication in all three types of viruses (A, B and C) and catalyses the cleavage of sialic acid residues. Neuraminidase catalysed cleavage allows newly formed viruses connected to the surface of the host cell to escape and to infect other cells after the budding process is complete. Inhibition of the enzyme causes the new virus particle to remain chemically attached to the cell surface and effectively stops the infective cycle. The structure neuraminidase enzyme was elucidated in the 90's and the structural data allowed the rational design of inhibitors that mimic the electrophilic species formed during the cleavage process (Scheme 8.1).

Retrosynthesis in the Manufacture of Generic Drugs: Selected Case Studies, First Edition, Pedro Paulo Santos and William Heggie.

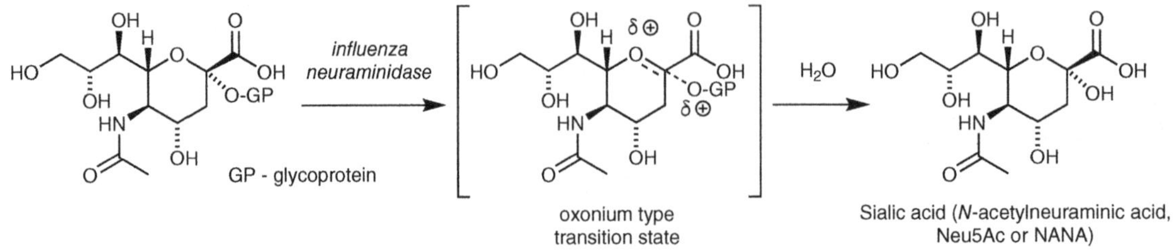

Scheme 8.1 Proposed role of neuraminidase in the replication of flu virus

Among the large number of compounds produced, Zanamivir and Oseltamivir emerged as forerunners (Figure 8.2). Zanamivir was first synthesised at Monash University in collaboration with Peter Colman who discovered the three-dimensional structure of neuraminidase. Zanamivir is administered by inhalation, whereas Oseltamivir phosphate has high bioavailability orally. Oseltamivir is a prodrug, the active compound being produced by *in vivo* ester hydrolysis. After a quick development phase (2.5 years) and an accelerated review, Tamiflu® was launched in November 1999, only four years after its discovery.

The urgency to find new lifesaving treatments to contain the H5N1 flu quickly brought Oseltamivir and Zanamivir into the spotlight worldwide. Laninamivir (Inavir®) and Peramivir (Rapivab®) are other recent compounds that belong to this class of drugs.

8.1 WORLDWIDE SALES AND PATENTS STATUS

The following graphs illustrate how research carried out in the pharmaceutical industry towards the production of the API and worldwide sales of the finished drug product evolved.

- US basic patent term 27/06/2017 (US5763483, Gilead1995b), 02/08/2016 (US5866601, Gilead1995a)
- Annual production[1]: 2019, 46,798 kg; 2018, 35,258 kg; 2017, 24,846 kg.
- Annual WW sales[1]: 2019, 1134.1M USD; 2018, 1517.7M USD; 2017, 1267.8M USD.

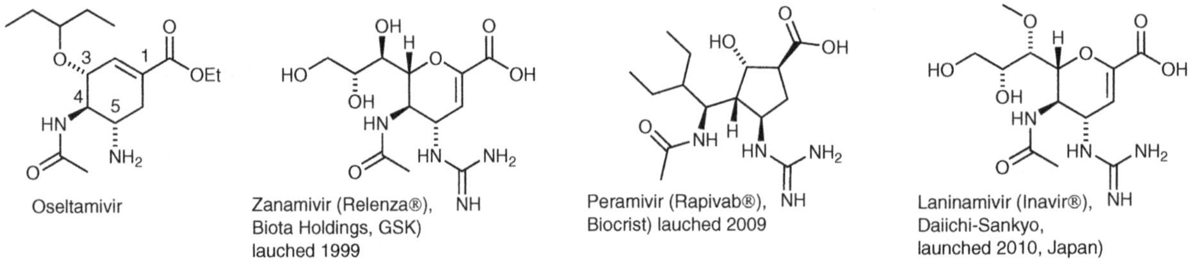

Figure 8.2 Chemical structure of selected neuraminidase inhibitors

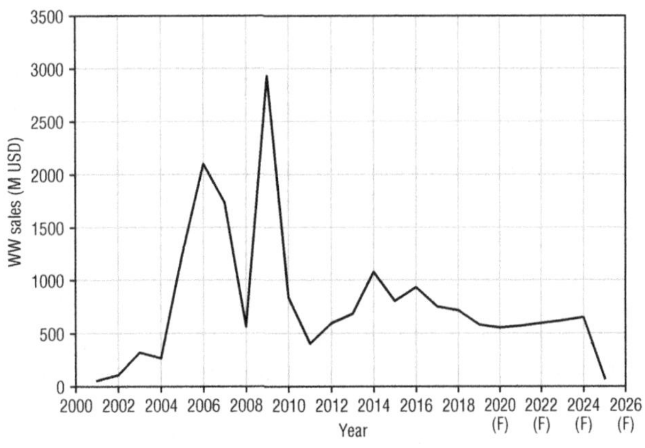

Figure 8.3 Oseltamivir WW sales evolution (in M USD)

Source: Data from GlobalData [Online] https://globaldata.com/

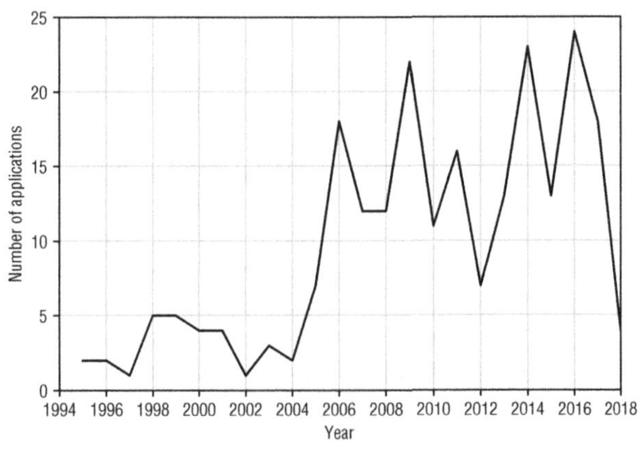

Figure 8.4 New Patent applications on Oseltamivir filed by year

1 Newport Modular Platform – Clarivate Analytics [Online] https://clarivate.com/cortellis/ [2020, April]

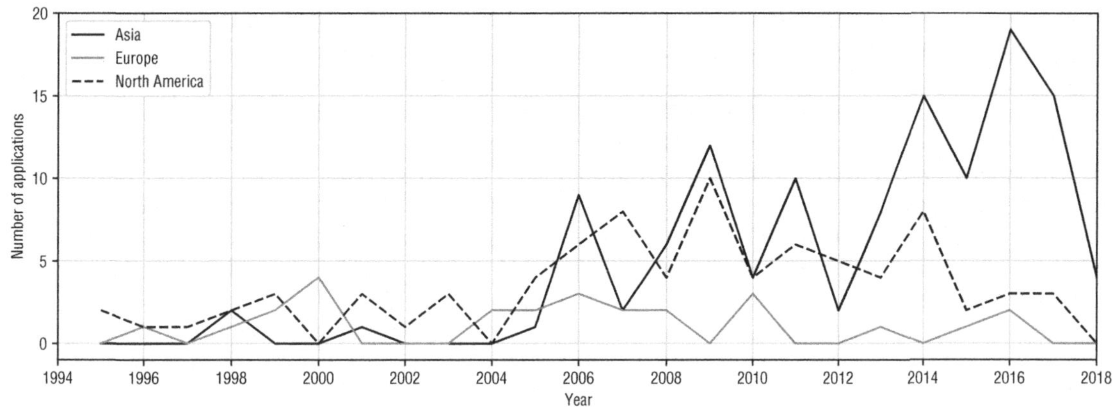

FIGURE 8.5 PATENT APPLICATIONS ON OSELTAMIVIR FILED BY REGION

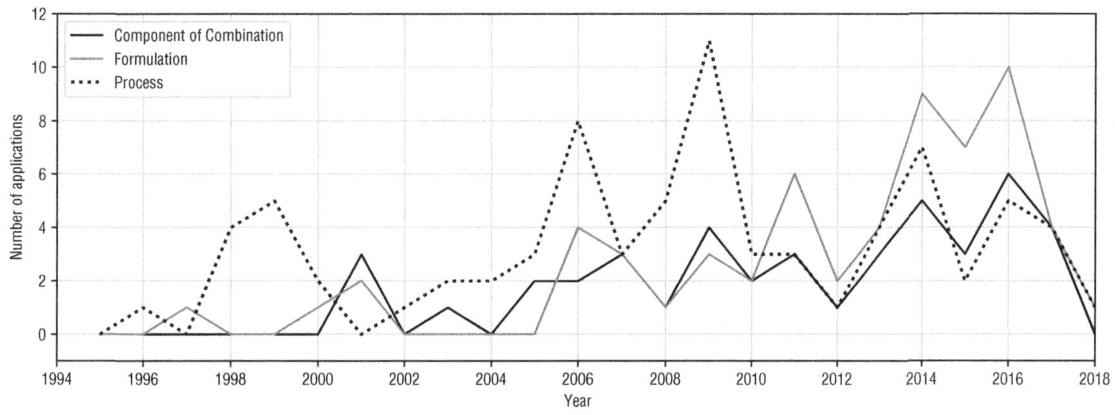

FIGURE 8.6 PATENT APPLICATIONS ON OSELTAMIVIR FILED BY TYPE OF PATENT

8.2 OSELTAMIVIR STRUCTURE AND GENERAL RETROSYNTHETIC ANALYSIS

Oseltamivir (Figure 8.7) is a fairly small but complex molecule and presents a tough challenge for the synthetic industrial chemists. The development of a reliable, safe and robust industrial process which can be scaled up is not an easy task. Some of the basic features of the molecule are:

- It is a 3,4,5-trisubstituted cyclohex-1-enecarboxylic acid ester possessing three contiguous chiral carbon atoms in the C6 ring[2]
- The position of the double bond is critical for drug activity (JACS1997) thus making the regiocontrol of its construction crucial during the synthesis
- The drug substance is a pure enantiomer, the three contiguous chiral sp^3 carbons each having the adjacent substituent in a *trans* configuration (α,β,α) in the C6 ring. In the most stable conformer of the cyclohexene ring the nitrogen substituents are equatorial and the ether pseudoequatorial (Figure 8.7). Hereinafter, the substituents are labelled as α when pointing backwards and β when pointing upwards when the molecule is drawn as in Figure 8.7. Throughout the text the intermediates will be presented in the same way as the final product to enable the reader to discern the stereochemistry of the diverse groups on the ring that appear in the intermediates

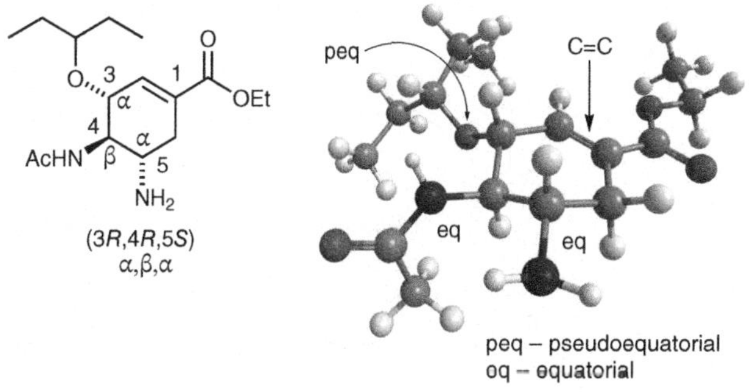

FIGURE 8.7 OSELTAMIVIR CHEMICAL STRUCTURE

2 Note that, throughout this chapter, the carbons atoms of intermediates will regularly be labelled according to the numbering that they will assume in the final Oseltamivir molecule. This will help the reader to identify each carbon atom throughout the analysis and description of the synthetic process.

- The 3-pentylether at C3 is a fairly bulky substituent and this may cause some difficulties for its introduction or, when already in place, may hinder subsequent reactions at adjacent positions
- Selective introduction of the acyl group at C4N over the C5N is required and migration of this acyl group from C4N to C5N has to be avoided
- The highly functionalized cyclohexene has the potential to aromatize and may cause issues during some synthetic steps, necessitating the use of mild conditions for various transformations.

8.2.1 Starting material targeted retrosynthesis

The first laboratory synthesis of Oseltamivir used (–)-shikimic acid as starting material. Even after Oseltamivir was finally identified as the target molecule for clinical development, its structural resemblance to (–)-shikimic acid made it clear that this material continued to be an especially favourable starting material for industrial synthesis. An examination of the (–) shikimic acid molecule indicates that the carbocyclic system is already present, the double bond is in the correct position and all the carbon atoms of the six-member ring are at the correct oxidation state, which signifies that no oxidation or reduction steps are required in the synthesis (Scheme 8.2). Yet another closely related potential starting material is (–)-quinic acid. In this case, a regioselective dehydration is required.

The introduction of the ring substituents predictably would use bimolecular displacements (S_N2 reactions). After an examination of the relative configuration at C3, C4 and C5 of shikimic acid and Oseltamivir, a synthetic plan can be devised (Scheme 8.3).

It is expected that the several hydroxyl groups present have to be selectively activated towards the displacements indicated in Scheme 8.3. With the application of some ingenuity, the use of a tri-*O*-mesyl intermediate may well be a good candidate for selective substitution reactions (Scheme 8.4), although at first sight, this might be considered somewhat improbable. Nonetheless, Process Chemists tend to explore these situations with a view to developing simpler and more streamlined processes.

It is feasible that advantage can be made of the minor chemical and stereochemical differences between the three oxygen substituents to promote regioselective reactions. Theoretically, the C3 carbon is the more reactive toward a nucleophile as it is allylic and is less sterically hindered, the C4 carbon being the most hindered.

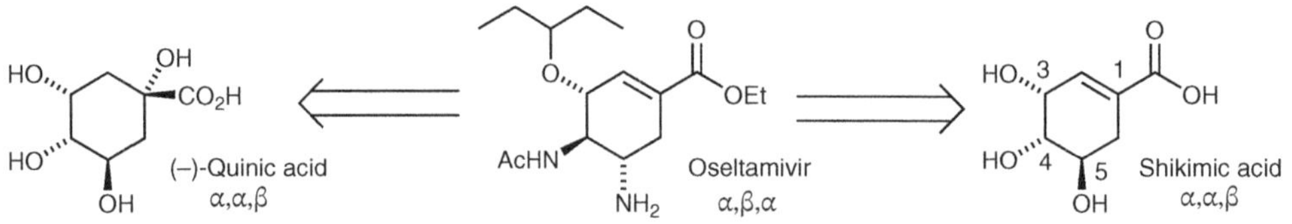

Scheme 8.2 Oseltamivir - structure and potential starting materials

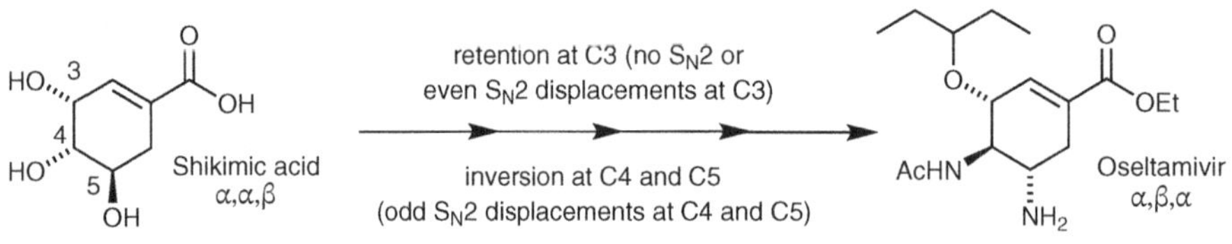

Scheme 8.3 Synthetic requirements to produce Oseltamivir from shikimic acid

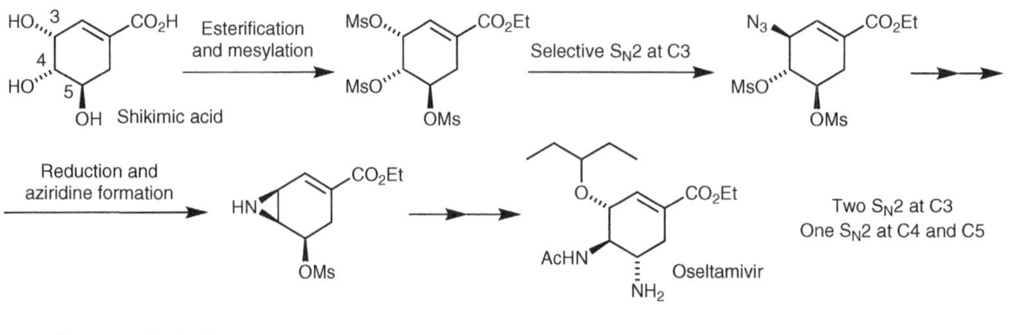

Scheme 8.4 Oseltamivir synthesis by direct functionalization transformation of a tri-*O*-mesyl shikimic acid derivative

Of note is the fact that an S_N2 reaction at the C3 carbon inverts configuration and a second S_N2 reaction is therefore necessary to produce the desired configuration. The use of a NH_2^- synthon for the first S_N2 reaction followed by an aziridine ring closure sets the system up for ring opening at the most reactive C3 carbon atom. Thus, the nitrogen is initially introduced at C3 and migrated to C4 in this sequence. This strategy allows for the facile introduction of the correct C3 and C4 substituents. The primary amine at C5 can then be introduced by an S_N2 reaction with a NH_2^- synthon.

This direct, selective transformation of a reactive intermediate is desirable for any industrial synthesis as it permits an abridged route wherein protective group chemistry is avoided.

Other, more traditional strategies can be elaborated by selectively activating the hydroxyl groups in shikimic acid towards displacement. In this case, protective groups are, in all probability, required. The *syn*-diol (C3/C4) is perfectly set up for protection as a ketal. Also of note is that the C5 hydroxyl group is comparatively more accessible than the hindered C4 hydroxyl group. An outline of the main retrosynthetic pathways to shikimic acid is shown in Scheme 8.5. Bimolecular substitution reaction ought to play a key role in the synthesis. Epoxide and aziridine intermediates are also of high importance.

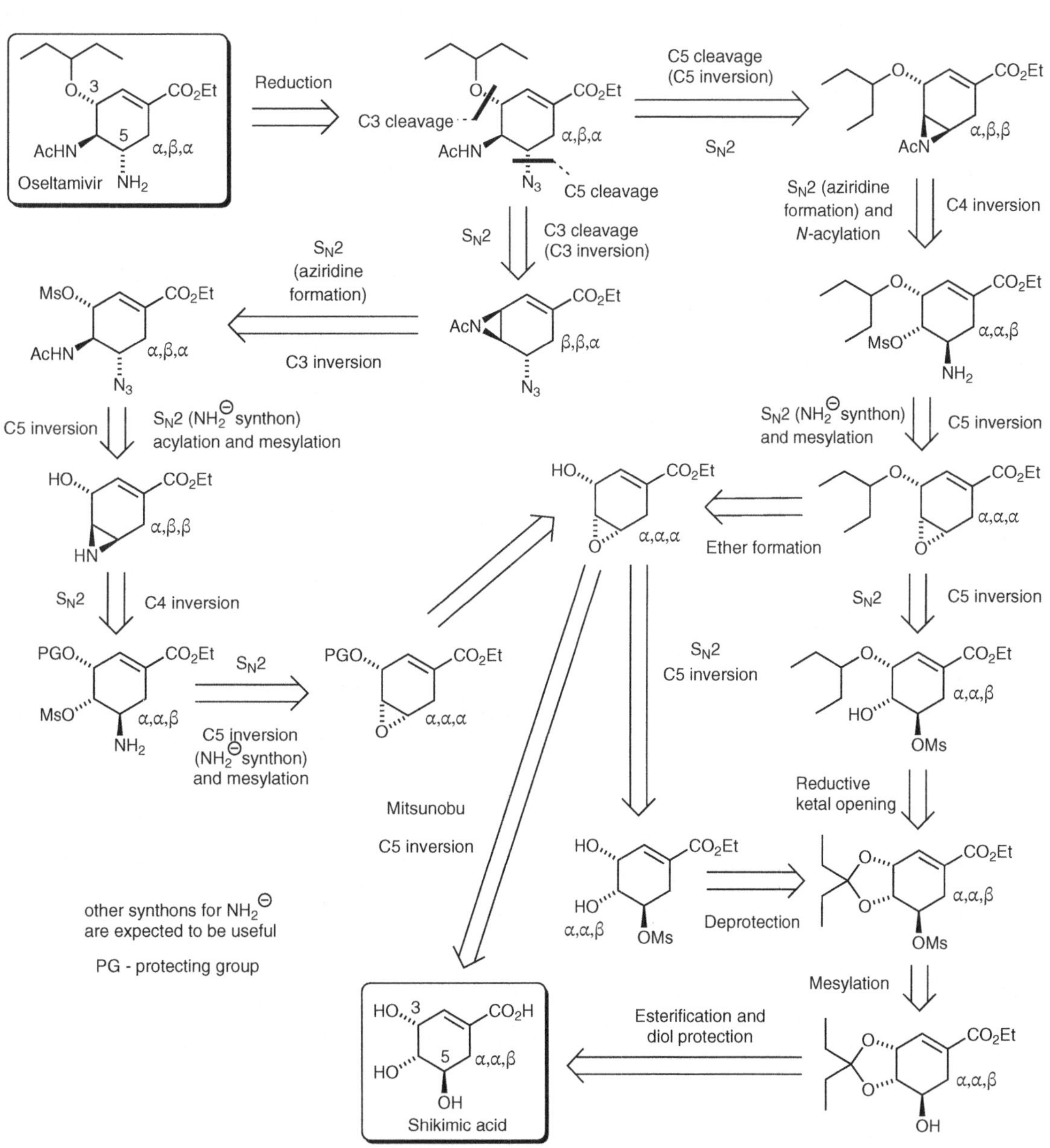

SCHEME 8.5 MAIN RETROSYNTHETIC ROUTES TO SHIKIMIC ACID

A possible option is the initial removal of the C3 side chain. The C3 carbon in shikimic acid has the same configuration as the C3 hydroxyl substituent in Oseltamivir and ether formation by nucleophile alkylation of C3 hydroxyl group with a 3-halopentane (or equivalent) gives the correct final stereochemistry. However, even if this transformation is expected to be feasible it is should not be particularly favourable because of steric hindrance. A better alternative employs the neighboring C4-nitrogen to facilitate the procedure *via* a ring opening of a C3/C4 aziridine. The desired regioselective ring opening, i.e. attack at C3, is favourable as C3 is allylic and also is less sterically hindered than C4. In the process, the C3 carbon suffers two S_N2 substitutions hence, the final carbon configuration remains unchanged (Scheme 8.5, left branch). Direct replacement of the primary amine at C4 looks less feasible. A possible strategy is to introduce a nitrogen at C5 *via* the intermediacy of a C4/C5 epoxide. Formation of a C4, C5 aziridine transfers the nitrogen to C4 and an α nitrogen function at C5 is introduced *via* aziridine ring opening with a nitrogen electrophile. It can be seen that, under these circumstances, α-epoxides which are required for the correct final stereochemistry are expected to be privileged intermediates in the synthesis of Oseltamivir. Notably, in this indirect procedure, the C4 carbon atom only suffers one S_N2. Thus, with one inversion of configuration the desired stereochemistry is obtained.

Formation of the reactive α epoxides ought to be more or less straightforward. The *syn*-α-diol in shikimic acid is protected as a ketal and the C5 hydroxyl mesylated. Thereafter, hydrolysis of the ketal produces the epoxide directly. During the entire process the C5 carbon is inverted three times.

In another likely approach, the order of the events is reversed, the C3 substituent being put in place earlier in the synthesis while the groups at the C4 and C5 carbons are introduced towards the end of the sequence (Scheme 8.5, right branch).

A reductive opening of diethyl ketal is yet another option to produce the C3 pentyloxy side chain followed by introduction of the C4 and C5 nitrogen functionality, wherein different nucleophiles can be used as NH_2 synthons, azide, allyl amine and benzylamine being some of the more common possibilities.

Several different routes have potential for development around the main strategic retrosynthesis shown above. Due to the close similarity between the target molecule and shikimic acid short and efficient routes are expected to be feasible and hence, are anticipated to be highly favourable for industrial scale production.

Notwithstanding, other retrosynthetic approaches can be elaborated to build the cyclohexene core from simpler starting materials.

8.2.2 Retrosynthetic strategies of the Oseltamivir cyclohexene ring

Alternative retrosynthesis of the cyclohexene core of Oseltamivir can be elaborated. In some cases, chirality in the staring material already exists and in others, the chirality has to be introduced during the synthesis. The unsaturated cyclohexene ring can be obtained from an aromatic ring or by a classical [4+2] cycloaddition (Diels-Alder) reaction. The CC double bond of Oseltamivir is a good handle to initiate several disconnections. Aldol, Dieckmann and similar reactions (for example Horner-Wadsworth--Emmons Olefination) as well as metathesis can all be applied to form the C6 ring. A retrosynthetic outline for Oseltamivir is shown in scheme A.

8.2.2.1 Producing the Oseltamivir core from an aromatic ring

Total reduction (i.e. *via* hydrogenation) of an appropriate aromatic ring followed by elimination is a good entry point. A retrosynthetic approach to a symmetric intermediate is an attractive option with phenol derivatives being starting materials of choice (Scheme 8.6). A desymmetrization step is key in making this strategy meaningful. For example, enzymatic hydrolysis of a symmetrical diester offers an effective solution.

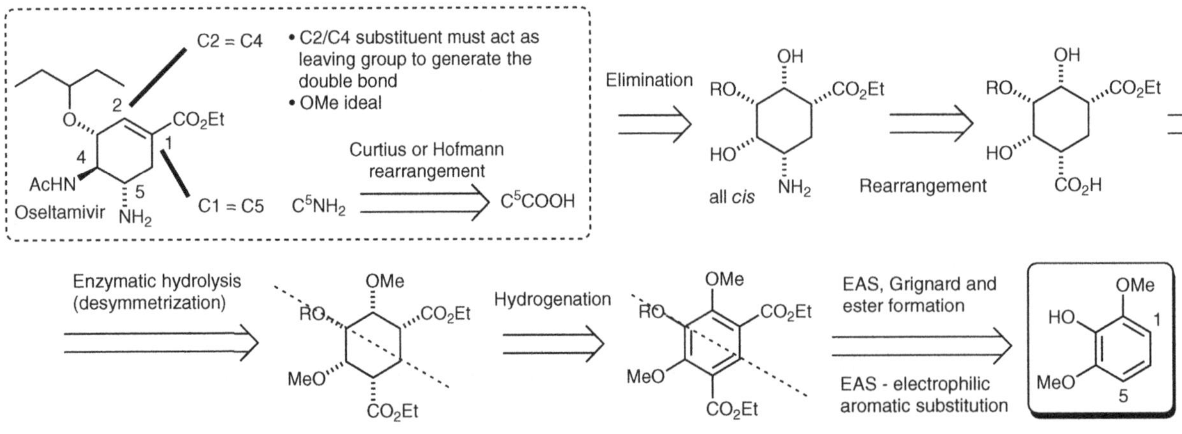

Scheme 8.6 Retrosynthesis of Oseltamivir to a symmetric aromatic starting material

Chiral metabolites obtained from fermentation of aromatic substrates are interesting starting materials and are freely available in large quantities (Scheme 8.7).

The double bond that does not have the electron-withdrawing group X attached is the more nucleophilic and can be functionalized selectively *via* electrophilic reactants (Scheme 8.8). The *cis* diol allows the formation of a bulky ketal that induces asymmetry on reaction of the more reactive, adjacent double bond.

In spite of the fact that several transformations are required to construct the Oseltamivir molecule, the starting diol possesses the appropriate framework to generate a viable synthetic route.

8.2.2.2 Forming Oseltamivir ring by a Diels-Alder cycloaddition

Construction of the C6 ring using a [4+2] cycloaddition is another potential option. An outline of two possible strategies *via* bicyclic intermediates is shown in Scheme 8.9. The relative position to the α,β-unsaturated ester and the C5 substituent makes it possible to generate the cyclohexene core from a C2-C5 bridged intermediate. The bicyclic cage structures are used to control *cis* functionalization of the carbon-carbon double bond formed in the Diels-Alder reaction, thus assuring the relative stereochemistry at C3 and C4. Since an appropriate asymmetric Diels-Alder reaction may not be feasible, a

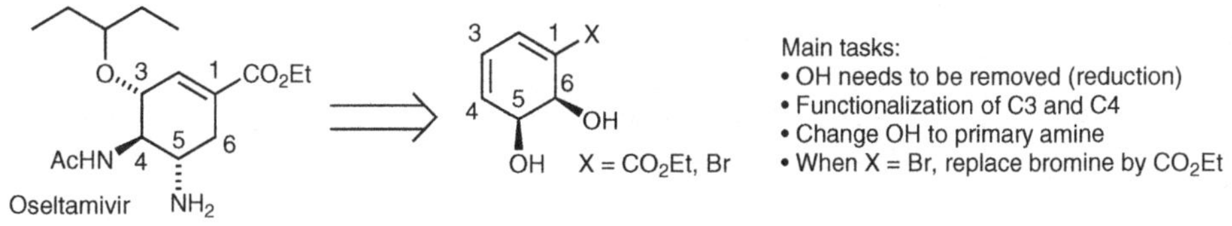

SCHEME 8.7 RETROSYNTHESIS OF OSELTAMIVIR TO A CHIRALLY PURE CYCLOHEXA-3,5-DIENE-1,2-DIOL DERIVATIVE

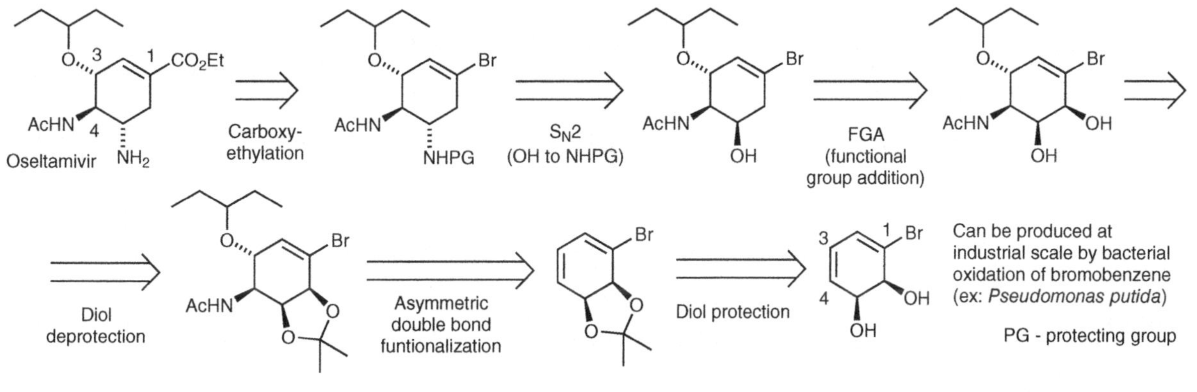

SCHEME 8.8 RETROSYNTHESIS OF OSELTAMIVIR TO (1*S*,2*S*)-3-BROMOCYCLOHEXA-3,5-DIENE-1,2-DIOL

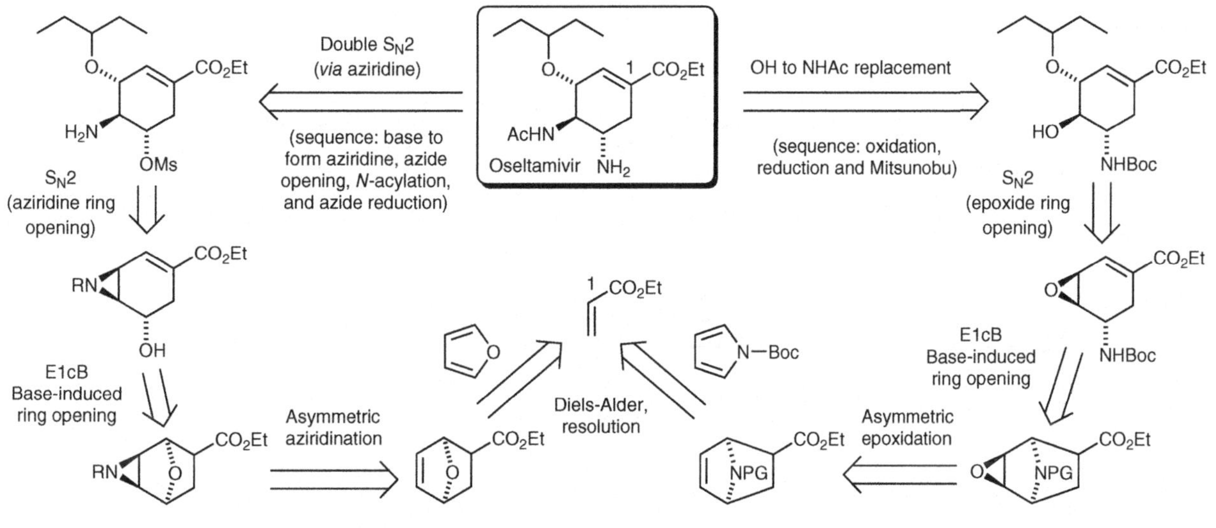

SCHEME 8.9 RETROSYNTHESIS OF OSELTAMIVIR *VIA* A RETRO-DIELS-ALDER DISCONNECTION 1

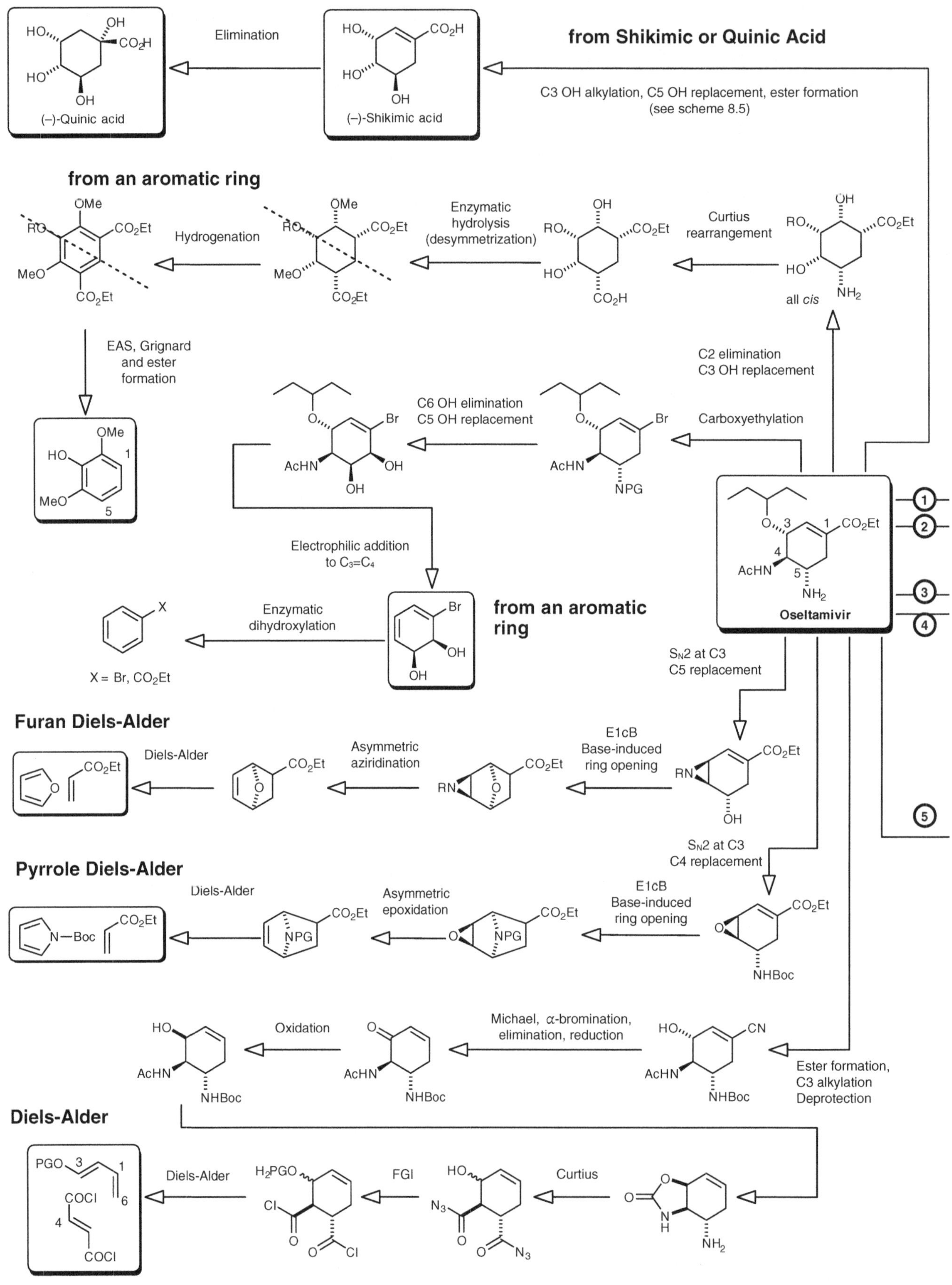

Scheme A Retrosynthetic outline for oseltamivir

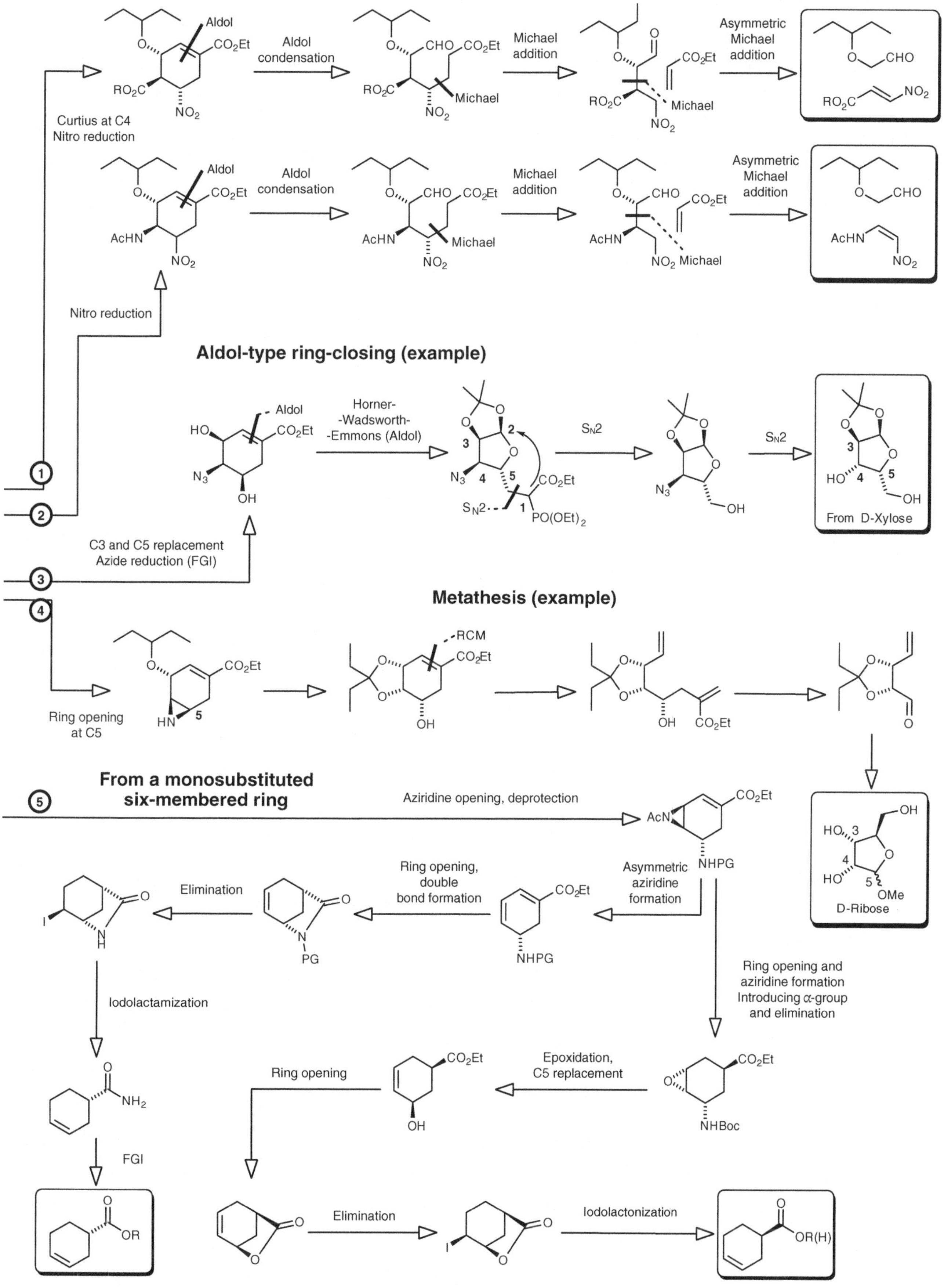
Aldol
Aldol condensation
Michael addition
Asymmetric Michael addition
Michael
Curtius at C4
Nitro reduction
Nitro reduction
Aldol-type ring-closing (example)
Horner-Wadsworth-Emmons (Aldol)
SN2
SN2
From D-Xylose
C3 and C5 replacement
Azide reduction (FGI)
Metathesis (example)
RCM
Ring opening at C5
From a monosubstituted six-membered ring
Aziridine opening, deprotection
D-Ribose
Asymmetric aziridine formation
Ring opening, double bond formation
Elimination
Iodolactamization
Ring opening and aziridine formation
Introducing α-group and elimination
Epoxidation, C5 replacement
Ring opening
FGI
Elimination
Iodolactonization

resolution step becomes necessary, thus reducing the efficiency of this strategy for industrial application.

A further Diels-Alder-based strategy is shown in Scheme 8.10. The cyclohexene sub-structure is obtained directly in the cycloaddition. The vicinal diamine can be obtained from the dicarboxylic acid derivative by a double Curtius rearrangement, whilst the relative configuration of the C4 and C5 carbon atoms is also produced during the ring formation step from the *trans*-ene. Again, this strategy would benefit from an asymmetric Diels-Alder reaction to avoid a resolution step.

Yet another possible approach is shown in Scheme 8.11, the key intermediate being a chiral diene generated by removal of the C3 and C4 groups (highlighted intermediate), their synthetic placement depending on an asymmetric functionalization of the more nucleophilic CC double bond, as above. The chiral diene can be generated from a chiral lactam but chiral bicyclic lactones and are just as effective as intermediates. Both can be produced from chiral cyclohex-3-ene-1-carboxylic acid derivatives, which in turn can be produced from asymmetric Diels-Alder reactions.

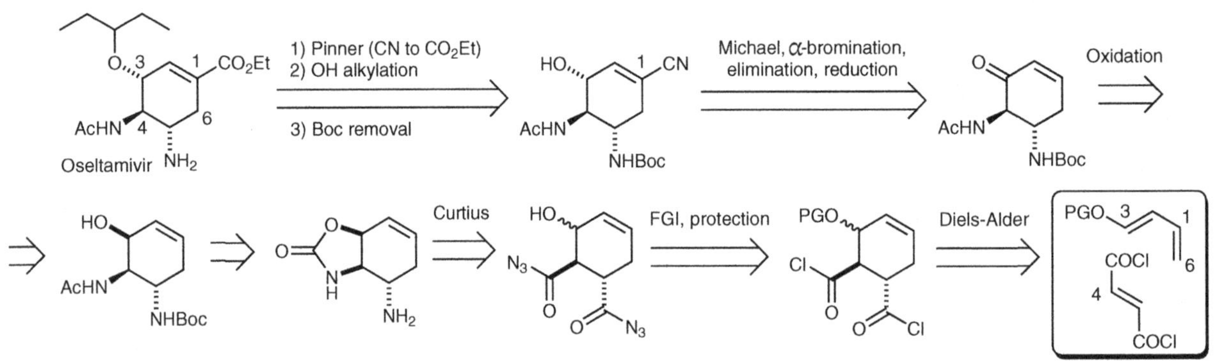

Scheme 8.10 Retrosynthesis of Oseltamivir *via* a retro-Diels-Alder disconnection 2

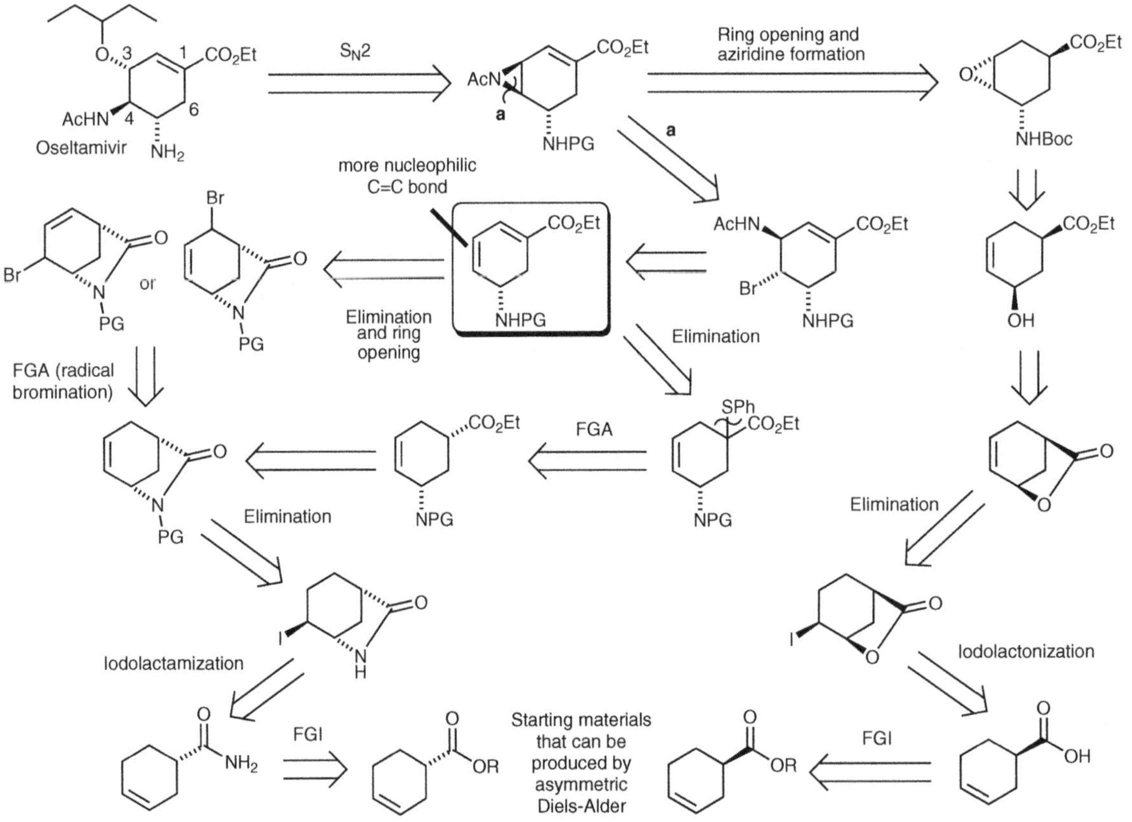

Scheme 8.11 Retrosynthesis of Oseltamivir *via* a chiral diene intermediate and bicyclic lactam and lactones

8.2.2.3 Forming Oseltamivir by CC double bond forming ring closure

The unsaturated carbon-carbon double bond of Oseltamivir can be used as a handle to open the C6 cycle. Aldol-type disconnections or ring-closing metathesis (RCM) are two possibilities (Scheme 8.12).

A chiral pool approach is a viable possibility and, in principle, all the chiral centres of the drug can be obtained from easily accessible carbohydrates. A basic retrosynthesis to a C5 and C6 carbohydrate is outlined in Scheme 8.13, using an intramolecular Horner-Wardsworth-Emmons (HWE) reaction to promote the ring closure. Protecting groups are necessary and the C1 and carboxylic acid atoms are added using triethyl phosphonoacetate that promotes the HWE reaction. When a C6 carbohydrate is used as starting material, the terminal carbon atom containing the primary hydroxyl group is removed and the phosphonate ester carbanion added to the C5 intermediate.

Other aldol-based processes starting from raw materials that do not contain all three of the chiral carbon atoms are also viable. In these cases, asymmetric syntheses are inferred to put the remaining groups in place in a controlled fashion. The use of the HWE cyclization avoids any ambiguity in the ring-closure.

A noteworthy retrosynthesis of Oseltamivir, that does not rely on deriving the chirality from a carbohydrate, builds up an open chain intermediate prior to the formation of the double bond by ring-closure is shown in Scheme 8.14. Firstly, the C5 amine is retro-transformed into a nitro group. Retrosynthesis continues by a double retro-Michael addition. Good control of the stereochemical transformations is key in forming the three new chiral centres in this relatively simple route if the synthesis is to be useful. The first Michael adduct can be produced *via* asymmetric catalysis but requires investigation to ascertain the best conditions. The Michael-aldol cascade that ends with the formation of

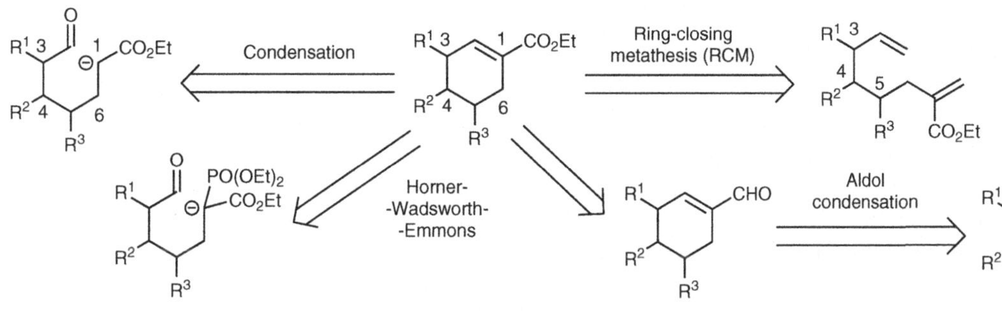

SCHEME 8.12 MAIN OPTIONS FOR RING-OPENING IN OSELTAMIVIR

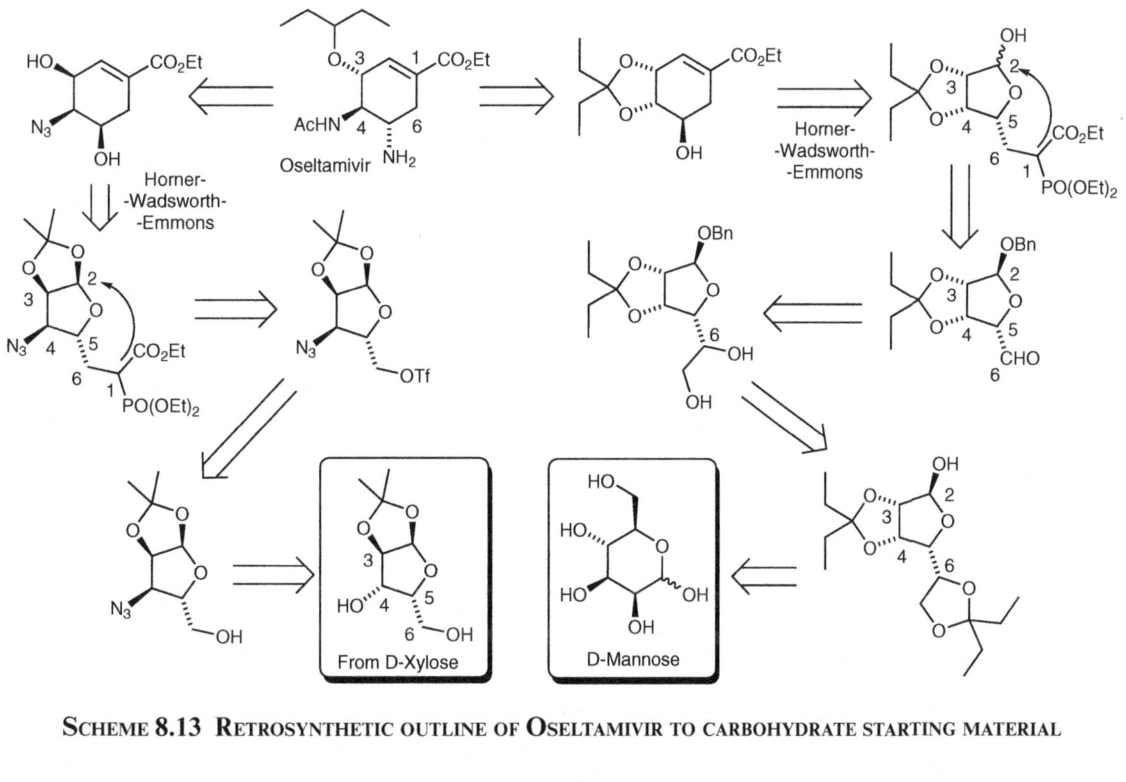

SCHEME 8.13 RETROSYNTHETIC OUTLINE OF OSELTAMIVIR TO CARBOHYDRATE STARTING MATERIAL

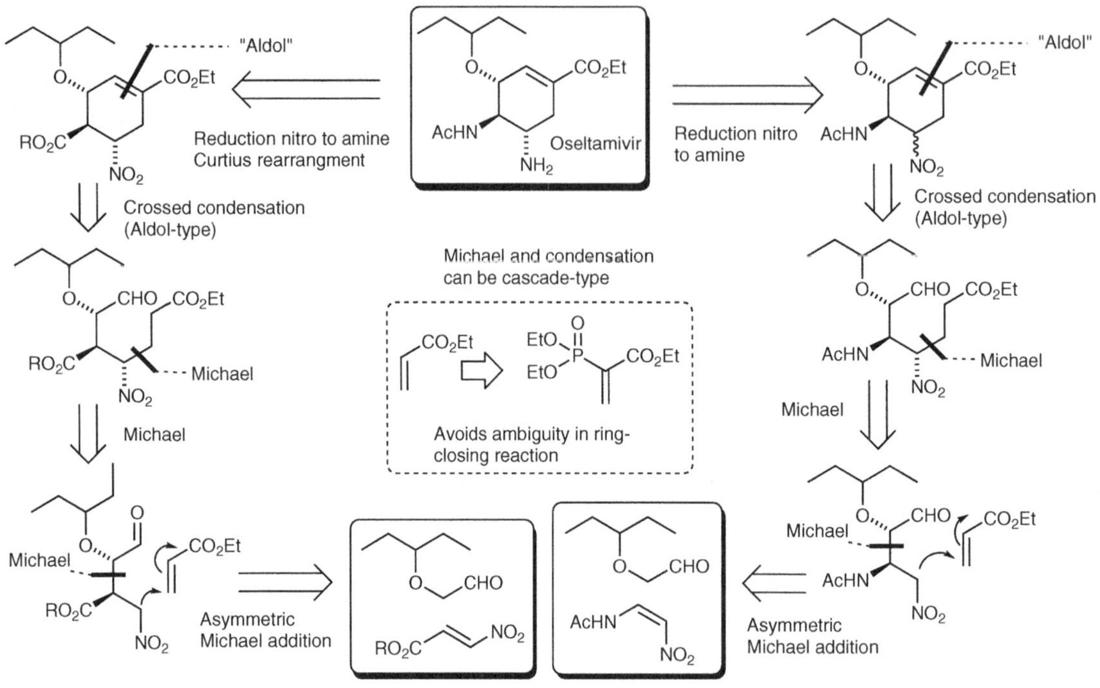

SCHEME 8.14 RETROSYNTHESIS OF OSELTAMIVIR BY A RETRO-ALDOL AND DI-RETRO-MICHAEL REACTIONS

the cyclohexene ring, in addition, sets the configuration at the C5 carbon. From an examination of a three-dimensional model of the final ring it can be seen that the C5-α isomer is more stable than the C5-β isomer (compare Figure 8.7) when the nitro group occupies an equatorial position. An equilibration process using base or acid catalysis can correct the stereochemistry when kinetic control rather than thermodynamic control produces the wrong (β) isomer.

The application of olefin metathesis for the synthesis of complex organic molecules was first reported at the beginning of the 1990s and has since then become a reliable and widely used synthetic methodology. As has been seen above for aldol-based processes, all of the stereochemical centres can be obtained from a suitable carbohydrate or produced by asymmetric induction before the ring is closed *via* metathesis (RCM). In all cases, a di-terminal alkene is cyclized with the release of ethylene (Scheme 8.15).

8.2.3 Retrosynthetic analysis summary

From the discussion above it can be concluded that various chemical strategies can be used to produce Oseltamivir. These retrosynthetic routes are presented in scheme A. Note that only the most important retrosynthetic routes were outlined, although each of these has several feasible variations. Hence, a large number of routes have the potential for development. The following sections describe the chemistry that was developed and reported for the retrosynthetic routes portrayed above.

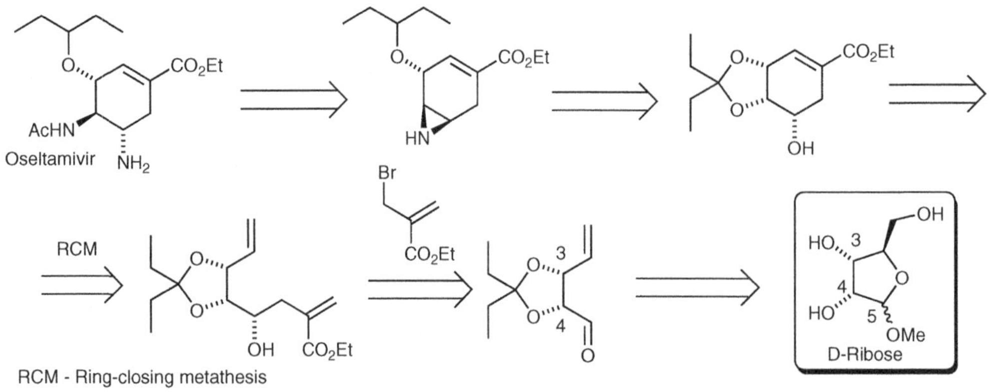

SCHEME 8.15 RETROSYNTHESIS OF OSELTAMIVIR BY A RING-CLOSING METATHESIS REACTION

A cursory analysis indicates that processes starting from shikimic acid are shorter and simpler, the chiral centres already being present in the starting material. Construction of the cyclohexene ring *via* the diverse processes depicted looks less promising because the synthetic routes are lengthier: the chiral centres having to be produced by asymmetric synthesis or because extensive protection/deprotection chemistry has to be used when using a chiral pool approach.

8.3 THE FIRST REPORTED OSELTAMIVIR SYNTHESIS

Initially Gilead identified the highly active free carboxylic acid of Oseltamivir as the drug candidate (Chimia2004, JACS1997). Natural (–)-shikimic or (–)-quinic acids were used as starting materials for its initial synthesis. Due to its higher oral bioavailability, the ethyl ester analogue (a pro-drug) was eventually selected as the candidate to market.

The first syntheses basically followed the retrosynthetic outline shown above in Scheme 8.5. The C4/C5 α-epoxide is obtained from shikimic acid in three ways (Scheme 8.16, Gilead1995a). The methyl ester of shikimic acid is subjected to a direct Mitsunobu reaction. Alternatively, the *syn*-diol is protected as a ketal, the third hydroxyl group mesylated and the protecting group removed, forming the epoxide in the process. A process from quinic acid was developed but suffered from a low yield because the elimination of water gives a mixture of double bond regioisomers. However, the unwanted isomer was easily aromatization making its removal possible without the use of chromatography. This allowed batches of several hundreds of grams to be carried out. The routes *via* a 5 mesylate or tosylate avoid the relatively uneconomical Mitsunobu reaction to form the epoxide.

The transformation of the epoxide into Oseltamivir is shown in Scheme 8.17 (Gilead1995a). The C4 β amine is produced by first introducing a nitrogen at C5 (β) by a regioselective ring opening of an epoxide (step A) and migrating the nitrogen *via* ring opening of an aziridine derivative (steps B and C). Opening of this aziridine intermediate by azide ion also inserts the C5 nitrogen atom with the correct stereochemistry. A second aziridine was instrumental in the introduction of the 3-pentylether group. In summary, configuration was inverted once at C4, twice at C3 and three times at C5. The critical regioselective openings of the aziridine and epoxide intermediates are mainly due to stereochemical hindrance. This discovery process of Oseltamivir was also the subject of a scientific article (JACS1997).

One of the motivations that led to the discovery of Oseltamivir was to find an orally active drug, less polar

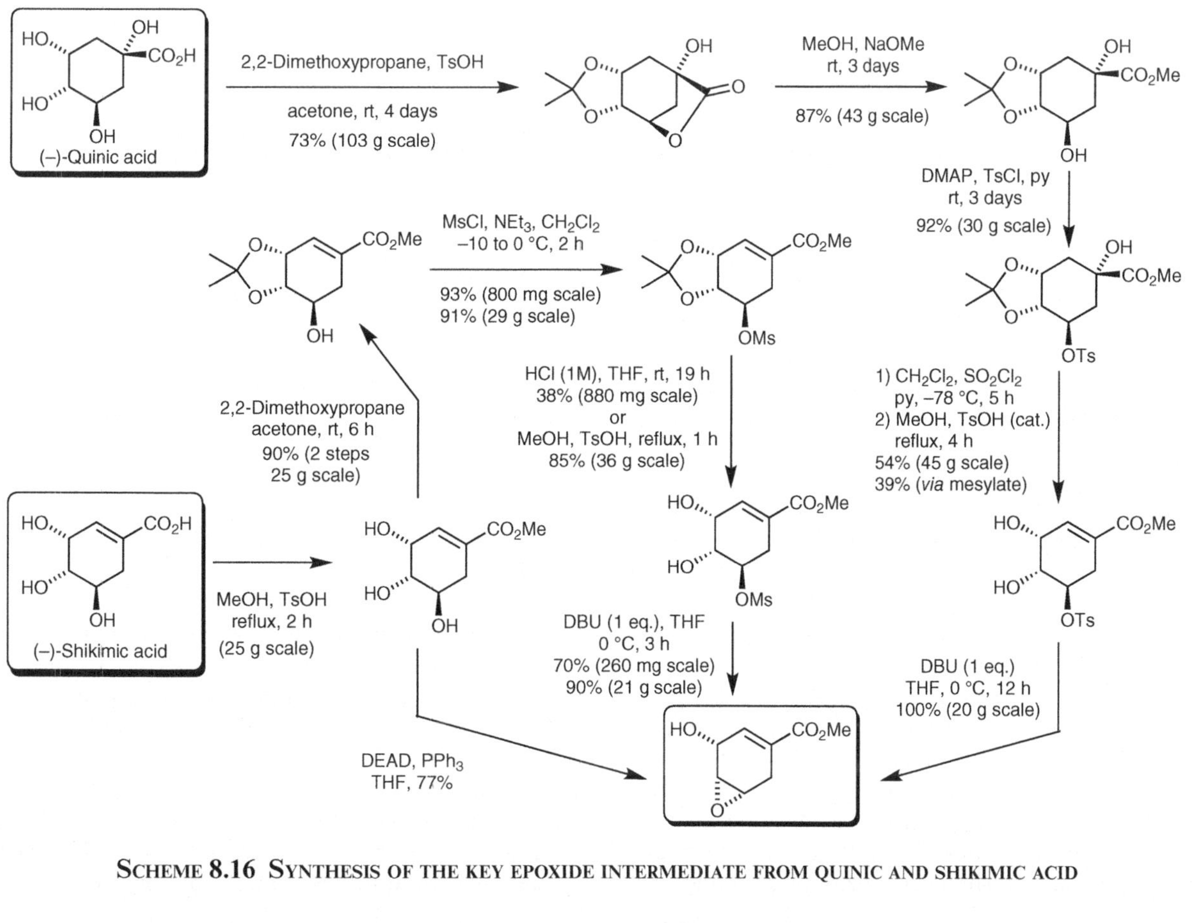

SCHEME 8.16 SYNTHESIS OF THE KEY EPOXIDE INTERMEDIATE FROM QUINIC AND SHIKIMIC ACID

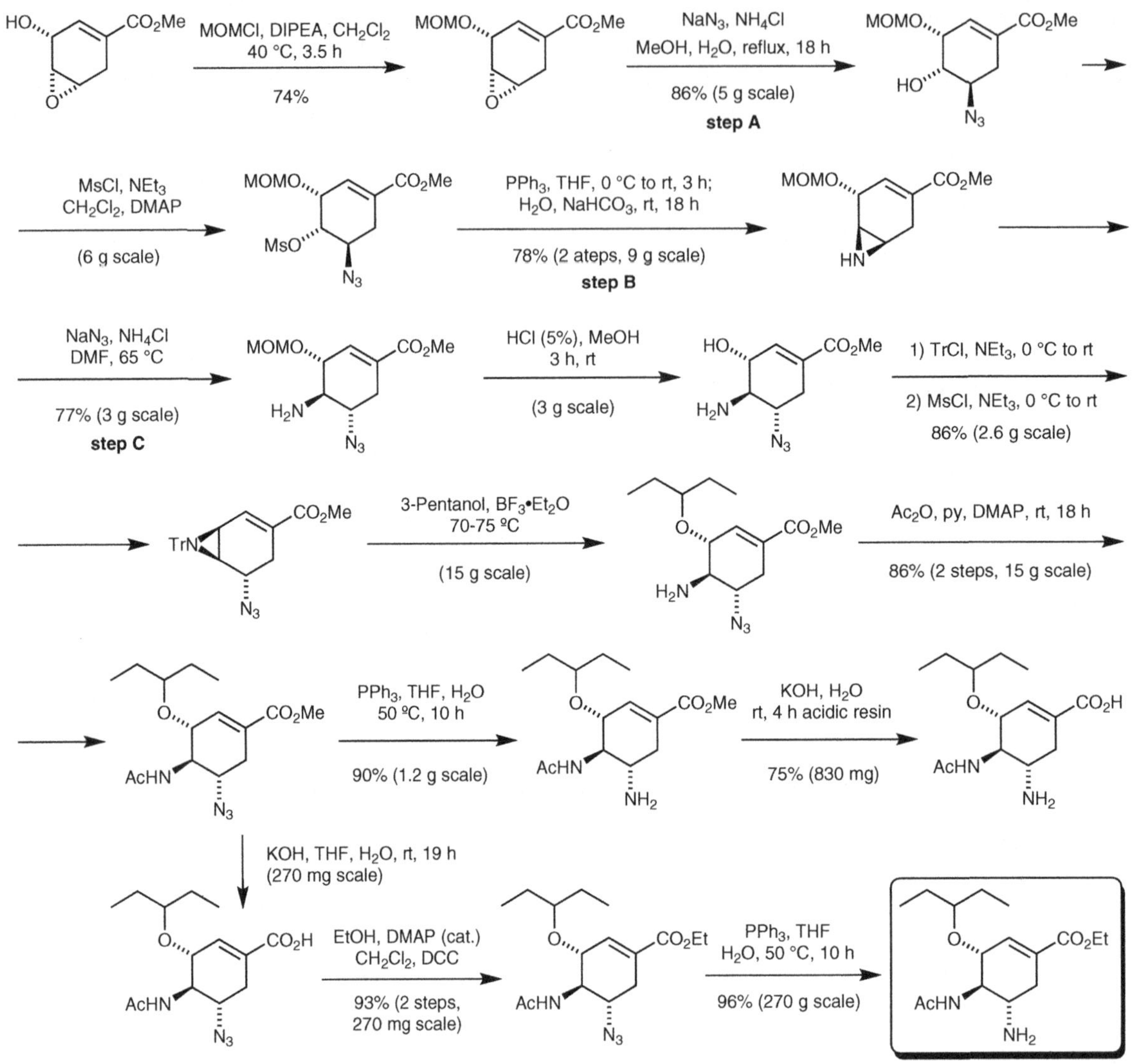

SCHEME 8.17 CONVERSION OF THE KEY EPOXIDE INTERMEDIATE INTO OSELTAMIVIR

Source: Gilead1995a, JACS1997

than NANA and Zanamivir (Scheme 8.1, Figure 8.2). The basic idea was to substitute the polar glycerol side chain for a spatially similar, less-polar group. The discovery route synthesis just described, fulfills this objective as a series of molecules can be efficiently prepared for screening from one intermediate. Hence, ring opening of the final aziridine with different alcohols gave this new class of biologically active materials and allowed selection of the most potent representative. Notably, variations in the ether side chain led to tremendous consequences for *in vivo* activity (JACS1997).

The synthesis, although well adapted to its original purpose and giving a reasonable accumulated yield (around 15%), is not suitable for large-scale supply of material. The discovery route, carried out on a gram scale, took 19 steps from quinic acid and 18 steps from shikimic acid. The use of azide-based chemistry, along with the stability of the various azide intermediates, requires careful evaluation. A side reaction, where azide adds in a conjugate fashion to the α,β -unsaturated ester, gives a β-azido ester, which is responsible for a positive Ames test in the final drug substance, even at low levels. The one Mitsunobu and two Staudinger reactions produce large amounts of undesirable waste material and several protecting groups are also employed.

8.4 ROUTES FROM SHIKIMIC ACID OR QUINIC ACID

Both shikimic and quinic acids are structurally close to the drug substance making a semi-synthesis from either an excellent choice for implementation on a large scale. Gilead, as well as several other companies, recognized this and invested in this strategy. These routes, which have many elements in common with the discovery syntheses, are summarized in schemes B to E. C4C5 epoxide intermediates clearly play a central role. Scheme B summarizes the routes from quinic or shikimic acid to the epoxides. Schemes C and D summarize the routes from epoxides to Oseltamivir. Other published quinic- or shikimic acid-base routes are summarized in scheme E.

8.4.1 First developments from shikimic acid or quinic acid

Starting from an advanced intermediate is recognized, in an industrial synthesis, to have clear advantages and improvements on the discovery route were the subject of the first studies.

An improved process from quinic acid using an elegant and novel regioselective reduction of a ketal to insert the ether side chain at C3 was developed (JOC1998, Gilead1996, 1997a, 1998, Scheme 8.18). Thus, the C3 group placement is re-deployed to earlier in the synthesis. As only an acetonide ketal produced crystalline intermediates, the initial transformations were carried out on this material and a ketal exchange was required before reductive ring opening. The isomeric alkene was removed *via* selective reaction of this impurity with pyrrolidine by an S_N2 displacement at the allylic mesylate of the unwanted isomeric akene, followed by an acidic aqueous extraction of the basic product. Hence, chromatography was avoided. Three crystalline intermediates were isolated in this synthesis, thus allowing removal of impurities and further avoidance of chromatography.

The epoxide was obtained in about 20% hence, the overall yield of this 12-step synthesis from (–)-quinic acid is still rather low (4.4%) but allowed implementation in standard pilot plant equipment at a kilogram scale. This route has several issues: potentially toxic and dangerous azide reagents and intermediates in addition to hazardous reaction conditions. Cost and availability of trimethylsilyl trifluoromethanesulfonate (TMSOTf) is problematic and the strong, unpleasant odour of PMe_3, characteristic of alkylphosphines, requires good containment. Low yields were recorded in two reactions due to the partial selectivity in the formation of the endocyclic double bond and in the ketal ring opening, whereas all the other reactions proceeded in reasonably high yields.

Improved processes to obtain an epoxide intermediate from shikimic or quinic acid, following the basic route

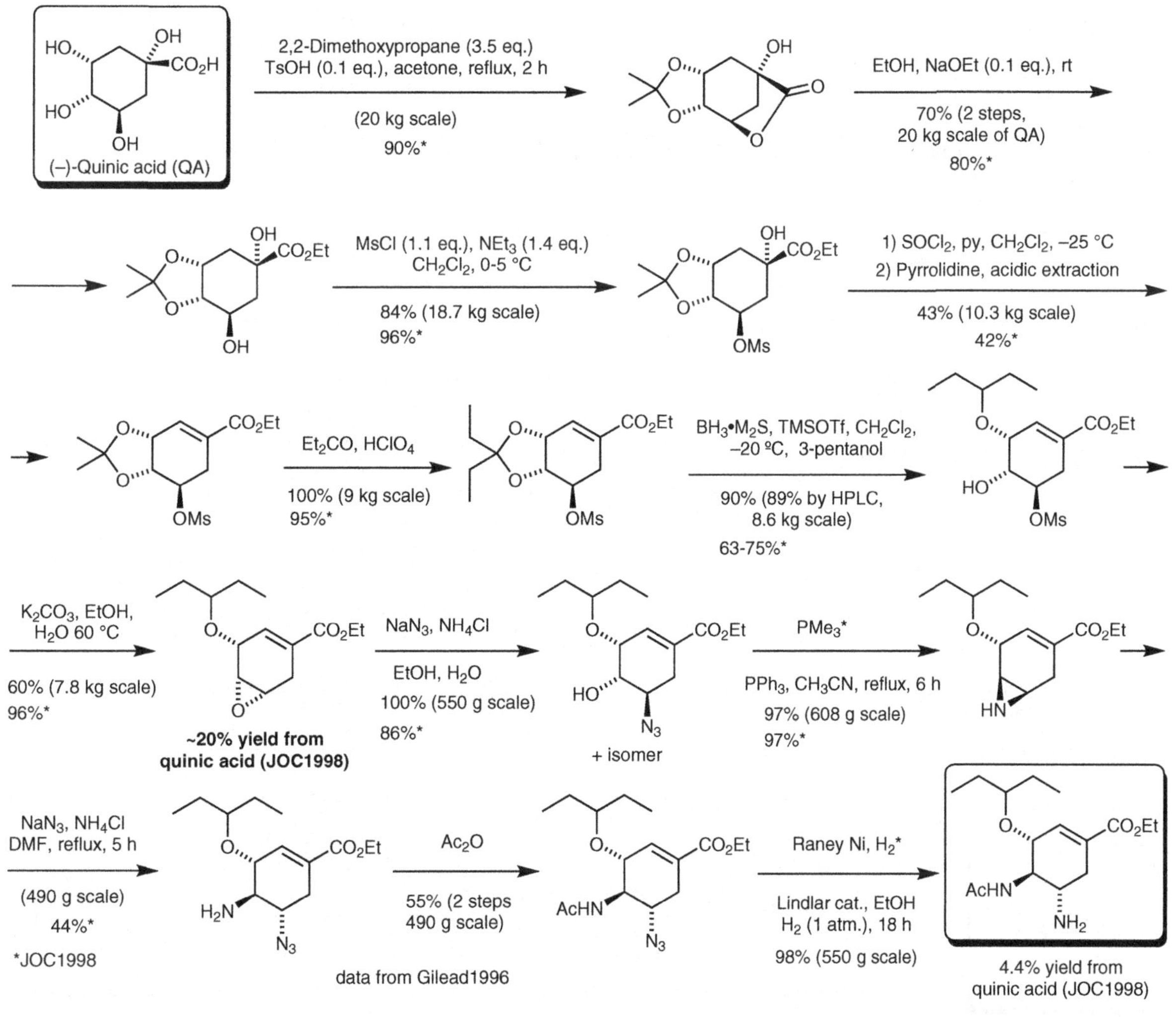

SCHEME 8.18 FIRST SCALABLE PROCESS FOR OSELTAMIVIR USING A REDUCTIVE KETAL OPENING AS A KEY STEP

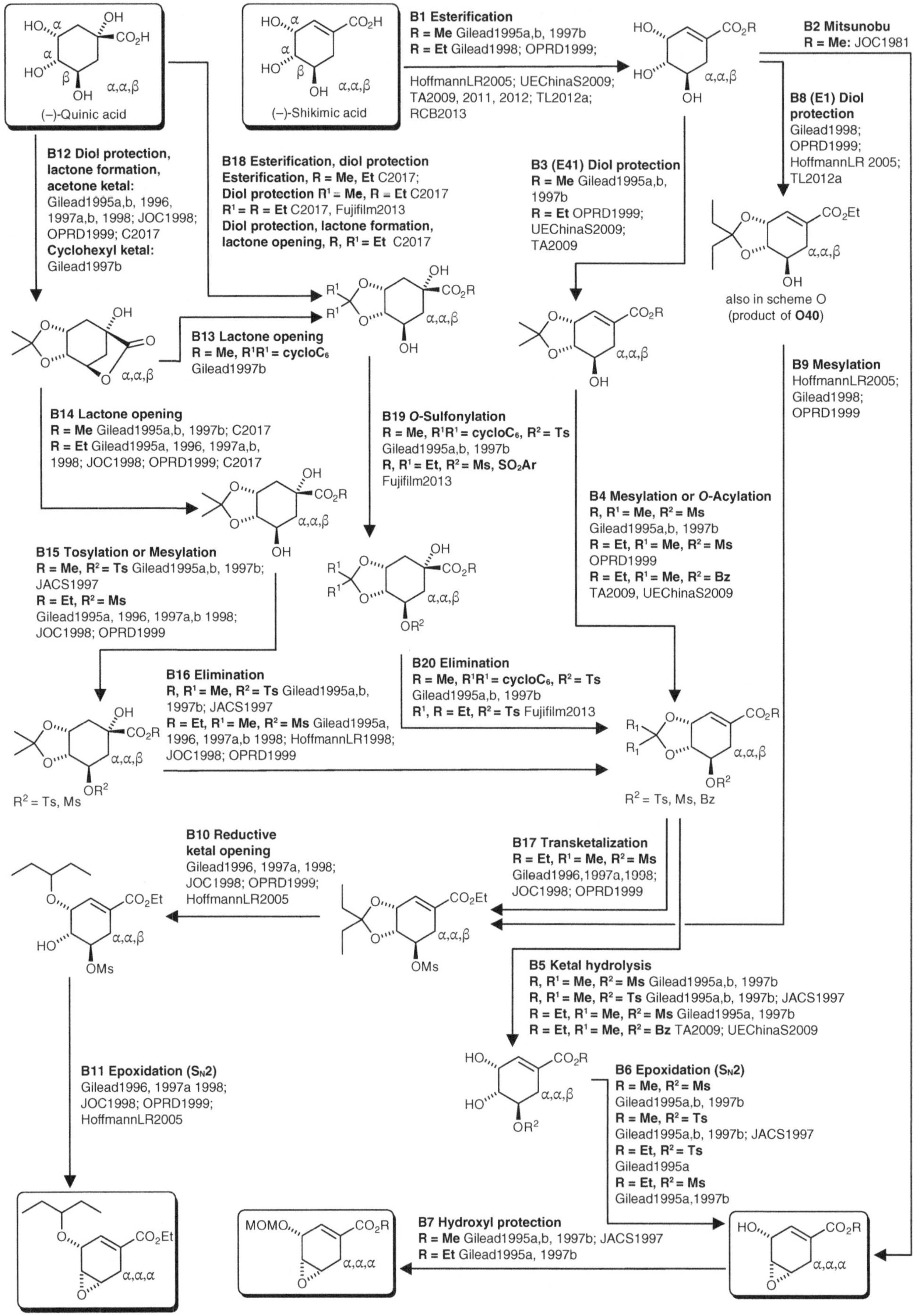

SCHEME B Synthesis of epoxide intermediates from (–)-shikimic acid or (–)-quinic acid (for reaction conditions see page 378)

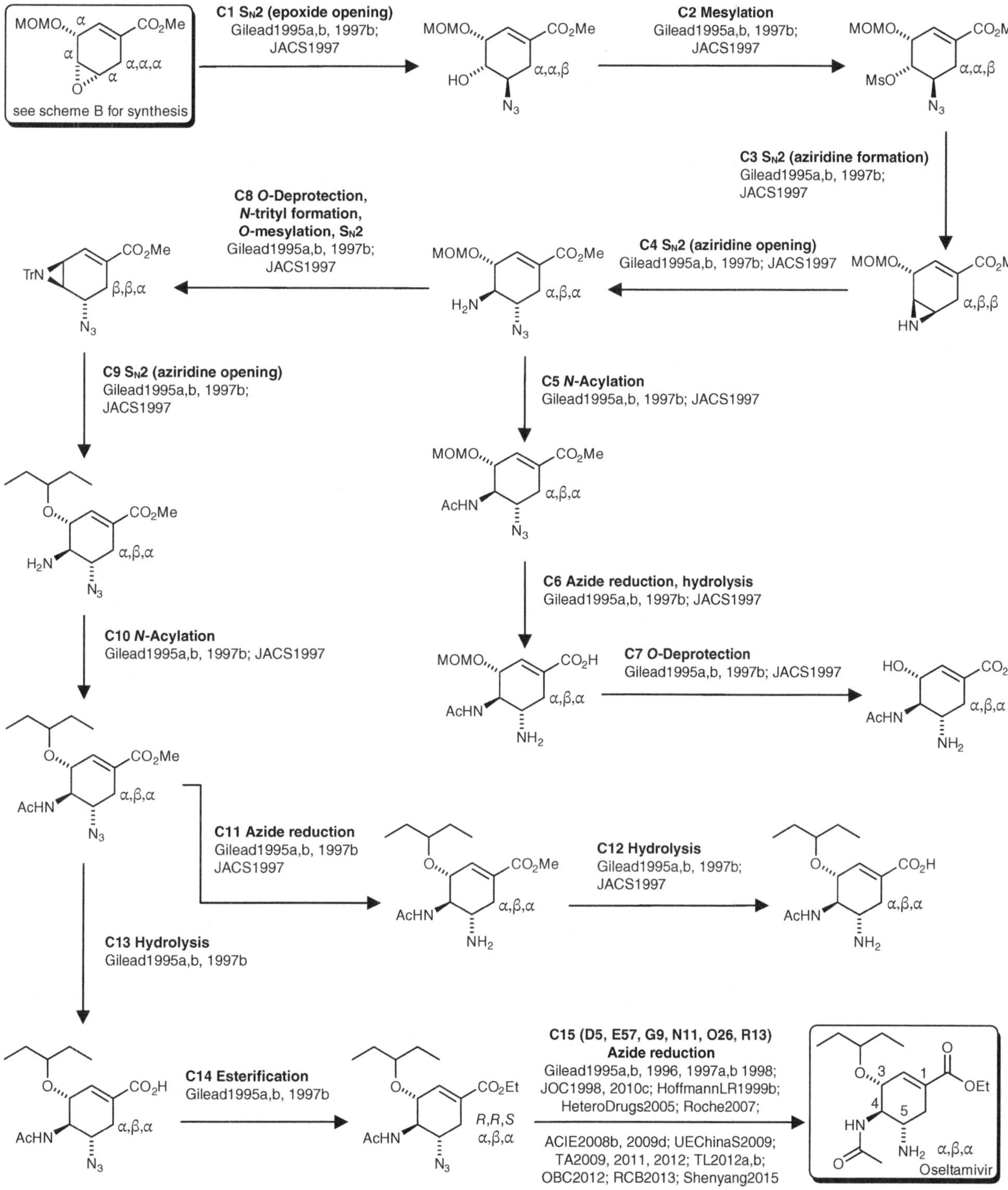

SCHEME C CONVERSION OF PROTECTED EPOXIDE INTO OSELTAMIVIR (FOR REACTION CONDITIONS SEE PAGE 379)

described in the previous scheme, were developed at kilogram scale (OPRD1999, Scheme 8.19). Consequently, the yield of the epoxide was doubled and simplification of the process achieved, the number of unit operations was reduced and a crystallisation and isolation of the alcohol-acetonide intermediate removed major impurities. A Vilsmeier salt assisted the dehydration step (Hoffmann LR1998) and reduction with Et_3SiH was employed for the ketal ring openin wherein the selectivity of the ring opening was much improved. The yield from quinic acid to the epoxide was raised to 35-38% and from shikimic acid to 63-65%. Of note was the fact that the impurity profile of the epoxide was the same when either starting material was used and all structurally related impurities were below 0.1% after crystallization of the epoxide. These findings are highly significant as raw materials derived from natural sources can vary considerably in both purity and the number and quantity of impurities presents. In the present case, the starting materials vary in purity from 85% to 99%, in addition to having a variable impurity profile. This improved process assures a constant quality and impurity profile of final product, thus making it suitable for implementation industrially and guaranteeing consistency of the product in the pharmacy.

However suitable this process is for implementation at industrial scale, the major issues associated with the use of azide anion remain to be addressed. Hence, substitution of azide in the synthesis continued to be a desirable objective.

8.4.2 Azide free routes

The concerns linked to the use of azide is a major concern for industrial operation, and consequently, several alternatives were explored (summarized in scheme D).

In the routes described above, benzylamine (**D9**), allylamine (**D10**) and *t*-butylamine (**D15**) are appropriate substitutes for the azide as nucleophile as all possess an alkyl chain that can be removed to give a primary amine. Benzylamine (using $Yb(OTf)_3$ catalyst) and allylamine gave regioselective epoxide ring opening (4% regioisomer in the case of the allylamine). Conditions for hydrogenolysis of the benzyl group resulted in concomitant reduction of the endocyclic CC double bond but in the case of allylamine the correct amine was produced (**D11**, Scheme 8.20).

To introduce the C4 and C5 nitrogens with the correct stereochemistry an interesting cascade-type process was initiated wherein allylamine is used twice, firstly to open the epoxide and the second time to open the aziridine ring (scheme D, **D12+D13+D14+D6** and Scheme 8.20). On removal of the allyl group (**D11**) after epoxide ring opening (**D10**), the α C5 amine is protected as an imine (**D12**)

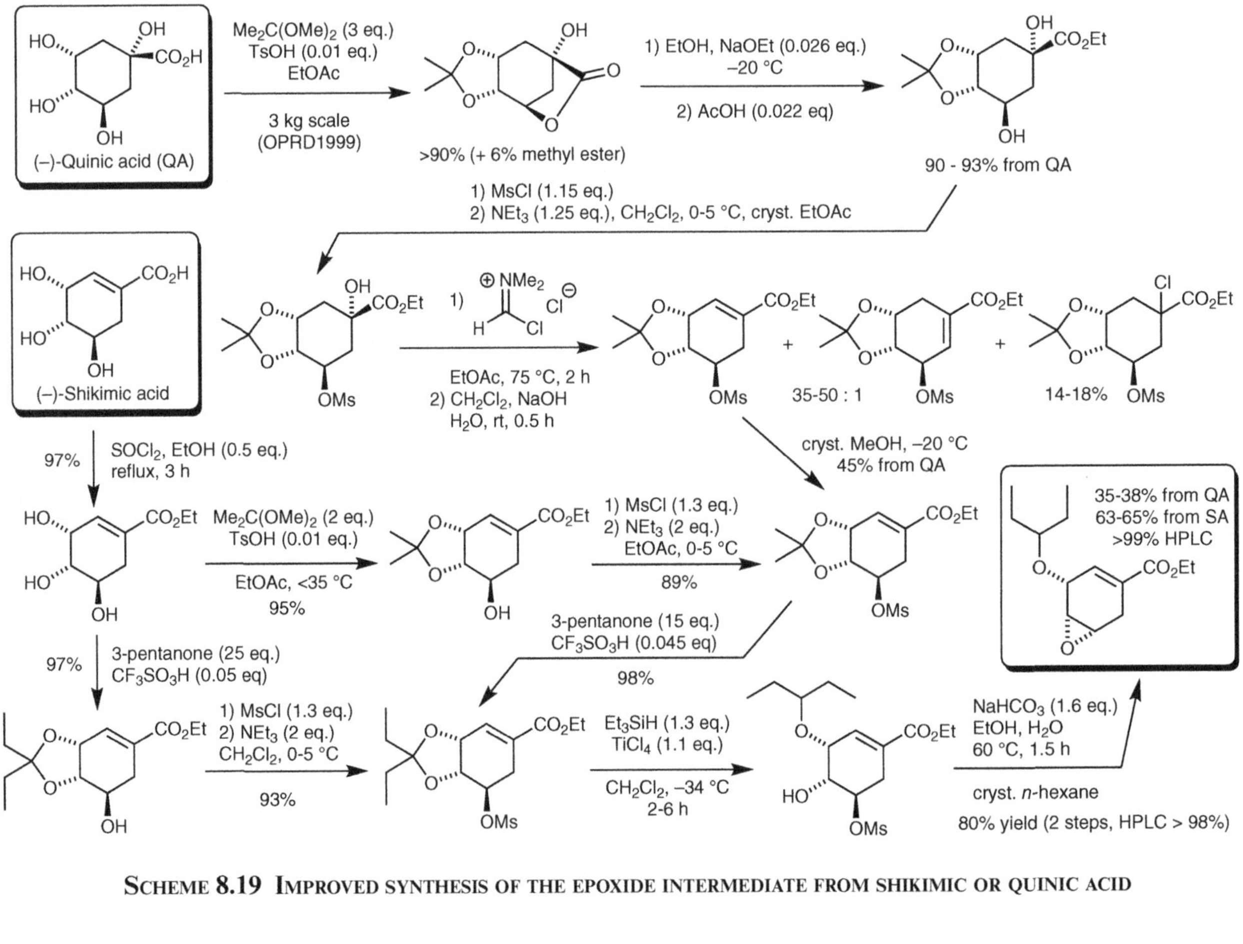

Scheme 8.19 Improved synthesis of the epoxide intermediate from shikimic or quinic acid

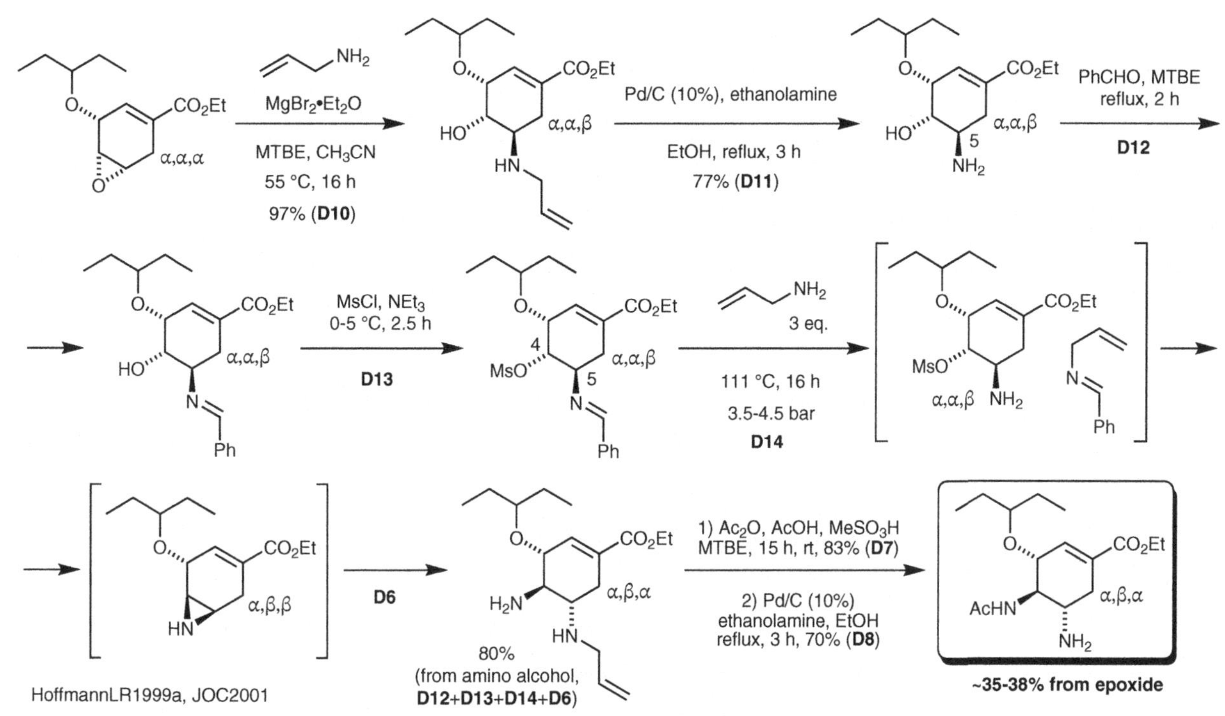

Scheme 8.20 The allylamine route from epoxide to Oseltamivir

thus allowing specific formation of the *O*-mesylate at C4 (**D13**). De-protection, aziridine ring formation and opening with allyl amine (**D14** + **D6**), C4 amine acetylation (**D7**) followed by carefully controlled de-allylation (**D8**), to avoid acyl migration, gives Oseltamivir The overall yield for this azide and chromatography free route was 35-38% compared to 27-29% yield for the previously best azide route (multigram scale, HoffmannLR1999a, JOC2001).

Another similar cascade-type route uses *t*-butylamine for epoxide ring opening (scheme D, **D15**) and passes through a *t*-butylaziridine intermediate (Scheme 8.21). Once more an allylamine (di-allylamine in this case) is used to ring open the aziridine ring (HoffmannLR2003, OPRD2004, **D17**). C4 *N*-acetylation under forcing conditions (**D18**, the other amine moiety is a tertiary di-allylamine and is not acetylated) gives a crystalline hydrochloric salt, which allows a clean-up point between the starting epoxide and Oseltamivir. After the acetylation of the C4 amine, the *t*-butyl group is removed under acidic conditions (**D19**) and the allyl groups by a η^3-allyl palladium-mediated transfer with 1,3-dimethylbarbituric acid as the nucleophilic scavenger (**D20**). The overall yield for this, once more, chromatography free route from epoxide was 61% (6 steps) and is a marked improvement on the previous routes developed at Gilead (27-29%) and Hoffman-La Roche (35-38%).

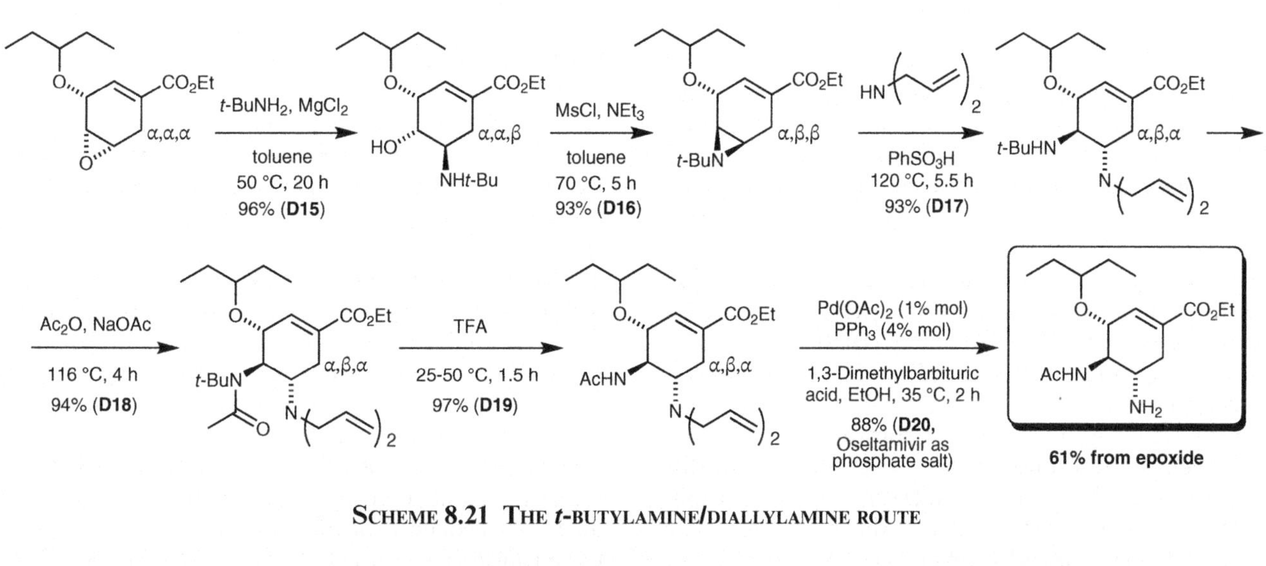

Scheme 8.21 The *t*-butylamine/diallylamine route

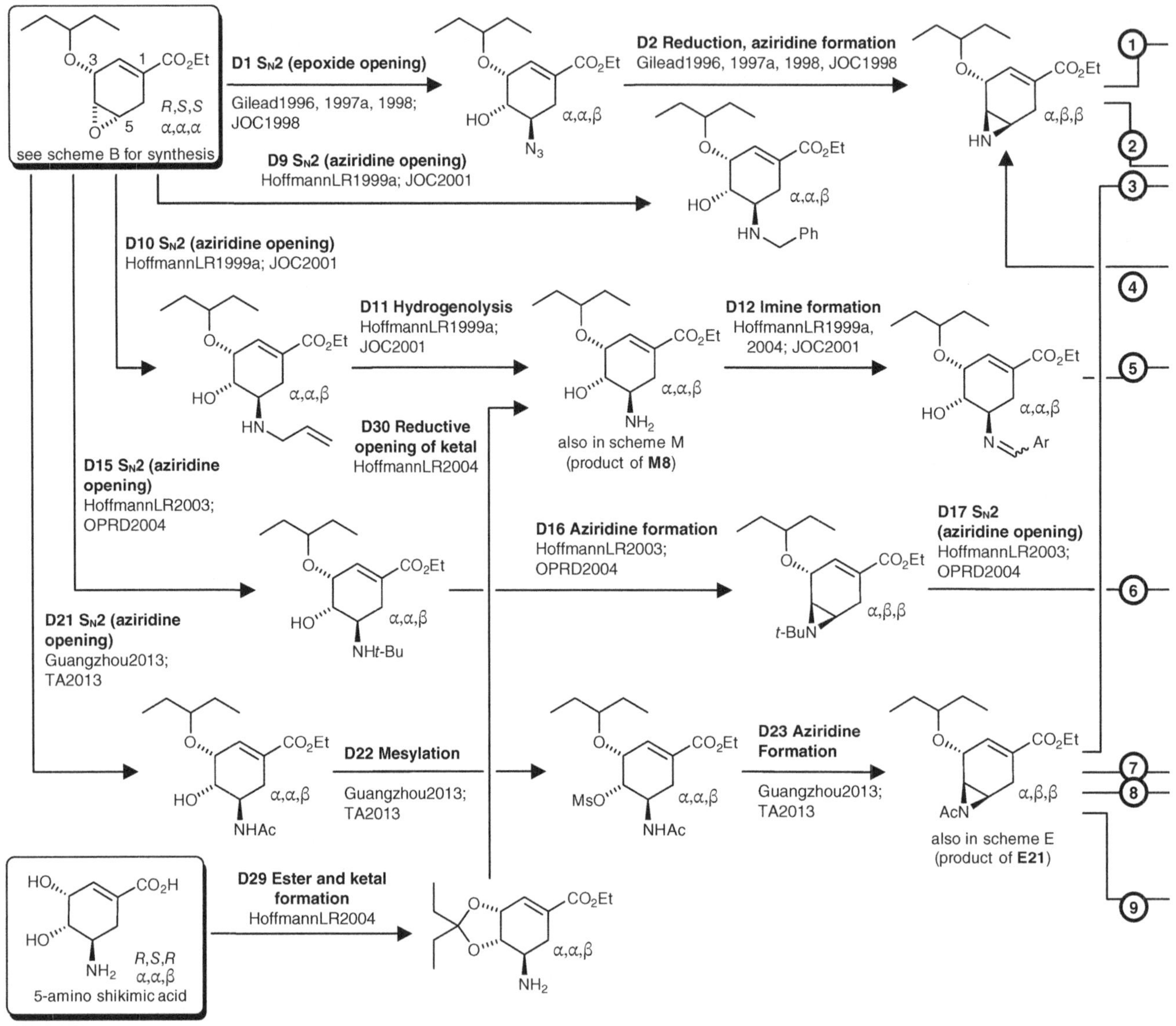

Scheme D Conversion of epoxide intermediate and 5-amino shikimic acid to Oseltamivir (for reaction conditions see page 380)

More recently, four novel routes using a Lewis acid catalysed ring opening of the epoxide with acetonitrile, which allow an early attachment of the acetyl group and accompanies the nitrogen transferred from C5 to C4, have been reported (scheme D, routes starting with reaction **D21**). After the epoxide ring opening step, conversion to Oseltamivir follows the traditional strategy *via* an acetylaziridine, ring opening with azide (**D31**), *t*-butylamine (**D24**), isopropyl 2,2,2-trichloroacetimidate (**D26**) or diallyl amine (**D28**). The yield of these short syntheses is high, 55%, 62%, 65% and 69% respectively.

A shorter, high yield process starting from 5-amino shikimic acid and following similar chemistry was disclosed (Scheme D, starting with **D29**). The low availability of the starting material makes this route unfeasible industrially (HoffmannLR2004).

8.4.3 Other routes from (–)-shikimic acid

A series of azide-routes that do not pass through a C4C5 epoxide were developed (summarized in scheme E). These were designed to be short and to provide high overall yield.

Route **Ea** first of all introduces the C3 ether group *via* reductive ketal ring opening (**E3**) followed by direct insertion of the C4 nitrogen *via* a Mitsunobu reaction with hydrazoic acid (**E4**). (Note that hydrazoic acid is a liquid, which is explosive at room temperature and pressure). The C5 amine was also introduced applying the same methodology (9 steps route with a 27% yield).

Route **Eb** opts to produce a C3/C4 epoxide in two efficient steps from the shikimic acid ethyl ester (**E8+E9**). A three-step tandem process (**E11**) produces a C3/C4 aziridine intermediate with the appropriate stereochemistry to be transformed into Oseltamivir. This highly efficient

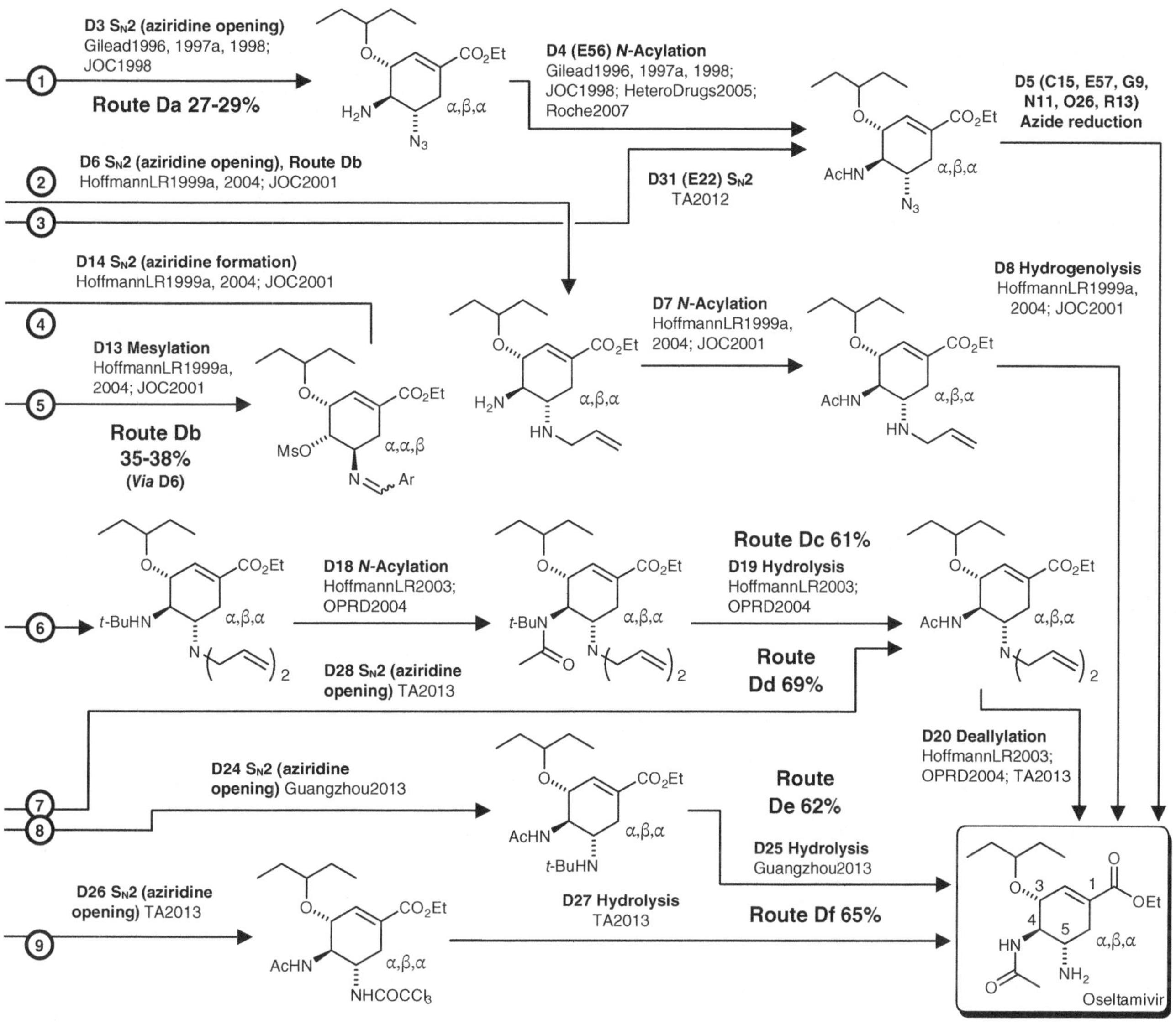

12 step synthesis, is azide free, allylamine being used to introduce the C5 amine group (45% yield).

All of the other processes opted to introduce an azide group at C3 during the initial stages of the synthesis, thus taking advantage of the fact that an allylic substituents on C3 is both more reactive towards S_N2 displacement and is less sterically hindered.

Routes **Ec** to **Ef** employ a cyclic 3,4-sulphite to first introduce an azide at C3 with inversion of stereochemistry, which is then transformed into a β-C3C4-aziridine. Ring opening of the aziridines with 3-pentanol gives the desired nitrogen group at C4. Further chemical manipulation produces the C5 primary amine. Route **Ec** (TA2012) has a 55% overall yield for the 11-steps while route **Ed** has 39% yield over a 9-step synthesis. *N*-acyl aziridines with the C5 hydroxyl protected by an acyl group were used in routes **Ee** and **Ef**.

The remaining routes opted to use a di- (**Eg**) and tri-*O*-mesyl derivative (**Eh**, **Ei**, **Ej**), respectively for the key reaction with azide at the more reactive C3 carbon. In **Eg**, the C5-hydroxyl was protected with a benzoyl group and the di-*O*-mesyl intermediate reacted with azide (12 steps, 39% yield). A rudimentary flash chromatography was used, still allowing scale up for this process.

In the tri-*O*-mesyl approach (routes **Eh**, **Ei** and **Ej**), the ability to differentiate between the mesylate groups is crucial for success. Three methods were reported to convert the tri-*O*-mesyl intermediate into Oseltamivir. The allylic mesylate group was selectively displaced with azide with inversion of stereochemistry at C3. This reaction requires low temperatures to avoid elimination and aromatization. The routes then follow a 3,4 aziridine strategy to introduce the C4 nitrogen. Displacement of the last mesylate group (C5) with azide is carried out at higher temperatures, since this position is less reactive. Other nitrogen nucleophiles were tested in the displacement of the mesylate groups (ammonia, benzylamine, allylamine, *t*-butylamine) but these failed. Route **Eh** has high reported yields and hence, would appear to be viable options for an industrial scale process (8 steps and 47% overall yield). Routes **Eh**, **Ei**, **Ej** to Oseltamivir have minimal protecting group use but used azide twice.

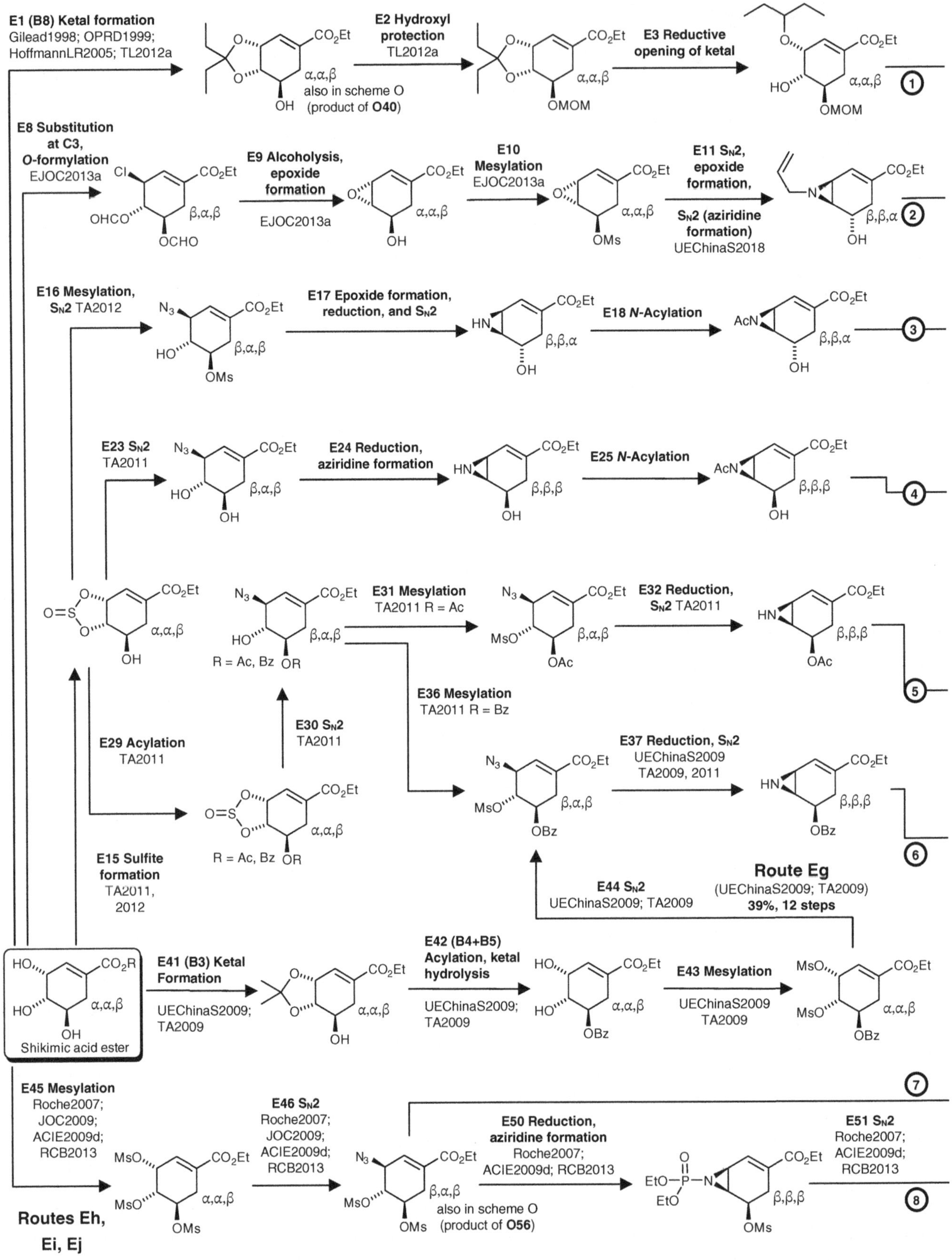

Scheme E Synthetic routes from (−)-shikimic acid that do not have a C4C5 epoxide as intermediate (for reaction conditions see page 380)

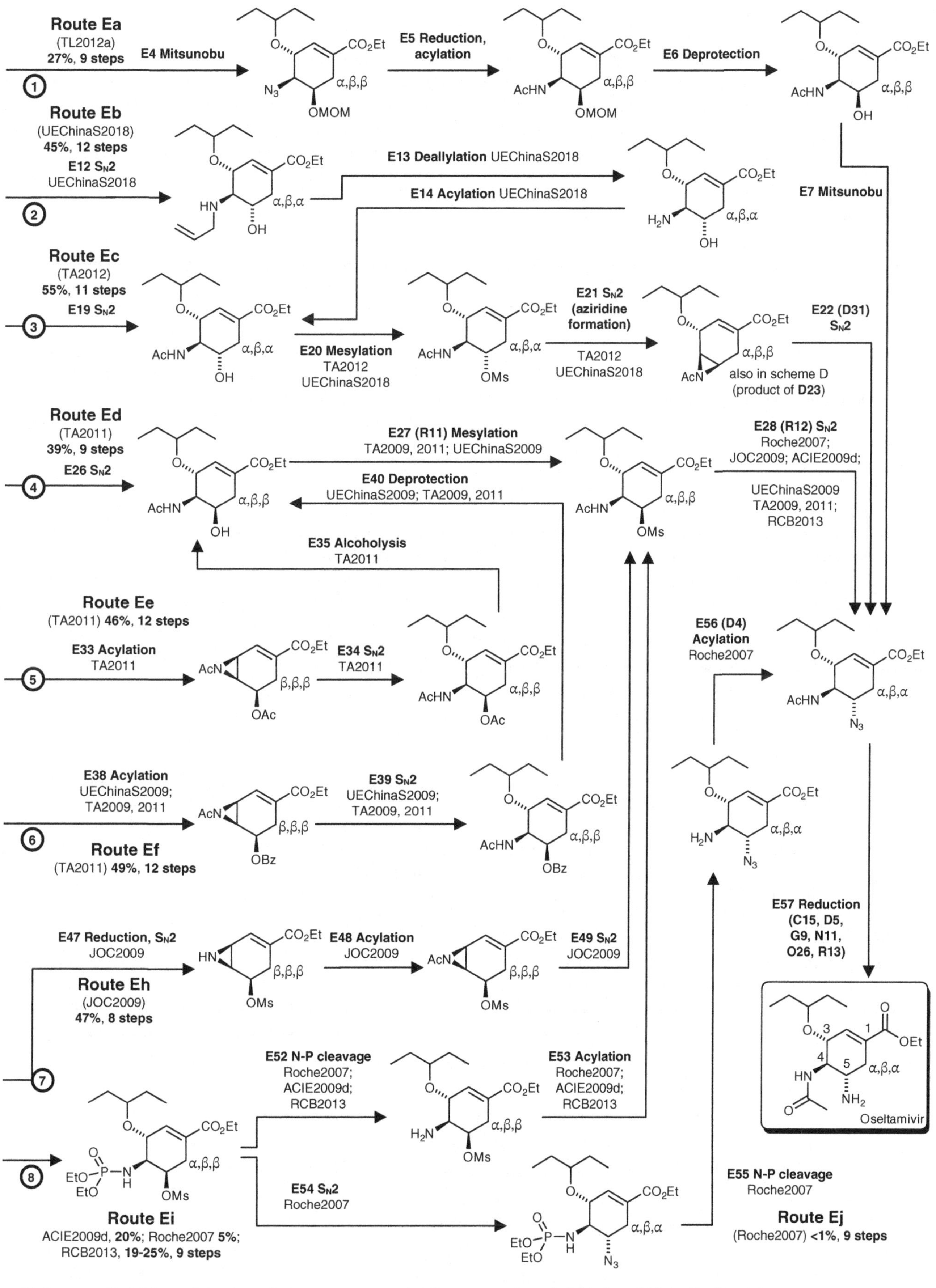

Note: overall yields and number of steps shown are for the conversion of shikimic acid into Oseltamivir.
A 97% yield for the formation of the shikimic acid ester was used for yield calculations.

8.5 THE SUPPLY PROBLEM

Shikimic and quinic acids are natural products and are intermediates in the biosynthesis of essential aromatic amino acids. They accumulate to some extent in various plants from which they can be extracted. Star anise, *Illicium verum*, indigenous to Northern Vietnam and Southern China, is a rich natural source of shikimic acid, 1 kg being obtained from 30 kg of dried fruits. However, both shikimic and quinic acids are expensive and have a limited supply. As a result, Gilead was forced to seek alternative synthetic routes from accessible and inexpensive materials. Attention was also focused on new sources and synthetic methods to obtain a reliable supply of shikimic and quinic acid.

8.5.1 Quinic and shikimic acid supply

At the time Oseltamivir went to market, the world supply of shikimic acid, the natural product starting material, was somewhat limited as these materials were derived from plant extracts. Although total chemical synthesis existed at the time, these procedures were incompatible with large-scale commercial production.

In early 2005, Roche announced a Tamiflu production shortage, due to insufficient shikimic acid supply. The shortness of supply is reflected in the following story. In 2006 the US government alone made plans to acquire and stockpile 300 million doses of Tamiflu (ACIE2006). This required 22.5 tons of the API. If a process yield of 36% overall is assumed, this translates into a requirement for 64 metric tons of shikimic acid. Before the threat of the global spread of the highly pathogenic H5N1 Asian flu strain in birds in 2006, shikimic acid was available at the price of \$40/kg but increased to \$400–500 per kilogram afterwards.

Improved methods for the extraction and purification of shikimic acid from Chinese star anise (*Illicium verum*) the main and preferred source of shikimic acid. Another possible sources is the Japanese star anise from the shikimi-no-ki shrub, *Illicium anisatum*, albeit also known to contain highly toxic compounds. Alternative sources and isolation procedures were developed and patented (AustinSU2005, Ind2006a,b, JiangnanU2006, Oregon2006, OL2015). A 5% weight yield of shikimic acid was obtained by aqueous isopropanol or ethanol extraction of dried powdered star anise (Ind2006b, OL2015) but values around 10% have been reported. Notably, extraction of *Illicium griffithii* fruits (seeds and pericarps) were reported to give a 12-18% weight yield (CSIRIndia2010, CS2015). A recent review on the plant sources of shikimic acid was published (JCB2012).

Genetically modified strains of *E. coli* have been used for the production of shikimic acid (MichiganSU1999a,c, Monsanto2007, JACS1999) and quinic acid (MichiganSU1999b, JACS1999) using glucose as feed stock. Several of these procedures were the subject of patent applications. One genetically engineered *E. coli* strain, deficient in both shikimate kinase isozymes (which effectively stops any further biotransformation of the shikimic acid) was reported to be capable of producing shikimic acid with a titer of 84 g/L and a yield of 33% from glucose (MichiganSU1999a). Today, fermentation approaches supply at least 30% of present requirements.

Processes for the recovery from a fermentation broth (DSM2000) and methods for producing high purity (greater than 92% by wt.) crystalline shikimic acid from an aqueous solution (Biogaia2002) have been claimed. Products obtained from fermentation typically exhibit fluctuations in quality resulting in large variations from batch to batch of final product even with most accommodating process. It is therefore, important to have adequate purification processes for raw materials in place to assure consistent quality.

The industrial production of shikimic acid has been reviewed (AMB2013, BA2012, CS2015).

A biosynthetic approach to the production of 5-aminoshikimate, an interesting starting material as it already has a primary amine group at C5, was also investigated by fermentation (Scheme 8.22). The major products were kanosamine and shikimic acid and, although the yield of the 5-aminoshikimate was very low (5%, OL2004), it was easily be isolated using a cation exchange resin (OL2004, MichiganSU2006).

The increasing demand for Tamiflu led Roche (Gilead's partner) and others, including those in the academic community, to develop new routes to Oseltamivir. The principal object of these investigations was to use cheap and widely available starting materials and to circumvent the use of either shikimic or quinic acid.

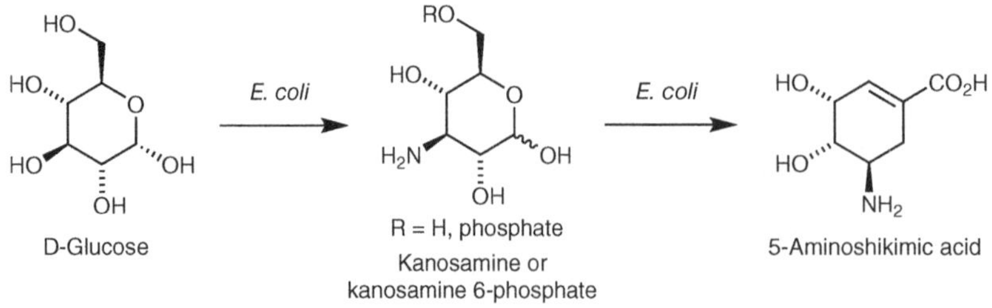

Scheme 8.22 Microbiologic production of 5-aminoshikimic acid

8.6 FIRST NON-SHIKIMIC OR QUINIC ACID SYNTHETIC ROUTES

Two different synthetic approaches that did not depend on natural products were patented by Hoffmann-La Roche in 2000. It is probable that these syntheses were designed to serve as back-up in case of any supply shortages. The first of these syntheses was based on a Diels-Alder approach whilst the second used aromatic compounds as starting materials.

8.6.1 The Diels-Alder furan approach

Diels–Alder reaction of inexpensive raw materials, furan and ethyl acrylate, produced a mixture of bicyclic compounds, thermodynamic control generating the racemic *exo* adduct as the major product (*exo*/*endo* 9:1, HoffmannLR2000a, Scheme 8.23). Enzymatic esterase kinetic resolution, in a biphasic system at 5% concentration, gave chirally pure *exo* adduct (6% yield, 93% ee). Aziridination of the CC double bond placed the nitrogen on the least hindered side of the bicycle i.e. on the opposite side to the oxygen bridge. Base treatment resulted in elimination of the oxygen β to the carboxylic ester, thus opening the bicycle, placing the double bond in the correct position and installing a hydroxyl group at C5. Subsequent steps follow similar chemistry to that applied in the shikimic acid route. Although this route eliminates the use of shikimic and quinic acids and has the resolution step early on in the synthesis it suffers from several issues, amongst which the impractical high dilution and low yield of the enzymatic resolution step (6%), which can seriously affect productivity.

8.6.2 Oseltamivir by total hydrogenation of an aromatic ring

An alternative methodology (HoffmannLR2000b, JOC2008, Scheme 8.24) started from inexpensive dimethyl pyrogallol 1,3 dimethyl ether, which was transformed into Oseltamivir in a series of high yielding reactions. The strategy exploited the C2 symmetry of the intermediates during the first part of the synthesis. After introduction of carboethoxy groups at positions 1 and 5 to give the symmetrically substituted aromatic isophthalic acid diester, a symmetric all-*cis* substituted cyclohexane was obtained *via* ruthenium-catalyzed hydrogenation. Subsequent demethylation of the ethers and desymmetrization of the diester with inexpensive pig-liver esterase (PLE) at relatively high concentration (10%), gave the C5 mono carboxylic acid in high yield (96%) and enantiomeric excess (96–98% ee). The C5 nitrogen was introduced by a Curtius rearrangement and the C4 nitrogen by an S_N2 displacement with azide. The hydrogen and hydroxyl group for the E1cB elimination, to produce the CC double bond, are in the ideal *trans*-configuration. The synthesis seems robust, does not require any chromatographic purification and has an excellent overall yield of 30% from 1,6-dimethoxy phenol, matching shikimic acid-based routes. However, the use of azide, DPPA and the Curtius rearrangement are of concern.

For the reasons discussed above, a flurry of activity from both industry and academia, resulted in many alternative routes being disclosed from 2003 onwards. A large number of these are, however, not suitable for large scale production and were not the subject of patent applications. The following sections on these alternative routes are organized according to the type of chemistry used.

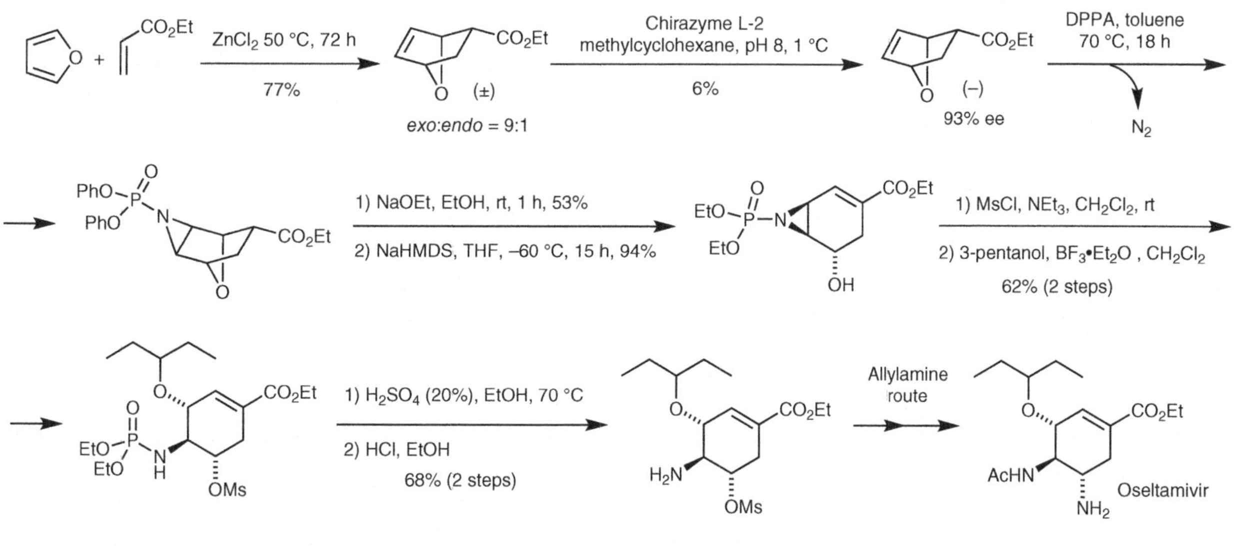

Scheme 8.23 Synthesis of Oseltamivir by a furan [4+2] cycloaddition

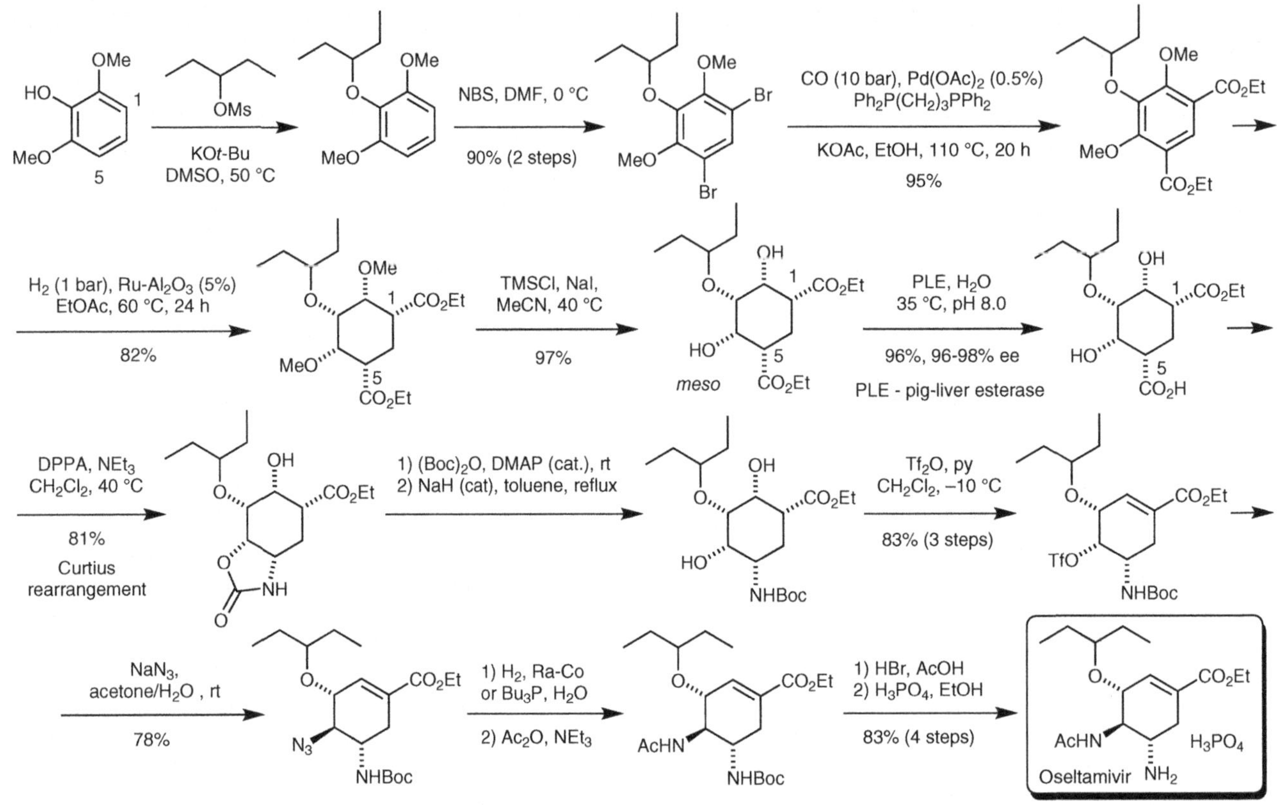

Scheme 8.24 Oseltamivir from aromatic compound by a desymmetrization approach

8.7 SYNTHESIS OF OSELTAMIVIR FROM CYCLOHEXADIENE DERIVATIVES

In addition to aromatic benzene derivatives, see Scheme 8.24 above, cyclohexadienes are also starting materials in several Oseltamivir synthesis.

8.7.1 *Meso* aziridines desymmetrization routes

A *meso*-aziridine, produced from cheap and widely available 1,4-cyclohexadiene, was the starting material in several routes (summarized in scheme F). The key step was a catalytic asymmetric ring opening of the *meso*-aziridine with trimethylsilyl azide ($TMSN_3$) in 96% yield and 91% ee (**F1**). Painstaking investigations were necessary to ascertain reaction conditions for this key step. An yttrium catalyst, prepared from $Y(OiPr)_3$ and a chiral phosphine oxide ligand (see structure: **F1** chiral ligand) were used. This sequence defines two of the asymmetric centres, C4 and C5 and the third, at C3, is produced by asymmetric induction.

After aziridine ring-opening the routes diverge. In the first (scheme F, route **Fa**, TokyoU2006a, JACS2006b), a functional group at C3 was inserted *via* an allylic oxidation (**F6**) and Michael addition to the resulting α,β-unsaturated ketone with TMSCN introduced a cyanide group (**F7**), which eventually became the carboxylic acid ester. The starting point of this sequence is a dicarbamate (di-NHBoc cyclohexene) with C2 symmetry and hence, both allylic positions are equivalent resulting in a unique unsaturated ketone. After further manipulation a 3,4-aziridination is produced (in **F10**) which is followed by nucleophilic attack by 3-pentanol introduced the ether side chain (**F11**). This 17 step route suffers from very low yield (1.4%) whilst several other issues also exist: extensive protecting group chemistry; the oxidation is problematic (use of stoichiometric amounts of toxic selenium dioxide and its disposal, **F6**) azide (**F1**) and cyanide (**F7**) chemistry are included, the Mitsunobu reaction (**F10**) is not easy to carry out and produces large quantities of waste.

Route **Fb** (TokyoU2006a, OL2007) followed a related strategy. The asymmetric ring opening was optimized and the catalyst recovered. The oxidation of the allylic position was avoided by introducing the C3/C4 double bond by an intramolecular iodo-lactonization followed by elimination (**F17+F18**). Instead of the Michael addition, direct addition to the carbonyl placed the cyanide at C1 (**F20**). The following allylic rearrangement (**F21**), *via* an intramolecular S_N2' allylic substitution, produced the C3/C4/C5 substituted cyclohexene that was converted to Oseltamivir using chemistry already described. Although the number of protecting groups is less, the synthesis still has 19 steps and a yield of only 8.4% on a milligram scale. Mitsunobu and Dess-Martin periodinane reactions were used once and twice, respectively. In the same publication a more concise route (12 steps) using the same general strategy was disclosed (route **Fc**, OL2007, the Dess-Martin reaction being used only once (5.0% yield, 1 g scale).

The α,β-unsaturated ketone, product of reaction **F6**, route **Fa**, was also obtained by a Diels-Alder cycloaddition (see Scheme 8.25 and scheme H, **H6**, BMCL2008, TokyoU2006c). However, chiral HPLC was used to obtain the desired enantiomer of the α,β-unsaturated ketone, thus making this route unrealistic at industrial scale.

Yet another route to this α,β-unsaturated ketone has been reported in the patent literature and uses a Ramberg--Bäcklund reaction to form a cyclohexene derivative (CSIRIndia2013, Scheme 8.26).

Apart from other safety concerns (use of: isocyanate; Dess Martin oxidation and MCPBA), the Meyer's modification of the Ramberg-Bäcklund reaction as applied here, requires highly toxic carbon tetrachloride to produce the α-halosulfone, the true precursor in the rearrangement.

8.7.2 Using iron diene carbonyl complexes

A methodology, based on cationic iron carbonyl complexes formed from a conjugated cyclohexadiene, was developed (Scheme 8.27, CC2007). These cationic species can react with several nucleophiles and after an oxidative de-complexation, the diene is still available for further transformation. Racemic iron complex was resolved by preparative HPLC of a diastereomeric mixture of an ether at C5. The resulting pure chiral cationic iron carbonyl complex was subjected to a stereoselective amination with Boc-amine. Inert, anhydrous conditions were critical to obtain a good yield. After oxidative de-complexation, a chiral diene (C5 substituted) suitable for further transformation into Oseltamivir was obtained. Note that the regioselectivity of the epoxidation reaction was assured by the presence of the carbamate group.

This 14 steps synthesis has an accumulated yield of 5% but the HPLC separation step is a major drawback to its use. Azide and MCPBA are used, both of which create safety concerns on an industrial scale.

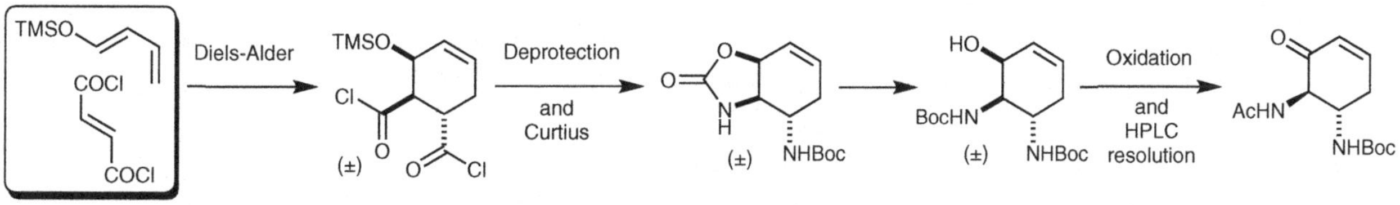

Scheme 8.25 Synthesis of key intermediate by a Diels-Alder approach

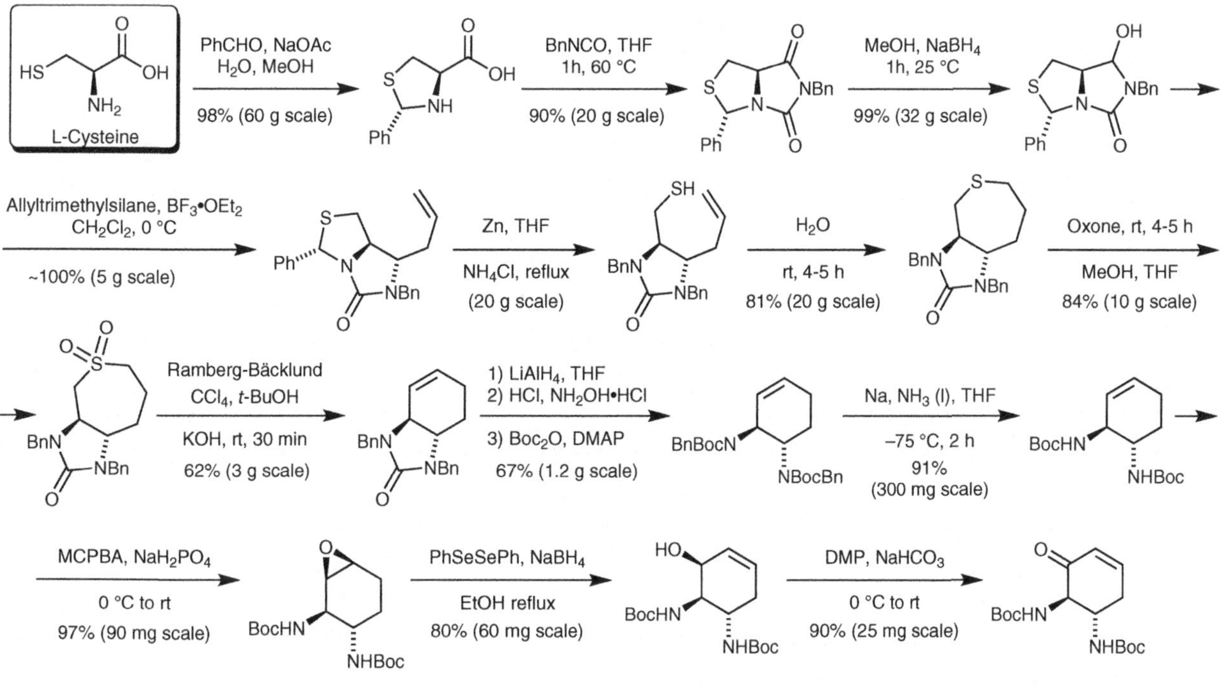

Scheme 8.26 Synthesis of key α,β-unsaturated ketone intermediate using a Ramberg-Backlund reaction as a key step

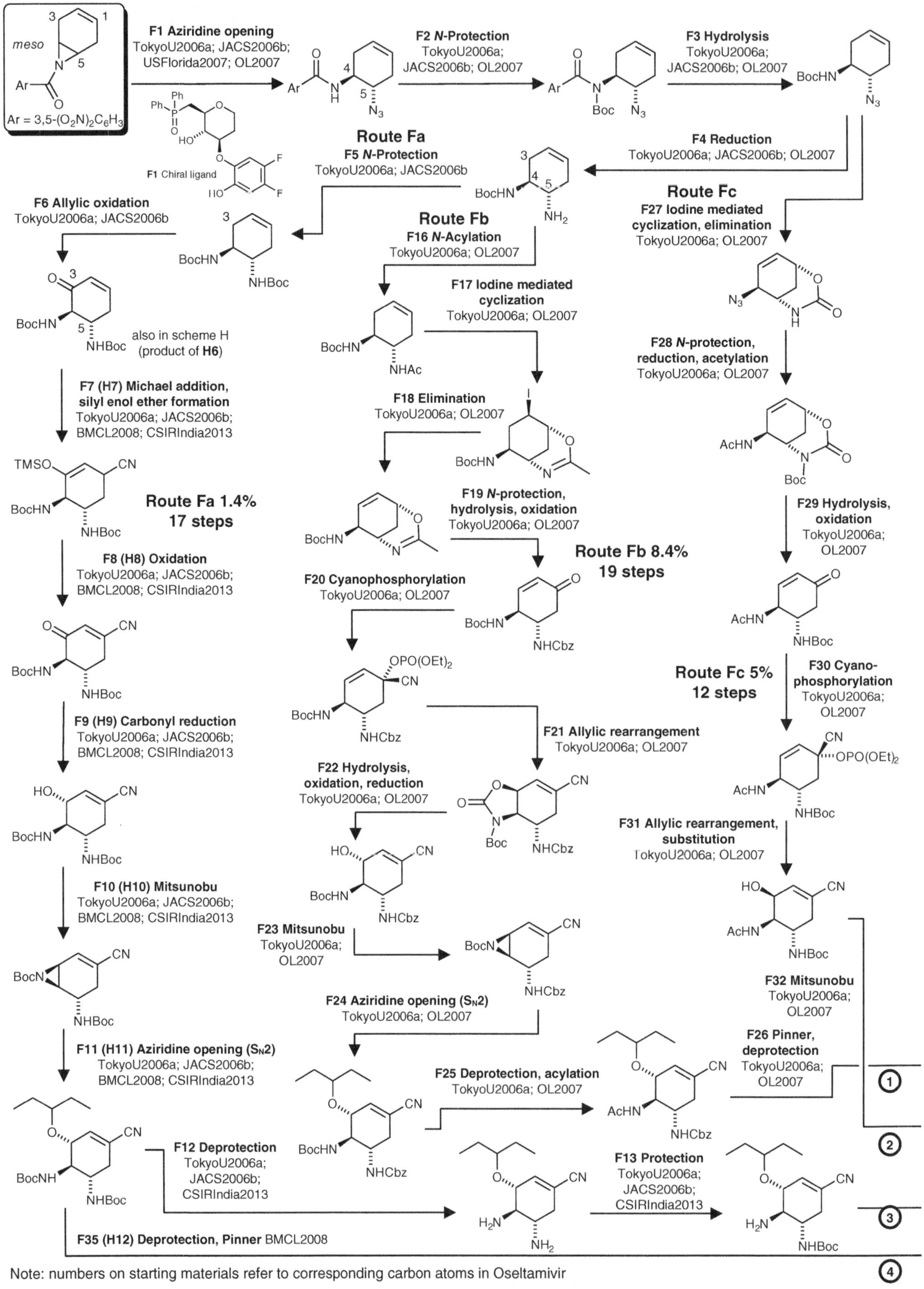

Scheme F Syntheses of Oseltamivir using an aziridine desymmetrization approach as a key step (for reaction conditions see page 381)

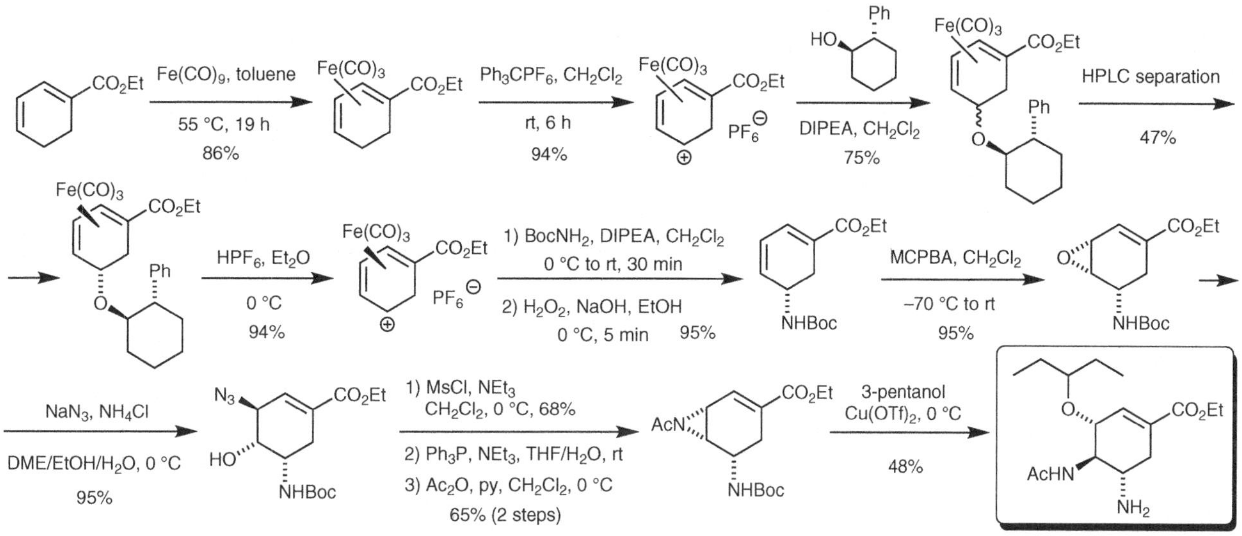

SCHEME 8.27 OSELTAMIVIR SYNTHESIS BASED ON CATIONIC IRON CARBONYL COMPLEXES

8.7.3 From cyclohexa-3,5-diene-1,2-diol derivatives

Enantiopure substituted cyclohexa-3,5-diene-1,2-diols are compounds of potential interest as starting materials for the manufacture of Oseltamivir. Several raw materials of this kind are commercially available and are produced on a large scale by microbial oxidation of aromatic derivatives. Syntheses of Oseltamivir have been published starting from bromo and carboxyethyl derivatives and are summarized in scheme G. The major transformations that are necessary on a bromodiol to produce Oseltamivir are shown in Scheme 8.28.

Two routes starting from this bromo diol have been published. Both of these routes, **Ga** and **Gb**, arrive at the same penta-substituted cyclohexene and a subsequent metal-mediated ethoxycarbonylation (**G8** and **G12**) is key to replace the bromine with the ester function.

In route **Ga** (ACIE2008b), after protection of the diol, electrophilic addition creates a bromonium ion at more nucleophilic double bond on the less hindered face of the ring (**G1**). Thereafter, attack of acetamide from the opposite face (second stage of reaction **G1**) produces the functionality and chirality at C3 and C4. Nucleophilic introduction of the ether at C3 *via* a

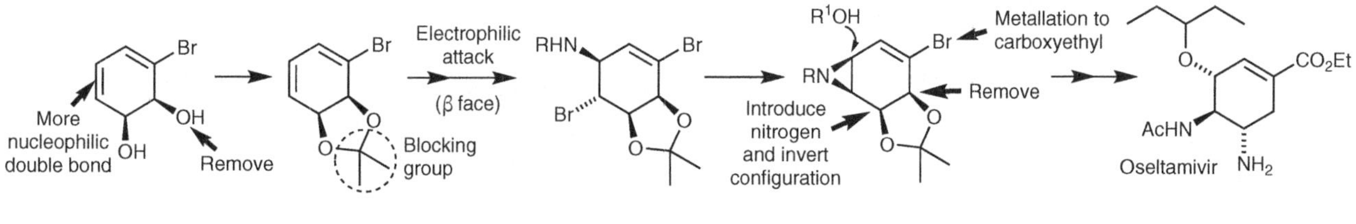

SCHEME 8.28 MAIN TASKS TO ACCOMPLISH FOR CONVERSION OF (1*S*,2*S*)-3-BROMOCYCLOHEXA-3,5-DIENE-1,2-DIOL INTO OSELTAMIVIR

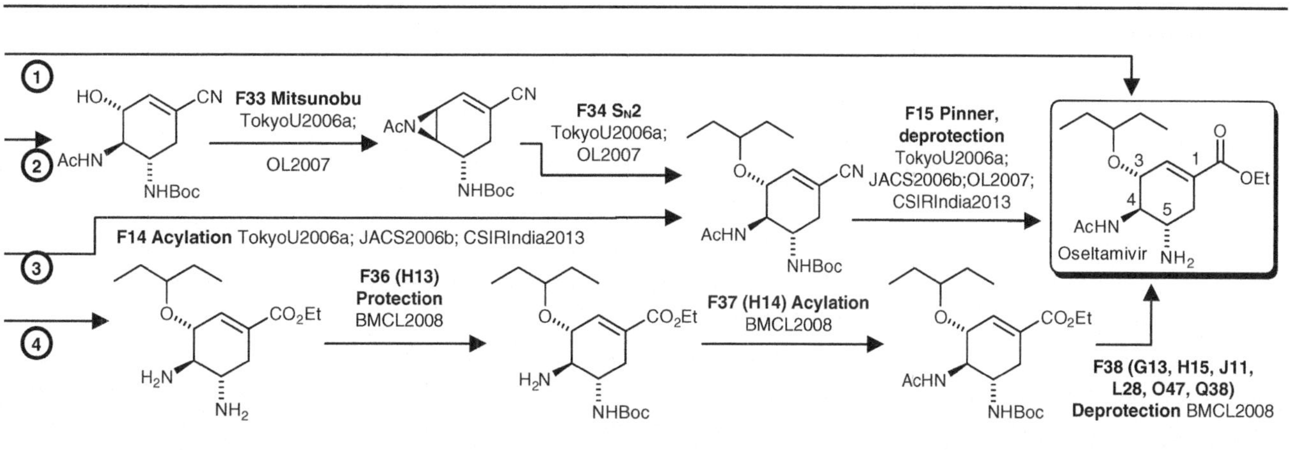

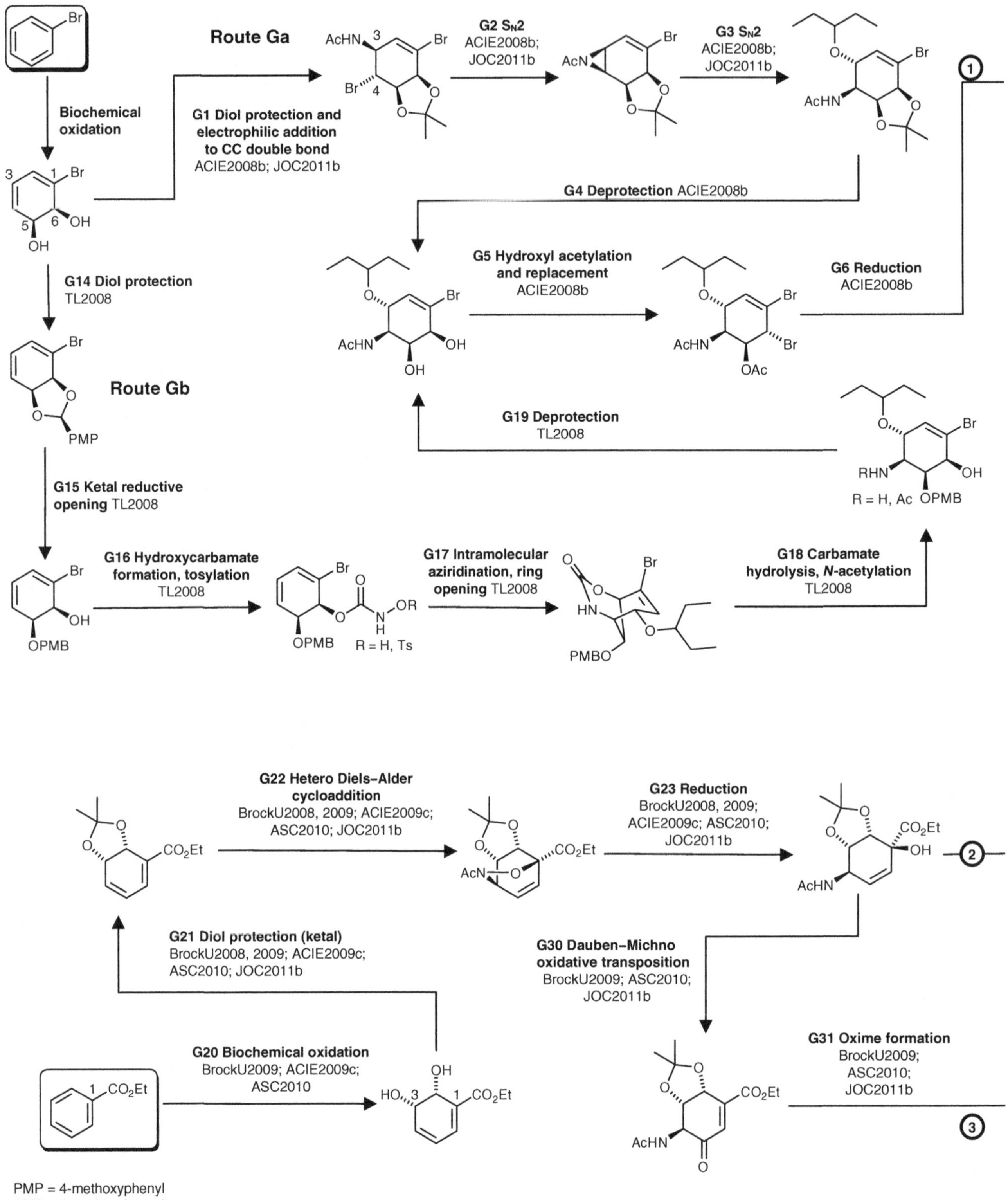

PMP = 4-methoxyphenyl
PMB = 4-methoxybenzyl

Note: numbers on starting materials refer to corresponding carbon atoms in Oseltamivir

Scheme G Synthesis of Oseltamivir from cyclohexa-3,5-diene-1,2-diol derivatives (for reaction conditions see page 382)

(for reaction conditions see page 382)

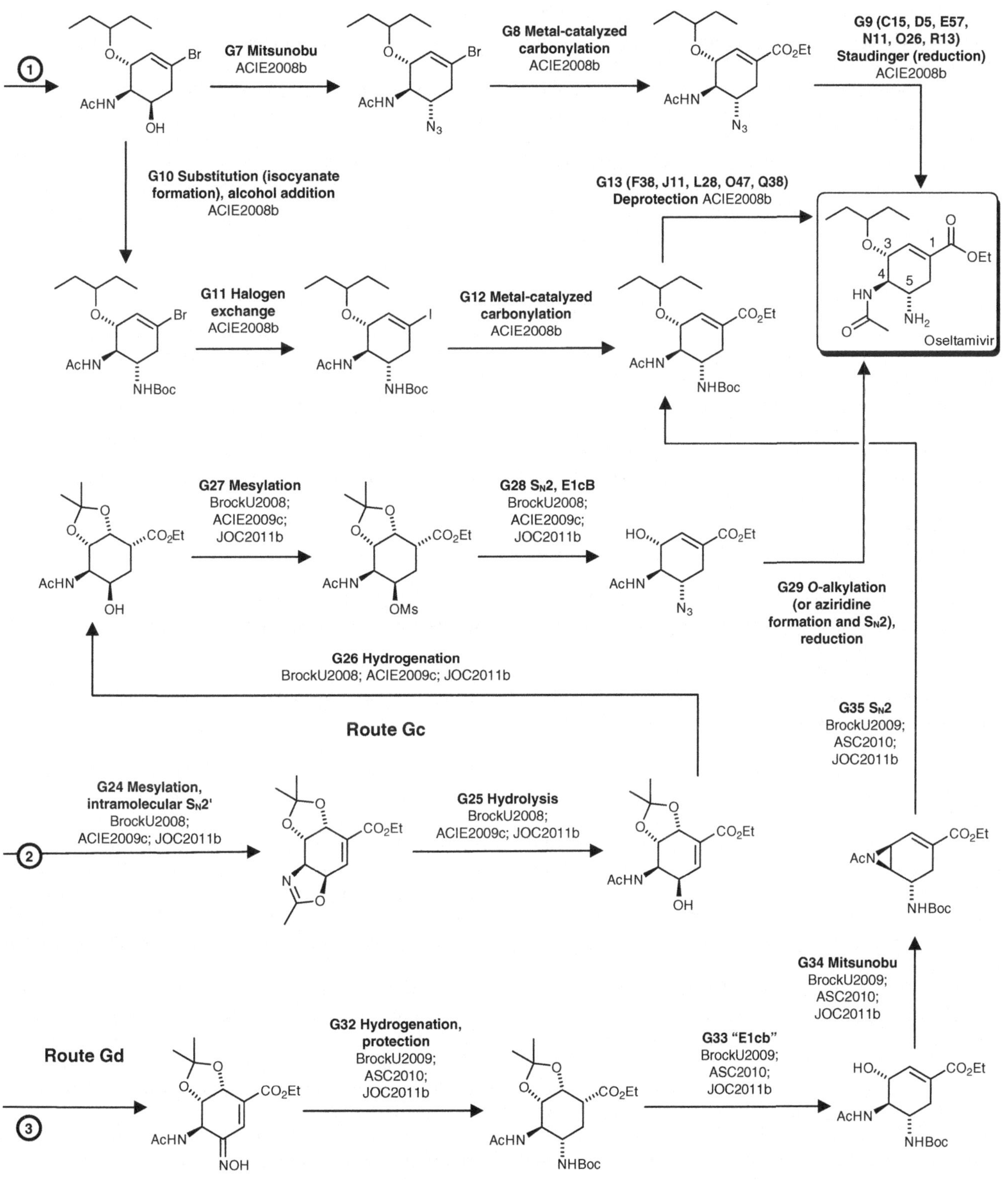

Route Ga (ACIE2008b) **26% (*via* G7), 11 steps; 22% (*via* G10), 13 steps**
Route Gb (TL2008) **6.2%, 14 steps**
Route Gc 10-12%, 10 steps (to G28 product)
Route Gd 4.4-10%, 12 steps

3,4-aziridine was not completely stereoselective (**G3**), probably because of the formation of a stable allylic cation allowing attack from both faces and chromatography was necessary for removal of the unwanted diastereomer. The C6 allylic hydroxyl group is converted to hydrogen by exchange with bromine and reduction (**G5+G6**). The other steps follow similar chemistry as described earlier. The synthesis is short (11 steps) and the yield is good (26%). High purity of the final product is made possible as some of the late intermediates can be isolated and re-crystallised. However, azide chemistry is used giving incentive to develop an azide-free version (scheme G, reactions **G10** to **G13**). Reaction of the C5 alcohol with tetrabutylammonium cyanate in the presence of triphenylphosphine and DDQ gives an isocyanate which is then treated with *t*-butanol (reaction **G10**). Carbonylation was only feasible after conversion of the bromide to the corresponding iodide (**G11+G12**). The yield of this azide-free process is 22%.

When carbonylation was carried out earlier in the sequence (JOC2011b) it was found that the presence of the ester caused problems during elimination of the C6 hydroxyl group, the main product being an isomer of Oseltamivir with a C1C6 double bond. Other unsuccessful variations have also been reported (JOC2011b).

Another route, starting with the bromodiol has been disclosed (TL2008, scheme G, route **Gb**). Intramolecular functionalization of the C3C4 double bond using a tethered electrophilic nitrogen atom gave a strained aziridine intermediate, which was readily opened with 3-pentanol (**G17**). This sequence introduced the C3 and C4 groups with the intended configuration. Chemical manipulation to give Oseltamivir follows previously reported chemistry. This 14-step route has an accumulate yield of 6% and reaches an intermediate of route **Ga** in 8 steps and 12% yield. This is in comparison with 5 steps and 51% yield of the **Ga** route to the same intermediate.

The intervention of the nickel-mediated ethoxycarbonylation at a late stage in the sequences above is disadvantageous because of the restrictions imposed on the presence of heavy metals in the final product: Therefore, the diol derived from ethyl benzoate was tested as starting material. For this third-generation synthesis to have success, the translocation of the double bond to an acrylic type ester has to be carried out (**G24** in route **Gc**, **G30** in **Gd** route). The synthesis uses a hetero-Diels–Alder cycloaddition wherein an acyl nitroso group adds to the polarized diene ketal of the diol (note the inverse electron demand, **G22**), resulting in the formation of two more chiral centres in a stereo and regioselective manner. Of these two routes (**Gc** and **Gd** in scheme G), the second one is azide free.

Significantly, the fourth-generation approach (route **Gd**) provided an especially efficient synthesis of the allylic alcohol, 52% overall yield from starting diol (steps **G21** to **G23**). The synthesis is short (10 chemical steps from the diol) and due to the crystallinity of the intermediates, no chromatography is presumed to be necessary at large scale. However, the oxidative transposition (**G30**) looks to be problematic as several oxidizing agents to replace toxic chromium trioxide were tested without success. Note that a direct, high yielding alkylation of the C3 hydroxyl group should avoid two late synthetic steps in the process (**G34** and **G35**).

Another proposal starts with a protected 4-hydroxycyclohexenone and introduces functionalization sequentially in the C6 rings (OCF2017, Scheme 8.29). Stereoselective aziridination of the double bond was a key step and the resulting aziridine was resolved enzymatically in high yield and enantiomeric excess.

8.8 BUILDING THE CARBOXYLIC RING OF OSELTAMIVIR BY A DIELS-ALDER REACTION

The Diels-Alder cycloaddition is frequently applied to prepare cyclohexenes and it is expected to play a significant role in constructing the core ring of Oseltamivir. The first of these processes to be developed was shown previously (Scheme 8.23). That process used furan as diene and resolution was enzymatic.

8.8.1 Using a [4+2] cycloaddition to produce a trisubstituted six-membered ring adduct

The initial attempts to use fumaryl chloride in a Diels-Alder approach produced a 2:1 mixture of *endo*/*exo* adducts (scheme H, **H1**), only the *endo* adduct being carried forward. Acyl azide formation (**H2**) followed by a double Curtius rearrangement (**H3**) gives a 3,4 cyclic carbamate which, after further transformation (**H4+H5**), gives a di Boc protected diamine α,β-unsaturated ketone. Preparative chiral HPLC gives material with the correct chirality at C4 and C5 (routes **Ha** and **Hb**, respectively **H6** and **H18**). Michael addition of nitrile then introduces the precursor for the ester function (**H7**, **H19**). The C3 group is eventually put in place by asymmetric induction by a bulky reducing agent, aziridination and ring opening with 3-pentanol. This route is relatively short but the accumulated yield is quite low due to the inefficient Diels-Alder reaction and the impracticable HPLC separation. The use of azide and a Curtius rearrangement (carried out in refluxing *t*-butanol) are problematic on an industrial scale.

This methodology (route **Hc**) was improved by application of a novel asymmetric Diels–Alder reaction catalyzed by a barium complex (TokyoU2006c, ACIE2009a, **H25**). The correct quirality at C4 and C5 was obtained in 91% yield with 95% ee along with a favourable diastereometric ratio of 1:5 at C3, appropriate for later closure to the *cis* cyclic carbamate.

Both amine groups were installed, once more, *via* a double Curtius rearrangement (**H28**). An S_N2' displacement with a *t*-butyldimethylsilyl or acetyl protected hydroxy malononitrile in the presence of palladium complexes introduces the functionality at C1 (**H31** and **H34**) and avoids oxidation to an enone seen in routes **Ha** and **Hb**.

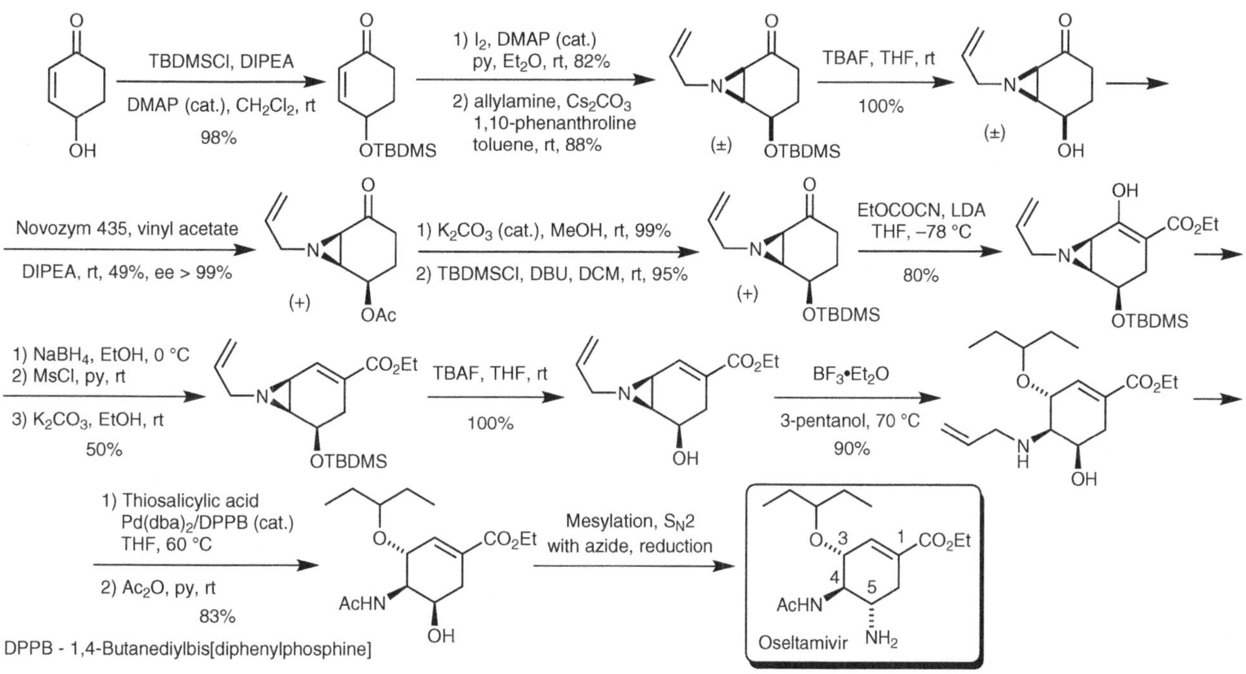

SCHEME 8.29 ENANTIOSELECTIVE SYNTHESIS OF OSELTAMIVIR STARTING FROM 4-HYDROXYCYCLOHEXENONE

Two approaches were used to introduce the C3 α group. In route **Hc1** anchimeric assistance of the C4 acetamide group directs epoxidation of C2C3 double bond to the β face (**H31**). As a result, after ring opening with simultaneous introduction of the desired 1,2 double bond a double Mitsunobu is necessary to produce the 3,4 aziridine (**H32**). In route **Hc2**, Ru-mediate dihydroxylation attacks the C–C double bond from the sterically less congested face (**H35**), producing the required configuration at C3. Compared to the previous routes, the overall yield of **Hc** routes is much improved but azide chemistry and the double Curtius are still present.

In a more recent proposal, a gram-scale, enantioselective synthesis was reported (Scheme 8.30, JACS2018). The asymmetry was produced by an enzymatic kinetic resolution followed by a key iron catalyzed stereoselective olefin diazidation step to introduce the C4 and C5 nitrogens. Using a [4+2] cycloaddition to produce a bicyclic adduct, as well as furan (see above, Scheme 8.23), hydropyridines and pyrrole derivatives have been employed in the Diels-Alder reaction to produce useful bicyclic adducts.

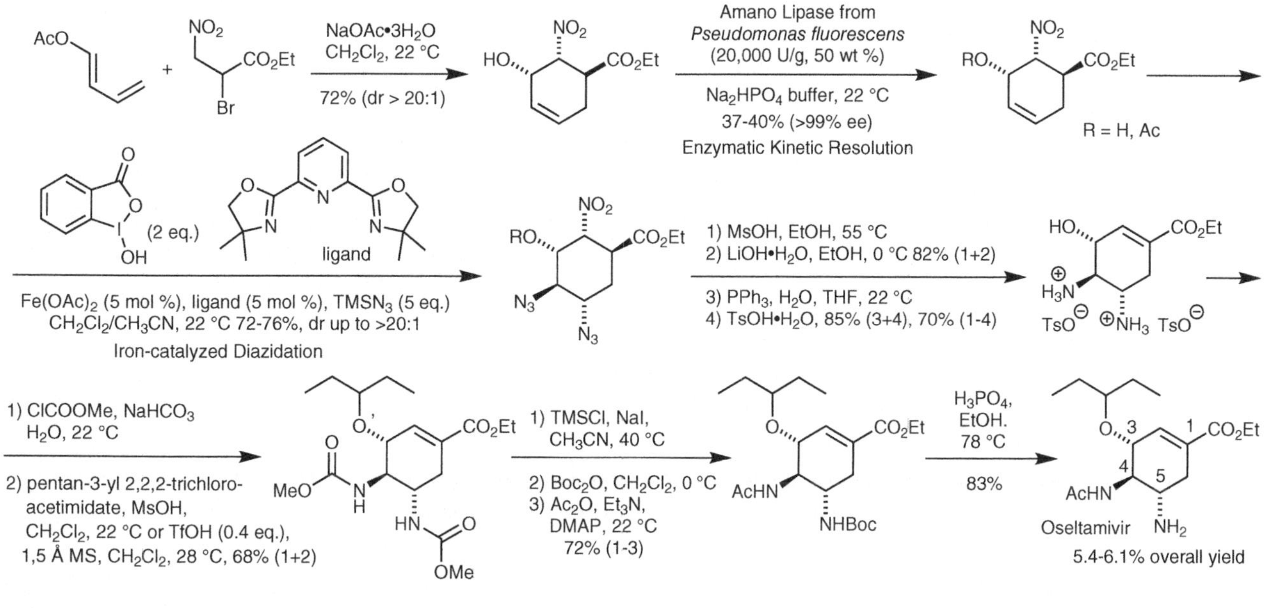

SCHEME 8.30 ENANTIOSELECTIVE SYNTHESIS OF OSELTAMIVIR *VIA* IRON-CATALYZED STEREOSELECTIVE OLEFIN DIAZIDATION

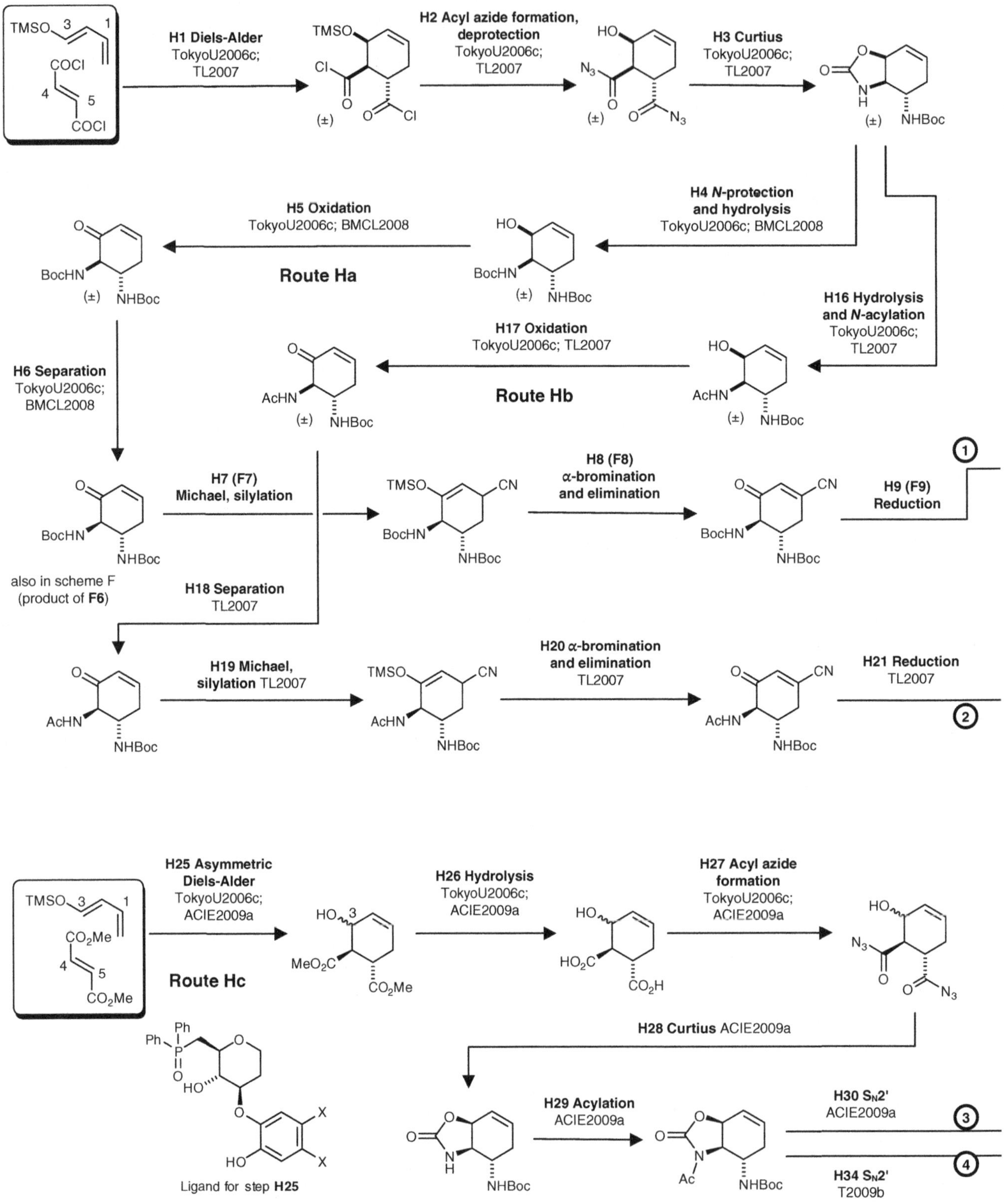

Note: numbers on starting materials refer to corresponding carbon atoms in Oseltamivir

Scheme H Synthesis of Oseltamivir ring using a fumaric acid derivative in a Diels-Alder reaction (for reaction conditions see page 383)

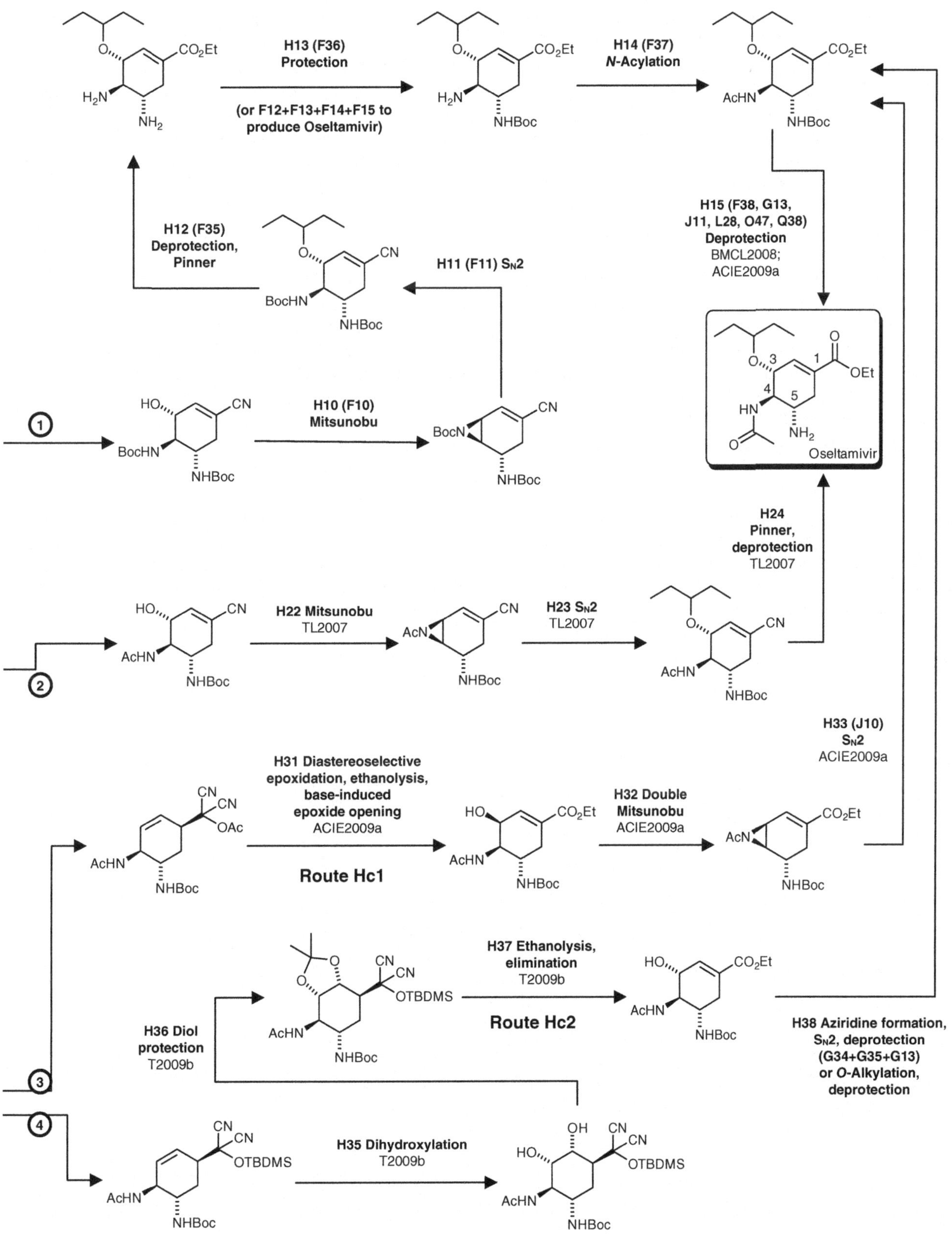
H13 (F36)
Protection
(or F12+F13+F14+F15 to produce Oseltamivir)
H14 (F37)
N-Acylation
H12 (F35)
Deprotection,
Pinner
H11 (F11) SN2
H15 (F38, G13, J11, L28, O47, Q38)
Deprotection
BMCL2008;
ACIE2009a
H10 (F10)
Mitsunobu
Oseltamivir
H24
Pinner,
deprotection
TL2007
H22 Mitsunobu
TL2007
H23 SN2
TL2007
H33 (J10)
SN2
ACIE2009a
H31 Diastereoselective epoxidation, ethanolysis, base-induced epoxide opening
ACIE2009a
Route Hc1
H32 Double Mitsunobu
ACIE2009a
H37 Ethanolysis, elimination
T2009b
Route Hc2
H36 Diol protection
T2009b
H38 Aziridine formation, SN2, deprotection (G34+G35+G13) or O-Alkylation, deprotection
H35 Dihydroxylation
T2009b

8.8.1.1 Diels-Alder with dihydropyridines as dienes

N-carboxybenzyl dihydropyridine and acrolein have been used in an asymmetric Diels-Alder reaction (Scheme 8.31, TokyoU2006b, ACIE2007, T2009a CC2010, JOC2014). The mixture of compounds from this reaction was oxidized and treated with bromine to produce the desired bromolactone as the only isolated product (26% yield, 99% ee, TokyoU2006b, ACIE2007, T2009a). No chromatography was used. After an exchange of protecting groups, the bicyclic lactone was subjected to Ru-mediated oxidation yielding a bicyclo[2.2.2]lactam. A Hofmann rearrangement replaced the carboxylic acid derivative by an allyloxycarbony protected amine linked to the 6-membered ring. Base mediated ring opening of the bicyclic lactam gives an aziridine, which can be converted to Oseltamivir in a small number of steps.

The reported accumulated yield was 5.6% from dihydropyridine (14 steps) and 22% yield from the first bromolactone intermediate. The yield to this crystalline bromolactone was only 26% but its ease of separation and purification was notable. The necessity to use group protection chemistry in addition to the nature of some reactants and reaction conditions are of concern (for example: bromine, Hofmann rearrangement). The costly ruthenium was recovered, thus giving an opportunity to contain costs. This route is azide-free and if the yield of the catalytic asymmetric Diels–Alder reaction could be improved, its practicality in carrying out the chemistry may make it feasible at scale.

8.8.1.2 Diels-Alder with pyrrole derivatives as dienes

A Diels–Alder reaction employing a pyrrole derivative can be a powerful tool to kick off a synthesis of Oseltamivir. However, since catalytic asymmetric cycloaddition is quite difficult to accomplish, it is not surprising that the feasibility of many published routes was tested by synthesising racemic Oseltamivir (scheme I, routes **Ia**, **Ib** and **Ic**). In a similar fashion to the dihydropyridine syntheses above, further manipulation relied heavily on the cage structure to control stereoselectivity up until a base-mediated ring opening reaction discloses the definitively substituted cyclohexene ring system (step **I6** for route **Ia** and **I12** for other routes). The presence of the carboxylic ester is crucial as removal of the acidic α-proton triggers ring opening. The more electron-rich double bond was epoxidised (route **Ia**, **I2**) or dihydroxylated (routes **Ib** and **Ic**, **I11**) thus introducing the C3 and C4 substituents. In **Ib** and **Ic**, the 3,4-diol was converted in several steps to a 3,4-aziridine intermediate *en route* to racemic Oseltamivir.

Chirally pure Oseltamivir was synthesised (route **Id**) and patented using the sequence of route **Ic**. Resolution was carried out by preparation of a diastereomeric mixture and selective crystallisation (**I22**, Shanghai IOC2009a). As alluded to before, a catalytic asymmetric Diels–Alder might make these routes viable for industrial use.

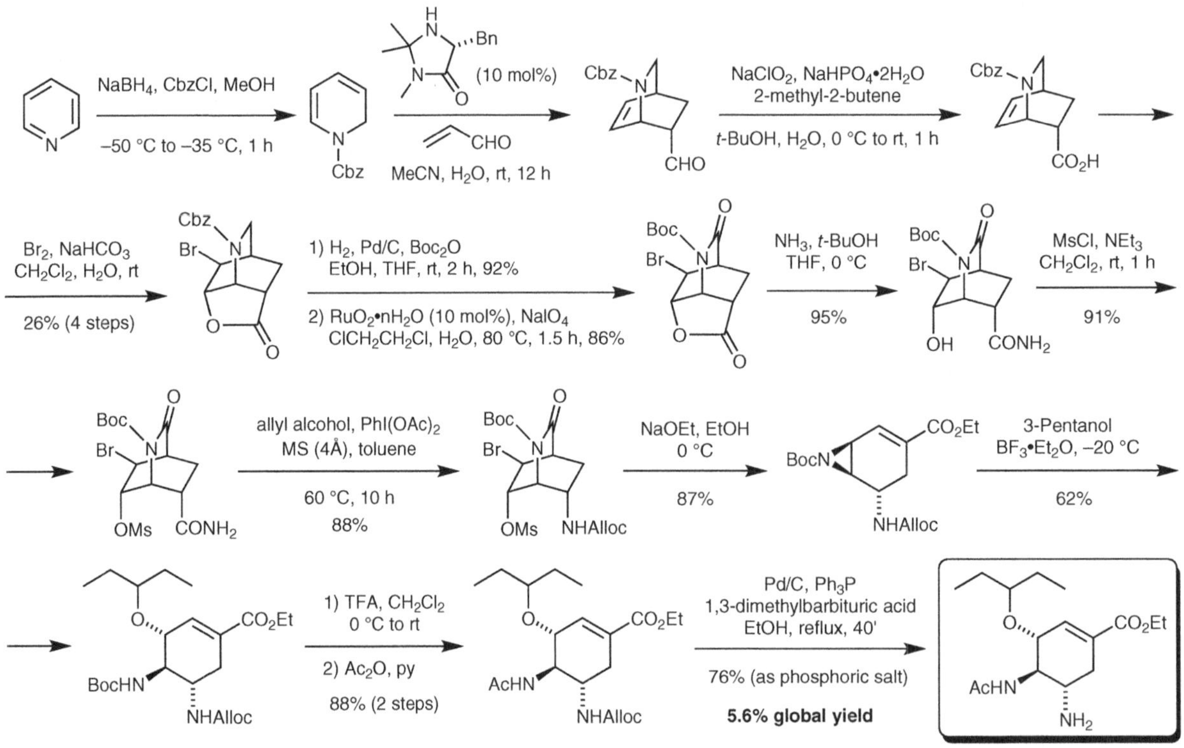

SCHEME 8.31 SYNTHESIS OF OSELTAMIVIR *VIA* A CYCLOADDITION WITH A DIHYDROPYRIDINE DIENE

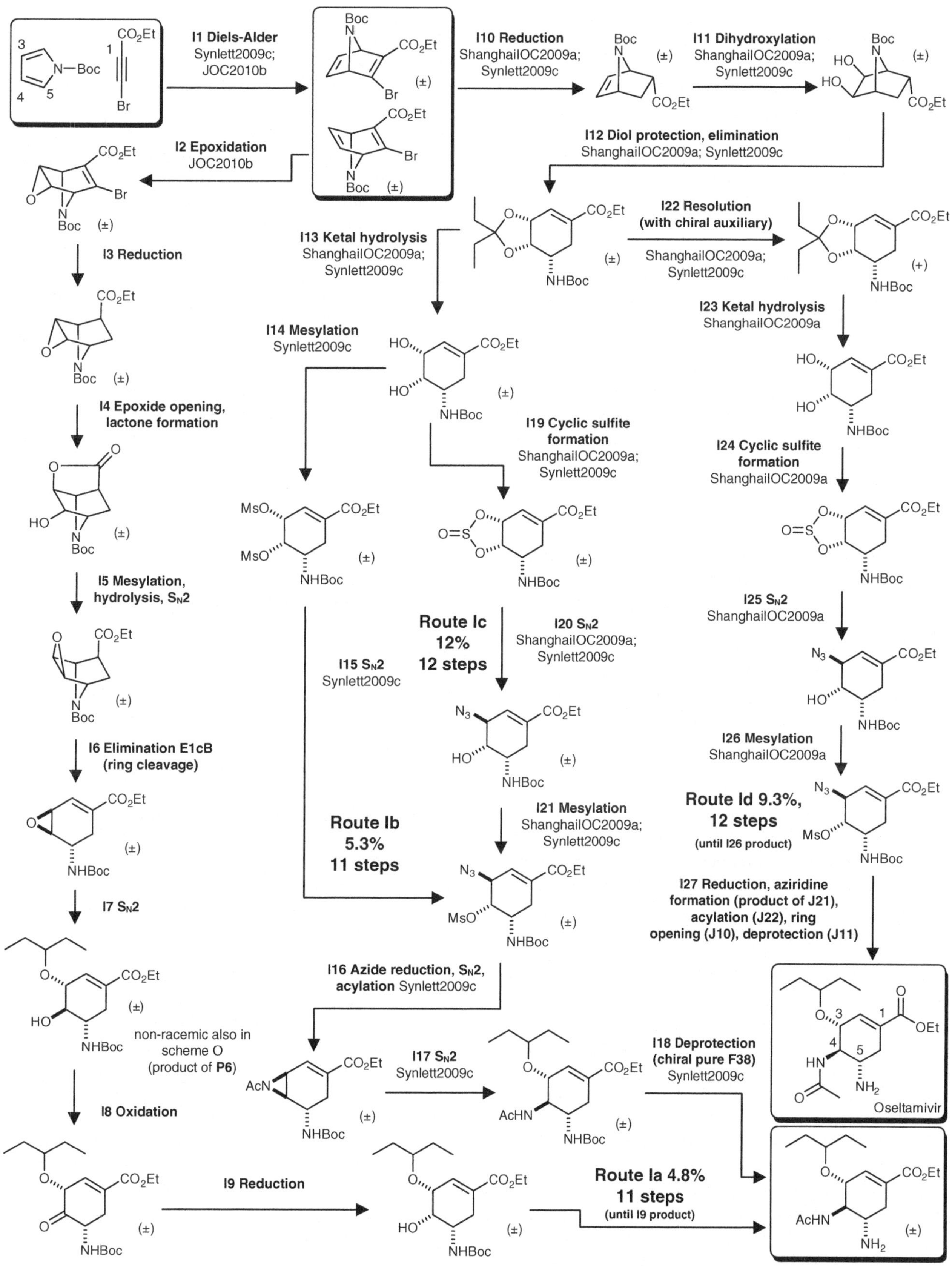

Note: numbers on starting materials refer to corresponding carbon atoms in Oseltamivir

Scheme I Syntheses of Oseltamivir using a Diels-Alder with a pyrrole derivative as diene (for reaction conditions see page 383)

8.8.2 Using a [4+2] cycloaddition to produce a monosubstituted six-membered ring adduct

An elegant and concise azide-free synthesis, starting with a catalytic enantioselective Diels–Alder reaction was disclosed in 2006 by Corey's group (scheme J, route **Ja**). The cycloaddition step (**J1**) is easily carried out at room temperature using a chiral oxazaborolidine catalyst (10% mol). The reaction was carried out on a multigram scale and gave excellent yield (97%) and with very high ee (97%). Recovery of the chiral ligand is simple and efficient. The resulting chiral cyclohexene carboxylic acid is the starting point for a series of sequential stereoselective reactions. An initial iodolactamization (**J3**) produces a bicycle and introduces the amine group at C5, followed by further manipulation to produce the key chiral disubstituted cyclohexadiene (highlighted in scheme J, result of **J7**). Subsequently, asymmetric bromoamidation (**J8**), aziridine formation (**J9**), aziridine ring opening with 3-pentanol (**J10**) and removal of the Boc protecting group (**J11**) completes synthesis of Oseltamivir.

Notably this short, high yielding (27% overall yield), 11 step synthesis was not patented. To make this synthesis viable for industrial manufacture some issues have to be tackled, for instance, an alternative for CCl_4 as solvent in the radical allylic bromination has to be found. Likewise, some of the materials used are either expensive or not universally available. An important outcome of the success of this synthesis was that the cyclohexadiene (formed in reaction **J7** and highlighted) was recognised as a effective intermediate in the synthesis. As a result, other researchers developed alternative methods to prepare it (summarized in scheme K).

Route **Jb** (T2011) started from (*R*)-3-cyclohexene-1-carboxylic acid (>99% ee) obtained readily on a large scale by an asymmetric Diels-Alder reaction. Iodolactonization (**J12**), similar to the iodolactamization used in step **J3** route **Ja**, followed by elimination of HI (**J13**), ring opening (**J14**) and Mitsunobu reaction (**J15**) places the amino group at C5 with the desired configuration. An asymmetric epoxidation (**J17**), anchimerically assisted by the NHBoc group introduces the chirality at C3 and C4. Chemical manipulations, including aziridine formation and ring opening as well as functionalization of the carbon α to the carboxylic ester and elimination, were required to complete the synthesis. In comparison with the other related synthesis, this route is longer, requires more chemical manipulation and greater use of protecting groups.

A recently developed approach used a Trost asymmetric allylic alkylation reaction with the commercially available lactone of racemic *cis* 3-hydroxyl cyclohex-4-ene carboxylic acid (see starting material - Oseltamivir numbering - for route **Jc**) to attach the cyclohexene skeleton to *N*-(trimethylsilyl)phthalimide (**J23**). Hence, the requisite stereochemistry for the entire Oseltamivir synthesis was set up in this single transformation (ACIE2008a, CEJ2011, route **Jc**), the configuration of the amine at C5 amine already having been established at this early stage. An oxidative sequence introduces a disubstituted cyclohexadiene similar to that in route **Ja**. Asymmetry at C3 and C4 was achieved *via* rhodium-catalysed aziridination (**J26**). Subsequent reactions, as seen previously, installed the 3-pentyl ether and the acetamide groups. The aziridination sequence proved to be more difficult than predicted and extensive optimization was required. A 2-(trimethylsilyl)ethanesulfonyl (SES) protecting group was found successful in reducing the competing ring opening reaction with attack of the alcohol at C4. Whilst the process is short (8 steps) and has good overall yield (30%), cost and availability of reagents as well as reaction conditions for the two crucial catalytic steps lactone, lactone desymmetrization (**J23**) and the aziridination (**J26**), need to be undertaken and evaluated. The high toxicity of hydrazine, used to remove the phthalyl protecting group is a safety concern. Microwave radiation, used to increase the reaction rate of the acylation step (reaction **J28**), is still difficult to apply on a large scale with present day technology.

8.8.3 Alternative syntheses for the Corey intermediate

The highlighted cyclohexadiene in route **Ja** was recognized as important intermediate for the synthesis of Oseltamivir and several approaches to its synthesis have been disclosed (summarized in scheme K). It should be noted that the original route (**Ja**) to this key diene was very efficient: only 7 synthetic steps with an overall yield of 73% being reported.

To complement this synthesis and improve on its performance any new approach must bring about: more overall simplicity; ease of operation and use freely available cheap reactants.

A brief synthesis to a racemic intermediate uses a Diels-Alder cycloaddition of *N*-nosyl-3-hydroxy-2-pyridone with ethyl acrylate as a starting point (33% yield, OL2008 route **Ke**). However, for this method to be useful an asymmetric Diels-Alder or a resolution is required.

Four other routes to the diene have been published (routes **Ka** to **Kd**), three of them using a chiral pool approach. Route **Ka** employs a catalytic asymmetric three-component reaction (**K1**) to set up the chirality of the carbon which will become C5 in Oseltamivir.

These routes use diverse chemistries to build the carbon cycle: a Dieckmann condensation for routes **Ka** and **Kb** (step **K3**) and an olefin metathesis ring closure in **Kc** and **Kd** (steps **K22** and **K32**).

Despite their ingenuity and the fact that all the raw materials are readily available, these routes suffer from their length and the extensive chemical manipulation that has to be used, making them less attractive than the original route (**Ja**).

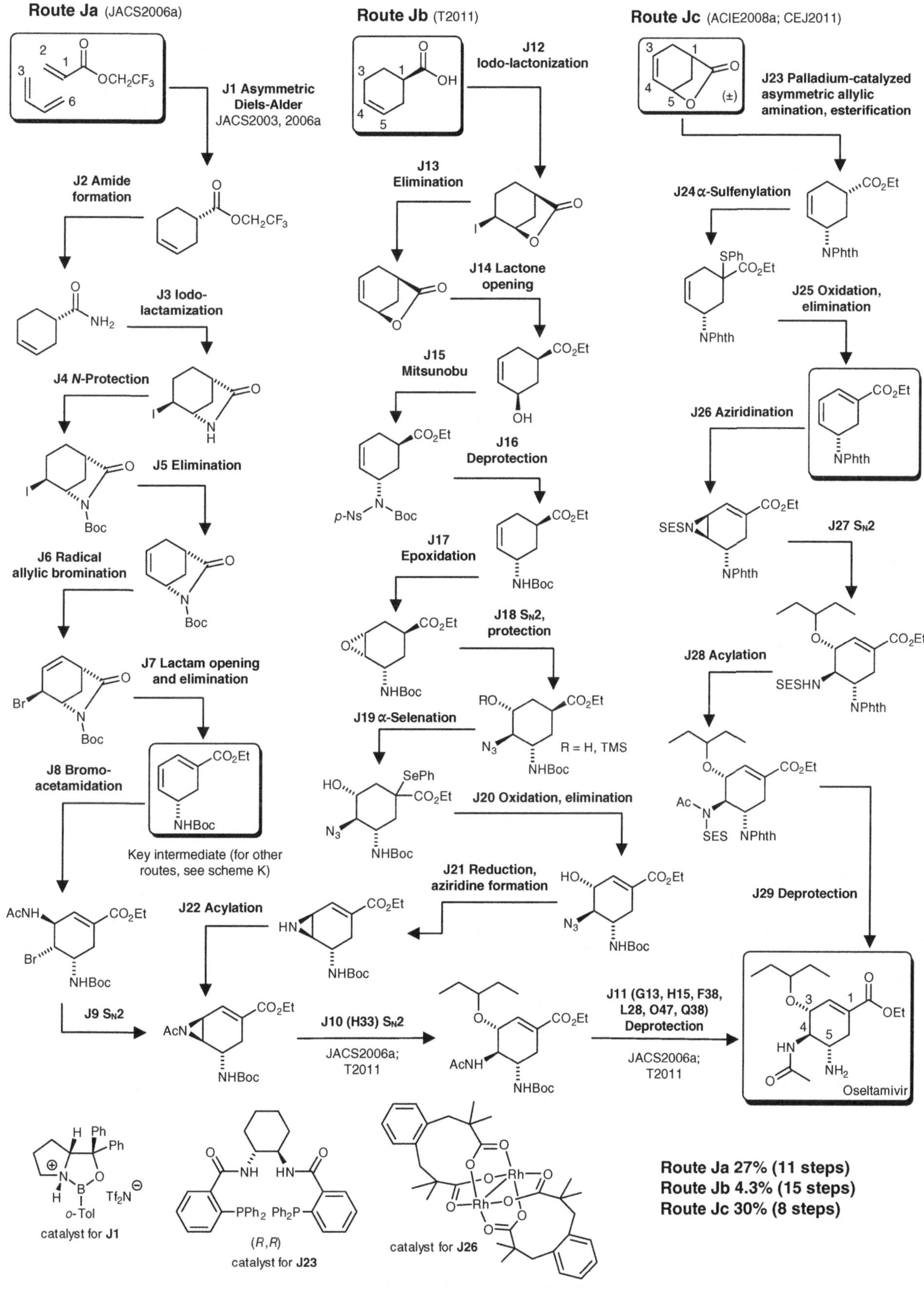

SCHEME J SYNTHESIS OF OSELTAMIVIR USING A CHIRAL CYCLOHEXENE-CARBOXYLIC ACID DERIVATIVE AS KEY INTERMEDIATE (FOR REACTION CONDITIONS SEE PAGE 384)

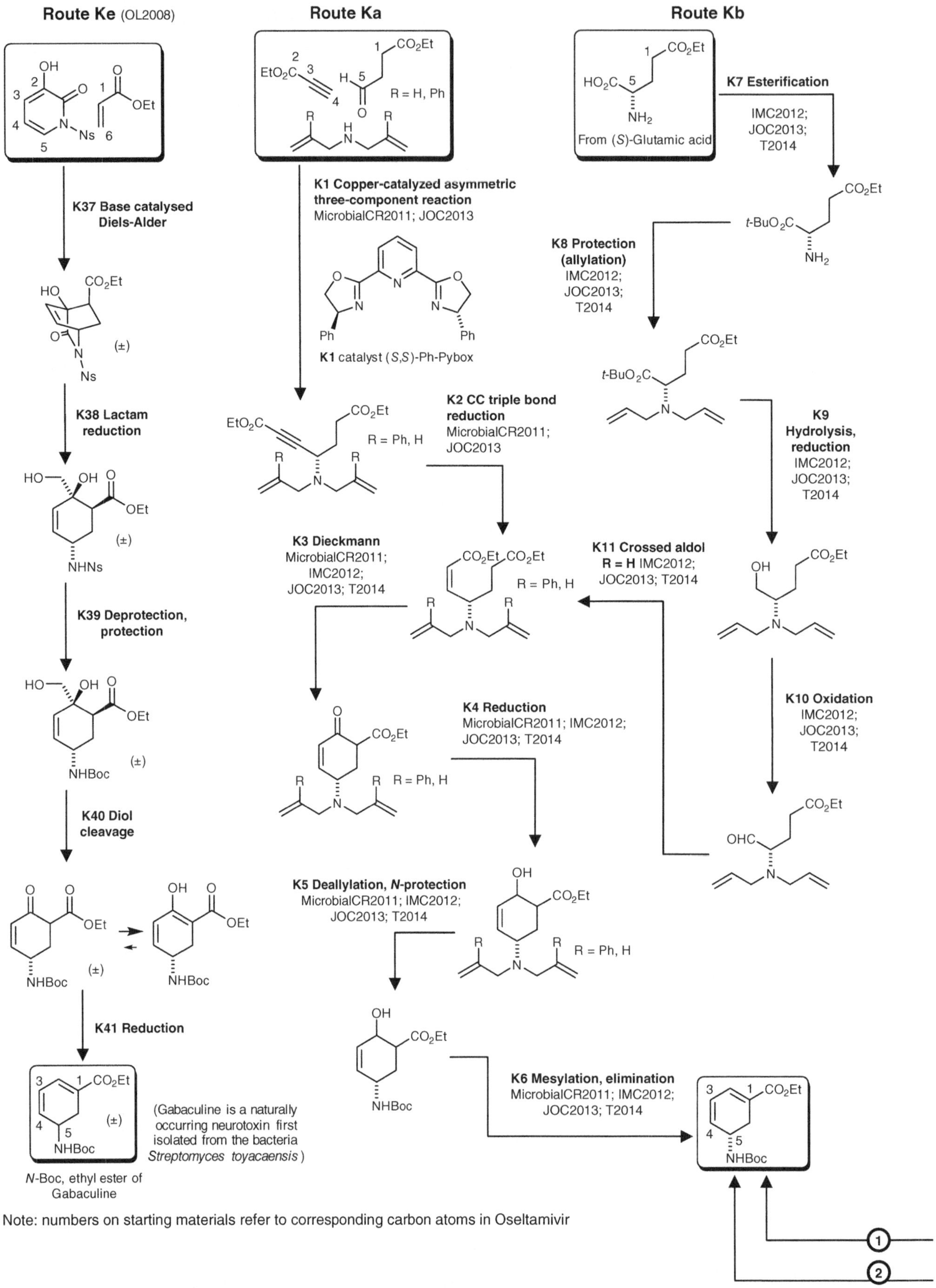

Scheme K Alternative syntheses for the Corey diene intermediate (for reaction conditions see page 384)

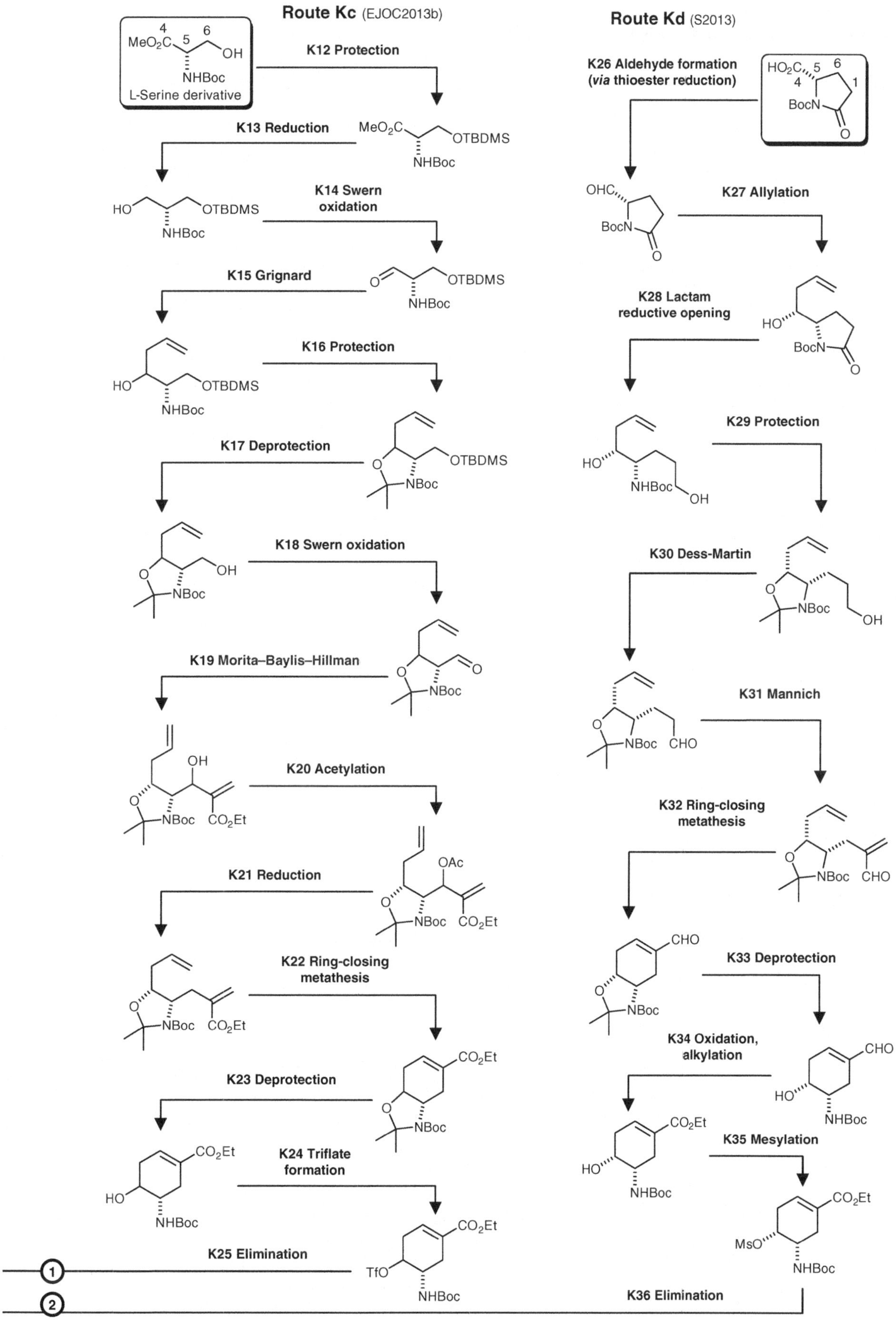
Route Kc (EJOC2013b)
L-Serine derivative
K12 Protection
K13 Reduction
K14 Swern oxidation
K15 Grignard
K16 Protection
K17 Deprotection
K18 Swern oxidation
K19 Morita–Baylis–Hillman
K20 Acetylation
K21 Reduction
K22 Ring-closing metathesis
K23 Deprotection
K24 Triflate formation
K25 Elimination
Route Kd (S2013)
K26 Aldehyde formation (via thioester reduction)
K27 Allylation
K28 Lactam reductive opening
K29 Protection
K30 Dess-Martin
K31 Mannich
K32 Ring-closing metathesis
K33 Deprotection
K34 Oxidation, alkylation
K35 Mesylation
K36 Elimination
1
2

8.9 OSELTAMIVIR RING BY A [3,3] SIGMATROPIC REARRANGEMENT

A high temperature concerted [3,3] sigmatropic rearrangement of a 5-vinyl-3,4-dihydropyran, prepared from readily available D-(+)-glucal, gives a pivotal cyclohexene, which is transformed into Oseltamivir in 22 steps and in about 3% overall yield (Nanyang2007, Scheme 8.32).

8.10 OSELTAMIVIR SYNTHESIS BY A MICHAEL ADDITION-HORNER-WADSWORTH-EMMONS CASCADE SEQUENCE

In 2008 an innovative protocol to assemble a highly functionalized ethyl cyclohexenecarboxylate core *via* a cascade nitro-Michael/Horner-Wadsworth-Emmons sequence was developed (scheme L, routes **La** and **Lb**) allowing a simple and high yielding route to Oseltamivir (Sumitomo2008).

The cyclohexane ring was built up by consecutive reactions, carried out in the same "pot" (**L1** and **L2**). The first reaction is an asymmetric conjugate addition of an aldehyde to a nitro-Michael acceptor (*E* isomer) in the presence of (*R*)-diphenylprolinol trimethylsilyl ether as catalyst (5-20% catalyst, reaction **L1**). This gives a mixture of *syn* and *anti* nitroaldehydes (*syn*, major 96% ee; *anti*, minor 87% ee) in quantitative yield. The second stage of the cycle involves domino reactions: a Michael addition with vinylphosphonate (ethyl 2-(diethoxyphosphoryl)acrylate) followed by an intramolecular Horner-Wadsworth-Emmons reaction to form the cyclohexene carboxylate ester (**L2**). Base treatment removes unwanted impurities giving a final mixture of 5*R* and 5*S*-diastereomers. The major isomer is the undesired 5*R* epimer. Fortunately, Michael addition of *p*-toluenethiol in ethanol, catalysed by Cs_2CO_3 at low temperature gives

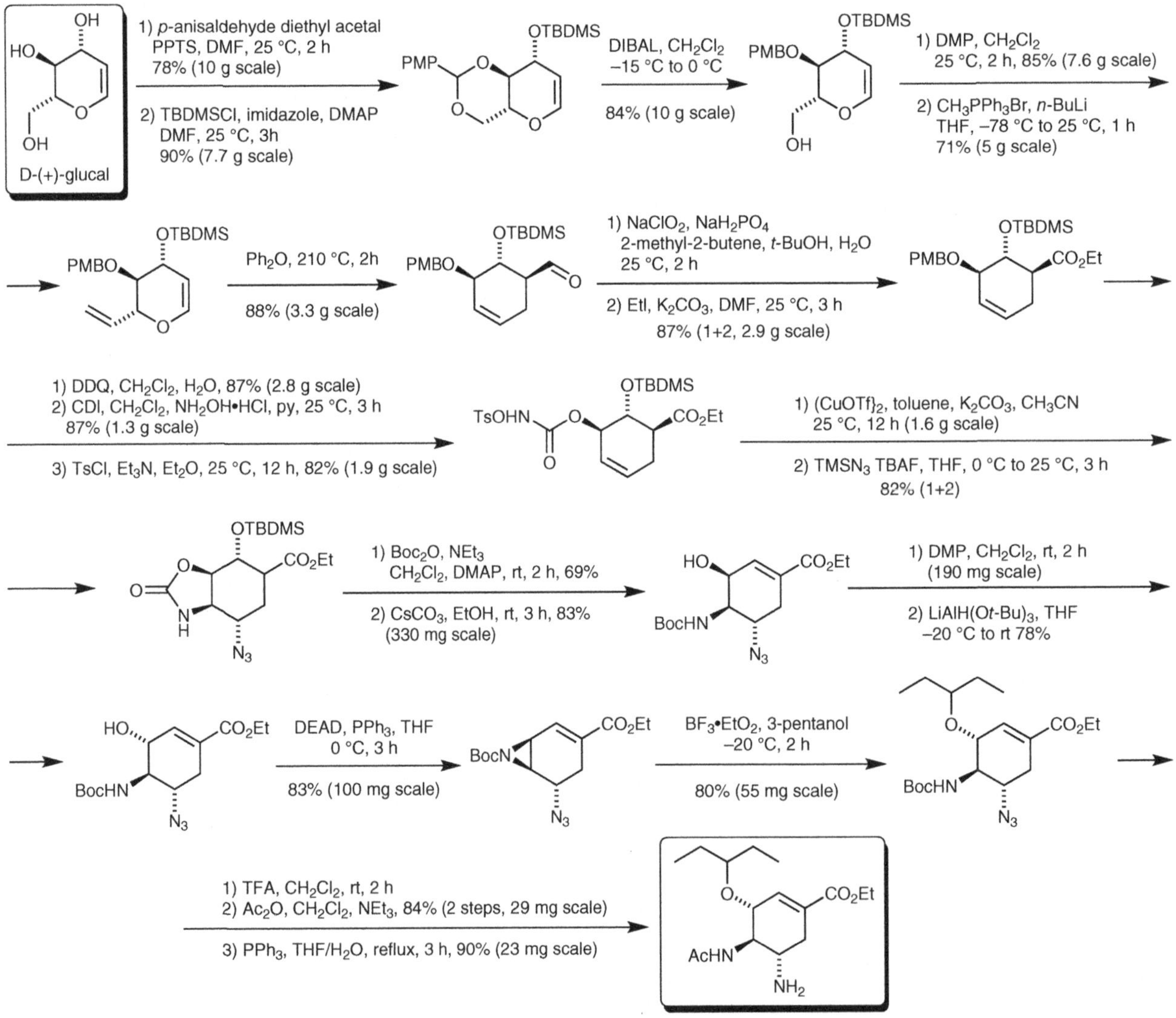

SCHEME 8.32 PRODUCING OSELTAMIVIR CYCLOHEXENE CORE BY A CLAISEN-TYPE REARRANGEMENT

the saturated cyclohexane in which the nitro group prefers to occupy an equatorial position (**L3**), brought about by base catalysed epimerization, thus quantitatively converting the *R* isomer into the desired *S* isomer. Note that simple acidic treatment, as well as with base, also promotes isomerization but a mixture of C5 epimers was always obtained.

To complete the synthesis, the three main tasks that remain are carried out in a one-pot procedure: zinc-mediated reduction of the nitro group (**L4**), elimination of *p*-toluenethiol on treatment with K_2CO_3 (**L5**) and a Curtius rearrangement in acetic anhydride to place the acetamide group at C4 (**L7**). The route **Lb** variation is best and the Curtius rearrangement is described as being possible without heating. The synthesis is reported to be straightforward and minimal purification of intermediates is required. One chromatography step was required. This 10-step route has an excellent overall yield (57%) and is simple to operate.

Route **Lb** was streamlined as follows: the catalyst loading was reduced to 1% mol and the chromatographic purification avoided (Scheme 8.33, EJOC2011). A safe protocol for the Curtius rearrangement by using trimethylsilyl azide as an azide source in microflow reaction was established. A single recrystallization step of a late-stage intermediate is able to guarantee constant quality of the final product. An accumulated yield of 60% is reported and two one-pot sequences are used. However, the preparation of the starting materials, (*E*)-*t*-butyl 3-nitroacrylate and (3-pentyl)oxyacetaldehyde requires five additional steps (3+2) (Scheme 8.34).

Apart from the safety of the Curtius rearrangement, scale-up of this process would require additional issues to be addressed: the handling and reactions of

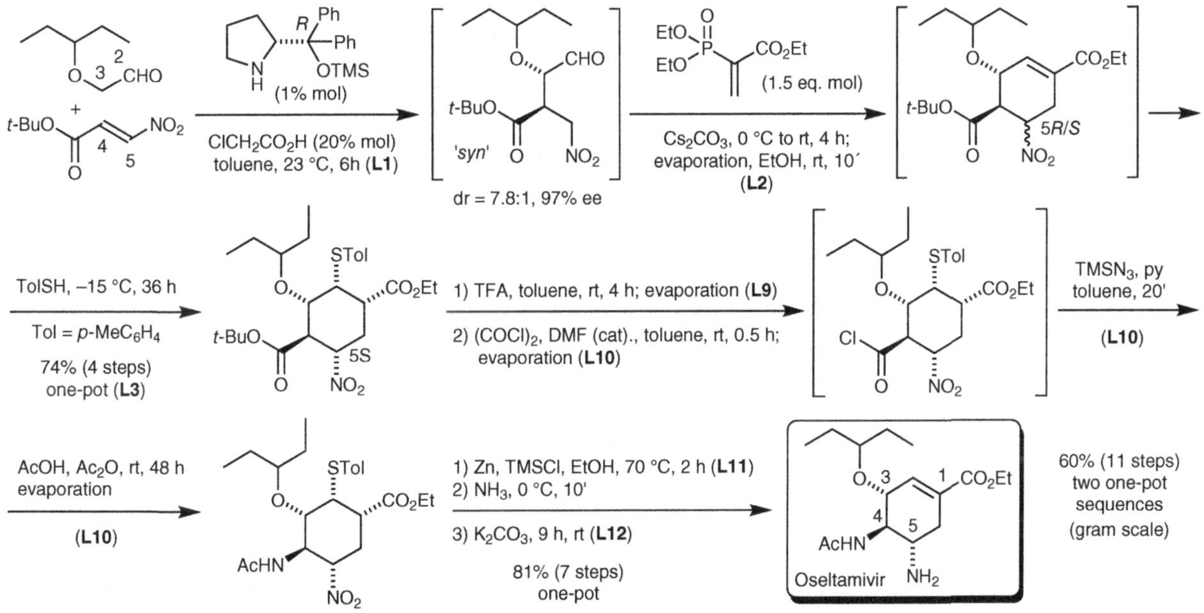

Scheme 8.33 Streamlined Oseltamivir synthesis by cascade nitro-Michael/Horner-Wadsworth-Emmons sequence (route Lb)

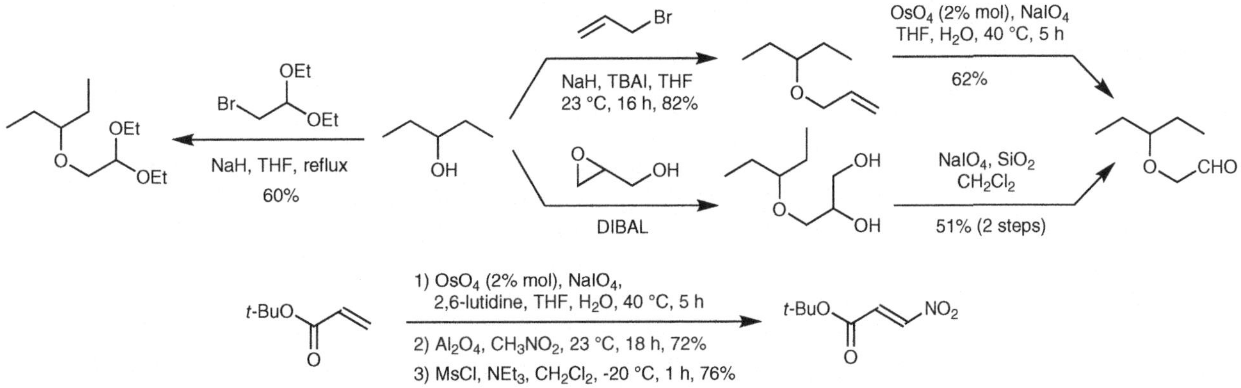

Scheme 8.34 Selected synthesis of starting materials for the Michael addition/Horner-Wadsworth-Emmons cascade sequence

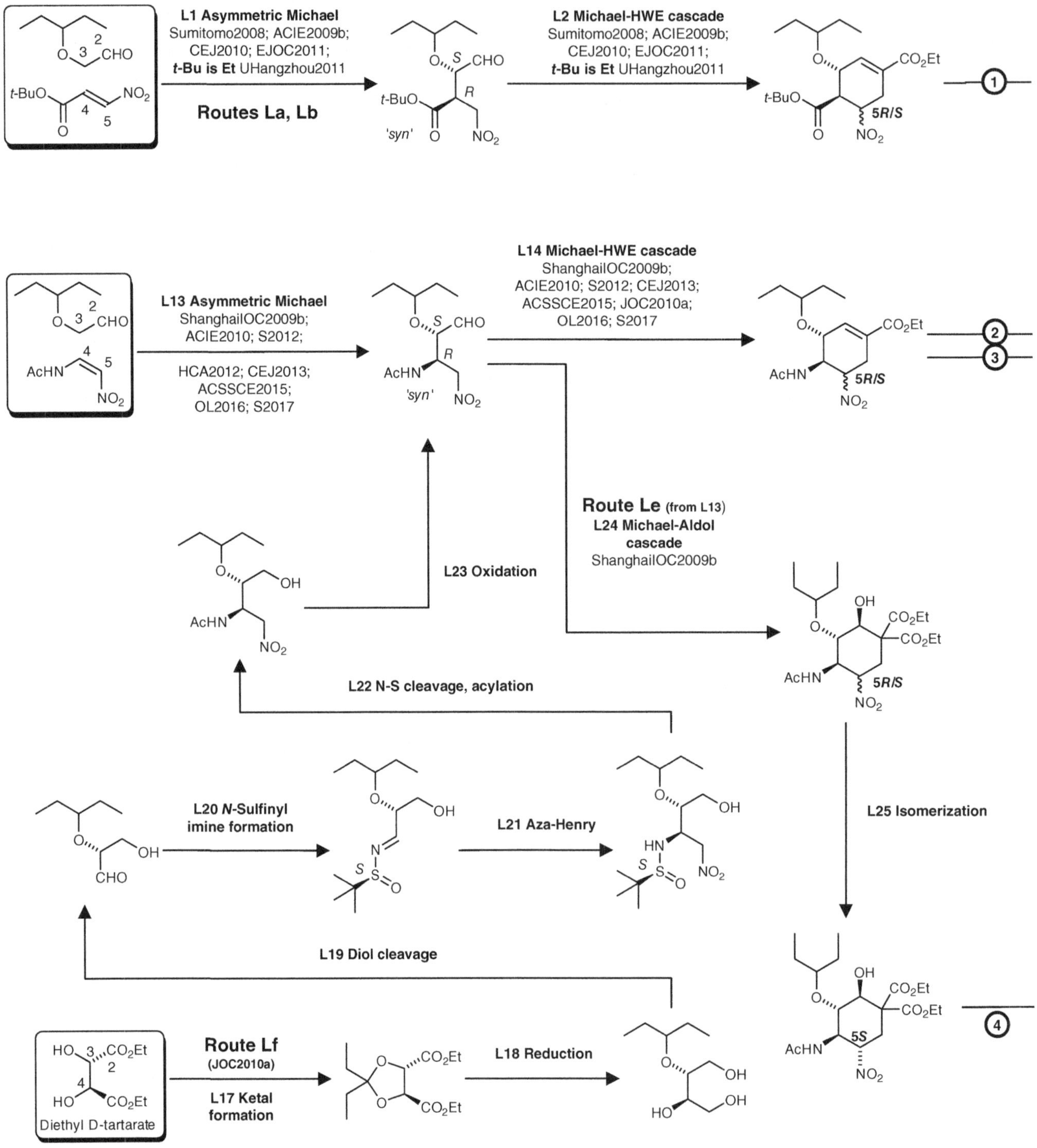

Note: numbers on starting materials refer to corresponding carbon atoms in Oseltamivir

Scheme L Synthesis of Oseltamivir by a Michael/Horner-Wardsworth-Emmons (HWE) domino reaction (for reaction conditions see page 385)

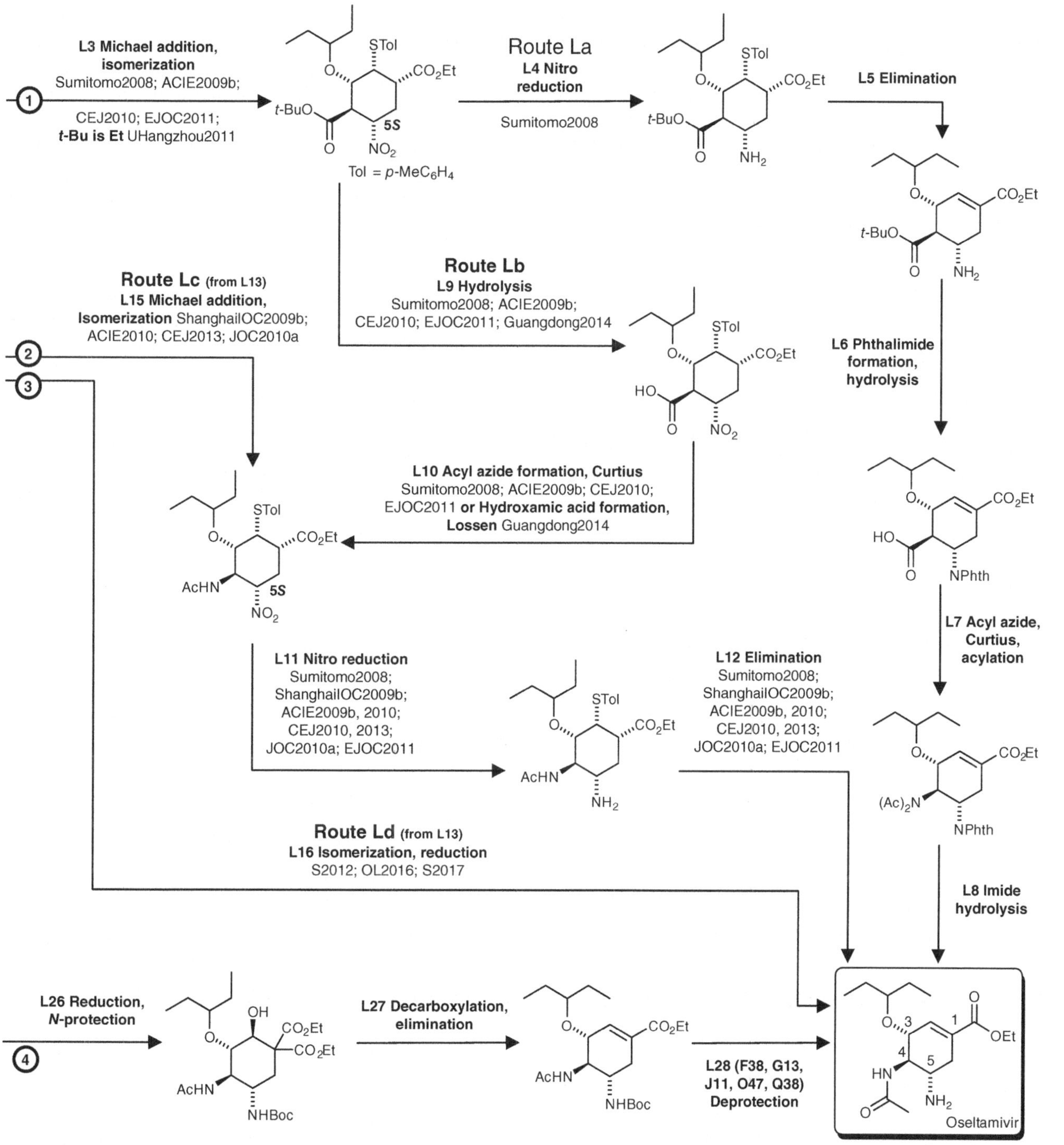

La 43% 13 steps (*via* L4)
Lb 60% 10 steps (2 *one-pot* sequences) (*via* L9)
Lc 28-36% 7 steps (all in *one-pot* sequence) (*via* L15)
Ld 13-15% 5 steps (all in *one-pot* sequence) (*via* L16)
Le 35% 8 steps (*via* L24)
Lf 13 steps 21% (*via* L17, L14 and L11)

L1 and **L13** catalyst

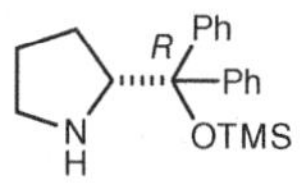

(other similar catalysts also used in UHangzhou201)

L13 catalyst

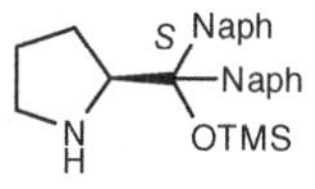

Naph = 2-Naphtyl

compounds containing nitro groups; zinc mediated reduction of the nitro group; the toxicity and odour of *p*-toluenethiol and the two highly toxic osmium tetroxide catalysed dihydroxylations. An osmium free process for the preparation of the aldehyde component was subsequently developed (S2012) and a diethylketal protected aldehyde used as an alternative in the first Michael addition (ACSSCE2015).

When a nitro substituted (*Z*)-enamide, stabilized by an intramolecular hydrogen bonding, is the Michael acceptor (**L13**) the Curtius rearrangement can be avoided (route **Lc**). Following an analogous sequence as before, this sets up an acetamido group at the future C4 atom in the cyclohexene ring. Since the rate, as well as diastereo- and enantioselectivity, are increased in this reaction, further investigation using this approach was undertaken (**L13**, CCC2012, S2012, CEJ2013, HCA2012, OL2016). Yields to the penta-substituted cyclohexane (**L13** to **L15**) were reported to have increased to 54% (96% ee) and thus, the overall yield for the process was 28-36% using a one-pot procedure (Scheme 8.35, CEJ2013).

The intervention of the *p*-toluenothiol can be avoided by performing the cyclization under microwave irradiation (route **Ld**, S2012), separating the C5 epimeric mixture by classical chromatography and epimerizing the isolated 5*R* isomer back to the intended 5*S* isomer (by base treatment followed by crystallisation). It is reported that the pure 5*R* isomer can be epimerized to a 96:4 to 98:2 *S*/*R* mixture in a 96% yield. The intervention of microwave radiation (S2012) greatly accelerates the ring forming reaction rate but the isomer separation by chromatography is a serious drawback for the use of this separation/isomerization approach at industrial scale.

This strategy was employed in the development a total synthesis of Oseltamivir, carried out in one pot and completed in approximately one hour and which avoided the use of *p*-toluenethiol (OL2016, Scheme 8.36). Hence cyclisation under the influence of microwave radiation and in the presence of *t*-BuOK gave eventually, a 50:50 mixture of *S*/*R* 5-nitro compound passing through nitronate salt and the trimethylsilyl nitronate. Oseltamivir was isolated after a zinc reduction and a flash chromatographic

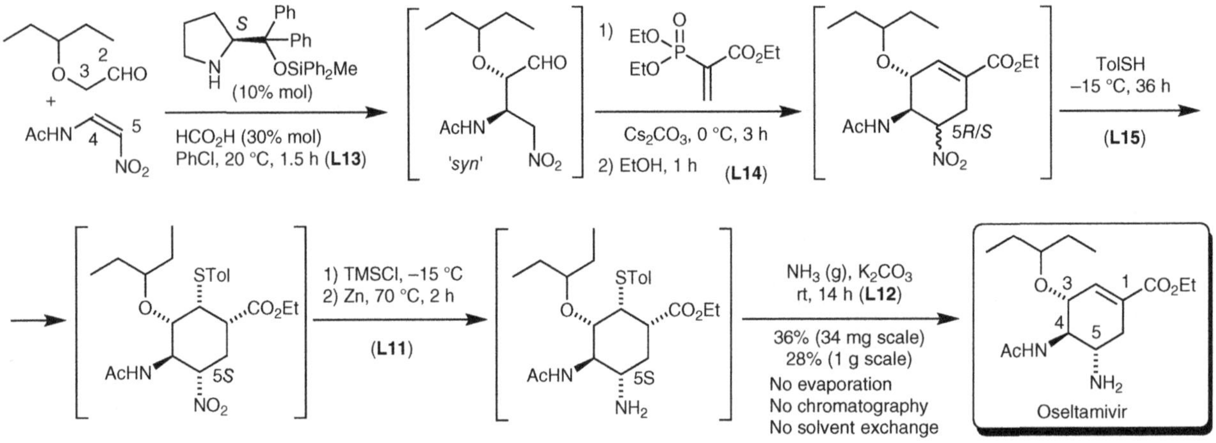

Scheme 8.35 Streamlined Oseltamivir synthesis by cascade nitro-Michael/ Horner-Wadsworth-Emmons sequence (route Lc)

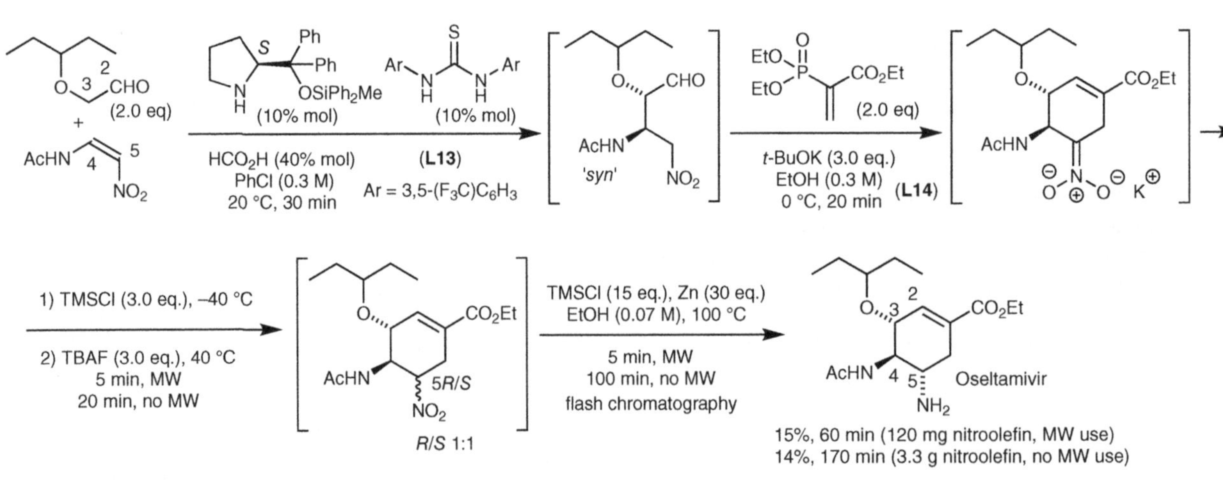

Scheme 8.36 A time economical and expedient total synthesis of Oseltamivir

separation. This five step, one-pot procedure, uses a single solvent and gives the API in one hour and in an overall yield of 15% (120 mg of the starting nitro olefin). A gram scale synthesis without microwave activation was completed in 170 minutes giving a 14% yield (3.3 g of the nitro olefin produced 1.1 g of Oseltamivir). Continuous flow methodology uses five contiguous flow units giving Oseltamivir in 13% total yield with a residence time of 310 minutes (S2017).

Instead of the toxic and potentially carcinogenic ethyl 2-(diethoxyphosphoryl)acrylate, diethyl 2-methylenemalonate has also been used as Michael acceptor in the ring forming reaction (ShanghaiIOC2009b, route **Le**). A lower overall yield was reported (35%).

In a chiral pool approach, diethyl D-tartrate is used as the source of the chirality of the C3 carbon atom (JOC2010a, Route **Lf**). Addition of nitromethane to a sulfinylimine in an aza-Henry reaction gave a mixture of diastereomers in a ratio of 10:1 (**L21**). Removal of the sulfinyl auxiliary, acetylation (**L22**) and oxidation (**L23**) gave the key nitro-aldehyde. This novel approach requires 13 steps to reach Oseltamivir in 21% overall yield. More steps are required to reach the key nitro-aldehyde in comparison with Route **Lc** and this, taken together with the use of nitromethane, is a drawback for its implementation in any industrial process.

8.11 ALDOL TYPE CONDENSATIONS AS KEY RING-FORMING STEP

Some alternatives that share the methodology for ring closure described in the previous section are the focus of this section (schemes M, N, O and P). Most of these routes follow a chiral pool approach, starting from widely available and inexpensive raw materials, selected from carbohydrates and amino acids and their derivatives.

A common practice is to modify the initial starting material to create the appropriate relative configurations of the incipient groups at C3, C4 and C5 before the ring closure. Inspiration from the biosynthesis of shikimic acid allowed investigators to recognize the potential of carbohydrates as starting materials for the synthesis of Oseltamivir.

The processes shown in scheme M use a Henry reaction (nitro-aldol type) to close the ring by forming the C4C5 bond (**M7**). Several chiral centres are present in the starting materials but only one of them is preserved during the synthesis (C3). All these processes pass through an aldehyde (at C2) that is used in a Morita-Baylis-Hillman reaction with ethyl acrylate to introduce the C1-CO_2Et unit of the final product (**M3**, **M11** and **M19**). Further chemical manipulation, addition of a nitro methyl group and elimination of a carbon atom, is necessary to set up C4 and C5.

The chiral centres of Oseltamivir are derived directly from the structure of the raw material when D-xylose (scheme N), *N*-acetyl-D-glucosamine and D-mannose (scheme O, routes **Oc**, **Od** and **Oe**) are used as staring materials. The C1 carbon and the ester group are added by nucleophilic substitution or condensation. In the case of xylose (scheme N), after activation of the primary hydroxyl group reaction with a phosphoryl-stabilized carbanion sets the stage for the Horner-Wardsworth-Emmons cyclization after opening the acetal ring. Further transformation to Oseltamivir is similar to that already described above.

When a C6 carbohydrate is used as starting material (*N*-acetyl-D-glucosamine and D-mannose, scheme O), the terminal carbon atom containing the primary hydroxyl group is removed prior to addition of the phosphonate ester carbanion to the resulting C5 intermediate.

In all other routes (O and P), one or more of the chiral centres has to be introduced by an asymmetric synthesis (routes **Oa** and **Ob** from mannitol and routes **Pa** and **Pb** from (*S*)-glutamic acid and L-methionine, respectively).

Despite the fact that these routes start from inexpensive chiral materials, most are long, require extensive functional group manipulation and chromatographic purification. In addition, they have only been demonstrated at a milligram scale and many intermediates are reported to be oils which are not conveniently manipulated at large scale.

8.12 METATHESIS AS KEY-RING FORMING STEP

The power of metathesis reactions to effect ring-closure attracted a great deal of attention and novel syntheses based on this strategy have been reported. Two of these routes designed to access the Corey diene (scheme K, routes **Kc** and **Kd**) have been discussed above. Compared with the first reported synthesis of the diene by classical chemistry (Scheme J, route **Ja**), all of these routes are longer and less efficient.

When the synthesis starts from with D- or L-serine (scheme Q), the chiral atom from the amino acid becomes C4 of Oseltamivir. The chirality in each of these staring materials is used to introduce the asymmetry of the other chiral atoms as they are added. In the cases where D-ribose, D-glucose or D-manitol are used as starting materials (schemes Q and R, routes **Qc**, **Qd**, **Rc** and **Rd**), the chirality of all atoms is directly derived from the structure of the carbohydrate. Routes **Ra** and **Rb** start from a simple chiral aziridine or epoxide, both derived from *cis*-2-butene-1,4-diol, to provide the asymmetry at C3 and C4. The three routes, **Rb**, **Rc** and **Rd**, all use similar aziridine-dienes for the RCM. Notably, all of these routes are long (14 to 25 steps), employ chromatographic separations to purify intermediates and have low overall yields (below 10%). Route **Qc** is exceptional, giving an intermediate structurally close to Oseltamivir in 9 steps in 23% overall yield (OL2010). Although this route uses azide as a nucleophile in two reactions, inexpensive and easily available D-ribose is the starting material producing a 5-epi-shikimic acid ketal in a concise and practical one-pot process, thus proving the usefulness of ring closure by a metathesis reaction. This sequence, given in more detail in Scheme 8.37, could be significant in circumventing any shortage of shikimic acid.

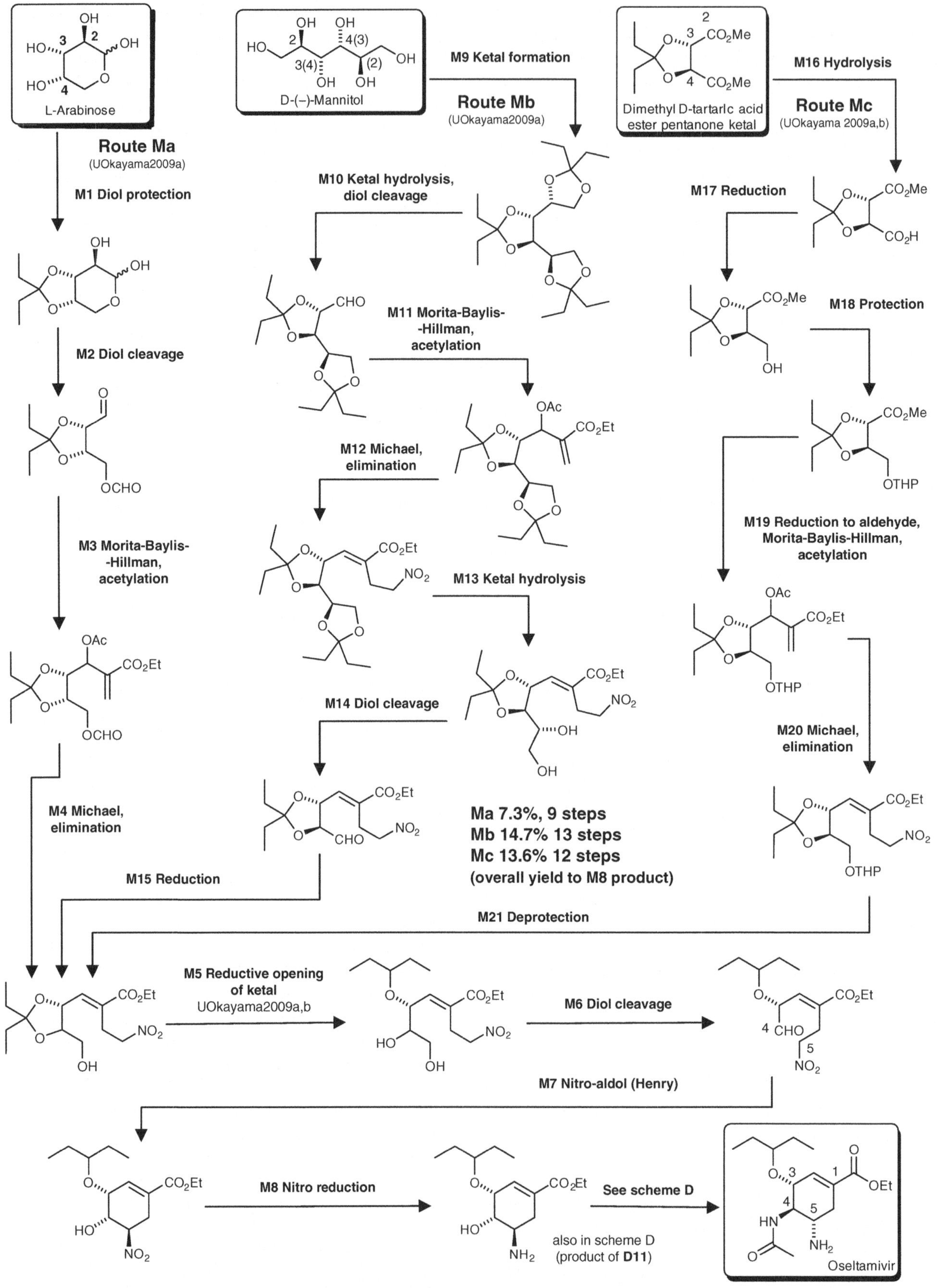

Note: numbers on starting materials refer to corresponding carbon atoms in Oseltamivir

Scheme M Synthesis of Oseltamivir by means of a nitro-aldol reaction to form the C6 ring (for reaction conditions see page 386)

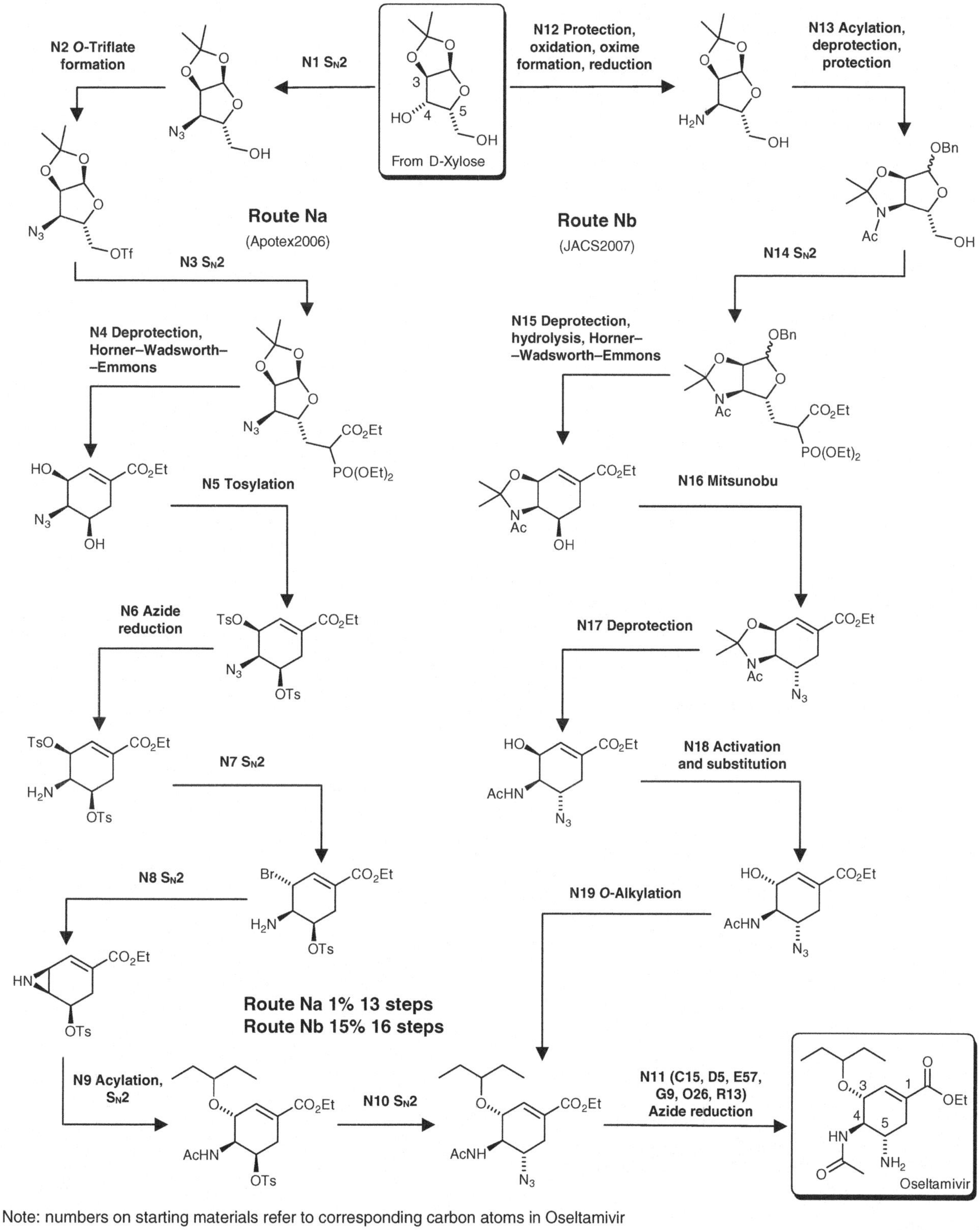

SCHEME N SYNTHESIS OF OSELTAMIVIR USING AN ALDOL-TYPE CONDENSATION FOR RING FORMATION (PART 1) (FOR REACTION CONDITIONS SEE PAGE 387)

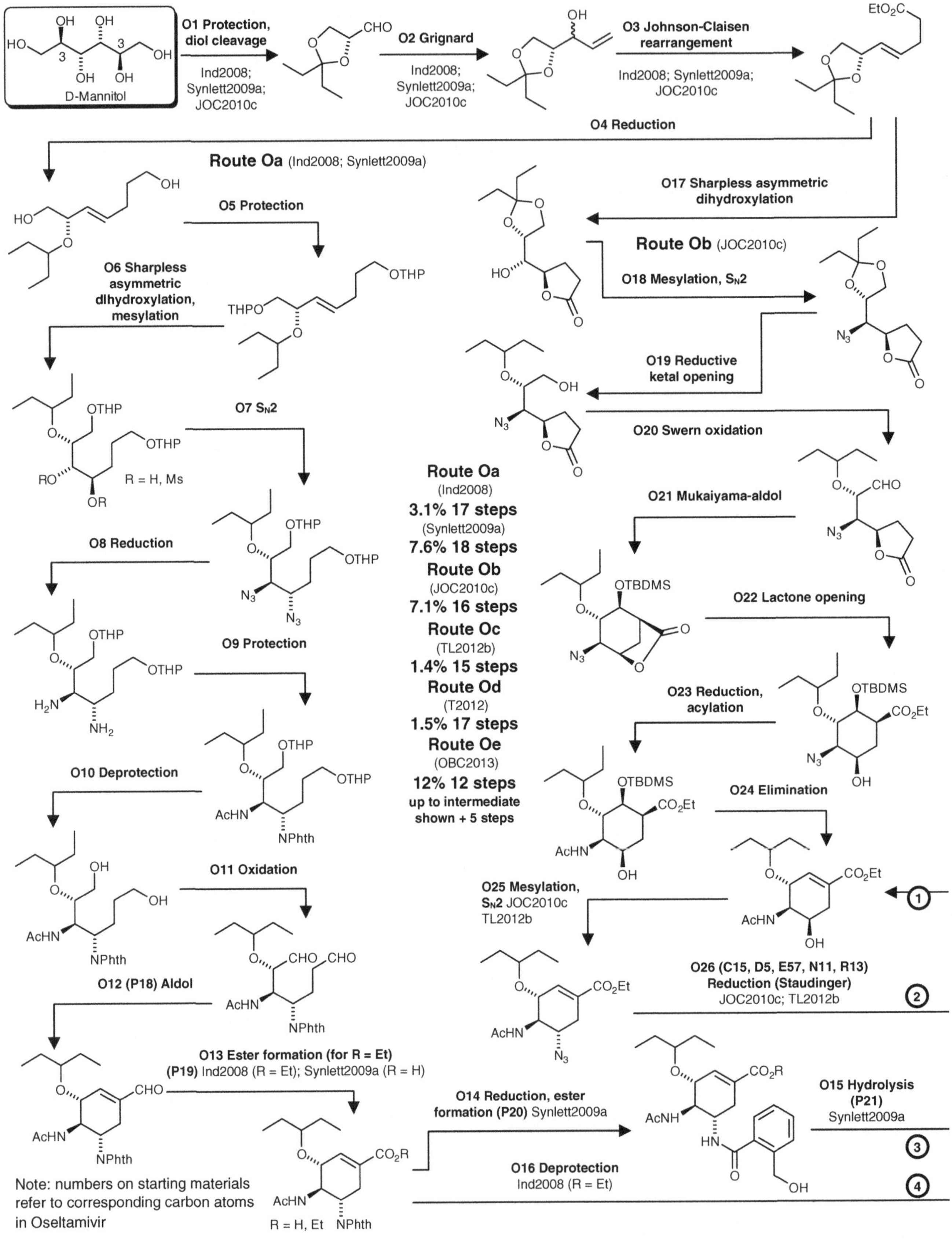

Scheme O Synthesis of Oseltamivir using an aldol-type condensation for ring formation (part 2) (for reaction conditions see page 387)

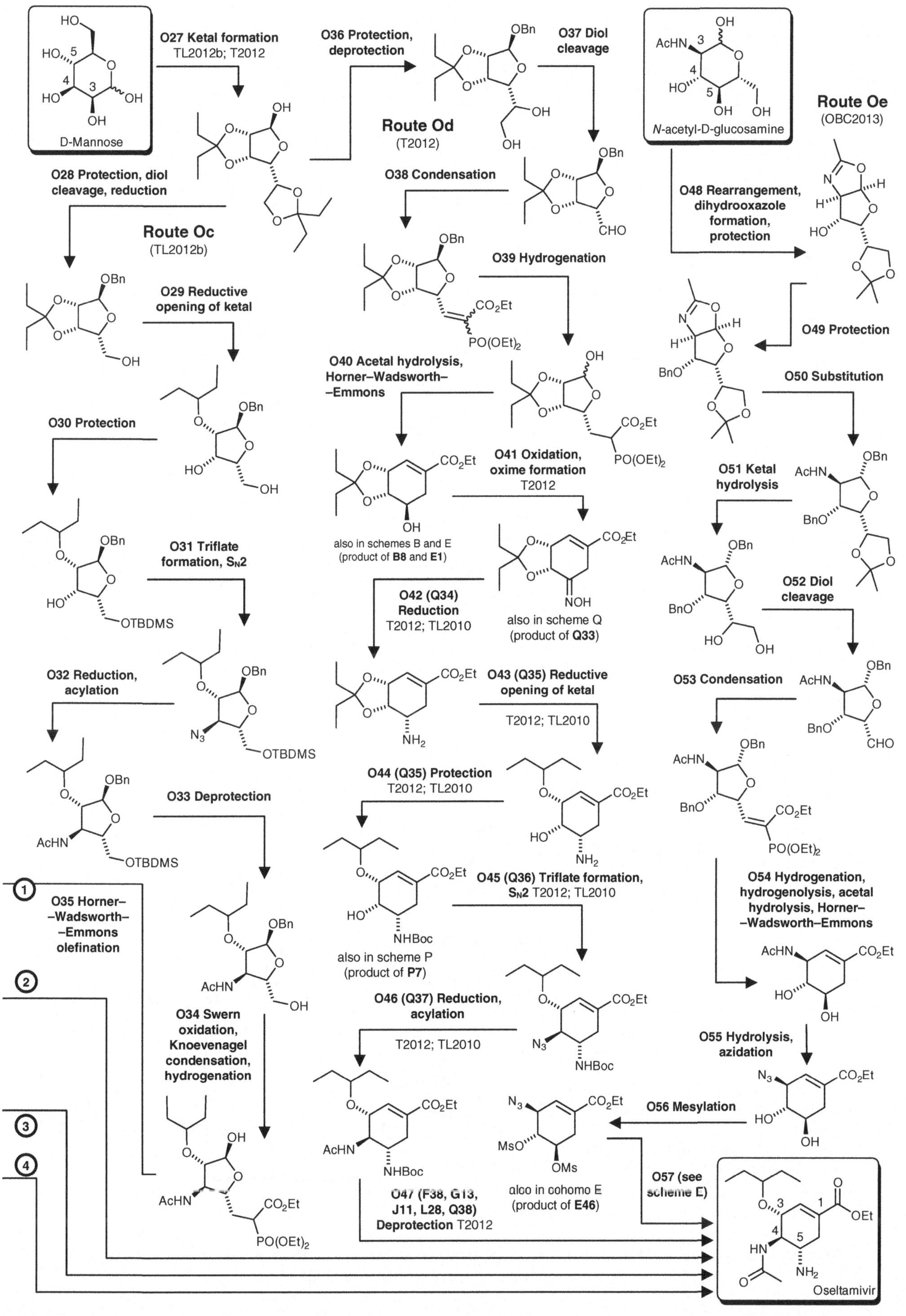

D-Mannose
O27 Ketal formation TL2012b; T2012
O28 Protection, diol cleavage, reduction
Route Oc (TL2012b)
O29 Reductive opening of ketal
O30 Protection
O31 Triflate formation, SN2
O32 Reduction, acylation
O33 Deprotection
O34 Swern oxidation, Knoevenagel condensation, hydrogenation
O35 Horner–Wadsworth–Emmons olefination
O36 Protection, deprotection
O37 Diol cleavage
Route Od (T2012)
O38 Condensation
O39 Hydrogenation
O40 Acetal hydrolysis, Horner–Wadsworth–Emmons
also in schemes B and E (product of B8 and E1)
O41 Oxidation, oxime formation T2012
also in scheme Q (product of Q33)
O42 (Q34) Reduction T2012; TL2010
O43 (Q35) Reductive opening of ketal T2012; TL2010
O44 (Q35) Protection T2012; TL2010
also in scheme P (product of P7)
O45 (Q36) Triflate formation, SN2 T2012; TL2010
O46 (Q37) Reduction, acylation T2012; TL2010
O47 (F38, G13, J11, L28, Q38) Deprotection T2012
also in scheme E (product of E46)
N-acetyl-D-glucosamine
Route Oe (OBC2013)
O48 Rearrangement, dihydrooxazole formation, protection
O49 Protection
O50 Substitution
O51 Ketal hydrolysis
O52 Diol cleavage
O53 Condensation
O54 Hydrogenation, hydrogenolysis, acetal hydrolysis, Horner–Wadsworth–Emmons
O55 Hydrolysis, azidation
O56 Mesylation
O57 (see scheme E)
Oseltamivir

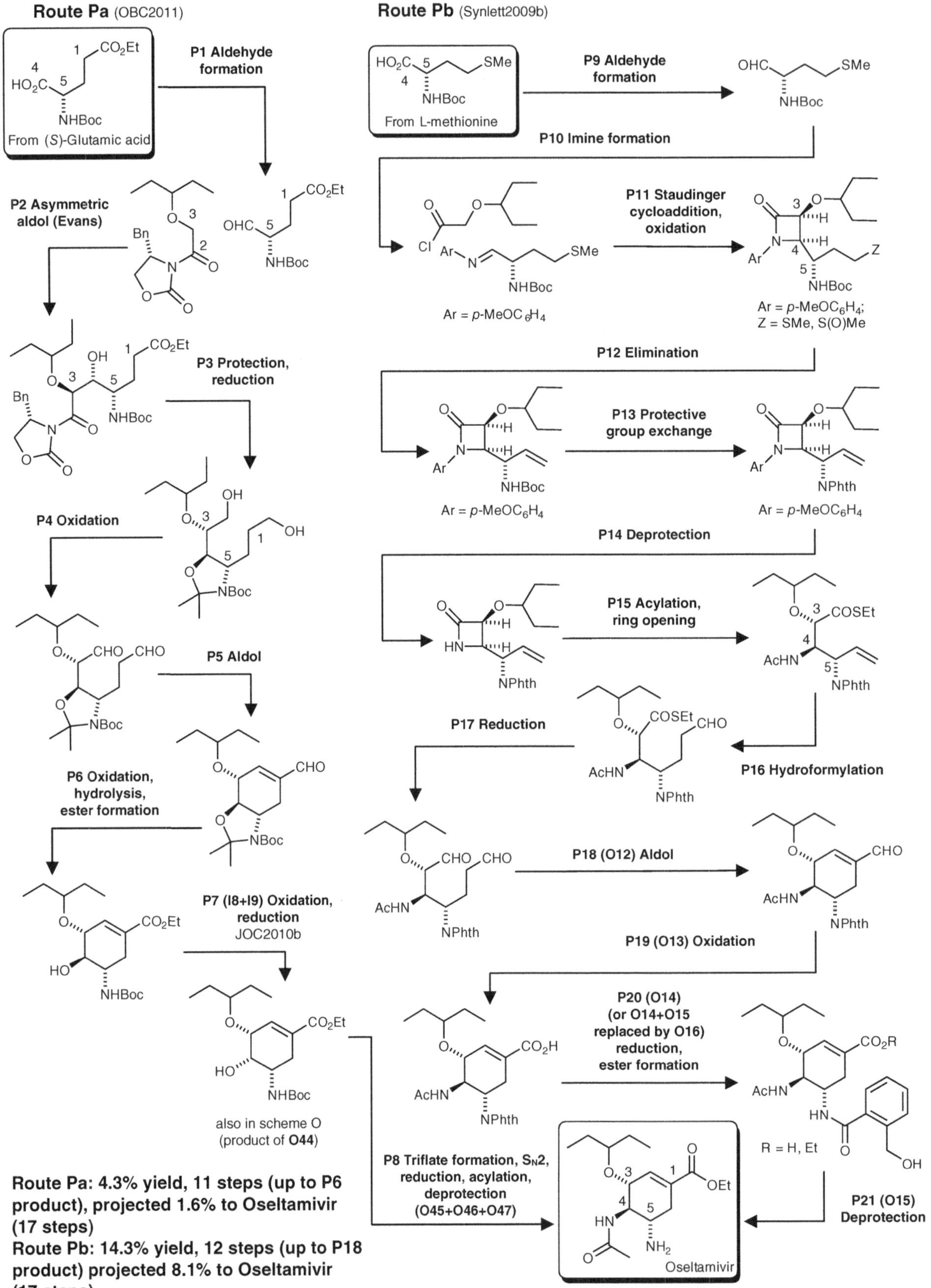

Route Pa: 4.3% yield, 11 steps (up to P6 product), projected 1.6% to Oseltamivir (17 steps)
Route Pb: 14.3% yield, 12 steps (up to P18 product) projected 8.1% to Oseltamivir (17 steps)

Note: numbers on starting materials refer to corresponding carbon atoms in Oseltamivir

SCHEME P SYNTHESIS OF OSELTAMIVIR USING AN ALDOL-TYPE CONDENSATION FOR RING FORMATION (PART 3) (FOR REACTION CONDITIONS SEE PAGE 388)

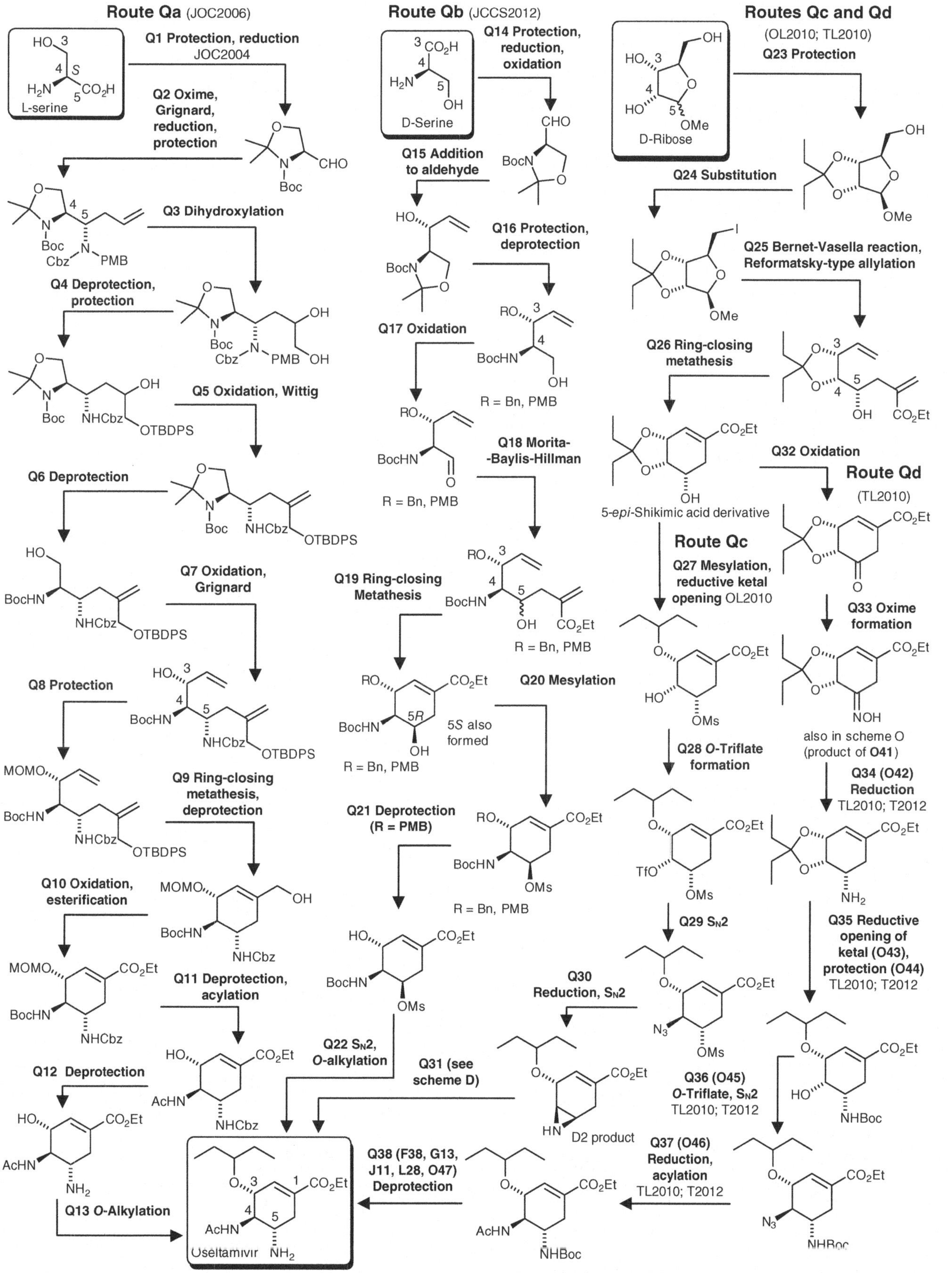

Note: numbers on starting materials refer to corresponding carbon atoms in Oseltamivir

SCHEME Q SYNTHESIS OF OSELTAMIVIR USING A RING-CLOSING METATHESIS REACTION AS A KEY STEP (PART 1) (FOR REACTION CONDITIONS SEE PAGE 388)

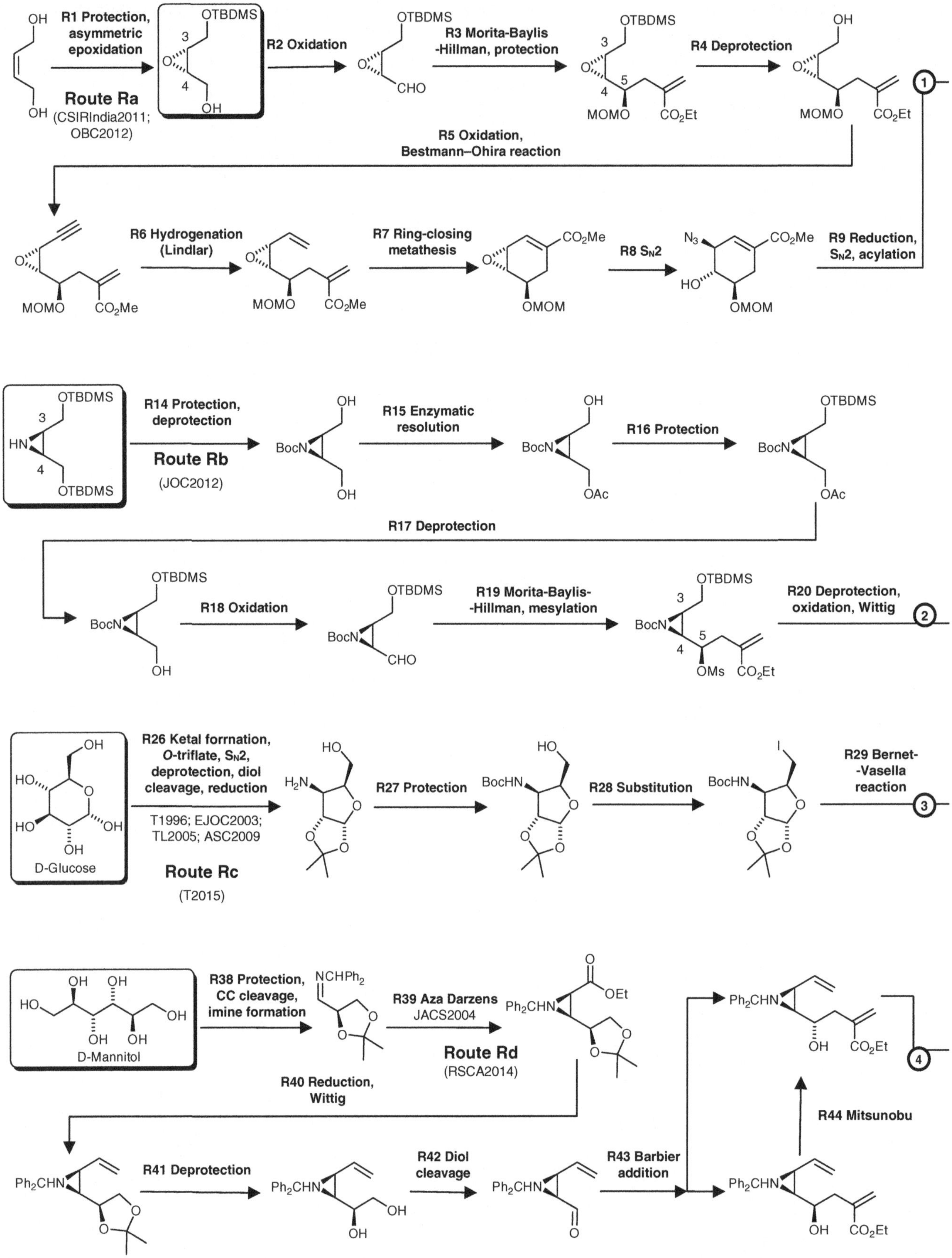

Note: numbers on starting materials refer to corresponding carbon atoms in Oseltamivir

Scheme R Synthesis of Oseltamivir using a ring-closing metathesis reaction as a key step (part 2) (for reaction conditions see page 389)

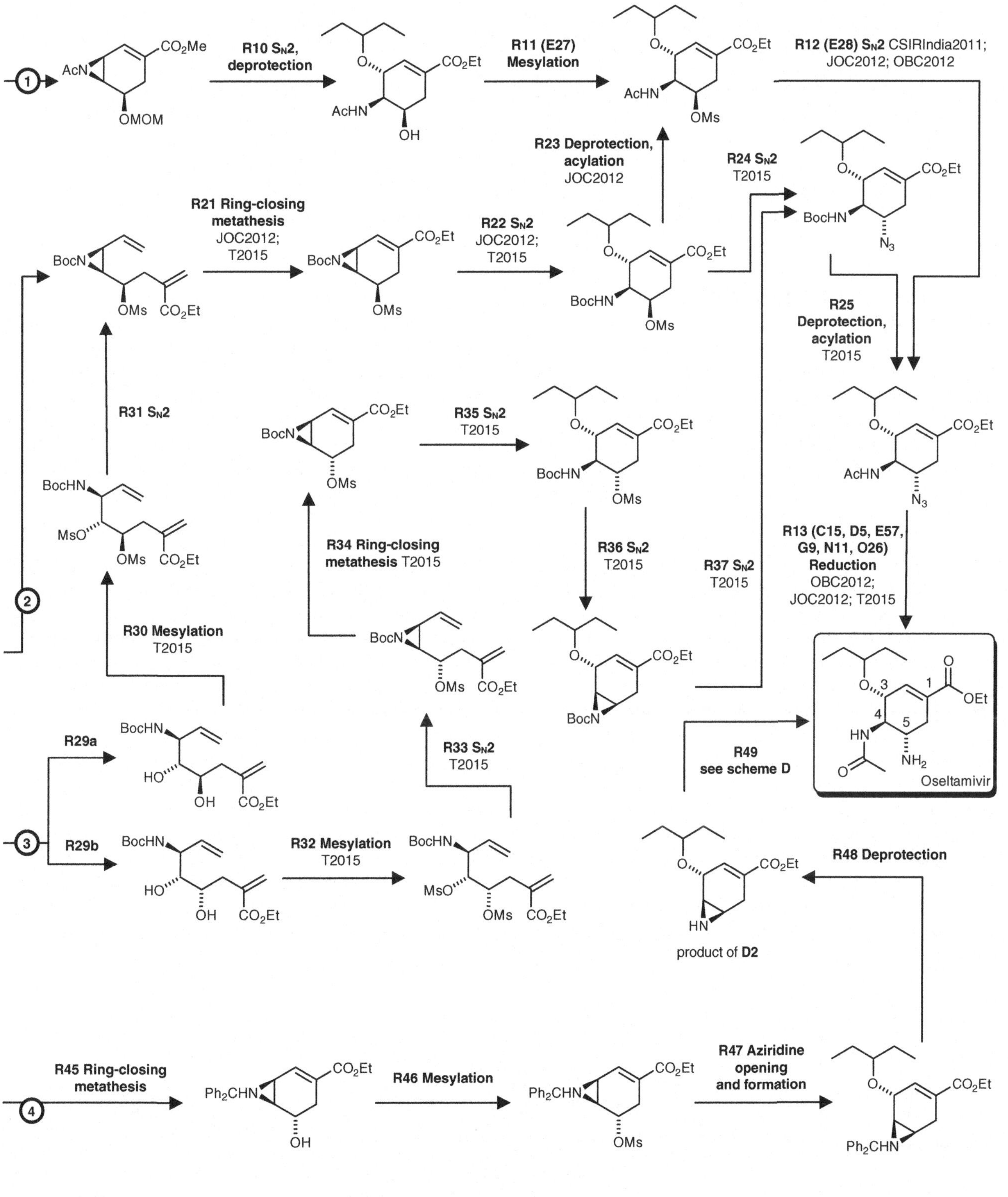

Route Ra: 5-5.7%, 18 steps; Route Rb: 1.5-2.1%, 17 steps; Route Rc: 4.6% (global of both branches from R27) 18 steps, Route Rd: 14% up to R47 product, 16 steps

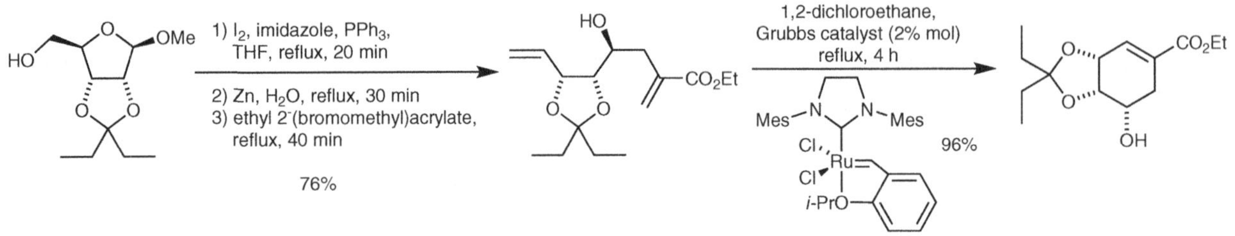

Scheme 8.37 Synthesis of 5-epi-shikimic acid derivative from D-ribose

8.13 STRATEGY COMPARISON AND CONCLUSIONS

Oseltamivir is a small molecule and its synthesis has revealed itself to be a challenge for chemists, both in industry and academia. As exemplified above, a large number of approaches have been described (for reviews see Chimia2007, EJOC2008, CR2009, IJC2011).

The complex structure of Oseltamivir and the alarm caused by the threat of a flu pandemic made it an attractive synthetic target for the academic community. Notwithstanding, Process Chemists were obliged to find viable industrial processes so that commercial quantities could be made available. The motivation of academic investigators, in comparison with those in industry, meant that many of the disclosed synthetic routes were not patent-protected.

Peer journal publications and the patent literature reveal fundamental differences. Patents must demonstrate both novelty and superiority over existing literature and their main focus is to discover processes for industrial implementation[3]. Academics seek new approaches and once the end point is reached generally do not make improvements. Patents focus on protecting an invention whilst revealing the minimum of information (really significant results can be hidden in an avalanche of secondary information). Hence, an analysis of either type of literature to ascertain applicability for industrial application is challenging. These pitfalls have been mentioned previously in the introduction.

In the case in hand, the initial shikimic acid/quinic acid synthesis set high standards, which were difficult to beat.

Apart from restrictions on the raw material, shikimic acid looks to be the best starting point for an industrial route because of its closeness structurally to Oseltamivir. All three stereogenic carbon atoms in Oseltamivir can be derived directly from shikimic acid.

A key intermediate in several of the reported routes was the epoxide, shown in Scheme 8.38. Shikimic acid is clearly preferred over quinic acid for its preparation as the introduction of the double bond in the latter case is reported to proceed with a yield of slightly over 50%. Quinic acid would only become a valid substitute for shikimic acid if it was available at much lower cost or perhaps, more easily available.

Other strategies have also been used to produce this desirable epoxide. Its transformation into Oseltamivir is shown in Scheme 8.39). Short syntheses have been developed, which are azide free and are scalable. These are preferred industrially. Various combinations of different nucleophiles have been employed to introduce the two vicinal nitrogens in the cyclohexene ring. Yields are reported to be similar and the ideal combination is not easily arrived at without extensive process develop-

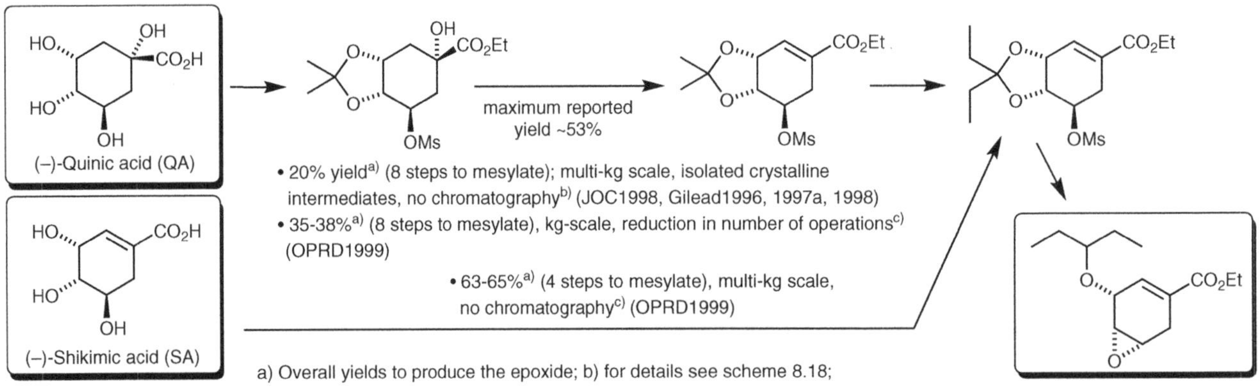

Scheme 8.38 Summary of synthetic routes from quinic and shikimic acid to the key epoxide intermediate

3 In many cases, companies will patent a viable industrial route, not with the objective of putting it into practice, but to stop competitors from using it to produce a competing product.

ment. The last four combinations in the list depicted in Scheme 8.39 all appear to have high potential for industrial application.

Additional routes from shikimic acid which do not depend on the formation of the epoxide have been published (scheme E). Some of these are short and give high yields, 8 to 12 steps and up to 55% overall yield. These routes are, *a priori*, good candidates for large scale processing. However, they all use the azide nucleophile in two steps in combination with a Staudinger reduction. A further drawback is the necessity for a relatively high temperature (80-90 °C) and excess azide (2-4 eq.) to introduce the nitrogen at C5, giving rise to safety concerns at industrial scale (TA2009, 2011, 2012). The alternative, a Mitsunobu reaction with explosive hydrazoic acid, is also not a realistic proposition (TL2012a).

Taken together, the above considerations lead to the conclusion that the best route to Oseltamivir starts from shikimic acid and follows the epoxide pathway (Scheme 8.38).

In the early days, the shortage of natural (–)-shikimic acid was a major concern. Today it is available in ton quantities by extraction of Chinese star anise and by fermentation using a genetically engineered *E. coli* (Roche asserts 84 g/L and a 33% yield from glucose). Given these now reliable sources of shikimic acid, chemical synthesis would appear not to be competitive although 5-epi-shikimic acid can be produced in a concise and practical one-pot process *via* ring-closing metathesis (Scheme 8.37, OL2010).

Hence, it appears that the most favourable route to Oseltamivir successfully applies traditional organic synthesis and, when the shikimic acid is not obtained from a natural source, biotechnology (Scheme 8.40).

It should be noted that Oseltamivir can be produced by a short process (6-7 steps) in good overall yielded (18-22%) starting from 5-aminoshikimic acid (schemes D and Scheme 8.41). If an alternative to, or improvement of, the low yielding (5%) fermentation/biocatalytic process known at present can be found, this could become an important option in the future.

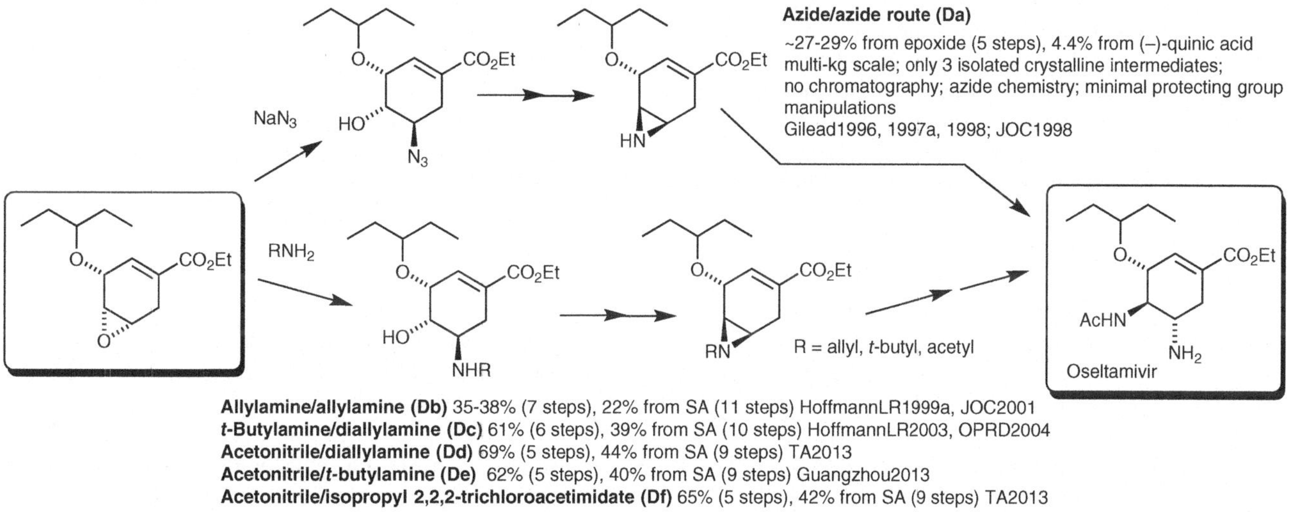

SCHEME 8.39 SUMMARY OF THE ROUTES TO CONVERT THE EPOXIDE INTERMEDIATE INTO OSELTAMIVIR

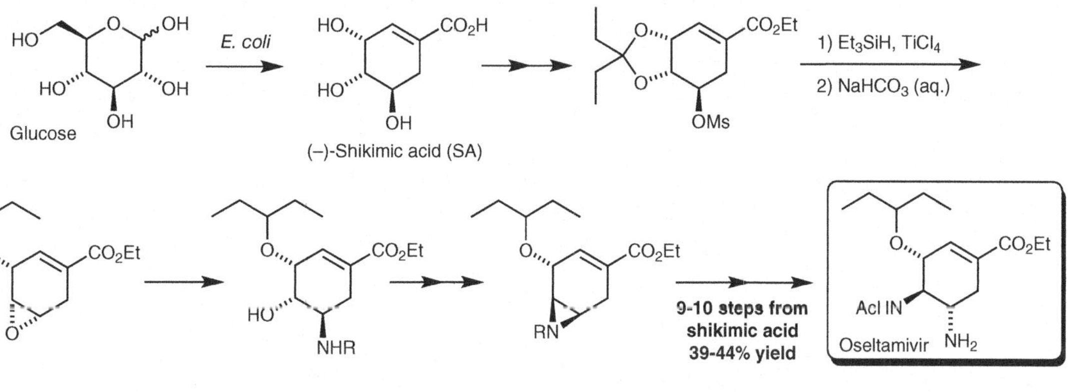

SCHEME 8.40 BEST INDUSTRIAL ROUTE TO OSELTAMIVIR

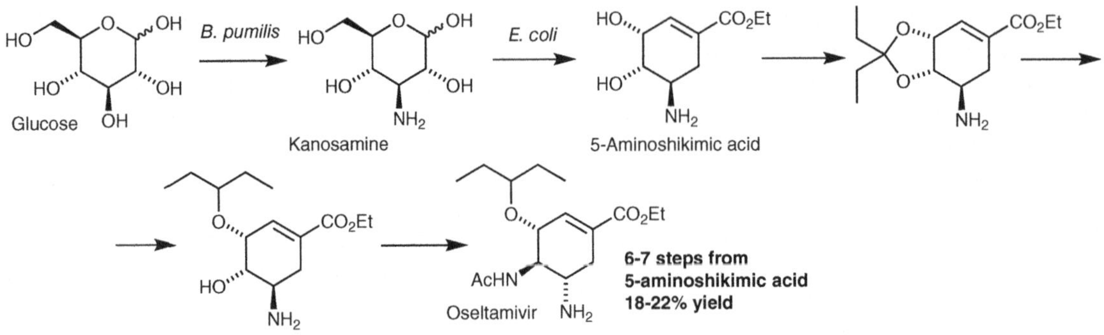

Scheme 8.41 Oseltamivir synthesis based on 5-aminoshikimic acid

Interestingly the Principles of Green Chemistry could help future route selection. Following the second principle – Atom Efficiency – six industrial syntheses and nine approaches by academic groups have been reviewed using an algorithm to determine their overall material efficiency (OPRD2009, JOC2011b). Not surprisingly, confirming the conclusions reached above, shikimic acid-based routes are amongst the most efficient.

As discussed above, numerous non-shikimic or quinic acid routes were devised, mainly by academic groups. Many of these can be excluded from consideration for application at an industrial level because of low yields, reduced throughput, chromatographic purifications and oily intermediates. Others may warrant further investigation (Scheme 8.42).

Amongst these are routes starting from chiral 3-bromocyclohexa-3,5-diene-1,2-diol, which is readily available in large quantities (scheme G). One such route (lower branch of **Ga** following **G10/G11/G12/G13**) is azide free, has crystalline intermediates and has a yield of 22% over 13 steps. The metalation step in the final stages and the chromatography used to remove an epimer after the ring opening of the aziridine with 2-pentanol, are not ideal.

Scheme J contains two of the most promising non-shikimic acid routes (**Ja** and **Jc**, 27% yield/11 steps and 30% yield/8 steps, respectively). Both are azide-free, use inexpensive starting materials and are asymmetric.

However, as discussed above, although some features of this sequence are favourable, there are various aspects that would have to be addressed before it can be put into practice industrially.

The one-pot cascade nitro-Michael/Horner-Wadsworth-Emmons reaction in scheme L, route **Lb**, also has merit and is perhaps the most attractive candidate of these routes for industrial application. The high overall yield of 57-60% reported and low catalyst loading are extremely attractive (Scheme 8.35 and Scheme 8.43).

Given that this route warrants more attention, additional work has been carried in which the following benefits were consolidated:

- Short, high yielding and simple to operate;
- Catalyst for asymmetric synthesis easily accessible and relatively inexpensive;
- One-pot procedures for sequential reactions (ACIE2011);
- Isolable, crystalline intermediate serves as a clean-up point;
- Viability proven at gram scale.

However, as mentioned before, availability of raw materials, odour of *p*-toluenethiol, azide chemistry, OsO_4 and CH_3NO_2, are all issues.

Finally, routes **Lc** and **Ld** deserve mention. These are similar to route **Lb**, employing a one-pot cascade nitro-Michael/Horner-Wadsworth-Emmons to close the C6 ring. The Curtius rearrangement can be avoided in both which, fortuitously, shortens the synthesis. The lower yields may render this route less appealing (yield for route **Lc** is 28-36%). The modification that eliminated the use of *p*-toluenethiol resulted in a one-pot operation where the complete transformation to Oseltamivir took minimal time (**Ld**). However, the low yield (13-15%) argues against its use at industrial scale.

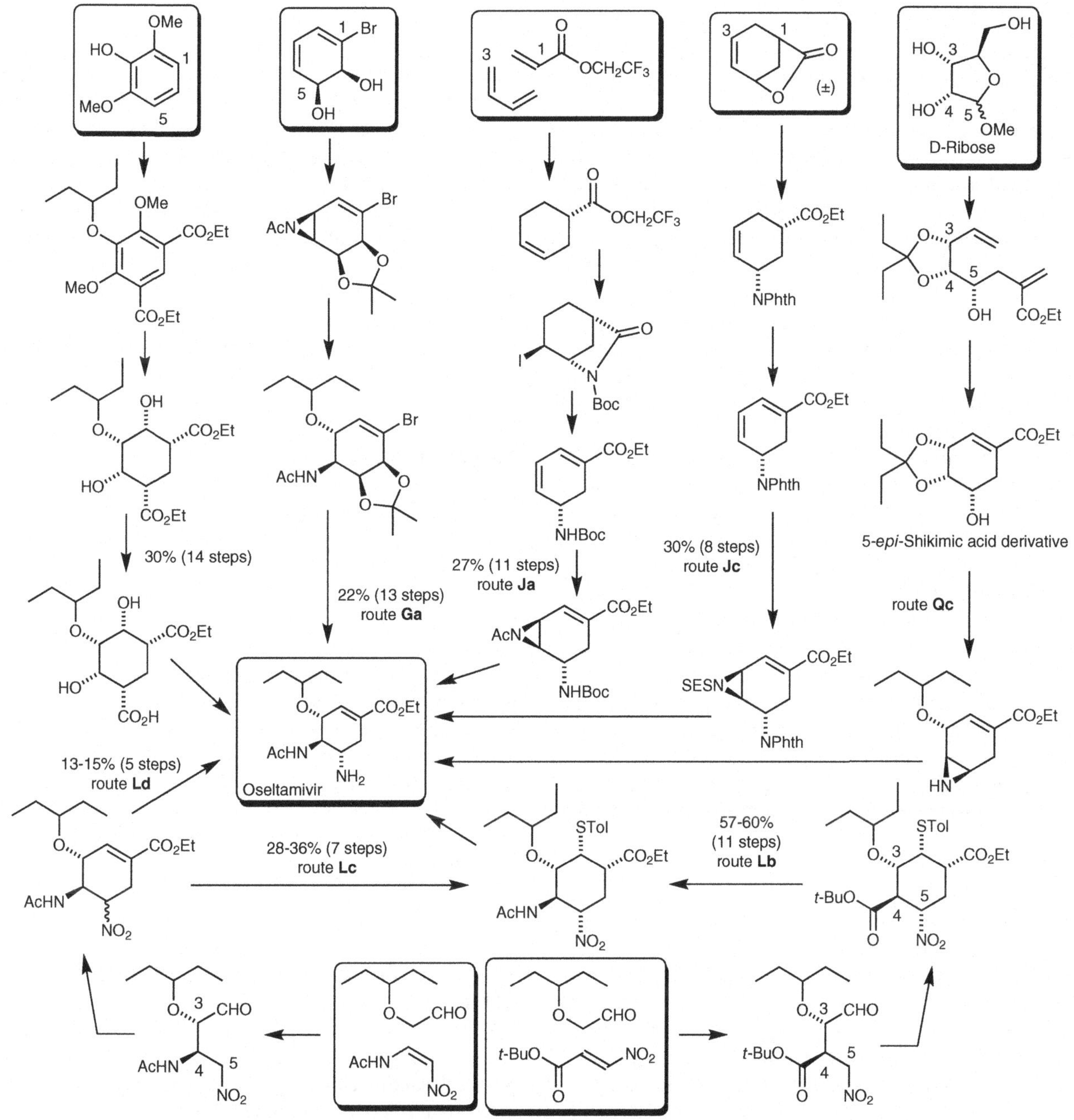

SCHEME 8.42 MOST PROMISING ROUTES THAT DO NOT START FROM SHIKIMIC ACID

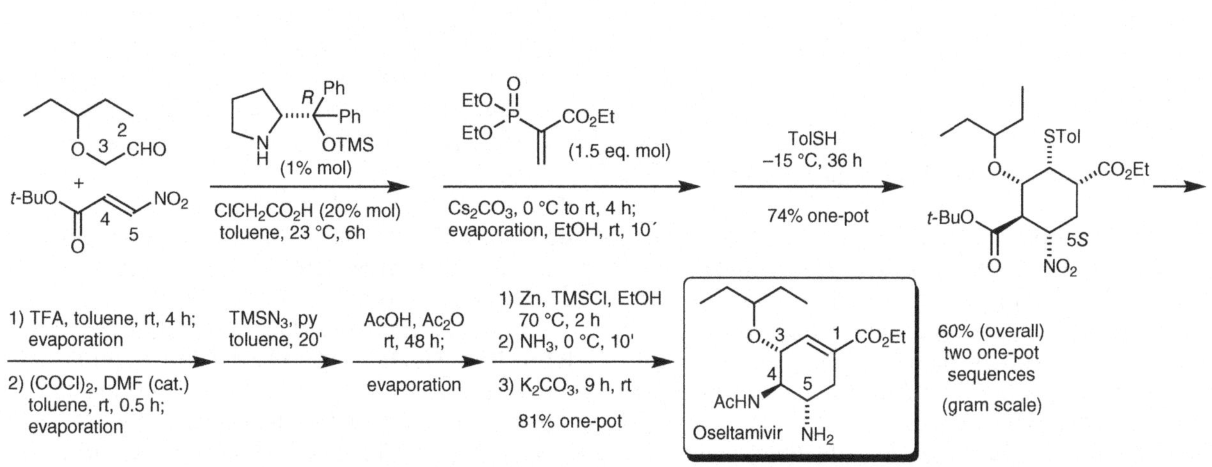

SCHEME 8.43 THE BEST SYNTHETIC ROUTE THAT DOES NOT START FROM SHIKIMIC ACID

8.14 LESSONS FROM THE OSELTAMIVIR CASE

Availability / cost of starting materials

Unquestionably, the unavailability of raw materials is an absolute showstopper in the production of any final product. Limited **availability** places constraints on the quantities that can be produced and tends to drive up prices. High raw material prices are reflected in the final cost of goods and these costs are passed on to the end user, who has to pay premium prices for the finished product. In many instances, when a drug is lifesaving, the consumer, the Health Insurance Company or the National Health Service accepts these substantial expenditures. In the case of Oseltamivir, which was required in large enough quantities to respond to a worldwide flu pandemic, this is not a **sustainable proposition**. The limited availability of shikimic and quinic acids, which were the preferred starting materials for industrial production, confronted the Pharmaceutical Industry with an enormous challenge, resulting in a flood of activity. The general interest from both industry and academia stimulated numerous alternative syntheses. The use of biological means to augment the supply of shikimic acid was successful, notably **without significant changes in quality**, which is a frequent outcome of altered/increased production. In parallel, alternative chemical routes, suitable for large scale production, were developed in a relatively short space of time, thus testifying to the resilience and ingenuity of chemists when faced with a humanitarian problem of this type.

Semisynthesis / total synthesis

The Pharmaceutical Industry chemically transforms many natural products, usually of high structural complexity and with a plethora of chiral centres, into drug substances. These semisynthetic procedures take advantage of the fact that nature easily produces complex starting materials that would otherwise be difficult to produce efficiently by total synthesis. In the case of Oseltamivir, neither the natural product starting material, shikimic acid, nor the final product, is of such a high degree of complexity that a total synthesis is unfeasible.

Biocatalysis as a tool for synthetic organic chemistry

Biocatalytic methods are key elements to some of the synthetic routes to Oseltamivir (see Scheme 8.24 as an example). These highly selective, chemo, regio, enantio and diastereospecific transformations complement traditional organic synthetic methods. They can be carried out under mild reaction conditions and side reactions are normally minimal.

Desymmetrization/asymmetric induction that rely on biochemical catalysts have been widely used and reported in the synthesis of Oseltamivir.

Purity of natural starting materials

The absolute purity and the impurity profile of natural products varies greatly according to many factors: time of harvest; degree of hydration; availability of neutrients; temperature during the biosynthesis, etc. Hence, different batches from different suppliers, or even the same supplier, may have widely ranging purities and dissimilar impurity levels. Shikimic acid is reported to have a purity ranging from 85 to 99% and a highly variable impurity profile. The Process Chemist's task, to assure that the industrial process produces final product of constant purity, is even more demanding in cases where natural products of high variability are the starting materials.

Validity of various approaches

Oseltamivir clearly demonstrates that there is more than one valid approach to resolve the problems that the synthesis poses. Use of:

- a chiral natural product as an advance intermediate
- chiral pool starting materials
- resolution to set the correct chirality
- asymmetric organocatalysis
- biocatalysis
- desymmetrization
- asymmetric induction

are amongst the techniques that were applied in the Oseltamivir case.

8.15 REACTION CONDITIONS FOR SCHEMES B TO R

8.15.1 Reaction conditions for scheme B (kg, g or mg after yield refers to reaction scale)

B1 R = Me: MeOH, TsOH, reflux, 100% (25 g, Gilead1995a,b, 1997b). **R = Et:** EtOH, $SOCl_2$, reflux, 97-100% (Gilead 1998, 2 kg, OPRD1999, UEChinaS2009, TA2009, 2011, 2012). EtOH, TsOH or $PhSO_3H$, reflux, YNR (50 g, HoffmannLR2005, TL2012a). EtOH (anh.), KU28 cationite, reflux, 93% (50 g, RCB2013).

B2 PPh_3, DEAD, THF, 0 °C, 77% (220 mg).

B3 R = Me: Acetone, 2,2-dimethoxypropane, TsOH, rt, 90% (25 g, Gilead1995a,b, 1997b). **R = Et:** 2,2-dimethoxypropane, TsOH, EtOAc, <35 °C, 95-96% (2 kg, OPRD1999, UEChinaS2009, TA2009).

B4 R, R^1 = Me, R^2 = Ms: MsCl, CH_2Cl_2, NEt_3, –10 °C to 0 °C, 91-93% (0.87-29 g, Gilead1995a,b, 1997b). **R = Et. R^1 = Me. R^2 = Ms:** EtOAc, MsCl, NEt_3, 20 °C 82% (2.5 kg, OPRD1999). **R = Et, R^1 = Me, R^2 = Bz:** BzCl, NEt_3, DMAP, CH_2Cl_2, 0 °C to rt, 98% (6.4 g, TA2009, EastChinaS2009).

B5 R = R^1 = Me, R^2 = Ms: HCl, THF, H_2O, 38% (1 g, Gilead1995a,b, 1997b). TsOH, MeOH, reflux, 85% (36 g, Gilead1995a,b, 1997b). **R = R^1 = Me, R^2 = Ts:** TsOH, MeOH, reflux, 54% (B16+B5, Gilead1995a,b, 1997b, JACS1997); **R = Et, R^1 = Me, R^2 = Ms:** AcOH, EtOH, reflux, 39% (B16+B5, Gilead1995a, 1997b). **R = Et, R^1 = Me, R^2 = Bz:** HCl, H_2O, EtOAc, rt, 94% (9 g, TA2009; UEChinaS2009).

B6 R = Me, R^2 = Ms: DBU, THF, 0 °C, 70-90% (0.26-21 g, Gilead1995a,b, 1997b). **R = Me, R^2 = Ts:** DBU, THF, 0 °C, 100% (20 g, Gilead1995a,b, 1997b, JACS1997). **R = Et, R^2 = Ts:** DBU, THF, 0 °C, 100% (Gilead1995a). **R = Et; R^2 = Ms:** DBU, THF, rt, 100% (Gilead1995a, 1997b).

B7 R = Me: $MeOCH_2Cl$, DIPEA, CH_2Cl_2, reflux, 74-100% (4 g, Gilead1995a,b, 1997b, JACS1997). **R = Et:** $MeOCH_2Cl$, DIPEA, CH_2Cl_2, reflux, 90% (12 g, Gilead1995a, 1997b).

B8 3-pentanone, TsOH, YNR (Gilead1998). 3-pentanone, TFA, 97% (OPRD1999). 3-pentanone, $CH(OEt)_3$, $PhSO_3H$, EtOH, 98% (25 g, HoffmannLR2005). 3-pentanone, $CH(OEt)_3$, 94% (B1+B8, TL2012a).

B9 MsCl, NEt_3, CH_2Cl_2, 80% (B1+B8+B9), 93% (OPRD1999), in *i*-PrOAc 86% (B1+B8+B9, 152 g, HoffmannLR2005).

B10 $BH_3{\bullet}Me_2S$, $CF_3SO_3Si(Me)_3$, CH_2Cl_2, –30 °C to –20 °C, 80% (8.6 kg, Gilead1996, 1997a, 1998), 75% (10.3 kg, Gilead1998, JOC1998). Et_3SiH, $TiCl_4$, CH_2Cl_2, –24 °C, 87% (2.6 kg, OPRD1999), YNR (145 g, HoffmannLR2005).

B11 $KHCO_3$, EtOH, H_2O, 60 °C, 60% (7.8kg, Gilead 1996, 1997a, 1998), 97% (16.4 kg, Gilead 1998, JOC1998), with $NaHCO_3$, 80% (B10+B11, 2.3 kg, OPRD1999), 83% (B10+B11, 146 g), 72% from SA (HoffmannLR2005).

B12 2,2-dimethoxypropane, TsOH, acetone, 73% (103 g, Gilead1995a,b, 1997b), 70% (Gilead 1996, 1997a, 1998), 70-90% (20 kg, JOC1998), YNR (3 kg, OPRD1999), 83% (0.2 g C2017). 2,2-dimethoxypropane, acetone, reflux, sulfonic acid catalyst, reflux, 86-91% (0.2 g C2017). **Cyclohexylketal:** cyclohexanone, TsOH, benzene, reflux, 96% (60 g, Gilead1997b).

B13 R = Me, R^{1}R^1 = cycloC$_6$: NaOMe, MeOH, rt, 80% (12.7 g).

B14 R = Me: NaOMe, MeOH, rt, 87% (43.5 g, Gilead1995a,b, 1997b). MeOH, TBD-PS [1,5,7-triazabicyclo[4.4.0]dec-5-ene (TBD), supported on polystyrene (TBD-PS)], 0 °C, 96% (C2017). **R = Et:** NaOEt, EtOH, 83% (Gilead1995a, 1997b), 70-80% (20 kg of QA, Gilead 1996, 1997a, 1998, JOC1998, 3 kg of QA, OPRD1999). EtOH, TBD-PS, 0 °C 86-90% (10-14% lactone, 0.2 g C2017).

B15 R = Me, R^2 = Ts: TsCl, py, DMAP, rt, 92% (30 g, Gilead1995a,b, 1997b, JACS1997). **R = Et, R^2 = Ms:** MsCl, DMAP, py, 100% (Gilead1995a, 1997b). MsCl, NEt_3, CH_2Cl_2, 84% (18.7 kg, Gilead1996, 1997a, 23 kg, Gilead1998), 69% from QA (3.5 kg, OPRD1999).

B16 R = R^1 = Me, R^2 = Ts: 1) SO_2Cl_2, py, CH_2Cl_2, –78 °C; 2) MeOH, TsOH, reflux, NI (45 g, Gilead1995a,b, 1997b, JACS1997). **R = Et, R^1 = Me, R^2 = Ms:** 1) SO_2Cl_2, py, CH_2Cl_2, –20 °C, 43% (10 kg, Gilead 1996, 1997a, 1998), 30% from QA (24 kg, Gilead 1998, JOC1998). $ClHC{=}NMe_2Cl$ (Vilsmeier reagent), EtOAc, 70-75 °C, 59% (45% from QA, 135 g, HoffmannLR1998), 65-68%, from QA (3kg of QA, OPRD1999). $POCl_3$, DMF, rt to 60 °C, 55% (80 g, HoffmannLR1998).

B17 R^1 = Me, R = Et, R^2 = Ms: 3-pentanone, $HClO_4$ (70%), 100% (8.9 kg, Gilead 1996, 1997a, 1998), 95% (10 kg, Gilead 1998, JOC1998). 3-pentanone, CF_3SO_3H, 40 °C, 98% (7.5 kg, OPRD1999).

B18 Esterification, R = Me, Et: ROH, sulfonic acid catalyst, reflux, 72-94% (0.2 g, C2017). **Diol protection, R^1 = Me, R = Et:** Acetone, sulfonic acid catalyst, rt, 49-83% (0.2 g, C2017). **R^1 = R = Et:** 3-pentanone, sulfonic acid catalyst, rt, 41-79% (0.2 g, C2017). 3-pentanone, 3,3-diethoxypentane, H_2SO_4, EtOH, 0 °C, 72% (34 g, Fujifilm2013). **R, R^1 = Et, Diol protection, lactone formation, lactone opening:** 1) 3-pentanone, 3,3-diethoxypentane, sulfonic acid catalyst, reflux, 91-100% (0.2 g C2017); 2) EtOH, TBD-PS 1,5,7-triazabicyclo[4.4.0]dec-5-ene (TBD), supported on polystyrene (TBD-PS), 0 °C, 84% (C2017).

B19 R = Me, R^{1}R^1 = cycloC$_6$, R^2 = Ts: TsCl, DMAP, py, rt, 89% (5.8 g, Gilead1995a,b, 1997b). **R = R^1 = Et, R^2 = (*i*-Pr)$_3$C$_6$H$_2$SO$_2$:** $(i\text{-Pr})_3C_6H_2SO_2Cl$, DMAP, py, 70 °C, 77% (10 g, Fujifilm2013); **R = R^1 = Et, R^2 = Ms:** MsCl, NEt_3, THF, rt, 65% (11 g, Fujifilm2013).

B20 R = Me, R^{1}R^1 = cycloC$_6$, R^2 = Ts: $POCl_3$, py, rt, 83% (2 isomers), 55% corrected yield, (0,44g, Gilead1995a,b, 1997b). TsCl, DMAP, py, rt, 89% (5.8 g scale, Gilead1995a,b). **R = R^1 = Et, R^2 = (*i*-Pr)$_3$C$_6$H$_2$SO$_2$:** 1) $p\text{-}O_2NC_6H_4SO_2Cl$, NEt_3, DMAP, CH_2Cl_2, rt, 56% (2.2 g); 2) NaI, imidazole, EtOAc, 60 °C, YNR; 3) NaOH, EtOAc 19-72% (0.2 g, Fujifilm2013). **R = R^1 = Et, R^2 = Ms:** 1) $MeP(OMe)_3I$, imidazole, EtOAc, rt; 2) NaI; 3) NaOH 84% (0.18 g, Fujifilm2013).

8.15.2 Reaction conditions for scheme C (kg, g or mg after yield refers to reaction scale)

C1 NaN_3, NH_4Cl, MeOH/H_2O, reflux, 86% (4.9 g, Gilead 1995a,b, 1997b, JACS1997).

C2 MsCl, NEt_3, DMAP, CH_2Cl_2, 0 °C, YNR (6.5 g, Gilead1995a,b, 1997b), 99% (6.5 g, JACS1997).

C3 1) PPh_3, THF, 0 °C to rt; 2) $NaHCO_3$ or NEt_3, H_2O, rt, 77-78% (C1+C2+C3, Gilead1995a,b, 1997b), (8.6 g, JACS1997).

C4 NaN_3, NH_4Cl, DMF, 65 °C, YNR (Gilead1995a,b), 86% (330 mg, Gilead1997b), 86% (3.2 g, Gilead 1995a,b, 1997b, JACS1997).

C5 AcCl, py, CH_2Cl_2 76% (C4+C5, 330 mg, Gilead 1995a,b, 1997b), (3 g, JACS1997).

C6 1) H_2 (1 bar), Lindlar cat. EtOH, 92% (39 mg); 2) THF, H_2O, KOH, 64% (33 mg).
C7 TFA (40%), CH_2Cl_2, 0 °C to rt, 100% (4 mg).
C8 1) HCl, MeOH, rt, 99%; 2) TrCl, NEt_3, CH_2Cl_2, 0 °C to rt; 3) MsCl, NEt_3, CH_2Cl_2, 0 °C to rt, 86% (2+3, 2.6 g).
C9 3-pentanol, BF_3•OEt_2, 75 °C, NI (0.1-15 g).
C10 Ac_2O, DMAP, py, rt, 66-69% (C9+C10, 15 g, Gilead1995a, 1997b, JACS1997), 53% (C9+C10, 0.1 g, Gilead1995b).
C11 Ph_3P, THF/H_2O, 50 °C, 80-90% (1 g).
C12 KOH, THF, rt, 75% (0.83 g, Gilead1995a, 1997b, JACS1997), 48% (31 mg, Gilead1995b).
C13 KOH, THF, H_2O, rt, YNR (0.27 g).
C14 DCC, DMAP or DIC, EtOH, CH_2Cl_2, 93-96%.
C15 PPh_3, THF, H_2O, 91-98% (0.3-0.9 g, Gilead1995a,b, 1997b, UEChinaS2009, JOC2010c, TA2011, 2012). PPh_3, THF, H_2O, reflux, 52% (8.5 g, HeteroDrugs2005). PPh_3, NEt_3, THF, H_2O, rt, 81% (34 mg, TL2012b). PPh_3, AcOH, THF, H_2O, rt to 45 °C, 84% (14 g, RCB2013). *n*-Bu_3P, EtOH, H_2O, AcOH, 5 °C to rt, 97% (50 g, HoffmannLR1999b, Roche2007). *n*-Bu_3P, EtOH, rt, 92% (ACIE2009d). H_2 (1 bar), Lindlar catalyst, EtOH (abs), 90-98% (0.1-0.5g, Gilead1996, 1997a, 1998, HoffmannLR1999b, TA2009, TL2012a, ACIE2008b). H_2 (1 bar), Raney Nickel, EtOH, 71% (0.33 kg, Gilead1998, JOC1998). H_2, Pd/C (5%), EtOH, rt, 63% (5 g, RCB2013). H_2S (g), py, 30 °C (8.5 g, HeteroDrugs2005).

8.15.3 Reaction conditions for scheme D (kg, g or mg after yield refers to reaction scale)

D1 NaN_3, NH_4Cl, EtOH, H_2O, 70-75 °C, 86-90% (550 g-1 kg).
D2 PPh_3, THF (anh.) or CH_3CN (anh.), reflux, 97% (608 g-2 kg).
D3 NaN_3, NH_4Cl, DMF, 70-85 °C, YNR (490 g-1.7 kg).
D4 Ac_2O, $NaHCO_3$, CH_2Cl_2, rt, 44-55% (D3+D4, 490 g-2 kg, Gilead1996, 1997a, 1998, JOC1998). Ac_2O, NEt_3, CH_2Cl_2, 25 °C, 74% (14 g, HeteroDrugs2005). Ac_2O, EtOAc, $NaHCO_3$, H_2O, rt, 70% (120 mg, Roche2007).
D5 see C15.
D6 see D14. Under reaction conditions reported in D14, D14+D6 occur in a cascade sequence.
D7 Ac_2O, AcOH, MsOH, MTBE, 20 °C, 83% (278 g).
D8 Pd/C (10%) EtOH, $H_2NCH_2CH_2OH$, reflux, 70% (176g).
D9 $BnNH_2$ (2 eq.), THF, 65 °C, $Yb(OTf)_3$ (0.2 eq.) or $MgBr_2$•OEt_2 (0.2 eq.), 80% (+15% regioisomer, 5g).
D10 1) $AllNH_2$, $MgBr_2$•OEt_2 (0.2 eq.), MTBE, CH_3CN (9:1), 55 °C; 2) $(NH_4)_2SO_4$, H_2O, 97% with 7% isomer, 89% corrected yield (254 g).
D11 1) Pd/C, EtOH, $H_2NCH_2CH_2OH$, reflux; 2) H_2SO_4, H_2O, 77% with 4% isomer, 73% corrected yield (312 g).
D12 PhCHO, MTBE, NI (271 g, HoffmannLR1999a, 2004, JOC2001). PhCHO, toluene, 110 °C, 100% (270 mg, JOC2001). ArCHO, EtOH, reflux, 99% (27 g, HoffmannLR1999a, Ar = *o*-TsO).
D13 MsCl, NEt_3, MTBE, YNR (HoffmannLR1999a, 2004, JOC2001). MsCl, EtOAc, NEt_3, 94% (360 mg, JOC2001, Ar = Ph). MsCl, NEt_3, EtOAc, 91% (9.2 g, HoffmannLR1999a, Ar = *o*-TsO).
D14 1) $AllNH_2$, 112 °C; 2) HCl, H_2O (under reaction conditions D14/D6 occurs in a cascade sequence), 80% (Ar = Ph, D12+D13+D14+D6, 271 g, HoffmannLR1999a, JOC2001), 89% (Ar = Ph, D12+D13+D14+D6, 271 g, HoffmannLR2004). $H_2NCH_2CH_2NH_2$, EtOH, reflux, 82% (Ar = Ph, 0.87 g, HoffmannLR1999a).
D15 *t*-$BuNH_2$, $MgCl_2$, toluene, 50 °C, 96% (105 g, OPRD2004), 98% (30 g, HoffmannLR2003).
D16 MsCl, NEt_3, toluene, 70 °C, 93% (163 g, OPRD2004), 100% (38 g, HoffmannLR2003).
D17 $(All)_2NH$, $PhSO_3H$, 120 °C, 93% (132 g, OPRD2004). $(All)_2NH$, CSA, 120 °C, 94% (36 g, HoffmannLR2003).
D18 Ac_2O, NaOAc (anh.), 110-116 °C, 94% (157 g, OPRD2004), 100% (45 g, HoffmannLR2003). Ac_2O, py, 100 °C, 88% (44 g, HoffmannLR2003).
D19 1) EtOH (anh.), HCl (anh.)/heptane, 15 °C, 92% (175 g); 2) TFA, 50 °C, 97% (142 g, OPRD2004), 94% (50 g, HoffmannLR2003). TFA, 25 °C, 72% (43 g, HoffmannLR2003).
D20 1,3-Dimethylbarbituric acid, PPh_3 (4% mol), $Pd(OAc)_2$ (1% mol), EtOH (anh.), 35 °C, 88-90% (28-32 g, OPRD2004, HoffmannLR2003). 1,3-Dimethylbarbituric acid, $Pd(PPh_3)_4$ (0.4% mol), THF, 89% (3.9 g, HoffmannLR2003).
D21 1) BF_3•OEt_2, CH_3CN, rt; 2) CH_3CN, H_2O, 95% (1+2, 7.5 -10 g).
D22 MsCl, NEt_3, EtOAc, DMAP, 0 °C, 95-98% (5-6 g).
D23 NaH, DMF, 89% (3 g, Guangzhou2013). NaH, CH_2Cl_2, DMSO, 91% (2 g, TA2013).
D24 *t*-$BuNH_2$, reflux, 90% (3 g).
D25 TFA, 80 °C, 86% (3 g).
D26 Isopropyl 2,2,2-trichloroacetimidate (1.5 eq.), BF_3•OEt_2, CH_2Cl_2, 15 °C, 94% (0.6 g).
D27 1) Cs_2CO_3, DMSO, 80 °C; 2) AcOH, H_2O, DMSO, 80 °C, 82% (0.46 g).
D28 $(All)_2NH$, reflux, 93% (2 g).
D29 **Ester formation:** EtOH, MsOH, reflux, 100% (13.7 g). $(EtO)_2SO$, $SOCl_2$, EtOH, reflux, 78%. **Ketal formation:** 3-pentanone, MsOH, reflux, 87% (0.9 g).
D30 Et_3SiH, $TiCl_4$, CH_2Cl_2, –70 °C, 51% (10.8 g).
D31 (E22) LiCl (4.2 eq.), NaN_3 (4 eq.), 80 °C, DMF, 92%.

8.15.4 Reaction conditions for scheme E (g or mg after yield refers to reaction scale)

E1 (B8) $CH(OCH_3)_3$, 3-pentanone, 94% from SA (TL2012a). See also B8.
E2 MOMCl, DIPEA, DMAP, CH_2Cl_2, 94% (4.5 g).
E3 Et_3SiH, $TiCl_4$, CH_2Cl_2, 70% (1.9 g).
E4 PPh_3, DEAD, HN_3, THF, 82% (2.1 g).
E5 AcSH, 2,6-lutidine, $CHCl_3$, 76% (1.2 g).

E6 $ZrCl_4$, *i*-PrOH, 92% (0.7 g).
E7 PPh_3, DEAD, HN_3, THF, 84% (0.7 g).
E8 $SOCl_2$, DMF, 0 °C to rt, 91% (5 g).
E9 K_2CO_3, EtOH, rt, 95% (10 g).
E10 MsCl, NEt_3, CH_2Cl_2, 0 °C, 95% (5 g).
E11 $AllNH_2$ (3 eq.), EtOH, H_2O, 75 °C, 93% (5 g).
E12 3-pentanol, BF_3•OEt_2 (1.5 eq.), rt to 70 °C, 90% (6 g).
E13 1,3-dimethylbarbituric acid, $Pd(OAc)_2$, PPh_3 (cat.), EtOH, 35 °C, 93% (5 g).
E14 Ac_2O, Na_2CO_3, H_2O, 0 °C, 95% (4 g).
E15 1) $SOCl_2$, NEt_3, 0-5 °C; 2) EtOAc, rt, 98% (10 g).
E16 1) MsCl, NEt_3, DMAP, EtOAc, 0 °C, 96% (10 g); 2) NaN_3 (1-6 eq.), NH_4Cl, DMF, H_2O, rt, 96% (10 g).
E17 1) Ph_3P (1 eq.), K_2CO_3 (1 eq.), EtOH, rt; 2) H_2O, EtOH, reflux, YNR (2 g).
E18 Ac_2O (1.5 eq.), NEt_3 (2 eq.), CH_2Cl_2, H_2O, 0 °C, 90% (E17+E18).
E19 3-pentanol, BF_3•OEt_2 (0.2 eq.), –8 °C, 93% (1 g).
E20 MsCl (1.5 eq.), NEt_3 (2 eq.), DMAP, EtOAc, 0 °C, 96% (2 g, TA2012), YNR (4 g, UEChinaS2018).
E21 NaH (4 eq.), THF, rt, 95% (2 g, TA2012), YNR (2 g, UEChinaS2018).
E22 (D31) LiCl (4.2 eq.), NaN_3 (4 eq.), DMF, 80 °C, 92% (1 g, TA2012).
E23 NaN_3 (2 eq.), EtOH (anh.), reflux, 93% (10 g).
E24 PPh_3 (1 eq.), Et_2NH•TsOH (0.1 eq.), rt; 2) 60 °C, 62% (2.1 g).
E25 Ac_2O (1.2 eq.), NEt_3 (2 eq.), CH_2Cl_2, 0 °C, 95% (0.8 g).
E26 3-pentanol, BF_3•OEt_2 (1.5 eq.), –8 °C, 93% (0.9 g).
E27 MsCl, NEt_3, CH_2Cl_2, 0 °C, 95% (1-2.5 g).
E28 NaN_3, DMSO, EtOH, 90 °C, 21% (340 mg, Roche 2007), 84% (1.2 g, TA2009), in DMF, 90 °C, 92% (1.2 g, TA2011). NaN_3, EtOH/H_2O, reflux, 88% (JOC2009), 72% (24 g, RCB2013). NaN_3, DMSO, EtOH, 90 °C, 66% (3.1 g, ACIE2009d). NaN_3, DMF, H_2O, 0 °C, 84% (2.5 g, UEChinaS2009).
E29 R = Bz: BzCl (1.2 eq.), NEt_3 (2 eq.), DMAP, CH_2Cl_2, 0 °C to rt, 98% (10 g). **R = Ac:** Ac_2O (1.5 eq.), NEt_3 (2.0 eq.), DMAP, CH_2Cl_2, rt, 98% (5 g).
E30 NaN_3 (2.5 eq.), DMF, H_2O, rt, 95% (5 g, R = Bz), 91% (5 g, R = Ac).
E31 MsCl (1.5 eq.), NEt_3 (1.2 eq.), DMAP, EtOAc, 0 °C, 95% (4 g).
E32 PPh_3 (1.1 eq.), NEt_3 (3 eq.), THF, H_2O, rt, 87% (2 g).
E33 Ac_2O (1.5 eq.), NEt_3 (2 eq.), EtOAc, 0 °C, 98% (2 g).
E34 3-pentanol, BF_3•OEt_2 (1.5 eq.), –5 °C, 91% (2 g).
E35 K_2CO_3 (1.1 eq.), EtOH (anh.), rt, 94% (2 g).
E36 MsCl (1.5 eq.), NEt_3 (1.2 eq.), DMAP, EtOAc, 0 °C, 96% (5 g).
E37 PPh_3 (1.1 eq.), THF, rt; 2) NEt_3 (3 eq.), THF, H_2O, rt, 88% (2-5 g).
E38 Ac_2O (2 eq.), NEt_3 (3 eq.), EtOAc, 0 °C, 98% (1.4-3.1 g).
E39 3-pentanol, BF_3•OEt_2 (1.5 eq.), –5 °C, 92% (1.6-3.5g).
E40 K_2CO_3 (1.1 eq.), EtOH, rt, 90% (1.7-4.1 g, TA2011, UEChinaS2009), 71% (E37+E38+E39+E40, TA2009).
E41 2,2-dimethoxypropane, TsOH, EtOAc, <35 °C, 95% (16.9 g, UEChinaS2009), 92% from SA (TA2009).
E42 1) BzCl, NEt_3, DMAP, CH_2Cl_2, 0 °C to rt, 98% (6.4 g); 2) HCl (cat.), EtOAc, H_2O, rt, 94% (9 g).
E43 MsCl (5 eq.), NEt_3 (4 eq.), EtOAc, DMAP, 0 °C, 97% (7.5 g).
E44 NaN_3 (4 eq.), DMF, H_2O, –5 °C, 95% (11.5 g).
E45 MsCl, NEt_3, DMAP, EtOAc, 0 °C, 91-97% (11.3-20 g, JOC2009, ACIE2009d, Roche 2007), 84% (55 g, RCB2013).
E46 NaN_3, acetone, H_2O, 0 °C, 92% (10 g, JOC2009). NaN_3, DMSO, rt, 70-100% (0.4-10.6 g, ACIE2009d, Roche2007). NaN_3, NH_4Cl, MeOH, 10 °C to rt, 98% (60 g, RCB2013).
E47 1) PPh_3, THF, rt; 2) NEt_3, H_2O, rt, 84% (8.1 g).
E48 Ac_2O, NEt_3, EtOAc, 0 °C, 98% (4.5 g).
E49 3-pentanol, BF_3•OEt_2, –8 °C to 0 °C, 86% (5 g).
E50 $P(OEt)_3$ (1.1 eq.), toluene, reflux, YNR (9.3 g, ACIE2009d), 22% (383 mg, Roche2007). PR_3, toluene, rt to reflux, YNR (47 g, RCB2013).
E51 3-pentanol, BF_3•OEt_2 (1.2 eq.), rt, 45% from SA (9.9 g, ACIE2009d), 62% (1.9 g, Roche2007), 66% (E50+E51, for R = Et, RCB2013), YNR (for R = Me, *i*-Pr, RCB2013).
E52 H_2SO_4 (conc.), EtOH, reflux, NI (5.4-40 g, ACIE2009d, RCB2013), 64% (120 mg, Roche2007).
E53 Ac_2O, EtOAc, rt, 73-82% (E52+R53, ACIE2009d, RCB2013). Ac_2O, $NaHCO_3$, EtOAc, H_2O, 91% (190 mg, Roche2007).
E54 NaN_3, EtOH, DMSO, 90 °C, 15% (104 mg).
E55 EtOH, H_2SO_4, 80 °C, 47% (300 mg).
E56 (D4) Ac_2O, EtOAc, H_2O, $NaHCO_3$, rt, 70% (118 mg).
E57 see C15.

8.15.5 Reaction conditions for scheme F (g or mg after yield refers to reaction scale)

F1 $Y(Oi\text{-}Pr)_3$ (2% mol), chiral ligand (4% mol), $TMSN_3$ (1.5 eq), EtCN, rt, 96%, 91% ee, recrystallization from *i*-PrOH, 72%, 99% ee (70-90 mg, JACS2006b, TokyoU2006a), 87%, 92% ee (1.5 g, JACS2006b). $Y(OiPr)_3$ (1% mol), catalyst (2% mol), $TMSN_3$ (1.5 eq), 2,6-dimethylphenol, EtCN, rt, 94%, 89% ee, recrystallization from *i*-PrOH 72%, 99% ee (30 g, OL2007).
F2 Boc_2O (1.5 eq.), DMAP (0.5 eq.), CH_3CN, rt, YNR (69 mg-11 g).
F3 NaOH (4 M), rt, 98-100% (69 mg-11 g, F2+F3).
F4 1) Ph_3P (2 eq.), CH_3CN, 50-70 °C; 2) H_2O, 40 °C, NI (1 g, JACS2006b). 1) Ph_3P (1 eq.), THF, 60-70 °C; 2) H_2O, 40 °C, NI (0.4-6.5 g, TokyoU2006a, OL2007).
F5 Boc_2O (2 eq.), NEt_3 (5 eq.) or py, CH_2Cl_2, rt, 87-90% (F4+F5, 0.4-1 g).
F6 1) SeO_2 (1 eq.), DMP (1.5 eq.), 1,4-dioxane, 80 °C; 2) DMP, CH_2Cl_2, 4 °C, recrystallization from *i*-PrOH/hexane, 62-68%, >99% ee (460 mg).
F7 $Ni(cod)_2$ (10% mol), cod (10% mol), TMSCN (3 eq.), THF, 60 °C (20 mg, JACS2006b,

CSIRIndia2013). Ni(cod)$_2$ (1 eq.), cod (1 eq.), TMSCN (1.5 eq.), THF, 40 °C, NI (TokyoU2006a, 820 mg). Ni(cod)$_2$ (20% mol), TMSCN (BMCL2008).

F8 1) NBS (1 eq.), THF 2) NEt$_3$ (14 eq.), 4 °C, 71% (20 mg, F7+F8, JACS2006b), 94% (20 mg, F8, JACS2006b), 71% (20 mg, F8, CSIRIndia2013). Pd(OAc)$_2$ (1.5 eq.), CH$_3$CN, 50 °C, 49% (F7+F8, 800 mg, TokyoU2006a). NBS, NEt$_3$, 66% (F7+F8, BMCL2008).

F9 LiAlH(O*t*-Bu)$_3$ (5 eq.), THF, 4 °C, 60% (20 mg, F7+F8+F9, JACS2006b, CSIRIndia2013), 94% (F9, 50 mg, TokyoU2006a, CSIRIndia2013). LiAlH(O*t*-Bu)$_3$, 85% (BMCL2008).

F10 DEAD (2.5 eq.), Ph$_3$P (2.5 eq.), THF, 4 °C, 83-87% (120 mg, JACS2006b, TokyoU2006a, CSIRIndia2013). PPh$_3$, DEAD, 60% (BMCL2008).

F11 3-Pentanol, BF$_3$•OEt$_2$ (1.5 eq.), 4 °C, 52-55% (23 mg, JACS2006b, TokyoU2006a, CSIRIndia2013). 3-Pentanol, BF$_3$•OEt$_2$, 60% (BMCL2008).

F12 TFA, CH$_2$Cl$_2$, rt, YNR (7-96 mg).

F13 Boc$_2$O (1.1 eq.), NEt$_3$, CH$_2$Cl$_2$, 4 °C, 63% (F12+F13, 7-96 mg).

F14 Ac$_2$O (2 eq.), DMAP, py, rt, 84% (47 mg).

F15 1) HCl (4 M), EtOH, 60 °C; 2) H$_2$O, 4 °C, 50% (26 mg, JACS2006b, CSIRIndia2013). HCl, EtOH, rt, 60% (OL2007, TokyoU2006a, 23 mg).

F16 Ac$_2$O, py, CH$_2$Cl$_2$, rt, 93-99% (F4+F16, 6.5 g).

F17 1) NIS, CH$_2$Cl$_2$/CHCl$_3$, 40-60 °C, NI (800 mg).

F18 DBU, CH$_2$Cl$_2$/CHCl$_3$, rt, YNR (800 mg).

F19 1) CbzCl, NaHCO$_3$, CH$_2$Cl$_2$/H$_2$O, rt, 85% (800 mg, F17+F18+F19 step 1); 2) K$_2$CO$_3$ or NaOMe, MeOH, rt, 99% (6 g OL2007); 3) DMP, CH$_2$Cl$_2$, rt, 96%, 99% ee (5.6 g, OL2007), 72% (1 g, TokyoU2006a, F19 step 2+ step 3).

F20 (EtO)$_2$P(O)CN (1 eq.), LiCN (17% mol), THF, –20 °C, dr = 20:1, NI (150 mg).

F21 1) Toluene, sealed tube, 150 °C (150 mg); 2) Boc$_2$O, DMAP, py, rt, 72% (F20+F21, 150 mg).

F22 1) Cs$_2$CO$_3$ (10-20% mol), MeOH, rt, 97% (1.4 g); 2) DMP (1.5 eq.), CH$_2$Cl$_2$, rt, 80-94% (48 mg-1.2 g, OL2007), 96% (step 1 + 2, TokyoU2006a); 3) LiAlH(O*t*-Bu)$_3$, THF, –20 °C to 0 °C, 89-91% (1 g, OL2007).

F23 DEAD, Ph$_3$P, THF, 0 °C, 80-87% (20-46 mg).

F24 3-Pentanol, BF$_3$•OEt$_2$, –20 °C, 51% (0.3 g, OL2007).

F25 1) TFA, CH$_2$Cl$_2$, 0 °C to rt; 2) Ac$_2$O, NEt$_3$ or py/DMAP, CH$_2$Cl$_2$, 0 °C to rt, 81% (130 mg, OL2007).

F26 1) HCl, EtOH, 60 °C; 2) NH$_3$(25% aq.), rt, 74% (22 mg, OL2007).

F27 1) I$_2$ (3 eq.), K$_2$CO$_3$ (3 eq.), KI (1 eq.), CHCl$_3$, 60 °C; 2) DBU (2 eq.), CH$_2$Cl$_2$, rt, 85% (1.4 g).

F28 1) Boc$_2$O (1.5 eq.), NEt$_3$ (2 eq.), DMAP (0.1 eq.), CH$_2$Cl$_2$, rt, 99% (0.9 g); 2) AcSH (1 eq.), 2,6-lutidine (10 eq.), CHCl$_3$, 60 °C, 83% (1.4 g).

F29 1) Cs$_2$CO$_3$ (30% mol), *n*-BuOH, 4 °C to rt, 86% (0.9 g); 2) DMP (1.2 eq.), CH$_2$Cl$_2$, 4 °C to rt, 90% (0.8 g).

F30 (EtO)$_2$P(O)CN (1 eq.), LiCN (1.3 eq.), THF, –20 °C, YNR (0.2 g).

F31 1) Toluene, sealed tube, 140 °C; 2) NH$_4$Cl, H$_2$O, rt, 78% (F30+F31, 0.2-0.8 g).

F32 DEAD (2 eq.), Ph$_3$P (2 eq.), *p*-O$_2$NC$_6$H$_4$CO$_2$H (2 eq.), THF, 4 °C; 2) LiOH (1 M), H$_2$O/THF, –20 °C 80% (0.8-1 g).

F33 DEAD (2 eq.), Ph$_3$P (2 eq.), THF, –20 °C, 66% (0.4 g).

F34 3-pentanol, BF$_3$•OEt$_2$ (1.5 eq.), 5 Å MS, –20 °C, 56% (120 mg).

F35 HCl, EtOH, NH$_4$OH, 89% (BMCL2008).

F36 Boc$_2$O, NEt$_3$, 80% (BMCL2008).

F37 AcCl (1.2 eq.), NEt$_3$ (3 eq.), THF, 55 °C, NI (20 mg, BMCL2008).

F38 HCl (4 M), EtOAc, 82% (F37+F38, 20 mg, BMCL2008).

8.15.6 Reaction conditions for scheme G (g or mg after yield refers to reaction scale)

G1 1) 2,2-dimethoxypropane, TsOH, acetone, 0 °C to rt; 2) SnBr$_4$ (cat.), *N*-bromoacetamide (1.2 eq.), CH$_3$CN, H$_2$O, –40 °C, 75% (5 g, ACIE2008b), 76% (400 mg, JOC2011b).

G2 LiHMDS, THF, –10 °C to 0 °C, YNR (11.7 g, ACIE2008b). KHMDS (1.1 eq.), *n*-Bu$_4$NBr (1.1 eq.), DME, 0 °C, 67% (JOC2011b, 18 g).

G3 3-Pentanol, BF$_3$•OEt$_2$, –10 °C to 0 °C, 73% (G2+G3, ACIE2008b, 8.3 g). 3-Pentanol, Cu(OTf)$_2$ (0.1 eq.), 0 °C, 76% (JOC2011b, 8 g).

G4 HCl, MeOH, 50 °C, 94% (8.8 g).

G5 AcOCMe$_2$COBr, THF, 0 °C to rt (7.4 g).

G6 LiBHEt$_3$, THF, 0 °C to rt, 82% (G5+G6, 9 g).

G7 DPPA, DIAD, PPh$_3$, THF, 40 °C, 84% (5.8 g).

G8 [Ni(CO)$_2$(PPh$_3$)$_2$], DIPEA, EtOH, THF, 80 °C, 81% (0.35 g).

G9 1) H$_2$, Lindlar catalyst; 2) H$_3$PO$_4$, EtOH, 50 °C, 91% (ACIE2008b).

G10 1) DDQ, PPh$_3$, *n*-Bu$_4$NOCN, CH$_3$CN, rt; 2) *t*-BuOH, reflux, 78% (1.4 g).

G11 CuI, KI, *N,N'*-dimethylethylenediamine, *n*-BuOH, 120 °C (2.1 g).

G12 Pd(OAc)$_2$, CO, NaOAc, EtOH, rt, 82% (G11+G12, 2.1 g).

G13 H$_3$PO$_4$, EtOH, 50 °C, 81% (1.2 g, ACIE2008b).

G14 4-Methoxybenzaldehyde dimethyl acetal, (+)-CSA, toluene, 0 °C, YNR (5.1 g).

G15 DIBAL, NEt$_3$, toluene, –78 °C to –30 °C, 85% (G14+G15).

G16 1) CDI, CH$_3$CN, 0 °C; 2) NH$_2$OH•HCl, imidazole, 0 °C to 18 °C, 56% (R = H, 88% conversion, 6.5 g); 3) TsCl, NEt$_3$, Et$_2$O, 10-18 °C, 79% (R = Ts, 3.2 g).

G17 3-pentanol, Cu(MeCN)$_4$PF$_6$, K$_2$CO$_3$, CH$_3$CN, 0-18 °C, 43% (130 mg).

G18 1) LiOH, 1,4-dioxane, H$_2$O, 100 °C 85% (R = H, 0.5 g); 2) AcCl, NEt$_3$, 0-18 °C, 99% (R = Ac, 82 mg).

G19 HCl, MeOH, 35 °C, 90% (78 mg).

G20 *E. coli* JM109 (pDTG601a), 48% (24 g, BrockU2009, ASC2010).

G21 2.2-dimethoxypropane, TsOH, acetone, rt, NI (5 g, BrockU2008, 2009, ASC2010).

G22 $CH_3CONHOH$ (1 eq.), $NaIO_4$ (1 eq.), MeOH, rt, 70-79% (5 g, G21+G22, BrockU2008, 2009), 88% (5 g, G21+G22, ASC2010).
G23 $Mo(CO)_6$ (1 eq.), CH_3CN, H_2O, 80 °C, 75% (1 g, ACIE2009c, BrockU2008, 2009). $Mo(CO)_6$, CH_3CN, reflux, 87% (1 g, ACS2010, JOC2011b).
G24 MsCl, NEt_3 (3 eq.), DMAP (cat.), CH_2Cl_2, rt, 54% (0.3-0.4 g).
G25 $CaCO_3$ (2 eq.), EtOH, H_2O, reflux, 72% (800 mg).
G26 Rh/Al_2O_3 (5% mol), H_2 (4 bar), EtOH/H_2O, 95% (150 mg).
G27 Ms_2O (1.9 eq.), NEt_3 (3.9 eq.), CH_2Cl_2, rt, 73% (50 mg).
G28 1) NaN_3 (10 eq.), H_2O, acetone, rt, 86% (25 mg); 2) DBU (1.5 eq.), CH_2Cl_2, rt, 85% (25 mg).
G29 Conditions not shown.
G30 CrO_3 (2.6 eq.), Ac_2O, CH_2Cl_2, 4 °C, NI (1 g).
G31 NH_2OH•HCl (10 eq.), EtOH, py, 75-82% (1 g, G30+G31).
G32 Rh/Al_2O_3 (5%), H_2 (4 bar), EtOH, H_2O, Boc_2O (2-3 eq.), 93% (73 mg), stepwise procedure, 50% (0.4 g).
G33 EtONa (0.3 eq.), EtOH, 94% (0.5 g, ASC2010, BrockU2009).
G34 $PPhMe_2$ (2.2 eq.), DIAD (2.2 eq.), CH_2Cl_2, 4 °C, 70-78% (84 mg).
G35 3-Pentanol, $Cu(OTf)_2$ (0.16 eq.), rt, 60% (16 mg).

8.15.7 Reaction conditions for scheme H (g or mg after yield refers to reaction scale)

H1 Toluene or THF, rt, NI (0.1 g diene, TokyoU2006c, 10.6 g diene, TL2007).
H2 1) $TMSN_3$ (2.1 eq), DMAP (0.1 eq.), toluene, rt; 2) HCl (1 N), 54% (H1+H2, 0.1 g, TokyoU2006c). 1) $TMSN_3$, DMAP, THF; 2) HCl (1 N), 55% (H1+H2, TL2007).
H3 1) Benzene, 80 °C, 96% (412 mg); 2) *t*-BuOH, reflux, 99% (0.3 g, TokyoU2006c). *t*-BuOH, reflux, 100% (8.6 g, TL2007).
H4 1) Boc_2O (1.2 eq.), NEt_3, DMAP (0.1 eq.), CH_2Cl_2, rt, 91% (5 mg); 2) Cs_2CO_3 (0.1 eq.), MeOH, rt, 100% (1.8 mg, TokyoU2006c). 1) Boc_2O; 2) Cs_2CO_3, MeOH, 84% (BMCL2008).
H5 DMP (1.5 eq.), CH_2Cl_2, rt, 36% (1.6 mg, TokyoU2006c). $(Me_2CHCO)_2O$, DMSO, 97% (BMCL2008).
H6 Chiralpak ADH, YNR.
H7 (F7) see F7.
H8 (F8) see F8.
H9 (F9) see F9.
H10 (F10) see F10.
H11 (F11) see F11.
H12 (F35) see F35.
H13 (F36) see F36.
H14 (F37) see F37.
H15 (F38) HCl (4 M), EtOAc, YNR (BMCL2008). TFA, H_3PO_4, 73% (ACIE2009a).
H16 1) LiOH, (3 eq.), H_2O, MeOH, 50 °C; 2) Ac_2O (1 eq.), NEt_3 (2 eq.), CH_2Cl_2, rt, 85% (0.1 g, TokyoU2006c). 1) LiOH; 2) Ac_2O, NEt_3, YNR (TL2007).
H17 MnO_2 (5 eq.), CH_2Cl_2, rt, 55% (13 mg, TokyoU2006c). Ac_2O, DMSO, *i*-PrOAc, 80 °C, 63% (14 mg, TokyoU2006c). $(Me_2CHCO)_2O$, DMSO, 53% (H3+H16+H17, TL2007).
H18 Chiral HPLC Daicel Chiralpak AD-H, YNR.
H19 TMSCN, $Ni(cod)_2$ (0.5 eq.), cod (0.5 eq.), YNR.
H20 NBS, NEt_3, YNR.
H21 $LiAl(Ot\text{-}Bu)_3H$, 44% (H19+H20+H21).
H22 DEAD, PPh_3, 66%.
H23 3-pentanol, BF_3•OEt_2, 56%.
H24 1) HCl, EtOH; 2) NaOH, H_2O, 60%.
H25 1) $Ba(OiPr)_2$, ligand (2% mol, X = H, F, Cl), CH_2Cl_2, –20 °C; 2) TBAF, AcOH, THF, 20 °C, 48%, *endo*/*exo* = 13.3:1, 86% ee (X = F), 34%, *endo*/*exo* = 4.6:1, 88% ee (X = F), 44%, *endo*/*exo* = 11.5:1, 90% ee (X = Cl) (43 mg, TokyoU2006c). $Ba(Oi\text{-}Pr)_2$ (2.5% mol), ligand (2.5% mol, X = F), CsF (2.5% mol), THF, –20 °C, 91%, 5:1, 95% ee (68 g, ACIE2009a).
H26 NaOH, MeOH, H_2O, 55-60 °C, YNR.
H27 DPPA, NEt_3, 1,4-dioxane, rt, 36% (H26+H27, 5 mg, TokyoU2006c). DPPA (3 eq.), NEt_3 (3 eq.), THF, 0 °C, 95% (H26+H27, ACIE2009a).
H28 *t*-BuOH, 80 °C, YNR.
H29 Ac_2O (2 eq.), NEt_3 (4 eq.), DMAP, CH_2Cl_2, rt, 98% (H28+H29), 95% ee. Recrystallization, 80%, >99% ee.
H30 $AcOCH(CN)_2$ (1.2 eq.), $[Pd_2(dba)_3]$•$CHCl_3$ (2% mol), dppf (4% mol), toluene, 60 °C, 85%.
H31 1) TFA (10 eq.), urea, H_2O_2 (20 eq.), Na_2HPO_4 (15 eq.), CH_2Cl_2, 48 °C; 2) K_2CO_3 (5 eq.), EtOH, rt, YNR.
H32 1) DEAD (2 eq.), PPh_3 (2 eq.), *p*-$O_2NC_6H_4CO_2H$ (2 eq.), THF, –20 °C; 2) LiOH (3 eq.), EtOH, –20 °C 65% (H31+H32.1+H32.2); 3) DIAD (2.1 eq.), Me_2PPh (2.1 eq.), NEt_3 (0.3 eq.), CH_2Cl_2, 48 °C, 76%.
H33 3-pentanol, BF_3•OEt_2 (1.5 eq.), –20 °C, 75%.
H34 $TBDMSOCH(CN)_2$, $[Pd_2(dba)_3]$•$CHCl_3$ (2% mol), dppf (8% mol), 88% (20 mg).
H35 $NaIO_4$, $RuCl_3$ (0.5% mol), H_2SO_4 (0.2 eq., 50 mg).
H36 2,2-dimethoxypropane, TsOH, 56% (H35+H26, 63 mg).
H37 NEt_3, HF, EtOH, DBU, 76% (28 mg).

8.15.8 Reaction conditions for scheme I (g or mg after yield refers to reaction scale)

I1 90 °C, 57-60% (11.7 g, JOC2010b).
I2 MCPBA, Et_2O, 64% (16 g).
I3 H_2, Pd/C, MeOH, 2-methylpyridine (2 eq.), 71% *endo*:*exo* 4:1 (0.16 g).
I4 1) NaOH (10%), H_2O; 2) H_2SO_4 (3 N, aq.), 89% (3.3 g).
I5 1) MsCl, NEt_3, DMAP, CH_2Cl_2, 95% (1 g); 2) KOH, DMF; 3) EtI, 85% (2+3, 0.1 g).
I6 LDA, HMPA, –78 °C, 62% (0.3 g).
I7 3-pentanol, BF_3•OEt_2, 54% (0.8 g).
I8 DMP, CH_2Cl_2, 87% (38 mg).

I9 $NaBH_4$, 91%, dr = 3:1 (67 mg).
I10 $NaBH_4$. DMSO, 49% (*endo*), 25% (Synlett2009c), 74% *endo*/*exo* 2:1 (0.5 g, ShanghaiIOC2009a).
I11 OsO_4, NMO, *t*-BuOH, H_2O, 96% (0.1 g).
I12 1) 3-pentanone, TsOH, 94%; 2) LiHMDS, THF, –78 °C to 0 °C, 80% (432 mg).
I13 TsOH, MeOH, 80% (142 mg).
I14 MsCl, NEt_3, 60%.
I15 NaN_3, DMF, 67%.
I16 1) PPh_3, NEt_3, THF/H_2O, rt; 2) Ac_2O, py, CH_2Cl_2, 0 °C, not exemplified. In a chiral pure enantiomer, 65% (CC2007)
I17 3-pentanol, $Cu(OTf)_2$, 0 °C, not exemplified. In a chiral pure enantiomer, 48% (CC2007)
I18 H_3PO_4, EtOH, not exemplified. In a chiral pure enantiomer, see F38.
I19 $SOCl_2$, NEt_3, 100% (Synlett2009c). THF, $SOCl_2$, NEt_3, 0 °C to rt, 100% (70 mg, ShanghaiIOC2009a).
I20 NaN_3, DMF, rt, 95%.
I21 MsCl, NEt_3, CH_2Cl_2, 95-97% (43 mg).
I22 1) LiOH, THF, H_2O, 96-98% (72 mg); 2) (*R*)-4-benzyloxazolidine-2-thione, EDC, DMAP, CH_2Cl_2, 96% (125 mg); 3) separation by crystallization, 50%; 4) DMAP, EtOH, CH_2Cl_2, rt, 73% (83 mg).
I23 MeOH, TsOH, rt, 80% (37 mg).
I24 $SOCl_2$, THF, NEt_3, 0 °C to rt, 96% (20 mg).
I25 NaN_3, DMF, rt, 73% (22 mg).
I26 MsCl, NEt_3, CH_2Cl_2, 0 °C, 97% (15 mg).

8.15.9 Reaction conditions for scheme J (g or mg after yield refers to reaction scale)

J1 Oxazaborolidine catalyst (10% mol), neat, 23 °C, 97%, >97% ee (12.2 g acrylate).
J2 NH_3, CF_3CH_2OH, 40 °C, 100% (15.7 g).
J3 1) TMSOTf, NEt_3, pentane; 2) I_2, Et_2O/THF, 84% (9.6 g).
J4 $(Boc)_2O$, NEt_3, DMAP, CH_2Cl_2, 99% (15.2 g).
J5 DBU, THF, reflux, 96% (29.1 g).
J6 NBS, AIBN (cat.), CCl_4, reflux, 95% (10.7 g).
J7 Cs_2CO_3, EtOH, 100% (8.5 g).
J8 $SnBr_2$ (5% mol), *N*-bromoacetamide, CH_3CN, –40 °C, 75% (0.5 g).
J9 KHMDS, *n*-Bu_4NBr, DME, –20 °C, 82% (0.16 g).
J10 3-pentanol, $Cu(OTf)_2$ (cat.), 0 °C, 61% (0.1 g, JACS2006a). 3-pentanol, BF_3•OEt_2, 70% (71 mg, T2001).
J11 H_3PO_4, EtOH (JACS2006a). TFA, CH_2Cl_2, 62% (62 mg, TA2011).
J12 I_2, KI, $NaHCO_3$, H_2O, 93% (0.86 g).
J13 DBU, toluene, 92% (1.6 g).
J14 K_2CO_3, EtOH, 90% (0.72 g).
J15 BocNHNNs-*p*, DEAD, PPh_3, toluene, 89% (0.89 g).
J16 DBU, $HSCH_2CH_2OH$, acetone, 91% (2.1 g).
J17 MCPBA, CH_2Cl_2, 84% (1.1g).
J18 1) $TMSN_3$, $Ti(OiPr)_3$, benzene (anh.), 5 °C to rt, 65%, 22% of regioisomer formed (580 mg); 2) TMSCl, CH_2Cl_2, NEt_3, 0 °C, 95% (430 mg). The regioisomer is converted to the *N*-acyl aziridine shown on the scheme by a similar sequence to the following one: 1) TMSCl, CH_2Cl_2, NEt_3, 0 °C, 94%; 2) LDA, PhSeBr, THF, 58%; 3) H_2O_2, py, CH_2Cl_2, 68% (1:3 isomers); 4) PPh_3, toluene, 71% (2 isomers); 5) DBU, toluene, 58%.
J19 LDA, PhSeBr, THF, 74% (0.5 g).
J20 1) H_2O_2, py, CH_2Cl_2, 51%, 17% of regioisomer formed (0.3 g); 2) DBU, toluene, 65% (SM is the mixture of isomers from step 1, R = H).
J21 PPh_3, toluene, 83% (97 mg).
J22 Ac_2O, NEt_3, DMAP, CH_2Cl_2, 87% (72 mg).
J23 1) $[(\eta^3\text{-}C_3H_5PdCl)_2]$ (2.5% mol), (*R*,*R*)-catalyst (7.5% mol), trimethylsilylphthalimide (1.5 eq.), THF, 40 °C; 2) TsOH, EtOH, reflux, 84%, 98% ee (5 g).
J24 KHMDS (1.5 eq.), $PhSSO_2Ph$ (1.8 eq.), THF, –78 °C to rt, 94% (4.9 g).
J25 1) MCPBA, $NaHCO_3$, 0 °C; 2) DBU, 60 °C, toluene, 85% (6.1 g)
J26 Catalyst (2% mol), $SESNH_2$ (1.1 eq.), MgO (2.3 eq.), $PhI(OCOCMe_3)_2$ (1.3 eq.), PhCl, 0 °C to rt, 86% (0.15 g).
J27 3-pentanol, BF_3•OEt_2 (1-5 eq.), 75 °C, 65% (0.1 g).
J28 Ac_2O, DMAP, py, MW, 150 °C, 84% (0.1 g).
J29 1) TBAF (2 eq.), THF, rt 95% (0.2 g); 2) NH_2NH_2 (5 eq.), EtOH, 68 °C, 100% (44 mg).

8.15.10 Reaction conditions for scheme K (g or mg after yield refers to reaction scale)

K1 CuBr (5-10% mol), (*S*,*S*)-Ph-Pybox (5-20% mol), alkyne (2 eq.), aldehyde (1 eq.), amine (2 eq.), 4 Å MS, toluene, −15 °C to 0 °C (26–50 mg aldehyde), 89%, 34% ee (R = H), 94%, 76% ee (R = Ph).
K2 $Pd_2(dba)_3$•$CHCl_3$ (2.5% mol), P(*o*-tol)$_3$ (10% mol), $Me_2HSiOSiHMe_2$ (1 eq.), AcOH (1 eq.), toluene, 40 °C, 36% (R = H, 1.25 g), 84% (R = Ph, 4.3 g).
K3 LHMDS (3 eq.), THF, −40 °C, NI (R = Ph, 0.34-3.6 g, JOC2013, MicrobialCR2011, IMC2012). LHMDS or LDA, THF, −40 °C, 62% (R = H, T2013).
K4 $NaBH_4$ (2 eq.), MeOH, −20 °C, 62% (K3+K4, 0.3 g, MicrobialCR2011), NI (JOC2013, T2014).
K5 1) $Pd(PPh_3)_4$ (10% mol), *N*,*N*-dimethylbarbituric acid (6 eq.), CH_2Cl_2, rt; 2) Boc_2O (5 eq.), $NaHCO_3$, H_2O, CH_3CN, rt, 55% (K3+K4+K5, 1.4 g, JOC2013, T2014), 91% (0.18 g, MicrobialCR2011, IMC2012).
K6 1) MsCl (1.1 eq.), NEt_3 (2 eq.), CH_2Cl_2, 4 °C; 2) DBU (3.1 eq.), CH_2Cl_2, rt, 87-100% (57-170 mg).
K7 $HClO_4$ (60%, 1.1 eq.), *t*-BuOAc, rt, 77% (15 g, JOC2013, IMC2012), with H_2SO_4, 62% (T2014).
K8 Allyl bromide, CH_3CN, K_2CO_3, *n*-Bu_4NI, 70 °C, 77% (3 g, IMC2012), in EtOH, reflux, 80-84% (4.6 - 9.4 g, JOC2013, IMC2012, T2014).
K9 1) HCO_2H, 80 °C; 2) $ClCO_2Et$ (3 eq.), NEt_3 (5 eq.), THF, −10 °C; 3) $NaBH_4$ (6 eq,), MeOH, 0 °C, 80% (1.4 g, JOC2013). 1) HCl, 1,4-dioxane, 50 °C, 97%; 2) $ClCO_2Et$, NEt_3, toluene, –20 °C; 3) $NaBH_4$, MeOH, –20 °C, 74% (2+3) (6.2 g, T2014). 1) HCO_2H, 60 °C (1.4 g); 2) $ClCO_2Et$, NEt_3, THF, –10 °C; 3) $NaBH_4$, MeOH, –10 °C, 80% (1.4 g, IMC2012).

K10 SO_3•py (4 eq.), NEt_3 (3.6 eq.), DMSO, 0 °C, NI (0.1 g, JOC2013, IMC2012). TEMPO (0.2 eq.), *N*-Methylimidazole (0.4 eq.), 2,2'-bipyridyl (0.2 eq.), $[Cu(CH_3CN)_4]PF_6$ (0.2 eq.), CH_3CN, 91% (4.8 g, T2014).
K11 $(CF_3CH_2O)_2P(O)CH_2CO_2Et$ (1.1 eq.), KHMDS (1.2 eq.), 18-crown-6 (5 eq.), THF, −78 °C, 80% (K10+K11, 0.15 g, JOC2013, IMC2012). $(CF_3CH_2O)_2P(O)CH_2CO_2Et$, DBU, NaI, THF, –40 °C, 93% (*Z*/*E* 7:1, 6.2 g, T2014), 62% (K10+K11, 0.1 g, IMC2012).
K12 TBDMSCl, imidazole, CH_2Cl_2, rt, 95% (5 g).
K13 $LiAlH_4$, THF, 0 °C, 92% (7.2 g).
K14 $(COCl)_2$, DMSO, NEt_3, CH_2Cl_2, –78 °C, YNR (6 g).
K15 AllMgBr, THF, –15 °C to rt, 85% (K14+K15, 7 g).
K16 2,2-dimethoxypropane, CSA (cat.), 0 °C to rt, 82% (5.8 g).
K17 TBAF, THF, 0 °C to rt, 96% (5.3 g).
K18 $(COCl)_2$, DMSO, NEt_3, CH_2Cl_2, –78 °C, YNR (3.5 g).
K19 Ethyl acrylate, DABCO, rt, 82% (K18+K19, 4.1 g).
K20 AcCl, py, 0 °C to rt, 96% (3.8 g).
K21 $NaBH_4$, DABCO, dioxane, H_2O, rt, 89% (3.9 g).
K22 Hoveyda-Grubbs II catalyst (10% mol), toluene, 60 °C, 71% (2.9 g).
K23 PPTS, $NaBH_4$, dioxane, H_2O, rt, 82% (1.9 g).
K24 Tf_2O, py, CH_2Cl_2, –78 °C, YNR (1.3 g).
K25 DBU, CH_2Cl_2, rt, 95% (K24+K25, 1.3 g).
K26 1) *i*-BuOCOCl, NEt_3, THF, 0 °C; 2) EtSH, NEt_3, 0 °C to rt, 90% (1+2, 1.1 g); 3) Et_3SiH, Pd/C, acetone, rt, 94% (1.1 g).
K27 In, allyl bromide, THF, rt, 82% (1.3 g).
K28 $NaBH_4$, THF, H_2O, rt, YNR (0.75 g).
K29 2,2-dimethoxypropane, TsOH, CH_2Cl_2, 95% (K28+K29, 0.6 g).
K30 DMP, CH_2Cl_2, rt, 85% (0.55 g).
K31 $Me_2N{=}CH_2I$, NEt_3, CH_2Cl_2, rt, 87% or CH_2O, H_2O, propanoic acid (cat.), pyrrolidine (cat.), *i*-PrOH, rt, 66% (380 mg).
K32 Grubbs 2nd generation Ru-catalyst (cat.), CH_2Cl_2, 35 °C, 92% (299 mg).
K33 LiCl, AcOH, H_2O, rt, 70% (54 mg).
K34 1) Oxone, DMF, rt; 2) EtI, K_2CO_3, DMSO, rt, 46% (83 mg).
K35 MsCl, NEt_3, CH_2Cl_2, 89% (14 mg).
K36 DBU, CH_2Cl_2, 87% (2 mg).
K37 NaOH, rt, 83% (325 mg).
K38 $NaBH_4$, THF, 0 °C, 77% (294 mg).
K39 1) PhSH, K_2CO_3, CH_3CN, rt, 55% (338 mg); 2) Boc_2O, H_2O, rt, YNR.
K40 1) $NaIO_4$, THF, H_2O ; 2) Boc_2O, NI.
K41 1) $NaBH_4$, EtOH; 2) MsCl, NEt_3, DMAP, CH_2Cl_2, 33%.

8.15.11 Reaction conditions for scheme L (g or mg after yield refers to reaction scale)

L1 $PhCO_2H$ (40% mol), (*R*)-diphenylprolinol TMS ether (20% mol), CH_2Cl_2, –20 °C (frequently NI), 100% *syn*:*anti* = 5:1, *syn* 95-96% ee, *anti* 87-96% ee (56-222 mg aldehyde, 50-200 mg nitro, 1.5:1.0 molar ratio, Sumitomo2008). $ClCH_2CO_2H$ (20% mol), (*R*)-diphenylprolinol TMS ether (5% mol), CH_2Cl_2, 23 °C, (frequently NI) 100%, *syn*:*anti* = 5:1, *syn* 96% ee, *anti* 87% ee, (25-200 mg nitro, aldehyde 1.5 eq., Sumitomo2008, ACIE2009b). $ClCH_2CO_2H$ (20% mol), (*R*)-diphenylprolinol TMS ether (1% mol), toluene, 23 °C, 100% (or NI), dr = 7.8:1, *syn* 97% ee, *anti* 81% ee (200 mg nitro; aldehyde 1.5 eq., CEJ2010, EJOC2011). (*R*)-*N*,*N*-dimethylbenzylamine prolinol TMS ether (5% mol), *o*-$O_2NC_6H_4CO_2H$ or $ClCH_2CO_2$(20% mol), CH_2Cl_2, 25 °C, NI (225 g nitro; 168 g aldehyde, 1.5 eq., *t*-Bu is Et, UHangzhou2011), with (*R*)-*N*,*N*-dimethylbenzylamine prolinol TBDPS ether or TBDMS ether or TES ether (5% mol), $ClCH_2CO_2H$, NI (168 g aldehyde, 225g nitro, *t*-Bu is Et, UHangzhou2011).
L2 Ethyl 2-(diethoxyphosphoryl)acrylate (1.5 eq.), Cs_2CO_3, 0 °C, EtOH, rt, 69% (L1+L2), dr (5*S*:5*R*) = 1:4 (L1 with $PhCO_2H$, 200 mg nitro, Sumitomo2008), 73% (L1+L2), dr (5*S*:5*R*) = 1:4, (L1 with $ClCH_2CO_2H$, 25 mg nitro, Sumitomo2008). Ethyl 2-(diethoxyphosphoryl)acrylate (1.5 eq.), benzene, 23 °C, 97% (L1 with $PhCO_2H$, 50 mg nitro, Sumitomo2008). 1) Ethyl 2-(diethoxyphosphoryl) acrylate, Cs_2CO_3, 0 °C; 2) Evaporation, EtOH, rt, NI (200 mg nitro, ACIE2009b). 1) Ethyl 2-(diethoxyphosphoryl)acrylate, Cs_2CO_3, toluene, 23 °C; 2) Evaporation, EtOH, rt, 71% dr (5*S*:5*R*) = 4:1 (62 mg, CEJ2010), 74% (L1+L2+L3, 1g, CEJ2010, EJOC2011). Ethyl 2-(diethoxyphosphoryl) acrylate (1.5 eq), Cs_2CO_3, 0 °C, EtOH, 25 °C, NI (168 g nitro, *t*-Bu is Et, UHangzhou2011).
L3 TolSH, NEt_3, EtOH, 23 °C, 71% 5*S*, 29% 5*R* (200 mg nitro, Sumitomo2008), 70% (L1+L2+L3, L1 with $ClCH_2CO_2H$, 200 mg nitro, Sumitomo2008). TolSH, Cs_2CO_3, EtOH, –40 °C to –15 °C, 90%, 5*S*, 100% ee (14 mg, 5*S*:5*R* 1:4, Sumitomo2008). TolSH, EtOH, –15 °C, 70-74% (L1+L2+L3, 0.2-1.1 g nitro, ACIE2009b), 90% (L3, 14 mg, CEJ2010, EJOC2011). TolSH, EtOH, –15 °C, 75% (L1+L2+L3), 94% ee (168 g nitro, *t*-Bu is Et UHangzhou2011), when intermediate is produced with (*R*)-*N*,*N*-dimethylbenzylamine prolinol TMS ether and *o*-$O_2NC_6H_4CO_2H$; with (*R*)-*N*,*N*-dimethylbenzylamine prolinol TMS ether/$ClCH_2CO_2H$ 75% (L1+L2+L3), 94% ee; with (*R*) *N*,*N*-dimethylbenzylamine prolinol TBDPS ether (5% mol), $ClCH_2CO_2H$, 70%, 90% ee (225g nitro, *t*-Bu is Et UHangzhou2011); with (*R*)-*N*,*N*-dimethylbenzylamine prolinol TBDMS ether (5% mol), $ClCH_2CO_2H$, 70%, 92% ee (225 g nitro, *t*-Bu is Et, UHangzhou2011); with (*R*)-*N*,*N*-dimethylbenzylamine prolinol TES ether (5% mol), $ClCH_2CO_2H$, 70%, 94% ee (225g nitro, *t*-Bu is Et, UHangzhou2011). Note: *R* isomer may be converted back to previous intermediate by: 1) MCPBA, CH_2Cl_2, rt; 2) Cs_2CO_3, EtOH, rt, 70% (1+2, 5 mg, Sumitomo2008).

L4 Zn, NH_4Cl, H_2O, THF, 90 °C, NI (19 mg, Sumitomo2008).
L5 Cs_2CO_3, EtOH, 23 °C, 97% (L4+L5, 19 mg, Sumitomo2008).
L6 1) Phthalic anhydride, NEt_3, toluene, reflux; 2) TFA, CH_2Cl_2, 23 °C, 90% (11 mg, Sumitomo2008).
L7 1) $(COCl)_2$, DMF, CH_2Cl_2; 2) NaN_3, H_2O, acetone; 3) Ac_2O, 90 °C, 74% (1+2+3, 14 mg, Sumitomo2008).
L8 NH_2NH_2•H_2O, EtOH, 50 °C, 96% (30 mg, Sumitomo2008).
L9 1) TFA, CH_2Cl_2, 23 °C, 88% (180 mg, Sumitomo2008), NI (90 mg, ACIE2009b), 93% (50 mg, CEJ2010). MeOH, HCl, rt, 98% (510 mg, Guangdong2014).
L10 1) $(COCl)_2$, DMF (cat.), CH_2Cl_2, 23 °C, 98% (11 mg); 2) NaN_3, H_2O, acetone, 0 °C, 96% (1+2, 40 mg); 3) Ac_2O, 90 °C, 86% (L9+L10.1.2.3, 80 mg); 4) NH_2NH_2, EtOH, 23 °C, 78% (4, 15 mg, Sumitomo2008). 1) $(COCl)_2$, DMF (cat.), toluene, 100% (109 mg); 2) NaN_3, acetone, 0-20 °C, 100% (83-215 mg, CEJ2010); 3) AcOH, Ac_2O, 23 °C, NI (90 mg, ACIE2009b), 95-100% (4-17 mg, CEJ2010), 92% (L9+L10, 122 mg, CEJ2010). $TMSN_3$, py, toluene, 23 °C NI (122 mg, CEJ2010, EJOC2011). **Hydroxamic acid formation, Lossen:** 1) $ClCO_2Et$, NEt_3, CH_2Cl_2, 0 °C to rt; 2) NH_2OH•HCl, MeOH, 92%; 3) Ac_2O, NEt_3, CH_2Cl_2, rt, 99% (450 mg); 4) DBU, THF, H_2O, reflux, 86% (510 mg, Guangdong2014).
L11 Zn, EtOH, HCl (2 N), 23 °C, 86% (30 mg, Sumitomo2008, CEJ2010). Zn, TMSCl, EtOH, 70 °C, NI (17 mg-0.26 g, ACIE2009b, CEJ2010).
L12 1) K_2CO_3, EtOH, 23 °C, 91% (9 mg, Sumitomo2008, CEJ2010), 83% (L9+L10+L11+L12, 215 mg, Sumitomo2008). 1) NH_3; 2) K_2CO_3, EtOH, 81-85% (L9+L10+L11+L12, 90 mg, ACIE2009b), 82% (L9+L10+L11+L12, 3 g, CEJ2010, EJOC2011), 81% (L11+L12, 0.2 g, CEJ2010), 85% (L11+L12, 47mg, ACIE2010), 82% (L15+L11+L12, JOC2010a), 28-36% (L13+L14+L15+L11+L12, 34 mg-1 g, CEJ2013). 1) NH_3, –20 °C; 2) K_2CO_3, rt, 92% (L11+L12, 0.26 g, ShanghaiIOC2009b).
L13 (*R*)-dinaphthylprolinol TMS ether (10% mol), $PhCO_2H$ (30% mol), $CHCl_3$, –5 °C, 81%, dr = 5:1, 96% ee (0.13 g nitro, ShanghaiIOC2009b), NI (1.3 g nitro, ACIE2010). (*R*)-diphenylprolinol TMS ether (10% mol), $ClCH_2CO_2H$ (20% mol), $CHCl_3/H_2O$, 0 °C, 88%, 98% ee, dr (*syn*:*anti*) = 81:19 (70 mg nitro, S2012). (*R*)-diphenylprolinol DPMS ether (10% mol), HCO_2H (30% mol), C_6H_5Cl, 20 °C, NI (34 mg-1 g of nitro, CEJ2013). Study of the acidic additive, most efficient: chloro- and bromoacetic acid, *rac*-mandelic acid, and 2,4-dinitrophenol, HCA2012. From aldehyde diethyl acetal: 1) [bmim]HSO_3 (1 eq.), H_2O, MW, 130 °C (0.25 g), 85%; 2) nitroolefin, HCO_2H (30% mol), $CHCl_3/H_2O$, catalyst (10% mol), 0 °C, NI (ACSSCE2015, OL2016).
L14 Ethyl 2-(diethoxyphosphoryl)acrylate (1 eq.), Cs_2CO_3 (3 eq.), CH_2Cl_2, rt, NI (0.26 g, ShanghaiIOC2009b). Ethyl 2-(diethoxyphosphoryl) acrylate Cs_2CO_3, 0 °C, NI (1.3 g nitro, ACIE2010). 1) Ethyl 2-(diethoxyphosphoryl)acrylate, Cs_2CO_3, 0 °C; 2) EtOH, NI (CEJ2013). Ethyl 2-(diethoxyphosphoryl) acrylate, DBU, LiCl, CH_3CN, –15 °C to 0 °C, 61% (L23+L14), 5*S*/5*R*, 3:2 (0.9 g, JOC2010a), can be upgraded to 96:4 *S*/*R* by K_2CO_3 (293 mg, S2012). Ethyl 2-(diethoxyphosphoryl)acrylate, K_2CO_3, 18-crown-6, CH_2Cl_2, 70 °C, MW, 66%, 5*S*/5*R* 2:3 (193 mg-4.5 g, S2012). Ethyl 2-(diethoxyphosphoryl)acrylate (1 eq.), K_2CO_3 (3 eq.), 18-crown-6, CH_2Cl_2, rt, 70 °C, 56% dr = 1:2 (ACSSCE2015, OL2016), in $CHCl_3$, 43%, dr = 1:2 (0.26 g, ACSSCE2015, OL2016).
L15 TolSH, –15 °C, 54%, 96% ee (L13+L14+L15, 1.3 g nitro, ACIE2010, JOC2010a, CEJ2013). TolSH, EtOH, 70% (L14+L15, 0.26 g, ShanghaiIOC2009b).
L16 Zn, HCl, EtOH, rt, 81% (45 mg, S2012, OL2016).
L17 3,3-dimethoxypentane, TsOH, toluene, reflux, 96%.
L18 $LiAlH_4$, $AlCl_3$, Et_2O, CH_2Cl_2, –30 °C to 0 °C, 88% (32 g).
L19 $NaIO_4$, THF, H_2O, rt, 95%.
L20 (*S*)-2-methylpropane-2-sulfinamide, $CuSO_4$, CH_2Cl_2, rt, 73% (9 g).
L21 CH_3NO_2, NaOH, 4 Å MS, rt, 86% (6.5 g).
L22 1) MeOH, HCl, rt; 2) Ac_2O, MeOH, rt, 83% (1 g).
L23 2-Iodoxybenzoic acid, EtOAc, 80 °C, YNR.
L24 Ethyl 2-(diethoxyphosphoryl)acrylate (1.2 eq.), CH_3CN, K_2CO_3 (1 eq.), 0°C to rt, 43%, 5*S*, 17%, 5*R* (0.26 g).
L25 5*R* isomer is converted to 5*S* by: CH_3CN, K_2CO_3, rt, 100%.
L26 1) Zn, AcOH, EtOH, 70 °C, 90% (0.1 g); 2) Boc_2O, CH_3CN, rt, 83% (1+2, 30 mg).
L27 LiCl (1 eq.), DMSO, 190 °C, 79% (16 mg).
L28 TFA, CH_2Cl_2, rt, 92% (16 mg).

8.15.12 Reaction conditions for scheme M (g or mg after yield refers to reaction scale)

M1 3,3-dimethoxypentane (1.5 eq.), DMF, TsOH, rt, 56% (3 g).
M2 KIO_4 (1.5 eq.), THF, H_2O, rt, 90% (3.3 g).
M3 1) Ethyl acrylate (3 eq.), DABCO (1.5 eq.), rt; 2) Ac_2O (3 eq.), NEt_3 (5 eq.), THF, 0 °C, 41% (0.42 g).
M4 CH_3NO_2 (20 eq.), KOH (3 eq.), EtOH, H_2O, rt, 68% (0.14 g).
M5 BH_3•SMe_2 (1M, 1.1 eq.), BF_3•OEt_2 (1.1 eq.), CH_2Cl_2, 0 °C, 93% (0.3 g).
M6 HIO_4•$2H_2O$ (1.2 eq.), THF, 0 °C, 100% (38 mg).
M7 $NaHCO_3$, THF, H_2O, rt, 56% (33 mg).
M8 Zn (6 eq.), HCl (aq. 1 N), EtOH, rt, 100% (75 mg).
M9 1) 3-pentanone (5 eq.), $CH(OEt)_3$ (10 eq.), THF, DL-CSA (0.1 eq.), rt to 60 °C; 2) D-(–)-mannitol, 60-70 °C, 97% (5 g).
M10 1) Ethylene glycol (1.1 eq.), MsOH (cat.), THF, rt; 2) HIO_4•$2H_2O$ (1.2 eq.), rt, 70% (404 mg).
M11 1) Ethyl acrylate (2 eq.), DABCO (1 eq.), 30 °C; 2) Ac_2O, 70% (255 mg).
M12 CH_3NO_2 (30 eq.), KOH (1.2 eq.), EtOH, rt, 75% (226 mg).
M13 DL-CSA (0.1 eq.), EtOH, rt, 98% (99 mg).

M14 HIO_4•$2H_2O$ (1.2 eq.), THF, rt, 85% (81 mg).
M15 $NaBH_4$ (1 eq.), THF, rt, 95% (62 mg).
M16 KOH (0.5 M aq.), MeOH, rt, 94% (1g).
M17 BH_3•SMe_2 (3 eq.), THF, 0 °C to rt, 87% (0.2 g).
M18 3,4-dihydro-2*H*-pyran (2 eq.), (+)-CSA (0.05 eq.), CH_2Cl_2, rt, 99% (2.3 g).
M19 1) DIBAL (1.1 eq.), toluene, –48 °C; 2) Ethyl acrylate (4.6 eq.), DABCO (1 eq.), rt; 3) Ac_2O (2 eq.), NEt_3 (3 eq.), DMAP (0.3 eq.), THF, 0 °C, 53% (1+2+3, 0.73 g).
M20 CH_3NO_2 (15 eq.), EtOH, KOH (1.1 eq.), 0 °C, 66% (233 mg).
M21 HCl (1 N), EtOH, H_2O, rt, 92% (27 mg).

8.15.13 Reaction conditions for scheme N (g or mg after yield refers to reaction scale)

N1 1) Tf_2O, py, –10 °C, 95% (200 g); 2) NaN_3, DMF, *n*-Bu_4NCl, 50 °C, 48% (145 g).
N2 Tf_2O, NEt_3, CH_2Cl_2, –20 °C, 67% (1.1 g).
N3 $EtO_2CCH_2PO(OEt)_2$, NaH, 15-crown-5, DMF, rt, 53% (4 g).
N4 1) TFA, rt to 30 °C, 96% (2 g); 2) THF, NaH, 0 °C, 66% (1+2, 1.8 g).
N5 TsCl, NEt_3, CH_2Cl_2, rt, 98% (0.1 g).
N6 PMe_3, toluene, CH_3CN, H_2O, 45 °C, 57% (1.1 g).
N7 LiBr, EtOH, rt, 84% (0.5 g).
N8 NEt_3, CH_2Cl_2, 35 °C, 95% (0.4 g).
N9 1) AcCl, NEt_3, CH_2Cl_2, rt, 53% (20 mg); 2) 3-pentanol, BH_3•SMe_2, 0 °C to rt, 93% (80 mg).
N10 NaN_3, EtOH, 75 °C, 74% (15 mg).
N11 H_2, Lindlar catalyst, EtOH, rt, 66% (20 mg, Apotex2006), 85% (JACS2007).
N12 1) PivCl, py, 0 °C, 89%; 2) PDC, Ac_2O, reflux; 3) NH_2OH•HCl, py, 60 °C, 82% (2+3); 4) $LiAlH_4$, THF, 0 °C to reflux, 88%.
N13 1) Ac_2O, py, 25 °C; 2) BnOH, toluene, 0-25 °C, 85% (1+2); 3) 2,2-dimethoxypropane, TsOH, toluene, 80 °C, 90%.
N14 1) Tf_2O, py, CH_2Cl_2, –15 °C; 2) $EtO_2CCH_2PO(OEt)_2$, NaH, 15-crown-5, DMF, 25 °C, 80% (1+2).
N15 1) H_2, Pd/C, EtOH, 25 °C; 2) NaH, THF, 25 °C, 83% (1+2).
N16 DPPA, DIAD, PPh_3, THF, 25 °C, NI.
N17 HCl, EtOH, reflux, 83% (N16+N17).
N18 1) Tf_2O, py, CH_2Cl_2, –15 °C to –10 °C; 2) KNO_2, 18-crown-6, DMF, 40 °C, 70% (1+2).
N19 $Cl_3CC(=NH)OCHEt_2$, CF_3SO_3H, CH_2Cl_2, 25 °C, 78%.

8.15.14 Reaction conditions for scheme O (g or mg after yield refers to reaction scale)

O1 1) 3,3-dimethoxypropane, CSA, DMF, rt; 2) KIO_4, $KHCO_3$, THF, rt, 61% (1+2, 25-45 g).
O2 Vinylmagnesium bromide, THF, toluene, 0 °C, 88% (25 g, Ind2008 Synlett2009a), 44% (O1+O2, 44.8 g, JOC2010c).
O3 Triethyl orthoformate (5.0 eq.), $EtCO_2H$, 132 °C, 85-95% (21-24 g).
O4 DIBAL (5.0 eq.), toluene, 0 °C to rt, 85% (7.5 g).
O5 DHP (2.5 eq.), PPTS (cat.), CH_2Cl_2, rt, 100% (5.2 g).
O6 1) $MsNH_2$ (1.0 eq.), AD-mix-β, *t*-BuOH, H_2O, 0 °C to rt, 97% (8.3 g, Ind2008); 2) MsCl (1.2 eq.), py, 0 °C to rt 71% (7.8 g, Synlett2009a), 97% (8.3 g, Ind2008).
O7 NaN_3 (4-6 eq.), DMSO, 80 °C, 79-85 °C (11 g).
O8 $LiAlH_4$ (1.1 eq.), THF, rt, 92% (7.1 g).
O9 1) $PhthNCO_2Et$ (0.9 eq.), NEt_3, DMAP, THF, 0 °C; 2) Ac_2O, py (2.0 eq,),rt, NI (5.8 g).
O10 PPTS, CSA, MeOH, rt, 36% (O9+O10, 5.8 g, Ind2008). CSA (10% mol), MeOH, rt, 32% (O4 to O10, Synlett2009a).
O11 TEMPO (5-10% mol), KBr (10-20% mol), NaOCl, $NaHCO_3$, CH_2Cl_2, H_2O, 91% (1.2 g).
O12 Bn_2NH•TFA (1.1 eq.), benzene, 50 °C, 80% (0.42 g, Ind2008), in toluene, 82% (O11+O12, Synlett2009a), 62% (O11+O12, Synlett2009b).
O13 MnO_2 (20 eq.), NaCN (6 eq.), EtOH, AcOH, 76% (0.2 g, R = Et, Ind2008). $NaClO_2$ (3.0 eq.), NaH_2PO_4, 2-methylbut-2-ene (10 eq.), *t*-BuOH, THF, H_2O, 0 °C to rt, 86% (Synlett2009a, R = H).
O14 1) $NaBH_4$ (5.0 eq.), *i*-PrOH, H_2O, rt, 93%; 2) K_2CO_3 (1.1 eq.), EtOH, H_2O, rt, NI; 3) EtI (5.0 eq.), DMSO, rt, 85% (2+3).
O15 HCl (4 M), 1,4-dioxane (5.0 eq.), EtOH, rt, 83%.
O16 NH_2NH_2•H_2O, EtOH, rt, 47% (60 mg).
O17 AD-mix-β, H_2O, *t*-BuOH, 93% (15.4 g).
O18 1) MsCl, NEt_3, 100%; 2) NaN_3, DMF, 73% (3.1 g).
O19 BH_3•SMe_2, TMSOTf, CH_2Cl_2, THF, 94% (1.6 g).
O20 $(COCl)_2$, DMSO, 92% (2.3 g).
O21 TBDMSOTf, DIPEA, CH_2Cl_2, 76% (0.9 g).
O22 LiBr, DBU, EtOH, 96% (2.2 g).
O23 H_2, Pd/C, Ac_2O, NEt_3, 96% (0.76 g).
O24 $LiClO_4$, DBU, EtOH, 62% (0.15 g).
O25 1) MsCl, NEt_3, 97-98%; 2) NaN_3, DMF, 78% (0.27 g, JOC2010c), in EtOH, H_2O, reflux, 91% (49 mg, TL2012b).
O26 PPh_3, THF, H_2O, 98% (0.22 g, JOC2010c). PPh_3, THF, H_2O, NEt_3, rt, 81% (34 mg scale, TL2012b).
O27 3-pentanone, DMF, H_2SO_4, rt, 65% (3.6 g).
O28 1) BnCl, NaH, DMF, rt (4.2 g); 2) MeOH, HCl, H_2O, rt (3.8 g); 3) $NaIO_4$, H_2O, EtOH, rt (2.2 g); 4) $NaBH_4$, EtOH, rt (1.9 g), 30% (O27+O28).
O29 BH_3•THF, $Cu(OTf)_2$, NEt_3, MeOH 75% (5 g). Et_3SiH, $TiCl_4$, CH_2Cl_2, –78 °C, 40%.
O30 TBDMSCl, imidazole, DMAP, CH_2Cl_2, 82% (310 mg).
O31 1) Tf_2O, py, CH_2Cl_2, 0 °C; 2) NaN_3, acetone, H_2O, rt, 89% (1+2, 350 mg).
O32 1) PPh_3, THF, rt; 2) H_2O, NEt_3, rt; 3) Ac_2O, py, 0 °C to rt, 70% (1+2+3, 330 mg).
O33 TBAF, THF, 0 °C to rt, 95% (175 mg).
O34 1) $(COCl)_2$, DMSO, NEt_3, CH_2Cl_2, –78 °C, 90% (2 g); 2) $(EtO)_2(O)PCH_2CO_2Et$, $TiCl_4$, NEt_3, CH_2Cl_2, 0 °C (1.7 g); 3) 10% Pd/C, EtOH, rt, 33% (3.3 g, 1+2+3).
O35 NaH, THF, 0 °C, 55% (400 mg).

O36 1) NaH, BnCl, DMF, rt, 70% (4 g); 2) HCl, MeOH, H_2O, 0 °C to rt, 70% (4 g).
O37 $NaIO_4$, EtOH, H_2O, rt, 30% (O27+O36+O37, 2.2 g).
O38 $TiCl_4$, $(EtO)_2(O)PCH_2CO_2Et$, NEt_3, CH_2Cl_2, 0 °C, 51% (1 g).
O39 H_2, Pd/C, EtOH, rt (300 mg).
O40 NaH, THF, 0 °C to rt, 52% (O39+O40, 300 mg).
O41 1) TEMPO, TCCA, CH_2Cl_2; 2) NH_2OH•HCl, py, EtOH, 72% (1+2, 2 g).
O42 $NaBH_4$, MoO_3, MeOH, rt, 92% (500 mg).
O43 Et_3SiH, $TiCl_4$, CH_2Cl_2, –78 °C to –10 °C (2 g).
O44 Boc_2O, MeOH, rt, 66% (O43+O44, 2 g).
O45 1) Tf_2O, py, CH_2Cl_2, –10 °C (620 mg); 2) NaN_3, acetone, H_2O, rt, NI.
O46 AcSH, 2,6-lutidine, $CHCl_3$, reflux, 71% (O45+O46).
O47 1) TFA, CH_2Cl_2, rt; 2) H_3PO_4, EtOH, 55 °C, 75% (100 mg).
O48 BF_3•OEt_2, acetone, reflux, 59% (21 g).
O49 BnBr, NaH, DMF, 0 °C to rt, 91% (3.4 g).
O50 TFA, BnOH, rt, 80% (4.2 g).
O51 AcOH, H_2O, rt, 95% (7.5 g).
O52 $NaIO_4$, acetone, H_2O, 0 °C, 96% (6.5 g).
O53 $EtO_2CCH_2PO(OEt)_2$, $TiCl_4$, NEt_3, THF, 0 °C, 76% (356 mg).
O54 1) Pd/C, H_2, EtOH, rt (2.3 g); 2) NaH, THF, 0 °C, 62% (O53+O54, 1.5 g).
O55 1) HCl, EtOH, reflux (210 mg); 2) Imidazole-1-sulfonyl azide, K_2CO_3, $CuSO_4$•$5H_2O$, EtOH, rt, 81% (1+2, 174 mg).
O56 MsCl, NEt_3, EtOAc, 0 °C, 85% (59 mg).

8.15.15 Reaction conditions for scheme P (g or mg after yield refers to reaction scale)

P1 1) $ClCO_2CH_2CH(Me)_2$; 2) EtSH, NEt_3 80% (1+2, 1g); 3) Et_3SiH, Pd/C, 2,6-lutidine, 81% (2.5 g).
P2 1) n-Bu_2BOTf, NEt_3, CH_2Cl_2; 2) Oxazolidinone intermediate, –78 °C to –20 °C; 3) H_2O_2, MeOH, 45% (1+2+3, 0.26 g).
P3 1) $(Me)_2C(OMe)_2$, CH_2Cl_2, 73%; 2) $NaBH_4$, THF, H_2O, 83% (88 mg).
P4 DMP, CH_2Cl_2, NI.
P5 Bn_2NH•TFA, toluene, rt, 52% (P4+P5, 0.5 g).
P6 1) Oxone, DMF, rt; 2) K_2CO_3, EtOH, H_2O; 3) EtI, DMSO, rt, 54% (1+2+3, 63 mg).
P7 see I8+I9.
P8 see O45+O46+O47.
P9 DIBAL (2.0 eq.), toluene, –70 °C, 91%.
P10 p-$MeOC_6H_4NH_2$ (0.9 eq.), $MgSO_4$, CH_2Cl_2, 0 °C to rt, 100%.
P11 1) AcCl, DIPEA, CH_2Cl_2, –15 °C to rt, 55% (Z = SMe); 2) NCS, CH_3CN, EtOH, H_2O, rt, 100% (Z = SOMe).
P12 Alpha-pinene–decalins, $NaHCO_3$, 155 °C, 81%.
P13 1) TFA, CH_2Cl_2, 0 °C to rt; 2) $PhthNCO_2Et$ (1.1 eq.), NEt_3, THF, 50 °C, 86% (1+2).
P14 CAN (3.0 eq.), CH_3CN, H_2O, 0 °C, 80%.
P15 1) LiHMDS (2.0 eq.), THF, –78 °C; 2) AcCl (2.0 eq.), –78 °C; 3) EtSH, NEt_3, –78 °C to –20 °C, 94% (1+2+3).
P16 Rh(*acac*)$(CO)_2$ (5% mol), BIPHEPHOS (10% mol), CO, H_2 (1:1, 6 bar), THF, 65 °C, 88%.
P17 Pd/C (10%), Et_3SiH (5.0 eq.), CH_2Cl_2, rt.
P18 (O12) Bn_2NH•TFA (1.1 eq.), toluene, 50 °C, 62% (P17+P18, Synlett2009b).
P19 see O13.
P20 see O14.
P21 see O15.

8.15.16 Reaction conditions for scheme Q (g or mg after yield refers to reaction scale)

Q1 Several syntheses of the Garner aldehyde are available, up to 85% yield. For a review on synthesis and use see BJOC2013. 1) HCl, MeOH, reflux, 100%; 2) Boc_2O, NEt_3, THF, 88%; 3) 2,2-dimethoxypropane, BF_3•OEt_2, acetone, 94%; 4) $LiAlH_4$, THF, –10 °C, 98%; 5) DMSO, $(COCl)_2$, NEt_3, CH_2Cl_2.
Q2 1) N-(4-methoxybenzyl)-hydroxylamine, $MgSO_4$, CH_2Cl_2, 74%; 2) $ZnBr_2$, AllMgBr, THF, Et_2O, –70 °C, 87%; 3) Zn•$Cu(OAc)_2$, AcOH/H_2O, 70 °C; 4) CbzCl, EtOAc, $NaHCO_3$, H_2O, 85% (3+4).
Q3 OsO_4, NMO, acetone, H_2O, 89%.
Q4 1) $Pd(OH)_2$, H_2, MeOH, 35 °C; 2) CbzCl, $NaHCO_3$, H_2O, EtOAc, 86% (1+2); 3) TBDPSCl, imidazole, CH_2Cl_2, rt, 96%.
Q5 1) $(COCl)_2$, DMSO, CH_2Cl_2, NEt_3, –78 °C; 2) Ph_3PCH_3Br, n-BuLi, THF, –78 °C to rt, 86% (1+2).
Q6 $BiBr_3$ (20% mol), CH_3CN, rt, 89%.
Q7 1) $(COCl)_2$, DMSO, CH_2Cl_2, NEt_3, –78 °C; 2) $H_2C{=}CHMgBr$ (3.0 eq.), $ZnBr_2$ (1.0 eq.), THF, –78 °C to –30 °C, 56% (19% diastereomer).
Q8 MOMCl, DIPEA, CH_2Cl_2, rt, 98% (1.3 g).
Q9 1) Grubs II catalyst (10% mol), CH_2Cl_2, rt, 98% (0.5 g); 2) TBAF, THF, rt, 96% (2.0 g).
Q10 1) PCC, 4 Å MS, CH_2Cl_2, rt (377 mg); 2) $NaClO_2$, K_2HPO_4, 2,3-dimethylbuta-1,3-diene, t-BuOH, THF, H_2O, 10 °C to rt, 88% (1+2, 380 mg); 3) EtOH, EDC, DIPEA, HOBt, CH_2Cl_2, rt, 85% (1.0 g).
Q11 1) HCl, EtOH, 0 °C to rt (0.4 g); 2) AcCl, Na_2CO_3, EtOH, 0 °C to rt, 83% (1+2).
Q12 $Pd(OAc)_2$, Et_3SiH, NEt_3, CH_2Cl_2, 0 °C to rt, 92% (64 mg).
Q13 Not shown.
Q14 1) AcCl, MeOH, reflux; 2) $(Boc)_2O$, NEt_3, THF, reflux; 3) 2,2-dimethoxypropane, acetone, BF_3•OEt_2, 25 °C; 4) $LiAlH_4$, THF, 25 °C, 99% (1+2+3+4); 5) $(COCl)_2$, DMSO, NEt_3, CH_2Cl_2, –78 °C, 94% (1+2+3+4+5).
Q15 $(CH_2{=}CH)_4Sn$, MeLi, $ZnBr_2$, Et_2O, –78 °C to 25 °C, 90%, *syn*/*anti* = 5:1 (4.6 g).
Q16 1) NaH, $PhCH_2Br$ or p-$MeOC_6H_4CH_2Br$, DMF, 25 °C; 2) TsOH, MeOH, 25 °C, 76% (R = Bn, 10 g), 70% (R = PMB, 1 g).
Q17 $(COCl)_2$, DMSO, NEt_3, CH_2Cl_2, –78 °C to 0 °C, 85% (R = Bn), 92% (R = PMB, 0.5 g).

Q18 In, $H_2C{=}C(CH_2Br)CO_2Et$, EtOH, H_2O, 25 °C, 76% (R = Bn), 78% (R = PMB, 1 g).
Q19 Grubbs catalyst II, CH_2Cl_2, reflux, 43% *S*, 38% *R* (R = Bn, 4 g) 23% *S*, 14% *R* (R = PMB, 0.3 g).
Q20 MsCl, NEt_3, CH_2Cl_2, 25 °C, 82% (5*S*, R = Bn, 70 mg), 80% (5*R*, R = Bn), 96% (5*S*, R = PMB, 154 mg), 93% (5*R*, R = PMB).
Q21 DQQ, CH_2Cl_2, H_2O, 25 °C, 71% (R = PMB).
Q22 Not shown.
Q23 3-pentanone, MeOH, HCl, $CH(OMe)_3$, reflux, 89% (0.5 g, OL2010), 98% (5 g, TL2010).
Q24 I_2, imidazole, PPh_3, CH_3CN, toluene, reflux, 90% (19 g, OL2010), 79% (5 g, Tl2010).
Q25 1) Zn, THF, H_2O, reflux; 2) $H_2C{=}C(CH_2Br)CO_2Et$, reflux, 78%, dr = 5.2:1 (0.5 g, OL2010), with In replacing Zn, 70% (1 g, TL2010).
Q26 Grubbs-Hoveyda II catalyst (2% mol), 1,2-dichloroethane, reflux, 99% (0.8 g, OL2010). Grubbs-II catalyst (10% mol), CH_2Cl_2, 40 °C, 60% (60 mg, TL2010).
Q27 **Mesylation, reduction:** 1) MsCl, py, 98% (OL2010); 2) $TiCl_4$, Et_3SiH, CH_2Cl_2, –34 °C, 25% (+ 75% regioisomer, OL2010). **Reduction, mesylation:** 1) $AlCl_3$, Et_3SiH, $CHCl_3$, MW, –50 °C to 0 °C, 67% (216 mg); 2) MsCl, py, –20 °C, 92% (96 mg).
Q28 Tf_2O, py, –10 °C to 0 °C, YNR (88 mg).
Q29 NaN_3, acetone, H_2O, 86% (Q28+Q29).
Q30 1) *n*-Bu_3P, THF; 2) NEt_3, H_2O, 84% (28 mg).
Q31 Not shown (see conversion of D2 product into Oseltamivir).
Q32 TEMPO, TCCA, CH_2Cl_2, 0 °C, YNR (2 g).
Q33 $NH_2OH{\bullet}HCl$, EtOH, py, rt, 71% (Q30+Q31).
Q34 $NaBH_4$, MoO_3, MeOH, rt, 93% (511 mg).
Q35 1) Et_3SiH, $TiCl_4$, CH_2Cl_2, –78 °C to –10 °C; 2) Boc_2O, MeOH, rt, 66% (2 g, TL2010).
Q36 1) Tf_2O, py, CH_2Cl_2, –19 °C; 2) NaN_3, acetone, H_2O, rt (0.6 g, TL2010).
Q37 AcSH, 2.8-lutidine, $CHCl_3$, reflux, 44% (Q36+Q37, 0.82 g, TL2010).
Q38 TFA, CH_2Cl_2, rt, 75% (0.1 g, TL2010).

8.15.17 Reaction conditions for scheme R (g or mg after yield refers to reaction scale)

R1 1) TBDMSCl, imidazole, CH_2Cl_2, 0 °C, 73% (20 g); 2) (+)-DET, Ti(O*i*-Pr)$_4$, Me_3COOH, 4 Å MS, 93% (20 g).
R2 TEMPO, $PhI(OAc)_2$, CH_2Cl_2, 95% (15 g).
R3 1) Ethyl-2-(bromomethyl)acrylate, Zn, NH_4Cl (aq.), THF, 64%, dr = 4:1 (4 g); 2) MOMCl, DIPEA, 90% (3g).
R4 TBAF, THF, 0 °C, 88% (1.1 g).
R5 1) DMP, DMSO, 25 °C (1.4 g); 2) K_2CO_3, MeOH, diethyl 1-diazo-2-oxopropylphosphonate (Bestmann–Ohira reagent, 1.5 eq.), 82%.
R6 H_2, Lindlar catalyst, py, 1-octene, EtOAc, 95% (0.24 g).
R7 Grubbs II, CH_2Cl_2, reflux, 90% (0.4 g).
R8 NaN_3, NH_4Cl, DMF, EtOH, H_2O, 83% (0.1 g).
R9 1) PPh_3, toluene, reflux (0.15 g); 2) Ac_2O, DMAP, NEt_3, CH_2Cl_2, 81% (1+2).
R10 1) 3-pentanol, $BF_3{\bullet}OEt_2$; 2) HCl (2 N), EtOH (0.16 g), 64% (1+2).
R11 MsCl, NEt_3, CH_2Cl_2, 0 °C, NI (0.3 g).
R12 NaN_3, DMF, H_2O, 80-90 °C, YNR (40 mg - 0.3 g).
R13 H_2, Lindlar catalyst, EtOH, rt, 64-72% (R11+R12+R13, 0.16 g, OBC2012), 80% (R12+R13, 12 mg, JOC2012), 88% (R13, 55 mg, T2015).
R14 1) Boc_2O, NEt_3, CH_2Cl_2, 88% (5.3 g); 2) TBAF, THF, 76% (6.1 g).
R15 Amino Lipase PS, vinyl acetate, *n*-hexane, THF, 68% (1 g).
R16 TBDMSCl, imidazole, CH_2Cl_2, 94% (0.8 g).
R17 K_2CO_3, MeOH, rt, 87% (1.4 g).
R18 DMP, CH_2Cl_2, rt, 87%.
R19 1) Ethyl-2-(bromomethyl)acrylate, Zn, NH_4Cl (aq.), THF, 71% (4.4 g); 2) MsCl, NEt_3, CH_2Cl_2, 84% (1.4 g).
R20 1) TBAF, THF, 0 °C, 72% (1.3 g); 2) DMP, CH_2Cl_2, 91% (0.74 g); 3) $PPh_3{\bullet}CH_3Br$, KHMDS, THF, 63% (0.67 g).
R21 Grubbs II, 40 °C, 49-55% (30-44 mg).
R22 3-pentanol, $BF_3{\bullet}OEt_2$, –10 °C, 84-95% (17-60 mg).
R23 1) TFA, CH_2Cl_2, 0 °C (51 mg); 2) Ac_2O, NEt_3, CH_2Cl_2, 93%.
R24 NaN_3 (9 eq.) DMF, H_2O, 90 °C, 68% (12 mg),
R25 1) TFA, CH_2Cl_2; 2) Ac_2O, 78% (82 mg).
R26 1) 2,2-dimethyoxypropane, TsOH; 2) Tf_2O, py; 3) DMF, NaN_3, 86–93%; 4) AcOH, H_2O; 5) $NaIO_4$, $NaHCO_3$, H_2O, 85–91%; 6) $NaBH_4$, EtOH; 7) PPh_3, THF, H_2O, 51–82%.
R27 Boc_2O, MeOH, rt, 83% (0.64 mg, T2015).
R28 I_2, PPh_3, imidazole, toluene, CH_3CN, reflux, 64% (2 g).
R29 1) Zn, AcOH, THF, H_2O, MW, 40 °C, 43%; 2) In, ethyl-2-(bromomethyl)acrylate, 40 °C, 39% (0.75 g).
R30 MsCl, NEt_3, CH_2Cl_2, 0 °C, 65% (133 mg).
R31 NaH (2 eq.), CH_2Cl_2, DMSO, rt, 44% (198 mg).
R32 MsCl, NEt_3, CH_2Cl_2, 0 °C, 60% (240 mg).
R33 NaH (2 eq.), CH_2Cl_2, DMSO, rt, 88% (208 mg).
R34 Grubbs II, CH_2Cl_2, 40 °C, 60% (127 mg).
R35 3-pentanol, $BF_3{\bullet}OEt_2$, –10 °C, 91% (45 mg).
R36 NaH (2 eq.), CH_2Cl_2, DMSO, rt, 87% (44 mg).
R37 NaN_3, NH_4Cl, DMF, 90 °C, 89% (5 mg).
R38 1) TsOH, DMF, 2,2-dimethoxypropane, heat, 80% (TL2005); 2) $NaIO_4$, $NaHCO_3$, CH_2Cl_2, 20 °C, 98% (EJOC2003); 3) $PhCH_2NH_2$, $MgSO_4$, Et_2O, 99% (T1996).
R39 N_2CHCO_2Et, TfOH, EtCN, –78 °C, 83%.
R40 1) DIBAL, CH_2Cl_2, –78 °C (10 g); 2) Ph_3PMeBr, *t*-BuOK, toluene, rt, 65% (R39+R40, RSCA2014).
R41 TMSOTf, CH_2Cl_2, 0 °C, 85% (6 g).
R42 $NaIO_4$, CH_2Cl_2, rt, NI (3.3 g).
R43 Ethyl-2-(bromomethyl)acrylate, THF, NH_4Cl, Zn, 94% (3.1 g, R42+R43, yield for the two isomers).
R44 1) *p*-O_2NPhCO_2H, DEAD, PPh_3, toluene, rt; 2) NaOEt, EtOH, –20 °C, 69% (0.5 g).
R45 Grubbs II, CH_2Cl_2, Ti(O*i*-Pr)$_4$, reflux, 74% (0.5 g).
R46 MsCl, NEt_3, DMAP, CH_2Cl_2, 0 °C to rt, 79% (150 mg).
R47 1) $BF_3{\bullet}Et_2O$, 3-pentanol; 2) NEt_3, rt, 80% (50 mg).
R48 HCO_2H, Pd/C.
R49 see scheme D.

8.16 REFERENCES

ACIE2006 Farina, V.; Brown, J. D. *Angewandte Chemie International Edition* **2006**, *45*, 7330-7334.

ACIE2007 Satoh, N.; Akiba, T.; Yokoshima, S.; Fukuyama, T. *Angewandte Chemie International Edition* **2007**, *46*, 5734-5736.

ACIE2008a Trost, B. M.; Zhang, T. *Angewandte Chemie International Edition* **2008**, *47*, 3759-3761.

ACIE2008b Shie, J. J.; Fang, J. M.; Wong, C. H. *Angewandte Chemie International Edition* **2008**, *47*, 5788-5791.

ACIE2009a Yamatsugu, K.; Yin, L.; Kamijo, S.; Kimura, Y.; Kanai, M.; Shibasaki, M. *Angewandte Chemie International Edition* **2009**, *48*, 1070-1076.

ACIE2009b Ishikawa, H.; Suzuki, T.; Hayashi, Y. *Angewandte Chemie International Edition* **2009**, *48*, 1304-1307.

ACIE2009c Sullivan, B.; Carrera, I.; Drouin, M.; Hudlicky, T. *Angewandte Chemie International Edition* **2009**, *48*, 4229-4231.

ACIE2009d Karpf, M.; Trussardi, R. *Angewandte Chemie International Edition* **2009**, *48*, 5760-5762.

ACIE2010 Zhu, S.; Yu, S.; Wang, Y.; Ma, D. *Angewandte Chemie International Edition* **2010**, *49*, 4656-4660.

ACIE2011 Vaxelaire, C.; Winter, P.; Christmann, M. *Angewandte Chemie International Edition* **2011**, *50*, 3605-3607.

ACSSCE2015 Tisovský, P.; Pe aška, T.; Me iarová, M.; Šebesta, R. *ACS Sustainable Chemistry & Engineering* **2015**, *3*, 3429-3434.

AMB2013 Rawat, G.; Tripathi, P.; Saxena, R. *Applied Microbiology and Biotechnology* **2013**, *97*, 4277-4287.

Apotex2006 Radatus, B.; Murthy, K.; Weeratunga, G.; Horne, S.; Kothakonda, K.; Wolf, E.; Wang, Z. (Apotex Pharmachem Inc) US2011040120 (prior. 10/07/2006, US).

ASC2009 Raluy, E.; Pàmies, O.; Diéguez, M. *Advanced Synthesis & Catalysis* **2009**, *351*, 1648-1670.

ASC2010 Werner, L.; Machara, A.; Hudlicky, T. *Advanced Synthesis & Catalysis* **2010**, *352*, 195-200.

AustinSU2005 Li, S.; Yuan, W.; Wang, P.; Zhang, Z.; Zhang, W.; Ownby, S. (Austin State Univ) US2007149805-A1 (prior. 23/12/2005, US).

BA2012 Ghosh, S.; Chisti, Y.; Banerjee, U. C. *Biotechnology Advances* **2012**, *30*, 1425-1431.

Biogaia2002 Malmberg, M.; Westrup, B. (Biogaia Fermentation Ab) WO2003057655, US2003138920, US6794164, EP1474375 (prior. 07/01/2002, US).

BJOC2013 Passiniemi, M.; Koskinen, A. M. P. *Beilstein Journal of Organic Chemistry* **2013**, *9*, 2641-2659.

BMCL2008 Morita, M.; Sone, T.; Yamatsugu, K.; Sohtome, Y.; Matsunaga, S.; Kanai, M.; Watanabe, Y.; Shibasaki, M. *Bioorganic & Medicinal Chemistry Letters* **2008**, *18*, 600-602.

BrockU2008 Hudlicky, T.; Sullivan, B. (Brock Univ) WO2009137916, US2011201577 (prior. 12/05/2008, US).

BrockU2009 Hudlicky, T.; Werner, L.; Machara, A. (Brock Univ) WO2011047466, US2012252890, US8716333 (prior. 23/10/2009, US).

C2017 Fraile, J. M.; Saavedra, C. J. *Catalysts* **2017**, *7*, 393.

CC2007 Bromfield, K. M.; Graden, H.; Hagberg, D. P.; Olsson, T.; Kann, N. *Chemical Communications* **2007**, 3183-3185.

CC2010 Nakano, H.; Osone, K.; Takeshita, M.; Kwon, E.; Seki, C.; Matsuyama, H.; Takano, N.; Kohari, Y. *Chemical Communications* **2010**, *46*, 4827-4829.

CCC2012 Weng, J.; Li, Y. B.; Wang, R. B.; Lu, G. *Chem Cat Chem* **2012**, *4*, 1007-1012.

CEJ2010 Ishikawa, H.; Suzuki, T.; Orita, H.; Uchimaru, T.; Hayashi, Y. *Chemistry – A European Journal* **2010**, *16*, 12616-12626.

CEJ2011 Trost, B. M.; Zhang, T. *Chemistry – A European Journal* **2011**, *17*, 3630-3643.

CEJ2013 Mukaiyama, T.; Ishikawa, H.; Koshino, H.; Hayashi, Y. *Chemistry – A European Journal* **2013**, *19*, 17789-17800.

Chimia2004 Abrecht, S.; Harrington, P.; Iding, H.; Karpf, M.; Trussardi, R.; Wirz, B.; Zutter, U. *CHIMIA International Journal for Chemistry* **2004**, *58*, 621-629.

Chimia2007 Abrecht, S.; Federspiel, M. C.; Estermann, H.; Fischer, R.; Karpf, M.; Mair, H.-J.; Oberhauser, T.; Rimmler, G.; Trussardi, R.; Zutter, U. *CHIMIA International Journal for Chemistry* **2007**, *61*, 93-99.

CR2009 Magano, J. *Chemical Reviews* **2009**, *109*, 4398-4438.

CS2015 Borah, J. C. *Current Science* **2015**, *109*, 1672-1679.

CSIRIndia2010 Bordoloi, M.; Borah, J.; Roy, D.; Dutta, S.; Baruah, N.; Rao, P. (Council of Scientific and Industrial Research India) IN201000575, WO2011110927, US2013137895, US8912357 (prior. 12/03/2010, IN).

CSIRIndia2011 Rawat, V.; Dey, S.; Arumugam, S. (Council of Scientific and Industrial Research India) WO2013061340, US2014243537, US9150498, EP2771311 (prior. 25/10/2011, IN).

CSIRIndia2013 Prataprao, C. S.; Narsing, C. P. (Council of Scientific and Industrial Research India) IN2956DE2013, US2016222029, WO2015049700, US9522922, EP3052473 (prior. 04/10/2013, IN).

DSM2000 Van Der Does, T.; Booij, J.; Kers, E.; Leenderts, E.; Sibeijn, M.; Agayn, V. (DSM Nv) WO200206203 (prior. 14/07/2000, EP).

EJOC2003 Sugisaki, Claudia H.; Ruland, Y.; Baltas, M. *European Journal of Organic Chemistry* **2003**, *2003*, 672-688.

EJOC2008 Shibasaki, M.; Kanai, M. *European Journal of Organic Chemistry* **2008**, *2008*, 1839-1850.

EJOC2011 Ishikawa, H.; Bondzic, B. P.; Hayashi, Y. *European Journal of Organic Chemistry* **2011**, *2011*, 6020-6031.

EJOC2013a Quan, N.; Nie, L.-D.; Zhu, R.-H.; Shi, X.-X.; Ding, W.; Lu, X. *European Journal of Organic Chemistry* **2013**, *2013*, 6389-6396.

EJOC2013b Bhowmik, S.; Batra, S. *European Journal of Organic Chemistry* **2013**, *2013*, 7145-7151.

Fujifilm2013 Uemura, M.; Sato, K. (Fujifilm Corp) JP6072247, WO2014208753 (prior. JP, WO).

Gilead1995a Bischofberger, N. W.; Kim, C. U.; Lew, W.; Liu, H.; Williams, M. A. (Gilead Sci Inc) WO9626933, US5866601, EP759917, EP976734 (prior. 27/02/1995, US).

Gilead1995b Liu, H.; Bischofberger, N.; Williams, M.; Kim, C.; Lew, W. (Gilead Sci Inc) US5763483 (prior. 29/12/1995, US).

Gilead1996 Kent, K. M.; Kim, C. U.; McGee, L. R.; Munger, J. D.; Prisbe, E. J.; Postich, M. J.; Rohloff, J. C.; St John, D. E.; Williams, M. A.; Zhang, L. (Gilead Sci Inc) WO9807685, US2002058823, US2002156300, US5859284, US6057459, US6204398, US6518438, EP920410, EP920410 (prior. 23/08/1996, US).

Gilead1997a McGee, L. R.; Munger, J. D.; Kim, C. U.; Postich, M. J.; Kelly, D. E.; Rohloff, J. C.; Zhang, L.; Williams, M. A.; Kent, K. M.; Prisbe, E. J. (Gilead Sci Inc) US5886213 (prior. 22/08/1997, US).

Gilead1997b Bischofberger, N. W.; Dahl, T. C.; Hitchcock, M. J. M.; Kim, C. U.; Lew, W.; Liu, H.; Mills, R. G.; Williams, M. A.; Bischofberger; Graham, M. R.; Hitchcock; Matthew, A. W.; Michael, J. M.; Wolfgang, N. (Gilead Sci Inc) WO9914185, US2004053999, EP1015417 (prior. 17/09/1997, US).

Gilead1998 Becker, M. W.; Chapman, H. H.; Kelly, D. E.; Kent, K. M.; Lew, W.; Louie, M. S.; McGee, L. R.; Prisbe, E. J.; Postich, M. J.; Rohloff, J. C.; Schultze, L. M.; Yu, R. H.; Zhang, L. (Gilead Sci Inc) WO9955664 (prior. 24/04/1998, US).

Guangdong2014 Rajiv, K. S.; Li, Y. (Guangdong Hec Pharmaceutical) CN104447451 (prior. 13/11/2014, CN).

Guangzhou2013 Zheng, Q.; Huang, A. (Guangzhou Tongjuan Pharmaceutical Technology Co Ltd) CN103304437 (prior. 16/05/2013, CN).

HCA2012 Hajzer, V.; Latika, A.; Durmis, J.; Šebesta, R. *Helvetica Chimica Acta* **2012**, *95*, 2421-2428.

HeteroDrugs2005 Parthasaradhi Reddy, B.; Rathnakar Reddy, K.; Raji Reddy, R.; Muralidhara Reddy, D.; Subash Chander Reddy, K. (Hetero Drugs Ltd) WO2007060681, US2009054682, US2010137632, US7687658, EP1951654, EP2204359 (prior. 25/11/2005, EP, WO, IN).

HoffmannLR1998 Mair, H. (Hoffmann La Roche & Co Ag F, Hoffmann La Roche Inc) US6130354, EP957078 (prior. 13/05/1998, EP).

HoffmannLR1999a Karpf, M.; Trussardi, R. (Hoffmann La Roche & Co Ag F) US2002095040, US6437171, US6939986, EP1059283 (prior. 11/06/1999, EP).

HoffmannLR1999b Mair, H. (Hoffmann La Roche & Co Ag F) US6462226, EP1112999 (prior. 03/12/1999, EP)

HoffmannLR2000a Abrecht, S.; Karpf, M.; Trussardi, R.; Wirz, B. (Hoffmann La Roche & Co Ag F) US2002006960, US6403824, EP1127872 (prior. 22/02/2000, EP).

HoffmannLR2000b Iding, H.; Wirz, B.; Zutter, U. (Hoffmann La Roche & Co Ag F) US2001036653, US6518048, EP1146036 (prior. 10/04/2000, EP).

HoffmannLR2003 Brown, J.; Harrington, P.; Hughes, R. (Hoffmann La Roche & Co Ag F, Roche Colorado Corp) WO2004080944, US2004180933, US7122684, EP1606242 (prior. 13/03/2003, US).

HoffmannLR2004 Karpf, M.; Trussardi, R. (Hoffmann La Roche & Co Ag F, Hoffmann La Roche Inc) WO2006024436, US2006047002, US7514580, EP1791808 (prior. 02/09/2004, EP).

HoffmannLR2005 Gabel, R.; Groaning, M.; Johnston, D. (Hoffmann La Roche & Co Ag F, Hoffmann La Roche & Co Ag F, Roche Palo Alto Llc) WO2007074091, EP2178819, US2007197807, US7473798 (prior. 28/12/2005, US).

IJC2011 Masakatsu, S.; Motomu, K.; Kenzo, Y. *Israel Journal of Chemistry* **2011**, *51*, 316-328.

IMC2012 Shibazaki, M.; Watanabe, T.; Kumagai, N.; Yamatsugu, K.; Kaliyamoorthy, A.; Furudate, M. (Institute Microbial Chemistry) JP6106452 (prior. 05/12/2012, JP).

Ind2006a Li, S.; Yuan, W.; Wang, P.; Zhang, Z.; Zhang, W.; Ownby, S. (Individual) US2007161818 (prior. 06/01/2006, US).

Ind2006b Iyer, S. V.; Pejakala, V.; Karabasanagouda; Wagle, S.; Balaya, L.; Kanaka, M.; Hiremath, M. (Individual) WO2007138607 (prior. 25/05/2006, WO, IN).

Ind2008 Bandai, T.; Oshitari, T. (Individual) JP2009215239, JP5192856 (prior. 11/03/2008, JP).

JACS1997 Kim, C. U.; Lew, W.; Williams, M. A.; Liu, H.; Zhang, L.; Swaminathan, S.; Bischofberger, N.; Chen, M. S.; Mendel, D. B.; Tai, C. Y.; Laver, W. G.; Stevens, R. C. *Journal of the American Chemical Society* **1997**, *119*, 681-690.

JACS1999 Draths, K. M.; Knop, D. R.; Frost, J. W. *Journal of the American Chemical Society* **1999**, *121*, 1603-1604.

JACS2003 Ryu, D. H.; Corey, E. J. *Journal of the American Chemical Society* **2003**, *125*, 6388-6390.

JACS2004 Williams, A. L.; Johnston, J. N. *Journal of the American Chemical Society* **2004**, *126*, 1612-1613.

JACS2006a Yeung; Hong, S.; Corey, E. J. *Journal of the American Chemical Society* **2006**, *128*, 6310-6311.

JACS2006b Fukuta, Y.; Mita, T.; Fukuda, N.; Kanai, M.; Shibasaki, M. *Journal of the American Chemical Society* **2006**, *128*, 6312-6313.

JACS2007 Shie, J.-J.; Fang, J.-M.; Wang, S.-Y.; Tsai, K.-C.; Cheng, Y.-S. E.; Yang, A.-S.; Hsiao, S.-C.; Su, C.-Y.; Wong, C.-H. *Journal of the American Chemical Society* **2007**, *129*, 11892-11893.

JACS2018 Li, H. Z.; Shen, S. J.; Zhu, C. L.; Xu, H. *Journal of the American Chemical Society* **2018**, *140*, 10619-10626.

JCB2012 Bochkov, D. V.; Sysolyatin, S. V.; Kalashnikov, A. I.; Surmacheva, I. A. *Journal of Chemical Biology* **2012**, *5*, 5-17.

JCCS2012 Lai, J.; Fang, J.-M. *Journal of the Chinese Chemical Society* **2012**, *59*, 426-435.

JiangnanU2006 Tang, L.; Li, Y.; Xiang, H.; Jiang, R. (Jiangnan Univ) CN101318893 (prior. 25/08/2006, CN).

JOC1981 McGowan, D. A.; Berchtold, G. A. *The Journal of Organic Chemistry 1981*, *46*, 2381-2383.

JOC1998 Rohloff, J. C.; Kent, K. M.; Postich, M. J.; Becker, M. W.; Chapman, H. H.; Kelly, D. E.; Lew, W.; Louie, M. S.; McGee, L. R.; Prisbe, E. J.; Schultze, L. M.; Yu, R. H.; Zhang, L. *The Journal of Organic Chemistry* **1998**, *63*, 4545-4550.

JOC2001 Karpf, M.; Trussardi, R. *The Journal of Organic Chemistry* **2001**, *66*, 2044-2051.

JOC2004 Cong, X.; Liao, Q.-J.; Yao, Z.-J. *The Journal of Organic Chemistry* **2004**, *69*, 5314-5321.

JOC2006 Cong, X.; Yao, Z.-J. *The Journal of Organic Chemistry* **2006**, *71*, 5365-5368.

JOC2008 Zutter, U.; Iding, H.; Spurr, P.; Wirz, B. *The Journal of Organic Chemistry* **2008**, *73*, 4895-4902.

JOC2009 Nie, L.-D.; Shi, X.-X.; Ko, K. H.; Lu, W.-D. *The Journal of Organic Chemistry* **2009**, *74*, 3970-3973.

JOC2010a Weng, J.; Li, Y.-B.; Wang, R.-B.; Li, F.-Q.; Liu, C.; Chan, A. S. C.; Lu, G. *The Journal of Organic Chemistry* **2010**, *75*, 3125-3128.

JOC2010b Kamimura, A.; Nakano, T. *The Journal of Organic Chemistry* **2010**, *75*, 3133-3136.

JOC2010c Ko, J. S.; Keum, J. E.; Ko, S. Y. *The Journal of Organic Chemistry* **2010**, *75*, 7006-7009.

JOC2011a Tanaka, T.; Tan, Q.; Kawakubo, H.; Hayashi, M. *The Journal of Organic Chemistry* **2011**, *76*, 5477-5479.

JOC2011b Werner, L.; Machara, A.; Sullivan, B.; Carrera, I.; Moser, M.; Adams, D. R.; Hudlicky, T.; Andraos, J. *The Journal of Organic Chemistry* **2011**, *76*, 10050-10067.

JOC2012 Oh, H.-S.; Kang, H.-Y. *The Journal of Organic Chemistry* **2012**, *77*, 8792-8796.

JOC2013 Alagiri, K.; Furutachi, M.; Yamatsugu, K.; Kumagai, N.; Watanabe, T.; Shibasaki, M. *The Journal of Organic Chemistry* **2013**, *78*, 4019-4026.

JOC2014 Kohari, Y.; Okuyama, Y.; Kwon, E.; Furuyama, T.; Kobayashi, N.; Otuki, T.; Kumagai, J.; Seki, C.; Uwai, K.; Dai, G.; Iwasa, T.; Nakano, H. *The Journal of Organic Chemistry* **2014**, *79*, 9500-9511.

MichiganSU1999a Frost, J.; Frost, K.; Knop, D. (Univ Michigan State) WO200044923, US6613552, EP1151126 (prior. 29/01/1999, US).

MichiganSU1999b Frost, J.; Frost, K. (Univ Michigan State) WO200044924, EP1151125, US2002142409, US2003203460, US2006003429, US2010279368, US6600077, US6620602, US7002047, US7642083, US8080397 (prior. 29/01/1999, US).

MichiganSU1999c Frost, J.; Frost, K.; Knop, D. (Univ Michigan State) WO200229078, US6472169 (prior. 29/01/1999, US)

MichiganSU2006 Frost, J.; Guo, J. (Univ Michigan State) US2007190621, US7977077 (prior. 30/01/2006, US).

MicrobialCR2011 Shibasaki, M.; Yamatsugu, K. (Microbial Chem Res Found, Biseibutsu Kagaku Kenkyusho) JP2012188356, WO2012121079, US2014046087, US8940919, EP2684866, EP3235796, EP3235797, EP3235806 (prior. 08/03/2011, JP).

Monsanto2007 Bogosian, G.; O'Neil, J. P.; Frantz, N. (Monsanto Technology Llc) WO2008128076, US2011020885, US8435769, EP2145010 (prior. 13/04/2007, US).

Nanyang2007 Liu, X.; Ma, J. (Univ Nanyang Technological) WO2009078813, US2011021762, US8304553 (prior. 19/12/2007, US).

OBC2011 Trajkovic, M.; Ferjancic, Z.; Saicic, R. *Organic & Biomolecular Chemistry* **2011**, *9*, 6927-6929.

OBC2012 Rawat, V.; Dey, S.; Sudalai, A. *Organic & Biomolecular Chemistry* **2012**, *10*, 3988-3990.

OBC2013 Chen, C.-A.; Fang, J.-M. *Organic & Biomolecular Chemistry* **2013**, *11*, 7687-7699.

OCF2017 Silva, S.; Maycock, C. D. *Org. Chem. Front.* **2017**, *4*, 236-240.

OL2004 Guo, J.; Frost, J. *Organic Letters* **2004**, *6*, 1585-1588.

OL2007 Mita, T.; Fukuda, N.; Roca, F.; Kanai, M.; Shibasaki, M. *Organic Letters* **2007**, *9*, 259-262.

OL2008 Kipassa, N.; Okamura, H.; Kina, K.; Hamada, T.; Iwagawa, T. *Organic Letters* **2008**, *10*, 815-816.

OL2010 Osato, H.; Jones, I.; Chen, A.; Chai, C. *Organic Letters* **2010**, *12*, 60-63.

OL2015 Just, J.; Deans, B.; Olivier, W.; Paull, B.; Bissember, A.; Smith, J. *Organic Letters* **2015**, *17*, 2428-2430.

OL2016 Hayashi, Y.; Ogasawara, S. *Organic Letters* **2016**, *18*, 3426-3429.

OPRD1999 Federspiel, M.; Fischer, R.; Hennig, M.; Mair, H.-J.; Oberhauser, T.; Rimmler, G.; Albiez, T.; Bruhin, J.; Estermann, H.; Gandert, C.; Göckel, V.; Götzö, S.; Hoffmann, U.; Huber, G.; Janatsch, G.; Lauper, S.; Röckel-Stäbler, O.; Trussardi, R.; Zwahlen, A. G. *Organic Process Research & Development* **1999**, *3*, 266-274.

OPRD2004 Harrington, P.; Brown, J.; Foderaro, T.; Hughes, R. *Organic Process Research & Development* **2004**, *8*, 86-91.

OPRD2009 Andraos, J. *Organic Process Research & Development* **2009**, *13*, 161-185.

Oregon2006 Anderson, K. (Univ Oregon State) WO2008027570, US2008058210, US8203020 (prior. 30/08/2006, US).

RCB2013 Kalashnikov, A. I.; Sysolyatin, S. V.; Sakovich, G. V.; Sonina, E. G.; Shchurova, I. A. *Russ. Chem. Bull. 2013*, *62*, 163-170.

Roche2007 Trussardi, R. (Roche Palo Alto, Hoffmann La Roche & Co Ag F) WO2009037137, US2009076296, US7531687, EP2193116 (prior. 18/09/2007, EP).

RSCA2014 Chavan, S.; Chavan, P.; Khairnar, L. *RSC Advances* **2014**, *4*, 11417-11419.

S2012 Rehák, J.; Hu ka, M.; Latika, A.; Brath, H.; Almássy, A.; Hajzer, V.; Durmis, J.; Toma, Š.; Šebesta, R. *Synthesis* **2012**, *44*, 2424-2430.

S2013 Trajkovic, M.; Ferjancic, Z.; Saicic, R. *Synthesis* **2013**, *45*, 389-395.

S2017 Ogasawara, S.; Hayashi, Y. *Synthesis* **2017**, *49*, 424-428.

ShanghaiIOC2009a Lin, Y.; Sun, H.; Wu, Y. (Shanghai Organic Chem Inst) CN101475573 (prior. 21/01/2009, CN).

ShanghaiIOC2009b Ma, D.; Wang, Y.; Zhu, S.; Yu, S.; Zhou, Q. (Shanghai Organic Chem Inst) CN102127003, WO2011076086, US2014221662, US9040738 (prior. 23/12/2009, CN).

Shenyang2015 Tian, Y.; Wang, K.; Li, L. (Univ Shenyang Pharmaceutical) CN105439884 (prior. 27/10/2015, CN).

Sumitomo2008 Hayashi, Y.; Ishikawa, H.; Yujiro, H.; Hayato, I. (Sumitomo Chem Co Ltd, Univ Tokyo Sci Educational Found) JP2010514536, WO2009145263, US2011082302, US2013296600, US8501980, US9139514, EP2301911 (prior. 30/05/2008, JP).

Synlett2009a Mandai, T.; Oshitari, T. *Synlett* **2009**, *2009*, 783-786.

Synlett2009b Oshitari, T.; Mandai, T. *Synlett* **2009**, *2009*, 787-789.

Synlett2009c Sun, H.; Lin, Y.-J.; Wu, Y.-L.; Wu, Y. Synlett 2009, **2009**, 2473-2476.

T1996 Junquera, F.; Merchán, F. L.; Merino, P.; Tejero, T. *Tetrahedron* **1996**, *52*, 7045-7052.

T2009a Satoh, N.; Akiba, T.; Yokoshima, S.; Fukuyama, T. *Tetrahedron* **2009**, *65*, 3239-3245.

T2009b Yamatsugu, K.; Kanai, M.; Shibasaki, M. *Tetrahedron* **2009**, *65*, 6017-6024.

T2011 Raghavan, S.; Babu, V. S. *Tetrahedron* **2011**, *67*, 2044-2050.

T2012 Chuanopparat, N.; Kongkathip, N.; Kongkathip, B. *Tetrahedron* **2012**, *68*, 6803-6809.

T2014 Furutachi, M.; Kumagai, N.; Watanabe, T.; Shibasaki, M. *Tetrahedron* **2014**, *70*, 9113-9117.

T2015 Kongkathip, B.; Akkarasamiyo, S.; Kongkathip, N. *Tetrahedron* **2015**, *71*, 2393-2399.

TA2009 Nie, L.-D.; Shi, X.-X. *Tetrahedron: Asymmetry* **2009**, *20*, 124-129.

TA2011 Nie, L.-D.; Shi, X.-X.; Quan, N.; Wang, F.-F.; Lu, X. *Tetrahedron: Asymmetry* **2011**, *22*, 1692-1699.

TA2012 Nie, L.-D.; Ding, W.; Shi, X.-X.; Quan, N.; Lu, X. *Tetrahedron: Asymmetry* **2012**, *23*, 742-747.

TA2013 Nie, L.-D.; Wang, F.-F.; Ding, W.; Shi, X.-X.; Lu, X. *Tetrahedron: Asymmetry* **2013**, *24*, 638-642.

TL2005 Ghosh, S.; Rao, R. V.; Shashidhar, J. *Tetrahedron Letters* **2005**, *46*, 5479-5481.

TL2007 Yamatsugu, K.; Kamijo, S.; Suto, Y.; Kanai, M.; Shibasaki, M. *Tetrahedron Letters* **2007**, *48*, 1403-1406.

TL2008 Matveenko, M.; Willis, A.; Banwell, M. *Tetrahedron Letters* **2008**, *49*, 7018-7020.

TL2010 Wichienukul, P.; Akkarasamiyo, S.; Kongkathip, N.; Kongkathip, B. *Tetrahedron Letters* **2010**, *51*, 3208-3210.

TL2011 Osato, H.; Jones, I.; Goh, H.; Chai, C. L.; Chen, A. *Tetrahedron Letters 2011*, *52*, 6352-6354.

TL2012a Kim, H.-K.; Park, K.-J. J. *Tetrahedron Letters* **2012**, *53*, 1561-1563.

TL2012b Chuanopparat, N.; Kongkathip, N.; Kongkathip, B. *Tetrahedron Letters* **2012**, *53*, 6209-6211.

TohokuU2009 Nakano, H.; Takeshita, M.; Matsuyama, H.; Kwon, E. S. (Univ Tohoku, Univ Muroran) JP2011068587 (prior. 25/09/2009, JP).

TokyoU2006a Shibasaki, M.; Kanai, M.; Mita, T.; Fukuda, N.; Fukuta, Y. (Univ Tokyo) JP2008502731, WO2007099843 (prior. 23/02/2006, JP).

TokyoU2006b Fukuyama, T.; Yokoshima, S.; Satoh, N.; Sato, N.; Vite, G. D. (Univ Tokyo) JP2008050336, WO2008013269, US2010010225, EP2050752 (prior. 28/07/2006, JP).

TokyoU2006c Shibazaki, M.; Kanai, M.; Kamijo, M.; Yamaji, K. (Univ Tokyo) JP2008081489 (prior. 31/08/2006, JP)

UEChinaS2009 Shi, X.; Nie, L. (Univ East China Sci & Technology) CN101538221 (prior. 19/03/2009, CN)

UEChinaS2018 He, Y.; Shi, X.; Nie, L. (Univ East China Science & Tech) CN108484467 (prior. 09/04/2018, CN).

UHangzhou2011 Jiang, J.; Lai, G.; Jiang, K.; Shen, H.; Xu, L.; Deng, Y.; Shi, Z. (Univ Hangzhou Normal) CN102180821 (prior. 10/03/2011, CN).

UOkayama2009a Ishikawa, T.; Saito, S.; Kudoh, T.; Sudo, K.; Nishiuchi, H.; Okada, M.; Taniguchi, M.; Suto, K. (Univ Okayama) JP2011518594, WO2010143734, US2012130092, US8524940, EP2441749 (prior. 09/06/2009, JP).

UOkayama2009b Ishikawa, A.; Kudo, T.; Nishiuchi, H.; Okada, M.; Saito, S.; Suto, K.; Taniguchi, M. (Univ Okayama) JP4496351, EP2441749 (prior. 01/09/2009, JP).

USFlorida2007 Antilla, J.; Rowland, E. (Univ South Florida) WO2009005775, US2009030212, US7799934 (prior. 29/06/2007, US).

9 A Final Word

It should be obvious from the descriptions of each case study that there is no one, unique method to manufacture a drug substance. Because of the plethora of chemical reactions available to the process development chemist many synthetic routes to the final Pharmaceutical Active Ingredient can be envisioned. The ultimate choice of route has to take into consideration many factors. Whether the chemistry can be done at scale is most certainly the most important, but even "scale" in this context has to be interpreted as many tonnes of product can be produced *via* plug flow reactors where only small amounts are actually being transformed at any one time.

Cost factors are also of the highest importance for medicines that have high volume demands. There are indeed drugs such as Soliris, Naglazyme and Cinryze, which cost much more than 250,000 USD per year, but these are rare and are used in patient populations of a few thousand or less. Typically, the cost of treatment should be in the 10,s to 100,s USD and perhaps as high as 1,000,s USD for life threatening conditions. What is clear is that there is a consensus that drugs are cheaper than, for example, hospitalisation to treat illness and the Pharmaceutical Industry has an obligation to maintain this state of affairs.

Individual companies have in-house technologies and expertise along with specific manufacturing trains that influence their choice of route to a drug substance. What could be a costly option for one company may be a cheap and efficient route for another.

In today´s world, a high priority for any responsible manufacturer must be environmental considerations. In keeping with these concerns, the chemical community has embraced the concept of Sustainability and Green Chemistry. This provides a framework whose aim is to design and improve materials, products, processes and systems, thus minimising their impact on the environment. These principles should take on a key role in the development and selection the synthetic route to an API.

The 12 principles of Green Chemistry are:

1. Prevent waste
2. Improve atom economy
3. Design less hazardous synthesis
4. Design benign chemicals
5. Use benign solvents and auxiliaries
6. Design for energy efficiency
7. Use renewable feed stocks
8. Reduce derivatives
9. Promote catalytic transformations (*versus* stoichiometric
10. Design for degradation
11. Real-time analysis for pollution prevention
12. Use of inherent benign chemistry for accident prevention

Hopefully, the industry will apply these more actively in the future.

Retrosynthesis in the Manufacture of Generic Drugs: Selected Case Studies, First Edition, Pedro Paulo Santos and William Heggie.

NOTES

Index

Wiley The manufacturer's authorized representative according to the EU
General Product Safety Regulation is Wiley-VCH GmbH, Boschstr. 12,
69469 Weinheim, Germany, e-mail: Product_Safety@wiley.com.

Printed and bound by CPI Group (UK) Ltd, Croydon, CR0 4YY
15/09/2026
02217284-0001